# Mechanics of Engineering Materials
## Second edition

## P. P. BENHAM

Professor Emeritus and formerly Professor of Aeronautical Engineering,
The Queen's University of Belfast

## R. J. CRAWFORD

Professor of Mechanical and Manufacturing Engineering
and Director of the School of Mechanical and Process Engineering,
The Queen's University of Belfast

## C. G. ARMSTRONG

Reader in the Department of Mechanical and Manufacturing Engineering,
The Queen's University of Belfast

PEARSON
Prentice
Hall

Harlow, England • London • New York • Boston • San Francisco • Toronto • Sydney • Singapore • Hong Kong
Tokyo • Seoul • Taipei • New Delhi • Cape Town • Madrid • Mexico City • Amsterdam • Munich • Paris • Milan

**Pearson Education Limited**
Edinburgh Gate
Harlow
Essex CM20 2JE
England

and Associated Companies throughout the world

*Visit us on the World Wide Web at:*
http://www.pearsoned.co.uk

First Published 1987
Second Edition 1996.

**British Library Cataloguing in Publication Data**
A catalogue entry for this title is available from the British Library
ISBN-10: 0-582-25164-8
ISBN-13: 978-0-582-25164-9

**Library of Congress Cataloging in Publication Data**
A catalog entry for this title is available from the Library of
Congress

17  16  15  14
09  08  07  06

Set by 6 in Monotype Enrhardt Roman
Printed in China
EPC/14

# Mechanics of
# Engineering Materials
## Second edition

# Contents

# Preface to the Second Edition

The historical development of this text was outlined in the Preface to the previous edition. During the nine years since its publication there have been no major changes in the theory of the subject but there have been advances in some of the manipulative tools which can be used to solve problems related to the strength of materials. In addition, with experience and hindsight one can always improve the presentation of material to make it more digestible to the reader.

In this new edition we have engaged the services of a new author Dr Cecil Armstrong to assist in a complete update of the text. New features of the book include the use of versatile solution techniques based on spreadsheets, the use of colour to enhance diagrams and provide emphasis of key concepts, an introduction to matrix methods as a powerful tool in engineering analysis, and the incorporation of additional practical problems to illustrate the application of the theory.

The essential structure and content of Chapters 1–9 has not been changed although regular users of the previous text will see subtle changes in the mode of presentation to make this important fundamental information more meaningful to students. Chapter 10 from the original text has been removed because it was felt that modern computer programs provide powerful tools for the solutions of problems in structures. In a general text of this type it is impossible to do justice to the wide range of methods available to analyse structures. Hence, we have moved the important fundamental information on tension coefficients, energy methods, etc. to other chapters and omitted the superficial and to some extent outdated solution methods based on scale drawings.

The remaining chapters have benefited from a complete overhaul and the chapter on Finite Elements has been almost totally re-written. Throughout the text the reader is introduced to the concept of creating a spreadsheet representation of a problem in strength of materials. Once such a solution has been formulated it is then a relatively simple matter to study the effects of changes in variables, to plot graphs of key variables and to optimize solutions. This is shown to be a very powerful solution methodology which provides a completely new insight to each problem and open up a whole new approach to problem solving in many subject areas.

The authors derived a great deal of pleasure in working on this new version of the text. However, although he participated fully in the revision of the book, regrettably Peter Benham died before he was able to see the new book in print. We know that the many young people who studied under him, and the readers all over the world who benefited from his books, were greatly saddened by his death. It is our hope that this new edition, about which he was very excited and proud, will be a fitting tribute to his memory.

R. J. CRAWFORD
C. G. ARMSTRONG
*10 November 1995*

# Preface to the First Edition

The S.I. edition of *Mechanics of Solids and Structures* by P. P. Benham and F. V. Warnock was first published in 1973. It appears to have been very well received over the ensuing period. This preface is therefore written both for those who are familiar with the past text and also those who are approaching this subject for the first time. Although the subject matter is still basically the same today as it has been for decades, there are a few developing topics which have been introduced into undergraduate courses such as finite element analysis, fracture mechanics and fibre composite materials. In addition, style of presentation and illustrations in engineering texts have changed for the better and certain limitations of the previous edition, e.g. the number of problems and worked examples, needed to be rectified. Professor Warnock died in 1976 after a period of happy retirement and so it was left to the other author and the publisher to take the initiative to construct a new textbook.

In order to provide fresh thinking and reduce the time of rewriting Dr Roy Crawford, Reader in Mechanical Engineering at the Queen's University of Belfast, was invited, and kindly agreed, to join the project as a co-author. Although the present text might be regarded as a further edition of the original book the new authorship team preferred to make a completely fresh start. This is reflected in the change of title which is widely used as an alternative to *Mechanics of Solids*. The dropping of the reference to structures does not imply any reduction in that topic as will be seen in the contents.

In order that the book should not become any larger with the proposed expansion of material in some chapters, it was decided that the previous three chapters on experimental stress analysis should be omitted as there are several excellent texts in this field. The retention of that part of the book dealing with mechanical properties of materials for design was regarded as important even though there are also specialized texts in this area.

The main part (Ch. 1–18) of this new book of course still deals with the basic subject of Solid Mechanics, or Mechanics of Materials, whichever title one may prefer, being the study of equilibrium and displacement systems in engineering components and structures to enable designs to be effected in terms of stress and strain and the selection of materials. These eighteen chapters cover virtually all that is required in the three-year syllabus of a university or polytechnic degree course in engineering, or the examinations of the engineering Council, C.N.A.A. etc.

Although there is a fairly natural ordering of the material there is some scope for variation and lecturers will have their own particular detailed preferences.

As in the previous text, the first eleven chapters are concerned with forces and displacements in statically-determinate and indeterminate components and structures, and the analysis of uniaxial stress and strain due to various forms of loading such as bending, torsion, pressure and temperature change. The basic concepts of strain energy (Ch. 9) and elastic stability (Ch. 11) are also introduced. In Chapter 12 a study is made of two-dimensional states of stress and strain with special emphasis on principal stresses and the analysis of strain measurements using strain gauges. Chapter 13 combines two chapters of the previous book and brings together the topics of yield prediction and stress concentration which are of such importance in design.

Also included in these two chapters is an elementary introduction to the stress analysis and failure of fibre composite materials. These relatively new advanced structural materials are becoming increasingly used, particularly in the aerospace industry, and it is essential for engineers to receive a basic introduction to them. These thirteen chapters constitute the bulk of the syllabuses covered in first and second-year courses.

Four of the next five chapters appeared in the previous text and deal with more advanced or specialized topics such as thick-walled pressure vessels, rotors, thin plates and shells and post-yield or plastic behaviour, which will probably occupy part of final-year courses.

One essential new addition is an introductory chapter on finite element analysis. It may seem presumptuous even to attempt an introduction to such a broad subject in one chapter, but it is an attempt to provide initial encouragement and confidence to proceed to the complete texts on finite elements.

Chapters 19 to 22 cover much the same ground as in the previous text, but have been brought up to date particularly in relation to fracture mechanics. Since these chapters have such importance in relation to design, a number of worked examples have been introduced, together with problems at the end of each chapter. Bibliographies have still been included for further reading as required.

The first Appendix covers the essential material on properties of areas. The second deals with the simple principles of matrix algebra. A useful table of mechanical properties is provided in the third Appendix.

One of the recommendations of the Finniston Report to higher education was that theory should be backed up by more practical industrial applications. In this context the authors have attempted to incorporate into the worked examples and problems at the end of each chapter realistic engineering situations apart from the conventional examination-type applications of theory.

There had been a number of enquiries for a solutions manual for the previous text and this can be very helpful to both lecturer and student. Consequently this text is accompanied by another volume which contains worked solutions to nearly 300 problems. The manual should be used alongside the main text, so that steps in each solution can be referred back to the appropriate development in the relevant chapter. It is most important not to approach solutions on the basis of plucking the 'appropriate formula' out of the text, inserting the numbers, and manipulating a calculator!

Every effort has been made by the authors to ensure accuracy of text and solutions, but lengthy experience demonstrates human fallibility in this respect. When errors subsequently come to light they will be corrected at the next reprinting and readers' patience and comments will be appreciated!

Some use has been made of data and diagrams from other published literature and, in addition to the individual references, the authors wish to make grateful acknowledgement to all persons and organizations concerned.

P. P. BENHAM
R. J. CRAWFORD
*1987*

# Notation

| | |
|---|---|
| $\alpha$ | angle, coefficient of thermal expansion |
| $\beta$ | angle |
| $\gamma$ | shear strain, surface energy per unit area |
| $\delta$ | deflection, displacement |
| $\epsilon$ | direct strain |
| $\eta$ | efficiency, viscosity |
| $\theta$ | angle, angle of twist, co-ordinate |
| $\lambda$ | lack of fit |
| $\nu$ | Poisson's ratio |
| $\rho$ | radius of curvature, density |
| $\sigma$ | direct stress |
| $\tau$ | shear stress |
| $\phi$ | angle, co-ordinate, stress function |
| $\omega$ | angular velocity |
| | |
| $A$ | area |
| $C$ | complementary energy |
| $D$ | diameter |
| $E$ | Young's modulus of elasticity |
| $F$ | force |
| $G$ | shear or rigidity modulus of elasticity, strain energy release rate |
| $H$ | force |
| $I$ | second moment of area, product moment of area |
| $J$ | polar second moment of area |
| $K$ | bulk modulus of elasticity, fatigue strength factor, stress concentration factor, stress intensity factor |
| $L$ | length |
| $M$ | bending moment |
| $N$ | number of stress cycles, speed of rotation |
| $P$ | force |
| $Q$ | shear force |
| $R$ | force, radius of curvature, stress ratio |
| $S$ | cyclic stress |
| $T$ | temperature, torque |
| $U$ | strain energy |
| $V$ | volume |
| $W$ | weight, load |
| $X$ | body force |
| $Y$ | body force |
| $Z$ | body force, section modulus |
| | |
| $a$ | area, distance, crack length |
| $b$ | breadth, distance, crack length |
| $c$ | distance |
| $d$ | depth, diameter |
| $e$ | eccentricity, base of Napierian logarithms |
| $g$ | gravitational constant |
| $h$ | distance |

| | |
|---|---|
| $j$ | number of joints |
| $k$ | diameter ratio of cylinder |
| $l$ | length |
| $m$ | mass, modular ratio, number of members |
| $n$ | number |
| $p$ | pressure |
| $q$ | shear flow |
| $r$ | co-ordinate, radius, radius of gyration |
| $s$ | length |
| $t$ | thickness, time |
| $u$ | displacement in the $x$- or $r$-direction |
| $v$ | deflection, displacement in the $y$- or $\theta$-direction, velocity |
| $w$ | displacement in the $z$-direction, load intensity |
| $x$ | co-ordinate, distance |
| $y$ | co-ordinate, distance |
| $z$ | co-ordinate, distance |

It should be noted that a number of these symbols have also been used to denote constants in various equations.

# 1 Statically Determinate Force Systems

> Structural and solid-body mechanics are concerned with analysing the effects of applied loads. These are *external* to the material of the structure or body and result in *internal* reacting forces, together with deformations and displacements, conforming to the principles of Newtonian mechanics. Hence a familiarity with the principles of statics, the cornerstone of which is the concept of *equilibrium of forces*, is essential.
>
> A force system is said to be statically determinate if the internal forces can be calculated by considering only the forces acting on the system.
>
> Forces result in four basic forms of deformation or displacement of structures or solid bodies and these are *tension, compression, bending,* and *twisting*.
>
> The equilibrium conditions in these situations are discussed so that the forces may be determined for simple engineering examples.

## 1.1 Revision of statics

A particle is in a state of equilibrium if the resultant force and moment acting on it are zero. This hypothesis can be extended to clusters of particles that interact with each other with equal and opposite forces but have no overall resultant. Thus it is evident that solid bodies, structures, or any subdivided part, will be in equilibrium if the resultant of all external forces and the resultant of all moments are zero. This may be expressed mathematically in the following six equations which relate to Cartesian co-ordinate axes $x$, $y$ and $z$.

$$\left.\begin{array}{l} \Sigma F_x = 0 \\ \Sigma F_y = 0 \\ \Sigma F_z = 0 \end{array}\right\} \qquad [1.1]$$

where $F_x$, $F_y$ and $F_z$ represent the components of force vectors in the co-ordinate directions.

$$\left.\begin{array}{l} \Sigma M_x = 0 \\ \Sigma M_y = 0 \\ \Sigma M_z = 0 \end{array}\right\} \qquad [1.2]$$

where $M_x$, $M_y$ and $M_z$ are components of moment vectors caused by the external forces acting about the axes $x$, $y$, $z$.

The above six equations are the necessary and sufficient conditions for equilibrium of a body.

If the forces all act in one plane, say $z = 0$, then

$$\Sigma F_z = \Sigma M_x = \Sigma M_y = 0$$

are automatically satisfied and the equilibrium conditions to be satisfied in a two-dimensional system are

$$\left.\begin{array}{l} \Sigma F_x = 0 \\ \Sigma F_y = 0 \\ \Sigma M_z = 0 \end{array}\right\} \qquad [1.3]$$

Forces and moments are vector quantities and may be resolved into components; that is to say, a force or a moment of a certain magnitude and direction may be replaced and exactly represented by two or more components of different magnitudes and in different directions.

Considering firstly the two-dimensional case shown in Fig. 1.1, the force $F$ may be replaced by the two components $F_x$ and $F_y$ provided that

$$\left.\begin{array}{l} F_x = F\cos\alpha \\ F_y = F\sin\alpha \end{array}\right\} \qquad [1.4]$$

**Fig. 1.1**

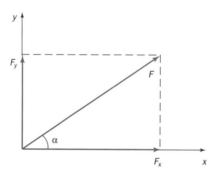

Note that throughout this book, externally applied forces will be shown coloured. Internal forces, that is those within a structural element, will be shown in black.

If the force $F$ were arbitrarily oriented with respect to three axes $x$, $y$, $z$ as in Fig. 1.2, then it could be replaced or represented by the following components:

$$\left.\begin{array}{l} F_x = F\cos\alpha \\ F_y = F\cos\beta \\ F_z = F\cos\gamma \end{array}\right\} \qquad [1.5]$$

**Fig. 1.2**

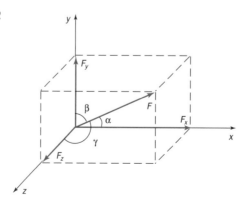

**Fig. 1.3**

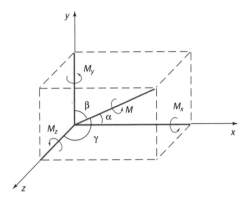

A couple or moment vector about an axis can similarly be resolved into a representative system of component vectors about other axes, as shown in Fig. 1.3 and represented by the following equations:

$$\left.\begin{array}{l} M_x = M\cos\alpha \\ M_y = M\cos\beta \\ M_z = M\cos\gamma \end{array}\right\} \tag{1.6}$$

**Example 1.1**

**A set of concurrent forces $F_i$ are defined by their components $F_x$, $F_y$, $F_z$ in kN as**

|       | $F_x$ | $F_y$ | $F_z$ |
|-------|-------|-------|-------|
| $F_1$ | 80    | −20   | −40   |
| $F_2$ | −40   | 60    | −80   |
| $F_3$ | 100   | −20   | 30    |
| $F_4$ | −30   | 10    | 40    |

**Calculate the magnitude and direction of the resultant of this force system.**

$$\bar{F}_x = \Sigma F_x = 80 - 40 + 100 - 30 = 110\,\text{kN}$$

$$\bar{F}_y = \Sigma F_y = -20 + 60 - 20 + 10 = 30\,\text{kN}$$

$$\bar{F}_z = \Sigma F_z = -40 - 80 + 30 + 40 = -50\,\text{kN}$$

The magnitude of the resultant force is

$$F_R = \sqrt{(\bar{F}_x^2 + \bar{F}_y^2 + \bar{F}_z^2)} = \sqrt{(110^2 + 30^2 + (-50)^2)} = 124\,\text{kN}$$

to three significant figures. The angles between the resultant force and the axes shown in Fig. 1.2 can now be found using

$$\cos\alpha = \frac{\bar{F}_x}{F_R} = \frac{110}{124} = 0.884, \text{ so } \alpha = 27.9°$$

$$\cos\beta = \frac{\bar{F}_y}{F_R} = \frac{30}{124} = 0.241, \text{ so } \beta = 76.1°$$

$$\cos \gamma = \frac{\bar{F}_z}{F_R} = \frac{-50}{124} = -0.402, \text{ so } \gamma = 113.7°$$

Since all the forces are concurrent, none of them have any moment about the common point through which they all act and there is therefore no resultant moment about that point.

**Spreadsheet solution**  In engineering practice there are few investigations where the problem is as clearly defined and straightforward as that given above. There is usually a degree of uncertainty about the supplied information, some data may be missing and the numerical computations are significantly more complicated. Thus laborious calculations are needed and the sensitivity of the results to variations in the given or assumed information has to be evaluated. This has lead many engineers to develop computer programs in languages such as Fortran, Basic or C to automate the required calculations.

In this book, selected numerical problems will be solved using commercial programming and modelling tools called **spreadsheets**. Probably the best known of these are Lotus 1–2–3, Borland Quattro and Microsoft Excel. A computer shreadsheet can be considered as a grid of cells which can contain text labels, numbers or formulae which may refer to numbers or numerical results in other cells. If the cell contains a formula, and a number is changed in any cell to which that formula refers, then the formula is automatically recalculated and the updated result is displayed. A range of alternatives can be quickly evaluated by entering different numbers

**Fig. 1.4**

|  | A | B | C | D | E |
|---|---|---|---|---|---|
| 1 |  | Fx | Fy | Fz | Magnitude |
| 2 | F1 | 80 | -20 | -40 |  |
| 3 | F2 | -40 | 60 | -80 |  |
| 4 | F3 | 100 | -20 | 30 |  |
| 5 | F4 | -30 | 10 | 40 |  |
| 6 |  |  |  |  |  |
| 7 | FR | +B2+B3+B4+B5 | @SUM(C2..C5) | @SUM(D2..D5) | @SQRT(B7^2+C7^2+D7^2) |
| 8 |  |  |  |  |  |
| 9 | Direction Cosines |  |  |  |  |
| 10 |  | alpha | beta | gamma |  |
| 11 |  | +B7/E7 | +C7/E7 | +D7/E7 |  |

|  | A | B | C | D | E |
|---|---|---|---|---|---|
| 1 |  | Fx | Fy | Fz | Magnitude |
| 2 | F1 | 80 | -20 | -40 |  |
| 3 | F2 | -40 | 60 | -80 |  |
| 4 | F3 | 100 | -20 | 30 |  |
| 5 | F4 | -30 | 10 | 40 |  |
| 6 |  |  |  |  |  |
| 7 | FR | 110 | 30 | -50 | 124.5 |
| 8 |  |  |  |  |  |
| 9 | Direction Cosines |  |  |  |  |
| 10 |  | alpha | beta | gamma |  |
| 11 |  | 0.884 | 0.241 | -0.402 |  |

in a given cell. Modern packages have sophisticated facilities for the graphical display of results, 'What-if' evaluation of a range of alternatives, querying databases and optimization. Although these programs were initially developed for financial modelling, many engineers and students now use these packages as sophisticated programmable calculators.

Figure 1.4(*a*) shows the data and formulae required to solve this problem. These were entered in a Quattro spreadsheet but they will also work with Lotus 1–2–3 or Microsoft Excel. Figure 1.4(*b*) shows the resulting display in the spreadsheet. Normally the formulae will not be visible and only the result will be displayed. Throughout this book, the cells in which the user should input data have been highlighted. The other cells contain either formulae or text labels and should not be overwritten.

As can be seen in Fig. 1.4(*a*), the total force in the *x*-direction is found in cell B7 by adding up the contents of cells B2, B3, B4 and B5. An even more convenient technique for summing up the *y*-components is shown in cell C7, where the built-in function @SUM is used to total the contents of cells C2 to C5. Within the spreadsheet program the Edit, Copy and Paste commands can be used to copy the same formula to cell D7, which then sums the *z*-components. Individual cells or groups of cells can be identified with arrow keys or mouse clicks – it is not necessary to type the cell references. The resultant force is found using the three-dimensional equivalent of eqn. [1.7] in cell E7, using the built-in function @SQRT to calculate the square root and $^\wedge$ to indicate that a number is to be raised to a power.

Once these formulae are entered, any change to any component of the four forces will cause an immediate recalculation of the magnitude of the resultant force and the directions. If only three forces are to be summed, deleting all the components of one force will give the correct answer. If a larger number of forces is to be summed, extra rows can be inserted above row 6 and the SUM formula in the current row 7 adjusted to the new range of cells.

**1.2 Resultant force and moment**

It is sometimes more convenient to replace a system of applied forces by a resultant which of course must have the same effect as those forces. Considering a two-dimensional case as illustrated in Fig. 1.5, then the most general solution is obtained by choosing any point A through which the resultant can act. Then the total force components in the co-ordinate directions are

$$\left.\begin{aligned} \bar{F}_x = \Sigma\, F_x \\ \bar{F}_y = \Sigma\, F_y \end{aligned}\right\} \qquad [1.7]$$

and the resultant force is given by

$$F_R = \sqrt{(\bar{F}_x^2 + \bar{F}_y^2)} \qquad [1.8]$$

However, this is not sufficient in itself since the moment due to the forces must be represented. This is done by having a couple acting about A such that

$$\bar{M} = \Sigma\, M_z \qquad [1.9]$$

**Fig. 1.5**

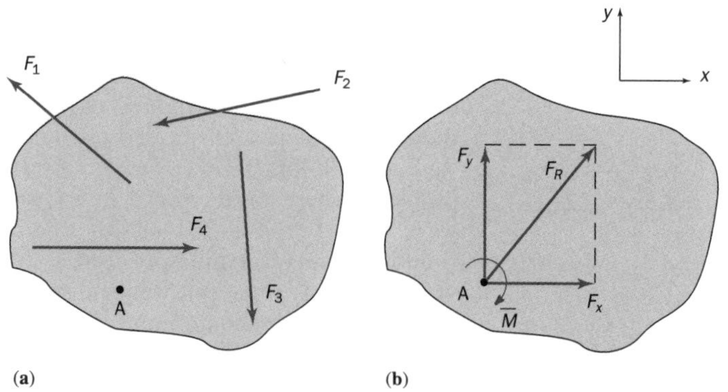

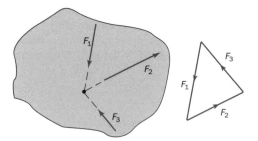

(a)                          (b)

In general, then, any system of forces can be replaced by a resultant force through, and a couple about, any chosen point.

The equivalent solution for a three-dimensional system of forces is similarly a couple and a resultant force whose direction is parallel to the axis of the couple.

One of the most useful constructions in force analysis is termed the *triangle of forces*. If a body is acted on by three forces then, for equilibrium to exist, these must act through a common point or else there will exist a couple about the point causing the body to rotate. In addition the magnitude and direction of the three force vectors must be such as to form a closed triangle as shown in Fig. 1.6.

**Fig. 1.6**

### 1.3 Types of structural and solid-body components

Structures are made up of a series of members of regular shape that have a particular function for load carrying. The shape and function are, through usage, implied in the name attached to the member. The first group is concerned with carrying loads parallel to a longitudinal axis. Examples are shown in Fig. 1.7. A member which prevents two parts of a structure from moving apart is subjected to a pull at each end, or tensile force, and is termed a *tie* (a). Conversely a slender member which prevents parts of a structure moving towards each other is under compressive force and is termed a *strut* (b). A vertical member which is perhaps not too slender and supports some of the mass of the structure is called a *column* (c). A *cable* (d) is a generally recognized term for a flexible string under tension which connects two bodies. It cannot supply resistance to bending action.

One of the most important of structural members is that which is frequently supported horizontally and carries transverse loading. This is

Fig. 1.7

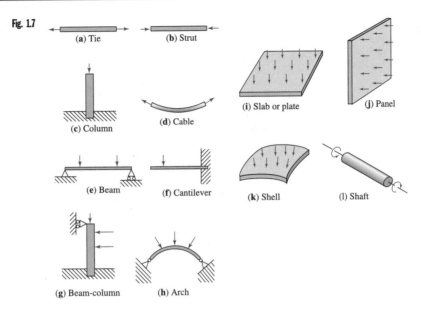

(a) Tie  (b) Strut

(c) Column  (d) Cable  (i) Slab or plate  (j) Panel

(e) Beam  (f) Cantilever  (k) Shell  (l) Shaft

(g) Beam-column  (h) Arch

known as a *beam* (*e*), a common special case of which is termed a *cantilever* (*f*), where one end is fixed and provides all the necessary support. A *beam–column* (*g*), as the name implies, combines the separate functions already described. The *arch* (*h*) has the same function as the beam or beam–column, but is curved in shape.

The filling-in and carrying of load over an area or space are achieved by *flat slabs* or *plates* (*i*), by *panels* (*j*) and also by *shells* (*k*), which are curved versions of the former.

The transmission of torque and twist is achieved through a member which is frequently termed a *shaft* (*l*).

The members described above can have a variety of cross-sectional shapes depending on the particular type of loading to be carried. Some typical cross-sections are illustrated in Fig. 1.8. Sections of this type are generally readily available from stock.

Fig. 1.8

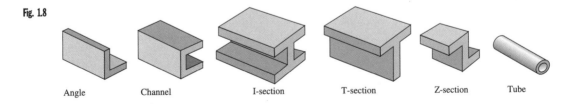

Angle  Channel  I-section  T-section  Z-section  Tube

## 1.4 Types of support and connection for structural components

The applied loading on a framework, beam or column is transmitted to the supports which will provide the required reacting forces to maintain overall equilibrium. Examples of supports of various kinds suitable to react to loading in a plane (two dimensions) are shown in Fig. 1.9. In the accompanying table the possible displacement and reacting force components are indicated.

**Fig. 1.9**

| Type of support | Equivalent force system |

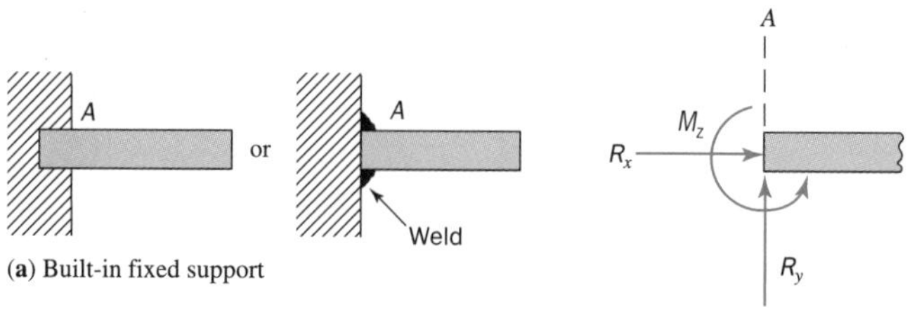

(a) Built-in fixed support

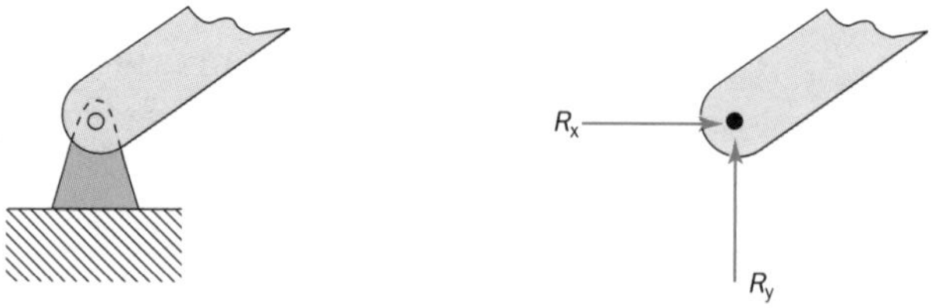

(b) Pin connection

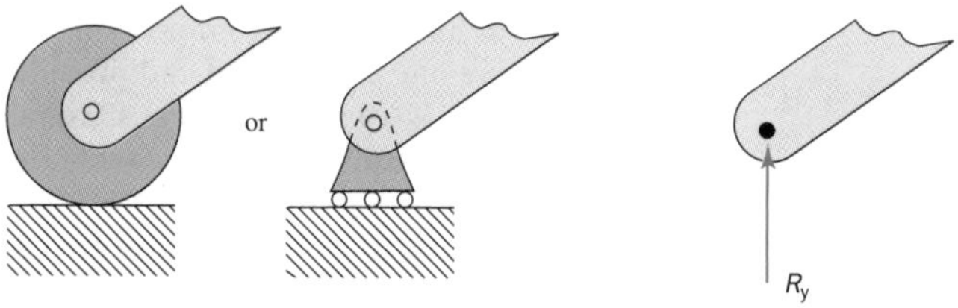

(c) Roller support

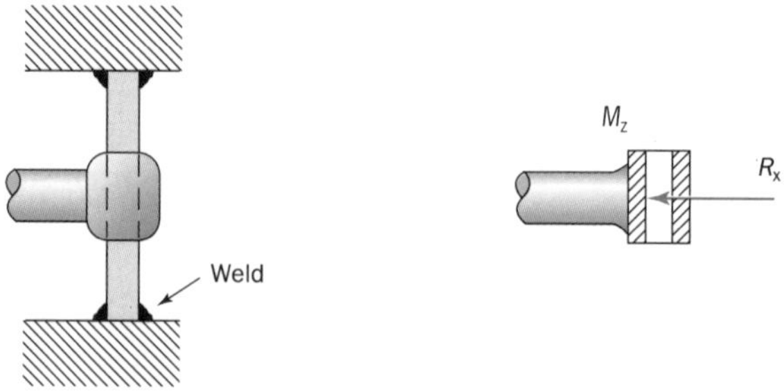

(d) Sliding support

|  | Displacement | | | Reacting force | | |
|---|---|---|---|---|---|---|
|  | $x$ | $y$ | $\theta$ | $R_x$ | $R_y$ | $M_z$ |
| (a) Built–in or fixed support |  |  |  | √ | √ | √ |
| (b) Pin connection |  |  | √ | √ | √ |  |
| (c) Roller support | √ |  | √ |  | √ |  |
| (d) Sliding support |  | √ |  | √ |  | √ |

**Fig. 1.10**

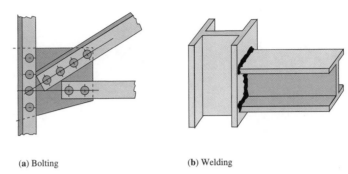

(a) Bolting                (b) Welding

The separate members of a structural framework are joined together by bolting, riveting or welding, two examples of which are shown in Fig. 1.10. Now, if these joints were ideally stiff, when the members of the framework were deformed under load, the angles between the members at the joint would not change. This would also imply that the joint was capable of transmitting a couple. However, calculations for a complete structure on this basis would become rather involved and tedious.

It is found in practice that there is some degree of rotation between members at a joint owing to the elasticity of the system. Furthermore, it has been shown that it is not unreasonable, for purposes of calculation, to assume that these joints may be represented by a simple ball and socket or pin in a hole. Even with this arrangement, which of course cannot transmit a couple or bending moment (other than by friction, which is ignored), deformations of the members are relatively small. Consequently changes in angle at the joints are also small, which is why this approximation is not unreasonable when applied to the actual joints. Thus it is common practice when calculating the forces in the members of a framework to assume that all joints are pinned.

## 1.5 Statical determinacy

In general, structural or solid-body mechanics involves determination of unknown forces within the structure or body. The approach taken in this analysis depends initially on whether the system under consideration is 'statically determinate' or 'statically indeterminate'.

If the number of equations available from statements of equilibrium is the same as the number of unknown forces (including reactions) then the problem is *statically determinate*. If the number of unknown reactions or internal forces in the structure or component is greater than the number of equilibrium equations available, then the problem is said to be *statically indeterminate*.

In order to solve a statically indeterminate problem it is necessary to consider additional equations relating to the displacement or deformation of the body.

The above statements are quite general and apply throughout this text.

## 1.6 Free-body diagrams

When commencing to analyse any force system acting on a component or structure it is essential firstly to have a diagram showing the forces acting. If the structure or part of it is separated from its surroundings and the possible reactions are inserted then a diagram of this system is called a *free-body* diagram. Examples of this are shown in Fig. 1.11.

**Fig. 1.11**

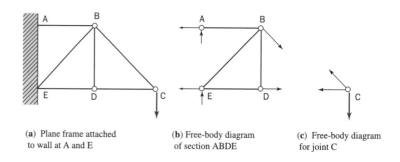

(a) Plane frame attached
to wall at A and E

(b) Free-body diagram
of section ABDE

(c) Free-body diagram
for joint C

## 1.7 Determinacy criteria for structures

The principles of statical determinacy will now be considered in relation to plane and space-frame structures. There are three classes of frame or truss, in concept, although one is not of practical interest:

(a) *Under-stiff.* If there are more equilibrium equations than unknown forces or reactions the system is unstable and is not a structure but a mechanism.

(b) *Just-stiff.* This is the *statically determinate* case for which there are the same number of equilibrium equations as unknown forces. If any member is removed then a part or the whole of the frame will collapse.

(c) *Over-stiff.* This is the *statically indeterminate* case in which there are more unknown forces than available equilibrium equations. There is at least one member more than is required for the frame to be just stiff.

Figure 1.12 shows an example of each of the three conditions. Remembering that each joint is pinned, then it is clear that in (a) the central square of members is not stable unless there is one diagonal member inserted. The number and arrangement of the members in case (b) is quite correct for the 'just-stiff' or statically determinate condition. In case (c) in contrast to (a) the central square has two diagonal braces one or other of which is unnecessary and hence the frame is 'over-stiff' or statically indeterminate.

It is useful to express the three cases in the form of mathematical criteria. Let the number of joints, including support points, in a frame be $j$, the number of members $m$, and the number of reactions $r$. Now, for a space frame there are three equilibrium equations applicable to each joint, namely $\Sigma F_x = 0$, $\Sigma F_y = 0$, $\Sigma F_z = 0$; hence there are $3j$ equations to determine $m + r$ unknown forces and reactions, and the statically determinate case is

**Fig. 1.12**

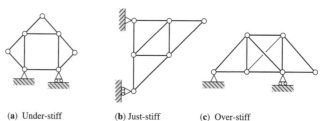

(a) Under-stiff

(b) Just-stiff

(c) Over-stiff

represented by

$$m + r = 3j \qquad [1.10]$$

When $m + r < 3j$ the members form a mechanism, and for $m + r > 3j$ the frame is over-stiff, or redundant, and therefore statically indeterminate.

When determining the reactions there are six equations for overall equilibrium of a framework:

$$\Sigma F_x = \Sigma F_y = \Sigma F_z = 0$$

$$\Sigma M_x = \Sigma M_y = \Sigma M_z = 0$$

Therefore the minimum value for $r$, when using the above criteria to allow for any general loading system, is six.

For frames lying only in one plane there are only two equilibrium equations at each joint and so the relationships comparable with the above are

$$\left. \begin{array}{c} m + r < 2j \\ m + r = 2j \\ m + r > 2j \end{array} \right\} \qquad [1.11]$$

The minimum value for $r$ in these expressions must be three for general forms of loading.

The above criterion for a just-stiff frame is a necessary but not a sufficient condition, since the *arrangement* of the members might still not provide the required stiffness.

**Example 1.2**

Examine the plane frames illustrated in Fig. 1.13. State the class of each and where members should be inserted or removed to make each statically determinate. Also indicate any redundant reactions.

Fig. 1.13

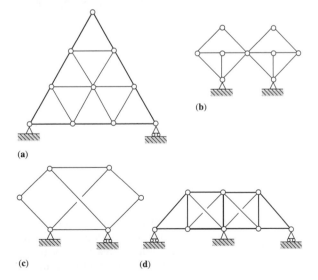

(a)

(b)

(c)

(d)

(a)  $m = 18$, $r = 3$, $j = 10$: hence $m + r > 2j$ and the frame is over-stiff or redundant. Any member may be removed from the central hexagon structure. None of the members may be removed from the apexes.

(b)  $m = 14$, $r = 4$, $j = 9$: hence $m + r = 2j$ and the frame is just stiff and statically determinate.

(c)  $m = 8$, $r = 3$, $j = 6$: hence $m + r < 2j$, which constitutes a mechanism. To make statically determinate insert a member between any pair of unconnected joints.

(d)  $m = 15$, $r = 4$, $j = 8$: hence $m + r > 2j$, which is redundant both in members and reactions. Hence remove either the left or right support and a diagonal member from each square.

## 1.8 Determination of axial forces by equilibrium statements

The members of pin-jointed plane or space frameworks can only carry axial forces, i.e. tension or compression (Fig. 1.14). These may be determined by considering the equilibrium of various parts of the structure as 'free bodies'.

### Plane pin-jointed frames

*Equilibrium at joints*

At any joint in a plane frame two equilibrium equations apply, namely

$$\Sigma F_x = 0 \quad \text{and} \quad \Sigma F_y = 0 \text{ (for a frame lying in the } z\text{-plane)}$$

Hence only two unknown forces can be determined. The method therefore entails making a free-body diagram centred on each joint at which there are *only two* unknowns. The forces are then resolved in the $x$- and $y$-directions so that the above two equations can be applied. It will probably be necessary at first to determine support reactions by considering equilibrium of the whole frame. This generally gives at least one joint where there is a known reaction and only two members having unknown forces from which to start the analysis. The method will be illustrated by the following example.

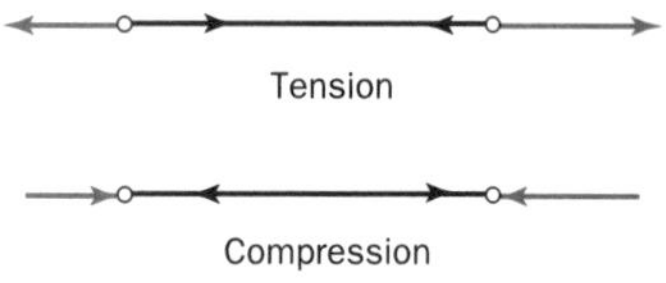

Tension

Compression

**Fig. 1.14**

## Example 1.3

**Determine the magnitude and the type of force in each member of the plane pin-jointed frame shown in Fig. 1.15.**

**Fig. 1.15**

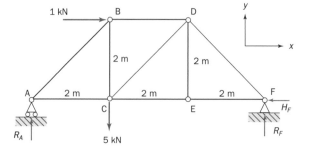

In Fig. 1.15, in order to save the space and having to draw the complete frame twice, once on its supports and a second time as a free body away from the supports with the reactions in place, the possible reactions have been placed directly onto the configuration diagram.

Considering the equilibrium of the whole frame and taking moments about joint F,

$$\Sigma M_z = (4 \times 5000) - 6R_A - (2 \times 1000) = 0 \quad \text{so that}$$

$$R_A = 3000\,\text{N}$$

For horizontal equilibrium,

$$\Sigma F_x = 0 = 1000 - H_F \quad \text{so that} \quad H_F = 1000\,\text{N}$$

For vertical equilibrium,

$$\Sigma F_y = 0 \quad \text{gives} \quad R_A + R_F - 5000 = 0$$

Hence

$$R_F = 2000\,\text{N}$$

It may be seen from Fig. 1.14 that members which are in tension have internal forces which act away from the joint. It is recommended that it is *assumed initially that each member is subjected to tension*. If the analysis shows that the internal force is negative this simply means that the member is in compression.

The only joints at which there are two unknowns are A and F. Start, say, at A and proceed as follows:

*Joint A*　The free-body diagram for this joint is as shown in Fig. 1.16(a).

$$\Sigma F_y = 0 \quad \text{gives} \quad F_{AB} \sin 45° + 3000 = 0$$

**Fig. 1.16**

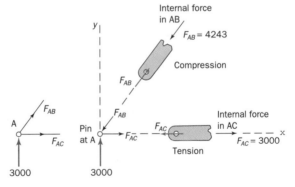

(a) Assumed forces on pin at A　(b) Actual forces on pin at A and members AB and AC

Hence

$$F_{AB} = -4243\,\text{N}$$

The negative sign shows that AB is in compression and not tension.

$$\Sigma F_x = 0 \quad \text{gives} \quad F_{AB} \cos 45° + F_{AC} = 0$$

Hence

$$F_{AC} = 4243 \cos 45° = 3000\,\text{N}$$

The next step *has* to be at joint B (rather than C).

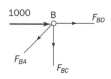

**Fig. 1.17**

*Joint B*   (Fig. 1.17)

$$\Sigma F_y = F_{BA} \cos 45° + F_{BC} = 0$$

$$F_{BC} = 4243 \cos 45° = 3000\,\text{N}$$

$$\Sigma F_x = F_{BD} + 1000 - F_{BA} \sin 45° = 0$$

$$F_{BD} = -4243 \sin 45° - 1000 = -4000\,\text{N}\quad\text{(compression)}$$

*Joint C*   There are now only two unknowns at this joint, $F_{CD}$ and $F_{CE}$. (Fig. 1.18)

$$\Sigma F_y = F_{CB} - 5000 + F_{CD} \cos 45° = 0$$

$$F_{CD} = (-3000 + 5000)\sqrt{2} = 2828\,\text{N}$$

$$\Sigma F_x = F_{CE} - F_{CA} + F_{CD} \cos 45° = 0$$

$$F_{CE} = 3000 - 2828 \cos 45° = 1000\,\text{N}$$

**Fig. 1.18**

Continuing the above process for joints E and D gives

$$F_{ED} = 0 \quad F_{EF} = 1000\,\text{N} \quad F_{DF} = -2828\,\text{N}$$

The final force distribution in the frame is shown in Fig. 1.19.

**Fig. 1.19**

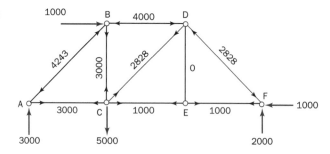

As the force in DE is zero one might ask whether that member is required. With this particular applied load, in fact, DE could be removed quite safely, since the two members CE and EF really act as one continuous member CEF in tension. However, if the applied load at C acted upwards the loads in all the members would have the same magnitude as before but would be reversed in sense and then CE and EF would be in compression. While ED was in position even without any force in it, it would keep CE and EF in line, whereas if ED were removed the joint E would move vertically due to the slightest misalignment and result in collapse.

**Matrix solutions**   An alternative and very powerful technique for solving problems of this type involves the use of matrix methods, which are described in Appendix B. These techniques are central to computer-aided analysis techniques such as

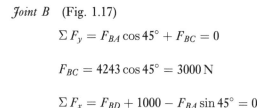

**Table 1.1**

| Joint | $F_{AB}$ | $F_{AC}$ | $F_{BC}$ | $F_{BD}$ | $F_{CD}$ | $F_{CE}$ | $F_{DE}$ | $F_{DF}$ | $F_{EF}$ | $R_A$ | $H_F$ | $R_F$ | External load |
|---|---|---|---|---|---|---|---|---|---|---|---|---|---|
| A | 0.707 | 1 | | | | | | | | | | | |
|   | 0.707 | | | | | | | | | 1 | | | |
| B | −0.707 | | | 1 | | | | | | | | | 1000 |
|   | −0.707 | | −1 | | | | | | | | | | |
| C | | −1 | | | 0.707 | 1 | | | | | | | |
|   | | | 1 | | 0.707 | | | | | | | | −5000 |
| D | | | | −1 | −0.707 | | | 0.707 | | | | | |
|   | | | | | −0.707 | | −1 | −0.707 | | | | | |
| E | | | | | | −1 | | | 1 | | | | |
|   | | | | | | | 1 | | | | | | |
| F | | | | | | | | −0.707 | −1 | | −1 | | |
|   | | | | | | | | 0.707 | | | | 1 | |

the finite element method (Chapter 17), but a brief description of the solution of Example 1.3 will be given here. Those unfamiliar with matrix methods may wish to skip this section.

The equations of joint equilibrium in Figs. 1.15 to 1.19 can be summarized in tabular form in Table 1.1. The twelve rows represent the $x$ and $y$ equations of force equilibrium for each of the six joints A–F. The columns contain the coefficients multiplying the nine unknown member forces and three reactions, plus one column for the known external loads.

An alternative way of writing these equations is as a *matrix equation* where a coefficient matrix $[A]$ is multiplied by a column vector $\{x\}$ containing the unknown member forces and reactions. The resulting forces are summed with the known external loads $\{y\}$ as

$$[A]\{x\} + \{y\} = \{0\}$$

To find $\{x\}$, which is a column vector containing the unknowns $F_{AB}$, $F_{AC}$, etc., the equivalent to a division is required. This is known as the matrix inverse; it can be found using standard numerical techniques and matrix multiplication and inversion routines are available in spreadsheets. Given the inverse matrix $[A]^{-1}$, the unknowns can be found by multiplying the inverse matrix by the known quantities, i.e.

$$\{x\} = [A]^{-1}\{-y\}$$

Members which are not connected to a given pin have zero coefficients multiplying the relevant member force. Members which are connected to a given pin have non-zero coefficients multiplying the member force. For horizontal equilibrium these are the cosine of the angle at which the member leaves the joint. For vertical equilibrium the coefficient is the sine of the member angle. The right-hand side of each equation is the negative of the component of any external force acting at that joint.

By this technique the series of equations at each joint described in Figs 1.16 to 1.18 become the matrix equation

$$
\begin{bmatrix}
0.707 & 1 & 0 & 0 & 0 & 0 & 0 & 0 & 0 & 0 & 0 & 0 \\
0.707 & 0 & 0 & 0 & 0 & 0 & 0 & 0 & 0 & 1 & 0 & 0 \\
-0.707 & 0 & 0 & 1 & 0 & 0 & 0 & 0 & 0 & 0 & 0 & 0 \\
-0.707 & 0 & -1 & 0 & 0 & 0 & 0 & 0 & 0 & 0 & 0 & 0 \\
0 & -1 & 0 & 0 & 0.707 & 1 & 0 & 0 & 0 & 0 & 0 & 0 \\
0 & 0 & 1 & 0 & 0.707 & 0 & 0 & 0 & 0 & 0 & 0 & 0 \\
0 & 0 & 0 & -1 & -0.707 & 0 & 0 & 0.707 & 0 & 0 & 0 & 0 \\
0 & 0 & 0 & 0 & -0.707 & 0 & -1 & -0.707 & 0 & 0 & 0 & 0 \\
0 & 0 & 0 & 0 & 0 & -1 & 0 & 0 & 1 & 0 & 0 & 0 \\
0 & 0 & 0 & 0 & 0 & 0 & 1 & 0 & 0 & 0 & 0 & 0 \\
0 & 0 & 0 & 0 & 0 & 0 & 0 & -0.707 & -1 & 0 & -1 & 0 \\
0 & 0 & 0 & 0 & 0 & 0 & 0 & 0.707 & 0 & 0 & 0 & 1
\end{bmatrix}
\begin{Bmatrix}
F_{AB} \\ F_{AC} \\ F_{BC} \\ F_{BD} \\ F_{CD} \\ F_{CE} \\ F_{DE} \\ F_{DF} \\ F_{EF} \\ R_A \\ H_F \\ R_F
\end{Bmatrix}
=
\begin{Bmatrix}
0 \\ 0 \\ -1000 \\ 0 \\ 0 \\ 5000 \\ 0 \\ 0 \\ 0 \\ 0 \\ 0 \\ 0
\end{Bmatrix}
$$

This equation is effectively $[A]\{x\} = -\{y\}$ and so the solution for the unknown forces in column vector $\{x\}$ is given by

$$\{x\} = [A]^{-1}\{-y\}$$

In expanded form, the solution is

$$
\begin{Bmatrix}
F_{AB} \\ F_{AC} \\ F_{BC} \\ F_{BD} \\ F_{CD} \\ F_{CE} \\ F_{DE} \\ F_{DF} \\ F_{EF} \\ R_A \\ H_F \\ R_F
\end{Bmatrix}
=
\begin{Bmatrix}
-4243 \\ 3000 \\ 3000 \\ -4000 \\ 2828 \\ 1000 \\ 0 \\ -2828 \\ 1000 \\ 3000 \\ 1000 \\ 2000
\end{Bmatrix}
$$

*Equilibrium of frame sections*
Returning to the analysis of the framework in Fig. 1.15, the alternative to considering a joint as a free body in equilibrium is to consider the equilibrium of a single member or group of members. This may have advantages in effort compared with analysing the equilibrium at every joint. In Fig. 1.20 cutting along the line XX provides either a free body of member AB to the left of the section or a free body of the remainder of the frame to the right of the section. Section YY splits the frame into two parts either of which could be regarded as a free body the equilibrium of which would give the forces in BD, CD and CE. In isolating part of the frame by a section it must be remembered that there are only three equilibrium equations that can be written $(\Sigma F_x = \Sigma F_y = \Sigma M_z = 0)$; hence only three unknown forces can be found for a particular 'section' and free body. Considering

Fig. 1.20

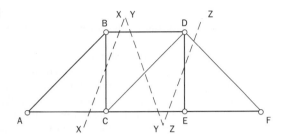

section XX then, as shown in Fig. 1.21, insert a force vector at the cut end of each member in the tensile sense.

Equilibrium for the free body requires that

$$\Sigma F_x = 0 \quad \Sigma F_y = 0 \quad \Sigma M_z = 0$$

Hence

$$\Sigma F_y = 3000 - F_{BC} = 0$$

$$F_{BC} = 3000 \text{ N}$$

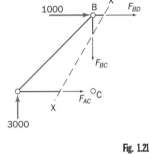

Fig. 1.21

Moments about A give $\Sigma M_z = (2 \times 1000) + 2F_{BD} + 2F_{BC} = 0$

$$F_{BD} = -4000 \text{ N} \quad \text{(compression)}$$

Finally, $\Sigma F_x = F_{BD} + F_{AC} + 1000 = 0$

$$F_{AC} = 3000 \text{ N}$$

These forces are the same as those obtained by taking equilibrium at each joint or using the matrix solution. The same results would have been

**Fig. 1.22** Free-body diagrams of parts of a fork-lift mechanism

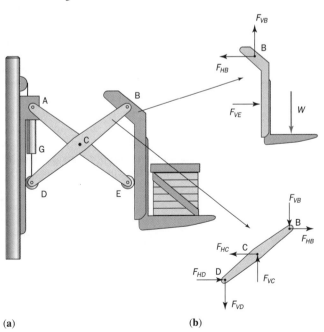

(a)                           (b)

obtained if the three equilibrium equations had been written for the remainder of the frame to the right of section XX.

**Other solid bodies and structures**

The principle of sectioning off a 'free body' and writing equilibrium statements, which was demonstrated for plane pin-jointed frames, is equally applicable in other engineering problems. Consider for example the front end of the fork-lift truck shown in Fig. 1.22(a). The individual elements of this may be analysed quite easily as illustrated in Fig. 1.22(b) so that each may be designed to withstand the forces acting on it.

**1.9  Forces by the method of tension coefficients**

This method was developed as a convenient way of solving for the forces in space frames, and it is of course also applicable to plane frames. It amounts to a shorthand notation for writing equilibrium equations of resolved force components at each joint.

The *tension coefficient* for a member is defined as the force, $F$, in the member divided by its length, $L$. Consider the member AB shown in Fig. 1.23; the resolved components of the force $F_{AB}$ in the co-ordinate directions $x$ and $y$ will be $F_{AB} \cos \alpha$ and $F_{AB} \sin \alpha$. Now, $\cos \alpha$ and $\sin \alpha$ can be expressed as the ratio of the length of the member projected onto the $x$- and $y$-axes divided by the length, $L_{AB}$, of the member; thus

$$\cos \alpha = \frac{x_{AB}}{L_{AB}} \quad \text{and} \quad \sin \alpha = \frac{y_{AB}}{L_{AB}}$$

Therefore

$$F_{AB} \cos \alpha = F_{AB} \frac{x_{AB}}{L_{AB}} = t_{AB} x_{AB}$$

and

$$F_{AB} \sin \alpha = F_{AB} \frac{y_{AB}}{L_{AB}} = t_{AB} y_{AB}$$

where $t_{AB}$ is the *tension coefficient* for the member.

Since the majority of configuration diagrams give dimensions which are related to the co-ordinate directions, it is simpler to express resolved components of force in terms of the tension coefficient and the projected length, rather than sines and cosines of angles. All members are assumed to be in tension initially and hence have a positive coefficient. If in the solution a tension coefficient turns out to be negative then this shows that the member is actually in compression. There is no need in this method to work out reactions at supports in advance as they can simply be included as unknowns in the equilibrium equations.

**Fig. 1.23**

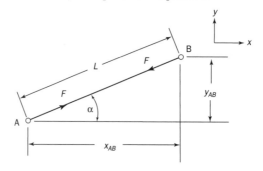

The first step is to choose a set of reference co-ordinate axes and directions and then to insert all the necessary reactions at the support points in the co-ordinate directions. The arrowhead direction is quite arbitrary. Now, commencing at any joint and considering one co-ordinate direction, write down the equilibrium equation. This will consist of the sum of the products of the tension coefficient $t$ and the projected length, say $x$ (for that co-ordinate direction), for each member at that joint plus any reaction or applied force at that point. It is important to remember that, taking the origin of the co-ordinate axes referenced at the joint, the projected lengths of the member *must* be given a sign appropriate to the positive and negative directions of the co-ordinate axes. Similarly, reactions and applied loads *must* also be given the appropriate sign relative to the axes. In the case of a space frame the above procedure must be repeated for each of the two other co-ordinate directions and then in turn for each joint.

Having obtained all the required equations these are solved for all the unknown tension coefficients and reactions. Each tension coefficient is then multiplied by the actual length of the member to give the force in the member. It will perhaps now be evident that, for clarity and ease of checking, a tabular system of solution is most desirable. The method will now be illustrated in the following example of a space frame.

**Example 1.4**

Use the method of tension coefficients to determine the reactions and the forces in the space frame shown in Fig. 1.24.

**Fig. 1.24**

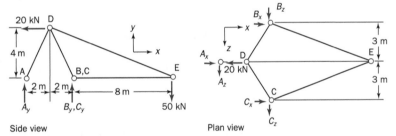

Side view          Plan view

The equilibrium equations for each joint are given in the following table.

| Eqn. no. | Joint/direction | Equilibrium equation |
|---|---|---|
| 1 | A/x | $A_x + 2t_{AD} = 0$ |
| 2 | A/y | $4t_{AD} + A_y = 0$ |
| 3 | A/z | $A_z = 0$ |
| 4 | B/x | $B_x - 2t_{BD} + 8t_{BE} = 0$ |
| 5 | B/y | $B_y + 4t_{BD} = 0$ |
| 6 | B/z | $B_z + 3t_{BD} + 3t_{BE} = 0$ |
| 7 | C/x | $C_x - 2t_{CD} + 8t_{CE} = 0$ |
| 8 | C/y | $C_y + 4t_{CD} = 0$ |
| 9 | C/z | $C_z - 3t_{CD} - 3t_{CE} = 0$ |
| 10 | D/x | $-20 - 2t_{AD} + 2t_{BD} + 2t_{CD} + 10t_{DE} = 0$ |
| 11 | D/y | $-4t_{DA} + 4t_{DB} - 4t_{DC} - 4t_{DE} = 0$ |
| 12 | D/z | $3t_{DC} - 3t_{DB} = 0$ |
| 13 | E/x | $-8t_{EB} - 8t_{EC} - 10t_{ED} = 0$ |
| 14 | E/y | $4t_{ED} - 50 = 0$ |
| 15 | E/z | $-3t_{EB} + 3t_{EC} = 0$ |

These equations may then be solved simultaneously to give the following forces and reactions:

| Member | Tension coefficient | Length (m) | Force (kN) | Reaction (kN) |
|--------|--------------------|-----------|-----------|---------------|
| AD | +20 | 4.48 | +89.6 | $A_x = -40$ |
| BD | −16.25 | 5.39 | −87.5 | $A_y = -80$ |
| CD | −16.25 | 5.39 | −87.5 | $A_z = 0$ |
| BE | −7.8 | 8.55 | −66.6 | $B_x = 30$ |
| CE | −7.8 | 8.55 | −66.6 | $B_y = 65$ |
| DE | +12.5 | 10.78 | +134.8 | $B_z = 72.4$ |
|  |  |  |  | $C_x = 30$ |
|  |  |  |  | $C_y = 65$ |
|  |  |  |  | $C_z = -72.4$ |

Earlier, in the analysis of plane frames, the concept of a matrix approach was introduced. Problems involving the method of tension coefficients may also be solved using matrix methods, although, once again, those unfamiliar with matrices may wish to skip this section. In this case the variables to be solved include the tension coefficient for each member and the unknown reactions. For the example above the matrix equation to be solved is

$$
\begin{bmatrix}
2 & 0 & 0 & 0 & 0 & 0 & 1 & 0 & 0 & 0 & 0 & 0 & 0 & 0 & 0 \\
4 & 0 & 0 & 0 & 0 & 0 & 0 & 1 & 0 & 0 & 0 & 0 & 0 & 0 & 0 \\
0 & 0 & 0 & 0 & 0 & 0 & 0 & 0 & 1 & 0 & 0 & 0 & 0 & 0 & 0 \\
0 & -2 & 8 & 0 & 0 & 0 & 0 & 0 & 0 & 1 & 0 & 0 & 0 & 0 & 0 \\
0 & 4 & 0 & 0 & 0 & 0 & 0 & 0 & 0 & 0 & 1 & 0 & 0 & 0 & 0 \\
0 & 3 & 3 & 0 & 0 & 0 & 0 & 0 & 0 & 0 & 0 & 1 & 0 & 0 & 0 \\
0 & 0 & 0 & -2 & 8 & 0 & 0 & 0 & 0 & 0 & 0 & 0 & 1 & 0 & 0 \\
0 & 0 & 0 & 4 & 0 & 0 & 0 & 0 & 0 & 0 & 0 & 0 & 0 & 1 & 0 \\
0 & 0 & 0 & -3 & -3 & 0 & 0 & 0 & 0 & 0 & 0 & 0 & 0 & 0 & 1 \\
-2 & 2 & 0 & 2 & 0 & 10 & 0 & 0 & 0 & 0 & 0 & 0 & 0 & 0 & 0 \\
-4 & -4 & 0 & -4 & 0 & -4 & 0 & 0 & 0 & 0 & 0 & 0 & 0 & 0 & 0 \\
0 & 3 & 0 & -3 & 0 & 0 & 0 & 0 & 0 & 0 & 0 & 0 & 0 & 0 & 0 \\
0 & 0 & -8 & 0 & -8 & -10 & 0 & 0 & 0 & 0 & 0 & 0 & 0 & 0 & 0 \\
0 & 0 & 0 & 0 & 0 & 4 & 0 & 0 & 0 & 0 & 0 & 0 & 0 & 0 & 0 \\
0 & 0 & 3 & 0 & -3 & 0 & 0 & 0 & 0 & 0 & 0 & 0 & 0 & 0 & 0
\end{bmatrix}
\begin{Bmatrix}
t_{AD} \\ t_{BD} \\ t_{BE} \\ t_{CD} \\ t_{CE} \\ t_{DE} \\ A_x \\ A_y \\ A_z \\ B_x \\ B_y \\ B_z \\ C_x \\ C_y \\ C_z
\end{Bmatrix}
=
\begin{Bmatrix}
0 \\ 0 \\ 0 \\ 0 \\ 0 \\ 0 \\ 0 \\ 0 \\ 0 \\ 20 \\ 0 \\ 0 \\ 0 \\ 50 \\ 0
\end{Bmatrix}
$$

This equation has the solution

$$\begin{Bmatrix} t_{AD} \\ t_{BD} \\ t_{BE} \\ t_{CD} \\ t_{CE} \\ t_{DE} \\ A_x \\ A_y \\ A_z \\ B_x \\ B_y \\ B_z \\ C_x \\ C_y \\ C_z \end{Bmatrix} = \begin{Bmatrix} 20 \\ -16.3 \\ -7.81 \\ -16.3 \\ -7.81 \\ 12.5 \\ -40 \\ -80 \\ 0 \\ 30 \\ 65 \\ 72.19 \\ 30 \\ 65 \\ -72.2 \end{Bmatrix}$$

It should be remembered that this equation must satisfy the requirements for a statically determine structure, eqn [1.10]. To assemble a matrix of this size manually is time consuming and error prone, but special purpose computer programs are available which can do this automatically, given a specification of the co-ordinates of the joints, the connectivity of the members, the restraints and the applied loads.

## 1.10 Bending of slender members

In the force analysis of frameworks the members were only subjected to axial force, namely tension or compression. The next step is to consider the effect of transverse loads acting on slender members. The deformation that results is termed *bending* and is of course very common in structures and machines – floor joists, railway axles, aeroplane wings, leaf springs, etc. External applied loads which cause bending give rise to internal reacting forces and moments. These have to be determined before it is possible to calculate stress and deflection.

The transverse externally applied load on a beam or bar can take one of two forms, concentrated or distributed. The former is illustrated in Fig. 1.25(*a*) in which the load acts on the surface of the beam along a line

Fig. 1.25

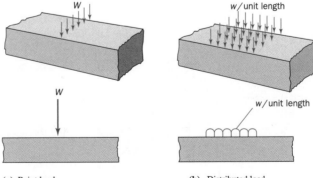

(a) Point load  (b) Distributed load

Fig. 1.26

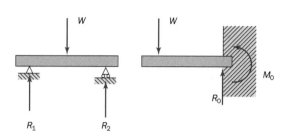

perpendicular to the longitudinal axis. This, of course, is an idealization, and in practice a concentrated load will cover a very short length of the beam.

A load which is distributed is shown in Fig. 1.25(b) and occupies a length of beam surface. The load intensity is always taken as constant across the beam thickness but may be uniformly or non-uniformly distributed along part or the whole length of the beam. In practice the particular conditions of force and displacement at beam supports may vary considerably. Theoretical solutions of beam problems generally employ two simplified forms of support. These are termed *simply supported* and *built-in* or *fixed*. The former is illustrated in Fig. 1.26(a) in which the beam rests on knife-edges or rollers. When the beam is loaded the support prevents vertical movement and there is a corresponding reaction, $R$, but since rotation at the supports is not prevented there is no restraining moment. Hence the deflection at the support is zero and the beam is free to take up a slope dictated by the applied load. The built-in support shown in Fig. 1.26(b) reacts with a transverse force and a moment. Thus both deflection and slope are fully restrained. The particular example illustrated of a beam built-in at one end and free at the other is termed a *cantilever*.

The number and type of supports also has a further important bearing on a beam solution by making it either *statically determinate* or *statically indeterminate* (Fig. 1.27). In the former the support reactions can be found simply from force and moment equilibrium equations. This applies, for example, to beams on two simple supports or one built-in support and no other. The two equilibrium equations for $F_y$ and $M_z$ are insufficient to find the reactions at the supports of a statically indeterminate beam owing to the presence of redundant forces. In this case it is necessary to consider also the

Fig. 1.27

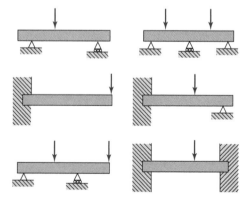

(a) Statically determinate          (b) Statically indeterminate

deflection of the beam in order to obtain additional equations to solve for the reactions.

## 1.11 Force and moment equilibrium during bending

A slender curved bar is shown in Fig. 1.28(*a*) subjected to various transverse loads. In general, to maintain equilibrium a force and a couple will be required at each end of the bar. The force can be resolved into two components, one perpendicular and the other parallel to each end cross-section. The internal forces are obtained by cutting the bar to give two free bodies and inserting at the cut sections the necessary forces and moments for equilibrium, as shown in Fig. 1.28(*b*). The couple, $M$, is termed a *bending moment*, the transverse force $Q$ is called a *shear force*, and $P$ is a *longitudinal force*. The most important stresses and deflections that occur during bending are due to $M$, rather than $Q$ or $P$.

**Fig. 1.28**

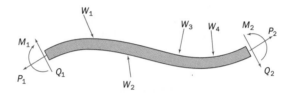

(**a**) Slender beam subjected to external forces and moments

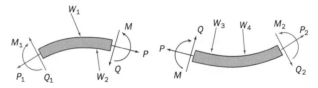

(**b**) Sections of beam showing internal forces and moments

## 1.12 Sign convention for bending

It is important that the analysis of internal forces in bending shall be consistent within and between various problems. This is best achieved by adopting a sign convention for loading, bending moment, shear force and distance/length. There is no standardized convention for bending, although there tends to be 'common practice' in, for example, civil engineering structural design. There is a fair degree of uniformity amongst textbooks and the convention used in this text (as in previous editions) follows the general pattern.

For horizontal beams, positive values are as shown in Fig. 1.29. There are several points worth noting; for example, loading $W$ and $w$ are taken positive downwards since for most horizontal members applied loading is in that sense. (If loading *upwards* was taken as positive there would be a profusion of negative signs.)

For horizontal beams, distance $x$ is *always* taken positive from left to right along the beam. Also the choice of $y$ positive downwards is again largely convenience and practice because most deflections tend to be downwards.

It is very important to remember that definitions of bending moment and shear force are illustrated as vector *pairs* in the positive (or negative) sense.

**Fig. 1.29**

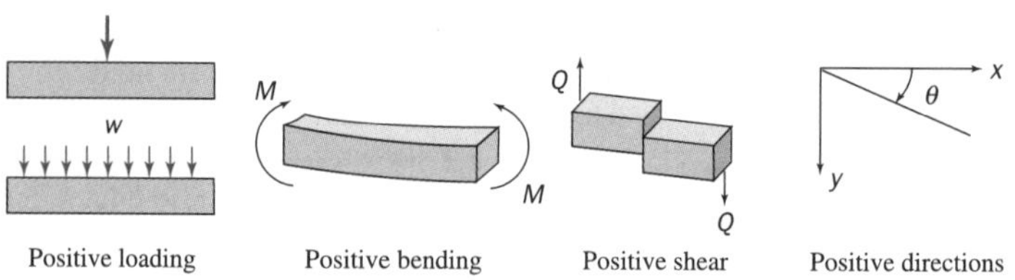

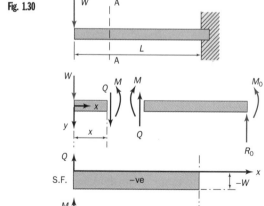

| | | | |
|---|---|---|---|
| Positive loading | Positive bending | Positive shear | Positive directions |

For example in the case of shear force, $Q$, the right-hand vector is taken as positive because it points in the positive $y$-direction and acts on a face which has its normal in the positive $x$-direction. The corresponding shear force on the other face is positive because it points in the negative $y$-direction and acts on a face whose normal is in the negative $x$-direction. It is important to note carefully this sign convention relating the nature of the shear force to the chosen $x$- and $y$-directions because the beam may not always be horizontal. A similar argument applies to the moment vectors, $M$.

The most important point to remember is *not* to change the sign convention (for some apparent convenience) in the middle of a problem solution.

### 1.13 Shear-force and bending-moment diagrams

Both stresses and deflections during bending are directly related to S.F. (Shear Force) and B.M. (Bending Moment); it is therefore desirable to know the distribution along the member and hence where maximum or minimum values occur. To this end S.F. and B.M. are computed for a number of cross-sections along the beam and diagrams plotted showing the distribution and magnitude. A few basic examples now follow and further illustrations of shear-force and bending-moment distributions appear at the start of Chapter 6 on bending stresses.

### 1.14 Cantilever carrying a concentrated load

This example is illustrated in Fig. 1.30. The first step is to choose any section AA at a distance $x$ from the left-hand end, cut the beam at this position and draw the free-body diagram, inserting the required internal

**Fig. 1.30**

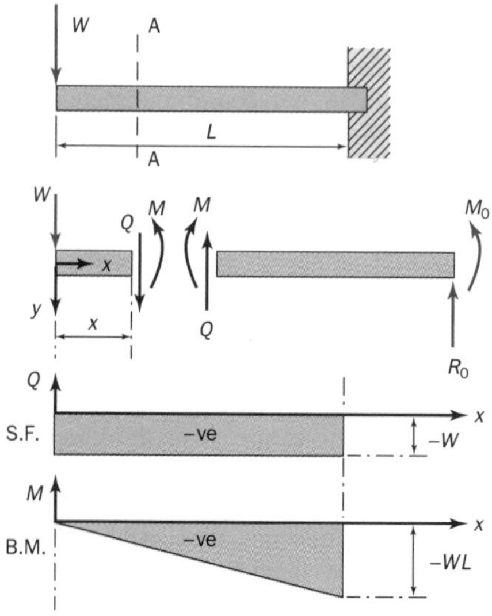

forces and moments in the positive sense according to the sign convention. It is advisable not to balance mentally the free body and put on the forces in what seems the 'right' sense. Note that in this first example, the free-body diagrams are shown for the cut sections to each side of AA. However, in the solution only the section to the left of AA is required.

*Shear force*  Vertical equilibrium for the left-hand free body gives

$$W + Q = 0$$

where $Q$ denotes the shearing force at a distance $x$ from the left end; hence

$$Q = -W$$

It is fairly obvious that this value of shear force is obtained at whatever distance the beam is cut between $x = 0$ and $x = L$.

*Bending moment*  Taking moments about the mid-point of section AA will give a moment equilibrium equation for the left-hand free body:

$$Wx + M = 0 \quad \text{or} \quad M = -Wx$$

This is a linear variation from $M = 0$ at $x = 0$ to $M = -WL$ at $x = L$, as shown in the diagram.

The reactions at the built-in end are, from the above,

$$R_0 = W \quad \text{and} \quad M_0 = -WL$$

Figure 1.30 illustrates the S.F. and B.M. diagrams which, in this example, are in the negative area below each zero base line.

If this problem had been set up so that the beam was supported at the left and the load applied at the right, then it would have been necessary to begin by determining $R_0$ and $M_0$ from overall equilibrium of the beam.

## 1.15  Cantilever carrying a uniformly distributed load

*Shear force*  Considering the cut section AA, the total downward load is $wx$ and vertical equilibrium of the free body in Fig. 1.31 is obtained as

$$wx + Q = 0 \quad \text{so that} \quad Q = -wx$$

**Fig. 1.31**

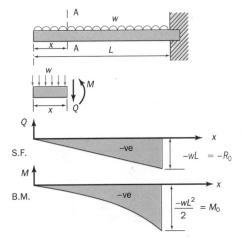

The shear-force diagram is therefore a linear variation from 0 to $-wL$ as shown in the figure.

*Bending moment*   The moment equilibrium equation is obtained by taking moments at AA; thus

$$wx\frac{x}{2} + M = 0 \quad \text{or} \quad M = -\frac{wx^2}{2} \tag{1.12}$$

which gives a parabolic shape of B.M. diagram varying from $M = 0$ at $x = 0$ to $M = -wL^2/2$ at $x = L$.

The support reactions are therefore

$$R_0 = wL \quad \text{and} \quad M_0 = -\frac{wL^2}{2}$$

### 1.16 Simply supported beam carrying a uniformly distributed load

In this case, Fig. 1.32, the reaction at the left-hand end has to be determined before the values of S.F. can be expressed. From vertical equilibrium and symmetry of the whole beam,

$$R_1 = R_2 = +\frac{wL}{2}$$

*Shear force*   For the equilibrium of the free-body section,

$$+Q - \frac{wL}{2} + wx = 0$$

$$Q = -wx + \frac{wL}{2}$$

This equation shows that the S.F. is zero at mid-span and equal to the reactions, $wL/2$, at $x = 0$ and $L$.

**Fig. 1.32**

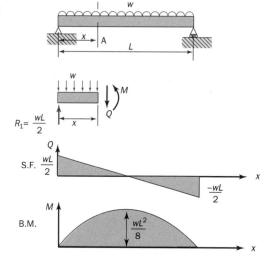

*Bending moment*    Moment equilibrium gives

$$M - \frac{wL}{2}x + wx\frac{x}{2} = 0$$

$$M = \frac{wLx}{2} - \frac{wx^2}{2} \qquad [1.13]$$

This equation represents a moment distribution which is parabolic, being zero at each support and having a maximum value of $wL^2/8$ at $x = L/2$.

## Example 1.5

**Sketch the S.F. and B.M. diagrams for the simply supported beam shown in Fig. 1.33.**

$$-R_A - R_B + 5000 + 10\,000 = 0 \quad \text{(force equilibrium)}$$

$$-5R_A + (5000 \times 4) - 5000 \times 2 \times 1 = 0 \quad \text{(moment equilibrium)}$$

$$R_A = 2000\,\text{N} \quad R_B = 13\,000\,\text{N}$$

Fig. 1.33

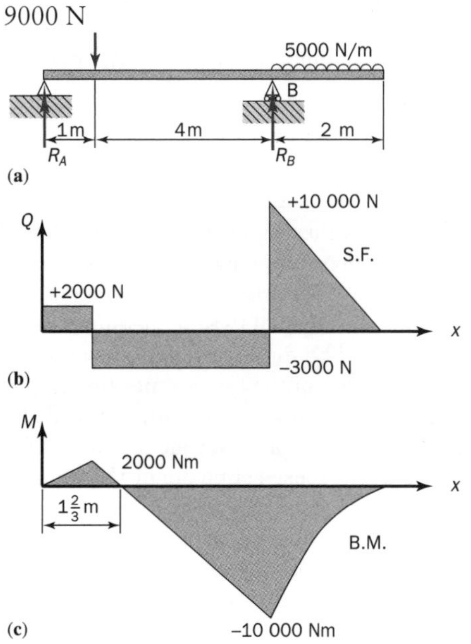

(a)

(b)

(c)

*Shear force*

| | | |
|---|---|---|
| $0 < x < 1$ : | $+Q - 2000 = 0$ | $Q = +2000\,\text{N}$ |
| $1 < x < 5$ : | $+Q - 2000 + 5000 = 0$ | $Q = -3000\,\text{N}$ |
| $5 < x < 7$ : | $+Q - 2000 + 5000 - 13\,000 + 5000(x - 5) = 0$ | $Q = -5000x + 35\,000\,\text{N}$ |

*Bending moment*

$0 < x < 1:$    $-2000x + M = 0$              $M = 2000x$

$1 < x < 5:$    $-2000x + 5000(x-1) + M = 0$      $M = 5000 - 3000x$

$5 < x < 7:$    $-2000x + 5000(x-1) - 13\,000(x-5)$

$$+5000(x-5)\frac{(x-5)}{2} + M = 0 \qquad M = -2500x^2 + 35\,000x - 122\,500$$

The S.F. and B.M. diagrams are shown in Fig. 1.33(*b*) and (*c*).

### 1.17 Point of contraflexure

In the above example it will be seen that the B.M. changes sign through a zero value and this is termed a *point of contraflexure*. Its position is determined by putting the bending-moment expression equal to zero. In real situations these points are of importance to the designer because, for example, supports placed at such points will not be subjected to bending moments.

### 1.18 Torsion of members

Torsion is the engineering word used to describe the process of twisting a member about its longitudinal axis as illustrated in Fig. 1.34. The angle of rotation of one section relative to another at a distance *l* along the member is termed the angle of twist $\theta$, measured in radians. The twist per unit length is thus $\theta/l$. The forces *F* required to cause this twist are shown at section B. The product of the forces times the distance to the axis is $2(F \cdot d/2)$ and this results in a moment about the axis called a *torque T*. For equilibrium of the member the applied torque at each end must be equal in magnitude and opposite in sense acting about the longitudinal axis.

The situation shown in Fig. 1.34 is described as pure torsion, i.e. the member is only subjected to torque and no other resultant lateral or longitudinal forces exist. However, in practice applied forces which set up torque frequently also induce bending and possibly end load on the component.

**Fig. 1.34**

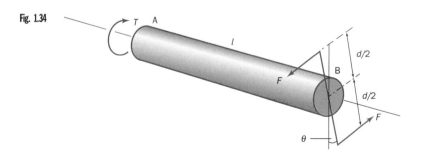

## 1.19 Members subjected to axial force, bending moment and torque

This situation was referred to at the end of the previous section and will now be considered in more detail.

Let us take a practical example with which many people have been personally involved, that of changing a wheel on a car. The wheel nuts have to be removed with a wheel brace or wrench. There are, of course, various designs to achieve the same objective which is to be able to apply enough torque to the nuts to loosen them and at a later stage to retighten them. A typical wheel brace is shown in Fig. 1.35 in which both hands apply force, one in the direction necessary to loosen the nut and the other as a reacting support. The torque set up on the nut is simply the force $F$ times the perpendicular distance between the plane in which the force is acting and the plane containing the nut. In addition, the brace is acting as a beam with a kink in it, i.e. simply supported at each end with a 'concentrated' load at the centre. In fact each section of the brace will have to transmit some combination of bending and torsion. In this particular case once the brace has been pushed on to the nut there will not thereafter be any axial force along the axis of the brace. To calculate the bending moment, shearing force and torque in each section of the brace, free-body diagrams must be drawn. If these are done correctly it is then a simple matter to determine the equilibrium conditions for each section. The following example will illustrate the method for slightly more complex internal reactions produced by the application of a single force.

Fig. 1.35

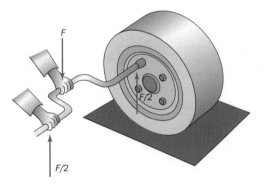

### Example 1.6

A length of pipe ABCD is connected rigidly to a pump unit at A and is bent through two right angles at B and C as shown in Fig. 1.36. A force $F$ has to be applied at D in order to connect the pipe to the next unit which is slightly out of alignment. Determine the system of forces, moments and torques which are set up in the pipe in relation to the co-ordinate axes given. AB is in the $x$-direction, BC in the $z$-direction and CD in the $y$-direction and the force $F$ lies in the $xz$-plane.

The first step is to resolve the force into the $x$- and $z$-directions, giving in each case $F\cos 45°$ or $F/\sqrt{2}$. To avoid confusion we shall temporarily call these $P$ and $R$ respectively. The next step is to draw the series of free-body diagrams as shown in Fig. 1.37, which should be self-explanatory, commencing with CD and successively following on with the segments CB and BA.

**Fig. 1.36**

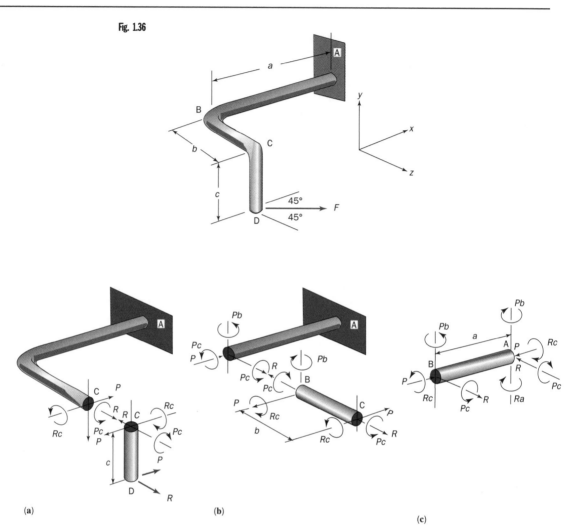

**Fig. 1.37**

It is seen that in each free body the transfer of the forces $P$ and $R$ to the next end of the segment requires a moment and/or torque to maintain equilibrium about the respective axes. This process is an *essential* first step towards the analysis of bending and torsional stresses and deflections which will be derived in Chapters 5, 6 and 7.

## 1.20 Torque diagram

In the analysis of torsion problems there is an analogous diagram to the shear-force and bending-moment diagrams for bending and this is a torque diagram. An example is shown in Fig. 1.38, which illustrates the variation in torque which may occur along a shaft depending on the various inputs and outputs of power transmission.

The values of $T_{AB}$, $T_{BC}$, etc., in the diagram may be determined from the free-body diagrams shown in Fig. 1.39.

A full analysis of torsion in shafts is given in Chapter 5.

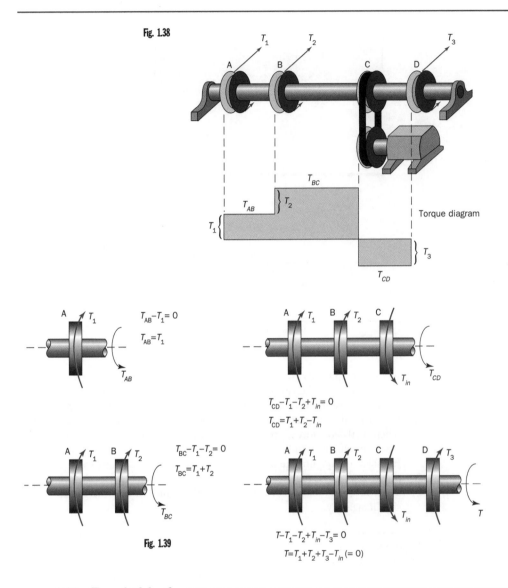

Fig. 1.38

$T_{AB} - T_1 = 0$

$T_{AB} = T_1$

$T_{BC} - T_1 - T_2 = 0$

$T_{BC} = T_1 + T_2$

Fig. 1.39

$T_{CD} - T_1 - T_2 + T_{in} = 0$

$T_{CD} = T_1 + T_2 - T_{in}$

$T - T_1 - T_2 + T_{in} - T_3 = 0$

$T = T_1 + T_2 + T_3 - T_{in} \,(= 0)$

## 1.21 The principle of superposition

Consider the beam problem in Fig. 1.40($a$). Although this would not be difficult to analyse for B.M. in its present form, imagine it split up into the two separate cases as in Fig. 1.40($b$) for which the B.M. diagrams for the two parts are as shown. Now, bending moment is always a first-order function of applied load, i.e. there are never terms in $W^2$ or $w^2$ etc.; therefore at any section along the beam the sum of the bending moments due to the loads acting separately would be exactly the same as the bending moment due to the combined load. Similarly the reactions at the supports due to the separate loads may be summed to give the reactions caused by the combined loading. The reader may also like to verify that the S.F. diagram could be derived in exactly the same manner.

The solution of the present example is obtained by summing the ordinates of bending moment at every section along the beam. This may also

**Fig. 1.40**

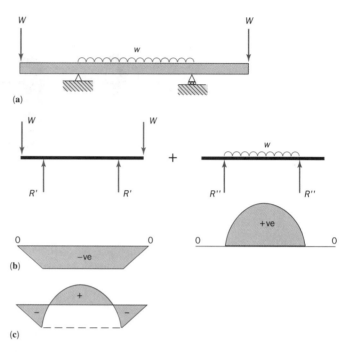

be done graphically by placing the two B.M. diagrams together as shown in Fig. 1.40(*c*) and the resultant is obtained as the shaded area.

This most important technique is known as the *principle of superposition*. It will be demonstrated at regular stages throughout subsequent chapters when analysing force, stress, strain and displacement systems. It is valid when the quantity to be determined is a linear function of the applied load. In effect, the quantity to be determined may be found by analysing each load or group of loads acting separately and then the results may be superimposed to obtain the total value due to all the loads acting simultaneously.

### 1.22  Summary

This chapter has served several purposes, the primary one being to emphasize the importance and give demonstrations of the development of *equilibrium situations* in solid bodies and structures. To this end it is also vital to be able to draw representative *free-body diagrams* with all the required force components. An appreciation of the meaning of *statical determinacy* and how to recognize such situations is also most important. The key equations for equilibrium are

$$\Sigma F_x = \Sigma F_y = \Sigma F_z = 0$$

$$\Sigma M_x = \Sigma M_y = \Sigma M_z = 0$$

Engineering components are subjected largely to *tension, compression, bending* and *torsion*, either separately or in a variety of combinations. It is essential, therefore, to understand how these various modes of deformation arise and the associated equilibrium requirements for internal and applied forces.

The *principle of superposition* has been shown to be a most valuable means of solution, and this will be illustrated further in succeeding chapters.

Finally, the use of spreadsheets has been introduced as a means of obtaining versatile, general solutions to engineering problems. This approach will be developed further throughout the book. For problems with a large number of variables, the convenience of matrix solutions has also been illustrated.

**Problems**

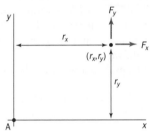

Fig. 1.41

1.1    A force $F$ has $x$ and $y$ components $F_x$ and $F_y$ respectively. The distances from a point A to any point on the line of action of $F$ and $r_x$ and $r_y$ in the $x$ and $y$ directions, Fig. 1.41. The moment about the $z$ axis of $F$ is therefore

$$M_z = r_x.F_y - r_y.F_x$$

Develop a spreadsheet where, given the $x$ and $y$ components of force and distance to a point on the line of action on the force, the resulting moment $M_z$ is calculated.

Using the formulae developed, find the componnts of the resultant force and moment of the system of forces in Fig. 1.42. Find the magnitudes of the resultant force and moment.

Fig. 1.42

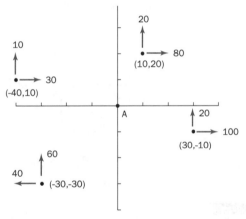

1.2    Modify the spreadsheet in Problem 1.1 so that the magnitude of the force and the angle it makes with the $x$ axis are entered and the $x$ and $y$ components are automatically calculated.

1.3    Using the formulae of Problem 1.2, find the resultant force and moment of the 15 kN force in Fig. 1.43. Find the moment due to a unit force acting at F3. From this find the magnitude of the force needed at F3 in order to balance the moment due to the 15 kN force.

Fig. 1.43

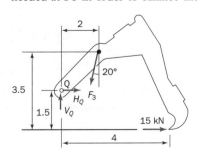

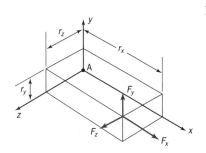

**Fig. 1.44**

1.4 A force $F$ has $x$, $y$ and $z$ components $F_x$, $F_y$ and $F_z$ respectively. The distances from a point A to any point on the line of action of $F$ are $r_x$, $r_y$ and $r_z$ in the $x$, $y$ and $z$ directions, Fig. 1.44. The moments about the $x$, $y$ and $z$ axes are therefore

$$M_x = r_y.F_z - r_z.F_y$$

$$M_y = r_z.F_x - r_x.F_z$$

$$M_z = r_x.F_y - r_y.F_x$$

Develop a spreadsheet where, given the $x$, $y$ and $z$ components of force and distance to a point on the line of action on the force, the resulting moments $M_x$, $M_y$ and $M_z$ are calculated. Use the spreadsheet to caculate the moments at point A in Fig. 1.36, when $F = 10\,\text{kN}$, $a = 1\,\text{m}$, $b = 0.6\,\text{m}$ and $c = 0.5\,\text{m}$.

*Note*: the sign of the displacements to go from A to a point on the line of action of $F$ is important.

the moments given in Fig. 1.37 are the reactions to force $F$ and are therefore opposite in sign to the moments caused by $F$.

1.5 Use the 3D moment spreadsheet from the previous question to find the moment of the applied forces at A and C about point B in Fig. 1.45. Sum the forces and moments to verify that the resultant force and moment of the forces at A, B and C is zero.

**Fig. 1.45**

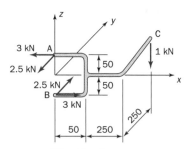

Dimensions in mm.

**Fig. 1.46**

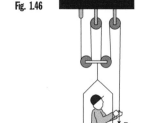

1.6 A maintenance cradle in a factory is supported by the pulley system in Fig. 1.46. A fitter weighing 82 kg can be raised (*a*) by pulling on the rope R himself or (*b*) by someone else pulling on the rope. Calculate the pull needed on the rope in each case. Friction and the weight of the cradle and pulleys may be ignored.

1.7 Figure 1.47 shows a tyical design for a vehicle weighbridge. Each of the levers A, B and C have *a:b* in the ratio of 1:10. If a balance load of

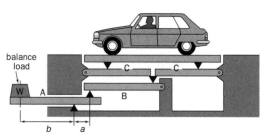

**Fig. 1.47**

12 N is required on the lever A then calculate the weight of the vehicle.

1.8   A certain design of brake drum is as shown in Fig. 1.48. If the drum rotates anticlockwise and generates a torque of 210 Nm, calculate the force needed in the cylinder C in order to stop the drum. The coefficient of friction between the drum and show is 0.35.

**Fig. 1.48**

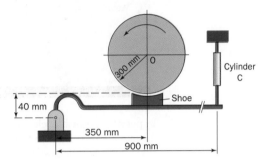

1.9   A wooden beam (SG = 0.6) is 15 cm by 15 cm by 4 m long and is hinged at A, as shown in Fig. 1.49. At what angle $\theta$ will the beam float in the water?

**Fig. 1.49**

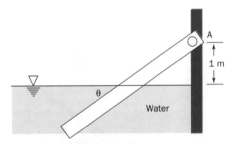

1.10  A 4 m wide gate for controlling water levels is shown in Fig. 1.50. For the levels shown calculate the magnitude and direction of the force in the hydraulic cylinder A. The density of the water is 1000 kg/m³.

**Fig. 1.50**

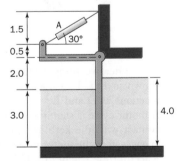

1.11  The door AB retains water in a channel as shown in Fig. 1.51. Determine the magnitude and direction of the force on the ground at B due to the pressure of the water. The weight of the door may be ignored.

**Fig. 1.51**

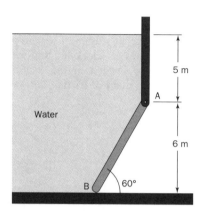

1.12  A special type of bicycle calliper brake is shown in Fig. 1.52. If it is pin-jointed at A, B, C and D and the cyclist applies a force of 300 N at A when the bicycle is stationary on a 1 in 7 slope, calculate whether or not the bicycle would move down the slope. The combined weight of the bicycle and cyclist is 102 kg and the coefficient of friction between the wheel and the brake block is 0.7. The outside diameter of the tyre is 690 mm and the diameter of the line of action of the brake block is 640 mm.

**Fig. 1.52**

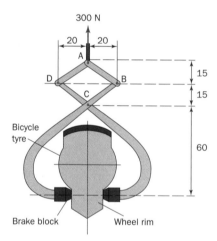

1.13  A cylinder weighing 800 N is supported between two rigid rods as shown in Fig. 1.53. Determine the tension in the cable AB and the reactions at D and E.

1.14  A motor cycle is standing on horizontal ground with its front wheel in contact with the kerb as shown in Fig. 1.54. When the rider slowly engages gear the motor cycle starts to rise over the kerb when the rear wheel torque reaches 0.3 kNm. For the instant when the front wheel just starts to leave the road calculate (a) the value of the coefficient of friction between the rear type and the road to avoid wheel slip, and (b) the magnitude and direction of the force on the front wheel. The wheel diameters are both 600 mm and the combined weght of the

Fig. 1.53

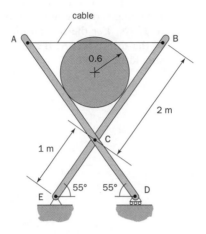

Fig. 1.54

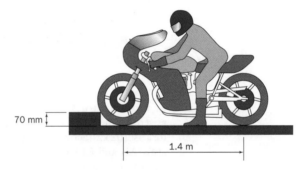

rider and the motor cycle is 285 kg acting through a point midway between the wheels.

1.15  At the moment of start-up, a power hacksaw is in the configuration shown in Fig. 1.55. If the torque input at the pulley drive is 150 Nm calculate the magnitude and direction of the force on the workpiece. The weight of the blade saddle is 45.9 kg and the weight of the support arm may be ignored. Friction between the blade saddle and the support arm may also be ignored.

Fig. 1.55

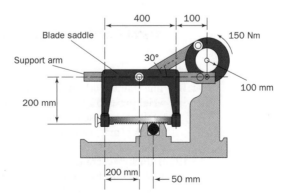

1.16 When the hydraulically-operated digger shown in Fig. 1.56 is removing earth the force on the bucket is horizontal wth a magnitude of 15 kN. For the arm positions indicated, calculate the force in each of the three rams A, B and C. The weight of the arms of the digger should be ignored.

**Fig. 1.56**

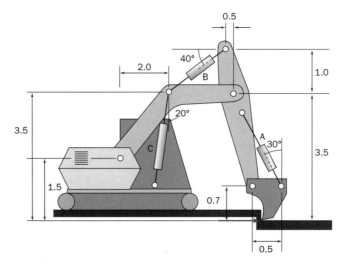

1.17 In a fork–life truck of the type shown in Fig. 1.57 the load of 3 kN is raised by means of the rotating screw AC. If the weight of the front fork is 102 kg (acting through its centroid G) and the weights of the arms AB and CD are 51 kg each, calculate the force in the screw AC and the reactions at A for the configuration shown.

**Fig. 1.57**

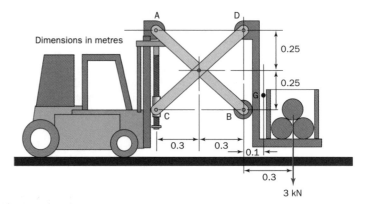

1.18 The frame work shown in Fig. 1.58 is used to support a steel car body weighing 200 kg. When the car body is suspended in (*a*) air and (*b*) totally immersed in a plating bath containing a liquid of density 1000 kg/m$^3$, calculate the support reactions and the forces in members AB, BF and FE. Density of steel = 7800 kg/m$^3$.

1.19 A dockyard crane may be considered to be a plane pin-jointed frame as shown in Fig. 1.59. Determine the support reactions and the forces in the members of the framework by resolution at joints.

Fig. 1.58

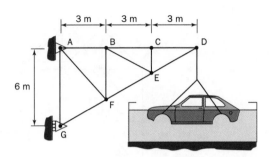

Fig. 1.59

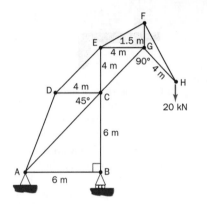

1.20   A plane pin-jointed frame is shown in Fig. 1.60. Determine the force
       in member DE when the frame is subjected to the forces shown.

Fig. 1.60

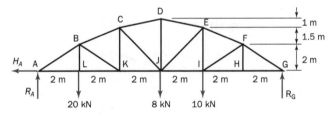

1.21   A crane with the dimensions shown in Fig. 1.61(*a*) supports a weight
       of 1000 kg. Use the method of sections to calculate the forces in
       members HG and HJ. If an alternative design is used as in (*b*) in
       which the cable is attached to the frame at F, calculate the force which
       this would cause in HJ.

1.22   Figure 1.62 shows a simple pin-jointed truss which must support a
       load $P$. Other known variables are the dimensions $b$ and $h$, the
       allowable stress in the bars $\sigma_0$ and the density of the bar material $\rho$.
           Program cells in a spreadsheet to calculate the following quantities,
       given numerical values for the variables above.
       (*a*)   Determine the forces in members AB and BC.
       (*b*)   If the members can support an allowable stress (force/unit area)
               of $\sigma_0$, determine the minimum cross-sectional area of each bar.
       (*c*)   Calculate the total weight of the truss, if both bars have a density
               $\rho$.

**Fig. 1.61**

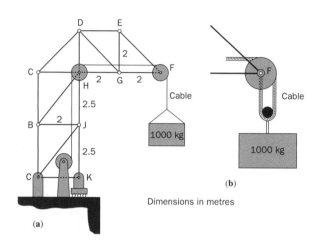

Cable

Cable

1000 kg

1000 kg

**(b)**

Dimensions in metres

**(a)**

**Fig. 1.62**

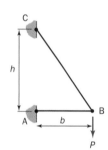

(d) Use the spreadsheet to evaluate numerical answers for parts (a)–(c), given $b = 1$ m, $h = \sqrt{2}$ m, $P = 1$ kN, $\sigma_0 = 100$ MN/m$^2$ and $\rho = 7800$ kg/m$^3$.

(e) Evaluate the total weight of the structure for values of h in the range 0.5–3 m to determine the minimum weight configuration.

1.23 Repeat problem 1.22 but use the same cross section for both members. In other words find the bigger cross sectional area and use that for both members in the subsequent calculation of weights. Hint, there is a spreadsheet function @ MAX which finds the largest value in a block of cells.

1.24 Calculate the forces in each of the members of the pin-jointed space frame shown in Fig. 1.63.

**Fig. 1.63**

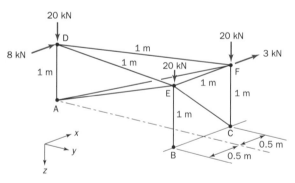

1.25 Use the method of tension coefficients to calculate the forces in the pin-jointed space frame shown in plan view in Fig. 1.64. The pinned supports A, B and C are at the same level, and DE is horizontal and at a height of 4 m above the supports.

1.26 Using the method of tension coefficients, determine the forces in each of the members of the framework illustrated in Fig. 1.65.

1.27 A semi-circular beam is subjected to a force of 2 kN as shown in Fig. 1.66. Draw diagrams to show the variation of axial force, shear force and bending moment in the beam and state the maximum values of each.

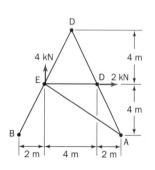

**Fig. 1.64**

**Fig. 1.65**

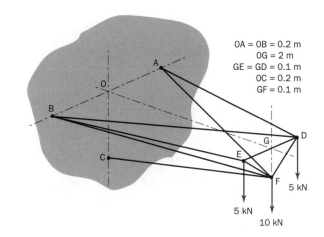

OA = OB = 0.2 m
OG = 2 m
GE = GD = 0.1 m
OC = 0.2 m
GF = 0.1 m

5 kN

5 kN

10 kN

**Fig. 1.66**

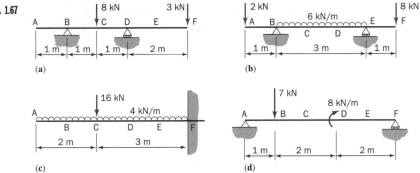

200 mm

α

2 kN                     2 kN

1.28 Sketch the shear-force and bending-moment diagrams, inserting the principal numerical values for the beams illustrated in Fig. 1.67. Also establish any positions of contraflexure.

**Fig. 1.67**

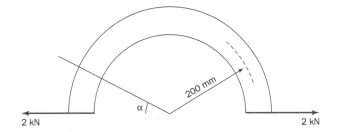

(a)

(b)

(c)

(d)

1.29 A 10 m × 6 m advertisement board is supported by three posts as in Fig. 1.68. Construct the shear-force and bending-moment diagrams for the vertical posts if the wind loading on the board is 981 N/m².

1.30 The bracket ABC is freely pivoted on the vertical rod shown in Fig. 1.69. Determine the forces, bending moments and torque transmitted on all parts of the bracket when a 1 kN vertical force is applied at C. Also calculate the reactions on the wall at A and B.

Fig. 1.68

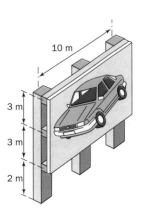

Fig. 1.69

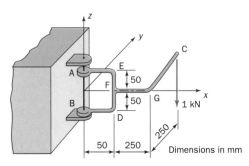

1.31 A power transmission system is illustrated in Fig. 1.70. The belt drives at A, C and D require 15 kW, 25 kW and 10 kW of power respectively at a shaft speed of 500 rev/min. Draw a diagram to show the torque in each section of the drive shaft.

Fig. 1.70

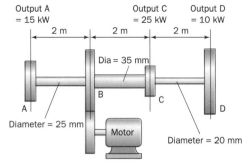

1.32 Use the principle of superposition to determine the B.M. diagram for the beam loaded as shown in Fig. 1.71.

Fig. 1.71

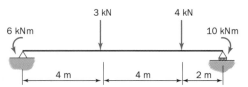

# 2 Statically Determinate Stress Systems

The effects of external applied forces can now be measured in terms of the internal reacting forces in a solid body or the members of a framework, as described in the previous chapter. However, at that stage no mention was made of the *cross-sectional size and shape* of the members. This aspect had no effect on the forces in the members, but conversely one should be able to describe quantitatively the way in which two members of different cross-sectional size would react to a particular value of force. This is done through the concept of *stress* and this chapter shows how stresses can be determined for simple engineering situations.

## 2.1 Stress: normal, shear and hydrostatic

Consider the member shown in Fig. 2.1(a) subjected to an external force, $F$, represented by the arrow at each end parallel to the longitudinal axis. The arrow simply represents a force *resultant* on the end faces and obviously the force is not actually applied solely along the line of the arrow. Similarly the internal force reaction does not act along just a single line but is transmitted throughout the bulk of material from grain to grain. If part of the member is cut off to give a free body as in Fig. 2.1(b) then equilibrium will be maintained by appropriate components of internal reacting force such as $\Delta F$ acting on elements of area $\Delta A$.

For equilibrium $\Sigma \Delta F = F$ and $\Sigma \Delta A = A$ the total cross-sectional area.

The internal *force per unit area* is called the stress, $\sigma$, and is given by

$$\sigma = \frac{\Delta F}{\Delta A}$$

In the limit,

$$\sigma = \lim_{\Delta A \to 0} \left( \frac{\Delta F}{\Delta A} \right) = \frac{\mathrm{d}F}{\mathrm{d}A}$$

So,

$$\mathrm{d}F = \sigma \, \mathrm{d}A$$

$$\int_0^F \mathrm{d}F = \int_{area} \sigma \, \mathrm{d}A$$

If it is assumed that the stress is constant over the area then

$$F = \sigma \int_{area} \mathrm{d}A = \sigma A$$

so that stress $\sigma = F/A$.

Note that this equation for direct or normal stress will only apply if the force acts through the centroid of the cross-section. If the applied force, $F$,

**Fig. 2.1**

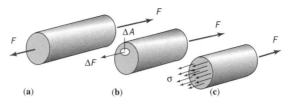

（a）　　　　　　（b）　　　　　　（c）

is offset then bending stresses will be set up along with the normal stress and the stress will no longer be constant across the material. This type of situation is considered in Chapter 6.

*Normal stress*
In the simple case in Fig. 2.1 the *average* direct stress is $F/A$ and will be denoted by the symbol $\sigma$ as in Fig. 2.1($c$). Normal or direct stress acts perpendicular to a plane and when acting outwards from the plane is termed *tensile stress* and given a *positive* sign. Stress acting towards a plane is termed *compressive stress* and is *negative* in sign. In order to denote the direction of a stress with respect to co-ordinate axes, a suffix notation is used, so that $\sigma_x$, $\sigma_y$, $\sigma_z$ represent the components of normal stress in the $x$-, $y$- and $z$-directions as shown in Fig. 2.2($a$).

**Fig. 2.2**

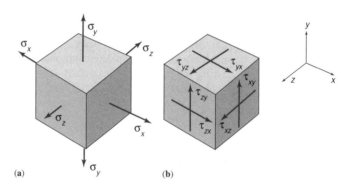

（a）　　　　　　　　　　　　（b）

*Shear stress*
When a man who is running wishes to slow down he applies his 'brakes'. This is achieved through mounting pressure between the soles of his shoes and the ground and thus increased frictional force parallel to the ground. This concept of a force applied tangential or parallel to a surface is termed a shear force. If internal reacting shear force is expressed as a force per unit area then it is termed a *shear stress*. It also acts parallel to any associated plane within the material and is denoted by the symbol $\tau$. A double suffix notation is required to define shear stresses with respect to co-ordinate axes. The first suffix gives the direction of the normal to the plane on which the stress is acting, and the second suffix indicates the direction of the shear stress component. Thus $\tau_{xy}$ is a shear stress acting on the $yz$-plane (the normal in the $x$-direction) and pointing in the $y$-direction. The sign convention associated with shear stress is defined as positive when the direction of the stress vector and the direction of the normal to the plane are both in the positive sense or both in the negative sense in relation to the directions of the co-ordinate axes. If the directions of shear stress and the normal to the

**Fig. 2.3**

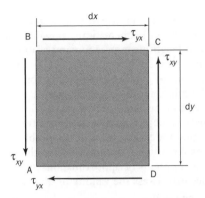

plane are opposite in sign then the shear stress is negative. There are twelve possible shear stress components in a three-dimensional stress system as indicated in Fig. 2.2(*b*) (those on the obscured faces have been omitted for clarity).

There is a further condition of shear stresses which always exists and is explained as follows. Consider the element of unit thickness in Fig. 2.3 subjected to shearing stresses along its edges $\tau_{xy}$ and $\tau_{yx}$. Vertical and horizontal force equilibrium shows that $\tau_{xy}$ on face AB is equal to $\tau_{xy}$ on face CD. Also, $\tau_{yx}$ on face AD is equal to $\tau_{yx}$ on face BC. The shear *forces* along the sides AB and CD are $(\tau_{xy} \times 1 \times \mathrm{d}y)$ and along AD and BC are $(\tau_{yx} \times 1 \times \mathrm{d}x)$. Taking moments about a $z$-axis through the centre of the element, for equilibrium

$$2(\tau_{xy} \times 1 \times \mathrm{d}y) \times \frac{\mathrm{d}x}{2} - 2(\tau_{yx} \times 1 \times \mathrm{d}x) \times \frac{\mathrm{d}y}{2} = 0$$

or

$$\tau_{xy} = \tau_{yx}$$

These are termed *complementary shear stresses*. Thus a shear stress on one plane is *always* accompanied by a complementary shear stress of the same sign and magnitude on a perpendicular plane.

*Hydrostatic stress*
This is a special state of direct stress which should be mentioned now although its importance will become more evident at later stages. Hydrostatic stress may be represented by the stress set up in a body immersed at a great depth in a fluid. The external applied pressure $p$ being equal at all points round the body gives rise to internal reacting compressive force and hence compressive stress which is equal in all directions, i.e. $\sigma_x = \sigma_y = \sigma_z = -p$.

## 2.2 A statically determinate stress system

If the stresses within a body can be calculated purely from the conditions of equilibrium of the applied loading and the internal forces then the problem is said to be statically determinate. There are very few examples of this nature; however, they do give further illustration of the application of equilibrium, and the remainder of the chapter will be devoted to solutions in this category.

## 2.3 Assumptions and approximations

Exact solutions for stress, displacement, etc., in real engineering problems are not always mathematically possible and even those that are possible can involve lengthy computation and advanced mathematical techniques which are not necessarily justifiable. This is because we seldom know the exact conditions of applied loading on a component or structure for its expected working life, and the materials used are not wholly predictable in behaviour. It therefore becomes necessary and desirable in most engineering problems to make some simplifying approximations and assumptions which, while not changing the basic nature of the problem, will allow a simpler solution and an answer which is not too far from the truth. It is important, however, that any assumptions or approximations are clearly stated at the start so that the reader may assess the validity of the answer in respect of what might be the exact solution.

Some of the problems to follow are in general not statically determinate but, with some realistic geometrical limitations, they can be solved purely from equilibrium conditions to give answers which, although not exact, are reasonably accurate for the purposes of engineering design.

## 2.4 Tie bar and strut (or column)

These are the simplest examples of statically determinate stress situations, since the equilibrium condition is simply that the external force at the ends of the member must be balanced by the internal force, which is the average stress multiplied by the cross-sectional area.

The case of the tie bar is illustrated in Fig. 2.4. The bar at ($a$), subjected to tension, has been cut perpendicular to the axis into two free bodies to show the normal stress $\sigma_x$. This, multiplied by the area on which it acts, must be in equilibrium with the applied force, $F$; hence

$$\sigma_x A = F \quad \text{or} \quad \sigma_x = \frac{F}{A} \quad \text{(tensile stress)}$$

**Fig. 2.4**

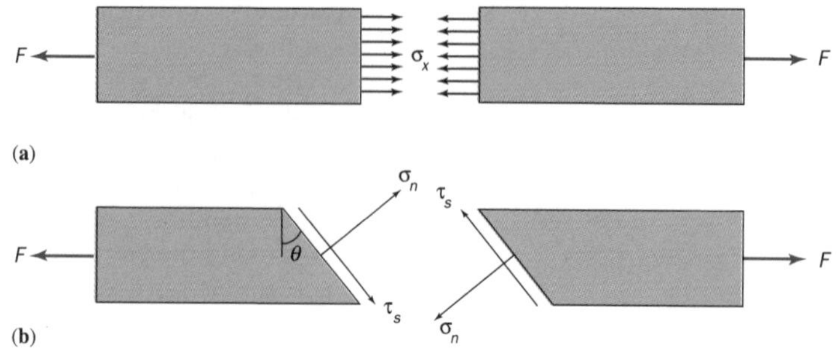

(a)

(b)

In Fig. 2.4($b$) it is seen that if the bar is cut into two free bodies at an angle to the axis then there will be two components of stress; one is normal to the plane $\sigma_n$, and the other is parallel to the plane $\tau_s$. The significance of these stress components is dealt with in detail in Chapter 12.

A similar reasoning to that above applies to the compression situation in the strut (column).

**Example 2.1**

**Figure 2.5(a) shows the general arrangement of an engine piston and connecting rod. The top surface of the piston is loaded by gas pressure. The resulting force is transmitted through the pin-jointed connecting rod to a crankshaft creating power. A heavy flywheel on the crankshaft smooths the variations in torque due to the rapidly changing gas pressure so that the angular velocity $\omega$ of the crank is effectively constant. The aim is to determine the minimum cross-sectional area of the connecting rod so that the maximum stress does not exceed a specified value.**

**Fig. 2.5**

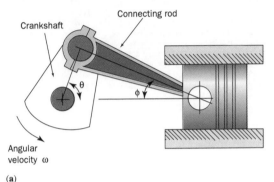

(a)

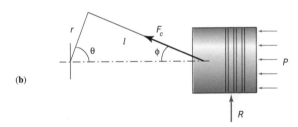

(b)

The pressure $p$ on the surface of a piston of bore $b$ creates a resultant force $F_g$ of

$$F_g = pA = \frac{p\pi b^2}{4}$$

The gas force is resisted by the force in the connecting rod $F_c$ and the side reaction $R$ from the cylinder wall, Fig. 2.5($b$). Horizontal force equilibrium requires that

$$-F_c \cos \phi - \frac{p\pi b^2}{4} = 0$$

so the conrod force is

$$F_c = -\frac{p\pi b^2}{4 \cos \phi}$$

If a stress of magnitude $\sigma_0$ is not to be exceeded, the cross–sectional area required is

$$A = \frac{\text{magnitude}(F_c)}{\sigma_0}$$

**Fig. 2.6**

|  | A | B | C | D | E | F | G |
|---|---|---|---|---|---|---|---|
| 1 | r | l | omega |  | BORE |  |  |
| 2 | (m) | (m) | (rpm) | (rad/s) | (m) |  |  |
| 3 | 0.015 | 0.055 | 15000 | +C3*2*@P/60 | 0.04 |  |  |
| 4 |  |  |  |  |  |  |  |
| 5 |  |  |  |  |  |  |  |
| 6 |  |  |  | Gas pressure | Gas force | Conrod force |  |
| 7 | theta | theta | phi | p | Fg | Fc |  |
| 8 | (degrees) | (radians) | (radians) | (Pa) | (N) | (N) |  |
| 9 |  |  |  |  |  |  |  |
| 10 | 0 | @RADIANS(A10) | @ASIN($A$3*@SIN(B10/$B$3) | 1145900 | +D10*@PI*$E$3^2/4 | –E10/@COS(C10) |  |
| 11 | 30 | @RADIANS(A11) | @ASIN($A$3*@SIN(B11/$B$3) | 1717800 | +D11*@PI*$E$3^2/4 | –E11/@COS(C11) |  |
| 12 | 60 | @RADIANS(A12) | @ASIN($A$3*@SIN(B12/$B$3) | 660100 | +D12*@PI*$E$3^2/4 | –E12/@COS(C12) |  |
| 13 | 90 | @RADIANS(A13) | @ASIN($A$3*@SIN(B13/$B$3) | 317800 | +D13*@PI*$E$3^2/4 | –E13/@COS(C13) |  |
| 14 | 120 | @RADIANS(A14) | @ASIN($A$3*@SIN(B14/$B$3) | 130900 | +D14*@PI*$E$3^2/4 | –E14/@COS(C14) |  |
| 15 | 150 | @RADIANS(A15) | @ASIN($A$3*@SIN(B15/$B$3) | 25900 | +D15*@PI*$E$3^2/4 | –E15/@COS(C15) |  |
| 16 | 180 | @RADIANS(A16) | @ASIN($A$3*@SIN(B16/$B$3) | 0 | +D16*@PI*$E$3^2/4 | –E16/@COS(C16) |  |
| 17 | 210 | @RADIANS(A17) | @ASIN($A$3*@SIN(B17/$B$3) | 700 | +D17*@PI*$E$3^2/4 | –E17/@COS(C17) |  |
| 18 | 240 | @RADIANS(A18) | @ASIN($A$3*@SIN(B18/$B$3) | 14000 | +D18*@PI*$E$3^2/4 | –E18/@COS(C18) |  |
| 19 | 270 | @RADIANS(A19) | @ASIN($A$3*@SIN(B19/$B$3) | 71400 | +D19*@PI*$E$3^2/4 | –E19/@COS(C19) |  |
| 20 | 300 | @RADIANS(A20) | @ASIN($A$3*@SIN(B20/$B$3) | 188900 | +D20*@PI*$E$3^2/4 | –E20/@COS(C20) |  |
| 21 | 330 | @RADIANS(A21) | @ASIN($A$3*@SIN(B21/$B$3) | 553700 | +D21*@PI*$E$3^2/4 | –E21/@COS(C21) |  |
| 22 | 360 | @RADIANS(A22) | @ASIN($A$3*@SIN(B22/$B$3) | 1145900 | +D22*@PI*$E$3^2/4 | –E22/@COS(C22) |  |
| 23 |  |  |  |  |  |  |  |
| 24 |  |  |  |  |  | @MAX(F(10 .. F22) | Maximum Conrod force |
| 25 |  |  |  |  |  | @MIN(F(10 .. F22) | Minimum Conrod force |
| 26 |  |  |  |  |  |  |  |
| 27 |  |  |  |  |  | 200000000 | Max stress |
| 28 |  |  |  |  |  |  |  |
| 29 |  |  |  |  |  | @ABS (F24/F27) | Area required for max force |
| 30 |  |  |  |  |  | @ABS (F25/F27) | Area required for min force |

(**a**) Data and cell formulae

|  | A | B | C | D | E | F | G |
|---|---|---|---|---|---|---|---|
| 1 | r | l | omega |  | BORE |  |  |
| 2 | (m) | (m) | (rpm) | (rad/s) | (m) |  |  |
| 3 | 0.015 | 0.055 | 15000 | 1571 | 0.04 |  |  |
| 4 |  |  |  |  |  |  |  |
| 5 |  |  |  |  |  |  |  |
| 6 |  |  |  | Gas pressure | Gas force | Conrod force |  |
| 7 | theta | theta | phi | p | Fg | Fc |  |
| 8 | (degrees) | (radians) | (radians) | (Pa) | (N) | (N) |  |
| 9 |  |  |  |  |  |  |  |
| 10 | 0 | 0.000 | 0.000 | 1145900 | 1440 | -1440 |  |
| 11 | 30 | 0.524 | 0.137 | 1717800 | 2159 | -2179 |  |
| 12 | 60 | 1.047 | 0.238 | 660100 | 830 | -854 |  |
| 13 | 90 | 1.571 | 0.276 | 317800 | 399 | -415 |  |
| 14 | 120 | 2.094 | 0.238 | 130900 | 164 | -169 |  |
| 15 | 150 | 2.618 | 0.137 | 25900 | 33 | -33 |  |
| 16 | 180 | 3.142 | 0.000 | 0 | 0 | 0 |  |
| 17 | 210 | 3.665 | -0.137 | 700 | -1 | 1 |  |
| 18 | 240 | 4.189 | -0.238 | 14000 | 18 | -18 |  |
| 19 | 270 | 4.712 | -0.276 | 71400 | 90 | -93 |  |
| 20 | 300 | 5.236 | -0.238 | 188900 | 235 696 1440 | -242 |  |
| 21 | 330 | 5.760 | -0.137 | 553700 |  | -702 |  |
| 22 | 360 | 6.283 | -0.000 | 1145900 |  | -1440 |  |
| 23 |  |  |  |  |  |  |  |
| 24 |  |  |  |  |  | 1 | Maximum Conrod force |
| 25 |  |  |  |  |  | -2179 | Minimum Conrod force |
| 26 |  |  |  |  |  |  |  |
| 27 |  |  |  |  |  | 200000000 | Max stress |
| 28 |  |  |  |  |  |  |  |
| 29 |  |  |  |  |  | 4.43970163E-09 | Area required for max force |
| 30 |  |  |  |  |  | 1.08950278E-05 | Area required for mi x force |

(**b**) Spreadsheet display

The angle $\phi$ can be determined as a function of crank angle $\theta$ from

$$r \sin \theta = l \sin \phi$$

or

$$\phi = \sin^{-1}\left(\frac{r}{l} \sin \theta\right)$$

In a small two-stroke chain-saw engine, the variation of gas pressure with crank angle is shown in the fourth and first columns respectively of a spreadsheet, Fig. 2.6(*a*). The appropriate formulae for the calculation of the angle $\phi$, gas pressure force and conrod force have been programmed into the first row underneath the appropriate titles. These formulae have then been copied to the rows beneath to calculate the quantities of interest for other angles $\theta$.

The only special point to note is that the location of the cell containing the cylinder bore, cell E3, should be the same in all rows when calculating the gas force. Normally when a cell is copied the cell addresses are automatically adjusted, so that if a formula +D10*@PI*E3^2/4 in cell E10 is copied to cell E11 the result is +D11*@PI*E4^2/4. In this case the reference to cell E3 should remain the same in all rows. This is indicated by marking it as an *absolute address* by pressing an appropriate function key when entering the cell, or by typing the address as $E$3.

The resulting spreadsheet display is shown in Fig. 2.6(*b*). From this data, it is very straightforward to create an *XY* graph of the variation of conrod force with angle $\theta$ as shown in Fig. 2.7. It can be seen that for the stroke $r$ and control length $l$ chosen here, the angle $\phi$ remains quite small and the gas pressure force and the conrod force are of similar magnitude. The maximum and minimum values of conrod force can be extracted with the appropriate functions @MIN and @MAX and the required cross-sectional areas calculated. Using the spreadsheet, the effect of varying any of the parameters such as cylinder bore, cylinder stroke or conrod length can immediately be evaluated.

This analysis ignores the inertia forces required to accelerate and decelerate the piston as it moves up and down the bore. For a chain-saw

**Fig. 2.7**

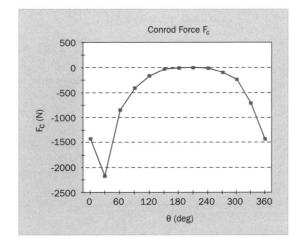

engine under no load the crank speed can approach $14\,000$ rev/min and inertial effects are very important.

### 2.5 Suspension-bridge cables

A common form of loading on a cable, for example in suspension bridges, is shown in Fig. 2.8(a). The loading, $w$ per unit length, is distributed uniformly on a *horizontal base*, the weight of the cable being neglected. In this particular example, the ends A and B are set at different heights above

**Fig. 2.8**

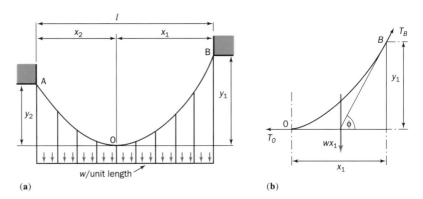

(a)                                                    (b)

the lowest point. It is useful for the analysis to cut the cable at O, to insert a reaction at that point, and to consider the equilibrium of the right-hand part of the cable. The free-body diagram in Fig. 2.8(b) shows that equilibrium is satisfied by the triangle of forces $T_B$, $T_O$ and $wx_1$. The position of the lowest point O and hence the distance $x_1$ is not known. The distance $x_1$ can be determined by equilibrium of moments of the forces for either part of the cable. Thus, taking moments about B, and noting that the force $wx_1$ can be taken as acting through the mid-point of $x_1$, then

$$T_O y_1 - wx_1 \frac{x_1}{2} = 0 \qquad [2.1]$$

You should now draw the free-body diagram for the left-hand part. Taking moments about A gives

$$T_O y_2 - w(l - x_1)\frac{(l - x_1)}{2} = 0 \qquad [2.2]$$

Inspection of eqn. [2.1] shows that, for any point on the cable having co-ordinates $(x, y)$ relative to O,

$$y = \frac{wx^2}{2T_O} \qquad [2.3]$$

which is an equation for a parabola.

Returning to eqns. [2.1] and [2.2] and eliminating $T_O$ gives

$$\frac{y_1}{y_2} = \frac{x_1^2}{(l - x_1)^2}$$

from which

$$x_1 = \frac{l(y_1/y_2)^{1/2}}{1 + (y_1/y_2)^{1/2}}$$   [2.4]

If the ends A and B are at the same level then $y_2 = y_1$ and $x_1 = l/2$.

The maximum tension in the cable will naturally occur at end B since this side of the cable supports the greater proportion of the load. The force $T_B$ can be determined from equilibrium of the triangle of forces $T_O$, $wx_1$ and $T_B$. Vertical equilibrium:

$$T_B \sin \phi - wx_1 = 0$$

But, from the geometry of the triangle,

$$\sin \phi = \frac{y_1}{[y_1^2 + (x_1/2)^2]^{1/2}}$$

Therefore

$$T_B = wx_1 \frac{[y_1^2 + (x_1/2)^2]^{1/2}}{y_1}$$   [2.5]

Similarly,

$$T_A = w(l - x_1) \frac{\{y_2^2 + [(l - x_1)/2]^2\}^{1/2}}{y_2}$$   [2.6]

where $x_1$ is given by eqn. [2.4].

The minimum tension in the cable is $T_O$, which is obtained from eqn. [2.1], by substituting the values of $x_1$ and $y_1$. Any required stress values are determined by dividing the appropriate tension by the cross-sectional area, $a$, of the cable, since it is assumed that the stress is uniformly distributed across the section and thus $\sigma = T/a$.

## 2.6  Thin ring or cylinder rotating

If a cylinder or ring is rotating at constant velocity, Fig. 2.9(a), then an *inward* radial component of force is required to provide the centripetal acceleration on each element of material. This *inward* force may be resolved into the tangential direction at each end of a typical element as shown in Fig. 2.9(b), which is thus subjected to circumferential tensile stress. This may be determined from a consideration of dynamical equilibrium. Alternatively we can reduce this dynamical situation to a static one by applying D'Alembert's principle, in which forces to accelerate masses are replaced by equal and opposite static forces. In this case the centripetal force is replaced by what is often termed a centrifugal force acting radially outwards, and we can then apply statical equilibrium.

Consider the rim of a wheel (no spokes) or a slice through a thin-walled cylinder each rotating about a central axis at velocity $\omega$. If this problem is to be statically determinate then the diameter of the ring or cylinder must be large, say greater than ten times the cross-sectional dimensions of the rim. It is then possible to assume a uniform distribution of stress over the cross-section in the circumferential direction, and in the radial and axial directions the stresses can be taken as zero or negligible.

**Fig. 2.9**

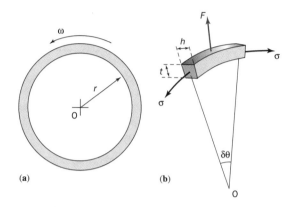

(a)    (b)

A small element of arc of the ring of cross-sectional area $A$ $(= th)$ rotating at uniform velocity $\omega$ is shown in Fig. 2.9($b$). The forces acting on the element are the radial inertia force $F$ and the circumferential tensile stress $\sigma$ acting over the area $A$. The resolved component of the radial forces inwards is

$$2\sigma th \ \sin\frac{\delta\theta}{2} \simeq 2\sigma th\frac{\delta\theta}{2} \quad \text{for small values of } \delta\theta$$

The mass of element is $\rho thr\delta\theta$ where $\rho$ is the mass per unit volume.

The radial centrifugal force, $F = (\rho thr\,\delta\theta)\omega^2 r$. For radial equilibrium, the resolved inward radial force must balance the outward force $F$; hence

$$2\sigma th\frac{\delta\theta}{2} - F = 0$$

$$\sigma th\,\delta\theta - \rho thr^2\omega^2\delta\theta = 0$$

and

$$\sigma = \rho\omega^2 r^2$$

$$= \rho v^2 \tag{2.7}$$

where $v$ is the tangential velocity of the central point of the element cross-section.

It should be noted that the tensile stress is independent of the shape and area of the cross-section of the rim.

An important practical example of the above effect of centrifugal action is the stress set up in the blades of gas turbine rotors which rotate at very high speed.

## 2.7 Thin shells under pressure

Thin-walled shells are used extensively in engineering in two principal forms: ($a$) made of reinforced concrete to form part of a civil engineering structure and stressed partly by self-weight and partly by the environment, namely wind, snow, etc.; ($b$) made of metal and used generally in engineering as storage containers for liquid, powder, gas, etc. Stresses will arise due to, say, uniform internal liquid or gas pressure, e.g. in a steam boiler, or pressure due to weight of liquids or solids contained.

In general, a shell of arbitrary wall thickness subjected to pressure contains a three-dimensional stress system. The stresses are perpendicular to the thickness and in two principal orthogonal directions tangential to the surface geometry. Each of these stresses has a non-uniform distribution through the thickness of material, and the solution is statically indeterminate. If, however, the wall thickness is less than about one-tenth of the principal radii of curvature of the shell, the variation of tangential stresses through the wall thickness is small and the radial stress may be neglected. The solution can then be treated as statically determinate.

In the present treatment the following additional simplifying assumptions will be made: that the shell acts as a membrane and does not provide bending resistance so that there are only uniform direct stresses present, which are often called *membrane* stresses; that the shell is formed by a surface of revolution and there are no discontinuities or sharp bends.

Consider first a general axi-symmetrical shell, Fig. 2.10(*a*), from which is cut an element bounded by two meridional lines and two lines perpendicular to the meridians as at (*b*). The notation is as follows:

$\sigma_1$ = tensile stress in meridional direction, meridional stress
$\sigma_2$ = tensile stress along a parallel circle, hoop stress
$r_1$ = meridional radius of curvature
$r_2$ = radius of curvature perpendicular to the meridian
$t$ = wall thickness

The forces on the edges of the element are, in the meridional direction, $\sigma_1 t \, ds_2$ and, in the perpendicular direction, $\sigma_2 t \, ds_1$. These forces have components inwards towards their centres of curvature. The radial components are $2\sigma_1 t \, ds_2 \sin(d\theta_1/2)$ and $2\sigma_2 t \, ds_1 \sin(d\theta_2/2)$. For small values of $d\theta$ the total radial force becomes

$$\sigma_1 t \, ds_2 d\theta_1 + \sigma_2 t \, ds_1 d\theta_2$$

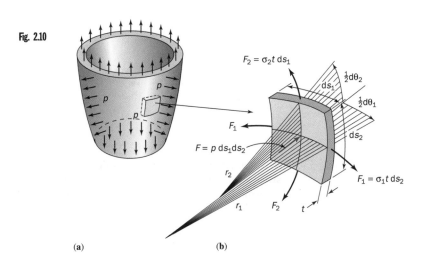

Fig. 2.10

(a)  (b)

or

$$\sigma_1 t \, ds_2 \frac{ds_1}{r_1} + \sigma_2 t \, ds_1 \frac{ds_2}{r_2}$$

The radial force due to the pressure $p$ acting over the surface area of the element is

$$p \, ds_1 ds_2$$

Thus, for equilibrium of the element,

$$\sigma_1 t \, ds_2 \frac{ds_1}{r_1} + \sigma_2 t \, ds_1 \frac{ds_2}{r_2} = p \, ds_1 ds_2$$

and, simplifying,

$$\frac{\sigma_1}{r_1} + \frac{\sigma_2}{r_2} = \frac{p}{t} \qquad\qquad [2.8]$$

In general $\sigma_1$ and $\sigma_2$ are both different and unknown and so another equilibrium equation has to be formulated relevant to the particular problem, in addition to eqn. [2.8], in order to solve for the stresses.

*Thin sphere*
The simplest cases to which eqn. [2.8] may be applied is the sphere. The symmetry about any axis implies that $\sigma_1 = \sigma_2 = \sigma$, and of course $r_1 = r_2 = r$; therefore the circumferential stress in any direction is $\sigma = pr/2t$. This stress may also be obtained by cutting the sphere across any diameter (Fig. 2.11) and considering the equilibrium of either of the free-body hemispheres in the same manner as shown in the next section for the cylinder.

**Fig. 2.11**

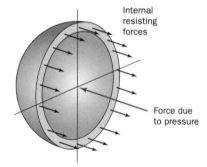

Internal resisting forces

Force due to pressure

*Thin cylinder*
When a thin-walled cylinder is subjected to internal pressure, there are axial stresses, $\sigma_x$, and a hoop (or circumferential) stress, $\sigma_y$. The latter can be determined using eqn. [2.8] but not the former. This is because $r_2 = \infty$.
    Using eqn. [2.8]

$$\frac{\sigma_y}{r} + \frac{\sigma_x}{\infty} = \frac{p}{t}$$

$$\sigma_y = \frac{pr}{t} \qquad\qquad [2.9]$$

As a further illustration of the use of free-body diagrams, this problem can also be solved from first principles.

*Axial equilibrium*    The force acting on each closed end of the cylinder owing to the internal pressure $p$, Fig. 2.12($b$), is obtained from the product of the pressure and the area on which it acts. Thus

$$\text{Axial force} = p\pi r^2$$

**Fig. 2.12**

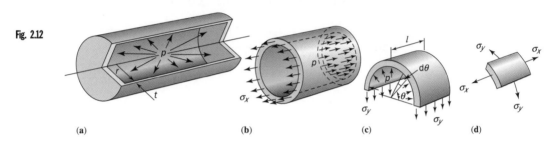

(a)  (b)  (c)  (d)

The part of the vessel shown in the free-body diagram, Fig. 2.12($b$), is in axial equilibrium simply under the action of the axial force above and the axial stress, $\sigma_x$, in the material. The radial pressure shown has no axial resultant force. The cross-sectional area of material is approximately $2\pi r t$, and therefore the reacting force is $\sigma_x \times 2\pi r t$, and, for equilibrium,

$$2\pi r t \sigma_x = \pi r^2 p$$

or

$$\sigma_x = \frac{pr}{2t} \qquad\qquad [2.10]$$

*Circumferential equilibrium*
Considering the equilibrium of part of the vessel, if the cylinder is cut across a diameter as in the free-body diagram in Fig. 2.12($c$) the internal pressure acting outwards must be in equilibrium with the circumferential stress, $\sigma_y$, as shown. Consider the length of cylinder $l$ and the small arc of shell subtending an angle $d\theta$ shown in the diagram. The radial component of force on the element is $p \times l \times r\,d\theta$; hence the vertical component is $p \times l \times r\,d\theta\,\sin\theta$. Therefore the total vertical force due to pressure is

$$\int_0^\pi prl\,\sin\theta\,d\theta = 2prl$$

It is useful to note that the vertical force can also be found by considering the pressure acting on the projected area at the diameter, which is $p(2rl)$.

The internal force required for equilibrium is obtained from the stress $\sigma_y$ acting on the two ends of the strip of shell. Hence the internal force is

$\sigma_y(2tl)$. For equilibrium, $2tl\sigma_y = 2rlp$, or

$$\sigma_y = \frac{pr}{t} \qquad [2.11]$$

This may be seen to be the same as the value obtained using eqn. [2.8].

Comparing eqns. [2.10] and [2.11] it is seen that the circumferential stress is twice the axial stress. Figure 2.12(d) shows a small element of the shell subjected to the axial and circumferential (hoop) stresses.

**Example 2.2**

A concrete dome is 250 mm thick, has a radius of 30 m and subtends an angle of 120° at the support ring. Calculate the stresses at the supports due to self-weight. The density for contrete is 2.3 Mg/m³. The dome is illustrated in Fig. 2.13.

**Fig. 2.13**

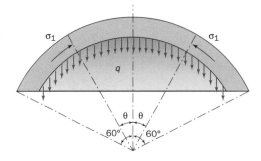

Consider the vertical equilibrium of a circular segment containing an angle $2\theta$; then if $q$ is the force per unit area due to weight of concrete,

$$(2\pi r \ \sin \ \theta \times t\sigma_1 \ \sin \ \theta) + 2\pi r(r - r \ \cos \ \theta)q = 0$$

$$\sigma_1 = -\frac{qr}{t}\frac{(1 - \cos \ \theta)}{\sin^2 \theta} = -\frac{qr}{t}\frac{1}{(1 + \cos \ \theta)}$$

Since self-weight, unlike applied pressure, does not act radially everywhere, eqn. [2.8] has to be modified to take account of the radially resolved component of weight, giving

$$\frac{\sigma_1}{r_1} + \frac{\sigma_2}{r_2} = -\frac{q}{t} \ \cos \ \theta$$

Substituting for $\sigma_1$,

$$\sigma_2 = \frac{qr}{t}\left(\frac{1}{1 + \cos \ \theta} - \cos \ \theta\right)$$

The stresses at the supports are obtained by putting $\theta = 60°$ and $q = 2.3 \times 9.81 \times 250/1000 = 5.65 \text{ kN/m}^2$:

$$\sigma_1 = -\frac{5.65 \times 30}{0.25} \times \frac{1}{1\frac{1}{2}} = -452 \text{ kN/m}^2$$

$$\sigma_2 = \frac{5.65 \times 30}{0.25}(\tfrac{2}{3} - \tfrac{1}{2}) = 113 \text{ kN/m}^2$$

**2.8 Shear in a coupling**

A pinned coupling for a tow bar is illustrated in Fig. 2.14($a$). Tensile force $F$ is transmitted across the joint being carried by shearing action on sections XX and YY of the pin.

From the earlier comments regarding complementary shear stresses, it is apparent that the shear stress at the cut section in Fig. 2.14($b$) cannot be constant across the section. This is because the shear stress at the outside surface is zero and hence the internal shear stress close to the surface must be zero. However, this type of problem is only statically determinate if we assume that the shear stress is uniform over the cross-section of the pin. It is important to remember therefore that $\tau$ is the average shear stress on the cross-section.

$$\text{Total internal shear force} = 2\tau A$$

Fig. 2.14

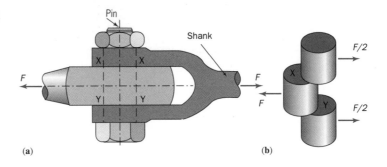

(a)                                                           (b)

For equilibrium,

$$F = 2\tau A$$

so the shear stress at the two sections of the pin is

$$\tau = \frac{F}{2A} \qquad\qquad [2.12]$$

**2.9 Torsion of a thin circular tube**

In general the twisting or torsion of solid and hollow circular-section members is statically indeterminate and is discussed fully in Chapter 5. However, as with thin-walled shells, if the radius of a circular tube is large (say over ten times) compared with the wall thickness then the stress can reasonably be taken as uniform through the thickness and the problem can be treated as statically determinate.

In Fig. 2.15 a tube of mean radius $r$ and wall thickness $t$ is subjected to opposing torques at each end. The deformation action is that of twisting,

Fig. 2.15

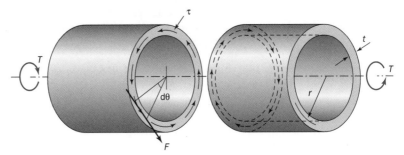

and considering the equilibrium of part of the tube, by cutting it at some section away from the end effects and perpendicular to the axis, an internal reacting torque will be required for equilibrium and to maintain the twist. This reaction torque will take the form of a 'uniform' shearing stress acting on the cut face shown. The shear stress does not vary around the tube as this would imply a variation in the complementary longitudinal shear stress and thus a resultant axial force when there is none.

Considering a small element of the tube wall subtending an angle $d\theta$, it has an area of $tr\,d\theta$. The tangential force $F$ shown in Fig. 2.15 is given as the shear stress acting over the element wall, so that $F = \tau tr\,d\theta$.

This force exerts a moment about the central axis of $Fr$ and the total reacting torque is obtained by summing all such moments around the periphery, giving

$$\int_0^{2\pi} \tau tr^2\,d\theta = 2\pi tr^2\tau$$

and this is in equilibrium with the applied torque. Hence $T = 2\pi tr^2\tau$ and the magnitude of the shear stress is

$$\tau = \frac{T}{2\pi tr^2} \tag{2.13}$$

## 2.10  Joints

All branches of engineering have to be able to connect together pieces of material which may then have to carry working loads through the connection or joint. The most common ways of constructing joints are by holding the segments firmly together with bolts, rivets or welds. The forces which have to be transmitted through a joint generally subject the connectors, e.g. bolts, to shear, and the holes in the material through which the connectors pass to tensile and compressive stresses. The determination of most of these stresses may be regarded as statically indeterminate, owing to the complex nature of load transfer through rows of bolts or rivets. Hence for a thorough treatment the reader is referred to the texts specifically concerned with engineering design. A few simple mechanical joints can be treated as statically determinate and the following example illustrates the application of the principles of equilibrium to such problems.

**Example 2.3**

A closed-ended cylindrical steel pressure vessel is 2 m in internal diameter. It is made up from 10 mm thick steel sheet formed into two semi-cylinders, which are riveted along longitudinal single-lap seams (as shown in Fig. 2.16) with 12 mm diameter rivets at 36 mm pitch. The dished ends are also single-lap riveted onto the cylinder at each end and the circumferential joints are made with 8 mm diameter rivets at 30 mm pitch. If the maximum shear stress in any rivet is not to exceed 60 MN/m² calculate the maximum allowable internal pressure and the resulting longitudinal and circumferential stresses in the wall of the cylinder.

*Longitudinal seam*   Maximum shear force per rivet is

$$60 \times 10^6 \times \frac{\pi}{4} \times \frac{12^2}{10^6} = 6.8\,\text{kN}$$

**Fig. 2.16**

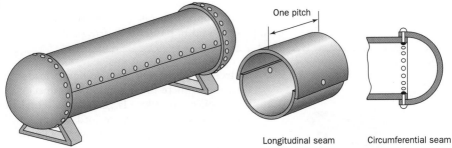

One pitch

Longitudinal seam    Circumferential seam

Separating force at joints due to pressure over one pitch length is

$$0.036 \times 2 \times p$$

The equilibrium condition is

$$0.036 \times 2 \times p = 2 \times 6.8$$

so that $p = 189 \, \text{kN/m}^2$.

*Circumferential seam*    Maximum shear force per rivet is

$$60 \times 10^6 \times \frac{\pi}{4} \times \frac{8^2}{10^6} = 3.02 \, \text{kN}$$

Number of rivets required is

$$\pi \times \frac{2000}{30} = 209$$

Total force in shear allowable $= 630 \, \text{kN}$

This must be in equilibrium with the longitudinal force due to pressure, which is $\pi/4 \times 2^2 \times p$.

Hence

$$\pi p = 630 \quad \text{and} \quad p = 201 \, \text{kN/m}^2$$

Therefore the maximum allowable pressure is $189 \, \text{kN/m}^2$.

$$\text{Circumferential stress}, \sigma_1 = \frac{pr}{t} = \frac{189 \times 1}{0.01} = 18.9 \, \text{MN/m}^2$$

$$\text{Axial stress}, \sigma_2 = \frac{18.9}{2} = 9.45 \, \text{MN/m}^2$$

**2.11  Summary**

The purpose of this chapter has been to introduce the concepts of normal and shearing *stresses* which represent the internal reacting forces per unit area within the material. Engineering design of components to establish size and shape depends on allowable values of stress which the material can tolerate when subjected to applied loads or forces. In a few cases it is possible to establish the stress for a given size or decide on the required size for an allowable stress by consideration *only* of the *equilibrium* of the system of external applied forces and internal reacting forces. These design examples are said to be *statically determinate*.

**Problems**

2.1 A steel rod of varying cross-section is loaded as shown in Fig. 2.17. Determine where the maximum stress occurs.

2.2 It is required to make a large concrete foundation block which, when supporting a comprehensive load together with its self-weight, will have the same compressive stress at all cross-sections. Determine a suitable profile.

2.3 The two parabolic cables of a suspension bridge are subjected to a horizontal uniformly-distributed load of $80\,kN/m$ as shown in Fig. 2.18. Calculate the required area of the cables at each end if their maximum permissible stress is $200\,MN/m^2$. What is the compressive load in the vertical columns?

2.4 A suspension footbridge spanning a ravine is constructed with twin cables and carries a horizontal uniformly-distributed loading of $2\,kN/m$ of span, which is $300\,m$. The lowest point of the cables is $50\,m$ below one cliff support, which is $10\,m$ below the higher cliff support. Determine a suitable cross-sectional area for each cable using a safety factor of 2 and a tensile stress of $300\,MN/m^2$.

**Fig. 2.17**

**Fig. 2.18**

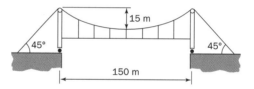

2.5 A cable is freely suspended from two points which are at the same horizontal level. If the cable is subjected to a uniformly-distributed loading of $w$ per unit length (self-weight, snow, birds), derive an expression for the maximum tension in the cable.

2.6 A small boat is anchored as shown in Fig. 2.19. When the tide causes a horizontal force of $1\,kN$ on the boat the steel rope is tangential at the anchor point. If the rope diameter is $20\,mm$ calculate (*a*) the distance between the boat and the anchor point, and (*b*) the maximum stress in the rope. The density of the steel is $7800\,kg/m^3$ and the density of the water is $1000\,kg/m^3$.

**Fig. 2.19**

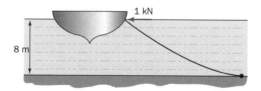

2.7 A chemical reaction process is carried out in a thin-walled steel cylinder of internal diameter $400\,mm$ with closed ends rotated about a longitudinal axis at a speed of $5000\,rev/min$. Whilst it is rotating, it is subjected to an internal pressure of $4\,MN/m^2$. If the maximum allowable tensile stress in any direction in the material is 175 MN/m², calculate a suitable shell thickness. Density of steel = 7.83 Mg/m³.

2.8 The pineline reducer shown in Fig. 2.20 has a uniform wall thickness of 3 mm. If the pipeline carries a fluid at a pressure of 0.7 Mg/m³, calculate the axial and hoop stresses in the reducer at a point halfway along its length. Assuming that the coupling at the large end of the reducer takes all the axial thrust, calculate the stress in each of the six 10 mm diameter retaining bolts.

Fig. 2.20

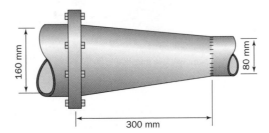

160 mm    80 mm

300 mm

2.9 A thin spherical steel vessel is made up of two hemispherical portions bolted together at flanges. The inner diameter of the sphere is 300 mm and the wall thickness of 6 mm. Assuming that the vessel is a homogeneous sphere, what is the maximum working pressure for an allowable tensile stress in the shell of 150 MN/m²?

If twenty bolts of 16 mm diameter are used to hold the flanges together, what is the tensile stress in the bolts when the sphere is under full pressure?

2.10 A thin wall conical shell supports a weight $W$ as shown in Fig. 2.21. Derive an expression for the stress in the wall of the cone.

Fig. 2.21

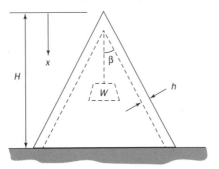

$H$    $x$    $\beta$    $W$    $h$

2.11 A conical storage tank has a wall thickness of 20 mm and an apex angle of 60°. If the vessel is filled with water to a depth of 3 m, calculate the maximum meridional and circumferential stresses. The water loading is 9.81 kN/m³.

2.12 In the mechanical digger shown in Fig. 2.22 the combined weight of the bucket and its contents is 1000 kg and the centre of gravity is at $G$. For the position shown in which the arm NKM is horizontal, calculate the shear stress in the pins at N and M. The inset sketch shows the joint arrangement in each case, and the pin diameters at N and M are 10 mm and 15 mm respectively.

**Fig. 2.22**

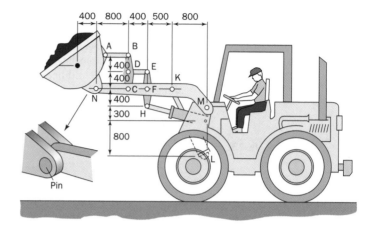

2.13  A solid circular rod of 10 mm diameter is coupled to a metal tube using a bonded rubber cylinder as shown in Fig. 2.23. If an axial pull of 10 kN is applied to the rod, calculate (*a*), the shear stress between the rod and the rubber, (*b*) the shear stress between the rubber and the metal tube, and (*c*) the axial stress in the tube.

**Fig. 2.23**

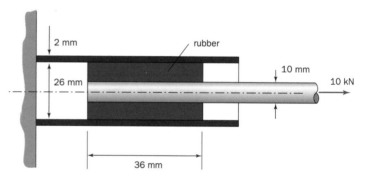

2.14  The hub of a pulley may be fastened to a 25 mm diameter shaft either by a square key or by a pin, as shown in Fig. 2.24. Determine the torque that each connection can transmit if the average shear stress in the key or pin is not to exceed 70 MN/m$^2$.

**Fig. 2.24**

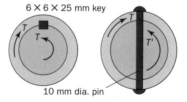

2.15  A splined shaft connection as shown in Fig. 2.25 is 50 mm long and is used to permit axial movement of the shaft relative to the hub during torque transmission. In order to facilitate axial movement in the connection it is to be desgned so that the side pressure on the splines does not exceed 7 MN/m$^2$. Calculate the power that could be

**Fig. 2.25**

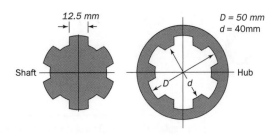

12.5 mm

D = 50 mm
d = 40mm

Shaft

Hub

transmitted by the shaft at 2000 rev/min and the shear stress in the splines at this power.

2.16 Extend the spreadsheet of Example 2.1 to calculate the torque about the crankshaft due to the gas pressure.

2.17 A thin-walled circular tube of 50 mm mean radius is required to transmit 300 kW at 500 rev/min. Calculate a suitable wall thickness so that the shear stress does not exceed 80 MN/m². 

2.18 (*a*) Construct a spreadsheet in which, given the applied torque on a thin tube, its radius and wall thickness, the resulting shear stress is calculated.

(*b*) Solve Problem 2.17 by trial and error. In other words try different values of wall thickness until the stress under the applied torque is approximated 80 MN/m².

2.19 A torque tube consists of two sections which are riveted together, as in Fig. 2.26, by 50 rivets of 4 mm diameter pitched uniformly and the radius of the mating surface of the tubes is 100 mm. If the limiting shear stress for the rivets is 180 MN/m² determine the maximum torque that can be transmitted through the joint.

**Fig. 2.26**

*T*

*T*

2.20 (*a*) Construct a spreadsheet to solve Problem 2.19, given any values for allowable stress, number of rivets, rivet diameter and tube radius.

(*b*) If a torque of 20 kN m must be transmitted, how many rivets are needed.

# 3 Stress–Strain Relations

As explained previously a statically indeterminate problem cannot be solved from the conditions of equilibrium alone; additional equations are required to find all the unknowns. These equations are obtained by studying the *geometry of deformation* of the component or structure and the *load–deformation* or *stress–strain relationship* for the material. These topics are dealt with in this chapter, but the detailed application in worked examples is carried out in Chapter 4.

## 3.1 Deformation

Deformations may occur in a material for a number of reasons, such as external applied loads, change in temperature, tightening of bolts, irradiation effects, etc. Bending, twisting, compression, torsion and shear or combinations of these are common modes of deformation. In some materials, e.g. rubber, plastics, wood, the deformations are quite large for relatively small loads, and readily observable by eye. In metals, however, the same loads would produce very small deformations requiring the use of sensitive instruments for measurement.

Stress values do not always provide the limiting factor in design, for although a component may be safe and employ material economically with regard to stress, the deformations accompanying that stress need to be considered. For example, large deflections are highly desirable in the case of springs or cushions. On the other hand, too large a deflection of an aeroplane wing can result, among other things, in a detrimental change in aerodynamic characteristics. A lathe bed which was not sufficiently rigid would not permit the required tolerances in machining.

In this and succeeding chapters there will be many problems in which the analysis of displacements will be considered specifically in addition to the determination of stress magnitude.

## 3.2 Strain

As explained in Chapter 2 the effect of a force applied to bodies of different size can be compared in terms of stress, i.e. the force per unit area. Likewise the deformation of different bodies subjected to a particular load is a function of size, and therefore comparisons are made by expressing deformation as a non-dimensional quantity given by the change in dimension per unit of original dimension, or in the case of shear as a

Fig. 3.1

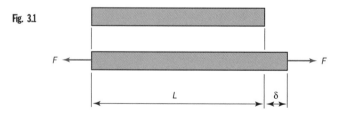

change in angle between two initially perpendicular planes. The non-dimensional expression of deformation is termed *strain*.

*Direct or normal strain*

Consider the uniform bar shown in Fig. 3.1 subjected to an axial tensile load $F$. If the resulting extension of the bar is $\delta$ and its unloaded length is $L$, then the direct tensile strain is

$$\varepsilon = \frac{\delta}{L}$$

If two bars identical in material, length $L$ and cross-sectional area were each subjected to a tensile load $F$, then the extension $\delta$ of each would be the same and the strain in both would be $\delta/L$. If the bars are now joined end on end and the same tensile force $F$ is applied, the overall extension of the combined bar will be $2\delta$. However, since the original length is now $2L$, the strain will be $2\delta/2L$, i.e. the same as for the separate bars.

Direct strain is defined as the increase in length per unit original length. If the bar had been subjected to a compressive force $F$ causing a reduction in length of $\delta$, the strain would be

$$\varepsilon = -\frac{\delta}{L}$$

The change in length is $-\delta$, so compressive strains are negative.

In the description above, the force $F$ was applied to a bar of constant cross-sectional area, so the stress was uniform at every point along the bar and so in fact is the strain. If, however, the stress varies along the bar, because for example the cross-sectional area varies, then so also will the strain and it is necessary to define strain at a point by considering a small element of undeformed length $\Delta x$, Fig. 3.2.

Fig. 3.2

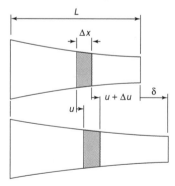

When the bar is loaded, the left-hand end of the section will be displaced to the right by an amount $u$. Owing to the strain in the element of length $\Delta x$, the right-hand end of the shaded section will be displaced by an amount $u + \Delta u$. The increase in length per unit length in the section is therefore

$$\varepsilon = \frac{\Delta u}{\Delta x}$$

In the limit as $\Delta x$ tends to zero, the strain at a point is

$$\varepsilon = \frac{du}{dx}$$

The total increase in length $\delta$ of the bar is given by

$$\delta = \int du = \int_0^L \varepsilon \, dx$$

The same suffix notation is used for strains as for stresses. $\varepsilon_x$ is the strain of a line measured in the $x$-direction and $\varepsilon_y$ is the strain of a line in the $y$-direction.

*Shear strain*

An element which is subjected to shear stress experiences deformation as shown in Fig. 3.3. The tangent of the angle through which two adjacent sides rotate relative to their initial position is termed *shear strain*. In many cases the angle is very small and the angle itself is used, expressed in radians, instead of the tangent, so that

$$\gamma = \angle AOB - \angle A'OB' = \phi$$

**Fig. 3.3**

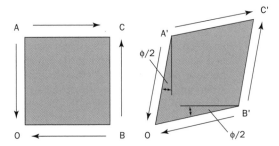

When $\angle A'OB' < \angle AOB$, then $\gamma$ is defined as a positive shear strain, and when $\angle A'OB' > \angle AOB$, $\gamma$ is termed a negative shear strain.

The shear strains corresponding to the shear stresses on the $xy$, $yz$ and $zx$ planes are $\gamma_{xy}$, $\gamma_{yz}$, $\gamma_{zx}$, $\gamma_{yx}$, $\gamma_{xz}$ and $\gamma_{zy}$.

*Volumetric strain*

The term 'hydrostatic stress' was used in Chapter 2 to describe a state of tensile or compressive stress equal in all directions within or external to a body. Hydrostatic stress causes a change in volume of the material which, if expressed per unit of original volume, gives a volumetric strain, or dilatation, denoted by $e$.

Volumetric strain may be expressed in terms of the three co-ordinate direct strains. Let a cuboid of material have sides initially of length $\Delta x$, $\Delta y$ and $\Delta z$. If the material is deformed, the new lengths of the sides will be $\Delta x(1 + \varepsilon_x)$, $\Delta y(1 + \varepsilon_y)$ and $\Delta z(1 + \varepsilon_z)$ respectively. The new volume is therefore

$$\Delta x \Delta y \Delta z (1 + \varepsilon_x)(1 + \varepsilon_y)(1 + \varepsilon_z)$$

Neglecting products of small quantities, for small strains

$$\text{New volume} = \Delta x \Delta y \Delta z (1 + \varepsilon_x + \varepsilon_y + \varepsilon_z)$$

$$\text{Original volume} = \Delta x \Delta y \Delta z$$

Therefore

$$\text{Volumetric strain } e = \frac{\text{change in volume}}{\text{original volume}}$$

$$e = (\varepsilon_x + \varepsilon_y + \varepsilon_z) \tag{3.1}$$

i.e. volumetric strain is given by the sum of the three direct co-ordinate strains.

**3.3 Elastic load–deformation behaviour of materials**

Studies of material behaviour made by Robert Hooke in 1678 showed that up to a certain limit the extension $\delta$ of a bar subjected to an axial tensile loading $F$ was often directly proportional to $F$, as in Fig. 3.4(a). This behaviour in which $\delta \propto F$ is known as *Hooke's law*. It is similarly found that for many materials uniaxial compressive load and compressive deformation

Fig. 3.4

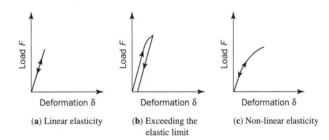

(a) Linear elasticity

(b) Exceeding the elastic limit

(c) Non-linear elasticity

are proportional up to a certain limit of load. A cylindrical bar which is twisted about its axis by opposing torques applied at each end is also found to have a linear torque–twist relationship up to a certain point. The maximum load up to which Hooke's law is applied is termed the *limit of proportionality*. If in each of the above cases at any particular load the same deformation exists both with increasing and decreasing load, and if after completely unloading the body it returns to exactly its original size, then it is said to exhibit the property of *elasticity*. This behaviour exists only over a certain range of load and deformation, the end point being termed the *elastic limit*. In general, the limit of proportionality is a shade lower than the elastic limit. If the elastic limit is exceeded it is found that some permanent deformation remains after removal of the load, as illustrated in Fig. 3.4(b). The stress in the material at the elastic limit is called the yield stress, $\sigma_y$. In most practical situations it is important to ensure that the stress in the component does not exceed $\sigma_y$. The situations which arise when yielding occurs are dealt with in Chapters 12 and 15.

**3.4 Elastic stress–strain behaviour of materials**

If the load, $F$, in Fig. 3.4(a) is divided by the original cross-sectional area of the bar, $A$, and the extensions on the abscissa are divided by the original

**Table 3.1**

| Material | $E$ $(GN/m^2)$ | $G$ $(GN/m^2)$ |
|---|---|---|
| Steels | 190–207 | 77–83 |
| Copper | 110–120 | 37–46 |
| Aluminium | 69–70 | 24–28 |
| Glass | 50–80 | 20-35 |

length of the bar, $L$, a graph of stress against strain is obtained. Since $A$ and $L$ are constants the stress–strain behaviour is also linear in the elastic range.

The slope of the line is constant and may be expressed as

$$\frac{F}{A} \bigg/ \frac{\delta}{L} = \frac{\sigma}{\varepsilon} = E$$

where $E$ is a constant for the material, and is called *Young's modulus of elasticity*. Since $\varepsilon$ is non-dimensional, $E$ has the dimensions of stress, i.e. force per unit area. Some typical values of $E$ are given for a few materials in Table 3.1.

A relationship between shear stress and shear strain may be derived from a torsion test on a cylindrical bar in which applied torque and angular twist are measured. The connections between torque and shear stress and twist and shear strain will be derived in Chapter 5. It is sufficient to state here that shear stress is proportional to shear strain within the elastic limit. Hence $\tau/\gamma = \text{constant} = G$.

The constant of proportionality, $G$, is known as the *modulus of rigidity*, or *shear modulus*, and has the dimensions of force per unit area. Typical values of $G$ for various materials are given in Table 3.1.

It can also be demonstrated experimentally that within the elastic range volumetric strain is proportional to the hydrostatic stress, $\sigma$, as defined in Chapter 2. The constant relating those two quantities is termed the *bulk modulus* and is denoted by the symbol $K$. Thus

$$\frac{\sigma}{e} = K$$

It will be shown in Chapter 11 that the elastic constants, $E$, $G$ and $K$ are related to one another.

**Example 3.1**

Calculate the overall change in length of the tapered rod shown in Fig. 3.5. It carries a tensile load of 10 kN at the free end, and at the step change in section a compressive load of 2 MN/m evenly distributed around a circle of 30 mm diameter. $E = 208 \text{ GN/m}^2$.

Firstly consider the general case of a tapered rod fixed at one end and subjected to a tensile load $F$ at the other end as in Fig. 3.6. The mean radius of any arbitrary slice at a distance $x$ from the upper end is

$$r = r_0 - (r_0 - r_1)\frac{x}{L}$$

Fig. 3.5

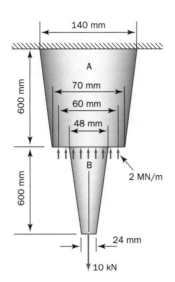

Fig. 3.6

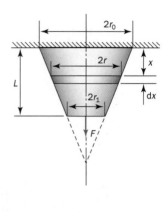

So the cross-sectional area of the slice is

$$A_x = \pi\left(r_0 - (r_0 - r_1)\frac{x}{L}\right)^2$$

If the slice extends an amount $du$ under load then its strain is

$$\varepsilon = \frac{du}{dx} = \frac{F}{A_x}\frac{1}{E}$$

The total extension of the rod is

$$u = \int_0^L \frac{F}{A_x E}\,dx$$

$$= \frac{F}{\pi E}\int_0^L \frac{dx}{[r_0 - (r_0 - r_1)(x/L)]^2}$$

$$= \frac{FL}{E\pi r_0 r_1}$$

Returning now to the stepped taper rod in Fig. 3.5, the extension of the lower part will be

$$u_B = \frac{10\,000 \times 0.6}{208 \times 10^9 \times \pi \times 0.024 \times 0.012} = +0.0319\,\text{mm}$$

The compressive load on the upper part will be treated as an axial concentrated load of magnitude

$$2 \times \pi \times 0.03 = 0.06\pi\,\text{MN} = 188.5\,\text{kN}$$

Resultant load acting on part A $= -188.5 + 10 = -178.5\,\text{kN}$

$$\text{Compression of A} = -\frac{178.5 \times 10^3 \times 0.6}{208 \times 10^9 \times \pi \times 0.07 \times 0.035}$$

$$= -0.0669\,\text{mm}$$

Therefore

$$\text{Overall deformation of rod} = -0.0669 + 0.0319$$

$$= -0.035\,\text{mm}$$

## 3.5 Lateral strain and Poisson's ratio

If a bar is subjected to, say, longitudinal tensile stress then it will extend in the direction of the stress and contract in the transverse or lateral directions, Fig. 3.7(a). If the member were subjected to uniaxial compressive stress then an expansion would occur in the lateral directions. It is found that the lateral strain is proportional to the longitudinal strain, and the constant of proportionality is termed *Poisson's ratio* denoted by the symbol $\nu$. Hence

$$\text{Lateral strain} = -\nu \times \text{direct strain (due to stress)}$$

For most metals $\nu$ is in the range from 0.28 to 0.32. It is important to remember that lateral strain can occur without being accompanied by lateral stress.

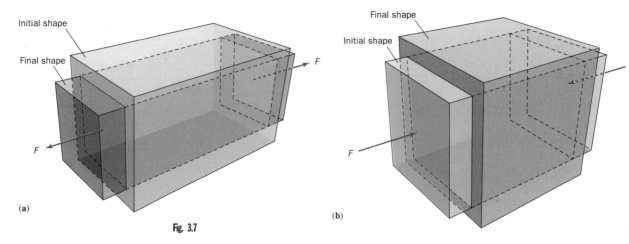

(a)      (b)

Fig. 3.7

## 3.6 Thermal strain

The effect of a change of temperature on a piece of material is a small change in size and hence strain. This can, in some circumstances, induce considerable stresses. The dependence of size on temperature variation is measured in terms of the basic quantity known as the *coefficient of linear thermal expansion*, denoted by $\alpha$. This defines the change in length per unit length for unit increase in temperature.

A rod of length $L$ has its temperature changed from $T_0$ to $T$ and the accompanying change in length is

$$\delta = \alpha L(T - T_0)$$

This may be expressed as a thermal strain:

$$\varepsilon_T = \frac{\delta}{L} = \alpha(T - T_0)$$

Increasing temperature causes expansion and thus a positive strain, while decreasing temperature results in contraction and negative strain. An important feature about this behaviour is that if there is no restraint on the material there can be strain unaccompanied by stress. However, if there is any restriction on free change in size then a *thermal stress* will result.

The total strain in a body experiencing thermal stress may be divided into two components, the strain associated with the stress, $\varepsilon_\sigma$, and the strain resulting from temperature change, $\varepsilon_T$. Thus

$$\varepsilon = \varepsilon_\sigma + \varepsilon_T$$

Hence

$$\varepsilon = \frac{\sigma}{E} + \alpha(T - T_0)$$

which is a more general form of the simple uniaxial stress–strain law.

## 3.7 General stress–strain relationships

Consider an element of material as in Fig. 3.8(*a*) subjected to a uniaxial stress, $\sigma_x$; the corresponding strain system is shown at (*b*). In the $x$-direction the strain is $\varepsilon_x$ and in the $y$- and $z$-directions the strains are $-\nu\varepsilon_x$ and $-\nu\varepsilon_x$, respectively. These strains may be written in terms of stress as $\varepsilon_x = \sigma_x/E$ and $\varepsilon_y = \varepsilon_z = -\nu\sigma_x/E$, the negative sign indicating contraction.

The element in Fig. 3.9(*a*) is subjected to triaxial stresses $\sigma_x$, $\sigma_y$ and $\sigma_z$. The total strain in the $x$-direction is therefore composed of a strain due to $\sigma_x$, a lateral strain due to $\sigma_y$ and a further lateral strain due to $\sigma_z$. Using the principle of superposition, the resultant strain in the $x$-direction is therefore the sum of the separate strains as shown at (*b*); hence

$$\varepsilon_x = \frac{\sigma_x}{E} - \frac{\nu\sigma_y}{E} - \frac{\nu\sigma_z}{E}$$

**Fig. 3.8**

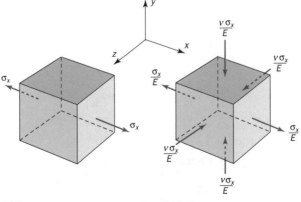

(**a**) Stress          (**b**) Strain

or

$$\varepsilon_x = \frac{\sigma_x}{E} - \frac{\nu}{E}(\sigma_y + \sigma_z)$$

Similarly

$$\varepsilon_y = \frac{\sigma_y}{E} - \frac{\nu}{E}(\sigma_z + \sigma_x)$$          [3.2]

and

$$\varepsilon_z = \frac{\sigma_z}{E} - \frac{\nu}{E}(\sigma_x + \sigma_y)$$

**Fig. 3.9**

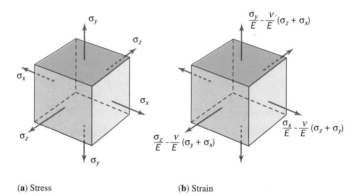

(a) Stress                                        (b) Strain

Equation [3.2] represents three equations for strain in terms of three stress components. These can be solved to give stresses in terms of the three strains as

$$\sigma_x = \frac{\nu E}{(1+\nu)(1-2\nu)}(\varepsilon_x + \varepsilon_y + \varepsilon_z) + \frac{E}{(1+\nu)}\varepsilon_x$$

$$\sigma_y = \frac{\nu E}{(1+\nu)(1-2\nu)}(\varepsilon_x + \varepsilon_y + \varepsilon_z) + \frac{E}{(1+\nu)}\varepsilon_y$$          [3.3]

$$\sigma_z = \frac{\nu E}{(1+\nu)(1-2\nu)}(\varepsilon_x + \varepsilon_y + \varepsilon_z) + \frac{E}{(1+\nu)}\varepsilon_z$$

Equation [3.3] above is sometimes written as

$$\sigma_x = \lambda e + 2\mu\varepsilon_x$$

$$\sigma_y = \lambda e + 2\mu\varepsilon_y$$

$$\sigma_z = \lambda e + 2\mu\varepsilon_z$$

where $\lambda$ and $\mu$ are the *Lamé elastic constants* of the material, $\lambda = \nu E/[(1+\nu)(1-2\nu)]$, $2\mu = E/(1+\nu)$ and $e = (\varepsilon_x + \varepsilon_y + \varepsilon_z)$ is the volumetric strain. It will be demonstrated in Chapter 11 that $\mu$ is also the shear modulus of the material.

There is no lateral strain associated with shear strain; hence the shear-stress/shear-strain relationships are

$$\gamma_{xy} = \frac{\tau_{xy}}{G} \quad \gamma_{yz} = \frac{\tau_{yz}}{G} \quad \gamma_{zx} = \frac{\tau_{zx}}{G}$$

**Plane stress**  In many practical situations the stress component in the $z$-direction is zero and this is referred to as a *plane stress* condition. The above equations may be applied with $\sigma_z = 0$, but note that the strain in the $z$-direction is not zero.

Usually $\varepsilon_z$ is not of interest and the direct strains in the $x$–$y$ plane can be found from the stresses using

$$\left.\begin{aligned} \varepsilon_x &= \frac{\sigma_x}{E} - \frac{\nu \sigma_y}{E} \\ \varepsilon_y &= \frac{\sigma_y}{E} - \frac{\nu \sigma_x}{E} \end{aligned}\right\} \qquad [3.4a]$$

Alternatively, if the direct strains are known, these equations may be solved to find the direct stresses, giving

$$\left.\begin{aligned} \sigma_x &= \frac{E}{1 - \nu^2}\left(\varepsilon_x + \nu\varepsilon_y\right) \\ \sigma_y &= \frac{E}{1 - \nu^2}\left(\varepsilon_y + \nu\varepsilon_x\right) \end{aligned}\right\} \qquad [3.4b]$$

**Plane strain**  If the strain in the $z$-direction is zero, then this condition is referred to as *plane strain* and the above equations may be applied using $\varepsilon_z = 0$. However, zero strain in the $z$-direction does not imply zero stress in that direction. This may be confirmed quite simply by considering the sample of material

**Fig. 3.10**

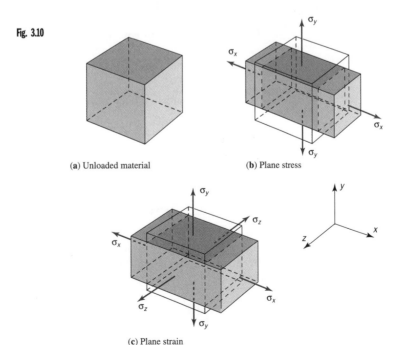

(a) Unloaded material

(b) Plane stress

(c) Plane strain

in Fig. 3.10($b$) subjected to tensile stresses in the $x$- and $y$-directions. Owing to the Poisson's ratio effect there would be a change in dimensions in the $z$-direction. To keep $\varepsilon_z = 0$ it would be necessary to have a stress in the $z$-direction, as shown in Fig. 3.10($c$).

**Thermal strains**

If in addition to strain due to stress there is also thermal strain due to change in temperature, then eqn. [3.2] has a thermal strain term added as in eqn. [3.5]. There is no thermal strain contribution to shear strain.

$$
\left.
\begin{aligned}
\varepsilon_x &= \frac{\sigma_x}{E} - \frac{\nu}{E}(\sigma_y + \sigma_z) + \alpha(T - T_0) \\
\varepsilon_y &= \frac{\sigma_y}{E} - \frac{\nu}{E}(\sigma_z + \sigma_x) + \alpha(T - T_0) \\
\varepsilon_z &= \frac{\sigma_z}{E} - \frac{\nu}{E}(\sigma_x + \sigma_y) + \alpha(T - T_0)
\end{aligned}
\right\}
\tag{3.5}
$$

**3.8 Strains in a statically determinate problem**

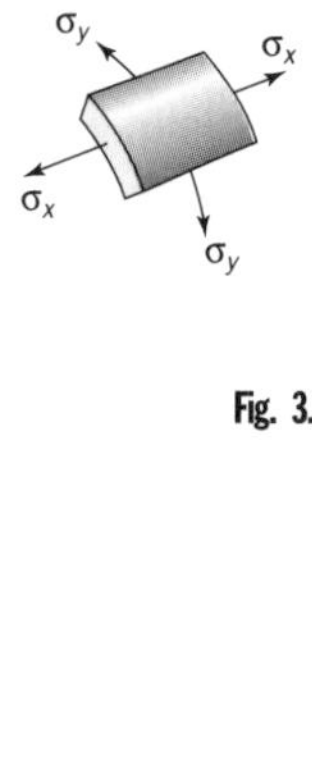

The stresses in a thin-walled cylinder under internal pressure were found in Chapter 2 as a statically determinate problem, and now that stress–strain relationships have been developed the strains in the cylinder can be found.

From eqns. [2.9] and [2.10] the axial and hoop stresses shown in Fig. 3.11 are given by

$$
\sigma_x = \frac{pr}{2t} \quad \text{and} \quad \sigma_y = \frac{pr}{t}
$$

**Fig. 3.11**

It will be shown later that $\sigma_z$ is equal to $-p$ at the inside surface and is zero at the outside surface. Therefore $\sigma_z$ is negligible in comparison with $\sigma_x$ and $\sigma_y$.

From the stress–strain equation [3.4$a$] the axial strain is

$$
\varepsilon_x = \frac{pr}{2tE} - \frac{\nu pr}{tE} = \frac{pr}{2tE}(1 - 2\nu)
\tag{3.6}
$$

and the circumferential or hoop strain is

$$
\varepsilon_y = \frac{pr}{tE} - \frac{\nu pr}{2tE} = \frac{pr}{2tE}(2 - \nu)
\tag{3.7}
$$

Taking a value for $\nu$ of 0.3 it is found that the ratio of the hoop to the axial strain is

$$
\frac{\varepsilon_y}{\varepsilon_x} = \frac{1.7}{0.4} = 4.25
$$

whereas the ratio for the hoop to axial stresses was 2.0.

In the thin sphere there is only circumferential stress and strain. In this case $\sigma_x = \sigma_y = \sigma$, so that

$$
\varepsilon = \frac{\sigma}{E} - \frac{\nu\sigma}{E}
$$

and since

$$\sigma = \frac{pr}{2t}$$

$$\varepsilon = \frac{pr}{2tE} \, (1 - v) \qquad\qquad [3.8]$$

## 3.9 Elastic strain energy

When a piece of material is deformed in simple tension, compression, bending or torsion, etc., within its elastic range, work is done by the applied loading. On removal of the loading the material returns to its undeformed state owing to the release of stored energy. This is termed *elastic strain energy* and has the same magnitude as the external work done. It is the release of strain energy in a stretched rubber band that enables a pellet to be projected from a catapult. The release of strain energy in a metal loaded and then unloaded in the elastic range is less obvious. The best illustration of it is a metal spring, the performance of which depends on the energy stored when the wire is bent and twisted during loading. The term *resilience* often associated with springs has in fact a more general meaning as the strain energy stored per unit volume.

## 3.10 Strain energy from normal stress

Consider the load–extension diagram shown in Fig. 3.12.

The work done during a small increment of extension d$u$ is $F \, \mathrm{d}u$. Therefore the total work done up to point N is

$$\text{Work done} = \int_0^\delta F \, \mathrm{d}u = \int_0^\delta \frac{AEu}{L} \mathrm{d}u$$

$$= \frac{AE}{2L} \delta^2$$

but

$$F_{max} = \frac{AE\delta}{L}$$

so

$$\text{Work done} = \tfrac{1}{2} F_{max} \delta$$

**Fig. 3.12**

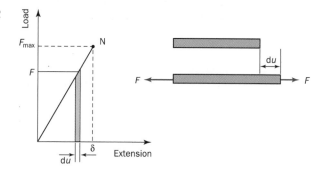

It should be noted that this is the area under the load–extension graph. This is the stored energy in the specimen. If we divide by $AL$ then the stored energy per unit volume, $U$, is given by

$$U = \frac{1}{2}\frac{F_{max}\delta}{AL} = \frac{1}{2}\sigma\varepsilon$$

Since

$$\varepsilon = \sigma/E,$$

$$U = \frac{1}{2}\frac{\sigma^2}{E} \quad \text{per unit volume} \qquad [3.9]$$

**3.11 Strain energy from shear stress**

If a piece of material is subject to pure shear then the strain energy stored per unit volume is represented by the area under the shear-stress/shear-strain curve shown in Fig. 3.13.

Hence $U = \frac{1}{2}\tau\gamma$ per unit volume, and since $\gamma = \tau/G$,

$$U = \frac{\tau^2}{2G} \quad \text{per unit volume} \qquad [3.10]$$

This expression only applies if the shear stress is uniform over the element of material.

**Fig. 3.13**

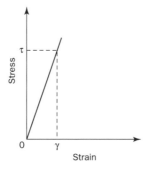

**Example 3.2**

**A car bumper is to be covered with a layer of energy-absorbing material so that in a 2 m/s collision the material will absorb the kinetic energy of the car and spring back without permanent deformation afterwards. Which one of the candidate materials below would give the cheapest solution?**

| Material | $\sigma_y$ (MN/m$^2$) | $E$ (GN/m$^2$) | Density (kg/m$^3$) | Relative cost/kg |
|---|---|---|---|---|
| Mild steel | 200 | 200 | 7800 | 1.0 |
| Polypropylene | 50 | 1.2 | 910 | 1.25 |
| Nylon | 75 | 2 | 1150 | 5.0 |

Assume that the kinetic energy of the car at $2\,\text{m/s}$ is a fixed quantity, as is the available area of the bumper. If the material must spring back after the impact, the largest stress to which it must be subjected is its yield stress $\sigma_y$. The largest energy/unit volume that can be absorbed is therefore

$$U = \frac{1}{2}\frac{\sigma_y^2}{E}$$

The total energy that can be absorbed is therefore

$$U_{car} = \frac{1}{2}\frac{\sigma_y^2}{E}At$$

where $A$ is the bumper area and $t$ the thickness of the energy-absorbing layer.

The required thickness of the energy-absorbing layer is therefore

$$t = \frac{U_{car}}{A}\bigg/\frac{1}{2}\frac{\sigma_y^2}{E}$$

The cost $C$ of the bumper covering is

$$C = c_w \rho A t$$

where $c_w$ is the cost for unit weight of the material and $\rho$ is its density. Substituting in for the required thickness

$$C = \frac{c_w \rho E}{\sigma_y^2} 2U_{car}$$

The best material from a cost point of view is therefore the one in which the combination of material properties $c_w \rho E/\sigma_y^2$ is least.

Table 3.2 shows that polypropylene is therefore the best material for this application. Note that only relative costs per unit weight have been quoted here. Absolute and relative prices fluctuate significantly over time.

Table 3.2

| Material | $C_w \rho E/\sigma_y^2$ |
| --- | --- |
| Mild steel | 0.039 |
| Polypropylene | 0.00055 |
| Nylon | 0.00204 |

## 3.12 Plastic stress–strain behaviour of materials

With relatively few exceptions in the design of structures and machines, stresses and strains are limited to the elastic range of a material. However, it is important to appreciate how a material behaves beyond the elastic range in what is termed the *plastic range* of stress and strain. As there is some elementary analysis involving yielding and plasticity in Chapters 12 and 15, the basic concept will be briefly introduced here.

When a metal specimen is subjected to uniaxial tension to fracture, the measurements of stress and strain will result in a diagram typically as shown

Fig. 3.14

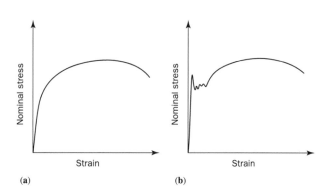

(a)                                    (b)

in Fig. 3.14(*a*) or (*b*). When the material has passed through the elastic range and enters the plastic range it is said to be *yielding*. Stress continues to increase with strain, but at a slower rate than in the elastic range, until a maximum value of nominal stress (load divided by original cross-sectional area) is reached which is termed the *tensile strength*. Thereafter the specimen enters a failure range terminating in complete fracture at one cross-section. The plastic range of strain can be perhaps 300 times the elastic range of strain, and demonstrates what is described as *ductility*, namely the ability to deform plastically.

## 3.13 Viscoelastic stress-strain behaviour of materials

Some materials, notably plastics and rubbers, do not exhibit a linear elastic range like metals, but show an interdependence of stress and strain with time. In addition these materials have a 'memory' in the sense that current strain or stress is always dependent on the loading history and after unloading considerable recovery of residual strain can occur at zero load. The above behaviour is termed *viscoelasticity*.

The simplest representation of viscoelasticity is the combined features of a Hookean solid and a Newtonian liquid. The former provides an elastic component, and the latter a viscous component (Fig. 3.15(*a*) and (*b*)). From the latter, stress is proportional to strain rate, $\dot{\varepsilon}$, and the constant of proportionality is $\eta$, the *coefficient of viscosity*. Hence the simplest possible relationship between stress, strain and time for linear viscoelastic behaviour

Fig. 3.15

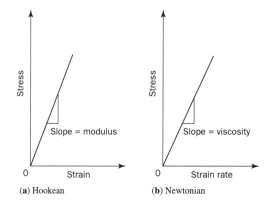

(a) Hookean                            (b) Newtonian

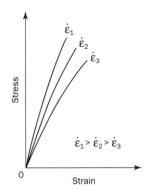

Fig. 3.16

is

$$\sigma = E\varepsilon + \eta\dot{\varepsilon} \qquad [3.11]$$

Because of the dependence on time and the behaviour known as *creep* (see Chapter 21) experimentally determined stress–strain relationships demonstrate a strain-rate dependence and non-linearity as shown in Fig. 3.16. As a result, the modulus of a viscoelastic material is not as constant as it is for metals, but depends on the magnitude and nature of the loading applied to the material. Design analysis for viscoelastic materials, e.g. the thermoplastic components, although still following the basic steps of equilibrium of forces and geometry of deformation, which are independent of the nature of the material, is somewhat more complex owing to the time-dependent stress–strain relationship. This matter is dealt with further in Chapter 21, which specifically relates to creep and viscoelasticity.

### 3.14 Summary

In this chapter the concept of *strain* was introduced together with the relationships that exist between *stress* and *strain* for engineering materials. The 'stage' is now set for the analysis of most engineering design problems involving a knowledge of *strength* and *stiffness*. The next three chapters apply these basic concepts to the design of important, widely used engineering components. An early appreciation of the importance of *strain energy* is essential for all engineers, since it has a major part to play through energy theorems in design analysis and further in relation to the mechanics of yielding and fracture.

### Problems

3.1 Determine the overall change in length for the steel rod shown in Problem 2.1 (Fig. 2.17). The length of the upper section is 300 mm and the two lower sections are each 400 mm long. $E_s = 200\,\mathrm{GN/m^2}$.

3.2 State whether the components, illustrated in Fig. 3.17 are in a state of plane stress or plane strain.
   (*a*) A grinding wheel rotating at high speed.
   (*b*) A plate of steel being cold-rolled.
   (*c*) A long thick-walled cylinder containing a fluid pressurized by two end pistons.

Fig. 3.17

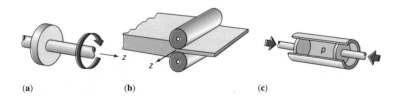

(a)        (b)        (c)

3.3 A 60 mm diameter mild-steel sphere has parallel flats machined on it 20 mm each side of the central axis. If a compressive load of 5 MN is applied perpendicular to the flats, calculate the decrease in length along the loading axis. The modulus of steel is $207\,\mathrm{GN/m^2}$.

3.4 A truncated cone with a base radius, $R$, and a height, $H$, is attached to a horizontal surface by its base and truncated at height $h$. If the material of the truncated cone has density $\rho$ and modulus $E$, show that its extension as a result of its own weight is given by

$$\delta = \rho g H^2 \, (H + 3h) \, / \, 6E(H + h)$$

3.5 A copper band 20 mm wide and 2 mm thick is a snug fit on a 100 mm diameter steel bar which may be assumed to be rigid. Determine the stress in the copper if its temperature is lowered by 50°C. $\alpha = 18 \times 10^{-6}/°C$, $E = 105 \, \text{GN/m}^2$.

3.6 What are the shear strain and angle of twist per unit length for the tube in Problem 2.17. $G = 85 \, \text{GN/m}^2$.

3.7 Express the stresses $\sigma_x$, $\sigma_y$, $\sigma_z$ in terms of the three co-ordinate strains and the elastic constants. Obtain similar expressions for the cases of plane stress, $\sigma_z = 0$ and plane strain, $\varepsilon_z = 0$.

3.8 Program the formulae of eqns [3.2] and [3.3] into a spreadsheet so that, given the material properties $E$ and $\nu$, and the stresses on an element of material, all the strains are calculated.

3.9 For Plane Stress loading, construct a spreadsheet whereby, given the Young's modulus and Poisson's ratio
(a) $\sigma_x$ and $\sigma_y$ are entered and $\varepsilon_x$ and $\varepsilon_x$ are calculated.
(b) $\varepsilon_x$ and $\varepsilon_x$ are entered and $\sigma_x$ and $\sigma_y$ are calculated.

3.10 Show that the volumetric strain, $e$ in an element subjected to triaxial stresses $\sigma_x$, $\sigma_y$ and $\sigma_z$, is given by

$$e = \frac{1 - 2\nu}{E} \, (\sigma_x + \sigma_y + \sigma_z)$$

3.11 Determine the maximum strain and change in diameter of the cylinder in Problem 2.7 if $E = 208 \, \text{GN/m}^2$ and $\nu = 0.3$.

3.12 A rectangular steel plate of uniform thickness has a strain gauge rosette bonded to one surface at the centre as shown in Fig. 3.18. It is placed in a test rig which can apply a biaxial force system along the edges of the plate. If the measured strains are $+0.0005$ and $+0.0007$ in the $x$- and $y$-directions, determine the corresponding stresses set up in the plate and the strain through the thickness. $E = 208 \, \text{GN/m}^2$ and $\nu = 0.3$.

**Fig. 3.18**

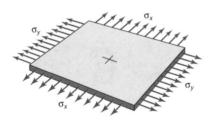

3.13 A trapeze artist weighs 50 kg and is balanced at the centre of a 3 mm diameter wire tightrope of 20 m length. There is an initial stress of $100 \, \text{MN/m}^2$ in the tightrope before the artist balances on it. Determine the strain energy stored in the wire. $E = 208 \, \text{GN/m}^2$.

3.14 Determine the shear-strain energy stored in the torsion tube of Fig. 2.13. $G = 85 \, \text{GN/m}^2$.

3.15 A block of material is subjected to strains $\varepsilon_x$, $\varepsilon_y$ and $\gamma_{xy}$.

(a) Derive a spreadsheet to calculate the new position of the corners of a unit square of material in the $xy$ plane i.e. points (0,0), (1,0), (1,1) and (0,1). Assume point (0,0) does not move.

(b) Normally the strains in engineering materials are very small. Add a 'strain magnification factor' which increases the displacements by a given factor.

(c) Link the input cells in which strains are specified to the output of the calculation of strains from stresses in Problem 3.9(a), so that the deformation in response to an applied stresses can be calculated.

(d) Create an $x$–$y$ graph of the original and magnified deformed shape of the unit square. Observe how the deformed shape changes in response to different stress values.

# 4 Statically Indeterminate Stress Systems

> The subject has now been developed sufficiently so that statically indeterminate problems involving uniform normal tensile or compressive stress can be solved. The principles of solution are fundamental to the mechanics of solids and structures and require the writing of:
> 1. Equations of *equilibrium of forces* – external (applied); internal (as a function of stress and area).
> 2. Equations describing the *geometry of deformation* or *compatibility of displacements*.
> 3. Relationships between *load–deformation* or *stress-strain* for the materials.
>     It is of the utmost importance to remember the above principles and apply them in a logical manner to all future problems. This chapter is devoted to illustrating the application of these principles to a variety of problems encountered in design. The solutions have been deliberately worked in letter symbols rather than numerical values to enable the steps to be followed more easily.

## 4.1 Interaction between components of different stiffness

A series of examples will now be studied involving different materials and also different component configurations. It is important to note the differences in equilibrium and deformation relationships that arise in these two component arrangements.

**Example 4.1**

A bimetallic rod is subjected to a compressive force, *F*, as shown in Fig. 4.1. Determine the overall change in length.

**Fig. 4.1**

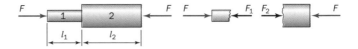

The two materials will be denoted by subscripts 1 and 2.

*Equilibrium*   It should be clear that the same force is carried through each material, so that

$$F_1 = F_2 = F \tag{4.1}$$

*Geometry of deformation*   The overall change in length is the sum of the changes in the two parts of the rod, so that

$$\delta = \delta_1 + \delta_2 \tag{4.2}$$

*Stress–strain relations*   Since it is a simple uniaxial stress system

$$\frac{\sigma_1}{\varepsilon_1} = E_1 \qquad\qquad [4.3]$$

$$\frac{\sigma_2}{\varepsilon_2} = E_2 \qquad\qquad [4.4]$$

Let the cross-sectional areas be $A_1$ and $A_2$. Then eqns. [4.3] and [4.4] can be re-expressed as

$$\frac{F_1}{A_1} = E_1 \frac{\delta_1}{l_1} \quad \text{and} \quad \frac{F_2}{A_2} = E_2 \frac{\delta_2}{l_2} \qquad\qquad [4.5]$$

Substituting the values of $\delta_1$ and $\delta_2$ into eqn. [4.2] gives

$$\delta = \frac{F_1 l_1}{A_1 E_1} + \frac{F_2 l_2}{A_2 E_2} = F\left(\frac{l_1}{A_1 E_1} + \frac{l_2}{A_2 E_2}\right)$$

For a series of $n$ bars

$$\delta = F \sum_{i=1}^{n} \frac{l_i}{A_i E_i} \qquad\qquad [4.6]$$

**Example 4.2**

**Two components of different materials are arranged concentrically and loaded through rigid end plates as shown in Fig. 4.2. Determinate the force carried by each component.**

**Fig. 4.2**

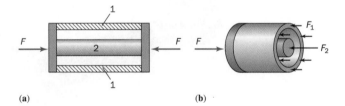

(a)                    (b)

*Equilibrium*   In this case the load is *shared* in some unknown proportions between the two parts, so that

$$F_1 + F_2 = F \qquad\qquad [4.7]$$

*Geometry of deformation*   If the unloaded lengths, $l$, are initially the same, then they will remain the same under load; hence

$$\delta_1 = \delta_2 = \delta \qquad\qquad [4.8]$$

*Stress–strain relations*   Again, for a simple uniaxial stress situation,

$$\frac{\sigma_1}{\varepsilon_1} = E_1 \quad \text{and} \quad \frac{\sigma_2}{\varepsilon_2} = E_2 \qquad\qquad [4.9]$$

From eqns. [4.8] and [4.9]

$$F_1 = E_1 A_1 \frac{\delta}{l} \quad \text{and} \quad F_2 = E_2 A_2 \frac{\delta}{l} \qquad [4.10]$$

Substituting in the equilibrium equation [4.7],

$$E_1 A_1 \frac{\delta}{l} + E_2 A_2 \frac{\delta}{l} = F$$

Thus

$$\delta = \frac{Fl}{E_1 A_1 + E_2 A_2} \qquad [4.11]$$

$$F_1 = \frac{FE_1 A_1}{E_1 A_1 + E_2 A_2} \quad \text{and} \quad F_2 = \frac{FE_2 A_2}{E_1 A_1 + E_2 A_2} \qquad [4.12]$$

**Example 4.3**

Figure 4.3 shows the connecting rod of an internal combustion engine. At the big end, where the rod is attached to the crankshaft of the engine, the bearing is split and the bearing cap is secured by two bolts which engage in threads in the rod as shown in Fig. 4.4. During assembly, the bolts will be tightened sufficiently to bring the mating surfaces into contact and then either a prescribed torque will be applied with an instrumented torque wrench or the bolt heads will be turned through a prescribed angle. In either case the tension in the bolt will be close to the yield stress of the material before any external load is applied. Why should this be done?

**Fig. 4.3**   Connecting rod

The analysis of Example 4.2 can be extended to components where there is an initial mismatch in size so that internal stresses are created even when no external load is applied. This analysis is useful in the design of bolted joints, especially in those cases where there is a risk of failure due to 'fatigue' or repetitive loading, a phenomenon which is described in Chapter 20.

Owing to the torque applied to the bolt heads there will initially be tension in the securing bolt which is opposed by an equal and opposite compression in the bearing cap. A general method of dealing with this situation is to treat the bolt and the surrounding material in the bearing cap as being subject to uniaxial stress. The two components can be represented as two springs in parallel as shown in Fig. 4.4($b$), but owing to the preload in the system the bolt (spring $k_1$) is stretched by an extra amount $\delta_0$.

**Fig. 4.4**

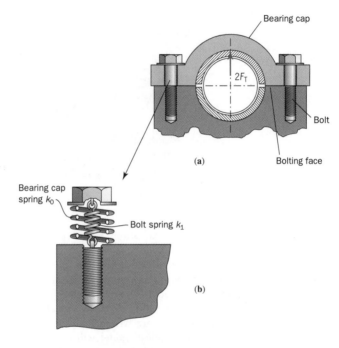

Bearing cap

$2F_T$

Bolt

(a)            Bolting face

Bearing cap spring $k_0$

Bolt spring $k_1$

(b)

*Equilibrium*    The total load supported is the sum of the load in each component:

$$F_T = F_1 + F_2$$

*Geometry of deformation*    The bolt is stretched by an extra amount $\delta_0$; hence

$$\delta_1 = \delta + \delta_0$$

$$\delta_2 = \delta$$

*Stress–strain relations*    The bolt is in simple uniaxial tension. The bearing cap has a varying cross-sectional area between the mating face and the underside of the bolt head. In reality the stress distribution can be quite complex in this component but by far the largest stress is the direct stress in the direction of the bolt axis. The simplest model of the loading is to treat the bearing cap as a simple member loaded in uniaxial compression. Its modulus and length are known. The cross-sectional area could be integrated over the length as in Example 3.1, but the simplest approach to obtaining an answer of the right order of magnitude is to take the area of the equivalent member as the area of the mating surface. As can be seen in Fig. 4.5, this is around four times the area of the bolt.

**Fig. 4.5**   Conrod bearing cap

The approximations introduced above are typical of those necessary in engineering practice to estimate likely stress levels in components during the early stages of design or to check the results of other more detailed calculations. Developing confidence and competence in choosing the best analytical model of a real engineering component is one of the most difficult tasks to teach and to learn. A useful principle is that it is better to have answers which are wrong by a factor of 10 than to have no estimate at all! Even the simplest model will give an appreciation of how the component works and the likely constraints on improving performance.

Assuming uniaxial stress in both components

$$F_1 = \frac{E_1 A_1}{L}(\delta + \delta_0) = k_1(\delta + \delta_0)$$

$$F_2 = \frac{E_2 A_2}{L}\delta = k_2\delta$$

The total force is therefore

$$F_T = F_1 + F_2 = k_1(\delta + \delta_0) + k_2\delta$$

Consider first the case when no external load is acting on the joint, i.e. $F_T = 0$. The deflection $\delta_{preload}$ is found from

$$k_1(\delta_{preload} + \delta_0) + k_2\delta_{preload} = 0$$

or

$$\delta_{preload} = -\frac{k_1}{k_1 + k_2}\delta_0$$

From this deflection, the preload force in the joint is

$$F_{preload} = F_1 = -F_2 = -k_2\delta_{preload} = \frac{k_1 k_2}{k_1 + k_2}\delta_0$$

When the applied load $F_T$ is not zero, then

$$F_T = k_1\delta + k_1\delta_0 + k_2\delta$$

The resulting deflection is

$$\delta = \frac{F_T - k_1\delta_0}{k_1 + k_2}$$

From this deflection the force in either component of the joint can be found. For the bolt

$$F_1 = k_1\delta + k_1\delta_0$$

$$= \frac{k_1}{k_1 + k_2}F_T - \frac{k_1^2}{k_1 + k_2}\delta_0 + k_1\delta_0$$

$$= \frac{k_1}{k_1 + k_2}F_T + \frac{k_1 k_2}{k_1 + k_2}\delta_0$$

$$= \frac{k_1}{k_1 + k_2}F_T + F_{preload}$$

The significance of this result is that if $k_2 \gg k_1$ and the externally applied load $F_T$ is varying, the load $F_1$ on the bolt has a large static component $F_{preload}$ but only a small varying component due to the varying $F_T$. If the cross-sectional area of the bearing cap is four times that of the bolt, then the variation in bolt force is only one-fifth of the variation in total load. This greatly reduces the possibility of a fatigue failure of the bolt due to repeated loading, as discussed in Chapter 20.

For the connecting rod material surrounding the bolt the load is

$$F_2 = \frac{k_2}{k_1 + k_2}F_T - F_{preload}$$

This can only be negative, i.e. compressive, since the conrod and the bearing cap will separate rather than carry a tensile force and a *fretting* failure due to relative motion may occur. The limiting case of a zero load occurs when

$$F_{preload} = \frac{k_2}{k_1 + k_2}F_T$$

In the present example, a preload of at least four-fifths of the maximum total load is required.

**Example 4.4**

A rigid member AB, the weight of which can be neglected, is supported horizontally at the pin joints A and C and by the spring at B as shown in Fig. 4.6(a). The stiffness of member CD is 2 kN/m and of the spring is 5 kN/m. Calculate the force in CD, which is initially unstressed, and the reaction at A when a vertical load of 10 kN is applied at B.

**Fig. 4.6**

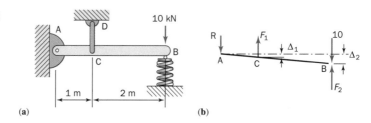

(a)                                                    (b)

*Equilibrium*    Let the reaction at A be $R$ and the forces in CD and in the spring be $F_1$ and $F_2$ respectively.

$$\textit{Vertical}: \quad 10 + R - F_1 - F_2 = 0 \tag{4.13}$$

$$\textit{Moments about B}: \quad 3R - 2F_1 = 0 \tag{4.14}$$

*Geometry of deformation*    The deformation is as shown in Fig. 4.6(*b*), so that from similar triangles

$$\frac{\Delta_1}{\Delta_2} = \frac{1}{3} \tag{4.15}$$

*Load–deformation relations*

$$\frac{F_1}{\Delta_1} = 2\,\text{kN/m} \quad \text{and} \quad \frac{F_2}{\Delta_2} = 5\,\text{kN/m} \tag{4.16}$$

From eqns. [4.15] and [4.16]

$$F_1 = \frac{2}{15} F_2 \tag{4.17}$$

From eqns. [4.13] and [4.14], eliminating $R$,

$$F_1 + 3F_2 = 30 \tag{4.18}$$

Hence the force in CD is $F_1 = 1.276\,\text{kN}$ and the reaction at A is $R = \frac{2}{3} F_1 = 0.851\,\text{kN}$.

## 4.2 Restraint of thermal strain

When the temperature of a piece of material is changed, its size will also change, and when expressed non-dimensionally this is termed *thermal strain* (see Chapter 3). If the thermal strain is not restricted in any manner, then at the new steady temperature a previously unstressed component will remain unstressed. Clearly if there is any form of restraint to free change in size then *thermal stress* will result. The following two examples will illustrate this situation.

**Example 4.5**

The bimetallic component illustrated in Fig. 4.2 consists of a steel rod of cross-sectional area 600 mm$^2$ coaxially surrounded by a copper tube of cross-sectional area 1200 mm$^2$. It is not subjected to any external load but its temperature is changed from 20 °C to 100 °C. Determine the axial stresses set up in the copper and the steel.

$$E_s = 205 \text{ GN/m}^2, \ E_c = 115 \text{ GN/m}^2,$$

$$\alpha_s = 11 \times 10^{-6}/°\text{C}, \ \alpha_c = 16 \times 10^{-6}/°\text{C}.$$

*Equilibrium*  Since there is no applied external force, the sum of the internal forces in the copper and steel must be zero. Therefore

$$F_c + F_s = 0 \qquad\qquad\qquad\qquad\qquad [4.19]$$

or

$$\sigma_c A_c + \sigma_s A_s = 0 \qquad\qquad\qquad\qquad\qquad [4.20]$$

*Geometry of deformation*  Since the two materials are initially stress-free and their ends are fixed together the total strain must be the same for each. Therefore

$$\varepsilon_c = \varepsilon_s \quad \text{or} \quad (\varepsilon_\sigma + \varepsilon_T)_c = (\varepsilon_\sigma + \varepsilon_T)_s \qquad\qquad [4.21]$$

where $\varepsilon_\sigma$ = strain due to stress, $\varepsilon_T$ = strain due to temperature change.

*Stress–strain relation*

$$\varepsilon_c = \frac{\sigma_c}{E_c} + \alpha_c(T - T_0) \qquad\qquad\qquad\qquad [4.22]$$

$$\varepsilon_s = \frac{\sigma_s}{E_s} + \alpha_s(T - T_0) \qquad\qquad\qquad\qquad [4.23]$$

Equating $\varepsilon_c$ and $\varepsilon_s$ from eqns. [4.22] and [4.23],

$$\frac{\sigma_c}{E_c} + \alpha_c(T - T_0) = \frac{\sigma_s}{E_s} + \alpha_s(T - T_0)$$

Using eqn. [4.20] to eliminate $\sigma_s$ gives

$$\frac{\sigma_c}{E_c} + \alpha_c(T - T_0) = -\frac{\sigma_c}{E_s}\frac{A_c}{A_s} + \alpha_s(T - T_0)$$

$$\sigma_c\left(\frac{1}{E_c} + \frac{A_c}{E_s A_s}\right) = (T - T_0)(\alpha_s - \alpha_c)$$

or

$$\sigma_c = \frac{A_s E_s E_c (T - T_0)(\alpha_s - \alpha_c)}{A_s E_s + A_c E_c}$$

and

$$\sigma_s = -\frac{A_c E_c E_s (T - T_0)(\alpha_s - \alpha_c)}{A_s E_s + A_c E_c}$$

The negative sign for $\sigma_s$ does not necessarily indicate a compressive stress but simply that it is opposite in sign to $\sigma_c$. The type of stress in each material is determined by the numerical values of the quantities, $T_0$, $T$, $\alpha_s$ and $\alpha_c$. Substituting the numerical values in the above equations gives

$$\sigma_c = -21.7\,\text{MN/m}^2$$

and

$$\sigma_s = -\frac{\sigma_c A_c}{A_s} = -\frac{-21.7 \times 1200}{600} = +43.4\,\text{MN/m}^2$$

Thus for an *increase* in temperature, $\alpha_c$ being greater than $\alpha_s$, the copper is prevented from expanding as much as if it were free and is put into compression. The steel is forced to expand more than it would if free and is therefore in tension.

If there was the situation of both an applied load as in Fig. 4.2 and a change in temperature then the solution could conveniently be obtained by using the principle of superposition and adding together the two separate results.

**Example 4.6**

A copper ring having an internal diameter of 150 mm and external diameter of 154 mm is to be shrunk onto a steel ring, of the same width, having internal and external diameters of 140 mm and 150.05 mm respectively.

What change in temperature is required in the copper ring so that it will just slide on to the steel ring?

What will be the uniform circumferential stress in each ring and also the interface pressure when assembled and back at room temperature?

Assume that there is no stress in the width direction, $E_s = 205\,\text{GN/m}^2$, $E_c = 100\,\text{GN/m}^2$, $\alpha_c = 18 \times 10^{-6}/°\text{C}$.

The circumferential length of the copper ring has to be increased by heating till it is fractionally larger than the circumferential length of the steel ring.

$$\text{Minimum required change in circumference} = \pi d_s - \pi d_c$$

$$= \pi \times 0.05$$

$$\text{Change in circumference due to heating} \quad = \pi d_c \times \alpha(T - T_0)$$

$$= \pi \times 150 \times 18 \times 10^{-6}(T - T_0)$$

Therefore

$$T - T_0 = \frac{0.05}{2700 \times 10^{-6}} = 18.5\,°\text{C}$$

When the assembly has returned to ambient temperature assume that the circumferential stresses in the copper and steel are uniformly distributed over each cross-section.

*Equilibrium*    Let the width of each ring be $w$; then, with thicknesses of 2 and 5 mm respectively,

$$(w \times 2)\sigma_c + (w \times 5)\sigma_s = 0$$

$$\sigma_c = -2.5\sigma_s \qquad [4.24]$$

*Geometry of deformation*    The circumferential strains in the copper and steel must be the same at the mating surface, so

$$\varepsilon_s = \varepsilon_c \qquad [4.25]$$

*Stress–strain relations*

$$\varepsilon_s = \sigma_s/E_s \quad \text{and} \quad \varepsilon_c = \sigma_c/E_c + \alpha_c \Delta T \qquad [4.26]$$

since it is only the copper ring that has the thermal strain component. From eqns. [4.25] and [4.26]

$$\frac{\sigma_s}{E_s} = \frac{\sigma_c}{E_c} + \alpha_c \Delta T \qquad [4.27]$$

Using eqn. [4.24],

$$\sigma_s \left( \frac{1}{E_s} + \frac{2.5}{E_c} \right) = -18 \times 10^{-6} \times 18.5$$

The negative sign is due to $\Delta T$ being a reduction in temperature. Substituting for $E_s$ and $E_c$,

$$\sigma_s = -11.15 \text{ MN/m}^2 \quad \text{and} \quad \sigma_c = +27.9 \text{ MN/m}^2$$

The radial pressure at the interface between the two rings may be treated as a thin cylinder under internal or external pressure, so that

$$p = \sigma_c t / r = 27.9 \times 2/75 = 745 \text{ kN/m}^2$$

---

## 4.3  Volume changes

The following problem analyses the change in volume of a vessel subjected to pressure and makes use of the relationship between hydrostatic stress and volume strain.

---

**Example 4.7**

A thin, spherical, steel shell with a mean diameter of 3 m and a wall thickness of 6 mm, is just filled with water at 20 °C at atmospheric pressure. Find the rise in gauge pressure if the temperature of the water and shell rises to 50 °C, and then determine the volume of water that would escape if a small leak developed at the top of the vessel.

**Steel: Young's modulus, $E = 200\,\text{GN/m}^2$**
**Coefficient of linear thermal expansion $\alpha = 11 \times 10^{-6}/°\text{C}$**
**Poisson's ratio= 0.3**
**Water: Bulk modulus, $K = 2.2\,\text{GN/m}^2$**
**Coefficient of volumetric thermal expansion $\alpha_v = 0.207 \times 10^{-3}/°\text{C}$**

*Equilibrium* Let the gauge pressure in the sphere after the rise in temperature be $p$; then from Chapter 2 the equilibrium condition is

$$\sigma = \frac{pr}{2t} = 125p \tag{4.28}$$

*Geometry of deformation* If there is to be a pressure at all then the water and sphere must remain in overall contact, and hence

Change in volume of sphere = change in volume of water

or

$$e_{sphere} = e_{water} \tag{4.29}$$

since the original volume is the same for each.

*Stress–strain relations* For the water the total volumetric strain is the sum of that due to pressure and that due to thermal strain:

$$e_{water} = -(p/K) + \alpha_v(T - T_0) \tag{4.30}$$

For the sphere the total volumetric strain is a function of strain due to stress (from pressure) and thermal strain:

$$e_{sphere} = e_{stress} + e_{thermal} \tag{4.31}$$

From eqn. [4.30],

$$e_{water} = -\frac{p}{2.2 \times 10^9} + (0.207 \times 10^{-3} \times 30)$$

$$= -(0.445p \times 10^{-9}) + (6210 \times 10^{-6})$$

The change in the internal capacity or volume of the sphere is

$$\tfrac{4}{3}\pi(r + \delta r)^3 - \tfrac{4}{3}\pi r^3$$

which gives, neglecting products of the small quantity $\delta r$,

$$\tfrac{4}{3}\pi \times 3r^2 \delta r$$

Expressing this as a volumetric strain,

$$\frac{\tfrac{4}{3}\pi \times 3r^2 \delta r}{\tfrac{4}{3}\pi r^3} = 3\frac{\delta r}{r}$$

It will now be shown that $\delta r/r$ is the linear or circumferential strain in the material of the sphere.

Change in circumference $= 2\pi(r + \delta r) - 2\pi r = 2\pi\delta r$

Therefore,

$$\text{Circumferential strain} = \frac{2\pi\delta r}{2\pi r} = \frac{\delta r}{r}$$

so that

$$\text{Volumetric strain of vessel} = 3 \times \text{circumferential strain}$$

Now, the hoop strain is given by eqn. (3.5):

$$\varepsilon = \frac{\sigma}{E} - \frac{\nu\sigma}{E} + \alpha(T - T_0)$$

$$= \frac{125p}{200 \times 10^9}(1 - 0.3) + (11 \times 10^{-6} \times 30)$$

Therefore the total volumetric strain is

$$e_{sphere} = 3\left(\frac{125 \times 0.7p}{200 \times 10^9} + 330 \times 10^{-6}\right)$$

Hence, using eqn. [4.29],

$$(-0.445p \times 10^9) + (6210 \times 10^{-6}) = 3\left(\frac{125 \times 0.7p}{200 \times 10^9} + (330 \times 10^{-6})\right)$$

from which

$$p = 2.97 \, \text{MN/m}^2$$

The volume of water which escapes through the leak is simply the difference of the *free* thermal expansions of the water and the vessel, since obviously there is no pressure present to affect the issue.

Volume of water escaping

$$= [(6210 \times 10^{-6}) - (3 \times 330 \times 10^{-6})] \times \tfrac{4}{3}\pi \times 1500^3$$

$$= 7.4 \times 10^{-3} \, \text{m}^3$$

## 4.4 Constrained material

Examples 4.1 and 4.2 introduced the concept of two materials reacting against each other in a simple *uniaxial* load and stress situation. There are some engineering components in which material is constrained on a two- or three-dimensional basis, so that more complex stress and strain distributions result. This type of situation is covered in detail in Chapter 14. The following example, although somewhat contrived, provides an illustration of the use of the general stress–strain relationships (see eqn. [3.2]).

### Example 4.8

A cylindrical block of concrete is encompassed by a close-fitting thin-walled steel tube of inside radius $r$ and wall thickness $t$ as shown in Fig. 4.7. If the concrete is subjected to a uniform pressure $p_1$ on its ends, determine the required ratio of $r/t$ so that the steel walls exert a lateral pressure $p_2$ on the concrete of at least $0.2p_1$. Assume that there is no friction between the concrete and steel. The moduli and

Poisson's ratio for concrete and steel are $E_c$, $\nu_c$ and $E_s$, $\nu_s$ respectively. Assume that there is no friction between the steel and the concrete. $E_s = 15E_c$, $\nu_c = 0.25$.

**Fig. 4.7**

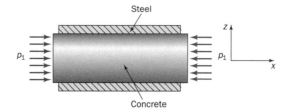

The physical nature of the problem is that as the concrete is compressed it expands laterally on to the tube. The resulting internal pressure in the tube is resisted by hoop stresses in the steel.

Superscripts $c$ and $s$ will be used to annotate the stresses and strains in the concrete and steel. The stresses in the concrete are $\sigma_x^c = -p_1$ and $\sigma_y^c = -p_2$. Also, by symmetry, $\sigma_z^c = -p_2$. In the steel, the hoop stress $\sigma_y^s = p_2 r/t$. For a thin-walled tube, $r/t \gg 1$, this is the only significant stress and so $\sigma_x^s = \sigma_z^s \approx 0$.

Since the tube is close-fitting, both steel and concrete will expand laterally by the same amount and the circumferential strains are equal i.e.

$$\varepsilon_y^s = \varepsilon_y^c$$

Using the stress–strain relations, eqn. [3.2], this implies

$$\frac{p_2 r}{t} \frac{1}{E_s} = -\frac{p_2}{E_c} + \frac{\nu_c}{E_c} p_1 + \frac{\nu_c}{E_c} p_2$$

$$\therefore \quad \frac{\nu_c}{E_c} p_1 = p_2 \left[ \frac{r}{t} \frac{1}{E_s} + \frac{(1 - \nu_c)}{E_c} \right]$$

$$\therefore \quad p_2 = p_1 \nu_c / \left[ \frac{r}{t} \frac{E_c}{E_s} + (1 - \nu_c) \right]$$

when $p_2 = \frac{1}{5} p_1$

$$\frac{1}{5} = \frac{\nu_c}{\left[ \frac{r}{t} \frac{E_c}{E_s} + (1 - \nu_c) \right]}$$

Taking $E_s = 15E_c$, $\nu_c = 0.25$

$$\frac{r}{t} = \frac{E_s}{E_c} [6\nu_c - 1] = 15 \cdot \frac{1}{2} = 7.5$$

A more general analysis involving interference between thick-walled cylinders is illustrated in Example 14.3.

**4.5   Maximum stress due to a suddenly applied load**

In Chapter 3, in the section dealing with the elastic strain energy stored under uniaxial stress, the load was applied gradually so that the work done

Fig. 4.8

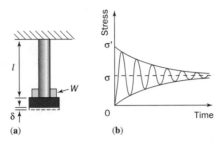

(a)          (b)

was the *average* load times the distance moved at the point where the load was applied. Now suppose that a bar fixed at the top with a flange at the lower end, as in Fig. 4.8(*a*), has a mass, *m*, suddenly released on to the flange. Let the momentary maximum extension, strain and stress in the bar be $\delta'$, $\varepsilon'$ and $\sigma'$ respectively. In addition, if the masses of the bar and flange are small compared with the mass, *m*, then a reasonable approximation to the behaviour is made by neglecting the effect of the former. Because the *full* load moves through the extension, $\delta'$,

$$\text{Work done} = W\delta' = mg\delta'$$

The strain energy stored per unit volume momentarily is $\frac{1}{2}\sigma'\varepsilon'$, so that

$$U = \tfrac{1}{2}\sigma'\varepsilon'\,al = \tfrac{1}{2}\sigma'a\delta'$$

Since the work done is equal to the strain energy,

$$mg\delta' = \tfrac{1}{2}\sigma'a\delta'$$

Thus $\sigma' = 2mg/a$; but $mg/a = \sigma$, the stress due to a gradually applied load, so that

$$\sigma' = 2\sigma \qquad\qquad [4.39]$$

or the momentary maximum stress due to a suddenly applied load is twice the stress for a gradually applied load. The bar will subsequently oscillate about the statical equilibrium position while the stresses and deformations rapidly die away, as shown in Fig. 4.8(*b*), to the value obtained for a gradually applied load. However, the momentary stress intensification by a factor of 2 might have serious consequences on a component.

## 4.6 Maximum stress due to impact

An extension of the above problem is the case where the mass *m* is dropped on to the flange from a height *h*, causing a momentary extension of the bar $\delta'$. The total potential energy is $mg(h + \delta')$, and the momentarily stored strain energy is

$$U' = \tfrac{1}{2}\sigma'a\delta'$$

Then neglecting the mass of the bar and flange and assuming no losses of energy during impact,

$$\tfrac{1}{2}\sigma'a\delta' = mg(h + \delta')$$

or

$$\tfrac{1}{2}\sigma'\delta' = \frac{mgh}{a} + \frac{mg}{a}\,\delta' \qquad\qquad [4.40]$$

Now $\delta' = (\sigma'/E)l$, and $mg/a = \sigma$ is the final steady stress; substituting into eqn. [4.40],

$$(\sigma')^2 - 2\sigma\sigma' - 2\sigma\frac{Eh}{l} = 0$$

Therefore

$$\sigma' = \sigma + \left(\sigma^2 + 2\sigma\frac{Eh}{l}\right)^{1/2} \qquad\qquad [4.41]$$

It will be seen from this equation that there are two practical situations to consider:

(a)  as $h$ tends to zero then $\sigma' = 2\sigma$, which is the result obtained in the previous section; and

(b)  for larger values of $h$, the term $2\sigma Eh/l$ is large relative to $\sigma^2$ and so eqn. [4.41] becomes

$$\sigma' = \left(\frac{2\sigma Eh}{l}\right)^{1/2}$$

The true situation for the stress during impact of one body on another is more complicated than is indicated in this approximate analysis. In practice the deformation and stress imposed on the bar at the point of impact take time to propagate along the length of the bar. The *stress wave*, as it is called, on reaching the fixed end of the bar will be reflected towards the point of initiation. Nevertheless, the above solution provides a good approximation in most practical situations.

**Example 4.9**

The lower part of a child's pogo stick is illustrated in Fig. 4.9 when the child and stick are just about to descend to the ground and the spring is undeformed. Determine the momentary maximum stress in the steel compression tube on impact with the ground, and compare this with the final steady stress. It may be assumed that the outer sleeve and supports are rigid and that the ground does not deform. The weights of the various parts may be neglected. $E = 208\ \text{GN/m}^2$; spring stiffness $= 18\ \text{kN/m}$.

**Fig. 4.9**

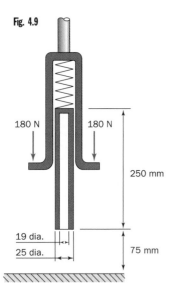

Let $\delta$ be the compression of the spring, and $x$ the compression of the tube. If the force in the spring on impact is momentarily $F$ then the strain energy stored in the spring and tube is

$$\frac{F\delta}{2} + \frac{\sigma^2}{2E} \times \text{volume}$$

$$\text{Tube volume} = 250\frac{\pi}{4}(25^2 - 19^2) \times 10^{-9} = 51.8 \times 10^{-6}\ \text{m}^3$$

Potential energy lost on impact $= 2 \times 180(0.075 + \delta + x)$

and

$$\delta = \frac{F}{18\,000} \, \text{m} \qquad x = \frac{Fl}{AE} = \frac{F \times 250 \times 10^{-3}}{208 \times 10^{-6} \times 208 \times 10^{9}} \, \text{m}$$

Equating potential and strain energies,

$$360\left(0.075 + \frac{F}{18\,000} + \frac{250F \times 10^{-6}}{208 \times 208}\right)$$

$$= \frac{F^2}{36\,000} + \frac{F^2 \times 51.8 \times 10^{-6}}{2 \times (208 \times 10^{-6})^2 \times 208 \times 10^{9}}$$

$$(F^2 \times 0.0278 \times 10^{-3}) - (F \times 20 \times 10^{-3}) - 27 = 0$$

$$F^2 - 720F - (970 \times 10^3) = 0$$

from which $F = 1410\,\text{N}$.

$$\sigma_{max} = \frac{1410}{208 \times 10^{-6}} = \textbf{6.78 MN/m}^2$$

Final steady stress, $\sigma = \dfrac{360}{208 \times 10^{-6}} = 1.73\,\text{MN/m}^2$

## 4.7 Summary

The importance of this chapter centres on the three requirements of principle set out in the introduction. If a problem cannot be solved by equilibrium statements alone, it is then described as statically indeterminate and we have to assess the geometry of deformation and also link stress and strain through the modulus and Poisson's ratio for the material. The bulk of the chapter has been devoted to a series of illustrative examples which set out the above three steps as appropriate to each case. It is therefore very important to work through each example to achieve a full understanding of the formulation of the equations, since from that point the solution is merely manipulative computation. The foregoing principles will be extensively used in the next two chapters on torsion and bending and thus this chapter must be thoroughly understood.

**Problems** 4.1 A composite shaft consists of a brass bar 50 mm in diameter and 200 mm long, to each end of which are concentrically friction-welded steel rods of 20 mm diameter and 100 mm length. During a tensile test to check the welds on the composite bar, at a particular stage the overall extension is measured as 0.15 mm. What are the axial stresses in the two parts of the bar? $E_{\text{brass}} = 120\,\text{GN/m}^2$, $E_{\text{steel}} = 208\,\text{GN/m}^2$.

4.2 A rigid bar AB is supported by 2 cables and subjected to a 10 kN force as shown in Fig. 4.10. Calculate the vertical displacement at the point of force application. $E_s = 200\,\text{GN/m}^2$.

**Fig. 4.10**

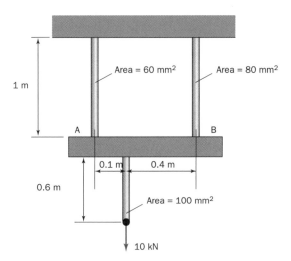

4.3    A spring-loaded buffer stop is illustrated in Fig. 4.11. The spring, which has a stiffness of 6 kN/mm, is located on the end of a steel tube of inner and outer diameters 21 mm and 29 mm respectively and 150 mm length. Determine accurately the total axial displacement of the system under a load of 30 kN. What would be the simple approximate solution? $E = 207\,\text{GN/m}^2$.

**Fig. 4.11**

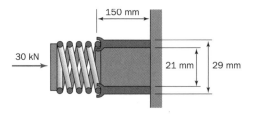

4.4    (*a*)  Create a spreadsheet to calculate the extension due to a load F of a number of bars in series, given the modulus *E*, cross sectional area *A* and length *L* of each bar. Calculate also the stress in each bar.

    (*b*)  Use the spreadsheet to find the extension of the bars in Problem 4.1 under a unit load. Use trial and error (or the spreadsheet 'Solve-for' facilities) to find the load to cause a total extension of $0.15 \times 10^{-3}\,\text{m}$.

    (*c*)  Use the spreadsheet to solve Problem 4.3 by replacing a bar stiffness $EA/L$ with the spring stiffness.

4.5    The spring shown in Fig. 4.12 has an unstretched length of 200 mm and a stiffness of 500 kN/m. If it is compressed to 150 mm, placed over the aluminium bar AB and then released, calculate the forces which will be exerted on each of the two side walls. Before the spring is released there is a gap of 0.15 mm between the end B and the right hand wall. $E_{\text{al}} = 70\,\text{GN/m}^2$.

4.6    The steel bolt shown in Fig. 4.13 has a thread pitch of 1.6 mm. If the nut is initially tightened up by hand so as to cause no stress in the copper spacing tube calculate the axial stresses in the copper and the

**Fig. 4.12**

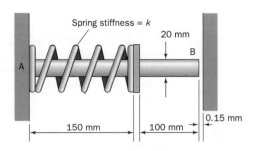

**Fig. 4.13**

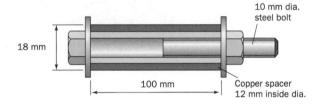

bolt if a spanner is then used to turn the nut through 90°. $E_c = 100\,\text{GN/m}^2$, $E_s = 208\,\text{GN/m}^2$.

4.7 Figure 4.14 shows a series of bars of modulus $E_i$, cross-sectional area $A_i$, length $L_i$ which are loaded in parallel by a force $F$.

The bars are slightly mismatched in length so that each bar has to be stretched by an amount $d_o$ to match its neighbours.

(a) Derive an expression for how the force in a given bar depends on $d$ and $d_{oi}$.

(b) By summing up forces on all bars, determine how the final deflection $d$ depends on the applied total force $F$.

(c) Set up a spreadsheet for an assembly of 2 bars. Calculate the stress in each bar once the deflection $d$ has been found.

(d) Use this to solve Problem 4.6. Note that the external load is zero in this case.

**Fig. 4.14**

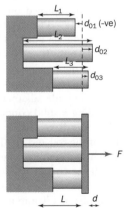

4.8　A hydraulic cylinder of 80 mm inside diameter and 4 mm wall thickness is welded to rigid end plates as shown in Fig. 4.15. The end plates are tied together by four rods of 8 mm diameter symmetrically arranged around the cylinder. Calculate the stresses in the rods and the cylinder at the cylinder design pressure of 20 MN/m². The cylinder and tie rods are made from steel with a Poisson's ratio value of 0.3.

**Fig. 4.15**

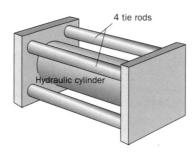

4.9　A rigid chute 2 m in length is supported horizontally, at a height of 0.33 m above a hopper, by a spring at one end of stiffness 30 kN/m and a second spring at mid-length of stiffness 20 kN/m (Fig. 4.16). Determine the position on the chute which a component of weight 2.5 kN reaches when the unsupported end of the chute just touches the edge of the hopper.

**Fig. 4.16**

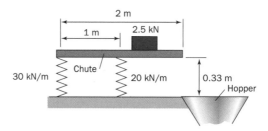

4.10　An elastic packing piece is bolted between a rigid rectangular plate and a rigid foundation by two bolts pitched 300 mm apart and symmetrically placed on the long centre-line of the plate, which is 450 mm long. The tension in each bolt is initially 120 kN, the extension of each bolt is 0.015 mm and the compression of the packing piece is 0.6 mm. If one bolt is further tightened to a tension of 150 kN, determine the tension in the other bolt.

4.11　A bimetallic temperature-sensitive component consists of a short steel tube of outside diameter 70 mm and inside diameter 60 mm, surrounding a solid copper rod of 50 mm diameter. At 20°C the rod and cylinder have exactly the same length. If a 100 kN load is placed on top of the rod and cylinder, calculate the forces in the two materials if the whole assembly is heated to 60°C. Calculate also the temperature at which the copper would take all the force. $E_s = 208 \, \text{GN/m}^2$, $E_c = 104 \, \text{GN/m}^2$, $\alpha_s = 12 \times 10^{-6}$ per deg C, $\alpha_c = 18.5 \times 10^{-6}$ per deg C.

4.12　A steel tube of 150 mm internal diameter and 8 mm wall thickness in a chemical plant is lined internally with a well-fitting copper sleeve of

2 mm wall thickness. If the composite tube is initially unstressed, calculate the circumferential stresses set up, assumed to be uniform through the wall thickness, in a unit length of each part of the tube due to an increase in temperature of 100°C. Neglect any temperature effect in the axial direction. For steel $\alpha = 11 \times 10^{-6}$ per deg C, $E = 208 \, \text{GN/m}^2$; for copper $\alpha = 18 \times 10^{-6}$ per deg C, $E = 104 \, \text{GN/m}^2$.

4.13 For a hydraulic test a steel tube of 80 mm internal diameter, 2 mm wall thickness and 1.2 m in length is fitted with end plugs and filled with oil at a pressure of $2 \, \text{MN/m}^2$. Determine the volume of oil leakage which would cause the pressure to fall to $1.5 \, \text{MN/m}^2$. Bulk modulus for the oil $= 2.8 \, \text{GN/m}^2$; for the steel $E = 208 \, \text{GN/m}^2$, $\nu = 0.29$.

4.14 A drop-weight shearing device consists of a vertical rod with a cross-sectional area of $125 \, \text{mm}^2$ which has an end collar which supports a spring of stiffness 150 kN/m, as shown in Fig. 4.17. If the shear tool of 10 kg mass is dropped through a height of 500 mm on to the spring, calculate (*a*) the initial instantaneous extension of the rod, (*b*) the maximum stress in the rod, and (*c*) the initial instantaneous compression of the spring. $E = 208 \, \text{GN/m}^2$.

Fig. 4.17

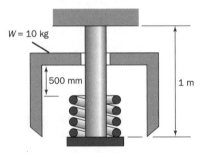

4.15 A drop-hammer used for forging metal is illustrated in Fig. 4.18. If the hammer is dropped through a height of 1 m on to the workpiece, calculate the resulting compression of the workpiece. Compare the force transmitted to the foundation for the system shown with that transmitted if the workpiece were resting on the foundation before the hammer was dropped 1 m on to it. Press: Mass of hammer $= 12\,000 \, \text{kg}$, mass of anvil $= 5000 \, \text{kg}$, spring stiffness $= 15 \, \text{MN/m}$. Workpiece: 30 mm dia. $\times$ 30 mm tall, modulus, $E = 208 \, \text{GN/m}^2$.

Fig. 4.18

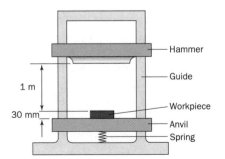

# 5 Torsion

One of the common engineering modes of deformation is that of torsion, in which a solid or tubular member is subjected to torque about its longitudinal axis resulting in twisting deformation. A design analysis is required in order to estimate the shear-stress distribution and angular twist for solid and hollow shafts of circular cross-section and thin-walled closed and open non-circular cross-sections. Engineering examples of the above are obtained in shafts transmitting power in machinery and transport, structural members in aeroplanes, springs, etc.

## 5.1 Torsion of a thin-walled cylinder

The thin cylinder of mean radius $r$, thickness $t$ and length $L$ shown in Fig. 5.1 is subjected to an axial torque $T$ at each end which causes the cylinder to twist about its longitudinal axis. In Chapter 2 it was shown that as a statically determinate problem only circumferential uniform shear stress in the wall of the cylinder was set up as a reaction to the torque $T$.

Fig. 5.1

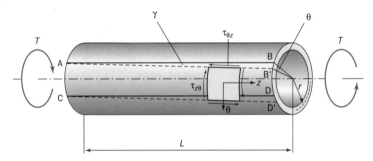

*Equilibrium*  Referring to Fig. 5.1, the shear stress $\tau_{z\theta}$ acting on an element of wall $tr\,d\theta$ gives a shear force

$$F = \tau_{z\theta}tr\,d\theta$$

This will provide a reacting moment about the central axis:

$$Fr = \tau_{z\theta}tr^2\,d\theta$$

Hence the total reacting torque will be

$$\int_0^{2\pi} \tau_{z\theta}tr^2\,d\theta$$

which is in equilibrium with the applied torque $T$. Therefore

$$T = \tau_{z\theta}tr^2 2\pi$$

or

$$\tau_{z\theta} = \frac{T}{2\pi r^2 t} \tag{5.1}$$

It should also be noted that the circumferential shear stress $\tau_{z\theta}$ is associated with a complementary shear stress $\tau_{\theta z}$ in the longitudinal direction in the wall. As there are no other shear stresses present, for simplicity the $z\theta$ suffices will be omitted.

*Geometry of deformation* The rotation of one end of the cylinder relative to the other through an angle $\theta$ results in a change in angle $\gamma$ between a cross-section and a longitudinal generator on the cylinder as in Fig. 5.1. The angle $\gamma$ is the shear strain associated with the shear stress. The displacement of B to B′ may be expressed both as $r\theta$ and $\gamma l$ and therefore

$$\gamma l = r\theta$$

and

$$\gamma = \frac{r\theta}{l} \tag{5.2}$$

*Stress–strain relationship* The stress–strain relationship in shear is expressed as

$$\frac{\tau}{\gamma} = G \tag{5.3}$$

From eqns. [5.1], [5.2] and [5.3] we can define the interrelationships for the cylinder as

$$\tau = \frac{Gr\theta}{l} = \frac{T}{2\pi r^2 t} \tag{5.4}$$

## 5.2 Torsion of a solid circular shaft

In the case of the thin-walled cylinder, in the previous section the shear stress was assumed to be constant throughout the wall, but in the case of a solid cylinder the shear stress varies over the cross-section. Firstly we shall require that:

(*a*) The shaft is straight and of uniform cross-section over its length.
(*b*) The torque is constant along the length of the shaft.

Further we note that the longitudinal and transverse symmetry of the shaft in relation to the applied torque enables the following deductions to be made:

1. Cross-sections which are plane before twisting remain plane during twisting.
2. Radial lines remain radial during twisting.
3. Deformation is by rotation of one cross-sectional plane relative to the next and planes remain normal to the axis of the shaft.

## 5.3 Relation between stress, strain and angle of twist

*Geometry of deformation*    The cylindrical shaft of length $L$ and outer radius $r_0$ subjected to torque $T$ may be regarded as being built up of a large number of thin-walled tubes just fitting inside each other. They are all twisted through the same angle of rotation $\theta$; therefore for any arbitrary tubes of radius $r_p$ and $r_q$ experiencing shear strain $\gamma_p$ and $\gamma_q$ from [5.2] we may write

$$\theta = \frac{\gamma_p L}{r_p} = \frac{\gamma_q L}{r_q}$$

or

$$\frac{\gamma_p}{r_p} = \frac{\gamma_q}{r_q} = \text{constant}$$

which demonstrates that at the centre of the shaft where $r$ is zero, $\gamma$ is zero, and that at the surface $\gamma$ is maximum and the variation is linear. Therefore

$$\frac{\gamma}{r} = \frac{\theta}{L} \tag{5.5}$$

*Stress–strain relation*    The shear-stress–shear-strain relation in terms of the shear modulus is

$$\frac{\tau}{\gamma} = G \tag{5.6}$$

From eqns. [5.5] and [5.6] we have

$$\frac{\tau}{r} = \frac{G\theta}{L} \tag{5.7}$$

Thus shear stress has a linear distribution across the shaft diameter, being zero at the centre and maximum at the outer surface, as shown in Fig. 5.2.

**Fig. 5.2**

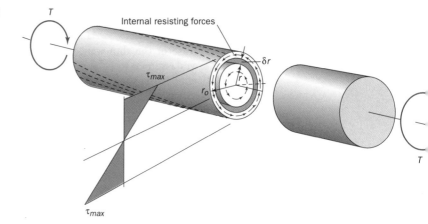

## 5.4 Relation between torque and shear stress

Referring again to any one of the thin tubes of radius $r$ and thickness $\delta r$ on which the shear stress is $\tau$.

*Equilibrium*

Force per unit length $= \tau\,\delta r$

Torque per unit length of tube about shaft axis $= \tau r\,\delta r$

Resisting torque on whole tube $= \tau r\,\delta r 2\pi r$

Resisting torque for whole cross-section $= \int_0^{r_o} \tau 2\pi r^2\,dr$

This is equal to the applied torque; therefore

$$T = \int_0^{r_o} \tau 2\pi r^2\,dr \qquad\qquad [5.8]$$

Using eqn. [5.7] to substitute for $\tau$,

$$T = \int_0^{r_o} \frac{G\theta}{L}2\pi r^3\,dr = \frac{G\theta}{L}\int_0^{r_o} 2\pi r^3\,dr \qquad [5.9]$$

The integral function

$$\int_0^{r_o} 2\pi r^3\,dr = \int r^2\,dA \quad \left(= \frac{\pi r_o^4}{2} \text{ for solid shaft}\right)$$

is the polar second moment of area of the section (see Appendix A) denoted by $\mathcal{J}$. Therefore,

$$T = \frac{G\theta}{L}\mathcal{J} \quad \text{or} \quad \frac{T}{\mathcal{J}} = \frac{G\theta}{L} \qquad\qquad [5.10]$$

Hence, from eqn. [5.7]

$$\tau = \frac{Tr}{\mathcal{J}}$$

which relates the shear stress to the torque and the geometry of the cross-section.

## 5.5 The torsion relationship

Combining eqns. [5.7] and [5.10] gives the fundamental relationship between shear stress, torque and geometry:

$$\frac{T}{\mathcal{J}} = \frac{\tau}{r} = \frac{G\theta}{L} \qquad\qquad [5.11]$$

The quantities concerned and their units are

$T =$ Torque, Nm

$\mathcal{J} =$ Polar second moment of area, m$^4$

$\tau =$ Shear stress, N/m$^2$ at radius $r$, m

$G =$ Shear modulus, N/m$^2$

$\theta =$ Angle of twist, radians (not degrees), over length $L$, m

## Example 5.1

Calculate the size of shaft which will transmit 40 kW at 2 rev/s. The shear stress is to be limited to 50 MN/m$^2$ and the twist of the shaft is not to exceed 1° for each 2 m length of shaft. The shear modulus $G$ is 77 GN/m$^2$.

Firstly converting the power to be transmitted into a torque,

$$T = \frac{40000}{2\pi \times 2} = 3183 \, \text{Nm}$$

From eqn. [5.11]

$$\tau_{max} = \frac{Tr_o}{J} = \frac{T}{\frac{1}{2}\pi r_o^3}$$

$$r_o^3 = \frac{2 \times 3183 \times 10^9}{\pi \times 50 \times 10^6} = 40.6 \times 10^3 \, \text{mm}^3$$

$$r_o = 34.4 \, \text{mm} \quad \text{on a stress basis}$$

Considering the twist criterion, from eqn. [5.10],

$$r_o^4 = \frac{2TL}{\pi G \theta}$$

$$= \frac{2 \times 3183 \times 2 \times 57.3 \times 10^{12}}{\pi \times 77 \times 10^9 \times 1}$$

$$= 302 \times 10^4 \, \text{mm}^4$$

$$r_o = 41.7 \, \text{mm} \quad \text{on a twist basis}$$

This is the governing criterion, and therefore the shaft diameter is 83.4 mm.

## 5.6 Torsion of a hollow circular shaft

The above analysis for the solid shaft is similarly applicable to the hollow shaft (thick-walled tube). Thus the torsion relationship equation [5.11] also expresses the conditions of equilibrium and compatibility for a hollow circular shaft. However, the radial boundaries are now $r = r_1$ and $r = r_2$, the outer and inner radii respectively, and thus the polar second moment of area is

$$J = \int_{r_2}^{r_1} 2\pi r^3 \, dr$$

$$= \frac{\pi}{2}(r_1^4 - r_2^4)$$

The shear stress varies linearly from $Tr_2/J$ at the bore to $Tr_1/J$ at the outer surface, as shown in Fig. 5.3.

The hollow shaft is more efficient in its use of stressed material than the solid shaft because the core of a solid shaft has relatively low stresses as compared with the outer layers. However, hollow shafts are not used widely in practice owing to the cost of machining, unless saving of weight is at a premium or it is necessary to pass services down the centre of the shaft.

**Example 5.2**

Compare the torque that can be transmitted by a hollow shaft with that of a solid shaft, as shown in Fig. 5.3, of the same material, weight, length and allowable stress.

**Fig. 5.3**

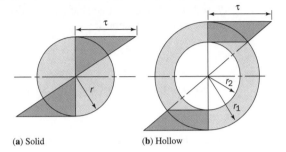

(a) Solid     (b) Hollow

Let $r_1$ and $r_2$ be the outer and inner radii of the hollow shaft and let $r$ be the radius of the solid shaft. Then, for the same maximum shear stress $\tau$,

$$T_{hollow} = \frac{\tau}{r_1} \frac{\pi}{2} (r_1^4 - r_2^4)$$

$$T_{solid} = \frac{\tau}{r} \frac{\pi}{2} r^4$$

Eliminating $\tau$ we have

$$\frac{T_{hollow}}{T_{solid}} = \frac{r_1^4 - r_2^4}{r_1 r^3} \qquad [5.12]$$

Since, for shafts of the same weight, $\pi r^2 = \pi(r_1^2 - r_2^2)$ per unit length, eqn. [5.12] can be simplified to

$$\frac{T_{hollow}}{T_{solid}} = \frac{r_1^2 + r_2^2}{r_1 r} = \frac{r_1}{r}\left(1 + \frac{1}{n^2}\right) \qquad [5.13]$$

where $n = r_1/r_2$

Now $r^2 = r_1^2 - r_2^2$, and putting $r_2$ in terms of $r_1/n$ gives

$$r^2 = r_1^2 - \left(\frac{r_1}{n}\right)^2 \quad \text{or} \quad \left(\frac{r_1}{r}\right) = \frac{n}{\sqrt{(n^2 - 1)}}$$

Therefore

$$\frac{T_{hollow}}{T_{solid}} = \frac{n^2 + 1}{n\sqrt{(n^2 - 1)}} \qquad [5.14]$$

It is common practice to take $n = 2$, which gives

$$\frac{T_{hollow}}{T_{solid}} = \frac{5}{2\sqrt{3}} = 1.44$$

Thus the hollow shaft can carry 44% greater torque than the solid shaft for the same weight, etc.

**Example 5.3**

**Compare the torsional stiffness of a thin-walled hollow tube with a solid shaft of the same material, torque capacity and allowable stress.**

For a thin tube, when mean radius $r$ is much less than wall thickness $t$, $\mathcal{J} = 2\pi r^3 t$. When the two shafts are subjected to equal torque, the maximum stress will be the same if

$$2\pi r^2 t = \frac{\pi a^3}{2}$$

where $a$ is the radius of the solid shaft. Alternatively

$$a = (4r^2 t)^{1/3}$$

The ratio of the stiffnesses will then be

$$k = \frac{T_{hollow}}{\theta_{hollow}} \Big/ \frac{T_{solid}}{\theta_{solid}} = \frac{G\mathcal{J}_{hollow}}{G\mathcal{J}_{solid}} = \frac{2\pi r^3 t}{\pi a^4/2} = \frac{4r^3 t}{a^4} = \left(\frac{r}{4t}\right)^{1/3}$$

The stiffest tube will be thin walled with the largest possible ratio of wall thickness to radius. However, when the tube becomes very thin it will fail, not by exceeding the maximum allowable shear stress in the material, but because the thin walls collapse by buckling or becoming unstable. A rolled up sheet of paper is very stiff in torsion when very small loads are applied, but it can not carry a useful amount of torque without buckling. An introduction to the analysis of buckling is given in Chapter 10.

It is usual that several possible modes of failure have to be considered in a design – there is a design goal, perhaps to maximize the shaft stiffness, subject to constraints such as maximum stress or the prevention of buckling.

## 5.7 Torsion of non-uniform and composite shafts

In certain shaft arrangements, the complete shaft is continuous but not of uniform diameter (Fig. 5.4), or the arrangement may be such that one shaft is hollow with another shaft arranged coaxially (Fig. 5.5). In each case it is necessary to investigate the conditions of both the torque and the angle of twist in each part of the system in order to obtain a sufficient number of equations for the solution.

### Continuous shaft having two different diameters

In this case (Fig. 5.4) the total torque $T$ is transmitted by each portion of the shaft; thus:

**Fig. 5.4**   Shafts in series

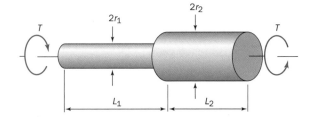

**Fig. 5.5** Shafts in parallel

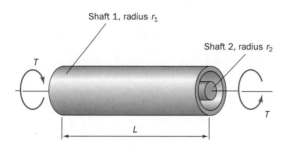

Shaft 1, radius $r_1$

Shaft 2, radius $r_2$

*Equilibrium*

$$T = T_1 = T_2 \qquad\qquad [5.15]$$

*Geometry of deformation*  The total deformation $\theta$ is due to $\theta_1$ over length $L_1$, plus $\theta_2$ over length $L_2$, so that

$$\theta = \theta_1 + \theta_2 \qquad\qquad [5.16]$$

*Stress–Strain Relation*  Substituting for $T_1$ and $T_2$ in eqn. [5.15], from [5.11]

$$\theta_1 = \frac{TL_1}{G\mathcal{J}_1} \quad \text{and} \quad \theta_2 = \frac{TL_2}{G\mathcal{J}_2} \qquad\qquad [5.17]$$

Hence,

$$\theta = \frac{T}{G}\left(\frac{L_1}{\mathcal{J}_1} + \frac{L_2}{\mathcal{J}_2}\right) \qquad\qquad [5.18]$$

If the shafts are of different materials then $G_1$ and $G_2$ may be used in the above analysis.

For a series of $n$ shafts in series, we can write

$$\theta = T \sum_{i=1}^{n} \frac{L_i}{G_i \mathcal{J}_i} \qquad\qquad [5.19]$$

**Concentric shafts**  In Fig. 5.5 the shafts have a common axis and are joined at the ends so that the total torque $T$ is made up of that carried by the hollow shaft and that carried by the solid shaft, these being $T_1$ and $T_2$ respectively:

*Equilibrium*

$$T = T_1 + T_2 \qquad\qquad [5.20]$$

*Geometry of deformation*  Both shafts twist through the same angle $\theta$ since their ends are rigidly connected; hence

$$\theta = \theta_1 = \theta_2$$

Using [5.11]

$$T = \frac{G\mathcal{J}\theta}{L}$$

In [5.20]

$$T = \frac{G_1 \mathcal{J}_1 \theta}{L_1} + \frac{G_2 \mathcal{J}_2 \theta}{L_2}$$

$$T = \frac{\theta}{L}(G_1 \mathcal{J}_1 + G_2 \mathcal{J}_2) \qquad [5.21]$$

This may be used to get the angle of twist as a result of any applied torque, or vice versa.

The maximum shear stresses in each shaft are

$$\tau_1 = \frac{Tr_1}{\mathcal{J}_1} \quad \text{and} \quad \tau_2 = \frac{Tr_2}{\mathcal{J}_2}$$

Hence

$$\frac{\tau_1}{\tau} = \frac{r_1}{r} \qquad [5.22]$$

so that the ratio of the maximum shear stresses is the same as the ratio of the outer diameters.

### Example 5.4

A solid alloy shaft of 50 mm diameter is to be friction welded concentrically to the end of a hollow steel shaft of the same external diameter (Fig. 5.6). Find the internal diameter of the steel shaft if the angle of twist per unit length is to be 75% of that of the alloy shaft.

What is the maximum torque that can be transmitted if the limiting shear stresses in the alloy and the steel are 50 MN/m$^2$ and 75 MN/m$^2$ respectively? $G_{steel} = 2.2G_{alloy}$.

**Fig. 5.6**

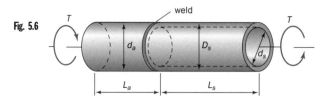

*Equilibrium*

$$T_{alloy} = T_{steel} = T \qquad [5.23]$$

*Geometry of deformation*

$$\frac{\theta_s}{L_s} = 0.75 \frac{\theta_a}{L_a} \qquad [5.24]$$

Hence

$$\frac{T_s}{\mathcal{J}_s G_s} = 0.75 \frac{T_a}{\mathcal{J}_a G_a}$$

Since

$$T_s = T_a \quad \text{and} \quad G_s = 2.2 G_a$$

$$\mathcal{J}_a = 2.2 \times 0.75 \mathcal{J}_s$$

$$\frac{\pi d_a^4}{32} = 2.2 \times 0.75 \frac{\pi}{32} (D_s^4 - d_s^4)$$

$$50^4 = (2.2 \times 0.75 \times 50^4) - (2.2 \times 0.75 \times d_s^4)$$

$$d_s^4 = \frac{0.65 \times 50^4}{2.2 \times 0.75} \qquad d_s = 39.6 \, \text{mm}$$

The torque that can be carried by the alloy is

$$T = \frac{\pi d^3}{16} \tau = \frac{\pi \times 50^3}{16 \times 10^9} \times 50 \times 10^6 = 1227 \, \text{Nm}$$

The torque that can be carried by the steel is

$$T = \frac{\pi}{16} \frac{(50^4 - 39.6^4)}{50 \times 10^9} \times 75 \times 10^6 = 1120 \, \text{Nm}$$

Hence the maximum allowable torque is 1120 Nm.

**Example 5.5**

A composite shaft of circular cross-section 0.5 m long is rigidly fixed at each end, as shown in Fig. 5.7. A 0.3 m length of the shaft is 50 mm in diameter and is made of bronze to which is joined the remaining 0.2 m length of 25 mm diameter made of steel. If the limiting shear stress in the steel is 55 MN/m² determine the maximum torque that can be applied at the joint. What is then the maximum shear stress in the bronze? $G_{steel} = 82$ GN/m²; $G_{bronze} = 41$ GN/m².

*Equilibrium* The free-body diagram Fig. 5.7(*b*) shows that the torque $T$ is shared between the two parts of the shaft. Therefore this is another

**Fig. 5.7**

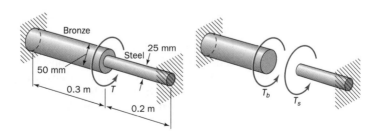

arrangement of the concentric shaft problem illustrated in the previous section.

$$T = T_s + T_b$$

The torque that can be carried by the steel is

$$T_s = \frac{2 \times 55 \times 10^6}{0.025} \times \frac{\pi \times 0.025^4}{32} = 169 \,\text{Nm}$$

*Geometry of deformation*    The angle of twist must be the same for each part at the joint; therefore

$$\theta_s = \theta_b$$

and the angle of twist for the steel is given by

$$\theta_s = \frac{169}{(\pi/32) \times 0.025^4} \times \frac{0.2}{82 \times 10^9} = 0.0108 \,\text{rad}$$

Therefore $\theta_b = 0.0108 \,\text{rad}$ and so

$$T_b = \frac{41 \times 10^9 \times 0.0108}{0.3} \times \frac{\pi \times 0.05^4}{32} = 906 \,\text{Nm}$$

The total torque that can be applied at the joint is

$$T_b + T_s = 906 + 169 = 1075 \,\text{Nm}$$

The maximum shear stress in the bronze is

$$\tau_b = \frac{41 \times 10^9 \times 0.0108}{0.3} \times \frac{25}{10^3} = 36.8 \,\text{MN/m}^2$$

## 5.8  Torsion of a thin tube of non-circular section

The torsion of a solid shaft of non-circular section is a complex problem, but in the case of a thin hollow shaft, or tube, a simple theory can be developed even if the tube thickness is not constant.

The thin-walled tube shown in Fig. 5.8 is assumed to be of constant cross-section throughout its length. The wall thickness is variable but at any point is taken to be $t$. Assume the applied torque $T$ to act about the longitudinal axis XX and to introduce shearing stresses over the end of the tube; these stresses have a direction parallel to that of a tangent to the tube at a given point. A shearing stress of magnitude $\tau$ at any point in the circumference has a complementary shear stress of the same magnitude acting in a longitudinal direction.

Consider the small portion ABCD of the tube and assume that the shearing stress $\tau$ is constant throughout the wall thickness $t$. The shearing force along the thin edge AB is $\tau t$ per unit length, and for longitudinal equilibrium of ABCD this force must be equal to that on the thin edge CD.

Now, since ABCD was an arbitrary choice, it follows that $\tau t$ is constant for all parts of the tube. The value $\tau t = q$ is called the *shear flow* and is an internal shearing force per unit length of the circumference of the section of the thin tube.

**Fig. 5.8**

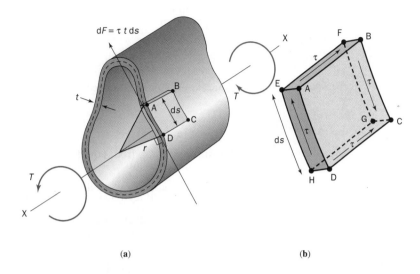

dF = τ t ds

(a)                                   (b)

The force, $dF$, acting in a tangential direction on an element of the perimeter of length ds is $\tau t$ ds, and if $r$ is the perpendicular distance from the tangent to the tube axis, then the moment of this force about the axis XX is $\tau tr$ ds and the total torque on the cross-section of the tube is

$$T = \oint \tau tr \, ds \qquad\qquad [5.25]$$

The integration extends over the whole circumference. Since $\tau t = q$ is constant, we may write

$$T = \tau t \oint r \, ds = q \oint r \, ds \qquad\qquad [5.26]$$

Now, $r$ ds is twice the shaded area shown, and $\oint r \, ds$ for the whole circumference is therefore equal to $2A$, where $A$ is the area enclosed by the centre-line of the wall of the tube (shown dotted); therefore

$$T = 2Aq \qquad\qquad [5.27]$$

Thus at any point the shearing stress is given by

$$\tau = \frac{q}{t} = \frac{T}{2At} \qquad\qquad [5.28]$$

Owing to the variation of shear stress around the circumference of the tube it is not possible to predict the deformations, and hence the angle of twist, in the simple manner used for the circular shaft or tube. The angle of twist $\theta$ can be determined, however, from the strain energy stored in the tube. Considering an axial strip along which the shear stress is constant, then from eqn. [3.9] the shear-strain energy per unit volume is

$$U_s = \frac{\tau^2}{2G}$$

If the strip is of length $l$, thickness $t$ and width $ds$, then the energy stored in the strip is

$$U_s = \frac{\tau^2}{2G} \times lt\,ds$$

Therefore the energy stored in the complete tube is

$$U_s = \oint \frac{\tau^2}{2G} lt\,ds$$

Substituting for $\tau$ from eqn. [5.28],

$$U_s = \oint \frac{T^2}{8A^2t^2G} lt\,ds = \frac{T^2l}{8A^2G} \oint \frac{ds}{t} \qquad [5.29]$$

But the stored energy is equal to the work $\frac{1}{2}T\theta$, since in the elastic range the torque is proportional to the angle of twist $\theta$; therefore

$$\frac{1}{2}T\theta = \frac{T^2l}{8A^2G} \oint \frac{ds}{t}$$

so that

$$\theta = \frac{Tl}{4A^2G} \oint \frac{ds}{t} \qquad [5.30]$$

If the tube is of constant thickness $t$ around the circumference $s$, then

$$\theta = \frac{Tls}{4A^2Gt}$$

or from eqn. [5.28] the angle of twist, $\theta$, may be expressed as a function of the shear stress, $\tau$.

$$\theta = \frac{\tau ls}{2AG} \qquad [5.31]$$

**Example 5.6**

The light-alloy stabilizing strut of a high-wing monoplane is 2 m long and has the cross-section shown in Fig. 5.9. Determine the torque that can be sustained and the angle of twist if the maximum shear stress is limited to 28 MN/m². G = 27 GN/m².

**Fig. 5.9**

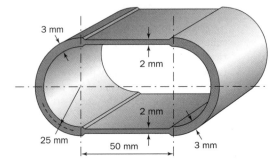

3 mm

2 mm

2 mm

25 mm    50 mm    3 mm

The area enclosed by the median line of the wall thickness is

$$A = (\pi \times 25^2) + (50 \times 50) = 4460 \, \text{mm}^2$$

The allowable torque is obtained using the minimum wall thickness:

$$T = 2At\tau$$

$$= 2 \times \frac{4460}{10^6} \times \frac{2}{10^3} \times 28 \times 10^6 = 500 \, \text{Nm}$$

The angle of twist is obtained from eqn. [5.30]:

$$\theta = \frac{Tl}{4A^2G} \oint \frac{ds}{t}$$

Therefore

$$\theta = \frac{500 \times 2}{4 \times 4460^2 \times 10^{-12} \times 27 \times 10^9} \left[ \frac{100}{2} + \frac{50\pi}{3} \right]$$

$$= 0.0476 \, \text{rad} = 2.73°$$

## 5.9 Torsion of a thin rectangular strip

An approximate solution for the torsion of a strip of rectangular cross-section whose thickness is small compared with the width, Fig. 5.10(a), can be obtained by considering the strip to be built up of a series of thin-walled concentric tubes which all twist by the same amount. One of these tubes is shown in Fig. 5.10(b), and the area enclosed by the median line is

$$A = (b - 2h)2h + \pi h^2$$

If $b$ is large compared with $h$ then the terms in $h^2$ can be ignored and

$$A \approx 2bh$$

If the tube is subjected to a torque $\delta T$ then, from eqn. [5.28], the shear stress is

$$\tau = \frac{\delta T}{4bh \, \delta h} \qquad \text{[5.32]}$$

**Fig. 5.10**

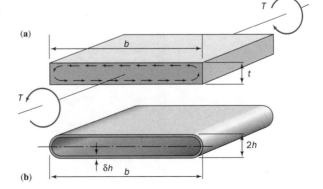

(a)

(b)

As the tube becomes infinitely thin,

$$\frac{\mathrm{d}T}{\mathrm{d}h} = 4bh\tau$$

and the torque carried by the strip is

$$T = \int_0^{t/2} 4bh\tau\,\mathrm{d}h \qquad\qquad [5.33]$$

From eqn. [5.31],

$$\tau = \frac{G\theta}{l}\frac{2A}{s} = \frac{G\theta}{l}\frac{4bh}{2b} \qquad\qquad [5.34]$$

Substituting in eqn. [5.33] for $\tau$,

$$T = \int_0^{t/2} \frac{G\theta}{l}\,8bh^2\,\mathrm{d}h$$

$$= \frac{1}{3}\,bt^3\,\frac{G\theta}{l} \qquad\qquad [5.35]$$

The quantity $bt^3/3$ is termed the *torsion constant*, but it is *not* the polar second moment of area for the section.

Equation [5.34] shows that the shear stress parallel to the long edge of the cross-section is proportional to the distance $h$ from the central axis. The maximum shear stress occurs at the outer surface and is

$$\tau_{max} = \frac{tG\theta}{l} \qquad\qquad [5.36]$$

**Example 5.7**

Determine the angle of twist per unit length and the maximum shear stress in the aluminium channel section shown in Fig. 5.11 when subjected to a pure torque of 20 Nm. Shear modulus = 27 GN/m².

**Fig. 5.11**

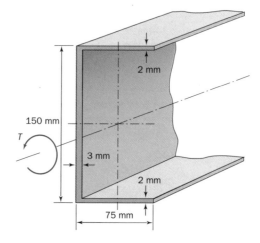

The channel may be analysed as three rectangular strips, the two flanges and the web, and the above solution will be used for each part. Let the proportions of the torque carried by the flanges and web be $T_1$ and $T_2$ respectively; then from eqn. [5.35],

$$T_1 = \frac{1}{3} \times \frac{75}{10^3} \times \left(\frac{2}{10^3}\right)^3 \times 27 \times 10^9 \frac{\theta}{l}$$

$$= 5.4 \frac{\theta}{l}$$

$$T_2 = \frac{1}{3} \times \frac{150}{10^3} \times \left(\frac{3}{10^3}\right)^3 \times 27 \times 10^9 \frac{\theta}{l}$$

$$= 36.5 \frac{\theta}{l}$$

But $2T_1 + T_2 = 20$; therefore

$$47.3 \frac{\theta}{l} = 20$$

and

$$\frac{\theta}{l} = 0.422 \, \text{rad/m} = \mathbf{24.2°/m}$$

The maximum shear stress in the flanges, from eqn. [5.36], is

$$\tau_{max} = \frac{2}{10^3} \times 27 \times 10^9 \times 0.422 = 22.8 \, \text{MN/m}^2$$

and in the web is

$$\tau_{max} = \frac{3}{10^3} \times 27 \times 10^9 \times 0.422 = 34.2 \, \text{MN/m}^2$$

**Example 5.8**

Compare the torsional strength and stiffness of a thin-walled tube of circular cross-section of mean radius *R* and thickness *t*, with the values for a similar tube having a thin slit cut along its full length, as shown in Fig. 5.12.

**Fig. 5.12**

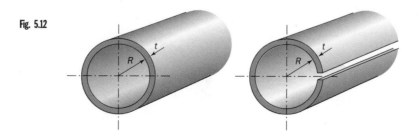

             Closed                   Open

(a) Strength:

$$\frac{T}{J} = \frac{\tau}{R}$$

           Treating open tube as a thin rectangular strip:

$$\tau = \frac{TR}{\pi R^3 t} = \frac{T}{\pi R^2 t} \qquad\qquad \tau = t \cdot \frac{G\theta}{L} = t \cdot \frac{3T}{bt^3} = \frac{3T}{\pi R t^2}$$

$$\frac{\text{Open}}{\text{Closed}} = \frac{3T\pi R^2 t}{\pi R t^2 T} = \frac{3R}{t}$$

(b) Stiffness:

$$\frac{T}{\theta} = \frac{GJ}{L} = \frac{G\pi R^3 t}{L} \qquad\qquad \frac{T}{\theta} = \frac{1}{3}bt^3 \frac{G}{L}$$

$$\frac{T}{\theta} = \frac{G}{L}(\pi R^3 t) \qquad\qquad\qquad \frac{T}{\theta} = \frac{G}{L}(\tfrac{1}{3}\pi R t^3)$$

$$\frac{\text{Open}}{\text{Closed}} = \frac{\tfrac{1}{3}\pi R^3 t}{\pi R^3 t} = \frac{t^2}{3R^2}$$

It may be seen from this example that a tube with a longitudinal slit has a very small torsional stiffness compared with that of a complete tube.

### 5.10 Effect of warping during torsion

In the previous section and the above example dealing with a thin-walled open section subjected to torsion, the pure torque was supposed to be applied to each end of the member in such a way that there was no axial restraint. Owing to the variation in transverse shear stress, for example in the flanges of the above channel or the I-section shown in Fig. 5.13, there is also a variation in longitudinal complementary shear stress which results in the axial movement of one flange with respect to the other. Therefore, cross-sections which were initially plane do not remain so during torsion and there is warping of any cross-section. If one or more sections of a member are constrained in some manner to remain plane during torsion then warping is restrained.

**Fig. 5.13**

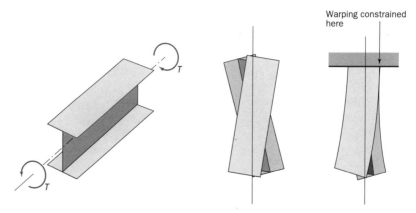

Warping constrained here

The warping constraint has most significance near the ends of the section. It decays with distance from the constraint so that after several section depths the influence of warping is negligible.

Resisting torque in the section is supplied in two ways, by simple torsion and also by torque set up through the restraint of warping. Thus an applied torque will cause a smaller angle of twist than when the section is free to warp, and torsional stiffness may be considerably increased if warping is restrained. Non-circular closed and solid-section members also exhibit warping, but the effect is much smaller in comparison with the open section. Further discussion of warping behaviour can be found in books such as that by Megson[1].

**5.11 Torsion of solid rectangular and square cross-sections**

The solution in section 5.9 is only applicable for a rectangular section in which the longer side is much greater than the other. Unfortunately the solution for torsion of general rectangular and square sections, although of some engineering interest, is very complex and beyond the scope of this text. (The reader who is interested in the full solution should consult a text such as *Theory of Elasticity* by Timoshenko and Goodier[2].) The result in terms of torque, maximum shear stress at the centre of the long side and angle of twist may be expressed as

$$T = \alpha b t^2 \tau_{max} = \beta b t^3 G \frac{\theta}{l} \qquad [5.37]$$

where $b$ is the longer and $t$ the shorter side, and $\alpha$, $\beta$ are factors dependent on the geometry, as given in Table 5.1. It may be seen that as $b/t$ becomes very large, $\beta \rightarrow \frac{1}{3}$ as shown in eqn. [5.35].

**Table 5.1**

| $b/t$ | 1 | 1.5 | 2 | 2.5 | 3 | 4 | 6 | 10 | $\infty$ |
|---|---|---|---|---|---|---|---|---|---|
| $\alpha$ | 0.208 | 0.231 | 0.246 | 0.256 | 0.267 | 0.282 | 0.299 | 0.312 | 0.333 |
| $\beta$ | 0.141 | 0.196 | 0.229 | 0.249 | 0.263 | 0.281 | 0.299 | 0.312 | 0.333 |

**5.12 Summary**

The analysis of the thin-walled circular tube highlights the basic steps of equilibrium of internal and external forces, the geometry of deformation of the tube and the elastic shear-stress/shear-strain relationship for the material. A full appreciation of the foregoing makes the understanding of the solution for the solid and hollow circular shafts relatively simple, leading to the basic relationship

$$\frac{T}{J} = \frac{G\theta}{L} = \frac{\tau}{r}$$

from which we see that shear stress and strain are linearly distributed across the section, being maximum at the outer surface. In the case of composite shafts the additional assessed information to the above is the manner in which the applied torque or the angle of twist is shared between different components.

Non-circular thin tubes and other open sections play an important part in engineering, and, whereas the stresses can be determined from the equilibrium state, we have to evaluate the shear-strain energy to determine the amount of twist.

Finally the effect of warping of the cross-section and its restraint have a considerable effect on the stiffness of short, non-circular torsion members and this must be taken into account.

**References**

1. Megson, T. H. G. (1990) *Aircraft Structures for Engineering Students*, 2nd edition, Edward Arnold, London.
2. Timoshenko, S. P. and Goodier, J. N. (1970) *Theory of Elasticity* 3rd edition, McGraw-Hill, New York.

**Problems**

5.1 A hollow steel shaft with an external diameter of 150 mm is required to transmit 1 MW at 300 rev/min. Calculate a suitable internal diameter for the shaft if its shear stress is not to exceed 70 MN/m².

Compare the torque-carrying capacity of this shaft with a solid steel shaft having the same weight per unit length and limiting shear stress. $G = 80\,\text{GN/m}^2$.

5.2 Given the applied torque on a shaft, its length and the shear modulus of the material from which it is made, develop a spreadsheet to calculate:

(*a*) the angle of twist of the shaft
(*b*) the maximum shear stress in the shaft

for

(*i*) a hollow circular tube
(*ii*) a thin rectangular strip.

5.3 Two shafts AB and CD are coupled by gear wheels as shown in Fig. 5.14. If the pitch circle diameter (PCD) of the gear wheel B is 180 mm and the PCD of gear wheel C is 60 mm, calculate the angle of twist of D relative to A when a torque of 3 kNm is applied at D. The shear modulus of the shaft material is 80 GN/m². Shaft AB is 0.15 m long and 30 mm diameter whilst shaft CD is 0.2 m long and 25 mm diameter.

**Fig. 5.14**

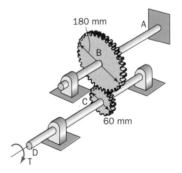

5.4 The gearbox in Fig. 5.15 is required to supply an output torque of 300 Nm at a shaft speed of 100 rev/min. The gear ratios are such that shaft 1 rotates at three times the speed of shaft 2, and shaft 2 rotates at five times the speed of shaft 3. Calculate the input power required from the motor and the diameters of shafts 2 and 3 if the allowable

**Fig. 5.15**

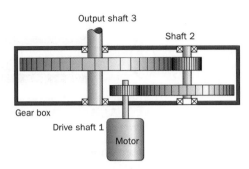

Output shaft 3

Shaft 2

Gear box

Drive shaft 1

Motor

shear stress of the shaft material is $400 \, \text{MN/m}^2$. Neglect losses in the system and assume a safety factor of 2.

5.5 A steel shaft has to transmit 1 MW at 240 rev/min so that the maximum shear stress does not exceed $55 \, \text{MN/m}^2$ and there is not more than 2° twist on a length of 30 diameters. Determine the required diameter of shaft. $G = 80 \, \text{GN/m}^2$.

5.6 A shaft carries five pulleys, A, B, C, D and E, and details of shaft diameters, lengths and pulley torques are given in Table 5.2. Determine in which sections the maximum shear stress and angle of twist occur. $G = 80 \, \text{GN/m}^2$.

**Table 5.2**

| Pulley | Torque (Nm) | Direction | Shaft | Length (m) | Diameter (mm) |
|--------|-------------|---------------|-------|-----------------|---------------|
| A | 60 | Clockwise | AB | 1 | 38 |
| B | 900 | Anticlockwise | BC | $\frac{1}{2}$ | 100 |
| C | 300 | Clockwise | CD | $\frac{1}{2}$ | 75 |
| D | 640 | Clockwise | DE | $1\frac{1}{2}$ | 75 |
| E | 100 | Anticlockwise | | | |

5.7 A hollow steel drive shaft with an outside diameter of 50 mm and an inside diameter of 40 mm is fastened to a coupling using using six 8 mm diameter bolts on a pitch circle diameter of 150 mm. If the shear stress in the shaft or bolt material is not to exceed $170 \, \text{MN/m}^2$ calculate the maximum power which the system could transmit at a rotational speed of 100 rev/min. What would be the effect of only using three of the bolts?

5.8 A torsional vibration damper consists of a hollow steel shaft fixed at one end and to the other end is attached a solid circular steel shaft which passes concentrically along the inside of the hollow shaft, as shown in Fig. 5.16. Determine the maximum torque $T$ that can be applied to the free end of the solid shaft so that the angle of twist where the torque is applied does not exceed 5°. Local effects where the two parts are connected may be ignored. Shear modulus, $G = 80 \, \text{GN/m}^2$.

5.9 Two shafts are connected end-to-end by means of a coupling in which there are twelve bolts on a pitch circle diameter of 250 mm. The

**Fig. 5.16**

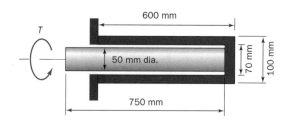

maximum shear stress is limited to $36 \, \text{MN/m}^2$ in the shafts and $16 \, \text{MN/m}^2$ in the bolts. If one shaft is solid, 50 mm diameter, and the other hollow, 60 mm external diameter, calculate the internal diameter of the latter and the bolt diameter so that both shafts and coupling are equally strong.

5.10 A compound drive shaft consists of a solid circular steel bar B surrounded for part of its length by an aluminium tube A as shown in Fig. 5.17. The contacting surfaces are smooth and the collar and the shaft are both rigidly fixed to a machine at D. Pin C fills a hole drilled completely through a diameter of the collar and shaft. The shearing deformation in the pin and the bearing deformation between the pin and shaft can be neglected. Calculate the maximum torque $T$ which can be applied to the steel shaft as shown, without exceeding an average shearing stress of $8 \, \text{MN/m}^2$ on the cross sectional area of the pin at C, the interface between the shaft and collar. $G_{steel} = 80 \, \text{GN/m}^2$; $G_{aluminium} = 28 \, \text{GN/m}^2$.

**Fig. 5.17**

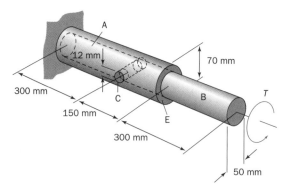

5.11 A hollow rectangular section tube with the dimensions shown in Fig. 5.18 is subjected to a torque of 20 Nm. Calculate the shear stresses in the walls of the tube and the angle of twist per unit length. $G = 27 \, \text{GN/m}^2$. Compare the results with those of the similar open section in Example 5.7.

5.12 An aluminium-alloy strut for a light aircraft having the cross-section shown in Fig. 5.19 is 3 m in length. If the shear stress is not to exceed $30 \, \text{MN/m}^2$ and the applied torque is 134 Nm, determine the required thickness $t$ of metal. What is the angle of twist? $G = 28 \, \text{GN/m}^2$.

5.13 A torsional member used for stirring a chemical process is made of a circular tube to which are welded four rectangular strips as shown in

**Fig. 5.18**

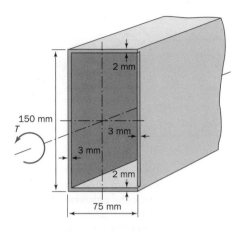

**Fig. 5.19**

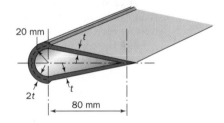

**Fig. 5.20**

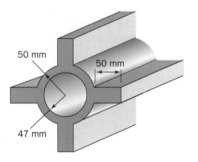

Fig. 5.20. The tube has inner and outer diameters of 94 mm and 100 mm respectively, each strip is 50 mm by 18 mm, and the stirrer is 3 m in length. If the maximum shearing stress in any part of the cross-section is limited to 56 MN/m², neglecting any stress concentration, calculate the maximum torque which can be carried by the stirrer and the resulting angle of twist over the full length. $\alpha = 0.264$, $\beta = 0.258$, $G = 83$ GN/m².

5.14 An I-beam has a width of 100 mm and depth of 150 mm, as shown in Fig. 5.21. Calculate the maximum pure torque which could be applied to this beam if the yield shear stress is 240 MN/m². Assume a safety factor of 3.

5.15 Using the spreadsheet formulae developed in Problem 5.2, sum the contributions from the individual thin rectangular strips to derive the torsional stiffness of an open section composed of 3 strips. From this

**Fig. 5.21**

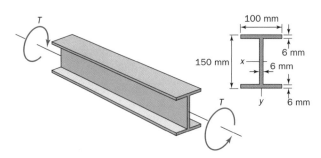

calculate the angle of twist for a shaft of a given length and the maximum shear stress in each strip. Use the resulting spreadsheet to check the answers for the channel section of Example 5.7 and the I-beam of Problem 5.14.

5.16 Using the spreadsheet formulae developed in Problem 5.2, sum the total angle of twist for two shafts in series, given the shear modulus $G$, section polar moment $J$ and length $L$. Calculate the maximum shear stress in each bar.

   Hint: the spreadsheet will be very similar to that of Problem 4.4 for the extension of a series of bars.

   (a) Use the resulting spreadsheet to confirm the results of Example 5.4. In other words, plug in the values obtained for the shaft diameters and torques, then check that the twist of the steel shaft is 75% of that in the alloy shaft and that the stress constraints are satisfied.

   (b) Use the spreadsheet to check the results of Example 5.5. Plug in the torques obtained for each shaft and confirm that the total twist is zero and that the maximum shear stresses are correct.

# 6 Bending: Stress

One of the most common modes of deformation of engineering structures and components is that of bending. The simplest form is the slender member, often termed a beam, subjected to transverse loading, which is the subject of this chapter. In Chapter 16 the more complex states of bending in plates and shells are studied.

Beams are designed both for strength and deflection, and this chapter is concerned with the former. The stresses that are set up in bending are tension and compression along the length and shear across the section of the beam. These stresses derive directly from the internal reactions of bending moment and shear force introduced in Chapter 1 in the applications of equilibrium. The importance of being able to establish shear-force and bending-moment distributions cannot be overemphasized, since the correct determination of a design based on allowable stresses and deflection depends upon that first step. Because of this, the chapter commences with a number of illustrations of the method of calculating shear-force and bending-moment distributions. This is then followed by analyses of bending stresses and shear stresses for various beam configurations.

## 6.1 Shear-force and bending-moment distributions

The reader should refer back to Chapter 1, if necessary, to revise the basic notation, sign conventions and equilibrium principles for establishing shear-force and bending-moment distributions. The following worked examples should help to reinforce the basic understanding of the procedure.

**Simply supported beam carrying concentrated loads**

Referring to the loading shown in Fig. 6.1, the left-hand reaction is first required and as the beam is statically determinate the reactions can be found from the equations of force and moment equilibrium.

$$-R_A - R_D + W_1 + W_2 = 0$$

$$-R_A L + W_1(L - a) + W_2(L - b) = 0$$

from which

$$R_A = \frac{W_1(L - a) + W_2(L - b)}{L}$$

The two concentrated loads cause mathematical discontinuities and so separate free-body diagrams and equilibrium equations have to be expressed for each different section, as shown in Fig. 6.1 and in the following equations respectively:

*Shear force*

$$\text{AB} \qquad 0 < x < a \qquad +Q - R_A = 0 \qquad Q = R_A$$

$$\text{BC} \qquad a < x < b \qquad +Q - R_A + W_1 = 0$$

$$Q = +R_A - W_1$$

**Fig. 6.1**

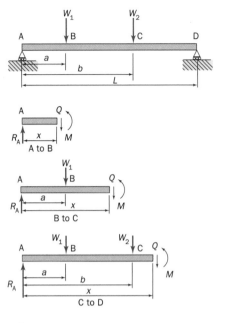

**Fig. 6.2**

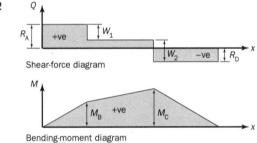

Shear-force diagram

Bending-moment diagram

$$\text{CD} \qquad b < x < L \qquad +Q - R_A + W_1 + W_2 = 0$$

$$Q = R_A - W_1 - W_2$$

*Bending moment*

$$\text{AB} \qquad 0 < x < a \qquad M - R_A x = 0 \qquad M = R_A x$$

$$\text{BC} \qquad a < x < b \qquad M - R_A x + W_1(x - a) = 0$$

$$M = R_A x - W_1(x - a)$$

$$\text{CD} \qquad b < x < L \qquad M - R_A x + W_1(x - a) + W_2(x - b) = 0$$

$$M = R_A x - W_1(x - a) - W_2(x - b)$$

The diagrams resulting from the above equations are shown in Fig. 6.2. It should be noted that

(a) The S.F. diagram *changes* in value at a support or concentrated load by the amount of the reaction or load.
(b) The B.M. is *always zero* at the ends of a beam which are either unsupported or on a simple support.

**Simply supported beam with an applied couple**   Apart from transverse loads, bending can be caused by a couple applied to the beam at a cross-section, as shown in Fig. 6.3. The reactions at the supports are obtained from moment equilibrium of the whole beam, from which

$$R_A L = R_B L = \bar{M} \quad \text{or} \quad R_A = R_B = \frac{\bar{M}}{L}$$

*Shear force*   The shear force is constant along the length of the beam and of value $Q = -\bar{M}/L$.

Fig. 6.3

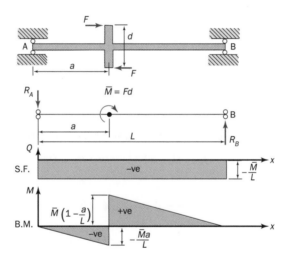

*Bending moment*

$$\text{For } x < a: \quad M + R_A x = 0; \quad \text{hence} \quad M = -\frac{\bar{M}x}{L}$$

$$\text{For } x > a: \quad M + R_A x - \bar{M} = 0; \quad \text{hence} \quad M = \bar{M} - \frac{\bar{M}x}{L}$$

$$= \frac{\bar{M}}{L}(L - x)$$

The S.F. and B.M. diagrams shown in Fig. 6.3 were derived from these equations.

**Example 6.1**    Sketch a B.M. diagram for the curved member shown in Fig. 6.4(a), giving the principal numerical values.

**Fig. 6.4**

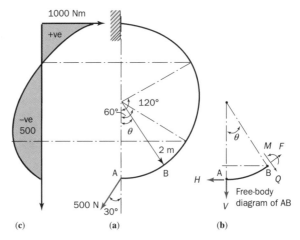

(c)                    (a)                         (b)

The determination of bending-moment distribution along a slender curved member follows exactly the same procedure as for straight beams.

It is more convenient to resolve the applied load into vertical and horizontal components. These are

$$V = 500 \ \cos \ 30° = 250\sqrt{3} \qquad \text{and} \qquad H = 500 \ \sin \ 30° = 250$$

A portion of the bar is cut off to give a free body and the forces and moment are inserted as shown in Fig. 6.4(b).

*Bending moment*    Taking moments about B

$$M - 500(1 - \ \cos \ \theta) + 500\sqrt{3} \ \sin \ \theta = 0$$

At

$$\theta = 0 \qquad M = 0$$

$$\theta = \pi/2 \qquad M = -365 \ \text{N/m}$$

$$\theta = \pi \qquad M = +1000 \ \text{N/m}$$

Also, if $M = 0 = 500(1 - \ \cos \ \theta - \sqrt{3} \ \sin \ \theta)$, then

$$\cos \ \theta = -\tfrac{1}{2} \qquad \text{and} \qquad \theta = 120°$$

which gives a point of contraflexure, where the bending moment changes sign. It may be seen that this coincides with the point at which the line of action of the applied force intersects the beam.

When $dM/d\theta = 0$,

$$\sin \ \theta - \sqrt{3} \ \cos \ \theta = 0 \qquad \theta = 60°$$

$$M_{60°} = 500\left(1 - \tfrac{1}{2} - \sqrt{3}\frac{\sqrt{3}}{2}\right) = -500 \ \text{Nm}$$

The B.M. diagram is plotted for convenience along the vertical axis as shown in Fig. 6.4(*c*).

**Example 6.2**

**Sketch the S.F. and B.M. diagrams for the cantilever shown in Fig. 6.5.**

Fig. 6.5

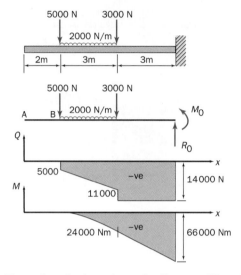

Even though there is no loading on AB and therefore no shear force or bending moment, the distance $x$ will still be measured from A rather than B.

*Shear force*

$$0 < x < 2 \quad Q = 0$$

$$2 < x < 5 \quad Q = -5000 - 2000(x - 2)$$

$$5 < x < 8 \quad Q = -5000 - 6000 - 3000 = -14\,000\,\text{N}$$

*Bending moment*

$$0 < x < 2 \quad M = 0$$

$$2 < x < 5 \quad M = -5000(x - 2) - 2000(x - 2)\frac{(x - 2)}{2}$$

$$= -1000(x - 2)(x + 3)\,\text{Nm}$$

$$\text{(parabolic distribution)}$$

$$5 < x < 8 \quad M = -5000(x - 2) - 6000(x - 3\tfrac{1}{2}) - 3000(x - 5)$$

$$= -14\,000x + 46\,000\,\text{Nm} \quad \text{(linear distribution)}$$

The diagrams and principal values are shown in Fig. 6.5.

**6.2   Relationships between loading, shear force and bending moment**

The beam shown in Fig. 6.6(*a*) carries distributed loading which varies in an arbitrary manner. Consider the free body, Fig. 6.6(*b*), of a small slice of length d*x* for which the loading may be regarded as uniform, *w*.

**Fig. 6.6**

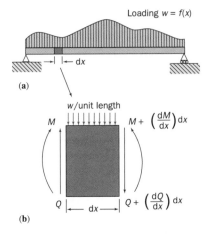

(a)

(b)

*Vertical equilibrium*

$$-Q + w\,\mathrm{d}x + \left(Q + \frac{\mathrm{d}Q}{\mathrm{d}x}\mathrm{d}x\right) = 0$$

$$w + \frac{\mathrm{d}Q}{\mathrm{d}x} = 0$$

$$w = -\frac{\mathrm{d}Q}{\mathrm{d}x} \qquad [6.1]$$

From eqn. [6.1] it follows that, between any two sections denoted by 1 and 2,

$$\int_1^2 \mathrm{d}Q = \int_1^2 -w\,\mathrm{d}x$$

or

$$Q_2 - Q_1 = \int_1^2 -w\,\mathrm{d}x \qquad [6.2]$$

Thus the *change* in shear force between any two cross-sections may be obtained from the area under the load distribution curve between those sections. The value of the shear force at point 2 is given by

$$Q_2 = Q_1 - \int_1^2 w\,\mathrm{d}x$$

The use of this equation is illustrated in Example 6.3.

*Moment equilibrium* Taking moments about the mid-point of the right-hand edge,

$$-M - Q\,dx + w\,dx\frac{dx}{2} + \left(M + \frac{dM}{dx}dx\right) = 0$$

Neglecting the term multiplied by $(dx)^2$,

$$-Q + \frac{dM}{dx} = 0$$

$$Q = \frac{dM}{dx} \qquad [6.3]$$

When $Q = 0$, $dM/dx = 0$, i.e. the S.F. is zero when the *slope* of a B.M. diagram is zero and not simply when $M$ is a maximum, although this also is implied. It is possible to have $M = M_{max}$ when $dM/dx \neq 0$ (see Fig. 6.4).

From eqn. [6.3], considering two sections, 1 and 2,

$$\int_1^2 dM = \int_1^2 Q\,dx$$

or

$$M_2 - M_1 = \int_1^2 Q\,dx$$

Thus the *change* in bending moment between any two sections is found from the area under the shear-force diagram between those sections. The value of the bending moment at point 2 is given by

$$M_2 = M_1 + \int_1^2 Q\,dx \qquad [6.4]$$

The use of this equation is illustrated in the following example.

**Example 6.3**

**Determine the position of zero shear force and the value of the maximum bending moment for the simply supported beam shown in Fig. 6.7. It carries a distributed load which varies linearly in intensity from zero at the left to w at the right-hand end. Plot the S.F. and B.M. diagrams.**

**Fig. 6.7**

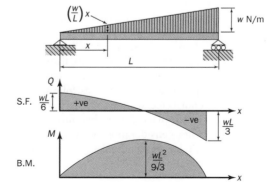

By similar triangles, the load intensity at $x$ from the left end is $(w/L)x$. From eqn. [6.2] the shear force is given by

*Shear force*

$$Q = Q_0 - \int \frac{wx}{L} \, \mathrm{d}x = Q_0 - \frac{wx^2}{2L}$$

*Bending moment*     From eqn. [6.4], and substituting for $Q$,

$$M = M_0 + \int Q \, \mathrm{d}x = M_0 + Q_0 x - \frac{wx^3}{6L}$$

Now, $M = 0$ when $x = 0$ and $L$. Hence

$$M_0 = 0 \quad \text{and} \quad Q_0 = \frac{wL}{6}$$

Thus

$$Q = -\frac{wx^2}{2L} + \frac{wL}{6}$$

and when $Q = 0$, $x = L/\sqrt{3}$. Now,

$$M = -\frac{wx^3}{6L} + \frac{wLx}{6}$$

Since $\mathrm{d}M/\mathrm{d}x = Q = 0$ at $x = L/\sqrt{3}$ is a possible solution for $M_{max}$,

$$M_{max} = -\frac{w}{6L}\left(\frac{L}{\sqrt{3}}\right)^3 + \frac{wL}{6}\frac{L}{\sqrt{3}} = \frac{wL^2}{9\sqrt{3}}$$

The S.F. and B.M. diagrams are as shown in Fig. 6.7.

## 6.3 Stress distribution in pure bending
### Assumptions

The determination of stress distributions in pure bending is a statically indeterminate problem, and hence all three basic principles stated in the introduction in Chapter 4 will need to be employed. However, the steps will be considered in a different order since at this stage there is insufficient information about the stress distribution to enable an equilibrium equation to be formulated. The approach initially will be to consider a prismatic beam of symmetrical cross-section subjected to pure bending, from which the *geometry of deformtion* can be studied and the strain distribution determined. The *stress–strain relations* will give the stress distribution which can be related to forces and moments through an *equilibrium condition*.

Before commencing the analysis it is necessary to make some assumptions and these are as follows:

1. Transverse sections of the beam which are plane before bending will remain plane during bending.
2. From consideration of symmetry during bending, transverse sections will be perpendicular to circular arcs having a common centre of curvature.

3. The radius of curvature of the beam during bending is large compared with the transverse dimensions.
4. Longitudinal elements of the beam are subjected only to simple tension or compression, and there is no lateral stress.
5. Young's modulus for the beam material has the same value in tension and compression.

## 6.4 Deformations in pure bending

### Longitudinal deformation

Consideration of the beam subjected to pure bending shown in Fig. 6.8 indicates that the lower surface stretches and is therefore in tension and the upper surface shortens and thus is in compression. Hence there must be an

**Fig. 6.8**

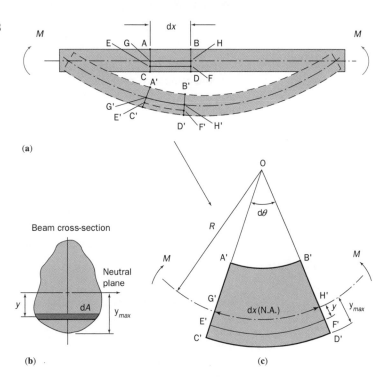

(a)

(b)    (c)

Beam cross-section

xz-plane in between in which longitudinal deformation is zero. This is termed the *neutral plane*, and a transverse axis lying in the neutral plane is the *neutral axis*. Consider the deformations between two sections AC and BD, a distance $dx$ apart, of an initially straight beam. A longitudinal fibre EF at distance $y$ below the neutral axis will have initially the same length as the fibre GH at the neutral axis. During bending EF stretches to become E'F', but GH, being at the neutral axis, is unstrained when it becomes G'H'. Therefore, if $R$ is the radius of curvature of G'H',

$$G'H' = GH = dx = R\,d\theta$$

$$E'F' = (R + y)\,d\theta$$

and the longitudinal strain in fibre $E'F'$ is

$$\varepsilon_x = \frac{E'F' - EF}{EF}$$

But $EF = GH = G'H' = R\,d\theta$; therefore

$$\varepsilon_x = \frac{(R+y)d\theta - R\,d\theta}{R\,d\theta}$$

Hence

$$\varepsilon_x = \frac{y}{R} \qquad\qquad [6.5]$$

and since $R = dx/d\theta$, therefore also

$$\varepsilon_x = y\frac{d\theta}{dx} \qquad\qquad [6.6]$$

From eqn. [6.5] it will be seen that strain is distributed linearly across the section, being zero at the neutral surface and having maximum values at the outer surfaces. It is important to note here that eqn. [6.5] is entirely independent of the type of material, whether it is in an elastic or plastic state and linear or non-linear in stress and strain.

*Transverse deformation*

Regarding deformations in the $y$- and $z$-directions, it is apparent that changes in length of the beam will result in changes in the transverse dimensions. For example, the fibres in compression will be associated with an increase in thickness, whereas the region in tension will show a decrease in beam thickness. The transverse strains will be

$$\varepsilon_z = \varepsilon_y = -\frac{\nu\sigma_x}{E}$$

and a beam of initially rectangular cross-section will take up the shape shown in Fig. 6.9. This can be easily demonstrated by bending an eraser. The neutral surface, instead of being plane, will be curved. This behaviour is termed *anticlastic curvature*. The deformations are extremely small and do not affect the solution for longitudinal strains.

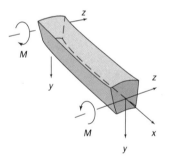

**Fig. 6.9**

**6.5 Stress–strain relationship**

Consider the segmental length of beam illustrated in Fig. 6.10 the cross-section of which, for simplicity at this stage of analysis, has symmetry about the vertical axis $y$, but not about a horizontal axis $z$.

As it has been assumed that $\sigma_y = \sigma_z = 0$ then the stress–strain relationship that is applicable for *linear-elastic bending* is

$$\varepsilon_x = \frac{\sigma_x}{E}$$

and, from eqn. [6.5]

$$\varepsilon_x = \frac{\sigma_x}{E} = \frac{y}{R}$$

**Fig. 6.10**

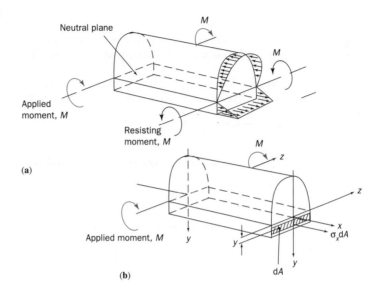

or

$$\frac{\sigma_x}{y} = \frac{E}{R} \tag{6.7}$$

Thus bending stress is also distributed in a linear manner over the cross-section, being zero where $y$ is zero, that is at the neutral plane, and being maximum tension and compression at the two outer surfaces where $y$ is maximum, as shown in Fig. 6.10(a).

## 6.6  Equilibrium of forces and moments

### Position of the neutral plane and axis

Consider an element of area, $dA$, at a distance $y$ from some arbitrary location of the neutral surface, Fig. 6.10(b).

The force on the element in the $x$-direction is

$$dF_x = \sigma_x \, dA$$

Therefore the total longitudinal force on the cross-section is

$$F_x = \int_A \sigma_x \, dA$$

where $A$ is the total area of the section.

Since there is no external axial force in pure bending, the internal force resultant must be zero; therefore

$$F_x = \int_A \sigma_x \, dA = 0$$

Using eqn. [6.7] to substitute for $\sigma_x$,

$$\frac{E}{R} \int_A y \, dA = 0$$

Since $E/R$ is not zero, the integral must be zero, and as this is the first moment of area about the neutral axis, it is evident that the centroid of the section must coincide with the neutral axis. (The first moment of area of a section about its centroid is zero; see Appendix A.)

**Internal resisting moment**

Returning to the element in Fig. 6.10(*b*), the moment of the axial force about the neutral surface is $y\,dF_x$. Therefore the total *internal resisting moment is*

$$\int_A y\,dF_x = \int_A y\sigma_x\,dA$$

This must balance the external applied moment $M$, so that for equilibrium

$$\int_A y\sigma_x\,dA = M \qquad [6.8]$$

or, substituting for $\sigma_x$,

$$M = \frac{E}{R}\int_A y^2\,dA$$

Now $\int_A y^2\,dA$ is the *second moment of area* (see Appendix A) of the cross-section about the neutral axis and will be denoted by $I$. Thus

$$M = \frac{EI}{R} \qquad \text{or} \qquad \frac{1}{R} = \frac{M}{EI} \qquad [6.9]$$

Using eqn. [6.7]

$$\frac{M}{I} = \frac{\sigma_x}{y}$$

and hence $\sigma_x = My/I$ which relates the stress to the moment and the geometry of the beam.

**6.7 The bending relationship**

Combining eqns. [6.7] and [6.9] gives the fundamental relationship between bending stress, moment and geometry:

$$\frac{M}{I} = \frac{\sigma}{y} = \frac{E}{R} \qquad [6.10]$$

The quantities concerned and their units are

$$M = \text{bending moment, N\,m}$$
$$I = \text{second moment of area, m}^4$$
$$\sigma_x = \text{stress, N/m}^2$$
$$y = \text{distance from neutral axis to point in question, m}$$
$$E = \text{Young's modulus, N/m}^2$$
$$R = \text{radius of curvature of neutral axis, m}$$

**6.8 The general case of bending**

The bending relationship, eqn. [6.10], is only an exact solution for the case of pure bending; however, in practice many beam problems involve bending moment and shearing force which vary along the length. In these cases it has been shown that eqn. [6.10], even if not exact, provides a solution which is quite accurate for engineering design, except for cross-sections at support points and concentrated loads where stress concentration may occur (see Chapter 12).

## 6.9 Section modulus

Using eqn. [6.10], for the outer surfaces of the beam, the maximum stresses will be

$$\sigma_{t\,max} = \frac{M y_{t\,max}}{I} = \frac{M}{Z_t}$$

$$\sigma_{c\,max} = \frac{M y_{c\,max}}{I} = \frac{M}{Z_c} \qquad [6.11]$$

where the subscripts denote tension and compression. The quantities $I/y_{t\,max}$ and $I/y_{c\,max}$ are functions of geometry only; they are termed the *section moduli* and are denoted by $Z_t$ and $Z_c$.

Values of $I$ and $Z$ for standard sections are given in manufacturers' handbooks.

### Example 6.4

Table 6.1 shows an extract from a steel designer's handbook giving some of the section properties of universal beams. Select a section to carry a 20 kN load which is cantilevered out 2 m from the support. The structural steel used has a yield stress of 250 MN/m², but the allowable bending stress is usually taken as 0.66 × 250 = 165 MN/m².

The maximum bending moment occurs at the support and is $M = 20 \times 2 = 40$ kN m. The bending stress due to this load is

$$\sigma = \frac{MD}{2I} = \frac{M}{Z} < \sigma_{allowable}$$

where $Z$ is the section modulus of the beam. Note that in handbooks this will often be called the elastic modulus. This is nothing to do with the Young's modulus of the material; it is a geometric property only and relates stress and internal moment during elastic, or recoverable, bending.

**Table 6.1**

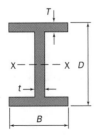

| Designation | | Depth of | Width of | Thickness | | Area of | Second moment of area for | Section modulus for |
|---|---|---|---|---|---|---|---|---|
| Serial size (mm) | Mass per metre (kg) | section $D$ (mm) | section $B$ (mm) | Web $t$ (mm) | Flange $T$ (mm) | section (cm²) | axis X–X (cm⁴) | axis X–X (cm³) |
| 254 × 102 | 28 | 260.4 | 102.1 | 6.4 | 10.0 | 36.2 | 4004 | 307.6 |
| | 25 | 257.0 | 101.9 | 6.1 | 8.4 | 32.1 | 3404 | 264.9 |
| 203 × 133 | 30 | 206.8 | 133.8 | 6.3 | 9.6 | 38.0 | 2880 | 278.5 |
| | 25 | 203.2 | 133.4 | 5.8 | 7.8 | 32.3 | 2348 | 231.1 |

Given the applied moment and the allowable stress, the necessary section can be found by rearranging eqn. [6.11] as

$$Z > \frac{M}{\sigma_{allowable}} > \frac{40 \times 10^3}{165 \times 10^6} > 242 \times 10^{-6} \, \text{m}^3$$

i.e. $Z > 242 \, \text{cm}^3$

From the table, a $203 \times 133$ section of $30 \, \text{kg/m}$ has a section modulus of $278 \, \text{cm}^3$, which should be sufficient.

Increasing the length of the cantilever, for example to allow more clearance beneath it, will cause a proportional increase in bending moment and stress on a given section. Bending moment also increases linearly with applied load.

An additional property, the plastic modulus of the section, is often given. This is used to relate the material yield stress and the largest, or ultimate, moment the section can carry when permanent deformation is allowed. Permanent, or plastic, deformation is covered in Chapter 15.

**Example 6.5**

A T-section bar has dimensions as shown in Fig. 6.11(b). The bar is used as a simply supported beam of span 1.5 m, the flange being horizontal as shown in Fig. 6.11(a). Calculate the uniformly distributed load which can be applied if the maximum tensile stress in the material is not to exceed 100 MN/m². What is then the greatest bending stress in the flange?

**Fig. 6.11**

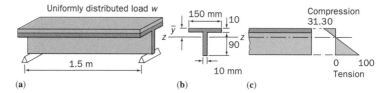

(a)                                        (b)          (c)

The first step is to find the position of the centre of area, which will also give the neutral axis. Taking first moments of area about the top surface,

$$(240 \times 10)\bar{y} = (1500 \times 5) + (900 \times 55)$$

$$\bar{y} = 23.8 \text{mm}$$

To find the second moment of area we can use the parallel axes theorem (see Appendix A):

$$I_z = \frac{150 \times 10^3}{12} + (150 \times 10 \times 18.8^2)$$

$$+ \frac{10 \times 90^3}{12} + (10 \times 90 \times 31.2^2) = 2028 \times 10^3 \, \text{mm}^4$$

$$M_{max} = \frac{w \times 1.5^2}{8} = 0.281 w \, \text{Nm}$$

The maximum tensile stress will occur on the bottom edge of the web where $y = y_{max} = 76.2$ mm. Therefore, using the allowable tensile stress of $100$ MN/m$^2$,

$$\frac{100 \times 10^6}{0.0762} = \frac{0.281w}{2028 \times 10^{-9}}$$

$$w = \frac{202.8}{0.0762 \times 0.281} = 9.5 \text{ kN/m}$$

Since stress is proportional to distance from the neutral axis, as shown in Fig. 6.11($c$).

$$\frac{\sigma_c}{y_c} = \frac{\sigma_t}{y_t}$$

Therefore the greatest compressive stress in the flange is

$$\sigma_c = 100 \times 10^6 \times \frac{23.8}{76.2} = 31.3 \text{ MN/m}^2$$

## 6.10 Beams made of dissimilar materials

In some circumstances it may be necessary or desirable to construct a beam such that the cross-section contains two different materials. Usually the object is for one material to act as a reinforcement to the other material. There would be a number of reasons (cost, weight, size, etc.) why the whole beam could not be made from the stronger material. The positioning of the reinforcement material might not be symmetrical with respect to the centroid of the cross-section and it could be embedded within or fixed in some manner to the outside of the main bulk material.

The arguments which were applied to the analysis of simple bending of a homogeneous beam also apply to the composite beam since the two materials constrain each other to deform in the same manner, e.g. to an arc of a circle for pure bending.

**Symmetrical sections**  Consider a beam cross-section consisting of a central part of, say, plastic or timber with reinforcing plates firmly bonded (no sliding) to the upper and lower surfaces along the length of the beam as shown in Fig. 6.12. The section is symmetrical about the centroid and neutral surface.

**Fig. 6.12**

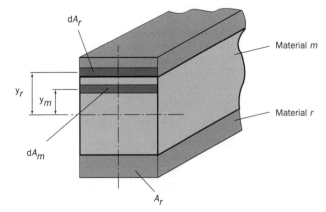

*Equilibrium*

(a) Since there is no *external* end load, longitudinal equilibrium requires that

$$\int_{A_m} \sigma_m \, dA_m + \int_{A_r} \sigma_r \, dA_r = 0 \qquad [6.12]$$

where the subscripts $m$ and $r$ refer to the main and reinforcing materials, respectively, as shown in Fig. 6.12.

(b) The sum of the internal resisting moments must be in equilibrium with the external applied moment

$$M_m + M_r = M$$

Hence, using eqn. [6.8] for each of the materials,

$$\int_{A_m} \sigma_m y_m \, dA_m + \int_{A_r} \sigma_r y_r \, dA_r = M \qquad [6.13]$$

The above equations are independent of whether the material is in an elastic or a plastic condition.

*Geometry of deformation*　Both materials deform to the same circular arc and, for a linear strain distribution, from eqn. [6.5]

$$\frac{1}{R} = \frac{\varepsilon_m}{y_m} = \frac{\varepsilon_r}{y_r} \qquad [6.14]$$

These relationships are purely a function of geometry and therefore are independent of the material and its properties.

*Stress–strain relations*　For linear elastic materials under uniaxial stress

$$\sigma_m = E_m \varepsilon_m$$

$$\sigma_r = E_r \varepsilon_r \qquad [6.15]$$

From eqns. [6.14] and [6.15], $\sigma_m = y_m E_m/R$ and $\sigma_r = y_r E_r/R$. Substitution in the equilibrium equation [6.13] gives

$$\left(\frac{E_m}{R}\right) \int_{A_m} y_m^2 \, dA_m + \left(\frac{E_r}{R}\right) \int_{A_r} y_r^2 \, dA_r = M$$

where the integrals are the second moments of area of the main and reinforcing materials, $I_m$ and $I_r$ respectively, about the neutral surface. Therefore

$$\frac{E_m I_m + E_r I_r}{R} = M$$

Substituting for $1/R$ gives

$$\frac{\sigma_m}{E_m y_m} = \frac{\sigma_r}{E_r y_r} = \frac{M}{E_m I_m + E_r I_r}$$

or

$$\sigma_m = \frac{ME_m y_m}{E_m I_m + E_r I_r}$$

$$\sigma_r = \frac{ME_r y_r}{E_m I_m + E_r I_r} \qquad\qquad [6.16]$$

**Example 6.6**

A timber beam of depth 100 mm and width 50 mm is to be reinforced with steel plates on each side, as shown in Fig. 6.13. The composite section will be subjected to a maximum bending moment of 6 kN/m. If the maximum stress in the timber is not to exceed 12 MN/m², calculate the required thickness for the steel plates, which are 100 mm in depth. What is the maximum stress in the steel? $E_{steel} = 205\,GN/m^2$, $E_{timber} = 15\,GN/m^2$.

$$I_{timber} = \frac{50 \times 100^3}{12} = 4.17 \times 10^6 \ mm^4$$

$$I_{steel} = \frac{t}{12} \times 100^3 = 0.0833 \times 10^6 t \ mm^4$$

**Fig. 6.13**

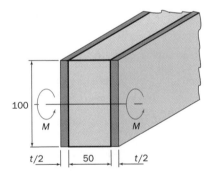

From eqn. [6.16] for limiting stress in the timber, at the top or bottom surface,

$$12 \times 10^6 = \frac{6000 \times 0.05 \times 15 \times 10^9 \times 10^{12}}{(15 \times 10^9 \times 4.17 \times 10^6) + (205 \times 10^9 \times 0.0833 \times 10^6 t)}$$

from which $t = 18.3$ mm; and since this is the total thickness of steel, the plates are each 9.15 mm thick.

The maximum stress in steel plates also occurs at the top and bottom edges,

$$\sigma = \frac{6000 \times 0.05 \times 205 \times 10^9 \times 10^{12}}{(15 \times 10^9 \times 4.17 \times 10^6) + (205 \times 10^9 \times 0.0833 \times 10^6 \times 18.3)}$$

$$= 164 \ MN/m^2$$

**Equivalent sections**   In beam sections which are symmetrical geometrically, but unsymmetrical with respect to location of the different materials, as, for example, in Fig. 6.14(a), the neutral axis no longer coincides with the centroid of the section. The problem can still be solved using eqns. [6.12]–[6.15], but the arithmetic is more tedious since the neutral axis has first to be found.

**Fig. 6.14**

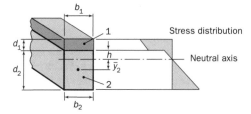

A more convenient approach, which is also valid, is to transform the composite section into an equivalent (from the view of resisting forces and moments) section of only one of the two (or more) materials. The solution is then of course a simple routine.

Consider the beam section shown in Fig. 6.14(a), in which the unknown neutral axis has been placed at a distance $h$ from the interface between the two materials. Assuming no sliding, the strain at the interface must be the same for each material; therefore, at the interface,

$$\varepsilon = \frac{\sigma_1}{E_1} = \frac{\sigma_2}{E_2} \qquad \qquad [6.17]$$

Also, rearranging eqns. [6.16],

$$M = \frac{\sigma_1}{h}\left(I_1 + \frac{E_2}{E_1}I_2\right) = \frac{\sigma_2}{h}\left(I_2 + \frac{E_1}{E_2}I_1\right) \qquad [6.18]$$

where $I_1$ and $I_2$ are the respective second moments of area about the neutral axis. Now,

$$\frac{E_2}{E_1}I_2 = \frac{E_2}{E_1}\left(\frac{b_2 d_2^3}{12} + b_2 d_2 \bar{y}_2^2\right) = b_1'\left(\frac{d_2^3}{12} + d_2 \bar{y}_2^2\right)$$

where $b_1' = (E_2/E_1)b_2$, so exactly the same resisting moment will exist if material 2 is replaced by material 1 having the *same* depth $d_2$ but a new width of $(E_2/E_1)b_2$. This forms the equivalent section shown in Fig. 6.15(a), which is entirely made of material 1 and is shown for the case of

**Fig. 6.15**

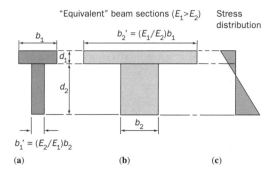

$E_1 > E_2$. In the converse manner, material 1 can be replaced by material 2 if the width is made $(E_1/E_2)b_1$ to give the equivalent section as in Fig. 6.15(*b*). The neutral axis can be found quite simply (from either equivalent section) as it now coincides with the centroid. The stress distribution in either equivalent section, Fig. 6.15(*c*), can be determined and transposed to that actually occurring as in Fig. 6.14(*b*) by using eqn. [6.17] for the condition at the interface.

**Example 6.7**

A timber beam of rectangular section 100 mm × 50 mm and simply supported at the ends of a 2 m span has a 30 mm × 10 mm steel strip securely fixed to the top surface as shown in Fig. 6.16(a) to protect the timber from trolley wheels. When the trolley is exerting a force at mid-span of 2 kN, determine the stress distribution at that section. $E_{steel} = 20E_{wood}$.

Fig. 6.16

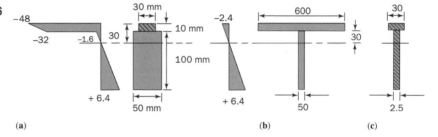

(a)          (b)          (c)

Firstly consider the equivalent section made entirely of timber as in Fig. 6.16(*b*). The 30 mm width of steel becomes $30 \times 20 = 600$ mm width of timber equivalent. The position of the centroid and hence of the neutral axis from the top surface is given by

$$\bar{y}(6000 + 5000) = (6000 \times 5) + (5000 \times 60)$$

$$\bar{y} = 30 \text{ mm}$$

$$I = \frac{600 \times 10^3}{12} + (6000 \times 25^2) + \frac{50 \times 100^3}{12}$$

$$+ (5000 \times 30^2)$$

$$= 1246.6 \times 10^4 \text{ mm}^4$$

Maximum bending moment,

$$M = WL/4 = (2000 \times 2)/4 = 1000 \text{ Nm}$$

At lower surface,

$$\sigma = \frac{1000 \times 0.08}{1246.6 \times 10^{-8}} = 6.4 \text{ MN/m}^2$$

At interface in timber, multiplying the stress by the ratio of the respective $y$-values,

$$\sigma = 6.4 \times -20/80 = -1.6 \text{ MN/m}^2$$

Since $E_{steel} = 20E_{wood}$, then from eqn. [6.17], $\sigma_{steel} = 20\sigma_{wood}$. Hence at interface in steel,

$$\sigma = -1.6 \times 20 = -32 \, \text{MN/m}^2$$

At top surface,

$$\sigma = 6.4 \times -\frac{30}{80} \times 20 = -48 \, \text{MN/m}^2$$

The distribution in the equivalent section is shown in Fig. 6.16(*b*) and the actual distribution in Fig. 6.16(*a*).

If the equivalent section is made out of steel then this is represented in Fig. 6.16(*c*) where the width of the timber is made into an equivalent width of steel by dividing 50 mm by 20, the modulus ratio, to give 2.5 mm.

The centroid is at

$$\bar{y}(300 + 250) = (300 \times 5) + (250 \times 60)$$

$$\bar{y} = 30 \, \text{mm}$$

which is the same as before since the proportions are the same. As only width dimensions have been changed by a factor of 20 compared with the previous equivalent section, there is no need to recalculate $I$, since

$$I_s = \frac{I_w}{20} = \frac{1246.6 \times 10^4}{20} = 623.3 \times 10^3 \, \text{mm}^4$$

At top surface, $\sigma = \dfrac{1000 \times (-0.03)}{623.3 \times 10^{-9}} = -48 \, \text{MN/m}^2$

At interface in steel, $\sigma = -48 \times \dfrac{20}{30} = -32 \, \text{MN/m}^2$

At interface in timber, $\sigma = -32 \times \dfrac{1}{20} = -1.6 \, \text{MN/m}^2$

At lower surface, $\sigma = 48 \times \dfrac{80}{30} \times \dfrac{1}{20} = +6.4 \, \text{MN/m}^2$

**Reinforced concrete sections**     Perhaps the most common example of a composite beam is one using steel bars to reinforce concrete. The steel is always embedded in the concrete on the tension side of the beam owing to the weakness of concrete in tension, but reinforcement may also be included on the compression side to reduce the amount of concrete required if it were not reinforced.

Consider the case illustrated in Fig. 6.17 and make the conventional assumption that the concrete takes all the compression and the reinforcing bars take all the tension. All the required relationships have been derived above, and it is only necessary now to solve for the unknown quantities as required.

**Fig. 6.17**

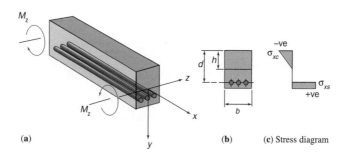

(a)  (b)  (c) Stress diagram

Let the distance of the neutral axis from the outer surface in compression be $h$, Fig. 6.17($b$), and the ratio of the elastic moduli

$$\frac{E_{steel}}{E_{concrete}} = m$$

From eqns. [6.14] and [6.15],

$$\sigma_c = \frac{y_c E_c}{R} \qquad \text{and} \qquad \sigma_s = \frac{y_s E_s}{R}$$

Substituting the stresses into eqn. [6.12] for longitudinal equilibrium gives

$$\frac{E_c}{R} \int_{A_c} y_c \, dA_c + \frac{E_s}{R} \int_{A_s} y_s \, dA_s = 0 \qquad [6.19]$$

For the steel reinforcement the tensile stress is considered constant over the cross-sectional area $A_s$ and concentrated at $y = (d - h)$ as in Fig. 6.17($c$), since

$$\int_{A_s} y_s \, dA_s = (d - h)A_s \qquad \text{and} \qquad \int_{A_c} y_c \, dA_c = -\frac{h}{2}(bh)$$

so that eqn. [6.19] becomes

$$-E_c \frac{bh^2}{2} + E_s(d - h)A_s = 0 \qquad [6.20]$$

and, substituting $E_s/E_c = m$ and rearranging gives

$$h = \left[ \left( \frac{mA_s}{b} \right)^2 + \frac{2mA_s d}{b} \right]^{1/2} - \frac{mA_s}{b} \qquad [6.21]$$

which gives the position of the neutral axis.

Substituting $\sigma_c$ and $\sigma_s$ for $\sigma_m$ and $\sigma_r$ in eqn. [6.13] for equilibrium of moments,

$$\frac{E_c}{R} \int_{A_c} y_c^2 \, dA_c + \frac{E_s}{R} \int_{A_s} y_s^2 \, dA_s = M$$

or

$$\frac{E_c}{R} \frac{bh^3}{3} + \frac{E_s}{R} A_s(d - h)^2 = M$$

Now,

$$\frac{1}{R} = \frac{\sigma_c}{y_c E_c} = \frac{\sigma_s}{y_s E_s}$$

so that substituting for $1/R$ gives

$$\sigma_c = \frac{Mh}{\frac{1}{3}bh^3 + mA_s(d-h)^2}$$

$$\sigma_s = \frac{M(d-h)m}{\frac{1}{3}bh^3 + mA_s(d-h)^2} \qquad [6.22]$$

**Example 6.8**

A reinforced concrete T-beam, Fig. 6.18, has a flange 1.5 m wide and 100 mm deep. The reinforcement is placed in the web 380 mm from the upper edge of the flange. The beam is designed so that the neutral axis coincides with the lower edge of the flange. The limits of stress are 110 MN/m² for steel and 4 MN/m² for concrete. The modulus ratio $E_{steel}/E_{concrete}$ is 15. Calculate (a) the area of the reinforcement, (b) the maximum moment the beam can resist, (c) the actual maximum stress in the steel and the concrete.

**Fig. 6.18**

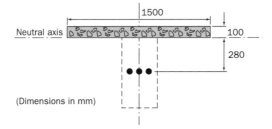

(Dimensions in mm)

(a) The area of steel can be calculated using eqn. [6.20]:

$$-\frac{1.5 \times 0.1^2}{2} + 15(0.38 - 0.1)A_s = 0$$

whence

$$A_s = 1790\,\text{mm}^2$$

(b) The denominator in eqns. [6.22] is given as

$$\tfrac{1}{3} \times 1.5 \times 0.1^3 + [15 \times 0.001\,79(0.38 - 0.1)^2] = 2.61 \times 10^{-3}\,\text{m}^4$$

Assuming the steel reaches the maximum stress, then from eqns. [6.22],

$$M = \frac{2.61 \times 10^{-3} \times 110 \times 10^6}{0.28 \times 15} = 68.3\,\text{kN/m}$$

Alternatively, if the concrete reaches the limiting stress,

$$M = \frac{2.61 \times 10^{-3} \times 4 \times 10^6}{0.1} = 104.5\,\text{kNm}$$

Therefore the maximum moment the beam can resist is 68.3 kN/m.
(c)  At this maximum moment the steel has reached its limiting stress of 110 MN/m$^2$.

$$\text{Actual maximum stress in concrete, } \sigma_c = \frac{68.3 \times 10^3 \times 0.1}{2.61 \times 10^{-3}}$$

$$= 2.62 \, \text{MN/m}^2$$

## 6.11 Combined bending and end loading

A number of situations arise in practice where a member is subjected to a combination of bending and longitudinal load. Problems of this type can be most easily dealt with by superposition of the individual components of stress to give resultant values.

Consider the rectangular-section beam shown in Fig. 6.19, which is subjected to bending moments $M$ about the $z$-axis, and an axial load $P$ in the $x$-direction.

If the end load acted alone, there would be a longitudinal stress, as shown in Fig. 6.20(a)

**Fig. 6.19**

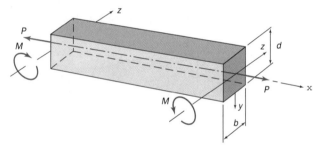

$$\sigma_x = \frac{P}{A}$$

If the moment acted alone, then, as shown in Fig. 6.20(b), the longitudinal bending stress would be

$$\sigma_x = \frac{My}{I}$$

By superposition the resultant stress due to $P$ and $M$ is

$$\sigma_x = \frac{P}{A} + \frac{My}{I} = \frac{P}{bd} + \frac{12My}{bd^3} \qquad [6.23]$$

The resultant distribution of $\sigma_x$ over the cross-section is shown in Fig. 6.20(c) and is obtained by superposition of the bending stresses (a) and the direct stress (b). An interesting feature is that the neutral surface no longer passes through the centroid of the cross-section since

$$\int_A \sigma_x \, dA \neq 0$$

**Fig. 6.20**

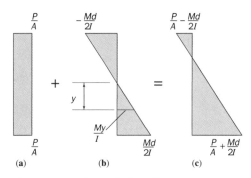

(a)    (b)    (c)

as in the case of simple bending.

Remember that $y$ in eqn. [6.23] is measured from the centroidal axis of the section and *not* from the new neutral axis.

## 6.12 Eccentric end loading

If the end load $P$ does not act at the centroid of the cross-section then it will itself set up bending moments about the principal axes. For example, in Fig. 6.21(*a*) a bar is subjected to a load $P$, which is eccentric from the $z$- and $y$-axes by amounts $m$ and $n$ respectively. The equivalent equilibrium system is with the load $P$ acting at the centroid and moments $Pm$ and $Pn$ acting about the $z$- and $y$-axes, as in Fig. 6.21(*b*).

**Fig. 6.21**

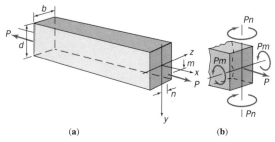

(a)    (b)

The direct stress due to $P$ alone is therefore

$$\sigma'_x = \frac{P}{A}$$

that due to bending about the $y$-axis is

$$\sigma''_x = \frac{Pnz}{I_y}$$

and that due to bending about the $z$-axis is

$$\sigma'''_x = \frac{Pmy}{I_z}$$

Therefore the resultant longitudinal stress is by superposition

$$\sigma_x = \sigma'_x + \sigma''_x + \sigma'''_x$$

$$= \frac{P}{A} + \frac{Pnz}{I_y} + \frac{Pmy}{I_z} \qquad [6.24]$$

Substituting $I_y = bd^3/12$ and $I_z = db^3/12$ gives

$$\sigma_x = \frac{P}{bd}\left(1 + \frac{12nz}{b^2} + \frac{12my}{d^2}\right) \qquad [6.25]$$

where $z$ lies between $\pm b/2$ and $y$ between $\pm d/2$.

A practical example of the use of this analysis is in civil engineering where short concrete beams are subjected to a compressive load which may be eccentric. In materials which are strong in compression but weak in tension, such as concrete, it is necessary to limit the eccentricities $m$ and $n$ so that no tensile stress is set up. The condition for no tension is that the compressive stress due to $P$ is greater than or equal to the maximum tensile bending stresses set up by the moments $Pm$ and $Pn$. Therefore, from eqn. [6.25],

$$1 - \frac{6n}{b} - \frac{6m}{d} \geq 0 \qquad [6.26]$$

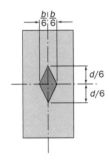

**Fig. 6.22**

This relationship defines the locus of maximum eccentricity as shown by the shaded area in Fig. 6.22. When $P$ is applied within the shaded area then no tensile stress will be set up anywhere in the cross-section. The limits on the $z$- and $y$-axes are $\pm b/6$ and $\pm d/6$, which has resulted in what is known as the *middle-third rule* for no tension. Typical distributions for various amounts of eccentricity along *one* principal axis are shown in Fig. 6.23.

**Fig. 6.23**

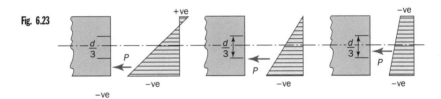

---

**Example 6.9**

In a tensile test within the elastic range on a specimen of circular cross-section an extensometer is being used which will only measure deformation on one side of the specimen. Determine how much eccentricity of loading will give rise to a 5% difference between the surface stress derived from the extensometer and the average stress over the cross-section.

Let the average stress be $\sigma$; then for a 5% error due to non-axial loading the resultant stress on one edge of the specimen will be $0.95\sigma$, and at the opposite end of the diameter $1.05\sigma$. From eqn. [6.25],

$$0.95\sigma = \sigma\left(1 - \frac{Amy_{max}}{I}\right)$$

Therefore

$$m = \frac{0.05I}{Ay_{max}}$$

For a circular cross-section $I = \pi d^4/64$. Therefore

$$m = \frac{0.05\pi d^4/64}{(\pi d^2/4)d/2} = 0.006\,25d$$

Thus in a tensile test on a specimen of 10 mm diameter, the eccentricity of loading must be *less* than 0.063 mm to avoid surface stresses being more than 5% greater than the average direct stress.

**Example 6.10**

A slotted machine link 6 mm thick, illustrated in Fig. 6.24(a), is subjected to a tensile load of 40 kN acting along the centre-line of the end faces. Find the stress distribution for a section through the slot such as AA.

**Fig. 6.24**

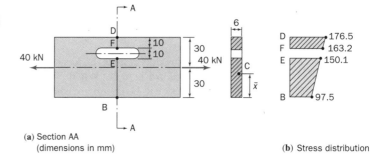

(a) Section AA
(dimensions in mm)

(b) Stress distribution

We must first find the centroid C of the section AA and by taking the moment of area about the bottom:

$$[(40 \times 6) + (10 \times 6)]\bar{x} = (40 \times 6 \times 20) + (10 \times 6 \times 55)$$

so that

$$\bar{x} = 27\,\text{mm}$$

Hence the load is acting eccentrically with respect to the centroid of the slot cross-section AA by an amount

$$e = 30 - 27 = 3\,\text{mm}$$

This eccentricity gives rise to a moment

$$M = 40\,000 \times 0.003 = 120\,\text{N/m}$$

which will cause tensile bending stress along the edge at D and compressive bending stress along the edge at B.

For bending only, the neutral axis passes through C and the greatest bending stresses occur at B and D; thus $y_{max} = 27$ and 33 mm and the second moment of area is $91.28 \times 10^3\,\text{mm}^4$. Therefore

$$\sigma_B = -\frac{120 \times 0.027}{91.28 \times 10^{-9}} = -35.5\,\text{MN/m}^2$$

$$\sigma_D = +\frac{120 \times 0.033}{91.28 \times 10^{-9}} = +43.5\,\text{MN/m}^2$$

$$\text{Direct stress} = \frac{40\,000}{0.05 \times 0.006} = +133\,\text{MN/m}^2$$

The resultant stresses will be

At B $\qquad +133 - 35.5 = 97.5\,\text{MN/m}^2$

At E $\qquad +133 + 17.1 = 150.1\,\text{MN/m}^2$

At F $\qquad +133 + 30.2 = 163.2\,\text{MN/m}^2$

At D $\qquad +133 + 43.5 = 176.5\,\text{MN/m}^2$

The distribution of stress is shown in Fig. 6.24($b$).

### 6.13 Asymmetrical or skew bending

The previous analysis has been concerned with bending about an axis of symmetry. However, many occasions arise in practice where bending will occur either of a section which does not have any axes of symmetry or of a symmetrical section about an asymmetrical axis. In order to express the conditions of equilibrium in asymmetrical bending a knowledge of first and second moments of area about arbitrary axes through the centroid of the section is required. It is therefore necessary to study certain properties of areas before embarking on the analysis of stress distribution in asymmetrical bending. The reader is here referred to Appendix A.

Fig. 6.25

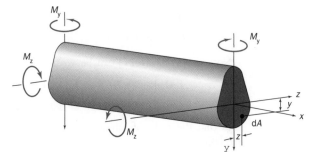

Asymmetrical pure bending of a beam is shown in Fig. 6.25 for positive external moments $M_y$ and $M_z$ applied about an arbitrary set of centroidal axes. Considering first bending in the $xy$-plane only, the equilibrium equations are

$$\int_A \sigma_x \, \mathrm{d}A = 0 \qquad \text{(neutral surface through the centroid)}$$

$$\int_A \sigma_x y \, \mathrm{d}A = M_z \qquad \text{and} \qquad \int_A \sigma_x z \, \mathrm{d}A = M_y$$

But $\sigma_x = yE/R_y$, where $R_y$ is the radius of curvature in the $xy$-plane; hence

$$M_z = \frac{E}{R_y} \int_A y^2 \, \mathrm{d}A = \frac{EI_z}{R_y}$$

and

$$M_y = \frac{E}{R_y} \int_A yz \, dA = \frac{EI_{yz}}{R_y}$$

$I_{yz}$ is called the *product moment of area* (see Appendix A).

For bending only in the $xz$-plane a procedure similar to the above gives

$$M_y = \frac{E}{R_z} \int z^2 \, dA = \frac{EI_y}{R_z}$$

$$M_z = \frac{E}{R_z} \int zy \, dA = \frac{EI_{yz}}{R_z}$$

For simultaneous bending in the $xy$- and $xz$-planes the above relationships may be superimposed to give

$$M_y = \frac{EI_y}{R_z} + \frac{EI_{yz}}{R_y}$$

$$M_z = \frac{EI_z}{R_y} + \frac{EI_{yz}}{R_z} \qquad [6.27]$$

The radii of curvature are obtained from the above equations as

$$\frac{1}{R_y} = \frac{M_z I_y - M_y I_{yz}}{E(I_y I_z - I_{yz}^2)}$$

$$\frac{1}{R_z} = \frac{M_y I_z - M_z I_{yz}}{E(I_y I_z - I_{yz}^2)}$$

Note that eqns. [6.27] can be written in matrix form as

$$\begin{Bmatrix} M_y \\ M_z \end{Bmatrix} = \begin{bmatrix} EI_y & EI_{yz} \\ EI_{yz} & EI_z \end{bmatrix} \begin{Bmatrix} 1/R_z \\ 1/R_y \end{Bmatrix} \qquad [6.28]$$

so that the radii of curvature can be determined from the inverse matrix relationship

$$\begin{Bmatrix} 1/R_z \\ 1/R_y \end{Bmatrix} = \frac{1}{E} \begin{bmatrix} I_y & I_{yz} \\ I_{yz} & I_z \end{bmatrix}^{-1} \begin{Bmatrix} M_y \\ M_z \end{Bmatrix}$$

$$= \frac{1}{E(I_y I_z - I_{yz}^2)} \begin{bmatrix} I_z & -I_{yz} \\ -I_{yz} & I_y \end{bmatrix} \begin{Bmatrix} M_y \\ M_z \end{Bmatrix} \qquad [6.29]$$

The resulting bending stress is therefore the sum of the components for bending in each of the $xy$- and $xz$-planes:

$$\sigma_x = \frac{yE}{R_y} + \frac{zE}{R_z}$$

$$= \frac{y(M_z I_y - M_y I_{yz}) + z(M_y I_z - M_z I_{yz})}{I_y I_z - I_{yz}^2} \qquad [6.30]$$

The neutral surface, where $\sigma_x = 0$, is defined by the plane

$$y(M_z I_y - M_y I_{yz}) + z(M_y I_z - M_z I_{yz}) = 0 \qquad [6.31]$$

Establishing the position of this plane enables one to determine the points which are furthest from the neutral plane and hence subjected to the maximum stresses. If bending is about the principal axes of the section for which $I_{yz} = 0$, then

$$\sigma_x = \frac{M_z y}{I_z} + \frac{M_y z}{I_y} \qquad [6.32]$$

If either $M_z$ or $M_y$ is in the opposite sense to that shown in Fig. 6.25, then the appropriate negative value must be used in eqn. [6.30]. This is illustrated in the following example.

If there is only a single external applied moment $M$ about an axis $ss$ inclined at $\theta$ to one of the principal axes, as in Fig. 6.26, then the moment vector $M$ can be resolved into components $M_y$ and $M_z$ about the $y$- and $z$-axes, so that

$$M_y = -M \sin \theta \quad \text{and} \quad M_z = M \cos \theta$$

**Fig. 6.26**

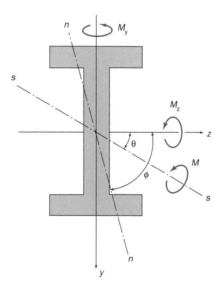

and substituting in eqn. [6.32],

$$\sigma_x = \frac{-Mz \sin \theta}{I_y} + \frac{My \cos \theta}{I_z} \qquad [6.33]$$

The neutral plane will no longer be perpendicular to the plane of bending, as in the symmetrical problem, but can be determined by putting $\sigma_x = 0$ above; then

$$\frac{-z \sin \theta}{I_y} + \frac{y \cos \theta}{I_z} = 0$$

and

$$\frac{y}{z} = \frac{I_z}{I_y} \tan \theta = \tan \phi$$

where $\phi$ is the inclination of $nn$, the neutral surface, to the $z$-axis.

The neutral surface is perpendicular to the plane of bending if either $I_z = I_y$ or $\theta = 0$.

**Example 6.11**

The angle section shown in Fig. 6.27 is subjected to a bending moment of 2 kN m about the z-axis. Determine the bending stress distribution.

**Fig. 6.27**

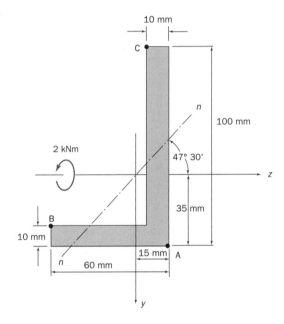

Considering the first moments of area about vertical and horizontal edges respectively,

$$\bar{z} = \frac{(90 \times 10 \times 5) + (60 \times 10 \times 30)}{(90 + 60)10} = 15 \,\text{mm}$$

$$\bar{y} = \frac{(90 \times 10 \times 55) + (60 \times 10 \times 5)}{(90 + 60)10} = 35 \,\text{mm}$$

which gives the co-ordinates of the centroid.

$$I_z = \frac{10 \times 90^3}{12} + (10 \times 90 \times 20^2) + \frac{60 \times 10^3}{12} + (60 \times 10 \times 30^2)$$

$$= 151.2 \times 10^4 \,\text{mm}^4$$

$$I_y = \frac{90 \times 10^3}{12} + (90 \times 10 \times 10^2) + \frac{10 \times 60^3}{12} + (10 \times 60 \times 15^2)$$

$$= 41.3 \times 10^4 \, \text{mm}^4$$

$$I_{zy} = [90 \times 10 \times 10 \times (-20)] + [60 \times 10 \times (-15) \times 30]$$

$$= -45 \times 10^4 \, \text{mm}^4$$

The position of the neutral axis may now be found using eqn. [6.31], from which, with $M_y = 0$,

$$y(2000 \times 41.3 \times 10^{-8}) + z(-2000 \times (-45) \times 10^{-8}) = 0$$

$$\frac{y}{z} = -1.09 \quad \text{and} \quad \phi = -47°30'$$

The maximum tensile stress will occur at A, and using eqn. [6.30], in which $M_y$ will be zero,

$$\sigma_A = \frac{2000[(0.035 \times 41.3 \times 10^{-8}) - (+0.015 \times (-45) \times 10^{-8})]}{[(41.3 \times 151.2) - 45^2]10^{-16}}$$

$$= 101 \, \text{MN/m}^2$$

At B,

$$\sigma_B = \frac{2000[(0.025 \times 41.3 \times 10^{-8}) - (-0.045 \times (-45) \times 10^{-8})]}{[(41.3 \times 151.2) - 45^2]10^{-16}}$$

$$= -47 \, \text{MN/m}^2$$

The maximum compressive stresses occurs at C, where

$$\sigma_C = \frac{2000[(-0.065 \times 41.3 \times 10^{-8}) - (0.005 \times (-45) \times 10^{-8})]}{[(41.3 \times 151.2) - 45^2]10^{-16}}$$

$$= -116.6 \, \text{MN/m}^2$$

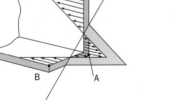

C

Neutral axis

B    A

**Fig. 6.28**    The distribution is illustrated in Fig. 6.28.

Figure 6.29(*b*) shows a spreadsheet for this problem. The second and product moments of area of the section have been derived from the co-ordinates of the points defining the cross-section as described in Appendix A. These properties plus the co-ordinates of the points relative to the section centroid are listed. Given this information, the stress at each point can be written economically in matrix form as

$$\sigma_x = [z \quad y] \frac{1}{(I_y I_z - I_{yz}^2)} \begin{bmatrix} I_z & -I_{yz} \\ -I_{yz} & I_y \end{bmatrix} \begin{bmatrix} M_y \\ M_z \end{bmatrix} \qquad [6.34]$$

|  | A | B | C | D | E | F | G | H | I |
|---|---|---|---|---|---|---|---|---|---|
| 1 |  |  |  |  |  | Iy | Iz | Izy | det |
| 2 |  |  |  |  |  | 1512500 | 413000 | −450000 | (F2*G2−H2^2) |
| 3 |  |  |  |  |  |  |  |  |  |
| 4 |  |  |  |  |  |  |  |  |  |
| 5 | Point | z | y | stress |  | Matrix |  | Moments | a |
| 6 |  |  |  |  |  |  |  |  | b |
| 7 | A | 15 | 35 | +B7*$I$7+C7*$I$8 |  | +F2/I2 | −H2/I2 | 0 | +F7*H7+G7*H8 |
| 8 |  | −45 | 35 | +B8*$I$7+C8*$I$8 |  | −H2/I2 | +G2/I2 | 2000*1000 | +F8*H7+G8*H8 |
| 9 | B | −45 | 25 | +B9*$I$7+C9*$I$8 |  |  |  |  |  |
| 10 |  | 5 | 25 | +B10*$I$7+C10*$I$8 |  |  | tan | −I7/I8 |  |
| 11 | C | 5 | −65 | +B11*$I$7+C11*$I$8 |  |  | theta | @ATAN(H10) | rads |
| 12 |  | 15 | −65 | +B12*$I$7+C12*$I$8 |  |  |  | @DEGREES(H11) | degrees |

(a) Data and cell formulae

|  | A | B | C | D | E | F | G | H | I |
|---|---|---|---|---|---|---|---|---|---|
| 1 |  |  |  |  |  | Iy | Iz | Izy | det |
| 2 |  |  |  |  |  | 1512000 | 413000 | −450000 | 4.22E+11 |
| 3 |  |  |  |  |  |  |  |  |  |
| 4 |  |  |  |  |  |  |  |  |  |
| 5 | Point | z | y | stress |  | Matrix |  | Moments | a |
| 6 |  |  |  |  |  |  |  |  | b |
| 7 | A | 15 | 35 | 100.51 |  | 3.58E-06 | 1.07E-06 | 0 | 2.13 |
| 8 |  | −45 | 35 | −27.47 |  | 1.07E-06 | 9.79E-06 | 2000000 | 1.96 |
| 9 | B | −45 | 25 | −47.04 |  |  |  |  |  |
| 10 |  | 5 | 25 | 59.60 |  |  | tan | −1.09E+00 |  |
| 11 | C | 5 | −65 | −116.58 |  |  | theta | −8.28E-0.01 | rads |
| 12 |  | 15 | −65 | −95.25 |  |  |  | −4.75E+01 | degrees |

(b) Spreadsheet Display

**Fig. 6.29**　For given moments on the section this can be simplified to

$$\sigma_x = \begin{bmatrix} z & y \end{bmatrix} \begin{bmatrix} a \\ b \end{bmatrix} \qquad [6.35]$$

where $a$ and $b$ are determined from above. Note that because the co-ordinates of the section points are given in mm, the moment has been entered in N mm so that the stresses appear in N/mm². Dimensions, moments and stresses can be calculated in any consistent set of units.

## 6.14 Shear stresses in bending

The presence of shear force indicates that there must be shear stress on transverse planes in the beam. It is not possible to make use of the conditions of geometry of deformation and the stress–strain relationships except in the development of an exact solution. However, from the assumptions about the validity of the bending-stress distribution, it is possible to estimate the transverse and longitudinal shear-stress distributions in the beam by using only the condition of equilibrium.

Firstly, consider the bending-stress distribution in the short section of beam of length $dx$ shown in Fig. 6.30. The bending moment increases from $M$ on GK to $M + (dM/dx)dx$ on HJ; therefore the bending stress on any arbitrary fibre must increase from

$$\sigma = \frac{My}{I} \qquad \text{on} \qquad \text{GK} \qquad [6.36]$$

**Fig. 6.30**

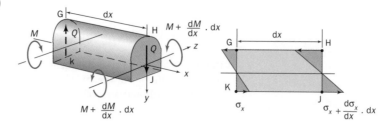

to

$$\sigma + \frac{\mathrm{d}\sigma}{\mathrm{d}x}\mathrm{d}x = \left(M + \frac{\mathrm{d}M}{\mathrm{d}x}\mathrm{d}x\right)\frac{y}{I} \qquad \text{on} \qquad \text{HJ} \qquad [6.37]$$

at each value of $y$.

Now consider the strip of beam below an $xz$-plane CDEF, Fig, 6.31, at a distance $y_1$ from the neutral surface. Each fibre in this strip will have an increase in bending stress $\mathrm{d}\sigma$ along the length $\mathrm{d}x$, so that taken over the hatched area $A$ there will be a resultant axial force equal to

$$\int_A \mathrm{d}\sigma\,\mathrm{d}A = P \qquad [6.38]$$

**Fig. 6.31**

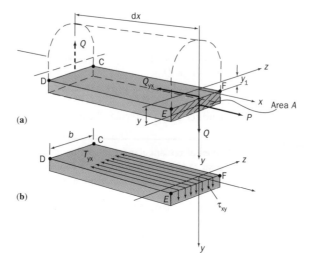

**(a)**

**(b)**

Equilibrium of the strip is maintained by a shear force $Q_{yx}$ acting on the surface of the plane CDEF. Therefore

$$Q_{yx} = \int_A \mathrm{d}\sigma\,\mathrm{d}A \qquad [6.39]$$

Now subtracting eqn. [6.36] from eqn. [6.37] gives

$$\mathrm{d}\sigma = \mathrm{d}M\frac{y}{I}$$

Hence, from eqn. [6.39],

$$Q_{yx} = \int_A \mathrm{d}M \frac{y}{I} \mathrm{d}A \qquad [6.40]$$

If the width of the strip at the plane $y = y_1$ is $b$ and this is small compared with the depth of the beam, then the shear stress on CDEF is almost uniformly distributed and can therefore be expressed as

$$\tau_{yx} = \frac{Q_{yx}}{b\,\mathrm{d}x}$$

Note that this is a *positive* shear stress, being the product of negative directions of the normal to the plane and the stress direction.

Substituting for $Q_{yx}$ in eqn. [6.40] gives

$$\tau_{yx} = \frac{\mathrm{d}M}{\mathrm{d}x} \frac{1}{bI} \int_A y\,\mathrm{d}A \qquad [6.41a]$$

But, from eqn. [6.3], $\mathrm{d}M/\mathrm{d}x = Q$ the vertical shear force on the section and the integral is the first moment of area $A$ about the neutral surface; therefore eqn. [6.41a] becomes

$$\tau_{yx} = \frac{Q}{bI} \int_A y\,\mathrm{d}A = Q\frac{A\bar{y}}{bI} \qquad [6.41b]$$

Using the principle of complementary shear stresses, Fig. 6.31(b), it is then evident that the vertical shear stress, $\tau_{xy}$, is also given by

$$\tau_{xy} = Q\frac{A\bar{y}}{bI} \qquad [6.42]$$

Again we note that $\tau_{xy}$ is *positive* and so is correctly related to the *positive* shear force $Q$.

This solution is only exact for a constant shear force along the beam; however, if the cross-section is small compared with the span, the error introduced by a varying shear force is quite small.

**Example 6.12**

**A beam of rectangular cross-section, depth $d$, thickness $b$, is simply supported over a span of length $l$, and carries a concentrated load $W$ at mid-span. Determine the distribution and maximum value of the transverse shear stress.**

Although changing sign at the centre of the span, the shear force $Q$ is constant in magnitude along the whole span and is equal to $W/2$.

Considering the cross-section shown in Fig. 6.32, the transverse shear stress on some arbitary line EF at a distance $y$ from the neutral surface is

**Fig. 6.32**

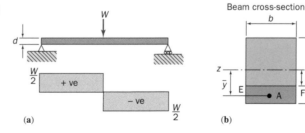

(a)

(b)

given by eqn. [6.42], where $A$ is the shaded area below EF and $\bar{y}$ is the distance of the centroid of $A$ from the neutral surface. Therefore

$$A\bar{y} = b\left(\frac{d}{2} - y\right) \times \frac{1}{2}\left(\frac{d}{2} + y\right) = \frac{b}{2}\left[\left(\frac{d}{2}\right)^2 - y^2\right]$$

The vertical shear stress is therefore

$$(\tau_{xy})_{EF} = \frac{Q}{bI} \times \frac{b}{2}\left[\left(\frac{d}{2}\right)^2 - y^2\right]$$

$$= \frac{W}{4I}\left[\left(\frac{d}{2}\right)^2 - y^2\right] \qquad [6.43]$$

The above expression shows that the distribution of vertical shear stress down the depth of the section is parabolic. The shear stress is zero at the outer fibres where $y = \pm d/2$, as it must be since the complementary shear stress in the longitudinal direction must be zero at a free surface. The maximum value is at the neutral surface where $y = 0$; therefore

$$\tau_{xy\,max} = \frac{Wd^2}{16I} = \frac{3W}{4bd}$$

If uniformly distributed the shear stress would be given by the shear force divided by the area, or

$$\tau_{xy\,mean} = \frac{W}{2bd}$$

Hence the maximum shear stress is 1.5 times the mean value.

**Example 6.13**

A box beam is built up of plate material riveted together as shown by the cross-section, Fig. 6.33. It is simply supported at each end of a 3 m span and carries a concentrated load of 12 kN at 1 m from one end. Estimate a suitable rivet diameter if the rivets are to be pitched at about 100 mm intervals. The shear stress is not to exceed 50 MN/m² in each rivet.

Fig. 6.33

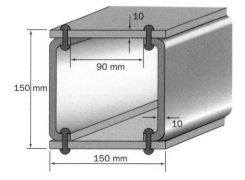

The maximum shear force will be equal to the larger of the two reactions, $12 \times 2/3 = 8\,\text{kN}$.

Now, the force tending to shear the rivets is due to the variation of bending stress along the length of the beam and is given by eqn. [6.41$b$] if slightly rearranged as follows:

$$\text{Shear force per unit length} = \tau_{yx}b = \frac{QA\bar{y}}{I}$$

If the rivet pitch is $p$ and the cross-sectional area $a$, then the allowable shear force per unit length is $\hat{\tau}a/p$. Hence

$$\frac{\hat{\tau}a}{p} = \frac{QA\bar{y}}{I}$$

$$\text{Required area, } a = \frac{QA\bar{y}p}{\hat{\tau}I}$$

$$I \approx \frac{0.15 \times 0.15^3}{12} - \frac{0.13 \times 0.11^3}{12} - (2 \times 0.09 \times 0.01 \times 0.06^2)$$

$$\approx 0.2132 \times 10^{-4}\,\text{m}^4$$

At the interface between the outer plates and the side plates where shearing would occur,

$$A\bar{y} = 150 \times 10 \times 70 = 105 \times 10^3\,\text{mm}^3$$

Therefore

$$a = \frac{8000 \times 105 \times 10^{-6} \times 0.1}{50 \times 10^6 \times 0.2132 \times 10^{-4}} = 7.89 \times 10^{-5}\,\text{m}^2$$

$$= 78.9\,\text{mm}^2$$

As there are two rivets at each interface resisting shear, the diameter of each is

$$d = \left(\frac{78.9}{2} \times \frac{4}{\pi}\right)^{\frac{1}{2}} = 7.1\,\text{mm}$$

The best compromise is to use 7 mm rivets at 97 mm pitch, giving 31 pitches in the length of the beam.

## 6.15 Bending and shear stresses in I-section beams

The I-section beam shown in Fig. 6.34($a$) is widely used in the construction of buildings, bridges, etc. The shape is efficient to resist both bending and shear. The latter is carried almost entirely by the web, and the flanges are located where the bending stress is highest. The practical I-section is usually idealized for ease of calculation into the rectangular shapes shown in Fig.

Fig. 6.34

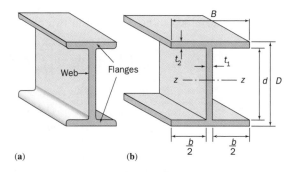

(a)　　　　　　　(b)

6.34($b$). The second moment of area of the section is given by

$$I = \frac{BD^3}{12} - \frac{(b/2)d^3}{12} - \frac{(b/2)d^3}{12} = \frac{BD^3 - bd^3}{12}$$

The bending stress distribution is then given by

$$\sigma = \frac{12My}{BD^3 - bd^3} \qquad\qquad [6.44]$$

which is illustrated in Fig. 6.35.

Fig. 6.35

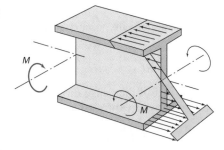

It can be shown that for a typical I-section the flanges carry approximately 80% of the total bending moment on the cross-section. In designing an I-section against failure, consideration must be given, in addition to the strength of the tension flange, to the avoidance of buckling (Chapter 12) in the compression flange.

**Shear stresses in web and flanges**　　The shear-stress distributions in I-section beams are rather more complex than in a rectangular section. Referring to Fig. 6.36, firstly we will examine the distribution of vertical shear $\tau_{xy}$ parallel to the axis $yy$.

For the web, from eqn. [6.41$b$], the shear stress at a distance $y'$ from the neutral axis is given by

$$\tau_{xy} = \frac{Q}{It_1} \int_{y'}^{D/2} y \, dA$$

Because the section has a different width for the flange and the web this integral has to be expressed in two parts:

$$\tau_{xy} = \frac{Q}{It_1}\left(\int_{y'}^{d/2} t_1 y \, dy + \int_{d/2}^{D/2} By \, dy\right)$$

$$= \frac{Q}{2I}\left[\left(\frac{d}{2}\right)^2 - y^2\right] + \frac{QB}{8It_1}(D^2 - d^2) \qquad [6.45]$$

This expression gives a parabolic distribution superimposed on a constant value as shown in Fig. 6.36. For sections in which the wall thickness is small relative to the overall dimensions, the shear stress in the web may be approximated by dividing the shear force by the area of the web.

**Fig. 6.36**

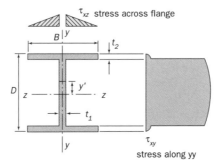

The maximum value occurs at the neutral axis, where $y = 0$, and is

$$\tau_{xy} = \frac{Q}{8I}\left(d^2 + \frac{B}{t_1}(D^2 - d^2)\right) \qquad [6.46]$$

In that part of the flange directly above and below the web the vertical shear stress can be expressed as

$$\tau_{xy} = \frac{Q}{BI}\int_{y'}^{D/2} By \, dy = \frac{Q}{2I}\left[\left(\frac{D}{2}\right)^2 - y'^2\right]$$

However, in those parts of the flange on each side of the web the top and bottom surfaces are 'free' from the load, and therefore the longitudinal and complementary vertical shear stresses must be zero. Thus the distribution is parabolic as for a rectangular section.

**Fig. 6.37**

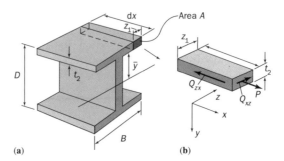

(a)  (b)

It is evident that in the flanges the vertical shear stress and its complementary component contribute little to balancing the longitudinal variation in bending stress. However, this may be achieved, as illustrated in Fig. 6.37, by means of a shear force $Q_{zx}$ lying in an $xy$-plane which cuts off a segment of the flange. A complementary shear force $Q_{xz}$ then occurs in a $yz$-plane. The net end load, $P$, is given by

$$P = \int_A d\sigma \, dA$$

(from eqn. [6.38]) and is equal to the shear force $Q_{zx}$, which is the shear stress $\tau_{zx}$ multiplied by the area $t_2 \, dx$, i.e.

$$P = Q_{zx} = \tau_{zx} t_2 \, dx$$

Therefore

$$\tau_{zx} t_2 \, dx = \int_A d\sigma_x \, dA$$

$$= \frac{dM}{I} \int_A y \, dA$$

$$\tau_{zx} = \frac{Q}{t_2 I} A\bar{y} \qquad [6.47]$$

where $Q$ is the *vertical* shear force on the section and $\bar{y}$ is still measured from the neutral axis to the centroid of the area.

$$A\bar{y} = z t_2 \frac{D - t_2}{2}$$

From eqn. [6.47] and the complementary (positive) shear-stress condition,

$$\tau_{zx} = \tau_{xz} = Q_z \frac{D - t_2}{2I} \qquad [6.48]$$

This is a linear distribution of shear stress in the $z$-direction, being zero at the outer edges of the flanges and a maximum at the joint with the web. The distribution is shown in Fig. 6.36, in which the maximum value is

$$\tau_{xz} = \frac{Q(D - t_2)(B - t_1)}{4I} \qquad [6.49]$$

**Example 6.14**

The vertical steel column of 5 m height and rolled I-section shown in Fig. 6.38 is built-in at the lower end and subjected to a transverse force of 4 kN at the free end. Calculate the bending- and shear-stress distributions at the fixed-end cross-section.

At the base

Bending moment $= 4 \times 5 = 20 \, \text{kNm}$

Shear force $= 4 \, \text{kN}$

Fig. 6.38

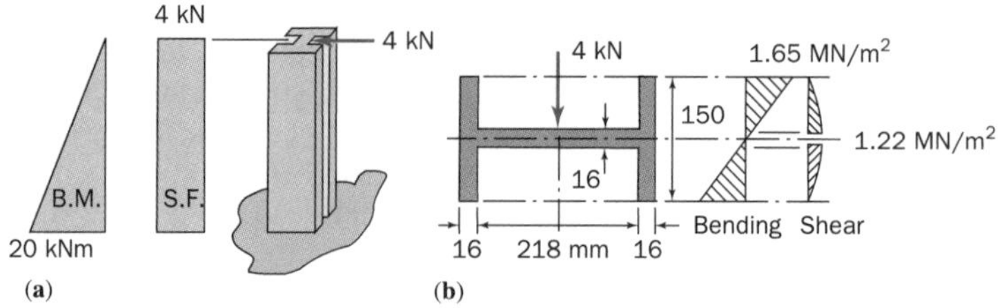

$$\text{Second moment of area} = \frac{32 \times 150^3}{12} + \frac{218 \times 16^3}{12}$$

$$= 9.075 \times 10^6 \, \text{mm}^4$$

$$y_{max} = \pm 0.075 \, \text{m}$$

Therefore

$$\sigma_{max} = \pm \frac{20 \times 10^3 \times 0.075}{9.075 \times 10^{-6}} = \pm 165 \, \text{MN/m}^2$$

In the flanges the shear stress in the $y$-direction is given by

$$\tau_{xy} = \frac{4000}{2 \times 0.016 \times 9.075 \times 10^{-6}} \int_y^{75} 2 \times 0.016 y \, dy$$

$$= \frac{440 \times 10^6}{2} (0.075^2 - y^2)$$

This is a parabolic distribution varying from zero at the outer surfaces to

$$\tau_{xy} = 220 \times 10^6 (0.075^2 - 0.008^2) = 1.22 \, \text{MN/m}^2$$

at the section where the flanges join the web.

In the web itself, since the width, which appears in the denominator for shear stress above, is 250 mm, it is evident that the shear stress in it can be neglected compared with that in the flanges.

A consideration of the geometry of the section indicates that shear stresses of the $\tau_{xz}$ type are also insignificant.

The distributions of bending and shear stresss are shown in Fig. 6.38.

## 6.16 Shear stress in thin-walled open sections and shear centre

There are a number of beam sections widely used, particularly in aircraft construction, in which the thickness of material is small compared with the overall geometry and there is only one or no axis of symmetry. These members are termed *thin-walled open sections*, and some common shapes are shown in Fig. 6.39

The arguments applied to the shear-stress distribution in the flanges of I-sections may also be applied in the above cases; however, there is one important difference owing to the lack of symmetry in the latter.

If the external applied forces, which set up bending moments and shear forces, act through the centroid of the section, then in addition to bending,

**Fig. 6.39**

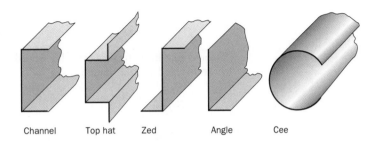

Channel     Top hat     Zed     Angle     Cee

twisting of the beam will generally occur. To avoid twisting, and cause only bending, it is necessary for the forces to act through a particular point, which may not coincide with the centroid. The position of this point is a function only of the geometry of the beam section; it is termed the *shear centre*.

The following examples illustrate how the position of the shear centre may be found. Referring to Fig. 6.40, in which the channel section is loaded

**Fig. 6.40**

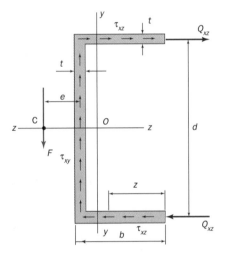

by a vertical force $F$, the $y$- and $z$-axes are principal axes and hence $I_{yz} = 0$. For the shear stress in the flanges the analysis is similar to that for the I-section.

$$\tau_{xz} = \frac{Q}{tI_z} \int_A y \, \mathrm{d}A$$

where $Q$ is the vertical shear force on the section caused by the applied force $F$. At a distance $z$ from the free edge of the flange,

$$\tau_{xz} = \frac{Q}{tI_z} \int_0^z \frac{d}{2} t \, \mathrm{d}z = \frac{Q}{tI_z} \times \frac{tzd}{2} \tag{6.50}$$

The shear stress varies linearly with $z$ from zero at the left to a maximum at the centre-line of the web:

$$\tau_{xz\,max} = \frac{Qbd}{2I_z}$$

The average shear stress is $Qbd/4I_z$, and therefore the horizontal shear force in the top and bottom flange is

$$Q_{xz} = \frac{Qb^2td}{4I_z}$$

The couple about the $x$-axis of these shear forces which would cause twisting of the section is

$$Q_{xz}d = \frac{Qb^2d^2t}{4I_z}$$

Twisting of the section is avoided if there is an opposing couple of equal magnitude. Let the vertical force $F$ act through a point C, the shear centre, at a distance $e$ from the middle of the web as shown in Fig. 6.40, so that $Fe$ balances $Q_{xz}d$. Now $Q$ must equal $F$ for equilibrium; therefore

$$Qe = Q_{xz}d$$

and hence

$$Qe = \frac{Qb^2d^2t}{4I_z}$$

or

$$e = \frac{b^2d^2t}{4I_z} \qquad\qquad [6.51]$$

which locates the position of the shear centre and is only a function of the geometry of the section. The vertical shear stress $\tau_{xy}$ in the web may be found in the same way as was that for the I–section.

**Example 6.15**

**A thin-walled tube of circular cross-section has inner and outer diameters of 50 and 70 mm respectively. If it is slit longitudinally on one side, at what position must a**

**Fig. 6.41**

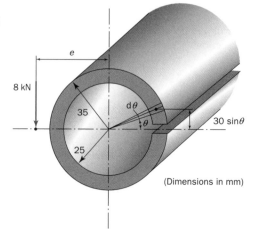

8 kN

35

25

$d\theta$

$\theta$

30 sin$\theta$

(Dimensions in mm)

vertical force of 8 kN be applied so that there is only bending and no twisting of the section? Calculate the maximum shear stress in the section.

Referring to Fig. 6.41, since the slit is narrow, the second moment of area, $I$, for the section may be taken as

$$I = \frac{\pi}{64}(70^4 - 50^4) = 872 \times 10^3 \, \text{mm}^4$$

For the element shown, the shear stress must be zero on the face of the slit and increases with distance from this free edge as

$$\tau = \frac{Q}{tI} \int_{\bar{A}} y \, dA$$

where $\bar{A}$ is the area of the segment of thin wall between the slit and the point of interest. For a thin-walled section $y = r \sin \theta$ and $dA = tr \, d\theta$, therefore,

$$\tau = \frac{Q}{tI} \int_0^\phi tr^2 \sin \theta \, d\theta = \frac{Qr^2}{I}(1 - \cos \phi)$$

The maximum shear stress occurs when $\phi = \pi$, when

$$\tau = \frac{2Qr^2}{I} = \frac{2 \times 8000 \times 30^2}{872 \times 10^3} = 16.5 \, \text{MN/m}^2$$

The torque set up by the shear-stress distribution is given by

$$T = \int_0^{2\pi} r \cdot \tau tr \, d\theta = \frac{Qr^4 t}{I} \int_0^{2\pi} (1 - \cos \theta) d\theta$$

$$= \frac{Q}{I} 2\pi r^4 t = 46.7 \times 10^4 \, \text{Nmm}$$

Let the shear centre be at a distance $e$ from the centre of the tube. Then equilibrium of torques gives

$$Qe = \frac{Q}{I} 2\pi r^4 t$$

$$e = \frac{2\pi r^4 t}{I} = 58.4 \, \text{mm}$$

Note that for a thin-walled tube, $I \sim \pi r^3 t$ and $e \sim 2r$. For a circular tube with no slit, it can be shown that the shear centre lies on the axis and the maximum shear stress is half the value for the slit tube.

The analysis of the slit tube and the channel section described earlier was relatively simple, since the section had one axis of symmetry about which bending was made to occur. The more general case is that of an asymmetric open section subjected to bending which is not a principal axis. The determination of the shear stresses and shear centre in this case is described in Megson[1]. Alternatively if the bending is considered relative to the principal axes of the section, where $I_{yz} = 0$, the analysis above can be used for each principal axis in turn. The shear stress arising from the component of shear force in each principal direction can then be combined by superposition.

## 6.17 Bending of initially curved bars

The theory of bending developed so far has been related to initially straight bars and beams. The analysis will now be extended to include members which are initially curved.

The geometry of curved bars has an important bearing on the bending-stress distribution. If the depth of the cross-section is small compared with the radius of curvature, then the stress distribution is linear as for straight beams. On the other hand, if the depth of section is of the same order as the radius of curvature, then a non-linear stress distribution occurs during bending.

Similar assumptions are made for curved beams as for straight beams, plane cross-sections remaining plane, etc., although a few of the assumptions are not strictly accurate for the case of a bar with a small radius of curvature.

**Fig. 6.42**

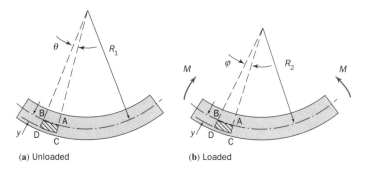

(a) Unloaded                    (b) Loaded

Consider the curved bar shown unloaded in Fig. 6.42(a) and subjected to pure bending $M$ (Fig. 6.42(b)) with initial and final radii of the neutral axis $R_1$ and $R_2$ respectively. The strain in a small element CD at a distance $y$ from the neutral axis is derived as for the straight beam and is

$$\varepsilon_{CD} = \frac{(R_2 + y)\phi - (R_1 + y)\theta}{(R_1 + y)\theta} \qquad [6.52]$$

but for an element AB at the neutral surface there is no change in length, so that $R_1\theta = R_2\phi$. Therefore

$$\varepsilon_{CD} = \frac{y(\phi - \theta)}{(R_1 + y)\theta} \qquad [6.53]$$

Making the substitution $\phi = R_1\theta/R_2$ gives

$$\varepsilon_{CD} = \frac{y[(R_1/R_2) - 1]}{R_1 + y} = \frac{y(R_1 - R_2)}{R_2(R_1 + y)} \qquad [6.54]$$

For the slender beam, $y$ can be neglected compared with $R_1$ and

$$\varepsilon = y\left(\frac{1}{R_2} - \frac{1}{R_1}\right) \qquad [6.55]$$

For $R_1$ infinite, i.e. a straight beam, the expression reduces to that found previously.

By using the same concept as for the straight beam it can be shown that for no applied end load the centroidal axis and the neutral axis coincide, and for equilibrium of the bending moment with the internal resisting moment

$$\frac{M}{I} = \frac{\sigma}{y} = E\left(\frac{1}{R_2} - \frac{1}{R_1}\right) \qquad [6.56]$$

## 6.18 Beams with a small radius of curvature

For the beam in which $y$ is not negligible compared with $R_1$, the strain at distance $y$ from the neutral axis is given by eqn. [6.54]. This is no longer a linear distribution of strain across the section as for the slender beam, and

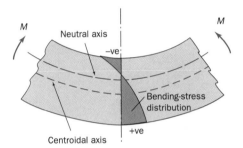

**Fig. 6.43**

hence *the distribution of stress is non-linear*, as indicated in Fig. 6.43, and *the centroidal and neutral axes no longer coincide*, as will now be shown.

Assuming that there is no applied end load,

$$\int_A \sigma dA = 0$$

and from eqn. [6.54]

$$\sigma = E\varepsilon = \frac{Ey(R_1 - R_2)}{R_2(R_1 + y)} \qquad [6.57]$$

Therefore

$$\frac{E(R_1 - R_2)}{R_2} \int_A \frac{y}{R_1 + y} dA = 0$$

since the integral must be zero, and this is not the first moment of area about the centroid; therefore the centroidal and neutral axes do not coincide.

For equilibrium of internal and external moments,

$$M = \int_A \sigma y \, dA$$

$$M = \frac{E(R_1 - R_2)}{R_2} \int_A \frac{y^2}{R_1 + y} dA \qquad [6.58]$$

The integral term may be re-expressed as

$$\int_A \frac{y^2}{R_1 + y}\, \mathrm{d}A = \int_A y\, \mathrm{d}A - R_1 \int_A \frac{y}{R_1 + y}\, \mathrm{d}A$$

$$= \int_A y\, \mathrm{d}A$$

since the second integral is zero as shown above. Now let the distance between the centroidal and neutral axes be $n$; then, for the cross-section of a curved beam shown in Fig. 6.44, $y = y' + n$; hence

$$\int_A y\, \mathrm{d}A = \int_A y'\, \mathrm{d}A + \int_A n\, \mathrm{d}A = nA$$

**Fig. 6.44**

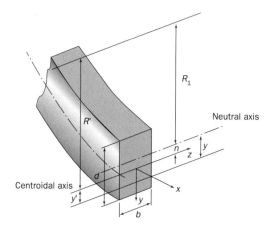

since $\int_A y'\, \mathrm{d}A$ is zero, being the first moment of area about the centroid. Thus

$$\int_A \frac{y^2}{R_1 + y}\, \mathrm{d}A = nA$$

Substituting into eqn. [6.58],

$$\frac{M}{nA} = \frac{E(R_1 - R_2)}{R_2}$$

and further substituting from eqn. [6.57],

$$\frac{M}{nA} = \frac{\sigma}{y}(R_1 + y)$$

or

$$\frac{\sigma}{y} = \frac{M}{nA(R_1 + y)} \qquad [6.59]$$

which is in a form similar to the bending-stress relationship for slender beams, the second moment of area term, $I$, being replaced by $nA(R_1 + y)$.

In order to determine the magnitude of bending stress it is first necessary to find the values of $n$ and $R_1$ for the particular shape of cross-section. From the condition that

$$\int_A \frac{y}{R_1 + y} \, dA = 0$$

it follows that

$$\int_A \frac{y' + n}{R' + y'} \, dA = 0$$

or

$$\int_A \left( \frac{R' + y'}{R' + y'} - \frac{R'}{R' + y} + \frac{n}{R' + y'} \right) dA = 0$$

$$\int_A dA - \int_A \frac{R'}{R' + y'} \, dA + \int_A \frac{n}{R' + y'} \, dA = 0$$

Hence

$$n = R' - \int_A \frac{1}{R' + Y'} \, dA \qquad\qquad [6.60]$$

and

$$R_1 = R' - n \qquad\qquad [6.61]$$

For rectangular and circular sections the values of $n$ and $R_1$ are obtained from eqns. [6.60] and [6.61].

*Rectangular section*

$$n = R' - \frac{d}{\log_e[(R' + d/2)/(R' - d/2)]} \qquad\qquad [6.62]$$

*Circular section of radius* r

$$n = R' - \frac{r^2}{2[R' - (R'^2 - r^2)^{1/2}]} \qquad\qquad [6.63]$$

**Example 6.16**

**A crane hook as illustrated in Fig. 6.45(a) is designed to carry a maximum force of 12 kN. Calculate the maximum tensile and compressive stresses set up on the cross-section AB shown at (b).**

The position of the centroid is given by

$$d_1 = \frac{[48 + (2 \times 24)]}{72} \frac{54}{3} = 24 \, \text{mm}$$

**Fig. 6.45**

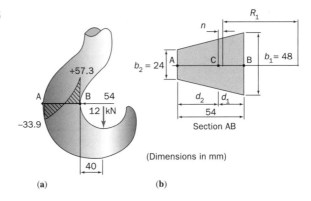

(Dimensions in mm)

(a)                   (b)

Hence

$$d_2 = 54 - 24 = 30 \, \text{mm}$$

The value of $n$ is given as follows:

$$n = R'$$

$$-\frac{\frac{1}{2}(b_1 + b_2)d}{\{b_2 + [(b_1 - b_2)/d](R' + d_2)\}\log_e[(R' + d_2)/(R' - d_1)] - (b_1 - b_2)}$$

$$= 78 - \frac{36 \times 54}{72 \log_e 2 - 24} = 3 \, \text{mm}$$

and

$$R_1 = 78 - 3 = 75 \, \text{mm}$$
$$A = 36 \times 54 = 1944 \, \text{mm}^2$$
$$M = -12\,000 \times 0.064 = -768 \, \text{Nm}$$
$$\text{(negative since curvature is reduced)}$$
$$\hat{y}_1 = -24 + 3 = -21 \, \text{mm}$$
$$\hat{y}_2 = +30 + 3 = +33 \, \text{mm}$$

$$\text{Direct stress on section AB} = \frac{12\,000}{0.001\,944}$$

$$= +6.2 \, \text{MN/m}^2$$

$$\text{Bending stress at B} = \frac{-768 \times -0.021}{0.003 \times 0.001\,944(0.075 - 0.021)}$$

$$= +51.1 \, \text{MN/m}^2$$

$$\text{Total stress at B} = +57.3 \, \text{MN/m}^2$$

$$\text{Bending stress at A} = \frac{-768 \times 0.033}{0.003 \times 0.001\,944(0.075 + 0.033)}$$

$$= -40.1\,\text{MN/m}^2$$

$$\text{Total stress at A} = -33.9\,\text{MN/m}^2$$

The higher value of stress at B illustrates the reason for having the trapezoidal cross-section, i.e. more material is available at B.

## 6.19 Summary

Many important practical problems of members subjected to bending have been analysed in this chapter and if this summary is to be succinct it must pick out the most important principles. Firstly, little progress can be made unless there is a clear understanding of the calculation of *shear-force and bending-moment distributions*. Then the *properties of areas* are always present in order to find the position of the neutral axis and the second moment of area of the cross-section. From this point in order to analyse a wide range of situations involving bending the same basic procedure is required for every case: (i) to write the *equilibrium statement* relating applied forces and internal reactions (as a function of stress), (ii) with appropriate assumptions to decide on *the geometry of deformation* of a suitable free body of the beam, and (iii) to link (i) and (ii) through the elastic *stress–strain relationship*. The final solution will also entail the use of the specific *boundary conditions* of the problem.

## Reference

1.  Megson, T. H. G. (1990) *Aircraft Structures for Engineering Students*, 2nd edition, Edward Arnold, London.

## Problems

6.1  Draw the bending-moment and shear-force diagrams for the beam loaded as shown in Fig. 6.46 and insert the principal values on each diagram.

**Fig. 6.46**

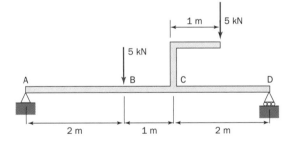

6.2  A gear wheel is mounted on a 25 mm diameter shaft which is driven through a 600 mm diameter pulley as in Fig. 6.47. During operation the gear wheel is subjected to a horizontal force of 3 kN and a vertical force of 5 kN. Calculate the value of the maximum bending moment on the shaft and sketch the bending-moment diagram.

6.3  A trolley of weight $W$ is supported on wheels as shown in Fig. 6.48. It can move over the full length of the beam AB. Show that the

**Fig. 6.47**

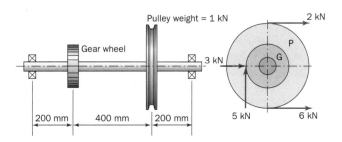

maximum bending moment occurs in the beam when a wheel of the trolley is at a distance $c$ from either support, where $c$ is given by

$$c = \frac{L}{4}\left(2 - \frac{d}{L}\right)$$

**Fig. 6.48**

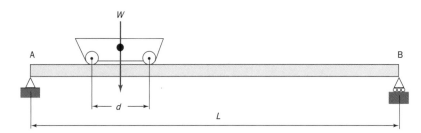

6.4    The portal frame shown in Fig. 6.49 is pinned at A, C and E. Sketch the bending-moment diagrams for the sides and top of the frame and insert the principal values.

**Fig. 6.49**

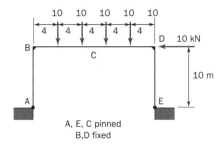

A, E, C pinned
B,D fixed

6.5    A dam 12 m in length retains water at a depth of 3 m as shown in Fig. 6.50. Sketch the bending-moment and shear-force diagrams for the dam. The density of water may be taken as $1000\,\text{kg/m}^3$.

6.6    A load of 50 kN is supported over a length of 4 m on the rigid beam AB as shown in Fig. 6.51. Plot the shear-force and bending-moment diagrams for the beam, stating the values of bending moment and shear force at key points along the beam. Calculate the magnitude and position of the maximum bending stress in the beam. $I = 30 \times 10^6 \text{mm}^4$.

6.7    A floor joist 6 m in length is simply-supported at each end and carries a varying distributed load of grain over the whole span. The loading

Fig. 6.50

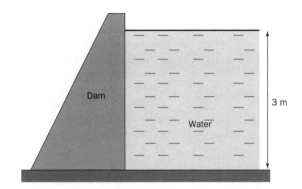

Fig. 6.51

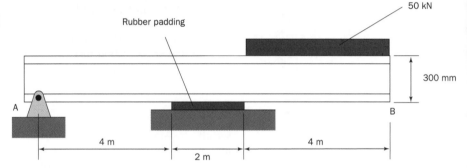

distribution is represented by the equation $w = ax^2 + bx + c$, where $w$ is the load intensity in kN/m at a distance $x$ along the beam, and $a$, $b$ and $c$ are constants. The load intensity is zero at each end and has its maximum value of 4 kN/m at mid-span.

Apply the differential equations of equilibrium relating load, shear force and bending moment to obtain the maximum values and distributions of shear force and bending moment.

6.8 A horizontal beam, the weight of which may be ignored, is loaded as shown in Fig. 6.52. The distributed load varies linearly from zero at the left-hand end to 6 kN/m at a distance of 3 m from the left-hand end. Sketch the shearing-force and bending-moment diagrams for the loading shown and insert the principal values of each on these diagrams.

6.9 A cantilevered balcony which projects 2 m out from a wall is constructed of timber joists at 0.33 m spacing supporting a boarded floor 12 mm thick. The design loading on the floor is 4.5 kN/m² and the self-weight of a joist and its associated boarding may be neglected.

Fig. 6.52

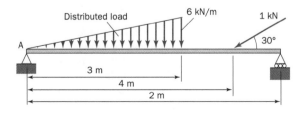

If each joist is to be 120 mm deep, determine the required width so that the maximum tensile bending stress does not exceed $10 \, \text{MN/m}^2$. For the purpose of calculating the second moment of area, it may be assumed that the neutral axis of the combined joist cross-section and its associated strip of boarding is at the mid-depth of the whole section.

Also find where the neutral axis is correctly loaded when the joist width is determined.

6.10 A channel which carries water is made of sheet metal 3 mm thick and has a cross-section of 400 mm width and 200 mm depth. If the maximum allowable depth of water is 150 mm determine the maximum simply-supported span. The maximum bending stress (tension or compression) is not to exceed $35 \, \text{MN/m}^2$, and self-weight may be neglected. Loading due to water, $9.81 \, \text{kN/m}^3$.

6.11 A $203 \times 133$ Universal Beam with the section properties given in the Table below is loaded as shown in Fig. 6.53. Determine (i) the largest bending stress which occurs under this loading and (ii) indicate where on the cross-section this stress occurs.

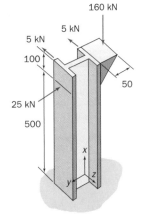

|  |  | Thickness | | | |
|---|---|---|---|---|---|
| Serial size (mm) | Mass per metre (kg) | Depth D (mm) | Width B (mm) | Web $t$ (mm) | Flange $T$ (mm) | Area cm² |
| $203 \times 133$ | 30 | 206.8 | 133.8 | 6.3 | 9.3 | 38.0 |

**Fig. 6.53**

| Second moment of area | | Elastic modulus | | Plastic modulus | |
|---|---|---|---|---|---|
| Axis $z$–$z$ (cm⁴) | Axis $y$–$y$ (cm⁴) | Axis $z$–$z$ (cm³) | Axis $y$–$y$ (cm³) | Axis $z$–$z$ (cm³) | Axis $y$–$y$ (cm³) |
| 2880 | 354 | 278.5 | 52.85 | 312.6 | 83.7 |

6.12 A beam with the cross-section shown in Fig. 6.54 is to be bent about the $x$–$x$ axis. Determine the optimum value of $h$ in order to minimize the outer fibre stress for a fixed bending moment and beam width, $B$. The second moment of area of a triangle about its base is given by

$$I = 1/12 \times \text{base} \times (\text{height})^3$$

6.13 Given the applied moment on a beam and the dimensions of its cross-section as shown in Fig. 6.55, develop a spreadsheet to calculate the maximum bending stress in the beam for
(i) a hollow circular section
(ii) a rectangular section

**Fig. 6.54**

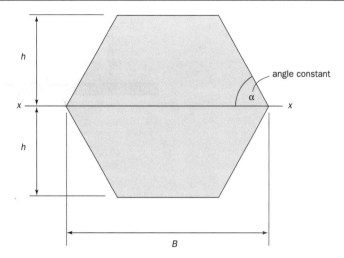

**Fig. 6.55**

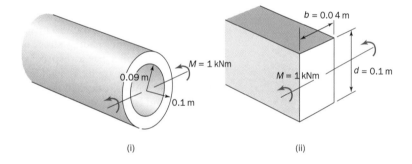

(i)                                    (ii)

6.14  The roof of a petrol station is made up of two main beams and eleven cross-beams (purlins) as shown in Fig. 6.56. The main beams have an I-section 200 mm wide, 600 mm deep with a web thickness of 10 mm and a flange thickness of 15 mm. The purlins also have an I-section 125 mm wide, 250 mm deep with a web thickness of 6 mm and a flange thickness of 10 mm. Calculate the maximum stresses in each type of beam if there is a snow loading of 120 kg/m² on the roof. The density of the beam steel is 7850 kg/m³ and you should allow for the weight of the beams.

6.15  A tapered shaft of length $l$ is built in at the larger end of diameter $d_2$, and is free at the smaller end of diameter $d_1$. A force $W$ is applied at the free end perpendicular to the axis of the shaft. Show that the maximum bending stress at any section distant $x$ from the free end of the shaft is given by

$$\frac{Wx}{(\pi/32)\{d_1 + (d_2 - d_1)(x/l)\}^3}$$

**Fig. 6.56**

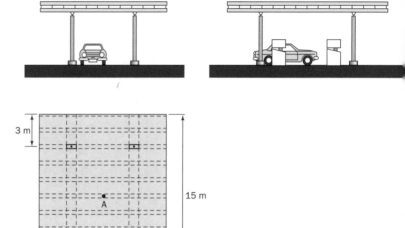

Hence determine the distance $x$ at which point the greatest value of the maximum bending stress occurs.

6.16  A small trailer has a suspension system as shown in Fig. 6.57. If the weight of the trailer is 4 kN and its centre of gravity is 0.5 m forward of the wheels, calculate the bending moments and torques in sections AB and BC. Calculate also the maximum bending and shear stresses in these sections. Ignore any effects at corners or changes in section.

6.17  A timber beam 80 mm wide by 160 mm deep is to be reinforced with two steel plates 5 mm thick. Compare the resisting moments for the

**Fig. 6.57**

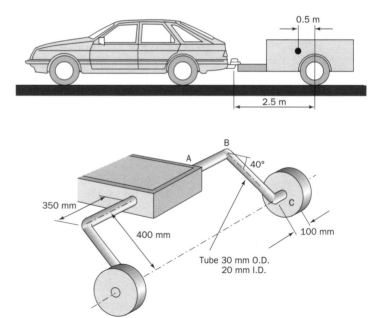

same value of the maximum bending stress in the timber when the plates are: (i) 80 mm wide and fixed to the top and bottom surfaces of the beam, and (ii) 160 mm deep and fixed to the vertical sides of the beam. $E$ for steel $= 20 \times E$ for timber.

6.18 A composite beam is to be made up of a U-shaped steel sheet with a wooden board glued on top as in Fig. 6.58. What depth must the wood be to cause the neutral axis in pure bending to be at the horizontal diameter of the semicircle? Calculate the maximum bending moment which may be applied to the beam if the maximum stresses in the steel and wood are not to exceed $280\,\text{MN/m}^2$ and $7\,\text{MN/m}^2$, respectively. The moduli for steel and wood are $210\,\text{GN/m}^2$ and $7\,\text{GN/m}^2$ respectively.

**Fig. 6.58**

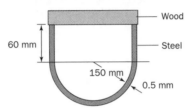

6.19 Extend the spreadsheet of Problem 6.13 to find the stress in a beam section made of multiple materials, given the modulus and dimensions of each part. Assume that the centroid of all the individual pieces lies on the neutral axis. Use the spreadsheet to check the results of Example 6.6.

6.20 Extend the spreadsheet above to find the stresses in a beam section of multiple materials where the distribution of materials is not symmetric. Use the spreadsheet to check the results of Example 6.7.

6.21 A reinforced-concrete beam has a rectangular cross-section 500 mm deep and 250 mm wide. The area of steel reinforcement is $1100\,\text{mm}^2$, and it is placed at 50 mm above the tension face. Calculate the resisting moment of the section and the stress in the steel if the compressive stress in the concrete is not to exceed $4.2\,\text{MN/m}^2$ and the modular ratio is 15.

6.22 A horizontal beam of rectangular cross-section 100 mm deep and 50 mm wide is simply supported at each end of a 1.5 m span. Vertical loads of 5 kN are applied at 0.5 m and 1 m from one end, and a horizontal tension of 40 kN is applied at the ends 25 mm below the upper surface. Determine and plot the distribution of longitudinal stress across the section at mid-span.

What eccentricity of end load is required so that there is just no resultant compressive stress at the outer surface?

6.23 The cross-section through a concrete dam is illustrated in Fig. 6.59. Calculate the required width of the base AB so that there is just no tensile stress at B. What is the resultant compressive stress at A? Loading due to water $= 9.81\,\text{kN/m}^3$. Weight of concrete $= 22.7\,\text{kN/m}^3$.

6.24 A concrete cooling tower may be assumed to consist of two truncated cones as in Fig. 6.60. If the estimated maximum horizontal wind pressure is $1.5\,\text{kN/m}^2$, calculate the wall thickness of the tower in order to avoid tensile stresses in the concrete. The density of the

**Fig. 6.59**

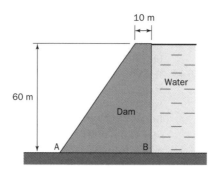

concrete is $2400 \, \text{kg/m}^3$. The volume of a truncated cone $= (1/3)\pi h \{R^2 + r^2 + Rr\}$.

**Fig. 6.60**

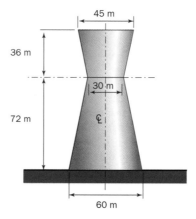

6.25 Derive an expression for the shear-stress distribution across a solid circular section rod of radius $R$, subjected to bending. Calculate the ratio of the maximum shear stress to the average shear stress on this section.

6.26 A channel-section beam is 50 mm wide and 50 mm deep with a 5 mm wall thickness. It is simply-supported over a length of 1 m and carries a uniformly-distributed load of 50 kN/m over its whole length. It also has a line load of 50 kN at mid-span. Sketch the shear stress distribution across the beam section 0.25 m from one of the supports and indicate the important values.

6.27 A circular tube of mean radius $r$ and wall thickness $t(\ll r)$ is subjected to a transverse shear force $Q$ during bending. Show, by derivation, that the maximum shear stress, $\tau$, occurs at the neutral axis and is equal to $Q/\pi r t$.

6.28 A wooden beam section can be made up by one of the two methods shown in Fig. 6.61. If the shear force in the beam is constant at 3 kN, calculate which design is preferable to keep the shearing forces in the nails to a minimum. If the nails can withstand a shear force of 400 N, determine the maximum permissible spacing of the nails along the beam for the design selected.

6.29 A beam is made up of four 50 mm × 100 mm pieces of timber glued to a 25 mm × 500 mm web of the same wood as shown in Fig. 6.62.

**Fig. 6.61**

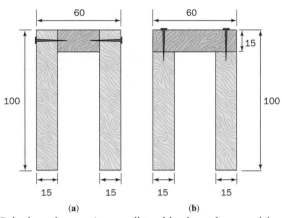

(a)          (b)

Calculate the maximum allowable shear force and bending moment that this section can carry. The maximum shearing stresses in the wood and glued joist must not exceed 500 and 250 kN/m² respectively, and the maximum permissible normal stress is 1 MN/m².

**Fig. 6.62**

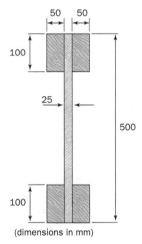

(dimensions in mm)

6.30 A metal bar AD is 10 mm wide and is bent into the shape shown in Fig. 6.63. If a vertical force of 80 N is applied at D, calculate the depth of the bar if both bending stresses and shear stresses in the metal must not exceed 60 MN/m².

6.31 A thin walled hollow square tube of sidelength $b$ is to be used in an aircraft wing. It is known from design handbooks that the side walls of the tube will locally buckle if the shear stress exceeds

$$\tau = 7.4 \frac{E}{1-\nu^2} \left(\frac{t}{b}\right)^2$$

where $E = 70$ GN/m², $\nu = 0.3$, $b = 100$ mm and $t$ is the thickness of the sheet from which the tube is fabricated. If a shear force of 10 kN must be supported: (a) find the shear stress in the side wall of the tube on the neutral axis when the tube is fabricated from 1.5 mm sheet; (b) find the minimum sheet thickness to prevent failure by local buckling.

**Fig. 6.63**

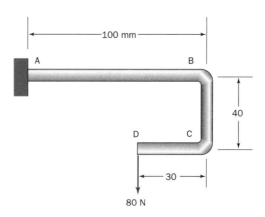

6.32  Extend the spreadsheet of Example 6.11 so that the additional stress due to an axial force of 10 kN is included.

6.33  Plot the shear stress distribution in the I-beam of Fig. 6.38 using a spreadsheet and eqn. [6.45]. Assume the section is loaded with a shear force of 4 kN along the line of the web, i.e. perpendicular to the shear force of Example 6.14. Compare the results with the average shear stress calculated by assuming the shear force is carried by the web only.

6.34  A Z-section beam is 2 m long and is supported as a cantilever with a 1 kN load at the free end. The direction of the 1 kN relative to the beam section is as shown in Fig. 6.64. Calculate the magnitude and position of the maximum tensile and compressive stresses on the section.

6.35  A 100 mm × 100 mm angle section as shown in Fig. 6.65 is built in at one end of its 1 m length and subjected to a point load of 3 kN at the free end. The point load is applied at an angle of 20° to the vertical axis as indicated. Calculate the stresses at L, M and N and the orientation of the neutral axis.

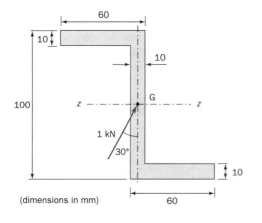

(dimensions in mm)

**Fig. 6.64**

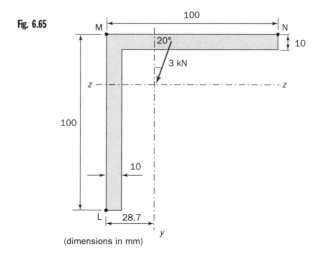

**Fig. 6.65**

(dimensions in mm)

6.36 A channel-section beam 1 m long is built in at one end and subjected to a point load of 1 kN at the free end. If the direction of the load is as in Fig. 6.66, calculate the direction of the neutral axis and the maximum values of the tensile and compressive stresses on the section.

Fig. 6.66

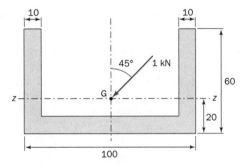

6.37 Determine the location of the shear centre for the beam cross-section shown in Fig. 6.67. Also calculate maximum values of the horizontal and vertical shear stresses in a flange and the web respectively for a vertical load of 500 kN applied at the shear centre.

Fig. 6.67

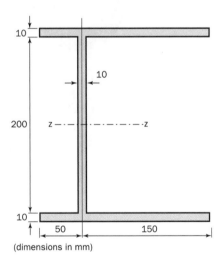

(dimensions in mm)

6.38 (i) Determine the shear centre of the three sections shown in Fig. 6.68. All are formed from plates of a constant thickness $t$, which is small relative to the other dimensions of the section. (ii) For the section of Fig. 6.6(a), calculate the maximum shear stress in the section when a shear force of 27 kN is applied vertically through the shear centre. Assume $b = 200$ mm, $d = 250$ mm and $t = 15$ mm.

6.39 A proving ring, used to calibrate a testing machine, has a mean diameter of 500 mm and a rectangular section 76 mm wide and

**Fig. 6.68**

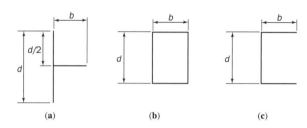

(a)            (b)            (c)

12.7 mm thick. If the maximum permitted stress under diametral compression is 55 MN/m², determine the maximum calibration load.

6.40   A U-shaped bar with a square cross-section (40 mm × 40 mm) is subjected to forces as shown in Fig. 6.69. Sketch the stress distribution at the section AB and insert the principal values.

**Fig. 6.69**

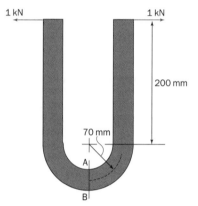

6.41   A chain coupling is made up of a 20 mm diameter steel rod bent into an 'S' shape as shown in Fig. 6.70. If the coupling is subjected to a tensile load of 1 kN as indicated, calculate the maximum stress in the steel.

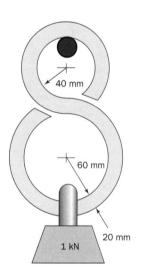

**Fig. 6.70**

# 7 Bending: Slope and Deflection

Having studied the stresses set up in bending, we now turn to the equally important aspect of beam stiffness. In many structural elements, such as floor joists or aircraft wings, the limiting constraint on the design is stiffness. Any design which is stiff enough will be strong enough. It is important that we should be able to calculate the deflection of a beam of given section, since for given conditions of span and load it would be possible to adopt a section which would meet a strength criterion but would give an unacceptable deflection.

The total deflection of a beam is due to a very large extent to the deflection caused by bending, and to a very much smaller extent to the deflection caused by shear. Various methods are available for determining the slope and deflection of a beam due to elastic bending, and examples of the use of each method will be found in this chapter. That part of the total deflection caused by shear will be discussed in Chapter 9.

## 7.1 The curvature–bending-moment relationship

In Fig. 7.1, $\theta$ is the angle which the tangent to the curve at C makes with the $x$-axis, and $(\theta - \mathrm{d}\theta)$ that which the tangent at D makes with the same axis. The normals to the curve at C and D meet at O. The point O is the centre of curvature and $R$ is the radius of curvature of the small portion CD of the deflection curve of the neutral axis.

Numerically $\mathrm{d}s = R\,\mathrm{d}\theta$ and $1/R = \mathrm{d}\theta/\mathrm{d}s$. Using the sign convention for bending described in Chapter 1 it will be seen that positive increments of $\mathrm{d}s$ from left to right are associated with a negative change in $\mathrm{d}\theta$. Thus, when signs are taken into account, the last equation becomes

$$\frac{1}{R} = -\frac{\mathrm{d}\theta}{\mathrm{d}s} \qquad [7.1]$$

Deflections of the neutral axis are denoted by the symbol $v$, measured positive downwards, and are assumed to be relatively small, giving a flat form of deflection curve; therefore no error is introduced in assuming that $\mathrm{d}s \approx \mathrm{d}x$, that $\theta \approx \tan\theta = \mathrm{d}v/\mathrm{d}x$, and hence that $\mathrm{d}\theta/\mathrm{d}s = \mathrm{d}^2v/\mathrm{d}x^2$. Therefore

$$\frac{1}{R} = -\frac{\mathrm{d}^2v}{\mathrm{d}x^2} \qquad [7.2a]$$

Fig. 7.1

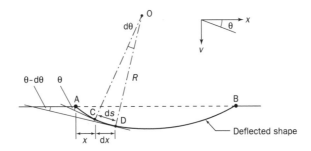

Note that from eqn. [6.5] the strain, $\varepsilon_x$, will be given by

$$\varepsilon_x = \frac{y}{R} = -y\frac{d^2v}{dx^2} \tag{7.2b}$$

Using eqn. [6.9], when elastic bending occurs,

$$\frac{1}{R} = \frac{M}{EI}$$

Therefore

$$\frac{d^2v}{dx^2} = -\frac{M}{EI} \tag{7.3}$$

This is the differential equation of the deflection curve. If the variation of $M$ with $x$ is known then this equation can be integrated twice to give the deflection, $v$.

**7.2 Slope and deflection by the double-integration method**

The first integration of eqn. [7.3] gives the slope of the beam at a distance $x$ along its length when the origin is taken at A. Therefore

$$\theta = \frac{dv}{dx} = \int \frac{-M}{EI}\,dx + C \tag{7.4}$$

The second integration gives the deflection of the beam at the above point, or

$$v = \int \frac{dv}{dx}\,dx = \int\left(\int \frac{-M}{EI}\,dx\right)dx + Cx + C_1 \tag{7.5}$$

$C$ and $C_1$, the constants of integration, can be evaluated from the known conditions of slope and deflection at certain points, usually at the supports. Equations [7.4] and [7.5] are widely used for determining the slope and deflection of a beam at a given point. Examples of their use will be found in the following paragraphs. In each case the sign convention used to obtain eqn. [7.3] will be adopted.

**Beam simply supported with distributed loading**

This problem is illustrated in Fig. 7.2. From eqn. [1.13], the bending moment at D is

$$M = \frac{wL}{2}x - \frac{wx^2}{2}$$

and

$$EI\frac{d^2v}{dx^2} = -\frac{wL}{2}x + \frac{wx^2}{2} \tag{7.6}$$

and

$$EI\frac{dv}{dx} = -\frac{wL}{4}x^2 + \frac{wx^3}{6} + C$$

**Fig. 7.2**

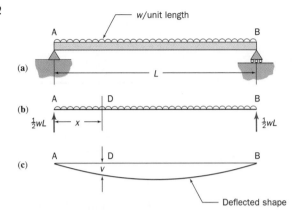

From symmetry $dv/dx = 0$, at $x = \frac{1}{2}L$; therefore $C = wL^3/24$. Hence

$$\frac{dv}{dx} = -\frac{w}{2EI}\left(\frac{Lx^2}{2} - \frac{x^3}{3} - \frac{L^3}{12}\right) \qquad [7.7]$$

The slopes at the ends of the beam are given by $wL^3/24EI$ at A, where $x = 0$, and $-wL^3/24EI$ at B, where $x = L$.

$$v = -\frac{w}{2EI}\left(\frac{Lx^3}{6} - \frac{x^4}{12} - \frac{L^3x}{12}\right) + C_1$$

At $x = 0$, $v = 0$; therefore $C_1 = 0$ and

$$v = -\frac{w}{12EI}\left(Lx^3 - \frac{x^4}{2} - \frac{L^3x}{2}\right) \qquad [7.8]$$

The maximum deflection occurs at mid-span, where $x = \frac{1}{2}L$. Therefore

$$v_{max} = -\frac{w}{12EI}\left(\frac{L^4}{8} - \frac{L^4}{32} - \frac{L^4}{4}\right) = \frac{5}{384}\frac{wL^4}{EI} \qquad [7.9]$$

**Simply supported beam with uniform bending moment**

The simply supported beam shown in Fig. 7.3 is acted on by a force system which sets up the moment, $\bar{M}$, at each end. The bending moment at any point, D, along the beam is $\bar{M}$, and with the previous sign convention,

$$EI\frac{d^2v}{dx^2} = -\bar{M} \qquad [7.10]$$

Therefore

$$EI\frac{dv}{dx} = -\bar{M}x + C$$

and

$$EIv = -\frac{\bar{M}x^2}{2} + Cx + C_1$$

**Fig. 7.3**

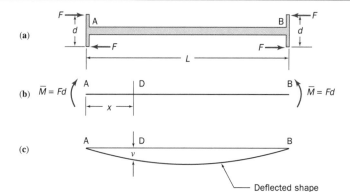

At $x = 0$, $v = 0$; therefore $C_1 = 0$; and at $x = L$, $v = 0$; therefore $C = \frac{1}{2}\bar{M}L$, and the slope $dv/dx$ is given by

$$\frac{dv}{dx} = \frac{1}{EI}\left(-\bar{M}x + \frac{\bar{M}L}{2}\right) \tag{7.11}$$

from which the slopes at the ends are $+\bar{M}L/2EI$ at A, $-\bar{M}L/2EI$ at B and at mid-span is zero. The deflection at any point is given by

$$v = \frac{1}{EI}\left(-\frac{\bar{M}x^2}{2} + \frac{\bar{M}L}{2}x\right) \tag{7.12}$$

The maximum deflection occurs at mid-span, and hence

$$v_{max} = \frac{\bar{M}L^2}{8EI} \tag{7.13}$$

**Cantilever with uniformly distributed loading**

The bending moment at D in Fig. 7.4 is

$$M = R_A x - M_A - \frac{wx^2}{2}$$

where $R_A = wL$ and $M_A = \frac{1}{2}wL^2$, the fixing moment at the support. Therefore

$$EI\frac{d^2v}{dx^2} = -M = -wLx + \frac{wL^2}{2} + \frac{wx^2}{2} \tag{7.14}$$

$$EI\frac{dv}{dx} = -\frac{wL}{2}x^2 + \frac{wL^2}{2}x + \frac{wx^3}{6} + C$$

At $x = 0$, $dv/dx = 0$; therefore $C = 0$. Hence

$$\frac{dv}{dx} = \frac{1}{EI}\left(-\frac{wL}{2}x^2 + \frac{wL^2}{2}x + \frac{wx^2}{6}\right) \tag{7.15}$$

At the free end, $x = L$. Therefore

$$\text{Slope at free end} = \frac{wL^3}{6EI} \tag{7.16}$$

**Fig. 7.4**

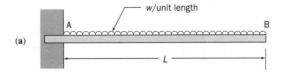

(a)

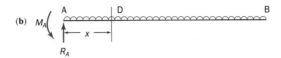

(b)

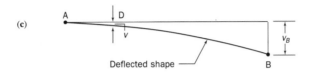

(c)

Deflected shape

$$v = \frac{1}{EI}\left(-\frac{wL}{6}x^3 + \frac{wL^2}{4}x^2 + \frac{wx^4}{24}\right) + C_1$$

and at $x = 0$, $v = 0$; therefore $C_1 = 0$. Hence

$$v = \frac{1}{EI}\left(-\frac{wL}{6}x^3 + \frac{wL^2}{4}x^2 + \frac{wx^4}{24}\right) \qquad [7.17]$$

The deflection at the free end B, where $x = L$, is given by

$$v_B = \frac{1}{EI}\left(-\frac{wL^4}{6} + \frac{wL^4}{4} + \frac{wL^4}{24}\right) = \frac{1}{8}\frac{wL^4}{EI} \qquad [7.18]$$

The reader may wish to check that this value for the deflection at the free end may also be obtained by analysing in the same way the cantilever in the configuration shown in Fig. 1.31 (i.e. fixed at the right-hand end rather than the left).

**Cantilever carrying a concentrated load**

The bending moment at D in Fig. 7.5 is given by

$$M = -M_A + R_A x$$

where $M_A = Wl$, the fixing moment at the support, and $R_A = W$. Therefore

$$M = -Wl + Wx$$

$$EI\frac{\mathrm{d}^2v}{\mathrm{d}x^2} = -M = Wl - Wx \qquad [7.19]$$

and

$$EI\frac{\mathrm{d}v}{\mathrm{d}x} = Wlx - \frac{Wx^2}{2} + C$$

**Fig. 7.5**

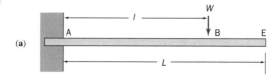

(a)

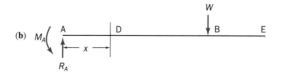

(b)

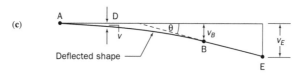

(c)

At $x = 0$, $\mathrm{d}v/\mathrm{d}x = 0$; therefore $C = 0$. Hence

$$\frac{\mathrm{d}v}{\mathrm{d}x} = \frac{Wlx}{EI} - \frac{Wx^2}{2EI} \qquad [7.20]$$

At $x = l$,

$$\frac{\mathrm{d}v}{\mathrm{d}x} = \frac{Wl^2}{2EI} \qquad [7.21]$$

$$v = \frac{Wlx^2}{2EI} - \frac{Wx^3}{6EI} + C_1$$

and at $x = 0$, $v = 0$; therefore $C_1 = 0$, and

$$v = \frac{W}{2EI}\left(lx^2 - \frac{x^3}{3}\right) \qquad [7.22]$$

which is the equation for the deflection curve for the beam.

For the deflection under the load we substitute $x = l$ in eqn. [7.22] and

$$v_B = \frac{1}{3}\frac{Wl^3}{EI} \qquad [7.23]$$

Either from eqn. [7.5] or from Fig. 7.5(c) and noting the fact that the moment and curvature are zero from B to E, it may be seen that

Deflection at free end E = deflection at B + (slope at B) × $(L - l)$

Therefore

$$v_E = \frac{1}{3}\frac{Wl^3}{EL} + \frac{Wl^2}{2EI}(L - l)$$

$$= \frac{Wl^2}{2EI}\left(L - \frac{l}{3}\right) \qquad [7.24]$$

**Cantilever with concentrated load at free end**

The slope and deflection under the load are obtained by substituting $L$ for $l$ in eqns. [7.21] and [7.23], and thus

$$v_{max} = \frac{1}{3}\frac{WL^3}{EI} \quad \text{and} \quad \theta = \frac{WL^2}{2EI} \qquad [7.25]$$

**Example 7.1**

A horizontal beam AB, simply supported at each end, carries a load which increases, at a uniform rate, from zero at one end. Determine the position of, and the value of, the maximum deflection.

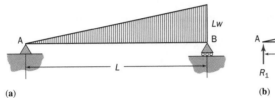

(a)

(b)

**Fig. 7.6**

Let $w$ be the intensity of loading at unit distance from A; then the intensity at C, Fig. 7.6, is $xw$ and that at B is $Lw$.

Taking moments about B,

$$R_1 L - Lw \cdot \frac{L}{2} \cdot \frac{L}{3} = 0$$

$$R_1 = \frac{L^2 w}{6}$$

Bending moment at $C = R_1 x - \frac{1}{2}x^2 w\left(\frac{x}{3}\right)$

$$= \frac{L^2 wx}{6} - \frac{x^3 w}{6}$$

$$EI\frac{\mathrm{d}^2 v}{\mathrm{d}x^2} = -\frac{L^2 wx}{6} + \frac{x^3 w}{6}$$

$$EI\frac{\mathrm{d}v}{\mathrm{d}x} = -\frac{L^2 w}{6}\frac{x^2}{2} + \frac{x^4 w}{24} + C$$

$$EIv = -\frac{L^2 w}{36}x^3 + \frac{x^5 w}{120} + Cx + C_1$$

The boundary conditions are, where $x = 0$, $v = 0$; therefore $C_1 = 0$. And at $x = L$, $v = 0$; therefore

$$0 = -\frac{L^5 w}{36} + \frac{L^5 w}{120} + CL$$

$$C = \frac{7}{360}L^4 w$$

and

$$EI\frac{dv}{dx} = -\frac{L^2 w}{12}x^2 + \frac{x^4 w}{24} + \frac{7}{360}L^4 w$$

$$= \frac{1}{360}(-30L^2 wx^2 + 15x^4 w + 7L^4 w)$$

At maximum deflection $dv/dx = 0$. Therefore

$$x^2 = \frac{30L^2 \pm \sqrt{(900L^4 - 420L^4)}}{30} = \frac{30L^2 - 21.9L^2}{30}$$

$$= 0.27L^2$$

so that

$$x = 0.52L$$

$$EIv = -\frac{L^2 w}{36}x^3 + \frac{w}{120}x^5 + \frac{7}{360}L^4 wx$$

$$= \frac{w}{360}(-10L^2 x^3 + 3x^5 + 7L^4 x)$$

$$v_{max} = \frac{w}{360}[-10L^2(0.52L)^3 + 3(0.52L)^5 + 7L^4(0.52L)]\frac{1}{EI}$$

$$= 0.00654\frac{L^5 w}{EI}$$

**Example 7.2**

A beam 4 m long is simply supported at its ends and carries a varying distributed load over the whole span. The equation to the loading curve is $w = ax^2 + bx + c$, where $w$ is the load intensity in kN/m, at a distance $x$ along the beam, measured from an origin at the left-hand support, and $a$, $b$ and $c$ are constants. The load intensity is zero at each end of the beam and reaches a maximum value of 100 kN/m at the centre of the span. Calculate the slope of the beam at each support and the deflection at the centre. $E = 208\,\text{GN/m}^2$, $I = 405 \times 10^{-6}\,\text{m}^4$.

The loading conditions are such that at $x = 0$, $w = 0$; therefore $c = 0$. Also, at $x = 4$, $w = 0$; therefore

$$0 = 16a + 4b \qquad \text{and} \qquad b = -4a$$

At $x = 2$, $w = 100$; therefore

$$100 = 4a + 2b \qquad \text{and} \qquad 100 = 4a - 8a$$

Hence

$$a = -25 \qquad \text{and} \qquad b = 100$$

Therefore

Loading distribution, $w = -25x^2 + 100x\,\text{kN/m}$

$$\text{Total load on beam} = \int_0^4 w\,\mathrm{d}x = \int_0^4 (-25x^2 + 100x)\mathrm{d}x = 267\,\text{kN}$$

Therefore the support reactions are each 133.5 kN, by symmetry.
   The shear-force distribution is given by

$$Q = -\int w\,\mathrm{d}x = -\int(-25x^2 + 100x)\mathrm{d}x = +25\frac{x^3}{3} - \frac{100x^2}{2} + A$$

At $x = 0$, $Q = R_1 = 133.5$; therefore $A = +133.5$.

$$\text{Bending moment, } M = \int Q\,\mathrm{d}x = \int\left(+\frac{25x^3}{3} - \frac{100x^2}{2} + 133.5\right)\mathrm{d}x$$

$$= +\frac{25x^4}{12} - \frac{100x^3}{6} + 133.5x + B$$

At $x = 0$, $M = 0$; therefore $B = 0$.

$$\text{Slope} = -\frac{1}{EI}\int M\,\mathrm{d}x = -\frac{1}{EI}\int\left(+\frac{25x^4}{12} - \frac{100x^4}{6} + 133.5x\right)\mathrm{d}x$$

$$= \frac{1}{EI}\left(-\frac{25x^5}{60} + \frac{100x^4}{24} - \frac{133.5x^2}{2} + C\right)$$

At $x = 2$, $\theta = 0$; therefore $C = 213$.
   When $x = 0$ and 4 m, $\theta = \pm0.002\,53\,\text{rad}$.
   The deflection is given by

$$v = \int\theta\,\mathrm{d}x = \frac{1}{EI}\int\left(-\frac{25x^5}{60} + \frac{100x^4}{24} - \frac{133.5x^2}{2} + 213\right)\mathrm{d}x$$

$$= \frac{1}{EI}\left(-\frac{25x^6}{360} + \frac{100x^5}{120} - \frac{133.5x^3}{6} + 213x + D\right)$$

At $x = 0$, $v = 0$; therefore $D = 0$. At mid-span, $x = 2$ and $v = 3.2\,\text{mm}$.

## 7.3 Discontinuous loading: Macaulay's method

When considering the bending-moment distribution for a beam with discontinuous loading (e.g. Fig. 1.33 or Fig. 6.1), a separate bending-moment expression has to be written for each part of the beam. This means that in deriving slope and deflection a double integration would have to be performed on each bending-moment expression and two constants would result for each section of the beam. A further example of discontinuous loading is shown in Fig. 7.7(*a*); in this case there would be three bending-

**Fig. 7.7**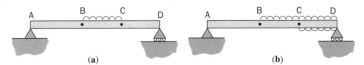

(a)                                    (b)

moment equations and thus six constants of integration. There are
apparently only two boundary conditions, those of zero deflection at each
end. However, at the points of discontinuity, B and C, both slope and
deflection must be continuous from one section to the next, so that

$$\text{At B} \qquad \left(\frac{\mathrm{d}v}{\mathrm{d}x}\right)_{AB} = \left(\frac{\mathrm{d}v}{\mathrm{d}x}\right)_{BC} \qquad \text{and} \qquad v_{AB} = v_{BC}$$

$$\text{At C} \qquad \left(\frac{\mathrm{d}v}{\mathrm{d}x}\right)_{BC} = \left(\frac{\mathrm{d}v}{\mathrm{d}x}\right)_{CD} \qquad \text{and} \qquad v_{BC} = v_{CD}$$

The above four conditions together with the two conditions of zero
displacement at each end enable the six constants of integration to be
determined. The derivation of the deflection curve by the above approach is
rather tedious; it is therefore an advantage to use the mathematical technique
termed a *step function*, commonly known as Macaulay's method when
applied to beam solutions. This approach requires one bending-moment
expression to be written down for a point close to the right-hand end to
cover the bending-moment conditions for the whole length of beam, and
hence, on integration, only two unknown constants have to be determined.

The step function is a function of $x$ of the form $f_n(x) = [x - a]^n$ such that
for $x < a, f_n(x) = 0$ and for $x > a, f_n(x) = (x - a)^n$. Note the change in the
form of brackets used: the square brackets are particularly chosen to indicate
the use of a step function, the curved brackets representing normal
mathematical procedure. The important features when using the step
function in analysis are that, if on substitution of a value for $x$ the quantity
inside the square brackets becomes negative, it is omitted from further
analysis. Square bracket terms must be integrated in such a way was to
preserve the identity of the bracket, i.e.

$$\int [x - a\,]^2\, \mathrm{d}x = \tfrac{1}{3}\,[x - a]^3$$

Also, for mathematical continuity, distributed loading which does not
extend to the right-hand end, as in Fig. 7.7(a), must be arranged to continue
to $x = l$, whether starting from $x = 0$ or $x = a$. This may be effected by the
superposition of loadings which cancel each other in the required portions of
the beam as shown in Fig. 7.7(b).

An applied couple $M_0$ must be expressed as a step function in the form
$M_0[x - a]^0$ so that the bracket can be integrated correctly.

The three common step functions for bending moment are shown in
Fig. 7.8 and several illustrative examples now follow.

**Beam simply supported**
**with concentrated load**

Taking moments about one end, the reactions at the supports in Fig. 7.9 are

$$R_1 = \frac{W(L - a)}{L} \qquad \text{and} \qquad R_2 = \frac{Wa}{L}$$

**Fig. 7.8**

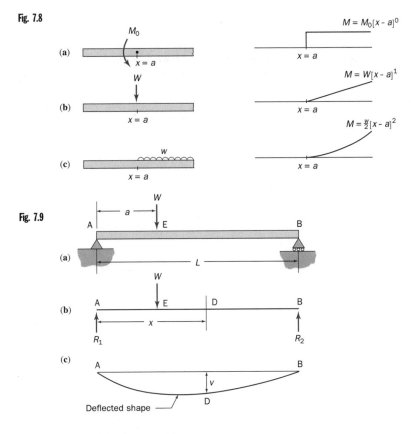

**Fig. 7.9**

When $x < a$ the moment at any distance $x$ from the left end of the beam is given by $M = R_1 x$.

When $x > a$ the moment is given by

$$M = R_1 x - W(x - a)$$

Using the step function concept to write one equation for the whole beam, then

$$M = R_1 x - W[x - a]$$

Note that when $x < a$, the quantity in the square brackets would become negative and so $W[x - a]$ is taken as zero. This leaves $M = R_1 x$ as above. Using the step function moment expression in eqn. [7.3]

$$EI \frac{d^2 v}{dx^2} = -M = -R_1 x + W[x - a] \qquad [7.26]$$

Integrating eqn. [7.26],

$$EI \frac{dv}{dx} = -\frac{R_1 x^2}{2} + \frac{W}{2}[x - a]^2 + C \qquad [7.27]$$

and

$$EI v = -\frac{R_1 x^3}{6} + \frac{W}{6}[x - a]^3 + Cx + C_1 \qquad [7.28]$$

If we omit the term inside the square brackets on the right-hand side of eqns. [7.27] and [7.28] when $x < a$, the equations are then of the correct form for the portion AE of the beam, and since the second term on the right-hand side of these equations vanishes for a value of $x = a$, then when these equations are used for the whole beam, both $dv/dx$ and $v$ will be continuous at the point E.

When $x = 0$, $v = 0$, and since the term inside the square brackets is omitted, $C_1 = 0$. For $x = L$, $v = 0$; therefore

$$0 = -\frac{R_1 L^3}{6} + \frac{W}{6}(L - a)^3 + CL$$

and

$$C = \frac{R_1 L^2}{6} - \frac{W}{6L}(L - a)^3$$

$$= \frac{W(L - a)L}{6} - \frac{W}{6L}(L - a)^3$$

$$= \frac{Wa}{6L}(L - a)(2L - a) \qquad [7.29]$$

Substituting the values of $C$ and $R_1$ in eqn. [7.28] and rearranging,

$$v = \frac{Wx}{6EI}\frac{L - a}{L}(2aL - a^2 - x^2) + \frac{W}{6EI}[x - a]^3 \qquad [7.30]$$

This equation gives the deflection at any point along the beam if the last term on the right-hand side is rejected when it becomes negative, i.e. for $x < a$. For the particular case when $x = a$, the deflection under the load is given by

$$v_E = \frac{Wa^2(L - a)^2}{3EIL} \qquad [7.31]$$

It should be noted that the maximum deflection occurs where $dv/dx = 0$ which is not, in fact, where the load is applied.

If $W$ is placed at mid-span so that $a = \frac{1}{2}L$, the deflection under the load is also the maximum deflection and is

$$v = \frac{WL^3}{48EI} \qquad [7.32]$$

**Beam with distributed load on part of the span**

As explained above (Fig. 7.7(*b*)), for mathematical continuity the loading must be continued to the right-hand end, and to maintain equilibrium, upward loading must be inserted from D to B as shown in Fig. 7.10.

At point E between D and B,

$$M = R_1 x - \frac{w}{2}[x - a]^2 + \frac{w}{2}[x - (a + b)]^2 \qquad [7.33]$$

**Fig. 7.10**

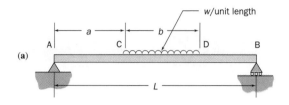

(a)

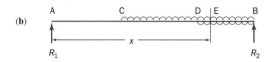

(b)

(c)

Deflected shape

the second and third terms being rejected when $x < a$ and the third term when $x < (a + b)$.

$$EI\frac{d^2v}{dx^2} = -R_1 x + \frac{w}{2}[x - a]^2 - \frac{w}{2}[x - (a + b)]^2$$

and

$$EI\frac{dv}{dx} = -\frac{R_1 x^2}{2} + \frac{w}{6}[x - a]^3 - \frac{w}{6}[x - (a + b)]^3 + C \qquad [7.34]$$

$$EIv = -\frac{R_1 x^3}{6} + \frac{w}{24}[x - a]^4 - \frac{w}{24}[x - (a + b)]^4 + Cx + C_1 \qquad [7.35]$$

The values of $C$ and $C_1$ are found from the conditions of $v = 0$ when $x = 0$ and $x = L$, the terms inside the square brackets being rejected when negative.

**Beam simply supported with an applied moment**

From moment equilibrium in Fig. 7.11,

$$R_1 = R_2 = \frac{\bar{M}}{L}$$

The bending moment at D, where $x > a$, is

$$M_D = R_1 x - \bar{M} = \frac{\bar{M}}{L}x - \bar{M}$$

A convenient way of dealing with a couple by Macaulay's method is to introduce a term $[x - a]^0$ which is in fact unity, but allows for subsequent integration in the correct manner:

$$EI\frac{d^2v}{dx^2} = -\frac{\bar{M}}{L}x + \bar{M}[x - a]^0$$

**Fig. 7.11**

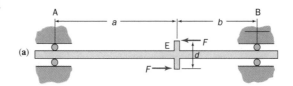

(a)

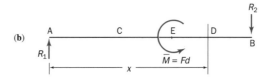

(b)

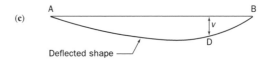

(c)

Deflected shape

and on integration we may write

$$EI\frac{dv}{dx} = -\frac{\bar{M}x^2}{2L} + \bar{M}[x-a]^1 + C$$

where the second term is integrated with respect to $(x - a)$. Therefore

$$EIv = -\frac{\bar{M}x^3}{6L} + \frac{\bar{M}}{2}[x-a]^2 + Cx + C_1 \qquad [7.36]$$

When $x < a$, the square-bracketed term on the right-hand side of the equation becomes negative and is rejected.

At $x = 0$, $v = 0$; therefore $C_1 = 0$. And at $x = L$, $v = 0$; therefore

$$0 = -\frac{\bar{M}L^2}{6} + \frac{\bar{M}}{2}[L-a]^2 + CL$$

and

$$C = \frac{\bar{M}}{6L}(-2L^2 + 6aL - 3a^2) \qquad [7.37]$$

Hence

$$v = \frac{1}{EI}\left(-\frac{\bar{M}}{6L}x^3 + \frac{\bar{M}}{2}[x-a]^2 - \frac{\bar{M}}{6L}(2L^2 - 6aL + 3a^2)x\right) \qquad [7.38]$$

At E, where $x = a$, the deflection is given by

$$EIv_E = -\frac{\bar{M}}{6L}a^3 - \frac{\bar{M}}{6L}[2L^2 - 6aL + 3a^2]a$$

or, putting $L = (a + b)$,

$$EIv_E = -\frac{\bar{M}}{6L}[a^3 - a^3 - 2ab(a-b)]$$

$$v_E = \frac{\bar{M}}{3EI} \frac{(a-b)ab}{L} \tag{7.39}$$

**Example 7.3**

A simply supported beam is subjected to the loading shown in Fig. 7.12. Calculate the deflection at a section 1.8 m from the left-hand end. $E = 70\,\text{GN/m}^2$, $I = 832\,\text{cm}^4$.

**Fig. 7.12**

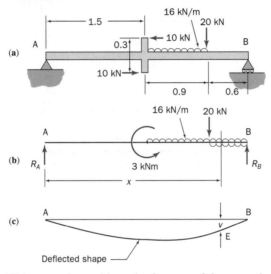

This example combines the features of the cases above, and so, to satisfy the Macaulay conditions, the distributed load must be extended to B and an equivalent negative load inserted to restore the correct resultant load distribution. Then

$$M = R_A x - 3[x - 1.5]^0 - 16\frac{[x-1.5]^2}{2} - 20[x-2.4]$$

$$+16\frac{[x-2.4]^2}{2}$$

$M = 0$ when $x = 3$; therefore $R_A = 10\,\text{kN}$.

$$EI\frac{\mathrm{d}v}{\mathrm{d}x} = -\frac{10x^2}{2} + 3[x-1.5] + \frac{16[x-1.5]^3}{6} + \frac{20[x-2.4]^2}{2}$$

$$-\frac{16[x-2.4]^3}{6} + C$$

$$EIv = -\frac{10x^3}{6} + \frac{3[x-1.5]^2}{2} + \frac{16[x-1.5]^4}{24} + \frac{20[x-2.4]^3}{6}$$

$$-\frac{16[x-2.4]^4}{24} + Cx + C_1$$

When $x = 0$, $v = 0$; therefore $C_1 = 0$. And when $x = 3$, $v = 0$; therefore

$$0 = -\frac{10(3)^3}{6} + \frac{3(1.5)^2}{2} + \frac{16(1.5)^4}{24} + \frac{20(0.6)^3}{6} - \frac{16(0.6)^4}{24} + 3C$$

from which $C = 12.54$. When $x = 1.8$ the third and fourth square-bracketed terms are omitted, and

$$EIv = -\frac{10 \times 1.8^3}{6} + \frac{3 \times 0.3^2}{2} + \frac{16 \times 0.3^4}{24} + (12.54 \times 1.8)$$

$$= 13 \, \text{kN/m}^3$$

$$v = \frac{13 \times 10^3 \times 10^3}{70 \times 10^9 \times 832 \times 10^{-8}} = 22.3 \, \text{mm}$$

**Example 7.4**

Calculate the position and magnitude of the maximum deflection for the beam shown in Fig. 7.13. $EI = 1000 \, \text{kNm}^2$.

**Fig. 7.13**

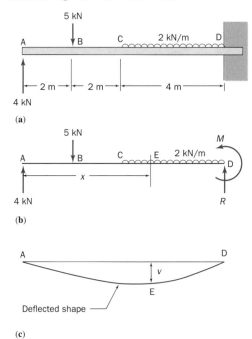

$$EI\frac{d^2v}{dx^2} = -4x + 5[x - 2] + \frac{2}{2}[x - 4]^2$$

$$EI\frac{dv}{dx} = -\frac{4x^2}{2} + \frac{5}{2}[x - 2]^2 + \frac{2}{6}[x - 4]^3 + C$$

$$EIv = -\frac{4x^3}{6} + \frac{5}{6}[x - 2]^3 + \frac{2}{24}[x - 4]^4 + Cx + C_1$$

The boundary conditions are $dv/dx = 0$ at $x = 8$, and $v = 0$ at $x = 8$; therefore

$$0 = -(2 \times 64) + (5 \times 18) + \frac{64}{3} + C$$

$$C = +16.7$$

$$0 = -\left(4 \times \frac{8^3}{6}\right) + (5 \times 36) + \frac{16^2}{12} + (16.7 \times 8) + C_1$$

$$C_1 = -6.6$$

At the left-hand end the deflection is obtained when $x = 0$; therefore

$$EIv = -6.6\,\text{kNm}^3$$

$$v = -\frac{6.6 \times 10^3}{1000} = -6.6\,\text{mm}$$

This may not be the maximum deflection and we must check elsewhere in the span. However, it is not sufficient merely to equate $dv/dx$ to zero since square-bracketed terms would then be included which might not be appropriate, depending on where $v_{max}$ occurred. The best way is to make a sensible guess as to the section where the maximum deflection is likely to occur and to determine the slope at each end of that section. The slopes will be of opposite sign if the guess was correct. If not, then an adjacent section must be treated in the same way. For example, assuming $dv/dx = 0$ occurs between B and C, then

$$\text{At B} \qquad EI\frac{dv}{dx} = -\left(4 \times \frac{2^2}{2}\right) + 16.7 = +8.7$$

$$\text{At C} \qquad EI\frac{dv}{dx} = -\left(4 \times \frac{4^2}{2}\right) + \left(\frac{5}{2} \times 2^2\right) + 16.7 = -5.3$$

and the assumption was correct. Therefore, using the condition that zero slope occurs between B and C,

$$\frac{dv}{dx} = 0 = -\frac{4x^2}{2} + \frac{5}{2}(x - 2)^2 + 16.7$$

Thus

$$x^2 - 20x + 53.4 = 0$$

from which $x = 3.17\,\text{m}$.

The deflection at this point is given by

$$EIv = -\left(\frac{4 \times 3.17^3}{6}\right) + \left(\frac{5}{6} \times 1.17^3\right) + (16.7 \times 3.17) - 6.6$$

$$= +26.44\,\text{kNm}^3$$

$$v = +\frac{26.44 \times 10^3}{1000} = +26.44 \, \text{mm}$$

Hence the maximum deflection occurs at 3.17 m from A and is downwards.

## 7.4 Superposition method

The principle of superposition which was introduced in Chapter 1 states that the effect of a given combined loading on a structure may be obtained by determining separately the effects of the various loads and then combining the results obtained. This type of superposition is only possible if each effect is linearly proportional to the load which produces it. Also, the deformation resulting from any given load should be small enough that it does not affect the conditions of application of the other loads.

In the case of beam deflections, the principle of superposition can be applied to give the total deflection of a beam which carries individual loads $W_1$, $W_2$, $W_3$, etc., or distributed loads $w_1$, $w_2$, $w_3$, etc. Let the bending moments at a section of the beam caused by each load when acting separately on the beam be $M_1$, $M_2$, $M_3$, etc., and the corresponding deflections be $v_1$, $v_2$, $v_3$, etc. Then the total bending moment is

$$M = M_1 + M_2 + M_3 + \ldots \tag{7.40}$$

But

$$M = -EI\frac{d^2v}{dx^2}$$

Therefore

$$v = -\frac{1}{EI}\int\left(\int M\,dx\right)dx$$

$$= -\frac{1}{EI}\int\left(\int (M_1 + M_2 + M_3 + \ldots)dx\right)dx$$

$$= -\frac{1}{EI}\int\left(\int M_1 dx\right)dx + \int\left(\int M_2 dx\right)dx + \int\left(\int M_3 dx\right)dx + \ldots$$

$$= v_1 + v_2 + v_3 + \ldots \tag{7.41}$$

Thus the deflection at a section of a beam subjected to complex loading can be obtained by the summation of the deflections caused at that section by the individual components of the loading.

## Example 7.5

Use the principle of superposition to determine the deflections at the ends and centre of the beam shown in Fig. 7.14. $EI = 500 \, \text{kN m}^2$.

This problem may be split into three components as shown in Fig. 7.15($a$), ($b$) and ($c$). It should be noted that when breaking a problem down into elements, it is essential that the same boundary conditions are employed in each case. For example, it would be incorrect to utilize a solution for beam

**Fig. 7.14**

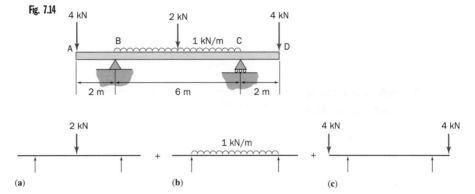

**Fig. 7.15**

deflection from Table 7.1 (at the end of the chapter) which did not use simple supports at B and C.

The respective deflections for cases (a), (b) and (c) are:

(a)
$$\delta_1 = \frac{Wl^3}{48EI} = +\frac{2000 \times 6^3 \times 10^3}{48 \times 500 \times 10^3} = +18\,\text{mm}$$

(b)
$$\delta_2 = \frac{5wl^4}{384EI} = +\frac{5 \times 1000 \times 6^4 \times 10^3}{384 \times 500 \times 10^3} = +33.8\,\text{mm}$$

(c) This may be treated as a beam subjected to couples at B and C of 8 kNm magnitude:

$$\delta_3 = \frac{Ml^2}{8EI} = -\frac{8000 \times 6^2 \times 10^3}{8 \times 500 \times 10^3} = -72\,\text{mm}$$

Resultant deflection $= +18 + 33.8 - 72 = -20.2\,\text{mm}$

To find the deflection at A or D it is necessary to know the slope in each case at B or C. Then

(a)
$$\delta_{1A} = \theta_{1B}l_{AB} = \frac{Wl_{BC}^2}{16EI}l_{AB} = -\frac{2000 \times 6^2 \times 2 \times 10^3}{16 \times 500 \times 10^3}$$

$$= -18\,\text{mm}$$

(b)
$$\delta_{2A} = \theta_{2B}l_{AB} = \frac{wl_{BC}^3}{24EI}l_{AB} = -\frac{1000 \times 6^3 \times 2 \times 10^3}{24 \times 500 \times 10^3}$$

$$= -36\,\text{mm}$$

(c)
$$\delta_{3A} = \theta_{3B}l_{AB} + \frac{W_A l_{AB}^3}{3EI}$$

$$= \frac{Ml_{BC}l_{AB}}{2EI} + \frac{W_A l_{AB}^3}{3EL}$$

$$= \frac{10^3}{500 \times 10^3}\left(\frac{8000 \times 6 \times 2}{2} + \frac{4000 \times 8}{3}\right)$$

$$= 117.3\,\text{mm}$$

$$\text{Resultant deflection at A or D} = -18 - 36 + 117.3$$

$$= +63.3 \, \text{mm}$$

## 7.5 Deflections due to asymmetrical bending

In asymmetrical bending, as in symmetrical bending, the deflection will occur perpendicular to the neutral plane. At any section along the beam the deflection may be calculated using the deflection formulae developed in this chapter. For example, for a cantilever of length $L$, carrying a load $W$ at the free end, the maximum deflection of the centroid of the cross-section will be given by

$$\delta = \frac{W_{NA}L^3}{3EI_{NA}}$$

where $W_{NA}$ is the component of the load perpendicular to the neutral axis and $I_{NA}$ is the second moment of area about the neutral axis.

The most convenient way to obtain $I_{NA}$ is to use the co-ordinates $I_z$, $I_{yz}$ and $I_y$, $-I_{yz}$ to construct a Mohr's circle for moments of area as illustrated in Appendix A. This then allows the second moment of area at any angle to the $y$- or $z$-direction to be determined.

### Example 7.6

**Calculate the maximum deflection of the centroid of the beam section shown in Fig. 6.27, if the beam is a cantilever of length 1 m. Young's modulus for the beam material is 210 GN/m².**

The beam deflection at the free end is perpendicular to the neutral axis (N.A.) shown in Fig. 6.28 and is given by

$$\delta = \frac{W_{NA}L^3}{3EI_{NA}}$$

The solution to Example 6.11 shows that the direction of the neutral axis is 47.45° anticlockwise from the $z$-axis.

The *vertical* end load on the beam is 2 kN, so

$$W_{NA} = (2000 \, \cos \, 47.45°) \, \text{N}$$

The second moment of area about the neutral axis is obtained from the Mohr's circle construction shown in Fig. 7.16. This gives

**Fig. 7.16** $I_v$ and $I_u$ are second moments of area about the principal axes (i.e. those were $I_{yz} = 0$)

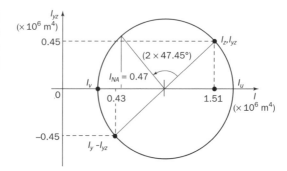

$I_{NA} = 0.46 \times 10^{-6} \, \text{m}^4$; therefore

$$\delta = \frac{(2000 \ \cos \ 47.45°)(1)^3}{3 \times 210 \times 10^9 \times 0.47 \times 10^{-6}} = 4.57 \, \text{mm}$$

This deflection is downwards to the right perpendicular to the neutral axis. The vertical and horizontal deflections of the centroid are given by

$$\text{Vertical deflection} = 4.57 \ \cos \ 47.45° = 3.1 \, \text{mm} \downarrow$$

$$\text{Horizontal deflection} = 4.57 \ \sin \ 47.45° = 3.4 \, \text{mm} \rightarrow$$

## 7.6 Beams of uniform strength

All of the beams analysed so far have had a uniform cross-section. This is convenient for manufacturing purposes and the dimensions of the cross-section will have been chosen on the basis that the maximum stress in the critical sections of the beam should equal the design stress for the material. As a consequence of this, the non-critical sections of the beam will be overdesigned, i.e. they will have too much material for the stresses they are carrying. A more economical design would be one in which the stress is the same at all sections. This would effectively be a beam of uniform strength.

The stress at any section in a beam may be obtained from the bending equation $\sigma = My/I$. The condition, therefore, of uniform strength is that $My/I$ shall be constant, i.e. the section modulus shall be proportional to the bending moment. The value of $I/y$ may be varied, in the case of rectangular beams, by altering the depth or altering the breadth.

**Beam having constant breadth**

$$\sigma = \frac{M_1 y_1}{I_1} = \frac{My}{I}, \quad \text{or} \quad \frac{M_1}{\frac{1}{6}bD_1^2} = \frac{M}{\frac{1}{6}bd^2}$$

Therefore

and

$$\frac{D_1}{d} = \sqrt{\frac{M_1}{M}}$$

$$\frac{M}{I} = \frac{M_1}{I_1}\frac{D_1}{d} = \frac{M_1}{I_1}\sqrt{\frac{M_1}{M}}$$

where $D_1$ is the depth at mid-span and $d$ the depth at a point distance $L$ from the support. In the case of a beam resting on supports, then, as before, considering one-half of the span,

$$v = \int_0^{L/2} \frac{M}{EI} x \, dx$$

$$= \frac{M_1^{3/2}}{EI_1} \int_0^{L/2} M^{-1/2} x \, dx \qquad [7.42]$$

For a concentrated load at mid-span, $M = \frac{1}{2}Wx$ and $M_1 = \frac{1}{4}WL$. Therefore

$$v = \frac{WL^3}{24EI_1}$$

Other cases may be solved by substituting the value of $M$ in eqn. [7.42] and integrating. Some examples of beams of uniform strength are shown in Fig. 7.17.

**Fig. 7.17** Uniform strength beam

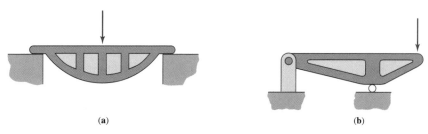

(a)                                 (b)

Table 7.1   Principal slope and deflection for beams with basic loading

| | | Slope | Deflection |
|---|---|---|---|
| (a) | | $-\dfrac{Ml}{EI}$ at B | $-\dfrac{Ml^2}{2EI}$ at B |
| (b) | | $-\dfrac{Ml}{12EI}$ at C $+\dfrac{Ml}{24EI}$ at A, B | 0 at C |
| (c) | | $\pm\dfrac{Ml}{2EI}$ at A, B | $+\dfrac{Ml^2}{8EI}$ at C |
| (d) | | $+\dfrac{Wl^2}{2EI}$ at B | $+\dfrac{Wl^3}{3EI}$ at B |
| (e) | | $\pm\dfrac{Wl^2}{16EI}$ at A, B | $+\dfrac{Wl^3}{48EI}$ at C |
| (f) | | 0 at A, B, C | $+\dfrac{Wl^3}{192EI}$ at C |
| (g) | | $+\dfrac{wl^3}{6EI}$ at B | $+\dfrac{wl^4}{8EI}$ at B |
| (h) | | $\pm\dfrac{wl^3}{24EI}$ at A, B | $+\dfrac{5wl^4}{384EI}$ at C |
| (i) | | 0 at A, B, C | $+\dfrac{wl^4}{384EI}$ at C |

## 7.7 Summary

All solutions for the slope and deflection of beams depend on the relationship between bending-moment distribution and curvature (eqn. [7.3]). It is therefore vital that the correct bending-moment expression can be stated. Although the double integration is relatively simple, the constants of integration can only be found by applying the correct boundary conditions at the supports. Equally, the successful application of Macaulay's method for loading discontinuities does depend on following the simple rules associated with the use of step functions. The superposition method is a valuable alternative to the double-integration method and choice depends on the nature of the problem. For example, it is obviously an advantage to use superposition if the case can be broken down into simple elements the solutions for which are readily available and can then be superposed. A final reminder may be made of the need to recognize the statically determinate nature of the problem; if this is not the case, then the treatment given in Chapter 8 is required. The principal values for the slope and deflection for various basic support and loading conditions are given in Table 7.1.

## Problems

7.1 A cantilevered deck is built in at the left end and is supported on a wall at 8 m from the end. The deck extends a further 4 m beyond this wall. The loading on each of the beams supporting the deck including self-weight is 20 kN/m and the flexural rigidity ($EI$) of each beam section is 200 MNm$^2$. Determine the level of the top of the wall (assumed rigid) relative to the fixed support so that the bending moment is zero at that support.

7.2 A horizontal beam is subjected to the loading shown in Fig. 7.18. Calculate the beam deflection under the 8 kN force. The flexural rigidity ($EI$) of the beam is 1 MNm$^2$.

**Fig. 7.18**

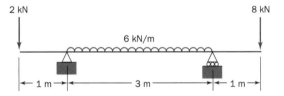

7.3 A beam 10 m long is simply supported at each end and carries the loading shown in Fig. 7.19. Calculate the position and the magnitude of the maximum deflection. The flexural rigidity ($EI$) of the beam is 100 MNm$^2$.

**Fig. 7.19**

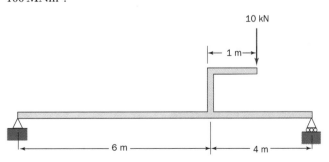

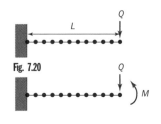

**Fig. 7.20**

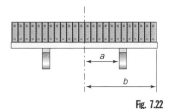

**Fig. 7.21**

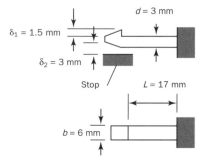

**Fig. 7.22**

**Fig. 7.23**

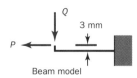

7.4   A cantilever beam is subjected to a point load $Q$ at its tip, as shown in Fig. 7.20. Create a spreadsheet to calculate the deflection of the beam at 10 points equally spaced along its length. Provide cells in which the length $L$, second moment of area $I$ and material modulus $E$ can be entered. Create an $x$–$y$ graph of the negative of deflection, $-v$, versus distance along the beam, $x$.

7.5   Extend the spreadsheet of Problem 7.4 so that a moment $M$ can also be applied to the tip and the deformed shape for the superimposed moment and shear force is calculated (Fig. 7.21). Observe how the deformed shape changes as different bending moments and shear forces are applied and compare the calculated shapes with the behaviour of a flexible ruler. (Try $E = I = I = W = 1$, $L = 10$, $M = 6$).

7.6   A bookshelf is to be supported by two wall brackets positioned symmetrically (Fig. 7.22). Where should the brackets be positioned if: (a) the maximum stress in the shelf is to be minimized; (b) the largest deflection of the shelf is to be minimized.

     Note that the deflection at both the end and the centre will need to be checked.

     A spreadsheet can be used to evaluate moment and deflection for a range of values of $a/b$.

7.7   Calculate the deflection of the roof of the petrol station in Problem 6.14 at points A, B and C.

7.8   A flexible mounting pad 3 m in length and of flexural stiffness $0.55\,MNm^2$ is simply-supported at one end and rests on a spring of stiffness $100\,N/mm$ at the other end. Determine the overall displacement of a load of $10\,kN$ placed on the pad at 2 m from the spring-supported end.

7.9   The plastic clip shown in Fig. 7.23 is injection moulded from a nylon material with modulus $2000\,MN/m^2$. The maximum allowable stress in the material is $100\,MN/m^2$. Each half of the clip must support an axial force $P$ of $250\,N$, which may be considered to act at the lip shown. The design constraints are:

(*a*) the force $Q$ to open the clip should not exceed $25\,\text{N}$

(*b*) the allowable stress should not be exceeded when: (i) the maximum axial force $P$ is applied at the point shown; (ii) the clip arm is pushed fully open against the stop in the centre of the clip.

Set up a spreadsheet to calculate the opening force and the stresses for any set of dimensions and load. Confirm that the clip shown in Fig. 7.23 meets the design requirements.

Note that the spreadsheet optimization facilities could be used to find the values of $b$ and $h$ which give the minimum volume of this material.

7.10 Part of a bridge is being assembled as shown in Fig. 7.24. The central section is hanging from a crane and is to be bolted to the left- and right-hand cantilever sections at A and B. In raising the section into position and before it is properly aligned it fouls at A and B, causing an upward force $W$ exerted by the crane, on the beams. Show that the deflection of the beam at the crane hook, if $E$ and $I$ are the same for

**Fig. 7.24**

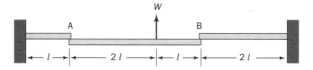

each of the beam sections, is

$$v = -\frac{15}{9}\frac{Wl^3}{EI}$$

7.11 A sight-screen for a cricket field is illustrated in Fig. 7.25. The cross-beams AB and CD can be assumed to be rigid and act as simple supports to the screen itself. When the screen is subjected to uniform wind pressure the location of the cross-beams is to be such that the horizontal deflections of the top and bottom edges are the same as the

**Fig. 7.25**

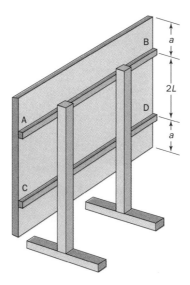

horizontal deflection of the centre line. Hence determine the relationship between $L$ and $a$.

7.12 A horizontal shaft 1 m in length is simply-supported in bearings at 100 mm from each end. It carries loads of 0.3 kN at each end and 1.2 kN at the centre. Find the deflection under the loads and the slope at the bearings. Check the solution by the method of superposition. $EI = 7 \, \text{kNm}^2$.

7.13 Calculate the vertical and horizontal components of deflection at the free end of the Z-section cantilever described in Problem 6.34. $E = 208 \, \text{GN/m}^2$.

7.14 A horizontal beam is simply-supported at each end of a 4 m span. It carries a distributed loading varying from 2 kN/m at the left end to 3 kN/m at the right-hand end. Find the position and magnitude of the maximum deflection. $E = 208 \, \text{GN/m}^2$; $I = 2 \times 10^{-4} \, \text{m}^4$.

7.15 A floor beam 6 m in length is simply supported at each end and carries a varying distributed load of grain over the whole span. The loading distribution is represented by the equation $w = ax^2 + bx + c$, where $w$ is the load intensity of the grain in kN/m at a distance $x$ along the beam, and $a$, $b$ and $c$ are constants. The load intensity is zero at each end and has its maximum value of 4 kN/m at mid-span. Calculate the slope at each end of the beam if $E = 208 \, \text{GN/m}^2$ and $I = 10^8 \, \text{mm}^4$.

7.16 A horizontal beam of rectangular section 25 mm wide by 50 mm deep is 1 m long and simply supported on (*a*) rigid rollers, (*b*) springs of stiffness 300 kN/m. A load of 15 kg falls 50 mm onto the beam at mid-span. Calculate the instantaneous maximum deflections and bending stresses. $E = 208 \, \text{GN/m}^2$.

# 8 Statically Indeterminate Beams

The design of beams depends initially on the evaluation of shear-force and bending-moment distributions in order to calculate stresses and deflections. A prerequisite is the calculation of support reactions, and in the case of statically determinate beam situations there are only two unknown reactions, which are found from the two equilibrium equations ($\Sigma M = 0$ and $\Sigma F = 0$). Thus a beam which is supported in such a way as to produce three or more reaction forces or moments is statically indeterminate. Some typical examples are shown in Fig. 8.1. The principal methods which are used for analysis are (i) double integration with Macaulay's method, (ii) superposition, (iii) moment–area. The application of these methods will be illustrated in a number of worked examples.

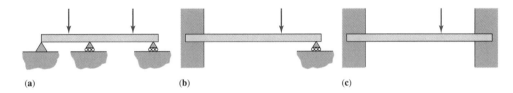

(a)                              (b)                              (c)

Fig. 8.1

## 8.1 Double-integration method

This method was first developed in Chapter 7. To reiterate briefly, the curvature is expressed in terms of bending moment at any point along the beam, this equation is then integrated twice, and the constants of integration are found from known boundary conditions of slope and deflection at the supports, or elsewhere. Whereas in the case of the statically determinate beam the reactions could be found *prior* to the above procedure, this is not possible for the indeterminate beam and the reactions must be carried through the analysis as unknown quantities. Since there are *always* enough boundary conditions to determine all the unknowns in the equations, the reactions can be evaluated together with the constants of integration. A number of worked cases now follow.

**Beam fixed at each end with uniformly distributed loading**

The problem is illustrated in Fig. 8.2, and it is evident that, owing to symmetry, $M_A = M_B$ and $R_A = R_B = wL/2$. However, we cannot determine $M_A$ and $M_B$ from a moment equilibrium equation. The next step is to write the equation for bending-moment distribution as a function of $x$ and equate this to $EI(\mathrm{d}^2v/\mathrm{d}x^2)$ (see eqn. [7.3]).

$$EI\frac{\mathrm{d}^2v}{\mathrm{d}x^2} = -M = -\frac{wLx}{2} + M_A + \frac{wx^2}{2} \qquad [8.1]$$

$$EI\frac{\mathrm{d}v}{\mathrm{d}x} = -\frac{wLx^2}{4} + M_Ax + \frac{wx^3}{6} + A \qquad [8.2]$$

$$EIv = -\frac{wLx^3}{12} + \frac{M_Ax^2}{2} + \frac{wx^4}{24} + Ax + B \qquad [8.3]$$

**Fig. 8.2**

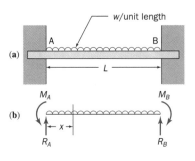

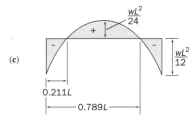

The boundary conditions are that when $x = 0$, $\mathrm{d}v/\mathrm{d}x = 0$ and $v = 0$ from which, respectively, $A = 0$ and $B = 0$; and when $x = L$, $\mathrm{d}v/\mathrm{d}x = 0$ and $v = 0$.

Either of these conditions can now be used to solve for $M_A$, which is found to be $M_A = M_B = wL^2/12$. The fact that $M_A$ and $M_B$ have worked out positive confirms that the directions chosen for them in Fig. 8.2(*b*) were correct.

At mid-span

$$M = \frac{wL^2}{24}$$

The deflection is a maximum at mid-span and is

$$v_{max} = \frac{wL^4}{384EI}$$

The bending-moment diagram, Fig. 8.2, shows two points of contra-flexure (where B.M. changes sign), occurring at $x = 0.211L$ and $0.789L$.

**Beam fixed at each end carrying a point load**

There are four unknown reactions illustrated in Fig. 8.3(*b*). Vertical force equilibrium gives

$$-R_A - R_B + W = 0 \qquad [8.4]$$

Taking moments about B gives

$$-R_A L + W(L - a) + M_A - M_B = 0 \tag{8.5}$$

**Fig. 8.3**

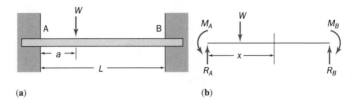

(a)                                         (b)

In this problem, when determining the moment at $x$ from the left-hand side, it is necessary to use the step function technique (Macaulay's) as there is a discontinuity of bending moment at the load, $W$.

$$EI \frac{\mathrm{d}^2 v}{\mathrm{d}x^2} = -R_A x + W[x-a] + M_A \tag{8.6}$$

$$EI \frac{\mathrm{d}v}{\mathrm{d}x} = -\frac{R_A x^2}{2} + \frac{W}{2}[x-a]^2 + M_A x + A \tag{8.7}$$

$$EIv = -\frac{R_A x^3}{6} + \frac{W}{6}[x-a]^3 + M_A \frac{x^2}{2} + Ax + B \tag{8.8}$$

The boundary conditions are: (i) when $x = 0$, $v = 0$ and $\mathrm{d}v/\mathrm{d}x = 0$; (ii) when $x = L$, $v = 0$ and $\mathrm{d}v/\mathrm{d}x = 0$.
    From (i)

$$A = 0 \quad \text{and} \quad B = 0$$

From (ii) and solving for $R_A$ and $M_A$,

$$R_A = \frac{W}{L^3}(L - a)^2(L + 2a)$$

$$M_A = \frac{Wa}{L^2}(L - a)^2$$

Using eqn. [8.5],

$$M_B = \frac{Wa^2}{L^2}(L - a)$$

From eqn. [8.8] the deflection under the load is

$$v = \frac{Wa^3(L - a)^3}{3EIL^3} \tag{8.9}$$

For the particular case when $a = L/2$,

$$R_A = R_B = \frac{W}{2} \quad \text{and} \quad M_A = M_B = \frac{WL}{8}$$

$$v_{max} = \frac{1}{192}\frac{WL^3}{EI} \tag{8.10}$$

**Cantilever with a prop**
**at the free end**

This situation is illustrated in Fig. 8.4, in which the prop, considered as a simple support, is, for generality, assumed to be at a level $\Delta$ above the fixed end. The unknown reactions are $P$, $R_B$ and $M_B$.

**Fig. 8.4**

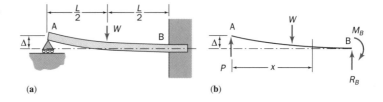

(a)                                                    (b)

Vertical force equilibrium gives

$$-P - R_B + W = 0 \qquad [8.11]$$

Taking moments about B gives

$$-PL + \frac{WL}{2} - M_B = 0 \qquad [8.12]$$

Using the step function method to deal with the discontinuous loading as in Chapter 7,

$$EI\frac{d^2v}{dx^2} = -Px + W\left[x - \frac{L}{2}\right] \qquad [8.13]$$

$$EI\frac{dv}{dx} = -\frac{Px^2}{2} + \frac{W}{2}\left[x - \frac{L}{2}\right]^2 + A \qquad [8.14]$$

$$EIv = -\frac{Px^3}{6} + \frac{W}{6}\left[x - \frac{L}{2}\right]^3 + Ax + B \qquad [8.15]$$

The boundary conditions are: (i) when $x = 0$, $v = -\Delta$; (ii) when $x = L$, $v = 0$ and $dv/dx = 0$.
    For (i), from eqn. [8.15]

$$B = -EI\Delta$$

(The term in square brackets has to be omitted as it is negative.)
    For (ii), from eqn. [8.14],

$$0 = -\frac{PL^2}{2} + \frac{WL^2}{8} + A \qquad [8.16]$$

and from eqn. [8.15],

$$0 = -\frac{PL^3}{6} + \frac{WL^3}{48} + AL - EI\Delta \qquad [8.17]$$

Solving eqns. [8.16] and [8.17],

$$P = \frac{5}{16}W + \frac{3EI\Delta}{L^3} \qquad [8.18]$$

and

$$A = +\frac{WL^2}{32} + \frac{3EI\Delta}{2L}$$

Substituting for $P$ in eqns. [8.11] and [8.12],

$$R_B = \frac{11}{16}W - \frac{3EI\Delta}{L^3}$$

and

$$M_B = \frac{3}{16}WL - \frac{3EI\Delta}{L^2}$$

Any required S.F., B.M., slope or deflection can now be determined.

**Continuous beam on multiple simple supports**

A fairly general example is illustrated in Fig. 8.5 in which $\Delta_B$ and $\Delta_C$ are known displacements due to the supports not being at the same level.

$$-R_A - R_B - R_C - R_D + W + wL = 0 \qquad [8.19]$$

**Fig. 8.5**

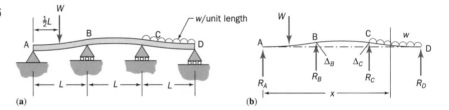

(a)          (b)

$$-3R_AL - 2R_BL - R_CL + \frac{5}{2}WL + \frac{wL^2}{2} = 0 \qquad [8.20]$$

$$EI\frac{d^2v}{dx^2} = -R_Ax + W\left[x - \frac{L}{2}\right] - R_B[x - L] - R_C[x - 2L]$$

$$+ \frac{w}{2}[x - 2L]^2 \qquad [8.21]$$

$$EI\frac{dv}{dx} = -\frac{R_Ax^2}{2} + \frac{W}{2}\left[x - \frac{L}{2}\right]^2 - \frac{R_B}{2}[x - L]^2 - \frac{R_C}{2}[x - 2L]^2$$

$$+ \frac{w}{6}[x - 2L]^3 + A \qquad [8.22]$$

$$EIv = -\frac{R_Ax^3}{6} + \frac{W}{6}\left[x - \frac{L}{2}\right]^3 - \frac{R_B}{6}[x - L]^3 - \frac{R_C}{6}[x - 2L]^3$$

$$+ \frac{w}{24}[x - 2L]^4 + Ax + B \qquad [8.23]$$

The boundary conditions are: (i) $x = 0$, $v = 0$; (ii) $x = L$, $v = -\Delta_B$; (iii) $x = 2L$, $v = -\Delta_C$; (iv) $x = 3L$, $v = 0$.

From (i),

$$B = 0$$

From (ii),

$$A = -\frac{WL^2}{48} + \frac{R_A L^2}{6} - \frac{EI\Delta_B}{L}$$

From (iii),

$$EI\Delta_C = R_A L^3 - \frac{25WL^3}{48} + \frac{R_B L^3}{6} + 2EI\Delta_B$$

From (iv),

$$0 = -4R_A L^4 + \frac{122WL^3}{48} - \frac{8R_B L^3}{6} - \frac{R_C L^3}{6} + \frac{wL^4}{24} - 3EI\Delta_B$$

Although perhaps somewhat laborious, these latter two equations together with eqns. [8.19] and [8.20] can be solved to give values for $R_A$, $R_B$, $R_C$ and $R_D$. From this stage any required aspect of this beam problem can be evaluated.

## 8.2 Superposition method

The principle of superposition can be very useful in finding redundant reactions, particular if a problem can be split up into 'standard' cases (see Table 7.1).

### Cantilever with a prop at the free end

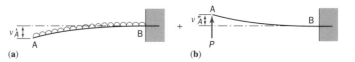

Fig. 8.6

This problem is illustrated in Fig. 8.6 and can be represented by superposition of the two parts shown in Fig. 8.7(a) and (b). If there were no support at A there would be a downward deflection due to the distribution load. If there were no distributed load and the reaction, $P$, at the support was considered as a force which could cause an upward deflection of the beam, then the necessary boundary condition is that the sum of these deflections must be zero.

Due to loading, $w$,

$$v'_A = +\frac{wL^4}{8EI}$$

Due to reaction force, $P$,

$$v''_A = -\frac{PL^3}{3EI}$$

But

$$v'_A + v''_A = 0$$

Fig. 8.7

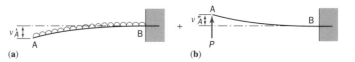

(a)          (b)

Therefore

$$+\frac{wL^4}{8EI} - \frac{PL^3}{3EI} = 0 \qquad [8.24]$$

Hence

$$P = \frac{3}{8}wL$$

From vertical and moment equilibrium,

$$R_B = \frac{5}{8}wL \qquad \text{and} \qquad M_B = \frac{wL^2}{8}$$

**Alternative superposition**

The case of Fig. 8.6 can also be split up in the manner shown in Fig. 8.8(*a*) and (*b*). The superposition must now satisfy the boundary condition of zero slope at B. Thus

$$\theta_B' + \theta_B'' = 0$$

Now,

$$\theta_B' = -\frac{wL^3}{24EI} \qquad \text{and} \qquad \theta_B'' = +\frac{M_BL}{3EI}$$

Therefore

$$-\frac{wL^3}{24EI} + \frac{M_BL}{3EI} = 0 \qquad [8.25]$$

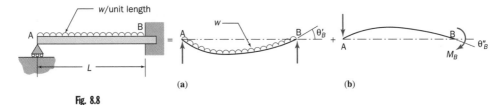

w/unit length

**(a)**     **(b)**

**Fig. 8.8**

$$M_B = \frac{wL^2}{8}$$

Knowing $M_B$ we can find $P$ and $R_B$ from the two equilibrium equations.

**Beam fixed horizontally at each end**

Before considering any particular form of applied loading, we will examine the effect of the fixing moments alone. Taking the general case of, say, $M_B > M_A$ as illustrated in Fig. 8.9(*a*), this may itself be put into the two parts in Fig. 8.9(*b*) and (*c*), giving slopes at each end of

$$\theta_A = -\frac{M_AL}{2EI} \qquad \text{and} \qquad \theta_B = +\frac{M_AL}{2EI}$$

$$\theta_A = -\frac{(M_B - M_A)L}{6EI} \qquad \text{and} \qquad \theta_B = \frac{(M_B - M_A)L}{3EI}$$

**Fig. 8.9**

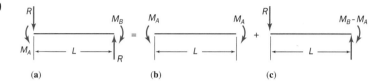

(a)          (b)          (c)

The resultant slopes are therefore

$$\theta_A = -\frac{(M_B + 2M_A)L}{6EI} \quad \text{and} \quad \theta_B = +\frac{(2M_B + M_A)L}{6EI} \qquad [8.26]$$

**Fixed beam with uniformly distributed loading**

This problem may be represented as in Fig. 8.10. The slopes at the ends for the simply supported part are $\pm wL^3/24EI$. Using the condition of zero slope at each end and the results in eqn. [8.26],

$$+\frac{wL^3}{24EI} - \frac{(M_B + 2M_A)L}{6EI} = 0 \qquad [8.27]$$

and

$$-\frac{wL^3}{24EI} + \frac{(2M_B + M_A)L}{6EI} = 0 \qquad [8.28]$$

**Fig. 8.10**

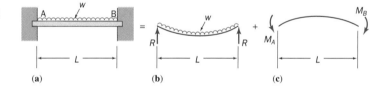

(a)          (b)          (c)

from which

$$M_A = M_B = \frac{wL^2}{12}$$

**Fixed beam with ends not at same level**

This is an important structural situation since a considerable bending moment can be set up by the ends not being at the same level, even without any applied loading to the beam.

Let $v$ be the difference in level between the ends of the beam, Fig. 8.11. A point of contraflexure occurs at mid-span, owing to symmetry, and each half of the beam may be taken as a cantilever, the free end of which is caused to deflect $\frac{1}{2}v$ by a force $P$ at the free end. Then, using the basic solution for a cantilever carrying a load at the free end,

$$\frac{v}{2} = \frac{1}{3}\frac{P(\frac{1}{2}L)^3}{EI} \quad \text{so that} \quad P = \frac{12EIv}{L^3}$$

**Fig. 8.11**

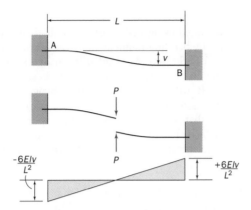

The bending-moment diagram on each half, due to $P$, is triangular, the maximum ordinate being

$$M_A = M_B = \pm \frac{12EIv}{L^3} \frac{L}{2} = \pm \frac{6EIv}{L^2}$$

as shown in Fig. 8.11.

If, say, a distributed load $w$ is now applied to the beam, then if the ends were at the same level the bending-moment diagram would be as shown in Fig. 8.12.

**Fig. 8.12**

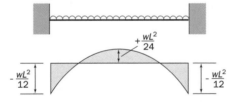

By superposition of the bending moments for the fixing moments and for the distributed loading, the total bending moment distribution may be obtained.

$$M_A = \frac{wL^2}{12} + \frac{6EIv}{L^2} \qquad \text{and} \qquad M_B = \frac{wL^2}{12} - \frac{6EIv}{L^2} \qquad [8.29]$$

The combined diagram of bending moment, due to the distributed load and to the fixing of the ends, is shown in Fig. 8.13.

**Fig. 8.13**

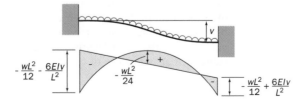

### 8.3 Moment–area method

In the previous sections, the slope and deflection of a loaded beam have been obtained using mathematical methods based on a knowledge of the variation of the bending moment along the beam. It is also possible to obtain the slope and deflection of a loaded beam by examining the geometric properties of the elastic curve of the deflected beam.

It will be shown that the change in slope or deflection between two points on a beam is related to the area under the bending-moment diagram for that section of the beam. This approach is referred to as the *moment–area method*. It may be used to determine the slope and deflection of statically determinate or statically indeterminate beams.

### Slope related to area of bending-moment diagram

In Fig. 8.14 a portion of length AB of a beam has a bending-moment diagram of area $A$ represented by CDEF. The distance of the centre of area G of the diagram from any chosen reference line HH is $\bar{x}$. An exaggerated view of the deflected beam is shown below the bending-moment diagram.

**Fig. 8.14**

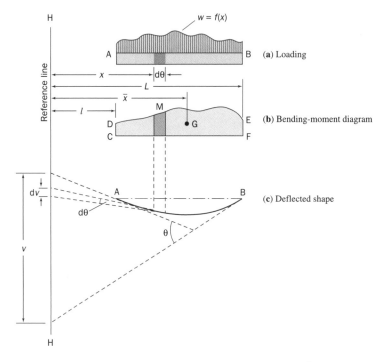

(a) Loading

(b) Bending-moment diagram

(c) Deflected shape

Consider a small piece of the beam of length $dx$ over which the bending moment may be assumed to be constant and equal to $M$. The *change* of slope over the small piece $dx$ is given by $d\theta$, where $d\theta$ is the small angle included between tangents drawn at each extremity of $dx$. Let $R$ be the radius of curvature of the small length $dx$ when deflected; now $d\theta$ is also the angle subtended at the centre of curvature of the element $dx$. Therefore

$$R\,d\theta = dx$$

and substituting for $R$ we get

$$d\theta = \frac{1}{EI} M \, dx$$

and integrating between the limits $L$ and $l$, the distances from the chosen reference line HH,

$$\theta = \int_l^L \frac{1}{EI} M \, dx = \frac{1}{EI} \int_l^L M \, dx \qquad [8.29a]$$

when the cross-section of the beam is constant.

Now $\int_l^L M \, dx$ is the area $A$ of the bending-moment diagram CDEF over the length AB. Therefore

$$\theta = \frac{A}{EI} \qquad [8.29b]$$

$\theta$ being the *change of slope* over the length AB.

Thus we have the important relationship that, over any portion of a loaded beam, the change of slope is equal to the area of the bending-moment diagram divided by $EI$. If the beam varies in cross-section, then $I$ must be retained within the integral in eqn. [8.29a].

**Deflection related to area of bending-moment diagram**

The next stage is to develop an expression for the deflection of the portion of beam AB. The intercept $dv$ on HH can be represented as

$$dv \approx x \, d\theta = x \frac{M}{EI} \, dx$$

Therefore

$$v = \frac{1}{EI} \int_l^L Mx \, dx$$

where $I$ is constant. Now $\int_l^L Mx \, dx$ is the first moment of the area $A$ of the bending-moment diagram on AB about the axis HH; and since $\bar{x}$ is the distance of the centre of area of $A$ from HH, then

$$v = \frac{A\bar{x}}{EL} \qquad [8.29c]$$

Thus the distance between the intercepts, on any chosen reference line, of the tangents drawn to the ends of any portion of a loaded beam is equal to the product of the area of the bending-moment diagram, over that portion of the beam, and the distance of the centre of area of this diagram from the reference line, divided by $EI$.

In Fig. 8.15, ACB is the deflected form of a loaded beam of length $l$ greatly exaggerated and we wish to find the deflection of any arbitrary point P. The slopes at A and B are $\theta_A$ and $\theta_B$ respectively.

The reference intercept lines are AK and BL at each end of the beam, and FE is a horizontal tangent at C.

**Fig. 8.15**

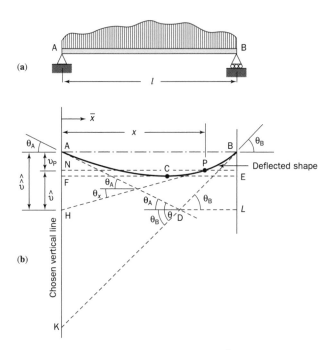

The deflection at P is $v_P$, where $v_P = \hat{\hat{v}} - \hat{v}$. Now

$$\hat{\hat{v}} = \frac{A_{PN}\bar{x}_P}{EI}$$

where $A_{PN}$ is the area of the bending moment diagram on PN and

$$\hat{v} \approx \text{PN} \cdot \theta_x$$

$$\therefore \quad v_P = \frac{A_{PN}\bar{x}_P}{EI} - \text{PN} \cdot \theta_x$$

since $\theta_A + \theta_x = \theta_P$. Therefore $\theta_x = \theta_P - \theta_A$, and

$$v_P = \frac{A_{PN}\bar{x}_P}{EI} - \text{PN}(\theta_P - \theta_A)$$

Now

$$\theta_A = \theta - \theta_B = \frac{A}{EI} - \theta_B$$

and

$$\theta_B l = AK = \frac{A\bar{x}}{EI}$$

therefore

$$\theta_A = \frac{A}{EI} - \frac{A\bar{x}}{EIl} = \frac{A}{EIl}(l - \bar{x})$$

Hence

$$v_P = \frac{A_{PN}\bar{x}_P}{EI} - \text{PN}\left\{\frac{A_{PN}}{EI} - A\frac{(l-\bar{x})}{EIl}\right\}$$

$$v_P = \frac{A_{PN}\bar{x}_P}{EI} - \frac{\text{PN}A_{PN}}{EI} + \text{PN}\frac{A(l-\bar{x})}{EIl}$$

The application of this method will now be illustrated using some worked examples.

**Fixed beam with irregular loading**

In the case of a horizontally fixed beam, $\theta = 0$ for the complete span, and since the product $EI$ is not zero, it follows from eqn. [8.29b] that $A$, the resultant area of the moment diagram for the beam, must be zero. This enables the deflection and slope at any point on the beam to be evaluated.

Referring to Fig. 8.16 let $A_1$ be the area of the 'free' bending-moment diagram AEFHB due to the distributed load on simple supports and let $\bar{x}_1$ be the distance of its centroid from A; also let $A_2$ be the area of the moment diagram ACDB due to the 'fixing' moments only, and $\bar{x}_2$ the distance of its centroid from A. Then

$$A_2 = \frac{(M_A + M_B)L}{2}$$

It is required that $A_1 + A_2 = 0$; thus

$$M_A + M_B = -\frac{2A_1}{L}$$

**Fig. 8.16**

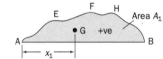

(a) Fixed beam with irregular loading

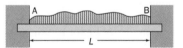

(b) Free moment diagram

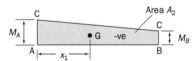

(c) Fixing moment diagram

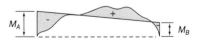

(d) Combined moment diagram

since there is no relative deflection of each end of the beam, then from eqn. [8.29$c$]

$$\frac{A_1 \bar{x}_1}{EI} + \frac{A_2 \bar{x}_2}{EI} = 0$$

Taking moments about A, for the areas under the fixing moment diagram

$$A_2 \bar{x}_2 = \frac{M_B L^2}{2} + \frac{(M_A - M_B)}{2} L \frac{L}{3}$$

$$= \frac{M_B L^2}{3} + \frac{M_A L^2}{6} \tag{8.31}$$

Therefore

$$-A_1 \bar{x}_1 = \frac{M_B L^2}{3} + \frac{M_A L^2}{6}$$

Hence

$$2M_B + M_A = -\frac{6A_1 \bar{x}_1}{L^2} \tag{8.32}$$

From eqns. [8.30] and [8.32],

$$M_A = -\frac{4A_1}{L} + \frac{6A_1 \bar{x}_1}{L^2} \tag{8.33}$$

and

$$M_B = -\frac{6A_1 \bar{x}_1}{L^2} + \frac{2A_1}{L} \tag{8.34}$$

**Fixed beam with concentrated central load**

By combining the positive free bending-moment diagram ABC and the negative fixing moment diagram, EABD, the resultant diagram AEFCGDB is obtained as shown in Fig. 8.17. Since area ABC+EABD must be zero for no change in slope at A compared with B

$$M_A L + \frac{WL^2}{8} = 0$$

Hence

$$M_A = -\frac{WL}{8} = M_B$$

and the points of contraflexure will be at G and F, distance $x$ from each end, where $x = \frac{1}{4}L$.

The value of $M$ at mid-span is numerically equal to that of $M$ at each end, but is positive. Considering half the span, and taking the chosen reference line through the left-hand support, we can obtain the deflection at mid-span using eqn. [8.29$c$]:

$$v = \frac{A\bar{x}}{EI}$$

**Fig. 8.17**

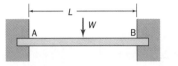

(a) Fixed beam with central loading

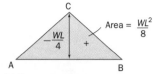

(b) Free moment diagram

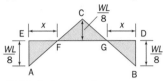

(d) Resultant moment diagram

(c) Fixing moment diagram

where $A$ and $\bar{x}$ are the values for the free bending-moment diagram and the fixed bending-moment diagram from the left end of the beam to the mid-span. Therefore

$$v_{max} = \frac{1}{EI}\left[\left(\frac{WL}{4} \times \frac{L}{4} \times \frac{2}{3}\frac{L}{2}\right) + \left(-\frac{WL}{8} \times \frac{L}{2} \times \frac{L}{4}\right)\right]$$

$$= \frac{WL^3}{EI}\left(\frac{1}{48} - \frac{1}{64}\right)$$

$$= \frac{1}{192}\frac{WL^3}{EI} \qquad [8.35]$$

It may be noted that this is the same result as was obtained earlier by the double-integration method (eqn. [8.10]).

**Continuous beam on multiple supports**

A beam resting on more than two supports is said to be *continuous*. Such a beam is represented by Fig. 8.18(*a*). Changes of curvature occur in each span, owing to negative bending moments at the supports. In the case represented, the supports are assumed to be at different levels, being displaced $v_0$, $v_1$ and $v_2$ from a horizontal line AB. Suppose the loading to be such that the 'free' (positive) and 'fixing' (negative) bending-moment diagrams are as shown at (*b*) and (*c*), the resultant diagram being shown at (*d*). The area of the resultant diagram on the span $l_1$ is $A_1$, and the distance of its centroid $G_1$ from the chosen reference line through the left-hand support is $\bar{x}_1$; also the area of the resultant diagram on the span $l_2$ is $A_2$, and $\bar{x}_2$ is the distance of its centroid $G_2$ from the other chosen reference lines through the right-hand support.

*(i) Bending moments at supports*
Draw CD, a common tangent at the point of contact of the central support, and let $\alpha$ which is actually a small angle, be its inclination to the horizontal. Taking intercepts, between tangents to the deflected beam, on a vertical line as positive when measured downwards, and vice versa as negative upwards, the same convention as for deflections, where the intercepts on the left-hand and right-hand reference lines are $z_0$ and $z_2$ respectively,

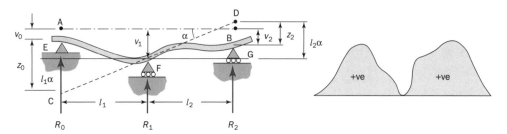

(**a**) Continuous beam                    (**b**) Free moment diagram

**Fig. 8.18**

(**c**) Fixing moment diagram

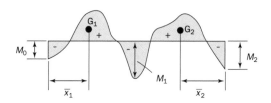

(**d**) Resultant moment diagram

$$z_0 = \frac{A_1\bar{x}_1}{EI} = v_1 - v_0 + l_1\alpha$$

$$-z_2 = \frac{A_2\bar{x}_2}{EI} = -l_2\alpha + (v_1 - v_2)$$

and

$$\frac{A_1\bar{x}_1}{EIl_1} - \frac{v_1 - v_0}{l_1} = \alpha = -\frac{A_2x_2^-}{EIl_2} + \frac{v_1 - v_2}{l_2}$$

Therefore

$$\frac{A_1\bar{x}_1}{EIl_1} + \frac{A_2\bar{x}_2}{EIl_2} = \frac{v_1 - v_0}{l_1} + \frac{v_1 - v_2}{l_2}$$

and

$$\frac{A_1\bar{x}_1}{l_1} + \frac{A_2\bar{x}_2}{l_2} = \left(\frac{v_1 - v_0}{l_1} + \frac{v_1 - v_2}{l_2}\right)EI \qquad [8.36]$$

When the supports are all at the same level, $v_0 = v_1 = v_2$, and

$$\frac{A_1 \bar{x}_1}{l_1} + \frac{A_2 \bar{x}_2}{l_2} = 0 \qquad [8.37]$$

Let the areas of the 'free' bending-moment diagrams be $S_1$ and $S_2$ and the distance of the centroid of $S_1$ from the reference line through the left-hand support be $x_1$, and the corresponding distance of the centroid of $S_2$ from the reference line through the right-hand support be $x_2$; then, the sum of the moments of the 'free' and 'fixing' moment diagrams are respectively

$$A_1 \bar{x}_1 = S_1 x_1 + \left( M_0 l_1 \frac{l_1}{2} + \frac{(M_1 - M_0)}{2} l_1 2 \frac{l_1}{3} \right)$$

and

$$A_2 \bar{x}_2 = S_2 x_2 + \left( M_2 l_2 \frac{l_2}{2} + \frac{(M_1 - M_2)}{2} l_2 2 \frac{l_2}{3} \right)$$

Substituting into eqn. [8.37] and rearranging gives

$$\frac{S_1 x_1}{l_1} + \frac{S_2 x_2}{l_2} + \frac{M_0 l_1}{6} + \frac{M_1}{3}(l_1 + l_2) + \frac{M_2 l_2}{6} = 0$$

and

$$M_0 l_1 + 2M_1(l_1 + l_2) + M_2 l_2 = -6\left( \frac{S_1 x_1}{l_1} + \frac{S_2 x_2}{l_2} \right) \qquad [8.38]$$

This is Clapeyron's *theorem of three moments*, and by taking the spans in pairs, sufficient equations are obtained to solve for the bending moments at the supports.

Useful forms of $6Sx/L$ are shown in Fig. 8.19:

(a)  for a point load,

$$\frac{6Sx}{L} = \frac{Wa}{L}(L^2 - a^2)$$

(b)  for a uniformly distributed load,

$$\frac{6Sx}{L} = \frac{wL^3}{4}$$

Fig. 8.19

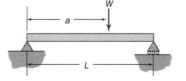

(a) For a point load

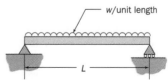

(b) For a uniformly distributed load

*(ii) Reactions at supports*

In order to calculate the support reactions, consider the beam in Fig. 8.18 and let the reactions at E, F and G be $R_0$, $R_1$, $R_2$ respectively. In Fig. 8.20 we have split the beam into two separate free bodies which have reactions $R_0'$, $R_1''$ and $R_1'$, $R_2''$ and bending moments at the supports, $M_0$, $M_1$ and $M_2$. Let the loading on each span be $f(W_1, w_1)$ and $f(W_2, w_2)$ and their centroidal distances from E and G be $\bar{x}_1$ and $\bar{x}_2$ respectively. Then taking moments about E for span EF

$$R_1'' l_1 + M_1 - M_0 - f(W_1, w_1)\bar{x}_1 = 0$$

$$R_1'' = \frac{1}{l_1}[f(W_1, w_1)\bar{x}_1 - (M_1 - M_0)] \qquad [8.39]$$

**Fig. 8.20**

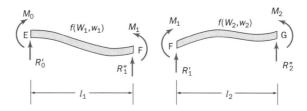

Similarly for span FG take moments about G:

$$R_1' l_2 + M_1 - M_2 - f(W_2, w_2)\bar{x}_2 = 0$$

$$R_1' = \frac{1}{l_2}[f(W_2, w_2)\bar{x}_2 - (M_1 - M_2)] \qquad [8.40]$$

Now

$$R_1 = R_1'' + R_1' \qquad [8.41]$$

Therefore

$$R_1 = \frac{f(W_1, w_1)\bar{x}_1}{l_1} + \frac{f(W_2, w_2)\bar{x}_2}{l_2} - \frac{M_1 - M_0}{l_1} - \frac{(M_1 - M_2)}{l_2}$$

$$[8.42]$$

$R_0'$ and $R_2''$ may be found from vertical equilibrium for each of the spans EF and FG and the above procedure can then be repeated for other adjacent spans to determine all the reactions. It then becomes a simple matter to plot the shear-force diagram for the beam.

**Example 8.1**

Draw the bending-moment and shearing-force diagrams for a continuous beam which is supported at three points at the same level, but free at its extremities. The spans are 15.2 m and 10.6 m; the 5.2 m span supports two loads of values 8900 N and 4450 N distance 6 m and 12 m reespectively from a free end, and the 10.6 m span is loaded uniformly with 1459 N per metre run.

**Fig. 8.21**

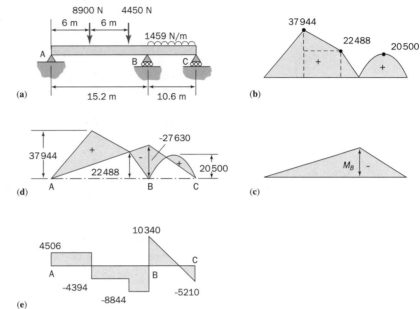

(a)

(b)

(d)

(c)

(e)

The maximum ordinate of the free bending-moment diagram (Fig. 8.21($b$)) for span BC is

$$\frac{wL^2}{8} = \frac{1459 \times 10.6 \times 10.6}{8} = 20\,500\,\text{N m}$$

Considering the span AB, in order to draw the free bending-moment diagram we must first determine the reaction at A due to the two concentrated loads only.

Taking moments about B for the free span AB

$$-15.2R'_A + (8900 \times 9.2) + (4450 \times 3.2) = 0$$

$$R'_A = 6324$$

Bending moment at 8900 N load $= 6324 \times 6$

$$= 37\,944$$

Bending moment at 4450 N load $= (6324 \times 12) - (8900 \times 6)$

$$= 22\,488$$

Referring to Fig. 8.21 $M_A = 0$ and $M_C = 0$, since the ends are free. Then eqn. [8.38],

$$M_A l_1 + 2M_B(l_1 + l_2) + M_C l_2 = -6\left(\frac{S_1 x_1}{l_1} + \frac{S_2 x_2}{l_2}\right)$$

becomes

$$2M_B(l_1 + l_2) = -6\left(\frac{S_1 x_1}{l_1} + \frac{S_2 x_2}{l_2}\right)$$

The free bending-moment diagram for AB is divided into four areas as shown in Fig. 8.21($b$) in order to calculate $S_1 x_1/l_1$:

$$\frac{S_1 x_1}{l_1} = \frac{1}{15.2}\left[\left(\frac{37\,944 \times 6}{2} \times \frac{6 \times 2}{3}\right) + (22\,488 \times 6 \times 9)\right.$$

$$\left. + \left(\frac{15\,465 \times 6 \times 8}{2}\right) + \left(\frac{22\,488 \times 3.2}{2} \times 13.07\right)\right]$$

$$= 165\,190\,\text{N m}^2$$

$$\frac{S_2 x_2}{l_2} = \frac{1}{10.6}\left(\frac{20\,500 \times 10.6 \times 2}{3} \times 5.3\right) = 72\,433\,\text{N m}^2$$

$$2 \times 25.8 M_B = -6(165\,190 + 72\,433)$$

$$M_B = -27\,630\,\text{N m}$$

The resultant bending-moment diagram is shown in Fig. 8.21($b$). Next we must find the reactions at A, B and C. Taking moments about B for span AB

$$-15.2 R_A + (8900 \times 9.2) + (4450 \times 3.2) - 27\,630 = 0$$

$$R_A = 4506\,\text{N}$$

Taking moments about B for span BC

$$+10.6 R_C - (15\,465 \times 5.3) + 27\,630 = 0$$

$$R_C = 5126\,\text{N}$$

Vertical equilibrium for the whole beam gives

$$-4506 - R_B - 5126 + 8900 + 4450 + 15\,465 = 0$$

$$R_B = 19\,183\,\text{N}$$

Note that this could also have been obtained by taking moments about A for span AB to get $R_B''$ and taking moments about C for BC to get $R_B'$ and then $R_B = R_B'' + R_B'$.

The shearing-force diagram can now be determined and is shown in Fig. 8.21($e$).

**Fig. 8.22**

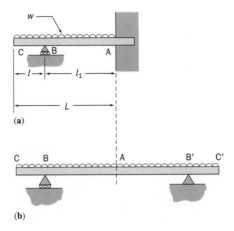

(a)

(b)

**Example 8.2**

A cantilever carries a uniformly distributed load w over its span L. Find the correct location for a prop which is to carry half the total load and to have the point of support at the same level as the fixed end as shown in Fig. 8.22.

Let the constraint be A, the prop B and the free end C. Then, imagining a similar span to the right of A, we can get the value of $M_A$:

$$M'_B = M_B = -\frac{wl^2}{2}$$

where $l = BC$ and $W$ is the evenly distributed load. Then, if $AB = l_1$, $AC = l_1 + l = L$. In eqn. [8.38],

$$S_1 = S_2 = \frac{wl_1^2}{8} \times \frac{2}{3}l_1 \qquad \text{and} \qquad x_1 = x_2 = \frac{l_1}{2}$$

Therefore

$$M_Bl_1 + 2M_A(l_1 + l_1) + M'_Bl_1 = -6\left[\frac{1}{l_1}\left(\frac{wl_1^2}{8} \times \frac{2}{3}l_1 \times \frac{l_1}{2}\right)\right]2$$

Substituting for $M_B$ and $M'_B$,

$$-wl^2 + 4M_A = -\frac{wl_1^2}{2}$$

Therefore

$$M_A = -\left(\frac{wl_1^2}{2} - wl^2\right)\frac{1}{4}$$

Now $R_B$ is equal to half the total load, $w(l_1 + l)$, and also, from eqn. [8.42],

$$R_B = \frac{wl_1}{2} - \frac{(M_B - M_A)}{l_1} + \frac{wl}{2} - \frac{(M_B - M_C)}{l} = \frac{w(l_1 + l)}{2}$$

so that $M_A - M_B(l + l_1) = 0$, since $M_C = 0$. Substituting for $M_A$ and $M_B$ and simplifying gives

$$l_1^2 - 4ll_1 - 6l^2 = 0$$

Therefore

$$l_1 = 5.16l = 0.840L$$

## 8.4 Summary

The first consideration in relation to this chapter is the recognition of a statically indeterminate beam situation, i.e. the appreciation that the support reactions cannot be determined from the two equilibrium equations for the beam. The next step is to choose one of the methods of solution that have been presented. This will depend on the nature of the problem, the availability of computation, etc. Double integration can be used for any case, but, for example, superposition is best suited to a convenient breakdown into simple separate elements. Moment–area can be quite a convenient method for some continuous beam situations as an alternative to double integration. Another important aspect is the correct application of the boundary conditions at support points. It is essential *in practice* to realize that supports which are supposed to be at the same level may not be so, and hence additional bending stresses can be induced. A so-called fixed or built-in support may not ensure zero deflection, or zero slope for that matter. It is also quite useful to remember that if points of contraflexure (zero B.M.) occur, then it may be more efficient to make a beam which has hinged connections at these points, e.g. in a multi-span bridge such as the Forth Bridge.

In structures that contain more than a few beams, the amount of calculation may become excessive. In these cases, computer packages based on matrix stiffness or finite element techniques (Chapter 17) are usually employed.

## Problems

8.1 An I-section beam of length 5 m is built in horizontally and at the same level at each end. If the maximum allowable stress is 90 MN/m$^2$, determine what could be the maximum uniformly-distributed load. The depth of section is 400 mm, and the second moment of area is $3 \times 10^{-4}$ m$^4$.

8.2 A horizontal beam of length $L$ is fixed at one end and simply-supported at the other. A uniformly-distributed load $w$ extends from the fixed end to mid-span. Determine all the reactions and the deflection at mid-span.

8.3 A bar of length $L$ and flexural stiffness $EI$ is built in horizontally and at the same level at each end. A clockwise couple $M$ is applied at mid-span. Find the slope at this point and the deflection curve for the bar.

8.4 A beam of 20 m length is fixed at the left end and simply-supported at the right end, where a clockwise couple of 210 kN-m is applied. A uniformly distributed load of 4 kN/m extends from the fixed support to mid-span, and a concentrated load of 60 kN is applied at 5 m from the right-hand end. Calculate the maximum deflection. $EI = 140$ MNm$^2$.

8.5 A beam is fixed horizontally at the left-hand end, A, and is simply-supported at the same level at B and C, distant 4 m and 6 m from A. A uniformly-distributed load of 2.5 kN/m is carried between B and C. Determine the fixing moments and reactions by the double-integration method.

8.6 A beam in a small bridge deck has been damaged at a particular point and temporarily a prop is to be placed beneath that point to carry half the concentrated load occurring at that position. The beam is 4 m long and has both ends built in at the same level as shown in Fig. 8.23 and the concentrated load $F$ occurs at 3 m from the left wall. The prop is to be a circular bar. Calculate its diameter so that, as stated, the beam and column each carry half the applied load. The second moment of area for the beam is $30 \times 10^7 \text{ mm}^4$ and the modulus of the beam material is three times the modulus of the column material.

Fig. 8.23

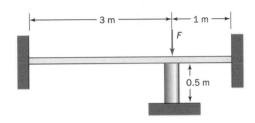

8.7 Use the principle of superposition to find the wall reactions and moments: (*a*) in Problem 8.3; (*b*) in Problem 8.2.

8.8 Draw the bending-moment diagram for the beam loaded as shown in Fig. 8.24.

Fig. 8.24

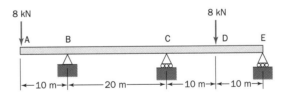

8.9 The beam AE is supported at four points as in Fig. 8.25. Draw the bending-moment and shear-force diagrams for the beam (*a*) if the beam is pinned at A and (*b*) if the beam just rests on A.

8.10 A continuous beam having three spans each of 10 m has the four simple supports A, B, C and D at the same level. Span AB carries a uniform load of 5 kN/m, and a concentrated load of 20 kN acts at 4 m to the left of D. Sketch the shear-force and bending-moment diagrams.

8.11 A horizontal beam is built in at one end and supported at three points as in Fig. 8.26. If it is subjected to point loads $W$ at the middle of the

**Fig. 8.25**

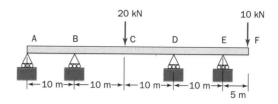

three spans AB, BC and CD, calculate the length of each span so that

$$M_A = M_B = M_C = \frac{WL_1}{8}.$$

**Fig. 8.26**

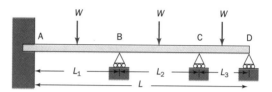

# 9 Energy Methods

The concept of stored elastic strain energy, introduced in Chapter 3, will now be developed further. It will be shown that, by considering the balance between internal strain energy and the external work done on a structure, the deflections or stresses at a particular point in a structure can be found more conveniently than before. These ideas will be demonstrated first in some simple structures such as helical springs where only a single load or moment is acting. The concepts of virtual work, strain energy and complementary energy will be introduced, since these provide the basis for a number of advanced techniques in the analysis of both statically determinate and statically indeterminate structures. Castigliano's hypothesis, which is especially useful for structures with zero or a small number of redundancies, will be used to solve some problems.

## 9.1 Work done by a single load

Consider the load–extension diagram in Fig. 3.12: the external work done during a small increment of deflection $du$ is $F \, du$. The total external work done $W$ when the extension is gradually increased from zero to a maximum value $\delta$ is

$$W = \int_0^\delta F \, du \qquad [9.1]$$

In other words, the work done is equal to the area under the load–displacement graph. If the material of the structure being loaded is linear elastic, the load–displacement relationship is linear and the work done is

$$W = \tfrac{1}{2} F \delta \qquad [9.2]$$

where $F$ is the maximum load at deflection $\delta$.

## 9.2 Work done by a single moment

Beams and shafts can support moments in bending ($M$) and torsion ($T$). In Chapter 1, it was shown that a moment $M$ can be replaced by an equivalent force couple $F.d$. The work done by a moment can be derived from the work done by the equivalent force couple. For a small rotation $d\theta$ about the axis of the moment, the work done by the two couple forces is as shown in Fig. 9.1.

$$W = 2 \int_0^\theta F \cdot \frac{d}{2} \, d\theta = \int_0^\theta M \, d\theta \qquad [9.3]$$

Again, if the structure behaves in a linear-elastic manner, the total work done when a moment is applied is

$$W = \tfrac{1}{2} M \theta \qquad [9.4]$$

**Fig. 9.1**

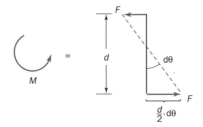

where $\theta$ is the total angle of rotation about the moment axis. The equivalent expression for the work done on a shaft in torsion is

$$W = \tfrac{1}{2} T\theta \qquad\qquad [9.5]$$

**Fig. 9.2**

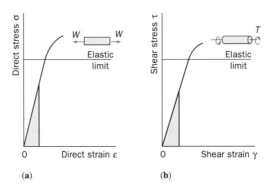

(a)                    (b)

### 9.3 Elastic strain energy: normal stress and shear stress

In Chapter 3 expressions were derived for the elastic strain energy stored in a material per unit volume under the action of direct stress or shear stress. These represented the area under the appropriate stress–strain curve, as shown in Fig. 9.2(a) and (b), and were given by

$$U = \frac{1}{2}\frac{\sigma^2}{E} \quad \text{per unit volume for normal stress}$$

and

$$U = \frac{1}{2}\frac{\tau^2}{G} \quad \text{per unit volume for shear stress}$$

Conservation of energy requires that the internal elastic strain energy must equal the work done by all the external forces and moments on the structure. By integrating the above expressions for strain energy density over the volume of the structure and equating these to the external work done, we can often obtain solutions for the deflections and stresses in a structure in a very economical way.

### 9.4 Strain energy in torsion

For a circular shaft subject to a torque $T$, the external work is $\tfrac{1}{2}T\theta$. The shear stress in a circular shaft is, from eqn. [5.11],

$$\tau = \frac{Tr}{J}$$

where $\mathcal{J} = \int_A r^2 \, dA$ is the polar moment of area of the shaft. The internal strain energy in a uniform shaft of length $L$ is then

$$U = \int_V \frac{1}{2} \frac{\tau^2}{G} \, dV = \frac{T^2}{2G\mathcal{J}^2} \int_0^L \int_A r^2 \, dA \, dx = \frac{T^2 L}{2G\mathcal{J}} \qquad [9.6]$$

Equating the external work and the internal strain energy,

$$\frac{T\theta}{2} = \frac{T^2 L}{2G\mathcal{J}}$$

From this, the angle of twist of the shaft is

$$\theta = \frac{TL}{G\mathcal{J}} \qquad [9.7]$$

## 9.5 Strain energy in bending

For a beam subjected to bending, the direct stress is

$$\sigma = \frac{My}{I}$$

The internal strain energy in a short length $dx$ of the beam is therefore

$$U = \int_V \frac{1}{2} \frac{\sigma^2}{E} \, dV = \int_0^L \int_A \frac{M^2}{2EI^2} y^2 \, dA \, dx = \int_0^L \frac{M^2}{2EI} \, dx \qquad [9.8]$$

This may be used to find, for example, the rotation of a beam due to a particular moment loading. Consider a cantilever with a moment at the free end as shown in Fig. 9.3. The external work done is

$$W = \frac{M\theta}{2}$$

**Fig. 9.3**

The internal strain energy is

$$U = \int_0^L \frac{M^2}{2EI} \, dx$$

Since moment is constant along the full length of the beam

$$U = \frac{M^2 L}{2EI}$$

Equating external work and internal strain energy, the rotation $\theta$ of the end of the beam where the moment is applied is found to be

$$\theta = \frac{ML}{EI} \qquad [9.9]$$

This method is further illustrated in the following examples.

**Beam simply supported with load at mid-span (Fig. 9.4)**

Since the bending-moment relationship is discontinuous over the length of the beam, it is necessary to split the integral of eqn. [9.8] into two parts. Thus for $0 < x < \frac{1}{2}L$, $M = \frac{1}{2}Fx$, and

$$\text{Strain energy up to the load} = \int_0^{L/2} \frac{F^2 x^2}{8EI}\, dx \qquad [9.10]$$

**Fig. 9.4**

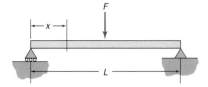

When $\frac{1}{2}L < x < L$, $M = \frac{1}{2}Fx - F(x - \frac{1}{2}L) = \frac{1}{2}F(L - x)$, and

Strain energy stored in second portion of beam

$$= \int_{L/2}^{L} \frac{F^2}{8EI}(L - x)^2\, dx$$

Therefore the total strain energy is

$$U = \int_0^{L/2} \frac{F^2 x^2}{8EI}\, dx + \int_{L/2}^{L} \frac{F^2(L - x)^2}{8EI}\, dx$$

$$U = \frac{F^2 L^3}{192EI} + \frac{F^2 L^3}{192EI}$$

$$U = \frac{F^2 L^3}{96EI} \qquad [9.11]$$

In this particular problem, since the load is at mid-span, the total strain energy could have been obtained by doubling the first integral (see eqn. [9.10]).

Now, the total strain energy is also equal to the external work done, so

$$\text{Work done, } W = \frac{1}{2}Fv_{max}$$

Therefore, from [9.11], $v_{max} = FL^3/48EI$.

**Deflection of a beam under impact loading**

Let a load $F$ strike a beam of span $L$ simply supported at its ends, at mid-span. If $h$ is the distance fallen by $W$ and $\delta$ is the deflection produced, then the work done is $F(h + \delta)$.

If $F_1$ is the equivalent static load applied at mid-span to produce the deflection $\delta$ then the work done by $F_1$ is given by $\frac{1}{2}F_1\delta$; therefore

$$\tfrac{1}{2}F_1\delta = F(h + \delta)$$

But the central deflection is given by the solution to the previous example,

$$\delta = \frac{F_1 L^3}{48EI}$$

Thus

$$\frac{48EI\delta^2}{2L^3} = F(h+\delta)$$

$$\delta^2 - \frac{FL^3\delta}{24EI} - \frac{FL^3h}{24EI} = 0$$

$$\delta = \frac{FL^3}{48EI} + \frac{1}{2}\sqrt{\left[\left(\frac{FL^3}{24EI}\right)^2 + \frac{FL^3h}{6EI}\right]} \qquad [9.12]$$

**Example 9.1**

**A beam of 3 m length is simply supported at each end and is subjected to a couple of 9 kNm at a point B, 2 m from the left end as shown in Fig. 9.5. Determine the slope at B. $EI = 30$ kNm².**

The reactions at A and C are $\bar{M}/L = 9000/3 = 3$ kN.

$$\text{When } 0 < x < 2 \quad M = \frac{\bar{M}x}{L}$$

$$\text{When } 2 < x < L \quad M = \frac{\bar{M}}{L}(x - L)$$

**Fig. 9.5**

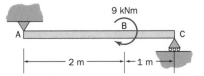

9 kNm

B

A        C

2 m        1 m

The strain energy stored is

$$U = \frac{1}{2EI}\int_0^2 \left(\frac{\bar{M}x}{L}\right)^2 dx + \frac{1}{2EI}\int_2^3 \left(\frac{\bar{M}}{L}(x-L)\right)^2 dx$$

$$= \frac{1}{60 \times 10^3}\left(\frac{9000}{3}\right)^2 \left\{\left[\frac{x^3}{3}\right]_0^2 + \left[\frac{(x-3)^3}{3}\right]_2^3\right\}$$

$$= \frac{3}{60 \times 10^3}\left(\frac{9000}{3}\right)^2$$

The work done at B is

$$\tfrac{1}{2}\bar{M}\theta = \frac{9000}{2}\theta$$

Therefore

$$\frac{9000}{2}\theta = \frac{3}{60 \times 10^3}\left(\frac{9000}{3}\right)^2$$

and

$$\theta = 0.1 \, \text{rad}$$

## 9.6  Helical springs

Springs are directly concerned with the theories of torsion, bending and strain energy and form an excellent example of their application. There are comparatively few machines which do not incorporate a spring to assist in their operation. The principal function of a spring is to absorb energy, store it for a long or short period, and then return it to the surrounding material. Two extremes of operation are found in a watch and on an engine valve. In the former case energy is stored for a long period and in the latter case the process is very rapid. The force required to produce a unit deformation of a spring is called the *stiffness*.

**Fig. 9.6**

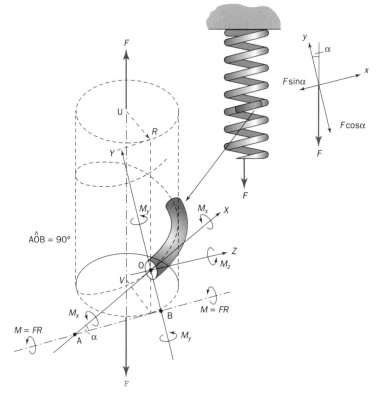

The geometry of a helical spring is shown in Fig. 9.6. The centre-line of the wire forming the spring is a helix on a cylindrical surface such that the helix angle is $\alpha$. Helical springs are designed and manufactured in two categories: close coiled and open coiled. In the former the helix angle $\alpha$ is very small and the coils almost touch each other. In the second case the helix angle is larger and the coils are spaced farther apart.

**Axial load on open-coiled spring**

The most common form of loading on a spring is a force $F$ acting along the central axis. Since this force acts at a distance $R$, the coil radius, from the axis of the wire, there will be torque and bending moment set up about the mutually perpendicular axes $OX$, $OY$ and $OZ$ on the cross-section at O in Fig. 9.6.

$$M_x = FR \cos \alpha \text{ (causing torsion on the wire)} = T$$
$$M_y = FR \sin \alpha \text{ (causing bending about the } y\text{-axis and a}$$
$$\text{change in } R)$$
$$M_z = 0$$

In addition, cross-sections of the wire are subjected to a transverse shear force $F \cos \alpha$ and an axial force $F \sin \alpha$. The stresses due to these forces are considerably smaller than those due to torsion and bending and are generally neglected.

The work done in deflecting the spring $\delta$ by the axial load $F$ is $\frac{1}{2} F\delta$, and the stored energies due to torsion and bending are $\frac{1}{2} T\theta$ and $\frac{1}{2} M\phi$, where $\theta$ is the angular twist of the wire and $\phi$ is the change in slope of the wire. Therefore

$$\tfrac{1}{2} F\delta = \tfrac{1}{2} T\theta + \tfrac{1}{2} M\phi$$

$$= \tfrac{1}{2}(FR \cos \alpha)\theta + \tfrac{1}{2}(FR \sin \alpha)\phi$$

hence, using eqns. [9.7] and [9.9] for $\theta$ and $\phi$ where $\theta$ is the rotation due to the torque and $\phi$ is the rotation due to the bending moment,

$$\delta = R\left(\cos \alpha \frac{TL}{JG} + \sin \alpha \frac{ML}{EI}\right)$$

$$= FR^2 L\left(\frac{\cos^2 \alpha}{JG} + \frac{\sin^2 \alpha}{EI}\right) \tag{9.13}$$

The length of wire in the spring is $2\pi nR \sec \alpha$, where $n$ is the number of complete coils.

$$J = \frac{\pi d^4}{32} \quad \text{and} \quad I = \frac{\pi d^4}{64}$$

where $d$ is the wire diameter: thus

$$\delta = \frac{64FR^3 n \sec \alpha}{d^4}\left(\frac{\cos^2 \alpha}{G} + \frac{2 \sin^2 \alpha}{E}\right) \tag{9.14}$$

**Axial load on close-coiled spring**

The helix angle $\alpha$ is very small in the close-coiled spring so that $\sin \alpha \to 0$ and $\cos \alpha \to 1$, and eqn. [9.14] reduces to

$$\delta = \frac{64FR^3 n}{Gd^4} \tag{9.15}$$

**Example 9.2**

A close-coiled helical spring is to have a stiffness of 1 kN/m compression, a maximum load of 50 N, and a maximum shearing stress of 120 MN/m². The solid length of the spring, i.e. when the coils are touching, is to be 45 mm. Find the diameter of the wire, the mean radius of the coils and the number of coils required. Shear modulus $G = 82\,GN/m^2$.

$$\text{Stiffness} = \frac{F}{\delta} = 1000 = \frac{Gd^4}{64R^3 n}$$

$$\text{Maximum torque } T = \frac{\pi d^3}{16} = 120 \times 10^6 \frac{\pi d^3}{16} = 50R$$

Hence

$$R = \frac{120 \times 10^6}{800} \pi d^3$$

$$\text{Closed length of spring } = nd = 0.045$$

Substituting for $n$ and $R$ in the equation for stiffness above.

$$1000 = \frac{Gd^4 \times d}{64 \times (15 \times 10^4 \pi d^3)^3 \times 0.045}$$

$$d^4 = \frac{82 \times 10^9}{64 \times (15\pi)^3 \times 10^{12} \times 45} = 273 \times 10^{-12}$$

Thus $d = 0.004\,06\,m = 4.06\,mm$; $R = 31.6\,mm$; and $n = 11$ coils.

**Example 9.3**

Compare the stiffness of a close-coiled spring with that of an open-coiled spring of helix angle $30°$. The two springs are made of the same steel and have the same coil radius, number of coils and wire diameter, and are subjected to axial loading $E = 2.5\,G$.

From eqn. [9.15], the stiffness of the close-coiled spring is

$$k_{cc} = \frac{F}{\delta} = \frac{Gd^4}{64R^3 n}$$

From eqn. [9.14] the stiffness of the open-coiled spring is

$$k_{oc} = \frac{F}{\delta} = \frac{d^4}{64R^3 n \, \sec\, \alpha(\cos^2 \alpha/G + 2 \, \sin^2 \alpha/E)}$$

$$\frac{k_{cc}}{k_{oc}} = G \, \sec\, \alpha \left( \frac{\cos^2 \alpha}{G} + \frac{2 \, \sin^2 \alpha}{E} \right)$$

$$= \sec\, \alpha \left( \cos^2 \alpha + \frac{2 \, \sin^2 \alpha}{2.5} \right)$$

$$= \frac{2}{\sqrt{3}} \left( \frac{3}{4} + \frac{1}{5} \right) = 1.1$$

**Axial torque on open-coiled spring**

The other type of loading on a spring which is of interest is that where the spring is subjected to a torque about the central axis. The resolved components of the axial torque $T_O$ about any cross-section are $T_O \sin \alpha$ about the axis of the wire and $T_O \cos \alpha$ changing the curvature of the coils. If one end of the spring moves round the longitudinal axis an amount $\psi$ relative to the other end owing to the torque $T_O$ then the work done is $\frac{1}{2} T_O \psi$ and the stored energies are $\frac{1}{2} T_O \theta \sin \alpha$ and $\frac{1}{2} T_O \phi \cos \alpha$, so that

$$\psi = \theta \sin \alpha + \phi \cos \alpha$$

$$= \frac{TL}{JG} \sin \alpha + \frac{ML}{EI} \cos \alpha$$

$$= T_O \pi n R \sec \alpha \left( \frac{\sin^2 \alpha}{JG} + \frac{\cos^2 \alpha}{EI} \right)$$

$$= \frac{128 T_O n R \sec \alpha}{d^4} \left( \frac{\sin^2 \alpha}{2G} + \frac{\cos^2 \alpha}{E} \right) \qquad [9.16]$$

**Axial torque on close-coiled spring**

Putting $\sec \alpha = \cos \alpha = 1$ and $\sin \alpha = 0$ in eqn. [9.16] gives

$$\psi = \frac{128 T_O R n}{E d^4} \qquad [9.17]$$

## 9.7 Shear deflection of beams

A deflection other than that due to bending moment occurs in beams owing to the shearing forces on transverse sections. This deflection may be found approximately from strain energy principles and by making use of the equation for shear stress at a point in the transverse section of a beam. For the majority of beams, where the span $L$ is large compared with the cross-section of the beam, it will be seen that the deflection due to shear is negligible in comparison with that due to bending.

**Cantilever with load at free end (Fig. 9.7)**

Assume the section to be rectangular, of breadth $b$ and depth $d$, and the total length of the beam to be $L$. If $v_s$ is the deflection, due to shear, at the free end, then

$$\text{Work done by load} = \tfrac{1}{2} F v_s$$

**Fig. 9.7**

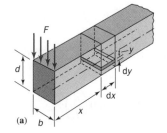

(a)

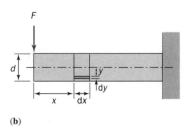

(b)

The shear stress at a distance $y$ from the neutral axis is obtained from eqn. [6.42]

$$\tau = \frac{6F}{bd^3} \left( \frac{d^2}{4} - y^2 \right)$$

where the shear force $Q = F$. Also, if $dy$ is the height of the strip in the direction of the depth of the beam, and we consider a small portion of the beam of length $dx$ and section $b\,dy$, we have

$$\text{Strain energy in strip} = \frac{1}{2} \frac{\tau^2}{G} dx\, b\, dy$$

and substituting for $\tau$, from above

$$\text{Strain energy} = \frac{1}{2G} dx \frac{36F^2}{b^2 d^6} \left( \frac{d^4}{16} - \frac{d^2 y^2}{2} + y^4 \right) b\, dy$$

Therefore the total strain energy for the piece of beam of length $dx$ is

$$U = \frac{18F^2 dx}{bd^6 G} \int_{-d/2}^{+d/2} \left( \frac{d^4}{16} - \frac{d^2 y^2}{2} + y^4 \right) dy$$

$$= \frac{3}{5} \frac{F^2 dx}{(bd)G} \tag{9.18}$$

The strain energy for the whole beam of length $L$ is

$$U_t = \frac{3}{5} \frac{F^2}{bdG} \int_0^L dx = \frac{3}{5} \frac{F^2 L}{bdG}$$

Equating the strain energy to the work done by $F$,

$$\tfrac{1}{2} F v_s = \frac{3}{5} \frac{F^2 L}{bdG}$$

Therefore

$$v_s = \frac{6}{5} \frac{FL}{bdG} \tag{9.19}$$

Thus the total deflection at the free end due to bending and shear is

$$v = v_b + v_s$$

$$= \frac{1}{3} \frac{FL^3}{EI} + \frac{6}{5} \frac{FL}{bdG}$$

For many materials, $E \simeq 3G$. With $I = \tfrac{1}{12} bd^3$, this equation may be rewritten as

$$v = \frac{FL^3}{3EI} \left[ 1 + 0.9 \left( \frac{d}{L} \right)^2 \right]$$

Hence, for a typical beam where $(d/L)$ is less than 0.1, the percentage error in neglecting the shear effect is less than 0.9%.

## 9.8 Virtual work

The principle of *virtual work* is one of the most powerful tools in structural analysis. The principle may be stated as follows:

> if a system of forces acts on a particle which is in statical equilibrium and the particle is given any virtual displacement then the net work done by the forces is zero.

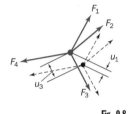

Fig. 9.8

A virtual displacement is any arbitrary displacement which is mathematically conceived and does not actually have to take place, but must be geometrically possible. The forces not only must be in equilibrium but are assumed to remain constant and parallel to their original lines of action.

If a body is subjected to the system of forces shown in Fig. 9.8 and is then given an arbitrary virtual displacement for which the corresponding displacements in the directions of the forces are $u_1$, $u_2$, $u_3$ and $u_4$, then static equilibrium exists for the system of forces if

$$F_1 u_1 + F_2 u_2 + F_3 u_3 + F_4 u_4 = 0$$

This expresses in mathematical form the principle of virtual work for the system considered, i.e. the sum of the work done by each virtual displacement is zero.

Let the resultant force of the above system be $P$ and assume a virtual displacement $\Delta$ in the direction of $P$; then for static equilibrium $P\Delta$ must be zero. Since $\Delta$ need not be zero, it follows that $P$ must be. Thus the resultant of a system of forces in equilibrium is zero.

**Application of virtual work method**

We can now make use of the above principle in deriving energy solutions for deflections. Consider the simple framework in Fig. 9.9 acted upon by forces $F_1$ and $F_2$. Let the displacements at the joints in the direction of the forces be $u_1$ and $u_2$. The internal reactions and deformations of the members of the frame are $P_1$, $P_2$, etc., and $\Delta_1$, $\Delta_2$, etc., respectively. Then, from the principle of virtual work,

$$F_1 u_1 + F_2 u_2 = P_1 \Delta_1 + P_2 \Delta_2 + \ldots + P_n \Delta_n$$

or

$$\sum_j Fu = \sum_n P\Delta \qquad [9.20]$$

Fig. 9.9

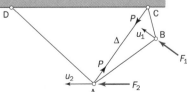

Thus the work done by external forces at the joints equals the internal work resulting from the tensions in the members. The summations in eqn. [9.20] cover all joints $j$, and all members $n$, for static equilibrium.

### 9.9 Displacements by the virtual work method

#### Actual forces, extensions and displacements

If a plane frame is subjected to *one* external load only and the displacement of the loaded joint is required *in the direction* of the load, then eqn. [9.20] can be used with all real values. This is illustrated in the following example.

**Example 9.4**

Determine the vertical and horizontal displacements of the joint D in the plane pin-joined framework in Fig. 9.10. All members have a cross-sectional area of 1000 mm² and modulus of 200 GN/m².

**Fig. 9.10**

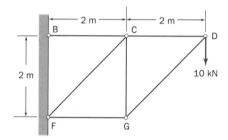

The first step is to calculate the forces in all the members by one of the methods described in Chapter 1. Next the change in length of each member is determined and it is convenient to tabulate the foregoing information:

| Member | Force $F$ (kN) | Length $L$ (m) | Change in length, FL/EA (mm) |
|--------|------|--------|--------|
| BC | +20 | 2 | +0.2 |
| CD | +10 | 2 | +0.1 |
| FG | −10 | 2 | −0.1 |
| FC | $-10\sqrt{2}$ | $2\sqrt{2}$ | −0.2 |
| GC | +10 | 2 | +0.1 |
| GD | $-10\sqrt{2}$ | $2\sqrt{2}$ | −0.2 |

Let the vertical displacement of joint D, which is where the load is applied and in the same direction, be $\delta$. Then, using numerical values from the table,

$$\sum_m P\Delta = (20 \times 0.2) + (10 \times 0.1) + (-10 \times -0.1)$$

$$+(-10\sqrt{2} \times -0.2) + (10 \times 0.1) + (-10\sqrt{2} \times -0.2)$$

$$= 12.66 \, \text{kNmm}$$

Therefore

$$10\delta = 12.66 \quad \text{and} \quad \delta = 1.266\,\text{mm}$$

The horizontal displacement of D cannot be found by the above method, since there is no force in the horizontal direction at D to give an external work term.

The above method of using the real forces and extensions cannot be used on a frame carrying several external loads, unless all the displacements at the loaded joints, except one, are known, since there would be several unknown $\delta$ values on the left-hand side of eqn. [9.20].

**Actual extensions and displacements with forces hypothetical**

The previous section demonstrated the use of eqn. [9.20] in which the work terms were composed of the real forces and real displacements. However, the equation is equally valid if the displacements are those actually occurring due to the applied loading, but the load and force terms are completely fictitious, so long as they form an equilibrium system. This feature enables one to determine the displacement of any joint in any direction, whether loaded or not, by the use of a 'dummy' load of unit magnitude.

Consider again Example 9.4 and the need to find the horizontal displacement at D. A dummy unit load is inserted horizontally at D. Then the equilibrium system of forces in the members *due to the unit load only* (the 10 kN is removed) can be found; the values are as given in the following table.

| Member | Actual extensions, $\Delta$, due to 10 kN load | Hypothetical forces, $P'$, due to dummy load | $P'\Delta$ |
|--------|------------------|------------------|--------|
| BC | +0.2 | 1 | +0.2 |
| FC | −0.2 | 0 | 0 |
| FG | −0.1 | 0 | 0 |
| CG | +0.1 | 0 | 0 |
| CD | +0.1 | 1 | +0.1 |
| GD | −0.2 | 0 | 0 |
| | | | $\Sigma = 0.3$ |

If the real horizontal displacement of D due to the actual load of 10 kN is $\delta$, then the external virtual work is $1 \times \delta$. The internal virtual work is given by the sum of the products of the actual changes in length, $\Delta$, of the members due to the actual load of 10 kN and the hypothetical equilibrium set of forces, $P'$, resulting from the unit dummy load. The simple computation is shown in the right-hand column of the above table. Applying eqn. [9.20] gives

$$1\delta = \sum P'\Delta \quad \text{or} \quad \delta = 0.3\,\text{mm}$$

**Example 9.5**

Determine the vertical displacement of joint G of the frame in Example 9.4.

Following the same procedure as above, a unit dummy load is placed vertically at G; the corresponding force system *due to the unit load only* is shown in the table below.

| Member | Actual extensions | Forces due to unit load | $P'\Delta$ |
|--------|-------------------|-------------------------|------------|
| BC | +0.2 | +1 | +0.2 |
| FC | −0.2 | −$\sqrt{2}$ | +0.2$\sqrt{2}$ |
| FG | −0.1 | 0 | 0 |
| CG | +0.1 | 1 | +0.1 |
| CD | +0.1 | 0 | 0 |
| GD | −0.2 | 0 | 0 |
| | | | $\Sigma = +0.5828$ |

Thus

$$1\delta_G = 0.5828 \quad \text{or} \quad \delta_G = 0.5828\,\text{mm}$$

## 9.10 Strain energy solutions for forces

Energy functions may be very usefully employed in the determination of deflections of frameworks, beams, shells, etc. The methods depend on the principle of virtual work as developed above. Now let us suppose that there is a small change in displacement at joint A of an amount $\delta u_1$, $\delta_2$ remaining constant, which results in changes $\delta\Delta_1$, $\delta\Delta_2$, etc., in the members for compatibility. Then

$$F_1(u_1 + \delta u_1) + F_2 u_2 = P_1(\Delta_1 + \delta\Delta_1) + P_2(\Delta_2 + \delta\Delta_2)$$

$$+ \ldots + P_n(\Delta_n + \delta\Delta_n)$$

$$= \sum_n P(\Delta + \delta\Delta) \tag{9.21}$$

Subtracting eqn. [9.20] from eqn. [9.21]

$$F_1\delta u_1 = \sum_n P\,\delta\Delta \tag{9.22}$$

But $P\,\delta\Delta$ is the increment of strain energy stored in a member of the system due to the increments of deformation $\delta\Delta$ caused by the change in displacement $\delta u_1$. Therefore for the system

$$\sum_n P\,\delta\Delta = \delta U$$

or

$$F_1\delta u_1 = \delta U \tag{9.23}$$

Thus, for an infinitely small change in displacement,

$$F_1 = \frac{\partial U}{\partial u_1} \tag{9.24}$$

By a similar argument we have that

$$F_2 = \frac{\partial U}{\partial u_2}$$

Thus the external force on a member is given by the partial derivative of the strain energy with respect to the displacement at the point of application and in the direction of the force.

This result can be illustrated in the solution of Example 4.2, where two bars were loaded in parallel so that both were subjected to the same extension $u$. The strain energy in each bar is

$$U = \int_V \frac{1}{2} \frac{\sigma^2}{E} \, \mathrm{d}V = \int_0^L \int_A \frac{P^2}{2EA^2} \, \mathrm{d}A \, \mathrm{d}x = \int_0^L \frac{P^2}{2EA} \, \mathrm{d}x = \frac{P^2 L}{2EA} \qquad [9.25]$$

Since $P$, the force in each component, is given by

$$P = \frac{EAu}{L} = ku \qquad [9.26]$$

where $k$ is the stiffness of the bar, the strain energy may be written in terms of the displacement of the end of the bars as

$$U_{total} = \tfrac{1}{2}(k_1 u^2 + k_2 u^2) \qquad [9.27]$$

where $k_1$ and $k_2$ are the stiffnesses of the two bars loaded in parallel. The external force required to cause the displacement $u$ is therefore

$$F = \frac{\partial U}{\partial u} = (k_1 + k_2)u \qquad [9.28]$$

or the displacement resulting from the applied force is

$$u = \frac{F}{k_1 + k_2} \qquad [9.29]$$

**9.11 Complementary energy solution for deflections**

We now return to the original proposition, and instead of changing the displacement $u_1$, we change the force $F_1$ by an amount $\delta F_1$, keeping $F_2$ constant; then there will be a reaction in the system causing changes $\delta P_1$, $\delta P_2$, etc., in the internal forces in the members. Now, by the principle of virtual work, we have

$$(F_1 + \delta F_1)u_1 + F_2 u_2 = (P_1 + \delta P_1)\Delta_1 + (P_2 + \delta P_2)\Delta_2$$

$$+ \ldots + (P_n + \delta P_n)\Delta_n$$

$$= \sum_n (P + \delta P)\Delta \qquad [9.30]$$

Subtracting eqn. [9.20] from eqn. [9.30]

$$\delta F_1 u_1 = \sum_n \delta P \, \Delta \qquad [9.31]$$

**Fig. 9.11**

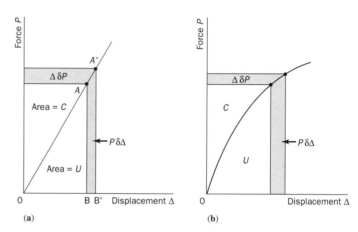

(a)                                                    (b)

Now considering Fig. 9.11($a$) or ($b$), the shaded area $P\delta\Delta$ is the increment of strain energy $\delta U$ below the load–deformation curve used in eqn. [9.22]. The shaded area above the load–deformation curve represents $\Delta\,\delta P$ in eqn. [9.31]; this is termed the *complementary energy* and is denoted by $C$; thus

$$\sum_{n}\delta P\,\Delta = \delta C$$

and therefore

$$\delta F_1 u_1 = \delta C \qquad\qquad [9.32]$$

For an infinitely small change in the force,

$$u_1 = \frac{\partial C}{\partial F_1} \qquad\qquad [9.33]$$

Thus the deflection at a point on a member in the direction of a force applied at that point is given by the partial derivative of the complementary energy with respect to the external force at the point.

The above energy theorems provide a most useful method of attack on many structural analysis problems. A further point of interest is illustrated in the load–deformation characteristics of Fig. 9.11($a$) and ($b$). The former illustrates linear elasticity for a member or system of members, while the latter represents non-linear elasticity, which can occur in certain structures and materials. In both cases the sum of the strain energy and complementary energy is given by the force times the deformation, i.e.

$$U + C = P\Delta \qquad\qquad [9.34]$$

but in the particular case of linear elasticity,

$$\delta U = \delta C = P\,\delta\Delta = \Delta\,\delta P$$

and

$$U = C = \tfrac{1}{2}P\,\Delta \qquad\qquad [9.35]$$

Because of this last relationship we can express displacements in terms of strain energy instead of complementary energy for linear-elastic systems. One of the earliest theorems of this form was due to Castigliano (1875) in which it was stated that for linear-elastic structures the partial derivative of the strain energy with respect to a force gives the displacement corresponding to that force, or

$$\frac{\partial U}{\partial P} = \Delta \qquad\qquad [9.36]$$

This relationship may be proved in the following way. Consider a force $P$ applied to a body giving a displacement $\Delta$. Then the work done or the stored strain energy is equal to OAB which equals $\frac{1}{2}P\Delta$ in Fig. 9.11($a$). If an additional force $\delta P$ is applied giving an additional deformation $\delta\Delta$, then the extra strain energy is

$$\text{BAA}'\text{B}' = P\,\delta\Delta + \tfrac{1}{2}\delta P\,\delta\Delta = \delta U \qquad\qquad [9.37a]$$

or $\delta U/\delta\Delta = P$, neglecting second-order products. Therefore,

$$\text{Total energy, OA}'\text{B}' = \tfrac{1}{2}P\Delta + P\,\delta\Delta + \tfrac{1}{2}\delta P\,\delta\Delta$$

If both forces had acted simultaneously, the stored strain energy would have been OA′B′ $= \frac{1}{2}(P + \delta P)(\Delta + \delta\Delta)$. Since work done is independent of the order of application of the forces, we have

$$\tfrac{1}{2}P\Delta + P\,\delta\Delta + \tfrac{1}{2}\delta P\,\delta\Delta = \tfrac{1}{2}(P + \delta P)(\Delta + \delta\Delta)$$

On simplifying, and neglecting small products, we find that

$$P\,\delta\Delta = \Delta\,\delta P \qquad\qquad [9.37b]$$

(i.e. $\delta U = \delta C$ for linear elasticity). Thus, substituting in eqn. [9.37a]

$$\delta U = \Delta\,\delta P + \tfrac{1}{2}\delta P\,\delta\Delta$$

Therefore

$$\frac{\delta U}{\delta P} = \Delta$$

neglecting the second-order term on the right. Hence

$$\frac{\partial U}{\partial P} = \Delta$$

which proves *Castigliano's hypothesis*.

The simplest applications of eqn. [9.36] are related to the deformations in tension and in torsion:

(i) in the case of a bar under a simple tensile force $F$,

$$U = \frac{F^2 L}{2AE}$$

and

$$\frac{\partial U}{\partial F} = \frac{FL}{AE} = \Delta, \text{ the extension of the bar;}$$

(ii)  and in torsion for a torque $T$,

$$U = \frac{T^2 L}{2G\mathcal{J}}$$

or

$$\frac{\partial U}{\partial T} = \frac{TL}{G\mathcal{J}} = \theta, \text{ the angle of twist.}$$

### 9.12 Bending deflection of beams

The complementary energy function can be used very conveniently to solve for beam deflections, since

$$\frac{\partial C}{\partial F} = \delta$$

or, using the notation for beams,

$$\frac{\partial C}{\partial W} = v$$

where $W$ is a concentrated load whose displacement (beam deflection) is $v$.

The complementary energy in bending of a small length of beam $dx$ is shown in Fig. 9.12(a), which is the moment–slope relationship. The shaded area is

$$dC = \theta \, dM$$

or

$$C = \int_0^M \theta \, dM \qquad\qquad [9.38]$$

For a linear-elastic beam, using eqn. [7.4]

$$\theta = \int_0^L \frac{M}{EI} \, dx$$

Therefore, from Fig. 9.12(b),

$$C = \int_0^M \int_0^L \frac{M}{EI} \, dx \, dM = \int_0^L \frac{M^2}{2EI} \, dx \qquad\qquad [9.39]$$

**Fig. 9.12**

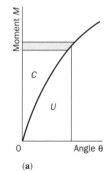

(a)

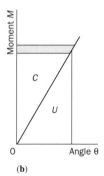

(b)

This result could also have been arrived at from the fact that $U = C$ in a linear-elastic system, and it has already been shown in eqn. [9.8] that the strain energy is

$$U = \int_0^L \frac{M^2}{2EI} \, dx$$

Given this expression for the strain energy, the deflection, $v$, under a concentrated load, $W$, may be determined from

$$v = \frac{\partial C}{\partial W} = \frac{\partial U}{\partial W}$$

We can solve for the deflection in one of two ways, either

$$v = \frac{\partial U}{\partial W} = \frac{\partial}{\partial W} \left( \int_0^L \frac{M^2}{2EI} \, dx \right) \tag{9.40}$$

or

$$v = \frac{\partial U}{\partial W} = \int_0^L \frac{M}{EI} \frac{\partial M}{\partial W} \, dx \tag{9.41}$$

It is merely a question of whether the bending-moment expression in terms of $W$ and $x$ is substituted and the integral evaluated, followed by partial differentiation with respect to $W$, or the latter is carried out first, substituted in the integral and then evaluated.

**Cantilever with concentrated load at free end**

At any distance $x$ from the free end, the bending moment is $M = Wx$; therefore

$$v = \frac{\partial}{\partial W} \int_0^L \frac{W^2 x^2}{2EI} \, dx = \frac{\partial}{\partial W} \left( \frac{W^2 L^3}{6EI} \right)$$

$$= \frac{WL^3}{3EI} \tag{9.42}$$

Alternatively, $\partial M / \partial W = x$; therefore

$$v = \int_0^L \frac{Wx}{EI} x \, dx = \int_0^L \frac{Wx^2}{EI} \, dx$$

$$= \frac{WL^3}{3EI}$$

**Beam simply supported with uniformly distributed load**

At any distance $x$ from the left-hand end the bending moment is

$$M = \frac{wL}{2} x - \frac{wx^2}{2}$$

and $\partial / \partial W$ of the above is zero, indicating no deflection, which is obviously not true. To get round this difficulty, we introduce an imaginary concentrated load $W$ at some point in the span, let us say mid-span for simplicity. Then, for $0 < x < L/2$,

$$M = \frac{W}{2}x + \frac{wL}{2}x - \frac{wx^2}{2}$$  [9.43]

and $\partial M/\partial W = \frac{1}{2}x$; therefore

$$v = 2 \int_0^{L/2} \frac{1}{EI}\left(\frac{W}{2}x + \frac{wL}{2}x - \frac{wx^2}{2}\right)\frac{x}{2}\,dx$$

$$= \frac{WL^3}{48EI} + \frac{5wL^4}{384EI}$$

Putting $W = 0$, we obtain

$$v_{max} = \frac{5}{384}\frac{wL^4}{EI}$$  [9.44]

If we require the deflection due to the point load only, we put $w = 0$; then

$$v = \frac{WL^3}{48EI}$$  [9.45]

**Example 9.6**

**A simply supported beam (Fig. 9.13) carries a concentrated load at a distance $a$ from the left-hand support and a distance $b$ from the other support. Determine the deflection of the beam underneath the load.**

$R_1 = Wb/L$ and the bending moment at C is $(Wb/L)x$. For the portion of beam AD, we have the complementary energy

$$C = \int_0^a \frac{(Wb/L)^2 x^2\,dx}{2EI} = \int_0^a \frac{W^2 b^2}{L^2 2EI}x^2\,dx = \frac{W^2 b^2 a^3}{6EIL^2}$$  [9.46]

Similarly, we may write that $C$ for the portion of beam DB is $W^2 a^2 b^3/6EIL^2$; therefore

Total value of $C$ for the beam

$$= \frac{W^2 b^2 a^3}{6EIL^2} + \frac{W^2 a^2 b^3}{6EIL^2} = \frac{W^2 a^2 b^2}{6EIL^2}(a + b)$$

$$= \frac{W^2 a^2 (L - a)^2}{6EIL}$$  [9.47]

**Fig. 9.13**

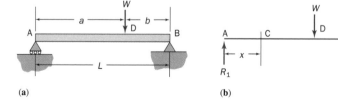

(a)

(b)

and

$$\text{Deflection underneath load} = \frac{\mathrm{d}C}{\mathrm{d}W} = \frac{Wa^2(L-a)^2}{3EIL} \qquad [9.48]$$

**Example 9.7**

**Determine the vertical and horizontal displacements of the end of the curved member shown in Fig. 9.14.**

**Fig. 9.14**

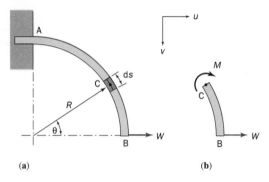

(a)                                          (b)

Considering first the displacement $u$ in the direction of the load.

$$U = \int_0^{\pi/2} \frac{M^2}{2EI} R \,\mathrm{d}\theta$$

$$u = \frac{\partial U}{\partial W} = \int_0^{\pi/2} \frac{M}{EI} \frac{\partial M}{\partial W} R \,\mathrm{d}\theta$$

$$M_c = WR \sin \theta, \qquad \frac{\partial M_c}{\partial W} = R \sin \theta$$

$$u = \int_0^{\pi/2} \frac{WR \sin \theta}{EI} (R \sin \theta) R \,\mathrm{d}\theta$$

$$= \int_0^{\pi/2} \frac{WR^3}{EI} \sin^2 \theta \,\mathrm{d}\theta$$

$$= \frac{\pi}{4} \frac{WR^3}{EI} \qquad [9.49]$$

To find the vertical displacement $v$, an imaginary vertical force $W_0$ is applied at B. The bending moment on C will be

$$M_c = WR \sin \theta - W_0 R(1 - \cos \theta)$$

and

$$\frac{\partial M_0}{\partial W_0} = -R(1 - \cos \theta)$$

Putting $W_0 = 0$ in the expression for $M_c$,

$$v = \frac{\partial U}{\partial W_0} = \int_0^{\pi/2} -\frac{WR}{EI} \sin \theta \cdot R(1 - \cos \theta) R \, d\theta$$

from which

$$v = -\frac{WR^3}{2EI} \qquad\qquad [9.50]$$

**Example 9.8**

**Determine the horizontal deflection of the member shown in Fig. 9.15.**

**Fig. 9.15**

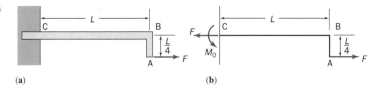

(a)                                                    (b)

The strain energy function is scalar; therefore the separate strain energy quantities for the two parts of the member can be added before proceeding to use Castigliano's theorem.

From A to B

$$M = +Fx \quad \text{and} \quad \frac{\partial M}{\partial F} = +x$$

From B to C

$$M = +\frac{FL}{4} \quad \text{and} \quad \frac{\partial M}{\partial F} = +\frac{L}{4}$$

$$\Delta = \frac{\partial U}{\partial F} = \frac{1}{EI} \int_0^{L/4} Fx \, x \, dx + \frac{1}{EI} \int_0^L \frac{FL}{4} \frac{L}{4} dx$$

$$= \frac{1}{EI} \left[ \frac{Fx^3}{3} \right]_0^{L/4} + \frac{1}{EI} \left[ \frac{FL^2 x}{16} \right]_0^L$$

$$= \frac{13FL^3}{192EI} \qquad\qquad [9.51]$$

**Example 9.9**

**Determine the bending moment for any cross-section of the slender ring shown in Fig. 9.16(a).**

In view of the symmetry of the ring, only one quadrant need be considered as shown in Fig. 9.16(b). Cutting the ring at any section the bending

**Fig. 9.16**

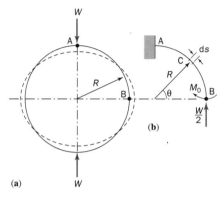

(a)

(b)

moment is given by

$$M_c = M_0 - \frac{WR}{2}(1 - \cos\,\theta) \qquad [9.52]$$

$M_0$ is unknown but may be obtained from the strain energy.

The strain energy is given by

$$U = \int_0^{\pi/2} \frac{1}{2EI}\left(M_0 - \frac{WR}{2}(1 - \cos\,\theta)\right)^2 R\,d\theta \qquad [9.53]$$

Owing to the symmetry at B, $\theta = 0 = \partial U/\partial M_0$. Therefore,

$$\frac{1}{EI}\int_0^{\pi/2}\left(M_0 - \frac{WR}{2}(1 - \cos\,\theta)\right)R\,d\theta = 0$$

from which

$$M_0 = WR\left(\tfrac{1}{2} - \frac{1}{\pi}\right) \qquad [9.54]$$

and

$$M_c = WR\left(\tfrac{1}{2}\cos\,\theta - \frac{1}{\pi}\right) \qquad [9.55]$$

## 9.13 The reciprocal theorem

In a linear structural system (Fig. 9.17), the deflection at point 1 due to forces $F_1$ at point 1 and $F_2$ at point 2 is, from the principle of superposition,

$$\Delta_1 = \Delta_{11} + \Delta_{12}$$

and the deflection at point 2 is

$$\Delta_2 = \Delta_{22} + \Delta_{21}$$

The deflections may be expressed in terms of *flexibility coefficients*, which are the displacements per unit force, as follows:

$$\Delta_1 = f_{11}F_1 + f_{12}F_2 \qquad [9.56]$$

$$\Delta_2 = f_{21}F_1 + f_{22}F_2 \qquad [9.57]$$

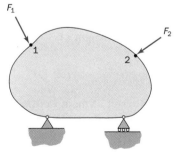

**Fig. 9.17**

If the strain energy of the system due to the application of these forces is $U$ then we may write

$$\Delta_1 = \frac{\partial U}{\partial F_1} \qquad\qquad [9.58]$$

and

$$\Delta_2 = \frac{\partial U}{\partial F_2} \qquad\qquad [9.59]$$

Partially differentiating eqns. [9.56] and [9.58] with respect to $F_2$,

$$\frac{\partial \Delta_1}{\partial F_2} = f_{12}$$

and

$$\frac{\partial^2 U}{\partial F_1 \, \partial F_2} = f_{12} \qquad\qquad [9.60]$$

Similarly, from eqns. [9.57] and [9.59],

$$\frac{\partial \Delta_2}{\partial F_1} = f_{21}$$

and

$$\frac{\partial^2 U}{\partial F_2 \, \partial F_1} = f_{21} \qquad\qquad [9.61]$$

From eqns. [9.60] and [9.61],

$$f_{21} = f_{12} \qquad\qquad [9.62]$$

This shows that the deflection at any point 1 due to a unit force at any point 2 is equal to the deflection at 2 due to a unit force at 1, providing the directions of the forces and deflections coincide in each of the two cases. This is termed the *reciprocal theorem*.

**Example 9.10**

**Determine the deflection at the tip of a cantilever beam of length $l$ when a load $F$ is applied at a distance $x$ from the fixed end.**

From eqn. [7.22], the deflection of a point at a distance $x$ from the fixed end and when a load $F$ is applied to the tip of a cantilever of length $l$ is

$$v = \frac{F}{2EI} \left( lx^2 - \frac{x^3}{3} \right)$$

By the reciprocal theorem, the same deflection will result at the tip of the cantilever when the load is applied at $x$. It can be verified that, with appropriate changes of variables, this is the same solution as is obtained for this problem in eqn. [7.24].

### 9.13 Summary

It can be seen from the diversity of applications, e.g. from coil springs to structures, that energy theorems have a very important part to play in engineering design analysis. Some of the concepts may be a little difficult to grasp at first, especially in relation to a physical appreciation. The principal advantage of energy methods is that the force, moment or deflection at a particular point in a structure (such as the end of a cantilever, or a node in a framework) can be found without investigating the details of the structural behaviour of every point.

---

**Problems**

9.1 Two possible designs of hydraulic cylinder are being considered as shown in Fig. 9.1(a) and (b). One involves the use of long bolts and the other uses short bolts of the same diameter. On the basis that the bolts are each required to absorb the same amount of strain energy, decide which of the two designs would be the best.

**Fig. 9.18**

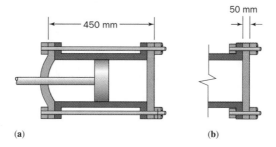

(a)                                        (b)

9.2 Two identical rectangular section steel members are loaded in the following way:
(a) one is subjected to an axial tensile force $W$;
(b) the other is supported as a cantilever and subjected to a load $W$ at its free end.
If each member is required to absorb the same strain energy, compare the maximum stresses which this would cause in each case.

9.3 Compare the strain energy stored in a beam simply-supported at each end and carrying a uniformly-distributed load, with that of the same beam carrying a concentrated load at mid-span and having the same value of maximum bending stress.

9.4 A solid shaft carries a flywheel of mass 150 kg and has a radius of gyration of 0.5 m. The shaft is rotating at a steady speed of 60 rev/min when a brake is applied at a point 4 m from the flywheel. Calculate the shaft diameter if the maximum instantaneous shear stress produced is 150 MN/m². Assume that the kinetic energy of the flywheel is taken up as torsional strain energy by the shaft. Neglect the inertia of the shaft. $G = 80$ GN/m².

9.5 A valve is controlled by two concentric close-coiled springs. The outer spring has twelve coils of 25 mm mean diameter, 3 mm wire diameter and 5 mm initial compression when the valve is closed. The free length of the inner spring is 6 mm longer than the outer. If the force required to open the valve 10 mm is 150 N, find the stiffness of the inner spring. If the diameter of the inner spring is 16 mm and the wide diameter is 2 mm, how many coils does it have? $G = 81$ GN/m².

9.6    When an open-coiled spring having ten coils is loaded axially the bending and torsional stresses are $140\,\text{MN/m}^2$ and $150\,\text{MN/m}^2$ respectively. Calculate the maximum permissible axial load and wire diameter for a maximum extension of $18\,\text{mm}$ if the mean diameter of the coils is eight times the wire diameter. $G = 80\,\text{GN/m}^2$, $E = 210\,\text{GN/m}^2$.

9.7    A bar of rectangular section $1\,\text{m}$ in length is simply-supported at each end and carries a uniformly distributed load. Determine the maximum depth of section so that the deflection due to shear shall not be greater than 2% of the total deflection. $E = 2.6\,G$.

9.8    A beam of $4\,\text{m}$ length carrying a concentrated load $W$ at mid-span is simply-supported at the right-hand end and is pin-jointed at the same level at the left-hand end to the free end of a horizontal cantilever of length $2\,\text{m}$. Use Castigliano's theorem to find the deflection under the load. Both beams have a flexural stiffness $EI$.

9.9    An anti-roll bar in a car suspension is mounted in bearings which allow rotation about the $z$ axis at points B and C but prevent any bending of section BC, Fig. 9.19. It is attached to opposite wheels at points A and D. If the bar is $20\,\text{mm}$ in diameter, what vertical force on the bar would be generated at point A when the wheel attached to point A raises by $1\,\text{mm}$ and point D falls by $1\,\text{mm}$? The bar is made of steel, with $E = 200\,\text{GN/m}^2$ and $\nu = 0.3$.

**Fig. 9.19**

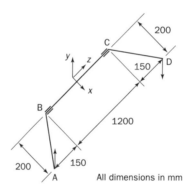

All dimensions in mm

9.10   A U-shaped pipe connecting two vessels has a radius $R$ and a leg-length $L$. Show that the deflection caused by forces $P$ applied at the free ends and perpendicular to the legs due to thermal expansion is

$$\Delta = \frac{P}{EI}\left\{\frac{2L^3}{3} + \pi L^2 R + \frac{\pi R^3}{2} + 4LR^2\right\}$$

9.11   A circular proving ring, used in the calibration of tensile testing machines, is shown in Fig. 9.20. (a) Show that the bending moment in the ring at any point 'C' is

$$M_c = M_o - \frac{WR}{2}(1 - \cos\,\theta)$$

(b) Given this moment distrbution, show that the reduction in diameter across AA' will be

$$\nu = \frac{WR^3}{EI}\left(\frac{\pi}{4} - \frac{2}{\pi}\right)$$

**Fig. 9.20**

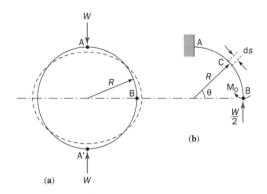

(a)          (b)

where $E$ is the Young's modulus of the ring material and $I$ is the second moment of area of the ring section.

9.12   A weight $W$ is supported by a semi-circular curved bar as shown in Fig. 9.21. If the supports at A and B to not move calculate (*a*) the reactions at the supports, (*b*) the position and magnitude of the maximum bending moment in the curved bar and (*c*) the vertical deflection at the weight $W$.

**Fig. 9.21**

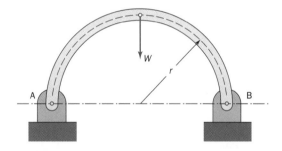

9.13   A childs pram uses four semicircular springs for its suspension as shown in Fig. 9.22. If the design weight of the pram and contents is $W$, derive expressions for the vertical and horizontal movement of the wheel axes relative to the body of the pram. The radius of each sring is $r$ and its second moment of areas is $I$. The modulus of the spring material is $E$.

9.14   A split ring of radius $R$ is used as a retainer on a machine shaft, and, in order to install it, it is necessary to apply outward tangential forces $F$

**Fig. 9.22**

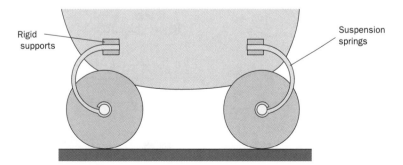

Rigid
supports

Suspension
springs

at the split to open up a gap $\delta$. If the flexural stiffness of the cross-section is $EI$, determine the required value of the forces.

9.15 A curved beam is in the form of a semicircle as shown in Fig. 9.23. The free end is pinned but is restrained to move in the horizontal direction only. If a horizontal force $F$ is applied at the free end show that the horizontal restraining force on the end of the beam is given by $4F/3\pi$.

**Fig. 9.23**

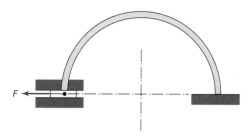

9.16 A curved beam is subjected to a load of 500 N at A, as shown in Fig. 9.24. If the product of Young's modulus and second moment of area for the beam section is given by $EI = 26\,\text{kN-m}^2$, calculate the vertical and horizontal deflections at A.

**Fig. 9.24**

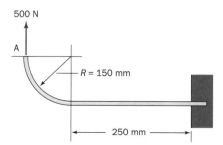

9.17 The frame ABCD in Fig. 9.25 is fixed at A and a horizontal force $F$ is applied at D. B and C are rigidly jointed and the lengths of the members are $AB = CD = h$ and $BC = L$. The relevant second moments of area are $I_1$ for AB and CD and $I_2$ for BC. Show that if D moves under load at an angle of $45°$ to the horizontal then

$$\frac{I_1}{I_2} = \frac{h}{3L}\left\{\frac{3L-4h}{2h-L}\right\}$$

Ignore the effects of axial load and shear force on the frame.

**Fig. 9.25**

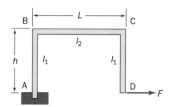

9.18 A circular section rod of diameter 20 mm has the shape of a quadrant of a circle of radius 300 mm (Fig. 9.26). If a vertical force 100 N is applied at the free end, determine the deflection in the direction of the applied force. The Young's modulus is 208 GN/m$^2$ and the shear modulus is 80 GN/m$^2$.

Fig. 9.26

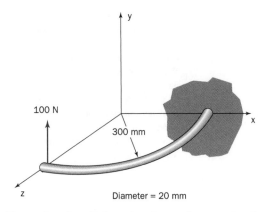

9.19 Show that for design situations where a circular section bar is to absorb an axial impact force, a uniform section rod of area A has better energy absorbing capacity than a stepped rod in which the cross-sectional areas are A and 2A. The overall lengths are assumed to be the same in each case.

9.20 The drive shaft from an electric motor has a diameter of 15 mm and a length of 100 mm. When the system is operating the pulley has a kinetic energy of 10 Nm. If the system comes to a sudden halt, calculate the angle of twist and the shear stress set up in the shaft. The shear modulus of the shaft material is 80 GN/m$^2$.

# 10 Buckling Instability

In earlier chapters basic analytical design procedures have been developed in which components and structural elements have been subjected to tension and compression forces, bending moment and torque. In this chapter we shall examine the specific effect of compressive forces in relation to the geometry and boundary conditions of members. Examples range from the compression force of combustion on the connecting rod of an internal combustion engine to the vertical columns used in structural steelwork to support all the vertical mass and forces in a building. In addition to axial and eccentrically aligned compression forces, columns or struts may be subjected to transverse loading which contributes to buckling and these cases will also be examined.

## 10.1 Stability of equilibrium

In previous chapters a fundamental condition in all the problems was the equilibrium of internal and external forces. Now, if the system of forces is disturbed owing to a small displacement of a body, two principal situations are possible: either the body will return to its original configuration owing to restoring forces during displacement, or the body will accelerate farther away from its original state owing to displacing forces. The former situation is termed *stable equilibrium* and the latter is termed *unstable equilibrium*.

Consider the simple case in Fig. 10.1(*a*) of a vertical bar pinned at the lower end and carrying an axial tensile force at the upper end. If there is a slight displacement from the vertical, the force will tend to restore the bar to its original position. In Fig. 10.1(*b*), however, the same bar subjected to a compressive load when displaced slightly from the vertical will accelerate towards a horizontal position, illustrating unstable equilibrium. A slightly more sophisticated case is shown in Fig. 10.1(*c*), where the bar is assisted in remaining vertical by the action of the horizontal springs. When the bar is displaced by an amount $x$ in either direction, there is a displacing moment $Px$ about O and a restoring moment $2KxL$, where $K$ is the stiffness of a

Fig. 10.1

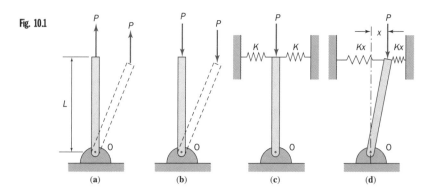

(a)        (b)        (c)        (d)

spring; hence we have

$$Px < 2KxL \rightarrow \text{stable}$$

$$Px > 2KxL \rightarrow \text{unstable}$$

The critical condition is when

$$Px = 2KxL \qquad \text{or} \qquad P_c = 2KL$$

and $P_c$ is termed the *critical load*, being the borderline between stable and unstable equilibrium.

**Elastic stability of slender members in compression**

The instability of structural members subjected to compressive loading may be regarded as a mode of failure, even though stress may remain elastic, owing to excessive deformation and distortion of the structure. This mode of failure is termed *buckling* and is prevalent in members for which the transverse dimension is small compared with the overall length.

A theory of buckling for slender columns under axial compression was developed by Leonhard Euler. In this analysis it will be assumed that the column is of uniform cross-section and either is ideally straight or has some defined initial curvature.

**Column with pinned ends**

The column shown in Fig. 10.2 is pin jointed at each end and it will be assumed to be straight when unloaded. An axial compressive load is applied with increasing magnitude until the column takes up the deformed shape as shown. It will be noted that although bending is taking place so that displacements are in the horizontal direction the notation used is the same as for a horizontal beam by rotating the column anticlockwise through 90°.

**Fig. 10.2**

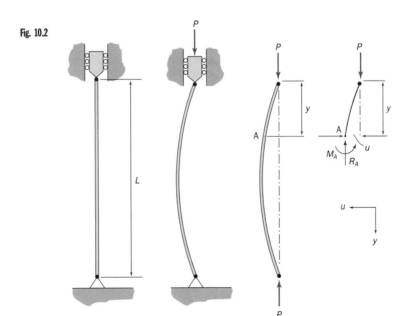

At a distance $y$ from the top joint the displacement is $u$ and the bending moment $M_A$ at section A is $Pu$ in the positive sense. Therefore, using the form of the equation derived for beams (eqn. [7.3]), we have

$$EI\frac{d^2u}{dy^2} = -Pu \qquad\qquad [10.1]$$

where $I$ is the *least* second moment of area of the cross-section. Therefore

$$\frac{d^2u}{dy^2} = -\frac{P}{EI}u = -k^2u$$

where $k = \sqrt{(P/EI)}$. Hence

$$\frac{d^2u}{dy^2} + k^2u = 0 \qquad\qquad [10.2]$$

The solution of this equation is

$$u = A\cos(ky) + B\sin(ky) \qquad\qquad [10.3]$$

The boundary conditions are that $u = 0$ at $y = 0$ and $L$; therefore

$$A = 0 \qquad \text{and} \qquad 0 = B\sin(kL)$$

Now the condition $B = 0$ merely gives the trivial case of the undeflected strut but the condition $\sin(kL) = 0$ leads to the solution $kL = n\pi$. Buckling first occurs for $n = 1$, from which the critical load $P_c$ is given by

$$\frac{P_c}{EI} = \frac{\pi^2}{L^2}$$

or

$$P_c = \frac{\pi^2 EI}{L^2} \qquad\qquad [10.4]$$

If $u = u_{max}$ at $y = L/2$, from symmetry, then $B = u_{max}$ and the deflection curve is given by

$$u = u_{max}\sin(ky) \qquad\qquad [10.5]$$

Note that the magnitude of $u_{max}$ cannot be determined from the boundary conditions, and it can in fact become arbitrarily large leading to *elastic instability* of the structure.

**Other end conditions**    Three other boundary conditions for the end restraint of columns are shown in Fig. 10.3.

Case ($a$) (*one end fixed, one end free*) may be treated as a pin-ended strut of equivalent length $2L$; hence

$$P_c = \frac{\pi^2 EL}{(2L)^2} = \frac{\pi^2 EI}{4L^2} \qquad\qquad [10.6]$$

Case ($b$) (*one end fixed, the other end only free to rotate*) does not have a readily assessed 'equivalent length' pin-ended strut and will be solved from

**Fig. 10.3**

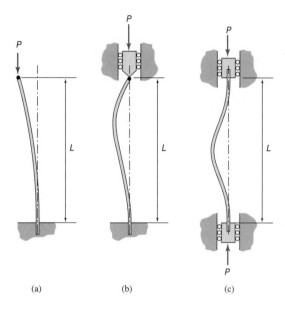

(a)    (b)    (c)

first principles. The free-body diagram is illustrated in Fig. 10.4($a$) and the bending moment resulting at point A is

$$M = Pu - Hy$$

from which

$$\frac{\mathrm{d}^2 u}{\mathrm{d}y^2} + \frac{P}{EI}u = \frac{Hy}{EI} \qquad\qquad [10.7]$$

Let $\sqrt{(P/EI)} = k$. The solution of eqn. [10.7] is

$$u = A\cos(ky) + B\sin(ky) + \frac{H}{P}y \qquad\qquad [10.8]$$

The boundary conditions are $u = 0$ at $y = 0$ and $\mathrm{d}u/\mathrm{d}y = 0$ at $y = L$, from which

$$A = 0 \qquad B = -\frac{H}{Pk}\sec(kL)$$

**Fig. 10.4**

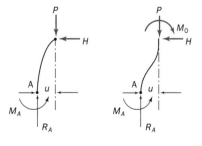

(a)    (b)

The deflection curve is thus

$$u = -\frac{H}{Pk}\frac{\sin(ky)}{\cos(kL)} + \frac{H}{P}y \qquad [10.9]$$

Finally $u = 0$ at $y = L$, and so

$$0 = -\frac{H}{Pk}\tan(kL) + \frac{H}{P}L$$

Therefore

$$\tan(kL) = kL \qquad [10.10]$$

The smallest value of $kL$ (other than $kL = 0$) to satisfy eqn. [10.10] is

$$kL = 4.49$$

or

$$P_c = 20\frac{EI}{L^2} \approx \frac{2\pi^2 EI}{L^2} \qquad [10.11]$$

This case is therefore approximately an 'equivalent length' of 0.7 that for the pin-ended column.

Case (c) (*both ends fixed*) is illustrated in Fig. 10.3(c) and the free-body diagram is shown in Fig. 10.4(b). It may be solved in a similar way to the previous cases using the boundary conditions that $u = 0$ at each end and $du/dy = 0$ at each end and at the middle. The solution is

$$P_c = \frac{4\pi^2 EI}{L^2} \qquad [10.12]$$

It may be seen that this situation is equivalent to a pin-ended strut of length $L/2$. Equations [10.6] and [10.12] contrast the differences in buckling load depending on the end conditions.

In general the critical buckling load for columns can be expressed as $P_c = \beta(EI/L^2)$, where $\beta$ has values as above or other values dependent on the end conditions. In practice these are rarely definable as 'pinned' or 'fixed' and the above formulae must only be applied with careful assessment.

## 10.2 Buckling characteristics for real struts

The load–deflection behaviour for an ideal Euler strut is illustrated in Fig. 10.5(a). For applied loads up to the critical value $P_c$ small transverse displacements $u$ can be maintained under load in a stable-equilibrium state.

**Fig. 10.5**

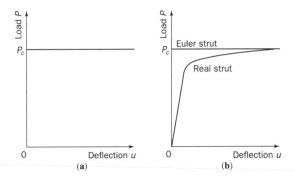

(a)            (b)

However, at $P_c$ and above the smallest transverse displacement is unstable and will rapidly grow to 'failure' of the strut. Obviously the strut can only accommodate a certain deflection up to its elastic limit; thereafter yielding and 'failure' would occur by plasticity.

In the case of a real strut, which incorporates some deficiency such as eccentricity of loading, deflection will occur from the moment when load is applied as shown in Fig. 10.5($b$). The curve becomes asymptotic to the Euler load at large deflection. Again, this situation will probably not be attained owing to yielding.

In order to appreciate the significance of stress during buckling behaviour we consider the Euler equation [10.4]

$$P_c = \frac{\pi^2 EI}{L^2} = \frac{\pi^2 EAr^2}{L^2}$$

where $A$ is the cross-sectional area and $r$ is the minimum *radius of gyration* of the section (see Appendix A). Therefore

$$\sigma_c = \frac{P_c}{A} = \frac{\pi^2 E}{(L/r)^2} \qquad [10.13]$$

The ratio $L/r$ is termed the *slenderness ratio*, and plotting this against $\sigma_c$ gives a curve known as the *Euler hyperbola*, as shown in Fig. 10.6.

Fig. 10.6

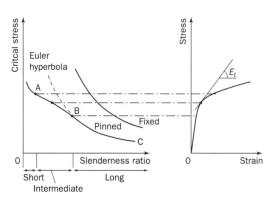

Buckling becomes the limiting mode of failure when the buckling stress given by equation [10.13] is less than or equal to the yield strength $\sigma_Y$ of the material, i.e.

$$\sigma_Y \geq \frac{\pi^2 E}{(L/r)^2} \qquad [10.14]$$

Alternatively, for a given material, the buckling stress will be less than the yield strength of the material for pin-jointed columns whose slenderness ratio exceeds a value which depends on the material properties

$$\frac{L}{r} \geq \pi \sqrt{\frac{E}{\sigma_Y}} \qquad [10.15]$$

This represents the range of slenderness ratios from B to C in Fig. 10.6. For a given material, a limiting slenderness ratio can be established below which the column will support at least the yield strength of the material and the design process is simplified. For structural steel ($\sigma_Y = 275\,\text{MN/m}^2$, $E = 200\,\text{GN/m}^2$), the critical slenderness ratio is 85. For columns with different end conditions, the appropriate equivalent length can be used, though the above expression is conservative. If the slenderness ratio is greater than the critical value, the limiting stress will be the buckling stress rather than the yield strength of the material.

For intermediate values of $L/r$, from A to B, instability will be accompanied by yielding. At a particular point on the stress–strain curve the stiffness is given by $E_t$, known as the *tangent modulus*. It is found that, in this elastic–plastic range, buckling loads can be predicted from the original Euler expression if the ordinary modulus $E$ is replaced by the tangent modulus $E_t$. For short columns, buckling instability does not occur but the stress may be sufficient to cause yielding.

Figure 10.6 also brings out the influence of end condition in relation to slenderness ratio.

**Example 10.1**

**A steel column is to be a fabricated I-section shown in Fig. 10.7. It is 6 m in height and is fixed at its lower end. It is to carry a design compressive load of 92 kN at the central axis O of the section. Determine the minimum dimensions of the column to resist buckling for (a) the upper end free, and (b) the upper end pinned but restrained in the horizontal direction by other structural members. What are the compressive stresses in each case? E = 200 GN/m².**

**Fig. 10.7**

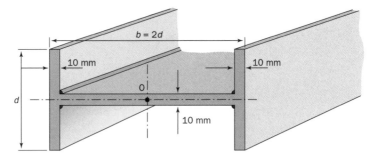

Case (*a*). The Euler buckling load is

$$P = \frac{\pi^2 EI}{4L^2}$$

$$I = \frac{4 \times 92 \times 10^3 \times 36}{\pi^2 \times 200 \times 10^9} = 6.72 \times 10^{-6}\,\text{m}^4$$

$I$ for the section in Fig. 10.7 is given by

$$\frac{2 \times 10 \times d^3}{12} + \frac{(2d - 20) \times 10^3}{12}$$

Therefore $(1.67d^3 + 167d - 1670) \times 10^{-12} = 6.72 \times 10^{-6}$,

$$d^3 + 100d \approx 4.03 \times 10^6$$

from which $d = 159\,\text{mm}$ and $b = 318\,\text{mm}$.

$$\text{Cross-sectional area} = 6160 \times 10^{-6}\,\text{m}^2$$

and

$$\text{Compressive stress} = \frac{92\,000}{6160} \times 10^6 = 14.9\,\text{MN/m}^2$$

Case (*b*). The Euler buckling load is, from eqn. [10.11],

$$P = 2\pi^2 \frac{EI}{L^2}$$

Therefore $I = 0.84 \times 10^{-6}\,\text{m}^4$ and, from the cubic equation for $d$ above, the solution is

$$d = 79\,\text{mm} \qquad \text{and} \qquad b = 158\,\text{mm}$$

$$\text{Cross-sectional area} = 2960 \times 10^{-6}\,\text{m}^2$$

and

$$\text{Compressive stress} = \frac{92\,000}{2960} \times 10^6 = 31.1\,\text{MN/m}^2$$

**Example 10.2**

A connecting rod in a high-speed mechanism must be 150 mm long and has to support a tensile load of 2 kN and a compressive load of 1.7 kN. The rod is connected through circular pins at each end. Decide on an appropriate size and shape of cross-section if the rod material has a yield stress of 275 MN/m² and a modulus of 200 GN/m².

The requirement to support a tensile load of 2 kN implies that the cross-sectional area $A$ is sufficient to prevent failure by yielding, i.e.

$$P_t \geq A\sigma_Y \qquad \qquad [10.16]$$

Failure due to buckling can occur by either of two mechanisms: buckling in a plane containing the axes of the pin joints at either end of the rod, or buckling in a plane perpendicular to this. The pins prevent rotation of the ends of the rod perpendicular to the pin axis, so for this mode of failure the rod effectively has fixed ends. The required buckling loads are therefore

$$P_{c-y} \geq \frac{\pi^2 EI_y}{L^2} \qquad \qquad [10.17]$$

and

$$P_{c-z} \geq \frac{4\pi^2 EI_z}{L^2} \qquad \qquad [10.18]$$

This implies that the optimum resistance to buckling occurs when $I_y = 4I_z$. A common section which can be dimensioned to meet this requirement is the H-section.

For a given cross-sectional area, the largest $I$-values will be obtained from a very large H with thin web and flanges. However, for rods manufactured by casting or forging, very thin sections are difficult to create. Rods fabricated from thin sheets of material are also liable to failure by local buckling or wrinkling, which will be described at the end of this chapter.

For the purposes of this exercise it will be assumed that the aspect ratio of the largest to the smallest dimension of the web or flange must be at most the given value $x$, Fig. 10.8

$$x = \frac{2B}{D-d} = \frac{d}{B-b} \qquad [10.19]$$

**Fig. 10.8**

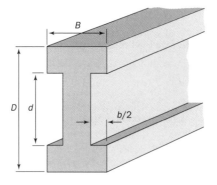

From this constraint, the minor dimensions $d$ and $b$ of the section can be found from $B$, $D$ and $x$ as

$$d = D - \frac{2B}{x} \qquad \text{and} \qquad b = B - \frac{d}{x} \qquad [10.20]$$

Figure 10.9 shows a spreadsheet in which the tensile and compressive loads which can be supported by given dimensions $B$ and $D$ can be evaluated. While these equations and constraints are very difficult to solve exactly, a trial and error variation of the $B$ and $D$ can be used to find a solution which meets the requirements.

**Fig. 10.9**

|  | A | B | C |
|---|---|---|---|
| 1 | Sy | 275 | |
| 2 | E | 200000 | |
| 3 | | | |
| 4 | L | 150 | |
| 5 | B | 2.49 | |
| 6 | D | 4.86 | |
| 7 | x | 3.20 | |
| 8 | | | |
| 9 | b | +B5-B10/B7 | |
| 10 | d | +B6-2*B5/B7 | |
| 11 | | | |
| 12 | A | +B5*B6-B9*B10 | |
| 13 | Iy | (B5*B6^3-B9*B10^3)/12 | |
| 14 | Iz | (2*(B6-B10)*B5^3+B10*B9^3)/12 | |
| 15 | | | Minimum |
| 16 | Pt | +B12*B1 | 2000 |
| 17 | Pc_y | @PI^2*B2*B13/B4^2 | 1700 |
| 18 | Pc_z | 4*@PI^2*B2*B14/B4^2 | +C17 |

(a) Data and cell formulae

|  | A | B | C |
|---|---|---|---|
| 1 | Sy | 275 | |
| 2 | E | 200000 | |
| 3 | | | |
| 4 | L | 150 | |
| 5 | B | 2.49 | |
| 6 | D | 4.86 | |
| 7 | x | 3.20 | |
| 8 | | | |
| 9 | b | 1.46 | |
| 10 | d | 3.30 | |
| 11 | | | |
| 12 | A | 7.27 | |
| 13 | Iy | 19.38 | |
| 14 | Iz | 4.84 | |
| 15 | | | Minimum |
| 16 | Pt | 2000 | 2000 |
| 17 | Pc_y | 1700 | 1700 |
| 18 | Pc_z | 1700 | 1700 |

(b) Spreadsheet display

Since a connecting rod is subjected to large accelerations it is desirable to minimize its weight, which for a rod of a given length and material is equivalent to minimizing the cross-sectional area. The problem can therefore be described as an *optimization* problem[1] where

the *goal* is to minimize the area;
subject to the *constraints* of tensile and buckling load being greater than or equal to the specified values;
by varying *design variables B* and *D*.

Most modern spreadsheets include facilities for optimization, and in fact the values of *B* and *D* in Fig. 10.9 were found by optimization. It can be verified by small changes in *B* and *D* that the dimensions shown give the minimum cross-sectional area for which the load constraints are met.

The minimum cross-sectional area which is necessary to satisfy both yielding and buckling constraints varies with aspect ratio as shown in Fig. 10.10. For large values of aspect ratio *x* a change in the aspect ratio makes no difference to the minimum weight, since the required cross-sectional area is the limiting constraint. It is undesirable to use an unnecessarily very thin-walled section, since the rod will occupy more space and be more difficult to manufacture.

**Fig. 10.10** Minimum cross-sectional area

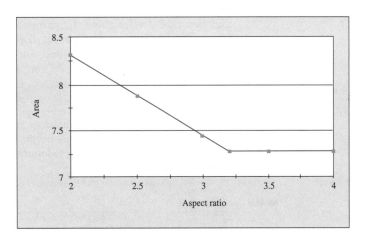

In Fig. 10.11 it can be seen that, for values of flange and web aspect ratio less than 3.2, Euler buckling of the rod is the limiting constraint and the tensile load which can be supported is larger than required. Thus the weight can be reduced if the area is distributed over a section with thinner walls. In Figs. 10.10 and 10.11 the smallest aspect ratio at which the minimum cross-sectional area is achieved was found by including aspect ratio as a design variable in the optimization and starting the search for the minimum from a small aspect ratio value.

Internal combustion engine connecting rods such as that shown in Fig. 4.3 are usually made with H-shaped cross-sections, even though their slenderness ratios are well below the value at which buckling becomes the critical mode of failure. One possible reason for this is that if the yielding load is exceeded, the buckling load becomes proportional to the tangent modulus, which falls rapidly with increasing plastic deformation. Thus it is

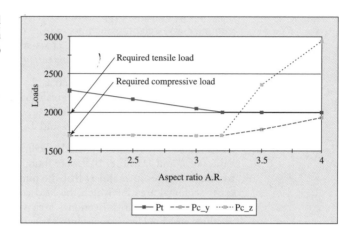

still advantageous to have a section where the ratio of the $I$-values is 4:1 to maximize the load at which the rod will collapse.

Thus, using an H-section, the rod should have sufficient cross-sectional area to prevent overstressing of the material under the largest of the tensile and compressive loads. It should have an $I$-value about the pin axis which is four times the $I$-value about the perpendicular axis. For the particular rod length and loading quoted above, the rod dimensions which allow a minimum cross-sectional area and the thickest web and flanges are: $D = 4.9\,\text{mm}$; $d = 3.3\,\text{mm}$; $B = 2.5\,\text{mm}$; $b = 1.5\,\text{mm}$.

## 10.3 Eccentric loading of slender columns

It is seldom in practice that a column or strut can be loaded exactly along its central axis as the Euler analysis implies. A general solution will now be developed for the case of a long column subjected to a load parallel to, but eccentric from, the central axis.

**Fig. 10.12**

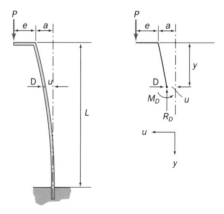

Consider the column illustrated in Fig. 10.12 displaced by the load $P$. Bending moment at D, $M_D = -P(e + a - u)$, where $e$ is the eccentricity, $a$ is the deflection at the free end and $u$ is the deflection at $y$ from the free end. Then

$$EI \frac{d^2u}{dy^2} = P(e + a - u)$$

$$\frac{d^2u}{dy^2} + \frac{P}{EI}u = \frac{P}{EI}(e + a) \qquad [10.21]$$

Let $\sqrt{(P/EI)} = k$. Then

$$u = A\cos(ky) + B\sin(ky) + e + a \qquad [10.22]$$

The boundary conditions are as follows:

At $y = 0$, $u = a$; therefore $A = -e$ and

$$u = -e\cos(ky) + B\sin(ky) + e + a \qquad [10.23]$$

At $y = L$, $du/dy = 0$; therefore $B = -e\tan(kL)$

At $y = L$, $u = 0$; therefore $a = e[\sec(kL) - 1]$

The deflection curve is therefore

$$u = e\sec(kL)\{1 - \cos[k(L - y)]\} \qquad [10.24]$$

The maximum bending stress occurs at the fixed end and is given by

$$\sigma = \pm \frac{Pec\sec(kL)}{Ar^2}$$

where $c$ is the half-depth of section in the plane of bending, $r$ the radius of gyration and $A$ the cross-section area. The maximum resultant compressive stress is

$$\sigma_c = -\frac{P}{A} - \frac{Pec\sec(kL)}{Ar^2} = -\frac{P}{A}\left(1 + \frac{ec\sec(kL)}{r^2}\right) \qquad [10.25]$$

Hence

$$P = -\frac{\sigma_c A}{1 + [ec\sec(kL)]/r^2} \qquad [10.26]$$

**Fig. 10.13**

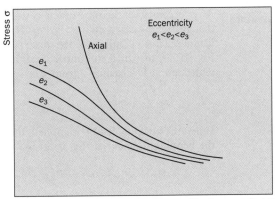

Note that $k$ is a function of $P$ and so $P$ exists in both sides of eqn. [10.26]. Therefore, a value for the critical buckling load can only be found by iteration.

The effect of eccentricity on maximum stress as a function of slenderness ratio is shown in Fig. 10.13.

**Example 10.3**

A vertical steel tube having 75 mm external and 62 mm internal diameters is 3 m long, fixed at the lower end and completely unrestrained at the upper end. The tube is subjected to a vertical compressive load parallel to, but eccentric by 6 mm from, the central axis. Determine the limiting value of the load so that there is no tensile stress at the base of the tube.

If the column had been loaded along its axis what would be the value of the Euler buckling load? $E = 208\,\mathrm{GN/m^2}$.

For zero tensile stress at the base

$$0 = -\frac{P}{A} + \frac{Pec\,\sec(kL)}{I}$$

Hence

$$\sec(kL) = \frac{I}{Aec} = \frac{0.815 \times 10^{-6}}{0.00138 \times 0.006 \times 0.0375} = 2.63$$

$$kL = 1.165$$

and

$$k = 0.388$$

$$P_c = 0.388^2 EI = 25.6kN$$

For the axially loaded column

$$P_c = \frac{\pi^2 EI}{4L^2} = \frac{\pi^2 \times 208 \times 10^9 \times 0.815 \times 10^{-6}}{4 \times 9}$$

$$= 46.5\ \mathrm{kN}$$

## 10.4 Struts having initial curvature

After eccentricity the next practical departure from the Euler idealization is that in some cases a column or strut may not be perfectly straight before loading. This will influence the onset of instability. The strut is illustrated in Fig. 10.14, in which the initial maximum deflection is $a_0$, the value of the deflection at D distance $y$ from P is $u_0$, and

$$u_0 = a_0 \sin\frac{\pi y}{L}$$

When the buckling load $P$ is applied, the deflection at $y$ is increased by $u$ and the bending moment at this point is $M_D$, where

$$M_D = P(u + u_0) = P\left(u + a_0 \sin\frac{\pi y}{L}\right)$$

**Fig. 10.14**

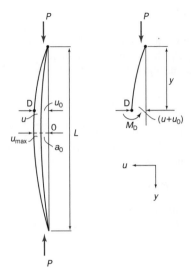

Hence

$$EI\frac{d^2u}{dy^2} = -P\left(u + a_0 \sin\frac{\pi y}{L}\right)$$

Therefore

$$\frac{d^2u}{dy^2} + k^2\left(u + a_0 \sin\frac{\pi y}{L}\right) = 0 \qquad [10.27]$$

where $k^2 = P/EI$; hence

$$u = A \cos(ky) + B \sin(ky) + \frac{k^2 a_0 \sin(\pi y/L)}{(\pi^2/L^2) - k^2}$$

The boundary conditions are that at $y = 0$ and $y = L$, $u = 0$; therefore $A = 0$ and $B \sin(kL) = 0$; and since $k$ is not zero, it follows that $B = 0$. Therefore

$$u = \frac{k^2 a_0 \sin(\pi y/L)}{(\pi^2/L^2) - k^2}$$

$$= \left(\frac{Pa_0}{\pi^2(EI/L^2) - P}\right) \sin\frac{\pi y}{L}$$

$$= \left(\frac{Pa_0}{P_e - P}\right) \sin\frac{\pi y}{L} \qquad [10.28]$$

where $P_e = \pi^2 EI/L^2$. Substituting for $u_0$,

$$u = \frac{u_0}{(P_e/P) - 1} \qquad [10.29]$$

Thus the effect of the end thrust $P$ is to increase the no-load maximum deflection $a_0$ by the multiplying factor $[(P_e/P) - 1]^{-1}$.

From eqn. [10.29] it will be observed that, as the value of $P$ approaches that of $P_e$, the basic Euler buckling load, the value of $u$ increases, tending to become infinite. At $y = \frac{1}{2}L$, the increased deflection is given by

$$u' = \frac{a_0}{(P_e/P) - 1}$$  [10.30]

Plotting values of $P$ and $u$, the curve shown in Fig. 10.15($a$) is obtained. Failure of the strut would occur before $P$ reached the theoretical value $P_e$. Equation [10.30] may be written in the form

$$(P_e/P)u' - u' = a_0$$  [10.31]

which shows that there is a linear relation between $u'$ and $u'/P$, Fig. 10.15($b$), the intercept on the axis of $u'$ being equal to $-a_0$.

**Fig. 10.15**

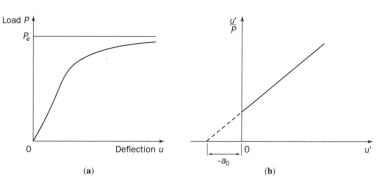

(a)                     (b)

Let $\sigma = P/A$ and $\sigma_e = P_e/A$, where $A$ is the cross-sectional area of the strut. Hence, from eqn. [10.28],

$$u = \left(\frac{\sigma}{\sigma_e - \sigma}\right) a_0 \sin \frac{\pi y}{L}$$  [10.32]

and total deflection at $y$ is given by

$$u + u_0 = \left(\frac{\sigma}{\sigma_e - \sigma}\right) a_0 \sin \frac{\pi y}{L} + a_0 \sin \frac{\pi y}{L}$$

$$= \left(\frac{\sigma_e}{\sigma_e - \sigma}\right) a_0 \sin \frac{\pi y}{L}$$  [10.33]

The maximum deflection occurs at O, where $y = \frac{1}{2}L$; therefore

$$u_{max} = \left(\frac{\sigma_e}{\sigma_e - \sigma}\right) a_0$$  [10.34]

Maximum bending moment is $a_0[\sigma_e/(\sigma_e - \sigma)]P$ and the maximum compressive stress is given by

$$\sigma_c = -\frac{Pa_0[\sigma_e/(\sigma_e - \sigma)]c}{I} - \frac{P}{A}$$  [10.35]

where $c$ is the distance from the neutral axis to the point of maximum compressive stress. If $r$ is the least radius of gyration of the section then, substituting $P/A = \sigma$,

$$\sigma_c = \sigma\left(\frac{\eta\sigma_e}{\sigma_e - \sigma} + 1\right) \qquad [10.36]$$

where $\eta = a_0 c/r^2$. Taking $\sigma_c$ equal to the yield stress in compression, $\sigma_Y$, we obtain a quadratic equation

$$\sigma^2 - \sigma[\sigma_Y + \sigma_e(\eta + 1)] + \sigma_Y\sigma_e = 0 \qquad [10.37]$$

from which the limiting value of $\sigma$ and hence $P$ can be determined.

## 10.5 Empirical formulae for design

In practice many strut or column designs do not fall in the category of slenderness ratio relevant to Euler theory. Consequently a number of empirical formulae have been devised for different classes of materials (metal, timber, concrete) to cover the range A to B in Fig. 10.6. There is not space here to discuss these in detail, but some are given in summary and the reader is referred to British Standards and other design codes as required.

### Rankine–Gordon

$$P = \frac{\sigma A}{1 + a(L/r)^2} \qquad [10.38]$$

where symbols have their previous meaning and $a$ is a constant dependent on end condition and material.

The formula applies for very short columns where buckling is not a factor, as well as for a range of larger slenderness ratios. Typical values for $a$ for pin-ended struts are 0.0001 for mild steel, 0.0006 for cast iron, and 0.0001 for timber.

For eccentric loading let the permissible load be $P'$; then for the elastic limiting condition

$$\sigma_Y = -\frac{P'}{A} - \frac{P'ec}{Ar_b^2}$$

where $r_b =$ radius of gyration in the plane of bending, and from eqn. [10.38] we can write

$$\sigma_Y = -\frac{P}{A} - \frac{aL^2P}{Ar^2}$$

Hence

$$P' = \frac{[1 + (aL^2/r^2)]P}{1 + (ec/r_b^2)} \qquad [10.39]$$

### Straight line

This represents the region A to B in Fig. 10.6 by a straight line expressed as

$$P = \sigma A[1 - c(L/r)] \qquad [10.40]$$

where $P =$ allowable load, $\sigma =$ allowable compressive stress and $c =$ a constant depending on the material and end restraint which is typically 0.005 for mild steel and 0.008 for cast iron.

**Parabolic**  This formula, which is intended to agree with the Euler formula for long columns, is

$$P = \sigma A[1 - c(L/r)^2] \qquad [10.41]$$

where $c$ is a constant; the other symbols have the same meanings as above. With pin ends and $L/r < 150$, then for mild steel $c$ may be taken as 0.000 023.

**10.6 Buckling under combined compression and transverse loading**

Since transverse loading of a slender member causes bending, the effect on instability under axial compression is similar to that of initial curvature of the strut, and consequently a more rapid rate of deflection occurs.

This situation is also described as a beam-column and one example has already been studied. The eccentrically loaded pin-ended column can be regarded as a beam–column carrying axial load $P$ together with end moments $P_e$. Before proceeding to specific transverse loading cases it is appropriate to derive the differential relationships as was done in Chapter 6 for bending without end load.

**Fig. 10.16**

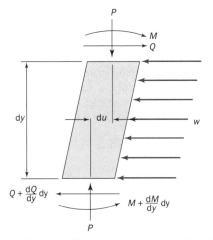

The equilibrium equations for the element shown in Fig. 10.16 are

$$w\,dy + \left(Q + \frac{dQ}{dy}dy\right) - Q = 0$$

and

$$-M - Q\,dy + w\,dy\frac{dy}{2} + \left(M + \frac{dM}{dy}dy\right) - P\,du = 0$$

These reduce to

$$\frac{dQ}{dy} = -w \qquad [10.42]$$

and

$$\frac{dM}{dy} - Q - P\frac{du}{dy} = 0 \qquad [10.43]$$

When $P = 0$ these equations are the same as [6.1] and [6.3]. Eliminating the shear force $Q$ between these equations

$$\frac{d^2M}{dy^2} - P\frac{d^2u}{dy^2} = -w$$

Substituting for $M$, the curvature $-EI(d^2u/dy^2)$ gives

$$EI\frac{d^4u}{dy^4} + P\frac{d^2u}{dy^2} = w$$

or

$$\frac{d^4u}{dy^4} + \frac{P}{EI}\frac{d^2u}{dy^2} = \frac{w}{EI} \qquad [10.44]$$

Equation [10.44] is the general differential equation for beam–column-type situations. If the lateral loading is zero then the equation is of the form that was used for pure column buckling cases.

The standard form of solution for eqn. [10.44] is

$$u = A\sin(ky) + B\cos(ky) + C_1wy^2 + C_2wy + C_3$$

where $k = \sqrt{(P/EI)}$ and $A$, $B$, $C_1$, $C_2$ and $C_3$ are constants related to the boundary conditions.

---

**10.7 Pin-ended strut carrying a uniformly distributed lateral load**

The bending moment at an arbitrary section A along the beam in Fig. 10.17 is

$$M_A = Pu + \frac{wL}{2}y - \frac{wy^2}{2} \qquad [10.45]$$

$$EI\frac{d^2u}{dy^2} = -Pu - \frac{wL}{2}y + \frac{wy^2}{2}$$

$$\frac{d^2u}{dy^2} + k^2u = \frac{wk^2y^2}{2P} - \frac{wLk^2y}{2P} \qquad [10.46]$$

The standard solution for this type of equation is

$$u = A\sin(ky) + B\cos(ky) + \frac{wy^2}{2P} - \frac{wL}{2P}y - \frac{w}{Pk^2} \qquad [10.47]$$

The boundary conditions are that at $y = 0$, $u = 0$; hence $B = w/Pk^2$; and at $y = L$, $u = 0$. Therefore

$$0 = A\sin(kL) + \frac{w}{Pk^2}\cos(kL) - \frac{w}{Pk^2}$$

**Fig. 10.17**

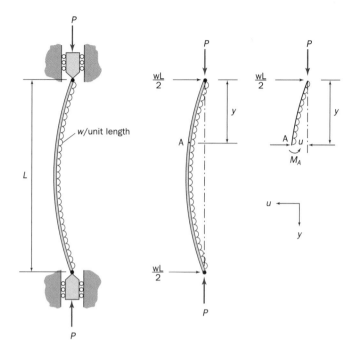

$$A = \frac{w}{Pk^2}\left[\operatorname{cosec}(kL) - \cot(kL)\right]$$

$$= \frac{w}{Pk^2}\,\tan\frac{kL}{2}$$

The deflection equation [10.47] becomes

$$u = \frac{w}{Pk^2}\left(\tan\frac{kL}{2}\,\sin(ky) + \cos(ky)\right) + \frac{wy^2}{2P} - \frac{wLy}{2P} - \frac{w}{Pk^2}$$

At $y = L/2, u = u_{max}$; therefore

$$u_{max} = \frac{w}{Pk^2}\left(\sec\frac{kL}{2} - 1\right) - \frac{wL^2}{8P} \qquad [10.48]$$

Now, $u_{max} \rightarrow \infty$ when $k(L/2) \rightarrow (n\pi/2)$. Hence

$$P_c = \frac{n^2\pi^2}{L^2}EI$$

for which the lowest value is

$$P_c = \frac{\pi^2 EI}{L^2}$$

which is the Euler load for a simple strut without transverse loading.

   Although theoretically the load–deflection relationship would become asymptotic to the Euler load, in fact 'failure' is governed by yielding at a lower load. Substituting the value of $u_{max}$ from eqn. [10.48] and $y = L/2$ in eqn. [10.45] gives

$$M_{max} = \frac{w}{k^2}\left(\sec\frac{kL}{2} - 1\right) \qquad [10.49]$$

from which, with a limiting value of $\sigma$, either an allowable value of $w$ can be determined for a particular end load $P$, or vice versa.

---

**Example 10.4**

**A part of a machine mechanism is illustrated in Fig. 10.18(a). The oscillating end portions AB and DE may be regarded as rigid, but can pivot freely at A, B, D and E. The central slender strut BCD has a simple restraint at C. What is the maximum load necessary to drive the mechanism?**

Fig. 10.18

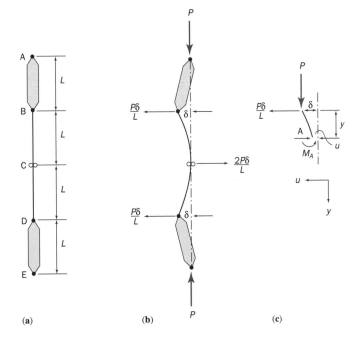

(a)          (b)          (c)

The deflected position of the members is shown in Fig. 10.18(b) with the appropriate forces acting on BCD. The basic differential equation is

$$EI\frac{d^2u}{dy^2} = P(\delta - u) + \frac{P\delta}{L}y$$

$$\frac{d^2u}{dy^2} + k^2u = k^2\delta\left(\frac{y}{L} + 1\right)$$

Differentiating,

$$\frac{d^3u}{dy^3} + k^2\frac{du}{dy} = k^2\frac{\delta}{L}$$

The solution of this equation is

$$u = A + B\sin(ky) + C\cos(ky) + \frac{\delta}{L}y$$

The boundary conditions are (i) $y = 0$, $M = 0$; (ii) $y = L$, $du/dy = 0$; (iii) $y = 0$, $u = \delta$; (iv) $y = L$, $u = 0$. From (iii) and (iv),

$$\delta = A + C \quad \text{and} \quad 0 = A + B\sin(kL) + C\cos(kL) + \delta$$

From (ii),

$$\frac{du}{dy} = Bk\cos(ky) - Ck\sin(ky) + \frac{\delta}{L}$$

$$0 = Bk\cos(kL) - Ck\sin(kL) + \frac{\delta}{L}$$

From (i)

$$\frac{d^2u}{dy^2} = -Bk^2\sin(ky) - Ck^2\cos(ky)$$

$$0 = -Ck^2$$

Hence $C = 0$, $A = \delta$, and

$$B = \frac{\delta}{kL\cos(kL)} = -\frac{2\delta}{\sin(kL)}$$

from which

$$\tan kL = 2kL$$

and

$$kL = 1.166$$

Hence the load required to drive the mechanism is

$$P_c = \left(\frac{1.166}{L}\right)^2 EI = 1.36\frac{EI}{L^2}$$

## 10.8 Other examples of instability

Instability is primarily a function of geometrical proportions, and in the case of columns this was expressed in terms of slenderness ratio in relation to compressive forces. Two other examples of members which become unstable with increasing load are shown in Fig. 10.19(a) and (b).

In the first case, because of the small width-to-depth ratio of the beam it is not possible to maintain the initial plane of bending, and the cross-section twists out of plane as shown. In the second case we again have a slenderness ratio influence in that the length of the bar under torsion is large compared with the diameter and it is not possible to maintain a straight axis of twist. Instability occurs through the shaft adopting a spiral axis when the torque reaches a critical level. There is not space here to analyse these cases but it is

Fig. 10.19

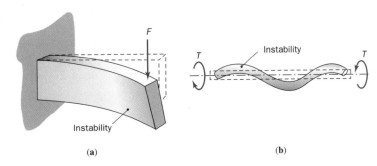

(a)     (b)

important for the designer to be aware of the possibility of these modes of deformation.

## 10.9 Local buckling

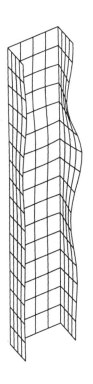

Fig. 10.20

In structures formed from thin sheets of material there is an additional possible mode of instability known as local buckling. When a thin rectangular sheet is subject to compressive stress it may buckle in a series of waves as illustrated in Fig. 10.20. The analysis of the critical buckling stresses is similar in principle to the buckling of a column, though the details are too complex to describe here.

Tables and formulae for the critical compressive stresses are given for a range of loading and boundary conditions in reference books such as *Roark's Formulas for Stress & Strain* by W.C. Young[2] or *Formulas for Stress, Strain, and Structural Matrices* by W. D. Pilkey[3]. Formulae for some simple cases where the sheet is very much longer in one direction than the other are given in Fig. 10.21. Using these, measures similar to the slenderness ratio of columns can be derived which indicate safe designs where buckling should not be the critical mode of failure.

For example, in an I-beam loaded in bending it is possible that the flange on the compression side may fail owing to local buckling before the yield strength of the material is approached. The critical buckling stress in a rectangular strip of material is

$$\sigma = K \frac{E}{1 - \nu^2} \left(\frac{t}{b}\right)^2 \qquad [10.50]$$

where $K$ is a constant which depends on the dimensions of the plate and the boundary conditions. Requiring that this stress is greater than the yield strength of the material implies that

$$\frac{b}{t} \leq \left(\frac{KE}{\sigma_y(1 - \nu^2)}\right)^{\frac{1}{2}} \qquad [10.51]$$

Treating half the flange as a long strip which is fixed at one long edge (to the web) and free at the other, $K = 1.21$. For structural steel with $E = 200$ GN/m$^2$ and $\sigma_y = 275$ MN/m$^2$ this implies that $b/t < 31$, i.e. the half-width of the flange should not be more than 31 times its thickness or the flange will locally buckle before the stress reaches the material yield strength. In fact local buckling is very sensitive to curvature or geometrical imperfections in the plate or residual stresses, so that the recommended ratio in the British steel design code is only 15 for an 'outstand' element of a

*Rectangular plate, $a \gg b \gg t$*

Compressive buckling stress

$$\sigma' = K \frac{E}{1 - \nu^2} \left( \frac{t}{b} \right)^2$$

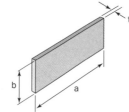

*Circular plate, $r \gg t$*

Uniform edge compression

$$\sigma' = 0.35 \frac{E}{1 - \nu^2} \left( \frac{t}{r} \right)^2$$

(a) fixed on two sides
$K = 5.7$

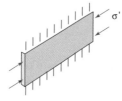

*Circular tube, $l \gg \sqrt{rt}$*

(b) fixed on one side
$K = 1.21$

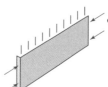

(a) longitudinal compression

$$\sigma' = \frac{1}{\sqrt{3}} \frac{E}{\sqrt{1 - \nu^2}} \frac{t}{r}$$

Shear buckling stress

$$\tau' = K \frac{E}{1 - \nu^2} \left( \frac{t}{b} \right)^2$$

(a) simply supported edges
$K = 4.4$

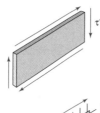

(b) torsion

$$\tau' = K \frac{E}{1 - \nu^2} \left( \frac{t}{l} \right)^2$$

where
$$K = 1.27 + \sqrt{9.64 + 0.466 H^{1.6}}$$
and
$$H = \sqrt{1 - \nu^2} \frac{l^2}{rt}$$

(b) clamped edges
$K = 7.4$

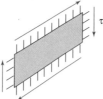

(c) external pressure

$$p' = \frac{1}{4} \frac{E}{1 - \nu^2} \frac{t}{r}$$

**Fig. 10.21** Local buckling stresses in a sheet

section in this material and 39 in an 'internal' element which is supported on both sides. Similar logic can be used to decide on minimum thicknesses of any material for any of the cases in Fig. 10.21 to ensure that buckling is not the limiting mode of failure.

Aircraft structures are perhaps the best example of structures where the prevention of local buckling is a critical design issue. The surface skins will be a millimetre or less in thickness, and in areas subjected to tensile loading there is no problem. However, all parts of the aircraft are subjected to some kind of bending action in flight and so some skins will be subject to compressive loading. The buckling tendency of the skin has to be limited by frequent stiffeners, e.g. ribs, stringers, etc. Even the web of a spar has to have a series of stiffeners in order to carry shear, since this gives rise to diagonal tension and compression, the latter causing wrinkling.

Thin sheet members subjected to compression or torsion may be stable in respect of the overall geometry, but will reveal local buckling instability characteristics as shown in Fig. 10.20.

**10.10 Summary**

Instability in structural elements has been shown to be an important factor in design as it is a mode of 'failure' dependent largely on compressive loading and geometrical proportions. The critical compressive load for the buckling of columns or struts can be expressed as

$$P_c = \frac{\beta EA}{(L/r)^2}$$

where $\beta$ is a constant depending on the material and end conditions. The idealized Euler theory is only applicable at large slenderness ratios and 'real' struts will 'fail' owing to exceeding the yield stress of the material long before attaining the Euler load. Eccentricity of loading, initial curvature and transverse loading can each contribute to a lowering of the allowable 'buckling' load as a function of the yield stress. Empirical formulae and design codes are now established for the design of columns for structural situations.

Local buckling of thin sheet material is also an important design consideration for which specialized texts such as *Theory of Elastic Stability* by Timoshenko and Gere[4], and design data sheets, should be consulted.

**References**

1. Arora, J. S. (1989) *Introduction to Optimum Design*, McGraw-Hill International, New York.
2. Young, W. C. (1989) *Roark's Formulas for Stress & Strain*, McGraw-Hill International, New York.
3. Pilkey, W. D. (1994) *Formulas for Stress, Strain, and Structural Matrices*, John Wiley, New York.
4. Timoshenko, S. P. and Gere, J. M. (1961) *Theory of Elastic Stability*, 2nd edition, McGraw-Hill, London.

**Problems**

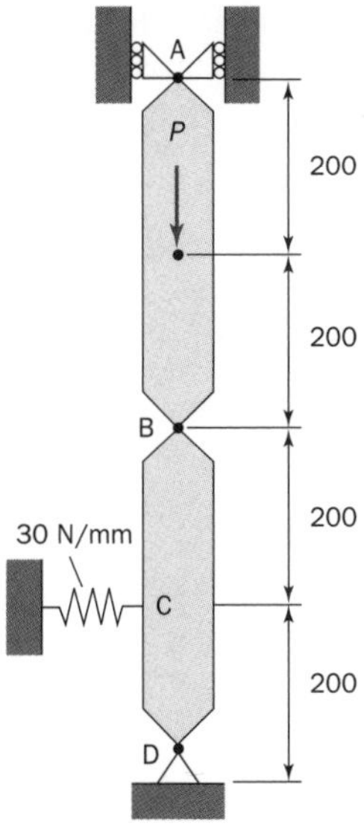

30 N/mm

200

200

200

200

200

Fig. 10.22

10.1 A machine mechanism consists of two rigid members each of length 400 mm connected by a frictionless hinge at B and pinned at A and D as illustrated in Fig. 10.22. A spring of stiffness 30 N/mm is attached to the lower member at C as shown. Determine the critical load, P, for the system.

10.2 A vertical mechanism linkage consists of a slender member of length 500 mm and stiffness 500 Nm$^2$, built in at the lower end and pinned at the upper end to a rigid member of length 250 mm. The upper end of the latter is pinned between rollers which are axially aligned with the whole strut. Determine the critical compressive load, when applied at the roller bearing, which will cause buckling.

10.3 A straight slender column of height 2.77 m is fixed at the lower end and is entirely free at the upper end. The design criterion is to limit the maximum compressive strain prior to buckling to 0.0008. Determine the required least radius of gyration.

10.4 A thin–walled square tube, Fig. 10.23(a) of length $L$ is to be designed to support a compressive load $P$. The lower end of the tube is fixed. In terms of the tube dimensions $L$, $h$ and $t$ and its material properties, write expressions for (i) the load at which the stress in the tube exceeds the yield stress $\sigma_y$ of the material, (ii) the load at which Euler buckling occurs (take the second moment of area $I$ for the cross-section as $h^3t / 2$), and (iii) the load at which local buckling of the tube wall occurs. Assume that the factor $K$ in eqn [10.50] is 4 in this case.

10.5    For a column 2 m high which must carry a load of 4000 N of structural steel ($E = 200$ GN/m$^2$, $\sigma_y = 200$ MN/m$^2$, $v = 0.3$), (i) calculate the minimum cross-sectional area of the column if the Euler buckling and yielding criteria are to be satisfied simultaneously, and (ii) calculate the minimum cross-sectional area if the local and Euler buckling conditions are to be met simultaneously.

10.6    Use a spreadsheet optimization to minimize cross-sectional area subject to the constraints that all the failure loads in Problem 10.4(i)–(iii) are greater than $P$ for the column of Problem 10.5.

**Fig. 10.23**

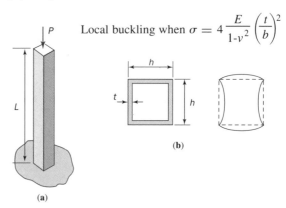

Local buckling when $\sigma = 4 \dfrac{E}{1-v^2}\left(\dfrac{t}{b}\right)^2$

10.7    A 4 m long strut has the cross-section shown in Fig. 10.24. Calculate the Euler buckling load for the strut if it has fixed ends. The Young's modulus for the strut material is 208 GN/m$^2$, $r_z = 43.6$ mm, $r_y = 36.7$ mm, where $r$ is the radius of gyration.

10.8    The column shown in Fig. 10.25 has pinned ends. The upper part, of length $(1 - n)L$, which is slender and of constant stiffness $EI$, is

**Fig. 10.24**

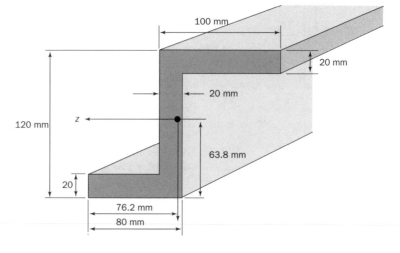

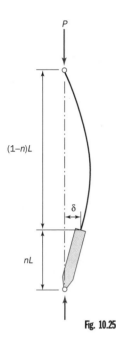

P

(1−n)L

δ

nL

**Fig. 10.25**

fixed to the lower part, of length $nL$, which is rigid. Show that, at instability, $\tan\{k(1 - n)L\} = -knL$. What is the critical load if the upper and lower parts are of equal length?

10.9  In a temperature-control device a copper strip measuring 8 mm × 4 mm and 100 mm long is pinned at each end. How much axial pre-compression is required so that buckling will occur after a temperature rise of 50°C? $E = 100$ GN/m², $\alpha = 18 \times 10^{-6}$ per deg C.

10.10  A strut 2 m long and pinned at both ends is subjected to an axial compressive force. If the strut cross-section is that shown in Fig. 10.26, calculate the Euler buckling load. $E = 207$ GN/m².

10.11  A circular steel column has a length of 2.44 m, an external diameter 101 mm and an internal diameter of 89 mm with its ends position fixed. Assuming that the centre-line is sinusoidal in shape with a maximum displacement at mid-length of 4.5 mm, determine the maximum stress due to an axial compressive load of 10 kN. $E = 205$ GN/m².

10.12  The compressive stress necessary to cause buckling in a long plate with one of its unloaded edges fixed so that it cannot rotate is

$$\sigma = -k\frac{Eh^2}{b^2}$$

where $h$ is the plate thickness and $b$ is its width, and $k = 1.32$ when the plate is very long.

(a)  Using this information, estimate the maximum width to thickness ratio $b/h$ that can be used in an angle section if the section must yield rather than locally buckle at maximum load, Fig. 10.27.

(b)  What will this ratio be for (i) aluminium alloy ($\sigma_y = 420$ MN/m², $E = 70$ GN/m²), (ii) structural steel ($\sigma_y = 350$ MN/m², $E = 200$ GN/m²).

(c)  Design an angle section of minimum weight in the aluminium alloy below to carry a compressive load of 10 kN which (i) will

**Fig. 10.26**

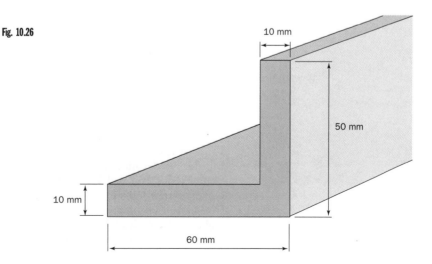

10 mm

50 mm

10 mm

60 mm

**Fig. 10.27**

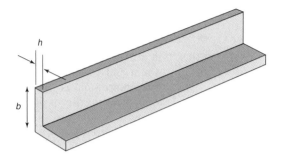

not yield, (ii) will not locally buckle, and (iii) has the largest second moment of area possible while still satisfying (i) and (ii).

10.13   A column is made up of two identical steel angle sections, as shown in Fig. 10.28 which are fixed at the lower end and free at the upper end. If the length of the column is 2.5 m and it is subjected to a compressive load of 8 kN calculate the safety factor in relation to buckling if:

(*a*)   the two angle sections are touching along AA but not connected in any way;

(*b*)   the two angle sections are fastened together along AA over the full length of the column.

The Young's modulus for steel is $207\,\text{GN/m}^2$ and the radius of gyration about $xx$ or $yy$ is 9.24 mm and about $zz$ is 5.92 mm.

**Fig. 10.28**

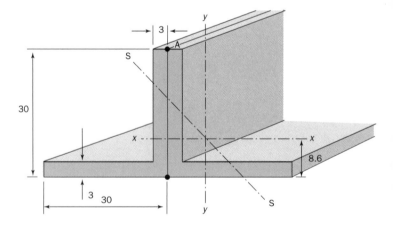

10.14   A tubular cast-iron column 5 m long has fixed ends and an external diameter of 250 mm. Calculate a suitable tube thickness if the column supports a load 1 MN. Assume a constant of 1/6400 in the Rankine formula and a stress of $80\,\text{MN/m}^2$.

10.15   A column of length $L$ is pinned at each end and is subjected to an axial compressive load $P$. A horizontal force $F$ is now applied at mid-height to the column. Show that the maximum bending-moment is obtained in the form

$$M_{max} = \frac{F}{2k} \tan\left(\frac{kL}{2}\right) \qquad \text{where} \qquad k = \sqrt{(P/EI)}$$

10.16 A horizontal beam of length 3.6 m is simply-supported at each end and is tubular, having internal and external diameters of 46 and 50 mm respectively. It is subjected to axial compression of 5 kN and uniformly-distributed transverse loading of 50 N/m. Determine the maximum surface stress. $E = 200\ \text{GN/m}^2$.

10.17 A pin-jointed truss is to support a load $W$ at a horizontal distance $b$ from a wall, Fig. 10.29. The member BC is a bar of circular section with radius $r$. Create a spreadsheet to determine the radius which is required to prevent bar BC from buckling. Given a material modulus $E$, evaluate the total volume of the bar for values of $h$ in the range 0.3–0.8, with $b = 1$, $W = 1000$, $E = 200\ \text{GN}/\text{m}^2$. Estimate the value of $h$ which gives a truss of minimum weight.

Fig. 10.29

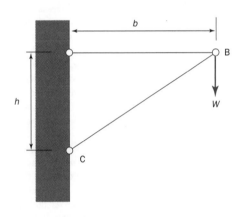

# 11 Stress and Strain Transformations

The preceding chapters have been concerned with 'one-dimensional' problems of stress and strain, i.e. in any particular example consideration has only been given to one type of stress acting in one direction. However, the majority of engineering components and structures are subjected to loading conditions, or are of such a shape that, at any point in the material, a complex state of stress and strain exists involving normal (tension, compression) and shear components in various directions. A simple example of this is a shaft which transmits power through a pulley and belt drive. An element of material in the shaft would be subjected to normal stress and shear stress due to bending action and additional shear stress from the torque required to transmit power. It is not necessarily sufficient to be able to determine the individual values of these stresses in order to select a suitable material from which to make the shaft because, as will be seen later, on certain planes within the element more severe conditions of stress and strain exist. The identification of these maximum stresses is an essential step in the design process.

In practice, however, it should be noted that stress cannot be measured directly. It can only be calculated from a knowledge of the strains in the material. These may be measured easily using electrical resistance strain gauges and generally a rosette of such gauges is used to measure strains in precise directions on the surface of the component. This chapter describes how the information from strain gauges may be analysed.

Finally the chapter introduces the way in which stresses and strains in fibre composite materials may be analysed. These materials are different to anything considered so far in the sense that they are anisotropic, i.e. they have different properties in different directions. However, it will be seen that the procedures developed in the chapter for the transformation of stresses and strains to different planes is directly applicable to fibre composites.

## 11.1 Symbols, signs and elements

Since conditions will be studied in which several different stresses occur simultaneously, it is essential to be consistent in the use of distinctive symbols, and a sign convention must be established and adhered to. A subscript notation will be used as follows:

$$\text{Direct stress} \quad \sigma_x, \; \sigma_y, \; \sigma_z$$

where the subscript denotes the direction of the stress.

$$\text{Shear stress} \quad \tau_{xy}, \; \tau_{yx}, \; \tau_{yz}, \; \tau_{zy}, \; \tau_{xz}, \; \tau_{zx}$$

where the first subscript denotes the direction of the normal to the plane on which the shear stress acts, and the second subscript the direction of the shear stress.

As in the previous work, tensile stress will be taken as positive and compressive stress negative. A shear stress is defined as positive when the

Fig. 11.1

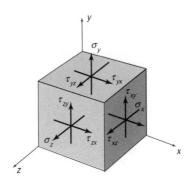

direction of the stress vector and the direction of the normal to the plane are both in the positive sense or both in the negative sense in relation to the co-ordinate axes. If the directions of the shear stress and the normal to the plane are opposed in sign, then the shear stress is negative. Pairs of complementary shear-stress components are therefore either both positive or both negative.

The angle between an inclined plane and a co-ordinate axis is positive when measured in the anticlockwise sense from the co-ordinate axis.

A general three-dimensional stress system is shown in Fig. 11.1.

**Plane stress**
If there are no normal and shear stresses on the two planes perpendicular to one of the co-ordinate directions, which implies that the complementary shear stresses are also zero, then this system is known as *plane stress* (see Section 3.7). It is a situation found, or approximately so, in a number of important engineering problems. The analysis of complex stresses which follows is only concerned with plane stress conditions.

## 11.2 Stresses on a plane inclined to the direction of loading

In preceding chapters the analysis has dealt with stress set up on a plane perpendicular (normal stress) or parallel (shear stress) to the direction of loading. However, if a piece of material is cut along a plane inclined at some angle $\theta$ to the direction of loading, then, in order to maintain equilibrium, a system of forces would have to be applied to the plane. This implies that there must be a stress system acting on that plane.

In Fig. 11.2 a bar is shown subjected to an axial tensile force $F$. The area of cross-section normal to the axis of the bar is denoted by $A$. If the bar is cut along the plane AB, inclined at an angle $\theta$ to the $y$-direction, then the

Fig. 11.2

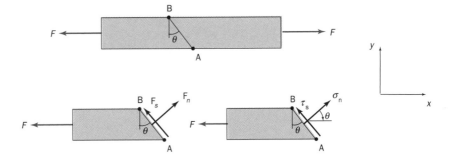

force $F$ applied to this portion of the bar can be reacted by component forces $F_n$ normal and $F_s$ tangential to the plane AB. Thus for equilibrium,

$$F_n = F \cos \theta \qquad \text{and} \qquad F_s = -F \sin \theta$$

The area of the plane AB will be the area of the bar normal to its axis multiplied by sec $\theta$.

Denoting on the plane AB the positive direct stress by $\sigma_n$ and the positive shear stress by $\tau_s$, then

$$\sigma_n = \frac{F_n}{A \sec \theta} = \frac{F \cos \theta}{A \sec \theta} = \frac{F}{A} \cos^2 \theta$$

and

$$\tau_s = \frac{F_s}{A \sec \theta} = \frac{-F \sin \theta}{A \sec \theta} = \frac{-F}{A} \sin \theta \cos \theta$$

$F/A$ is the direct stress normal to the axis of the bar and is equal to $\sigma_x$. Therefore

$$\sigma_n = \sigma_x \cos^2 \theta \qquad\qquad [11.1]$$

and

$$-\tau_s = \sigma_x \sin \theta \cos \theta = \tfrac{1}{2} \sigma_x \sin 2\theta \qquad\qquad [11.2]$$

Note that, in eqn. [11.1], when $\theta = 0$, $\sigma_n$ is a maximum and equal to $\sigma_x$, and when $\theta = 90°$, $\sigma_n$ is zero, indicating that there is no transverse stress in the bar.

Again, in eqn. [11.2], the magnitude of $\tau_s$ will be a maximum when $\sin 2\theta$ is a maximum, i.e. when $2\theta = 90°$ and $270°$, or $\theta = 45°$ and $135°$; the value of $\tau_s$ is then $\tfrac{1}{2}\sigma_x$ on the planes prescribed by $\theta = 45°$ and $135°$. This result is borne out in practice, and for materials whose shear strength is less than half the tensile strength, direct tensile loading results in failure along planes of maximum shear stress.

## 11.3 Element subjected to normal stresses

The rectangular element of material of unit thickness, Fig. 11.3, is subjected to tensile stresses in the $x$- and $y$-directions as shown.

Fig. 11.3

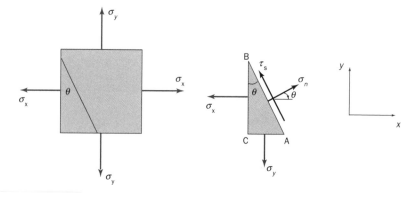

Considering a corner cut off the element by the plane AB inclined at $\theta$ to the $y$-axis, then positive normal and shear stresses will act on the plane as shown. For equilibrium of ABC, the forces on AB, BC and CA must also be in equilibrium. As the element is of constant unit thickness, the areas of the faces are proportional to the lengths of the sides of the triangle. Resolving *forces* normal to the plane AB,

$$\sigma_n AB - \sigma_x BC \cos\theta - \sigma_y AC \sin\theta = 0 \qquad [11.3]$$

Dividing through by AB,

$$\sigma_n - \sigma_x \frac{BC}{AB}\cos\theta - \sigma_y \frac{AC}{AB}\sin\theta = 0$$

Therefore

$$\sigma_n = \sigma_x \cos^2\theta + \sigma_y \sin^2\theta = \tfrac{1}{2}(\sigma_x + \sigma_y) + \tfrac{1}{2}(\sigma_x - \sigma_y)\cos 2\theta$$
$$[11.4]$$

Resolving *forces* parallel to AB,

$$\tau_s AB + \sigma_x BC \sin\theta - \sigma_y AC \cos\theta = 0 \qquad [11.5]$$

Dividing by AB,

$$\tau_s + \sigma_x \frac{BC}{AB}\sin\theta - \sigma_y \frac{AC}{AB}\cos\theta = 0$$

$$-\tau_s = \sigma_x \cos\theta \sin\theta - \sigma_y \sin\theta \cos\theta = \tfrac{1}{2}(\sigma_x - \sigma_y)\sin 2\theta \qquad [11.6]$$

If $\sigma_y$ is made zero, eqns. [11.4] and [11.6] reduce to eqns. [11.1] and [11.2] for the normal and shear stresses on a plane inclined to the direction of loading.

## 11.4 Element subjected to shear stresses

The rectangular element of the previous section is now considered with shear stresses on the faces instead of normal stresses. The plane AB inclined at $\theta$ to the $y$-axis, Fig. 11.4, is subjected to positive normal and shear stresses as previously. Consider the equilibrium of the triangular portion ABC. Resolving *forces* normal to the plane AB,

$$\sigma_n AB - \tau_{xy} BC \sin\theta - \tau_{yx} AC \cos\theta = 0 \qquad [11.7]$$

**Fig. 11.4**

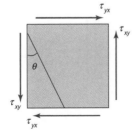
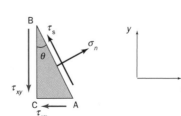

Dividing through by AB as before,

$$\sigma_n - \tau_{xy}\frac{BC}{AB}\sin\theta - \tau_{yx}\frac{AC}{AB}\cos\theta = 0$$

$$\sigma_n = \tau_{xy}\cos\theta\sin\theta + \tau_{yx}\sin\theta\cos\theta \qquad [11.8]$$

But, from a consideration of complementary shear stresses,

$$\tau_{xy} = \tau_{yx} \qquad [11.9]$$

Therefore

$$\sigma_n = 2\tau_{xy}\sin\theta\cos\theta = \tau_{xy}\sin 2\theta \qquad [11.10]$$

Resolving forces parallel to the plane AB,

$$\tau_s AB - \tau_{xy}BC\cos\theta + \tau_{yx}AC\sin\theta = 0 \qquad [11.11]$$

Dividing by AB,

$$\tau_s - \tau_{xy}\frac{BC}{AB}\cos\theta + \tau_{yx}\frac{AC}{AB}\sin\theta = 0$$

Therefore

$$-\tau_s = \tau_{yx}\sin^2\theta - \tau_{xy}\cos^2\theta = -\tau_{xy}\cos 2\theta \qquad [11.12]$$

## 11.5 Element subjected to general two-dimensional stress system

The general two-dimensional stress system shown in Fig. 11.5 may be obtained by a summation of the conditions of stress in Figs. 11.3 and 11.4. Hence the equations obtained for $\sigma_n$ and $\tau_s$ under normal stresses and shear stresses separately may be added together to give values for the normal and shear stress on the inclined plane AB, Fig. 11.5, in the general stress system. Therefore, from eqns. [11.4] and [11.10],

$$\sigma_n = \tfrac{1}{2}(\sigma_x + \sigma_y) + \tfrac{1}{2}(\sigma_x - \sigma_y)\cos 2\theta + \tau_{xy}\sin 2\theta \qquad [11.13]$$

and from eqns. [11.6] and [11.12]

$$\tau_s = -\tfrac{1}{2}(\sigma_x - \sigma_y)\sin 2\theta + \tau_{xy}\cos 2\theta \qquad [11.14]$$

**Fig. 11.5**

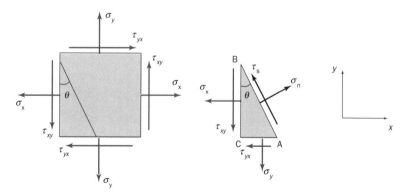

The validity of this method of superposition may be checked by considering the equilibrium of the triangular element of Fig. 11.5 and resolving forces perpendicular and parallel to the plane AB as was done in the previous sections.

It must be remembered that the signs in eqns. [11.13] and [11.14] are dependent on the directions chosen for the arrows (indicating stress) on the element in Fig. 11.5. If for any reason the directions of the stresses are different, e.g. compression instead of tension, then the stress components in eqns. [11.13] and [11.14] must be changed entered as negative quantities.

**Example 11.1**

A marine propeller shaft of 200 mm diameter is subjected to a torque of 126 kNm and a pure bending moment of 157 kNm. During inspection, a small surface crack is observed at 60° to the longitudinal axis of the shaft, Fig. 11.6. Determine the normal and shear stresses at the crack as it passes through positions A, B and C during rotation of the shaft.

**Fig. 11.6**

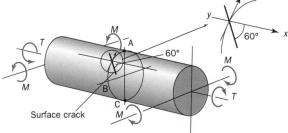

This is an example to illustrate the common engineering situation of shafts subjected to combined bending and torsion.

The second moments of area are as follows:

$$I = \frac{\pi d^4}{64} = \frac{\pi (0.2)^4}{64} = 78.54 \times 10^{-6} \, \text{m}^4$$

$$J = \frac{\pi d^4}{32} = 157.1 \times 10^{-6} \, \text{m}^4$$

Using the appropriate equations for bending and torsion,

$$\sigma = \frac{My}{I} = \pm \frac{157 \times 10^3 \times 0.1}{78.54 \times 10^{-6}} = \pm 200 \, \text{MN/m}^2$$

$$\tau = \frac{Tr}{J} = \frac{126 \times 10^3 \times 0.1}{157.1 \times 10^{-6}} = 80 \, \text{MN/m}^2$$

The stress components at A, B and C are

$$\sigma_x^A = +200 \, \text{MN/m}^2, \qquad \tau_{xy}^A = 80 \, \text{MN/m}^2$$

$$\sigma_x^B = 0, \qquad \tau_{xy}^B = 80 \, \text{MN/m}^2$$

$$\sigma_x^C = -200 \, \text{MN/m}^2, \qquad \tau_{xy}^C = 80 \, \text{MN/m}^2$$

Using eqns. [11.13] and [11.14] and noting that the correct value of $\theta$ on the element is $30°$ as shown in Fig. 11.7,

$$\sigma_n^A = 100 + 100\cos 60° + 80\sin 60°$$

$$= +219.3\,\text{MN/m}^2$$

$$\tau_s^A = -100\sin 60° + 80\cos 60°$$

$$\tau_s^A = -46.6\,\text{MN/m}^2$$

By a similar analysis we find

$$\sigma_n^B = +69.3\,\text{MN/m}^2 \qquad \tau_s^B = +40\,\text{MN/m}^2$$

$$\sigma_n^C = -80.7\,\text{MN/m}^2 \qquad \tau_s^C = +126.6\,\text{MN/m}^2$$

The stress components above are illustrated acting on the planes in Fig. 11.7. This example is extended in relation to principal stresses in Example 11.5.

**Fig. 11.7**

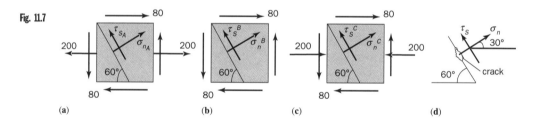

(a)          (b)          (c)          (d)

## 11.6 Mohr's stress circle

Considering eqns. [11.13] and [11.14] once more and rewriting,

$$\sigma_n - \tfrac{1}{2}(\sigma_x + \sigma_y) = \tfrac{1}{2}(\sigma_x - \sigma_y)\cos 2\theta + \tau_{xy}\sin 2\theta \qquad [11.15]$$

$$\tau_s = -\tfrac{1}{2}(\sigma_x - \sigma_y)\sin 2\theta + \tau_{xy}\cos 2\theta \qquad [11.16]$$

Squaring both sides and adding the equations,

$$[\sigma_n - \tfrac{1}{2}(\sigma_x + \sigma_y)]^2 + \tau_s^2 = \tfrac{1}{4}(\sigma_x - \sigma_y)^2 + \tau_{xy}^2 \qquad [11.17]$$

This is the equation of a circle of radius

$$\sqrt{[\tfrac{1}{4}(\sigma_x - \sigma_y)^2 + \tau_{xy}^2]}$$

and whose centre has the co-ordinates

$$[\tfrac{1}{2}(\sigma_x + \sigma_y),\, 0]$$

The circle represents all possible states of normal and shear stress on any plane through a stressed point in a material, and was developed by the German engineer Otto Mohr. The element of Fig. 11.5 and the corresponding Mohr diagram are shown in Fig. 11.8.

Fig. 11.8

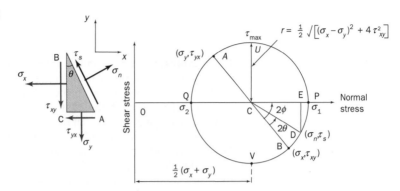

The sign convention used on the circle will be, for normal stress, positive to the right and negative to the left of the origin. Shear stresses which might be described as trying to cause a clockwise rotation of an element are plotted above the abscissa, and shear stresses tending to cause anticlockwise rotation are plotted below the axis.

It is important to remember that shear stress plotted, say, below the $\sigma$-axis, although being regarded as negative in the circle construction, may be either positive or negative on the physical element according to the shear-stress convention previously defined. Likewise, positive shear stress on the circle may be either positive or negative on the element.

The diagram is constructed as follows: using co-ordinate axes of normal stress and shear stress, *both to the same scale*, point B $(\sigma_x, \tau_{xy})$ is plotted representing the direct and shear stress acting on the plane BC of the element. Assuming in this case that $\sigma_y < \sigma_x$, the point A $(\sigma_y, \tau_{yx})$ is plotted to represent the stresses on the plane AC of the element. The normal stress axis bisects the line joining AB at C, and with centre C and radius AC a circle is drawn.

An angle equal to twice that in the element, i.e. $2\theta$, is set off from BC in the anticlockwise direction (the same sense as in the element), and the line CD then cuts the circle at the point whose co-ordinates are $(\sigma_n, \tau_s)$. These are then the normal and shearing stresses on the plane AB in the element.

The validity of the diagram is demonstrated thus:

$$\sigma_n = \mathrm{OC} + \mathrm{CE} = \tfrac{1}{2}(\sigma_x + \sigma_y) + r\cos(2\phi - 2\theta)$$

$$= \tfrac{1}{2}(\sigma_x + \sigma_y) + r\cos 2\theta \cos 2\phi + r\sin 2\theta \sin 2\phi$$

But

$$\cos 2\phi = \frac{\tfrac{1}{2}(\sigma_x - \sigma_y)}{r} \qquad \text{and} \qquad \sin 2\phi = \frac{\tau_{xy}}{r}$$

Therefore

$$\sigma_n = \tfrac{1}{2}(\sigma_x + \sigma_y) + \tfrac{1}{2}(\sigma_x - \sigma_y)\cos 2\theta + \tau_{yx}\sin 2\theta$$

$$\tau_s = \mathrm{DE} = r\sin(2\phi - 2\theta)$$

$$= r(\sin 2\phi \cos 2\theta - \cos 2\phi \sin 2\theta)$$

Substituting for $\cos 2\phi$ and $\sin 2\phi$,

$$\tau_s = -\tfrac{1}{2}(\sigma_x - \sigma_y)\sin 2\theta + \tau_{xy}\cos 2\theta$$

These expressions for $\sigma_n$ and $\tau_s$ are seen to be the same as those derived from equilibrium of the element. Thus, if at a point in a material the stress conditions are known on two planes, then the normal and shear stresses on any other plane through the point can be found using Mohr's circle.

Certain features of the diagram are worthy of note. The sides of the element AC and CB, which are $90°$ apart, are represented on the circle by AC and CB, $180°$ apart. A compressive direct stress would be plotted to the left of the shear-stress axis. The maximum shear stress in an element is given by the top and bottom points of the circle, i.e.

$$\tau_{max} = \pm\sqrt{[\tfrac{1}{4}(\sigma_x - \sigma_y)^2 + \tau_{xy}^2]} \tag{11.18}$$

and the corresponding normal stress is $\tfrac{1}{2}(\sigma_x + \sigma_y)$. The angle $\theta$ to the plane on which a maximum shear stress acts is obtained from the circle as

$$\tan 2\theta = \tan(90° + 2\phi) = -\cot 2\phi$$

Therefore

$$\tan 2\theta = -\left(\frac{\sigma_x - \sigma_y}{2\tau_{xy}}\right) \tag{11.19}$$

The second plane of maximum shear stress is displaced by $90°$ from that above.

**Example 11.2**

At a point in a complex stress field $\sigma_x = 40\,\text{MN/m}^2$, $\sigma_y = 80\,\text{MN/m}^2$ and $\tau_{xy} = -20\,\text{MN/m}^2$. Use Mohr's circle solution to find the normal and shear stresses on a plane at $45°$ to the $y$-axis.

**Fig. 11.9**

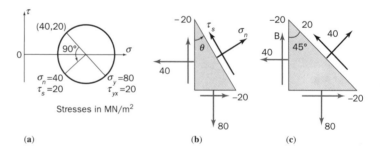

Stresses in MN/m²

(a)                    (b)                    (c)

The stresses on the element are shown in Fig. 11.9($b$) noting that the shear stresses are in the negative direction. The corresponding Mohr's circle is shown in Fig. 11.9($a$). From Fig. 11.9($a$) for $\theta = 45°$ on the element which is $90°$ on the circle,

$$\sigma_n = 40\,\text{MN/m}^2 \qquad \text{and} \qquad \tau_s = 20\,\text{MN/m}^2$$

Note that $\tau_s$ will be upwards to the left as shown in Fig. 11.9($c$) because it is below the axis on the circle (i.e. would turn the element anticlockwise). The values of $\sigma_n$ and $\tau_s$ may be checked by calculation using eqns. [11.13] and [11.14].

**Example 11.3**

Construct a Mohr's circle for the following point stresses: $\sigma_x = 60\,\text{MN/m}^2$, $\sigma_y = 10\,\text{MN/m}^2$ and $\tau_{xy} = +20\,\text{MN/m}^2$, and hence determine the stress components and planes in which the shear stress is a maximum.

The stresses on the element are as shown in Fig. 11.10(*b*) and the corresponding Mohr's circle is shown in Fig. 11.10(*a*).

Fig. 11.10

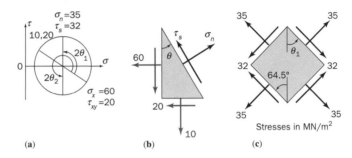

(a)    (b)    (c)

Stresses in MN/m²

From Fig. 11.10, the normal and maximum shear-stress components are

$$\sigma_n = 35\,\text{MN/m}^2 \qquad \text{and} \qquad \tau_s = 32\,\text{MN/m}^2$$

acting on the planes $\theta_1 = 64.5°$, $\theta_2 = 154.5°$. The orientation of the element experiencing these stresses is shown in Fig. 11.10(*c*).

**11.7 Principal stresses and planes**

It has been shown that Mohr's circle represents all possible states of normal and shear stress at a point. From Fig. 11.8 it can be seen that there are two planes, QC and CP, 180° apart on the diagram and therefore 90° apart in the material, on which the shear stress $\tau_s$ is zero. These planes are termed *principal planes* and the normal stresses acting on them are termed *principal stresses*. The latter are denoted by $\sigma_1$ and $\sigma_2$ at P and Q respectively, and are the maximum and minimum values of normal stress that can be obtained at a point in a material. The values of the principal stresses can be found either from eqn. [11.17] by putting $\tau_s = 0$, or directly from the Mohr diagram; hence

$$\sigma_1 = \frac{\sigma_x + \sigma_y}{2} + \frac{1}{2}\sqrt{[(\sigma_x - \sigma_y)^2 + 4\tau_{xy}^2]} \qquad [11.20]$$

$$\sigma_2 = \frac{\sigma_x + \sigma_y}{2} - \frac{1}{2}\sqrt{[(\sigma_x - \sigma_y)^2 + 4\tau_{xy}^2]} \qquad [11.21]$$

where $\sigma_1$ is the maximum and $\sigma_2$ the minimum principal stress. The planes are specified by

$$2\theta = 2\phi \qquad \text{and} \qquad 180° + 2\phi$$

or

$$\theta = \phi \quad \text{and} \quad 90° + \phi$$

But from geometry in Fig. 11.11(a)

$$\tan 2\phi = \frac{\tau_{xy}}{\frac{1}{2}(\sigma_x - \sigma_y)} \quad \left( \sin 2\phi = \frac{\tau_{xy}}{\sqrt{[\frac{1}{2}(\sigma_x - \sigma_y)^2 + \tau_{xy}^2]}} \right)$$

Therefore

$$\phi = \frac{1}{2} \tan^{-1}\left(\frac{2\tau_{xy}}{\sigma_x - \sigma_y}\right) \quad \text{and} \quad 90° + \frac{1}{2}\tan^{-1}\left(\frac{2\tau_{xy}}{\sigma_x - \sigma_y}\right) \quad [11.22]$$

Thus the magnitude and direction of the principal stresses at any point in a material depend on $\sigma_x$, $\sigma_y$ and $\tau_{xy}$ at that point, Fig. 11.11.

Fig. 11.11

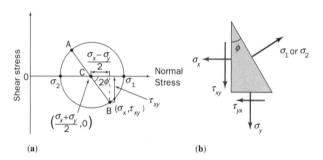

(a)                                              (b)

## 11.8 Maximum shear stress in terms of principal stresses

It was shown earlier, eqn. [11.18], that the maximum shear stress at a point is given by

$$\tau_{max} = \pm\frac{1}{2}\sqrt{[(\sigma_x - \sigma_y)^2 + 4\tau_{xy}^2]}$$

If expressions [11.20] and [11.21] for $\sigma_1$ and $\sigma_2$ are subtracted, then

$$\sigma_1 - \sigma_2 = \pm\sqrt{[(\sigma_x - \sigma_y)^2 + 4\tau_{xy}^2]} \quad [11.23]$$

and therefore

$$\tau_{max} = \frac{1}{2}(\sigma_1 - \sigma_2) \quad [11.24]$$

It should be noted that principal stresses are considered a maximum or minimum mathematically, e.g. a compressive or negative stress is less than a positive stress, irrespective of numerical value.

In Mohr's circle the principal planes PC and QC in Fig. 11.8 are at 90° to those of maximum shear stress, UC and VC, and therefore in the material the angles between these two sets of planes become 45°, or the maximum shear-stress planes bisect the principal planes.

## 11.9 General two-dimensional state of stress at a point

In the preceding paragraphs certain specific stress conditions which exist on planes in an element subjected to normal and shear stress have been derived, and these are shown on elements of the material in Fig. 11.12. Although it

Fig. 11.12

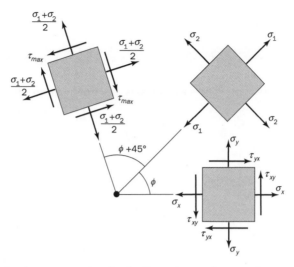

has been convenient in the foregoing analyses to draw an element of material it must be remembered that the stresses actually act at a *point* in the material.

**Example 11.4**

**Determine the principal stresses and maximum shear stresses at points A and B by calculation and at point C by a Mohr's circle construction for the I-section beam shown in Fig. 11.13.**

Fig. 11.13

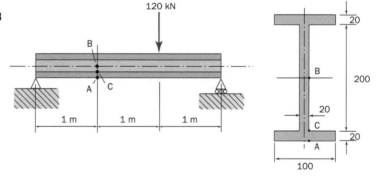

The reaction at the left-hand end is $40\,\mathrm{kN}$ and hence at the required cross-section of the beam

$$M = 40 \times 1 = 40\,\mathrm{kNm} \qquad Q = 40\,\mathrm{kN} \qquad I = 61\,867 \times 10^3\,\mathrm{mm}^4$$

$$y_A = +120\,\mathrm{mm}, \qquad y_B = 0, \qquad y_C = +100\,\mathrm{mm}$$

The bending stresses are, from eqn. [6.10],

$$\sigma_x^A = \frac{My_A}{I} = \frac{40 \times 10^3 \times (+0.12)}{61\,867 \times 10^{-9}} = 77.5\,\mathrm{MN/m}^2$$

$$\sigma_x^B = 0$$

$$\sigma_x^C = +77.5 \times \frac{0.1}{0.12} = +64.5\,\text{MN/m}^2$$

The shear stresses are, from eqn. [6.42]

$$\tau_{xy}^B = \frac{QA\bar{y}}{bI} = \frac{40 \times 10^3}{61\,867 \times 10^{-9}} \left( \frac{(0.2)^2}{8} + \frac{0.1}{8 \times 0.02} (0.24^2 - 0.2^2) \right)$$

$$= 10.34\,\text{MN/m}^2$$

$$\tau_{xy}^C = \frac{40 \times 10^3 \times 2000 \times 10^{-6} \times 0.11}{0.02 \times 61\,867 \times 10^{-9}} = 7.11\,\text{MN/m}^2$$

$$\tau_{xy}^A = \frac{40 \times 10^3 \times (0.24 - 0.02)(0.1 - 0.02)}{4 \times 61\,867 \times 10^{-9}} = 2.84\,\text{MN/m}^2$$

Using eqns. [11.20] and [11.21], for the principal stresses

$$\sigma_1^A = \frac{77.5}{2} + \tfrac{1}{2}(77.5^2 + 4 \times 2.84^2)^{1/2} = +77.6\,\text{MN/m}^2$$

$$\sigma_2^A = \frac{77.5}{2} - \tfrac{1}{2}(77.5^2 + 4 \times 2.84^2)^{1/2} = -0.1\,\text{MN/m}^2$$

At B, $\sigma_x = \sigma_y = 0$; hence

$$\sigma_1^B = +\tau_{xy}^B = +10.34\,\text{MN/m}^2, \qquad \sigma_2^B = -\tau_{xy}^B = -10.34\,\text{MN/m}^2$$

Using eqn. [11.24] for the maximum shear stresses,

$$\tau_{max}^A = \frac{77.6 - (-0.1)}{2} = 38.8\,\text{MN/m}^2$$

$$\tau_{max}^B = \frac{10.34 - (-10.34)}{2} = 10.34\,\text{MN/m}^2$$

**Fig. 11.14**

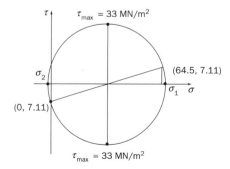

The Mohr's circle construction is shown in Fig. 11.14, for point C, from which the principal stresses and maximum shear stress are

$$\sigma_1^C = +65.3 \, \text{MN/m}^2, \quad \sigma_2^C = -0.77 \, \text{MN/m}^2, \quad \tau_{max}^C = 33 \, \text{MN/m}^2$$

**Example 11.5**

**Derive expressions, in terms of bending moment $M$ and torque $T$, for the magnitude and direction of the principal stresses at points A, B and C on the shaft in Example 11.1.**

The maximum bending and shear stresses occur at the outer surface of the shaft and are given by

$$\sigma_x = \frac{32M}{\pi d^3} \quad \text{and} \quad \tau_{xz} = \frac{16T}{\pi d^3}$$

where $d$ is the diameter of the shaft.

The stress components on an element of material at the surface points A, B and C are shown in Fig. 11.7. The shear stresses are the same in each; however, the bending stress is maximum tension at A, zero at B (the neutral plane) and maximum compression at C. The principal stresses at these three points are therefore

At A $\qquad \sigma_1, \sigma_2 = \frac{1}{2}\sigma_x \pm \frac{1}{2}\sqrt{(\sigma_x^2 + 4\tau_{xy}^2)}$

At B $\qquad \sigma_1 = -\sigma_2 = \tau_{xy}$

At C $\qquad \sigma_1, \sigma_2 = -\frac{1}{2}\sigma_x \pm \frac{1}{2}\sqrt{(\sigma_x^2 + 4\tau_{xy}^2)}$

At B, therefore, there is a state of pure shear.

The principal stresses can be expressed in terms of the bending moment and torque by substituting for $\sigma_x$ and $\tau_{xz}$:

At A $\qquad \sigma_1, \sigma_2 = \frac{16M}{\pi d^3} \pm \sqrt{\left[\left(\frac{16M}{\pi d^3}\right)^2 + \left(\frac{16T}{\pi d^3}\right)^2\right]}$

$$= \frac{16}{\pi d^3}[M \pm \sqrt{(M^2 + T^2)}] \qquad [11.25]$$

At B $\qquad \sigma_1 = -\sigma_2 = \frac{16T}{\pi d^3}$

At C the values are the same as at A but are negative.

The inclinations of the principal planes to the $z$-axis are

$$\theta = \frac{1}{2}\tan^{-1}\left(\frac{T}{M}\right) \quad \text{and} \quad 90° + \frac{1}{2}\tan^{-1}\left(\frac{T}{M}\right)$$

The maximum shear stress is given by

$$\tau = \frac{16}{\pi d^3}\sqrt{(M^2 + T^2)} \qquad [11.26]$$

## 11.10　Maximum shear stress in three dimensions

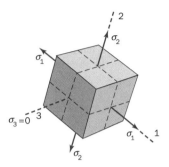

**Fig. 11.15**

So far the analysis of complex stress has been restricted to the two-dimensional situation. Therefore the value of the maximum shear stress obtained from the equations or the Mohr diagram represents the maximum shear stress in the $(x, y)$-plane. However, even for a state of plane stress this may not be the maximum shear stress in the material. To obtain the true maximum shear stress for use in design calculations (see Chapter 12) it is necessary to consider all three principal planes. The three-dimensional element subjected to the principal stresses is shown in Fig. 11.15. The principal stress $\sigma_3$ is zero in this particular case since we have been concerned only with plane stress situations. For convenience, consider the axis of the three principal stresses to be labelled 1, 2 and 3. Considering each of the three planes $(1, 2)$, $(2, 3)$ and $(3, 1)$ in turn it is possible to construct a Mohr diagram for each, as shown in Fig. 11.16. It is then apparent that a composite Mohr diagram could be constructed, as shown in Fig. 11.17, by superimposing these three diagrams. This Mohr diagram then enables the true maximum shear stress in the material to be determined.

It is evident from the composite Mohr diagram that if $\sigma_1$ and $\sigma_2$ are either both positive or both negative than $\tau_{max}$ in the $(1, 2)$-plane will not be the maximum shear stress in the material. It should also be noted that in the above superposition of the Mohr diagrams for the three planes, it is not essential that $\sigma_3 = 0$, so that in fact Mohr diagrams may be used generally for three-dimensional stress systems.

**Fig. 11.16**

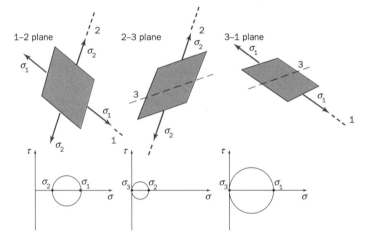

**Fig. 11.17**　Composite Mohr diagram for three-dimensional system

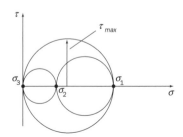

## 11.11  States of strain

In the following analysis all strains are considered to be small in magnitude. *Normal strains* are defined as the ratio of change in length to original length in a particular direction, and a subscript notation similar to that for stresses will be adopted, $\varepsilon_x$ and $\varepsilon_y$ being the direct strains in the $x$- and $y$-directions respectively, positive for tension and negative for compression. *Shear strain* is defined as the change in angle between two planes initially at right angles, and the symbol and subscripts $\gamma_{xy}$ will be used for shear referred to the $x$- and $y$-planes.

### Plane strain

*Plane strain* is the term used to describe the strain system in which the normal strain in, say, the $z$-direction, along with the shear strains $\gamma_{zx}$ and $\gamma_{zy}$, are zero. It should be noted that plane stress is not the stress system associated with plane strain. It will be evident from the definitions of plane stress and plane strain in Chapter 3 that plane strain, i.e. $\varepsilon_z = 0$, is associated with a three-dimensional stress system, and likewise plane stress is related to a three-dimensional strain system.

The stress system in Fig. 11.18(*a*) will give rise to a strain system combining direct and shear strains as shown in an exaggerated manner at (*b*). The object now is to determine the direct strain, $\varepsilon_n$, and shear strain, $\gamma_s$, for directions normal and tangential to a plane, inclined at $\theta$ to a co-ordinate direction, in terms of the direct strain, $\varepsilon_x$, $\varepsilon_y$, and shear strain, $\gamma_{xy}$, $\gamma_{yx}$, referred to the co-ordinate planes.

**Fig. 11.18**

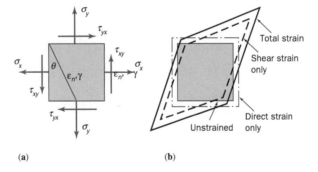

(a)    (b)

## 11.12  Normal strain in terms of co-ordinate strains

Referring to Fig. 11.19 $(A'C' - AC)/AC$ gives the strain normal to the plane FB related to the normal stress $\sigma_n$. Considering the triangles ACD and A'C'D' and with d$u$ as the increase in length from AD to A'D', and d$v$ the increase in length from CD to C'D', then

$$A'D' = AD + \mathrm{d}u = AD\left(1 + \frac{\mathrm{d}u}{AD}\right) = AD(1 + \varepsilon_x)$$

$$C'D' = CD + \mathrm{d}v = CD\left(1 + \frac{\mathrm{d}v}{CD}\right) = CD(1 + \varepsilon_y)$$

Similarly

$$A'C' = AC(1 + \varepsilon_n)$$

**Fig. 11.19**

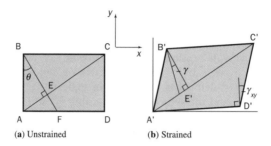

(a) Unstrained　　　(b) Strained

Now,

$$(A'C')^2 = (A'D')^2 + (C'D')^2 - 2A'D'.\,C'D'\cos(90° + \gamma_{xy})$$

or

$$(AC)^2(1 + \varepsilon_n)^2 = (AD)^2(1 + \varepsilon_x)^2 + (CD)^2(1 + \varepsilon_y)^2$$
$$+2AD(1 + \varepsilon_x)CD(1 + \varepsilon_y)\sin\gamma_{xy}$$

Since strains are assumed small, then $\sin\gamma_{xy} \approx \gamma_{xy}$ and second-order powers may be neglected. We have

$$(AC)^2(1 + 2\varepsilon_n) = (AD)^2(1 + 2\varepsilon_x) + (CD)^2(1 + 2\varepsilon_y)$$
$$+2AD.\,CD.\,\gamma_{xy}$$

which, with $(AC)^2 = (AD)^2 + (CD)^2$, reduces to

$$2\varepsilon_n(AC)^2 = 2\varepsilon_x(AD)^2 + 2\varepsilon_y(CD)^2 + 2AD.\,CD.\,\gamma_{xy}$$

Dividing through by $2(AC)^2$ and introducing $\sin\theta$ and $\cos\theta$,

$$\varepsilon_n = \varepsilon_x\cos^2\theta + \varepsilon_y\sin^2\theta + \gamma_{xy}\sin\theta\cos\theta \qquad [11.27]$$

Therefore

$$\varepsilon_n = \frac{\varepsilon_x + \varepsilon_y}{2} + \frac{\varepsilon_x - \varepsilon_y}{2}\cos 2\theta + \tfrac{1}{2}\gamma_{xy}\sin 2\theta \qquad [11.28]$$

## 11.13　Shear strain in terms of co-ordinate strains

Referring to Fig. 11.19, the shear strain $\gamma_s$ related to the shear stress $\tau_s$, Fig. 11.5, is given by the change in angle between EB and AE, or $\angle AEB - \angle A'E'B'$. Considering the triangles AEB and A'E'B', then, as before,

$$A'B' = AB(1 + \varepsilon_y)$$

$$A'E' = AE(1 + \varepsilon_n)$$

$$E'B' = EB(1 + \varepsilon_{n+90°})$$

$\varepsilon_{n+90°}$ is the normal strain in a direction at 90° to $\varepsilon_n$. Now

$$(A'B')^2 = (A'E')^2 + (E'B')^2 - 2A'E'.E'B'\cos(90° + \gamma_s) \quad [11.29]$$

or

$$(AB)^2(1 + \varepsilon_y)^2 = (AE)^2(1 + \varepsilon_n)^2 + (EB)^2(1 + \varepsilon_{n+90°})^2$$
$$-2AE(1 + \varepsilon_n)EB(1 + \varepsilon_{n+90°})\cos(90° + \gamma_s)$$

But $\cos(90° + \gamma_s) = -\sin\gamma_s \approx -\gamma_s$, and, neglecting the second order of small quantities,

$$(AB)^2(1 + 2\varepsilon_y) = (AE)^2(1 + 2\varepsilon_n) + (EB)^2(1 + 2\varepsilon_{n+90°})$$
$$+2AE.EB.\gamma_s$$

But $(AB)^2 = (AE)^2 + (EB)^2$, and dividing by $2(AB)^2$ gives

$$\varepsilon_y = \varepsilon_n \sin^2\theta + \varepsilon_{n+90°}\cos^2\theta + \gamma_s \sin\theta \cos\theta$$

Rewriting in terms of $2\theta$,

$$-\frac{1}{2}\gamma_s \sin 2\theta = \frac{\varepsilon_{n+90°} + \varepsilon_n}{2} + \frac{\varepsilon_{n+90°} - \varepsilon_n}{2}\cos 2\theta - \varepsilon_y \quad [11.30]$$

Now,

$$\varepsilon_{n+90°} = \frac{\varepsilon_x + \varepsilon_y}{2} + \frac{\varepsilon_x - \varepsilon_y}{2}\cos 2(\theta + 90°) + \frac{1}{2}\gamma_{xy}\sin 2(\theta + 90°)$$

$$= \frac{\varepsilon_x + \varepsilon_y}{2} - \frac{\varepsilon_x - \varepsilon_y}{2}\cos 2\theta - \frac{1}{2}\gamma_{xy}\sin 2\theta$$

and

$$\varepsilon_n = \frac{\varepsilon_x + \varepsilon_y}{2} + \frac{\varepsilon_x + \varepsilon_y}{2}\cos 2\theta + \frac{1}{2}\gamma_{xy}\sin 2\theta$$

Therefore

$$\frac{\varepsilon_{n+90°} + \varepsilon_n}{2} = \frac{\varepsilon_x + \varepsilon_y}{2}$$

and

$$\frac{\varepsilon_{n+90°} - \varepsilon_n}{2} = -\frac{\varepsilon_x - \varepsilon_y}{2}\cos 2\theta - \frac{1}{2}\gamma_{xy}\sin 2\theta$$

Substituting the above expressions in eqn. [11.30],

$$-\frac{1}{2}\gamma_s \sin 2\theta = \frac{\varepsilon_x + \varepsilon_y}{2} - \frac{\varepsilon_x - \varepsilon_y}{2}\cos^2 2\theta - \frac{1}{2}\gamma_{xy}\sin 2\theta\cos 2\theta - \varepsilon_y$$

$$= \frac{\varepsilon_x - \varepsilon_y}{2}(1 - \cos^2 2\theta) - \frac{1}{2}\gamma_{xy}\sin 2\theta\cos 2\theta$$

Dividing through by $-\sin 2\theta$,

$$\tfrac{1}{2}\gamma_s = -\left(\frac{\varepsilon_x - \varepsilon_y}{2}\right)\sin 2\theta + \tfrac{1}{2}\gamma_{xy}\cos 2\theta \qquad [11.31]$$

### 11.14 Mohr's strain circle

The graphical method of obtaining normal and shear stress on any plane by Mohr's circle can also be employed to determine normal and shear strains at a point. Considering eqns. [11.28] and [11.31] and rewriting, we have

$$\varepsilon_n - \frac{\varepsilon_x + \varepsilon_y}{2} = \frac{\varepsilon_x - \varepsilon_y}{2}\cos 2\theta + \tfrac{1}{2}\gamma_{xy}\sin 2\theta$$

and

$$\tfrac{1}{2}\gamma_s = -\frac{\varepsilon_x - \varepsilon_y}{2}\sin 2\theta + \tfrac{1}{2}\gamma_{xy}\cos 2\theta$$

Squaring each and adding produces the expression

$$\left(\varepsilon_n - \frac{\varepsilon_x + \varepsilon_y}{2}\right)^2 + (\tfrac{1}{2}\gamma_s)^2 = \left(\frac{\varepsilon_x - \varepsilon_y}{2}\right)^2 + (\tfrac{1}{2}\gamma_{xy})^2 \qquad [11.32]$$

which is the equation of a circle of radius $\tfrac{1}{2}\sqrt{[(\varepsilon_x - \varepsilon_y)^2 + \gamma_{xy}^2]}$, and with centre at $[\tfrac{1}{2}(\varepsilon_x + \varepsilon_y), 0]$ relating $\varepsilon_n$ and $\gamma_s$.

The Mohr's circle as shown in Fig. 11.20 is constructed in the same manner as for stresses. The correct position for plotting shear strain on the circle, i.e. above or below the $\varepsilon$-axis, may be found either by relating the deformation of the element to the corresponding shear-stress system, or by the convention that a positive shear strain in the element corresponds to the sides of the deformed element having positive slope in relation to the co-ordinate axes. On co-ordinate axes of normal strain $\varepsilon$ and semi-shear strain $\tfrac{1}{2}\gamma$, *each to the same scale*, the points $(\varepsilon_x, \tfrac{1}{2}\gamma_{xy})$ and $(\varepsilon_y, \tfrac{1}{2}\gamma_{yx})$ are set up, and a circle is drawn with the line joining these two points as diameter. The normal and semi-shear strain $\varepsilon_n$ and $\tfrac{1}{2}\gamma_s$ in a direction at $\theta$ to the $x$-direction are obtained from the intersection of a radius with the circle, at $2\theta$ (anticlockwise) from AB.

Fig. 11.20

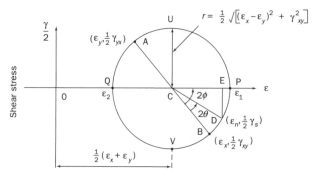

### 11.15 Principal strain and maximum shear strain

The maximum and minimum values of the normal strain at a point are given by P and Q in Fig. 11.20, whence

$$\varepsilon_1 = \frac{\varepsilon_x + \varepsilon_y}{2} + \frac{1}{2}\sqrt{[(\varepsilon_x - \varepsilon_y)^2 + \gamma_{xy}^2]} \quad \text{(maximum)} \qquad [11.33]$$

and

$$\varepsilon_2 = \frac{\varepsilon_x + \varepsilon_y}{2} - \frac{1}{2}\sqrt{[(\varepsilon_x - \varepsilon_y)^2 + \gamma_{xy}^2]} \quad \text{(minimum)} \qquad [11.34]$$

These are termed the *principal strains* and may be compared for similarity with the expressions for principal stresses. The former occur on mutually perpendicular planes making angles $2\theta = 2\phi$ and $180° + 2\phi$, or $\theta = \phi$ and $90° + \phi$ with the $x$-direction. From the diagram, when $2\theta = 2\phi$,

$$\theta = \frac{1}{2}\tan^{-1}\left(\frac{\gamma_{xy}}{\varepsilon_x - \varepsilon_y}\right) \qquad \text{and} \qquad 90° + \frac{1}{2}\tan^{-1}\left(\frac{\gamma_{xy}}{\varepsilon_x - \varepsilon_y}\right)$$

$$[11.35]$$

Either by substituting for $\theta$ in eqn. [11.31] or by reference to the circle, it is found that the shear strain is zero for the planes of principal strain.

Since $\tau_{xy} = G\gamma_{xy}$ when the shear strain is zero, so also must be the shear stress. But it was previously shown that the shear stress was zero on the principal stress planes, and therefore the planes of principal stress and principal strain must coincide.

The maximum shear strain occurs at the top and bottom points of the circle when $\varepsilon_n = \frac{1}{2}(\varepsilon_x + \varepsilon_y)$ and $\frac{1}{2}\gamma_s = \frac{1}{2}\sqrt{[(\varepsilon_x - \varepsilon_y)^2 + \gamma_{xy}^2]}$; therefore

$$\gamma_{max} = \sqrt{[(\varepsilon_x - \varepsilon_y)^2 + \gamma_{xy}^2]} \qquad [11.36]$$

Alternatively, this may be written in terms of the principal strains as

$$\frac{1}{2}\gamma_{max} = \frac{\varepsilon_1 - \varepsilon_2}{2} \qquad \text{or} \qquad \gamma_{max} = \varepsilon_1 - \varepsilon_2$$

## 11.16 Experimental stress analysis

In many practical situations, the geometry of a structure or part of it may be complex and yet there is a need to know accurately the levels of stress in the material. This may necessitate measurements rather than calculations but unfortunately it is not possible to measure force or stress directly. All that one can do is measure the displacements or deformations which arise as a result of the force or stress. The strain can then be related to the stress from a knowledge of the modulus of the material.

One of the most widely used methods of experimental stress analysis is based on the strain gauge. The principle of the electrical resistance strain gauge was discussed by Lord Kelvin when he observed that the electrical resistance of a wire changes when it is stretched.

Modern strain gauges are of metal foil construction as illustrated in Fig. 11.21. The gauge is bonded onto the surface where the strain is to be measured. It is then connected to an electrical circuit to monitor its resistance. When the material surface is deformed, the gauge also experiences the same strain and this can be quantified from the measurement of the change in its resistance and a knowledge of the calibration constant (gauge factor) for the strain gauge.

It will be seen from Fig. 11.21 that foil gauges come in a variety of patterns to measure strains in different directions. It should be noted that it is only possible to measure direct strains, not shear strains. The following example illustrates how the results from strain gauges may be analysed.

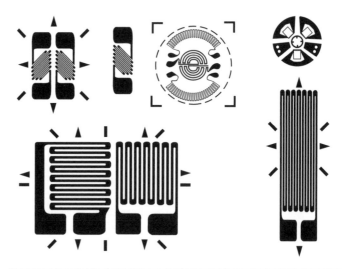

**Fig. 11.21** Typical foil strain gauges (published by courtesy of Measurement Group, Vishay)

**Example 11.6**

Two electrical resistance strain gauges are sited at $\pm 45°$ to the axis of a 75 mm diameter shaft. The shaft is rotating, and in addition to transmitting power, it is subjected to an unknown bending moment and a direct thrust. The readings of the gauges are recorded, and it is found that the maximum or minimum values for each gauge occur at 180° intervals of shaft rotation and are $-0.0006$ and $+0.0003$ for the two gauges at one instant and $-0.0005$ and $+0.0004$ for the same gauges 180° of rotation later. Determine the transmitted torque, the applied bending moments and the end thrust. Assume all the forces and moments are steady, i.e. do not vary during each rotation of the shaft.

$E = 208\,\text{GN/m}^2$, $\nu = 0.29$, $G = 80\,\text{GN/m}^2$. The shaft and strain gauges are illustrated diagrammatically in Fig. 11.22.

**Fig. 11.22**

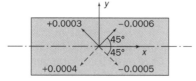

The simplest starting point is to find the torque. This may be found from the shear stress which is in turn related to shear strain. From eqn. [11.31] we have

$$\gamma_{xy} = -(\varepsilon_p - \varepsilon_q)\sin 2\theta + \gamma_{pq}\cos 2\theta$$

$$= -(0.0003 + 0.0006)\sin 90° + \gamma_{pq}\cos 90°$$

$$= -0.0009$$

$$\tau_{xy} = -0.0009 \times 80 = -72\,\text{MN/m}^2$$

Hence the torque is

$$T = \frac{\pi}{16} \times 0.075^3 \times 72 \times 10^6 = 5.97\,\text{kNm}$$

In order to find the bending moment and end thrust it is necessary to know the strains in the $x$-direction for the two specific rotational positions of the shaft. From eqn. [11.28] we have

$$\text{Top} \quad -0.0006 = \frac{\varepsilon_x^T + \varepsilon_y^T}{2} - 0.000\,45$$

and

$$+0.0003 = \frac{\varepsilon_x^T + \varepsilon_y^T}{2} + 0.000\,45$$

Hence

$$\varepsilon_x^T + \varepsilon_y^T = -0.0003$$

$$\text{Bottom} \quad -0.0005 = \frac{\varepsilon_x^B + \varepsilon_y^B}{2} - 0.000\,45$$

or

$$+0.0004 = \frac{\varepsilon_x^B + \varepsilon_y^B}{2} + 0.000\,45$$

Hence

$$\varepsilon_x^B + \varepsilon_y^B = -0.0001$$

To eliminate $\varepsilon_y$ from the above equations we use the relationship $\varepsilon_y = -\nu\varepsilon_x$:

$$\varepsilon_x^T = -\frac{0.0003}{1-\nu} = -0.000\,423$$

$$\varepsilon_x^B = -\frac{0.0001}{1-\nu} = -0.000\,141$$

These strains are the sum of the bending strain which reverses in sign for 180° rotation and the steady compressive strain (due to end thrust).

$$\text{Bending strain} = \pm\tfrac{1}{2}(\varepsilon_x^T - \varepsilon_x^B) = \pm0.000\,141$$

$$\text{Compressive strain} = -\tfrac{1}{2}(\varepsilon_x^T + \varepsilon_x^B) = -0.000\,282$$

Hence

$$\sigma_{xb} = \pm0.000\,141 \times 208 \times 10^9 = \pm29.3\,\text{MN/m}^2$$

from which

$$M = \frac{\pi}{32} \times 0.075^3 \times 29.3 \times 10^6 = 1.21 \, \text{kNm}$$

$$\sigma_{xc} = -0.000\,282 \times 208 \times 10^9 = -58.6 \, \text{MN/m}^2$$

and so the end thrust $F$ is given by

$$F = \frac{\pi}{4} \times 0.075^2 \times (-58.6) \times 10^6 = -259 \, \text{kN}$$

**11.17 Rosette strain computation and circle construction**

For the complete determination of strain at a point on the surface of a component, it is necessary to measure the strain in three directions at the point. This is achieved by cementing an electrical resistance rosette strain gauge to the surface.

**Fig. 11.23**

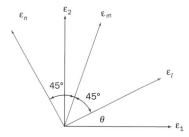

Let the three measured strains be $\varepsilon_l$, $\varepsilon_m$ and $\varepsilon_n$ and the angle between the directions $l$ and $m$, and $m$ and $n$, be 45° in each case. Then this arrangement is known as a 45° rosette as shown in Fig. 11.23. If the angle between $\varepsilon_l$ and the principal strain $\varepsilon_1$ is $\theta$, then from eqn. [11.27], the principal strains are related to the measured strains as follows:

$$\left. \begin{array}{l} \varepsilon_l = \varepsilon_1 \cos^2 \theta + \varepsilon_2 \sin^2 \theta \\ \varepsilon_m = \varepsilon_1 \cos^2(\theta + 45°) + \varepsilon_2 \sin^2(\theta + 45°) \\ \varepsilon_n = \varepsilon_1 \cos^2(\theta + 90°) + \varepsilon_2 \sin^2(\theta + 90°) \end{array} \right\} \qquad [11.37]$$

These equations may be rewritten as

$$\left. \begin{array}{l} \varepsilon_l = \tfrac{1}{2}(\varepsilon_1 + \varepsilon_2) + \tfrac{1}{2}(\varepsilon_1 - \varepsilon_2) \cos 2\theta \\ \varepsilon_m = \tfrac{1}{2}(\varepsilon_1 + \varepsilon_2) - \tfrac{1}{2}(\varepsilon_1 - \varepsilon_2) \sin 2\theta \\ \varepsilon_n = \tfrac{1}{2}(\varepsilon_1 + \varepsilon_2) - \tfrac{1}{2}(\varepsilon_1 - \varepsilon_2) \cos 2\theta \end{array} \right\} \qquad [11.38]$$

Solving the above equations simultaneously gives

$$\varepsilon_1 = \tfrac{1}{2}(\varepsilon_l + \varepsilon_n) + \frac{\sqrt{2}}{2} \sqrt{[(\varepsilon_l - \varepsilon_m)^2 + (\varepsilon_m - \varepsilon_n)^2]}$$

$$\varepsilon_2 = \tfrac{1}{2}(\varepsilon_l + \varepsilon_n) + \frac{\sqrt{2}}{2} \sqrt{[(\varepsilon_l - \varepsilon_m)^2 + (\varepsilon_m - \varepsilon_n)^2]} \qquad [11.39]$$

$$\tan 2\theta = \frac{2\varepsilon_m - \varepsilon_l - \varepsilon_n}{\varepsilon_l - \varepsilon_n} \qquad [11.40]$$

To obtain principal stresses from principal strains we use the stress–strain relationships,

$$\varepsilon_1 = \frac{\sigma_1}{E} - \frac{\nu\sigma_2}{E}$$

$$\varepsilon_2 = \frac{\sigma_2}{E} - \frac{\nu\sigma_1}{E} \qquad [11.41]$$

from which

$$\sigma_1 = \frac{E}{1 - \nu^2}(\varepsilon_1 + \nu\varepsilon_2)$$

$$\sigma_2 = \frac{E}{1 - \nu^2}(\varepsilon_2 + \nu\varepsilon_1) \qquad [11.42]$$

An alternative approach to the analysis of the strain rosette is to use the construction described below based on Mohr's circle.

Fig. 11.24

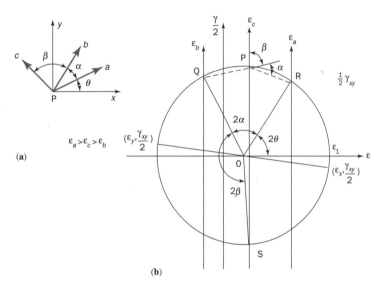

(a)

(b)

Consider the general case of three strain gauges, $a$, $b$ and $c$, having arbitrary orientation as shown in Fig. 11.24($a$) and strain readings $\varepsilon_a$, $\varepsilon_b$ and $\varepsilon_c$ in the given directions. As an example to illustrate the method assume $\varepsilon_a > \varepsilon_c > \varepsilon_b$. The procedure is as follows:

1. Set up a vertical axis to represent $\varepsilon = 0$ (which will subsequently be the semi-shear-strain axis).
2. Draw three lines parallel to the above axis at the appropriate distances representing the values (positive or negative) of $\varepsilon_a$, $\varepsilon_b$ and $\varepsilon_c$.
3. On the middle line of these three (representing the middle value of the three strains $\varepsilon_c$) mark a point P representing the origin of the rosette.
4. Draw the rosette configuration at the point P but *lining up gauge c* (in this particular example) *along its vertical ordinate*.
5. Project the directions of gauges $a$ and $b$ to cut their respective vertical ordinates at Q and R.

6. Construct perpendicular bisectors of PQ and PR; where these intersect is the centre of the strain circle, O.
7. Draw the circle on this centre, which of course should pass through the points P, Q and R. Insert the horizontal strain abscissa through O.
8. Join O to Q, R and S, where S is the other intersection of the circle with the middle vertical line.
9. The lines OQ, OR and OS represent the three gauges *on the circle* where $2\alpha$ and $2\beta$ are the angles between OR and OQ, and OQ and OS, respectively.
10. From the circle read off as required the principal strains $\varepsilon_1$, $\varepsilon_2$ or the chosen co-ordinate direction strains $\varepsilon_x$, $\varepsilon_y$, $\gamma_{xy}$.

**Example 11.7**

At a point on the surface of a component, a 60° rosette strain gauge positioned as shown in Fig. 11.25(a) measures strains of $\varepsilon_l = 0.000\,46$, $\varepsilon_m = 0.0002$ and $\varepsilon_n = -0.000\,16$. Use Mohr's strain circle to determine the magnitude and direction of the principal strains and hence the principal stresses. $E = 208\,\text{GN/m}^2$, $\nu = 0.29$.

**Fig. 11.25**

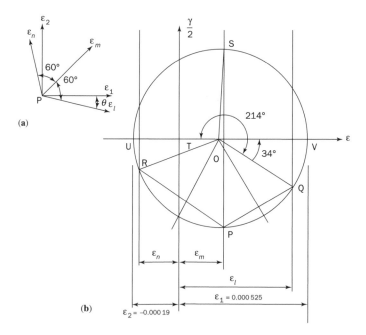

(a)

(b)

To construct the strain circle, Fig. 11.25(*b*), we use the procedure described above. The principal strain values are represented by TV and TU, and therefore

$$\varepsilon_1 = 0.000\,525 \qquad \text{and} \qquad \varepsilon_2 = -0.000\,19$$

The angle between $\varepsilon_l$ and $\varepsilon_1$ on the circle is 34° and between $\varepsilon_l$ and $\varepsilon_2$, 214°. Therefore

$$\theta_1 = 17° \qquad \text{and} \qquad \theta_2 = 107°$$

The principal stresses are given by eqns. [11.42]; thus

$$\sigma_1 = \frac{208 \times 10^9}{1 - 0.29^2} [0.000\,525 + 0.29 \times (-0.000\,19)] = 107\,\text{MN/m}^2$$

$$\sigma_2 = \frac{208 \times 10^9}{1 - 0.29^2} (-0.000\,19 + 0.29 \times 0.000\,525) = -9\,\text{MN/m}^2$$

**11.18 Spreadsheet solution for strain gauge rosette**

A general method for the analysis of a three-gauge rosette with the elements at arbitrary orientations $\theta_1$, $\theta_2$ and $\theta_3$ will now be demonstrated. Using eqn. [11.27], the strain in a rosette element at an angle $\theta$ is

$$\varepsilon_n = \varepsilon_x \cos^2 \theta + \varepsilon_y \sin^2 \theta + \gamma_{xy} \sin \theta \cos \theta$$

Given the strains $\varepsilon_a$, $\varepsilon_b$ and $\varepsilon_c$ in each element of the rosette lying at angles $\theta_1$, $\theta_2$ and $\theta_3$ respectively, three linear equations are obtained for the three unknowns $\varepsilon_x$, $\varepsilon_y$ and $\gamma_{xy}$ as

$$\begin{bmatrix} \cos^2\theta_1 & \sin^2\theta_1 & \sin\theta_1\cos\theta_1 \\ \cos^2\theta_2 & \sin^2\theta_2 & \sin\theta_2\cos\theta_2 \\ \cos^2\theta_3 & \sin^2\theta_3 & \sin\theta_3\cos\theta_3 \end{bmatrix} \begin{Bmatrix} \varepsilon_x \\ \varepsilon_y \\ \gamma_{xy} \end{Bmatrix} = \begin{Bmatrix} \varepsilon_a \\ \varepsilon_b \\ \varepsilon_c \end{Bmatrix} \qquad [11.43]$$

Here the three equations are written, as described in Appendix B, as a single matrix equation,

$$[A]\{x\} = \{y\} \qquad [11.44]$$

The solution can be found using the inverse matrix $[A]^{-1}$ as

$$\begin{Bmatrix} \varepsilon_x \\ \varepsilon_y \\ \gamma_{xy} \end{Bmatrix} = \begin{bmatrix} \cos^2\theta_1 & \sin^2\theta_1 & \sin\theta_1\cos\theta_1 \\ \cos^2\theta_2 & \sin^2\theta_2 & \sin\theta_2\cos\theta_2 \\ \cos^2\theta_3 & \sin^2\theta_3 & \sin\theta_3\cos\theta_3 \end{bmatrix}^{-1} \begin{Bmatrix} \varepsilon_a \\ \varepsilon_b \\ \varepsilon_c \end{Bmatrix} \qquad [11.45]$$

The inverse matrix can be found using the 'macro' or menu commands beginning in cell D16 of Fig. 11.26. In some spreadsheets such as Microsoft Excel, the inverse matrix can be inserted as a special array formula and the macro is not necessary. Once this matrix is evaluated for a given set of angles, then as soon as the element strains are entered in the highlighted cells H7..H9, the co-ordinate strains are calculated and appear in cells H11..H13. The principal strains, principal angles and maximum shear strain are then evaluated using the basic formulae of eqns. [11.33]–[11.36] and displayed in cells H16..H20.

Once the principal strains have been obtained the corresponding principal stresses can be found using the stress–strain relations. Since a strain gauge rosette will almost always lie on a free surface which has no normal or shear stress acting on it, a state of plane stress exists. For a state of plane stress, stresses can be calculated from the strains using

$$\begin{Bmatrix} \sigma_x \\ \sigma_y \end{Bmatrix} = \frac{E}{1 - \nu^2} \begin{bmatrix} 1 & \nu \\ \nu & 1 \end{bmatrix} \begin{Bmatrix} \varepsilon_x \\ \varepsilon_y \end{Bmatrix} \qquad [11.46]$$

**Fig. 11.26**

| | A | B | C | D | E | F | G | H |
|---|---|---|---|---|---|---|---|---|
| 1 | Rosette | calculations | | | | | | |
| 2 | | Enter rosette element angles and invert matrix D7 F9 | | | | | | |
| 3 | | Enter rosette element strains and get, eps_x, eps_y and gamma_xy. | | | | | | |
| 4 | | | | Principal Strains, gamma_max and Principal Angle | | | | |
| 5 | | | | | | | | |
| 6 | Element | theta (deg) | Theta (rad) | Matrix to invert | | | | Strain |
| 7 | a | 0 | @RADIANS(B7) | @COS(C7)^2 | @SIN(C7)^2 | @SIN(C7)*@COS(C7) | eps_a | 0.00046 |
| 8 | b | 60 | @RADIANS(B8) | @COS(C8)^2 | @SIN(C8)^2 | @SIN(C8)*@COS(C8) | eps_b | 0.0002 |
| 9 | c | 120 | @RADIANS(B9) | @COS(C9)^2 | @SIN(C9)^2 | @SIN(C9)*@COS(C9) | eps_c | −0.00016 |
| 10 | | | | Inverted matrix | | | | |
| 11 | | | | 1 | 0 | 0 | eps_x | +D11*H7+E11*H8+F11*H9 |
| 12 | | | | −0.33333333333 | 0.666666666666667 | 0.666666666666667 | eps_y | +D12*H7+E12*H8+F12*H9 |
| 13 | | | | 4.6983901144659E−16 | 1.1547005383793 | −1.1547005383793 | gamma_xy | +D13*H7+E13*H8+F13*H9 |
| 14 | | | | | | | | |
| 15 | | | | | | | | |
| 16 | | | | Macro to invert matrtix | | | eps_1 | 0.5*(H11+H12)+0.5*H20 |
| 17 | | | | {Invert Source A:D7..F9} | | | eps_2 | +H16−H20 |
| 18 | | | | {Invert Destination A:D11..f13} | | | theta_p | @DEGREES@(ASIN(H13/H20))/2 |
| 19 | | | | {Invert Go} | | | | +H18+90 |
| 20 | | | | | | | gamma_m | @SQRT(H11-H12)^2+H13^2) |

(**a**) Data and cell formulae

| | A | B | C | D | E | F | G | H |
|---|---|---|---|---|---|---|---|---|
| 1 | Rosette | calculations | | | | | | |
| 2 | | Enter rosette element angles and invert matrix D7 F9 | | | | | | |
| 3 | | Enter rosette element strains and get, eps_x, eps_y and gamma_xy. | | | | | | |
| 4 | | | | Principal Strains, gamma_max and Principal Angle | | | | |
| 5 | | | | | | | | |
| 6 | Element | theta (deg) | Theta (rad) | Matrix to invert | | | | Strain |
| 7 | a | 0.000 | 0.000 | 1.000 | 0.000 | 0.000 | eps_a | 0.000460 |
| 8 | b | 60.000 | 1.047 | 0.250 | 0.750 | 0.433 | eps_b | 0.000200 |
| 9 | c | 120.000 | 2.094 | 0.250 | 0.750 | −0.4.33 | eps_c | −0.000160 |
| 10 | | | | Inverted matrix | | | | |
| 11 | | | | 1.000 | 0.000 | 0.000 | eps_x | 0.000460 |
| 12 | | | | −0.333 | 0.667 | 0.667 | eps_y | −0.000127 |
| 13 | | | | 0.000 | 1.155 | −0.1.155 | gamma_xy | 0.000416 |
| 14 | | | | | | | | |
| 15 | | | | | | | | |
| 16 | | | | Macro to invert matrtix | | | eps_1 | 0.000526 |
| 17 | | | | {Invert Source A:D7..F9} | | | eps_2 | −0.000193 |
| 18 | | | | {Invert Destination A:D11..f13} | | | theta_p | 17.660 |
| 19 | | | | {Invert Go} | | | | 107.660 |
| 20 | | | | | | | gamma_m | 0.000719 |

(**b**) Spreadsheet Display

and

$$\tau_{xy} = \frac{E}{2(1+\nu)}\gamma_{xy}$$

[11.47]

Equation [11.47] is the same as eqn. [11.49], which will be derived below. A general purpose spreadsheet for calculating stresses given strains, or strains given stresses, under plane stress is shown in Fig. 11.27. The material

**Fig. 11.27**

|   | A | B | C | D | E | F | G | H |
|---|---|---|---|---|---|---|---|---|
| 1 | Stress strain relations (Plane Stress) | | | | | | | |
| 2 | | | | | | | | |
| 3 | Material Properties | | | | | | | |
| 4 | E | 2.08E+11 | | | | | | |
| 5 | nu | 0.29 | | | | | | |
| 6 | | | | | | | | |
| 7 | | Enter strains, get stresses | | | | | | |
| 8 | | | | Modulus matrix | | | Strains | Stresses |
| 9 | | | | +B4/(1-B5^2) | +D9*B5 | | 0.000525 | +D9*G9+E9*G10 |
| 10 | | | | +E9 | +D9 | | −0.00019 | +D10*G9+E10*G10 |
| 11 | | | | | | +B4/(2*(1+B5)) | 0 | +F11*G11 |
| 12 | | | | | | | | |
| 13 | | Enter stresses, get strains | | | | | | |
| 14 | | | | Compliance matrix | | | Stresses | Strains |
| 15 | | | | 1/B4 | −B5/B4 | | 107000000 | +D15*G15+E15*G16 |
| 16 | | | | −B5/B4 | 1/B4 | | −9000000 | +D16*G15+E16*G16 |
| 17 | | | | | | 2*(1+B5)/B4 | 0 | +F17*G17 |

(**a**) Data and cell formulae

|   | A | B | C | D | E | F | G | H |
|---|---|---|---|---|---|---|---|---|
| 1 | Stress strain relations (Plane Stress) | | | | | | | |
| 2 | | | | | | | | |
| 3 | Material Properties | | | | | | | |
| 4 | E | 2.08E+11 | | | | | | |
| 5 | nu | 0.29 | | | | | | |
| 6 | | | | | | | | |
| 7 | | Enter strains, get stresses | | | | | | |
| 8 | | | | Modulus matrix | | | Strains | Stresses |
| 9 | | | | 2.27E+11 | 6.59E+10 | | 5.25E−04 | 1.07E−08 |
| 10 | | | | 6.59E+10 | 2.27E+11 | | −1.90E−04 | −9E+06 |
| 11 | | | | | | 8.06E+10 | 0.00E+00 | 0.00E+00 |
| 12 | | | | | | | | |
| 13 | | Enter stresses, get strains | | | | | | |
| 14 | | | | Compliance matrix | | | Stresses | Strains |
| 15 | | | | 4.81E−12 | −1.39E−12 | | 1.07E+08 | 5.27E−04 |
| 16 | | | | −1.39E−12 | 4.81E−12 | | −9.00E+06 | −1.92E−04 |
| 17 | | | | | | 1.24E−11 | 0.00E+00 | 0.00E+00 |

(**b**) Spreadsheet Display

properties $E$ and $\nu$ are input in cells B4..B5, strains can be input in G9..G11 or stresses can be input in G15..G17.

## 11.19 Relationships between the elastic constants

In Chapter 3 four constants of elasticity were defined relating various conditions of stress and strain. These are Young's modulus, $E$, shear modulus, $G$, bulk modulus, $K$, and Poisson's ratio, $\nu$. It will now be shown that these constants are not independent of one another.

### Relationship between $K$, $E$ and $\nu$

It was shown in eqn. [3.1] that volumetric strain is given by the sum of the three linear strains along the axes of the element. Hence

$$e = \varepsilon_x + \varepsilon_y + \varepsilon_z$$

But from eqns [3.2]

$$\varepsilon_x = \frac{\sigma_x}{E} - \frac{\nu}{E}(\sigma_y + \sigma_z)$$

Also for hydrostatic stress, $\sigma_x = \sigma_y = \sigma_z = \sigma$; therefore

$$\varepsilon_x = \frac{\sigma}{E}(1 - 2\nu)$$

Similarly

$$\varepsilon_y = \frac{\sigma}{E}(1 - 2\nu) \qquad \text{and} \qquad \varepsilon_z = \frac{\sigma}{E}(1 - 2\nu)$$

Therefore, summing the above three strains, we have

$$e = \frac{3\sigma}{E}(1 - 2\nu)$$

or

$$E = 3\frac{\sigma}{e}(1 - 2\nu)$$

But $\sigma/e = K$, the bulk modulus; therefore

$$E = 3K(1 - 2\nu) \hspace{3cm} [11.48]$$

## Relationship between $E$, $G$ and $\nu$

The square element of unit thickness shown in Fig. 11.28(a) is acted on by pure shearing stresses. This system is equivalent to the system of direct stresses on the element of Fig. 11.28(b), and from equilibrium $\sigma_{n_1} = \sigma_{n_2} = \tau_s$. The strain along the diagonal AB in terms of the stresses is given by

$$\varepsilon_{n_1} = \frac{\sigma_{n_1}}{E} - \frac{\nu}{E}(-\sigma_{n_2})$$

i.e. the extension due to $\sigma_{n_1}$ plus the lateral expansion in the direction AB due to the compression $\sigma_{n_2}$. But

$$\sigma_{n_1} = \sigma_{n_2} = \tau_s$$

Fig. 11.28

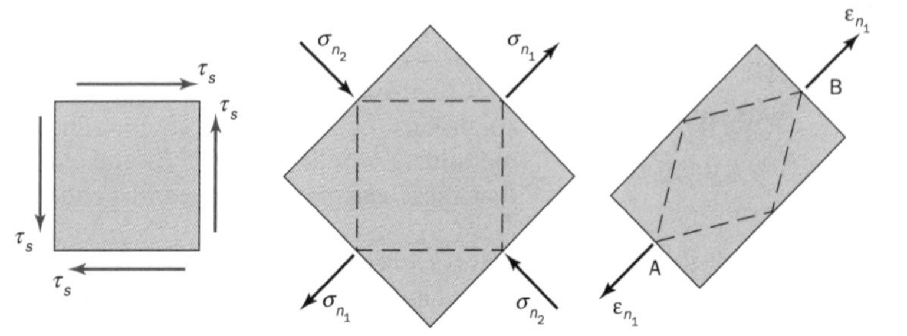

(a) Pure shear stress          (b) Equivalent direct stress          (c) Strain systems

Therefore

$$\varepsilon_{n_1} = \frac{\tau_s}{E}(1 + \nu)$$

Since for pure shear $\varepsilon_x$ and $\varepsilon_y$ are zero, then from eqn. [11.28],

$$\varepsilon_{n_1} = \tfrac{1}{2}\gamma_s \sin 2\theta$$

In this case $\theta = 45°$; therefore

$$\varepsilon_{n_1} = \tfrac{1}{2}\gamma_s$$

Equating the above expressions,

$$\tfrac{1}{2}\gamma_s = \frac{\tau_s}{E}(1 + \nu)$$

or

$$E = \frac{2\tau_s}{\gamma_s}(1 + \nu)$$

But $\tau_s/\gamma_s = G$, the shear modulus; therefore

$$E = 2G(1 + \nu) \qquad [11.49]$$

**Relationship between $K$, $G$ and $\nu$**

It follows from eqns. [11.48] and [11.49] that

$$3K(1 - 2\nu) = 2G(1 + \nu)$$

or

$$K = \frac{2G(1 + \nu)}{3(1 - 2\nu)} \qquad [11.50]$$

Thus if any two of the four constants are known, or can be measured, then the other two can be determined.

**11.20 Stress and strain transformations in composites**

In recent years there has been a rapid growth in the use of fibre-reinforced composites. The major advantage of such materials is that high strength and stiffness can be achieved at low weight. Products that have benefitted from the use of composites include aircraft, ships, automobiles, chemical vessels and sporting goods. In these industries the base material is usually metal or plastic and the fibres used include glass, carbon, aramid ('Kevlar'), boron and asbestos. In some cases short ('chopped') fibres are used and this provides a significant property enhancement over the base resin. However, by far the greatest improvement in properties is observed if the fibres are continuous. For example, if unidirectional carbon fibres are added to an epoxy resin, the modulus of the resulting composite is improved by a factor of about 60 (see Table 11.1) and the strength by a factor of about 30 compared with the unreinforced base resin. However, the composite is markedly anisotropic in that in the direction perpendicular to the fibre axis the modulus is only improved by a factor of about 2 and the strength is

**Table 11.1**  Typical properties of some materials

| Material | $E_x^*$ (GN/m²) | $E_y$† (GN/m²) | $G_{xy}$ (GN/m²) | $\nu_x$ | Density (kg/m³) |
|---|---|---|---|---|---|
| Carbon fibre/epoxy | 180 | 10 | 7.2 | 0.28 | 1600 |
| Kevlar/epoxy | 76 | 5.5 | 2.3 | 0.34 | 1460 |
| Glass/epoxy | 39 | 8.4 | 4.2 | 0.26 | 1800 |
| Spruce | 8.9 | | 2.5 | | 400 |
| Aluminium alloy | 70 | 70 | 27 | | 2770 |

*Parallel to fibres or grain.
†Perpendicular to fibres.

**Fig. 11.29**   Arrangement of fibre orientation in laminate

likely to be reduced. Therefore, in the aircraft industry, for example, in order to get property enhancement in all the required directions within the component, it is normal practice to build up a laminate structure where each layer has fibres arranged in the desired direction, as shown in Fig. 11.29.

As it is becoming increasingly likely that engineers and designers will at some stage have to become involved in the design of components made from fibre composites it is important that they should have an appreciation of the stages in the design process. In the following sections a brief introduction is given to the laminate theory involved in fibre-reinforced composites.

## 11.21 Analysis of a lamina

In this analysis it is necessary to consider both the local co-ordinates $(x, y)$ for the lamina and the global co-ordinates $(X, Y)$ for the applied stress system.

### On-axis properties

Consider first of all a single lamina in which the fibres are all aligned in the global $X$-direction as shown in Fig. 11.30. The lamina is thin in relation to its transverse dimensions and therefore it will be in a state of plane stress when forces are applied to it.

**Fig. 11.30**

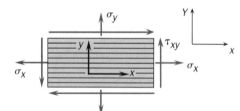

Recognizing that the lamina is anisotropic with modulus values of $E_x$ and $E_y$ in its $x$- and $y$-directions and corresponding Poisson's ratios $\nu_x$ and $\nu_y$, then the strains in the lamina may be written as

$$\varepsilon_X = \frac{\sigma_X}{E_x} - \frac{\nu_y \sigma_Y}{E_y}$$

$$\varepsilon_Y = \frac{\sigma_Y}{E_y} - \frac{\nu_x \sigma_X}{E_x} \qquad [11.51]$$

$$\gamma_{XY} = \frac{\tau_{XY}}{G_{xy}}$$

Note that it can be shown that $\nu_x/E_x = \nu_y/E_y$.

It is convenient for subsequent analysis to write these three equations in matrix form (a brief introduction to the use of matrices is given in Appendix B):

$$\left\{ \begin{array}{c} \varepsilon_X \\ \varepsilon_Y \\ \gamma_{XY} \end{array} \right\} = \left[ \begin{array}{ccc} 1/E_x & -\nu_y/E_y & 0 \\ -\nu_x/E_x & 1/E_y & 0 \\ 0 & 0 & 1/G_{xy} \end{array} \right] \left\{ \begin{array}{c} \sigma_X \\ \sigma_Y \\ \tau_{XY} \end{array} \right\} \qquad [11.52]$$

or in abbreviated form

$$\{\varepsilon\} = [S]\{\sigma\} \qquad [11.53]$$

where $[S]$ is referred to as the *compliance* matrix for the lamina.

At this point it is worth noting that in order to describe completely an anisotropic material subject to a triaxial stress system, the compliance matrix will have 36 terms.

$$\left\{ \begin{array}{c} \varepsilon_X \\ \varepsilon_Y \\ \varepsilon_Z \\ \gamma_{YZ} \\ \gamma_{ZX} \\ \gamma_{XY} \end{array} \right\} = \left[ \begin{array}{cccccc} S_{11} & S_{12} & S_{13} & S_{14} & S_{15} & S_{16} \\ S_{21} & S_{22} & S_{23} & S_{24} & S_{25} & S_{26} \\ S_{31} & S_{32} & S_{33} & S_{34} & S_{35} & S_{36} \\ S_{41} & S_{42} & S_{43} & S_{44} & S_{45} & S_{46} \\ S_{51} & S_{52} & S_{53} & S_{54} & S_{55} & S_{56} \\ S_{61} & S_{62} & S_{63} & S_{64} & S_{65} & S_{66} \end{array} \right] \left\{ \begin{array}{c} \sigma_X \\ \sigma_Y \\ \sigma_Z \\ \tau_{YZ} \\ \tau_{ZX} \\ \tau_{XY} \end{array} \right\}$$
$$[11.54]$$

In practice for isotropic materials the number of constants may be reduced if we assume the following:

1. Shear stresses do not affect normal strains and normal stresses do not affect shear strains. Hence

$$S_{14} = S_{15} = S_{16} = S_{24} = S_{25} = S_{26} = S_{34} = S_{35} = S_{36} = 0$$

2. Shear strains are only affected by shear stresses in the same plane. Hence

$$S_{45} = S_{46} = S_{54} = S_{56} = S_{65} = S_{64} = 0$$

3. The effect of $\sigma_X$ on $\varepsilon_X$ is the same as the effect of $\sigma_Y$ on $\varepsilon_Y$, etc. Hence

$$S_{11} = S_{22} = S_{33}$$

4. The effect of $\sigma_Y$ on $\varepsilon_X$ is the same as the effect of $\sigma_Z$ on $\varepsilon_X$, etc. Hence

$$S_{12} = S_{13} = S_{21} = S_{23} = S_{31} = S_{32}$$

5. The effect of $\tau_{XY}$ on $\gamma_{XY}$ is the same as the effect of $\tau_{YZ}$ on $\gamma_{YZ}$, etc. Hence

$$S_{44} = S_{55} = S_{66}$$

Hence for isotropic materials the matrix in eqn. [11.54] reduces to one in which there are only three constants $S_{11}$, $S_{12}$ and $S_{66}$, the values of which are

$$S_{11} = \frac{1}{E} \quad S_{12} = -\frac{\nu}{E} \quad S_{66} = \frac{1}{G}$$

It may be seen that eqn. [11.54] then reduces to eqns. [3.2] and [3.4]. When introducing the stress analysis of isotropic materials it is generally more convenient to use these simple equations, but it should be remembered that they are derived from a much more general situation.

However, in the analysis of laminates, which by their nature are anisotropic, the basis equations are a little more complex and it is generally found that their manipulation is simplified by the use of matrix algebra. For a lamina it may be shown (see Jones[1], for example) that eqn. [11.54] reduces to the form

$$[S] = \begin{bmatrix} S_{11} & S_{12} & S_{16} \\ S_{21} & S_{22} & S_{26} \\ S_{61} & S_{62} & S_{66} \end{bmatrix}$$

where, from eqn. [11.52]

$$S_{11} = \frac{1}{E_x} \quad S_{22} = \frac{1}{E_y} \quad S_{66} = \frac{1}{G_{xy}}$$

$$S_{12} = S_{21} = -\frac{\nu_x}{E_x} = -\frac{\nu_y}{E_y}$$

$$S_{16} = S_{61} = S_{26} = S_{62} = 0$$

If eqn. [11.51] is rearranged to give stresses in terms of strains, then

$$\sigma_X = (\varepsilon_X + \nu_y \varepsilon_Y) E_x / (1 - \nu_x \nu_y)$$

$$\sigma_Y = (\varepsilon_Y + \nu_y \varepsilon_X) E_y / (1 - \nu_x \nu_y) \qquad [11.55]$$

$$\tau_{XY} = G \gamma_{XY}$$

In matrix form this may be written as

$$\begin{Bmatrix} \sigma_X \\ \sigma_Y \\ \tau_{XY} \end{Bmatrix} = \begin{bmatrix} Q_{11} & Q_{12} & Q_{16} \\ Q_{21} & Q_{22} & Q_{26} \\ Q_{61} & Q_{62} & Q_{66} \end{bmatrix} \begin{Bmatrix} \varepsilon_X \\ \varepsilon_Y \\ \gamma_{XY} \end{Bmatrix} \qquad [11.56]$$

where $[Q]$ is called the *stiffness* matrix. It is symmetrical and the individual terms are

$$Q_{11} = E_x / (1 - \nu_y \nu_x)$$

$$Q_{22} = E_y / (1 - \nu_y \nu_x)$$

$$Q_{66} = G_{xy}$$

$$Q_{12} = Q_{21} = \nu_x E_y / (1 - \nu_x \nu_y) = \nu_y E_x / (1 - \nu_y \nu_x)$$

$$Q_{16} = Q_{61} = Q_{26} = Q_{62} = 0$$

Note that from eqns. [11.53] and [11.56] we may write

$$\{\sigma\} = [Q]\{\varepsilon\} = [S]^{-1}\{\varepsilon\}$$

Therefore it is apparent that the stiffness matrix is the inverse of the compliance matrix, so that

$$[Q] = [S]^{-1} \tag{11.57}$$

**Off-axis properties** Consider now a situation where the $x$–$y$ co-ordinates of the lamina do not coincide with the global $X$–$Y$ co-ordinates. This is shown in Fig. 11.31.

**Fig. 11.31** Fibres aligned at an angle to global X-direction

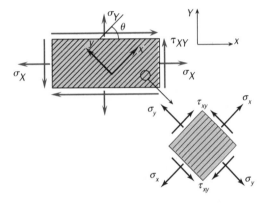

When stresses are applied to the lamina in the global ($X$–$Y$) co-ordinates these must be transformed to the local ($x$–$y$) axes for the lamina since it is in these directions that its properties are known. The transformation necessary has already been performed in Section 11.5. Therefore using eqns. [11.13] and [11.14] we may write

$$\sigma_x = \sigma_X \cos^2 \theta + \sigma_Y \sin^2 \theta + 2\tau_{XY} \sin \theta \cos \theta \tag{11.58}$$

$$\sigma_y = \sigma_X \sin^2 \theta + \sigma_Y \cos^2 \theta - 2\tau_{XY} \sin \theta \cos \theta \tag{11.59}$$

$$\tau_{xy} = -\sigma_X \sin \theta \cos \theta + \sigma_Y \sin \theta \cos \theta + \tau_{XY} (\cos^2 \theta - \sin^2 \theta) \tag{11.60}$$

Note that $\sigma_Y$ is obtained by letting $\theta = (\theta + 90°)$ in eqn. [11.13].

Again using matrix notation and letting $s = \sin \theta$, $c = \cos \theta$, eqns. [11.58], [11.59] and [11.60] may be written as

$$\begin{Bmatrix} \sigma_x \\ \sigma_y \\ \tau_{xy} \end{Bmatrix} = \begin{bmatrix} c^2 & s^2 & 2sc \\ s^2 & c^2 & -2sc \\ -sc & sc & (c^2 - s^2) \end{bmatrix} \begin{Bmatrix} \sigma_X \\ \sigma_Y \\ \tau_{XY} \end{Bmatrix} \tag{11.61}$$

or, in shorthand form,

$$\{\sigma\}_{xy} = [T]\{\sigma\}_{XY} \tag{11.62}$$

where $[T]$ is the transformation matrix which may also be inverted to give global stress components in terms of local stress components:

$$\{\sigma\}_{XY} = [T]^{-1}\{\sigma\}_{xy} \tag{11.63}$$

The inversion of $[T]$ gives $[T]^{-1}$ as

$$[T]^{-1} = \begin{bmatrix} c^2 & s^2 & -2sc \\ s^2 & c^2 & 2sc \\ sc & -sc & (c^2 - s^2) \end{bmatrix} \tag{11.64}$$

From the analysis in Section 11.14 it may be seen that the transformation of strain is similar to that of stress, so we may write

$$\left\{ \begin{array}{c} \varepsilon_x \\ \varepsilon_y \\ \frac{1}{2}\gamma_{xy} \end{array} \right\} = [T] \left\{ \begin{array}{c} \varepsilon_X \\ \varepsilon_Y \\ \frac{1}{2}\gamma_{XY} \end{array} \right\} \tag{11.65}$$

At this stage we have developed methods of transforming the values of stresses and strains independently from the loading axes to the fibre axis or vice versa. In practice it is much more important to be able to establish the contribution to overall stiffness which is made by a lamina in which the fibres are at an angle to the loading axis. This is because in the construction of laminates a number of lamina will be arranged at different orientations and bonded together. The stiffness of the laminate will be the sum of the contribution made by each of the individual lamina.

In order to determine the stiffness in the global directions for a lamina in which the fibres are at an angle $\theta$ to the global $X$-direction then the following three steps are involved:

1. Determine the strains in the local $(x, y)$-directions by transforming the applied strains through $\theta°$ from the global $(X, Y)$-directions.
2. Calculate stresses in local directions using the on-axis stiffness matrix $[Q]$.
3. Transform stresses back to the global directions through an angle of $-\theta°$.

Using matrix notation these steps are shown below.

*Step 1*

$$\left\{ \begin{array}{c} \varepsilon_x \\ \varepsilon_y \\ \gamma_{xy} \end{array} \right\} = \begin{bmatrix} c^2 & s^2 & sc \\ s^2 & c^2 & -sc \\ -2sc & 2sc & (c^2 - s^2) \end{bmatrix} \left\{ \begin{array}{c} \varepsilon_X \\ \varepsilon_Y \\ \gamma_{XY} \end{array} \right\}$$

Note the modification to the transformation matrix $[T]$ so that we may write $\gamma_{xy}$ instead of $\frac{1}{2}\gamma_{xy}$.

*Step 2*

$$\left\{\begin{array}{c} \sigma_x \\ \sigma_y \\ \tau_{xy} \end{array}\right\} = \left[\begin{array}{ccc} Q_{11} & Q_{12} & 0 \\ Q_{21} & Q_{22} & 0 \\ 0 & 0 & Q_{66} \end{array}\right] \left\{\begin{array}{c} \varepsilon_x \\ \varepsilon_y \\ \gamma_{xy} \end{array}\right\}$$

*Step 3*

$$\left\{\begin{array}{c} \sigma_X \\ \sigma_Y \\ \tau_{XY} \end{array}\right\} = \left[\begin{array}{ccc} c^2 & s^2 & -2sc \\ s^2 & c^2 & 2sc \\ sc & -sc & (c^2 - s^2) \end{array}\right] \left\{\begin{array}{c} \sigma_x \\ \sigma_y \\ \tau_{xy} \end{array}\right\}$$

Hence to express global stresses in terms of global strains we perform the following matrix multiplication:

$$\left\{\begin{array}{c} \sigma_X \\ \sigma_Y \\ \tau_{XY} \end{array}\right\} = \left[\begin{array}{ccc} c^2 & s^2 & -2sc \\ s^2 & c^2 & 2sc \\ sc & -sc & (c^2 - s^2) \end{array}\right] \left[\begin{array}{ccc} Q_{11} & Q_{12} & 0 \\ Q_{21} & Q_{22} & 0 \\ 0 & 0 & Q_{66} \end{array}\right]$$

$$\times \left[\begin{array}{ccc} c^2 & s^2 & sc \\ s^2 & c^2 & -sc \\ -2sc & 2sc & (c^2 - s^2) \end{array}\right] \left\{\begin{array}{c} \varepsilon_X \\ \varepsilon_Y \\ \gamma_{XY} \end{array}\right\}$$

which provides an overall stiffness matrix $[\bar{Q}]$ where

$$\left\{\begin{array}{c} \sigma_X \\ \sigma_Y \\ \tau_{XY} \end{array}\right\} = \left[\begin{array}{ccc} \bar{Q}_{11} & \bar{Q}_{12} & \bar{Q}_{16} \\ \bar{Q}_{21} & \bar{Q}_{22} & \bar{Q}_{26} \\ \bar{Q}_{61} & \bar{Q}_{62} & \bar{Q}_{66} \end{array}\right] \left\{\begin{array}{c} \varepsilon_X \\ \varepsilon_Y \\ \gamma_{XY} \end{array}\right\} \qquad [11.66]$$

$$\bar{Q}_{11} = \frac{1}{\lambda}[E_x \cos^4\theta + E_y \sin^4\theta + (2\nu_x E_y + 4\lambda G)\cos^2\theta \sin^2\theta]$$

$$\bar{Q}_{21} = \bar{Q}_{12} = \frac{1}{\lambda}[\nu_x E_y(\cos^4\theta + \sin^4\theta)$$
$$+ (E_x + E_y - 4\lambda G)\cos^2\theta \sin^2\theta]$$

$$\bar{Q}_{61} = \bar{Q}_{16} = \frac{1}{\lambda}[\cos^3\theta \sin\theta(E_x - \nu_x E_y - 2\lambda G)$$
$$- \cos\theta \, \sin^3\theta(E_y - \nu_x E_y - 2\lambda G)]$$

$$\bar{Q}_{22} = \frac{1}{\lambda}[E_y \cos^4\theta + E_x \sin^4\theta + \sin^2\theta \cos^2\theta(2\nu_x E_y + 4\lambda G)]$$

$$\bar{Q}_{62} = \bar{Q}_{26} = \frac{1}{\lambda}[\cos\theta \sin^3\theta(E_x - \nu_x E_y - 2\lambda G)$$
$$- \cos^3\theta \sin\theta(E_y - \nu_x E_y - 2\lambda G)]$$

$$\bar{Q}_{66} = \frac{1}{\lambda}[\sin^2\theta\cos^2\theta(E_x + E_y - 2\nu_x E_y - 2\lambda G)$$
$$+\lambda G(\cos^4\theta + \sin^4\theta)]$$

in which $\lambda = (1 - \nu_x\nu_y)$.

Note that for high-performance composites, $E_x$ is typically much larger than $E_y$ or $G$. As $\nu_x$ and $\nu_y$ are also relatively small the stiffness values may be approximated by

$$\bar{Q}_{11} \simeq E_x\cos^4\theta \qquad\qquad \bar{Q}_{22} \simeq E_x\sin^4\theta$$

$$\bar{Q}_{21} \simeq E_x\sin^2\theta\cos^2\theta \qquad \bar{Q}_{26} \simeq E_x\cos\theta\sin^3\theta$$

$$\bar{Q}_{61} \simeq E_x\cos^3\theta\sin\theta \qquad \bar{Q}_{66} \simeq E_x\sin^2\theta\cos^2\theta$$

By a similar analysis it may be shown that, for applied stresses (rather than applied strains) in the global directions, the overall compliance matrix $[\bar{S}]$ has the form

$$\left\{\begin{array}{c} \varepsilon_X \\ \varepsilon_Y \\ \gamma_{XY} \end{array}\right\} = \left[\begin{array}{ccc} \bar{S}_{11} & \bar{S}_{12} & \bar{S}_{16} \\ \bar{S}_{21} & \bar{S}_{22} & \bar{S}_{26} \\ \bar{S}_{61} & \bar{S}_{62} & \bar{S}_{66} \end{array}\right]\left\{\begin{array}{c} \sigma_X \\ \sigma_Y \\ \tau_{XY} \end{array}\right\} \qquad [11.67]$$

where

$$\bar{S}_{11} = S_{11}\cos^4\theta + S_{22}\sin^4\theta + (2S_{12} + S_{66})\cos^2\theta\sin^2\theta$$

$$\bar{S}_{21} = \bar{S}_{12} = (S_{11} + S_{22} - S_{66})\cos^2\theta\sin^2\theta + S_{12}(\cos^4\theta + \sin^4\theta)$$

$$\bar{S}_{61} = \bar{S}_{16} = (2S_{22} - 2S_{12} - S_{66})\cos^3\theta\sin\theta$$
$$-(2S_{22} - 2S_{12} - S_{66})\sin^3\theta\cos\theta$$

$$\bar{S}_{22} = S_{11}\sin^4\theta + S_{22}\cos^4\theta + (2S_{12} + S_{66})\cos^2\theta\sin^2\theta$$

$$\bar{S}_{62} = \bar{S}_{26} = (2S_{11} - 2S_{12} - S_{66})\cos\theta\sin^3\theta$$
$$-(2S_{22} - 2S_{12} - S_{66})\sin\theta\cos^3\theta$$

$$\bar{S}_{66} = 4(S_{11} - 2S_{12} + S_{66})\cos^2\theta\sin^2\theta + S_{66}(\cos^2\theta - \sin^2\theta)^2$$

These calculations, though laborious to perform by hand, are very straightforward to perform by computer using matrix methods. From the moduli and Poisson's ratios, eqn. [11.52] provides the lamina compliance $[S]$. The lamina stiffness is the inverse of the compliance matrix, i.e.

$$[Q] = [S]^{-1} \qquad\qquad [11.68]$$

and the off-axis lamina stiffness $[\bar{Q}]$ in the global $(X-Y)$-co-ordinate system is

$$[\bar{Q}] = [T_\sigma]^{-1}[Q][T_\varepsilon] \qquad [11.69]$$

where $[T_\sigma]$ and $[T_\varepsilon]$ are the transformation matrices of eqns. [11.61] and [11.65]. The off-axis compliance is the inverse of the stiffness, i.e.

$$[\bar{S}] = [\bar{Q}]^{-1} \qquad [11.70]$$

Given facilities for matrix inversion and multiplication, the detailed formulae for each of the terms in eqn. [11.67] are not needed. If the modulus values are required for a lamina with the fibres running at some angle $\theta$ to the global $X$-axis, they can be determined from inspection of the compliance matrix as

$$E_X = \frac{1}{\bar{S}_{11}}, \qquad E_Y = \frac{1}{\bar{S}_{66}}, \qquad G_{XY} = \frac{1}{\bar{S}_{66}}$$

## 11.22 Analysis of a laminate

At this stage we are in a position to describe the stress–strain behaviour in any co-ordinate direction for a lamina consisting of unidirectional fibres. Such a lamina is used in beams and tension/compression members where the excellent longitudinal properties can be used to good advantage. However, in many cases the low transverse properties could not be tolerated. For such applications it is usual to build up a laminate in which laminae are arranged at different orientations in order to achieve the desired overall properties in the laminate. The orientations of the individual lamina can be in any desired combination. Generally there are two broad categories of laminates – those which are symmetric about the mid-plane and those which are unsymmetric, as shown in Fig. 11.32.

**Fig. 11.32** Different types of laminate construction

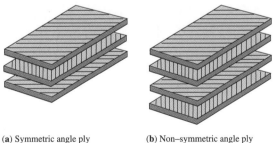

(a) Symmetric angle ply          (b) Non–symmetric angle ply

We will consider only those laminates which have mid-plane symmetry and the reader should refer to textbooks on laminate theory[1-3] for the unsymmetric cases. Laminates which are symmetric can be analysed by a simple extension of the theory developed in the previous section for unidirectional composites. In both cases their mechanical behaviour is described by three sets of elastic constants. Thus a symmetric laminate behaves like a homogeneous anisotropic plate and under uniaxial loading its effective modulus is simply the arithmetic average of the constituent laminae.

### 11.23 In-plane behaviour of a symmetric laminate

During the manufacture of a laminate the individual plies or laminae are bonded securely together so that when loaded they all experience the same strain. However, because the stiffnesses of the plies are all different, the stresses will not be the same in each case. This is illustrated in Fig. 11.33.

**Fig. 11.33** Stresses and strains in uniaxially loaded symmetrical laminate

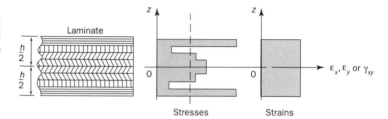

When defining the overall stress–strain behaviour of a laminate it is necessary to use average stresses. These are defined as

$$\bar{\sigma}_X = \frac{1}{h} \int_{-h/2}^{h/2} \sigma_X \, dZ \qquad [11.71]$$

$$\bar{\sigma}_Y = \frac{1}{h} \int_{-h/2}^{h/2} \sigma_Y \, dZ \qquad [11.72]$$

$$\bar{\tau}_{XY} = \frac{1}{h} \int_{-h/2}^{h/2} \tau_{XY} \, dZ \qquad [11.73]$$

or in matrix form

$$\left\{ \begin{array}{c} \bar{\sigma}_X \\ \bar{\sigma}_Y \\ \bar{\tau}_{XY} \end{array} \right\} = \frac{1}{h} \int_{-h/2}^{h/2} \left\{ \begin{array}{c} \sigma_X \\ \sigma_Y \\ \tau_{XY} \end{array} \right\} dZ \qquad [11.74]$$

$$= \frac{1}{h} \int_{-h/2}^{h/2} \left[ \begin{array}{ccc} \bar{Q}_{11} & \bar{Q}_{12} & \bar{Q}_{16} \\ & \bar{Q}_{22} & \bar{Q}_{26} \\ \text{sym.} & & \bar{Q}_{66} \end{array} \right] \left\{ \begin{array}{c} \varepsilon_X \\ \varepsilon_Y \\ \gamma_{XY} \end{array} \right\} dZ$$

Note that the stiffness matrix $[Q]$ is symmetric about its diagonal.
As the strains are independent of $Z$ they can be taken outside the integral:

$$\left\{ \begin{array}{c} \bar{\sigma}_X \\ \bar{\sigma}_Y \\ \bar{\tau}_{XY} \end{array} \right\} = \frac{1}{h} \int_{-h/2}^{h/2} \left[ \begin{array}{ccc} \bar{Q}_{11} & \bar{Q}_{12} & \bar{Q}_{16} \\ & \bar{Q}_{22} & \bar{Q}_{26} \\ \text{sym.} & & \bar{Q}_{66} \end{array} \right] dZ \left\{ \begin{array}{c} \varepsilon_X \\ \varepsilon_Y \\ \gamma_{XY} \end{array} \right\}$$

$$\left\{ \begin{array}{c} \bar{\sigma}_X \\ \bar{\sigma}_Y \\ \bar{\tau}_{XY} \end{array} \right\} = [A] \left\{ \begin{array}{c} \varepsilon_X \\ \varepsilon_Y \\ \gamma_{XY} \end{array} \right\} \qquad [11.75]$$

where, for example,

$$A_{11} = \frac{1}{h}\int_{-h/2}^{h/2} \bar{Q}_{11}\, dZ = \frac{2}{h}\int_{0}^{h/2} \bar{Q}_{11}\, dZ$$

Within a single lamina, such as the $i$th, the $\bar{Q}$-terms are constant so, referring to Fig. 11.34, the integral may be replaced by a summation.

**Fig. 11.34**

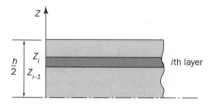

$$A_{11} = \frac{2}{h}\sum \bar{Q}_{11}^{i}\int_{Z_{i-1}}^{Z_{i}} dZ = \frac{2}{h}\sum \bar{Q}_{11}^{i} h_{i} \qquad [11.76]$$

$$A_{11} = \sum \bar{Q}_{11}^{i}\left(\frac{2h_{i}}{h}\right) \qquad [11.76]$$

Thus the stiffness matrix for a symmetric laminate may be obtained by adding, in proportion to the lamina thickness, the corresponding terms in the stiffness matrix for each of the laminae. Note also that the volume fraction of material in the $i$th lamina is given by

$$v_{i} = \left(\frac{2h_{i}}{h}\right)$$

and this is a convenient substitution to use when performing the summations indicated by eqn. [11.76]. The sum of all the $v_{i}$ terms for the different laminae will of course be 1.

Having obtained all the terms for the stiffness matrix $[A]$ as indicated above, this may then be inverted to give the compliance matrix $[a]$ and hence the modulus values for the laminate.

The approximate analysis for the off-axis lamina stiffness terms such as $Q_{11} \simeq E_{x}\cos^{4}\theta$ may be extended to a lamina giving

$$A_{11} \simeq E_{x}\sum v_{i}\cos^{4}\theta_{i}$$

where $v_{i}$ is the volume fraction of laminae at angle $\theta_{i}$ in the laminate.

For cross-ply laminates and balanced angle-ply laminates equation [11.75] will be simplified by the fact that $A_{16} = A_{61} = 0$; $A_{26} = A_{62} = 0$.

The following laminate properties may then be obtained from inspection of the compliance matrix (see eqn. [11.52]):

$$E_{X} = \frac{1}{a_{11}} \qquad E_{Y} = \frac{1}{a_{22}} \qquad G = \frac{1}{a_{66}}$$

$$\nu_{X} = \frac{-a_{12}}{a_{11}} \qquad \nu_{Y} = \frac{-a_{12}}{a_{22}}$$

Since diagonal terms in the laminate stiffness matrix, such as $A_{11}$, are usually large compared with off-diagonal terms, such as $A_{12}$, the laminate modulus $E_X$ is approximated by the relation

$$E_X \simeq \sum v_i E_{xi} \cos^4 \theta_i$$

where $v_i$ is the volume fraction of laminae at angle $\theta_i$, with longitudinal modulus $E_{xi}$.

**Example 11.8**

A filament-wound composite cylindrical pressure vessel is made up from ten plies of continuous carbon fibres in an epoxy resin. The arrangement of the plies is as shown in Fig. 11.35. There are two plies at $60°$, two plies at $-60°$ and the remainder are in the hoop direction. Calculate the maximum permissible pressure in the cylinder if the hoop strain is not to exceed 1%. At this pressure calculate the axial strain in the cylinder. The properties of the individual plies are $E_x$ = 180 GN/m², $E_y$ = 10 GN/m², $G_{xy}$ = 7 GN/m², $v_x$ = 0.28. For the cylinder $\frac{r}{t}$ = 350.

**Fig. 11.35** Composite cylindrical pressure vessel

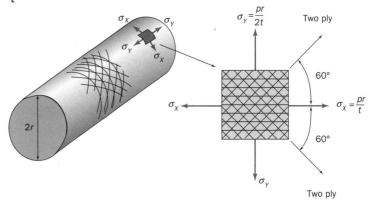

The first step in the solution is to get the stiffness matrix terms for each ply in the global co-ordinate directions. Thus from eqn. [11.66]:

$$(\bar{Q}_{11})_{60°} = \frac{1}{\lambda} [E_x \cos^4 60° + E_y \sin^4 60°$$
$$+ (2v_x E_y + 4\lambda G) \cos^2 60° \sin^2 60°] = 23.24 \, \text{GN/m}^2$$

Similarly,

$$(\bar{Q}_{11})_{-60°} = 23.24 \, \text{GN/m}^2 \qquad \text{and} \qquad (\bar{Q}_{11})_{0°} = 180.78 \, \text{GN/m}^2$$

Then, from eqn. [11.76],

$$A_{11} = 0.2(\bar{Q}_{11})_{60°} + 0.2(\bar{Q}_{11})_{-60°} + 0.6(\bar{Q}_{11})_{0°}$$

$$= 117.8 \, \text{GN/m}^2$$

In a similar way the other terms in the stiffness matrix for the laminate may be calculated to give

$$[A] = \begin{bmatrix} 117.8 & 14.6 & 0 \\ 14.6 & 49.5 & 0 \\ 0 & 0 & 18.8 \end{bmatrix}$$

This may then be inverted to give

$$[a] = \begin{bmatrix} 8.81 & -2.6 & 0 \\ -2.6 & 20.97 & 0 \\ 0 & 0 & 53.2 \end{bmatrix} \times 10^{-3}$$

From which

$$E_X = \frac{1}{a_{11}} = \frac{1}{8.81 \times 10^{-3}} = 113.5 \, \text{GN/m}^2$$

$$E_Y = \frac{1}{a_{22}} = \frac{1}{20.97 \times 10^{-3}} = 47.7 \, \text{GN/m}^2$$

$$G_{XY} = \frac{1}{a_{66}} = \frac{1}{53.2 \times 10^{-3}} = 18.8 \, \text{GN/m}^2$$

$$\nu_X = \frac{-a_{12}}{a_{11}} = \frac{2.6}{8.81} = 0.295$$

$$\nu_Y = \frac{-a_{12}}{a_{22}} = \frac{2.6}{20.97} = 0.124$$

Then, expressing the axial and hoop strains in terms of $\sigma_Y$ and $\sigma_X$,

$$\begin{Bmatrix} \varepsilon_X \\ \varepsilon_Y \\ \gamma_{XY} \end{Bmatrix} = \begin{bmatrix} 8.81 & -2.6 & 0 \\ -2.6 & 20.97 & 0 \\ 0 & 0 & 53.2 \end{bmatrix} 10^{-3} \begin{Bmatrix} \sigma_X \\ \sigma_Y \\ \tau_{XY} \end{Bmatrix}$$

$$= \begin{bmatrix} 8.81 & -2.6 & 0 \\ -2.6 & 20.97 & 0 \\ 0 & 0 & 53.2 \end{bmatrix} 10^{-3} \begin{Bmatrix} 350 \\ 175 \\ 0 \end{Bmatrix} p$$

where $p$ is the internal pressure.

Hoop strain $\varepsilon_X = (8.81 \times 350 - 2.6 \times 175)10^{-3}p = 0.01$

$$p = 3.8 \, \text{MN/m}^2$$

Also, at this pressure the axial strain

$$\varepsilon_Y = (-2.6 \times 350 + 20.97 \times 175)10^{-3} \times 3.8 \times 10^{-3} = 1.05\%$$

Note that the approximate formulae for the laminate modulus give

$$E_X \approx \sum v_i E_{xi} \cos^4 \theta_i = 180[0.2\cos^4 60 + 0.2\cos^4(-60)$$

$$+ 0.6\cos^4 0] = 112.5\,\text{GN/m}^2$$

$$E_Y \approx \sum v_i E_{xi} \cos^4(\theta_i + 90) = 180(0.2\cos^4 150 + 0.2\cos^4 30$$

$$+ 0.6\cos^4 90) = 40.5\,\text{GN/m}^2$$

which are quite close to the exact values calculated above.

## 11.24 Flexural behaviour of a symmetric laminate

When a laminate is subjected to flexure there will be a uniform strain gradient across the section but, as shown in Fig. 11.36, the stress variation will be non-linear owing to the different stiffnesses of the individual laminae.

**Fig. 11.36** Stresses and strains in laminate subjected to flexure

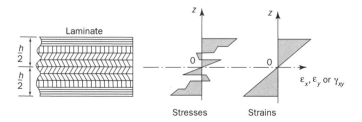

Using an analysis similar to that in the previous section it may be shown that the terms in the stiffness matrix for the laminate are given by

$$D_{11} = \sum \bar{Q}_{11}^i \left( \frac{I_i}{I_{COMP.}} \right) \tag{11.78}$$

where $I_i$ and $I_{COMP.}$ are the second moments of area for the $i$th lamina and the complete laminate respectively. Inversion of the matrix $[D]$ will then enable moduli values to be obtained as before.

## 11.25 Summary

The design of engineering components depends on the assessment of the critical stress levels occurring. In general these are the principal stresses, the importance of which will be demonstrated in the next chapter. It is therefore most important through either the analytical derivations or the Mohr's circle construction to be able to investigate the state of stress at a point in the material. It will be appreciated that for a component that already exists we cannot directly measure stress due to applied forces.

However, we can measure *in situ* displacements and strains, from which the associated stress system can be derived, using the stress–strain relationships. Therefore it is equally important to understand two-dimensional strain analysis for which the Mohr strain circle is most useful. Finally, although we seldom have to derive one elastic constant from two of the remainder, it is fundamental to the elastic behaviour of materials that the four constants are interrelated.

This chapter has also introduced the theory of fibre composite materials which by their nature are markedly anisotropic. It will be seen that the stresses and strains in a laminate, although apparently complex in nature, may be determined using the stress and strain transformations developed at the beginning of the chapter.

**References**

1. Jones, R. M. (1975) *Mechanics of Composite Materials*, Scripta Book Co., Washington, D.C.
2. Tsai, S. W. and Hahn, H. T. (1980) *Introduction to Composite Materials*, Technomic, Westport, CT.
3. Agarwal, B. D. and Broutman, L. J. (1980) *Analysis and Performance of Fiber Composites*, John Wiley, New York.

**Problems**    11.1    For the engineering components illustrated in Fig. 11.37 it is necessary to determine the normal and shear stresses at the points shown. Sketch an element in each case showing the magnitude and sense of the stresses on each face.

**Fig. 11.37**

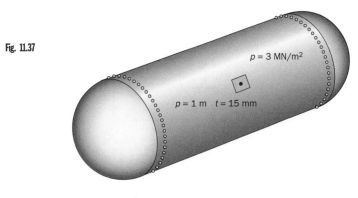

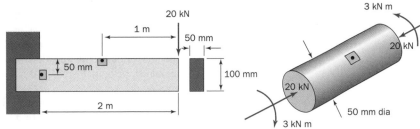

11.2    For the elements illustrated in Fig. 11.38 calculate the stress components on the inclined planes shown.

**Fig. 11.38**

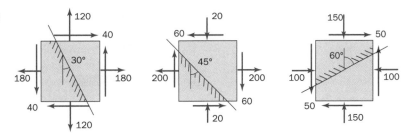

11.3    At a point in a boiler rivet the material of the rivet is undergoing the action of a shear stress of $50\,\text{MN/m}^2$ whilst resisting movement between the boiler plates and a tensile stress of $40\,\text{MN/m}^2$ due to the extension of the rivet. Find the magnitude of the tensile stresses at the same point acting on two planes making an angle of $80°$ to the axis of the rivet.

11.4    Construct Mohr's circles for the stress systems given in Problem 12.2 and check the solutions for the stresses on the inclined planes.

11.5    A cube is subjected to a hydrostatic pressure, ie the same pressure in all directions. Show by both calculation and Mohr's circle that the resulting maximum shear stress is zero.

11.6    At a point in the cross-section of a girder there is a tensile stress of $50\,\text{MN/m}^2$ and a positive shearing stress of $25\,\text{MN/m}^2$. Find the principal planes and stresses, and sketch a diagram showing how they act.

11.7    Draw the Mohr stress circles for the states of stress at a point given in Figs. 11.39 (a) and (b). For (a) determine and show the magnitude and orientation of the principal stresses. For (b) show the stress components on the inclined plane.

Fig. 11.39

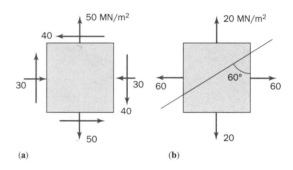

(a)                                    (b)

11.8    Determine the principal stresses, maximum shear stresses and their orientations for locations A, B and C of Example 11.1.

11.9    The loads applied to a piece of material cause a shear stress of $40\,\text{MN/m}^2$ together with a normal tensile stress on a certain plane. Find the value of this tensile stress if it makes an angle of $30°$ with the major principal stress. What are the values of the principal stresses?

11.10   The I-section beam shown in Fig. 11.40 is simply-supported over a length of $6\,\text{m}$ and subjected to a point load of $12\,\text{kN}$ at mid-span. Calculate the values of the principal stresses at a point on the cross-section $54.5\,\text{mm}$ above the neutral axis.

Fig. 11.40

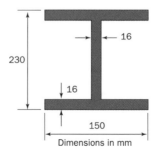

Dimensions in mm

11.11 A pulley of 250 mm diameter is keyed to the unsupported end of a 50 mm diameter shaft which overhangs 200 mm from a bearing. The pulley belt tension on the tight side is three times that on the slack side. Determine the largest values for these tensions if the maximum principal stress in the shaft is not to exceed 150 MN/m².

11.12 For a curved bar loaded as in Fig. 11.41 determine at what position the maximum principal stress has its greatest value. Calculate the latter and also the maximum shear stress for a split ring of radius 250 mm, bar diameter 50 mm and loaded with 5 kN at one end perpendicular to the plane of the ring.

**Fig. 11.41**

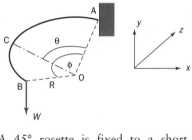

11.13 A 45° rosette is fixed to a short rectangular section pillar as illustrated in Fig. 11.42. If the gauges read $\varepsilon_a = 72 \times 10^{-6}$, $\varepsilon_b = 100 \times 10^{-6}$ and $\varepsilon_c = -240 \times 10^{-6}$ determine the values of the forces $F$ and $W$. The cross-sectional area of the pillar is 600 mm² and it is made from steel for which $E = 207$ GN/m² and $v = 0.3$.

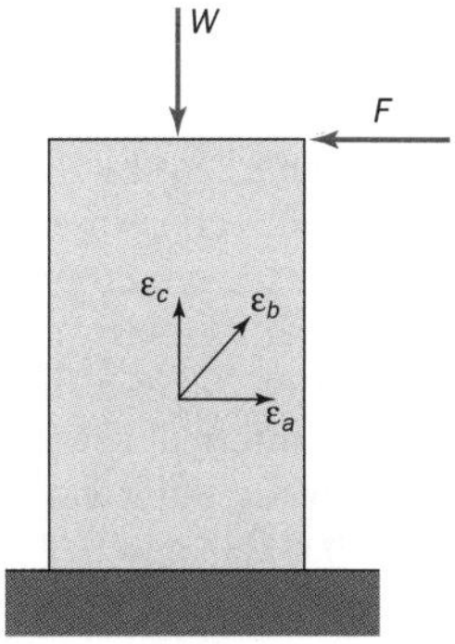

11.14 A state of two-dimensional strain is $\varepsilon_x = 0.0007$, $\varepsilon_y = -0.0006$; $\gamma_{xy} = \gamma_{yx} = 0.0003$. Calculate the principal strains in magnitude and direction and check the results using Mohr's circle construction.

11.15 At a particular point on the surface of a component the principal strain directions are known, but it is not convenient to attach electrical resistance strain gauges in these directions. However, it is possible to cement gauges at 30° and 60° anticlockwise from the major principal strain direction, and the readings from these gauges are +0.0009 and −0.0006 respectively. Construct the Mohr strain circle and find the value of the principal stresses and maximum shear stress. $E = 208$ GN/m², $\varepsilon = 0.29$.

11.16 The principal strains, $\varepsilon_1$ and $\varepsilon_2$, are measured at a point on the surface of a shaft which is subjected to bending and torsion. The values are $\varepsilon_1 = 0.0011$ and $\varepsilon_2 = -0.0006$, and $\varepsilon_1$ is inclined at 20° to the axis of the shaft. If the diameter of the shaft is 51 mm and the rigidity modulus for the material is 83 GN/m², determine analytically the applied torque and the maximum shear stress in the material at the point concerned and check graphically.

11.17 There is a method of experimental stress analysis called the brittle-coating method where the object to be analysed is coated with a thin coating of brittle material. When the component is loaded, the coating will crack perpendicular to the maximum tensile strain when a critical threshold strain is reached. You can assume both component and coating are subject to the same strain and the component behaves in a linear elastic manner.

An aluminium component ($E = 70$ GN/m², $v = 0.3$) is coated. At a particular point cracks in the coating first appear at half the

**Fig. 11.42**

design load. When the load is further increased to 0.8 times the design load a second set of cracks appears perpendicular to the first. The critical strain for the coating used is $500 \times 10^{-6}$.

(i)   What are the principal strains in the plane of the component surface at this point under the design load?

(ii)  What are the principal stresses under the design load? You can assume the component surface is in a state of plane stress.

(iii) If the first set of cracks was perpendicular to the $x$ axis, what would be the normal stress on a plane whose normal is at an angle of $30°$ to the $x$ axis under the design loading?

11.18   A 60 mm diameter solid shaft has a strain gauge mounted at $65°$ to the axis of the shaft. In service a torque is applied to the shaft and the strain gauge reads $200 \times 10^{-6}$. Calculate the value of the torque if the shaft is made from steel with $E = 207 \, \mathrm{GN/m^2}$ and $\nu = 0.3$.

11.19   Three strain gauges A, B and C are fixed to a point on the surface of a test plate at $120°$ intervals, and the strains recorded are $\varepsilon_A = +0.00108$, $\varepsilon_B = +0.00064$, $\varepsilon_C = +0.00090$. Draw Mohr's strain circle for this problem and determine the principal strains and the inclination of gauge A to the direction of the greater principal strain.

11.20   At a certain point in a steel structural element the directions of the principal stresses $\sigma_1$ and $\sigma_2$ are known. Measurements by strain gauges show that there is a tensile strain of 0.00083 in the direction of $\sigma_1$ and a compressive strain of 0.00052 in the direction of $\sigma_2$. Find the magnitudes of $\sigma_1$ and $\sigma_2$, stating whether tensile or compressive, and the maximum shear stress. $\nu = 0.28$, $E = 207 \, \mathrm{GN/m^2}$.

11.21   A thin-walled aluminium alloy pressure vessel of 200 mm diameter and 3 mm wall thickness is subjected to an internal pressure of $6 \, \mathrm{MN/m^2}$. Strain gauges which are bonded to the outer surface in the hoop and axial directions give readings of 0.00243 and 0.00057 at full pressure respectively. Determine the four elastic constants for the material.

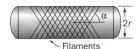
Filaments

**Fig. 11.43**

11.22   A chemical pressure vessel is to be manufactured from glass fibres in an epoxy matrix as illustrated in Fig. 11.43. If the optimum fibre orientation is that in which the fibres are subjected to tensile stresses with no transverse or shear stresses, determine the optimum value of $\alpha$.

11.23   A unidirectional fibre reinforced composite has a strength of $1500 \, \mathrm{MN/m^2}$ when loaded in the fibre direction, but only $100 \, \mathrm{MN/m^2}$ when loaded perpendicular to the fibres. When sheared parallel to the fibres it fails at a stress of $120 \, \mathrm{MN/m^2}$.

(a)   A specimen of the composite with fibres running at $30°$ to the loading axis is subjected to a uniaxial stress of $\sigma_0$, Fig. 11.44. Determine the normal stress parallel to the fibres, the normal stress perpendicular to the fibres and the shear stress parallel to the fibres.

(b)   At what value of $\sigma_0$ will the composite fail, and will it fail parallel or perpendicular to the fibres, or by shear? You can assume the material will fail when any one of the stresses reaches a critical value.

Fig. 11.44

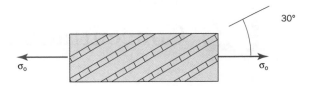

$\sigma_0$                                    $\sigma_0$

30°

11.24  A cylindrical pressure vessel is made up of continuous carbon fibres
in an epoxy matrix. The fibres are wound at ±45° from the cylinder
axis. Calculate the axial and hoop strains in the cylinder when the
internal pressure is $5\,\text{MN/m}^2$. The cylinder diameter is 1 m and the
wall thickness is 12 mm. Unidirectional carbon fibres in an epoxy
matrix have the following properties: $E_x = 130\,\text{GN/m}^2$, $E_y = 7\,\text{GN/m}^2$; $G_{xy} = 5.6\,\text{GN/m}^2$, $\nu_{xy} = 0.3$.

11.25  Part of the hull of a speedboat is in the form of a flat sheet of fibre-
reinforced polyester. The fibres are continuous glass strands and the
lay-up is such that there are four plies at 45°, four plies at −45° and
two plies at 0°. Calculate the in-plane stiffness of the sheet in the 0°
and 90° directions. A unidirectional glass-fibre composite has the
following properties: $\nu_{xy} = 0.3$, $E_x = 40\,\text{GN/m}^2$, $E_y = 9.8\,\text{GN/m}^2$,
$G_{xy} = 2.8\,\text{GN/m}^2$.

# 12 Yield Criteria and Stress Concentration

All the theoretical analysis of the previous chapters has made use of a linear stress–strain relationship. This is because Hooke's law established that metals have a linear-elastic stress–strain range. however, if a ductile metal is subjected to simple axial loading, it is found that beyond a certain point, stress is no longer proportional to strain, which results in there being a permanent deformation when the stress is removed, as illustrated in Fig. 12.1. The material is then said to have yielded. Knowing the stress at which yielding behaviour commenced, it would then be a simple matter to design a component from the same material to withstand a particular axial load without any yielding occurring. This example is simple as there is only one principal stress to consider.

The problem of designing a pressure vessel, rotating disc, or some component containing a complex principal stress system so that the material remains elastic, i.e. no yielding, when under full load is rather more complex. One could adopt a trial and error method of building a component and testing it to find when the deformations were no longer recoverable, but this would obviously be very uneconomical. It is therefore essential to find some criterion based on stresses, or strains, or perhaps strain energy in the complex system which can be related to the simple axial conditions mentioned above. If a theoretical criterion can be established which predicts complex material behaviour, it is then only necessary to establish experimentally the yield point in a simple tension or compression test.

A number of theoretical criteria for yielding have been proposed over the past century but only those now currently accepted and used for ductile and brittle materials will be discussed.

**Fig. 12.1**

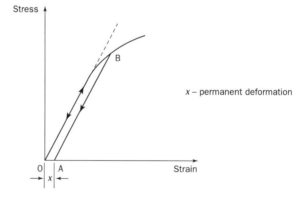

## 12.1 Yield criteria: ductile materials

Materials which exhibit 'yielding' followed by some plastic deformation prior to fracture as measured under simple tensile or compressive stress are termed *ductile*. This is a very important property as it provides a design reserve for materials if they should exceed the elastic range during service. It also has significance in relation to stress concentration as explained later in this chapter. It is therefore essential to have a method of designing to avoid the possibility of yielding under complex stress situations.

Plastic deformation is related to 'slip', as it is termed, in the grain structure which is dependent on shearing action. There are two yield criteria based on concepts of shear which are now accepted and used for ductile materials. There are known as the maximum shear-stress (Tresca) criterion and the shear-strain energy (von Mises) criterion. These will now be developed.

## Maximum shear-stress (Tresca) criterion

Theories of yielding are generally expressed in terms of principal stresses, since these completely determine a general state of stress. The element of material shown in Fig. 12.2 is subjected to three principal stresses and it will be taken that $\sigma_1 > \sigma_2 > \sigma_3$.

The French engineer Tresca, who proposed this theory, made the assumption that yielding is dependent on the maximum shear stress in the material reaching a critical value. This is taken as the maximum shear stress at yielding in a uniaxial tensile test. The maximum shear stress in the complex stress system will depend on the relative values and signs of the three principal stresses, always being half the difference between the maximum and the minimum. It should be remembered that the minimum stress can be zero or compressive, in which case it is negative in value.

For a general three-dimensional stress system, or in the two-dimensional case with one of the stresses tensile, one compressive and the third zero, the maximum shear stress is

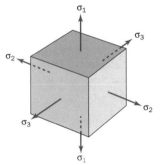

Fig. 12.2

$$\tau_{max} = (\sigma_1 - \sigma_3)/2$$

Under uniaxial tension there is only one principal stress, $\sigma_1 (\sigma_2 = \sigma_3 = 0)$, so that the maximum shear stress is

$$\tau_{max} = \sigma_1/2$$

and at yield this becomes $\tau_Y = \sigma_Y/2$.

Therefore the Tresca criterion states that yielding will occur when

$$(\sigma_1 - \sigma_3)/2 = \sigma_Y/2$$

or when the maximum principal stress difference equals the yield stress in simple tension, i.e.

$$\sigma_1 - \sigma_3 = \sigma_Y \qquad [12.1]$$

For the case when two of the principal stresses are of the same type, tension or compression, and the third is zero, then we have

$$\tau_{max} = (\sigma_1 - 0)/2 = \sigma_1/2$$

and yielding occurs when

$$\sigma_1/2 = \sigma_Y/2 \qquad \text{or} \qquad \sigma_1 = \sigma_Y \qquad [12.2]$$

## Shear-strain energy (von Mises) criterion

Huber in 1904 proposed that the total elastic strain energy stored in an element of material could be considered as consisting of energy stored due to change in volume and energy stored due to change in shape, i.e. distortion or shear. It was proposed that the latter contribution of stored strain energy could provide a viable criterion for complex yield conditions. The same criterion was also suggested independently by Maxwell, von Mises and Hencky, but is now generally referred to as the von Mises criterion.

**Fig. 12.3**

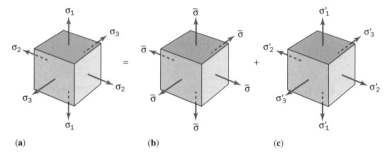

(a)                              (b)                              (c)

In order to show that the deformation of a material can be separated into change in volume and change in shape, consider the element in Fig. 12.3 subjected to the principal stresses $\sigma_1$, $\sigma_2$ and $\sigma_3$. These may be written in terms of the 'average' stresses in the element as follows:

$$\left.\begin{array}{l} \sigma_1 = \bar{\sigma} + \sigma_1' \\ \sigma_2 = \bar{\sigma} + \sigma_2' \\ \sigma_3 = \bar{\sigma} + \sigma_3' \end{array}\right\} \qquad [12.3]$$

where $\bar{\sigma}$ is the average or mean stress defined as

$$\bar{\sigma} = (\sigma_1 + \sigma_2 + \sigma_3)/3 \qquad [12.4]$$

and the $\sigma'$ represent the deviatoric stresses.

Now, when an element as in Fig. 12.3(b) is subjected to $\bar{\sigma}$ in all directions, this hydrostatic stress will produce a change in volume, but no distortion. Adding together eqns. (12.3) gives

$$\sigma_1 + \sigma_2 + \sigma_3 = 3\bar{\sigma} + \sigma_1' + \sigma_2' + \sigma_3'$$

but $\bar{\sigma} = \frac{1}{3}(\sigma_1 + \sigma_2 + \sigma_3)$; hence

$$\sigma_1' + \sigma_2' + \sigma_3' = 0 \qquad [12.5]$$

But from the stress–strain relationships,

$$\left.\begin{array}{l} \varepsilon_1' = \dfrac{\sigma_1'}{E} - \dfrac{\nu}{E}(\sigma_2' + \sigma_3') \\[2mm] \varepsilon_2' = \dfrac{\sigma_2'}{E} - \dfrac{\nu}{E}(\sigma_3' + \sigma_1') \\[2mm] \varepsilon_3' = \dfrac{\sigma_3'}{E} - \dfrac{\nu}{E}(\sigma_1' + \sigma_2') \end{array}\right\} \qquad [12.6]$$

Hence

$$\varepsilon_1' + \varepsilon_2' + \varepsilon_3' = e' = \frac{(1 - 2\nu)}{E}(\sigma_1' + \sigma_2' + \sigma_3') \qquad [12.7]$$

and since the sum of the three stresses is zero, eqn. [12.5],

$$\varepsilon_1' + \varepsilon_2' + \varepsilon_3' = e' = 0 \qquad [12.8]$$

Thus the deviatoric stress components cause no change in volume but only a change in shape.

We now turn to the determination of the strain energy quantities in the expression

$$U_T = U_V + U_S$$

where $U_T$ = total strain energy, $U_V$ = volumetric strain energy and $U_S$ = shear or distortion strain energy.

The total strain energy per unit volume is given by the sum of the energy components due to the three principal stresses and principal strains so that

$$U_T = \tfrac{1}{2}\sigma_1\varepsilon_1 + \tfrac{1}{2}\sigma_2\varepsilon_2 + \tfrac{1}{2}\sigma_3\varepsilon_3 \qquad [12.9]$$

Substituting for the principal strains from the stress–strain relationships and rearranging gives

$$U_T = \frac{1}{2E}(\sigma_1^2 + \sigma_2^2 + \sigma_3^2) - \frac{\nu}{2E}(2\sigma_1\sigma_2 + 2\sigma_2\sigma_3 + 2\sigma_3\sigma_1)$$

$$\text{per unit volume} \qquad [12.10]$$

The volumetric strain energy can now be determined from the hydrostatic component of stress, $\bar{\sigma}$.

$$U_V = \tfrac{1}{2}\bar{\sigma}e$$

$$= \tfrac{1}{2}\bar{\sigma}\frac{3\bar{\sigma}}{E}(1 - 2\nu)$$

Substituting for $\bar{\sigma}$ from eqn. [12.4] gives

$$U_V = \frac{1 - 2\nu}{6E}(\sigma_1 + \sigma_2 + \sigma_3)^2 \qquad \text{per unit volume} \qquad [12.11]$$

But $U_S = U_T - U_V$; therefore

$$U_S = \frac{1}{2E}[\sigma_1^2 + \sigma_2^2 + \sigma_3^2 - 2\nu(\sigma_1\sigma_2 + \sigma_2\sigma_3 + \sigma_3\sigma_1)]$$

$$- \frac{1 - 2\nu}{6E}(\sigma_1 + \sigma_2 + \sigma_3)^2$$

which reduces to

$$U_S = \frac{1 + \nu}{6E}[(\sigma_1 - \sigma_2)^2 + (\sigma_2 - \sigma_3)^2 + (\sigma_3 - \sigma_1)^2]$$

$$\text{per unit volume} \qquad [12.12]$$

or alternatively, using the relationship between $E$, $G$ and $\nu$,

$$U_S = \frac{1}{12G}[(\sigma_1 - \sigma_2)^2 + (\sigma_2 - \sigma_3)^2 + (\sigma_3 - \sigma_1)^2]$$

$$\text{per unit volume} \qquad [12.13]$$

Now, the shear or distortion strain energy theory proposes that yielding commences when the quantity $U_S$ reaches the equivalent value at yielding in simple tension. In the latter case $\sigma_2 = \sigma_3 = 0$ and $\sigma_1 = \sigma_Y$; therefore

$$U_S = \frac{\sigma_Y^2}{6G} \qquad \text{per unit volume} \qquad [12.14]$$

and

$$\frac{1}{12G}[(\sigma_1 - \sigma_2)^2 + (\sigma_2 - \sigma_3)^2 + (\sigma_3 - \sigma_1)^2] = \frac{\sigma_Y^2}{6G}$$

or

$$(\sigma_1 - \sigma_2)^2 + (\sigma_2 - \sigma_3)^2 + (\sigma_3 - \sigma_1)^2 = 2\sigma_Y^2 \qquad [12.15]$$

In the two-dimensional system, $\sigma_3 = 0$ and

$$\sigma_1^2 + \sigma_2^2 - \sigma_1\sigma_2 = \sigma_Y^2 \qquad [12.16]$$

for yielding to occur.

The above analysis has been directly aimed at establishing a yield criterion on an energy basis. However, from eqn. [12.15] one might equally well propose that yielding occurs as a function of the differences between principal stresses. On this hypothesis it is evident that eqn. [12.15] can also be obtained by considering the root mean square of the principal stress differences in the complex stress system in relation to simple tension. Thus

$$\{\tfrac{1}{3}[(\sigma_1 - \sigma_2)^2 + (\sigma_2 - \sigma_3)^2 + (\sigma_3 - \sigma_1)^2]\}^{1/2} = [\tfrac{1}{3}(2\sigma_Y^2)]^{1/2}$$

$$[12.17]$$

The right-hand side of the equation is obtained for simple tension by putting $\sigma_1 = \sigma_Y$ and $\sigma_2 = \sigma_3 = 0$. Equation [12.17] reduces to

$$(\sigma_1 - \sigma_2)^2 + (\sigma_2 - \sigma_3)^2 + (\sigma_3 - \sigma_1)^2 = 2\sigma_Y^2$$

which is the same as eqn. [12.15].

An alternative presentation of eqn. [12.15] is

$$\sigma_e = \frac{1}{\sqrt{2}}[(\sigma_1 - \sigma_2)^2 + (\sigma_2 - \sigma_3)^2 + (\sigma_3 - \sigma_1)^2]^{1/2}$$

where $\sigma_e$ is the von Mises equivalent stress. The basis of the von Mises yield criterion is that when $\sigma_e$ reaches $\sigma_Y$, the yield stress in simple tension, the material is deemed to have yielded.

Many experiments have been conducted under complex stress conditions to study the behaviour of metals and it has been shown that hydrostatic pressure, and by inference hydrostatic tension, does not cause yielding. Now any complex stress system can be regarded as a combination of hydrostatic stress and a function of the difference of principal stresses, and therefore a yield criterion such as that of Tresca or von Mises which is based on principal stress difference would seem to be the most logical.

**Yield envelope and locus**     For the case of three principal stresses, all non-zero, the shear-strain energy criterion, eqn. [12.15], is represented by a circular cylinder whose longitudinal axis is equally inclined to the three co-ordinate axes $\sigma_1$, $\sigma_2$, $\sigma_3$ (Fig. 12.4). The surface of the cylinder represents the envelope between an elastic stress system within the cylinder and a plastic stress state outside.

**Fig. 12.4**

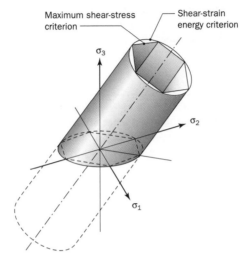

The maximum shear-stress criterion for three-dimensional states of stress is represented by a hexagonal cylinder lying within the circular cylinder as shown shaded in Fig. 12.4. Again, the hexagonal envelope divides elastic from plastic or yielded stress states.

If we wish to consider two-dimensional stress states in which, say, $\sigma_3 = 0$, then the yield boundary or locus is given by the intersection of the $\sigma_1\sigma_2$-plane with the two cylinders, as shown by the dashed lines in Fig. 12.4. The yield loci for the above two criteria for ductile materials subjected to two-dimensional principal stress states are illustrated in Fig. 12.5.

**Fig. 12.5**

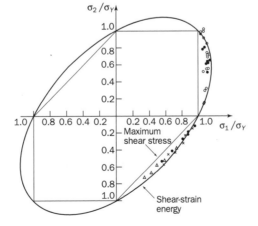

A number of experimental studies have been carried out on various ductile metals to establish the appropriate criterion for yielding under various combinations of principal stresses. There are several classical test methods for studying complex stress behaviour. These include a thin-walled tube subjected to internal pressure and tensile or compressive axial load, or a tube under combined torsion and axial load or bending. Suitable measurements of strain or deformation are recorded as applied loading is increased to a point where linear elasticity is no longer obtained. The principal stresses may then be calculated and plotted to represent yielding, as shown in Fig. 12.5. It is seen that there is close correlation with the von Mises shear-strain energy criterion and that the Tresca maximum shear-stress criterion is also satisfactory but somewhat more conservative.

**Example 12.1**

A mild steel shaft of 50 mm diameter is subjected to a bending moment of 1.9 kNm. If the yield point of the steel in simple tension is 200 MN/m², find the maximum torque that can also be applied according to: (a) the maximum shear stress; (b) the shear-strain energy theories of yielding.

The maximum bending stress occurs at the surface of the shaft and is given by

$$\sigma_x = \frac{32M}{\pi d^3} = \frac{32}{\pi} \times \frac{1900}{125 \times 10^{-6}} = 155 \, \text{MN/m}^2$$

The maximum shear stress at the surface is

$$\tau_{xy} = \frac{16T}{\pi d^3} = \frac{16}{\pi} \times \frac{T}{125 \times 10^{-6}} = 40.7 \times 10^3 \, T$$

*(a) Maximum shear-stress theory*

$$\tau = \frac{\sigma_1 - \sigma_2}{2} = \frac{200 \times 10^6}{2}$$

$$\frac{\sqrt{(\sigma_x^2 + 4\tau_{xy}^2)}}{2} = 200 \times 10^6$$

$$155^2 + [4 \times (0.0407\,T)^2] = 200^2$$

$$0.001\,66\,T^2 = 4000$$

$$T = 1.55 \, \text{kNm}$$

*(b) Shear-strain energy theory*

$$\sigma_1^2 + \sigma_2^2 - \sigma_1\sigma_2 = (200 \times 10^6)^2$$

Putting $\sigma_x^2 + 4\tau_{xy}^2 = A$,

$$\tfrac{1}{4}(\sigma_x + \sqrt{A})^2 + \tfrac{1}{4}(\sigma_x - \sqrt{A})^2 - \tfrac{1}{4}(\sigma_x - \sqrt{A})(\sigma_x + \sqrt{A})$$

$$= (200 \times 10^6)^2$$

$$\tfrac{1}{4}[\sigma_x^2 + 2\sigma_x\sqrt{A} + A + \sigma_x^2 - 2\sigma_x\sqrt{A} + A - \sigma_x^2 + A]$$

$$= (200 \times 10^6)^2$$

Simplifying gives

$$\sigma_x^2 + 3\tau_{xy}^2 = (200 \times 10^6)^2$$

$$155^2 + [3 \times (0.001\,66 T^2)] = 200^2$$

$$0.001\,66 T^2 = 5330$$

Therefore

$$T = 1.79\,\text{kNm}$$

## Example 12.2

**A thin-walled steel cylinder of 2 m diameter is subjected to an internal pressure of 2.5 MN/m². Using a safety factor of 2 and a yield stress in simple tension of 400 MN/m², calculate the wall thickness on the basis of the Tresca and von Mises yield criteria. It may be assumed that the radial stress in the wall is negligible.**

The stress system in the wall of the cylinder consists of three principal stresses, circumferential, axial and radial, of which the last may be neglected and will be taken as zero. Hence, using eqns. [2.10] and [2.11] we have

$$\sigma_1 = \frac{pr}{t} \quad \text{and} \quad \sigma_2 = \frac{pr}{2t}$$

*(a) Tresca criterion*   Since both axial and circumferential stresses are tension the maximum difference between principal stresses gives

$$\tau_{\max} = \frac{\sigma_1 - 0}{2} = \frac{\sigma_Y}{2} \quad \text{at yielding}$$

Hence

$$\sigma_1 = \sigma_Y$$

Therefore

$$\frac{2.5 \times 1000}{t} = \frac{400}{2}$$

$$t = 12.5\,\text{mm}$$

*(b) Von Mises criterion*   For $\sigma_3 \simeq 0$ we have

$$\sigma_1^2 + \sigma_2^2 - \sigma_1\sigma_2 = \sigma_Y^2$$

$$\frac{p^2 r^2}{t^2} + \frac{p^2 r^2}{4t^2} - \frac{p^2 r^2}{2t^2} = \sigma_Y^2$$

$$\frac{3}{4}\frac{p^2 r^2}{t^2} = \sigma_Y^2$$

$$t = \left(\frac{3}{4}\right)^{1/2} \cdot \frac{pr}{\sigma_Y} = \left(\frac{3}{4}\right)^{1/2} \cdot \frac{2.5 \times 1000}{200}$$

$$t = 10.8\text{mm}$$

The slightly larger plate thickness given by the Tresca criterion illustrates its more conservative characteristic compared with the von Mises criterion.

### 12.2 Fracture criteria: brittle materials

Brittleness in a material may be defined as an inability to deform plastically. Materials such as glass, some cast irons, concrete and some plastics, when subjected to tensile stress, will generally fracture at or just beyond the elastic limit. We are therefore not much concerned with a yield criterion as a fracture criterion for brittle materials under complex principal stress states. Traditionally the most widely used criterion has been that suggested by Rankine known as the *maximum principal stress theory*.

### Maximum principal stress (Rankine) criterion

This hypothesis, proposed by Rankine, which was also intended for use to predict yielding of a ductile material, states that 'failure' (i.e. fracture of a brittle material or yielding of a ductile material) will occur in a complex stress state when the maximum principal stress reaches the stress at 'failure' in simple tension. The two-dimensional locus for this theory is illustrated in Fig. 12.6. It will be noticed that in the first and third quadrants the boundary is the same as for the maximum shear-stress theory.

**Fig. 12.6**

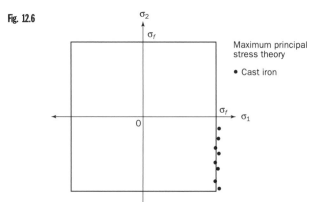

### Mohr fracture criterion

Some materials, such as case iron, have much greater strength in compression than in tension. Mohr proposed that, in the first and third quadrants of a 'failure' locus, a maximum principal stress theory was appropriate based on the ultimate strength of the material in tension or compression respectively. In the second and fourth quadrants where the two

Fig. 12.7

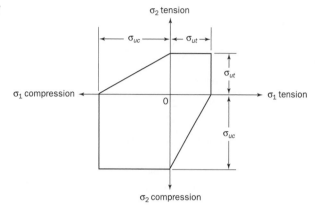

principal stresses are of opposite sign the maximum shear-stress theory should apply.

This results in a diagram as shown in Fig. 12.7.

**Fracture mechanics**    The modern approach to fracture in brittle materials is to recognize that all real materials contain defects which are capable of initiating failure without yielding in a brittle material. The likelihood of a specific defect causing failure under a particular stress system can be assessed using the procedures illustrated in Chapter 20. This may involve the use of statistical methods due to the random nature of inherent flaws in real materials.

**Example 12.3**

In a cast-iron component the maximum principal stress is to be limited to one-third of the tensile strength. Determine the maximum value of the minimum principal stress using the Mohr theory. What would be the values of the principal stresses associated with a maximum shear stress of 390 MN/m²? The tensile and compressive strengths of the cast iron are 360 MN/m² and 1410 MN/m² respectively.

Fig. 12.8

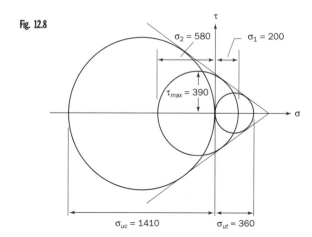

Maximum principal stress $= 360/3 = 120\,\text{MN/m}^2$ (tension). According to Mohr's theory, in the second and fourth quadrant

$$\frac{\sigma_1}{\sigma_{ut}} + \frac{\sigma_2}{\sigma_{uc}} = 1$$

Therefore

$$\frac{120}{360} + \frac{\sigma_2}{-1410} = 1 \qquad \text{and} \qquad \sigma_2 = -940\,\text{MN/m}^2$$

Mohr's stress circle construction for the second part of this problem is shown in Fig. 12.8. If the maximum shear stress is $390\,\text{MN/m}^2$, a circle is drawn of radius 390 units to touch the two envelope lines. The principal stresses can then be read off as $+200\,\text{MN/m}^2$ and $-580\,\text{MN/m}^2$.

## 12.3 Strength of laminates

In the previous sections we have been dealing with isotropic materials, so that it has been possible to develop a failure criterion on the basis of one limiting parameter (the yield strength). For fibre-reinforced composites, however, it is not possible to define failure in terms of a single parameter. It was shown in Chapter 11 that unidirectional composites have a longitudinal strength which is many times the transverse and shear strengths. Thus when a multi-axial stress system is applied it is necessary to consider the effect of this in relation to the various strength components of the composite. For laminates the picture is complicated still further by the different orientations of the individual lamina or ply in relation to the applied stress system. However, since a chain is only as strong as its weakest link, the strength of a laminate will be determined by the strength of the individual plies within the laminate.

One of the most popular failure criteria for laminates is the Tsai–Hill[1] criterion. This is based on the von Mises failure criterion which was expanded by Hill to anisotropic bodies and applied to composites by Tsai. The criterion may be expressed as

$$\left(\frac{\sigma_x}{T_x}\right)^2 - \left(\frac{\sigma_x \sigma_y}{T_x^2}\right) + \left(\frac{\sigma_y}{T_y}\right)^2 + \left(\frac{\tau_{xy}}{S_{xy}}\right)^2 = 1 \qquad [12.18]$$

where

$\sigma_x =$ the stress parallel to the fibres;
$\sigma_y =$ the stress perpendicular to the fibres;
$T_x =$ the tensile strength parallel to the fibres;
$T_y =$ the tensile strength perpendicular to the fibres;
$\tau_{xy}, S_{xy} =$ the shear stress and shear strength values.

The Tsai–Hill criterion would be used as follows. Consider a laminate subjected to in-plane stresses which are applied relative to the global $X$–$Y$-directions. The strains may then be calculated in the $X$–$Y$-directions using the compliance matrix $[a]$ (see Chapter 11)

$$\left\{ \begin{array}{c} \varepsilon_X \\ \varepsilon_Y \\ \gamma_{XY} \end{array} \right\} = [a] \left\{ \begin{array}{c} \sigma_X \\ \sigma_Y \\ \tau_{XY} \end{array} \right\} \qquad [12.19]$$

For example, if there is only a stress in the global $X$-direction, then

$$\varepsilon_X = a_{11}\sigma_X \quad \varepsilon_Y = a_{12}\sigma_X \quad \gamma_{XY} = 0$$

For in-plane loading the strains will be the same in all the plies so that these strain values may be used to get the stress in each ply. Using the stiffness matrix $[A]$ for each ply,

$$\left\{ \begin{array}{c} \sigma_X \\ \sigma_Y \\ \tau_{XY} \end{array} \right\}_{ply} = [A]_{ply} \left\{ \begin{array}{c} \varepsilon_X \\ \varepsilon_Y \\ \gamma_{XY} \end{array} \right\}_{ply/laminate} \quad [12.20]$$

From the terminology for $[A]$ in Chapter 11, this would give

$$\sigma_X = \bar{Q}_{11}\varepsilon_X + \bar{Q}_{12}\varepsilon_Y + \bar{Q}_{16}\gamma_{XY}$$

$$\sigma_Y = \bar{Q}_{21}\varepsilon_X + \bar{Q}_{22}\varepsilon_Y + \bar{Q}_{26}\gamma_{XY}$$

$$\tau_{XY} = \bar{Q}_{61}\varepsilon_X + \bar{Q}_{62}\varepsilon_Y + \bar{Q}_{66}\gamma_{XY}$$

These are the stresses in each ply in the global $X$–$Y$-directions. They would then need to be transferred to the local $x$–$y$-directions using the transformation matrix $[T]$.

$$\left\{ \begin{array}{c} \sigma_x \\ \sigma_y \\ \tau_{xy} \end{array} \right\} = [T] \left\{ \begin{array}{c} \sigma_X \\ \sigma_Y \\ \tau_{XY} \end{array} \right\} \quad [12.21]$$

At this point the Tsai–Hill criterion could be used to establish whether or not failure would be expected in the ply under consideration. To do this eqn. [12.18] could be applied or the alternative form shown below which is popular because it gives a safety factor S.F.:

$$\text{S.F.} = \frac{T_x}{\sqrt{[\sigma_x^2 - \sigma_x\sigma_y + (T_x^2/T_x^2)\sigma_y^2 + (T_x^2/S_{xy}^2)\tau_{xy}^2]}} \quad [12.22]$$

## 12.4 Concepts of stress concentration

In previous chapters the problems analysed have had stress distributions which were either uniform or varied smoothly and gradually over a significant area. However, in the vicinity of the point of application of a concentrated load there is a rapid variation in stress over a small area, in which the maximum value is considerably higher than the average stress in the full section of the material. This situation is known as a *stress concentration*.

The cause of stress concentration is perhaps most readily understood from consideration of analogous systems such as the flow of a fluid. In a simple strut subjected to an axial compressive force as shown in Fig. 12.9(a), the force is transmitted through the strut via the medium of the stresses exerted on every small element of material by its neighbouring elements. The lines of transmission of the stress are similar to the flow lines which would be observed if a fluid entered (at the point of application of the force) a channel of the same cross-section as the strut. The densely packed flow lines at the entry and exit points are representative of the concentration of stress at those points.

**Fig. 12.9**

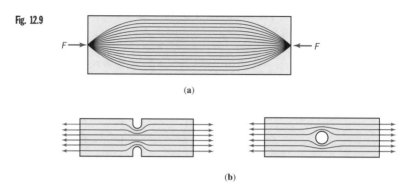

(a)

(b)

Another way in which a stress concentration can be produced is at a geometrical discontinuity in a body, such as a hole, keyway, or other sharp change in sectional dimensions. Figure 12.9(*b*) again uses the flow analogy to illustrate the stress concentration effect which occurs at notches and holes.

Two questions come to mind when considering the above effect. If, as is the case, all points of support, load application and geometric discontinuity disturb the uniformity of stress and cause stress concentration, it is surprising that one can obtain realistic results by, say, the simple bending and torsion theories considered earlier. This problem was studied theoretically by St Venant, who stated the following principle.

> If the forces acting on a small area of a body are replaced by a statically equivalent system of forces acting on the same area, there will be considerable changes in the local stress distribution, but the effect on the stresses at distances large compared with the area on which the forces act will be negligible.

For instance, in a bar gripped at each end and subjected to axial tension, the stress distribution at the ends will vary considerably according to whether gripping is by screw thread, button head, or wedge jaws. However, it has been shown that, at a distance of between one and two diameters from the ends, the stress distribution is quite uniform across the section. Similarly, it is immaterial how the couples are applied at the ends of a beam in pure bending. So long as the length is markedly greater than the cross-sectional dimensions, the assumptions and simple theory of bending will hold good at a distance of approximately one beam depth away from the concentrated force.

The second feature of interest is that, although a stress concentration is only effective locally (St Venant), the peak stress at this point is sometimes far in excess of the average stress calculated in the body of the component. Why is it then normal practice under static loading to base design calculations on the main field of stress, and not on the maximum stress concentration value where a load is applied? For instance, a point or line application of load would theoretically cause an infinite elastic stress in the material under the load. This obviously cannot occur in practice since a ductile material will reach a yield point and plastic deformation will occur under the point of application of the load. The effect of the plastic flow is to cause a local redistribution of stress, which relieves the stress concentration slightly so that the peak value of stress does not continue to increase with increasing load at the same rate as in the elastic range. Eventually, with still greater loading, general yielding in the body of the material will tend to catch up and encompass what was the stress concentration area.

Brittle materials have little or no capacity for plastic deformation and therefore the stress concentration is maintained up to fracture. Whether or not there is an accompanying reduction in nominal strength depends largely on the structure of the material. Those such as glass and some cast irons which have inherent internal flaws, which themselves set up stress concentration, show little reduction in strength over the unnotched condition. Others which have a homogeneous stress-free structure will show a considerable decrease in static strength for a severe notch.

From all of the foregoing it would appear that stress concentration does not present too serious a problem for components in service. However, there are two main aspects of material behaviour in which stress concentration plays the major part in causing failure. These are fatigue and brittle fracture (notch brittle reaction of a normally ductile metal), both of which topics are dealt with at length in later chapters.

The theoretical analysis of stress concentration is generally very complex by classical mathematics. Many theoretical solutions are due to Neuber, and a number of individual problems have been solved by other theoreticians and are available in published handbooks[3,7]. The development of finite element analysis in recent times has provided many more solutions.

The principal experimental method which has provided many simple and accurate solutions to problems of stress concentration is the technique of photoelasticity.

## 12.5 Concentrated loads and contact stresses

### Concentrated load on the edge of a plate

The local distribution of stress at the point of application of a concentrated load normal to the edge of an infinite plate was first studied in 1891 using photoelasticity. This led to the theoretical solutions a year later of Boussinesq and Flamant.

**Fig. 12.10**

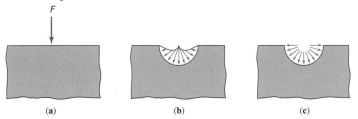

(a)  (b)  (c)

Consider the three systems of forces shown in Fig. 12.10 acting on the edge of an infinitely large plate of thickness $b$. The resultant force in each case is the same, and hence the systems are statically equivalent and therefore satisfy the principle of St Venant. Now, case $(a)$ is the one we wish to solve, but this will result in practice in a small volume of plastic flow as explained previously. To overcome this difficulty we replace the point load on the straight edge by a radial distribution of forces, as in $(b)$ or $(c)$, around a small semi-circular groove. Experiment has shown that the forces in $(c)$ give the better representation of the stress distribution due to a concentrated load on a straight edge. The solution by Flamant on this basis shows that the stress distribution is a simple radial one involving compression only. Using polar co-ordinates, and referring to Fig. 12.11, any element distance $r$ from O at an angle $\theta$ to the normal to the edge of the plate is subjected to a simple radial compression only of magnitude

$$\sigma_r = -\frac{2F\cos\theta}{\pi b r}$$

**Fig. 12.11**

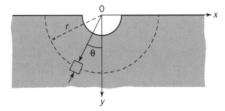

**Concentrated load bending a beam**

The cross-section of the beam at which the load is acting is subjected to a complex stress condition composed of the stress due to simple bending plus the stress due to the concentrated load itself.

**Fig. 12.12**

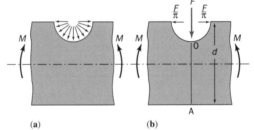

(a)                    (b)

Considering the radial pressure distribution on the small groove in Fig. 12.12(a), then the horizontal components give rise to forces $F/\pi$ acting parallel to the edge of the beam, so that the system of forces equivalent to the pressure distribution is as shown in Fig. 12.12(b). In this problem we are not considering an infinite plate, but a beam of finite depth, and consequently the horizontal forces, $F/\pi$, set up longitudinal tension and bending stresses. The former is given simply as load divided by area or $F/\pi \times (1/bd)$. The latter are determined by considering the bending moment about the axis of the beam given by $(F/\pi) \times \frac{1}{2}d$. The bending stresses are therefore

$$\sigma_x = \mp\frac{Fd}{2\pi}\frac{y}{I}$$

The total stress acting across the section OA of the beam is then obtained by the superposition of the various separate quantities:

$$\sigma_{x_{OA}} = \pm\frac{Fl}{4}\frac{y}{I} \mp \frac{Fd}{2\pi}\frac{y}{I} + \frac{F}{\pi bd}$$

$$= \pm\left(\frac{l}{4} - \frac{d}{2\pi}\right)\frac{12Fy}{bd^3} + \frac{F}{\pi bd}$$

**Fig. 12.13**

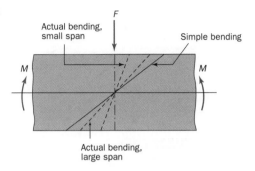

This expression is often referred to as the Wilson–Stokes solution.

The distribution of $\sigma_{x_{OA}}$ for long and short spans is compared with the simple bending distribution in Fig. 12.13, and it is seen that the more accurate solution gives rise to maximum longitudinal stresses which are *less* than those from simple bending theory.

Hence in this problem, although the stress concentration causes high normal compressive stresses, the tensile bending stress which would be expected to be the cause of failure is in fact reduced by the concentrated load.

**Contact stress**

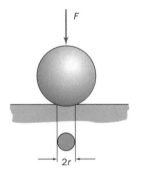

**Fig. 12.14**

Another important problem involving stress concentration is the condition of contact of two bodies under load. Typical examples may be found in the mating of gear teeth, in a shaft in a bearing, and in the balls and rollers in bearings. Solutions are too complex and lengthy to be considered here but a few useful results will be quoted.

Considering first the situation of a ball under loaded contact with a flat surface, Fig. 12.14, the point of contact when unloaded develops into a small spherical surface when under load, which is initially elastic deformation.

The radius of the circular contact area is given as

$$r = 0.88 \left( \frac{FD}{E} \right)^{1/3}$$

where $F$ is the contact force, $D$ is the diameter of the ball and $E$ is the modulus of the two materials (assumed the same in this instance).

The distribution of pressure over the contact area is such that a maximum value occurs at the centre of the circle equal to

$$\sigma_{max} = 0.62 \left( \frac{FE^2}{D^2} \right)^{1/3}$$

From the dimensions and pressure on the contact surface the stress distribution can be calculated along the axis normal to the contact. If maximum shear stress is taken as the criterion for yielding, it is found that the greatest value occurs not at the surface of contact but at a small depth below the surface in each body. It is generally at this point that failure of the material would originate if the loading were excessive.

For a roller in contact with a plane surface, Fig. 12.15(a), the contact area is rectangular of length $l$, and width $w$, with maximum pressure occurring at the centre of the rectangle.

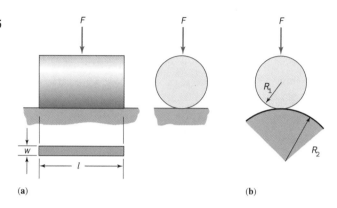

Fig. 12.15

(a)                                                                (b)

$$w = 2.15 \left( \frac{FD}{El} \right)^{1/2}$$

where $F$ is the contact force, $D$ the diameter of the roller and $l$ its length, and $E$ the modulus of the two materials, as before. The maximum pressure is given as

$$\sigma_{max} = 0.59 \left( \frac{FE}{lD} \right)^{1/2}$$

The contact dimensions and peak pressures for contact between two spheres of different diameter or two rollers of different diameter as shown in Fig. 12.15($b$) can be found from the above formulae if the sphere or roller diameter $D$ is replaced with twice the relative radius of curvature $R$. The relative radius of curvature between two surfaces of radii $R_1$ and $R_2$ is given by

$$\frac{1}{R} = \frac{1}{R_1} + \frac{1}{R_2} = \frac{R_1 + R_2}{R_1 R_2}$$

The contact of a ball of radius $R_1$ with a spherical seat is represented by taking $R_2$ as the negative of the seat radius. Contact between a roller and a cylindrical seat is treated similarly. It should be noted, however, that these analyses are only valid when the contact dimensions are small relative to the radii of curvature; they will not model accurately the contact between heavily loaded, closely conforming components.

## 12.6 Geometrical discontinuities

It was previously explained in the introduction that abrupt changes in geometry of a component give rise to stress concentration in a similar manner to those described in previous sections for loading. In most cases, the failure of a component can be attributed to some form of geometrical stress raiser, from either bad design or misfortune.

Typical examples of stress raisers are oil holes, keyways and splines, threads, and fillets at changes of section.

Figure 12.16($a$) illustrates a bar under tensile loading into which has been machined two grooves. The uniform stress is $\sigma$ at a distance from the discontinuity where the cross-sectional area is $A_1$. The mean or nominal

Fig. 12.16

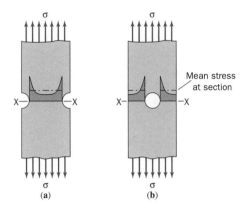

stress at section XX, where the area is $A_2$, will be $\sigma(A_1/A_2)$. However, it is seen that at the base of the grooves there is a peak value of $\sigma$ which is much higher than the average stress. There is a steep stress gradient to the lower levels of stress in the central region of the bar.

The area under this stress curve must be the same as the area under the mean stress lines, since the applied load is the same in each case.

**Stress concentration factor**

The stress concentration set up by such geometrical discontinuities is a function of the shape and dimensions of the discontinuity, and is expressed in terms of the *elastic stress concentration factor* denoted by $K_t$:

$$K_t = \frac{\text{maximum boundary stress at the discontinuity}}{\text{average stress at that cross-section of the body}}$$

This factor is constant within the elastic range of the material. Since this is dependent on the geometrical proportions and shape of the notch and the type of stress system (tension, torsion, etc.), it is readily appreciated that to cover a wide range of parameters would require a great deal more space than part of one chapter. Furthermore, the limitations on theoretical solutions for stress concentration at notches have resulted in many analyses being obtained experimentally (generally photoelastically) and presented as charts of $K_t$ against geometrical proportions. The subject has been treated very thoroughly by Neuber[2], Peterson[3], Frocht[4] and others. For complex three-dimensional stress concentrations, finite element techniques are frequently used nowadays.

A few cases of special interest will now be considered.

**Circular hole**

The distribution of axial and transverse stress at the hole cross-section is illustrated in Fig. 12.16($b$). The peak stress arises on plane XX at the edge of the hole and, for a small hole, in a thin infinite plate subjected to tension, the stress concentration factor $K_t = 3$. This is the highest value obtained and, for a circular hole in a finite width strip under axial loading $K_t$, it lies between 2 and 3. The relationship between $K_t$ and geometrical proportions is plotted in Fig. 12.17.

In pure bending of a finite width plate with a transverse hole, $K_t$ is a function of plate thickness as well as of radius of hole $r$ and width of plate $w$. For a very thick plate and small hole, the $K_t$ against $r/w$ curve is identical

**Fig. 12.17** Stress concentration factor $K_t$ for axial loading of a finite width plate with a transverse hole (Peterson[3]; by courtesy of John Wiley & Sons, Inc.)

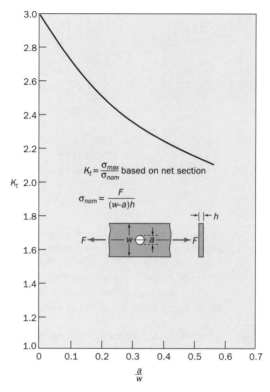

$$K_t = \frac{\sigma_{max}}{\sigma_{nom}} \text{ based on net section}$$

$$\sigma_{nom} = \frac{F}{(w-a)h}$$

with Fig. 12.17 for the tension strip. When the hole is large and the plate thin, $K_t$ varies from 1.85 to 1.1 for decreasing width.

The case of a shaft with a transverse hole subjected to tension, bending or torsion is a common one. It introduces the feature of a three-dimensional stress system as against plane stress in the plate case. The stress concentration factor in a triaxial stress field, although defined in the same way as for a biaxial stress system, will be denoted by $K_t'$. Frocht has studied photoelastically a circular bar with a transverse hole under tension and has

**Fig. 12.18** Summary of experimental results for shafts with transverse holes in tension (Frocht[4]; by courtesy of John Wiley & Sons, Inc.)

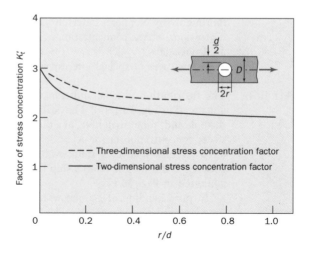

--- - Three-dimensional stress concentration factor
—— Two-dimensional stress concentration factor

obtained a curve of $K_t'$ against $r/d$ as shown in Fig. 12.18. Also plotted here for comparison is Howland's curve for $K_t$ for a plate under tension, and it is interesting to note that $K_t'$ is noticeably higher.

**Fillet radius**    Almost without exception cylindrical components do not have a uniform diameter from one end to the other. The journal bearings of a crankshaft have to mate into the web, and a motor shaft or railway axle requires shoulders to retain bearings or a wheel. At these changes of section a sharp internal corner would introduce an intolerable stress concentration which could lead to failure by fatigue. The problem is lessened by the introduction of a fillet radius, to blend one section smoothly into the next. Even a fillet radius will give rise to some stress concentration; however, this will be considerably less than with the sharp corner. The stress concentration at a fillet radius is not readily amenable to mathematical treatment and solutions have been obtained by photoelasticity and other experimental means.

Values of $K_t$ for various geometrical proportions have been obtained for shafts in tension, bending and torsion. The last two cases are probably the most common in practice and therefore only the charts for these, in Figs. 12.19 and 12.20, have been included, for reasons of space.

**Fig. 12.19** Stress concentration factor $K_t$ for the bending of a shaft with a shoulder fillet (Peterson[3]; by courtesy of John Wiley & Sons, Inc.)

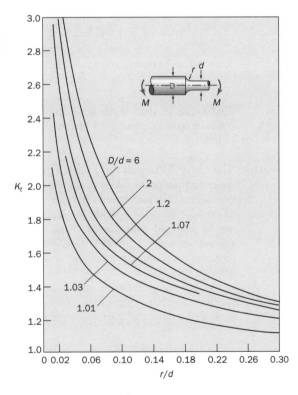

**Keyways and splines in torsion**    Another design feature which requires careful attention is the keyway or spline in a shaft subjected to torsion.

The rectangular keyway of standard form having root fillets has been solved mathematically by Leven, and the curve for $K_{ts}$ against the ratio of

**Fig. 12.20**   Stress concentration
factor $K_{ts}$ for the torsion of a shaft
with a shoulder fillet (Peterson[3];
by courtesy of John Wiley &
Sons, Inc.)

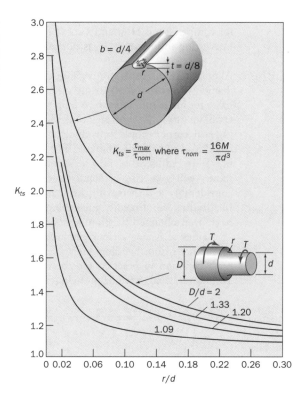

**Fig. 12.20**   Stress concentration factor $K_{ts}$ for the torsion of a shaft with a shoulder fillet (Peterson[3]; by courtesy of John Wiley & Sons, Inc.)

$$K_{ts} = \frac{\tau_{max}}{\tau_{nom}} \text{ where } \tau_{nom} = \frac{16M}{\pi d^3}$$

fillet radius to shaft diameter is shown in Fig. 12.20. The maximum shear stress occurs at a point 15° from the bottom of the keyway on the fillet radius.

**Gear teeth**   The stress distribution in a loaded gear tooth is a complex problem. Stress concentration at the point of mating of two teeth due to contact load varies in position with rotation of the teeth, and further stress concentration occurs at the root fillets of a tooth, the latter being the more serious. The former can be analysed with the aid of the expressions given in the section under contact stresses. The fillet stress concentration is a function of the load components (causing bending and direct stress) and the gear tooth geometry.

A very comprehensive investigation was conducted by Jacobson[5], who studied involute spur-gears (20° pressure angle) photoelastically and produced a series of charts of strength factor against the reciprocal of the

**Fig. 12.21**

number of teeth in the gear, which cover the whole possible range of spur-gear combinations. Figure 12.21 shows the photoelastic stress pattern and this clearly indicates the areas of stress concentration at the root radius and point of contact.

**Screw threads**     One of the most common causes of machinery or plant having to be shut down is the fatigue of bolts or studs. This is principally due to the high stress concentration at the root of the thread[6]. For a bolt and nut of conventional design, the load distribution along the screw (Fig. 12.22) is far from uniform and reaches a maximum intensity at the plane of the bearing face of the nut. This is due in part to the unmatched and opposing signs of the strains in the screw and nut. This can be overcome to a large extent by altering the design of the nut, principally so that the nut thread is in tension and so matching the strains more evenly with the screw.

**Fig. 12.22** Stress distribution in ordinary stud and nut (Brown and Hickson[6]; reprinted by permission of the Institution of Mechanical Engineers)

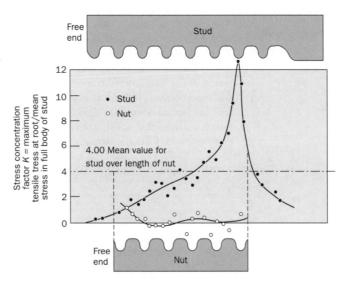

**Example 12.4**

A shaft is stepped from 48 mm diameter to 40 mm diameter through a fillet of 2.4 mm radius. A gear wheel is keyed to the larger diameter of the shaft and the radii at the internal corners of the keyway are each 1.5 mm. What is the maximum torque that can be transmitted if the steel has a shear stress limit of 200 MN/m²?

$$D/d = 48/40 = 1.2 \qquad \text{and} \qquad r/d = 2.4/40 = 0.06$$

From Fig. 12.20, $K_{ts} = 1.5$ at the fillet radius. At the keyway $r/d = 1.5/48 = 0.031$. From Fig. 12.20, $K_{ts} = 2.4$. Now

$$K_{ts} = \frac{\tau_{max}}{\tau_{nom}}$$

where

$$\tau_{max} = 200\,\text{MN/m}^2 \qquad \text{and} \qquad \tau_{nom} = \frac{16T}{\pi d^3}$$

At the fillet radius

$$\frac{16T}{\pi \times 0.04^3} = \frac{200 \times 10^6}{1.5}$$

$$T = 1676\,\text{N}\,\text{m}$$

At the keyway

$$\frac{16T}{\pi \times 0.048^3} = \frac{200 \times 10^6}{2.4}$$

$$T = 1810\,\text{N}\,\text{m}$$

The maximum torque that can be transmitted is $1676\,\text{N}\,\text{m}$ which is governed by the shear stress at the fillet radius.

## 12.7 Yield and plastic stress concentration factors

The usual definition of stress concentration factor, as given in the introduction, is in terms of the maximum stress at the discontinuity. There is of course a biaxial stress condition at the free surface of a notch, and therefore it might be more realistic to express stress concentration in terms of the maximum 'equivalent stress,' i.e. using one of the yield criteria discussed earlier. If the biaxial stresses can be determined theoretically at the free surface of a notch, then considering the shear-strain energy criterion, which seems the most appropriate for both static and fatigue stress conditions in a ductile material, we have for $\sigma_1 > \sigma_2$ and $\sigma_3 = 0$

$$\sigma_{equiv} = \sqrt{(\sigma_1^2 - \sigma_1\sigma_2 + \sigma_2^2)}$$

Then

$$K_e = \frac{\sigma_{equiv}}{\sigma_{nom}}$$

In the elastic range there is a constant ratio between $\sigma_1$ and $\sigma_2$ for any particular point on the surface; therefore

$$\sigma_1 = k\sigma_2$$

and

$$K_e = \frac{\sigma_2\sqrt{[1 - (1/k) + (1/k^2)]}}{\sigma_{nom}}$$

It is seen from the above that $K_e$ is always less than $K_t$, except when $k = 1$ and then $K_e = K_t$. This may partly explain why notched strength reduction in fatigue (Chapter 20) is nearly always less than would be expected from considerations of $K_t$.

If loading is increased to the point where yielding occurs at the notch, there is a redistribution of stress similar to that at a concentrated load. The stress concentration factor is still defined in the same way as in the elastic range but is now denoted by $K_p$, the *plastic stress concentration factor*, and this is now also a function of the degree of plastic deformation that has occurred. It is found in practice that, for ductile materials, as $\sigma_{max} \rightarrow \sigma_{ult}$, $K_p \rightarrow 1$, i.e. the stress concentration does not reduce the strength of a component loaded statically. However, in fatigue (Chapter 20) stress concentration only achieves a plastic state in a microscopic locality which is contained in a larger elastic stress concentration area. This is the cause in all cases of the initiation of the fatigue crack.

## 12.8 Summary

The latter part of this chapter demonstrates that although designers basically work in the elastic range of materials and indeed employ safety factors to make that more sure, the presence of stress concentration can still result in some local yielding. It is therefore very important to have acceptable yield criteria to apply to complex stress states. For ductile materials it is now well proven that the shear-strain energy and maximum shear-stress theories both give satisfactory predictions of the onset of yielding. Brittle materials, on the other hand, fracture rather than yield, and this situation can be adequately designed for on the basis of the maximum principal stress theory or the Mohr modified maximum shear-stress criterion.

There are plenty of data nowadays on stress concentration factors owing to the primary effects in fatigue failure and fracture mechanics which will be discussed in later chapters. From a design point of view we obviously cannot eliminate stress raisers entirely, but it is possible to minimize their effect by careful attention to the detail of points of load application, appropriate surface hardening and heat treatment and the avoidance of sharp internal corners and sudden changes of section.

## References

1.  Tsai, S. W. and Hahn, H. T. (1980) *Introduction to Composite Materials*, Technomic, Westport, CT.
2.  Neuber, N. (1946) *Theory of Notch Stress*, Edwardes, MI.
3.  Peterson, P. E. (1974) *Stress Concentration Factors*, John Wiley, New York.
4.  Frocht, M. M. *Photoelasticity*, Vol. I (1941), Vol. II (1948), John Wiley, New York.
5.  Jacobson, M. A. (1955) 'Bending stresses in spur gear teeth', *Proc. I.Mech.E.*, **169**, 587–694.
6.  Brown, A. F. C. and Hickson, V. M. (1952–1953) 'A photoelastic study of stresses in screw threads', *Proc. I.Mech.E.*, **16**, 605–612.
7.  Young, W. C. (1989) *Roark's Formulas for Stress & Strain*, McGraw-Hill International, New York.

## Problems

12.1   A shaft subjected to pure torsion yields at a torque of 1.2 kNm. A similar shaft is subjected to a torque of 720 Nm and a bending moment $M$. Determine the maximum allowable value of $M$ according to (*a*) maximum shear stress theory, (*b*) shear strain energy theory.

12.2   Show that the maximum shear stress in a helical spring subjected either to axial force or axial couple is independent of the helix angle of the coils.

An open-coiled helical spring has ten coils of 50 mm pitch and 76 mm mean diameter made from steel wire of 12.7 mm diameter. If the 0.1% proof stress for the steel is 840 MN/m², determine the allowable axial load according to the maximum shear stress criterion. $E = 206 \text{ GN/m}^2$, $\nu = 0.3$.

12.3    A circular steel cylinder of wall thickness 10 mm and internal diameter 200 mm is subjected to a constant internal pressure of 15 MN/m$^2$. Determine how much (*a*) axial tensile load, and (*b*) axial compressive load can be applied to the cylinder before yielding commences according to the maximum shear stress theory. The yield stress of the material in simple tension is 240 MN/m$^2$. Assume that the radial stress in the wall of the cylinder is zero. Sketch the plane yield stress locus for the maximum shear stress theory, and show the two points representing the cases above.

12.4    A thin walled cylindrical tank has a diameter of 2 m. If it is subjected to an internal pressure of 1 MN/m$^2$, calculate the required wall thickness for the tank if yielding is not to occur. The von Mises theory should be used to check for yielding. The yield stress in simple tension for the material is 280 MN/m$^2$ and a factor of safety of 1.5 should be used.

12.5    A hollow tube with an external diameter of 100 mm and a wall thickness of 5 mm is subjected to an axial force of 100 kN. If the tensile yield stress for the material is 280 MN/m$^2$, use the Tresca Criterion to establish whether or not a torque of 6 kNm could be applied without causing the tube to yield. A safety factor of 2 should be used to allow for stress concentrations.

12.6    Part of a supporting bracket for a machine consists of a steel rod of 12 mm diameter fixed at its lower end and containing a right-angle bend which lies in the *xy*-plane shown in Fig. 12.23. The force *P* applied at the free end lies in the *yz*-plane inclined at 30° to the *z*-axis. The 0.1% proof stress for yielding in simple tension of the steel is 200 MN/m$^2$. Calculate the value of *P* which will cause yielding at point A on the outer surface according to the shear-strain energy criterion.

12.7    In the pulley system shown in Fig. 12.24 the pulleys cause an additional load of 500 N each. Calculate a suitable shaft diameter so as to avoid failure by the maximum shear stress criterion. The tensile yield strength of the shaft material is 248 MN/m$^2$. The shaft weight may be neglected and the shaft bearings may be treated as simple supports.

12.8    A cast-iron tube 50 mm and 40 mm outside and inside diameters respectively, is being assembled into a structure. Owing to misalignment it is subjected to a torque about the longitudinal axis of 2.5 kNm and a tensile force of 50 kN and it fractures. It was

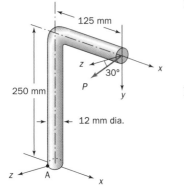

**Fig. 12.23**

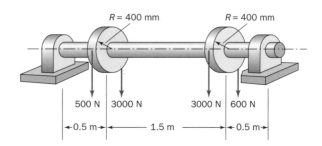

**Fig. 12.24**

discovered that the line of application of the tensile force was parallel to the axis of the tube but offset from it. Calculate the amount of eccentricity which must have occurred to cause failure of the tube according to the maximum principal stress theory. The failure stress of the cast iron in simple tension is $280\,MN/m^2$.

12.9   A cast-iron cylinder of 60 mm internal diameter and 5 mm wall thickness is to be used to check the Mohr theory of failure. The tensile and compressive strengths of the material have been measured as $400\,MN/m^2$ and $1200\,MN/m^2$ respectively. Determine (a) the internal pressure to cause failure, and (b) the axial compressive load to cause failure when combined with an internal pressure of $50\,MN/m^2$.

12.10  An aircraft fuselage spacing strut is made from carbon-fibre-reinforced epoxy and has the shape shown in Fig. 12.25. In the manufacture of the strut, six plies are laid in the loading direction and then two plies at $60°$ and two plies at $-60°$ from the loading axis. In service, the axial load causes a compressive stress of $70\,MN/m^2$ in the strut. Use the Tsai-Hill criterion to establish whether or not the composite would be expected to fail at this stress and, if not, determine the stress at which it would fail. The compressive strengths of a unidirectional carbon-fibre composite are $840\,MN/m^2$ in the fibre direction and $42\,MN/m^2$ in the transverse direction. The shear strength is $56\,MN/m^2$. Also for the unidirectional lamina, $E_x = 207\,GN/m^2$, $E_y = 7.7\,GN/m^2$, $G_{xy} = 4.9\,GN/m^2$, $\nu_{xy} = 0.3$.

Fig. 12.25

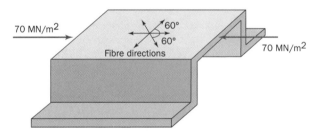

12.11  A filament-wound composite-pressure vessel consists of carbon fibre in an epoxy matrix. The vessel is cylindrical with a diameter of 600 mm and a wall thickness of 10 mm. The fibres are arranged with two plies at $45°$, two plies at $-45°$ and six plies at $0°$ to the axis of the cylinder. Use the Tsai-Hill criterion to estimate the internal pressure which would cause the vessel to fail. For a unidirectional composite using the fibre-matrix combination the following data applies: $E_x = 207\,GN/m^2$, $T_x = 1200\,MN/m^2$, $E_y = 7.7\,GN/m^2$, $T_y = 28\,MN/m^2$, $G_{xy} = 4.9\,GN/m^2$, $T_{xy} = 43\,MN/m^2$, $\nu_{xy} = 0.3$.

12.12  A steel beam of width 30 mm and of depth 90 mm is simply-suported at each end of a 2 m span. It is subjected to a concentrated load of 5 kN at mid-span. Determine the approximate depth of compressive yielding beneath the loading point. Taking into account the effect of the concentrated load, compare the maximum resultant tensile stress on the lower surface of the beam with that obtained by simple bending theory. Compressive yield stress $= 400 \, \text{MN/m}^2$.

12.13 A railway-wagon wheel is 500 mm in diameter and has an approximate contact width on the rail of 40 mm. If the compressive yield stress of the rail steel is $600 \, \text{MN/m}^2$ and the modulus is $208 \, \text{GN/m}^2$ determine the working load that can be carried per when using a load factor of 2.

12.14 A spherical steel pressure vessel is 3 m in diameter and contains a hole of 200 mm diameter to accommodate a safety valve. If the working pressure is $1.4 \, \text{MN/m}^2$ determine a suitable value for the shell thickness. The allowable tensile stress for the steel is $350 \, \text{MN/m}^2$. (*Hint*: Make a reasonable estimate for $K_t$ using Fig. 12.17.)

12.15 A shaft projects through a roller bearing from where it may be assumed to be cantilevered. It is 50 mm diameter for a length of 100 mm and then is stepped down to 25 mm diameter for a further 100 mm to the free end. At this point a load of 2.68 kN is applied. If the limiting design bending stress is $280 \, \text{MN/m}^2$ determine a suitable value for the filet radius at the change of section. What is the safety factor at the bearing housing? (*See* Fig. 12.19 for relevant data.)

CHAPTER 13 Variation of Stress and Strain

In Chapter 11 the conditions of stress and strain at a poingt in a material were considered. It is now necessary to take the analysis a stage further by examining the variation of stress between adjacent points and deriving suitable expressions for this variation. As in the earlier work, relationships for stresses may be found by considering the equilibrium of a small element of material. The solution of these equations of equilibrium must satisfy the boundary conditions of the problem as defined by the applied forces. However, it is not possible to obtain the individual components of stress directly from the above equations, owing to the statically indeterminate nature of the problem. It is necessary, therefore, to consider the elastic deformations of the material such that, in a continuous strain field, the displacements are compatible with the stress distribution. These relationships are termed the *equations of compatibility*. From this point it is only required to have a relationship between stress and strain, e.g. Hooke's law, to obtain a complete solution of the stress components in a body.

The equations of equilibrium and compatibility are quite general and may be derived in terms of various co-ordinate systems. The mathematical solution of a problem may often be simplified if an appropriate set of co-ordinates is chosen. With this in mind, and suitable illustrative applications in the following chapter, the various equations will be derived in two-dimensional Cartesian and cylindrical co-ordinates.

## 13.1 Equilibrium equations: plane stress Cartesian co-ordinates

Consider the equilibrium of a small rectangular element of dimensions $\delta x$, $\delta y$, $\delta z$, Fig. 13.1. Owing to the variation of stress through the material, $\sigma_{x_{AB}}$ is a little different from $\sigma_{x_{CD}}$, and likewise for the other stresses $\sigma_y$ and $\tau_{xy}$. The variation that must occur over any particular face may be neglected, as it cancels out when the force equilibrium on opposite pairs of faces is considered. On this occasion body forces arising from gravity, inertia, etc., will be taken into account, and these are shown as $X$ and $Y$ per unit volume.

Fig. 13.1

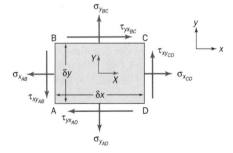

For equilibrium in the $x$-direction,

$$(\sigma_{x_{CD}} - \sigma_{x_{AB}})\delta y\,\delta z + (\tau_{yx_{BC}} - \tau_{yx_{AD}})\delta x\,\delta z + X\,\delta x\,\delta y\,\delta z = 0$$

[13.1]

Dividing by $\delta x\,\delta y\,\delta z$ gives

$$\frac{\sigma_{x_{CD}} - \sigma_{x_{AB}}}{\delta x} + \frac{\tau_{yx_{BC}} - \tau_{yx_{AD}}}{\delta y} + X = 0$$

In the limit, as $\delta x \rightarrow 0$ and $\delta y \rightarrow 0$ and the element becomes smaller and smaller, the terms become partial differentials with respect to $x$ and $y$, and thus

$$\frac{\partial \sigma_x}{\partial x} + \frac{\partial \tau_{yx}}{\partial y} + X = 0 \qquad [13.2]$$

Considering the $y$-direction,

$$(\sigma_{y_{BC}} - \sigma_{y_{AD}})\delta x\,\delta z + (\tau_{xy_{CD}} - \tau_{xy_{AB}})\delta y\,\delta z + Y\,\delta x\,\delta y\,\delta z = 0$$

$$[13.3]$$

Dividing by $\delta x\,\delta y\,\delta z$ and for $\delta x \rightarrow 0$ and $\delta y \rightarrow 0$,

$$\frac{\partial \sigma_y}{\partial y} + \frac{\partial \tau_{xy}}{\partial x} + Y = 0 \qquad [13.4]$$

It is often the case that the only body force is the weight of the component and that it can be neglected in comparison with the applied forces. Then

$$\frac{\partial \sigma_x}{\partial x} + \frac{\partial \tau_{yx}}{\partial y} = 0$$

$$\frac{\partial \sigma_y}{\partial y} + \frac{\partial \tau_{xy}}{\partial x} = 0$$

## 13.2 Equilibrium equations: plane stress cylindrical co-ordinates

As has been previously stated, there are certain cases such as cylinders, discs, curved bars, etc., in which it is rather more convenient to use $r$, $\theta$, $z$ co-ordinates. Consider the element ABCD, Fig. 13.2, which is bounded by radial lines OC and OD, subtending an angle $\delta\theta$ at the origin, and circular arcs AB and CD at radii $r$ and $r + \delta r$ respectively. The element is of thickness $\delta z$.

In the preceding section and Fig. 13.1, the stress variation on the element was represented symbolically by the different letter subscripts in order that the first analysis could be written down simply. However, it is quite usual to show stress variation in terms of partial derivatives. The body forces are shown as $R$ radially and $\Theta$ tangentially per unit volume.

Considering equilibrium along the radial centre-line of the element, there will be, in addition to forces from the radial stresses, the resolved components of force from the circumferential stress, $\sigma_\theta$, and the shear

**Fig. 13.2**

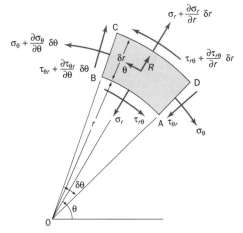

stresses $\tau_{r\theta}$ and $\tau_{\theta r}$. Hence

$$\left(\sigma_r + \frac{\partial \sigma_r}{\partial r} \delta r\right)(r + \delta r)\delta\theta\,\delta z - \sigma_r r\,\delta\theta\,\delta z$$

$$-\left(\sigma_\theta + \frac{\partial \sigma_\theta}{\partial \theta} \delta\theta\right)\delta r\,\delta z \sin\frac{\delta\theta}{2} - \sigma_\theta\,\delta r\,\delta z \sin\frac{\delta\theta}{2}$$

$$+\left(\tau_{\theta r} + \frac{\partial \tau_{\theta r}}{\partial \theta} \delta\theta\right)\delta r\,\delta z \cos\frac{\delta\theta}{2} - \tau_{\theta r}\,\delta r\,\delta z \cos\frac{\delta\theta}{2} + Rr\,\delta\theta\,\delta r\,\delta z = 0$$

$$[13.7]$$

As $\delta\theta \to \theta$, $\sin\frac{1}{2}\delta\theta \to \frac{1}{2}\delta\theta$ and $\cos\frac{1}{2}\delta\theta \to 1$. Also, neglecting second- and higher-order terms and dividing by $r\,\delta r\,\delta\theta\,\delta z$, the equation reduces to

$$\frac{\sigma_r}{r} + \frac{\partial \sigma_r}{\partial r} - \frac{\sigma_\theta}{r} + \frac{1}{r}\frac{\partial \tau_{\theta r}}{\partial \theta} + R = 0$$

or

$$\frac{\partial \sigma_r}{\partial r} + \frac{1}{r}\frac{\partial \tau_{\theta r}}{\partial \theta} + \frac{\sigma_r - \sigma_\theta}{r} + R = 0 \qquad [13.8]$$

Resolving in the tangential direction,

$$\left(\sigma_\theta + \frac{\partial \sigma_\theta}{\partial \theta} \delta\theta\right)\delta r\,\delta z \cos\frac{\delta\theta}{2} - \sigma_\theta\,\delta r\,\delta z \cos\frac{\delta\theta}{2}$$

$$+\left(\tau_{\theta r} + \frac{\partial \tau_{\theta r}}{\partial \theta} \delta\theta\right)\delta r\,\delta z \sin\frac{\delta\theta}{2} + \tau_{\theta r}\,\delta r\,\delta z \sin\frac{\delta\theta}{2}$$

$$+\left(\tau_{r\theta} + \frac{\partial \tau_{r\theta}}{\partial r} \delta r\right)(r + \delta r)\delta\theta\,\delta z - \tau_{r\theta} r\,\delta\theta\,\delta z$$

$$+\Theta r\,\delta r\,\delta\theta\,\delta z = 0 \qquad [13.9]$$

In the limit, as $\delta\theta \to 0$, neglecting the appropriate terms and dividing by $r\,\delta r\,\delta\theta\,\delta z$,

$$\frac{1}{r}\frac{\partial\sigma_\theta}{\partial\theta} + \frac{\partial\tau_{r\theta}}{\partial r} + 2\frac{\tau_{\theta r}}{r} + \Theta = 0 \qquad [13.10]$$

**Axial symmetry**     In certain cases, such as a ring, disc or cylinder, the body is symmetrical about a central axis, $z$, through O, Fig. 13.2. Then by symmetry the stress components depend on $r$ only, and $\sigma_\theta$ at any particular radius is constant. Also, from the consideration of symmetry, the shear-stress components $\tau_{\theta r}$ must vanish. Equation [13.10] no longer exists and the equilibrium equation [13.8] becomes

$$\frac{d\sigma_r}{dr} + \frac{\sigma_r - \sigma_\theta}{r} + R = 0 \qquad [13.11]$$

If the body force can be neglected, then

$$\frac{d\sigma_r}{dr} + \frac{\sigma_r - \sigma_\theta}{r} = 0 \qquad [13.12]$$

## 13.3 Strain in terms of displacement: Cartesian co-ordinates

With reference to a set of fixed axes the movement of an elastic body consists of displacement and rotation of the body combined with strain in the material. Consider a continuous strain field and the displacement of elements OA, of length $\delta x$, and OB, of length $\delta y$, referred to the axes O$x$ and O$y$, Fig. 13.3. The point O moves to O$'$ having co-ordinates $u$ and $v$, which are in general functions of $x$ and $y$. The rate of change of $u$ with respect to $x$ will be $\partial u/\partial x$. Therefore, since OA is of length $\delta x$, the point A will move to A$'$, where the displacement in the $x$-direction will be $u + (\partial u/\partial x)\delta x$. The change in length along this axis is thus $(\partial u/\partial x)\delta x$ and the strain is, in the limit as $\delta x \to 0$,

$$\varepsilon_x = \frac{(\partial u/\partial x)\delta x}{\delta x} = \frac{\partial u}{\partial x} \qquad [13.13]$$

**Fig. 13.3**

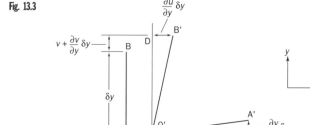

Strain in the $y$-direction is obtained from a consideration of the displacement of OB to O'B'. The rate of change of $v$ in the $y$-direction with respect to $y$ will be $\partial v/\partial y$, and therefore the point B' will have been displaced from B in the $y$-direction an amount $v + (\partial v/\partial y)\delta y$. The strain occurring will thus be

$$\varepsilon_y = \frac{(\partial v/\partial y)\delta y}{\delta y} = \frac{\partial v}{\partial y} \qquad [13.14]$$

The shearing strain in the element AOB will be given by the change from the original right angle to the new angle A'O'B'. Hence

$$\gamma_{xy} = \angle CO'A' + \angle DO'B'$$

For small displacements,

$$\angle CO'A' \approx \frac{CA'}{O'C} \qquad \text{and} \qquad DO'B' \approx \frac{DB'}{O'D}$$

Now, CA' is the rate of change of $v$ in the $x$-direction for an amount $\delta x$ or $(\partial v/\partial x)\delta x$. Similarly DB' is the rate of change of $u$ in the $y$-direction for a length $\delta y$, giving $(\partial u/\partial y)\delta y$. Therefore

$$\angle CO'A' \approx \frac{(\partial v/\partial x)\delta x}{\delta x} \approx \frac{\partial v}{\partial x}$$

and

$$\angle DO'B' \approx \frac{(\partial u/\partial y)\delta y}{\delta y} \approx \frac{\partial u}{\partial y}$$

Thus, for small displacements,

$$\gamma_{xy} = \frac{\partial v}{\partial x} + \frac{\partial u}{\partial y} \qquad [13.15]$$

In a two-dimensional strain field, the strains in terms of displacements are therefore

$$\varepsilon_x = \frac{\partial u}{\partial x} \quad \varepsilon_y = \frac{\partial v}{\partial y} \quad \gamma_{xy} = \frac{\partial v}{\partial x} + \frac{\partial u}{\partial y} \qquad [13.16]$$

## 13.4 Strain in terms of displacement: cylindrical co-ordinates

When dealing with the analysis of elements having a circular geometry (curved bars, discs, etc.) it is often more convenient to consider strain and displacements in terms of cylindrical co-ordinates as was done for the equilibrium equations. Consider the element ABCD, Fig. 13.4, subtending an angle $\delta\theta$, with AB at radius $r$ and CD at $r + \delta r$. This element is displaced to A'B'C'D' so that the radial and tangential movements to A are $u$ and $v$ respectively. The displacement of the point D to D' in the $r$-direction will be $u + (\partial u/\partial r)\delta r$, where $\partial u/\partial r$ is the rate of change of $u$ with respect to $r$. The change in length of AD is therefore $(\partial u/\partial r)\delta r$, and hence the strain in the

Fig. 13.4

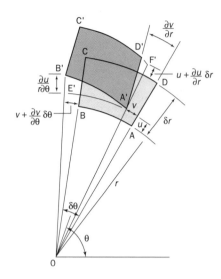

radial direction is

$$\varepsilon_r = \frac{(\partial u/\partial r)\delta r}{\delta r} = \frac{\partial u}{\partial r}$$ [13.17]

In the tangential direction there are two effects of displacement on strain. As B moves to B′ there is the change of $v$ with respect to $\theta$, giving $(\partial v/\partial \theta)\delta\theta$ as the increase in length and hence a strain of

$$\frac{(\partial v/\partial\theta)\delta\theta}{r\,\delta\theta} \qquad \text{or} \qquad \frac{1}{r}\cdot\frac{\partial v}{\partial\theta}$$

There is also the tangential strain due to the element moving out to the new radius, $r + u$. This part is

$$\frac{(r + u)\delta\theta - r\,\delta\theta}{r\,\delta\theta} = \frac{u}{r}$$

Therefore, the total tangential strain is

$$\varepsilon_\theta = \frac{1}{r}\frac{\partial v}{\partial\theta} + \frac{u}{r}$$ [13.18]

The shear strain, $\gamma_{r\theta}$, is given by the difference between $\angle$DAB and $\angle$D′A′B′, which is

$$\gamma_{r\theta} = \angle\text{D}'\text{A}'\text{F}' + \angle\text{B}'\text{A}'\text{E}'$$

Now, $\angle$D′A′F′ is the difference between the tangential displacement of D to D′, $(\partial v/\partial r)\delta r$, which as an angle is $\partial v/\partial r$, and the rigid-body rotation about O, $\angle$DOF′, which is $v/r$. Therefore

$$\angle\text{D}'\text{A}'\text{F}' = \frac{\partial v}{\partial r} - \frac{v}{r}$$

$\angle B'A'E'$ is due to variation of radial displacement $u$ in the tangential direction and is therefore $\partial u / r\, \partial\theta$. Hence

$$\gamma_{r\theta} = \frac{\partial v}{\partial r} - \frac{v}{r} + \frac{1}{r}\frac{\partial u}{\partial\theta} \qquad [13.19]$$

The three strains in terms of displacement are

$$\varepsilon_r = \frac{\partial u}{\partial r} \qquad [13.20]$$

$$\varepsilon_\theta = \frac{1}{r}\frac{\partial v}{\partial\theta} + \frac{u}{r} \qquad [13.21]$$

$$\gamma_{r\theta} = \frac{\partial v}{\partial r} + \frac{1}{r}\frac{\partial u}{\partial\theta} - \frac{v}{r} \qquad [13.22]$$

**Axial symmetry**

For problems which are symmetrical about a $z$-axis through O, there will be no tangential displacement, $v$, and since $u$ will not vary with $\theta$ there will be no shear strain, $\gamma_{r\theta}$, therefore $\tau_{r\theta}$ is zero, as stated in Section 13.2. The above equations then reduce to

$$\varepsilon_r = \frac{\partial u}{\partial r} = \frac{\mathrm{d}u}{\mathrm{d}r} \qquad [13.23]$$

$$\varepsilon_\theta = \frac{u}{r} \qquad [13.24]$$

**13.5 Compatibility equations: Cartesian co-ordinates**

Because the three strains of eqn [13.16] are expressed in terms of the two displacements, $u$ and $v$, there must be a relationship between them. This may be obtained by differentiating $\varepsilon_x$ twice with respect to $y$, $\varepsilon_y$ twice with respect to $x$, and $\gamma_{xy}$ with respect to both $x$ and $y$. Hence

$$\frac{\partial^2 \varepsilon_x}{\partial y^2} = \frac{\partial^3 u}{\partial x\, \partial y^2} \qquad \frac{\partial^2 \varepsilon_y}{\partial x^2} = \frac{\partial^3 v}{\partial y\, \partial x^2} \qquad \frac{\partial^2 \gamma_{xy}}{\partial x\, \partial y} = \frac{\partial^2}{\partial x\, \partial y}\left(\frac{\partial v}{\partial x} + \frac{\partial u}{\partial y}\right)$$

Eliminating $u$ and $v$ between these three equations provides the relationship

$$\frac{\partial^2 \varepsilon_x}{\partial y^2} + \frac{\partial^2 \varepsilon_y}{\partial x^2} = \frac{\partial^2 \gamma_{xy}}{\partial x\, \partial y} \qquad [13.25]$$

which is known as a *compatibility equation* in terms of strains. To obtain a compatibility relationship in terms of stresses, it is only necessary to substitute for the strains in eqn. [13.25], using the stress–strain relationships.

For the case of plane stress, $\sigma_z = 0$, the stress–strain relationships are

$$\varepsilon_x = \frac{\sigma_x}{E} - \frac{\nu\sigma_y}{E}$$

$$\varepsilon_y = \frac{\sigma_y}{E} - \frac{\nu\sigma_x}{E}$$

$$\gamma_{xy} = \frac{\tau_{xy}}{G} = \frac{2\tau_{xy}(1+\nu)}{E}$$

Substituting the above in eqn. [13.25],

$$\frac{1}{E}\frac{\partial^2\sigma_x}{\partial y^2} - \frac{\nu}{E}\frac{\partial^2\sigma_y}{\partial y^2} + \frac{1}{E}\frac{\partial^2\sigma_y}{\partial x^2} - \frac{\nu}{E}\frac{\partial^2\sigma_x}{\partial x^2} = \frac{2(1+\nu)}{E}\frac{\partial^2\tau_{xy}}{\partial x\,\partial y} \qquad [13.26]$$

Considering the equilibrium equations and neglecting the body force, if it is only the weight of the body, then differentiating eqn. [13.5] with respect to $x$ and eqn. [13.6] with respect to $y$ and adding we get

$$\frac{\partial^2\sigma_x}{\partial x^2} + \frac{\partial^2\sigma_y}{\partial y^2} = -2\frac{\partial^2\tau_{xy}}{\partial x\,\partial y} \qquad [13.27]$$

Eliminating $\tau_{xy}$ between eqns. [13.26] and [13.27],

$$\frac{1}{E}\frac{\partial^2\sigma_x}{\partial y^2} - \frac{\nu}{E}\frac{\partial^2\sigma_y}{\partial y^2} + \frac{1}{E}\frac{\partial^2\sigma_y}{\partial x^2} - \frac{\nu}{E}\frac{\partial^2\sigma_x}{\partial x^2}$$

$$= -\frac{(1+\nu)}{E}\left(\frac{\partial^2\sigma_x}{\partial x^2} + \frac{\partial^2\sigma_y}{\partial y^2}\right)$$

Simplifying,

$$\frac{\partial^2\sigma_x}{\partial x^2} + \frac{\partial^2\sigma_y}{\partial x^2} + \frac{\partial^2\sigma_x}{\partial y^2} + \frac{\partial^2\sigma_y}{\partial y^2} = 0$$

or

$$\left(\frac{\partial^2}{\partial x^2} + \frac{\partial^2}{\partial y^2}\right)(\sigma_x + \sigma_y) = 0 \qquad [13.28]$$

An analysis similar to that above can be used to show that the compatibility eqn. [13.28] also applies to the case of plane strain.

## 13.6 Compatibility equations: cylindrical co-ordinates

As in the case of Cartesian co-ordinates there must be a relationship between the strains in cylindrical co-ordinates. This may be obtained by eliminating $u$ and $v$ between the strain–displacement equations [13.20]–[13.22].

Differentiating eqn. [13.21] with respect to $r$ and eqn. [13.22] with respect to $\theta$, dividing the latter by $r$ and subtracting gives

$$\frac{\partial\varepsilon_\theta}{\partial r} - \frac{1}{r}\frac{\partial\gamma_{r\theta}}{\partial\theta} = \frac{1}{r}\frac{\partial u}{\partial r} - \frac{u}{r^2} - \frac{1}{r^2}\frac{\partial^2 u}{\partial\theta^2}$$

Multiplying by $r^2$ and substituting for $\partial u/\partial r$, we have

$$r^2 \frac{\partial \varepsilon_\theta}{\partial r} - r \frac{\partial \gamma_{r\theta}}{\partial \theta} = r\varepsilon_r - u - \frac{\partial^2 u}{\partial \theta^2} \qquad [13.29]$$

Differentiating this equation with respect to $r$, and eqn. [13.20] with respect to $\theta$ twice, and eliminating $u$ gives

$$2r \frac{\partial \varepsilon_\theta}{\partial r} + r^2 \frac{\partial^2 \varepsilon_\theta}{\partial r^2} - \frac{\partial \gamma_{r\theta}}{\partial \theta} - r \frac{\partial^2 \gamma_{r\theta}}{\partial \theta \, \partial r} = \varepsilon_r + r \frac{\partial \varepsilon_r}{\partial r} - \varepsilon_r - \frac{\partial^2 \varepsilon_r}{\partial \theta^2}$$

and simplifying,

$$\frac{\partial \gamma_{r\theta}}{\partial \theta} + r \frac{\partial^2 \gamma_{r\theta}}{\partial \theta \, \partial r} = \frac{\partial^2 \varepsilon_r}{\partial \theta^2} + r^2 \frac{\partial^2 \varepsilon_\theta}{\partial r^2} + 2r \frac{\partial \varepsilon_\theta}{\partial r} - r \frac{\partial \varepsilon_r}{\partial r} \qquad [13.30]$$

This is the compatibility equation in terms of strain. To obtain a similar relationship for stresses it is necessary to use the stress–strain relationships and equilibrium equations in cylindrical co-ordinates.

The stress–strain equations for plane stress

$$\varepsilon_r = \frac{\sigma_r}{E} - \frac{\nu \sigma_\theta}{E}$$

$$\varepsilon_\theta = \frac{\sigma_\theta}{E} - \frac{\nu \sigma_r}{E}$$

$$\gamma_{r\theta} = \frac{\tau_{r\theta}}{G} = \frac{2(1+\nu)}{E} \tau_{r\theta}$$

Substituting in eqn. [13.30],

$$\frac{2(1+\nu)}{E} \frac{\partial \tau_{r\theta}}{\partial \theta} + \frac{2(1+\nu)}{E} r \frac{\partial^2 \tau_{r\theta}}{\partial \theta \, \partial r} = \frac{1}{E} \frac{\partial^2 \sigma_r}{\partial \theta^2} - \frac{\nu}{E} \frac{\partial^2 \sigma_\theta}{\partial \theta^2}$$

$$+ \frac{r^2}{E} \frac{\partial^2 \sigma_\theta}{\partial r^2} - r^2 \frac{\nu}{E} \frac{\partial^2 \sigma_r}{\partial r^2} + \frac{2r}{E} \frac{\partial \sigma_\theta}{\partial r} - \frac{2r\nu}{E} \frac{\partial \sigma_r}{\partial r}$$

$$- \frac{r}{E} \frac{\partial \sigma_r}{\partial r} + \frac{r\nu}{E} \frac{\partial \sigma_\theta}{\partial r} \qquad [13.31]$$

Now the equilibrium equations [13.8] and [13.10] are used to eliminate $\tau_{r\theta}$. Multiply through eqn. [13.8] by $r$, and differentiate with respect to $r$; then

$$\frac{\partial \sigma_r}{\partial r} + r \frac{\partial^2 \sigma_r}{\partial r^2} + \frac{\partial^2 \tau_{r\theta}}{\partial r \, \partial \theta} + \frac{\partial \sigma_r}{\partial r} - \frac{\partial \sigma_\theta}{\partial r} = 0$$

Differentiating eqn. [13.10] with respect to $\theta$,

$$\frac{1}{r} \frac{\partial^2 \sigma_\theta}{\partial \theta^2} + \frac{\partial^2 \tau_{r\theta}}{\partial r \, \partial \theta} + \frac{2}{r} \frac{\partial \tau_{r\theta}}{\partial \theta} = 0$$

Multiplying each of the above equations by $r$, adding and simplifying, we get

$$2 \frac{\partial \tau_{r\theta}}{\partial \theta} + 2r \frac{\partial^2 \tau_{r\theta}}{\partial r \, \partial \theta} = - \frac{\partial^2 \sigma_\theta}{\partial \theta^2} - r^2 \frac{\partial^2 \sigma_r}{\partial r^2} - 2r \frac{\partial \sigma_r}{\partial r} + r \frac{\partial \sigma_\theta}{\partial r} \qquad [13.32]$$

Substituting for $\tau_{r\theta}$ in eqn. [13.31] and simplifying,

$$\frac{\partial^2 \sigma_r}{\partial r^2} + \frac{\partial^2 \sigma_\theta}{\partial r^2} + \frac{1}{r}\frac{\partial \sigma_r}{\partial r} + \frac{1}{r}\frac{\partial \sigma_\theta}{\partial r} + \frac{1}{r^2}\frac{\partial^2 \sigma_r}{\partial \theta^2} + \frac{1}{r^2}\frac{\partial^2 \sigma_\theta}{\partial \theta^2} = 0$$

or

$$\left(\frac{\partial^2}{\partial r^2} + \frac{1}{r}\frac{\partial}{\partial r} + \frac{1}{r^2}\frac{\partial^2}{\partial \theta^2}\right)(\sigma_r + \sigma_\theta) = 0 \qquad [13.33]$$

This is the equation of compatibility in terms of stresses. The complete analysis of stress distribution in a body may now be made using the equilibrium equations [13.8] and [13.10], the above compatibility equation, and the boundary conditions appropriate to the applied forces or displacements.

For cases of axial symmetry, since stress and displacement are independent of $\theta$, the equation becomes

$$\left(\frac{\partial^2}{\partial r^2} + \frac{1}{r}\frac{\partial}{\partial r}\right)(\sigma_r + \sigma_\theta) = 0 \qquad [13.34]$$

Multiplying out eqn. [13.34] gives

$$\frac{\partial^2 \sigma_r}{\partial r^2} + \frac{1}{r}\frac{\partial \sigma_r}{\partial r} + \frac{\partial^2 \sigma_\theta}{\partial r^2} + \frac{1}{r}\frac{\partial \sigma_\theta}{\partial r} = 0 \qquad [13.35]$$

Now, from eqn. [13.12],

$$\sigma_\theta = r\frac{\partial \sigma_r}{\partial r} + \sigma_r \qquad [13.36]$$

Therefore

$$\frac{\partial \sigma_\theta}{\partial r} = \frac{\partial \sigma_r}{\partial r} + r\frac{\partial^2 \sigma_r}{\partial r^2} + \frac{\partial \sigma_r}{\partial r}$$

and

$$\frac{\partial^2 \sigma_\theta}{\partial r^2} = 3\frac{\partial^2 \sigma_r}{\partial r^2} + r\frac{\partial^3 \sigma_r}{\partial r^3}$$

Substituting these expressions for $\sigma_\theta$ in eqn. [13.35] and gathering terms together gives

$$r\frac{\partial^3 \sigma_r}{\partial r^3} + 5\frac{\partial^2 \sigma_r}{\partial r^2} + \frac{3}{r}\frac{\partial \sigma_r}{\partial r} = 0 \qquad [13.37]$$

which is the general equation for $\sigma_r$ in an axially symmetric stress system with no body force. It can be verified by substitution that one particular solution of this equation is

$$\sigma_r = A - \frac{B}{r^2} \qquad [13.38]$$

and substituting for $\sigma_r$ and $\partial\sigma_r/\partial r$ in eqn. [13.36] gives

$$\sigma_\theta = A + \frac{B}{r^2} \qquad [13.39]$$

$A$ and $B$ are constants which are determined from the particular boundary conditions of the problem.

These expressions for $\sigma_r$ and $\sigma_\theta$ will be derived and used for specific examples in the next chapter.

## 13.7 Equilibrium in terms of displacement: plane stress Cartesian co-ordinates

The equations of stress equilibrium can be expressed in terms of the components of displacement. Substituting the strain-displacement relations, eqn. [13.16], into the plane stress–strain equations, eqn. [3.3b], gives

$$\sigma_x = \frac{E}{1-\nu^2}(\varepsilon_x + \nu\varepsilon_y) = \frac{E}{1-\nu^2}\left(\frac{\partial u}{\partial x} + \nu\frac{\partial v}{\partial y}\right)$$

$$\sigma_y = \frac{E}{1-\nu^2}(\varepsilon_y + \nu\varepsilon_x) = \frac{E}{1-\nu^2}\left(\frac{\partial v}{\partial y} + \nu\frac{\partial u}{\partial x}\right)$$

$$\tau_{xy} = \frac{E}{2(1+\nu)}\gamma_{xy} = \frac{E}{2(1+\nu)}\left(\frac{\partial u}{\partial y} + \frac{\partial v}{\partial x}\right) \qquad [13.40]$$

Substituting these equations into the plane stress equilibrium equations, eqns. [13.5] and [13.6], gives the equilibrium equations in terms of displacements as

$$2\frac{\partial^2 u}{\partial x^2} + (1+\nu)\frac{\partial^2 v}{\partial x\,\partial y} + (1-\nu)\frac{\partial^2 u}{\partial y^2} = 0$$

$$2\frac{\partial^2 v}{\partial y^2} + (1+\nu)\frac{\partial^2 u}{\partial x\,\partial y} + (1-\nu)\frac{\partial^2 v}{\partial x^2} = 0 \qquad [13.41]$$

Since the equilibrium equations are written in terms of the two displacement components $u$ and $v$, the compatibility equation [13.28] is automatically satisfied and does not need to be considered. A similar approach will be employed in Chapter 14 to derive the equilibrium equation for a thin rotating disc in terms of the radial displacement $u$ in cylindrical co-ordinates.

## 13.8 Summary

Many engineering design problems involve complex variations of the stress and strain fields within the component. However, we still must employ the three basic tenets of equilibrium of forces, compatibility of strain and displacements and the stress–strain relationships of elasticity. Now that we have derived the necessary equations in Cartesian and cylindrical co-ordinates we can proceed in the next chapter to apply these to some design examples.

**Problems** 13.1 Show that the equilibrium equation in the radial direction for a varying stress field in spherical co-ordinates $r$, $\theta$, $\psi$ with body force $R$ is of the following form:

$$R \sin \psi + \frac{\partial \sigma_r}{\partial r} \sin \psi + \frac{1}{r}\left(2\sigma_r \sin \psi - \sigma_\theta \sin \psi - \sigma_\psi \sin \psi \right.$$
$$\left. + \tau_{\psi r} \cos \psi + \frac{\partial \tau_{r\theta}}{\partial \theta} + \frac{\partial \tau_{\psi r}}{\partial \psi} \sin \psi \right) = 0$$

13.2 What are the strain-displacement relationships in spherical co-ordinates for an axi-symmetrical stress field?

Using the equilibrium equation in the previous question, simplified for an axi-symmetrical stress field without body force, show that the displacement at radius $r$, for a spherical shell, is given by

$$\frac{d^2 u}{dr^2} + \frac{2}{r}\frac{du}{dr} - \frac{2u}{r^2} = 0$$

13.3 For a particular problem the strain-displacement equations in cylindrical co-ordinates are

$$\varepsilon_r = \frac{\partial u}{\partial r} \qquad \varepsilon_\theta = \frac{u}{r} \qquad \varepsilon_z = \gamma_{r\theta} = \gamma_{\theta z} = \gamma_{zr} = 0$$

Show that the compatibility equation in terms of the stresses $\sigma_r$ and $\sigma_\theta$ is

$$r\nu\frac{\partial \sigma_r}{\partial r} - r(1-\nu)\frac{\partial \sigma_\theta}{\partial r} + \sigma_r - \sigma_\theta = 0$$

What is the problem?

13.4 Derive compatibility equations from the following strain-displacement relationships:

$$(a) \quad \gamma_{xy} = \frac{\partial u}{\partial y} + \frac{\partial v}{\partial x} \qquad \gamma_{xz} = \frac{\partial u}{\partial z} \qquad \gamma_{yz} = \frac{\partial u}{\partial z}$$

$$(b) \quad \varepsilon_z = \frac{\partial w}{\partial z} \qquad \gamma_{\theta z} = \frac{1}{r}\frac{\partial w}{\partial \theta}$$

13.5 Commencing with the six strain-displacement relationships in three-dimensional Cartesian co-ordinates, derive the six compatibility equations for three-dimensional states of strain.

# 14 Applications of the Equilibrium and Strain–Displacement Relationships

It has been explained in principle in the last chapter how the stress components may be determined in a body by use of equilibrium, compatibility and the particular boundary conditions of the problem. In a majority of cases, the solutions are complex but there are a few problems in beams and axi-symmetrical bodies in which a simpler analysis is possible using the equilibrium, strain–displacement and stress–strain relationships. As it is important to understand how to apply these principles, the present chapter commences with two simple beam-bending situations. These are followed by important engineering components, namely the thick-walled cylinder used typically in high-pressure chemical engineering and the rotating disc or rotor used in steam and gas turbines.

## 14.1 Boundary conditions

The first and most critical step in any stress analysis is specification of the correct boundary conditions. With the development of modern numerical and computer-aided techniques, the solution of a properly defined analysis can often be obtained with little or no effort on the part of the engineer. This is of little value, however, if the wrong problem is solved!

The boundary conditions for any analysis of the stress equilibrium and compatibility equations will be of two kinds:

1. Specified displacements. Over some part of the object boundary the displacement of the material may be known.
2. Specified surface tractions. Over the remainder of the object boundary the direct stress acting normal to the boundary, and/or the shear stress acting on the surface, may be known.

Fig. 14.1

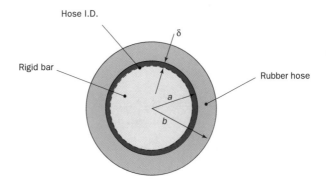

Consider for example a rubber hose which has been forced over a rigid bar whose outer diameter is larger than the bore of the hose, Fig. 14.1. The boundary conditions at the inner diameter of the rubber hose are that every point has been forced radially outward by an amount equal to the difference in radii. In terms of the radial and circumferential displacements $u$ and $v$ respectively, this implies

$$\text{at} \quad r = a, \qquad u = \delta$$

$$v = 0$$

Note that the pressure exerted on the hose at $r = a$ by forcing it over the tube must be found from the analysis. The radial displacement and normal stress cannot both be specified.

At the outer diameter of the hose the resulting displacement is not known; this must be found from the analysis. What is known, however, is that on that external surface, no normal or shear stress is acting. This implies

$$\text{at} \quad r = b, \qquad \sigma_r = \tau_{r\theta} = 0$$

Note that it is not possible to specify $\sigma_\theta$ as a boundary condition in this problem, since all the surfaces on which $\sigma_\theta$ acts are in the interior of the hose.

## 14.2 Shear stress in a beam

The distribution of transverse shear stress in a beam in terms of the shear force and the geometry of the cross-section was obtained in Chapter 6, using simple bending theory. An alternative approach will now be developed.

Consider the beam in Fig. 14.2, which, for simplicity of solution, is shown simply supported and carrying a uniformly distributed load, $w$ per unit length. The origin of Cartesian co-ordinates is taken on the neutral axis, with $x$ positive left to right and $y$ positive downwards.

This is treated as a two-dimensional problem with no variation of stress across the width of the beam, and therefore only two equations of equilibrium are applicable. Using eqns. [13.5] and [13.6],

$$\frac{\partial \sigma_x}{\partial x} + \frac{\partial \tau_{xy}}{\partial y} = 0 \tag{14.1}$$

$$\frac{\partial \sigma_y}{\partial y} + \frac{\partial \tau_{xy}}{\partial x} = 0 \tag{14.2}$$

**Fig. 14.2**

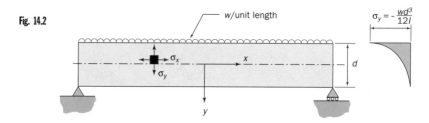

Making use of the exact solution for pure bending

$$\sigma_x = \frac{My}{I} \qquad [14.3]$$

then eqn. [14.2] is not required, neither is any strain–displacement relationship. This is because, in the derivation of eqn. [14.3], the geometry of deformation and the stress–strain relationship were included. Substituting for $\sigma_x$ in eqn. [14.1] gives

$$\frac{\partial(My/I)}{\partial x} + \frac{\partial \tau_{xy}}{\partial y} = 0$$

Therefore

$$\partial \tau_{xy} = -\frac{\partial M}{\partial x} \frac{y}{I} \partial y$$

But $\partial M / \partial x = Q$, the shear force on the section, so that

$$\partial \tau_{xy} = -\frac{Qy}{I} \partial y \qquad [14.4]$$

Integrating gives

$$\tau_{xy} = -\frac{Q}{I} \int y \, dy + C = -\frac{Qy^2}{2I} + C \qquad [14.5]$$

At the top and bottom free surface of the beam the shear stress must be zero; therefore

$$\tau_{xy} = 0 \quad \text{at} \quad y = \pm\frac{d}{2} \qquad \text{from which} \qquad C = \frac{Qd^2}{8I}$$

and

$$\tau_{xy} = -\frac{Qy^2}{2I} + \frac{Qd^2}{8I} \qquad [14.6]$$

At the neutral axis $y = 0$ and the shear stress has its maximum value

$$\tau_{xy} = \frac{Qd^2}{8I} \qquad [14.7]$$

This agrees with the value obtained in Chapter 6.

## 14.3 Transverse normal stress in a beam

A further example on the analysis of beams is that of the distribution of direct stress in the $y$-direction due to the application of a distributed load $w$. For the beam in Fig. 14.2 the equilibrium equation which is applicable is eqn [14.2]:

$$\frac{\partial \sigma_y}{\partial y} + \frac{\partial \tau_{xy}}{\partial x} = 0$$

Now, the shear stress, $\tau_{xy}$, was determined in eqn. [14.6], and substituting that value in eqn. [14.2] gives

$$\frac{\partial \sigma_y}{\partial y} + \frac{\partial}{\partial x}\left(-\frac{Qy^2}{2I} + \frac{Qd^2}{8I}\right) = 0$$

But from eqn [6.1] $\partial Q/\partial x = -w$. Therefore

$$\frac{\partial \sigma_y}{\partial y} + \frac{wy^2}{2I} - \frac{wd^2}{8I} = 0 \qquad [14.8]$$

or

$$\sigma_y = -\int\left(\frac{wy^2}{2I} - \frac{wd^2}{8I}\right)dy + C$$

$$= -\frac{w}{I}\left(\frac{y^3}{6} - \frac{d^2 y}{8}\right) + C \qquad [14.9]$$

Using the boundary condition that at the upper surface $y = -\frac{1}{2}d$, the compressive stress is $\sigma_y = -w/b$, where $b$ is the beam width; then

$$C = -\frac{w}{b} + \frac{w}{I}\left(-\frac{d^3}{48} + \frac{d^3}{16}\right)$$

Substituting

$$\frac{1}{b} = \frac{d^3}{12I}$$

$$C = -\frac{wd^3}{24I}$$

Therefore

$$\sigma_y = -\frac{w}{I}\left(\frac{y^3}{6} - \frac{d^2 y}{8} + \frac{d^3}{24}\right) \qquad [14.10]$$

The distribution of stress is illustrated in Fig. 14.2.

A check on this solution may be made by considering the condition at the lower free surface. Here $y = +\frac{1}{2}d$, from which $\sigma_y = 0$, which is correct.

## 14.4 Stress distribution in a pressurized thick-walled cylinder

This problem is of considerable practical importance in pressure vessels and gun barrels. It is an application of the cylindrical co-ordinate system, $r$, $\theta$, $z$.

A long hollow cylinder which is subjected to uniformly distributed internal and external pressure is shown in Fig. 14.3(a) and (b). The two methods of maintaining the pressure inside the cylinder are either by end caps which are attached to the cylinder as shown in Fig. 14.3(a) or by pistons in each end of the cylinder, Fig. 14.3(b). Considering a cross-sectional slice XX as shown in Fig. 14.4, the deformations produced are symmetrical about the longitudinal axis of the cylinder, and the small element of material in the wall supports the stress system shown. This is the

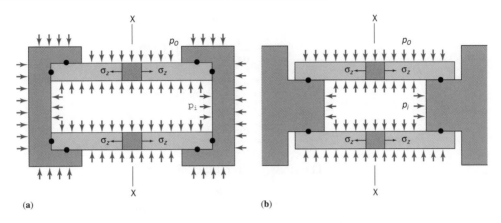

**(a)**                                                **(b)**

**Fig. 14.3**     same as in Fig. 13.2 for the general stress system except that for axial symmetry $\tau_{r\theta} = 0$ and $\sigma_\theta$ is constant at any particular radius. Hence $\sigma_\theta$ and $\sigma_r$ are principal stresses and additionally are quite independent of the method of end closure of the cylinder. Considering axial stress $\sigma_z$ and axial strain $\varepsilon_z$ then both of these occur in the case of end cap closures (Fig. 14.3(a)). The axial stress, $\sigma_z$, is constant and the axial strain, $\varepsilon_z$, is $dw/dz$.

For closure by pistons (Fig. 14.3(b)) it is evident that $\sigma_z = 0$ and $\varepsilon_z$ occurs only due to the Poisson's ratio effect of $\sigma_r$ and $\sigma_\theta$. From the symmetry of the system and for a long cylinder, we come to the conclusion that plane cross-sections remain plane when subjected to pressure and therefore axial deformation, $w$, across the section is independent of $r$ and $dw/dr = 0$.

**Fig. 14.4**

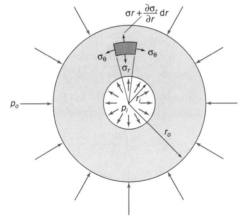

**Cylinder with end caps**     The equations of equilibrium for an element of material are

$$\frac{d\sigma_r}{dr} + \frac{\sigma_r - \sigma_\theta}{r} = 0 \qquad \text{(eqn. [13.12])} \qquad\qquad [14.11]$$

and since $\sigma_z$ is constant

$$\frac{d\sigma_z}{dz} = 0 \qquad\qquad [14.12]$$

The strain–displacement equations are

$$\varepsilon_r = \frac{du}{dr} \qquad \text{(eqn. [13.23])} \tag{14.13}$$

$$\varepsilon_\theta = \frac{u}{r} \qquad \text{(eqn. [13.24])} \tag{14.14}$$

$$\varepsilon_z = \frac{dw}{dz} \tag{14.15}$$

The stress–strain relationships are

$$\varepsilon_r = \frac{\sigma_r}{E} - \frac{\nu}{E}(\sigma_\theta + \sigma_z) = \frac{du}{dr} \tag{14.16}$$

$$\varepsilon_\theta = \frac{\sigma_\theta}{E} - \frac{\nu}{E}(\sigma_z + \sigma_r) = \frac{u}{r} \tag{14.17}$$

$$\varepsilon_z = \frac{\sigma_z}{E} - \frac{\nu}{E}(\sigma_r + \sigma_\theta) = \frac{dw}{dz} \tag{14.18}$$

Differentiating eqn. [14.17] with respect to $r$ gives

$$\frac{E}{r}\left(\frac{du}{dr} - \frac{u}{r}\right) = \frac{d\sigma_\theta}{dr} - \nu\frac{d\sigma_z}{dr} - \nu\frac{d\sigma_r}{dr}$$

Substituting for $du/dr$ and $u/r$ from eqns. [14.16] and [14.17] and simplifying,

$$\frac{1+\nu}{r}(\sigma_r - \sigma_\theta) = \frac{d\sigma_\theta}{dr} - \nu\frac{d\sigma_z}{dr} - \nu\frac{d\sigma_r}{dr} \tag{14.19}$$

Now, since $\varepsilon_z = $ constant, $d\varepsilon_z/dr = 0$ and differentiating eqn. [14.18] gives

$$\frac{d\sigma_z}{dr} = \nu\left(\frac{d\sigma_r}{dr} + \frac{d\sigma_\theta}{dr}\right) \tag{14.20}$$

Substituting into eqn. [14.19] for $d\sigma_z/dr$ from eqn. [14.20] and $(\sigma_r - \sigma_\theta)/r$ from eqn. [14.11] and simplifying gives

$$(1 - \nu^2)\left(\frac{d\sigma_\theta}{dr} + \frac{d\sigma_r}{dr}\right) = 0 \tag{14.21}$$

From eqns. [14.21] and [14.20] we see that $d\sigma_z/dr = 0$ and therefore $\sigma_z$ is constant through the wall thickness. Integrating eqn. [14.21] shows that

$$(\sigma_\theta + \sigma_r) = \text{constant} = 2A \tag{14.22}$$

Eliminating $\sigma_\theta$ between eqns. [14.22] and [14.11] gives

$$\frac{d\sigma_r}{dr} + \frac{2\sigma_r - 2A}{r} = 0 \tag{14.23}$$

from which, multiplying by $r^2$,

$$2Ar - 2r\sigma_r - r^2\frac{d\sigma_r}{dr} = 0$$

and

$$2Ar - \frac{\mathrm{d}}{\mathrm{d}r}(r^2 \sigma_r) = 0$$

By integration,

$$Ar^2 - r^2 \sigma_r = B$$

Hence

$$\sigma_r = A - \frac{B}{r^2} \qquad [14.24]$$

and from eqn. [14.22],

$$\sigma_\theta = A + \frac{B}{r^2} \qquad [14.25]$$

where $A$ and $B$ are constants which may be found using the boundary conditions. If will be noted that these equations are the same as eqns. [13.38] and [13.39], which confirms that they satisfy the equilibrium and compatibility conditions.

**Cylinder with pistons**    In this case, Fig. 14.3(*b*), $\sigma_z = 0$ and there is a condition of plane stress. This solution has been included to show that we can arrive at the same expressions for $\sigma_r$ and $\sigma_\theta$ by deriving a differential equation for displacement, $u$.

Putting $\sigma_z = 0$ in eqns. [14.16] and [14.17] and solving for $\sigma_r$ and $\sigma_\theta$ in terms of $u$ gives

$$\sigma_r = \left(\frac{\mathrm{d}u}{\mathrm{d}r} + \frac{\nu u}{r}\right)\frac{E}{(1-\nu^2)} \qquad [14.26]$$

$$\sigma_\theta = \left(\nu\frac{\mathrm{d}u}{\mathrm{d}r} + \frac{u}{r}\right)\frac{E}{(1-\nu^2)} \qquad [14.27]$$

From eqn. [14.26],

$$\frac{\mathrm{d}\sigma_r}{\mathrm{d}r} = \left(\frac{\mathrm{d}^2u}{\mathrm{d}r^2} + \nu\frac{\mathrm{d}u}{\mathrm{d}r} - \frac{\nu u}{r}\right)\frac{E}{(1-\nu^2)} \qquad [14.28]$$

Substituting eqns. [14.26], [14.27] and [14.28] into eqn. [14.11] and simplifying gives

$$\frac{\mathrm{d}^2u}{\mathrm{d}r^2} + \frac{1}{r}\frac{\mathrm{d}u}{\mathrm{d}r} - \frac{u}{r^2} = 0 \qquad [14.29]$$

This differential equation expresses radial equilibrium in terms of the displacement, $u$, in the cylinder wall.

The general solution of this equation is

$$u = Cr + \frac{C'}{r} \qquad [14.30]$$

Substituting for $u/r$ and $\mathrm{d}u/\mathrm{d}r$ in eqns. [14.26] and [14.27],

$$\sigma_r = \left( C(1+\nu) - \frac{C'}{r^2}(1-\nu) \right) \frac{E}{(1-\nu^2)} \qquad [14.31]$$

$$\sigma_\theta = \left( C(1+\nu) + \frac{C'}{r^2}(1-\nu) \right) \frac{E}{(1-\nu^2)} \qquad [14.32]$$

where $C$ and $C'$ are constants.

These equations may be rewritten with different constants as

$$\sigma_r = A - \frac{B}{r^2}$$

and

$$\sigma_\theta = A + \frac{B}{r^2}$$

which are the same as eqns. [14.24] and [14.25].

**Boundary conditions**  The next stage is the determination of the constants $A$ and $B$.

*1. Internal and external pressure*
The boundary conditions of this problem are: at $r = r_i$, $\sigma_r = -p_i$ (pressure being negative in sign); and at $r = r_o$, $\sigma_r = -p_o$,

$$-p_i = A - \frac{B}{r_i^2} \qquad \text{and} \qquad -p_o = A - \frac{B}{r_o^2}$$

from which, eliminating $A$, we get

$$B = \frac{(p_i - p_o)r_i^2 r_o^2}{r_o^2 - r_i^2} \qquad \text{and} \qquad A = \frac{p_i r_i^2 - p_o r_o^2}{r_o^2 - r_i^2}$$

Therefore the radial and hoop stresses become

$$\sigma_r = \frac{p_i r_i^2 - p_o r_o^2}{r_o^2 - r_i^2} - \frac{(p_i - p_o)r_i^2 r_o^2}{r^2(r_o^2 - r_i^2)}$$

$$\sigma_\theta = \frac{p_i r_i^2 - p_o r_o^2}{r_o^2 - r_i^2} + \frac{(p_i - p_o)r_i^2 r_o^2}{r^2(r_o^2 - r_i^2)} \qquad [14.33]$$

These equations were first derived by Lamé and Clapeyron in 1833.

Let the radius ratio $r_o/r_i = k$; then eqn. (14.33) may be written as

$$\left. \begin{aligned} \sigma_r &= \frac{1}{k^2 - 1}\left[ p_i\left(1 - \frac{r_o^2}{r^2}\right) - p_o k^2\left(1 - \frac{r_i^2}{r^2}\right) \right] \\[2mm] \sigma_\theta &= \frac{1}{k^2 - 1}\left[ p_i\left(1 + \frac{r_o^2}{r^2}\right) - p_o k^2\left(1 + \frac{r_i^2}{r^2}\right) \right] \end{aligned} \right\} \qquad [14.34]$$

It is important to note that the stresses depend on the $k$ ratio rather than on the absolute dimensions.

### 2. Internal pressure only

An important special case of the above is when the external pressure is atmospheric only and can be neglected in relation to the internal pressure. Then with $p_o = 0$,

$$\sigma_r = \frac{p_i r_i^2}{r_o^2 - r_i^2}\left(1 - \frac{r_o^2}{r^2}\right) = \frac{p_i}{k^2 - 1}\left(1 - \frac{r_o^2}{r^2}\right) \qquad [14.35]$$

$$\sigma_\theta = \frac{p_i r_i^2}{r_o^2 - r_i^2}\left(1 + \frac{r_o^2}{r^2}\right) = \frac{p_i}{k^2 - 1}\left(1 + \frac{r_o^2}{r^2}\right) \qquad [14.36]$$

At the inner surface, $\sigma_r$ and $\sigma_\theta$ each have their maximum magnitude so that at $r = r_i$,

$$\sigma_r = -p_i \quad \text{(radial compressive stress)}$$

It is appropriate at this point to note that the radial stress shown on the element in Fig. 14.4 in the positive sense, i.e. tension, is in fact in the opposite sense, i.e. compression.

The circumferential or hoop stress at $r = r_i$ is

$$\sigma_\theta = \frac{r_o^2 + r_i^2}{r_o^2 - r_i^2} p_i$$

$$= \frac{k^2 + 1}{k^2 - 1} p_i$$

At the outer surface, where $r = r_o$,

$$\sigma_r = 0 \quad \text{and} \quad \sigma_\theta = \frac{2p_i}{k^2 - 1}$$

**Stress distributions for $\sigma_\theta$ and $\sigma_r$**   To complete the basic analysis of the elastically deformed thick-walled pressure vessel, the variation of the two principal stresses $\sigma_\theta$ and $\sigma_r$ is shown plotted through the wall thickness in Fig. 14.5 for internal pressure and a $k$ ratio of 3.

**Fig. 14.5**

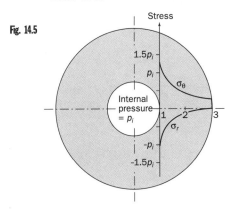

**Example 14.1**

The cylinder of a hydraulic jack has a bore (internal diameter) of 150 mm and is required to operate up to 13.8 MN/m². Determine the required wall thickness for a limiting tensile stress in the material of 41.4 MN/m².

The given boundary conditions are that at $r = 75 \times 10^{-3}$, $\sigma_r = -13.8 \times 10^6$ and $\sigma_\theta = 41.4 \times 10^6$, since the maximum tensile hoop stress occurs at the inner surface. Therefore

$$-13.8 \times 10^6 = A - \frac{B}{5600 \times 10^{-6}}$$

and

$$41.4 \times 10^6 = A + \frac{B}{5600 \times 10^{-6}}$$

Adding the two equations,

$$2A = 27.6 \times 10^6$$

$$A = 13.8 \times 10^6 \, \text{N/m}^2 \qquad \text{and} \qquad B = 154.5 \, \text{kN}$$

At the outside surface, $\sigma_r = 0$; therefore

$$0 = A - \frac{B}{r^2} = 13.8 \times 10^6 - \frac{154.5 \times 10^3}{r^2}$$

$$r^2 = 0.011\,2 \, \text{m}^2$$

$$r = 0.106 \, \text{m}$$

**Axial stress and strain**

Now that expressions have been developed for the radial and circumferential stresses within the cylinder, the next step is to consider what conditions of stress and strain can exist axially along the cylinder. These will depend on the boundary conditions at the ends of the cylinder.

*1. Cylinder with end caps but free to change in length*
In this case there must be equilibrium between the force exerted on the end cover by the internal pressure and the force of the axial stress integrated across the wall of the vessel. Therefore, from Fig. 14.3(a)

$$\sigma_z(\pi r_o^2 - \pi r_i^2) - p_i \pi r_i^2 = 0$$

so that

$$\sigma_z = \frac{p_i r_i^2}{r_o^2 - r_i^2} = \frac{p_i}{k^2 - 1} \tag{14.37}$$

and

$$\varepsilon_z = \frac{\sigma_z}{E} - \frac{\nu}{E}(\sigma_r + \sigma_\theta) = \frac{(1 - 2\nu)p_i}{E(k^2 - 1)}$$

## 2. Pressure retained by piston in each end of cylinder

Since there is no connection between the piston and the cylinder, the axial force due to pressure is reacted entirely by the pistons, and therefore there can be no axial stress in the wall of the cylinder. Thus

$$\sigma_z = 0 \qquad [14.38]$$

and

$$\varepsilon_z = -\frac{\nu}{E}(\sigma_r + \sigma_\theta) = -\frac{2\nu p_i}{E(k^2 - 1)}$$

## 3. Cylinder built-in between rigid end supports

For this case $\varepsilon_z = 0$; in other words, plane strain exists.
Therefore

$$\frac{\sigma_z}{E} - \frac{\nu}{E}(\sigma_r + \sigma_\theta) = 0$$

$$\sigma_z = \nu(\sigma_r + \sigma_\theta) \qquad [14.39]$$

Substituting for $\sigma_r$ and $\sigma_\theta$ from eqns. [14.35] and [14.36],

$$\sigma_z = \frac{2\nu p_i r_i^2}{r_o^2 - r_i^2} = \frac{2\nu p_i}{k^2 - 1} \qquad [14.40]$$

**Maximum shear stress in the cylinder**

Since the radial and circumferential stresses are principal stresses the maximum shear stress in the plane of the cross-section is given by

$$\tau_{max} = \frac{\sigma_\theta - \sigma_r}{2}$$

$$= \frac{p_i}{k^2 - 1}\left(\frac{r_o}{r}\right)^2 \qquad [14.41]$$

This applies for all three end conditions because in each case $\sigma_r < \sigma_z < \sigma_\theta$.

**Yielding in the cylinder**

Yielding will commence at the inner surface and using the Tresca criterion we have

$$\sigma_Y = \sigma_\theta - \sigma_r \qquad \text{at} \qquad r = r_i$$

Using eqn. [14.41],

$$\sigma_Y = \frac{2k^2 p_i}{k^2 - 1}$$

Hence the internal pressure to cause yielding is

$$p_i = \frac{(k^2 - 1)}{2k^2}\sigma_Y \qquad [14.42]$$

This applies for each of the end conditions considered previously.

If the von Mises criterion is employed, using eqn. [12.15] we have

$$(\sigma_\theta - \sigma_r)^2 + (\sigma_r - \sigma_z)^2 + (\sigma_z - \sigma_\theta)^2 = 2\sigma_Y^2$$

Substituting the expressions for the three principal stresses and simplifying gives the internal pressure to cause initial yielding at the bore.

For the three cases considered previously this gives

1.  *Cylinder with end caps but free to change in length*

$$p_i = \frac{(k^2 - 1)\sigma_Y}{k^2\sqrt{3}} \qquad\qquad [14.43a]$$

2.  *Pressure retained by piston in each end of cylinder*

$$p_i = \frac{(k^2 - 1)\sigma_Y}{\sqrt{(3k^4 + 1)}} \qquad\qquad [14.43b]$$

3.  *Cylinder built-in between rigid end supports*

$$p_i = \frac{(k^2 - 1)\sigma_Y}{\sqrt{(3k^4 - 4\nu + 4\nu^2 + 1)}} \qquad\qquad [14.43c]$$

## 14.5 Methods of containing high pressure

From Fig. 14.5 it will be observed that there is a marked variation in the stress in the wall of a thick cylinder subjected to internal pressure, and this situation gets worse when designing for even higher pressures. In order to secure a more uniform stress distribution, one method is to build up the cylinder by 'shrinking' one tube on the outside of another (Fig. 14.6(a)). The inner tube is subjected to hoop compression by the shrink fit of the external tube, which will therefore be subjected to pressure causing hoop tension. When the compound tube is subjected to working pressure, the resultant stresses are the algebraic sum of that due to the shrinking and that due to the internal pressure. The resultant tensile stress at the inner surface of the inner tube is not so large as if the cylinder were composed of one thick tube. The final tensile stress at the inner surface of the outer tube is larger than if the cylinder consisted of one thick tube. Thus a more even stress distribution is obtained.

Another technique used in special industrial pressure vessels is to wind around the outside of a tube a high-tensile-strength ribbon of a rectangular section with sufficient tension to bring the tube into a state of hoop compression (Fig. 14.6(b)). Subsequent internal pressure then has to overcome the hoop compression before tensile stress can be set up in the tube.

A further method for creating hoop compression at the bore of a cylinder is known as *autofrettage*. This consists in applying internal pressure to a single cylinder until yielding and a prescribed amount of plastic deformation occurs at the bore (Fig. 14.6(c)). Since strain increases along the radius of the cylinder, on the release of pressure, the elastic recovery of the material causes the bore of the cylinder to be subjected to compressive hoop stress.

Fig. 14.6

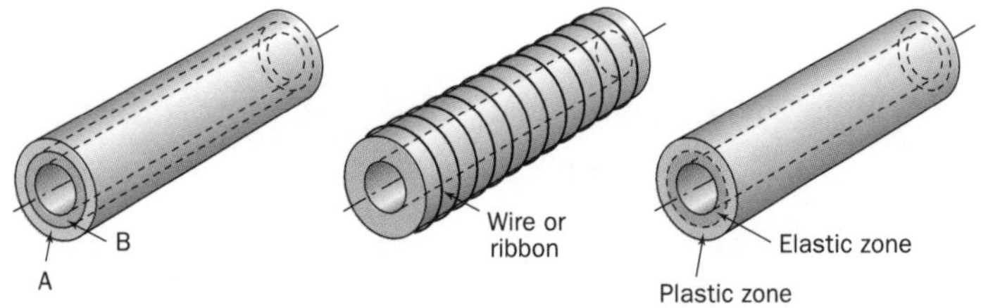

(**a**) Compound cylinder     (**b**) Wire-wound cylinder     (**c**) Autofrettaged cylinder

At the same time tensile hoop stress occurs in the outer material. This technique is discussed further in Chapter 15, in the section on residual stresses.

## 14.6 Stresses set up by a shrink-fit assembly

A shrink fit between two components is a very important and secure method of assembly. It consists, in the case of two cylindrical objects, of the inner diameter of the outer cylinder being slightly less (by a fraction of a millimetre) than the outer diameter of the inner cylinder. Consequently, when they are at the same temperature the outer cannot be passed over the inner. However, if the outer cylinder is heated and the inner cylinder is cooled then the thermal expansion and contraction can be made sufficient to allow one cylinder to pass over the other. On returning each to room temperature there is 'interference' at the mating surface since they cannot regain their original dimensions at the interface. The two components are locked firmly together and a system of radial and circumferential stresses are set up at the interface and through the wall of each cylinder. For elastic conditions the principle of superposition can be used to add the stresses due to shrink-fit interference to those due to internal pressure or rotation.

For two cylindrical components we require eqns. [14.24] and [14.25] for each component given by

$$\sigma_\theta = A + \frac{B}{r^2}, \qquad \sigma_r = A - \frac{B}{r^2} \qquad \text{(inner component)}$$

$$\sigma_\theta = C + \frac{D}{r^2}, \qquad \sigma_r = C - \frac{D}{r^2} \qquad \text{(outer component)} \qquad [14.44]$$

where the constants $A$, $B$, $C$ and $D$ are determined from the boundary conditions. For shrink-fit stresses only, the boundary conditions are that $\sigma_r$ is zero at the inside of the inner cylinder and outside of the outer cylinder, and at the mating surface $r_m$ the radial stress in each vessel must be the same; therefore

$$A - \frac{B}{r_m^2} = C - \frac{D}{r_m^2} \qquad [14.45]$$

Finally, at the mating surface the radial interference $\delta$ is the sum of the displacement of the inner cylinder inwards, $-u'$, and the outer cylinder outwards, $+u''$; thus

$$\delta = -u' + u'' = r_m(\varepsilon_\theta'' - \varepsilon_\theta') \qquad [14.46]$$

We next substitute into eqn. [14.46] the expressions for $\varepsilon'_\theta$ and $\varepsilon''_\theta$,

$$\varepsilon'_\theta = \frac{\sigma'_\theta}{E'} - \frac{\nu'}{E'}\sigma'_r$$

and

$$\varepsilon''_\theta = \frac{\sigma''_\theta}{E''} - \frac{\nu''}{E''}\sigma''_r$$

and thence the relationships (eqn. [14.44]) for $\sigma'_\theta$, $\sigma''_\theta$, $\sigma'_r$ and $\sigma''_r$ at $r = r_m$.

We now have sufficient equations to solve for the constants $A$, $B$, $C$ and $D$.

The following example illustrates the analytical process.

**Example 14.2**

A bronze bush of 25 mm wall thickness is to be shrunk onto a steel shaft 100 mm in diameter. If an interface pressure of 69 MN/m² is required, determine the interference between bush and shaft. Steel: $E = 207$ GN/m², $\nu = 0.28$; bronze: $E = 100$ GN/m², $\nu = 0.29$.

Using constants $A$ and $B$ for the shaft and $C$ and $D$ for the bush, then the radial stress for the shaft is

$$\sigma_{r_s} = A - \frac{B}{r^2}$$

At the centre of the shaft $r = 0$ and this might imply that $\sigma_{r_s}$ was infinite, but this cannot be so and therefore $B$ must be zero; hence $\sigma_{r_s} = A = \sigma_{\theta_s}$ at all points in the shaft. The boundary conditions are

$$\text{At the interface} \qquad \sigma_{r_s} = -69 \times 10^6 = A$$

$$\text{At } r_m = 50, \qquad \sigma_{r_b} = C - \frac{D}{0.0025} = -69 \times 10^6$$

$$\text{At } r_o = 75, \qquad \sigma_{r_b} = C - \frac{D}{0.0056} = 0$$

From which $D = 312 \times 10^3$ and $C = 55.5 \times 10^6$. So at $r_m = 50$ mm, $\sigma_{r_b} = -69$ MN/m², $\sigma_{\theta_b} = 180$ MN/m².

Now, the interference is

$$\delta = -u_s + u_b = r_m(\varepsilon_{\theta_b} - \varepsilon_{\theta_s})$$

where $r_m = 50$. Substituting the values for $\sigma_{\theta_b}$, $\sigma_{r_b}$, $\sigma_{\theta_s}$, $\sigma_{r_s}$,

$$\delta = 0.05\{[180 - 0.29(-69)]/100 - [-69 - 0.28(-69)]/207\}$$

$$= 0.112 \text{ mm}$$

which is the interference required at the nominal interface radius of 50 mm between the shaft and the bush.

**14.7 Stress distribution in a pressurized compound cylinder**

As was explained earlier in this chapter the containment of high internal pressure in, for example, chemical processes can be achieved more effectively by shrinking two or more cylinders, one over the other, to give a compound or multi-tube vessel. The analysis simply uses the basic thick cylinder equations for $\sigma_r$ and $\sigma_\theta$ together with the shrink fit and other boundary conditions.

The method and stress distribution is illustrated by the following worked example.

**Example 14.3**

A vessel is to be used for internal pressures up to 207 MN/m². It consists of two hollow steel cylinders which are shrunk one on the other. The inner tube has an internal diameter of 200 mm and a nominal external diameter of 300 mm, while the outer tube is 300 mm nominal and 400 mm for the inner and outer diameters respectively. The interference at the mating surface of the two cylinders is 0.1 mm. Determine the radial and circumferential stresses at the bores and outside surfaces. The axial stress in the cylinders is to be neglected. $E = 207$ GN/m². Compare the stress distributions with that for a single steel cylinder, having the same overall dimensions, subjected to the same internal pressure.

The boundary conditions are: inner tube, $r = 100$ mm, $\sigma_r = -207$ MN/m²; outer tube, $r = 200$ mm, $\sigma_r = 0$; at the mating surface or interface, $r = 150$ mm nominally, and the radial stresses in the inner and outer tubes are equal, $\sigma_{ri} = \sigma_{ro}$. Also the radial displacement, $u_i$, of the inner cylinder inwards plus the radial displacement, $u_o$, of the outer cylinder outwards due to the shrink fit must equal the interference value. Therefore

$$-u_i + u_o = 0.1 \text{ mm}$$

Using the above conditions and constants $A$, $B$ and $C$, $D$ for the inner and outer tubes respectively, we have four equations:

$$-207 \times 10^6 = A - \frac{B}{0.01} \tag{14.47}$$

$$0 = C - \frac{D}{0.04} \tag{14.48}$$

$$A - \frac{B}{0.0225} = C - \frac{D}{0.0225} \tag{14.49}$$

and

$$-\frac{u_i}{0.15} + \frac{u_o}{0.15} = \frac{0.0001}{0.15}$$

or

$$-\varepsilon_{\theta i} + \varepsilon_{\theta o} = \frac{0.0001}{0.15}$$

and substituting for the strains in terms of the stresses,

$$-\sigma_{\theta i} + \nu \sigma_{r_i} + \sigma_{\theta_o} - \nu \sigma_{r_o} = \frac{0.0001}{0.15} \times 207 \times 10^9$$

and since $\sigma_{r_i} = \sigma_{r_o}$,

$$\sigma_{\theta_o} - \sigma_{\theta i} = 138 \times 10^6$$

Table 14.1

| Radius (mm) | $\sigma_r$ (MN/m$^2$) | $\sigma_\theta$ (MN/m$^2$) | |
|---|---|---|---|
| 100 | −207 | +263 | } inner cylinder |
| 150 | −76.4 | +132.4 | |
| 150 | −76.4 | +272.6 | } outer cylinder |
| 200 | 0 | +196.4 | |

Therefore

$$C + \frac{D}{0.0225} - \left(A + \frac{B}{0.0225}\right) = 138 \times 10^6 \qquad [14.50]$$

The next step is to solve for the constants using eqns. [14.47] to [14.50] and the following values are obtained:

$$A = 28\,\text{MN/m}^2, \qquad B = 2.35\,\text{MN}$$

$$C = 98.2\,\text{MN/m}^2, \qquad D = 3.93\,\text{MN}$$

The required stresses are then computed from the basic equations with the values of the constants above, Table 14.1

If the cylinder had been made of one thick tube of the same overall dimensions as the compound vessel then at the bore $r = 100$ mm, $\sigma_r = -207\,\text{MN/m}^2$

$$-207 \times 10^6 = A - \frac{B}{0.01} \qquad [14.51]$$

and at $r = 200$ mm, $\sigma_r = 0$

$$0 = A - \frac{B}{0.04} \qquad [14.52]$$

Hence $A = 69\,\text{MN/m}^2$ and $B = 2.76\,\text{MN}$ from which $\sigma_\theta$ and $\sigma_r$ are as follows:

| Radius (mm) | $\sigma_r$ (MN/m$^2$) | $\sigma_\theta$ (MN/m$^2$) |
|---|---|---|
| 100 | −207 | +345 |
| 150 | −53.5 | +191.5 |
| 200 | 0 | +138 |

The values of $\sigma_\theta$ may be compared from the compound cylinder and the monobloc cylinder.

Figure 14.7 shows diagrammatically the distribution of radial and hoop stresses through the wall of the compound and single cylinders, and illustrates the more efficient use of material in the former case.

The stress distribution due to shrinkage only may be obtained directly as the difference between the curves for the single and compound cylinders as plotted above. This approach would only apply if the compound tube was made of the one type of material throughout.

**Fig. 14.7**
Compound cylinder ———;
single cylinder — — —;
shrinkage stress, —·—

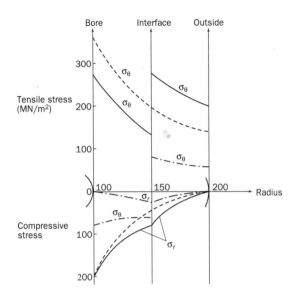

Shrinkage stresses alone could have been obtained by calculation from the four boundary conditions: $\sigma_r = 0$ at $r = 100$ and 200, and $\sigma_{r_i} = \sigma_{r_o}$ and $-u_i + u_o = 0.1$ at the interface radius $r = 150$, as described earlier.

If the two tubes are of different materials then the appropriate elastic constants have to be used where they occur in the various equations.

## 14.8 Spreadsheet solution for stress in a compound cylinder

The general case for the stress in a compound cylinder of two different materials subject to internal and external pressure will now be solved using a spreadsheet. The modulus and Poisson's ratio of the inner and outer cylinder will be taken as $E'$ and $\nu'$, $E''$ and $\nu''$ respectively. The necessary equations to solve for the four arbitrary constants are:

1. At the inner radius of the inner cylinder a known pressure is applied, i.e. at $r = a$

$$\sigma'_r = A - \frac{B}{r^2} = -p_a \qquad [14.53]$$

2. At the outer radius of the outer cylinder a known pressure is applied, i.e. at $r = c$

$$\sigma''_r = C - \frac{D}{r^2} = -p_c \qquad [14.54]$$

3. At the interface, there is continuity of direct stress, i.e. at $r = b$

$$\sigma'_r - \sigma''_r = A - \frac{B}{r^2} - C + \frac{D}{r^2} = 0 \qquad [14.55]$$

4.   At the interface, there is compatibility of deformation, i.e. at $r = b$

$$\frac{\delta}{b} = -\frac{u'}{b} + \frac{u''}{b}$$

$$= -\frac{\sigma'_\theta}{E} + \nu\frac{\sigma'_r}{E} + \frac{\sigma''_\theta}{E} - \nu\frac{\sigma''_r}{E}$$

$$= -\frac{(1-\nu')}{E'}A - \frac{(1+\nu')}{b^2 E'}B + \frac{(1-\nu'')}{E''}C + \frac{(1+\nu'')}{b^2 E''}D$$

[14.56]

These equations can be written in matrix form as

$$\begin{bmatrix} 1 & -1/a^2 & 0 & 0 \\ 0 & 0 & 1 & -1/c^2 \\ 1 & -1/b^2 & -1 & 1/b^2 \\ -(1-\nu')/E' & -(1+\nu')/b^2 E' & (1-\nu'')/E'' & (1+\nu'')/b^2 E'' \end{bmatrix}$$

$$\times \begin{bmatrix} A \\ B \\ C \\ D \end{bmatrix} = \begin{bmatrix} -p_a \\ -p_c \\ 0 \\ \delta/b \end{bmatrix}$$

[14.57]

This matrix equation can be solved using a spreadsheet by inverting the coefficient matrix and then multiplying the resulting inverse by the right-hand side (see Appendix B). A 'macro' (a series of menu commands in Quattro Pro) to perform this operation is shown in Fig. 14.8 starting at cell G17. Once the unknowns $A$, $B$, $C$, $D$ have been found the pressure at the interface can be found from the radial stress in either component at that radius, e.g.

$$\sigma'_r = A - \frac{B}{b^2}$$

[14.58]

Figure 14.8 shows the results for the data of Example 14.3. The only inputs (shown shaded) are the inner, interface and outer radii, the material properties, the amount of interference and the internal and external pressures. Once this information is entered and the macro to perform the matrix solution is executed, the radial and circumferential stresses are calculated automatically. The results corresponding to those in Table 14.1 are shown in cells A19..C22. Any consistent set of units can be used. If dimensions are given in m, then modulus, stress and pressure values must be given in $N/m^2$.

This spreadsheet can also be used to check the interface pressure generated by the interference calculated in Example 14.1, if a very small value (say 1.0E−6) is used for the inner radius to approximate a solid inner cylinder.

|  | A | B | C | D | E | F | G | H |
|---|---|---|---|---|---|---|---|---|
| 1 | a | b | c | E' | V' | E" | V" | Interference |
| 2 | 0.1 | 0.15 | 0.2 | 207000000000 | 0.28 | 207000000000 | 0.28 | 0.0001 |
| 3 |  |  |  |  |  |  |  |  |
| 4 | Matrix to invert |  |  |  |  |  |  |  |
| 5 | 1 | -1/A2^2 | 1 | 0 |  |  |  |  |
| 6 | 0 | 0 | 0 | -1/C2^2 |  |  |  |  |
| 7 | 1 | -1/B2^2 | 1 | -B7 |  |  |  |  |
| 8 | -(1-E2)/D2 | -(1+E2)/(B2^2*D2) | (1+G2)/F2 | (1+G2)/(B2^2*F2) |  |  |  |  |
| 9 |  |  |  |  |  |  |  |  |
| 10 | Inverse |  |  |  |  |  |  | A-D |
| 11 | -0.3333333333333 | 1.3333333333333 | 1.1233333333333 | -60375000000 |  | -207000000 | -Pa | 28750000 |
| 12 | -0.0133333333333 | 0.01333333333333 | 0.01123333333333 | -603750000 |  | 0 | -Pc | 2357500 |
| 13 | -0.3333333333333 | 1.3333333333333 | 0.48333333333333 | 43125000000 |  | 0 |  | 97750000 |
| 14 | -0.0133333333333 | 0.01333333333333 | 0.01933333333333 | 1725000000 |  | +H2/B2 |  | 3910000 |
| 15 |  |  |  |  |  |  |  |  |
| 16 |  |  |  |  |  |  | Macro to solve matrix equation |  |
| 17 |  |  |  |  |  |  | {Invert.Source a5 ..d8} |  |
| 18 | Radius | Sigma_r | Sigma_t |  |  |  | {Invert.Destination a11 ..d14} |  |
| 19 | +A2 | +H11-H12/A19^2 | +H11+H12/A19^2 | Inner Cylinder |  |  | {Invert.Go} |  |
| 20 | +B2 | +H11-H12/A20^2 | +H11+H12/A20^2 |  |  |  | {Multiply.matrix_1 a11 .d14} |  |
| 21 | +B2 | +H13-H14/A21^2 | +H13+H14/A21^2 | Outer Cylinder |  |  | {Multiply.matrix_2 f11 ..d14} |  |
| 22 | +C2 | +H13-H14/A22^2 | +H13+H14/A22^2 |  |  |  | {Multiply.Destination h11 ..h14} |  |
| 23 |  |  |  |  |  |  | {Multiply.Go} |  |

(**a**) Data and cell formulae

|  | A | B | C | D | E | F | G | H |
|---|---|---|---|---|---|---|---|---|
| 1 | a | b | c | E' | V' | E" | V" | Interference |
| 2 | 0.1 | 0.15 | 0.2 | 2.07E+11 | 0.28 | 2.07E+11 | 0.28 | 0.0001 |
| 3 |  |  |  |  |  |  |  |  |
| 4 | Matrix to invert |  |  |  |  |  |  |  |
| 5 | 1 | -100 | 0 | 0 |  |  |  |  |
| 6 | 0 | 0 | 1 | -25 |  |  |  |  |
| 7 | 1.00 | -44.44 | -1.00 | 44.44 |  |  |  |  |
| 8 | -3.48E-12 | -2.75E-10 | 3.48E-12 | 2.75E-10 |  |  |  |  |
| 9 |  |  |  |  |  |  |  |  |
| 10 | Inverse |  |  |  |  |  |  | A-D |
| 11 | -0.33 | 1.33 | 1.12 | -6.04E+10 |  | -2.07E+08 | -Pa | 28750000 |
| 12 | -0.01 | 0.01 | 0.01 | -6.04E+08 |  | 0.00E+00 | -Pc | 2357500 |
| 13 | -0.33 | 1.33 | 0.48 | 4.31E+10 |  | 0 |  | 97750000 |
| 14 | -0.01 | 0.01 | 0.02 | 1.72E+09 |  | 6.67E-04 |  | 3910000 |
| 15 |  |  |  |  |  |  |  |  |
| 16 |  |  |  |  |  |  |  |  |
| 17 |  |  |  |  |  |  | Macro to solve matrix equation |  |
| 18 | Radius | Sigma_r | Sigma_t |  |  |  | {Invert.Source a5 ..d8} |  |
| 19 | 0.1 | -2.07E+08 | 2.64E+08 | Inner Cylinder |  |  | {Invert.Destination a11 ..d14} |  |
| 20 | 0.15 | -7.60E+07 | 1.34E+08 |  |  |  | {Invert.Go} |  |
| 21 | 0.15 | -7.60E+07 | 2.72E+08 | Outer Cylinder |  |  | {Multiply.matrix_1 a11 .d14} |  |
| 22 | 0.2 | -0.00 | 1.96E+08 |  |  |  | {Multiply.matrix_2 f11 ..d14} |  |
| 23 |  |  |  |  |  |  | {Multiply.Destination h11 ..h14} |  |
| 24 |  |  |  |  |  |  | {Multiply.Go} |  |

(**b**) Spreadsheet display

**Fig. 14.8**

## 14.9 Stress distribution in a thin rotating disc

A simplified model of a component such as a gas turbine rotor is a uniformly thin disc which, when rotating at a constant velocity, is subjected to stresses induced by centripetal acceleration. This is a problem which produces deformations symmetrical about the rotating axis. If the disc is thin in section then it is assumed that plane stress exists, so the radial and hoop stresses are constant through the thickness, and there is no stress in the $z$–direction.

**Fig. 14.9**

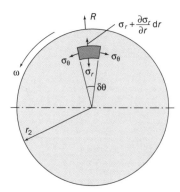

The equation of equilibrium of an element, Fig. 14.9, is that derived in Chapter 13 for the axially symmetric stress system, but in this case a body force term must be included which is determined from the centripetal acceleration. Hence, from eqn. [13.11],

$$\frac{\mathrm{d}\sigma_r}{\mathrm{d}r} + \frac{\sigma_r - \sigma_\theta}{r} + R = 0$$

where $R$ is the body force per unit volume.

In order to analyse the rotating disc as a static equilibrium problem, D'Alembert's principle is applied whereby the inward force on the element due to the centripetal acceleration is replaced with an outward centrifugal force given by

$$F = (\rho r\, \delta\theta\, \delta r\, z)\omega^2 r = Rr\, \delta\theta\, \delta r\, z$$

where $\rho$ is the density, $\omega$ is the steady rotational velocity in radians per second and $z$ is the thickness of the disc. Therefore

$$R = \rho\omega^2 r$$

and

$$\frac{\mathrm{d}\sigma_r}{\mathrm{d}r} + \frac{\sigma_r - \sigma_\theta}{r} + \rho\omega^2 r = 0 \qquad\qquad [14.59]$$

The strain–displacement equations for axial symmetry are

$$\varepsilon_r = \frac{\mathrm{d}u}{\mathrm{d}r}, \qquad \varepsilon_\theta = \frac{u}{r}$$

and the stress–strain relationships are

$$\varepsilon_r = \frac{\sigma_r}{E} - \frac{\nu\sigma_\theta}{E}, \qquad \varepsilon_\theta = \frac{\sigma_\theta}{E} - \frac{\nu\sigma_r}{E}$$

Using these four equations to obtain $\sigma_r$ and $\sigma_\theta$ in terms of $u$,

$$\sigma_r = \left(\frac{\mathrm{d}u}{\mathrm{d}r} + \frac{\nu u}{r}\right)\frac{E}{1 - \nu^2} \qquad\qquad [14.60]$$

$$\sigma_\theta = \left(\nu\frac{\mathrm{d}u}{\mathrm{d}r} + \frac{u}{r}\right)\frac{E}{1 - \nu^2} \qquad\qquad [14.61]$$

Substituting in eqn. [14.59] we obtain

$$\frac{d^2u}{dr^2} + \frac{1}{r}\frac{du}{dr} - \frac{u}{r^2} + \left(\frac{1-\nu^2}{E}\right)\rho\omega^2 r = 0$$

or

$$\frac{d^2u}{dr^2} + \frac{1}{r}\frac{du}{dr} - \frac{u}{r^2} = -\left(\frac{1-\nu^2}{E}\right)\rho\omega^2 r \qquad [14.62]$$

This is a linear differential equation of the second order. The general solution consists of the sum of two separate solutions known as the complementary function and the particular integral. The former is the solution of the left-hand side and the latter is obtained by considering the right-hand side of eqn. [14.62]. Thus the complementary function is

$$u = Cr + \frac{C'}{r}$$

and the particular integral is

$$u = -\left(\frac{1-\nu^2}{E}\right)\frac{\rho\omega^2 r^3}{8}$$

The complete solution is therefore

$$u = Cr + \frac{C'}{r} - \left(\frac{1-\nu^2}{E}\right)\frac{\rho\omega^2 r^3}{8} \qquad [14.63]$$

in which $C$ and $C'$ are constants to be determined from the boundary conditions.

Using eqn. [14.63] and substituting for $u$ and $(du/dr)$ in eqns. [14.60] and [14.61] and simplifying the various constant terms by inserting new ones, $A$ and $B$, we obtain

$$\sigma_r = A - \frac{B}{r^2} - \left(\frac{3+\nu}{8}\right)\rho\omega^2 r^2 \qquad [14.64]$$

$$\sigma_\theta = A + \frac{B}{r^2} - \left(\frac{1+3\nu}{8}\right)\rho\omega^2 r^2 \qquad [14.65]$$

The constants $A$ and $B$ are found from the appropriate boundary conditions of the problem.

**Solid disc with unloaded boundary**  If the disc is continuous from the centre to some outer radius $r = r_2$, then it is apparent that, unless $B = 0$, the stresses would become infinite at $r = 0$. To find $A$ it is only necessary to use the condition that

$$\sigma_r = 0 \qquad \text{at} \qquad r = r_2$$

from which

$$A = \frac{3+\nu}{8}\rho\omega^2 r_2^2$$

and

$$\sigma_r = \frac{3+\nu}{8}\rho\omega^2(r_2^2 - r^2) \qquad [14.66]$$

$$\sigma_\theta = \frac{3+\nu}{8}\rho\omega^2 r_2^2 - \frac{1+3\nu}{8}\rho\omega^2 r^2$$

$$= \frac{\rho\omega^2}{8}[(3+\nu)r_2^2 - (1+3\nu)r^2] \qquad [14.67]$$

**Fig. 14.10**

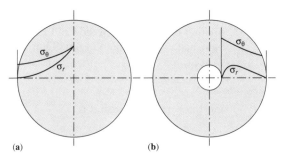

(a)                    (b)

The distributions of radial and hoop stresses are shown in Fig. 14.10(a). The maximum stress occurs at the centre, where $r = 0$, and then

$$\sigma_r = \sigma_\theta = \frac{3+\nu}{8}\rho\omega^2 r_2^2$$

**Disc with a central hole and unloaded boundaries**

In this case the boundary conditions are that the radial stress will be zero at $r = r_1$ and $r = r_2$, the radius of the hole and outside periphery of the disc respectively. Therefore

$$\sigma_r = A - \frac{B}{r_2^2} - \frac{3+\nu}{8}\rho\omega^2 r_2^2 = 0$$

and also

$$\sigma_r = A - \frac{B}{r_1^2} - \frac{3+\nu}{8}\rho\omega^2 r_1^2 = 0$$

Solving these two equations for the constants gives

$$B = \frac{3+\nu}{8}\rho\omega^2 r_1^2 r_2^2 \qquad \text{and} \qquad A = \frac{3+\nu}{8}\rho\omega^2(r_1^2 + r_2^2)$$

Therefore

$$\sigma_r = \frac{3+\nu}{8}\rho\omega^2\left(r_1^2 + r_2^2 - \frac{r_1^2 r_2^2}{r^2} - r^2\right) \qquad [14.68]$$

and

$$\sigma_\theta = \frac{3+\nu}{8}\rho\omega^2\left(r_1^2 + r_2^2 + \frac{r_1^2 r_2^2}{r^2} - \frac{1+3\nu}{3+\nu}r^2\right) \qquad [14.69]$$

The maximum value of the hoop stress, $\sigma_\theta$, is at $r = r_1$, and is given by

$$\sigma_{\theta\,max} = \frac{3+\nu}{4}\rho\omega^2\left(r_2^2 + \frac{1-\nu}{3+\nu}r_1^2\right) \tag{14.70}$$

$\sigma_r$ is a maximum when $d\sigma_r/dr = 0$ or $r = \sqrt{(r_1 r_2)}$; therefore

$$\sigma_{r\,max} = \frac{3+\nu}{8}\rho\omega^2(r_2 - r_1)^2 \tag{14.71}$$

The stress distributions of $\sigma_\theta$ and $\sigma_r$ for the disc with the hole are shown in Fig. 14.10(b).

**Disc with a loaded boundary**  In general, rotor discs will have mounted on the outer boundary a large number of blades. These will themselves each have a centrifugal force component which will have to be reacted at the periphery of the disc. Given the mass of each blade, its effective centre of mass and the number of blades we can compute the force due to each blade at a particular value of $\omega$. Multiplying by the number of blades gives the total force which may then be computed as a uniformly distributed load. Dividing this by the thickness of the outer boundary gives the required value of $\sigma_r$ to use as the boundary condition when evaluating $A$ and $B$.

**Disc shrunk onto a shaft**  The concept of a shrink fit between two components was developed earlier for the compound cylinder. The same principle is very valuable for locating a rotating disc onto a shaft and avoiding mechanical attachments, e.g. bolting, riveting. The stresses set up by the shrink-fit mechanism are quite independent when the disc is stationary. However, since at the mating surface the radial stress is compressive, when rotation commences there is a superposition of radial tensile stress. It is therefore necessary to ensure that the shrink-fit stress is always in excess of the rotational stress at the mating surfaces so that the disc does not become 'free' on the shaft.

## 14.10 Rotational speed for initial yielding

The design of rotating discs must take account of the limit of rotational speed which would induce initial yielding at some point in the disc.

**Solid disc**  The maximum stress occurs at $r = 0$ and is

$$\sigma_r = \sigma_\theta = \frac{3+\nu}{8}\rho\omega^2 r_2^2$$

Using the maximum shear-stress criterion (Tresca)

$$\sigma_Y = \sigma_\theta \qquad (\text{since } \sigma_z = 0)$$

Therefore

$$\sigma_Y = \frac{(3+\nu)}{8}\rho\omega_Y^2 r_2^2$$

$$\omega_Y = \frac{1}{r_2}\sqrt{\frac{8\sigma_Y}{(3+\nu)\rho}} \tag{14.72}$$

Using the shear-strain energy criterion (von Mises)

$$\sigma_Y^2 = \sigma_\theta^2 + \sigma_r^2 - \sigma_\theta \sigma_r$$

$$= \sigma_\theta^2 \quad \text{(since } \sigma_\theta = \sigma_r\text{)}$$

$$\sigma_Y = \sigma_\theta$$

and

$$\omega_Y = \frac{1}{r_2}\sqrt{\frac{8\sigma_Y}{(3+\nu)\rho}} \qquad [14.73]$$

which is the same value as for the Tresca criterion.

**Disc with central hole**  The maximum hoop stress occurs at $r = r_1$ and is

$$\sigma_\theta = \frac{3+\nu}{4}\rho\omega^2\left(r_2^2 + \frac{(1-\nu)}{(3+\nu)}r_1^2\right)$$

For the Tresca criterion

$$\sigma_Y = \sigma_\theta = \frac{3+\nu}{4}\rho\omega_Y^2\left(r_2^2 + \frac{(1-\nu)}{(3+\nu)}r_1^2\right)$$

$$\omega_Y = \left(\frac{4\sigma_Y}{\rho[(3+\nu)r_2^2 + (1-\nu)r_1^2]}\right)^{1/2} \qquad [14.74]$$

For the von Mises criterion since $\sigma_r = \sigma_z = 0$ at $r = r_1$ yield occurs when $\sigma_Y = \sigma_\theta$; hence the rotational speed at yield is the same as given by eqn. [14.74].

**Example 14.4**

A steel ring has been shrunk onto the outside of a solid steel disc and shaft. The interface radius is 250 mm and the outer radius of the assembly is 356 mm. If the pressure between the ring and the disc is not to fall below 34.5 MN/m², and the circumferential stress at the inside of the ring must not exceed 207 MN/m², determine the maximum speed at which the assembly can be rotated. What is then the stress at the centre of the disc? $\rho = 7.75$ Mg/m³, $\nu = 0.28$.

For the ring at $r = 356$ mm, $\sigma_r = 0$, and at $r = 250$ mm, $\sigma_r = -34.5 \times 10^6$; therefore

$$0 = A - \frac{B}{0.126} - \left(\frac{3+0.28}{8} \times 7.75\omega^2 \times 0.126 \times 10^3\right)$$

and

$$-34.5 \times 10^6 = A - \frac{B}{0.0625}$$

$$- \left(\frac{3+0.28}{8} \times 7.75\omega^2 \times 0.062.5 \times 10^3\right)$$

from which

$$B = (4280 + 0.025\omega^2)10^3 \quad \text{and} \quad A = (34\,000 + 0.6\omega^2)10^3$$

Also when $r = 250\,\text{mm}$, $\sigma_\theta$ must not exceed $207\,\text{MN/m}^2$; therefore

$$207 \times 10^6 = A + \frac{B}{0.0625}$$

$$-\left(\frac{1 + (3 \times 0.28)}{8} \times 7.75 \times 10^3\omega^2 \times 0.0625\right)$$

Substituting for $A$ and $B$,

$$207 \times 10^3 = 34\,000 + 0.6\omega^2 + 68\,500 + 0.4\omega^2 - 0.111\omega^2$$

From which

$$\omega = 343\,\text{rad/s} \quad = 3280\,\text{rev/min}$$

For the solid disc, using constants $C$ and $D$, as shown previously, $D$ must be zero; therefore

$$\sigma_r = C - \left(\frac{3 + \nu}{8}\right)\rho\omega^2 r^2$$

At $r = 250$, $\sigma_r = -34.5 \times 10^6$; therefore

$$-34.5 \times 10^6 = C - \left(\frac{3 + 0.28}{8} \times 7.75 \times 10^3 \times 117\,500 \times 0.0625\right)$$

and

$$C = (-34.5 \times 10^6) + (23.3 \times 10^6)$$

$$= -11.2 \times 10^6\,\text{N/m}^2$$

But at the centre of the disc, $r = 0$ and $\sigma_r = \sigma_\theta = C$; therefore

$$\sigma_r = \sigma_\theta = -11.2\,\text{MN/m}^2$$

This type of problem also lends itself to the type of spreadsheet solution illustrated in section 14.8.

**14.11 Stresses in a rotor of varying thickness with rim loading**

The thin uniform discs analysed in the previous section, although illustrating equilibrium and compatibility concepts, do not represent a very realistic design configuration. Because the hoop stress $\sigma_\theta$ is highest at the bore and reduces towards the periphery it is more economical to vary the thickness in similar proportions. A varying cross-section for a turbine rotor might be as shown in Fig. 14.11. A simple method of solution was devised by Donath for steam turbines which consisted of dividing up the cross-section into a number of constant thickness rings as shown.

The equations [14.64] and [14.65] for $\sigma_r$ and $\sigma_\theta$ can be applied to each ring observing the required conditions of equilibrium and compatibility at

**Fig. 14.11**

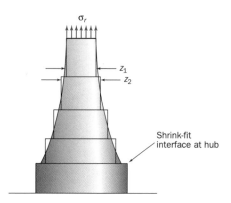

each. Let

$$S \text{ (sum)} = \sigma_\theta + \sigma_r = 2A - 4\frac{(1+\nu)}{8}\rho\omega^2 r^2 \qquad [14.75]$$

and

$$D \text{ (difference)} = \sigma_\theta - \sigma_r = 2\frac{B}{r^2} + 2\frac{(1-\nu)}{8}\rho\omega^2 r^2 \qquad [14.76]$$

Replacing $\omega r$ by the tangential velocity $V$ at radius $r$ gives

$$S = \frac{(1+\nu)}{2}\rho(-V^2 + C_1) \qquad [14.77]$$

$$D = \frac{(1-\nu)}{4}\rho\left(V^2 + \frac{C_2}{V^2}\right) \qquad [14.78]$$

Donath then constructed a chart of a series of $S$ and $D$ curves for various values of $C_1$ and $C_2$ and for a particular density and Poisson's ratio. A typical Donath chart is illustrated in Fig. 14.12.

If values of $S$ and $D$ are known at any radius then $\sigma_\theta$ and $\sigma_r$ can be found since

$$\sigma_\theta = \frac{S+D}{2} \qquad \text{and} \qquad \sigma_r = \frac{S-D}{2} \qquad [14.79]$$

The procedure is as follows having divided the disc into several rings of constant thickness:

1. Calculate the rim loading due to blades and thence $\sigma_{r_\theta}$ at the rim.
2. Assume a value of $\sigma_{\theta_\theta}$ at the rim and then calculate $S$ and $D$ at the rim.
3. Using the chart and the tangential velocity at the rim will give starting positions on the appropriate $S$ and $D$ curves.
4. Proceed along the $S$ and $D$ lines from the rim to the first interface between rings at the correct tangential velocity for that interface radius. Use these new values of $S$ and $D$ to calculate $\sigma_{r_1}$.
5. At this interface we must satisfy equilibrium for each side of the interface; hence

$$\sigma_{r_1} . z_1 = \sigma_{r_2} . z_2 \qquad [14.80]$$

Fig. 14.12

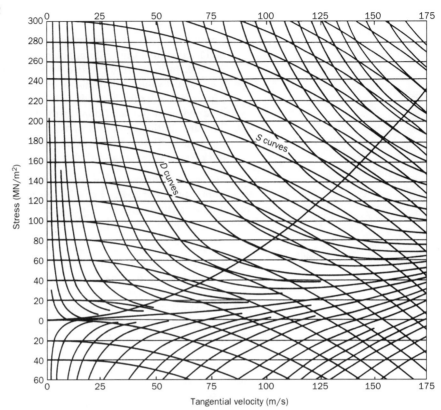

where $z_1$ and $z_2$ are the thicknesses of the adjacent rings.

Now $\Delta\sigma_r = \sigma_{r_1} - \sigma_{r_2}$

$$\Delta\sigma_r = \sigma_{r_1}\left(1 - \frac{\sigma_{r_2}}{\sigma_{r_1}}\right) = \sigma_{r_1}\left(1 - \frac{z_1}{z_2}\right) \qquad [14.81]$$

and we must also satisfy compatibility so that circumferential strains must be equal on each side of the step. Hence, $\varepsilon_{\theta_1} = \varepsilon_{\theta_2}$ and

$$\sigma_{\theta_1} - \nu\sigma_{r_1} = \sigma_{\theta_2} - \nu\sigma_{r_2} \qquad [14.82]$$

$$\Delta\sigma_\theta = \nu\Delta\sigma_r$$

$$\Delta\sigma_\theta = \nu\sigma_{r_1}\left(1 - \frac{z_1}{z_2}\right) \qquad [14.83]$$

Adding and subtracting eqns. [14.81] and [14.83],

$$\Delta\sigma_\theta + \Delta\sigma_r = \Delta S = (1+\nu)\Delta\sigma_r \qquad [14.84]$$

$$\Delta\sigma_\theta - \Delta\sigma_r = \Delta D = (\nu - 1)\Delta\sigma_r \qquad [14.85]$$

6. The new values of $S$ and $D$ for the other side of the interface are calculated using eqns. [14.81], [14.84] and [14.85].

7. Follow these curves on the chart to the next interface velocity $V$ and calculate $\sigma_r$ using the values of $S$ and $D$. Repeat calculations 5 and 6.

8. Continue for the remaining steps until reaching the centre or the inner radius. The final values should be $\sigma_r = \sigma_\theta$ for a solid disc and $\sigma_r = 0$ or a shrink-fit pressure for a disc with a central hole.

9. The correct values will probably not be obtained on the first run and an adjustment will then be made to the $\sigma_\theta$ value at the rim and the process repeated until the conditions at the bore or centre are correct.

10. The distribution of $\sigma_r$ and $\sigma_\theta$ may now be plotted by calculation from the values off the $S$ and $D$ curves at the radii for which each ring thickness equals the disc thickness.

The above analysis may seem rather outdated with the current availability of computers, but the object here is to demonstrate the principle of analysis and the Donath chart is merely a graphical aid. The following simplified worked example will help to clarify the solution.

**Example 14.5**

A steel rotor disc of 800 mm diameter and varying thickness, as shown in Fig. 14.11, rotates at 2860 rev/min. The outer periphery is subjected to radial and circumferential stresses of 17 and 33 MN/m² respectively. Evaluate the magnitudes and distributions of radial and circumferential stresses in the disc. The interface radii are 50, 100, 175 and 275 mm and the ring segments are of width 100, 65, 43 and 30 mm respectively. $\nu = 0.3$.

The first step is to calculate the velocities at the periphery and at each ring interface and these are given in the table below.

At the periphery

$$S = 33 + 17 = 50 \, \text{MN/m}^2$$

and

$$D = 33 - 17 = 16 \, \text{MN/m}^2$$

which gives the starting points on the chart, Fig. 14.12, at a velocity of 120 m/s. The remainder of the solution is shown in tabulated form. The final values of $\sigma_r$ and $\sigma_\theta$ are determined from the values of $S$ and $D$ half-way between each interface where the ring thickness equals the disc thickness. In this example further iteration is not required and the negative value of $\sigma_r$ is due to a shrink fit.

| $V$ (m/s) | $z_1$ | $z_2$ | $z_1/z_2$ | $(1 - z_1/z_2)$ | $S$ | $D$ | $\sigma_r$ | $\Delta\sigma_r$ | $\Delta S$ | $\Delta D$ | $\sigma_\theta$ | $\sigma_r$ |
|---|---|---|---|---|---|---|---|---|---|---|---|---|
| 120 | – | 30 | – | – | 50 | 16 | 17 | – | – | – | 33 | 17 |
| 82.4 | 30 | 43 | 0.7 | 0.3 | 90 | 0 | 45 | 13.5 | 17.5 | −9.5 | | |
| | | | | | | | | | | | 47 | 39 |
| 52.4 | 43 | 65 | 0.66 | 0.34 | 95 | 5 | 45 | 15.3 | 19.9 | −10.7 | | |
| | | | | | | | | | | | 49 | 27 |
| 30 | 65 | 100 | 0.65 | 0.35 | 80 | 38 | 21 | 7.3 | 9.5 | −5.1 | | |
| | | | | | | | | | | | 72 | 0 |
| 15 | 100 | – | – | – | 74 | 140 | −33 | – | – | – | 107 | −33 |

### 14.12 Summary

The application of the equilibrium and strain–displacement relationships has been demonstrated at length in relation to two important engineering components, namely the thick-walled pressure vessel and the rotating turbine rotor. Various design problems have been examined such as containing very high pressure by using a compound vessel. This leads to the concept of shrink fitting and the associated initial stresses prior to pressurizing. Numerical constants in the equations for radial and hoop stresses have to be determined from the boundary conditions which must be accurately assessed. The axial condition in the cylinder wall depends on the method of end closure and sealing and this is an important design consideration.

Stresses set up by rotation are a major design feature of turbine rotors and compressors. The analysis commenced with the thin uniform disc to enable a full understanding of the plane stress solution for $\sigma_\theta$ and $\sigma_r$ obtained from the equilibrium, compatibility and stress–strain equations, together with a variety of boundary conditions. It became a fairly straightforward process then to develop an analysis for the varying-thickness rotor.

Finally, we must remember that elastic design relies on the application of yield criteria limits such as those of Tresca and von Mises to enable us to determine a limiting pressure for a cylinder and limiting speed for a rotor. However, it will be seen in Chapter 15 that there can be good reasons for developing limited amounts of plastic deformation in a cylinder or rotor to induce favourable residual stresses which allow enhanced performance. In *all* cases the basic principles of equilibrium of forces, geometry of deformation and the elastic or plastic stress–strain relationship, together with the appropriate boundary conditions, must be followed.

### Problems

14.1   A beam of depth $d$ and length $l$ is simply-supported at each end and carries a uniformly distributed load over the whole span. Show that the maximum vertical direct stress $\sigma_y$ is $\frac{4}{3}(d/l)^2$ times the maximum bending stress $\sigma_x$ at mid-span, and therefore in most cases may be considered insignificant in relation to the latter.

14.2   Solve Problem 13.3 using the thick-walled cylinder relationships and compare the difference in axial loads.

14.3   Determine the $k$ ratio for a thick-walled cylinder subjected to an internal pressure of $80 \, \mathrm{MN/m^2}$ if the circumferential stress is not to exceed $140 \, \mathrm{MN/m^2}$. What are the maximum shear stresses at the inside and outside surfaces?

14.4   In a pressure test on a hydraulic cylinder of $120 \, \mathrm{mm}$ external diameter and $60 \, \mathrm{mm}$ internal diameter the hoop and longitudinal strains are measured by means of strain gauges on the outer surface and found to be $266 \times 10^{-6}$ and $69.6 \times 10^{-6}$ respectively for an internal pressure of $100 \, \mathrm{MN/m^2}$. Determine the actual hoop stress at the outer surface and compare this result with the calculated value. Determine also the safety factor for the cylinder according to the maximum shear stress theory. The properties of the cylinder material are as follows: $\sigma_y = 280 \, \mathrm{MN/m^2}$; $E = 208 \, \mathrm{MN/m^2}$; $\nu = 0.29$.

14.5    A cylinder of internal radius $a$ and external radius $b$ is sealed at atmospheric pressure and put into the sea. If the pressure due to the water is 10 atmospheres, calculate the maximum hoop stress in the cylinder in units of bars. Assume that the volume of the cylinder does not change.

14.6    Derive expressions for the radial, circumferential and axial strains at the inner and outer surfaces of a thick-walled cylinder of radius ratio $k$ with closed ends subjected to internal pressure $p$.

14.7    One method of determining Poisson's ratio for a material is to subject a cylinder to internal pressure and to measure the axial, $\varepsilon_z$, and circumferential, $\varepsilon_\theta$, strains on the outer surface. Show that

$$\nu = \frac{\varepsilon_\theta - 2\varepsilon_z}{2\varepsilon_\theta - \varepsilon_z}$$

The axial and circumferential strains on the outer surface of a closed-ended cylinder of diameter ratio 3 subjected to internal pressure were found to be $1.02 \times 10^{-4}$ and $4.1 \times 10^{-4}$ respectively. Calculate the internal pressure and the hoop strain at the bore. It may be assumed that

$$\sigma_z = \frac{\sigma_r + \sigma_\theta}{2} \qquad E = 207 \text{ GN/m}^2$$

14.8    Show that the pressure generated at the interface between two cylinders when they are shrunk together is given by

$$p = \left(\frac{E\delta}{r}\right) \left[\frac{k_2{}^2 + 1}{k_2{}^2 - 1} + \frac{k_1{}^2 + 1}{k_1{}^2 - 1}\right]^{-1}$$

where $\delta$ is the interference between the outer and inner cylinders and $r$ is the nominal interface radius. The suffices 1 and 2 refer to the inner and outer cylinders respectively.

14.9    A steel insert of 60 mm ID and 70 mm OD is shrunk with a diametral interference of 0.1 mm into an aluminium mould for forming plastic. The mould may be taken as a cylinder with an OD of 100 mm. What interference pressure will be generated? For steel $E = 200$ GN/m$^2$, $\nu = 0.33$; for aluminium $E = 70$ GN/m$^2$, $\nu = 0.33$.

**Fig. 14.13**

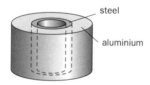

14.10   The ram of a hydraulic actuator is 50.00 mm diameter. It is located in a bearing with an internal diameter of 50.05 mm and an outer diameter of 75 mm, Fig. 14.14.

(*a*) If the ideal conditions for sliding require a clearance on the diameter of 0.025 mm, what is the optimum compressive load on the actuator ram? The bearing is unstressed.

(b)   At what compressive load will the bearing and ram start to interfere?

(c)   What will the interference pressure between ram and bearing be at a load of 2000 kN?

For both ram and bearing, take $E = 200$ GN/m$^2$, $\nu = 0.30$.

**Fig. 14.14**

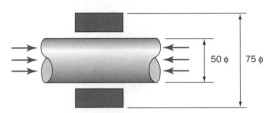

50 ϕ      75 ϕ

14.11   Figure 14.15 shows a dowel having an interference fit in a steel plate which is very large compared to the hole diameter. Both dowel and plate have the same mechanical properties: $E = 200$ GN/m$^2$; $\nu = 0.3$; $\sigma_y = 400$ GN/m$^2$. What is the maximum interference between dowel and hole that can be used without causing yielding of either component? What force would be necessary to pull the dowel out of the hole, given a coefficient of friction $\mu = 0.1$? (You can neglect the effect of any axial force on the interface pressure, i.e. assume plane stress). What minimum amount of interference would be necessary to ensure the pull-out force was at least half the maximum value?

If either component could be heat treated to increase its yield stress, which should it be?

**Fig. 14.15**

10 mm ϕ

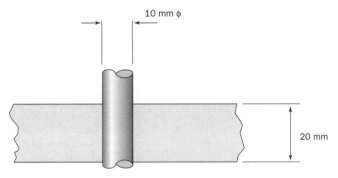

20 mm

14.12   A large circular saw, Fig. 14.16 may be considered as a 4 mm thick disc with a central hole bolted to a rigid inner disc rotating at 2500 rpm.

If the disc material is steel ($E = 200$ GN/m$^2$, $\nu = 0.3$, $\rho = 7800$ kg/m$^3$) estimate the shearing force that has to be carried by each one of the 24 securing bolts.

14.13   Figure 14.17 shows a rigid metal fitting attached to the end of a thick-walled rubber hydraulic hose. The seal is formed due to compression of the outer wall of the hose, which has a modulus $E = 100$ MN/m$^2$ and $\nu = 0.5$. The inner radius of the hose and the mating diameter of the fitting are both 6 mm. Assuming that there is no axial load on the hose, find the interface pressures between the

**Fig. 14.16**

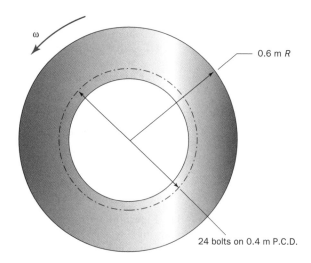

inner and outer radii of the hose and the metal fitting. From your results, estimate the hydraulic pressure at which the fitting would start to leak.

**Fig. 14.17**

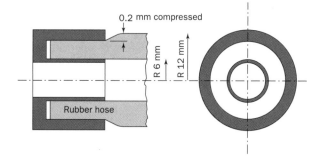

14.14 A gun barrel is formed by shrinking a tube of 224 mm external diameter and 168 mm internal diameter upon another tube of 126 mm internal diameter. After shrinking, the radial pressure at the common surface is $13.8 \, \text{MN/m}^2$. Determine the hoop stresses at the inner and outer surfaces of each tube. Plot diagrams to show the variation of the hoop and radial stresses with radius for both tubes.

14.15 A steel tube has an internal diameter of 25 mm and an external diameter of 50 mm. Another tube, of the same steel, is to be shrunk over the outside of the first so that the shrinkage stresses just produce a condition of yield at the inner surface of each tube. Determine the necessary difference in diameters of the mating surfaces before shrinking and the required external diameter of the outer tube. Assume that yielding occurs according to the maximum shear stress criterion and that no axial stresses are set up due to shrinking. Yield stress in simple tension or compression = $414 \, \text{MN/m}^2$, $E = 207 \, \text{GN/m}^2$.

14.16 A steel bush is to be shrunk on to a steel shaft so that the internal diameter is extended 0.152 mm above its original size. The inside and outside diameters of the sleeve are 203 mm and 305 mm respectively. Find (i) the normal pressure intensity between the bush and the shaft, (ii) the hoop stresses at the inner and outer surfaces of the bush. $E = 207 \text{ GN/m}^2$, $\nu = 0.28$.

14.17  A solid steel shaft of 0.2 m diameter has a bronze bush of 0.3 m outer diameter shrunk on to it. In order to remove the bush the whole assembly is raised in temperature uniformly. After a rise of 100°C the bush can just be moved along the shaft. Neglecting any effect of temperature in the axial direction, calculate the original interface pressure between the bush and the shaft. $E_{steel} = 208 \text{ GN/m}^2$; $\nu_{steel} = 0.29$; $\alpha_{steel} = 12 \times 10^{-6}$ per deg C; $E_{bronze} = 112 \text{ GN/m}^2$; $\nu_{bronze} = 0.33$; $\alpha_{bronze} = 18 \times 10^{-6}$ per deg C.

14.18  Commencing from the relationship derived in Problem 13.2 show that the radial and circumferential stresses in a thick-walled spherical shell may be expressed as

$$\sigma_r = A - 2\frac{B}{r^3} \qquad \text{and} \qquad \sigma_\theta = A + \frac{B}{r^3}$$

Determine the maximum shear stress at the inner and outer surfaces of a spherical shell, having a $k$ ratio of 1.5, for an internal pressure of 7 MN/m².

14.19  A steel shaft of radius 6 mm has been forced into a steel disc of thickness 1 mm and outer radius 12 mm. There was an interference of 0.02 mm on the diameters of disc and shaft. Assuming a state of plane stress and a coefficient of friction of 0.1, what torque would be required to turn the disc on the shaft? For steel $E = 200 \text{ GN/m}^2$, $\nu = 0.3$.

14.20  A thin disc of inner and outer radii 150 and 300 mm respectively rotates at 150 rad/sec. Determine the maximum radial and hoop stresses. $\nu = 0.304$, $\rho = 7.7 \text{ Mg/m}^3$.

14.21  (*a*)  A solid circular disc of uniform thickness is to be used as a flywheel to store kinetic energy in a city bus. Derive an expression for the maximum shear stress in the flywheel.

(*b*)  The energy $U$ stored in a rotating disc is given by

$$U = \tfrac{1}{2} \, I\omega^2$$

where $I = \tfrac{1}{2}\rho\pi r_0{}^4 t$, $\rho$ is the material density, $r_0$ is the disc radius, $t$ is its thickness and $\omega$ is the speed of angular rotation. Assuming that the thickness of the disc is fixed, but the radius is limited only by the strength of the material, which of the materials below would allow the disc to store most energy?

| Material | $\sigma_y$ (MN/m²) | $\rho$ (kg/m³) | $\nu$ |
|---|---|---|---|
| GFRP | 1000 | 1800 | 0.3 |
| Titanium Alloy | 1200 | 4500 | 0.3 |
| Mild Steel | 220 | 7800 | 0.3 |

14.22  A thin uniform disc with a central hole is pressed on a shaft in such a manner that when the whole is rotated at $n$ revolutions per minute the pressure at the common surface is $p$. Derive an expression for the hoop stress in the disc at the periphery, if the inside radius of the disc is $r_1$ and the outside radius $r_2$.

14.23   A solid steel disc 457 mm in diameter and of small constant thickness has a steel ring of outer diameter 610 mm and the same thickness shrunk on to it. If the interference pressure is reduced to zero at a rotational speed of 3000 rev/min, calculate the difference in diameters of the mating surfaces of the disc and ring before assembly and the interface pressure. $\nu = 0.29$, $\rho = 7.7\,\mathrm{Mg/m^3}$, $E = 207\,\mathrm{GN/m^2}$.

14.24   An aircraft window consists of a flat circular sheet of perspex 0.3 m in diameter. If the aircraft is pressurized, i.e. assuming sea level pressure is maintained even at high altitude, what thickness of perspex would be required to prevent yielding? Justify your assumed loading and boundary conditions. For perspex, $E = 4\,\mathrm{GN/m^2}$, $\nu = 0.5$, $\sigma_y = 60\,\mathrm{MN/m^2}$.

14.25   A steel rotor disc of uniform thickness 50 mm has an outer rim of diameter 750 mm and a central hole of diameter 150 mm. There are 200 blades each of weight 0.22 kg at an effective radius of 430 mm pitched evenly around the periphery. Determine the rotational speed at which yielding first occurs according to the maximum shear stress criterion. Yield stress in simple tension for the steel is 700 MN/m², $\nu = 0.29$, $\rho = 7.3\,\mathrm{Mg/m^3}$, $E = 207\,\mathrm{GN/m^2}$.

14.26   A disc is to be designed having uniform strength, that is, the radial and hoop stresses are the same at any point in the disc. Show that the required profile of thickness variation is given by

$$z = z_0 e^{-\rho \omega^2 r^2 / 2\sigma}$$

# 15 Elementary Plasticity

Engineering design is primarily concerned with maintaining machines and structures working within their elastic range. The analyses of Chapter 12 were specifically related to the assessment of the yield boundaries for components and the influence of stress concentration in possibly causing material to exceed the elastic range locally. However, it would be imprudent if designers knew nothing of what would happen to components that were, say, grossly overloaded to the point where marked yielding and plastic deformation occurred. Another important aspect is that in some circumstances enhanced performance can be achieved by prior plastic deformation resulting in favourable residual stresses as, for example, in a thick-walled pressure vessel or rotor disc. A further application of plasticity relates to forming metals, and although this is generally a subject outside the scope of this text we shall look at the simple elements of beams and shafts plastically deformed.

It is evident that the same principles must apply as for elastic deformation, namely equilibrium of forces, compatibility of deformations and a stress–strain relationship. It is the nature of this latter material behaviour which particularly dictates the final solution. Elastic–plastic stress–strain relationships are illustrated in Fig. 15.1(a), (b) and (c). The first is a typical curve for a real strain-hardening material and its linear to non-linear development causes some complication in analysis. Because of this, semi-idealized behaviour is often assumed as in Fig. 15.1(b) in which strain hardening occurs linearly from initial yield, while in Fig. 15.1(c) strain hardening is ignored and we have a linear-elastic non-hardening plastic relationship.

## 15.1 Plastic bending of beams: plastic moment

In considering the behaviour of beams subjected to pure bending which results in fibres being stressed beyond the limit of proportionality, the following assumptions will be made:

1. That the fibres are in a condition of simple tension or compression.
2. That any cross-section of the beam will remain plane during bending as in elastic bending. This means that the strain distribution will be linear even if the stress distribution is not.

**Fig. 15.1**

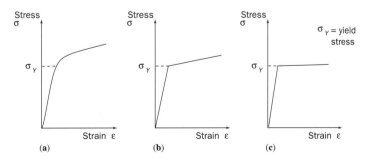

In elastic bending of a beam there is a linear stress distribution over the cross-section and when the extreme fibres reach the yield stress, the bending moment is given by

$$M_Y = \sigma_Y \frac{I}{y}$$    [15.1]

where $I$ is the second moment of area of the cross-section about the neutral axis, and $y$ is the distance from the neutral axis to an extreme fibre. The value of the yield bending moment $M_Y$ will be found for beams of various cross-section.

**Rectangular section**  From eqn. [15.1].

$$M_Y = \sigma_Y \frac{bd^2}{6}$$    [15.2]

and the stress distribution corresponding to this condition is shown in Fig. 15.2(*a*), all the fibres of the beam being in the elastic condition.

**Fig. 15.2**

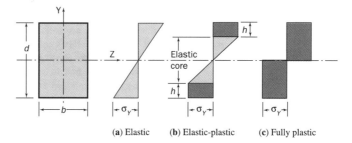

(**a**) Elastic    (**b**) Elastic-plastic    (**c**) Fully plastic

When the bending moment is increased above the value given in eqn. [15.2], some of the fibres near the top and bottom surfaces of the beam begin to yield and the appropriate stress diagram for a non-strain hardening material is given in Fig. 15.2(*b*). With further increase in bending moment, plastic deformation penetrates deeper into the beam. The total bending moment is obtained by consideration of both the plastic stress near the top and bottom of the beam and the elastic stress in the core of the beam. This moment is called the *elastic–plastic bending moment*.

From Chapter 6, the bending moment is given by

$$M = \int \sigma y \, dA$$

The elastic component is obtained from eqn. [15.2] in which the depth is now $(d - 2h)$, so

$$M = \sigma_Y \frac{b(d - 2h)^2}{6}$$

The plastic component of the moment, shown shaded heavily in Fig. 15.2, is given by

$$M = \sigma_Y bh(d - h)$$

Hence

$$M = \sigma_Y bh(d - h) + \sigma_Y \frac{b(d - 2h)^2}{6}$$

$$= \frac{\sigma_Y bd^2}{6}\left[1 + 2\frac{h}{d}\left(1 - \frac{h}{d}\right)\right] \qquad [15.3]$$

At a distance $(\frac{1}{2}d - h)$ from the neutral axis, the stress in the fibres has just reached the value $\sigma_Y$; then, if $R$ is the radius of curvature, we have

$$\sigma_Y = \frac{E(\frac{1}{2}d - h)}{R}$$

or

$$\frac{1}{R} = \frac{\sigma_Y}{E(\frac{1}{2}d - h)} \qquad [15.4]$$

The values of $M$ and $1/R$ calculated from eqns. (15.3) and (15.4) when plotted give the graph shown in Fig. 15.3. The connection between these quantities is linear up to a value of $M = M_Y$. Beyond this point the relationship is non-linear and the slope decreases with increase in depth, $h$, of the plastic state. When $h$ becomes equal to $\frac{1}{2}d$, the stress distribution becomes that shown in Fig. 15.2($c$) and the highest value of bending moment is reached.

This *fully plastic moment* is given by eqn. [15.3], putting $h = \frac{1}{2}d$, as

$$M_p = \frac{3}{2}\sigma_Y \frac{bd^2}{6} = \sigma_Y \frac{bd^2}{4} \qquad [15.5]$$

$$= \frac{3}{2}M_Y \qquad [15.6]$$

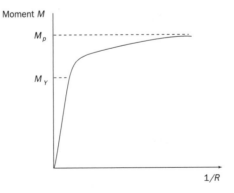

Fig. 15.3

The value of $M_p$ is shown in Fig. 15.3 and is the horizontal asymptote of the curve. Plastic collapse of the beam is shown, by eqn. (15.6), to occur at a bending moment of one and a half times that at initial yielding of the extreme fibres of the beam.

The ratio of $M_p/\sigma_Y$ is termed the *plastic section modulus*, $Z_{plastic}$ (analogous to the elastic section modulus, $Z$, discussed in Chapter 6.) For the rectangular section it is given by $bd^2/4$ from eqn. [15.5]. It may be noted that it is simply a function of the geometry of the cross-section and is often available in standard tables of data. When $Z_{plastic}$ and $\sigma_Y$ are known, the plastic moment is given by

$$M_p = \sigma_Y Z_{plastic}$$

**I-section**   When yielding is about to occur at the extreme fibres, the beam is still in the elastic condition and, for the dimensions in Fig. 15.4(*a*),

$$M_Y = \sigma_Y \left( \frac{bd^3}{12} - \frac{b_1 d_1^3}{12} \right) \frac{2}{d}$$   [15.7]

**Fig. 15.4**

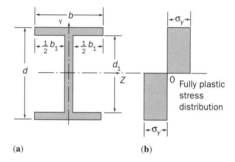

(a)                                    (b)

In the fully plastic condition, the stress diagram is shown in Fig. 15.4(*b*), and from eqn. [15.5] the fully plastic moment is given by

$$M_p = \sigma_Y \left( \frac{bd^2}{4} - \frac{b_1 d_1^2}{4} \right)$$   [15.8]

and the ratio $M_p/M_Y$ which is termed the *shape factor* is

$$\frac{M_p}{M_Y} = \frac{(bd^2/4 - b_1 d_1^2/4)}{(bd^3/12 - b_1 d_1^3/12)} \frac{d}{2}$$

or

$$\frac{M_p}{M_Y} = \frac{3}{2} \frac{(1 - b_1 d_1^2/bd^2)}{(1 - b_1 d_1^3/bd^3)}$$   [15.9]

In an I-beam, 100 mm × 300 mm, with flanges and web 14 mm and 98 mm thick respectively,

$$\frac{M_p}{M_Y} = \frac{3}{2} \frac{[1 - (91 \times 272^2)/(100 \times 300^2)]}{[1 - (91 \times 272^3)/(100 \times 300^3)]} = 1.16$$

This shape factor is fairly representative of standard rolled I-section beams, and the fully plastic moment is only 16% greater than that at which initial yielding occurs.

**Asymmetrical section**   In the two previous cases the neutral axis in bending of the section coincided with an axis of symmetry. If the cross-section is asymmetical about the axis

Fig. 15.5

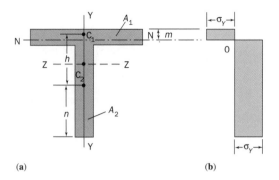

(a)          (b)

of bending, then the position of the neutral axis must be determined. Figure 15.5(a) shows a T-bar section in which YY is the only axis of symmetry and ZZ passes through the centroid of the section.

In the fully plastic condition the beam is bent about the neutral axis NN. If $A_1$ and $A_2$ are the areas of the cross-section above and below NN respectively, then since there can be no longitudinal resultant force in the beam during bending without end load,

$$A_1\sigma_Y = A_2\sigma_Y$$

or $A_1 = A_2 = \frac{1}{2}A$, where $A$ is the total area of the cross-section. Thus for the fully plastic state the neutral axis divides the cross-section into two equal areas and the stress diagram is shown in Fig. 15.5(b).

If $C_1$ is the centroid of the area $A_1$, $C_2$ the centroid of the area $A_2$, and $h$ the distance between $C_1$ and $C_2$, then the fully plastic moment is given by

$$M_p = \frac{1}{2}A\sigma_Y h = \frac{1}{2}\sigma_Y Ah \qquad [15.10]$$

This equation applies for any shape of cross-section. The reader may wish to check its use for a rectangular cross-section to obtain eqn. [15.5].

Tables on the properties of areas, such as elastic and plastic section modulus, moments of area, etc., for standard sections are widely available from material suppliers and trade organizations.

**Example 15.1**

The flange and web of the T-bar section in Fig. 15.5 are each 12 mm thick, the flange width is 100 mm, and the overall depth of the section is 100 mm. The centroid of the section is at a distance of 70.6 mm from the bottom of the web, and the second moment of area $I_z$ of the section about a line through the centroid and parallel to the flange is $2.03 \times 10^6$ mm$^4$. Determine the value of the shape factor.

Let $m$ be the distance of the neutral axis NN from the top of the flange (Fig. 15.5); then

$$A_1 = 100m \quad\text{and}\quad A_2 = 100(12 - m) + (88 \times 12)$$

$$100m = 100(12 - m) + (88 \times 12)$$

$$m = 11.3\,\text{mm}$$

If $n$ is the distance of the centroid of area $A_2$ from the bottom of the web, then

$$(100 \times 0.7 \times 88.35) + (12 \times 88 \times 44) = [(100 \times 0.7) + (12 \times 88)]n$$

$$n = 46.8 \, \text{mm}$$

Therefore $h$, the distance between $C_1$ (the centroid of $A_1$) and $C_2$ (the centroid of $A_2$), is

$$h = 88.7 - 46.8 + \frac{11.3}{2} = 47.55 \, \text{mm}$$

so that

$$M_p = \frac{12 \times 188}{2} \sigma_Y \times 47.55 = 53\,636\sigma_Y$$

$$M_Y = \frac{2.03 \times 10^6}{70.6} \sigma_Y = 28\,754\sigma_Y$$

and the shape factor

$$\frac{M_p}{M_Y} = 1.87$$

## 15.2 Plastic collapse of beams

The fully plastic bending moment developed in the preceding section was due to the application of pure bending. A beam would therefore become fully plastic at *all* cross-sections along the whole length once $M_p$ was reached. However, in practice pure bending rarely occurs and the bending-moment distribution varies depending on the loading conditions. The point of maximum bending moment along the beam will be the first cross-section which becomes fully plastic as the load magnitude increases. Cross-sections adjacent to the fully plastic section will have commenced yielding to various depths. For a beam simply supported at each end and carrying a central concentrated load the shape of the plastic zone associated with the central fully plastic cross-section is illustrated in Fig. 15.6(*a*). The boundary between elastic and plastic material is parabolic in shape.

The plastic zones for distributed loading are triangular in shape, as shown in Fig. 15.6(*b*). When a cross-section such as those shown reaches the fully plastic state it cannot carry any higher loading and the beam forms a hinge at that cross-section. This is termed a *plastic hinge* about which rotation of the two halves of the beam occurs, as shown in Fig. 15.7. When

**Fig. 15.6**

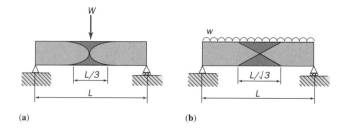

(a)                              (b)

**Fig. 15.7**

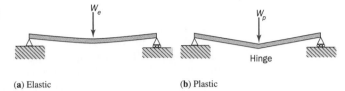

(**a**) Elastic                    (**b**) Plastic

one or more plastic hinges occur such that the beam or structure becomes a *mechanism* then this situation is described as *plastic collapse*.

In the example of Fig. 15.7 the maximum bending moment is at the centre and is $WL/4$. Therefore plastic collapse occurs for the *single* hinge formation and

$$M_p = \frac{W_p L}{4} \qquad \text{or} \qquad W_p = \frac{4M_p}{L} \qquad [15.11]$$

The next example is a cantilever propped at the free end and carrying a concentrated load at mid-span, as shown in Fig. 15.8(*a*). In situations like this, plastic collapse will only occur if multiple hinges form. The sequence of events will be as follows:

1.  Elastic behaviour occurs in a structural member until a plastic hinge is formed at a section.
2.  If rotation at this hinge results in diffusion of the load to other parts of the structure or supports then additional load may be carried until another plastic hinge is formed.
3.  As each hinge forms the moment remains constant at the fully plastic value irrespective of additional load or deformation.
4.  When there is no remaining stable portion able to carry additional load then collapse will occur.
5.  The structure as a whole, or in part, will form a simple mechanism at collapse.
6.  The collapse load may be calculated by statical equilibrium if the locations of the hinges can be identified.

**Fig. 15.8**

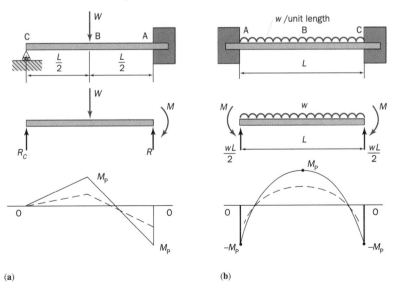

(**a**)                                    (**b**)

Considering the loading situation in Fig. 15.8(a), the elastic bending-moment diagram is shown dashed. At some value of load, $W_Y$, the beam will yield at A. The moment at A will be $M_Y$ at this point. As the load is increased beyond $W_Y$, yielding will also start to occur at B. As the load is further increased, a plastic hinge forms at A when the moment there reaches $M_p$. This will not cause the beam to collapse. As the load is further increased, the moment at A remains at $M_p$ and the beam behaves like a statically determinate system. Eventually a plastic hinge forms at B and the bending moment diagram is shown as the solid line in Fig. 15.8(a). At this point the beam will collapse.

The value of load $W_p$ which causes plastic collapse may be determined from a static equilibrium analysis of the beam because the moments at A and B are known to be equal.

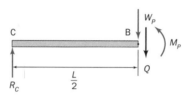

Taking moments about A

$$-M_p + \frac{W_p L}{2} - R_c L = 0$$

$$R_C = \frac{W_p}{2} - \frac{M_p}{L} \qquad [15.12]$$

where $R_C$ is the vertical reaction at C.

Taking a free-body diagram for section CB, moment equilibrium about B gives

$$M_p - \frac{R_c L}{2} = 0 \qquad [15.13]$$

Using this in [15.12] gives

$$\frac{6M_p}{L} = W_p$$

Note also that in Chapter 8, it was shown that the elastic moment at A is given by

$$M = \frac{3}{16} WL$$

At the yield condition $W = W_Y$ and $M = M_Y$ so substituting for $L$ from [15.13] enables the ratio of the plastic load to yield load to be compared:

$$\frac{W_p}{W_Y} = \frac{9}{8} \frac{M_p}{M_Y} \qquad [15.14]$$

Next consider the case of a beam fixed at both ends carrying a uniformly distributed load as shown in Fig. 15.8(*b*). The elastic bending-moment distribution is shown dashed. At the collapse condition, plastic hinges will form at A, B and C.

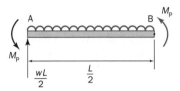

Taking a free body diagram of AB, moment equilibrium about A gives

$$M_p + M_p - \frac{w_p L}{2}\left(\frac{L}{4}\right) = 0$$

$$w_p = \frac{16M_p}{L^2} \tag{15.15}$$

Also, as shown in Chapter 8, the elastic moment at A is given by

$$M = \frac{wL^2}{12}$$

Letting $M = M_Y$ and $w = w_Y$ at the yield condition, then once again we may use eqn. [15.15] to get the ratio of the plastic load to yield load. This will be

$$\frac{w_p}{w_Y} = \frac{4}{3}\frac{M_p}{M_Y} \tag{15.16}$$

This method of obtaining the plastic load is illustrated further in the following example.

**Example 15.2**

**The beam illustrated in Fig. 15.9 is made of I-section mild steel having a shape factor of 1.15 and a yield stress of 240 MN/m². Using a load factor against collapse of 2 find the required section modulus.**

The elastic bending-moment diagram is such that peaks (of different magnitude) will occur at A, B and C. Plastic hinges will form when these peaks reach the value of $M_p$. Collapse will occur when the three hinges are formed as shown in Fig. 15.9(*b*).

**Fig. 15.9**

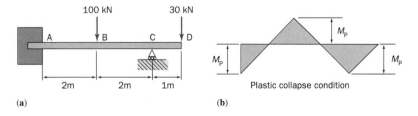

(a)                                                     (b)

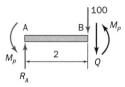

Statical equilibrium may be applied to the collapse condition. Taking a free-body diagram for AB and applying moment equilibrium about B gives

$$M_p + M_p - 2R_A = 0$$

$$M_p = R_A$$

Then taking a free-body diagram for AC and applying moment equilibrium about C gives

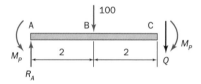

$$M_p - M_p + (2 \times 100) - 4R_A = 0$$

$$R_A = 50 \text{ kN}$$

Therefore, from above $M_p = 50 \text{ kN m}$.
     Also

$$\frac{M_p}{M_Y} = 1./15$$

Hence

$$M_Y = \frac{2 \times 50}{1.15} = 86.96 \text{ kN m}$$

having incorporated the load factor of 2.
     Now the elastic section modulus

$$Z = \frac{M_Y}{\sigma_Y} = \frac{86.96}{240} = 0.362 \text{ m}^3$$

For the hinge to form at C due to the load at D we have

$$M_p = WL = 30 \times 1 = 30 \text{ kN m}$$

This value would require a section modulus of

$$Z = \frac{2 \times 30}{1.15 \times 240} = 0.217 \text{ m}^3$$

Therefore the larger section modulus is required.

### 15.3 Plastic torsion of shafts: plastic torque

In the following discussion it will be assumed that we have an ideal stress–strain relationship for the material as shown in Fig. 15.10, that a plane cross-section of the shaft remains plane when in the plastic state, and that a radial line remains straight. The shearing strain $\gamma$ at a distance $r$ from the axis of the shaft will be given by $\gamma = r\theta/L$.

**Fig. 15.10**

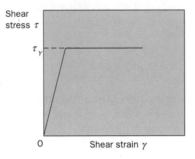

When the shaft has a torque applied in the elastic range, the shear stress, $\tau$, increases from zero at the shaft axis to a maximum value at the surface of the shaft, and from eqn. [5.11]

$$T = \frac{\tau\pi}{2}r^3$$

for a solid circular shaft. When the shear stress at the surface of the shaft has reached the value $\tau_Y$ the torque required to give this stress is

$$T_Y = \frac{\tau_Y\pi}{2}r^3 \tag{15.17}$$

If the torque is increased beyond this value, then plasticity occurs in fibres at the surface of the shaft and the stress diagram is as shown in Fig. 15.11. The torque carried by the elastic core is

$$T_1 = \frac{\tau_Y\pi}{2}r_Y^3 \tag{15.18}$$

**Fig. 15.11**

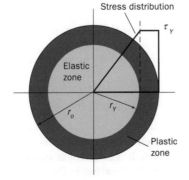

where $r_Y$ is the interface radius between elastic and plastic material, and that carried by the plastic zone is

$$T_2 = \int_{r_Y}^{r_o} 2\pi r^2 \tau_Y \, dr = \frac{2\pi}{3} \tau_Y (r_o^3 - r_Y^3) \qquad [15.19]$$

and the total torque, $T$, is

$$T_1 + T_2 = \tfrac{1}{2}\tau_Y \pi r_Y^3 + \tfrac{2}{3}\pi\tau_Y(r_o^3 - r_Y^3)$$

Therefore

$$T = \tfrac{2}{3}\pi r_o^3 \tau_Y \left(1 - \frac{r_Y^3}{4r_o^3}\right) \qquad [15.20]$$

When the fully plastic condition is reached, the shear stress at all points in the cross-section is $\tau_Y$, and it follows from eqn. [15.19] that the fully plastic torque is given by

$$T_p = \frac{2\pi}{3}\tau_Y r_o^3 \qquad [15.21]$$

and the ratio $T_p/T_Y$ is

$$\frac{T_p}{T_Y} = \frac{4}{3}$$

When the fibres at the outer surface of the shaft are about to become plastic, the angle of twist is given by eqn. [5.11]

$$\theta_Y = \frac{\tau_Y L}{G r_o} \qquad [15.22]$$

and when the shaft is in the elastic–plastic condition, the angle of twist of the elastic core is given by

$$\theta = \frac{\tau_Y L}{G r_y} \qquad [15.23]$$

Since we have assumed that radii remain straight, then the outer plastic region has the same angle of twist. From eqns. [15.22] and [15.23] it follows that

$$\frac{\theta_Y}{\theta} = \frac{r_Y}{r_o} \qquad [15.24]$$

It is evident that as the shaft approaches the fully plastic state, the angle of twist tends to infinity.

Equation [15.20] may be expressed in the form

$$T = \tfrac{2}{3}\pi r_o^3 \tau_Y \left[1 - \frac{1}{4}\left(\frac{\theta_Y}{\theta}\right)^3\right] \qquad [15.25]$$

**Example 15.3**

A mild steel shear coupling in a metal-working process is 40 mm in diameter and 250 mm in length. It is subjected to an overload torque of 1800 Nm which is known to have caused shear yielding in the shaft.

Determine the radial depth to which plasticity has penetrated and the angle of twist. $\tau_Y = 120\,\text{MN/m}^2$, $G = 80\,\text{GN/m}^2$.

Using eqn. [15.20] for the elastic–plastic torque

$$1800 = \tfrac{2}{3}\pi \times 0.02^3 \times 120 \times 10^6 \left(1 - \frac{r_Y^3}{4 \times 0.02^3}\right)$$

from which $r_Y = 15\,\text{mm}$.

Hence the depth of plastic deformation is 5 mm.

The shear strain at $r_Y = 15\,\text{mm}$ is

$$\gamma = \frac{120 \times 10^6}{80 \times 10^9} = 0.0015$$

but

$$\gamma L = r_Y \theta$$

Therefore

$$0.0015 \times 0.25 = 0.015\theta$$

and

$$\theta = 0.025\,\text{rad} = 1.43°$$

## 15.4 Plasticity in a pressurized thick-walled cylinder

The problems analysed in the previous sections only involved a uniaxial stress condition, and hence a simple tensile or shear yield stress was sufficient to define the onset of plastic deformation. In two- or three-dimensional stress systems it is necessary to use a yield criterion of the type discussed in Chapter 12 in order to determine the initiation of plastic flow. A thick-walled cylindrical pressure vessel is a good example of this type of situation in that there will be radial and hoop stresses and generally also axial stress. The process of inducing plastic deformation partly through the wall thickness is known as *autofrettage* and its purpose was described in Chapter 14.

The following analysis will only consider an ideal elastic–plastic material, since the problem for a strain-hardening material is beyond the scope of this text. Furthermore, in order to simplify the mathematics the maximum shear stress (Tresca) theory of yielding will be adopted.

For a thick cylinder under internal pressure, the maximum shear stress occurs at the inner surface (see Fig. 14.5) and therefore as the pressure is increased, plastic deformation will commence first at the bore and penetrate deeper and deeper into the wall, until the whole vessel reaches the yield condition.

At a stage when plasticity has penetrated partly through the wall, the vessel might be regarded as a compound cylinder with the inner tube plastic and the outer elastic. If the elastic–plastic interface is at a radius $a$ and the radial pressure there is $p_a$, then from eqns. [14.35] and [14.36],

$$\sigma_r = \frac{p_a a^2}{r_o^2 - a^2}\left(1 - \frac{r_o^2}{a^2}\right) \qquad [15.26]$$

$$\sigma_\theta = \frac{p_a a^2}{r_o^2 - a^2}\left(1 + \frac{r_o^2}{a^2}\right) \tag{15.27}$$

and

$$\tau_{max} = \frac{\sigma_\theta - \sigma_r}{2} = \frac{p_a r_o^2}{r_o^2 - a^2} \tag{15.28}$$

But at the interface, yielding has just been reached; therefore

$$\tau_{max} = \frac{\sigma_Y}{2}$$

and

$$\frac{p_a r_o^2}{r_o^2 - a^2} = \frac{\sigma_Y}{2}$$

Therefore

$$p_a = \frac{\sigma_Y}{2r_o^2}(r_o^2 - a^2) \tag{15.29}$$

From this value of $p_a$ the stress conditions in the elastic zone can be determined using eqns. [14.35] and [14.36]. It is now necessary to consider the equilibrium of the plastic zone in order to find the internal pressure required to cause plastic deformation to a depth of $r = a$. The equilibrium equation is [14.11].

$$r\frac{d\sigma_r}{dr} + \sigma_r - \sigma_\theta = 0 \tag{15.30}$$

Hence

$$r\frac{d\sigma_r}{dr} - 2\tau_{max} = 0$$

or

$$\frac{d\sigma_r}{dr} = \frac{\sigma_Y}{r}$$

Integrating this equation gives

$$\sigma_r = \sigma_Y \log_e r + C \tag{15.31}$$

Now, at $r = a$, $\sigma_r = -p_a$; therefore

$$-p_a = \sigma_Y \log_e a + C$$

or

$$C = -p_a - \sigma_Y \log_e a$$

Substituting for $C$ in eqn. [15.31],

$$\sigma_r = \sigma_Y \log_e r - p_a - \sigma_Y \log_e a \tag{15.32}$$

$$= -\sigma_Y \log_e \frac{a}{r} - p_a \tag{15.33}$$

Therefore, using eqn. [15.29],

$$\sigma_r = -\sigma_Y \log_e \frac{a}{r} - \frac{\sigma_Y}{2r_o^2}(r_o^2 - a^2) \qquad [15.34]$$

which gives the distribution of radial stress in the plastic zone; and at

$$r = r_i \qquad \sigma_r = -p_i$$

Therefore

$$p_i = +\sigma_Y \log_e \frac{a}{r_i} + \frac{\sigma_Y}{2}\left(1 - \frac{a^2}{r_o^2}\right) \qquad [15.35]$$

where $p_i$ is the internal pressure to cause yielding to a depth of $r = a$. The hoop stress is

$$\sigma_\theta = \sigma_Y + \sigma_r$$

Therefore

$$\sigma_\theta = \sigma_Y\left(1 - \log_e \frac{a}{r}\right) - \frac{\sigma_Y}{2}\left(1 - \frac{a^2}{r_o^2}\right) \qquad [15.36]$$

The internal pressure, $p_{max}$, required to cause yielding right through the wall is found by putting $a = r_o$ in eqn. [15.35]:

$$p_{max} = +\sigma_Y \log_e \frac{r_o}{r_i} \qquad [15.37]$$

and from eqns. [15.34] and [15.36]

$$\sigma_r = -\sigma_Y \log_e \frac{r_o}{r} \qquad [15.38]$$

$$\sigma_\theta = \sigma_Y\left(1 - \log_e \frac{r_o}{r}\right) \qquad [15.39]$$

The stress distribution of $\sigma_r$ and $\sigma_\theta$ for the cases considered above are illustrated in Fig. 15.12 in terms of $\sigma_Y$ for a vessel where $r_o = 2r_i$.

**Fig. 15.12**

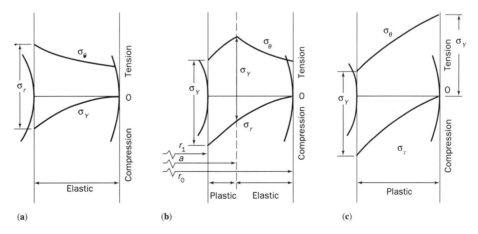

(a)  (b)  (c)

**Example 15.4**

A thick-walled cylindrical pressure vessel has to be autofrettaged prior to its use for a high-pressure chemical process. The radius ratio is 2.5 and 20% of the wall thickness has to be brought into the plastic range of the alloy steel which has a yield stress of 400 MN/m². Calculate the internal pressure required to achieve the specified plastic deformation. What are the values of hoop stress at the bore, elastic–plastic interface and the outer surface caused by that internal pressure? Determine the reserve factor against yielding right through the wall thickness.

To determine $a$, the radius of the elastic–plastic interface, in terms of $r_i$ we know that 20% of the wall thickness is in the plastic range so that

$$a = r_i + 0.2(r_o - r_i) = 1.3r_i$$

From eqn. [15.35],

$$p_i = 400 \log_e 1.3 + \frac{400}{2}\left[1 - \left(\frac{1.3r_i}{2.5r_i}\right)^2\right]$$

$$= 104.8 + 146 = 250.8 \, \text{MN/m}^2$$

At the bore

$$\sigma_r = -250.8 \, \text{MN/m}^2$$

and

$$\sigma_\theta = 400 - 250.8 = 149.2 \, \text{MN/m}^2$$

At the interface from eqn. [15.34] putting $r = a$,

$$\sigma_r = -\frac{400}{2}\left[1 - \left(\frac{1.3r_i}{2.5r_i}\right)^2\right] = -146 \, \text{MN/m}^2$$

and

$$\sigma_\theta = 400 - 146 = 254 \, \text{MN/m}^2$$

The value of $\sigma_r$ at the interface is also the value of $p_a$ in eqn. [15.29]. Therefore at $r = r_o$

$$\sigma_r = 0$$

and from eqn. [15.27]

$$\sigma_\theta = \frac{146 \times 2}{2.69} = 108.5 \, \text{MN/m}^2$$

The shape of the stress distribution is similar to that in eqn. [15.12].
    For yielding right through the wall we use eqn. [15.37]

$$p_{max} = 400 \log_e 2.5 = 366.4 \, \text{MN/m}^2$$

The reserve factor on pressure is

$$\frac{366.4}{250.8} = 1.46$$

## 15.5 Plasticity in a rotating thin disc

In a rotating disc with a central hole the maximum hoop stress occurs on the surface of the hole, and both hoop and radial stresses are positive throughout the disc, the former being the greater of the two. Therefore maximum shear stress is given by

$$\tau_{max} = \frac{\sigma_\theta - 0}{2} \tag{15.40}$$

since $\sigma_z = 0$.

Yielding occurs first at the hole where $r = r_1$ and gradually spreads outwards with increasing rotational speed. Using the maximum shear-stress criterion, the speed $\omega_Y$ at which yielding first commences is given by

$$\tau_{max} = \frac{\sigma_\theta}{2} = \frac{\sigma_Y}{2} \tag{15.41}$$

From eqn. [14.69] putting $r = r_1$,

$$\sigma_\theta = \frac{\rho\omega_Y^2}{4}[r_2^2(3 + \nu) + r_1^2(1 - \nu)] = \sigma_Y \tag{15.42}$$

Therefore initial yielding commences at a speed of

$$\omega_Y = \left(\frac{4\sigma_Y}{\rho[r_2^2(3 + \nu) + r_1^2(1 - \nu)]}\right)^{1/2} \tag{15.43}$$

Next we determine the rotational speed $\omega$ at which there is a plastic zone from $r = r_1$ to $r = c$. The equilibrium equation is

$$r\frac{d\sigma_r}{dr} + \sigma_r - \sigma_\theta + \rho\omega^2 r^2 = 0$$

and, since $\sigma_\theta = \sigma_Y$ in the plastic zone,

$$r\frac{d\sigma_r}{dr} + \sigma_r = \sigma_Y - \rho\omega^2 r^2 \tag{15.44}$$

Integrating this equation,

$$r\sigma_r = r\sigma_Y - \frac{\rho}{3}\omega^2 r^3 + K$$

Hence

$$\sigma_r = \sigma_Y - \frac{\rho}{3}\omega^2 r^2 + \frac{K}{r} \tag{15.45}$$

At $r = r_1$, $\sigma_r = 0$, so

$$0 = \sigma_Y - \frac{\rho}{3}\omega^2 r_1^2 + \frac{K}{r_1}$$

or

$$K = -r_1\sigma_Y + \frac{\rho}{3}\omega^2 r_1^3$$

Substituting for $K$ in eqn. [15.45]

$$\sigma_r = \sigma_Y - \frac{\rho}{3}\omega^2 r^2 - \frac{r_1}{r}\sigma_Y + \frac{\rho}{3}\omega^2 \frac{r_1^3}{r} \qquad [15.46]$$

At $r = c$, eqn. [15.46] becomes

$$\sigma_r = \frac{\sigma_Y}{c}(c - r_1) - \frac{\rho}{3}\frac{\omega^2}{c}(c^3 - r_1^3) \qquad [15.47]$$

But this value of $\sigma_r$ must be the same as $\sigma_r$ at $r = c$ for the elastic zone. It is firstly necessary to determine the constants $A$ and $B$ in eqns. [14.64] and [14.65]. The boundary conditions are: at $r = r_2$, $\sigma_r = 0$; and at $r = c$, $\sigma_\theta = \sigma_Y$. Therefore

$$0 = A - \frac{B}{r_2^2} - \frac{3+\nu}{8}\rho\omega^2 r_2^2$$

and

$$\sigma_Y = A + \frac{B}{c^2} - \frac{1+3\nu}{8}\rho\omega^2 c^2$$

from which

$$A = \frac{c^2\sigma_Y + \frac{1}{8}\rho\omega^2[(1+3\nu)c^4 + (3+\nu)r_2^4]}{c^2 + r_2^2}$$

$$B = \frac{c^2 r_2^2\sigma_Y + \frac{1}{8}\rho\omega^2 c^2 r_2^2[(1+3\nu)c^2 - (3+\nu)r_2^2]}{c^2 + r_2^2}$$

Therefore, for the elastic zone at $r = c$,

$$\sigma_r = \frac{c^2\sigma_Y + \frac{1}{8}\rho\omega^2[(1+3\nu)c^4 + (3+\nu)r_2^4]}{c^2 + r_2^2}$$

$$- \frac{r_2^2\sigma_Y + \frac{1}{8}\rho\omega^2 r_2^2[(1+3\nu)c^2 - (3+\nu)r_2^2]}{c^2 + r_2^2}$$

$$- \left(\frac{3+\nu}{8}\right)\rho\omega^2 c^2 \qquad [15.48]$$

Equating ens. [15.47] and [15.48] and simplifying,

$$\omega = \left(\frac{12\sigma_Y[2cr_2^2 - r_1(c^2 + r_2^2)]}{\rho[3(3+\nu)cr_2^4 - (1+3\nu)(2r_2^2 - c^2)c^3 - 4r_1^3(r_2^2 + c^2)]}\right)^{1/2} \qquad [15.49]$$

where $\omega$ is the angular speed to cause plasticity to a radial depth of $r = c$.

If $\omega_p$ is the speed at which the disc becomes fully plastic, then substituting $c = r_2$ in eqn. [15.49] and simplifying,

$$\omega_p = \left(\frac{3\sigma_Y}{\rho(r_2^2 + r_1 r_2 + r_1^2)}\right)^{1/2} \qquad [15.50]$$

The stress distributions of $\sigma_r$ and $\sigma_\theta$ for various degrees of plastic deformation are shown in Fig. 15.13 for $r_2 = 10r_1$.

Fig. 15.13

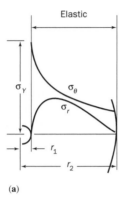
(a)

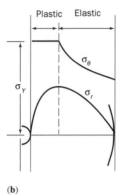
(b)

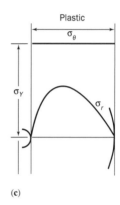
(c)

## 15.6 Residual stress distributions after plastic deformation

### Bending or torsion

When a beam is bent or a shaft twisted beyond the elastic limit, permanent deformation occurs which does not disappear when the load is removed. Those regions which have suffered permanent deformation prevent those which are elastically strained from recovering their initial dimension when the load is removed. Consequently the interaction produces what are termed *residual stresses*.

In a beam of rectangular cross-section, we will assume for simplicity that the beam has been bent to the fully plastic condition such that the stress distribution diagrams are the two rectangles Ocdb and Oeka, shown in Fig. 15.14(a). Assuming that when the material is stretched beyond the yield point and then unloaded it will be linear elastic during unloading, then the bending stresses which are superposed while the beam is being unloaded follow the linear law represented by the line $a_1b_1$. The shaded areas represent the stresses which remain after the beam is unloaded and are thus the residual stresses produced in the beam by plastic deformation.

The rectangular and triangular stress distributions represent bending moments of the same magnitude; hence the moment of the rectangle Ocdb about the axis eOc is equal to the moment of the triangle $Obb_1$ about the same axis. Since bd represents $\sigma_Y$ it follows for equal moments about the axis eOc that the stress represented by $bb_1$ is equal to $1\frac{1}{2}\sigma_Y$, and thus the maximum residual tension and compression stresses after loading and unloading are $db_1$, and $a_1k = \frac{1}{2}\sigma_Y$. Near the neutral axis the residual stresses are equal to $\sigma_Y$.

Fig. 15.14

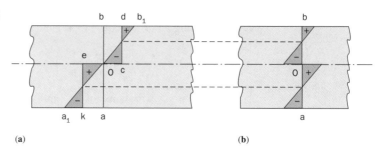

(a)                                                        (b)

In Fig. 15.14(*b*) the residual stresses are replotted on a conventional base of a cross-section, in order to show these residual tensile and compressive stresses more clearly. Diagrams showing the residual stresses in a partially yielded beam are shown in Fig. 15.15. In this case the material has yielded to such an extent that the stress distribution diagrams are represented by the areas Ocdb and Oeka in Fig. 15.15(*a*). Again assuming the material to follow Hooke's law during unloading, the bending stress during this operation will follow the linear law represented by $a_1b_1$, such that the moments of Ocdb and $Obb_1$ about the neutral axis are equal, and the shaded areas represent the residual stresses. These stresses are shown replotted on a conventional cross-section in Fig. 15.15(*b*).

Fig. 15.15

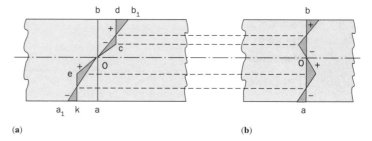

(a)                                                        (b)

The solution for residual stresses after plastic torsion follows the same arguments as above (see Example 15.7).

After any plastic deformation there will be some elastic recovery of the deformation. This 'springback' of the material causes difficulties in metal-forming applications such as sheet pressing and pipe bending, since the tools forming the shape have to be designed to apply sufficient additional deformation so that the final shape of the component after recovery is what is desired.

The amount of the recovered deformation can be estimated by an elastic analysis of the deformation due to the unloading moment or torque. The procedure is the same as for the calculation of residual stress above.

Example 15.5

**A flat bar 10 mm × 3 mm of mild steel ($E = 200$ GN/m$^2$, $\sigma_Y = 200$ MN/m$^2$) is formed by bending it round a circular former. What is the maximum change in curvature which will occur due to elastic springback?**

The maximum moment that could occur due to elastic unloading is the fully plastic moment, eqn. [15.5]

$$M_p = \sigma_Y \frac{bd^2}{4} \tag{15.51}$$

The curvature caused by an elastic moment of this magnitude would be

$$\frac{1}{R} = -\frac{M}{EI} = -\frac{\sigma_Y}{E}\frac{3}{d} \tag{15.52}$$

Assuming the bar is bent about the shorter dimension, i.e. $d = 3\,\mathrm{mm}$, then

$$\frac{1}{R} = -\frac{200 \times 10^6}{200 \times 10^9}\frac{3}{3 \times 10^{-3}} = -1\,\mathrm{m}^{-1} \tag{15.53}$$

from which the change in radius of curvature is

$$R = 1\,\mathrm{m} \tag{15.54}$$

The most useful point that this analysis illustrates is that the amount of elastic recovery is greatest in materials which have a large yield strain $\varepsilon_Y = \sigma_Y/E$. The amount of elastic recovery is also proportional to the amount of elastic deformation before yielding begins. Elastic recovery in torsion is analysed in Example 15.7.

**Axially symmetric components**

If plastic deformation is caused in one part of a body and not in the remainder, then, on removal of external load, there still exists a stress system in the body, owing to the strain gradient, and hence interaction between parts of the body not being able to return to the unstrained state. Residual stresses have important implications in engineering practice.

In the pressurized thick-walled cylinder it was assumed that plastic deformation had partly penetrated the wall. On release of pressure the elastic outer zone tries to return to its original dimensions, but is partly prevented by the permanent deformation of the inner plastic material. Hence the latter is put into hoop compression and the former is in hoop tension, such that equilibrium exists. The residual stress distributions are then as shown diagrammatically in Fig. 15.16. They may be calculated by using eqns. [14.35] and [14.36] to determine the elastic unloading stresses from the internal pressure $p_i$ (eqn. [15.35]) and superposing these onto the loaded stress distribution. The residual compressive hoop stress at the bore has to

Fig. 15.16

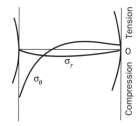

be nullified first on repressurizing, thus allowing a greater elastic range of hoop stress and therefore a greater internal pressure.

When a rotating disc is stopped after partial plasticity has occurred, a similar condition of residual stresses is obtained as for the thick cylinder above. Compressive hoop stress is obtained at the central hole, which may

**Fig. 15.17**

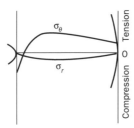

be calculated using the elastic stress equations and the appropriate rotational speed, followed by superposition onto the loaded stress pattern. The above action, known as *overspeeding*, helps to increase the elastic stress range, and hence the speed limit available under working conditions. The residual stress distributions for this case are shown in Fig. 15.17 for the same conditions as in Fig. 15.13(*b*).

**Example 15.6**

**After the autofrettage process is completed in Example 15.4 what are the residual stresses at the bore, interface and outer surface of the vessel assuming elastic unloading?**

The elastic unloading stress *range* may be found from the elastic equations for $\sigma_\theta$ and $\sigma_r$ using $p_i = 250.8\,\mathrm{MN/m^2}$.
    At $r = r_i$,

$$\sigma_\theta = \frac{250.8}{k^2 - 1}(1 + k^2)$$

$$= 250.8 \times \frac{7.25}{5.25} = 346.3\,\mathrm{MN/m^2}$$

Therefore the residual stress is

$$\sigma_\theta^R = +149.2 - 346.3$$

$$= -197.1\,\mathrm{MN/m^2}$$

At $r = a$,

$$\sigma_\theta = \frac{250.8}{5.25}(1 + 3.69) = 224\,\mathrm{MN/m^2}$$

The residual stress

$$\sigma_\theta^R = +254 - 224$$

$$= +30\,\mathrm{MN/m^2}$$

At $r = r_o$, $\sigma_\theta = 95.5\,\mathrm{MN/m^2}$, and the residual stress

$$\sigma_\theta^R = +108.5 - 95.5$$

$$= +13\,\mathrm{MN/m^2}$$

At $r = r_i$ and $r = r_o$ the residual radial stresses are zero and at $r = a$ the radial stress due to unloading elastically from $p_i = 250.8\,\text{MN/m}^2$ is $\sigma_r = -128.5\,\text{MN/m}^2$. Therefore the residual stress is

$$\sigma_r^R = -146 - (-128.5) = -17.5\,\text{MN/m}^2$$

The shape of the residual stress distribution is similar to that of Fig. 15.16.

**Example 15.7**

**Determine the residual shear-stress distribution after elastic unloading for the plastically deformed shear coupling in Example 15.3. What is the permanent twist?**

The elastic unloading torque of 1800 N m gives rise to a shear-stress range at the outer surface of

$$\tau = \frac{16 \times 1800}{\pi \times 0.04^3} = 143\,\text{MN/m}^2$$

The residual shear stress at the outer surface of the coupling is the value of $143\,\text{MN/m}^2$ less the yield stress value of $120\,\text{MN/m}^2$ which gives $23\,\text{MN/m}^2$.

The residual shear stress is zero at $r = 0$. It is also zero where the unloading stress range equals $120\,\text{MN/m}^2$; therefore from Fig. 15.18(a)

$$\frac{120}{143} = \frac{r}{20}$$

$$r = 16.8\,\text{mm}$$

**Fig. 15.18**

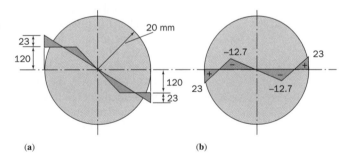

(a)                                                                 (b)

As shown in Example 15.3, the elastic–plastic boundary occurs at a radius of 15 mm where the elastic shear stress is

$$\tau = 143 \times \frac{15}{20}$$

$$= 107.3\,\text{MN/m}^2$$

and the *residual* shear stress is $107.3 - 120 = -12.7\,\text{MN/m}^2$.

The complete residual shear-stress diagram is shown in Fig. 15.18(b).

The elastic unloading twist or 'spring back' is given by

$$\theta = \frac{1800 \times 32}{\pi \times 0.4^4} \times \frac{0.25}{80 \times 10^9} = 0.0224\,\text{rad}$$

Permanent angle of twist $= 0.025 - 0.0224$

$$= 0.0026 \, \text{rad}$$

### 15.7  Summary

The occurrence of plastic deformation in engineering components is not common. Nevertheless an understanding of the behaviour is important for the analysis of modes of failure and in various metal-forming processes. Since equilibrium and strain–displacement equations are independent of the type or state of the material, the principal difference from elastic analysis is the form of the stress-strain relationship. For convenience this is often assumed to be elastic–ideally plastic, i.e. non-strain hardening, although a linear plastic strain-hardening relationship is a straightforward extension of the former.

Plastic bending and torsion represent elementary illustrations of the principles of analysis. Plasticity in pressurized cylinders and rotating discs can be seen to have important practical implications for performance through the influence of residual stresses and strains. The introduction of this latter concept is an opportunity to emphasize the role, both advantageous and detrimental, played by residual stresses in engineering components.

### Problems

15.1    Determine the ratio of the fully plastic to the maximum elastic moment for a beam of elastic-perfectly plastic material subjected to pure bending for the following shapes of cross-section: (*a*) solid circular; (*b*) solid square about a diagonal axis; (*c*) thin-walled circular tube; (*d*) thin-walled square tube about a centroidal axis parallel to one of the sides.

15.2    A steel beam of I-section, as shown in Fig. 15.19, and of length 5 m is simply-supported at each end and carries a uniformly-distributed load of 114 kN/m over the full span. Steel reinforcing plates 12 mm thick are welded to each flange and are made of elastic-ideally plastic material. Calculate the plate width such that yielding has just spread through each reinforcing plate at mid-span under the given load. Determine the positions along the reinforcing plates at which the outer surfaces have just reached the yield point. Yield stress $= 300 \, \text{MN/m}^2$, second moment of area $= 80 \times 10^{-6} \, \text{m}^4$.

15.3    The flange and web of a T-section are each 12 mm thick, the flange width is 100 mm and the overall depth is 100 mm. The beam is simply-supported over a length of 2 m and it is subjected to a point load $W$ at mid-span. Calculate the maximum value of $W$ if the beam is to be is to be designed such that yielding is permitted to penetrate the web to a depth of 20 mm. The yield strength of the beam material is 300 MN/m$^2$.

15.4    A short column of 0.05 m square cross-section is subjected to a compressive load of 0.5 MN parallel to but eccentric from the central axis. The column is made from elastic-perfectly plastic material which has a yield stress in tension or compression of 300 MN/m$^2$. Determine the value of the eccentricity which will result in the cross-section becoming just fully plastic.

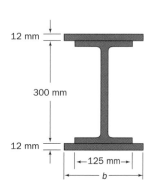

12 mm

300 mm

12 mm

|←125 mm→|

*b*

**Fig. 15.19**

15.5 Prove that for a beam simply-supported at each end and carrying a concentrated load at mid-span, which has developed full plasticity, the shape of the boundary between elastic and plastic material is parabolic.

15.6 A horizontal cantilever of length $L$ is simply-supported at the same level at the free end and is subjected to a uniformly-distributed load $w$ over the full span. Determine the location and magnitude of the collapse load.

15.7 Part of a small bridge deck is represented as shown in Fig. 15.20. What will be the mode of collapse and the value of the collapse load?

Fig. 15.20

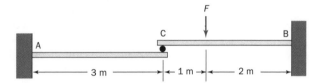

15.8 A steel shaft 100 mm in diameter and 1 m long is in an elastic–plastic state under a torque of 30 kNm. Determine the diameter of the elastic core of the shaft and the angle of twist. What is the value of the fully plastic torque for the shaft? $\tau_Y = 120$ MN/m$^2$, $G = 80$ GN/m$^2$.

15.9 A solid circular shaft is subjected to pure torsion and the material is elastic–perfectly plastic with a yield stress in shear of 152 MN/m$^2$. When the shear stress at one-third of the radius from the centre of the shaft reaches the yield stress, determine the shear strain on the outer surface. Also fnd the ratio of the torque carried in the above conditions to the maxmum elastic torque for the shaft. $G = 83$ MN/m$^2$.

15.10 A solid cylindrical composite shaft 1 m long consists of a copper core of 50 mm diameter surrounded by a well-fitting steel sleeve having an external diameter of 62 mm. If the steel has an elastic–perfectly plastic stress–strain relationship, determine the torque that can be applied to the shaft to cause yielding to develop just through to the inner surface of the sleeve. Neglect any stresses set up by the fit between the two parts of the shaft and assume there is no slipping at the copper–steel interface. $G$ (steel) = 83 GN/m$^2$, $G$ (copper) = 45 GN/m$^2$. $\tau_Y$ (steel) = 124 MN/m$^2$, $\tau_Y$ (copper) = 76 MN/m$^2$.

15.11 A thick-walled cylinder of radius ratio 2:1 and made of an elastic–perfectly plastic material is subjected to internal pressure. Plot a diagram of $(p_a/p_{max})$ against $a$, where $p_a$ is the internal pressure to cause yielding to a depth $a$ through the wall, and $p_{max}$ is the pressure which results in yielding right through the wall.

15.12 A metal disc of uniform thickness is 300 mm diameter and has a central hole of 50 mm diameter. Determine the increase in rotational speed over that for initial yielding at the hole necessary to cause plastic deformation throughout the disc. $\nu = 0.3$.

15.13 Calculate the residual stress at the outer surfaces of the column in Problem 15.4 after elastic unloading from the fully plastic condition.

15.14 A rectangular beam 30 mm wide and 50 mm deep is simply-supported over a length of 2 m. If it is subjected to a uniformly distributed load of 8 kN/m calculate the depth of penetration of plastically deformed material in the beam. If the load is removed calculate the force necessary at mid-span to straighten the beam. The yield stress of the beam material is 240 MN/m$^2$.

15.15   If the composite shaft in Problem 15.10 has the torque removed, calculate the residual shear stress at the outer surface of the steel sleeve and plot the residual stress distrbution.

15.16   A thin circular disc with a central hole has inner and outer diameters of 51 and 304 mm respectively. It is required to have a residual compressive hoop stress of 77 MN/m$^2$ at the hole when the disc is stationary. Assuming ideal elastic–plastic conditions, yield according to the maximum shear stress theory and elastic unloading, determine the rotational speed necessary to effect the required residual stress. By how much is this speed greater than the speed for initial yielding? Also find the depth of the plastic zone. $\sigma_Y = 340$ MN/m$^2$, $\rho = 7.83$ Mg/m$^3$, $\nu = 0.3$.

# 16 Thin Plates and Shells

The purpose of plates in engineering is to cover, generally, a rectangular or circular area and to support concentrated or distributed loading normal to the plane of the plate. A typical example is a pressure diaphragm, as a safety or control device, supported around its circular periphery and subjected to uniform pressure on one face and perhaps a central point load on the opposite face.

Some simple examples of thin shells were studied in Chapter 2 to illustrate statically determinate problems.

The engineering applications of thin shells include storage tanks for liquids or solids and pressure vessels for a variety of chemical processes, rocket motor casings, boiler drums, etc.

'Thin' is a relative term which indicates that the thickness of the material is small compared with the overall geometry, a ratio of 10:1 or greater being the usual criterion.

Solutions which are more accurate for 'thick' plates and shells are complex and may be found in specialized texts.

A further factor which affects the nature of the analysis is the range of deformation, again in qualitative terms referred to as 'small', or 'large' relative to the sheet thickness. This chapter will only consider *small* elastic deformations of plates and shells.

## 16.1 Assumptions for small deflection of thin plates

(a) No deformation in the middle plane of the plate, i.e. a neutral surface.
(b) Points in the plate lying initially on a normal to the middle plane of the plate remain on the normal during bending.
(c) Normal stresses in the direction transverse to the plate can be disregarded.

Assumption (a) does not of course hold if there are external forces acting in the middle plane of the plate. Assumption (b) disregards the effect of shear force on deflection.

The deflection, $w$, is a function of the two co-ordinates in the plane of the plate, the elastic constants of the material and the loading.

## 16.2 Relationships between moments and curvatures for pure bending
### Rectangular co-ordinates

The first important step in the analysis of plates is similar to that for beams and is to relate the bending moments to curvature from which slope and deflection are determined.

Consider an element of material as shown in Fig. 16.1 cut from a plate subjected to pure bending as in Fig. 16.2. The bending moments $M_x$ and $M_y$ *per unit length* are positive as drawn acting on the middle of the plate. This plane is undeformed and constitutes the *neutral surface*. The material above it is in a state of biaxial compression, and below it, in biaxial tension. The curvatures of the mid-plane in sections parallel to the $xz$- and $yz$-planes are denoted by $1/R_x$ and $1/R_y$ respectively. At a depth $z$ below the neutral surface the strains in the $x$- and $y$-directions of a lamina such as abcd are

$$\varepsilon_x = \frac{z}{R_x} \quad \text{and} \quad \varepsilon_y = \frac{z}{R_y} \qquad [16.1]$$

**Fig. 16.1**

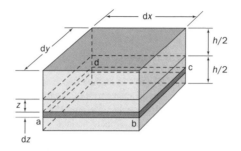

**Fig. 16.2**

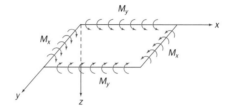

using the same approach as for beams in Chapter 6.

The stress-strain relationships are

$$\varepsilon_x = \frac{\sigma_x}{E} - \frac{\nu\sigma_y}{E} \quad \text{and} \quad \varepsilon_y = \frac{\sigma_y}{E} - \frac{\nu\sigma_x}{E}$$

Combining with eqn. [16.1] and rearranging gives

$$\left.\begin{aligned}
\sigma_x &= \frac{Ez}{1-\nu^2}\left(\frac{1}{R_x} + \frac{\nu}{R_y}\right) \\[2mm]
\sigma_y &= \frac{Ez}{1-\nu^2}\left(\frac{1}{R_y} + \frac{\nu}{R_x}\right)
\end{aligned}\right\} \qquad [16.2]$$

Equations [16.2] show that bending stresses are a function of plate curvatures and are proportional to distance from the neutral surface.

Next, we equate the required equilibrium between the internal moments, due to the bending stresses acting on the sides of the element, and the applied moments $M_x$ and $M_y$ per unit length:

$$\left.\begin{aligned}
\int_{-h/2}^{h/2} \sigma_x z \, \mathrm{d}y \, \mathrm{d}z &= M_x \, \mathrm{d}y \\[2mm]
\int_{-h/2}^{h/2} \sigma_y z \, \mathrm{d}x \, \mathrm{d}z &= M_y \, \mathrm{d}x
\end{aligned}\right\} \qquad [16.3]$$

Substituting from eqns. [16.2] for $\sigma_x$ and $\sigma_y$ in eqns. [16.3] and integrating gives

$$\left.\begin{aligned}
M_x &= \frac{Eh^3}{12(1-\nu^2)}\left(\frac{1}{R_x} + \frac{\nu}{R_y}\right) = D\left(\frac{1}{R_x} + \frac{\nu}{R_y}\right) \\[2mm]
M_y &= \frac{Eh^3}{12(1-\nu^2)}\left(\frac{1}{R_y} + \frac{\nu}{R_x}\right) = D\left(\frac{1}{R_y} + \frac{\nu}{R_x}\right)
\end{aligned}\right\} \qquad [16.4]$$

where

$$D = \frac{Eh^3}{12(1 - \nu^2)}$$

is termed the *flexural rigidity*.

Since the principal curvatures are given by

$$\frac{1}{R_x} = -\frac{\partial^2 w}{\partial x^2} \quad \text{and} \quad \frac{1}{R_y} = -\frac{\partial^2 w}{\partial y^2}$$

where $w$ is the deflection in the $z$-direction, the relationships between the applied moments and curvatures are

$$\left.\begin{array}{c} M_x = -D\left(\dfrac{\partial^2 w}{\partial x^2} + \nu\dfrac{\partial^2 w}{\partial y^2}\right) \\[3ex] M_y = -D\left(\dfrac{\partial^2 w}{\partial y^2} + \nu\dfrac{\partial^2 w}{\partial x^2}\right) \end{array}\right\} \qquad [16.5]$$

These equations can be written in matrix form as

$$\left\{\begin{array}{c} M_x \\ M_y \end{array}\right\} = -D\begin{bmatrix} 1 & \nu \\ \nu & 1 \end{bmatrix}\left\{\begin{array}{c} \partial^2 w/\partial x^2 \\ \partial^2 w/\partial y^2 \end{array}\right\} \qquad [16.6]$$

Alternatively, the curvatures arising from given moments on an element are

$$\left\{\begin{array}{c} \partial^2 w/\partial x^2 \\ \partial^2 w/\partial y^2 \end{array}\right\} = -\frac{1 - \nu^2}{D}\begin{bmatrix} 1 & -\nu \\ -\nu & 1 \end{bmatrix}\left\{\begin{array}{c} M_x \\ M_y \end{array}\right\} \qquad [16.7]$$

Two special cases of interest can be identified: one where only a single moment is acting, the other where the plate is curved in one direction only.

*Beam bending*
If a total moment $M$ is applied to opposite ends of a rectangular plate of width $b$, so that $M_x = M/b$, while the other edges are free, $M_y = 0$, eqn. [16.7] reduces to

$$\frac{\partial^2 w}{\partial x^2} = -\frac{(1 - \nu^2)}{D}M_x = -\frac{12M}{Eh^3 b} = -\frac{M}{EI} \qquad [16.8]$$

Therefore plate theory reduces to beam theory when only a single moment is applied. The other curvature is

$$\frac{\partial^2 w}{\partial y^2} = -\nu\frac{(1 - \nu^2)}{D}M_x = -\nu\frac{M}{EI} \qquad [16.9]$$

This 'anticlastic' curvature, mentioned in Chapter 6, arises from the Poisson's ratio effect. On the side of the beam in tension a lateral contraction occurs while on the side in compression lateral expansion occurs, Fig. 16.3. These opposite changes in width can only be accommodated if a curvature of the opposite sign occurs in the transverse direction.

Fig. 16.3

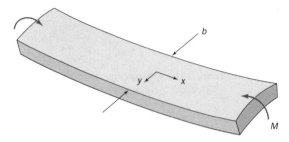

*Cylindrical bending*
If the plate is constrained so that displacement varies in only one direction, e.g. in a short, wide beam or the axi-symmetric shell described later in Example 16.3, then a state of cylindrical bending is said to occur, Fig. 16.4. Assuming that $w$ varies with $x$ only so that $\partial^2 w/\partial y^2 = 0$, the moment curvature relations become

$$M_x = -D\frac{\partial^2 w}{\partial x^2}$$

$$M_y = -\nu D\frac{\partial^2 w}{\partial x^2} = \nu M_x \qquad [16.10]$$

Fig. 16.4

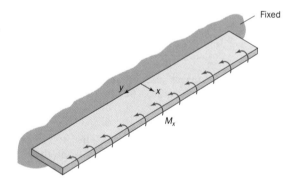

The plate is stiffer than the equivalent beam by a factor of $1/(1-\nu^2)$. Note that lateral bending moments of $\nu M_x$ are required, so that cylindrical bending cannot in fact occur near the free edges of a rectangular plate.

**Symmetrical bending of circular plates in cylindrical co-ordinates**    When the loading on the surface of a circular plate is symmetrical about a perpendicular central axis, the deflection surface is also symmetrical about that axis. Any diametral section may be used to indicate the deflection curve and the associated slope $\psi$ and deflection $w$ at any radius $r$, as shown in Fig. 16.5.

The curvature of the plate in the diametral plane $rz$ is

$$\frac{1}{R_r} = -\frac{d^2 w}{dr^2} \qquad [16.11]$$

Fig. 16.5

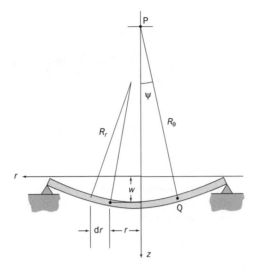

and for small values of $w$ (positive downwards) the slope at any point is

$$\psi = -\frac{\mathrm{d}w}{\mathrm{d}r}$$

The second principal radius of curvature $R_\theta$ is in the plane perpendicular to $rz$ and is represented by lines such as PQ which form a conical surface so that

$$\frac{1}{R_\theta} = \frac{\psi}{r} \qquad [16.12]$$

Now we shall consider an element of the plate subjected to bending moments along the edges $M_r$ per unit length and $M_\theta$ per unit length respectively, as shown in Fig. 16.6. This element can be analysed in the same manner as for rectangular co-ordinates. Thus eqns. [16.4] can be expressed for the circular plate as follows in terms of slope $\psi$:

$$\left.\begin{aligned} M_r &= \frac{Eh^3}{12(1-\nu^2)}\left(\frac{\mathrm{d}\psi}{\mathrm{d}r} + \nu\frac{\psi}{r}\right) \\[2mm] M_\theta &= \frac{Eh^3}{12(1-\nu^2)}\left(\frac{\psi}{r} + \nu\frac{\mathrm{d}\psi}{\mathrm{d}r}\right) \end{aligned}\right\} \qquad [16.13]$$

Fig. 16.6

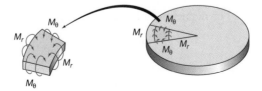

These equations can then be interpreted in terms of curvatures or deflection by further substitution to give

$$M_r = -D\left(\frac{\mathrm{d}^2 w}{\mathrm{d}r^2} + \frac{\nu}{r}\frac{\mathrm{d}w}{\mathrm{d}r}\right) = D\left(\frac{1}{R_r} + \frac{\nu}{R_\theta}\right)$$

$$M_\theta = -D\left(\frac{1}{r}\frac{\mathrm{d}w}{\mathrm{d}r} + \nu\frac{\mathrm{d}^2 w}{\mathrm{d}r^2}\right) = D\left(\frac{1}{R_\theta} + \frac{\nu}{R_r}\right)$$

$$\left.\right\} \qquad [16.14]$$

## 16.3 Relationships between bending moment and bending stress

We can determine bending stress as a function of bending moment by eliminating the curvatures between eqns. [16.2] and [16.4] so that for rectangular plates,

$$\sigma_x = \frac{12 M_x z}{h^3} \quad \text{and} \quad \sigma_y = \frac{12 M_y z}{h^3} \qquad [16.15]$$

Similar expressions apply for the bending stresses in circular plates as a function of $M_r$ and $M_\theta$.

## 16.4 Relationships between load, shear force and bending moment

If an element, Fig. 16.7, is taken from the plate then it must be in equilibrium under the action of uniformly distributed loading $p$ per unit area and the resulting shear forces $Q$ per unit length and bending moments $M$ per unit length. Owing to symmetry there are no shear forces on the radial sides of the element. For vertical equilibrium,

$$Qr\,\mathrm{d}\theta + pr\,\mathrm{d}r\,\mathrm{d}\theta - \left(Q + \frac{\mathrm{d}Q}{\mathrm{d}r}\mathrm{d}r\right)(r + \mathrm{d}r)\mathrm{d}\theta = 0$$

**Fig. 16.7**

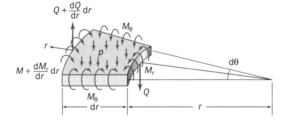

from which

$$\frac{\mathrm{d}Q}{\mathrm{d}r} + \frac{Q}{r} = p \qquad [16.16]$$

For moment equilibrium,

$$\left(M_r + \frac{\mathrm{d}M_r}{\mathrm{d}r}\mathrm{d}r\right)(r + \mathrm{d}r)\mathrm{d}\theta - M_r r\,\mathrm{d}\theta - 2M_\theta\,\mathrm{d}r\sin\frac{\mathrm{d}\theta}{2}$$

$$+ Qr\,\mathrm{d}\theta\,\mathrm{d}r = 0$$

which may be simplified to

$$r\frac{dM_r}{dr} + M_r - M_\theta + Qr = 0 \qquad [16.17]$$

## 16.5 Relationships between deflection, slope and loading

Next we substitute eqns. [16.13] into eqn. [16.17] and after simplifying obtain

$$\frac{d^2\psi}{dr^2} + \frac{1}{r}\frac{d\psi}{dr} - \frac{\psi}{r^2} = -\frac{Q}{D} \qquad [16.18]$$

which relates the slope at any radius to the shear force.

If eqns. [16.14] are substituted into eqn. [16.17] then

$$\frac{d^3w}{dr^3} + \frac{1}{r}\frac{d^2w}{dr^2} - \frac{1}{r^2}\frac{dw}{dr} = \frac{Q}{D} \qquad [16.19]$$

expresses the variation of deflection with radius.

Equations [16.18] and [16.19] can be expressed in a form which makes a solution by integration rather more obvious, as follows:

$$\frac{d}{dr}\left[\frac{1}{r}\frac{d}{dr}(r\psi)\right] = -\frac{Q}{D} \qquad [16.20]$$

and

$$\frac{d}{dr}\left[\frac{1}{r}\frac{d}{dr}\left(r\frac{dw}{dr}\right)\right] = \frac{Q}{D} \qquad [16.21]$$

The shear force $Q$ is a function of the applied loading $p$ and may be related by simple statical equilibrium or by integrating eqn. [16.16]. Multiplying through by $r\,dr$ gives

$$r\,dQ + Q\,dr = pr\,dr \quad \text{or} \quad d(Qr) = pr\,dr$$

$$Qr = \int_0^r pr\,dr$$

Substituting in eqn. [16.21]

$$r\frac{d}{dr}\left[\frac{1}{r}\frac{d}{dr}\left(r\frac{dw}{dr}\right)\right] = \frac{1}{D}\int_0^r pr\,dr \qquad [16.22]$$

If we know that $p$ is equal to $f(r)$ or is constant then eqn. [16.22] can be integrated to find the deflection at any radius.

The constants of integration are evaluated from the boundary conditions for the particular problem being solved.

Some typical loading and boundary situations will now be considered.

## 16.6 Plate subjected to uniform pressure

In this problem the right-hand side of eqn. [16.22] reduces to $pr^2/2$, since $p$ is constant; therefore

$$\frac{d}{dr}\left[\frac{1}{r}\frac{d}{dr}\left(r\frac{dw}{dr}\right)\right] = \frac{pr}{2D} \qquad [16.23]$$

Integrating,

$$\frac{1}{r}\frac{d}{dr}\left(r\frac{dw}{dr}\right) = \frac{pr^2}{4D} + C_1$$

Multiplying both sides by $r$ and integrating again,

$$r\frac{dw}{dr} = \frac{pr^4}{16D} + \frac{C_1 r^2}{2} + C_2$$

$$\frac{dw}{dr} = \frac{pr^3}{16D} + \frac{C_1 r}{2} + \frac{C_2}{r} \qquad [16.24]$$

Finally

$$w = \frac{pr^4}{64D} + \frac{C_1 r^2}{4} + C_2 \log_e r + C_3 \qquad [16.25]$$

**Clamped periphery**

For a plate of radius $a$ the boundary conditions are $dw/dr = 0$ at $r = 0$ and $r = a$, and $w = 0$ at $r = a$ (Fig. 16.8(a)).

Fig. 16.8

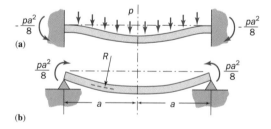

Hence

$$C_2 = 0 \quad \text{and} \quad \frac{pa^3}{16D} + C_1\frac{a}{2} = 0 \quad \text{so that} \quad C_1 = -\frac{pa^2}{8D}$$

From eqn. [16.25],

$$0 = \frac{pa^4}{64D} - \frac{pa^4}{32D} + C_3 \quad \text{or} \quad C_3 = \frac{pa^4}{64D}$$

The deflection curve is thus

$$w = \frac{p}{64D}(a^2 - r^2)^2 \qquad [16.26]$$

The maximum deflection occurs at the centre, so that

$$w_{max} = \frac{pa^4}{64D} \qquad [16.27]$$

Bending stress may now be determined from the moment–slope relationships, eqns. [16.13] and [16.24]:

$$M_r = \frac{p}{16}[a^2(1+\nu) - r^2(3+\nu)]$$

$$M_\theta = \frac{p}{16}[a^2(1+\nu) - r^2(1+3\nu)]$$

At the periphery $r = a$ and

$$M_r = -\frac{pa^2}{8} \quad \text{and} \quad M_\theta = -\nu\frac{pa^2}{8}$$

At the centre $r = 0$ and

$$M_r = M_\theta = \frac{pa^2}{16}(1+\nu) \qquad [16.28]$$

From eqns. [16.15] we have

$$\sigma_r = \frac{12M_r z}{h^3} \quad \text{and} \quad \sigma_\theta = \frac{12M_\theta z}{h^3}$$

The maximum stresses occur at $r = a$ and $z = \pm h/2$; hence

$$\sigma_{r_{max}} = \pm\frac{6M_{r_{max}}}{h^2} = \pm\frac{3}{4}\frac{pa^2}{h^2} \qquad [16.29]$$

$$\sigma_{\theta_{max}} = \pm\frac{3}{4}\nu pa^2/h^2$$

**Simply supported periphery** For this case the boundary conditions to determine three constants of integration are: $M_r = 0$, $r = a$; $dw/dr = 0$, $r = 0$; and $w = 0$, $r = a$. The problem will instead be tackled by an alternative method using the principle of superposition. We can use the solution from the fixed edge case combined with that for a plate simply supported and subjected to edge moments equal but opposite in sense to the fixing moment, as illustrated in Fig. 16.8(b).

Since the plate deforms to a spherical surface $M_r = M_\theta$, $R_r = R_\theta$ then from eqns. [16.13]

$$\frac{1}{R} = \frac{M}{D(1+\nu)}$$

The deflection at the centre of a spherical surface of radius $a$ and curvature $1/R$ is

$$w = \frac{a^2}{2R}$$

Therefore

$$w = \frac{Ma^2}{2D(1+\nu)} \tag{16.30}$$

But from the previous section the fixing moment is $M_r = -pa^2/8$; therefore

$$w = \frac{pa^4}{16D(1+\nu)} \tag{16.31}$$

The resultant deflection at the centre for the simply supported plate is, by superposition of eqns. [16.27] and [16.31],

$$w = \frac{pa^4}{64D} + \frac{pa^4}{16D(1+\nu)} = \frac{5+\nu}{64(1+\nu)D}pa^4 \tag{16.32}$$

The maximum bending moment occurs at the centre, and by superposition of eqn. [16.28] and $M = pa^2/8$ for the pure bending contribution,

$$M_r = M_\theta = \frac{pa^2}{16}(3+\nu) \tag{16.33}$$

The maximum stress occurs at the centre and is again obtained using eqn. [16.15]; thus

$$\sigma_{r_{max}} = \frac{3}{8}\frac{pa^2}{h^2}(3+\nu)$$

## 16.7 Plate with central circular hole

### Edge moments

A ring of moments is applied to the inner and outer boundaries as shown in Fig. 16.9. Since there is no shear force at the inner boundary, eqn. [16.20] reduces to

$$\frac{d}{dr}\left[\frac{1}{r}\frac{d}{dr}\left(r\frac{dw}{dr}\right)\right] = 0 \tag{16.34}$$

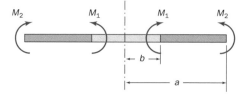

**Fig. 16.9**

Integrating twice and simplifying gives

$$\frac{dw}{dr} = \frac{C_1 r}{2} + \frac{C_2}{r} \tag{16.35}$$

and

$$w = \frac{C_1 r^2}{4} + C_2 \log_e r + C_3 \tag{16.36}$$

To find $C_1$ and $C_2$ we use the boundary conditions: at $r = b$, $M_r = M_1$; and at $r = a$, $M_r = M_2$; and by substituting eqn. [16.35] and its differential into eqns. [16.13] and solving, we obtain

$$C_1 = -\frac{2(a^2 M_2 - b^2 M_1)}{(1 + \nu)(a^2 - b^2)D}$$

$$C_2 = -\frac{a^2 b^2 (M_2 - M_1)}{(1 - \nu)(a^2 - b^2)D}$$

Since $w = 0$ at $r = a$ then, from eqn. [16.36],

$$C_3 = \frac{a^2 (a^2 M_2 - b^2 M_1)}{2(1 + \nu)(a^2 - b^2)D} + \frac{a^2 b^2 (M_2 - M_1) \log_e a}{(1 - \nu)(a^2 - b^2)D}$$

The deflection at any radius may now be determined by substituting the values of $C_1$, $C_2$ and $C_3$ into eqn. [16.36].

**Edge forces**  In this case uniformly distributed transverse forces $Q_0$ are applied at the inner and outer edges as shown in Fig. 16.10. The shear force per unit length at radius $r$ is

$$Q = \frac{2\pi b Q_0}{2\pi r} = \frac{b Q_0}{r}$$

Substituting in eqn. [16.20] and integrating gives

$$\frac{\mathrm{d}w}{\mathrm{d}r} = \frac{b Q_0}{4D}(r 2 \log_e r - 1) + \frac{C_1 r}{2} + \frac{C_2}{r} \tag{16.37}$$

$$w = \frac{b Q_0 r^2}{4D}(\log_e r - 1) + \frac{C_1 r^2}{4} + C_2 \log_e r + C_3 \tag{16.38}$$

**Fig. 16.10**

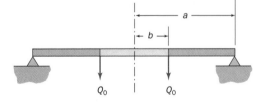

For a simply supported periphery the boundary conditions are: at $r = a$, $w = 0$ and $M_r = 0$; at $r = b$, $M_r = 0$.

By substituting eqn. [16.37] and its differential into eqns. [16.14] and solving, we obtain

$$C_1 = -\frac{b Q_0}{2D}\left(\frac{1 - \nu}{1 + \nu} + \frac{2(a^2 \log_e a - b^2 \log_e b)}{a^2 - b^2}\right)$$

$$C_2 = \frac{b Q_0}{2D}\left(\frac{(1 + \nu)}{(1 - \nu)}\frac{a^2 b^2}{a^2 - b^2}\log_e \frac{b}{a}\right)$$

Then, from eqn. [16.38],

$$C_3 = \frac{a^2 b Q_0}{4D}\left(1 + \frac{1}{2}\frac{(1-\nu)}{(1+\nu)} - \frac{b^2}{a^2 - b^2}\log_e\frac{b}{a} - \frac{2(1+\nu)}{(1-\nu)}\right.$$

$$\left. \times \frac{b^2}{a^2 - b^2}\log_e\frac{b}{a}\log_e a\right)$$

These constants may now be substituted into eqns. [16.37] and [16.38] to give the slope and deflection at any radius.

## 16.8 Solid plate with central concentrated force

### Simply supported edge

Considering the problem solved in the previous section and equating the total load around the inner periphery to the concentrated force $F$ so that

$$2\pi b Q_0 = F$$

and taking the limiting case when $b$ is infinitely small, $b^2\log_e(b/a)$ tends to zero and the above constants of integration become

$$C_1 = -\frac{F}{4\pi D}\left(\frac{1-\nu}{1+\nu} + 2\log_e a\right), \qquad C_2 = 0$$

$$C_3 = \frac{Fa^2}{8\pi D}\left[1 + \frac{1}{2}\left(\frac{1-\nu}{1+\nu}\right)\right]$$

Substituting these values into eqn. [16.38],

$$w = \frac{F}{8\pi D}\left(\frac{3+\nu}{2(1+\nu)}(a^2 - r^2) + r^2\log_e\frac{r}{a}\right) \qquad [16.39]$$

which is the deflection at any radius for a solid plate simply supported and subjected to a concentrated force at the centre.

### Fixed edge

For the fixed edge case we find the slope at the edge for the simply supported plate by differentiating eqn. [16.39]; then this slope has to be made zero by the superposition of the appropriate ring of edge moments. The deflection due to these moments may then be superposed onto the deflection given by eqn. [16.39] to obtain the total deflection.

### Example 16.1

A cylinder head valve of diameter 38 mm is subjected to a gas pressure of 1.4 MN/m². It may be regarded as a uniform, thin, circular plate simply supported around the periphery by the seat, as shown in Fig. 16.11. Assuming that the valve stem applies a concentrated force at the centre of the plate, calculate the movement of the stem necessary to lift the valve from its seat. The flexural rigidity of the valve is 260 N m and Poisson's ratio for the material is 0.3.

We have already derived solutions for a simply supported plate subjected to uniform loading, $p$, and a concentrated force, $F$, at the centre; hence the deflection at the centre is equal to the sum of the deflections due to the two separate load components. Therefore

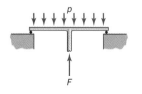

Fig. 16.11

$$w_{max} = \frac{Fa^2}{16}\frac{(3+\nu)}{(1+\nu)} + \frac{pa^4}{64D}\frac{(5+\nu)}{(1+\nu)}$$

but when the valve lifts from its seat $F = -\pi a^2 p$; therefore

$$w_{max} = -\frac{(7+3\nu)}{(1+\nu)}\frac{pa^4}{64D}$$

$$= -\frac{7.9 \times 1.4 \times 10^6 \times 0.019^4}{1.3 \times 64 \times 260}$$

$$= -0.0665\,\text{mm}$$

**16.9  Other forms of loading and boundary condition**

The solutions that have been obtained in the previous sections can be used to advantage with the principle of superposition to analyse a number of other plate problems.

**Concentric loading**

A plate which is uniformly loaded transversely along a circle of radius $b$, as illustrated in Fig. 16.12($a$), can be split into the two components shown at ($b$), and the separate solutions which have been obtained previously can be superposed using the appropriate boundary conditions, namely that there must be continuity of slope at radius $b$.

**Fig. 16.12**

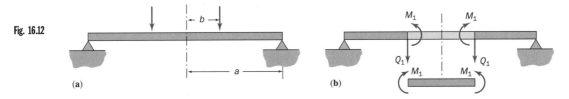

(a)            (b)

**Distributed loading on plate with central hole**

The situation illustrated in Fig. 16.13($a$) can be simulated by taking the deflection of the solid plate subjected to uniform loading and superposing that due to the appropriate moment and shear force at radius $b$, as in Fig. 16.13($b$).

**Fig. 16.13**

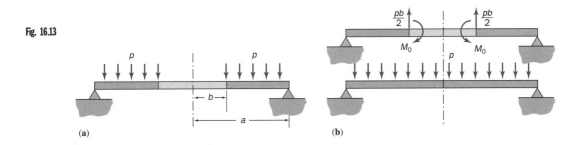

(a)            (b)

**Fixed inner boundary**

This edge condition can be associated with several types of loading and merely entails applying the appropriate moment to give zero slope and shear force as a function of the applied loading.

A variety of configurations and loadings are illustrated in Fig. 16.14, all of which can be dealt with easily by superposition of the required

**Fig. 16.14**

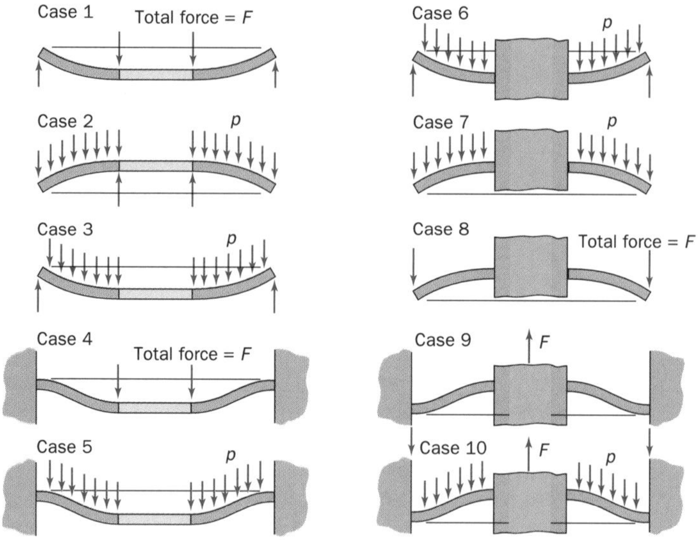

Case 1   Total force = F

Case 2   p

Case 3   p

Case 4   Total force = F

Case 5   p

Case 6   p

Case 7   p

Case 8   Total force = F

Case 9   F

Case 10   F   p

components which give the appropriate boundary conditions. However, in all these cases the maximum deflection can be represented by the following relations:

$$w_{max} = c' \frac{pa^4}{Eh^3} \quad \text{or} \quad w_{max} = c' \frac{Fa^2}{Eh^3}$$

where $c'$ is a factor involving the ratio $a/b$ and Poisson's ratio.

The maximum stresses can also be expressed by formulae as follows:

$$\sigma_{max} = c'' \frac{pa^2}{h^2} \quad \text{or} \quad \sigma_{max} = \frac{c'' F}{h^2}$$

where $c''$ is also a factor as defined above. Values of $c'$ and $c''$ for $\nu = 0.3$ and $a/b$ in the range $1\frac{1}{4}$ to 5 for the cases in Fig. 16.14 are given in Table 16.1.

**Table 16.1**   Coefficients $c'$ and $c''$ for the plate cases shown in Fig. 16.14

| $a/b =$ | 1.25 | | 1.5 | | 2 | | 3 | | 4 | | 5 | |
| Case | $c''$ | $c'$ | $c''$ | $c'$ | $c''$ | $c'$ | $c''$ | $c'$ | $c''$ | $c'$ | $c''$ | $c'$ |
| --- | --- | --- | --- | --- | --- | --- | --- | --- | --- | --- | --- | --- |
| 1 | 1.10 | 0.341 | 1.26 | 0.519 | 1.48 | 0.672 | 1.88 | 0.734 | 2.17 | 0.724 | 2.34 | 0.704 |
| 2 | 0.66 | 0.202 | 1.19 | 0.491 | 2.04 | 0.902 | 3.34 | 1.220 | 4.30 | 1.300 | 5.10 | 1.310 |
| 3 | 0.592 | 0.184 | 0.976 | 0.414 | 1.440 | 0.664 | 1.880 | 0.824 | 2.08 | 0.830 | 2.19 | 0.813 |
| 4 | 0.194 | 0.00504 | 0.320 | 0.0242 | 0.454 | 0.0810 | 0.673 | 0.172 | 1.021 | 0.217 | 1.305 | 0.238 |
| 5 | 0.105 | 0.00199 | 0.259 | 0.0139 | 0.480 | 0.0575 | 0.657 | 0.130 | 0.710 | 0.162 | 0.730 | 0.175 |
| 6 | 0.122 | 0.00343 | 0.336 | 0.0313 | 0.74 | 0.1250 | 1.21 | 0.291 | 1.45 | 0.417 | 1.59 | 0.492 |
| 7 | 0.135 | 0.00231 | 0.410 | 0.0183 | 1.04 | 0.0938 | 2.15 | 0.293 | 2.99 | 0.448 | 3.69 | 0.564 |
| 8 | 0.227 | 0.00510 | 0.428 | 0.0249 | 0.753 | 0.0877 | 1.205 | 0.209 | 1.514 | 0.293 | 1.745 | 0.350 |
| 9 | 0.115 | 0.00129 | 0.220 | 0.0064 | 0.405 | 0.0237 | 0.703 | 0.062 | 0.933 | 0.092 | 1.13 | 0.114 |
| 10 | 0.090 | 0.00077 | 0.273 | 0.0062 | 0.71 | 0.0329 | 1.54 | 0.110 | 2.23 | 0.179 | 2.80 | 0.234 |

## 16.10 Axi-symmetrical thin shells

The analysis of thin shells of revolution subjected to uniform pressure was treated as a statically determinate problem in Chapter 2. The fundamental relationship between the principal membrane stresses in the wall and the principal curvatures of the shell to the applied pressure and wall thickness was shown to be

$$\frac{\sigma_1}{r_1} + \frac{\sigma_2}{r_2} = \frac{p}{t} \qquad [16.40]$$

The simple applications of eqn. [16.40] to the cylinder and sphere under internal pressure were also dealt with in Chapter 2, together with an example on the self-weight of a concrete dome. The following example on liquid storage illustrates a further use of a thin shell.

### Example 16.2

The water storage tank illustrated in Fig. 16.15, of 20 mm uniform wall thickness, consists of a cylindrical section which is supported at the top edge and joined at the lower end to a spherical portion. An angle-section reinforcing ring of 5000 mm² cross-sectional area is welded into the lower joint as shown.

Calculate the maximum stresses in the cylindrical and spherical portions of the tank and the hoop stress in the reinforcing ring when the water is at the level shown. Density of water is 1000 kg/m³.

Fig. 16.15

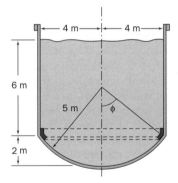

The total force due to the weight of water in the tank is

$$W = 9.81 \times 10^3 \left[ (\pi \times 6 \times 4^2) + \left( \frac{2\pi}{3} \times 5^3 \right) + \left( \frac{\pi \times 3^3}{3} \right) \right.$$

$$\left. - (\pi \times 5^2 \times 3) \right]$$

$$= 3.5\,\text{MN}$$

The axial stress in the cylindrical part of the tank is

$$\sigma_a = \frac{3.5}{2\pi \times 4 \times 0.02} = 6.95\,\text{MN/m}^2$$

and the hoop stress is

$$\sigma_{h_{max}} = \frac{6 \times 9.81 \times 10^3 \times 4}{0.02} = 11.8 \, \text{MN/m}^2$$

The maximum stress in the spherical portion of the tank occurs at the bottom, where the pressure is $9.81 \times 10^3 \times 8 \, \text{N/m}^2$; therefore

$$\sigma_{max} = \frac{9.81 \times 10^3 \times 8 \times 5}{2 \times 0.02} = 9.81 \, \text{MN/m}^2$$

For the reinforcing ring, the tangential force per unit length at the edge of the spherical part is $W/(2\pi \times 4 \sin \phi)$. The inward radial component of that force is

$$\frac{W}{2\pi \times 4 \sin \phi} \cos \phi$$

and therefore the compressive force in the ring is

$$\frac{W \cot \phi}{2\pi \times 4} \times 4$$

and the compressive stress is

$$\sigma_c = \frac{W \cot \phi}{2\pi} \times \frac{1}{5000 \times 10^{-6}}$$

$$= \frac{3.5 \times 10^6 \times \frac{3}{4}}{2\pi \times 5 \times 10^{-3}} = 83 \, \text{MN/m}^2$$

## 16.11  Local bending stresses in thin shells

Whenever there is a change in geometry of the shell, particularly for discontinuities in the meridian such as in the above example, the membrane stresses cause displacements which give rise to local bending in the wall. The resulting bending stresses may be significant in comparison with the membrane stresses. This was the reason for introducing the reinforcing ring in the above problem.

To illustrate the method of analysing local bending we will consider the elementary situation of a cylindrical vessel with hemispherical ends of the same thickness subjected to internal pressure as illustrated in Fig. 16.16.

The membrane stresses for the cylinder are

$$\sigma_1 = \frac{pr}{t} \quad \text{and} \quad \sigma_2 = \frac{pr}{2t}$$

**Fig. 16.16**

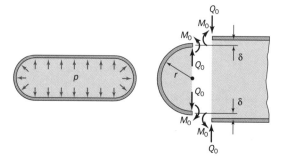

and for the hemisphere,

$$\sigma = \frac{pr}{t}$$

as calculated in Chapter 2.

The corresponding radial displacements for the cylinder and hemisphere, respectively, are

$$u_c = \frac{r}{E}(\sigma_1 - \nu\sigma_2) = \frac{pr^2}{2tE}(2 - \nu)$$

$$u_h = \frac{r}{E}(\sigma_1 - \nu\sigma_2) = \frac{pr^2}{2tE}(1 - \nu)$$

The difference in deformation radially is

$$\delta = \frac{pr^2}{2tE}[(2 - \nu) - (1 - \nu)] = \frac{pr^2}{2tE} \qquad [16.41]$$

In order to overcome this difference, shear and moment reactions are set up at the joint as shown in Fig. 16.16.

Since the cylindrical section is symmetrical with respect to its axis we may consider the reactions $Q_0$ and $M_0$ per unit length acting on a strip of unit width as illustrated in Fig. 16.17. The inward bending of the strip due to $Q_0$ sets up compressive circumferential strain. If the radial displacement is $v$ then

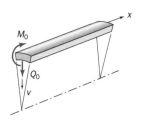

**Fig. 16.17**

$$\varepsilon_\theta = \frac{v}{r} \quad \text{and} \quad \sigma_\theta = \frac{Ev}{r}$$

The circumferential force due to this stress on the edge of the strip per unit length is $Evt/r$. The outward radial component of this force is

$$R = 2\frac{Evt}{r}\sin\frac{d\theta}{2} = \frac{Evt}{r}d\theta$$

$$= \frac{Evt}{r^2} \qquad [16.42]$$

This force opposes the deflection of the strip and is distributed along the strip, being proportional to the deflection $v$ at any point. This is a special case of the bending of a beam on an elastic foundation for which the deflection curve is

$$EI\frac{d^4v}{dx^4} = -w$$

where $w$ is the distributed loading function. For the present problem $EI$ is replaced by $D$, since the strip is restrained from distortion by adjacent material as for plates; therefore

$$D\frac{d^4v}{dx^4} = -\frac{Et}{r^2}v = -4D\beta^4 v \qquad [16.43]$$

The solution of this equation is

$$v = \exp(\beta x)(A \cos \beta x + B \sin \beta x)$$

$$+\exp(-\beta x)(C \cos \beta x + D' \sin \beta x) \qquad [16.44]$$

where $A$, $B$, $C$ and $D'$ are constants determined by the boundary conditions, and

$$\beta = \left(\frac{Et}{4Dr^2}\right)^{1/4} = \left(\frac{3(1-\nu^2)}{r^2t^2}\right)^{1/4}$$

As $x \to \infty$, $v \to 0$ and $M \to 0$, which gives $A = B = 0$. At $x = 0$, $M = M_0$ and $Q = Q_0$; therefore

$$D\frac{d^2v}{dx^2} = -M_0$$

and

$$D\frac{d^3v}{dx^3} = -Q_0$$

from which

$$C = \frac{1}{2\beta^3 D}(Q_0 - \beta M_0) \quad \text{and} \quad D' = \frac{M_0}{2\beta^2 D}$$

Substituting into eqn. [16.44], the deflection curve for the strip becomes

$$v = \frac{e^{-\beta x}}{2\beta^3 D}[Q_0 \cos \beta x - \beta M_0(\cos \beta x - \sin \beta x)] \qquad [16.45]$$

This is a rapidly damped oscillatory curve, and thus bending of the cylinder and head is local to the joint. The bending deflection and stresses decay over a characteristic length. For most materials,

$$\frac{1}{\beta} = \sqrt{(rt)}[3(1-\nu^2)]^{1/4} \simeq 1.25\sqrt{(rt)}$$

It can be seen that the bending stresses and deflections decay most rapidly in thin-walled tubes of small diameter.

When the cylinder and head are of the same material and wall thickness then the deflections and slopes at the joint produced by $Q_0$ are equal and $M_0 = 0$. Therefore the boundary condition is at $x = 0$, $v = \delta/2$, and from eqn. [16.45],

$$\frac{\delta}{2} = \frac{Q_0}{2\beta^3 D}$$

or

$$Q_0 = \delta\beta^3 D = \frac{pr^2}{2tE}\frac{Et}{4\beta r^2} = \frac{p}{8\beta} \qquad [16.46]$$

The deflection curve becomes

$$v = \frac{e^{-\beta x} \, p \cos \beta x}{16\beta^4 D} = \frac{pr^2}{4tE} \, e^{-\beta x} \cos \beta x \qquad [16.47]$$

By differentiating this equation twice the bending moment and hence the bending stress can be calculated for any cross-section. These have to be added to the membrane stresses to get the resultant stress.

If the wall thickness of the head and cylinder are different then $M_0 \neq 0$ and the boundary conditions would be that (a) the sum of the edge deflections must be zero, and (b) the rotation of the edges must be the same.

The above solution is equally applicable for other shapes of head.

**Example 16.3**

Calculate the local bending stresses in the vessel shown in Fig. 16.16 if $p = 1\,\text{MN/m}^2$, $r = 500\,\text{mm}$, $t = 100\,\text{mm}$, $\nu = 0.3$.

$$\beta = \left( \frac{3(1 - 0.3^2)}{0.5^2 \times 0.01^2} \right)^{1/4} = 18.2\,\text{m}^{-1}$$

$$Q_0 = \frac{p}{8\beta} = \frac{1 \times 10^6}{8 \times 18.2} = 6.87\,\text{kN/m}$$

Since the bending moment in the strip is $M = -D(\mathrm{d}^2 v/\mathrm{d} x^2)$, and from eqn. [16.45] with $M_0 = 0$,

$$v = \frac{Q_0}{2\beta^3 D} \, e^{-\beta x} \cos \beta x$$

Therefore

$$M = -\frac{Q_0}{\beta} \, e^{-\beta x} \sin \beta x$$

This expression takes the largest value for $\beta x = \pi/4$, which gives

$$M_{max} = 0.121\,\text{kN m/m}$$

This gives rise to a maximum bending stress of

$$\text{(axial)} \quad \sigma_b = \frac{6M_{max}}{t^2} = \frac{6 \times 0.121 \times 10^3}{0.01^2} = 7.26\,\text{MN/m}^2$$

The membrane stress is

$$\text{(axial)} \quad \sigma_m = \frac{pr}{2t} = \frac{1 \times 0.5}{2 \times 0.01} = 25\,\text{MN/m}^2$$

The total axial stress is $\sigma_a = 7.26 + 25 = 32.26\,\text{MN/m}^2$.

The bending of the strip also produces circumferential stresses: (a) since the strip displacement, $v$, varies only in the axial direction, the strip is loaded in cylindrical bending giving circumferential stresses of $\pm 6\nu M/t^2$; (b) stresses due to shortening of the circumference of $-Ev/r$.

Using the above values for $v$ and $M$ and summing gives

$$\text{(hoop)} \quad \sigma_b = \frac{Q_0 \exp(-\beta x)}{\beta t^2}\left(6\nu \sin \beta x - \frac{12(1 - \nu^2)}{2\beta^2 tr}\cos \beta x\right)$$

$$= 22.6 \exp(-\beta x)(0.3 \sin \beta x - 0.55 \cos \beta x)\,\text{MN/m}^2$$

The maximum value of this expression is $1.58\,\text{MN/m}^2$, which is small compared with the hoop membrane stress.

$$\text{(hoop)} \quad \sigma_m = \frac{pr}{t} = \frac{1 \times 0.5}{0.01} = 50\,\text{MN/m}^2$$

Hence local bending does not have a serious influence in this particular case.

## 16.12  Bending in a cylindrical storage tank

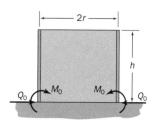

Fig. 16.18

An upright cylindrical storage tank, of radius $r$, uniform wall thickness $t$, and height $h$, is filled to the top with liquid of density $\rho$. The base of the tank is built into its foundation, Fig. 16.18, and we need to design for the maximum bending moment due to discontinuity in the shell at the base. Since $t \ll h$ or $r$ the shell may be regarded as infinitely long.

The governing equation is basically the same as eqn. [16.43] with an additional term which defines the variation of pressure loading due to the liquid:

$$D\frac{\mathrm{d}^4 v}{\mathrm{d}x^4} + 4D\beta^4 v = -\rho g(h - x) \quad\quad [16.48]$$

The particular integral part of the solution is $-\rho g(h - x)/4D\beta^4$ and the complete solution is

$$v = \exp(\beta x)(C_1 \cos \beta x + C_2 \sin \beta x)$$

$$+\exp(-\beta x)(C_3 \cos \beta x + C_4 \sin \beta x) - \frac{\rho g(h - x)}{4D\beta^4} \quad\quad [16.49]$$

The boundary conditions are as follows:
1.  The height of the cylinder can be regarded as 'infinite', so that $M \to 0$ and $v \to 0$, giving $C_1 = C_2 = 0$.
2.  At $x = 0$, $v = 0$ and $\mathrm{d}v/\mathrm{d}x = 0$; hence

$$C_3 = \frac{\rho g h}{4D\beta^4} \quad \text{and} \quad C_4 = \frac{\rho g}{4D\beta^4}\left(h - \frac{1}{\beta}\right)$$

Putting these values in eqn. [16.49] gives the deflection curve

$$v = -\frac{\rho g h}{4D\beta^4}\left[1 - \frac{x}{h} - e^{-\beta x}\cos \beta x - e^{-\beta x}\left(1 - \frac{1}{\beta h}\right)\sin \beta x\right]$$

$$[16.50]$$

But $M = -D(\mathrm{d}^2v/\mathrm{d}x^2)$, so that differentiating eqn. [16.50] twice and substituting gives

$$M = \frac{\rho g h}{2\beta^2}\left[-\mathrm{e}^{-\beta x}\sin\beta x + \left(1 - \frac{1}{\beta h}\right)\mathrm{e}^{-\beta x}\cos\beta x\right] \qquad [16.51]$$

Now, the maximum value of $M$ occurs at the discontinuity, where $x = 0$; therefore

$$M_{max} = \frac{\rho g h}{2\beta^2}\left(1 - \frac{1}{\beta h}\right) = \frac{\rho g}{2\beta^3}(\beta h - 1)$$

The resultant bending stresses can now be calculated as in Example 16.3, and as shown in Fig. 16.19. This illustrates the rapid decay in the localized bending as we move away from the restraint. This plot was generated using a spreadsheet 'what if' facility to calculate hoop and axial stress for a range of $x$-values.

**Fig. 16.19**

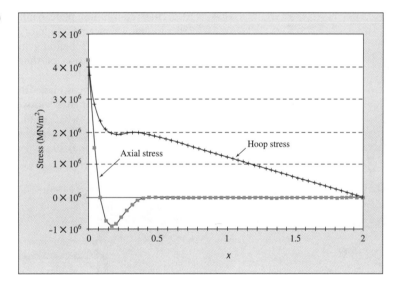

16.13 **Summary**

The basic principles for the analysis of plates have been developed involving equilibrium, geometry of deformation and the stress–strain relationships. These are essentially the two-dimensional development from the simple bending theory for beams. Thereafter, each plate solution is dependent on the particular boundary conditions of loading and support.

The design of thin shells depends principally on the magnitude of the general system of membrane stresses. However, attention must also be given to the effect of local bending stresses at regions of discontinuity in the shell. These stresses are of the same order of magnitude as the membrane stresses, but they decay rapidly with distance from the discontinuity in thin-walled shells. Somewhat surprisingly, reinforcement of a shell with rigid stiffeners increases the stress in a shell due to local bending.

**Bibliography**

Timoshenko, S, and Woinowsky-Kreiger, S. (1970) *Theory of Plates and Shells*, McGraw-Hill, New York.
Young, W. C. [1989] *Roark's Formulas for Stress and Strain*, McGraw-Hill International, New York.

**Problems**

16.1    Show that for a flat circular steel plate subjected to a uniform pressure on one surface, the maximum stress when the periphery is freely supported is 1.65 times that when the periphery is clamped. $\nu = 0.3$.

16.2    Calculate the ratio of (i) the radial stresses, and (ii) the tangential stresses at the edge and centre of a flat circular steel plate clamped at its periphery and subjected to a uniform pressure $p$. $\nu = 0.29$.

16.3    A circular thin steel diaphragm having an effective diameter of 200 mm is clamped around its periphery and is subjected to a uniform gas pressure of $180 \, kN/m^2$. Calculate a minimum thickness for the diaphragm if the deflection at the centre is not to exceed 0.5 mm. $E = 208 \, GN/m^2$, $\nu = 0.287$.

16.4    A circular aluminium plate 6 mm thick has an outer diameter of 250 mm and a concentric hole of 50 mm diameter. The edge of the hole is subjected to a bending moment of magnitude $900 \, Nm/m$. Determine the deflection of the inner edge relative to the outer. $E = 70 \, GN/m^2$, $\nu = 0.3$.

16.5    The end plate of a tube is made from 5 mm thick steel plate as in Fig. 16.20. If a 30 mm diameter rod welded to the end plate is subjected to a force of 10 kN what would be the movement of the rod? Calculate also the maximum stresses in the end plate. $E = 207 \, GN/m^2$, $\nu = 0.29$.

**Fig. 16.20**

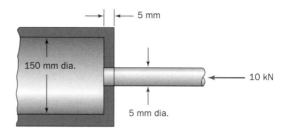

150 mm dia.

5 mm

5 mm dia.

10 kN

16.6    Figure 16.21 shows a long hydraulic cylinder and piston. Find an expression for the radial displacement of the cylinder along its length in terms of the supply pressure $p$, and the bending stresses in the cylinder at the locaton of the piston.

**Fig. 16.21**

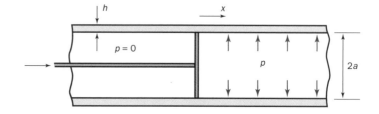

$h$

$x$

$p = 0$

$p$

$2a$

16.7 A circular plate 500 mm diameter and 2.5 mm thick is clamped around its edge and is subjected to a concentrated load of 900 N at its centre. Caculate the radial and tangential bending stresses at the fixed edge. $\nu = 0.29$.

16.8 The initially flat, circular plate shown in Fig. 16.22 is stepped so that the central disc has twice the flexural rigidity ($2D$) of the outer annulus ($D$). There is no external loading on the disc except at the edges, where the plate is clamped so that the slope at its edge is $\alpha$. Find the moment at the step in the plate. For the plate material take $\nu = 0.333$.

**Fig. 16.22**

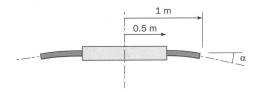

16.9 A pressure transducer uses a probe to measure the deflection of a thin circular diaphragm, clamped at its periphery and having a rigid seat at its centre to eliminate curvature in line with the probe as shown in Fig. 16.23. Calculate the thickness of the diaphragm if the probe can measure 0–0.25 mm and the range of the pressure transducer is to be 0–0.5 MN/m². Young's modulus and Poisson's ratio for the diaphragm material are 207 GN/m² and 0.3 respectively.

**Fig. 16.23**

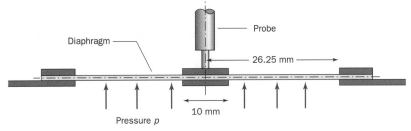

16.10 A 200 mm diameter circular plate is clamped around its periphery and subjected to a uniform pressure of 15 MN/m². If a rod supports the plate at its centre so that the central deflection of the plate is zero calculate the force in the rod.

16.11 A circular plate is to be simply-supported at its periphery and subjected to a load of 15 kN distrbuted around a circle which has the same centre as the plate. If the ratio of the plate diameter to the loading circle diameter is 2 calculate the thickness of the plate so that its maximumn stress does not exceed 200 MN/m². $\nu = 0.3$.

16.12 A circular steel plate of 304 mm diameter and 12 mm thick is clamped around the edge. A concentric ring of loading of 20 kN is applied uniformly on a circle of 152 mm diameter. Calculate the deflection at the centre of the plate. $E = 208$ GN/m², $\nu = 0.29$.

16.13   Determine the membrane stresses in a concrete hemispherical dome of radius 4 m and thickness 200 mm. The concrete has a density of 2.31 Mg/m³.

16.14   A conical water-storage tank has an included angle of 60° as shown in Fig. 16.24 and a vertical depth of water of 3 m. If the wall thickness is 5 mm determine the location and magnitude of the maximum hoop and meridional stresses. The water loading is 9.81 kN/m³.

**Fig. 16.24**

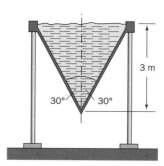

16.15   A toroidal pressure vessel has an interior diameter of 1 m and an exterior diameter of 2 m as shown in Fig. 16.25, with a wall thickness of 4 mm. Derive expressions for the principal membrane stresses and determine values at the horizontal section of symmetry for an internal pressure of 150 kN/m².

**Fig. 16.25**

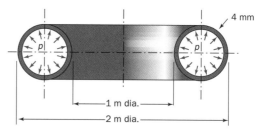

16.16   A steel pipe of outer diameter 200 mm and wall thickness 6 mm is to have a steel ring shrunk on to the outside. The inside diameter of the ring is 199 mm and it is 80 mm² in cross-sectional area. Determine the maximum local bending stress in the pipe due to the shrink pressure of the ring. $E = 208$ GN/m² and $\nu = 0.3$.

16.17   A stainless steel drum of outer diameter 400 mm and 6 mm wall thickness is closed at each end by steel discs which house central bearings. The drum has to rotate in service at 1000 rev/min. Determine the local bending stress in the drum where it is fixed to the discs. $E = 208$ GN/m², $\nu = 0.3$, $\rho = 7700$ kN/m³.

16.18   An upright cylindrical steel tank is built in to a concrete base. It is 3 m in diameter and has a wall thickness of 12 mm. If the height of water during a test is 2 m determine the maximum bending stress. $\rho = 1000$ kg/m³, $\nu = 0.3$.

CHAPTER **17** Finite Element Method

The finite element method for analysing structural parts has been around since the 1950s. The method was first developed for use in the aerospace and nuclear power industries. Here, the safety of the structures is critical: they involve large capital expenditure and the economic consequences of a failure are very severe, so the cost of the analysis is justified. Today the method is also extensively used in areas such as the automotive industry, where components are relatively cheap but are manufactured in large volumes. Furthermore, any small reduction in the safe weight of a component such as a connecting rod can lead to additional benefits in areas such as vibration reduction and fuel economy.

The growth in the usage of finite element methods is directly attributable to the rapid advances in computing technology in recent years. Today there are a number of large software companies developing and marketing finite element and associated modelling software. As a result, there exist commercial finite element packages capable of solving the most sophisticated problems, not just in structural or stress analysis, but for a wide range of phenomena such as steady and dynamic temperature distributions, fluid flow, and manufacturing processes such as injection moulding and metal forming.

Despite the proliferation and power of commercial software, it is still very important to have an understanding of the principles of the technique, so that an appropriate analysis model can be selected, correctly defined and interpreted. In this chapter the basic principles of the technique will be illustrated on some of the simplest types of finite element. Any solved problems are sufficiently small that the numerical details do not interfere with the interpretation of the results.

A full appreciation of the power and utility of the technique will require some experience and experimentation with commercial software. NAFEMS (the National Agency for Finite Element Methods and Standards) describe a series of benchmark problems for which answers are known. These are frequently used to develop the expertise of novices.

## 17.1 Principle of finite element method

If a truss of the type shown in Fig. 17.1(*a*) is being analysed then it is a straightforward exercise because it is formed from discrete members. The

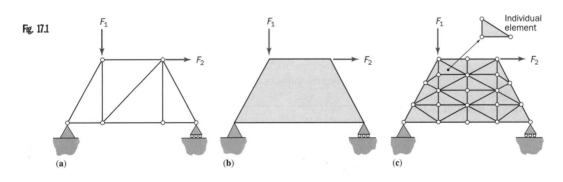

Fig. 17.1

(a)   (b)   (c)

paths of force transmission through the truss to the ground are readily apparent and the forces may be determined using equilibrium equations as illustrated in Chapter 1. If the truss had been statically indeterminate then both equilibrium equations and deformation compatibility would be necessary, but the method of solution is well established. If, however, a plate of the same shape as the truss (Fig. 17.1(*b*) is to be analysed then this is not so straightforward. The reason is that the plate is an elastic continuum and since the force transmission paths are not readily apparent the problem does not lend itself to simple mathematical analysis. Although, for the purpose of analysis, it might be tempting to consider the truss as being equivalent to the plate, this would not give accurate results since it ignores the restraining effect which all points in a continuum will experience and exert on neighbouring points. However, if the continuum was considered to be subdivided into a large number of triangular panels (Fig. 17.1(*c*)) it should be possible to develop a picture of the stress distribution in the whole plate by analysing each of the small panels in turn. To do this it would of course be necessary (*a*) to analyse the *equilibrium* of each of the triangular panels in relation to its neighbours and (*b*) to have available equations for the *geometry of deformation* and the *stress–strain relationships* for a triangular panel. This subdivision of a continuum into a large number of discrete elements is the basis of the finite element method of stress analysis. The triangular panels referred to in this example are the 'elements', but this is only one type of element. Others include a spring element (one dimensional), a plane rectangular element (two dimensional) and solid elements (three dimensional) as shown in Fig. 17.2.

**Fig. 17.2**   Types of finite elements

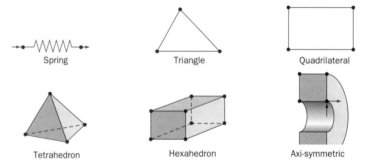

The accuracy of the solution depends on the number of subdivisions (elements); the more there are the greater the accuracy. However, although the analysis of each individual element is straightforward the analysis of a large number of elements becomes extremely tedious. For this reason finite element solutions to problems are generally carried out on computers and there are many commercial software packages available.

As an introduction to the finite element method it is convenient to consider the matrix analysis of skeletal structures using the stiffness method. An introduction to matrix algebra is given in Appendix B.

## 17.2  Analysis of uniaxial bars

### A single element

When a uniaxial bar such as a pin–jointed tie or strut is part of a structure, its ends will be able to move due to displacement of the structure and deformation of the member. This may be modelled by a spring element as shown in Fig. 17.3.

Fig. 17.3

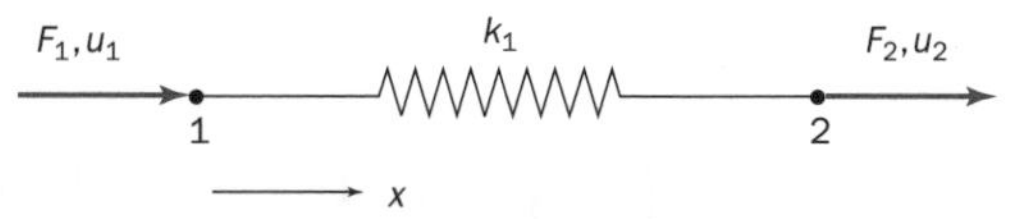

$F_1, u_1$  $k_1$  $F_2, u_2$

1  2

$\longrightarrow x$

The points of attachment of the element to other parts of the structure are called *nodes* and are indicated by the points 1 and 2 in Fig. 17.3. Here, $F$ and $u$ are the force and displacement values and their suffix indicates the node to which they apply. In situations where the spring element is used to model a tie or strut of length $L$ and area $A$, the stiffness of the spring $k_1$ will be given by

$$k_1 = \frac{AE}{L} \qquad [17.1]$$

where $E$ is Young's modulus for the material.

For the simple system illustrated in Fig. 17.3, using the sign convention that forces and displacements are positive in the $x$-direction, then the forces may be related to the displacements by the following equations:

$$F_1 = k_1(u_1 - u_2) = k_1 u_1 - k_1 u_2 \qquad [17.2]$$

$$F_2 = k_1(u_2 - u_1) = -k_1 u_1 + k_1 u_2 \qquad [17.3]$$

Equations [17.2] and [17.3] may be written in matrix form as

$$\begin{Bmatrix} F_1 \\ F_2 \end{Bmatrix} = \begin{bmatrix} k_1 & -k_1 \\ -k_1 & k_1 \end{bmatrix} \begin{Bmatrix} u_1 \\ u_2 \end{Bmatrix} \qquad [17.4]$$

In shorthand form this may be written as

$$\{F\} = [K^e]\{u\} \qquad [17.5]$$

where $[K^e]$ is referred to as the *stiffness matrix* for the spring element. An important property of the stiffness matrix for an element (and, as will be seen later, for a complete structure) is that it is symmetrical.

**An assembly of bar elements**

Consider now a system consisting of two bar elements as shown in Fig. 17.4.

Using eqn. [17.4] the force–displacement equation for each element may be written as

$$\begin{Bmatrix} F_1 \\ F_2 \end{Bmatrix} = \begin{bmatrix} k_1 & -k_1 \\ -k_1 & k_1 \end{bmatrix} \begin{Bmatrix} u_1 \\ u_2 \end{Bmatrix}$$

$$\begin{Bmatrix} F_2 \\ F_3 \end{Bmatrix} = \begin{bmatrix} k_2 & -k_2 \\ -k_2 & k_2 \end{bmatrix} \begin{Bmatrix} u_2 \\ u_3 \end{Bmatrix}$$

Fig. 17.4

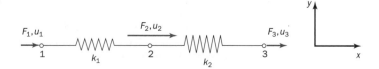

$F_1, u_1$  $F_2, u_2$  $F_3, u_3$  $y$

1  2  3  $x$

$k_1$  $k_2$

If each of these two equations is expanded so that they are in an equivalent form

$$\begin{Bmatrix} F_1 \\ F_2 \\ F_3 \end{Bmatrix} = \begin{bmatrix} k_1 & -k_1 & 0 \\ -k_1 & k_1 & 0 \\ 0 & 0 & 0 \end{bmatrix} \begin{Bmatrix} u_1 \\ u_2 \\ u_3 \end{Bmatrix} \quad \text{and}$$

$$\begin{Bmatrix} F_1 \\ F_2 \\ F_3 \end{Bmatrix} = \begin{bmatrix} 0 & 0 & 0 \\ 0 & k_2 & -k_2 \\ 0 & -k_2 & k_2 \end{bmatrix} \begin{Bmatrix} u_1 \\ u_2 \\ u_3 \end{Bmatrix}$$

The forces in the overall system are obtained by adding all the forces at each node. This may be obtained by adding the matrices to give

$$\begin{Bmatrix} F_1 \\ F_2 \\ F_3 \end{Bmatrix} = \begin{bmatrix} k_1 & -k_1 & 0 \\ -k_1 & k_1 + k_2 & -k_2 \\ 0 & -k_2 & k_2 \end{bmatrix} \begin{Bmatrix} u_1 \\ u_2 \\ u_3 \end{Bmatrix} \qquad [17.6]$$

or

$$\{F\} = [K]\{u\}$$

where $[K]$ is the stiffness matrix for the *structure*, i.e. the assembly of two spring elements, and $[F]$ is the vector of externally applied loads.

The influence of each term in the structural stiffness matrix may be visualized as follows. Imagine all the nodes in the structure except node $i$ are restrained so that all the $u_j$ terms ($j \neq i$) are zero. If then $u_i$ is given a unit value, i.e. $u_i = 1$, the force needed at all the other nodes to hold the displacements to zero at node $j$ is $K_{ij}$.

Thus in the example above, if $u_1 = u_3 = 0$ and $u_2 = 1$, the forces are

$$\begin{Bmatrix} F_1 \\ F_2 \\ F_3 \end{Bmatrix} = \begin{bmatrix} K_{11} & K_{12} & K_{13} \\ K_{21} & K_{22} & K_{23} \\ K_{31} & K_{32} & K_{33} \end{bmatrix} \begin{Bmatrix} 0 \\ 1 \\ 0 \end{Bmatrix} = \begin{Bmatrix} K_{12} \\ K_{22} \\ K_{32} \end{Bmatrix} = \begin{Bmatrix} -k_1 \\ k_1 + k_2 \\ -k_2 \end{Bmatrix}$$

$$[17.7]$$

Note that if nodes $i$ and $j$ are not connected by an element, $K_{ij} = 0$.

This approach to a solution is the basis of the finite element method and it is illustrated in the following example of a statically indeterminate force system.

**Example 17.1**

Three dissimilar materials are friction welded together and placed between rigid end supports as shown in Fig. 17.5. If forces of 50 kN and 100 kN are applied as indicated calculate the movement of the interfaces between the materials and the forces exerted on the end supports.

**Fig. 17.5**

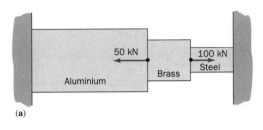

(a)

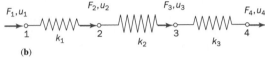

(b)

**Table 17.1**

| For aluminium | For brass | For steel |
|---|---|---|
| Area $= 400\,\mathrm{mm}^2$ | Area $= 200\,\mathrm{mm}^2$ | Area $= 70\,\mathrm{mm}^2$ |
| Length $= 280\,\mathrm{mm}$ | Length $= 100\,\mathrm{mm}$ | Length $= 100\,\mathrm{mm}$ |
| $E = 70\,\mathrm{GN/m}^2$ | $E = 100\,\mathrm{GN/m}^2$ | $E = 200\,\mathrm{GN/m}^2$ |

This system may be representeld by a model consisting of three spring elements as shown:

$$k_1 = \frac{A_1 E_1}{L_1} = \frac{400 \times 70 \times 10^3}{280} = 100\mathrm{kN/mm}$$

$$k_2 = 200\mathrm{kN/mm}$$

$$k_3 = 140\mathrm{kN/mm}$$

The force–deformation relationships for each element may then be written as

$$\begin{Bmatrix} F_1 \\ F_2 \end{Bmatrix} = \begin{bmatrix} 100 & -100 \\ -100 & 100 \end{bmatrix} \begin{Bmatrix} u_1 \\ u_2 \end{Bmatrix}$$

$$\begin{Bmatrix} F_2 \\ F_3 \end{Bmatrix} = \begin{bmatrix} 200 & -200 \\ -200 & 200 \end{bmatrix} \begin{Bmatrix} u_2 \\ u_3 \end{Bmatrix}$$

$$\begin{Bmatrix} F_3 \\ F_4 \end{Bmatrix} = \begin{bmatrix} 140 & -140 \\ -140 & 140 \end{bmatrix} \begin{Bmatrix} u_3 \\ u_4 \end{Bmatrix}$$

Using the two rules for the formation of the overall stiffness matrix,

$$\begin{Bmatrix} F_1 \\ F_2 \\ F_3 \\ F_4 \end{Bmatrix} = \begin{bmatrix} 100 & -100 & 0 & 0 \\ -100 & (100 + 200) & -200 & 0 \\ 0 & -200 & (200 + 140) & -140 \\ 0 & 0 & -140 & 140 \end{bmatrix} \begin{Bmatrix} u_1 \\ u_2 \\ u_3 \\ u_4 \end{Bmatrix}$$

This yields the equations

$$F_1 = 100u_1 - 100u_2 \qquad\qquad\qquad\qquad [17.8]$$

$$F_2 = -100u_1 + 300u_2 - 200u_3 \qquad\qquad\qquad [17.9]$$

$$F_3 = -200u_2 + 340u_3 - 140u_4 \qquad\qquad\qquad [17.10]$$

$$F_4 = -140u_3 + 140u_4 \qquad\qquad\qquad\qquad [17.11]$$

Using the boundary conditions that $u_1 = u_4 = 0$ and letting $F_2 = -50\,\text{kN}$, $F_3 = 100\,\text{kN}$ then eqns. [17.9] and [17.10] may be solved simultaneously to give $u_2 = 0.048\,\text{mm}$ and $u_3 = 0.323\,\text{mm}$. Using eqns. [17.8] and [17.11] then gives $F_1 = -4.8\,\text{kN}$ and $F_4 = -45.2\,\text{kN}$. These forces will both act to the left on the elements and to the right on the supports.

Note that in any finite element or stiffness analysis, sufficient displacement restraints must be applied to prevent rigid-body motion of the structure. This requirement may be illustrated by considering a single element as shown in Fig. 17.3 with a known force $P$ applied at node 2. The stiffness matrix gives

$$\begin{bmatrix} k & -k \\ -k & k \end{bmatrix} \begin{Bmatrix} u_1 \\ u_2 \end{Bmatrix} = \begin{Bmatrix} F_1 \\ P \end{Bmatrix} \qquad\qquad\qquad 17.12$$

which corresponds to the two scalar equations

$$\begin{aligned} ku_1 - ku_2 &= F_1 \\ -ku_1 + ku_2 &= P \end{aligned} \qquad\qquad\qquad [17.13]$$

It can be seen that

$$F_1 = -P \qquad\qquad\qquad\qquad [17.14]$$

i.e. equilibrium requires an equal and opposite force at node 1. The two equations are now

$$\begin{aligned} ku_1 - ku_2 &= -P \\ -ku_1 + ku_2 &= P \end{aligned} \qquad\qquad\qquad [17.15]$$

The second equation is the same as the first and there is no unique solution for $u_1$ and $u_2$. The determinant of the stiffness matrix is

$$\begin{vmatrix} k & -k \\ -k & k \end{vmatrix} = 0 \qquad\qquad\qquad [17.16]$$

and it can be shown that the inverse matrix does not exist. The system cannot therefore be solved to find the displacements. The physical reason for this is that if one end of the bar is not restrained the structure can 'float' back and forth on the $x$-axis. Equilibrium may be satisfied, but unless the displacement of one end of the bar is fixed, the displacement of the other end cannot be found.

In a one-dimensional analysis, one restraint is sufficient to prevent rigid-body motion. In a two-dimensional analysis, three restraints are required: two to prevent $x$- and $y$-translation, one to prevent rotation about the $z$-axis, Fig. 17.6. In a three-dimensional analysis, six restraints are required to

Fig. 17.6

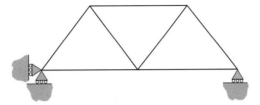

prevent rigid-body translation and rotation about each of the three co-ordinate axes.

Failure to prevent rigid-body motion is a very common cause of difficulty for inexperienced users of commercial finite element packages. Error or warning messages using phrases such as 'zero pivot' or 'zero tangent stiffness' often arise from this cause.

Some analyses, such as Example 17.1, have more than the minimum number of restraints. It is also possible to have a sufficient number of restraints, but fail to prevent rigid-body motion. In Fig. 17.7 the two-dimensional analysis has three restraints. Rotation about the $z$-axis and translation in the $y$-direction are prevented, but translation in the $x$-direction is not restrained.

Fig. 17.7

## 17.3 Analysis of frameworks

Although the bar elements considered so far have been collinear this need not be the case. The spring analogy could, for example, be used to determine the forces and deformation of a framework as shown in Fig. 17.8.

Fig. 17.8

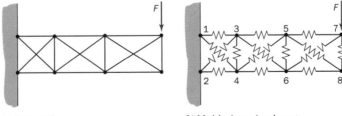

(**a**) Framework        (**b**) Model using spring elements

However, as the spring elements in the model are at different angles to one another it is necessary to express the forces and deformations for each element (calculated for its own *local* co-ordinate system $x$, $y$) to the deformation of the whole framework relative to the global co-ordinate system $X$, $Y$. For example, consider member 4–5 in Fig. 17.9.

**Fig. 17.9**

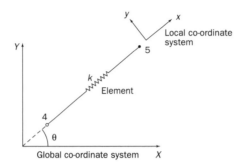

In the general case it is necessary to consider forces and displacements in both local co-ordinate directions at each node. For a pin-jointed member there will only be forces in the $x$-direction so the $F_y$-components will be zero. Thus eqn. [17.4] may be expanded to the form

$$\begin{Bmatrix} F_{x_4} \\ F_{y_4} \\ F_{x_5} \\ F_{y_5} \end{Bmatrix} = \begin{bmatrix} k & 0 & -k & 0 \\ 0 & 0 & 0 & 0 \\ -k & 0 & k & 0 \\ 0 & 0 & 0 & 0 \end{bmatrix} \begin{Bmatrix} \delta_{x_4} \\ \delta_{y_4} \\ \delta_{x_5} \\ \delta_{y_5} \end{Bmatrix} \qquad [17.17]$$

Referring to Fig. 17.9 the forces in the local co-ordinate directions may be related to the forces in the global co-ordinate directions as follows:

$$F_{x_4} = F_{X_4} \cos\theta + F_{Y_4} \sin\theta$$

$$F_{y_4} = -F_{X_4} \sin\theta + F_{Y_4} \cos\theta$$

$$F_{x_5} = F_{X_5} \cos\theta + F_{Y_5} \sin\theta$$

$$F_{y_5} = -F_{X_5} \sin\theta + F_{Y_5} \cos\theta$$

Using $s = \sin\theta$ and $c = \cos\theta$ as a convenient shorthand, these equations may be expressed in the matrix form

$$\begin{Bmatrix} F_{x_4} \\ F_{y_4} \\ F_{x_5} \\ F_{y_5} \end{Bmatrix} = \begin{bmatrix} c & s & 0 & 0 \\ -s & c & 0 & 0 \\ 0 & 0 & c & s \\ 0 & 0 & -s & c \end{bmatrix} \begin{Bmatrix} F_{X_4} \\ F_{Y_4} \\ F_{X_5} \\ F_{Y_5} \end{Bmatrix}$$

or

$$\{F\}_L = [T]\{F\}_G \qquad [17.18]$$

where $L$ refers to the local co-ordinate system and $G$ refers to the global co-ordinate system. Here $[T]$ is called the transformation matrix. A similar analysis will show that the local displacements are related to the global displacements by the following equation:

$$\{\delta\}_L = [T]\{\delta\}_G \qquad [17.19]$$

Now, by combining eqns. [17.5], [17.18] and [17.19] we may write

$$[T]\{F\}_G = [K^e][T]\{\delta\}_G$$

Premultiplying each side by $[T]^{-1}$ this becomes

$$\{F\}_G = [T]^{-1}[K^e][T]\{\delta\}_G \qquad [17.20]$$

or

$$\{F\}_G = [K^e]_G\{\delta\}_G \qquad [17.21]$$

where $[K^e]_G$ is the stiffness matrix for the element in global co-ordinates. Performing the matrix multiplication $[T]^{-1}[K^e][T]$ gives the value of $[K^e]_G$ as

$$[K^e]_G = k \begin{bmatrix} c^2 & cs & -c^2 & -cs \\ cs & s^2 & -cs & -s^2 \\ -c^2 & -cs & c^2 & cs \\ -cs & -s^2 & cs & s^2 \end{bmatrix} \qquad [17.22]$$

where $k = AE/L$.

The use of this method of analysing frameworks using spring elements is illustrated in the following example.

---

**Example 17.2**

**Determine the vertical and horizontal displacements at the loading point in the framework shown in Fig. 17.10. The value of _AE_ for each of the members is 200 MN.**

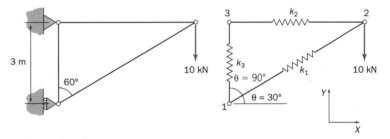

**Fig. 17.10**

From eqn. [17.22] the stiffness matrix for each of the elements in global co-ordinates will be

$$[K_1^e]_G = \begin{bmatrix} 25 & 14.4 & -25 & -14.4 \\ 14.4 & 8.33 & -14.4 & -8.33 \\ -25 & -14.4 & 25 & 14.4 \\ -14.4 & -8.33 & 14.4 & 8.33 \end{bmatrix}$$

$$[K_2^e]_G = \begin{bmatrix} 38.5 & 0 & -38.5 & 0 \\ 0 & 0 & 0 & 0 \\ -38.5 & 0 & 38.5 & 0 \\ 0 & 0 & 0 & 0 \end{bmatrix}$$

$$[K_3^e]_G = \begin{bmatrix} 0 & 0 & 0 & 0 \\ 0 & 66.7 & 0 & -66.7 \\ 0 & 0 & 0 & 0 \\ 0 & -66.7 & 0 & 66.7 \end{bmatrix}$$

**Table 17.2**

| Member | Length (m) | $\theta$ | $c (= \cos\theta)$ | $s (= \sin\theta)$ | $k$ (MN/m) |
|--------|-----------|----------|---------------------|---------------------|------------|
| 1-2 | 6 | 30 | 0.866 | 0.5 | 33.33 |
| 3-2 | 5.196 | 0 | 1 | 0 | 38.5 |
| 3-1 | 3 | 90 | 0 | 1 | 66.7 |

If these matrices are expanded with rows and columns of zeros so that they are in an equivalent form, they may then be added to give the stiffness matrix for the whole structure. This will give the following equation:

$$\begin{Bmatrix} F_{X_1} \\ F_{Y_1} \\ F_{X_2} \\ F_{Y_2} \\ F_{X_3} \\ F_{Y_3} \end{Bmatrix} = \begin{bmatrix} 25 & 14.4 & -25 & -14.4 & 0 & 0 \\ 14.4 & 75 & -14.4 & -8.33 & 0 & -66.7 \\ -25 & -14.4 & 63.5 & 14.4 & -38.5 & 0 \\ -14.4 & -8.33 & 14.4 & 8.33 & 0 & 0 \\ 0 & 0 & -38.5 & 0 & 38.5 & 0 \\ 0 & -66.7 & 0 & 0 & 0 & 66.7 \end{bmatrix} \times \begin{Bmatrix} \delta_{X_1} \\ \delta_{Y_1} \\ \delta_{X_2} \\ \delta_{Y_2} \\ \delta_{X_3} \\ \delta_{Y_3} \end{Bmatrix} \qquad [17.23]$$

Recognizing that $\delta_{X_1} = \delta_{X_3} = \delta_{Y_3} = 0$ and also $F_{Y_1} = F_{X_2} = 0$ then this set of simultaneous equations may be solved for $\delta_{Y_1}$, $\delta_{X_2}$ and $\delta_{Y_2}$.

$$\begin{bmatrix} 75 & -14.4 & -8.33 \\ -14.4 & 63.5 & 14.4 \\ -8.33 & 14.4 & 8.33 \end{bmatrix} \begin{Bmatrix} \delta_{Y_1} \\ \delta_{X_2} \\ \delta_{Y_2} \end{Bmatrix} = \begin{Bmatrix} 0 \\ 0 \\ -10^4 \end{Bmatrix} \qquad [17.24]$$

Hence

$$\delta_{Y_1} = -0.15 \text{ mm} \qquad \text{(i.e. downwards)}$$

$$\delta_{X_2} = -0.45 \text{ mm} \qquad \text{(i.e. to the right)}$$

$$\delta_{Y_2} = -2.13 \text{ mm} \qquad \text{(i.e. downwards)}$$

The values of the reactions at the supports may now be determined from eqn. [17.23]. This gives $F_{X_1} = 17.3 \text{ kN}$, $F_{Y_1} = 10 \text{ kN}$ and $F_{X_3} = -17.3 \text{ kN}$ which agrees with the values obtained from a simple equilibrium analysis of the structure.

The forces in each of the elements may be determined by reverting to the local co-ordinate system for each element. In the local co-ordinate system of Fig. 17.9, the force $f$ in the element is

$$f = AE\varepsilon = \frac{EA}{L}(\delta_{x_5} - \delta_{x_4})$$

$$= k(\delta_{x_j} - \delta_{x_i}) \qquad [17.25]$$

for an element connecting nodes $i$ and $j$ and $k = EA/L$ for that member. Using the displacement transformation, eqn. [17.19],

$$\delta_{x_i} = \begin{bmatrix} c & s \end{bmatrix} \begin{Bmatrix} \delta_{X_i} \\ \delta_{Y_i} \end{Bmatrix} \qquad [17.26]$$

and

$$\delta_{x_j} = [c \quad s] \begin{Bmatrix} \delta_{X_j} \\ \delta_{Y_j} \end{Bmatrix} \qquad [17.27]$$

so that the strain in the element connecting nodes $i$ and $j$ is

$$\varepsilon = \frac{1}{L}[-c \quad -s \quad c \quad s] \begin{Bmatrix} \delta_{X_i} \\ \delta_{Y_i} \\ \delta_{X_j} \\ \delta_{Y_j} \end{Bmatrix} \qquad [17.28]$$

and the force is

$$f = k[-c \quad -s \quad c \quad s] \begin{Bmatrix} \delta_{X_i} \\ \delta_{Y_i} \\ \delta_{X_j} \\ \delta_{Y_j} \end{Bmatrix} \qquad [17.29]$$

For example, for member 1–2, Fig. 17.10, the force is given by

$$f_{12} = 33.33[-0.866 \quad -0.5 \quad 0.866 \quad 0.5] \times \begin{Bmatrix} 0 \\ -0.15 \\ 0.45 \\ -2.12 \end{Bmatrix}$$

$$= -20 \text{ kN}$$

Similar $f_{32} = 17.3$ kN, and these values agree with the values obtained by taking a free-body diagram at the loading point.

The framework in this example was deliberately kept simple in order that the steps in the solution could be illustrated and the calculation performed manually. For a large plane or three-dimensional structure the individual steps in the solution are identical to those illustrated but, although the calculations are straightforward, they are so numerous that they are best left to a computer. Even on a computer the time taken to obtain a solution can be relatively long if a large complex structure is being analysed. This is where it can be beneficial to have some understanding of the nature of the calculations being performed so that data input can be rationalized to streamline the solution procedure. For example, one good way to achieve this in a structural analysis program is to ensure that the difference between the node numbers on each element is kept to the minimum. This has the effect of condensing the data into a band along the main diagonal of the overall stiffness matrix. This then reduces the subsequent computation and provides a valuable saving in computing time and disk space requirements.

This method can be used on either statically determine or statically indeterminate problems. For statically determinate problems, either the method of joints or the method of tension coefficients described in Chapter 1 is slightly more economical, since only the member forces have to be solved for. However, with the finite element method, the extra computation time is usually negligible in all but the largest structures, the displacements are obtained 'for free' and general purpose commercial packages can be used.

Before extending the analysis of the bar element to other more general elements, it is convenient to express the steps outlined so far in a more generalized form.

**Geometry of deformation**

In the one-dimensional bar, the displacement pattern may be expressed as a linear polynomial

$$u = \alpha_1 + \alpha_2 x \qquad [17.30]$$

where $\alpha_1$ and $\alpha_2$ are constants which may be determined from the nodal displacements and geometry of the element. At node 1, $x = 0$, so

$$u = u_1 = \alpha_1 \qquad [17.31]$$

whilst at node 2, $x = L$, so

$$u = u_2 = \alpha_1 + \alpha_2 L \qquad [17.32]$$

From this

$$\alpha_1 = u_1 \qquad [17.33]$$

and

$$\alpha_2 = \frac{u_2 - u_1}{L} \qquad [17.34]$$

The resultant variation in displacement over the element is

$$u = \alpha_1 + \alpha_2 x$$

$$= \left(1 - \frac{x}{L}\right) u_1 + \frac{x}{L} u_2$$

$$= \left[1 - \frac{x}{L} \quad \frac{x}{L}\right] \begin{Bmatrix} u_1 \\ u_2 \end{Bmatrix}$$

$$= [N]\{u\} \qquad [17.35]$$

Here $[N]$ represents what are called the *shape functions* of the element, which specify the form of the variation in displacement within the element, which in this case is linear.

The shape function associated with a particular displacement takes the value of one at that node and zero at any other node, see Fig. 17.11. The displacement at any point within an element can be found by multiplying the shape function matrix (which varies with position) by the nodal displacements (which are simply numbers). This applies to one-, two- and three-dimensional elements.

The only strain of interest in the element is

$$\varepsilon_x = \frac{du}{dx} = \frac{d}{dx}[N]\{u\}$$

$$= \frac{d}{dx}\left[1 - \frac{x}{L} \quad \frac{x}{L}\right] \begin{Bmatrix} u_1 \\ u_2 \end{Bmatrix} \qquad [17.36]$$

Fig. 17.11

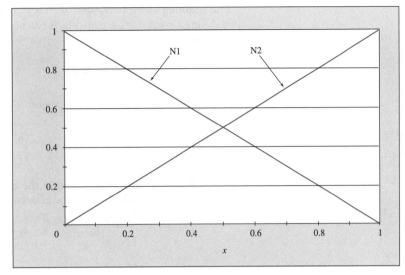

Only the shape functions vary with $x$, so

$$\varepsilon_x = \begin{bmatrix} -\dfrac{1}{L} & \dfrac{1}{L} \end{bmatrix} \begin{Bmatrix} u_1 \\ u_2 \end{Bmatrix}$$

$$= [B]\{u\} \qquad\qquad [17.37]$$

The matrix $[B]$ gives the strain at any point due to unit nodal displacement. Thus if node 1 is displaced by one unit, $\varepsilon_x = -1/L$ (compressive), whilst if node 2 is displaced by one unit, $\varepsilon_x = 1/L$ (tensile).

**Stress–strain relationships** The stress–strain relationship for a bar in uniaxial tension is just

$$\sigma = E\varepsilon = [E][B]\{u\}$$

$$= E \begin{bmatrix} -\dfrac{1}{L} & \dfrac{1}{L} \end{bmatrix} \begin{Bmatrix} u_1 \\ u_2 \end{Bmatrix} \qquad\qquad [17.38]$$

Here the matrix $[E]$ contains only a single value. For two- and three-dimensional elements, matrices with more elements will be necessary, but the procedure is just the same.

**Equilibrium of forces** The final step in the procedure is to relate the forces on each node of the element to the stresses within it. For the one-dimensional bar

$$F_1 = -A\sigma_x \qquad F_2 = A\sigma_x \qquad\qquad [17.39]$$

or

$$\begin{Bmatrix} F_1 \\ F_2 \end{Bmatrix} = \begin{bmatrix} -A \\ A \end{bmatrix} [E] \begin{bmatrix} -\dfrac{1}{L} & \dfrac{1}{L} \end{bmatrix} \begin{Bmatrix} u_1 \\ u_2 \end{Bmatrix}$$

$$= \dfrac{EA}{L} \begin{bmatrix} 1 & -1 \\ -1 & 1 \end{bmatrix} \begin{Bmatrix} u_1 \\ u_2 \end{Bmatrix} \qquad\qquad [17.40]$$

The resulting element stiffness matrix is exactly the same as eqn. [17.4].

Using virtual work arguments, it can be shown[1] that for any element, the stiffness matrix is given by

$$[K^e] = \int_V [B]^T [E][B] \, \mathrm{d}V \qquad [17.41]$$

where $V$ is the volume of the element. Thus for the one-dimensional bar element

$$[K^e] = \int_V \begin{bmatrix} -1/L \\ 1/L \end{bmatrix} [E] \begin{bmatrix} -\dfrac{1}{L} & \dfrac{1}{L} \end{bmatrix} \mathrm{d}V \qquad [17.42]$$

Since $[B]$ and $[E]$ are constant within the element then for this element, the matrices are just multiplied by the element volume $V = AL$ to give

$$[K^e] = \begin{bmatrix} -1/L \\ 1/L \end{bmatrix} [E] \begin{bmatrix} -\dfrac{1}{L} & \dfrac{1}{L} \end{bmatrix} AL$$

$$= \frac{EA}{L} \begin{bmatrix} 1 & -1 \\ -1 & 1 \end{bmatrix} \qquad [17.43]$$

which is the same as eqn. [17.4].

For higher-order elements, where the terms of $[B]$ vary within the element, the integration of eqn. [17.41] is usually performed numerically.

The general equation [17.40] can also be used to determine the stiffness matrix of the two-dimensional framework element. The matrix $[E]$ relating stress and strain is the same as for the one-dimensional element, but the matrix $[B]$ was shown, eqn. [17.28], to be

$$[B] = \frac{1}{L} \begin{bmatrix} -c & -s & c & s \end{bmatrix} \qquad [17.44]$$

Therefore

$$[K^e] = \int_V [B]^T [E][B] \, \mathrm{d}V = \frac{1}{L^2} \begin{bmatrix} -c \\ -s \\ c \\ s \end{bmatrix} [E] \begin{bmatrix} -c & -s & c & s \end{bmatrix} AL$$

$$= \frac{EA}{L} \begin{bmatrix} c^2 & cs & -c^2 & -cs \\ cs & s^2 & -cs & -s^2 \\ -c^2 & -cs & c^2 & cs \\ -cs & -s^2 & cs & s^2 \end{bmatrix} \qquad [17.45]$$

## 17.4 Analysis of beam elements

Although the stresses and deformation in beams have been extensively covered in Chapters 6–8, the amount of manual calculation required for statically indeterminate structures containing multiple members can become prohibitive. The development of a finite element representing a single beam will now be presented. Only the simplest one-dimensional beam which lies on the $x$-axis and is loaded in the $y$-direction will be described. Assemblies of these elements allow one-dimensional beams with multiple sections or loadings to be rapidly analysed using a computer, and the extension to two- or three-dimensional frameworks is straightforward.

**Geometry of deformation**

Figure 17.12 shows a beam bent in the $xy$-plane. The neutral axis undergoes no strain and planes normal to the neutral axis remain plane. The bending strain in the beam at any section is, eqn. [7.2b],

$$\varepsilon_x = -y\frac{d^2v}{dx^2} \tag{17.46}$$

**Fig. 17.12**

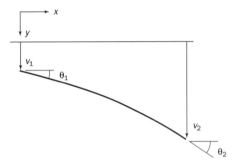

When we assemble a series of the elements together we must ensure that both displacement and slope are compatible between adjacent elements. In this case two variables, slope and displacement, are required at each end of the beam. The assumed form of the deflection of the beam must therefore contain four unknowns, i.e.

$$v = \alpha_1 + \alpha_2 x + \alpha_3 x^2 + \alpha_4 x^3 = \begin{bmatrix} 1 & x & x^2 & x^3 \end{bmatrix} \begin{Bmatrix} \alpha_1 \\ \alpha_2 \\ \alpha_3 \\ \alpha_4 \end{Bmatrix} \tag{17.47}$$

The slope of the beam at any point is then

$$\theta = \frac{dv}{dx} = \alpha_2 + 2\alpha_3 x + 3\alpha_4 x^2 = \begin{bmatrix} 0 & 1 & 2x & 3x^2 \end{bmatrix} \begin{Bmatrix} \alpha_1 \\ \alpha_2 \\ \alpha_3 \\ \alpha_4 \end{Bmatrix} \tag{17.48}$$

Substituting in the boundary conditions at each end of the beam, i.e.

at $x = 0$    $v = v_1 = \alpha_1$

$\theta = \theta_2 = \alpha_2$

at $x = L$    $v = v_2 = \alpha_1 + \alpha_2 L + \alpha_3 L^2 + \alpha_2 L^3$

$\theta = \theta_2 = \alpha_2 + 2\alpha_3 L + 3\alpha_2 L^2$

we obtain four equations in the four unknowns as

$$
\begin{Bmatrix} v_1 \\ \theta_1 \\ v_2 \\ \theta_2 \end{Bmatrix} =
\begin{bmatrix}
1 & 0 & 0 & 0 \\
0 & 1 & 0 & 0 \\
1 & L & L^2 & L^3 \\
0 & 1 & 2L & 3L^2
\end{bmatrix}
\begin{Bmatrix} \alpha_1 \\ \alpha_2 \\ \alpha_3 \\ \alpha_4 \end{Bmatrix}
\qquad [17.49]
$$

These equations can be solved either directly or by matrix inversion to give

$$
\begin{Bmatrix} \alpha_1 \\ \alpha_2 \\ \alpha_3 \\ \alpha_4 \end{Bmatrix} =
\begin{bmatrix}
1 & 0 & 0 & 0 \\
0 & 1 & 0 & 0 \\
-3/L^2 & -2/L & 3/L^2 & -1/L \\
2/L^3 & 1/L^2 & -2/L^3 & 1/L^2
\end{bmatrix}
\begin{Bmatrix} v_1 \\ \theta_1 \\ v_2 \\ \theta_2 \end{Bmatrix}
\qquad [17.50]
$$

**Fig. 17.13**

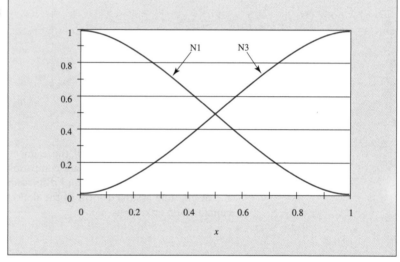

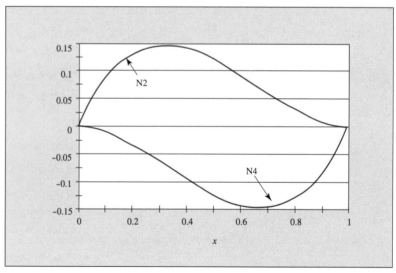

The equations for $\alpha_1 \ldots \alpha_4$ can be substituted back into the original equation for displacement to give

$$
v = \begin{bmatrix} 1 & x & x^2 & x^3 \end{bmatrix} \begin{bmatrix} 1 & 0 & 0 & 0 \\ 0 & 1 & 0 & 0 \\ -3/L^2 & -2/L & 3/L^2 & -1/L \\ 2/L^3 & 1/L^2 & -2/L^3 & 1/L^2 \end{bmatrix} \begin{Bmatrix} v_1 \\ \theta_1 \\ v_2 \\ \theta_2 \end{Bmatrix}
$$

$$
= \begin{bmatrix} 1 - 3\dfrac{x^2}{L^2} + 2\dfrac{x^3}{L^3} & x - 2\dfrac{x^2}{L} + \dfrac{x^3}{L^2} & 3\dfrac{x^2}{L^2} - 2\dfrac{x^3}{L^3} & -\dfrac{x^2}{L} + \dfrac{x^3}{L^2} \end{bmatrix}
$$

$$
\times \begin{Bmatrix} v_1 \\ \theta_1 \\ v_2 \\ \theta_2 \end{Bmatrix} = [N]\{u\} \tag{17.51}
$$

Here $[N]$ is the shape function matrix and $\{u\}$ is a vector of nodal displacements and rotations. The individual shape functions are shown in Fig. 17.13. They describe the deformed shape of the beam in response to unit displacement or rotation of the ends of the beam.

**Strain displacement relations**   The bending strain in the beam can be derived from the displacement using

$$
\varepsilon_x = -y\frac{\mathrm{d}^2 v}{\mathrm{d}x^2} = -y\frac{\mathrm{d}^2}{\mathrm{d}x^2}[N]\{u\}
$$

$$
= -y\begin{bmatrix} -\dfrac{6}{L^2} + \dfrac{12x}{L^3} & -\dfrac{4}{L} + \dfrac{6x}{L^2} & \dfrac{6}{L^2} - \dfrac{12x}{L^3} & -\dfrac{2}{L} + \dfrac{6x}{L^2} \end{bmatrix} \begin{Bmatrix} v_1 \\ \theta_1 \\ v_2 \\ \theta_2 \end{Bmatrix}
$$

$$
= [B]\{u\} \tag{17.52}
$$

The individual terms in the matrix $[B]$ describe the variation in strain within the element in response to unit displacement or rotation of the beam ends.

**Stress–strain relations**   The bending stress is obtained directly from the strain as

$$
\sigma = E\varepsilon_x = [E][B]\{u\} \tag{17.53}
$$

**Equilibrium**   The bending moment at either end of the beam can be obtained from

$$
M = \int_A \sigma y \, \mathrm{d}A \tag{17.54}
$$

From the bending moment, the moments acting at each end of the beam can be determined. Moment equilibrium can then be used to find the forces acting at the nodes, so that the nodal forces and moments resulting from a given set of imposed nodal displacements and rotations can be determined. This defines the element stiffness matrix.

An alternative technique is to use the general equation for the stiffness matrix, eqn. [17.41],

$$[K^e] = \int_V [B]^T [E][B] \, \mathrm{d}V$$

$$= \int_L \int_A E y^2 \, \mathrm{d}A \begin{bmatrix} -6/L^2 + 12x/L^3 \\ -4/L + 6x/L^2 \\ 6/L^2 - 12x/L^3 \\ -2/L + 6x/L^2 \end{bmatrix} [E]$$

$$\times \begin{bmatrix} -\dfrac{6}{L^2} + \dfrac{12x}{L^3} & -\dfrac{4}{L} + \dfrac{6x}{L^2} & \dfrac{6}{L^2} - \dfrac{12x}{L^3} & -\dfrac{2}{L} + \dfrac{6x}{L^2} \end{bmatrix} \mathrm{d}x$$

$$= \frac{EI}{L^3} \begin{bmatrix} 12 & 6L & -12 & 6L \\ 6L & 4L^2 & -6L & 2L^2 \\ -12 & -6L & 12 & -6L \\ 6L & 2L^2 & -6L & 4L^2 \end{bmatrix} \qquad [17.55]$$

The final stiffness equations are

$$\frac{EI}{L^3} \begin{bmatrix} 12 & 6L & -12 & 6L \\ 6L & 4L^2 & -6L & 2L^2 \\ -12 & -6L & 12 & -6L \\ 6L & 2L^2 & -6L & 4L^2 \end{bmatrix} \begin{Bmatrix} v_1 \\ \theta_1 \\ v_2 \\ \theta_2 \end{Bmatrix} = \begin{Bmatrix} Q_1 \\ M_1 \\ Q_2 \\ M_2 \end{Bmatrix} \qquad [17.56]$$

**Fig. 17.14**

The *nodal forces* and *nodal moments* corresponding to the nodal displacements and rotations are shown in Fig. 17.14. Note that these correspond to a positive displacement in the *y*-direction or a positive rotation about the *z*-axis, so that the force and moment contributions from adjacent elements can be added during element assembly. This is a different sign convention from that used for *shear force* and *bending moment* derived in Chapter 1.

**Example 17.3**

**Find the slope and deflection at the free end of a cantilever beam, loaded at the free end as shown in Fig. 17.15**

**Fig. 17.15**

The boundary conditions for this single-element problem are

$$\text{at } x = 0 \qquad v_1 = \theta_1 = 0$$

$$\text{at } x = L \qquad Q_2 = W \qquad M_2 = 0$$

Using $v_1 = \theta_1 = 0$, we can eliminate the first two rows and columns from the element stiffness matrix leaving

$$\frac{EI}{L^3} \begin{bmatrix} 12 & -6L \\ -6L & 4L^2 \end{bmatrix} \begin{Bmatrix} v_2 \\ \theta_2 \end{Bmatrix} = \begin{Bmatrix} W \\ 0 \end{Bmatrix} \qquad [17.57]$$

The second equation is

$$-6Lv_2 + 4L^2\theta_2 = 0 \qquad [17.58]$$

or

$$\theta_2 = \frac{3v_2}{2L} \qquad [17.59]$$

The first equation is

$$12v_2 - 6L\theta_2 = 3v_2 = \frac{WL^3}{EI} \qquad [17.60]$$

Therefore

$$v_2 = \frac{WL^3}{3EI} \qquad [17.61]$$

$$\theta_2 = \frac{WL^2}{2EI} \qquad [17.62]$$

These are the solutions listed for these problems in Table 7.1.

**Example 17.4**

**Find the deflection of a centrally loaded beam, built-in at both ends as shown in Fig. 17.16. Determine also the bending moments at the fixed ends.**

**Fig. 17.16**

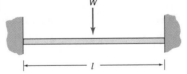

This is an example of a problem where symmetry can be used to reduce the size of the finite element problem to be solved, although in this case the reduction is from two elements to one! The stated problem is equivalent to analysing only half of the structure, carrying half the total load and with the symmetry boundary condition. (See Fig. 17.17.)

At node 1, the boundary conditions are $v_1 = \theta_1 = 0$. At node 2, $\theta_2 = 0$. The full stiffness relationship is

**Fig. 17.17**

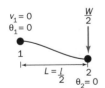

$$\frac{EI}{L^3} \begin{bmatrix} 12 & 6L & -12 & 6L \\ 6L & 4L^2 & -6L & 2L^2 \\ -12 & -6L & 12 & -6L \\ 6L & 2L^2 & -6L & 4L^2 \end{bmatrix} \begin{Bmatrix} 0 \\ 0 \\ v_2 \\ 0 \end{Bmatrix} = \begin{Bmatrix} Q_1 \\ M_1 \\ W/2 \\ M_2 \end{Bmatrix} \qquad [17.63]$$

where $L = l/2$. The third row of this matrix equation gives

$$\frac{EI}{L^3} \cdot 12v_2 = \frac{EI}{(l/2)^3} \cdot 12v_2 = \frac{W}{2}$$ [17.64]

Therefore

$$v_2 = \frac{Wl^3}{192EI}$$ [17.65]

which is the same result as is given in Table 7.1. The second row gives

$$\frac{EI}{L^3} \cdot -6Lv_2 = M_1$$ [17.66]

Therefore

$$M_1 = -\frac{WL}{4} = -\frac{Wl}{8}$$ [17.67]

**Example 17.5**

**Determine the deflection of the free end of a cantilever beam subject to a uniformly distributed load $w$ per unit length as shown in Fig. 17.18.**

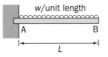

w/unit length

A      B

L

**Fig. 17.18**

The basic problem here is that external forces and moments can only be applied to the nodal points defining the element. The solution is to apply nodal forces and moments which do the same amount of virtual work as the distributed load over the element[1]. These 'kinematically equivalent loads' can be derived from

$$\{f\} = \int_A [N] q \, dA$$

where $[N]$ are the shape functions, $q$ is the pressure distributed over the element edge and $A$ is the area of the edge.

All commercial packages have facilities for applying distributed loads and calculating the correct equivalent nodal loads for internal use. These facilities should be used if at all possible, since the results are sometimes not obvious.

**Fig. 17.19**

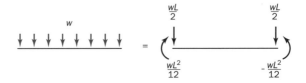

For a beam, the correct equivalent nodal forces and moments are shown in Fig. 17.19. Imposing the conditions $v_1 = \theta_1 = 0$, we are left with

$$\frac{EI}{L^3} \begin{bmatrix} 12 & -6L \\ -6L & 4L^2 \end{bmatrix} \begin{Bmatrix} v_2 \\ \theta_2 \end{Bmatrix} = \begin{Bmatrix} wL/2 \\ -wL^2/12 \end{Bmatrix}$$ [17.68]

which has the inverse matrix solution

$$\left\{ \begin{array}{c} v_2 \\ \theta_2 \end{array} \right\} = \frac{L}{12EI} \left[ \begin{array}{cc} 4L^2 & 6L \\ 6L & 12 \end{array} \right] \left\{ \begin{array}{c} wL/2 \\ -wL^2/12 \end{array} \right\}$$

$$= \frac{wL^4}{EI} \left\{ \begin{array}{c} 1/8 \\ 1/6 \end{array} \right\} \qquad [17.69]$$

These are again the same results as are obtained in Table 7.1. The further development of this type of element to allow the beam to be loaded while positioned at an angle $\theta$ in the $xy$-plane follows a similar procedure to that described for the bar element. Both element types can be further extended to allow their positioning at any orientation in three dimensions.

The beam and bar elements can be combined and extended to include torsion so that the behaviour of a three-dimensional assembly of long slender members subject to any combination of bending and torsional moments, and axial and shear forces, can be analysed in a straightforward manner. The analysis package only requires the nodal positions, the element topology (the nodes at either end of the element), the section properties (cross-sectional area, moments of inertia and torsional constant), material properties ($E$), loads and restraints.

## 17.5 Analysis of continua

Although the use of a simple linear spring element is a convenient way to introduce finite element methods, it is quite limited in its application. The major advantage of the finite element method is its ability to model complex two-dimensional and three-dimensional solids. In these cases the elements used may be of the types shown earlier in Fig. 17.2. However, the approach to a solution is still similar to the method illustrated for the linear spring element. In essence the solid continuum is modelled by a *mesh* of plane or three-dimensional elements which are joined to each other at their node points. The system of external loads acting on the actual solid must then be replaced by an equivalent system of forces acting at the node points. The type and number of elements used can be decided by the analyst. In general the accuracy of the solution will be greater if the number of elements is large. However, computer time (and cost) also increases with the number of elements chosen so it is generally wise only to use a dense concentration of elements in the critical areas of the solid which are likely to be of particular interest. Typical examples are shown in Fig. 17.20.

**Fig. 17.20**

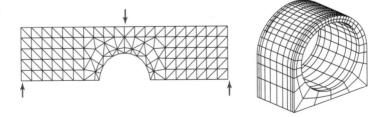

A computer program is then used to obtain the distribution of forces and displacements in the solid based on the stiffness matrix for the particular type of element chosen. Modern commercial packages include procedures for estimating the error arising from a given density of elements and adaptively refining the mesh where necessary, so that answers of known accuracy can be obtained.

## 17.6 Stiffness matrix for a triangular element

Figure 17.21 shows a two-dimensional triangular element typical of that used in plane stress problems.

**Fig. 17.21**

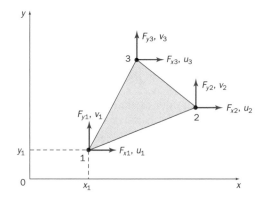

### Geometry of deformation

In general, two-dimensional displacement patterns may be expressed in terms of two linear polynomials

$$u = \alpha_1 + \alpha_2 x + \alpha_3 y$$
$$v = \beta_1 + \beta_2 x + \beta_3 y \qquad [17.70]$$

At node 1

$$u_1 = \alpha_1 + \alpha_2 x_1 + \alpha_3 y_1$$
$$v_1 = \beta_1 + \beta_2 x_1 + \beta_3 y_1 \qquad [17.71]$$

Similar expressions may be written for $u_2$, $v_2$, $u_3$ and $v_3$, and if each set of equations is written in matrix form so that they may be added then we get

$$
\begin{Bmatrix} u_1 \\ v_1 \\ u_2 \\ v_2 \\ u_3 \\ v_3 \end{Bmatrix}
=
\begin{bmatrix}
1 & x_1 & y_1 & 0 & 0 & 0 \\
0 & 0 & 0 & 1 & x_1 & y_1 \\
1 & x_2 & y_2 & 0 & 0 & 0 \\
0 & 0 & 0 & 1 & x_2 & y_2 \\
1 & x_3 & y_3 & 0 & 0 & 0 \\
0 & 0 & 0 & 1 & x_3 & y_3
\end{bmatrix}
\begin{Bmatrix} \alpha_1 \\ \alpha_2 \\ \alpha_3 \\ \beta_1 \\ \beta_2 \\ \beta_3 \end{Bmatrix}
\qquad [17.72]
$$

or in shorthand

$$\{\delta\} = [C]\{\alpha\}$$

This may be rearranged to give the coefficients $\{\alpha\}$ in terms of the displacements

$$\{\alpha\} = [C]^{-1}\{u\} \qquad [17.73a]$$

By matrix manipulation of $[C]$, $[C]^{-1}$ may be written as

$$[C]^{-1} = \frac{1}{2a} \begin{bmatrix} a_1 & 0 & a_2 & 0 & a_3 & 0 \\ b_1 & 0 & b_2 & 0 & b_3 & 0 \\ c_1 & 0 & c_2 & 0 & c_3 & 0 \\ 0 & a_1 & 0 & a_2 & 0 & a_3 \\ 0 & b_1 & 0 & b_2 & 0 & b_3 \\ 0 & c_1 & 0 & c_2 & 0 & c_3 \end{bmatrix} \qquad [17.73b]$$

where

$$\begin{array}{lll} a_1 = (x_2 y_3 - x_3 y_2) & b_1 = (y_2 - y_3) & c_1 = (x_3 - x_2) \\ a_2 = (x_3 y_1 - x_1 y_3) & b_2 = (y_3 - y_1) & c_2 = (x_1 - x_3) \\ a_3 = (x_1 y_2 - x_2 y_1) & b_3 = (y_1 - y_2) & c_3 = (x_2 - x_1) \\ 2a = (a_1 + a_2 + a_3) = 2(\text{area of element}) \end{array} \qquad [17.73c]$$

**Strain–displacement relations**    The next step is to get an expression for the strains in the element as a function of its geometry and nodal displacements. The strains in the element may be determined from the displacements as described in Chapter 13 as

$$\varepsilon_x = \frac{\partial u}{\partial x} = \alpha_2$$

$$\varepsilon_y = \frac{\partial v}{\partial y} = \beta_3$$

$$\gamma_{xy} = \frac{\partial u}{\partial y} + \frac{\partial v}{\partial x} = \beta_2 + \alpha_3 \qquad [17.74]$$

Expressing eqn. [17.74] in matrix form

$$\left\{ \begin{array}{c} \varepsilon_x \\ \varepsilon_y \\ \gamma_{xy} \end{array} \right\} = \begin{bmatrix} 0 & 1 & 0 & 0 & 0 & 0 \\ 0 & 0 & 0 & 0 & 0 & 1 \\ 0 & 0 & 1 & 0 & 1 & 0 \end{bmatrix} \left\{ \begin{array}{c} \alpha_1 \\ \alpha_2 \\ \alpha_3 \\ \beta_1 \\ \beta_2 \\ \beta_3 \end{array} \right\} \qquad [17.75]$$

or in shorthand form

$$\{\varepsilon\} = [H]\{\alpha\} \qquad [17.76]$$

Then combining eqns. [17.73a] and [17.76]

$$\{\varepsilon\} = [H][C]^{-1}\{u\} \qquad [17.77]$$

So, multiplying the respective matrices (see Appendix B) gives,

$$\{\varepsilon\} = [B]\{u\}$$

where

$$[B] = \frac{1}{2a} \begin{bmatrix} b_1 & 0 & b_2 & 0 & b_3 & 0 \\ 0 & c_1 & 0 & c_2 & 0 & c_3 \\ c_1 & b_1 & c_2 & b_2 & c_3 & b_3 \end{bmatrix} \qquad [17.78]$$

It may be seen that the terms in this matrix are known since they are a function of the co-ordinates of the nodes.

Furthermore all the terms of the matrix are constants, so the strains will all be constant within the element. This element is therefore known as the constant strain triangle. If the stresses vary within the part being analysed the true stresses will be approximated as a series of piecewise constant values across a given section.

**Stress–strain relationships**  For a two-dimensional plane stress element, the stresses and strains are related by the following equations:

$$\sigma_x = \left(\frac{E}{1-\nu^2}\right)(\varepsilon_x + \nu\varepsilon_y)$$

$$\sigma_y = \left(\frac{E}{1-\nu^2}\right)(\varepsilon_y + \nu\varepsilon_x)$$

$$\tau_{xy} = \frac{E}{2(1+\nu)}\gamma_{xy}$$

In matrix form these equations may be expressed as

$$\begin{Bmatrix} \sigma_x \\ \sigma_y \\ \tau_{xy} \end{Bmatrix} = \frac{E}{1-\nu^2} \begin{bmatrix} 1 & \nu & 0 \\ \nu & 1 & 0 \\ 0 & 0 & \frac{1}{2}(1-\nu) \end{bmatrix} \begin{Bmatrix} \varepsilon_x \\ \varepsilon_y \\ \gamma_{xy} \end{Bmatrix}$$ [17.79]

or in shorthand form

$$\{\sigma\} = [D]\{\varepsilon\}$$ [17.80]

where $[D]$ expresses the material properties of the element.

**Equilibrium of forces**  The effect of the external stress system on the triangular element is as shown in Fig. 17.22.

**Fig. 17.22**

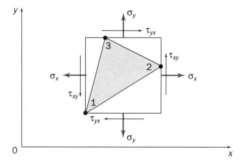

Assuming the thickness of the element to be $h$, the force on the left-hand side of the element is $-\sigma_x h(y_3 - y_1)$. This force is assumed to be shared equally by nodes 1 and 3, so making the earlier substitution of $b_2 = (y_3 - y_1)$ then

$$F'_{x_1} = F'_{x_3} = -\tfrac{1}{2}\sigma_x h b_2$$ [17.81]

Note, however, that this is just one component of the forces at nodes 1 and 3. There will be other components due to the direct stress on the right-hand side of the element and the shear stresses on the top and bottom of the element.

Considering the right-hand side of the element, over nodes 1–2 the force to be shared is $\sigma_x h(y_2 - y_1)$. Hence

$$F''_{x_1} = -\tfrac{1}{2}\sigma_x h b_3 \qquad [17.82]$$

The force at node 1 due to the shear stresses will comprise contributions from face 1–3 and face 1–2.

$$F'''_{x_1} = -\tfrac{1}{2}\tau_{xy}h(c_2 + c_3) = \tfrac{1}{2}\tau_{xy}hc_1 \qquad [17.83]$$

Combining eqns. [17.81], [17.82] and [17.83] gives

$$F_{x_1} = \tfrac{1}{2}h(b_1\sigma_x + c_1\tau_{xy}) \qquad [17.84]$$

Similar expressions may be obtained for the $y$-direction at node 1 and the $x$- and $y$-directions at nodes 2 and 3. The overall interrelationships between nodal forces and applied stresses may then be written in matrix form.

$$\begin{Bmatrix} F_{x_1} \\ F_{y_1} \\ F_{x_2} \\ F_{y_2} \\ F_{x_3} \\ F_{y_3} \end{Bmatrix} = \frac{h}{2} \begin{bmatrix} b_1 & 0 & c_1 \\ 0 & c_1 & b_1 \\ b_2 & 0 & c_2 \\ 0 & c_2 & b_2 \\ b_3 & 0 & c_3 \\ 0 & c_3 & b_3 \end{bmatrix} \begin{Bmatrix} \sigma_x \\ \sigma_y \\ \tau_{xy} \end{Bmatrix} \qquad [17.85]$$

Comparing eqn. [17.85] with [17.78], it can be seen that this is equivalent to

$$\{F\} = ah[B]^T\{\sigma\} \qquad [17.86]$$

We are now in a position to determine the stiffness matrix for the triangular element since the matrices $[B]$ and $[D]$ are available from eqns. [17.78] and [17.79]:

$$[K^e] = ah[B]^T[D][B] \qquad [17.87]$$

where $V = ah$ is the volume of the element. Since all the terms in $[B]$ and $[D]$ are constant within the element, this is equivalent to the general eqn. [17.41]

$$[K^e] = \int_V [B]^T[D][B]\,\mathrm{d}V \qquad [17.88]$$

The stiffness matrix for a triangular element is thus a $6 \times 6$ matrix. It can be appreciated that when a structure is modelled by a large number of triangular elements, the global stiffness matrix becomes extremely large and can only be handled by a computer. The following example illustrates the response of a single element to an imposed displacement.

**Example 17.6**

A single element of length *l*, width *w* as shown in Fig. 17.23 is fixed at nodes 1 and 3 while node 2 is displaced to the right by an amount $\delta$. Find the strain and stress in the material.

**Fig. 17.23**

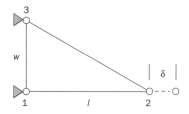

The strains can be determined from the nodal displacements, which are all known in this case, and the matrix $[B]$ defined by eqns. [17.73c] and [17.78] as

$$\left\{\begin{array}{c} \varepsilon_x \\ \varepsilon_y \\ \gamma_{xy} \end{array}\right\} = \frac{1}{wl}\begin{bmatrix} -w & 0 & w & 0 & 0 & 0 \\ 0 & -l & 0 & 0 & 0 & l \\ -l & -w & 0 & w & l & 0 \end{bmatrix}\left\{\begin{array}{c} 0 \\ 0 \\ \delta \\ 0 \\ 0 \\ 0 \end{array}\right\} = \left\{\begin{array}{c} \delta/l \\ 0 \\ 0 \end{array}\right\}$$

[17.89]

There is a strain of $\delta/l$ in the horizontal direction. Note that there is no lateral contraction $\varepsilon_y$ in the material arising from the extension in the $x$-direction since nodes 1 and 3 are fixed. Equally there is no shear strain, since the element edges 1–3 and 1–2 which were initially perpendicular remain so. The corresponding stress can be obtained by substituting in eqn. [17.79] as

$$\left\{\begin{array}{c} \sigma_x \\ \sigma_y \\ \tau_{xy} \end{array}\right\} = \frac{E}{1-\nu^2}\begin{bmatrix} 1 & \nu & 0 \\ \nu & 1 & 0 \\ 0 & 0 & (1-\nu)/2 \end{bmatrix}\left\{\begin{array}{c} \varepsilon_x \\ \varepsilon_y \\ \gamma_{xy} \end{array}\right\}$$

$$= \frac{E}{1-\nu^2}\frac{\delta}{l}\left\{\begin{array}{c} 1 \\ \nu \\ 0 \end{array}\right\}$$

[17.90]

## 17.7 Effect of mesh density

Since the simple triangular element studied here can only model a constant state of stress, we cannot expect to achieve accurate answers if the true stress is varying significantly over the area covered by the element. This effect is illustrated in the analysis of a rectangular region in pure bending in Fig. 17.24. These analyses were performed with a commercial finite element package. The shading within each element is proportional to the stress component $\sigma_x$.

In the first analysis, Fig. 17.24(a), only two elements were used through the thickness of the beam and large differences between the stress in adjacent elements can be seen. For a three-noded triangle, the 'stress jump' between adjacent elements is a measure of the error in the analysis, though improved estimates of the correct stress at a given node can be determined by averaging the stress at all the elements connected to it.

Fig. 17.24

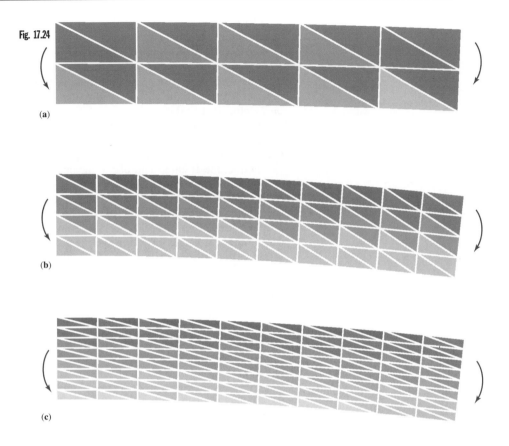

(a)

(b)

(c)

As the number of elements is increased, Fig. 17.24, the size of the stress discontinuities decreases and a more accurate estimate is obtained, though a very large number of elements are required for reasonable accuracy.

## 17.8  Other continuum elements

Owing to these accuracy limitations, the constant strain triangle element is not in very common use. In practice, elements which can model a quadratic variation in displacement or a linear variation in stress are usually preferred. These may be triangular or quadrilateral in shape for plane stress analyses. A further advantage of this type of element is that it can have curved sides to approximate curved boundaries. The results for a coarse mesh with these elements gives exact answers for the pure bending case, Fig. 17.25.

Fig. 17.25

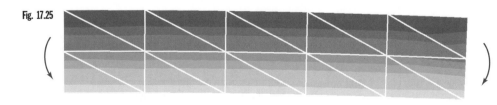

*Plane stress* elements such as the constant strain triangle allow the analysis of loading in the plane of a thin sheet of material. Plane stress elements can be converted to analyse *plane strain* or *axi-symmetric* problems with minimal effort. If a thin sheet is loaded perpendicular to its plane then *plate* elements must be used. The development of these is analogous to the beam element described earlier in the chapter. If the thin sheet of material is curved then *shell* elements must be used.

If the region of material to be analysed is long and slender then a one-dimensional *bar* or *beam* element will normally be used, since the stress at every point can be estimated by solving a small number of equations for only two nodes, one at either end of the member. This allows a very rapid and economical computer solution.

If the component to be analysed is not either long and slender or a thin sheet, then three-dimensional elements such as hexahedron or tetrahedron must be used, Fig. 17.2. The resulting analysis will be much more time consuming. A single brick element capable of modelling a quadratic variation in displacement has 20 nodes with three possible displacements in $x$-, $y$- and $z$-directions at each node. Thus 60 equations must be solved to determine the behaviour of a single element. Using three-dimensional elements for thin sheets or long slender bars is therefore uneconomic and can result in numerical problems if the element is too distorted. However, with these elements, accurate predictions can be obtained for the stress distribution in very complex geometries which would be impossible to analyse in any other way.

**Summary**

In the finite element method, the structure to be analysed is subdivided into a mesh of finite-sized elements of simple shape. Within each element, the variation of displacement is assumed to be determined by simple polynomial shape functions and nodal displacements. Equations for the strains and stresses are then developed in terms of the unknown nodal displacements. From this, the equations of equilibrium are assembled in a matrix form which can be easily programmed on a computer. After applying the appropriate boundary conditions, the nodal displacements are found by solving the matrix stiffness equation. Once the nodal displacements are known, element strains and stresses can be calculated.

A range of line, surface and solid element types are available for efficiently solving problems in long slender members, thin sheets and chunky solids respectively. Different types of line and surface element are used for in-plane and bending loads.

The power and generality of this method is demonstrated in excellent commercial packages, but the analyst or designer needs to have a basic understanding of the principles of the technique in order to apply it safely and with confidence.

**Reference**

1.    Zienckiewicz, O. C. (1989) *The Finite Element Vol. 1, Basic formulation and linear problems*, 4th edition, McGraw-Hill, New York.

**Bibliography**

Bathe, K. J. [1982] *Finite element procedures in engineering analysis*, Prentice Hall, London.

Bickford, W. B. [1994] *A First Course in the Finite Element Method*, 2nd edition, Irwin, Boston.

Fagan, M. J. (1992) *Finite Element Analysis. Theory and Practice*, Longman Scientific & Technical, Harlow.

NAFEMS (1986) *A Finite Element Primer*, Glasgow.

**Problems** 17.1 (*a*) Show that a finite element for analysing torsion of a straight bar (Fig. 17.26) has the stiffness matrix relationship

$$\frac{GJ}{L} \begin{bmatrix} 1 & -1 \\ -1 & 1 \end{bmatrix} \begin{Bmatrix} \phi_1 \\ \phi_2 \end{Bmatrix} = \begin{Bmatrix} T_1 \\ T_2 \end{Bmatrix}$$

where $G$ = shear modulus, $J$ = polar moment of inertia, $L$ = bar length, $T$ = torque and $\theta$ = angle of twist.

(*b*) Form an assembly of two of these bars in series and use this to check the results of Examples 5.4 and 5.5.

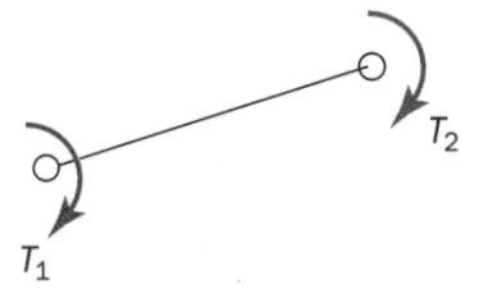

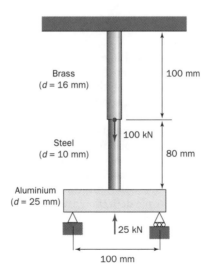

**Fig. 17.26**

17.2 Three dissimilar cylindrical rods are jointed together as indicated in Fig. 17.27. The brass cylinder is fixed to the rigid support at the top and the aluminium cylinder may be regarded as simply-supported with a point load at mid-span. Using the stiffness matrix approach, calculate the movement at each interface and the forces at the rigid supports when 25 kN and 100 kN forces are applied as indicated. $E_b = 100$ GN/m$^2$, $E_s = 200$ GN/m$^2$, $E_a = 70$ GN/m$^2$.

**Fig. 17.27**

Brass
(*d* = 16 mm)

100 mm

Steel
(*d* = 10 mm)

100 kN

80 mm

Aluminium
(*d* = 25 mm)

25 kN

100 mm

17.3 For the simple pin-jointed framework shown in Fig. 17.28 use the stiffness matrix approach to calculate the vertical deflection at the 20 kN load and the forces in each of the members. For each member the product $AE = 240$ MN.

17.4 Use the method of stiffness matrices to determine the vertical deflection at the loading point in the plane pin-jointed framework shown in Fig. 17.29. The product of cross-sectional area and modulus for each member is 400 MN.

Fig. 17.28

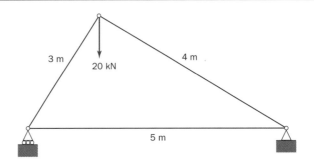

Fig. 17.29

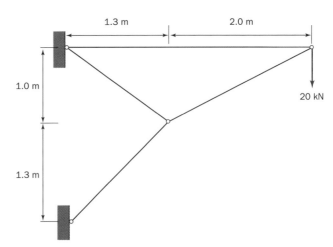

17.5   (a) Program the shape functions of Eqn. [17.51] into a spreadsheet so that they are calculated at an arbitrary point $x$ along the length of a beam. Calculate the displacement at $x$ when given the nodal displacements and rotations.

      (b) Plot a graph of the shape functions and displacement $u$ at 10 points along the length of the beam.

      (c) Insert the nodal displacements and rotations resulting from a load $W$ on the end of the beam as given in Table 7.1. View the graph of the deformed shape. Assume $W = E = I = L = 1$.

17.6   Use the stiffness matrix of a single beam finite element to determine the deflection of a simply-supported beam to (a) a concentrated load $W$ applied to the centre (Fig. 17.30(a)), (b) a distributed load of $w/$ unit length (Fig. 17.30(b)).

Fig. 17.30   17.7   The cantilever beam shown in Fig. 17.31 is to be represented by three triangular finite elements. Construct the master stiffness matrix for the beam.

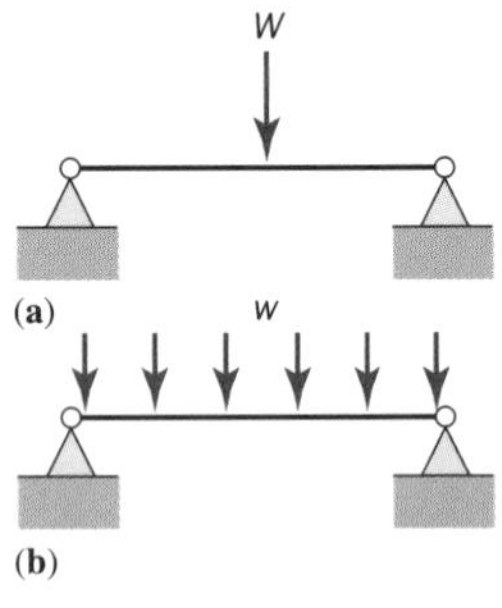

Fig. 17.31

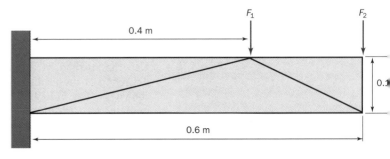

# 18 Tension, Compression, Torsion and Hardness

For the design engineer it is just as essential to have an understanding of material behaviour, to aid in the appropriate selection of type and condition, as it is to be able to calculate the stresses and strains which will have to be withstood by the material of a component or structure. This textbook is primarily devoted to the latter aspect and there are many excellent texts which deal fully with the engineering properties of materials. These remaining four chapters are intended therefore to provide a sufficient introduction to enable the reader to appreciate the significance of material response to stress and environmental conditions.

   This first chapter concentrates on the principal laboratory tests which are used to characterize materials from which British Standards (B.S.), other international standards and manufacturers' specifications of material properties are derived.

**18.1 Stress–strain response in a uniaxial tension test**

**Tensile testing of metals**

The principal concepts of elastic and plastic uniaxial tensile stress–strain behaviour were introduced in Chapter 3 and it is therefore only necessary briefly to reiterate certain key aspects.

   Figure 18.1(a) shows the typical shape of a flat or round bar specimen for tension testing (B.S. EN10002-1:1990). The enlarged ends are for gripping in the jaws of a testing machine and the reduced parallel portion contains the *gauge length*, across the ends of which is mounted an *extensometer*. This is a sensitive instrument which measures the very small longitudinal deformations that occur in the *elastic range*.

   If a second instrument is used to measure the lateral contraction of the gauge length as the load is increased then the ratio of lateral to longitudinal strains can be determined, which is Poisson's ratio $\nu$.

   Figure 18.1(b) is a stress–strain graph for the elastic range which is derived from the loads applied by the testing machine and the extensions of

**Fig. 18.1**

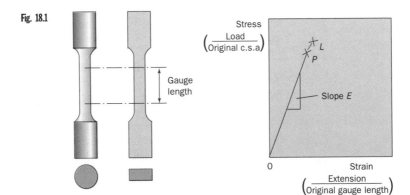

(a) Tensile specimens

(b) Elastic stress-strain curve

**Fig. 18.2**   Nominal and true tensile
stress–strain curves

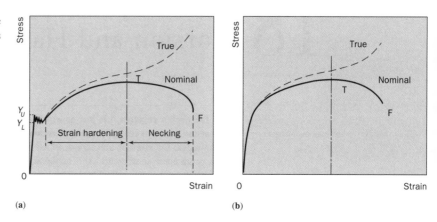

**Fig. 18.2**   Nominal and true tensile stress–strain curves

the gauge length measured by the extensometer. For metals it is a linear relationship, obeying Hooke's law, the slope of which is the Young's modulus of elasticity $E$. The end point of linearity is termed the *limit of proportionality* $P$, and a closely adjacent point is the *elastic limit L*.

When straining of a ductile metal is continued beyond the elastic limit, yielding commences and the material is in the plastic range. There are two characteristically different types of transition from elastic to plastic behaviour. These are illustrated in Fig. 18.2(*a*) and (*b*): the first is principally found in low- and medium-carbon steels and the second in alloy steels and non-ferrous metals. In the former, the point $Y_U$, at which there is a sudden drop in load with further strain, is termed the *upper yield point*. This is followed by $Y_L$, the *lower yield point*, from which there is a marked extension at almost constant load.

The upper yield point of low- and medium-carbon steels is a complex phenomenon, which is a function of strain rate, temperature, type of testing machine and geometry of the specimen. In some cases the upper yield point does not appear and the curve departs from the elastic range immediately into the horizontal lower-yield range. The latter, however, may be regarded as a material property and is used as the yield point stress for design purposes. It is this rather unique shape of stress–strain curve to which one approximates when using an ideal elastic-plastic relationship in plastic bending or torsion theory as in Chapter 15, since mild steel is widely used for structural work.

Many materials, particularly the light alloys, do not exhibit a clearly defined elastic limit, limit of proportionality or yield point, and several methods of indicating these stresses are in use. The most widely used is that involving a measure of permanent set and is called the *offset method*.

Figure 18.3 shows the stress–strain relationship for a material stressed beyond the limit of proportionality and then unloaded. The slope during the unloading stage is practically similar to that during the elastic range of loading. The stress for any given amount of inelastic deformation is easily obtained from the stress–strain diagram. The amount $x$ is set off on the strain or extension axis and a line AB is drawn parallel to the straight line portion of the loading curve, and the intersection at B gives the stress $\sigma_p$ for a permanent set of $x\%$ strain.

**Fig. 18.3**  Determination of proof stress

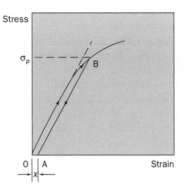

The elastic limit is taken to be that stress at which there is a permanent set of 0.02%; this is generally called the 0.02% elastic limit. The proof stress is that stress at which there is a permanent set of 0.1% and is called the 0.1% proof stress. The yield stress is that stress at which there is a permanent set of 0.2% of the gauge length.

On a 50 mm gauge length, the extensions corresponding to the above stresses are 0.01, 0.05 and 0.1 mm respectively, and thus extensometer measurements are required for the accurate production of the load–extension diagram.

Once the initial yield region has been passed in Fig. 18.2(*a*) and (*b*), the stress–strain curves have a common form. An increasing stress is required to cause continued straining, and this behaviour is known as *work hardening* or *strain hardening* and the metal does in fact become harder.

The property of work hardening is quite important and will be discussed further later. The specimen has a limit to which it can be work hardened uniformly in a simple tension test, and this is reached when the slope of the nominal stress-strain curve becomes zero as at T, in Fig. 18.2(*a*) and (*b*). This point is known as the *tensile strength* (in the past, the ultimate tensile strength U.T.S.) and is given by the maximum load carried by the specimen divided by the original cross-sectional area. This is a very important quantity as it is sometimes used in design in conjunction with a safety factor, and is always quoted when comparing metals and describing their mechanical properties. It is also an approximate guide to hardness and fatigue strength.

Up to the point T, the parallel gauge length of the specimen has reduced in cross-section quite uniformly; however, at or close to the tensile strength T one section is slightly weaker due to inhomogeneity than the rest and begins to thin more rapidly, forming a waist or *neck* in the gauge length. Further extension is now concentrated in the neck and a reducing load is required for continued straining, and thus the nominal stress–strain curve also falls off, until fracture occurs at F.

The lower stress–strain curves in Fig. 18.2 were based on nominal values of stress, i.e. the load at all points was divided by the original cross-sectional area. This is a convenient arrangement for most practical purposes and is the generally adopted procedure. However, as the specimen is strained, so the cross-sectional area reduces, and hence the true stress is higher than the nominal stress. The true stress is obtained by dividing the load by the current area of the specimen corresponding to that load. The current area of the bar may be obtained by direct measurement at various stages during the test or by calculation.

The shape of the true stress curve is shown in Fig. 18.2. It is only strictly valid up to the point where necking commences, since the change in geometry at the neck sets up a complex stress system which cannot be determined simply from the load divided by the area of the neck.

The foregoing has dealt principally with the property of 'strength' or stress; another property which is of almost equal importance is that of *ductility*, or the ability of a material to withstand plastic deformation. In a tensile test this is expressed in two ways, by the percentage elongation of the gauge length after fracture and by the percentage reduction in cross-sectional area referred to the neck or minimum section at fracture.

The second quantity is expressed algebraically as

$$\text{Reduction in area} = \frac{A_O - A_F}{A_O} \times 100\%$$

where $A_O$ and $A_F$ are the original and fracture areas respectively.

Elongation is the increase in the gauge length divided by the original gauge length, or algebraically

$$\text{Elongation} = \frac{L_F - L_O}{L_O} \times 100\%$$

where $L_O$ and $L_F$ are the original and final gauge lengths respectively. Elongation is only partly a material property since it is also dependent on the geometrical form of the test piece.

In cases where the specimen necks, the distribution of strain over the gauge length of a specimen can vary considerably from one metal to another, as illustrated in Fig. 18.4. This will depend on the grain size and microstructure.

**Fig. 18.4** Strain distribution along a tensile test piece

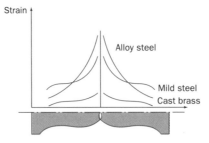

The elongation at fracture includes the necked region, and this is relatively independent of original gauge length and is more a function of the material and shape of cross-section.

Two basic types of fracture can be obtained in tension, depending on the material, temperature, strain rate, etc., and these are termed *brittle* and *ductile*. The main features of the former are that there is little or no plastic deformation, the plane of fracture is normal to the tensile stress and separation of the crystal structure occurs. In the second type, ductile failure is preceded by a considerable amount of plastic deformation and fracture is by shear or sliding of the crystal structure on microscopic or general planes at about 45° to the tensile stress. Probably the most well-known type of failure is called the *cup and cone* in the cylindrical bar.

A cylindrical bar, even though relatively ductile, may not produce a sufficient neck to set up marked triaxiality of stress, and failure is found to occur on a single shear plane right across the specimen. When the two parts of a fractured flat bar are placed together it is seen that there is a gap in the middle region, showing the initiation of fracture there, while the outer regions continued to extend.

## 18.2 Stress–strain response in a compression test

The mechanical properties of a ductile metal are generally obtained from a tension test. However, compression behaviour is of interest in the metal-forming industry, since most processes, rolling, forging, etc., involve compressive deformations of the metal, and also often of the forming equipment.

In compression an elastic range is exhibited as in tension and the elastic modulus, proportional limit and yield point or proof stresses have closely corresponding values for the two types of deformation. The real problem arises in a compression test when the metal enters the plastic range. The test piece has to be relatively short $(D/L > 4)$ to avoid the possibility of instability and buckling. The axial compression is accompanied by lateral expansion, but this is restrained at the ends of the specimen owing to the friction between the machine platens and the end faces, and consequently on a short specimen marked barrelling occurs as in Fig. 18.5. This causes a non-uniformity of stress distribution, and conical sections of material at each end are strained and hardened to a lesser degree than the central region. The effect on the load–compression curve, after the smaller values of plastic strain have been achieved, is a fairly rapid rise in the load required to overcome friction and cause further compression.

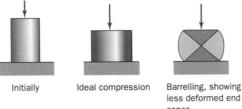

**Fig. 18.5** Deformation during a compression test

Initially        Ideal compression        Barrelling, showing less deformed end cones

Owing to the barrelling effect, only an average stress can be computed from the load–compression curve, based on an average area determined from considerations of constant volume.

Various methods have been attempted to overcome the effects of barrelling, none of which is completely successful. The most satisfactory appears to be the technique of using several cylinders of the same metal having different diameter-to-length ratios. Incremental compression tests are conducted on the set of cylinders at a series of loads of increasing magnitude, measuring the strain for each of the cylinders at each load. Extrapolation of curves of $D/L$ against strain with load as parameter to a value of $D/L = 0$, representing an infinitely long specimen where barrelling would be negligible, enables the true compressive stress–strain curve to be determined, Fig. 18.6. Failure of a ductile metal in compression only occurs owing to excessive barrelling causing axial splitting around the periphery.

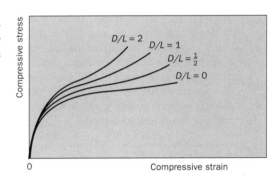

For brittle materials, such as flake cast iron, concrete, etc., which would
not normally be used in tension, the compression test is used to give
quantitative mechanical properties. Although end friction still occurs, which
affects the stress values somewhat, owing to the absence of ductility in these
materials the barrelling condition is barely achieved. The examples of failure
of cylindrical specimens shown in Fig. 18.7 illustrate that fracture takes
place on planes of maximum shear stress.

**Fig. 18.7** Modes of compression
failure in various materials

Concrete          Flake iron          Timber

## 18.3 Stress–strain response in a torsion test

The usual method of obtaining a relationship between shear stress and shear
strain for a material is by means of a torsion test. This may be conducted on
a circular-section solid or tubular bar. By applying a torque to each end of
the test piece by a testing machine and measuring the angular twist over a
specified gauge length, a torque–twist diagram can be plotted. This is the
equivalent in torsion to the load–extension diagram in tension. It was shown
theoretically in Chapter 14 that, under elastic conditions in torsion, the
applied torque is proportional to the angle of twist on the assumption that
shear stress is proportional to shear strain. This is found to be true
experimentally, and the linear torque–twist relationship obtained enables the
shear or rigidity modulus to be determined, since

$$G = \frac{T}{\theta} \frac{L}{J}$$

where the symbols are as specified in Chapter 5.

The torsion test is not like the tension or compression tests in which the
stress is uniform across the section of the specimen. In torsion there is a
stress gradient across the cross-section, and hence at the end of the elastic
range yielding commences in the outer fibres first while the core is still
elastic, whereas in the direct stress tests yielding occurs relatively evenly
throughout the material. With continued twisting into the plastic range,
more and more of the cross-section yields until there is penetration to the

Fig. 18.8

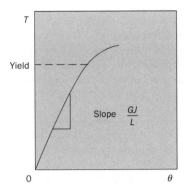

axis of the bar (see section 15.3). The torque–twist diagram, Fig. 18.8, appears of much the same form as a load–extension diagram, and work hardening will occur at a gradually decreasing rate as straining proceeds, but of course there is no fall-off in the curve, as in tension, since necking cannot take place. In fact, ductile metals can absorb extremely high values of shear strain (200%) before failure occurs.

Although the shear-stress/shear-strain relationship can be determined easily in the elastic range, difficulties are introduced for a solid bar in the plastic range owing to the stress variation mentioned above. One solution to this problem is to conduct the torsion test on a thin-walled tubular specimen in which the shear stress in the plastic range may be assumed to be constant through the wall thickness, and is given by

$$\tau = \frac{T}{2\pi r^2 t}$$

where $t$ is the wall thickness and $r$ is the mean radius. Shear strain is obtained in the plastic range from the same assumptions that apply in the elastic range; hence

$$\gamma = \frac{r\theta}{L}$$

(The tangent of the angle $\gamma$ must be used for large strains.)

The limitation on the torsion test of a thin-walled tube is the possibility of elastic or plastic instability. If the full shear-stress/shear-strain graph is required in the plastic range, there are two possible approaches. One is the construction developed by Nadai. This derives the stress–strain curve from the torque–twist curve based on a test on a solid bar. The other method utilizes the relationship between yield stress in simple tension and that in pure shear as derived from the von Mises criterion, eqn [12.15].

Putting $\sigma_2 = 0$ and $\sigma_1 = -\sigma_3 = \tau$ for pure shear we have

$$6\tau_Y^2 = 2\sigma_Y^2$$

or

$$\tau_Y = \sigma_Y/\sqrt{3}$$

Hence the yield stress in torsion is 0.577 times the yield stress in simple tension. It can also be shown that increments of plastic shear strain are equal to $\sqrt{3}$ times the increments of plastic tensile strain for a material so that it is possible to construct the *plastic* shear stress–strain curve from the *plastic* tensile stress–strain curve.

Fracture in torsion for ductile metals generally occurs in the plane of maximum shear stress perpendicular to the axis of the bar, whereas for brittle materials failure occurs along a 45° helix to the axis of the bar owing to tensile stress across that plane.

**18.4  Plastic overstrain and hysteresis**

The behaviour of materials in their plastic range is of particular interest in relation to the analytical treatment in Chapter 15 of plasticity of engineering components.

If a metal is taken into its plastic range in tension, compression or torsion, and at some point the load is removed, then the unloading line, although having slight curvature, approximates to the slope of the original elastic range. On reapplication of load the line diverges only slightly from that for unloading, and yielding occurs only when the load has reached the previous point of commencement of unloading. This effect is shown diagrammatically in Fig. 18.9 and is known as *overstraining*. It should be emphasized that this word does not mean damage to the static properties of the metal although it may convey that impression.

Overstrain is in fact a very useful means of obtaining a higher yield stress as shown in Fig. 18.9. Instances of the usefulness of overstraining on components are mentioned in Chapter 15.

**Fig. 18.9**  Hysteresis loops during overstrain

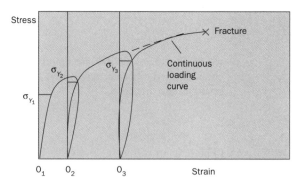

In Fig. 18.9 the 'loops' formed by the unloading and reloading lines during overstrain are caused by mechanical hysteresis, i.e. there is a lag between stress relaxation and strain recovery, and similarly on reloading. In unidirectional loading (tension only) hysteresis loops are generally quite narrow and in some metals, virtually non-existent. Hysteresis is also the term used to describe the loop obtained when reversed loading is conducted on a material, i.e. through yield in tension followed by compression or vice versa. Figure 18.10 shows the form of a tension–compression loop.

**18.5.  Hardness measurement**

Hardness of materials is a concept which is not directly employed in engineering design as, for example, yield stress may be. However, it is a 'property' which is intimately related to strength and ductility, as discussed

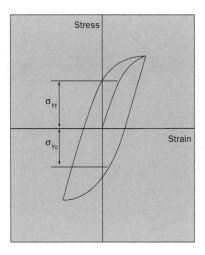

in the next section, and in that context it gives valuable quality-control
information.

The term *hardness* of a material may be defined in several different ways.
These are principally in relation to the resistance to permanent deformation
such as indentation, abrasion, scratching and machining. Although the last
three are of importance in certain circumstances, they have a limited
application in practice, and therefore discussion will be restricted to
hardness as measured by resistance to indentation. One of the first recorded
tests of this type was made by de Réaumur (1722) in which a piece of
material was indented by a tool made of the same material and the volume of
the resulting indentation measured. Since then there have been several
variations of the principle of this type of test used and widely adopted in
engineering practice. The most common are the Brinell, Vickers and
Rockwell methods and these are discussed in the following sections.

**Brinell method**   Brinell published the details of the indentation hardness test he devised in
1901. The principle involved is that a hardened steel ball is pressed under a
specified load into the surface of the metal being tested (B.S. 240: 1991).
The hardness, which is quoted as a number, is then defined as

$$\text{Hardness number} = \frac{\text{load applied to indenter in kg}}{\text{contact area of indentation in mm}^2}$$

or Brinell hardness number (B.H.N.) $= P/A$. It is noted that the number
has in fact units of pressure. The contact area $A$ is given by

$$A = \pi D h$$

$$= \frac{\pi D}{2}[D - \sqrt{(D^2 - d^2)}]$$

where $h$ = depth of impression in mm, $D$ = ball diameter in mm and $d$ =
surface projected diameter of impression in mm.

If the hardness number of a metal is determined at several different
values of load it is found that the number is not constant. The results give all
or part of a curve of the form shown in Fig. 18.11 depending on the

**Fig. 18.11** Variation of hardness with load in the Brinell test

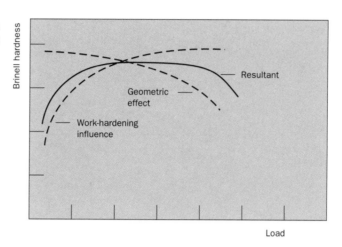

condition of the material. The curve is attributed to two separate effects. The rising portion with increase in load is caused by the non-proportional effect of work hardening on the size of the impression. Thus a soft metal will show a marked apparent rise in hardness while a heavily cold-worked material will show none.

The falling part of the curve in Fig. 18.11 with increasing indenter load is caused by geometrical dissimilarity between the spherical areas of successive impressions. This feature is very important and is worthy of further analysis. Consideration of Fig. 18.12 shows that similarity can only be obtained for different loads if different ball sizes are used, since the total angle subtended by the centre of a ball and the indentation must be equal in each case. Hence the condition for similarity is that

$$\frac{d_1}{D_1} + \frac{d_2}{D_2} = \text{constant}$$

**Fig. 18.12** Indentation geometry in the Brinell test

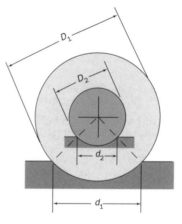

For a given angle of indentation the mean pressure is $P/\frac{1}{4}\pi d^2$, but since $d = \text{constant} \times D$ it follows that for similarity $P/D^2 = \text{constant}$.

The highest value on the curve of Fig. 18.11 is known as the *optimum hardness number* and is the figure quoted for a material when hardness is

**Table 18.1**  Typical hardness values

| Material | Condition | Hardness Brinell | Hardness Vickers | Tensile strength, (MN/m$^2$) |
|---|---|---|---|---|
| Pure aluminium | Annealed | 25 | (25) | 54 |
| | Cold rolled (hard) | 55 | (55) | 185 |
| Duralumin alloy (3.5–4.5 Cu, 0.4–0.8 Mg, 0.4–0.7 Mn, 0.7 Si) | Solution and precipitation treated | 120 | (120) | 433 |
| 6% Al–Zn, 4% Mg alloy | Solution and precipitation treated | 180 | (181) | 587 |
| Pure copper | Annealed | 42 | (42) | 221 |
| | Cold rolled (hard) | 119 | (119) | 371 |
| Brass (60–40) | Cold drawn | 178 | (179) | 659 |
| Mild steel (0.19 C) | Annealed | 127 | (127) | 451 |
| | Cold rolled | (192) | 595 | |
| Si–Mn spring steel | Quenched and tempered | (415) | 435 | 1180 |
| 4% Ni, 1.5% Cr, 0.3% C steel | Quenched and tempered | (434) | 460 | 1640 |
| Ball-bearing steel | – | – | 700 | – |
| Tungsten carbide | Sintered | – | 1200 | – |

*Note:* Equivalent values are in parentheses.

required. Brinell obtained the optimum hardness for steels using a 10 mm diameter ball at a load of 3000 kg, and these conditions have become a British Standard for hardness testing. Using the above values it is seen that $P/D^2 = 30$, and it is therefore possible to obtain comparable hardness numbers on a metal for different sizes of ball if the load is chosen to satisfy the above relationship. For softer metals and thin sheet it is found that other values are required for the constant equal to $P/D^2$ in order to give optimum hardness. Values of $P/D^2 = 10$ (non-ferrous alloys), 5 (copper, aluminium) and 1 (lead, tin) have been adopted as standard. The peak of the hardness curve generally occurs for a value of $d/D$ between 0.25 and 0.5 and this fact can be used to assist in choosing the correct value of $P/D^2$. It is important when quoting a hardness number to state the conditions used, ball size, load, etc. It is also essential to space successive impressions adequately and keep them clear of the edge of the material, owing to the plastic deformation caused in the area around the indentation ('ridging' for hard metals, 'sinking' for soft metals).

The size of the Brinell indentation is such that the test is generally employed for checking raw stock or unmachined components rather than finished products. Typical hardness values for various materials are given in Table 18.1.

Another very useful feature of the Brinell method is that an empirical relationship has been found to exist between hardness number and tensile strength for steels. Thus, $K \times$ B.H.N. (kg/mm$^2$) = tensile strength (MN/m$^2$), where $K$ lies between 3.4 and 3.9 for the majority of steels. Hence the Brinell test can be used to get an approximate value for the tensile strength of a metal. This can be useful as a non–destructive method for checking if the correct heat treatment has been carried out or to determine the properties of a failed component.

**Vickers method**

This test was devised about 1920 and employs a square-based diamond pyramid as the indenting tool (B.S. 427: 1990). The angle between opposite faces of the pyramid is 136° and this was chosen so that close correlation can be obtained between Vickers and optimum Brinell hardness numbers. The angle of 136° corresponds to the geometry of an impression given by a $d/D$ ratio of 0.375.

In the Vickers test, hardness number is defined in the same way as for Brinell, i.e. indenter load, kg, divided by the contact area of the impression, mm$^2$. If $l$ is the average length of the diagonal of the impression, Fig. 18.13, the contact area is given by

$$\frac{l^2}{2 \sin \frac{1}{2}(136)} = \frac{l^2}{1.854}$$

**Fig. 18.13**

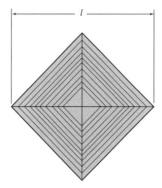

Therefore the Vickers pyramid number

$$\text{V.P.N.} = 1.854 \frac{P}{l^2}$$

There are two features of this test which are essentially different and advantageous over the Brinell method. Firstly, there is geometrical similarity between impressions under different indenter loads, and hardness number is virtually independent of load as shown in Fig. 18.14, except at very low loads where there is often a higher hardness owing to a 'skin' effect on the test piece. The standard loads recommended in B.S. 427 are 1, 2.5, 5, 10, 20, 30, 50 and 100 kg.

The second advantage of the Vickers test is that the upper limit of hardness number is controlled by the diamond, therefore allowing values up to 1500 to be determined, which is far in excess of that possible with the steel ball in the Brinell test.

The extremely small size of the impression necessitates a very good surface finish on the test sample, but means that it is advantageous in

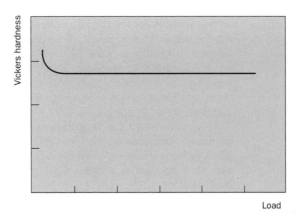

checking the hardness of finished components without leaving a noticeable mark.

**Rockwell method**    This test was introduced in the U.S.A. at about the same time as the Vickers test in England. It is quite popular as it has a wide range of versatility, is rapid and useful for finished parts.

Two types of penetrator are employed for different purposes, a diamond cone with rounded point for hard metals and a $\frac{1}{16}$ in (1.6 mm) diameter hardened steel ball for metals of medium and lower hardness values.

In this test hardness is defined in terms of the depth of the impression rather than the area, and the hardness number is read directly from an

**Fig. 18.15** Load application in the Rockwell test

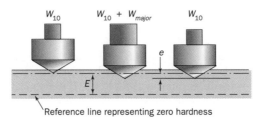

indicator on the machine having three scales, A, B and C.

The procedure for applying load to the specimen is rather different from the Vickers and Brinell methods and is illustrated in Fig. 18.15. Initially a minor load of 10 kg is applied, which is followed by a major load, being an additional 50, 90 or 140 kg depending on the indenter and type of metal. The major load is now removed, but the minor load is retained while the hardness number is read, where

$$H_R = E - e$$

and $H_R$ = Rockwell number, $e$ = depth of penetration due to the major load only but while the minor load is operating and $E$ = arbitrary constant which is dependent on the type of penetrator. The various relevant details of the test may be found in B.S. 891: 1989.

**Comparison of hardness values**

Owing to the wide use of the Brinell, Vickers and Rockwell methods and the varying preferences for any one of these tests, there are occasions when the same material or component is hardness tested by different methods in different laboratories. This has led to a demand for some correlation between hardness values determined by the three tests. It has been shown that there is no general relationship between the hardness scales, and empirical formulae only hold good for materials of closely similar composition and condition.

However, based on experimental results, the British Standards Institution has issued a table (B.S. 860: 1989) of *approximately* comparative values for the three tests, but it is emphasized that it is not intended that the table shall be used as a conversion system for standard values from one hardness scale to another.

## 18.6 Hardness of viscoelastic materials

Hardness measurements are also necessary for assessing non-metallic materials such as plastics, rubbers and composites. Indentation methods can be used, but owing to the much lower levels of hardness compared with metals very low indenter loads are used. Material thickness must be adequate in relation to the depth of indentation and a solid mounting plate must be used. The strain–time dependence of viscoelastic materials also has to be taken into account. In particular the depth of penetration of the indenter will increase with time under load and the geometry of the indent may change due to recovery after load removal. In general it has been found that conventional Brinell- and Vickers-type tests are not successful with plastics and rubbers because these methods require a clear impression on the indent after the indenter is removed. Using a ball indenter the image of the indent tends to be imprecise; with a pyramid indenter the indent is sharper, but in both cases the image is difficult to see owing to the poor reflection of light from the surface of the plastic or rubber. For plastics the most successful results are obtained by using a conical or ball indenter (B.S. 2782: Methods 365, B, D) or a pyramid indenter and measuring the *depth of penetration* of the indenter during load application. For rubber, a rigid ball indenter is used (B.S. 903: Part A26). Hardness testing of rubber is in fact very common in industry because there is a known relationship between the hardness number (I.R.H.D. – International Rubber Hardness Degrees) and Young's modulus for vulcanized rubber.

## 18.7 Factors influencing strength, ductility and hardness

The chemical composition of a material and the heat treatment to which it has been subjected have a great effect on the strength and ductility of the material. The mechanical properties of steel are very largely influenced by the amount of carbon in the steel. Its strength increases with increase in carbon content, but the ductility decreases as illustrated in Fig. 18.16.

A steel may be hardened by heating it to a high temperature and then rapidly cooling it in a cold liquid. The strength is greatly increased by this process, but its ductilty is reduced and it is in a brittle state. It may be brought to the required degree of strength and ductility by further heat treatment, or tempering. The effect of tempering on a nickel–chrome steel for use in highly stressed components in automobiles and aircraft is shown in Fig. 18.17. The steel was hardened by heating to 820 °C, soaked for $\frac{3}{4}$ hour and quenched in oil. It was then tempered for 1 hour at a range of temperatures and cooled in oil.

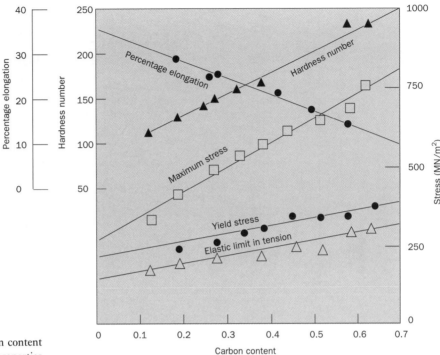

**Fig. 18.16** Effect of carbon content on mechanical properties

Similar influences on strength and ductility are obtained for non-ferrous alloys by the appropriate types of heat treatment.

The operating temperature of materials can range quite widely under atmospheric extremes quite apart from special engineering situations such as internal combustion, or liquid petroleum gas (L.P.G.) storage. It is therefore necessary for engineers to have data in this context. A few brief comments follow herewith and a more detailed treatment of high-temperature creep behaviour is given in Chapter 21.

At low temperatures, for example, the strength of a mild steel is increased; the material, however, may become very brittle under certain circumstances, its ductility being almost negligible (see Chapter 19).

With increase in temperature, the elastic limit decreases, and is accompanied by a corresponding decrease in the modulus of elasticity. The tensile strength of the material decreases until a temperature of approximately 150 °C is reached. From this temperature the strength increases, reaching a maximum value in the neighbourhood of 300 °C, after which it decreases as the temperature is further increased. Table 18.2 shows the effect of high temperature on the tensile strengths of various steels.

The ductility decreases with increase in temperature until a temperature of about 150 °C is reached. After this temperature the ductility increases with further increase of temperature. The increase is not regular, however, but takes place in an erratic manner.

Table 18.3 shows the effect of temperature on a cast- and wrought-aluminium alloy respectively.

One must conclude by emphasizing the importance of the interdependence of strength, ductility and hardness.

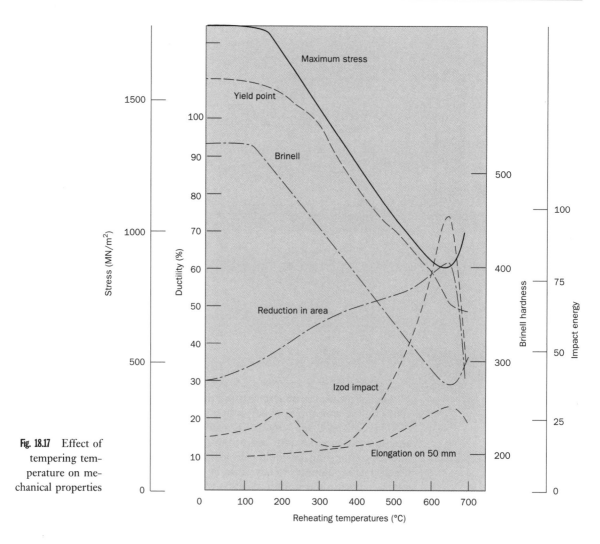

**Fig. 18.17** Effect of tempering temperature on mechanical properties

**Table 18.2**

| Steel | Tensile tests, cold | | | | Tensile strength, hot (°C), (MN/m²) | | | | | | |
| | Elastic limit (MN/m²) | Tensile strength (MN/m²) | Elongation on 50 mm (%) | Reduction of area (%) | 600 | 650 | 700 | 750 | 800 | 850 | 900 |
|---|---|---|---|---|---|---|---|---|---|---|---|
| 3% Nickel–chrome | 745 | 874 | 23.0 | 62.5 | 525 | 365 | 209 | 173 | 142 | 107 | 70 |
| Stainless steel | 726 | 746 | 24.0 | 58.1 | 374 | 264 | 158 | 278 | 94 | 87 | 121 |
| Silicon–chrome | 790 | 975 | 21.0 | 40.0 | 650 | 534 | 392 | 275 | 165 | 111 | 60 |
| Chrome steel | 650 | 835 | 24.5 | 55.0 | 560 | 464 | 294 | 201 | 108 | 111 | 113 |
| Cobalt–chrome | 650 | 896 | 13.0 | 22.0 | 695 | 472 | 383 | 210 | 155 | 90 | 125 |
| High-speed steel | 711 | 943 | 15.0 | 24.0 | 634 | 411 | 330 | 260 | 151 | 119 | 130 |
| High-nickel chrome | 650 | 1050 | 27.0 | 45.0 | 660 | 592 | 523 | 442 | 371 | 300 | 232 |

Table 18.3

| Temperature (°C) | Tensile strength (MN/m²) | | |
|---|---|---|---|
| | Wrought | Sand cast | Die cast |
| 20 | 433 | 286 | 371 |
| 200 | 332 | 248 | 340 |
| 250 | 301 | 216 | 302 |
| 300 | 201 | 201 | 240 |
| 350 | 124 | 136 | 139 |

In broad terms a high hardness value is associated with high tensile strength and medium to low ductility, while the converse is that a low hardness number relates to lower tensile strength and higher ductility. Hence hardness measurement is widely used as a quality-control test for materials during manufacture.

In applications it is almost always true that a combination of material properties is required. There will be several possible modes of failure which must be prevented, or the design must be optimized so that, for example, cost or weight is minimized.

**Material selection criteria**

A very powerful tool for identifying candidate materials during the development of new design concepts are the Materials Selection Charts developed by Ashby[1]. Figure 18.18 shows a plot of Young's modulus against density for a range of engineering materials. Materials with high modulus but low density are in the top-left corner of the chart. Quantitative criteria for selecting the best material for a bar of minimum weight with a given stiffness when loaded in tension $(E/\rho)$ or bending $(E^{\frac{1}{2}}/\rho)$ are derived by Ashby. These criteria can then be used to identify candidate materials from the chart.

Figure 18.19 shows modulus against strength for a similar range of materials. 'Springy' materials are those which can absorb a large amount of elastic strain energy per unit volume, $\frac{1}{2}\sigma^2/E$, Chapter 3. On the log–log scale shown all materials lying on a given line of slope 2 have the same value of this parameter and materials below and to the right of the line will be better than those above and to the left. Thus soft butyl rubber is a good choice at one extreme, while at the other end of the strength range ceramics and laminates of engineering composites are also good choices.

The same chart may also be used to choose materials for resistance to buckling. In Chapter 10, eqn. [10.15], it was shown that buckling will occur before yielding at a critical slenderness ratio

$$\frac{L}{r} \geq \pi \sqrt{\frac{E}{\sigma_Y}} \qquad [18.1]$$

All materials on the line at 45° in Fig. 18.19 will have the same critical slenderness ratio. Materials above and to the left of a given 45° line can be used in longer columns before buckling becomes the limiting mode of failure.

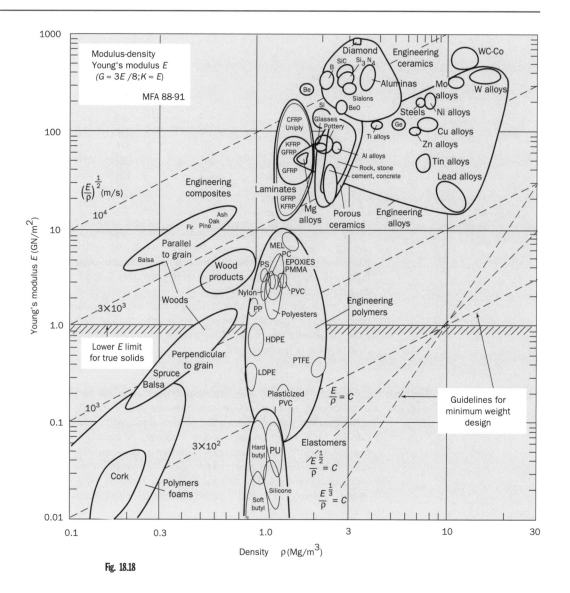

**Fig. 18.18**

## 18.8  Summary

The properties discussed in this chapter represent the basic information that the designer needs to have available in relation to the selection of materials for 'steady-load' components and structures.

The tensile, compression or shear moduli together with the yield (or proof) stress and Poisson's ratio are needed for linear–elastic design. The plastic properties of tensile or compressive strength and ductility are required to give a measure of reserve safety in the event of exceeding the yield level. The hardness of the material gives a check on its condition.

However, it is important not to get a false sense of security, since it will be seen in the next three chapters that although what appears to be a very suitable material in the required condition has been selected for 'static'

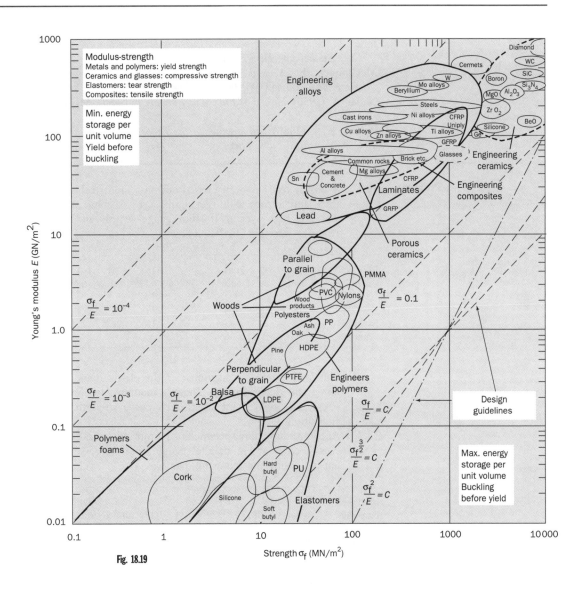

**Fig. 18.19**

working conditions other factors can have very serious consequences during the service life of a component or structure.

**Reference**

1. Ashby, M. F. (1992) *Materials Selection in Mechanical Design*, Pergamon Press, Oxford.

**Bibliography**

Ashby, M. F. and Jones, D. R. (1980) *Engineering Materials I: An Introduction to their Properties and Applications*, Pergamon Press, Oxford.
Ashby, M. F. and Jones, D. R. (1986) *Engineering Materials II: An Introduction to Microstructures, Processing and Design*, Pergamon Press, Oxford.

Dieter, G. E. (1976) *Mechanical Metallurgy*, McGraw-Hill, New York.

Jones, D. R. (1993) *Engineering Materials III: Mechanical Failure Analysis: Case Studies and Design Implications*, Pergamon Press, Oxford.

McClintock, F. A. and Argon, A. S. (1966) *Mechanical Behaviour of Materials*, Addison-Wesley, Reading.

Nadai, A. (1950) *Theory of Flow and Fracture of Solids*, McGraw-Hill, New York.

Pascoe, K. J. (1980) *Introduction to the Properties of Engineering Materials*, Van Nostrand, New York.

**Problems**   18.1   A tensile test has been carried out on a mild steel specimen 10 mm thick and 50 mm wide rectangular cross-section. An extensometer was attached over a 100 mm gauge length and load extension readings were obtained as follows:

| Load (kN)        | 16    | 32    | 64    | 96    | 128   | 136   | 144   | 152   | 158   |
|------------------|-------|-------|-------|-------|-------|-------|-------|-------|-------|
| Extension (mm)   | 0.016 | 0.032 | 0.064 | 0.096 | 0.128 | 0.137 | 0.147 | 0.173 | 0.605 |
| Load (kN)        | 154   | 168   | 208   | 222   | 226   | 216   | 192   | 185.4 |       |
| Extension (mm)   | 1.81  | 2.42  | 7.25  | 12.0  | 16.8  | 22.0  | 24.0  | Fracture |    |

Plot load–extension diagrams for the elastic range and the plastic range and determine: (i) Young's modulus; (ii) proportional limit stress; (iii) yield point stress; (iv) tensile strength; (v) percentage elongation.

18.2   An aluminium alloy specimen of 1.2 mm thickness and 25 mm width cross-section and a parallel gauge length of 50 mm is tested in tension giving the following data:

| Load (kN)        | 1.8    | 3.6    | 5.4   | 6.4   | 7.2   | 7.6   | 8.0   | 8.4   | 8.8   |
|------------------|--------|--------|-------|-------|-------|-------|-------|-------|-------|
| Extension (mm)   | 0.0443 | 0.0886 | 0.133 | 0.155 | 0.181 | 0.198 | 0.219 | 0.246 | 0.281 |
| Load (kN)        | 9.2    | 10.8   | 12.0  | 12.4  |       |       |       |       |       |
| Extension (mm)   | 0.332  | 0.645  | 1.05  | 1.94  |       |       |       |       |       |

Determine values for: (i) Young's modulus; (ii) 0.1% proof stress; (iii) 0.5% proof stress; (iv) tensile strength.

18.3   The following load–compression data has been obtained on four copper cylinders of 12 mm diameter and lengths 24, 12, 6, 4 mm. Construct the true compressive stress–strain curve for the material.

| $d_0/h_0$     | Load kN        | 13.5 | 22.5 | 29.0 | 34.0 | 40.0 |
|---------------|----------------|------|------|------|------|------|
| 3             | % Compression  | 3.3  | 11.5 | 19.5 | 26.7 | 37.1 |
| 2             | % Compression  | 3.8  | 12.6 | 21.4 | 29.5 | 41.5 |
| 1             | % Compression  | 4.0  | 13.8 | 23.3 | 32.2 | 45.8 |
| $\frac{1}{2}$ | % Compression  | 4.3  | 14.4 | 24.2 | 33.6 | 47.9 |

18.4 A Wallace-type of micro-hardness testing machine measures the depth of penetration of a standard Vickers indenter. During a test on polypropylene, using an applied weight of 300 g the instrument records $2.66 \times 10^2$ mm. From this information calculate the Vickers hardness number of the plastic.

18.5 A nickel–chrome alloy steel component fails in service and an investigation is required. A Brinell hardness test is carried out on the broken part giving a value of 320 B.H.N. An Izod impact sample cut from the part gives a fracture energy of 50 J. The design maximum stress for the component was 1075 MN/m². What is the main contributing factor for the failure? (*Hint*: Examine Fig. 18.17).

# CHAPTER 19 Fracture Mechanics

> The material properties discussed in the previous chapter principally relate to the quality control of materials and to initial material selection by a designer. We now need to recognize that in spite of carefully employing the design stress and strain analysis procedures to avoid failure by gross deformation, elastic instability or exceeding the yield stress, there are other factors to be taken into account in design. These 'factors' include straining rate, fluctuating stresses, stress concentration, metallurgical flaws, high and low temperatures, corrosion and other special effects. The designer needs to be aware that these variables can cause an engineering component to fail by *fracture* which may lead to a catastrophic disaster and, in the most serious cases, loss of life.
>
> This chapter and the remaining two will attempt to cover these topics in a manner which will provide an initial insight which can be built on as required through the specific texts in these areas.

## 19.1 Fracture concepts

Fracture is concerned with the *initiation* and *propagation* of a crack or cracks in the material until the extent of cracking is such that the applied loading can no longer be sustained by the component or structure. It is generally accepted that initiation of a crack is relatively difficult to design against and, in fact, most components or structures will contain some crack-like flaw/ defect by the time manufacturing is completed in spite of rigorous inspection procedures. We, therefore, must try to design for non-propagation or, at next best, controlled propagation. In the latter case normal in-service inspections should enable the presence of growing cracks to be detected. This tends to be the case in respect of the phenomenon known as *fatigue* which is fully discussed in the next chapter. Cracks which propagate in an uncontrolled manner at very high velocity through materials is a situation of greater danger in service and is the theme of this chapter.

From a metallurgical viewpoint there are only two paths for a crack passing through a metal, either transcrystalline or intercrystalline. The latter only occurs in a few particular circumstances, e.g. creep, stress, corrosion. The former is the more general mechanism of which there are two types related to the crystallographic planes known as *shear* and *cleavage*. Shear is the result of certain crystal planes sliding over one another, termed *slip*, and is associated with a great deal of macroscopic plastic deformation. Cleavage occurs on different crystallographic planes caused by a normal (tensile) stress and involves negligible plastic deformation. Shear fractures have a dull appearance, sometimes described as fibrous, but cleavage fractures reveal smooth reflecting planes described as bright and crystalline or granular. By relating these two modes broadly to stress–strain characteristics it is found that the ductile shear mode tends to be above-yield stress fracture with high energy absorption and hence *toughness*. On the other hand, a fully brittle cleavage mode would be associated with a low toughness, low energy absorption failure which apparently occurs below the yield stress. Of course there are the possibilities of mixed-mode fractures, depending on a combination of a number of factors, between the two extremes above.

One of the more important influences on the mode of fracture is the state of stress. In engineering components, as compared with simple laboratory uniaxial stress tests, a complex stress system generally exists. In a triaxial stress state where $\sigma_1 > \sigma_2 > \sigma_3$, the maximum shear stress is $\frac{1}{2}(\sigma_1 - \sigma_3)$, but as $\sigma_3 \to \sigma_1$, $\tau \to 0$. In the extreme case of hydrostatic tension and compression, $\sigma_1 = \sigma_2 = \sigma_3$ and $\tau = 0$. It is evident that shear cannot occur and hence cleavage fracture will result. The introduction of a discontinuity or notch into a piece of material causes a stress concentration and triaxiality of stress to a degree which depends on notch geometry and loading condition.

Most metals exhibit some temperature dependence of fracture over a range from, say, $-100\,^{\circ}\text{C}$ to $+100\,^{\circ}\text{C}$. Toughness is reduced by lowering temperature, and so this aspect of the working environment of engineering structures and components must be taken into account.

Unstable fracture manifested itself first as a serious engineering problem of nominally ductile low-carbon steels from the mid-1930s to the mid-1950s. Large welded structures such as ships, bridges and storage tanks failed in a catastrophic and apparently brittle manner. From this grew an extensive research programme into what was then called *brittle fracture*. Although factors such as stress concentration at 'notches', weld defects and low temperature contributed to the initiation of a crack, the principal controlling factors on fast propagation of the crack are the ability of the material to absorb energy, i.e. the toughness, and the existence of crack arrest barriers. In the latter context riveted or bolted plate structures were better than the 'continuous' all-welded structure if the welds were not of high quality.

The past 40 years have seen the development of ultra-high-strength alloys for rocket motors and space vehicles. Some of these low-ductility materials were found to be susceptible to unstable fracture from small defects owing to low toughness. This resulted in the development of the theory of linear-elastic fracture mechanics (L.E.F.M.) which was accompanied by the establishment of special tests to measure the fracture resistance of materials. The analytical and experimental techniques of fracture mechanics now have a major influence on design for crack growth which is (i) unstable, (ii) intermittent/cyclic (fatigue) and (iii) time dependent in metallic and non-metallic materials which are either homogeneous or fibre–matrix composites.

## 19.2 Linear-elastic fracture mechanics

Linear-elastic fracture mechanics (L.E.F.M.) developed from the early work of Griffith[1] who sought to explain why the observed strength of a material is considerably less than the theoretical strength based on the forces between atoms. He concluded that real materials must contain small defects and cracks which reduce their strength. These cracks cause stress concentrations but they cannot be allowed for by calculation of a linear-elastic stress concentration factor $K_t$. This is because an elliptical defect, Fig. 19.1, has its stress concentration factor defined by the equation

$$K_t = 1 + 2\left(\frac{a}{b}\right)$$  [19.1]

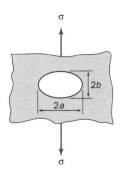

**Fig. 19.1** Elliptical defect in a stressed plate

As $b \to 0$ the defect becomes a crack, but $K_t \to \infty$ which would suggest that a material with a crack would not be able to withstand any applied forces. This is contrary to what is observed so Griffith developed a concept to explain how a stable crack could exist in a material. He postulated that a crack only becomes unstable if an increment of crack growth results in more stored energy being released than can be absorbed by the creation of the new crack surface.

Based on this premise and with subsequent refinements, principally by Irwin[2], L.E.F.M. has developed as an analytical approach to fracture. It relates the stress distribution in the vicinity of a crack tip to other parameters such as the nominal stress applied to the structure and the size, shape and orientation of the crack. Thus it permits representation of the material fracture properties, often in terms of a single parameter.

There have been two main approaches: (i) energy; (ii) stress intensity factor.

### 19.3 Strain energy release rate

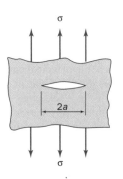

**Fig. 19.2** Sharp crack in a stressed infinite plane

For a through crack of length $2a$ in an infinite body of unit thickness, as shown in Fig. 19.2, the surface energy $U_s$ stored in the material due to the formation of the crack is given by

$$U_s = (2a)2\gamma \qquad [19.2]$$

where $\gamma =$ surface energy per unit area. In the context of the fracture of brittle materials this term is replaced by $\gamma = \frac{1}{2}G$, where $G$ is energy absorbed per unit area of crack (note that $G$ refers to the area of crack which will be half the new surface area).

Thus eqn. [19.2] may be written as

$$U_s = 2aG \qquad [19.3]$$

From the concept of elastic strain energy introduced in Chapter 3 the elastic energy $U_e$ released by the formation of the crack is given by

$$U_e = \frac{1}{2}\int_a \sigma(x).\Delta(x, a)\,\mathrm{d}x \qquad [19.4]$$

where $\sigma(x)$ is the stress distribution in the vicinity of the crack, and $\Delta(x, a)$ is the vertical opening of the crack.

It can be shown[3] that for the through crack of length $2a$ in an infinite plate, Fig. 19.3,

$$U_e = \frac{\pi\sigma^2 a^2}{E}k \qquad [19.5]$$

where $k = (1 - \nu^2)$ for plane strain and 1 for plane stress and $\nu$ is Poisson's ratio.

Thus the surface energy which is developed in the material is increasing linearly with crack length, whereas the energy released by the formation of the crack increases with (crack length)$^2$. This is illustrated in Fig. 19.4.

The net energy in the presence of the crack is thus the mathematical summation of the surface energy $U_s$ and the energy released $U_e$. Griffith proposed that the threshold between a stable crack and an unstable crack

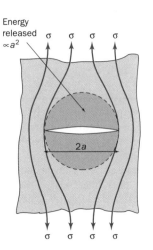

Energy released $\propto a^2$

**Fig. 19.3**

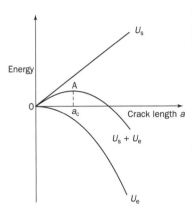

Energy

$U_s$

A

0

$a_c$

Crack length $a$

$U_s + U_e$

$U_e$

**Fig. 19.4** Variation of energy with crack length

occurs when an increment of crack growth causes more energy to be released than can be absorbed in the material. Thus the critical condition is $dU/da = 0$ which occurs at point A in Fig. 19.4 and hence the critical crack length $a_c$ is defined. From eqns. [19.3] and [19.5]

$$\frac{d}{da}\left(2aG_C - \frac{\pi\sigma^2 a^2}{E}k\right) = 0$$

$$2G_C = \frac{2\pi\sigma^2 a}{E}k$$

This then reduces to

$$(EG_C)^{1/2} = \sigma(\pi a)^{1/2} \qquad \text{(plane stress)}$$

$$\left(\frac{EG_C}{1-\nu^2}\right)^{1/2} = \sigma(\pi a)^{1/2} \qquad \text{(plane strain)} \qquad [19.6]$$

These equations are an expression of the conditions for fast fracture in a brittle material. It should be noted that $G_C$ is a material property which is referred to as the *critical strain energy release rate, toughness* or *crack extension force*, and has the units J/m$^2$.

A high value means that it is hard to propagate cracks in the material as, for example, in copper for which $G_C \simeq 10^3$kJ/m$^2$. This may be compared with a value for glass of approximately 0.01 kJ/m$^2$.

**Example 19.1**

Determine what size of defect will cause a fracture of glass before it reaches its ideal strength. Defect-free thin whiskers of glass can reach strengths of 3600 MN/m$^2$, but the toughness $G_c$ is only 0.01 kJ/m$^2$. The modulus of glass is 69 GN/m$^2$.

The largest defect which can be tolerated before a fracture occurs below the ideal strength $\sigma_f$ is when both modes of failure occur at the same stress, i.e.

$$(EG_C)^{1/2} = \sigma_f(\pi a)^{1/2} \qquad [19.7]$$

The size of defect $a$ at which this occurs is

$$a = \frac{EG_C}{\pi\sigma_f^2} = \frac{69 \times 10^9 \times 0.01 \times 10^3}{\pi(3600 \times 10^6)^2} = 1.7 \times 10^{-8}\,\text{m} \qquad [19.8]$$

Any defect which is larger than this extremely small value will reduce the apparent strength of the material, since it will fail by fracture long before its ideal strength is reached.

**Example 19.2**

The fracture stress of a large sheet of steel with a central crack of 40 mm is 480 MN/m$^2$. What is the fracture stress of a similar sheet with a crack of 100 mm?

Fracture occurs when

$$\sqrt{(EG_C)} = \sigma_1\sqrt{(\pi a_1)} = \sigma_2\sqrt{(\pi a_2)} \qquad [19.9]$$

Therefore, since the crack length $2a = 40$ mm in the first sheet and $100$ mm in the second, the stress to cause a fracture in the second sheet is

$$\sigma_2 = \frac{\sigma_1 \sqrt{(\pi a_1)}}{\sqrt{(\pi a_2)}} = \frac{480\sqrt{(20 \times 10^{-3})}}{\sqrt{(50 \times 10^{-3})}} = 304 \, \text{MN/m}^2 \qquad [19.10]$$

## 19.4 Stress intensity factor

Although Griffith put forward the original concept of L.E.F.M. he restricted his work to brittle materials (e.g. glass) and it was Irwin who developed the technique for metals. He examined the equations that had been developed for the stresses in the vicinity of a sharp crack in a large plate as illustrated in Fig. 19.5. The equations for the elastic stress distribution at the crack tip are as follows:

$$\left. \begin{aligned}
\sigma_x &= \frac{K}{(2\pi r)^{1/2}} \cos\left(\frac{\theta}{2}\right) \left[1 - \sin\left(\frac{\theta}{2}\right) \sin\left(\frac{3\theta}{2}\right)\right] \\
\sigma_y &= \frac{K}{(2\pi r)^{1/2}} \cos\left(\frac{\theta}{2}\right) \left[1 + \sin\left(\frac{\theta}{2}\right) \sin\left(\frac{3\theta}{2}\right)\right] \\
\tau_{xy} &= \frac{K}{(2\pi r)^{1/2}} \sin\left(\frac{\theta}{2}\right) \cos\left(\frac{\theta}{2}\right) \cos\left(\frac{3\theta}{2}\right)
\end{aligned} \right\} \qquad [19.11]$$

and

$$\sigma_z = \frac{2\nu K}{(2\pi r)^{1/2}} \cos\left(\frac{\theta}{2}\right) \qquad \text{(plane strain)}$$

or

$$\sigma_z = 0 \qquad \text{(plane stress)}$$

**Fig. 19.5** Stress distribution at crack tip in an infinite plate

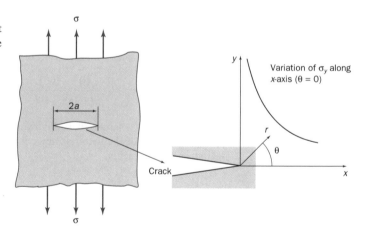

He observed that the stresses are proportional to $(\pi a)^{1/2}$, where $a$ is the half-length of the crack. On this basis, Irwin defined a *stress intensity factor K*

as

$$K = \sigma(\pi a)^{1/2} \qquad [19.12]$$

The stress intensity factor is a means of characterizing the elastic stress distribution near the crack tip but in itself has no physical reality. It has units of $MN\,m^{-3/2}$ and should not be confused with the elastic stress concentration factor $K_t$.

Thus the basis of the L.E.F.M. design approach is that:

(a)  All materials contain cracks or flaws.
(b)  The stress intensity value $K$ may be calculated for the particular loading and crack configuration.
(c)  Failure is predicted if $K$ exceeds the critical value $K_C$ for the material.

The *critical stress intensity factor* is sometimes referred to as the *fracture toughness* and will be designated $K_C$. By comparing eqns. [19.12] and [19.6] it may be seen that $K_C$ is related to $G_C$ by the following equations:

$$(EG_C)^{1/2} = K_C \qquad \text{(plane stress)} \qquad [19.13]$$

$$\left(\frac{EG_C}{1 - \nu^2}\right)^{1/2} = K_C \qquad \text{(plane strain)} \qquad [19.14]$$

In order to extend the applicability of L.E.F.M. beyond the case of a central crack in an infinite plate, $K$ is usually expressed in the more general form

$$K = Y\sigma(\pi a)^{1/2} \qquad [19.15]$$

where $Y$ is a geometry factor and $a$ is the half-length of a central crack or the full length of an edge crack.

The methods for evaluating stress intensity factors for particular loading situations are formidable and outside the scope of this book. The main theoretical methods include: boundary collocation, conformal mapping, numerical methods and analysis of stress functions. The reader should refer to other texts[3,4] for details of these methods.

Figure 19.6 shows some crack configurations of practical interest and the expressions for $K$ are as follows:

(a)  Central crack of length $2a$ in a sheet of finite width

$$K = \sigma(\pi a)^{1/2}\left(\frac{W}{\pi a}\tan\frac{\pi a}{W}\right)^{1/2} \qquad [19.16]$$

(b)  Edge cracks in a plate of finite width

$$K = \sigma(\pi a)^{1/2}\left(\frac{W}{\pi a}\tan\frac{\pi a}{W} + \frac{0.2W}{\pi a}\sin\frac{\pi a}{W}\right)^{1/2} \qquad [19.17]$$

(c)  Single edge cracks in a plate of finite width

$$K = \sigma(\pi a)^{1/2}\left[1.12 - 0.23\left(\frac{a}{W}\right) + 10.6\left(\frac{a}{W}\right)^2 - 21.7\left(\frac{a}{W}\right)^3\right.$$

$$\left. + 30.4\left(\frac{a}{W}\right)^4\right] \qquad [19.18]$$

**Fig. 19.6**   Typical crack configurations

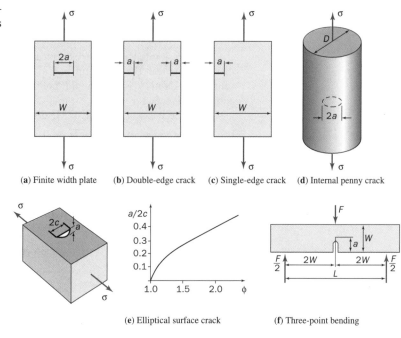

(a) Finite width plate   (b) Double-edge crack   (c) Single-edge crack   (d) Internal penny crack

(e) Elliptical surface crack                (f) Three-point bending

*Note:* In most cases $(a/W)$ is very small so $Y = 1.12$.

(*d*)  Penny-shaped internal crack

$$K = \sigma(\pi a)^{1/2}\left(\frac{2}{\pi}\right) \qquad\qquad [19.19]$$

assuming $a \ll D$.

(*e*)  Semi-elliptical surface flaw

$$K = \sigma(\pi a)^{1/2}\left(\frac{1.12}{\phi^{1/2}}\right) \qquad\qquad [19.20]$$

(*f*)  Three-point bending

$$K = \frac{3FL}{2BW^{3/2}}\left[1.93\left(\frac{a}{W}\right)^{1/2} - 3.07\left(\frac{a}{W}\right)^{3/2} + 14.53\left(\frac{a}{W}\right)^{5/2}\right.$$
$$\left. - 25.11\left(\frac{a}{W}\right)^{7/2} + 25.8\left(\frac{a}{W}\right)^{9/2}\right] \qquad [19.21]$$

or

$$K = \frac{F}{BW^{1/2}}\cdot f_1\left(\frac{a}{W}\right) \qquad\qquad [19.22]$$

A handbook of stress intensity factors for a range of geometrical
configurations and loadings has been prepared by Rooke and Cartwright[5].

## 19.5 Modes of crack tip deformation

So far it has been assumed that the loading plane is symmetrical with respect to the crack plane. This is probably the most common situation and is referred to as the opening mode (designated as mode I), Fig. 19.7(*a*). Therefore to be strictly correct the stress intensity factor should have the suffix I. There are in fact three deformation possibilities and the other two are shown in Fig. 19.6(*b*) and (*c*).

(*a*) Opening mode (mode I) having symmetry about the $(x, y)$- and $(x, z)$-planes.
(*b*) Sliding or shear mode (mode II) having antisymmetry about the $(x, z)$-plane and symmetry about the $(x, y)$-plane.
(*c*) Tearing mode (mode III) having antisymmetry about the $(x, y)$- and $(x, z)$-planes.

**Fig. 19.7** Cracking modes

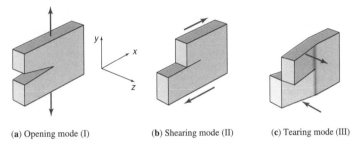

(**a**) Opening mode (I)   (**b**) Shearing mode (II)   (**c**) Tearing mode (III)

Thus the stress intensity factor should be referred to as $K_I$, $K_{II}$ or $K_{III}$ depending on the deformation mode.

## 19.6 Experimental determination of critical stress intensity factor

From the previous sections it is apparent that the stress intensity factor $K$ for a particular loading situation may be calculated from a knowledge of the nominal stress and the size, shape and orientation of the crack. The basis of L.E.F.M. is that fast fracture will occur when $K$ reaches the critical value for the material. This critical value $K_C$ may be determined experimentally from standardized tests[6,7] on samples of the material.

**Effect of size**

It has been found that the value of the critical stress intensity factor $K_C$ at which unstable crack growth occurs under static loading depends on the specimen thickness as illustrated in Fig. 19.8.

**Fig. 19.8** Variation of fracture toughness with plate thickness

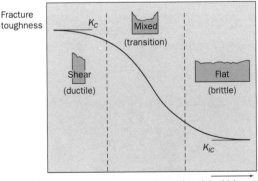

The limiting value of $K_C$ is observed in plane strain (which is the maximum constraint condition). It is designated $K_{IC}$ and is usually referred to as the fracture toughness. It is important not to confuse $K_I$ with $K_{IC}$. As shown earlier, $K_I$ depends on the configuration of the system, but $K_{IC}$ is a material property and is independent of the configuration of the system. Since $K_{IC}$ is the most conservative value of fracture toughness for the material, it is this value which is used in design calculations. Therefore the standardized test conditions are carefully chosen to ensure that it is $K_{IC}$ which is determined and, after testing, checks are made on the validity of the test conditions.

**Test methods**     The two most commonly used test methods involve bending or tensile loading of the test specimens shown in Fig. 19.9. The specimens contain a carefully machined notch which may have a plane front or a chevron profile. A chevron notch has been found to keep the crack in-plane and so the machining operation is not quite so critical as with a plane notch.

**Fig. 19.9**  Details of tensile and bending fracture toughness specimen

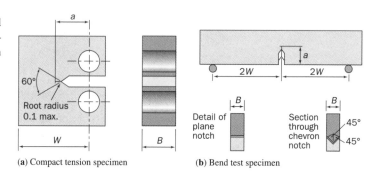

(a) Compact tension specimen      (b) Bend test specimen

In order to get a sharp crack the machined notch is extended by applying a cyclically varying force to the test piece. This cyclic stressing, in which the peak value of stress is chosen to ensure that $K_{max}$ is less than 70% of the critical value $K_{IC}$, causes the crack to grow by a fatigue mechanism (see

**Fig. 19.10**  Fracture toughness testing

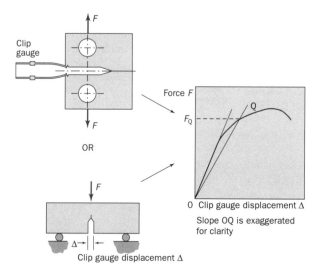

Chapter 20). The sharp crack thus produced should be at least 1.25 mm long. At this point the test specimen is removed from the fatigue machine and subjected to a static test. Typical forms of test are illustrated in Fig. 19.10.

The force is measured directly from the load cell on the testing machine and this is recorded automatically as the vertical axis on an $X–Y$ plotter. The horizontal axis is obtained from a displacement (clip) gauge attached to the test piece as indicated in Fig. 19.10.

From the graph so obtained a value of force, termed $F_Q$, is obtained so that an interim value of stress intensity factor $K_Q$ may be calculated. If the material is perfectly elastic up to fracture then the peak force is taken as $F_Q$. Assuming that the material exhibits some non-linearity as in Fig. 19.10, then the procedure for obtaining $F_Q$ is to draw a line OQ with a slope of 95% of the initial slope of the force–displacement graph. The value of force at the point where this line intersects the $F–\Delta$ characteristic is taken as $F_Q$.

The fracture surface is then examined so that the crack length $a$ may be determined as the average of $a_1$, $a_2$ and $a_3$ where $a_1$ and $a_3$ are mid-way between the centre-line of the specimen and its edge as shown in Fig. 19.11.

**Fig. 19.11** Crack length measurements

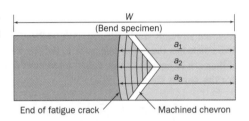

Using the values of $F_Q$ and $a$ thus obtained the following equations are used to calculate $K_Q$:

$$\text{Bending} \quad K_Q = \frac{F_Q}{BW^{1/2}} f_1\left(\frac{a}{W}\right) \qquad [19.23]$$

$$\text{Tension} \quad K_Q = \frac{F_Q}{BW^{1/2}} f_2\left(\frac{a}{W}\right) \qquad [19.24]$$

The values of $f_1(a/W)$ and $f_2(a/W)$ depend on the particular specimen geometry. (See Table 19.2, p. 529.)

Having thus obtained an interim value of stress intensity factor $K_Q$, checks are made that the following conditions have been satisfied:

$$B, a \geq 2.5\left(\frac{K_Q}{\sigma_Y}\right)^2, \qquad 0.45 < \left(\frac{a}{W}\right) < 0.55, \qquad F_Q \leq 1.1 F_{yield}$$

If these conditions are upheld then the test is a valid one and $K_Q$ is taken to be $K_{IC}$. Otherwise the result must be discarded. If desired $G_{IC}$ may then be calculated from eqn. [19.22]. Table 19.1 gives typical values of $K_{IC}$ and $G_{IC}$ for a range of materials.

**Example 19.3**

A pressure vessel is to be fabricated from plate steel which may be either: (i) a maraging (18% nickel) steel with $\sigma_Y = 1900 \, MN/m^2$, $K_{IC} = 82 \, MN \, m^{-3/2}$ or (ii) a

**Table 19.1**  Typical values of $K_{IC}$ and $G_{IC}$

| Material | $K_{IC}$ (MN m$^{-3/2}$) | $G_{IC}$ (kJ/m$^2$) |
|---|---|---|
| Mild steel | 100–200 | 50–95 |
| High-strength steel | 30–150 | 5–110 |
| Cast iron | 6–20 | 0.2–3.0 |
| Titanium alloys | 30–120 | 7–120 |
| Aluminium alloys | 22–45 | 7–30 |
| C.F.R.P. (uniaxial fibres)* | 20–45 | 2–30 |
| G.F.R.P.(uniaxial fibres)* | 10–100 | 3–60 |
| G.F.R.P. (laminate) | 10–60 | 5–100 |
| Wood* | 8–13 | 6–20 |
| Glass | 0.3–0.7 | 0.002–0.01 |
| Acrylic (P.M.M.A.) | 1.0–2.0 | 1.3–1.6 |
| Polycarbonate | 1.0–3.5 | 2.0–5.0 |
| Concrete | 0.2–0.4 | 0.03–1 |
| Epoxy | 0.5–0.7 | 0.08–0.34 |

*Perpendicular to fibre direction.

**medium strength steel with $\sigma_Y = 1000$ MN/m$^2$, $K_{IC} = 50$ MN m$^{-3/2}$. Which of these two steels has the better tolerance to defects and compare their fracture toughness if they are to have the same defect tolerance? A factor of safety of 2 should be used for the design stress.**

Assuming that eqn. [19.15] is appropriate for the steel plates then
(i) For the maraging steel the critical defect size is given by

$$a_c = \frac{K_{IC}^2}{\pi \sigma_d^2} \qquad \text{where} \qquad \sigma_d = \frac{\sigma_Y}{2}$$

$$= \frac{82^2}{\pi (950)^2} = 2.4 \, \text{mm}$$

So the critical crack length is 4.8 mm.
(ii) For the medium-strength steel

$$a_c = \frac{50^2}{\pi (500)^2} = 3.18 \, \text{mm}$$

So the critical crack length is 6.36 mm.
Hence the medium-strength steel is more tolerant of defects in the material. In order that the maraging steel would have the same tolerance its $K_{IC}$-value would need to be

$$K_{IC} = \sigma_d (\pi a_c)^{1/2} = 950(\pi \times 3.18 \times 10^{-3})^{1/2}$$

$$= 95 \, \text{MN m}^{-3/2}$$

This analysis can be extended into a general material selection chart as described in Ashby.[14] On the log–log plot of fracture toughness against strength of Fig. 19.12, all materials lying on a given 45° line have the same

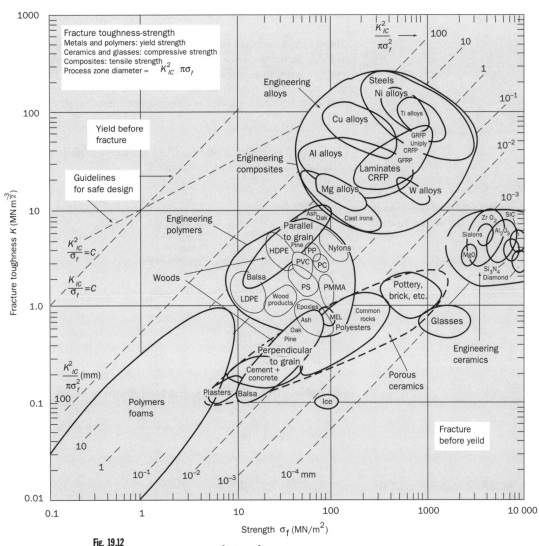

**Fig. 19.12**

value of $a_c = K_{IC}^2/(\pi\sigma_f^2)$. As noted in the figure, the strength $\sigma_f$ of metals and polymers can be taken as their yield stress, $\sigma_Y$. Materials which can tolerate a defect larger than the value of $a_c$ corresponding to a given line on the graph lie above and to the left of that line. If a defect of that size exists in the material it will yield before it fractures. Those materials with smaller critical defect sizes lie below and to the right of the line corresponding to a given value of $a_c$. If a defect of size $a_c$ exists in a structure made from this material, failure will occur by fracture before the material yield stress is reached.

## 19.7 Fracture mechanics for ductile materials

**Plastic zone correction**

The stress field equations given earlier show that the elastic stresses would become very large in the vicinity of the crack tip where $r \ll a$. In practice these large stresses do not occur because in a ductile material this region becomes plastically deformed. It would appear therefore that in such materials the formation of a plastic zone at the crack tip would invalidate the use of L.E.F.M.

However, Irwin has shown that when yielding occurs at the crack tip, L.E.F.M. techniques may still be applied if an equivalent crack length is used, i.e. a physical crack length plus an allowance for the extent of the plastic zone. This zone is generally represented by a circular boundary of radius $r_p$ at the crack tip and the equivalent crack length then becomes

$$a' = (a + r_p)$$ [19.25]

This effect is illustrated in Fig. 19.13 where the modification to the stress system due to local yielding is indicated.

**Fig. 19.13** Stress distribution at crack tip due to local yielding

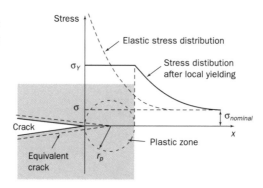

For plane stress situations $r_p$ has been shown to be given by

$$r_p = \frac{1}{2\pi}\left(\frac{K}{\sigma_Y}\right)^2$$ [19.26]

**Fig. 19.14** Variation of crack tip plastic zone across thickness of material

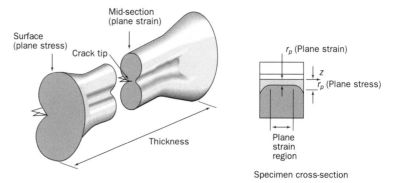

For plane strain cases the extent of the plastic zone is less as indicated in Fig. 19.14. In this case the radius of the plastic zone is given approximately by the following expression:

$$r_p = \frac{1}{6\pi}\left(\frac{K}{\sigma_Y}\right)^2$$ [19.27]

**Example 19.4**

A wide sheet of aluminium alloy has a central crack 25 mm long. If the fracture stress for the sheet is 200 MN/m² and the yield stress of the material is 400 MN/m², calculate the fracture toughness of the material (i) using L.E.F.M., and (ii) using the plastic zone correction.

(i) Using L.E.F.M.

$$K_{IC} = \sigma_{max}(\pi a)^{1/2}$$

$$= 200\left(\pi\left(\frac{0.025}{2}\right)\right)^{1/2}$$

$$K_{IC} = 39.6\,\text{MN}\,\text{m}^{-3/2}$$

(ii) Using correction for the plastic zone

$$K_{IC} = \sigma_{max}[\pi(a + r_p)]^{1/2}$$

$$= \sigma_{max}\left\{\pi a\left[1 + \frac{1}{2}\left(\frac{\sigma_{max}}{\sigma_Y}\right)^2\right]\right\}^{1/2}$$

$$= 200\left\{\pi\left(\frac{0.025}{2}\right)\left[1 + \frac{1}{2}\left(\frac{200}{400}\right)^2\right]\right\}^{1/2}$$

$$K_{IC} = 42\,\text{MN}\,\text{m}^{-3/2}$$

The difference between the values of $K_{IC}$ calculated by the elastic and plastic zone correction methods will increase as the ratio $(\sigma_{max}/\sigma_Y)$ increases, i.e. when the size of the plastic zone increases. In order to ensure that the plastic zone dimensions are small compared with other dimensions of the test piece, it is found that the more ductile the material the larger must be the test specimen in order to meet the criteria laid down for the validity of the test. In some cases this can lead to specimen handling problems as well as difficulty in getting a testing machine with enough capacity to break the specimen. For these reasons it would be advantageous to have an alternative fracture test method utilizing smaller specimens of tough materials.

This need is also borne out by the fact that for materials which exhibit extensive plasticity before fracture, even the use of the correct crack length approach becomes inappropriate. As a result of this a number of alternative test and analysis methods have been developed and the more important of these will now be described.

**Crack opening displacement (C.O.D.)** This method was proposed by Dugdale[8] and developed by Wells[9]. It is based on the fact that owing to plasticity at the crack tip, the crack opens in the direction of the applied stress as illustrated in Fig. 19.15.

The C.O.D. test is performed using the same type of specimen shown in Fig. 19.9 and the same procedure as described for determining $K_{IC}$[10]. The clip gauge measures the sample opening $V_g$ which is a measure of the resistance of the material to fracture initiation. The objective is to determine

**Fig. 19.15** Increased crack opening
due to yielding at crack tip

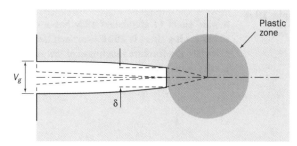

the critical crack opening at the onset of crack extension. A typical test
record is shown in Fig. 19.16($a$) and this indicates how the plastic
component $V_p$ of the clip gauge displacement is obtained.

**Fig. 19.16**

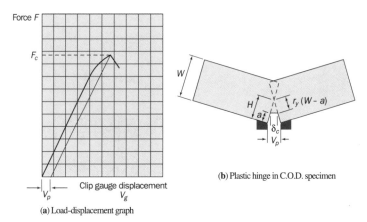

(a) Load-displacement graph

(b) Plastic hinge in C.O.D. specimen

For this type of characteristic the critical displacement is taken as the
value corresponding to the maximum applied force $F_c$. Occasionally other
shapes of characteristics are obtained and the British Standard[10] should be
referred to for details of how to obtain $V_p$ in these cases. The clip gauge
displacement $V_p$ must be converted to the crack opening displacement $\delta$
using

$$\delta = \frac{K^2(1 - \nu^2)}{2\sigma_Y E} + \frac{0.4(W - a)V_p}{0.4W + 0.6a + z} \qquad [19.28]$$

where

$$K = \frac{F}{BW^{1/2}} f_1\left(\frac{a}{W}\right)$$

and $z =$ thickness of knife edge. We obtain $f_1(a/W)$ from Table 19.2.

A basic requirement of toughness tests for critical C.O.D. is that they
should be carried out at full thickness. These tests do not have to satisfy
criteria related to plane strain conditions because they seek to determine a
toughness value relevant to the particular thickness of interest. For
application to welded structures it is necessary for tests to be carried out on
material representing different regions of the welded joints.

**Table 19.2**

| $a/W$ | 0.45 | 0.46 | 0.47 | 0.48 | 0.49 | 0.5 | 0.51 | 0.52 | 0.53 | 0.54 | 0.55 |
|---|---|---|---|---|---|---|---|---|---|---|---|
| $f_1(a/W)^*$ | 9.1 | 9.37 | 9.66 | 9.96 | 10.28 | 10.61 | 10.96 | 11.33 | 11.71 | 12.12 | 12.55 |
| $f_2(a/W)^{\dagger}$ | 8.34 | 8.57 | 8.81 | 9.06 | 9.32 | 9.6 | 9.9 | 10.21 | 10.54 | 10.89 | 11.26 |

*Three-point bending (for support span $= 4W$). †Tension.

From elastic–plastic analysis of the crack tip region using several simplifying assumptions, the critical crack tip opening displacement has been related to the critical values of fracture toughness[3]. The most commonly used relations are

$$\delta_c = \frac{K_C^2}{\lambda E \sigma_Y} \qquad \text{(plane stress)} \qquad [19.29]$$

$$\delta_c = \frac{K_{IC}^2(1 - \nu^2)}{\lambda E \sigma_Y} \qquad \text{(plane strain)} \qquad [19.30]$$

where $\lambda$ is a constant constraint factor which theoretical analyses have shown to be in the range 1–2 and which experimental measurements have shown to be approximately unity for both plane stress and plane strain situations.

**Example 19.5**

**A pressure vessel has a diameter of 2 m and a wall thickness of 10 mm. During routine inspection it is discovered that there is a crack 6.5 mm deep on the outside surface of the cylinder. The steel used in the vessel is known to be tough and a C.O.D. test on a sample gives the following results:**

**Sample width, $B = 25$ mm;    $V_p = 0.35$ mm**

**Sample depth, $W = 50$ mm;    $F_C = 65$ kN**

**Crack length, $a = 25$ mm;    $z = 2.5$ mm**

**Calculate the maximum internal pressure to which the vessel could be subjected. $E = 207$ GN/m², $\sigma_Y = 500$ MN/m², $\nu = 0.3$.**

From eqn. [19.28]

$$\delta = \frac{K^2(1 - \nu^2)}{2\sigma_Y E} + \frac{0.4(W - a)V_p}{0.4W + 0.6a + z}$$

where

$$K = \frac{F}{BW^{1/2}} f_1\left(\frac{a}{W}\right)$$

From Table 19.2, at $(a/W) = 0.5$, $f_1(a/W) = 10.61$

$$K = \frac{0.065 \times 10.61}{0.025(0.05)^{1/2}} = 123.4 \,\text{MN}\,\text{m}^{-3/2}$$

From eqn. 19.28

$$\delta_c = \frac{(123.4)^2(0.91)}{2 \times 500 \times 207 \times 10^3} + \frac{0.4 \times 25 \times 0.35 \times 10^{-3}}{0.4(50) + 0.6(25) + 2.5}$$

$$= 0.16 \times 10^{-3}\,\text{m}$$

Using eqn. [19.29], assuming plane stress conditions and $\lambda = 1$,

$$K_C = (E\sigma_Y\delta_c)^{1/2} = (207 \times 10^3 \times 500 \times 0.16 \times 10^{-3})^{1/2}$$

$$= 128.7\,\text{MN}\,\text{m}^{-3/2}$$

Using eqn. [19.15] with $Y = 1.16$ for the crack geometry,

$$\text{Hoop stress, } \sigma_\theta = \frac{K_C}{1.16(\pi a)^{1/2}} = \frac{128.7}{1.16(\pi \times 6.5 \times 10^{-3})^{1/2}}$$

$$= 776\,\text{MN/m}^2$$

$$\sigma_\theta = \frac{pd}{2t}$$

$$p = \frac{2t\sigma_\theta}{d} = \frac{2 \times 10 \times 776}{2000} = 7.8\,\text{MN/m}^2$$

**$J$-contour integral**    During the period when the C.O.D. method was being developed in the U.K. an alternative approach was being developed in the U.S.A.[11]. This has become known as the $J$-contour integral approach. It is based on the finding that for a two-dimensional crack situation, the sum of the strain energy density and the work terms along a path completely enclosing the crack tip are independent of the path taken. This is shown in Fig. 19.17.

**Fig. 19.17** Crack tip co-ordinate system and arbitrary line integral contour

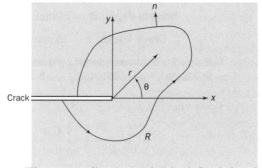

The energy line integral $J$ is defined for either the elastic or elastic–plastic behaviour as follows:

$$J = \int_p w\,dy - T\left(\frac{\partial u}{\partial x}\right)ds \qquad [19.31]$$

where $p$ = any contour path surrounding the crack tip (*note*: the integral is evaluated anticlockwise starting at the lower surface of crack and proceeding along any path to the top surface); $w$ = strain energy density = $\int_0^\varepsilon \sigma\,d\varepsilon$; $T =$

traction vector defined according to the outward normal $n$ along path $p$; $u =$ displacement vector; and $s =$ arc length.

From a more physical viewpoint $\mathcal{J}$ may be interpreted as the potential energy difference between two identically loaded bodies having crack sizes $(a)$ and $(a + \mathrm{d}a)$. In this context

$$\mathcal{J} = -\frac{1}{B}\frac{\partial U}{\partial a}$$

where $B =$ material width and $U =$ strain energy or work done (area under the load–displacement curve).

The $\mathcal{J}$-contour integral provides a means for describing the severity of conditions at a crack tip in a non-linear elastic material. Although it is a work absorption rate, $\mathcal{J}$ is equivalent to Griffith's strain energy release rate concept (note $\mathcal{J} = G$ for the linear elastic situation) and it is related to the stress intensity factor used in L.E.F.M. As with these other approaches, the critical value of $\mathcal{J}$ has been found to be dependent on thickness and test specimen geometry. However, recommended test procedures have been developed and using the suggested specimen thickness $B \geq 25\mathcal{J}_{IC}/\sigma_Y$ (or $50\mathcal{J}_{IC}/\sigma_Y$), the specimen thickness of, for example, a tough structural steel would be only about 10–20 mm compared with a required thickness of approximately 100 mm to conduct a valid L.E.F.M. test on such a material.

The most commonly used test methods for obtaining $\mathcal{J}_{IC}$ are based on compact tension or notch bend specimens as illustrated in Fig. 19.18(a) and (b). A special test-piece geometry is used to permit load-line displacements to be measured as shown in Fig. 19.18(c).

**Fig. 19.18** Test pieces used to determine $\mathcal{J}$

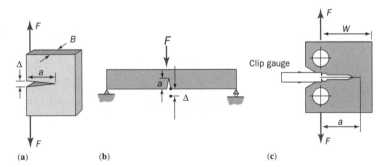

(a)  (b)  (c)

From these test specimens, $\mathcal{J}$ is calculated from

$$\mathcal{J} = \frac{2U}{B(W - a)} \qquad [19.32]$$

where $U =$ the total energy under the load–displacement diagram (using the load-line displacement $\Delta$), $W =$ the specimen width and $a =$ the crack depth (so $(W - a)$ is the remaining ligament of the test piece).

When using these tests methods it is important not to overlook the fact that the definition of $\mathcal{J}$ as a measure of work in fracturing the specimen is just a convenient simplification of the original $\mathcal{J}$-contour integral. It so happens that for the notched bend and compact tension specimens, the total energy represented by the load–displacement diagram is directly proportional to the rate of release of energy as the crack extends a small amount. It

**Fig. 19.19** Procedure for $J_{IC}$ measurement: (*a*) load test pieces to various displacements; (*b*) measure crack extension; (*c*) calculate $J$ for each test piece and plot $J$ against d$a$; (*d*) construct two curves for $J_{IC}$ measurement

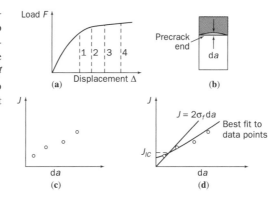

is this energy release rate which really justifies the linking of the work done to $J$. The steps in a typical test procedure are illustrated in Fig. 19.19 and summarized below.

(*a*) Load the test piece to different displacement values with a testing machine in displacement control.
(*b*) Unload and mark the extent of crack growth.
(*c*) Fracture each test piece at a low temperature (e.g.-150°C) and measure the crack extension.
(*d*) Calculate $J$-values from the load against load-line displacement record using eqn. [19.32] with $U$ being the area under the graph at the displacement of interest.
(*e*) Plot a curve of $J$ against crack extension d$a$.
(*f*) Draw a straight line $J = 2\sigma_f$ d$a$ to intersect the best-fit line through the data points. Here, $\sigma_f$ is taken as $\frac{1}{2}(\sigma_Y + \sigma_u)$, where $\sigma_u$ is the tensile strength of the material.
(*g*) The critical value of $J$ is at the intersection of the two lines.

The most common method of application of the $J$-contour integral approach is to convert experimentally determined values of $J_{IC}$ to equivalent values of $G_{IC}$ or $K_{IC}$. As would be expected from the similarity of the tests, it has also been shown that $J$ is related to the crack opening displacement (C.O.D.). The following equations are frequently used to relate the various fracture toughness parameters:

$$\frac{K_{IC}^2}{E'} = G_{IC} = J_{IC} = \lambda\sigma_Y\delta_c \qquad [19.33]$$

where

$$E' = E \qquad \text{for plane stress}$$

$$E' = E/(1 - \nu^2) \qquad \text{for plane strain}$$

**Example 19.6**

A bend test piece of the type shown in Fig. 19.9 is made from mild steel. The width $B$ is 20 mm and the depth $W$ is 25 mm. At the point of crack growth the crack length $a$ is 15.3 mm and the area under the load–deflection graph is 14.7 J. If the modulus of the mild steel is 207 GN/m², calculate its fracture toughness.

From eqn. [19.32],

$$J = \frac{2U}{B(W-a)}$$

$$= \frac{2 \times 14.7}{0.02(25 - 15.3)10^{-3}} = 151.5 \, \text{kJ/m}^2$$

From eqn. [19.33],

$$K_{IC} = \sqrt{(EJ)}$$

$$= \sqrt{(207 \times 10^9 \times 151.5 \times 10^3)} \, \text{N m}^{-3/2}$$

$$= 177 \, \text{MN m}^{-3/2}$$

*R* **curves**    Another method which has been developed to extend the techniques of L.E.F.M. into the regime of elastic–plastic fracture involves the use of crack extension resistance curves or *R* curves. An *R* curve is a plot of the crack growth resistance (*R*) in a material as a function of the actual or effective crack extension ($\Delta a$). Here, *R* represents the energy absorbed ($dU_s$) per increment of crack growth ($da$). It is therefore given by the value of $dU_s/da$ prior to unstable crack growth at the critical point; *R* has the same units as the stress intensity factor *K*.

It was shown earlier that $dU_s/da$, i.e. *R*, was independent of the crack length since $U_s$ varied linearly with crack length *a* in Fig. 19.3. This is approximately true for cracks under plane strain conditions, but in situations involving larger proportions of plane stress failure, *R* is no longer independent of crack length.

Figure 19.20 shows a typical variation of *R* (calculated using the effective crack length) with crack extension. A.S.T.M. E561 provides specific instructions for the measurement of *R*. If curves of *G* (corrected to allow for the size of the plastic zone) are superimposed on the *R* curves then the point of instability occurs at the point of tangency between the two types of curves. The value of *G* should be calculated from $G = Y^2 \sigma_Y^2 (a + r_p)/E$.

The solid curve shows the resistance to crack extension (*R* curve). The dashed lines show the driving force for crack extension. Both dashed lines

**Fig. 19.20**  Typical *G–R* curves

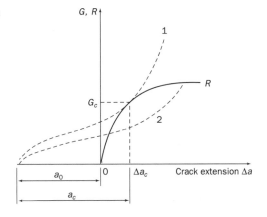

are defined by $EG = Y^2\sigma_Y^2 a$, but for curve 1 the stress is greater than for curve 2. The higher stress causes unstable crack growth when the crack length reaches the value $a_c$, shown in Fig. 19.20.

## 19.8 Toughness measurement by impact testing

In addition to the more recent methods of fracture toughness testing for design by fracture mechanics analysis, the traditional quality-control method of measuring the energy absorption of a material during fracture has been by impact bend tests.

The two forms of the test most widely used are the Charpy and Izod. The principle of the former is shown in Fig. 19.21. The test piece is a square bar of material, $10\,\text{mm} \times 10\,\text{mm} \times 55\,\text{mm}$, containing a notch cut in the

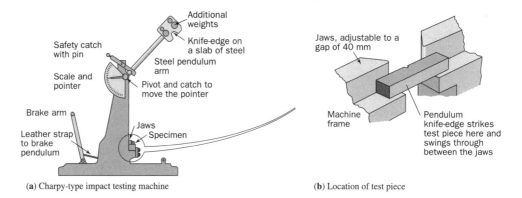

(a) Charpy-type impact testing machine        (b) Location of test piece

**Fig. 19.21**    middle of one face. The notch is a 45° vee, 2 mm deep, with a root radius of 0.25 mm. The test piece is simply supported at each end on anvils 40 mm apart. A heavy pendulum is supported at one end in a bearing on the frame of the machine, and a striker is situated at the other end. The pendulum in its initially raised position has an available energy of 300 J and on release swings down to strike the specimen immediately behind the notch, bending and fracturing it between the supports. A scale and pointer indicate the energy absorbed during fracture.

In the Izod test, Fig. 19.22, the specimen is of circular section, 11.43 mm in diameter and 71 mm long, or square section, $10\,\text{mm} \times 10\,\text{mm} \times 75\,\text{mm}$, and the Izod notch is a vee, as described above for the Charpy test, 3.33 mm and 2 mm deep for the round and square specimens respectively. The specimen

**Fig. 19.22**   Arrangement of test piece in the Izod test

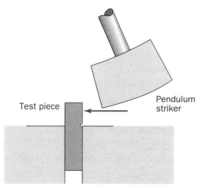

Table 19.3

| Material | Condition | Tensile strength (MN/m$^2$) | Reduction in area (%) | Izod impact energy at room temperature (J) |
|---|---|---|---|---|
| 0.1 C, 0.3 Mn steel | Annealed | 377 | 65 | 54.2 |
| 0.21 C, 0.82 Mn steel | Annealed | 505 | 58 | 40.6 |
| 0.5 C steel | Normalized | 787 | 63 | 29.8 |
| Ni–Cr–Mo steel | Quenched 840 °C, tempered 650 °C | 895 | 64 | 114.0 |
| | Quenched 840 °C, tempered 500 °C | 1472 | 47 | 29.8 |
| 3.0 Ni, 1.0 Cr steel | Quenched 840 °C | 1668 | 35 | 19.0 |
| | Quenched and tempered 550 °C | 865 | 60 | 93.5 |
| Stainless steel (18–8) | Cold rolled | 987 | – | 46.1 |

is supported as a vertical cantilever 'built-in' to jaws up to the notched cross-section. A pendulum and striker having an initial energy of 166 J is arranged to swing and strike the free end of the test piece with a velocity of 3–4 m/s on the same side as the notch and 22 mm above it. The energy absorbed by the test piece is again recorded on a scale.

It is seen that in both tests the notch is on the tension side of bending, thus initiating fracture. The Charpy and Izod tests have been adopted as British Standards and details of alternative types of specimen and other conditions are given in B.S. 131: 1982. There is no direct correlation between energy values given in each of these tests; however, experimental results on a wide range of steels show that there is a linear relationship over the range from 20 to 95 J.

Some typical values of strength, ductility and toughness for steels are given in Table 19.3. The most notable feature is the effect of the type of heat treatment given, whether normalizing, quenching or quenching and tempering. The reason for this is the difference in metallurgical structure in each case influencing the ductility of the material, and the ease with which a crack can propagate through the notched bar. The quenched structure is hard and brittle and can absorb little energy. In the quenched and tempered structure, a higher tempering temperature gives greater ductility and therefore higher impact energy. The importance of temperature effects on fracture mechanics can be easily demonstrated in the Charpy or Izod impact tests by prior heating or cooling of the specimen.

If the test temperature is varied from 'high' to 'low', then a metal such as mild steel, which exhibits the property of brittle fracture, will have a range of test temperature in which the mechanism of fracture changes from shear to cleavage. This is known as the *transition temperature* range, and within this range will be determined, according to some criterion, a transition temperature.

**Fig. 19.23**  Transition curves of brittle–ductile energy against temperature

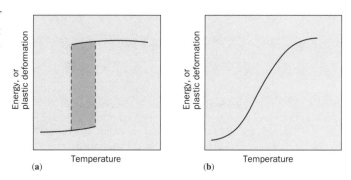

(a)             (b)

Typical transition curves are shown in Fig. 19.23(a) and (b) in which plastic deformation at, or energy to, fracture is plotted against test temperature. The first shows what is known as bimodal behaviour in which there are two distinct branches, one at high energy and temperature and the other at low energy and temperature, joined by a narrow region (10–20 °C) of scattered points at upper and lower energy values. The second diagram shows continuous behaviour where there is a gradual fall from high to low energy with decrease in temperature. The transition range in this case can be as much as 100 °C from complete shear to entire cleavage.

## 19.9 Relationship between impact testing and fracture mechanics

The major advantage of the Charpy and Izod types of impact tests is that they are quick and convenient tests to perform when compared with the much more demanding test schedule essential to obtain fracture toughness parameters. Hence there have been a number of attempts in recent years to correlate $K_{IC}$ with the Charpy impact strength $C_v$. No general rule has yet emerged but the research has met with some success. For high-strength steels a relationship has been found to exist between $K_{IC}$ and $C_v$ for the upper shelf region, and a variation of this can be applied to the transition temperature region[12]. Typical results are shown in Fig. 19.24 and it is felt that the type of relationships observed may have more general applicability to the lower-strength steels, particularly for the upper shelf $C_v$-values.

**Fig. 19.24**  Impact strength against critical stress intensity for maraging steel

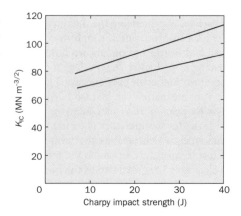

Thus, although Charpy testing says nothing about stress levels, which has in the past made it impossible to use Charpy toughness values in design calculations, the correlation between $K_{IC}$ and $C_v$ has encouraged attempts to provide a criterion for material selection based on $C_v$.

Finally, there has been extensive work done on the use of Charpy tests to obtain fracture mechanics parameters for polymers. In particular a very useful relationship has been developed[13] between the energy absorbed in breaking the precracked specimen $C_v$ (in J) and the critical strain energy release rate $G_{IC}$ (in $J/m^2$). This has the form

$$C_v = G_{IC}BD\phi \qquad [19.34]$$

where $B$ = specimen breadth, $D$ = specimen depth and $\phi$ = a geometry factor. This permits $G_{IC}$, which is a material property, to be determined and hence $K_{IC}$ (from eqn. [19.14]) although it should be noted that for polymers the modulus $E$ is not a constant.

## 19.10 Summary

The development of fracture mechanics on the basis that materials contain crack-like defects from which fast fracture may occur, has led to the emergence of new design concepts. In essence these assume that the critical value of the stress intensity factor $K_{IC}$ is a material property which may be used to calculate the maximum defect size for a given stress or the maximum permissible stress for a given intrinsic defect size. Linear elastic fracture mechanics may be applied with confidence in situations where fracture occurs under essentially elastic conditions. The theory can also be extended to include materials which exhibit a relatively small amount of yielding prior to fracture, and the quantity $(K_{IC}/\sigma_Y)^2$ provides a measure of the size of the plastic zone at the crack tip. However, for very tough materials which exhibit gross yielding prior to fracture the use of L.E.F.M. is no longer valid. In such cases recourse may be made to methods such as crack opening displacement (C.O.D.), $J$-contour integral or resistance curves which have emerged in the wake of the success achieved by L.E.F.M.

The traditional toughness tests such as the Charpy and Izod tests still have an important role as quality-control tests, but have the disadvantage that they do not provide information which can be used in design calculations.

## References

1. Griffith, A. A. (1921) 'The phenomena of rupture and flow in solids', *Phil. Trans. Ry. Soc.*, A221, 163–197.
2. Irwin, G. R. (1957) 'Fracture mechanics', *J. Appl. Phys.*, 24, 361.
3. Parker, A. P. (1981) *The Mechanics of Fracture and Fatigue*, E. & F. N. Spon, London.
4. Paris, P. C. and Sih, G. C. (1965) 'Fracture toughness and its applications', *ASTM STP* 381.
5. Rooke, D. P. and Cartwright, D. J. (1976) *Compendium of Stress Intensity factors*, HMSO, London.
6. B.S. 5447: 1977. *Plane Strain Fracture Toughness of Metallic Materials*.
7. A.S.T.M. Test Method E399-74. *Standard Method of Test for Plane Strain Fracture Toughness of Metallic Materials*.
8. Dugdale, D. S. (1960) 'Yielding of steel sheets containing slits', *J. Mech. Phys. Solids*, 8, 100–108.

9. Wells, A. A. (1961) 'Unstable crack propagation in metals', *R.Ae.S. Symposium on Crack Propagation, Cranfield*, p. 210–230.
10. B.S. 5762: 1979. *Crack Opening Displacement (C.O.D.) Testing.*
11. Rice, J. F. (1962) *Fracture*, Vol. II, ed. H. Liebowitz, Academic Press, New York.
12. Rolfe, S. T. and Barsom, J. M. (1977) *Fracture and Fatigue Control in Structures*, Prentice Hall, Englewood Cliffs, NJ.
13. Plati, E. and Williams, J. G. (1975) *Polymer*, **16**, 915.
14. Ashby, M. F. (1992) *Materials Selection in Mechanical Design*, Pergamon Press, Oxford.

**Bibliography**

Anderson, T. L. (1991) *Fracture Mechanics. Fundamentals and Applications*, CRC Press, Boca Raton, FL.

Hellan, K. (1984) *Introduction to Fracture Mechanics*, McGraw-Hill, London.

Hertzberg, R. W. (1989) *Deformation and Fracture Mechanics of Engineering Materials*, 3rd edition, John Wiley, New York.

Knott, J. F. and Withey, P. A. (1993) *Fracture Mechanics. Worked examples*, 2nd edition, Institute of Materials, London.

Parker, A. P. (1981) *The Mechanics of Fracture and Fatigue. An Introduction.* E. & F. N. Spon, London.

Powell, G. W. and Mahmoud, S. E. (1986) *Metals Handbook, 9th edition, Vol. 11. Fracture Analysis and Prevention*, American Society for Metals, Metals Park, OH.

Smith, R. N. L. (1991) *BASIC Fracture Mechanics*, Butterworth–Heinemann, Oxford.

**Problems**

19.1   Describe how the critical strain energy release rate, $G_{IC}$, might be obtained by experiment.

19.2   Three fracture toughness samples of an aluminium alloy have identical external dimensions and a thickness of 25 mm. During three tests on the samples the following information was obtained:

| Test | Sample crack length (mm) | Applied load (kN) | Sample elongation (mm) |
|------|--------------------------|-------------------|------------------------|
| 1 | 20 | 185 | Sample fractured |
| 2 | 19.5 | 120 | 0.26 |
| 3 | 20.5 | 120 | 0.263 |

If the Young's modulus and Poisson's ratio values for the aluminium are 70 GN/m$^2$ and 0.3 respectively, calculate the fracture toughness of the material.

19.3   A compact tension fracture mechanics specimen containing a crack 53 mm in length fails at a load of 59.1 kN. Estimate the fracture toughness of the material, given that the stress intensity $K_I$ in the specimen was

$$K_I = \frac{P}{BW^{1/2}} Y$$

where the specimen thickness $B = 50\,\text{mm}$, effective width $W = 100\,\text{mm}$ and $Y$ is given in the table below. $a$ is the length of crack in the specimen.

| $a/W$ | 0.50 | 0.51 | 0.52 | 0.53 | 0.54 |
|-------|------|------|------|------|------|
| $Y$   | 9.60 | 9.90 | 10.21 | 10.54 | 10.89 |

If the yield stress of the material is $400\,\text{MN/m}^2$, was this estimate a valid one?

19.4 Calculate the critical defect size for each of the following steels assuming they are each to be subjected to a stress of $\frac{1}{2}\sigma_Y$. Comment on the results obtained.

| Steel | Yield strength, $\sigma_Y$ ($\text{MN/m}^2$) | Fracture toughness ($\text{MN m}^{-3/2}$) |
|-------|------|------|
| Mild steel | 207 | 200 |
| Low-alloy steel | 500 | 160 |
| Medium–carbon steel | 1000 | 280 |
| High-carbon steel | 1450 | 70 |
| 18% Ni (maraging) steel | 1900 | 75 |
| Tool steel | 1750 | 30 |

19.5 In a fracture toughness test involving three-point bending, the following information was recorded: support span $= 180\,\text{mm}$; specimen thickness $= 22.8\,\text{mm}$; specimen width $= 44.72\,\text{mm}$; crack length $= 21.92\,\text{mm}$; fracture force $= 19.8\,\text{kN}$. Establish whether or not these data are suitable for measuring $K_{IC}$ for the material. The yield stress of the material is $350\,\text{MN/m}^2$.

19.6 The load–deflection graph from a tensile fracture toughness test on a 50 mm thick steel specimen is shown in Fig. 19.25. Also shown (to scale) is the fracture surface of the broken specimen. If the yield stress for the steel is $1650\,\text{MN/m}^2$, establish whether the test is valid for the determination of $K_{IC}$.

19.7 A sheet of glass 0.5 m wide and 18 mm thick is found to contain a number of surface cracks 3 mm deep and 10 mm long. If the glass is placed horizontally on two supports, calculate the maximum spacing of the supports to avoid fracture of the glass due to its own weight. For glass $K_{IC} = 0.3\,\text{MN m}^{-3/2}$ and density $= 2600\,\text{kg/m}^3$.

19.8 The accident report on a steel pressure vessel which fractured in a brittle manner when an internal pressure of $19\,\text{MN/m}^2$ had been applied to it shows that the vessel had a longitudinal surface crack 8 mm long and 3.2 mm deep. A subsequent fracture mechanics test on a sample of the steel showed that it had a $K_{IC}$ value of $75\,\text{MN m}^{-3/2}$. If the vessel diameter was 1 m and its wall thickness was 10 mm, determine whether the data reported are consistent with the observed failure.

19.9 A simple lifting crane is illustrated in Fig. 19.26. If the tie bar is found to have 3 mm long cracks at the pin-joint hole, calculate the maximum vertical force at the pulley to avoid fracture of the tie bar.

**Fig. 19.25**

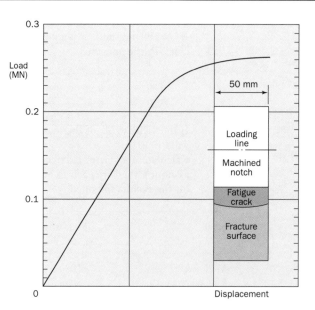

The fracture toughness of the tie-bar steel is $75\,\mathrm{MNm}^{-3/2}$. Comment on the suitability of the equation used to calculate $K_{IC}$.

**Fig. 19.26**

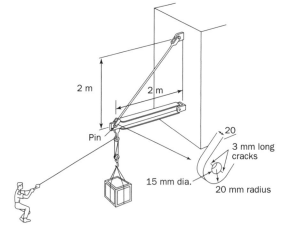

19.10 (a) Using an energy balance, show that the stress intensity for a long crack running in the axial direction of a thin walled pipe, Fig. 19.27 is given by

$$K = \frac{pR}{t}\sqrt{\pi R}$$

(b) Estimate the stress intensity for a short crack when $L \ll R$.

19.11 A large sheet of aluminium alloy is loaded in tension to a stress of 200 MPa. If its yield stress $\sigma_y = 400$ MPa and fracture toughness $K_{IC} = 42$ MN m$^{-3/2}$, what is the maximum tolerable defect size (i) using linear elastic fracture mechanics; (ii) using a plastic zone correction?

Fig. 19.27

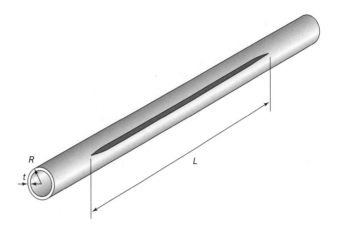

19.12 A vessel is protected from excessive pressure by a bursting disc, a flat circular plate clamped at its edges with a sharp notch machined across a diameter in the unpressurized side (Fig. 19.28). What depth of notch woud be required for a bursting pressure $q_c$ of $20\,MN/m^2$. The plate has radius $r = 40\,mm$ and its thickness $h$ is $4\,mm$. The plate material has the properties $E = 200\,GN/m^2$, $\sigma_Y = 1500\,MN/m^2$, $\nu = 0.333$ and critical COD $= 4\,\mu m$. You can assume that the stress intensity round the notch is given by $K = 1.12\sigma\sqrt{\pi a}$, and $K_C = \sqrt{E\sigma_Y \delta_c}$.

The maximum stress acting perpendicular to the notch can be derived from Eqn. [16.28]. It occurs at $r = 0$, when

$$\sigma_\theta = \sigma_r = \frac{3q_c r^2 (1 + \nu)}{8h^2}$$

Fig. 19.28

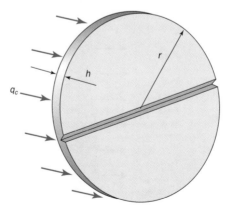

19.13 In a chemical plant, a thick walled cylindrical pressure wheel made from a titanium alloy is subjected to occasional surges in internal pressure of up to $100\,MN/m^2$. During a routine inspection a $2\,mm$

long crack oriented perpendicular to the hoop direction is found in the cylinder bore. Should the cylinder be taken out of service immediately or can this crack be tolerated? The internal radius of the vessel is 10 mm while the outer radius is 15 mm. The hoop stress in the vessel is given by Eqn. [14.36]. $K_{IC} = 106$ MN m$^{-3/2}$.

19.14  An aluminium alloy plate with a yield stress of 450 MN/m$^2$ fails in service at a stress of 110 MN/m$^2$. The conditions are plane stress and there is some evidence of ductility at the fracture. If a surface crack 20 mm long is observed at the fracture plane calculate the size of the plastic zone at the crack tip. Calculate also the percentage error likely if L.E.F.M. was used to obtain the fracture toughness of this material.

19.15  A large medium-carbon steel crane hook is thought to contain penny-shaped internal cracks. If the non-destructive test equipment used on the hook is not capable of detecting cracks smaller than 20 mm diameter, determine the fracture toughness required from this steel if the safety factor on stress is to be 2. The yield stress of the steel is 1050 MN/m$^2$.

19.16  A large plate carrying a tensile load contains a hole with two cracks propagating from the hole in a direction perpendicular to the applied load, Fig. 19.29.

(*a*)  Estimate the stress $\sigma_0$ at which the plate will fracture by both the following methods. (i) Assume the stress intensity in the cracks is the same as the stress intensity in the edge of an infinite plate whose average stress is the same as the peak stress round a hole in an uncracked plate. (ii) Assume that the only effect of the hole is to make the two cracks at opposite sides of the hole appear as one long continuous crack.

The stress intensity for a crack of length $a$ in the edge of a plate is $K = 1.12\sigma\sqrt{\pi a}$. For a central crack of length $2a$ in the middle of an infinite plate $K = \sigma\sqrt{\pi a}$.

(*b*)  At what ratio $r/L$ do both approaches give the same estimate? Which approach is more likely to be correct for $r \ll L$?

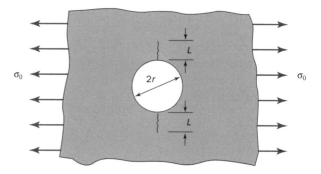

**Fig. 19.29**

19.17  As a result of a C.O.D. test on a medium-carbon steel, the following data were recorded: sample width = 25.03 mm; $V_p = 0.33$ mm; sample depth = 49.98 mm; $F_c = 80$ kN; crack length = 24.00 mm; z = 2.5 mm. Calculate (*a*) the fracture toughness of the material and (*b*) the percentage error in using L.E.F.M. to calculate this value. For the steel, $E = 207$ GN/m$^2$, $\sigma_Y = 950$ MN/m$^2$, $\nu = 0.3$.

19.18 In a three-point bend test on a mild steel sample the load at the point of crack growth and the final crack length were noted as 70 kN and 26 mm respectively. At the point of crack growth the area under the load–deflection graph was 32.6 J. If the specimen width (*W*) and thickness (*B*) were 49.98 mm and 25.05 mm respectively, calculate the fracture toughness of the mild steel. If a C.O.D. gauge (width $z = 2.5$ mm) had been attached to this specimen calculate the plastic component of the gauge reading which you would have expected to record. For the mild steel, $\sigma_Y = 240$ MN/m$^2$, $E = 207$ GN/m$^2$ and $\nu = 0.3$.

19.19 A bench-top impact testing machine for plastics as shown in Fig. 19.30. The pendulum weighs 4 kg and its centre of gravity is at A. When there is no specimen in position it is found that the pendulum swings through to an angle $\theta = 129°$. When there is a specimen in position the pendulum swings through to $\theta = 89°$. Calculate (*a*) the windage and friction losses in the machine and (*b*) the energy absorbed in breaking the specimen.

Fig. 19.30

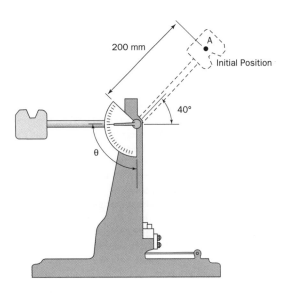

# 20 Fatigue

In the early part of the nineteenth century the failure of some mechanical components subjected to nominal stresses well below the tensile strength of the material aroused some interest among a few engineers of that time. The fact that puzzled these early engineers was that a component such as a bolt or a shaft made from a ductile material such as mild steel could fracture suddenly in what appeared to be a brittle manner. There was no obvious defect in workmanship or material, and the only feature common to these failures was the fact that the stresses imposed were not steady in magnitude, but varied in a cyclical manner. This phenomenon of failure of a material when subjected to a number of varying stress cycles became known as *fatigue*, since it was thought that fracture occurred owing to the metal weakening or becoming 'tired'. The first real attack on this problem was made by the German engineer Wöhler in 1858. Since then a great deal of research has been conducted on fatigue of metals, and in more recent times other materials also, and although this work has resulted in an ever-increasing understanding of the problem there is as yet no complete solution. What has been established is that the early theories that the metal becomes 'crystalline' or brittle under the action of the cyclic load are erroneous. It is now well known that a fatigue failure starts on a microscopic scale as a minute crack or defect in the material and this gradually grows under the action of the stress fluctuations until complete fracture occurs.

It has been estimated that at least 75% of all machine and structural failures have been caused by some form of fatigue. It is therefore evident that every engineer should be aware of this phenomenon, and have some idea of its mechanics and what can be done to minimize or avoid the risk of this type of failure. It is astonishing how many design engineers ignore warnings about the possibility of fatigue failures and include in their design geometrical shapes which cause stress concentrations and initiate fatigue failures. Indeed even when the danger of fatigue failure is recognized it is not always possible to avoid owing to the many stages of manufacture which a component may go through and which are outside the direct control of the designer. There have been reports of situations where the designer specified non-destructive testing of components to search for flaws that would initiate fatigue failures, and when none were found the test engineer used a metal stamp to mark the component as having passed inspection. Then in service fatigue cracks initiated from the stamp mark!

Figure 20.1(*a*) shows a typical situation where fatigue failures can arise owing to geometrical configurations, and Fig. 20.1(*b*) illustrates the appearance of the fracture surface in such cases.

## 20.1 Forms of stress cycle

Throughout the working life of a component subjected to cyclical stress the magnitude of the upper and lower limits of cycles may vary considerably as shown in Fig. 20.2. However, it has been general practice when considering fatigue behaviour to assume or employ a sinusoidal cycle having constant upper and lower stress limits throughout the life.

Fig. 20.1   Fatigue cracks in an
            engine crankshaft

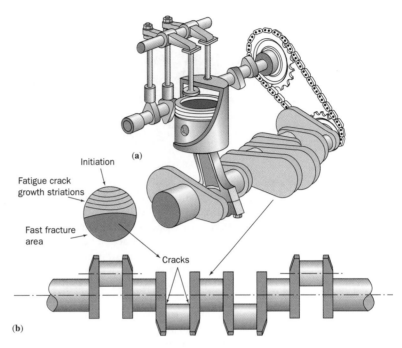

Fig. 20.2   Variable stress spectrum

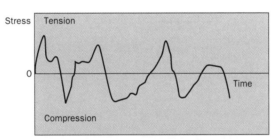

Figure 20.3 shows a general type of stress cycle, which is termed
*fluctuating*, in which an alternating stress is imposed on a mean stress. This
cycle can consist of any combination of upper and lower limits, within the
static strength, which are both positive, or both negative.

Fig. 20.3   Stress parameters

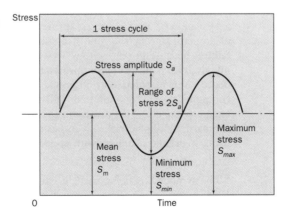

**Fig. 20.4** Typical stress–time cycles

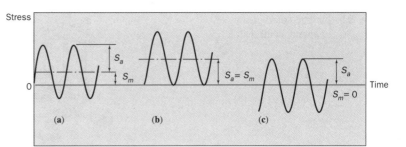

When one stress limit in a cycle is positive and the other negative as shown in Fig. 20.4(*a*) it is known as a reversed cycle. There are two particular cases of the fluctuating and reversed cycles which arise frequently in engineering practice. The first is a fluctuating cycle in which the mean stress is half the maximum, the minimum being zero, as shown in Fig. 20.4(*b*). The second is a symmetrical reversed cycle as shown in Fig. 20.4(*c*) in which the mean stress is zero and the upper and lower limits are equal positive and negative.

The relationship between the various stress values is of some importance. The mean stress, $S_m$,* is half the algebraic sum of the maximum stress, $S_{max}$, and the minimum stress, $S_{min}$. The range of stress, $2S_a$, is the algebraic difference between $S_{max}$ and $S_{min}$. The ratio of the minimum to the maximum stress is termed the *stress ratio*, $R$. Hence

$$S_m = \frac{S_{max} + S_{min}}{2} \qquad [20.1]$$

$$2S_a = S_{max} - S_{min} \qquad [20.2]$$

$$R = \frac{S_{min}}{S_{max}} \qquad [20.3]$$

It should be noted that the foregoing has only considered stresses as being positive or negative in a general sense. In practice, fatigue can be generated in direct stress due to axial loading or bending or shear stress due to cyclic torsion or any combinations of these.

## 20.2 Test methods

Fatigue failures occur most often in moving machinery parts, e.g. shafts, axles, connecting rods, valves, springs, etc. However, the wings and fuselage of an aeroplane or the hull of a submarine are also susceptible to fatigue failures because in service they are subjected to variations of stress. As it is not always possible to predict where fatigue failures will occur in service and because it is essential to avoid premature fractures in articles such as aircraft components, it is common to do full-scale testing on aircraft wings, fuselage, engine pods, etc. This involves supporting the particular aircraft section or submarine hull or car chassis in jigs and applying cyclically varying stresses using hydraulic cylinders with specially controlled valves.

* The symbol $S$ is being used for stress to conform with B.S. 3518: Part I: 1984; however, $\sigma$ is also a recommended symbol (I.S.O.).

**Fig. 20.5**   Rotating bending fatigue test

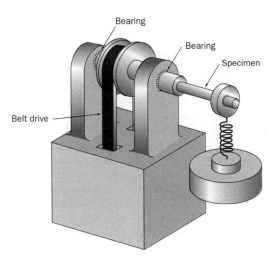

Laboratory tests are also carried out on particular materials to establish their fatigue characteristics and to study factors such as their susceptibility to stress concentrations. The earliest and still widely used method of fatigue testing of laboratory specimens is by means of rotation bending. A cylindrical bar is arranged either as a cantilever (Fig. 20.5) or as a beam in pure bending. It is then rotated while subjected to a bending moment and hence each fibre of the bar suffers cycles of reversed bending stress. A bending fatigue test can also be arranged, without rotation, by alternating bending in one plane. An advantage of this test over the former is that a mean bending stress can be introduced.

Axial load fatigue tests, although requiring more sophisticated and expensive equipment, have the significant advantage that they subject a volume of material to a uniform stress condition as opposed to the stress gradient in bending.

There are three main principles of operation for axial load fatigue machines. Firstly, there is the system whereby an electromagnet in series with the specimen is rapidly energized and de-energized, thus applying cyclical force. To reduce the amount of power input to the electromagnet to achieve the required load amplitude, a resonant vibration system is employed using springs. Another method of achieving a resonant condition of springs in series with a specimen is by mechanical means using a small out-of-balance rotating mass. The third arrangement for producing fluctuating forces on a specimen is by means of hydraulic pulsation.

Frequency is an important factor in fatigue testing, owing to the large number of cycles it is necessary to achieve at lower stress ranges, and consequent time factor in obtaining fatigue data. Bending fatigue machines generally run in the range from 30 to 80 Hz. The electrical excitation push–pull machines operate between 50 and 300 Hz, while the mechanical excitation is usually restricted to about 50 Hz. The hydraulic machines have a frequency, often variable, in the range from about 1 to 50 Hz. The frequency of an actual test is generally controlled by the stiffness of the specimen, i.e. high frequency can only be achieved for high stiffness and low amplitude of deformation. The testing capacity of axial load fatigue machines varies typically from 20 to 2000 kN.

Cyclic torsion or combined bending and torsion fatigue machines generally operate on the principle of direct mechanical displacement of the specimen by a variable eccentric, crank and connecting-rod system. Depending on the capacity of the machine, frequencies can vary from 16 to 50 Hz.

### 20.3 Fatigue data

#### S–N curves

The most readily obtainable information on fatigue behaviour is the relationship between the applied cyclic stress $S$ and the number of cycles to failure $N$. When plotted in graphical form the result is known as an $S$–$N$ curve. Figure 20.6 shows the three ways of plotting the variables: (a) $S$ against $N$; (b) $S$ against $\log_{10} N$, (c) $\log_{10} S$ against $\log_{10} N$. The number of cycles to failure at any stress level is termed the *endurance*, and this may vary between a few cycles at high stress and as much as 100 million cycles at low cyclic stresses, for a complete $S$–$N$ curve. It is immediately apparent that a linear scale for endurance $N$ over the whole range is impractical. It is for this reason that the semi-logarithmic or double-logarithmic plots (b) and (c) are employed. The former, (b), is the most widely used method of presentation.

**Fig. 20.6** Typical methods of presenting fatigue curves

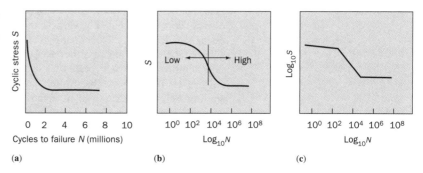

(a)                                    (b)                                    (c)

#### High-endurance fatigue

This relates to endurances from about $10^4$ cycles to 'infinity' (or $50 \times 10^6$ in terms of a laboratory test). The $S$–$N$ curve in Fig. 20.6(b) falls rapidly from $10^4$ cycles followed by a 'knee' after which, depending on the material, the curve either becomes parallel to the $N$-axis or continues with a steadily decreasing slope. Most steels and ferrous alloys exhibit the former types of curve, and the stress range at which the curve becomes horizontal is termed the *fatigue limit*. Below this value it appears that the metal cannot be fractured by fatigue. In general, non-ferrous metals do not show a fatigue limit and fractures can still be obtained even after several hundred million cycles of stress. It is usual, therefore, to quote what is termed an *endurance limit* for these metals, i.e. the stress range to give a specific large number of cycles, usually $50 \times 10^6$. Typical $S$–$N$ curves for an aluminium alloy and a steel tested in air for a plain condition, i.e. no stress concentration, under axial loading, are given in Fig. 20.7.

#### Low-endurance fatigue

For many years the low-endurance region ($<3000$ cycles) of fatigue was ignored because the bulk of engineering cyclic stress situations had to have a working life ranging from several millions to infinity before failure. Hence the important information was contained in the lower stress–higher cycles to failure part of the curve. However, a low cycle to failure does not necessarily mean a short lifetime because it is a function of cyclic frequency. Hence an

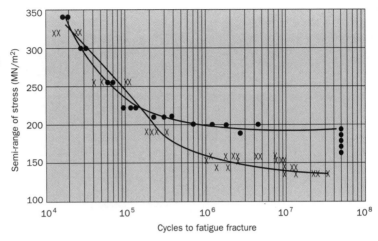

**Fig. 20.7** Aluminium alloy 24S-T3 reversed axial stress fatigue curve (×); mild steel reversed axial stress fatigue curve (■)

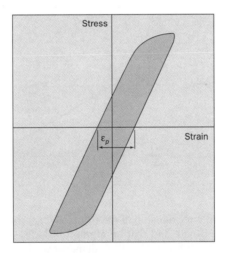

**Fig. 20.8** Plastic strain hysteresis loop

aircraft fuselage is only pressurized once every flight and so it may take years to accumulate 1000 cycles and a pressure vessel may work for 25 years before achieving 1000 cycles of cleaning and inspection.

Failure at low endurances results from stresses and strains which are high and will result in marked plastic deformation (hysteresis) in every cycle, Fig. 20.8. Owing to strain-hardening (or softening) effects it becomes of some significance whether the cycle is of controlled load (stress) amplitude or controlled strain amplitude, a problem which does not arise at long endurance and hence low stress. A good example of strain cycling, which came into prominence with the design of nuclear plants and aerospace vehicles, is the effect of repeated thermal changes in a component.

At low endurances the $S$–$N$ and $\varepsilon$–$N$ curves take the form shown in Fig. 20.9. It is usual to consider the $S$–$N$ curve as starting from a point at a quarter cycle representing a tension test or single pull to fracture. Likewise the $\varepsilon$–$N$ curve is started at a quarter cycle using the true fracture ductility in simple tension as the $\varepsilon$-value. Much work has been concentrated on strain cycle testing, and it has been found that, if the plastic strain range $\varepsilon_p$, or width of the hysteresis loop, is determined during the life of the specimen,

**Fig. 20.9** Typical stress cycle and strain cycle fatigue curves at low endurance

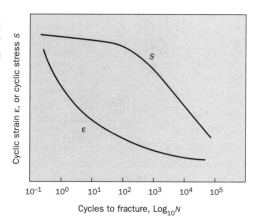

**Fig. 20.10** Relationship of log $\epsilon_p$ with log $N$

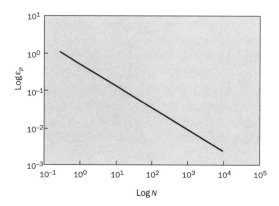

then a relationship of the form $\varepsilon_p N^\alpha = $ constant holds for most metals up to $10^5$ cycles, and that on a graph of $\log \varepsilon_p - \log N$ a straight line results, Fig. 20.10. It has further been demonstrated that $\alpha$ is between $-0.5$ and $-0.6$ for many metals at room temperature, and the constant is broadly related to the ductility of the material, i.e. the greater the latter, the larger the constant term.

Thermal strain cycling, i.e. cyclic strains induced by changes in temperature, appears to yield the same results as if the specimen or component were kept at a constant temperature of a value equal to the upper limit of the previous temperature cycle, and the cyclic strains induced mechanically.

**Statistical nature of fatigue**

Fatigue failure of materials has long been recognized as a random phenomenon. Thus, even in carefully controlled experiments, at any selected stress amplitude there is often a larger scatter in the number of cycles to failure as seen in Fig. 20.7. The reason for this is that, although on a macroscopic scale it is convenient to consider a material as being a homogeneous continuum, on a microscopic scale this is far from being the case. A metal, for example, is known to consist of a random distribution of internal defects such as microcracks, dislocations and inclusions contained within a network of grains which have randomly oriented slip planes and grain boundaries. Thus when the same cyclic stress amplitude is applied to nominally identical specimens it is extremely unlikely that each sample will fail after the same number of stress cycles. This is because, on a microscopic scale, the conditions which the fatigue cracks experience as they propagate through each specimen will be quite different.

Thus although fatigue data is usually presented as a single line on an $S$–$N$ curve (Fig. 20.7) it should be realized that this line does not predict the exact fatigue life at a particular stress amplitude: $S$–$N$ curves are usually plotted from a large number of fatigue test results which have been analysed by some type of statistical method. In the simplest case the line drawn represents the 50% probability of failure. This means that at stress $S_1$ in Fig. 20.11, 50% of the samples tested will have failed before cycles $N_1$. In some cases other probability lines will also be drawn. Figure 20.11 shows the 1% and 99% probabilities of failure and this type of graph is sometimes called a $P$–$S$–$N$ diagram.

**Fig. 20.11** Probability fatigue curve

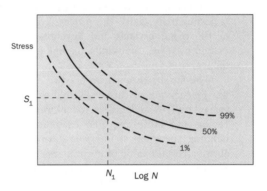

## Use of fatigue data for design

In general engineering, components subjected to cyclical stresses will be designed on an 'infinite' life basis. The design stress range will therefore be related to a choice of material having a fatigue or endurance limit, with a suitable safety factor, in excess of that working stress requirement.

There are a number of factors that influence the magnitude of the fatigue limit which will affect design stresses accordingly and these will be discussed later.

Limited or finite life fatigue relates to endurances principally in the range $10^5$–$10^7$ cycles and it is here that design can benefit from analysis of fatigue crack propagation. The micromechanisms of fracture and the fracture mechanics of crack growth in fatigue are introduced in the next two sections.

Finally low-endurance–high-strain fatigue is a rather specialized area of design involving cyclical plasticity which is beyond the scope of this short chapter.

### Example 20.1

**A mild steel shaft has a pulley mounted at each end and is supported symmetrically in between two bearing housings. When rotating under load the shaft will be subjected to a cyclical bending moment of $\pm 1.2$ kN m. Using a safety factor of 2 and the data in Fig. 20.7, determine a suitable diameter for the shaft for infinite fatigue life.**

From Fig. 20.7 the fatigue limit for infinite life is approximately $\pm190$ MN/m$^2$ and using a factor of 2 the design stress will be $\pm95$ MN/m$^2$.

The bending relationship is

$$\sigma = \frac{My}{I}$$

which for a shaft diameter of $D$ gives

$$\pm95 \times 10^6 = \pm\frac{1.2 \times 10^3 \times 32}{\pi D^3}$$

Hence $D^3 = 128.6 \times 10^{-6}$ m$^3$, or $D = 50.48$ mm.

## 20.4 Micromechanisms of fatigue: initiation and propagation

A great deal of research has been devoted to a study of the mechanism of fatigue, and yet there is still not a complete understanding of the phenomenon. It is not an easy problem to handle theoretically or experimentally, since the process commences within the atomic structure of the metal crystals and develops from the first few cycles of stress, extending over thousands or millions of subsequent cycles to eventual failure.

The fatigue mechanism has two distinct phases, *initiation* of a crack and the *propagation* of this crack to final rupture of the material. One of the earliest (1910) and classic metallurgical studies of initiation was made by Ewing and Humfrey, who examined the polished surface of a rotating bending specimen at intervals during its fatigue life. They observed that above a certain value of cyclic stress (the fatigue limit) some crystals on the surface of the specimen developed bands during cycling. These bands are the result of sliding or shearing of atomic planes within the crystal and are termed *slip bands*. With continued cyclic action these slip bands broaden and intensify to the point where separation occurs within one of the slip bands and a crack is formed as shown in Fig. 20.12(*a*). In the 1950s Forsyth discovered that a process of intrusion and extrusion at the surface could cause a crack to be formed as illustrated in Fig. 20.12(*b*) and (*c*). This crack initially develops along the slip plane of the grain in which it was formed

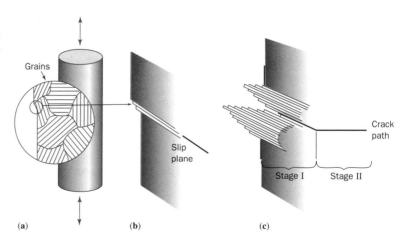

**Fig. 20.12** Stages in the development of a crack along a slip plane

**Fig. 20.13** How fatigue cracks grow (from Ashby and Jones[1])

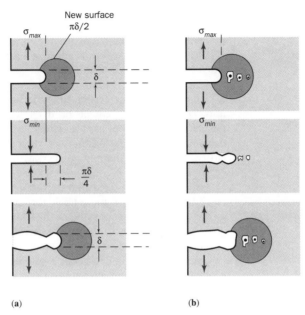

(a)                                    (b)

(Stage I growth) but eventually propagates across other grains. This Stage II growth occurs, not as a consequence of any progressive structural damage, but as a result of the stress concentration effect at the crack tip as it becomes sharp during unloading.

The mechanism of crack growth is shown in Fig. 20.13. In a material which is free from defects, the tensile stress produces a plastic zone which causes the crack tip to stretch open by an amount $\delta$. This creates new surface at the crack tip. During the compressive part of the cycle, the crack is squeezed shut and the new surface folds. This causes the crack to advance by an amount approximately equal to $\delta$. This process is then repeated during each cycle so that the crack growth rate $da/dN$ is approximately equal to $\delta$.[1] This mechanism is illustrated in Fig. 20.13(a).

**Fig. 20.14** Stages of crack growth

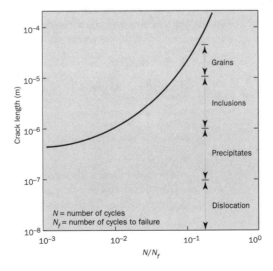

In a real engineering material there will always be microscopic defects. As a result of these, holes will form within the plastic zone. These will link up with one another and with the crack tip. This is illustrated in Fig. 20.13(*b*). The crack will thus advance a little faster than before because it is aided by the presence of the holes.

There have been a number of theories on the initiation and propagation of fatigue cracks. In recent times it has been suggested that fatigue is initiated by the movement of dislocations. A dislocation is a fault or misplacement in the atomic lattice of the metal. Microscopic plastic deformation allows dislocations or vacancies to 'move' through the atomic lattice, and it is thought that coagulation of dislocations forms the beginnings of a crack.

Others have suggested that dislocations are much too small to have any real effect and that it is more likely that cracks develop from intrinsic defects in the material. These may be of the order of 0.5 $\mu$m in size and Fig. 20.14 gives an idea of the scale of this relative to microstructural dimensions. It also indicates the rate at which a crack might propagate through a material. Propagation of a fatigue crack is a complex phenomenon depending on the geometry of the component, the material, the type of stressing, and the environment. Propagation can occupy as much as 90% of the total endurance; hence movement is relatively slow. Some experimental evidence reveals a discontinuous progress of the crack: it is moving for some cycles and stationary for others. Such investigations have also shown that there is a fatigue crack growth threshold below which cracks can exist in a material but will not propagate. This can now be explained in terms of the concept of the stress intensity factor $K$ which was introduced in Chapter 19, and the emergence of fracture mechanics has cast new light on fatigue crack growth phenomena. Fracture mechanics also permits life expectancy to be predicted for a material or structure containing a crack-like defect of known size—something not possible using the $S$–$N$ curve approach since this does not separate out the initiation and propagation phases.

## 20.5 Fracture mechanics for fatigue

Fracture mechanics can only be applicable to fatigue *after* the crack initiation phase to enable crack growth to be predicted. In Chapter 19 it was shown that the state of stress in the vicinity of a crack in an infinite body could be expressed in terms of the stress intensity factor, $K$, where

$$K = Y\sigma\sqrt{(\pi a)} \qquad\qquad [20.4]$$

In cyclic loading, $K$ varies over a stress intensity range $\Delta K$ where

$$\Delta K = K_{max} - K_{min}$$

$$\Delta K = Y(S_{max} - S_{min})\sqrt{(\pi a)} \qquad\qquad [20.5]$$

using $S$ in place of $\sigma$.

The simplest and most widely used expression which relates the range of stress intensity factor to the crack growth rate during cyclic loading is the Paris–Erdogan equation. This takes the form

$$\frac{da}{dN} = C(\Delta K)^m \qquad\qquad [20.6]$$

**Fig. 20.15** Crack growth data for an aluminium alloy

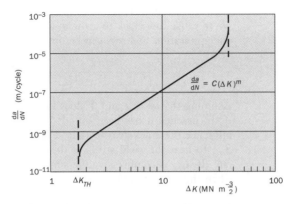

**Table 20.1** Fracture mechanics constants for a range of materials

| Material | $\Delta K_{TH}$ (MN m$^{-3/2}$) | $m$ | $C^*$ ($\times 10^{-11}$) |
|---|---|---|---|
| Mild steel | 3.2–6.6 | 3.3 | 0.24 |
| Structural steel | 2.0–5.0 | 3.85–4.2 | 0.07–0.11 |
| Structural steel in sea water | 1.0–1.5 | 3.3 | 1.6 |
| Aluminium | 1.0–2.0 | 2.9 | 4.56 |
| Aluminium alloy | 1.0–2.0 | 2.6–3.9 | 3–19 |
| Copper | 1.8–2.8 | 3.9 | 0.34 |
| Titanium | 2.0–3.0 | 4.4 | 68.8 |

*Units of $C$ will give $da/dN$ in m/cycle when $\Delta K$ is in MN m$^{-3/2}$.

where $C$ and $m$ are material constants although their values will depend on factors such as the nature of the environment and the level of residual stresses. Typical values of $C$ and $m$ for a range of materials are given in Table 20.1. The way in which the crack growth rate varies during cycling is shown in Fig. 20.15. This illustrates that at the lower end there is a threshold value of stress intensity range $\Delta K_{TH}$ below which the crack will not propagate. It has been found that for most metals $\Delta K_{TH}$ is approximately proportional to the elastic modulus. At the upper end, the crack growth rate tends towards an infinitely large value. As the crack grows, $K_{max}$ increases and failure will occur when this exceeds the fracture toughness of the material $K_{IC}$ or when the remaining ligament of material ahead of the crack tip fails by plastic collapse.

As eqn. [20.6] expresses the rate at which a crack grows under specified cyclic stress conditions, it is possible to use this to predict fatigue lifetimes.

From eqns. [20.5] and [20.2]

$$\Delta K = Y . S_R \sqrt{(\pi a)}$$

where $S_R$ is the range of variation of tensile stress.

Substituting in eqn. [20.6],

$$\frac{da}{dN} = C[Y . S_R \sqrt{(\pi a)}]^m \qquad [20.7]$$

Letting the initial crack size in the material be $a_0$, then the number of cycles $N_f$ to cause this crack to grow to a size $a_f$ at which failure would occur in one application of stress can be obtained by integrating eqn. [20.7]

$$N_f = \frac{2}{C(Y.S_R)^m \pi^{m/2}(2-m)}(a_f^{1-m/2} - a_0^{1-m/2}) \qquad [20.8]$$

assuming that $m \neq 2$ and $Y$ does not vary with $a$.

In practice it has been found[2,3] that the mean stress intensity will also influence the crack growth rate. A large mean stress which raises $S_{max}$ and $S_{min}$ to near the yield stress of the material will increase the crack growth rate. If $S_{min}$ is compressive the simplest analysis is that the crack will close when the stress is compressive and so $S_{min}$ should be taken as zero.

**Example 20.2**

A support bracket is welded to a backing plate as shown in Fig. 20.16. A fluctuating force in the coupling rod causes a stress variation of $\pm 50\,MN/m^2$ at the weld. Using the crack growth data in Fig. 20.15 calculate the maximum size of defect which could be tolerated in the weld.

Fig. 20.16

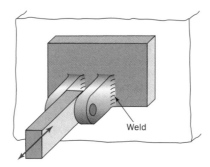

Weld

From Fig. 20.15 the threshold value of the stress intensity factor $\Delta K_{TH}$ is $1.65\,MN\,m^{-3/2}$:

$$\Delta K_{TH} = Y(\Delta S)\sqrt{(\pi a)}$$

In this case, assuming an edge crack in the weld, then $Y = 1.12$ and $\Delta S$ is taken as $50\,MN/m^2$ since it is only the tensile part of the cycle which causes fatigue crack growth:

$$1.65 = 1.12(50)\sqrt{(\pi a_0)}$$

$$a_0 = 0.27\,mm$$

**Example 20.3**

The blades of a turbine rotor are fitted into aluminium alloy discs on the rotor as shown in Fig. 20.17. If during the assembly of the system a 0.1 mm deep scratch is made in the surface of the disc as indicated, (a) calculate how many stress cycles the disc can withstand before fatigue failure occurs and (b) if the rotor is required to undergo at least 2000 cycles in service, by what factor should the rotation speed be increased in a test to verify that any cracks in the disc are too small to grow to a critical size in service? The rotation of the turbine causes a stress of $350\,MN/m^2$ at

Fig. 20.17

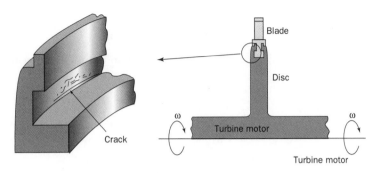

Blade

Disc

Crack

Turbine motor

$\omega$

$\omega$

Turbine motor

**the plane of the scratch and the crack growth data for the disc material is given in Fig. 20.15.**

(a) From Fig. 20.15 the following information can be obtained:

$$\frac{\mathrm{d}a}{\mathrm{d}N} = 4 \times 10^{-11} (\Delta K)^{3.54}$$

and

$$K_{IC} = 35 \, \mathrm{MN \, m}^{-3/2}$$

From $K_{IC}$ it is possible to calculate the size of crack which would cause failure in a single stress application. This will be the value of $a_f$ in eqn. [20.8]:

$$a_f = \frac{1}{\pi} \left( \frac{K_{IC}}{YS} \right)^2 = \frac{1}{\pi} \left( \frac{35}{1.12 \times 350} \right)^2 = 2.54 \, \mathrm{mm}$$

It is evident that as the actual crack depth is only 0.1 mm there is no danger of fast fracture during running of the rotor, even at full speed. However, repeated ON/OFF cycles for the rotor will cause the crack to grow. In this case the number of cycles to cause failure can be predicted from eqn. [20.8]:

$$N_f = \frac{2}{C(YS_R)^m \pi^{m/2}(2-m)} \left( a_f^{1-m/2} - a_0^{1-m/2} \right)$$

$$= \frac{2 \times [(2.54 \times 10^{-3})^{-0.77} - (0.1 \times 10^{-3})^{-0.77}]}{4 \times 10^{-11}(1.12 \times 350)^{3.54} \pi^{1.77}(-1.54)}$$

$$= 3117 \, \mathrm{cycles}$$

(b) The critical crack size for failure in service has been found to be 2.54 mm. A crack of length $a_0$ will grow to this length in 2000 cycles if

$$N_f = \frac{2}{C(Y.S_R)^m \pi^{m/2}(2-m)} \left( a_f^{1-m/2} - a_0^{1-m/2} \right)$$

or

$$2000 = -2.83[(2.54 \times 10^{-3})^{-0.77} - a_0^{-0.77}]$$

from which

$$a_0 = 0.168 \, \text{mm}$$

The stress necessary to cause fracture with a crack of this length is

$$S = \frac{K_{IC}}{Y\sqrt{(\pi a)}} = \frac{35}{1.12\sqrt{(\pi \times 1.68 \times 10^{-4})}} = 1361 \, \text{MN/m}^2$$

Since the stress in a rotating disc is proportional to (angular velocity)$^2$ (see Chapter 14), the required overspeed is a factor $x$ of

$$x = \sqrt{\frac{1361}{350}} = 2$$

### 20.6 Influential factors

#### Mean stress

It is quite common to think of fatigue in terms of the range of cyclic stress; however, the mean stress in the cycle has an important influence on fatigue behaviour. There are two obvious limiting conditions for the mean and range of stress. One is for the mean stress equal to the static strength in tension or compression, whence the range of stress must be zero. The other condition is for zero mean stress and a stress range equal to twice the fatigue limit for fully reversed stress. Between these boundaries there is an infinite number of combinations of mean and range of stress. It is obviously impossible to study the problem experimentally completely; consequently, empirical laws have been developed to represent the variation of mean and range of stress, in terms of static strength values and the fatigue curve for reversed stress (zero mean). This latter condition is the most widely used for obtaining experimental data, principally because of the simplicity of the rotating beam test.

**Fig. 20.18** Diagrams for mean stress against semi-range of stress

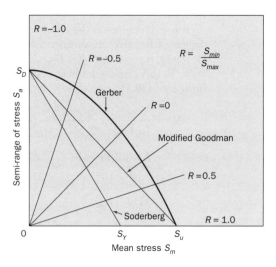

If a diagram is plotted of the semi-range of stress as ordinate and the mean stress as abscissa, known as an $S_a$–$S_m$ diagram, as in Fig. 20.18, then the limiting conditions are $S_m = 0$ and $S_a = S_D$, the fatigue limit for reversed stress, and $S_a = 0$ and $S_m = S_u$, the tensile strength of the material. Between these limits it is required to have a line which represents the locus of all combinations of $S_a$ and $S_m$ which result in the same fatigue endurance. A straight line joining these pairs of co-ordinates represents one empirical law known as the modified Goodman relationship. Algebraically this is given by

$$S_a = S_D\left(1 - \frac{S_m}{S_u}\right) \qquad [20.9]$$

Another relationship is obtained by joining the limiting co-ordinates by a parabola known as the Gerber parabola. This is expressed algebraically in the above symbols as

$$S_a = S_D\left[1 - \left(\frac{S_m}{S_u}\right)^2\right] \qquad [20.10]$$

A more conservative line for design purposes was proposed by Soderberg using a yield stress instead of the tensile strength in the Goodman relationship:

$$S_a = S_D\left(1 - \frac{S_m}{S_Y}\right) \qquad [20.11]$$

The simplest conclusion that can be reached from the foregoing is that an increase in tensile mean stress in the cycle reduces the allowable range of stress for a particular endurance. This applies similarly for direct stress or shear-stress (torsional) fatigue.

Compressive mean stresses appear to cause little or no reduction in stress range, and some materials even shown an increase; consequently the $S_a$–$S_m$ diagram is not symmetrical about the zero mean stress axis.

## Geometrical factors

### 1. Stress concentration

Probably the most serious effect in fatigue is that of stress concentration. It is virtually impossible to design any component in a machine without some discontinuity such as a hole, keyway, or change of section. These features are known as *stress raisers* or sources of stress concentration. This concept was introduced in Chapter 12, and it was explained that under static loading in the elastic range the local peak stress at a notch or discontinuity is raised in magnitude above the nominal stress on a cross-section away from the notch. The theoretical elastic stress concentration factor $K_t$ for a notch is defined as the maximum stress at the notch divided by the average stress on the minimum area of cross-section at the notch.

For ductile metals, static stress concentration does not reduce the strength owing to redistribution of stress when the material at the notch enters the plastic range. However, under fatigue loading the position is very different, since a fatigue limit tends to correspond with the static elastic limit of the material. Consequently, the fatigue limit for the material having the peak stress at the discontinuity would correspond to the static elastic limit, and hence the fatigue limit based on the nominal stress would be reduced by a factor dependent on the elastic stress concentration factor. In practice this is generally not so and the notched fatigue strength is rather better than the 'plain' fatigue strength divided by $K_t$. This has led to what is termed the *fatigue strength reduction factor*, which is defined as

$$K_f = \frac{\text{plain fatigue strength at } N \text{ cycles or fatigue limit}}{\text{notched fatigue strength at } N \text{ cycles or fatigue limit}}$$

and generally $K_t > K_f > 1$ as shown in Fig. 20.19.

**Fig. 20.19** Comparison of plain and notched fatigue curves

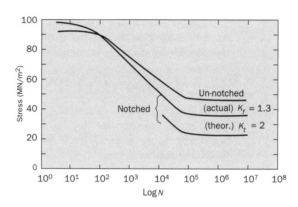

Another way of representing material reaction to stress concentration in fatigue is by means of a *notch sensitivity factor q*, which is defined in terms of the theoretical stress concentration factor and the fatigue strength reduction factor. Thus

$$q = \frac{K_f - 1}{K_t - 1} \qquad [20.12]$$

and for a component or specimen which yields $K_f$ in fatigue equal to $K_t$, the factor $q = 1$ shows maximum sensitivity. Where there is no strength reduction, $K_f = 1$ and $q$ becomes zero showing no sensitivity.

Geometrical discontinuities such as notches, holes, etc., have little or no strength reduction effect at low endurances for *stress cycling* conditions. When the cycles are low, the material fatigue strength is closer to the static strength. Consequently, if the notched static strength is higher than the plain tensile strength, the notched fatigue strength at the upper end of an *S–N* curve may be higher than the plain fatigue strength as shown in Fig. 20.19.

**Fig. 20.20** Size effect

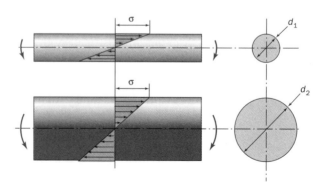

### 2. Size

It has been found that for geometrically similar components a large size has a slightly lower fatigue strength than a small size in cyclic bending or torsion. As this size effect is not found in uniaxial fatigue, it has been primarily attributed to the different gradients of stress and strain as shown in Fig. 20.20.

**Example 20.4**

The stepped shaft shown in Fig. 20.21 is subjected to a steady axial pull of 50 kN and a uniform bending moment *M*. If the yield strength of the rotor material is 300 MN/m² and the fatigue limit in reversed bending is 200 MN/m², calculate the maximum value of *M* to avoid fatigue failure in the rotor. $K_t$ for the fillet radius is 1.55 and the notch sensitivity factor *q*= 0.9.

**Fig. 20.21**

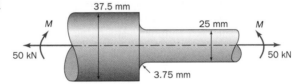

Mean stress,
$$S_m = \frac{F}{A} = \frac{50 \times 10^3 \times 4}{\pi(25)^2} = 101.8 \, \text{MN/m}^2$$

Alternating stress,
$$S_a = \frac{My}{I} = \frac{M \times 12.5 \times 64}{\pi(25)^4}$$

$$= M \times 6.52 \times 10^{-4} \, \text{MN/m}^2$$

Here, $K_f$ is obtained from eqn. [20.12],

$$0.9 = \frac{K_f - 1}{1.55 - 1}$$

from which $K_f = 1.5$.

Applying Soderberg's rule to allow for the mean stress,

$$\sigma_a = \frac{\sigma_\Delta}{K_f}\left(1 - \frac{\sigma_m}{\sigma_y}\right)$$

$$6.52 \times 10^{-4}M = \frac{200}{1.5}\left(1 - \frac{101.8}{300}\right)$$

$$M = 135100 \, \text{N mm}$$

## Environmental effects

### 1. Temperature

Many components are subjected to fatigue conditions while working at temperatures other than ambient. Components in aircraft at high altitude may experience many degrees of frost, while a steam or gas turbine will be running at several hundred degrees Celsius.

Tests on a number of aluminium and steel alloys ranging down to
$-50\,°C$, and lower in some cases, have shown that the fatigue strength is as
good as and often a little better than at normal temperature $(+20\,C)$.

On the other hand, fatigue tests at higher temperatures show little or no
effect up to about $300\,°C$, after which for steels to about $400\,°C$, there is an
increase in fatigue strength to a maximum value, followed by a rapid fall to
values well below that at $+20\,°C$. A further interesting feature is that a
material having a fatigue limit characteristic at ambient temperature will lose
this at high temperatures, and the $S–N$ curve will continue to fall slightly
even at high endurances, and an endurance limit has to be quoted.

Above a certain temperature there is interaction between fatigue and
creep effects (see Chapter 21), and it is found that up to, say, $700\,°C$ for
heat-resistant alloy steels, fatigue is the criterion of fracture, whereas at
higher temperatures, creep becomes the cause of failure. This has led to the
use of combined fatigue–creep diagrams, similar to Fig. 20.18 where the
cyclic stress is plotted against the steady stress which produces failure in a
specified endurance at a particular temperature.

### 2. Corrosion

Fatigue tests are generally conducted in air as a reference condition, but in
practice many components are subjected to cyclic stress in the presence of a
corrosive environment. Corrosion is essentially a process of oxidation, and
under static conditions a protective oxide film is formed which tends to
retard further corrosion attack. In the presence of cyclic stress the situation
is very different, since the partly protective oxide film is ruptured in every
cycle allowing further attack. A rather simplified explanation of the
corrosion fatigue mechanism is that the microstructure at the surface of the
metal is attacked by the corrosive, causing an easier and more rapid initiation
of cracks. The stress concentration at the tips of fissures breaks the oxide
film and the corrosive in the crack acts as a form of electrolyte with the tip of
the crack becoming an anode from which material is removed, thus assisting
the propagation under fatigue action. It has been shown that the separate
effects of corrosion and fatigue when added do not cause as serious a
reduction in strength as the two conditions acting simultaneously.

One of the important aspects of corrosion fatigue is that a metal having a
fatigue limit in air no longer possesses one in the corrosive environment, and
fractures can be obtained at very low stress after hundreds of millions of
cycles.

A particular form of corrosion fatigue may occur in situations which
involve the relative movement of contacting surfaces under the action of an
alternating load. This is known as fretting corrosion. Some materials are
more susceptible to it than others and hence there are preferred
combinations of materials in situations where it is likely to arise. The use
of antifretting compounds, the reduction of surface stresses and surface
hardening have also been found to alleviate the problem.

**Surface finish and
treatments**       Fatigue failures in metals almost always initiate at a free surface so that the
surface condition has a significant effect on fatigue endurance. Immediate
improvement can be effected by polishing a machined surface since this
reduces the mild stress concentration effect of a lathe-turned, milled or
ground surface.

Manufacturing processes also may introduce residual stresses and strain hardening. It is important therefore to appreciate the effect which these operations may have on fatigue endurance.

Surface coating of ferrous metals is often done to improve their wear and corrosion characteristics. In general it is found that such coating does not improve the fatigue strength of the substrate and, depending on the plating conditions, may cause a considerable reduction in fatigue life. Again the process of anodizing aluminium to improve its wear and corrosion resistance also has the effect of reducing its fatigue strength.

The introduction of residual compressive stresses on the surface of a metal has been shown to be a very successful method of improving fatigue endurance. Conversely, the presence of residual tensile stresses on the surface has a very detrimental effect. There are a number of metallurgical methods of introducing compressive residual stresses through hardening of the surface layers of the material. The three main methods are induction (or flame) hardening, carburizing and nitriding. It is interesting that the greatest improvement in fatigue strength is observed when stress concentrations are present.

There are also a number of methods of physically introducing compressive residual stresses at the surface. These include shot peening and skin rolling, both of which have the additional advantage that they remove any stress concentration marks left by machining operations.

## 20.7 Cumulative damage

Although most fatigue tests and some components are subjected to a constant amplitude of cyclic stress during the life to fracture, there are many instances of machine parts and structures which receive a load spectrum, i.e. the load and cyclic stress vary in some way under working conditions. To establish any difference between fatigue under varying and constant amplitude conditions, tests are conducted in which a certain number of

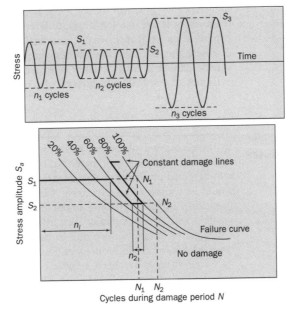

**Fig. 20.22** Diagrammatic representation of cumulative damage

cycles are done at one stress level followed by a number at a higher or lower stress, and this sequence is repeated till failure occurs. It is suggested that damage by fatigue action accumulates and that a certain total damage line is represented by the $S$–$N$ curve. One way of representing this algebraically was proposed by Miner. If $n_1$ cycles are conducted at a stress level $S_1$, at which the fracture endurance would be $N_1$, and if this is followed by $n_2$ cycles out of $N_2$ at a second stress level $S_2$, as in Fig. 20.22 and so on, then for $h$ such blocks

$$\frac{n_1}{N_1} + \frac{n_2}{N_2} + \frac{n_3}{N_3} + \ldots + \frac{n_h}{N_h} = 1$$

or

$$\sum_{q=1}^{h} \frac{n_q}{N_q} = 1 \qquad [20.13]$$

Test results often show the sum of the cycle ratios $(n/N)$ differing widely from the value of unity, generally covering a range from about 0.6 to 2.0 with, in a few cases, extreme values well outside this range.

The value of unity in eqn. [20.13] tends to be an overall average, but when using this approach in design it is prudent to use a lower value (such as 0.5) in order to allow for uncertainties in the hypothesis.

Other models for cumulative damage have been proposed in order to obtain more conservative estimates of the total fatigue life for multi-level sinusoidal stress histories. However, any slight improvements in accuracy are outweighed by the increased complexity of the analysis. The problem is that cumulative damage is dependent on stress history, mean stress, stress concentration, etc., and no simple model can be developed to predict accurately the fatigue life with such a wide range of variables. For example, it is often found that for a two-stress level test, in which one stress is applied for a number of cycles and then run to failure at a second stress, if $S_1 < S_2$, then $\Sigma(n/N) > 1$, and for $S_1 > S_2$, $\Sigma(n/N) < 1$. In addition, the variation from unity is greater for larger differences between $S_1$ and $S_2$. A further complication is that in many practical situations it is difficult to define precisely the nature of the loading pattern. For example, in the design of a vehicle axle it is unreasonable to expect the designer to anticipate all the extremes of stress which different driving styles will cause in the vehicle.

## 20.8 Failure under multi-axial cyclic stresses

In the discussion of fatigue so far, we have considered only the effects of cyclic uniaxial stresses on materials. However, in practice many components are subjected to biaxial or triaxial stress systems for which laboratory testing would be complex and expensive. Consequently, in a similar manner to the prediction of static yielding under complex stress by the use of a simple tensile yield stress (Chapter 12), attempts have been made to establish a criterion for fatigue failure due to complex cyclic stresses, in terms of the uniaxial stress fatigue limit for the material.

Some success has been achieved in the use of the shear-strain energy criterion to predict cyclical failure. Hence from eqn. [12.15] the yield stress in simple tension is replaced by the fatigue limit, $S_D$, in, say, rotating bending for the material, so that

$$(\sigma_1 - \sigma_2)^2 + (\sigma_2 - \sigma_3)^2 + (\sigma_3 - \sigma_1)^2 = 2S_D^2$$

The applicability of this criterion may be related to the static yield situation in that microyielding will occur at the crack tip as propagation proceeds. Fatigue cracks generally initiate at a free surface so that $\sigma_3$ will be zero in the above equation. If mean stresses are present then this must be allowed for by using the above equation as a function of both alternating stresses and mean stresses and using the appropriate values of $S_a$ and $S_m$, from uniaxial data, on the right-hand side. This is illustrated in Example 20.5.

More recent research has shown that the above criterion has a rather limited range of applicability and that two strain parameters are required instead of the single equivalent uniaxial cyclic stress. The parameters proposed are the largest strain circle created in a fatigue cycle and the position of that strain circle in strain space. These are represented as $\frac{1}{2}(\varepsilon_1 - \varepsilon_3)$ and $\frac{1}{2}(\varepsilon_1 + \varepsilon_3)$ respectively from Mohr's strain circles. The intermediate strain $\varepsilon_2$ has been shown to control the direction in which the crack, once initiated, will grow.

**Example 20.5**

A 25 mm diameter geared shaft, Fig. 20.23, is subjected to a fully reversed bending moment of $\pm 1094$ kN m as shown. The gear is a shrink fit onto the shaft and sets up radial and tangential stresses of 350 MN/m² on the surface of the shaft. If during operation the gear wheel is subjected to a fluctuating torque of $\pm 900$ N m estimate the fatigue life of the shaft. The shaft material has a yield stress of 925 MN/m² and its bending fatigue life may be predicted from

$$N_f = 5.2 \times 10^{56} \left( \frac{1}{S_f} \right)^{16.5}$$

where $S_f$ is the stress amplitude in MN/m².

**Fig. 20.23**

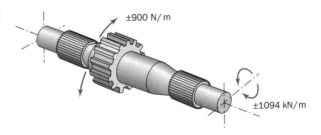

±900 N/m

±1094 kN/m

The bending stress is

$$\sigma_x = \frac{My}{I} = \frac{\pm 1094 \times 10^3 \times 12.5 \times 64}{\pi \times (25)^4} = \pm 713 \,\text{MN/m}^2$$

The shear stress is

$$\tau_{xy} = \frac{Tr}{J} = \frac{\pm 900 \times 10^3 \times 12.5 \times 32}{\pi \times (25)^4} = \pm 293 \,\text{MN/m}^2$$

**Fig. 20.24**

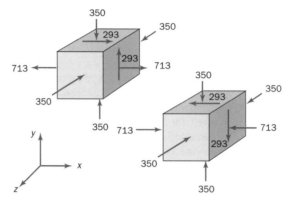

Figure 20.24 shows the stress systems and from Mohr's circle the principal stresses may be determined for the maximum and minimum limits of cyclic stress:

| Cyclic maximum (MN/m$^2$) | Cyclic minimum (MN/m$^2$) |
| --- | --- |
| $\sigma_1 = 800$ | $\sigma_1 = -870$ |
| $\sigma_2 = -420$ | $\sigma_2 = -190$ |
| $\sigma_3 = -350$ | $\sigma_3 = -350$ |

| Amplitudes (MN/m$^2$) | Mean (MN/m$^2$) |
| --- | --- |
| $\sigma_{1a} = 835$ | $\sigma_{1m} = -35$ |
| $\sigma_{2a} = 115$ | $\sigma_{2m} = -305$ |
| $\sigma_{3a} = 0$ | $\sigma_{3m} = -350$ |

Using the shear-strain energy criterion,

$$2S_{Da}^2 = (\sigma_{1a} - \sigma_{2a})^2 + (\sigma_{2a} - \sigma_{3a})^2 + (\sigma_{3a} - \sigma_{1a})^2$$

$$= (835 - 115)^2 + (115 - 0)^2 + (0 - 835)^2$$

$$S_{Da} = 784 \text{MN/m}^2$$

Similarly, $S_{Dm} = 295 \text{ MN/m}^2$

Using the Soderberg rule,

$$S_{Da} = S_f \left( 1 - \frac{S_{Dm}}{\sigma_Y} \right)$$

$$S_f = \frac{784}{1 - 295/925} = 1151 \text{ MN/m}^2$$

From the equation describing the fatigue life of shaft material

$$N_f = 5.2 \times 10^{56} \left( \frac{1}{1151} \right)^{16.5} = 1.61 \times 10^6 \text{ cycles}$$

## 20.9 Fatigue of plastics and composites

Reinforced and unreinforced plastics are susceptible to brittle fatigue failures in much the same way as metals, and the fracture mechanics approach has been used to predict crack growth rates. In addition, however, the high damping and low thermal conductivity of plastics means that under cyclic stressing there is also the possibility of short-term thermal-softening failures if precautions are not taken to dissipate the heat generated. Therefore, although in metals the fatigue strength is relatively unaffected by frequencies in the range 3–100 Hz, with thermoplastics the cyclic frequency is important since it has a pronounced effect on the temperature rise in the material.

This is illustrated in Fig. 20.25. At a relatively low cyclic frequency, $f_1$, the material will fail by thermal softening if the stress amplitude is greater than $\sigma_A$. At cyclic stresses below $\sigma_A$, the material fails by a normal crack initiation and propagation mechanism similar to that observed in metals. Thus at a cyclic frequency of $f_1$, the fatigue curve which would be seen is ABCFG.

At a higher cyclic frequency, $f_2$, a thermoplastic will experience thermal softening at lower stress amplitudes – down to $\sigma_B$ in Fig. 20.25. Below this stress level, conventional fatigue failures are once again observed. Hence, at this higher cyclic frequency, the fatigue behaviour which is observed is described by DEFG.

**Fig. 20.25** Fatigue failure behaviour of thermoplastics

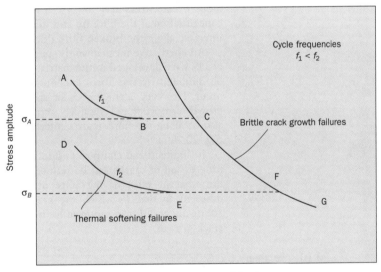

With non-reinforced moulded plastics, fatigue cracks usually initiate within the bulk of the material because the moulding operation tends to produce a protective skin which inhibits crack growth from the surface. However, plastics are not immune to the effect of stress concentrations and many fatigue cracks in moulded articles are initiated at holes, sharp corners, etc.

Fibre-reinforced plastics (composites) can have a high resistance to fatigue crack growth. However, the properties of many composites are

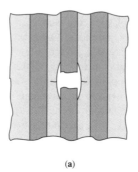
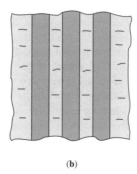
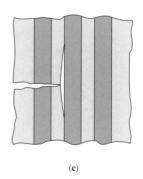

(a)                              (b)                              (c)

**Fig. 20.26**   Fatigue failure mechanisms in composites

anisotropic, i.e. direction dependent, and fatigue crack resistance may be low in certain directions. There are many types of fibre-reinforced composites, e.g. unidirectional fibres, bidirectional fibres, random short fibres, thermoplastic matrix, thermosetting matrix, etc. It would be inappropriate in this book to cover in detail the many types of fatigue behaviour which can be observed in the various permutations of matrix and fibre type. However, the following observations will give the reader a feel for the general type of behaviour to be expected. Those requiring additional information should refer to specialist texts such as that by Hertzberg[4].

In uniaxial fibre composites the material may fail by a number of mechanisms as illustrated in Fig. 20.26. In the left-hand diagram, the high, local, cyclic stress has broken a fibre. This causes a high shear-stress concentration at the fibre tip and this will lead to failure of the fibre–matrix interface along the broken fibre ('debonding').

An alternative mechanism is shown in Fig. 20.26(b). In this case, fatigue cracks have developed in the matrix. Initially the growth of these cracks will be inhibited by the fibres and in some cases no further propagation will occur. However, at higher cyclic stress levels, a fibre will break and crack growth can occur progressively across the section. Such crack growth can also be along the fibre–matrix interface in neighbouring fibres as shown in Fig. 20.26(c).

In bidirectional composites and general multi-ply laminates the rate of progression of cracks can be reduced owing to the constraint provided by the adjacent plies having fibre orientations along a direction which is different to that in which the crack is propagating. Thus the strategic arrangement of fibres can produce materials with a high resistance to fatigue crack growth.

### 20.10   Summary

It has been shown that the possibility of premature failure of a material by a fatigue mechanism is an extremely important design consideration in situations where fluctuating stresses are either applied directly or transmitted to the material. The majority of all failures which occur in practice can be attributed to fatigue and the greatest percentage of these are caused by bad design, usually ill-considered positioning and/or geometry of stress raisers such as holes, abrupt changes in section, keyways, etc. Other factors which affect fatigue endurance are the level of the mean stress, surface condition and nature of the environment.

Although the crack growth mechanism during fatigue is not completely understood it is known that when a component is subjected to a variety of stress levels the damage incurred is cumulative. It has also been found that fracture mechanics is a useful tool in predicting fatigue life provided crack growth data for the material is available.

**References**

1. Ashby, M. F. and Jones, D. R. H. (1980) *Engineering Materials*, Pergamon Press, Oxford.
2. Hellan, K. (1984) *Introduction to Fracture Mechanics*, McGraw-Hill, London.
3. Smith, R. N. L. (1991) *BASIC Fracture Mechanics*, Butterworth-Heinemann, Oxford.
4. Hertzberg, R. W. (1989) *Deformation and Fracture Mechanics of Engineering Materials*, 3rd edition, John Wiley, New York.

**Bibliography**

Duggan, T. V. and Byrne, J. (1977) *Fatigue as a Design Criterion*, Macmillan, London.

Forrest, P. G. (1962) *Fatigue of Metals*, Pergamon Press, Oxford.

Klesnil, M. and Likas, P. (1989) *Fatigue of Metallic Materials*, Elsevier, Amsterdam.

Madayag, A. F. (1969) *Metal Fatigue: Theory and Design*, John Wiley, New York.

Miller, K. J. (1991) *Metal Fatigue – Past, Current and Future*, 27th John Player Lecture. Preprint 3, Institution of Mechanical Engineers, London.

Osgood, C. G. (1982) *Fatigue Design*, Pergamon Press, Oxford.

Pook, L. P. (1983) *The Role of Crack Growth in Metal Fatigue*, The Metal Society, London.

**Problems**
20.1 A switching device consists of a rectangular cross-section metal cantilever 200 mm in length and 30 mm in width. The required operating displacement at the free end is $\pm 2.7$ mm and the service life is to be 100 000 cycles. To allow for scatter in life performance a factor of 5 is employed on endurance. Using the fatigue curves given in Fig. 20.7 determine the required thickness of the cantilever if made in (*a*) mild steel, (*b*) aluminium alloy. $E_{steel} = 208$ GN/m$^2$, $E_{aluminium} = 79$ GN/m$^2$.

20.2 (*a*) A fatigue fracture produced by cyclic uniaxial tension stresses exhibits circular striations which have their centre at the point of crack initiation (usually on the surface). Explain why the striations have this shape. (*b*) If a shaft is subjected to cyclic torsional stresses, on what planes would you expect the fatigue cracks to grow?

20.3 A pressure vessel support bracket is to be designed so that it can withstand a tensile loading cycle of 0–500 MN/m$^2$ once every day for 25 years. Which of the following steels would have the greater tolerance to intrinsic defects in this application: (i) a maraging steel ($K_{Ic} = 82$ MN/m$^{-\frac{3}{2}}$, $C = 0.15 \times 10^{-11}$, $m = 4.1$), or (ii) a medium-strength steel ($K_{Ic} = 50$ MN/m$^{-\frac{3}{2}}$, $C = 0.24 \times 10^{-11}$, $m = 3.3$)? For the loading situation a geometry factor of 1.12 may be assumed.

20.4 A series of crack growth tests on a moulding grade of polymethyl methacrylate gave the following results:

| $\mathrm{d}a/\mathrm{d}N$ (m/cycle) | $2.25 \times 10^{-7}$ | $4 \times 10^{-7}$ | $6.2 \times 10^{-7}$ | $11 \times 10^{-7}$ | $17 \times 10^{-7}$ | $29 \times 10^{-7}$ |
|---|---|---|---|---|---|---|
| $\Delta K$ (MN m$^{-3/2}$) | 0.42 | 0.53 | 0.63 | 0.79 | 0.94 | 1.17 |

If the material has a critical stress intensity factor of 1.8 MN/m$^{-3/2}$ and it is known that the moulding process produces defects 40 $\mu$m long, estimate the maximum repeated tensile stress which could be applied to this material for at least 10$^6$ cycles without causing fatigue failure.

20.5 As part of the mechanism of a machine a cam is used to cause a metal beam to oscillate, as shown in Fig. 20.27. If the design of the cam is such that the beam deflection varies between a maximum of 3 mm and a minimum of 1 mm relative to its undeflected position, calculate a suitable beam depth to avoid fatigue failure in the beam material, using the Gerber, modified Goodman and Soderberg methods to allow for the effect of mean stress. A fatigue strength reduction factor of 1.8 should be assumed. The tensile and yield strengths of the beam material are 350 MN/m$^2$ and 200 MN/m$^2$ respectively, and its fatigue strength in fully reversed cycling is 100 MN/m$^2$. Young's modulus for the steel is 207 GN/m$^2$.

**Fig. 20.27**

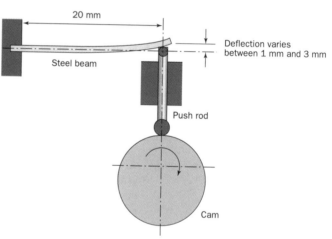

20 mm

Deflection varies between 1 mm and 3 mm

Steel beam

Push rod

Cam

20.6 A shaft of circular cross-section is subjected to a steady bending moment of 1500 Nm and simultaneously to an alternating bending moment of 1000 Nm in the same place (so that the total moment fluctuates between 2500 Nm and 500 Nm). Calculate the necessary diameter of the shaft if the factor of safety is to be 2.5. The yield stress of the material is 210 MN/m$^2$ and the fatigue limit in reversed bending is 170 MN/m$^2$. Calculate also the diameter of the shaft if stress concentrations are to be allowed for with a fatigue strength reduction factor of 2. Assume that the Soderberg rule applies.

20.7 A connecting-rod of circular cross-section with a diameter of 50 mm is subjected to an eccentric longitudinal load at a distance of 10 mm from the centre of the cross-section. The load varies from a value of $-F/2$ to $F$. The material used has a tensile strength of 420 MN/m$^2$ and a fatigue limit in fully reversed loading of 175 MN/m$^2$. Determine the limiting value of $F$ to avoid fatigue failure if the modified Goodman relation applies.

**Fig. 20.28**

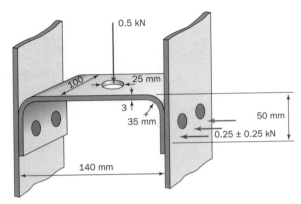

20.8 Part of the structure of an aircraft is shown in Fig. 20.28 The central hole in the inverted U-section supports a vertical force of 0.5 kN. During flight the upright section is subjected to a cyclical force which varies from 0 to 0.5 kN. If all the parts are made from an aluminium alloy with a yield strength of 392 MN/m² and fatigue strength of 270 MN/m², estimate where you expect fatigue failure to occur. Use the Soderberg rule to allow for the effect of mean stresses. The stress concentration factor at the curved portion of the U-channel may be taken as 1.85 and $K_t$ at the central hole may be obtained from Chapter 12.

20.9 A series of tensile fatigue tests on stainless steel strips containing a central through hole gave the following values for the fatigue endurance of the steel. If the steel strips were 100 mm wide, comment on the notch sensitivity of the steel.

| Hole diameter (mm) | No hole | 5 | 10 | 20 | 25 |
|---|---|---|---|---|---|
| Fatigue endurance (MN/m²) | 600 | 250 | 270 | 320 | 370 |

20.10 The fatigue endurances from the $S–N$ curve for a certain steel are:

| Stress (MN/m²) | Fatigue endurance (cycles) |
|---|---|
| 350 | 2 000 000 |
| 380 | 500 000 |
| 410 | 125 000 |

If a component manufactured from this steel is subjected to 600 000 cycles at 350 MN/m² and 150 000 cycles at 380 MN/m², how many cycles can the material be expected to withstand at 410 MN/m² before fatigue failure occurs, assuming that Miner's cumulative damage theory applies?

20.11 The analysis of the cyclic stresses on part of the landing gear of an aircraft shows that during each flight it is subjected to the following stress history:

100 000 cycles at ±50 MN/m²
10 000 cycles at ±100 MN/m²

**Fig. 20.29**

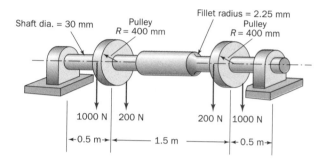

$$500 \text{ cycles at } \pm 150 \, \text{MN/m}^2$$
$$200 \text{ cycles at } \pm 180 \, \text{MN/m}^2$$
$$100 \text{ cycles at } \pm 185 \, \text{MN/m}^2$$
$$50 \text{ cycles at } \pm 200 \, \text{MN/m}^2$$
$$10 \text{ cycles at } \pm 210 \, \text{MN/m}^2$$

If the $S$–$N$ curve, for the part material is given by $S = 500 \, N^{-0.064}$, where $S$ is in $\text{MN/m}^2$, estimate how many flights this component can withstand before fatigue failure occurs.

20.12   During service a steel cylinder, 320 mm diameter, is to be subjected to an internal pressure which varies from 0 to $p$. If the wall thickness of the cylinder is 8 mm estimate the maximum permissible value of $p$ to avoid fatigue failure in the cylinder. The tensile strength of the steel is $440 \, \text{MN/m}^2$ and the fatigue endurance limit in fully reversed cycling is $210 \, \text{MN/m}^2$. A fatigue strength reduction factor of 1.8 may be assumed and the modified Goodman relationship should be used for the mean stress effect.

20.13   Part of a pulley power transmission system is shown in Fig. 20.29. Assuming a notch sensitivity factor of 1, calculate the required endurance limit in the shaft steel in order to avoid fatigue failure. The stress concentration factors for the shaft should be obtained from Chapter 12. The shaft bearings may be regarded as simple supports and the pulleys each provide additional load of 300 N.

CHAPTER
# 21  Creep and Viscoelasticity

For the majority of engineering designs the variation of the ambient temperature is not great and the stiffness and strength of the metal may be regarded as a constant. However, there are also engineering applications which occur at high temperature in fields such as steam plant, gas turbines, nuclear and chemical processes, kinetic heating of supersonic aircraft,[1] etc.

It was mentioned briefly in Chapter 18 that in general the effect of temperatures of up to several hundred degrees Celsius on metals is to lower their yield and tensile strengths by very considerable amounts compared with ambient conditions. Another factor which did not arise during the development of stress–strain solutions of engineering problems in earlier chapters was the effect of the length of time under which a component or structure was subjected to stress. It was assumed that the application of external loading would develop particular values of stress and strain and these would remain constant until the applied loading was removed. This is certainly so for the bulk of engineering alloys in the elastic range at room temperature. However, with increasing temperature it is possible for a material to have increasing strain with time even at *constant* applied load. The time dependence of strain is termed *creep*. Another complementary time-dependent response is termed *stress relaxation* which occurs when the strain or deformation of a component or structure is kept constant for a time period during which a reducing applied load (stress) is required to maintain the strain. These are most important properties in the design of any high-temperature component.

Nowadays plastics, both reinforced and unreinforced, play a major role as engineering materials. A brief reference was made in Chapter 3 to their stress–strain–time behaviour and this is known as *viscoelasticity*, being the interactive results of Newtonian viscosity and Hookean elasticity. Unreinforced thermosplastics are particularly susceptible to creep and room temperature is 'high' enough for plastics to exhibit the phenomenon, the actual rate of creep being a function of the imposed stress.

The first part of this chapter concentrates on the problem of creep and stress relaxation as a time-dependent high-temperature phenomenon affecting metals. The second part of the chapter deals with the basic considerations of viscoelastic data and creep design for plastics.

## 21.1 Stress–strain–time–temperature relationships

Creep manifests itself in metals at temperatures above about $0.3T_m$, where $T_m$ is the absolute melting temperature, and around $0.5T_m$ creep strain becomes considerable. Thus Andrade[2] commenced studies of creep behaviour in 1910 using lead, since this metal exhibits creep at room temperature. The majority of creep experiments are carried out under uniaxial constant loading conditions for a particular chosen test temperature which must be very accurately controlled. Measurements of extension are made at frequent intervals of time until the specimen fractures or the experiment is stopped after a sufficiently lengthy period. Typical curves of creep strain against time are plotted in Fig. 21.1 for various constant load (nominal stress) levels at a constant specimen temperature.

**Fig. 21.1**  Typical creep curves for
various stresses at constant temperature

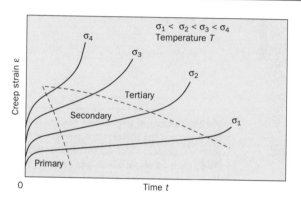

There are four principal aspects of each of the curves shown in Fig. 21.1 as follows:

(a) *initial strain*, which is elastic but may extend marginally into the plastic range due to the first application of load;

(b) *primary stage*, a period of decreasing creep rate during which strain hardening is occurring more rapidly than softening due to the high temperature;

(c) *secondary stage*, in which the creep rate is virtually constant through equilibrium between strain hardening and thermal softening;

(d) *tertiary stage*, an increasing rate of strain, due to microstructural instability from prolonged high temperature and to gradual increase in stress level and stress concentration at cracks in the grain boundaries, which leads to complete fracture of the specimen.

It is important to note that elongation to fracture in creep, even for a ductile metal, is only a fraction of that obtained for continuous loading to fracture at high temperature. The very small gradient of the secondary stage at low stresses suggests that there might be a limiting creep stress, below which $d\varepsilon/dt = 0$, having similar significance as the fatigue limit under cyclic stress. However, it has been shown that there is no reliable criterion of this form for creep, and a 'limiting creep stress' is based on a permissible creep strain after a given time.

The family of curves shown in Fig. 21.1 is based on stress as the parameter and temperature constant; however, a similar family of curves would be obtained for a particular constant stress with temperature as the parameter. Thus a complete picture of the creep behaviour of a metal necessitates the construction of several families of creep curves for various stresses and temperatures.

## 21.2  Empirical representations of creep behaviour

Typical of the standards of creep strength required for metals are the stresses to give minimum creep rates of 1% strain in 10 000 hours, or 1% strain in 100 000 hours.

A 10 000 hour test occupies approximately 1 year, and it is therefore evident that, although a few creep tests may be conducted over periods of this length or longer, it is a very slow and inconvenient process for obtaining a range of data at different stresses and temperatures. As a result, methods have been sought whereby long-life data can be extrapolated from short-term tests.

Creep strain $\varepsilon_c$, which is a function of stress, $\sigma$, time, $t$, and temperature, $T$, can be represented by three functions as follows:

$$\varepsilon_c = f_1(\sigma).f_2(t).f_3(T) \qquad [21.1]$$

(a) *Stress function*: the most commonly used functions are

$$f_1(\sigma) = A_1\sigma^{\eta} \quad f_1(\sigma) = A_2 \sin h\left(\frac{\sigma}{\sigma_0}\right)$$

$$f_1(\sigma) = A_3 \exp\left(\frac{\sigma}{\sigma_0'}\right) \qquad [21.2]$$

where $A_1$, $A_2$, $A_3$ are constants and $\sigma_0$ and $\sigma_0'$ are reference stresses.

(b) *Time function*: this is usually expressed as a polynomial, as first suggested by Andrade[2], one reasonably applicable form being

$$\varepsilon_c = \alpha t^{1/3} + \beta t + \gamma t^3 \qquad [21.3]$$

in which $\alpha$, $\beta$ and $\gamma$ are material constants, but are functions of stress and temperature, relating to the primary, secondary and tertiary stages respectively.

(c) *Temperature function*: the most generally used temperature function is

$$f_3(T) = \exp\left(-\frac{\Delta H}{RT}\right) \qquad [21.4]$$

where $\Delta H$ is the activation energy, $R$ is the universal gas constant and $T$ is the absolute temperature. This type of expression is fundamental to all rate processes.

When we come to designing for creep in components, interest centres principally on the secondary stage, where at low stresses the creep rate is constant for very long times producing the major contribution to the total creep strain to fracture. Consequently, in eqn. [21.3] the tertiary term is neglected and the primary is replaced by a constant strain $\varepsilon_0$, being the intercept of the extrapolated secondary stage back on to the strain axis (Fig. 21.2). Thus

$$\varepsilon_c = \varepsilon_0 + \left(\frac{d\varepsilon}{dt}\right)t \qquad [21.5]$$

**Fig. 21.2** Simplified creep curve

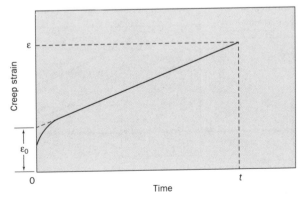

**Fig. 21.3** Curves of creep strain
against log time, extrapolated to a
specified strain

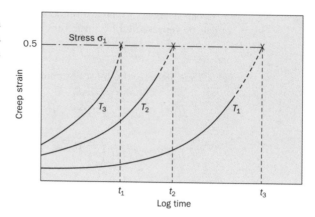

**Fig. 21.3** Curves of creep strain
against log time, extrapolated to a
specified strain

where $d\varepsilon/dt = \dot{\varepsilon}_c$ is the secondary-stage creep rate. This minimum creep
rate has been experimentally related to stress by the empirical expression,

$$\dot{\varepsilon}_c = B\sigma^n \qquad [21.6]$$

where $B$ and $n$ are material constants. The dependence on temperature can
then be included by writing

$$\dot{\varepsilon}_c = B\sigma^n \exp\left(-\frac{\Delta H}{RT}\right) \qquad [21.7]$$

A log–log plot of $\dot{\varepsilon}_c$ against $\sigma$ yields a straight line. However, extrapolation of
the data, particularly at high stress, can be unreliable owing to the
dependence of $n$ and $\Delta H$ on the particular stress/temperature regime.
Alternatively, combining eqns. [21.5] and [21.6], we can express the time to
reach a specified value of total creep strain in the secondary stage as follows:

$$t = \frac{\varepsilon_c - \varepsilon_0}{B\sigma^n} \qquad [21.8]$$

However, this approach can be uncertain in that tertiary creep may
commence before the predicted value of secondary creep is achieved.

**Fig. 21.4** Curves of temperature
against log time, extrapolated to
specified time

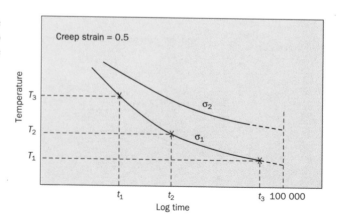

A safer method of predicting the stress and temperature permissible for a specified creep strain of say 0.5% in 100 000 hours is as follows. Creep tests are carried out at various stresses and temperatures to obtain a significant part of the secondary-stage creep in each test. Graphs of creep strain against log time for several temperatures at a particular stress are then extrapolated to the limiting creep strain of say 0.5% as in Fig. 21.3. Next we plot graphs of temperature against log time for various stresses at the strain limit of 0.5% as shown in Fig. 21.4. These curves are then extrapolated to a required life of say 100 000 hours. Although several extrapolations are involved in this method, they tend to be rather more reliable than a simple extrapolation of a creep–time curve.

## 21.3 Creep-rupture testing

As there is as yet no real substitute for a few long-term tests to ensure reliable creep knowledge, it is useful to have a quick sorting test to enable the best material from a group to be selected for long term tests.

The creep-rupture test is widely used for the above purpose and also as a guide to the rupture strength at very long endurances. The principle is to apply various values of stress, in successive tests at constant temperature, of a magnitude sufficient to cause rupture in times from a few minutes to several hundred hours. Plotting log stress against log time as in Fig. 21.5 yields a family of straight lines with temperature as parameter. These tend to extrapolate back to the respective hot tensile strengths. Extrapolation forward to longer times is possible, but care has to be exercised in that oxidation owing to the high temperature can cause a marked increase in the slope, and hence reduction in stress for a required life.

**Fig. 21.5** Log rupture/log time relation for chromium–molybdenum–silicon steel (adapted from Lessels[3]; by courtesy of John Wiley & Sons, Inc.)

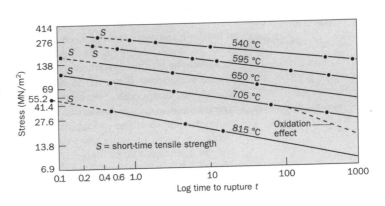

One of the physical approaches to the problem of creep suggests that viscous flow in fluids is analogous to secondary creep in metals and hence a rate-process theory is applicable, relating creep rate and temperature, using eqn. (21.4)

$$\dot{\varepsilon} = \frac{d\varepsilon}{dt} = A \exp(-\Delta H/RT) \qquad [21.9]$$

where $A$ is a material constant.

Larsen and Miller[4] have analysed the above relationship and put it in the form

$$\frac{\Delta H}{R} = T(\log_e A + \log_e t - \log_e \varepsilon) \qquad [21.10]$$

or

$$\left(\frac{\Delta H}{R}\right)_\sigma = T(a + \log_e t) \qquad [21.11]$$

for a given value of strain. Here $a$ is a constant for a given strain and $(\Delta H/R)_\sigma$ is a function of the stress level $\sigma$. The right-hand side of eqn. [21.11] is known as the Larsen–Miller[4] parameter, $T(a + \log_e t)$, and plotting $\log_e \sigma$ against the parameter often yields a family of straight lines for different creep strains, known as master creep curves, which correlate well over a wide range of times, temperatures and different metals. From the above curves a general relationship may be written in the form

$$\log_e \sigma = C_1 + C_2 T(a + \log_e t) \qquad [21.12]$$

where $C_1$ and $C_2$ are constants. Combining the strain-rate/stress equation

$$\dot{\varepsilon} = B\sigma^n$$

with that above for stress, temperature and time gives

$$\dot{\varepsilon} = C\sigma^{n'} \exp(-\alpha T) \qquad [21.13]$$

where $C$, $\alpha$ and $n'$ are material creep constants.

## 21.4 Tension creep test equipment

The most common form of creep test is conducted in simple tension, and since constant loading on the specimen is required over very long periods a dead-weight loading system is usually employed. A typical arrangement is where a specimen is housed in a furnace. In previous sections it has been stated that temperature variation has a great influence on the minimum creep rate; it is therefore essential to have uniform and constant temperature along the length of the specimen. The relevant British Standard specifies a maximum variation along the gauge length of 2 °C and a variation of mean temperature of not more than ±1 °C up to 600 °C and ±2 °C from 600 °C to 1000 °C. If a resistance furnace is used, it is usual to have three separately controlled resistance elements to compensate for non-uniform flow of heat through the furnace. Thermocouples and temperature indicators are used for measurement and control.

## 21.5 Creep during pure bending of a beam

In order to demonstrate the application of eqn. [21.6], the case of a beam subjected to pure bending at high temperature will be considered.

The assumptions that will be made in this problem are as follows:

(a) plane sections remain plane under creep deformation;
(b) longitudinal fibres experience only simple axial stress;
(c) creep behaviour is the same in tension as in compression.

If the radius of curvature of the neutral axis is $R$, then the strain at a distance $y$ from the axis is

$$\varepsilon = \frac{y}{R}$$

and for creep in the secondary stage,

$$\frac{d\varepsilon}{dt} = B\sigma^n$$

Therefore

$$\frac{d(y/R)}{dt} = B\sigma^n$$

or

$$\frac{y}{R} = B\sigma^n t$$

and

$$\sigma = \left(\frac{y}{RBt}\right)^{1/n} \tag{21.14}$$

If the bar is of width $b$ and depth $d$, equilibrium of the external and internal moments is given by

$$M = 2\int_0^{d/2} \sigma b y \, dy$$

Substituting for $\sigma$, using eqn. [21.14],

$$M = 2\int_0^{d/2} \left(\frac{y}{RBt}\right)^{1/n} by \, dy = \frac{2b}{(RBt)^{1/n}} \int_0^{d/2} y^{1+(1/n)} \, dy$$

Integration gives

$$M = \frac{2b}{(RBt)^{1/n}} \frac{n}{2n+1} \left(\frac{d}{2}\right)^{2+(1/n)}$$

Substituting for $(RBt)^{1/n}$ from eqn. [21.14] gives

$$M = \frac{2b\sigma}{y^{1/n}} \frac{n}{2n+1} \left(\frac{d}{2}\right)^{2+(1/n)}$$

Inserting the second moment of area $I = \frac{1}{12}bd^3$ and rearranging, we have the bending stress at distance $y$ from the neutral axis as

$$\sigma = \frac{My}{I} \frac{(2n+1)}{3n} \left(\frac{2y}{d}\right)^{(1/n)-1} \tag{21.15}$$

It has been put in this form to show the difference from the simple linear-elastic distribution of stress, $My/I$.

The stress distribution across the section is shown in Fig. 21.6 for $n = 1$, which corresponds to no creep and simple elastic conditions, and $n = 10$,

**Fig. 21.6**  Distribution of bending stress for the cases of zero creep and secondary-stage creep

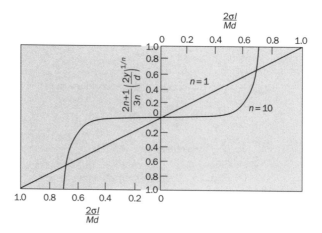

where secondary-stage creep is occurring. It is seen that the effect of creep is to relax the outer fibres' stress, but to increase the core bending stress. This is because minimum creep rate must be proportional to the distance from the neutral axis in order to satisfy eqn. [21.14], and this automatically adjusts the stresses to the distribution shown.

## 21.6  Creep under multi-axial stresses

Under the above heading we find typical engineering components such as thin-walled tubes subjected to internal pressure and combinations with bending or torsion, thick-walled cylinders under internal pressure, rotating discs and plates subjected to bending. The number of experimental investigations has been limited owing to the intricacies of high-temperature testing and complex loading. However, there have been sufficiently comprehensive studies to verify the analytical approach that is generally adopted.

In order to relate uniaxial creep data to a biaxial problem, some of the laws of plasticity are invoked, namely that (*a*) the principal strains and stresses are coincident in direction, (*b*) plastic deformation occurs at constant volume, (*c*) the maximum shear stresses and shear strains are proportional.

For constant volume the sum of the three principal strains is zero; therefore,

$$\varepsilon_1 + \varepsilon_2 + \varepsilon_3 = 0 \qquad [21.16]$$

Expressing maximum shear stress and strain in terms of the difference in principal stress and strains,

$$\frac{\varepsilon_1 - \varepsilon_2}{\sigma_1 - \sigma_2} = \frac{\varepsilon_2 - \varepsilon_3}{\sigma_2 - \sigma_3} = \frac{\varepsilon_3 - \varepsilon_1}{\sigma_3 - \sigma_1} = \beta \qquad [21.17]$$

Rearranging the above equations to give the individual principal strains in terms of the principal stresses,

$$\left.\begin{aligned} \varepsilon_1 &= \frac{2\beta}{3}[\sigma_1 - \tfrac{1}{2}(\sigma_2 + \sigma_3)] \\ \varepsilon_2 &= \frac{2\beta}{3}[\sigma_2 - \tfrac{1}{2}(\sigma_3 + \sigma_1)] \\ \varepsilon_3 &= \frac{2\beta}{3}[\sigma_3 - \tfrac{1}{2}(\sigma_1 + \sigma_2)] \end{aligned}\right\} \qquad [21.18]$$

These may be expressed as a constant creep rate by writing

$$\left.\begin{aligned} \dot{\varepsilon}_1 &= \alpha[\sigma_1 - \tfrac{1}{2}(\sigma_2 + \sigma_3)] \\ \dot{\varepsilon}_2 &= \alpha[\sigma_2 - \tfrac{1}{2}(\sigma_3 + \sigma_1)] \\ \dot{\varepsilon}_3 &= \alpha[\sigma_3 - \tfrac{1}{2}(\sigma_1 + \sigma_2)] \end{aligned}\right\} \qquad [21.19]$$

where $\alpha$ is a function relating the three principal stresses to the simple uniaxial stress creep condition.

Using the von Mises yield criterion to obtain the equivalent uniaxial stress $\sigma_e$, gives

$$\sigma_e = \frac{1}{\sqrt{2}}\sqrt{[(\sigma_1 - \sigma_2)^2 + (\sigma_2 - \sigma_3)^2 + (\sigma_3 - \sigma_1)^2]} \qquad [21.20]$$

From the simple secondary-stage creep law,

$$\dot{\varepsilon} = B\sigma_e^n$$

and for simple tension, $\sigma_2$ and $\sigma_3 = 0$ and $\sigma_e = \sigma_2$; therefore

$$\dot{\varepsilon} = \alpha\sigma_e$$

and hence

$$\alpha = B\sigma_e^{n-1} \qquad [21.21]$$

Therefore the three principal creep rates may be written as

$$\left.\begin{aligned} \dot{\varepsilon}_1 &= B\sigma_e^{n-1}[\sigma_1 - \tfrac{1}{2}(\sigma_2 + \sigma_3)] \\ \dot{\varepsilon}_2 &= B\sigma_e^{n-1}[\sigma_2 - \tfrac{1}{2}(\sigma_3 + \sigma_1)] \\ \dot{\varepsilon}_3 &= B\sigma_e^{n-1}[\sigma_3 - \tfrac{1}{2}(\sigma_1 + \sigma_2)] \end{aligned}\right\} \qquad [21.22]$$

**Example 21.1**

An Ni–Cr–Mo alloy steel tube of 100 mm, diameter and 3 mm wall thickness is to operate at 400 °C with internal pressure for a service life of 100 000 hours. Determine the allowable pressure for a creep strain limit of 0.5%. The constants in the minimum creep equation at 400 °C are $n = 3$ and $B = 1.45 \times 10^{-23}$ per hour per MN/m$^2$.

In the thin tube under internal pressure where $\sigma_1$ is the hoop stress and $\sigma_2$ the axial stress, $\sigma_1 = 2\sigma_2$ and $\sigma_3 = 0$. Hence from eqns. [21.20] and [21.22]

$$\sigma_e = \frac{\sqrt{3}}{2}\sigma_1$$

and

$$\dot{\varepsilon}_1 = \left(\frac{\sqrt{3}}{2}\right)^{n+1} B\sigma_1^n$$

$$\dot{\varepsilon}_2 = 0$$

$$\dot{\varepsilon}_3 = -\left(\frac{\sqrt{3}}{2}\right)^{n+1} B\sigma_1^n$$

where $\sigma_1 = pr/t$ (eqn. [2.10]).

It is interesting to observe that there is no creep in the axial direction, which has also been verified experimentally.

The allowable internal pressure is controlled by the circumferential or hoop strain rate $\varepsilon_1$; therefore

$$\varepsilon_1 = \left(\frac{\sqrt{3}}{2}\right)^{n+1} B\left(\frac{pr}{t}\right)^n t_h$$

where $t_h = $ time in hours.

Substituting the design values,

$$0.005 = \left(\frac{\sqrt{3}}{2}\right)^4 1.45 \times 10^{-23} \left(p \cdot \frac{50}{3}\right)^4 10^5$$

from which

$$p^4 = 794.4 \times 10^8$$

$$p = 530 \text{N/m}^2$$

If the problem had combined the internal pressure with axial tension or torsion, the solution would only entail the use of a different ratio of $\sigma_1$ to $\sigma_2$ to give the required expression for $\sigma_e$ above.

## 21.7 Stress relaxation

The chapter has dealt so far with creep in the form of time-dependent increase in strain at constant stress; an alternative manifestation of creep is a time-dependent decrease in stress at constant strain. A common example of this phenomenon is the relaxation of tightening stress in the bolts of flanged joints in steam and other hot piping, with the resulting possibility of leakage. Another important case is the component subjected to a cycle of thermal strain. The effect is illustrated by means of the stress–strain curve in Fig. 21.7, where thermal expansion has set up compressive stress followed by relaxation with time. On cooling down, a higher residual tensile stress is set up than if there had been no relaxation of compressive stress. This situation can lead to failure eventually in a metal such as a flake cast iron which is weak in tension. Thermal strain concentration around nozzle openings in pressure vessels is another example where reversal of stress after relaxation could in time lead to a thermal fatigue failure.

A curve of stress relaxation against time is similar to a mirror image of a curve of creep strain against time as shown in Fig. 21.8. Just as much care has to be taken with relaxation testing as with conventional creep testing

**Fig. 21.7** Hysteresis including stress relaxation

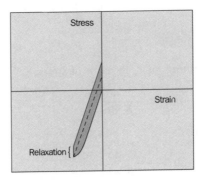

since results are very sensitive to temperature variation. The procedure usually adopted is to load the specimen to an initial stress which will give a specified strain of, say, 0.15%. The stress is then adjusted with time so that the specified strain is maintained.

In the Barr and Bardgett types of test the decrease in stress is noted after 48 hours for various initial stresses at constant temperature. If the decrease in stress is plotted against initial stress, the intercept of the curve on the latter axis gives the initial stress required for 'zero' decrease in stress, i.e. no relaxation. However, this is a very short-term test and can only safely be used for sorting materials.

**Fig. 21.8** Relaxation of stress with time at constant strain and temperature

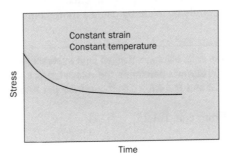

## 21.8 Stress relaxation in a bolt

Bolted flanged joints at high temperature represent an example of the problem of stress relaxation. Consider a bolt which is initially tightened to a stress $\sigma_0$, producing an elastic strain $\varepsilon_0$, and to simplify the problem it is assumed that the flange is not deformed by the bolt stress. After a period of time $t$ the effect of creep is to induce plastic deformation or creep strain, which allows a relaxation of stress $\sigma$ and elastic strain. Now, the total strain must remain the same if the flange is rigid; therefore,

$$\varepsilon_0 = \varepsilon_c + \frac{\sigma}{E}$$

or differentiating with respect to time,

$$0 = \frac{d\varepsilon_c}{dt} + \frac{1}{E}\frac{d\sigma}{dt}$$

or

$$\dot{\varepsilon}_c = -\frac{1}{E}\frac{d\sigma}{dt} \qquad [21.23]$$

Substituting for $\dot{\varepsilon}_c$ in terms of stress,

$$B\sigma^n = -\frac{1}{E}\frac{d\sigma}{dt}$$

Therefore

$$dt = -\frac{1}{EB}\frac{d\sigma}{\sigma^n} \qquad [21.24]$$

The time for relaxation of stress, from $\sigma_0$ initially to $\sigma_t$ at time $t$, is then obtained by integrating eqn. [21.24], and

$$t = -\frac{1}{EB}\int_{\sigma_0}^{\sigma_t}\frac{d\sigma}{\sigma^n}$$

Therefore

$$t = \frac{1}{EB}\frac{1}{(n-1)}\left(\frac{1}{\sigma_t^{n-1}} - \frac{1}{\sigma_0^{n-1}}\right) \qquad [21.25]$$

It is found in practice that relaxation of stress is more rapid than that given above owing to the effects of primary creep, and allowances for this can only be made by using a more complex creep-rate–stress–time function.

**Example 21.2**

The bolts holding a flanged joint in steam piping are tightened to an initial stress of 400 MN/m². Determine the relaxed stress after 10 000 hours. $E = 200\,\text{GN/m}^2$, $n = 3$ and $B = 4.8 \times 10^{-34}$ per hour per N/m².

From eqn. [21.25]

$$10\,000 = \frac{1}{200 \times 10^9 \times 4.8 \times 10^{-34} \times 2}$$

$$\times \left(\frac{1}{(\sigma_t \times 10^6)^2} - \frac{1}{(400 \times 10^6)^2}\right)$$

$$19.2 \times 10^{-7} = \frac{1}{\sigma_t^2} - \frac{1}{400^2}$$

and

$$\sigma_t = 349.8\,\text{MN/m}^2$$

**21.9 Creep during variable load or temperature**

Experiment and analysis has concentrated principally on constant load and constant temperature conditions during creep. There are some engineering applications in which the loading conditions change from time to time at high temperature as illustrated in Fig. 21.9. These are not necessarily

Fig. 21.9

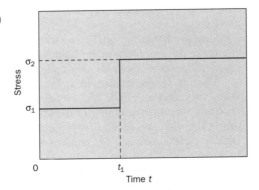

cyclical stress in the fatigue sense, as discussed in the next section, but result in different creep rates during each different load sequence.

Several hypotheses have been proposed for predicting creep strain related to changes of load which are constant before the change and constant after. It is beyond the scope of this chapter to develop these hypotheses, but two which are commonly quoted are *time-hardening* and *strain-hardening* theories. These take the following analytical forms:

$$\text{Time hardening:} \quad \dot{\varepsilon}_c = f(\sigma, t) \qquad\qquad [21.26]$$

$$\text{Strain hardening:} \quad \dot{\varepsilon}_c = g(\sigma, \varepsilon_c) \qquad\qquad [21.27]$$

The former implies that creep rate is a function only of the stress and the current time. The creep curve after the change of stress from $\sigma_1$ to $\sigma_2$ has the same shape as the constant stress curve from the time of change, i.e. the curve, $\sigma_2$, is moved vertically as shown in Fig. 21.10(a).

The latter implies that strain rate depends only on the stress and the current plastic strain. Again, as above, the same shape of curve is assumed, but now the appropriate portion of the curve, $\sigma_2$, from the time of change is moved horizontally as shown in Fig. 21.10(b).

Fig. 21.10

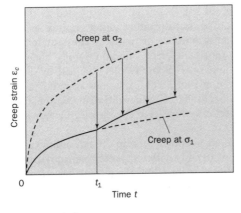

(a) Time hardening

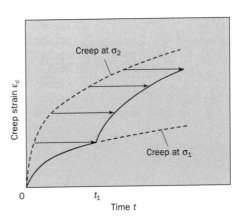

(b) Strain hardening

These hypotheses can also be applied in a variable strain–stress–relaxation context.

The predictions of total creep strain after load (stress) change are rather better in the case of the strain-hardening approach compared with the time-hardening method both for step-up load change and step-down load change.

Change of temperature in a component or structure, unless extremely slow, will always induce non-uniform thermal strain gradients and hence thermal stresses in addition to the stress due to applied load. It is evident, therefore, that prediction of creep rates or total strain is extremely difficult under variable temperature conditions.

## 21.10 Creep–fatigue interaction

Engineering developments such as gas and steam turbines, rockets and supersonic aircraft have involved the use of metals not only at very high temperatures but also with dynamic fluctuating stresses. In short, the problem is one in which the mean or steady component of stress can induce creep, and the alternating component of stress may lead to fatigue failure. The earliest investigations into this problem were made between 1936 and 1940 by various German investigators, and since then there have been many interesting studies both in this country and the U.S.A.

The problem of fatigue at high temperature was discussed in Chapter 20, and this phenomenon can be unaccompanied by creep for fully reversed or zero mean stress. Therefore, when considering a material for high-temperature service it is usual to think of the behaviour in terms of a diagram such as Fig. 21.11, in which fatigue failure is the criterion within certain stress and temperature limits, and beyond these creep is the predominant factor. If it is a question not of one or the other phenomenon acting on its own, but of both influences operating simultaneously, then the solution becomes rather more involved. Fatigue is essentially a cycle-dependent mechanism, whereas creep is time dependent. It is therefore both desirable and convenient to express fatigue behaviour at high temperature also in terms of time to rupture as suggested by Tapsell[5]. One of the reasons for this is because of the greater dependence of fatigue on cyclic frequency at high temperature.

The most useful way of presenting data for combined creep and fatigue conditions is in the form of a diagram of alternating stress against steady or mean stress, which is similar in most respects to the $S_a$–$S_m$ diagram in normal fatigue (Fig. 20.18). Test results are plotted as the combination of alternating and mean stress to produce either rupture or a specified creep strain after a particular number of hours at constant temperature. Points along the abscissa represent creep conditions only and points along the ordinate are for fatigue only. Some results obtained by Tapsell[5] on 0.26% carbon steel are given in Fig. 21.12 for various amounts of total creep strain occurring in 100 hours at 400 °C under different combinations of cyclic and steady stress. Theoretically predicted curves are also shown for creep strains of 0.002 and 0.005.

The influence of alternating stress on the minimum creep rate and time to rupture varies considerably with temperature, material and length of time. At higher temperatures or long life, the alternating stress appears to have little effect on creep rate; in fact, there are cases where creep strengthening has resulted. On the other hand, at lower temperatures or shorter rupture times, fatigue appears to play a more detrimental part, giving a higher creep

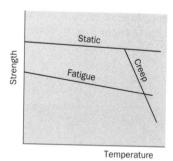

**Fig. 21.11**  Strength limitations with increasing temperature

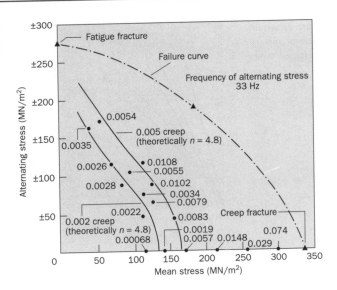

**Fig. 21.12** Total creep of 0.26% carbon steel occurring in 100 hours at 400°C

rate. The rupture strain is also somewhat reduced by the presence of cyclic stress.

## 21.11 Viscoelasticity

Because of the increasing use of plastics, both reinforced and unreinforced, in engineering load-bearing applications, it is important that the response of these materials to stress and environmental conditions should be appreciated. A brief mention was made of viscoelastic stress–strain–time behaviour in Chapter 3 and the discussion will be extended somewhat further, particularly in relation to creep, in the remainder of this chapter.

In a viscoelastic material the stress is a function of strain and time and so may be described by an equation of the form

$$\sigma = f(\varepsilon, t) \qquad [21.28]$$

This response is known as *non-linear* viscoelasticity, but as it is not amenable to simple analysis it is frequently approximated by the following form:

$$\sigma = \varepsilon . f(t) \qquad [21.29]$$

This response is the basis of *linear* viscoelasticity and simply indicates that, in a tensile test for example, for a fixed value of elapsed time the stress will be directly proportional to the strain. These different stress–strain–time responses are shown schematically in Fig. 21.13.

Viscoelastic materials invariably exhibit a time-dependent strain response to a constant strain, which is *relaxation*. In addition, when the applied stress is removed the materials have the ability to recover slowly over a period of time. These effects occur at ambient temperature and, therefore, are a principal design consideration as compared with metals for which creep and relaxation only occur in higher-temperature environments.

In Chapter 3 it was explained that tensile test characteristics for plastics are extremely sensitive to rate of straining. They are equally sensitive to tensile test temperature and, in some materials, the humidity condition. As a

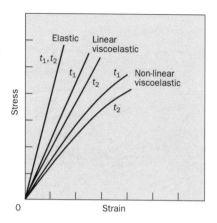

result of these special effects in plastics it is not reasonable to quote properties such as modulus, yield strength, etc, as a single value without qualifying these with details of the test condition.

## 21.12 Creep behaviour of plastics

Plastics exhibit a similar shape of creep curve of creep strain against time for constant stress and temperature as for metals (Fig. 21.1). However, one distinct difference is the ability of plastics to 'recover' strain after the removal of the applied load, and this effect is shown in Fig. 21.14. Plastics also have 'memory' and the current behaviour in creep or relaxation is dependent on all the past history of stress, strain and time on that sample (assuming constant temperature).

**Fig. 21.14** Typical creep and recovery behaviour

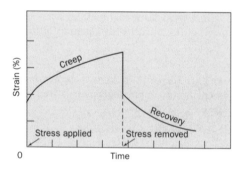

Attempts have been made to simulate polymer structure and its creep and recovery responses by mechanical-type modelling using two principal elements. These are a coil spring, which represents Hookean behaviour (linear load deformation), and a dashpot (a piston in an oil-filled container), which represents viscous Newtonian response (linear load–deformation rate). These elements may be coupled in two ways: (i) with the spring in series with the piston of the dashpot, and this is termed a Maxwell model; and (ii) with the spring in a 'parallel' location to the dashpot, so that applied load is 'shared' between the two elements, and this arrangement is known as the Kelvin–Voigt model. Because these models are individually quite inadequate to represent even linear viscoelasticity, more complex assemblies of the above units have been studied and, although somewhat more representative, they still do not give an adequate prediction of creep response which could be applied in design. This is principally because of the non-linear viscoelastic nature of polymers. The alternative to the empirical methods above is the use of experimental data obtained on the particular plastics for which design exercises have to be carried out.

Creep data is initially presented in the form of graphs of creep strain against log time, since linear time is inconvenient to encompass both short- and long-term tests. A family of creep curves is illustrated in Fig. 21.15(*a*) and two commonly used derivative graphs are shown in Fig. 21.15(*b*) and (*c*). The former is constructed by taking a constant strain section through the curves in Fig. 21.15(*a*) to give what is termed an *isometric* curve. A constant time section through the creep curves gives a stress–strain diagram as shown in Fig. 21.15(*c*) which is known as an *isochronous* curve.

**Fig. 21.15** Isometric and isochronous curves from deep curves

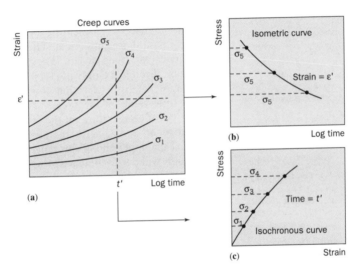

The isometric curve is often used as a good approximation of stress relaxation behaviour since this specific experimental method is less common than creep testing.

Isochronous curves can be developed independently without having to obtain a family of creep curves as above. The method involves a series of mini-creep and recovery tests in tension on a material. A stress is applied to a sample and the strain recorded after a time $t$ (typically 100 s), the stress is then removed and the specimen allowed to recover for a period of four times the loading time, i.e. $4t$ or 400 s. A larger stress is then applied to the same specimen and, after recording the strain at time $t$, this stress is removed and the material allowed to recover. This procedure is repeated until sufficient points have been obtained for the isochronous curve to be plotted. Obtaining isochronous curves by this direct experimental method is less time consuming and more economical than through creep experiments. In fact, one can, of course, derive creep curves from several isochronous experiments for different time intervals, e.g. $10^2$, $10^3$, $10^4$ s. Isochronous curves plotted on linear scales may not show up the slight non-linearity at low strains so an alternative plot on log–log scales is used so that any non-

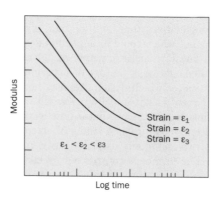

**Fig. 21.16** Typical variation of modulus with time

Modulus

Strain = $\varepsilon_1$
Strain = $\varepsilon_2$
Strain = $\varepsilon_3$

$\varepsilon_1 < \varepsilon_2 < \varepsilon_3$

Log time

linearity is demonstrated by the slope of the line being less than unity, i.e. less than 45° on the log–log paper.

Another method of representing long-term creep behaviour is by means of curves of modulus against time. These are shown in Fig. 21.16 for three values of constant creep strain. They were derived by taking a constant strain section through a family of creep curves and dividing the stress values by the strain to give relaxation moduli which are plotted against the respective time-value intersections.

The effect of temperature on the creep of plastics principally relates to a typical range of atmospheric temperatures such as −30 °C to +40 °C. However, at the upper end of this range of temperature, marked acceleration of creep rates will occur compared with the mid-range value and a family of isochronous curves for one elapsed time and several temperatures will be of the form shown in Fig. 21.17.

The prediction of creep response to step changes in stress is perhaps even more complex than for metals. It has been tackled by a variety of methods of superposition of parts of individual creep and recovery curves. The principle that is most frequently quoted is the Boltzmann-type superposition but, as with most methods, this only relates to linear viscoelasticity. Solutions for non-linear viscoelastic superposition are too intractable to be of practical use.

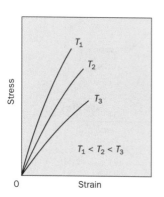

Stress

$T_1$
$T_2$
$T_3$

$T_1 < T_2 < T_3$

0     Strain

**Fig. 21.17** Isochronous curves for different temperatures at one elapsed time

## 21.13 Designing for creep in plastics

The design of metallic structures and components is generally not temperature or time dependent and is based on linear-elastic reversible stress–strain behaviour and small deformations. However, *any* load-bearing component to be made out of plastic has first and foremost to be designed for time-dependent deformation and, secondly, for time-dependent fracture.

Although the basic tenets of equilibrium of forces and compatibility of deformations apply in the design of a plastic component, the problem arises in relation to a suitable stress–strain–time law to link the foregoing. The more accurate methods that have been proposed have the drawback of being extremely complex and unattractive to the average designer. Perhaps the most acceptable approach devised has been called the *pseudo-elastic design method*. This involves the use of time-dependent 'elastic constant', moduli and contraction (Poisson's) ratio substituted into classical equations in place of the true elastic constants. The time-dependent value of modulus must be carefully determined to allow for the service life and limiting strain for the plastic component. The limiting strain value for the particular plastic should generally be decided in consultation with the material manufacturers and, typically, might be of the order of 1–2% strain. From this point the use of published experimental creep data for the material is quite straightforward in developing the component design. The following examples will illustrate the technique and the relevant creep data is given in Fig. 21.18.

**Fig. 21.18**  Creep curves for acetal at 20°C

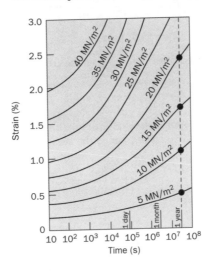

**Example 21.3**

A solid circular acetal rod, 0.15 m in length, is clamped horizontally at one end and the free end is subjected to a vertical load of 25 N. Determine a suitable diameter for the rod for a limiting strain of 2% in 1 year. What would be the maximum deflection at this time?

Using the creep curves in Fig. 21.18, a 1 year isochronous curve is plotted as shown in Fig. 21.19 from which an allowable stress of $17.1 \text{MN/m}^2$ is obtained at the 2% strain limit.

**Fig. 21.19**  A 1 year isochronous curve for acetal at 20°C

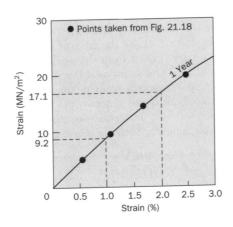

The maximum bending moment is $25 \times 0.15 = 3.75 \,\text{N m}$. Using the bending-stress relationship,

$$\sigma = \frac{My}{I} = \frac{32M}{\pi d^3}$$

$$d^3 = \frac{32 \times 3.75 \times 10^9}{\pi \times 17.1 \times 10^6} \,\text{mm}^3$$

$$d = 13.07 \,\text{mm}$$

The maximum deflection at the free end is given by

$$\delta = \frac{WL^3}{3EI}$$

The appropriate value of modulus may be obtained from the isochronous curve at 2% strain; hence the secant modulus

$$E(t) = \frac{17.1}{0.02} = 855 \,\text{MN/m}^2$$

Therefore,

$$\delta = \frac{25 \times 0.15^3 \times 64 \times 10^3}{3 \times 855 \times 10^6 \times \pi \times 0.013\,07^4} = 23 \,\text{mm}$$

An alternative way of obtaining the design stress would have been to plot a 2% isometric curve and read off the stress at 1 year.

**Example 21.4**

A circular acetal diaphragm is 2 mm thick and is clamped around its periphery giving a clear diameter of 100 mm. It is to be subjected to uniform pressure for a service life of 1 year with a material creep strain limitation of 1% and a maximum central deflection of 3 mm. Determine the allowable working pressure.

The central deflection of a clamped-edge circular plate subjected to uniform pressure was given in eqn. [16.27] as

$$w = \frac{12(1 - \nu^2)pa^4}{64Eh^3}$$

In this problem we shall have two time-dependent functions to consider, the modulus $E(t)$ and the creep or lateral contraction ratio, $\nu(t)$. The former may be determined from the isochronous curve of Fig. 21.19 for a strain limit of 1%. The secant modulus is given as $920 \,\text{MN/m}^2$. The data available for creep contraction ratio (the time-dependent equivalent of Poisson's ratio) is rather limited and generally lies between 0.3 and 0.4 but can rise to near 0.5 for 'rubbery' materials. For this problem we shall take a value of 0.35.
Rewriting the equation above gives

$$p = \frac{64Eh^3 w}{12(1 - \nu^2)a^4} = \frac{64 \times 920 \times 2^3 \times 3}{12(1 - 0.35^2) \times 50^4}$$

$$p = 21.4 \,\text{kN/m}^2$$

We could equally well have obtained the required modulus value by taking a 1% strain section through the creep curves and plotting modulus values against log time and extrapolating to 1 year.

## 21.14 Creep rupture of plastics

The creep rupture behaviour of metals was described earlier and a similar phenomenon also occurs with plastics. Under the sustained action of a constant load, plastics exhibit a failure mode associated with the creep deformation of the material. This type of failure is sometimes referred to as 'static fatigue', but the preferred engineering term is *creep rupture* or, perhaps more generally, *creep failure*. The more general terminology of creep failure is necessary for plastics because, although rupture of the material will eventually occur owing to creep, there may be earlier visible phenomena such as whitening, necking or crazing (crack-like features in glassy plastics) which, as far as the user is concerned, terminate the useful life of the component. Some typical creep failure data for plastics is shown in Fig. 21.20. Two important points should be noted from this. The first is that the creep failure data may be correlated with the isometric data obtained from a constant strain section across the creep curves. A second very important point is that, although the fracture as a result of creep is generally ductile in nature, there is a tendency towards embrittlement in some materials when subjected to constant loads for very long periods of time. This results in a sharp drop-off in the rupture line in Fig. 21.20. Such brittle failures can have serious consequences in practice since there is no prior warning of imminent fracture and there is no ductile tearing of the material to absorb the energy of the fracture. The possible 'knee' in the fracture line is something to be wary of when extrapolating short- or medium-term creep failure data to long lifetimes.

**Fig. 21.20** Creep rupture behaviour

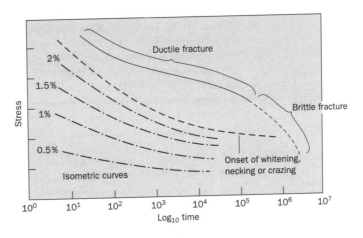

## 21.15 Summary

In the case of the design of metal components and structures, creep may be regarded as not a very common occurrence. However, in the case of the design of plastics components, it is of fundamental importance from the start. For either type of material there is no exact analytical method and heavy reliance must be placed on experimental data.

For metals the secondary stage of creep, which yields a constant minimum creep rate for very long periods at low stresses, provides the basis for design using an expression of the form $\varepsilon = B\sigma^n$. This relationship can also be used in the companion time-dependent phenomenon known as stress relaxation.

While situations of constant stress or strain can be handled reasonably, those involving changes in stress or strain at regular intervals of time are very difficult to treat quantitatively. Creep-rupture testing provides very valuable short-term data for the assessment and sorting of materials.

Creep and stress relaxation of plastics, while having many similarities with the behaviour of metals, has the added features of memory and recovery which make in-depth analysis very difficult. However, the interpretation of basic creep curve data into isometric and isochronous curves allows a convenient and acceptable approach to design by the pseudo-elastic design method as illustrated in the worked examples above.

**References**

1.  Pomeroy, C. D. (1978) *Creep of Engineering Materials*, Ch. 9, I.Mech.E., London.
2.  Andrade, E. N. da C. (1910) 'The viscous flow in metals and allied phenomena', *Proc. R. Soc.*, **A84**, 1.
3.  Lessels, J. M. (1954) *Strength and Resistance of Metals*, John Wiley, New York.
4.  Larsen, F. R. and Miller, J. A. (1952) 'Time–temperature relationship for rupture and creep stresses', *Trans. ASME*, **74**, 765.
5.  Tapsell, H. J. (1952) *Symposium on High Temperature Steels and Alloys for Gas Turbines*, Iron Steel Inst., London, p. 43.

**Bibliography**

Crawford, R. J. (1987) *Plastics Engineering*, Pergamon Press, Oxford.
Faupel, J. H. (1981) *Engineering Design*, Ch. 12, John Wiley, New York.
Finnie, I. and Heller, W. R. (1959) *Creep of Engineering Materials*, McGraw-Hill, New York.
*Metals Handbook*, 9th edition, Vol. 8, *Mech. Testing*, (1985) John R. Newley (co-ordinator), American Society for Metals, Metals Park, OH.
Penny, R. K. and Marriott, D. L. (1971) *Design for Creep*, McGraw-Hill, New York.
Pomeroy, C. D. (1978) *Creep of Engineering Materials*, I.Mech.E., London.

**Problems**  21.1  A series of creep tests on an austenitic high-temperature alloy gave the following results:

| Stress (MN/m$^2$) | $\epsilon_0$ (%) | Minimum creep rate (mm/mm/hr) |
|---|---|---|
| 70 | 0.041 | $27 \times 10^{-8}$ |
| 105 | 0.061 | $15.5 \times 10^{-6}$ |
| 140 | 0.081 | $27.5 \times 10^{-5}$ |
| 210 | 0.122 | $15.8 \times 10^{-3}$ |
| 280 | 0.162 | 0.281 |
| 350 | 0.203 | 2.62 |

Calculate how long would elapse before a steady stress of $125\,MN/m^2$ would cause a strain of 1% in this material.

21.2 The following table shows the creep data obtained for a metal using a stress of $100\,MN/m^2$ at a range of temperature. If the constant "a" in the Larsen–Miller parameter is 20, calculate the time to failure at a stress of $100\,MN/m^2$ when the temperature of the material is $700°C$. $(R = 8.314\,J\,mol^{-1}\,K)$.

| Temperature (°C) | 100 | 200 | 250 | 300 |
|---|---|---|---|---|
| $\epsilon_0$ (mm/mm/hr) | $2.1 \times 10^{-23}$ | $2.77 \times 10^{-17}$ | $4.25 \times 10^{-15}$ | $2.7 \times 10^{-13}$ |
| Temperature | 400 | 500 | 600 | |
| $\epsilon_0$ (mm/mm/hr) | $1.72 \times 10^{-10}$ | $2.06 \times 10^{-8}$ | $8.25 \times 10^{-7}$ | |

21.3 The following creep rupture data was recorded for an alloy steel when it was tested at a range of stresses and temperatures:

| Temperature (°C) | Stress (MN/m²) | Time to failure (hr) |
|---|---|---|
| 500 | 300 | 4724 |
| 600 | 200 | 570 |
| 700 | 100 | 297 |
| 800 | 60 | 95.3 |
| 1000 | 30 | 8.6 |

If a component made from this material is required to last at least 10 000 hours at a stress of $150\,MN/m^2$, what is its maximum permissible service temperature? The constant "a" in the Larsen–Miller parameter for the alloy is 20.

21.4 Figure 21.21 shows a delayed-action contact switch. When the pin is removed the compressed spring causes a tensile stress in the previously unstressed lead rod. Due to creep of the lead the gap between the contact points decreases steadily. Caculate the delay time if the free length of the spring is 40 mm and its stiffness is 10 N/mm. For the lead $\varepsilon_0 = 5 \times 10^{-10}\sigma^{7.5}$ mm/mm/hr.

Fig. 21.21

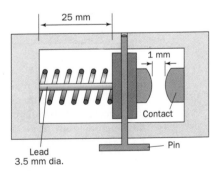

Lead
3.5 mm dia.

21.5 A sheet of high-temperature alloy is clamped in positon using a toggle clamp as illustrated in Fig. 21.22. When the lever A is in the vertical (clamp) position the sheet thickness is reduced by 10 $\mu$m. If the sheet is subjected to a pull of 5 kN, calculate how long the clamp could retain the sheet in position. The coefficient of friction between the clamp and the alloy is 0.6. Creep data for the alloy is given in Problem 21.1. $E = 207\,\text{GN/m}^2$.

**Fig. 21.22**

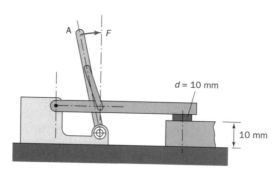

21.6    An aerosol container is to be moulded from an acetal copolymer, for which the creep curves are given in Fig. 21.18. The container diameter is 50 mm and it has a uniform wall thickness of 2 mm. The base of the container is designed with a "skirt" to prevent rocking when the bottom deforms under pressure (see Fig. 21.23). Calculate the depth of the skirt if the container is expected to be subjected to an internal pressure of $120\,\text{kN/m}^2$ (absolute) for 1 year. Poisson's ratio for acetal may be taken as 0.4.

**Fig. 21.23**

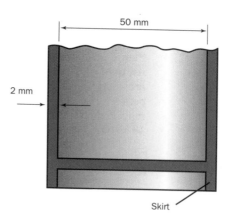

Skirt

21.7    A cylindrical acetal container is subjected to an internal pressure of $0.7\,\text{MN/m}^2$. For aesthetic reasons the strain in the container is not to exceed 2%. If the diameter and wall thickness of the container are 60 mm and 1 mm respectively calculate how long the container may be regarded as serviceable. Creep curves for the acetal are given in Fig. 21.18.

21.8    A plastic snap-fit connection is shown in Fig. 21.24. If the pin will slip out when the transverse clamping force exerted by the clasp is 33 N calculate (a) the clamping force when the pin is first inserted, and (b) the elapsed time before the pin would slip out. Use the creep curves in Fig. 21.18.

**Fig. 21.24**

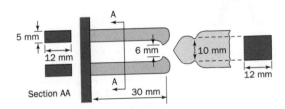

Section AA

21.9 A long thin-walled pipe constrained by end fittings made of polyvinylchloride is subjected to a steady internal pressure of $700 \, kN/m^2$ at 20°C. If a tensile stress of $17.5 \, MN/m^2$ is not to be exceeded and the internal radius is 100 mm, determine a suitable wall thickness. What will be the increase in diameter after 1000 hours? The mean creep contraction ratio $\nu_t$ is 0.45, and tensile creep curves provide the following values at 1000 hours:

| $\sigma(MN/m^2)$ | 6.9 | 13.8 | 20.7 | 27.6 | 34.5 |
|---|---|---|---|---|---|
| $\epsilon$ (%) | 0.2 | 0.48 | 0.92 | 1.72 | 3.38 |

# Properties of Areas

The analysis of stresses developed in symmetrical and unsymmetrical bending of beams (Chapter 6) depends on the shape and area of the cross-section of a beam. The reason for this is because internal *forces* and *moments* derive from stress acting on elements of area. To obtain the *total* shear-force or bending-moment effect on a cross-section we must sum up or integrate all the constituent elements and the final expressions involve integrals for the total area (shape) in question. It is these integrals and these solutions which we refer to as 'properties' of areas. There are direct comparisons with the 'properties' of masses which are used in engineering dynamics or mechanics of machines. The integrals which will need to be evaluated for any cross-sectional shape are described as the first moment, the second moment and the product moment of an area. It may seem incongruous to speak of the 'moment' of an area since a moment implies a mass or force multiplied by a distance. However, the integrals do consist of areas multiplied by distances and that is how the expression *moment of area* has become established (this is *not* to be confused with moment of inertia which relates to mass). Co-ordinate axes *y* and *z* are used throughout for beam cross-sections, since the *x*-axis is along the length of the beam.

## A.1   First moment of area

Referring to the plane figure in Fig. A.1, the first moment of the element of area d$A$ about the $z$-axis is $y$ d$A$; therefore the first moment of the whole figure is $\int_A y\, \mathrm{d}A$ about the $z$-axis, the suffix $A$ indicating summation over the whole area. The first moment of the whole figure about the $y$-axis is $\int_A z\, \mathrm{d}A$.

## A.2   Position of centre of area, or centroid

Let the co-ordinates of the centre of area C.A. be $\bar{z}$ and $\bar{y}$ as shown in Fig. A.1. Then the moment of the whole area about an axis is the same as the sum of the moments of all the elements of area about that axis, or

$$A\bar{y} = \int_A y\, \mathrm{d}A \quad \text{so that} \quad \bar{y} = \frac{1}{A}\int_A y\, \mathrm{d}A \qquad [A.1]$$

Similarly,

$$\bar{z} = \frac{1}{A}\int_A z\, \mathrm{d}A \qquad [A.2]$$

**Fig. A.1**

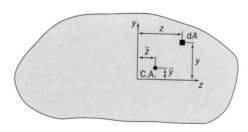

If either or both of the $z$- and $y$-axes pass through the centre of area then $\bar{z}$ or $\bar{y}$ or both are zero, and

$$\int_A y\,\mathrm{d}A = 0 \quad \text{and/or} \quad \int_A z\,\mathrm{d}A = 0$$

If a shape has one axis of symmetry then the centre of area will lie on that axis. If there are two axes of symmetry then their intersection will be the centre of area.

**Example A.1**

**Determine the location of the centre of area for the concrete beam cross-section shown in Fig. A.2.**

**Fig. A.2**

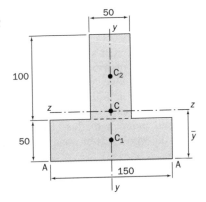

Since there is a vertical axis of symmetry the centre of area will lie somewhere on that axis as shown by C.

Take as a reference horizontal axis the lower edge of the section AA, and let the distance of C from AA be $\bar{y}$. The cross-section can be divided into two rectangles by the dashed line. The centre of area of the upper rectangle is at $C_2$ at a distance of 100 mm from AA. The centre of the area of the lower rectangle is at $C_1$ at a distance of 25 mm from AA.

The areas of the upper and lower rectangles are 5000 and 7500 mm$^2$ respectively and the total area of the figure is 12 500 mm$^2$.

Referring to eqn. [A.1] above we may write

$$12\,500\bar{y} = 5000 \times 100 + 7500 \times 25$$

from which

$$\bar{y} = 55\,\text{mm}$$

**A.3 Second moment of area**

If the first moment of an element $\mathrm{d}A$ about an axis is multiplied again by its respective co-ordinate we obtain the second moment of area, namely $y^2\,\mathrm{d}A$, or $z^2\,\mathrm{d}A$. The second moment of area of the whole figure about the $z$-axis is

$$\int_A y^2\,\mathrm{d}A \quad \text{denoted as} \quad I_z \hspace{3cm} [A.3]$$

and about the $y$-axis is

$$\int_A z^2 \, dA \quad \text{denoted as} \quad I_y \tag{A.4}$$

The *radius of gyration* of an area $A$ with respect to the $z$-axis is defined by the quantity $r_z$ which satisfies the relation

$$I_z = A r_z^2$$

from which we can write

$$r_z = \sqrt{\frac{I_z}{A}} \tag{A.5}$$

In a similar way

$$r_y = \sqrt{\frac{I_y}{A}} \tag{A.6}$$

**Second moment of area for a rectangle**   Common structural cross-sectional shapes are composed of rectangles and the second moment of area of a rectangle is obtained as follows.

Because of the double symetry the centre of area C is at the centre of the rectangle of width $b$ and depth $d$ shown in Fig. A.3.

Consider an element of area $b \, dy$ as shown.

The second moment of this element about the $z$ axis is $y^2(b \, dy)$. To obtain $I_z$ for the whole section we must integrate between the limits of $\pm d/2$ so that

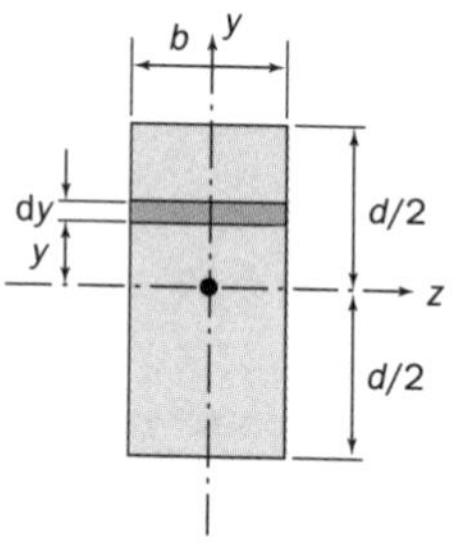

$$I_z = \int_{-d/2}^{+d/2} y^2 b \, dy = \left[ \frac{by^3}{3} \right]_{-d/2}^{+d/2}$$

$$= \frac{bd^3}{12}$$

**Fig. A.3**   By a similar analysis we can obtain

$$I_y = \frac{db^3}{12}$$

**Example A.2**

**Determine the second moment of area for a solid circular cross-section of 50 mm diameter about an axis through the centre.**

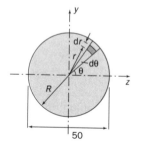

**Fig. A.4**

The element of area marked in Fig. A.4 is $dA = r \, d\theta \, dr$ and the second moment of this element about the $z$ axis is $(r \sin \theta)^2 r \, dr$. If we now integrate this expression between the limits of 0 to $2\pi$ we shall have the second moment of an annular element about the $z$ axis, which is

$$\int_0^{2\pi} (r \sin \theta)^2 r \, d\theta \, dr = \pi r^3 \, dr$$

and for the solid circle the second moment is

$$I_z = \int_0^R \pi r^3 \, dr = \frac{\pi R^4}{4} = \frac{\pi D^4}{64} = \frac{\pi \times 50^4}{64} = 30.7 \times 10^4 \text{ mm}^4$$

**A.4 Parallel axes theorem**

It is sometimes necessary to determine the second moment of area about axes parallel to the centroidal axes.

**Fig. A.5**

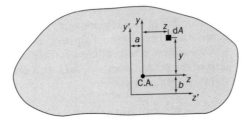

Referring to Fig. A.5, the second moment of the element $dA$ about the $z'$-axis is $(y + b)^2 \, dA$, and for the whole figure

$$I_{z'} = \int_A (y + b)^2 \, dA$$

$$= \int_A y^2 \, dA + 2b \int_A y \, dA + \int_A b^2 \, dA$$

but $\int_A y \, dA = 0$, since it is the first moment about a centroidal axis, so

$$I_{z'} = I_z + b^2 A \qquad\qquad [A.7]$$

and by a similar analysis

$$I_{y'} = I_y + a^2 A \qquad\qquad [A.8]$$

**Example A.3**

**Determine the second moments of area of the section in Example A.1 about its centroidal axes.**

Any section composed of rectangles can be broken up for analysis into its separate components. Therefore, in the case of the vertical $yy$-axis which passes through the centres of area $C_1$ and $C_2$, we do not need the parallel axes theorem, so

$$I_y = \frac{100 \times 50^3}{12} + \frac{50 \times 150^3}{12} = 15.1 \times 10^6 \, \text{mm}^4$$

In order to calculate the value of $I_z$ we need to apply the parallel axes theorem to both the top and bottom rectangles as follows:

$$I_{z_1} = \frac{150 \times 50^3}{12} + (150 \times 50)30^2 = 8.3 \times 10^6 \, \text{mm}^4$$

The first term is the $I$ about a horizontal axis through $C_1$ and the second term is the area of the rectangle multiplied by the square of the distance between $C_1$ and C.

$$I_{z_2} = \frac{50 \times 100^3}{12} + (100 \times 50)45^2 = 14.3 \times 10^6 \, \text{mm}^4$$

The total value of $I_z$ is the sum of the two parts above:

$$I_z = 8.3 \times 10^6 + 14.3 \times 10^6 = 22.6 \times 10^6 \, \text{mm}^4$$

## A.5 Product moment of area

The moment of area requirements for the analysis of *symmetrical* bending of beams have been covered up to this section. However, for *unsymmetrical sections* subjected to bending a further moment of area property is required termed the *product moment of area*. It is defined in relation to the plane figure shown in Fig. A.6 as

$$I_{zy} = \int_A zy \, dA \qquad\qquad [A.9]$$

**Fig. A.6**

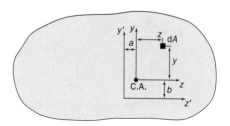

where the axes pass through the centre of area C.A. of the figure, Two important differences from the second moments of area $I_z$ and $I_y$ are that $I_{zy}$ can have either *positive* or *negative* values since $z$- and $y$-values can be positive and negative. Secondly, if either or both of the axes are an axis of symmetry then $I_{zy} = 0$.

For the case of the product moment of area related to parallel axes $z'y'$, it is straightforward to show that

$$I_{z'y'} = I_{zy} + abA \qquad\qquad [A.10]$$

## Example A.4

**Determine the product moment of area of the triangular section shown in Fig. A.7 in relation to the centroidal axes z, y.**

The equation of the diagonal of the triangle is $y = h/3 - hz/k$. Hence

$$I_{zy} = \int zy \, da = \int_{-k/3}^{2k/3} \int_{-h/3}^{h(k-3z)/3k} zy \, dz \, dy$$

$$= \frac{h^2}{6k^2} \int_{-k/3}^{2k/3} (3z^3 - 2kz^2) \, dz = -\frac{k^2 h^2}{72}$$

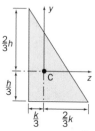

**Fig. A.7**

### A.6 Spreadsheet calculation of area and section properties of a polygon

The area of the trapezium underneath a straight line segment from $(z_i, y_i)$ to $(z_{i+1}, y_{i+1})$ in Fig. A.8 is given by $(z_i - z_{i+1})(y_i + y_{i+1})/2$. Note that if the order of the points was reversed a negative area would be computed.

#### Area under a line segment

Fig. A.8

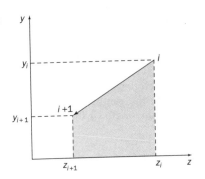

#### Area of a triangle

The area of a triangle defined by points $(z_1, y_1)$, $(z_2, y_2)$ and $(z_3, y_3)$ shown in Fig. A.9 is formed by the area under the line from 2 to 3 plus the area under line 3–1 minus the area under the segment 1–2. Since $z_1$ is less than $z_2$, the area under the segment between points 1 and 2 is negative and summing up the areas under the three segments will give the right answer, i.e.

$$A = \tfrac{1}{2}[(z_1 - z_2)(y_1 + y_2) + (z_2 - z_3)(y_2 + y_3)$$
$$+ (z_3 - z_1)(y_3 + y_1)] \qquad [A.11]$$

Fig. A.9

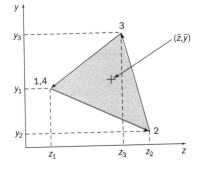

#### Area of a polygon

For a polygon of $n$ sides

$$A = \tfrac{1}{2}\sum_{i=1}^{n}(z_i - z_{i+1})(y_i + y_{i+1}) \qquad [A.12]$$

We can create a spreadsheet to calculate the area of any polygon by making a table of the $(z, y)$-co-ordinates of the polygon vertices, calculating the contributions of the individual line segments and then summing them. Enter the formula for one segment and then copy the formula for as many segments as are required.

There are two points to note in constructing the table of co-ordinates.

1. The first point has to be repeated at the end of the list to account for the line segment from the last point back to the first, forming a closed loop.
2. The points must be entered in anticlockwise order as you progress round the boundary of the polygon.

A ZY graph of the vertices can be easily constructed in the spreadsheet to check that we have correctly entered the data, though the relative scaling of the $z$- and $y$-directions may not be the same.

To calculate the centre of area and the second moments of area of a polygon about the origin $(0, 0)$ the equivalent formulae are[1]

$$\bar{z} = \frac{1}{6A} \sum_{i=1}^{n} (y_{i+1} - y_i)(z_i^2 + z_i z_{i+1} + z_{i+1}^2) \qquad [A.13]$$

$$\bar{y} = \frac{1}{6A} \sum_{i=1}^{n} (z_i - z_{i+1})(y_i^2 + y_i y_{i+1} + y_{i+1}^2) \qquad [A.14]$$

Fig. A.10a

| | A | B | C | D | E | F | G | H |
|---|---|---|---|---|---|---|---|---|
| 1 | z | y | Area | First Moments of Area | | Second Moments of Area | | |
| 2 | | | | Mz | My | Iz | Iy | Iyz |
| 3 | 0 | 0 | 0 | 0 | 0 | 0 | 0 | 0 |
| 4 | -60 | 0 | 0 | 0 | 0 | 0 | 0 | 0 |
| 5 | -60 | -10 | 1000 | -108000 | -15000 | -15000 | 200000 | 630000 |
| 6 | -10 | -10 | 0 | 0 | 0 | 0 | 0 | 0 |
| 7 | -10 | -100 | 2000 | -27000 | -300000 | -300000 | 4000000 | 1800000 |
| 8 | 0 | -100 | 0 | 0 | 0 | 0 | 0 | 0 |
| 9 | +A3 | +B3 | | | | | | |
| 10 | | | | zbar | ybar | | | |
| 11 | Totals | | @SUM(C3..C9)/2 | @SUM(D3..D8)/(6*C11) | @SUM(E3..E8)/(6*C11) | @SUM(F3..F8)/12 | @SUM(G3..G8)/12 | @SUM(H3..H8)/72 |
| 12 | | | | | | | | |
| 13 | | | | | | Second Moments of Area about centroid | | |
| 14 | | | | | | Izc | Iyc | Izyc |
| 15 | | | | | | +F11-$C$11*E11^2 | +G11-$C$11*D11^2 | +H11-C11*E11*D11 |
| 16 | | | | | | | | |
| 17 | Coordinates relative to centroid | | | | | | | |
| 18 | z | y | | | | | | |
| 19 | | | | | | | | |
| 20 | +A3-$D$11 | +B3-$E$11 | | | | | | |
| 21 | +A4-$D$11 | +B3-$E$11 | | | | | | |
| 22 | +A5-$D$11 | +B3-$E$11 | | | | | | |
| 23 | +A6-$D$11 | +B3-$E$11 | | | | | | |
| 24 | +A7-$D$11 | +B3-$E$11 | | | | | | |
| 25 | +A8-$D$11 | +B3-$E$11 | | | | | | |

(a) Data and cell formulae

Note that the formulae in C3..H8 are too long to display here. They correspond to the individual terms in the summations represented by eqn. ( A.12) - ( A.20)

Fig. A.10b

| | A | B | C | D | E | F | G | H |
|---|---|---|---|---|---|---|---|---|
| 1 | z | y | | First Moments of Area | | Second Moments of Area | | |
| 2 | | | Area | Mz | My | Iz | Iy | Iyz |
| 3 | 0 | 0 | 0 | 0 | 0 | 0 | 0 | 0 |
| 4 | −60 | 0 | 0 | −108000 | 0 | 0 | 0 | 0 |
| 5 | −60 | −10 | 1000 | 0 | −15000 | 200000 | 8640000 | 6300000 |
| 6 | −10 | −10 | 0 | −27000 | 0 | 0 | 0 | 0 |
| 7 | −10 | −100 | 2000 | 0 | 300000 | 4000000 | 360000 | 18000000 |
| 8 | 0 | −100 | 0 | 0 | 0 | 0 | 0 | 0 |
| 9 | 0 | 0 | | | | | | |
| 10 | | | | zbar | ybar | | | |
| 11 | Totals | | 1500 | −15 | −35 | 3350000 | 750000 | 337500 |
| 12 | | | | | | | | |
| 13 | | | | | | Second Moments of Area about centroid | | |
| 14 | | | | | | Izc | Iyc | Izyc |
| 15 | | | | | | 1512500 | 412500 | −450000 |
| 16 | | | | | | | | |
| 17 | Coordinates relative to centroid | | | | | | | |
| 18 | z | y | | | | | | |
| 19 | | | | | | | | |
| 20 | 15 | 35 | | | | | | |
| 21 | −45 | 35 | | | | | | |
| 22 | −45 | 25 | | | | | | |
| 23 | 5 | 25 | | | | | | |
| 24 | 5 | −65 | | | | | | |
| 25 | 15 | −65 | | | | | | |

(**b**) Spreadsheet display

$$I_z = \frac{1}{12} \sum_{i=1}^{n} (z_i - z_{i+1})(y_i^3 + y_i^2 y_{i+1} + y_i y_{i+1}^2 + y_{i+1}^3) \qquad [A.15]$$

$$I_y = \frac{1}{12} \sum_{i=1}^{n} (y_{i+1} - y_i)(z_i^3 + z_i^2 z_{i+1} + z_i z_{i+1}^2 + z_{i+1}^3) \qquad [A.16]$$

$$I_{zy} = \frac{1}{72} \sum_{i=1}^{n} (z_i - z_{i+1})[z_{i+1}(9y_{i+1}^2 + 6y_i y_{i+1} + 3y_i^2)$$

$$+ z_i(9y_i^2 + 6y_i y_{i+1} + 3y_{i+1}^2)] \qquad [A.17]$$

The second moments of area about the centroid can be determined using the parallel axes theorem from

$$I_{zc} = I_z - A\bar{y}^2 \qquad [A.18]$$

$$I_{yc} = I_y - A\bar{z}^2 \qquad [A.19]$$

$$I_{zyc} = I_{zy} - Azy \qquad [A.20]$$

The resulting spreadsheet for the section of Example 6.11 is shown in Fig. A.10.

**Fig. A.11**

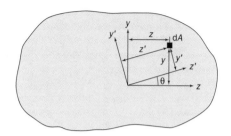

## A.7 Transformation of moments of area

In unsymmetrical bending of beams it is sometimes necessary to consider the nature of bending about a set of axes rotated through an angle $\theta$ with respect to a reference direction of axes as shown in Fig. A.11. Let the moments of area be $I_z$, $I_y$ and $I_{zy}$ with respect to the reference axes $z$, $y$ and $I_{z'}$, $I_{y'}$ and $I_{z'y'}$ with respect to different axes $z'$, $y'$ at an anticlockwise angle $\theta$ to the former.

$$I_{z'} = \int_A y'^2 \, \mathrm{d}A = \int_A (y\cos\theta - z\sin\theta)^2 \, \mathrm{d}A$$

$$= I_z \cos^2\theta + I_y \sin^2\theta - 2I_{zy}\sin\theta\cos\theta$$

$$= \tfrac{1}{2}(I_z + I_y) + \tfrac{1}{2}(I_z - I_y)\cos 2\theta - I_{zy}\sin 2\theta \qquad [A.21]$$

$$I_{z'y'} = \int_A z'y' \, \mathrm{d}A = \int (z\cos\theta + y\sin\theta)(y\cos\theta - z\sin\theta) \, \mathrm{d}A$$

$$= (\cos^2\theta - \sin^2\theta)I_{zy} + I_z\sin\theta\cos\theta - I_y\sin\theta\cos\theta$$

$$= \tfrac{1}{2}(I_z - I_y)\sin 2\theta + I_{zy}\cos 2\theta \qquad [A.22]$$

These two equations provide the relationships between moments of area about two sets of rectangular axes with a common origin. It is interesting to note the similarity of form between these equations and those for two-dimensional stress transformation, eqns. [11.13] and [11.14]. It does in fact suggest the existence of 'principal' second moments of area about axes for which the product moment of area is zero. From eqn. [A.22], putting $I_{z'y'} = 0$ we obtain

$$\tan 2\theta = \frac{2I_{zy}}{(I_z - I_y)} \qquad [A.23]$$

and this defines the axes about which maximum and minimum principal second moments of area $I_u$ and $I_v$ occur.

## A.8 Mohr's circle for moments of area

The simplest way of determining the principal second moments of area and indeed the moments of area about any set of axes is to construct a Mohr's circle in the same manner as that for stresses or strains.

If eqns. [A.21] and [A.22] are squared and added to each other to eliminate the angle $2\theta$ we obtain the following equation:

$$[I_{z'} - \tfrac{1}{2}(I_z + I_y)]^2 + I_{z'y'}^2 = \tfrac{1}{4}(I_z - I_y)^2 + I_{zy}^2 \qquad [A.24]$$

This represents a circle with axes of product moments of area $(I_{zy})$, as ordinate, and second moments of area $(I_z, I_y)$, as abscissa. The centre of the circle is located at

$$\left(\frac{I_z + I_y}{2}, 0\right)$$

and the radius of the circle is

$$[\tfrac{1}{4}(I_z - I_y)^2 + I_{zy}^2]^{1/2}$$

Since there are no negative values of second moments of area the circle is always to the right of the ordinate.

Figure A.12 shows the circle construction for the shape shown shaded top right and similarly top left. The circle is drawn from known, or calculated, values of $I_z$, $I_y$ and $\pm I_{zy}$. For required values about axes $z'y'$ at $\theta$ to axes $ZY$ we draw the diameter ECF at $2\theta$ anticlockwise from ABC. The values at E and F are those required. The principal second moments of area $I_u$ and $I_v$ are the values at U and V, where $I_{zy} = 0$. The directions of the principal axces are given by the chords UG and VG which are shown as axes O$u$ and O$v$ on the area (top left).

**Fig. A.12**

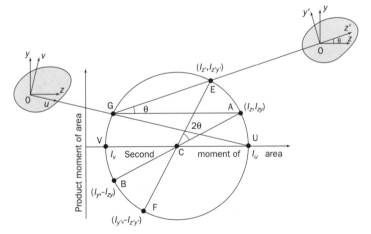

**Example A.5**

Use the circle construction to determine the principal second moments of area for the angle section of Example 6.11 in which $I_y = 41.3 \times 10^4$ mm$^4$, $I_z = 151.2 \times 10^4$ mm$^4$ and $I_{zy} = 45 \times 10^4$ mm$^4$.

The centre of the circle is located at $(96.5, 0)$ and with the values of $I_z$ and $I_{zy}$ the circle is drawn as shown in Fig. A.13. The maximum and minimum

**Fig. A.13**

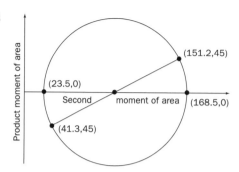

principal second moments of area are seen to be $168.5 \times 10^4$ and $23.5 \times 10^4\,\text{mm}^4$.

## A.9 Polar second moment of area

The second moment of area about an axis perpendicular to the plane of an area is termed the *polar second moment of area* and it is an essential part of the analysis of the shear stresses in the torsion of circular sections (Chapter 5). Referring to Fig. A.14 the polar second moment of area of $dA$ is $r^2\,dA$ and for the whole figure

$$\mathcal{J}(\text{or } I_p) = \int_A r^2\,dA$$

**Fig. A.14**

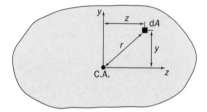

Also, since $r^2 = z^2 + y^2$,

$$\mathcal{J} = \int_A z^2\,dA + \int_A y^2\,dA$$

$$= I_y + I_z \qquad\qquad [A.25]$$

This is known as the *perpendicular axes theorem*.

## References

1. Cope, R. J., Sawko, F. and Tickell, R. G. (1982) *Computer Methods for Civil Engineers*, McGraw-Hill, London.

## APPENDIX B Introduction to Matrix Algebra

### B.1 Matrix definitions

A matrix is an array of terms as shown below:

$$[A] = \begin{bmatrix} a_{11} & a_{12} & a_{13} & \cdots & a_{1n} \\ a_{21} & a_{22} & a_{23} & \cdots & a_{2n} \\ a_{31} & a_{32} & a_{33} & \cdots & a_{3n} \\ \vdots & \vdots & \vdots & & \vdots \\ a_{m1} & a_{m2} & a_{m3} & \cdots & a_{mn} \end{bmatrix}$$

If $n = 1$ then we have a matrix consisting of a single column of terms and this referred to as a *column matrix*. If $m = 1$ then the matrix is called a *row matrix*.

If in the analysis of a problem there is a set of simultaneous equations then the use of matrices can be a very convenient shorthand way of expressing and solving the equations. For example, consider the following set of equations:

$$y_1 = a_{11}x_1 + a_{12}x_2 + a_{13}x_3 + \ldots + a_{1n}x_n$$

$$y_2 = a_{21}x_1 + a_{22}x_2 + a_{23}x_3 + \ldots + a_{2n}x_n$$

$$y_3 = a_{31}x_1 + a_{32}x_2 + a_{33}x_3 + \ldots + a_{3n}x_n$$

$$\vdots$$

$$y_m = a_{m1}x_1 + a_{m2}x_2 + a_{m3}x_3 + \ldots + a_{mn}x_n$$

These may be written in matrix form as follows:

$$\{y\} = [A]\{x\} \qquad\qquad [B.1]$$

where $\{y\}$ and $\{x\}$ are column matrices.

### B.2 Matrix multiplication

The matrix equation [B.1] involves the multiplication of the matrices $[A]$ and $\{x\}$. To do this one must apply the simple rules of *matrix multiplication*. These are:

(a) two matrices may only be multiplied if the number of columns in the first is equal to the number of rows in the second;

(b) the terms in the product matrix resulting from the multiplication of matrix $[A]$ with a matrix $[B]$ are given by

$$c_{ij} = \sum_{k=1}^{n} a_{ik}b_{kj} \qquad\qquad [B.2]$$

The use of these rules is illustrated in the following example:

$$\begin{bmatrix} a_{11} & a_{12} \\ a_{21} & a_{22} \end{bmatrix} \begin{bmatrix} b_{11} & b_{12} & b_{13} \\ b_{21} & b_{22} & b_{23} \end{bmatrix}$$

$$= \begin{bmatrix} (a_{11}b_{11} + a_{12}b_{21}) & (a_{11}b_{12} + a_{12}b_{22}) & (a_{11}b_{13} + a_{12}b_{23}) \\ (a_{21}b_{11} + a_{22}b_{21}) & (a_{21}b_{12} + a_{22}b_{22}) & (a_{21}b_{13} + a_{22}b_{23}) \end{bmatrix}$$

Suppose

$$[A] = \begin{bmatrix} 2 & 4 \\ 6 & 8 \end{bmatrix}, \quad [B] = \begin{bmatrix} 1 & 2 & 3 \\ 3 & 2 & -1 \end{bmatrix}$$

Then

$$[C] = [A][B] = \begin{bmatrix} 14 & 12 & 2 \\ 30 & 28 & 10 \end{bmatrix}$$

## B.3  Matrix addition and subtraction

Matrix algebra also involves the *addition and subtraction* of matrices. The rules for this are as follows:

(*a*)  matrices may only be added or subtracted if they are of the same order, i.e. they each contain the same number of rows and columns;

(*b*)  the terms in the resulting matrix are given by

$$c_{ij} = a_{ij} \pm b_{ij} \qquad\qquad [B.3]$$

The following example illustrates the use of these rules:

$$\begin{bmatrix} d_{11} & d_{12} & d_{13} \\ d_{21} & d_{22} & d_{23} \end{bmatrix} \pm \begin{bmatrix} b_{11} & b_{12} & b_{13} \\ b_{21} & b_{22} & b_{23} \end{bmatrix}$$

$$= \begin{bmatrix} (d_{11} \pm b_{11}) & (d_{12} \pm b_{12}) & (d_{13} \pm b_{13}) \\ (d_{21} \pm b_{21}) & (d_{22} \pm b_{22}) & (d_{23} \pm b_{23}) \end{bmatrix}$$

Suppose

$$[D] = \begin{bmatrix} -2 & 4 & 5 \\ 6 & 8 & -3 \end{bmatrix} \quad [B] = \begin{bmatrix} 1 & 2 & 3 \\ 3 & 2 & -1 \end{bmatrix}$$

Then

$$[E] = [D] + [B] = \begin{bmatrix} -1 & 6 & 8 \\ 9 & 10 & -4 \end{bmatrix}$$

$$[F] = [D] - [B] = \begin{bmatrix} -3 & 2 & 2 \\ 3 & 6 & -2 \end{bmatrix}$$

**B.4  Inversion of a matrix**

Referring back to the set of simultaneous equations at the beginning of this appendix, the objective is usually to solve these for the unknown $x$ terms. This is where the use of matrices has a major advantage because referring to eqn. [B.1] we may rewrite this as

$$\{x\} = [A]^{-1}\{y\} \qquad [B.4]$$

This equation expresses the solution to the set of simultaneous equations in that each of the unknown $x$ terms is now given by a new matrix $[A]^{-1}$ multiplied by the known $y$ terms. The new matrix is called the *inverse* of matrix $[A]$. The determination of the terms in the inverse matrix is beyond the scope of this brief introduction. Suffice to say that it may be obtained very quickly on a computer and hence the solution to a set of simultaneous equations is determined quickly using eqn. [B.4].

**B.5  Transpose of a matrix**

The *transpose* of a matrix $[A]$ is denoted by $[A]^T$. It is determined by exchanging the rows and columns in the original matrix. Thus referring to the matrix $[A]$ at the beginning of this appendix, then

$$[A]^T = \begin{bmatrix} a_{11} & a_{21} & a_{31} & \cdots & a_{m1} \\ a_{12} & a_{22} & a_{32} & \cdots & a_{m2} \\ a_{13} & a_{23} & a_{33} & \cdots & a_{m3} \\ \vdots & \vdots & \vdots & & \vdots \\ a_{1n} & a_{2n} & a_{3n} & \cdots & a_{mn} \end{bmatrix}$$

So if

$$[A] = \begin{bmatrix} 2 & 4 \\ 6 & 8 \end{bmatrix} \quad \text{then} \quad [A]^T = \begin{bmatrix} 2 & 6 \\ 4 & 8 \end{bmatrix}$$

**B.6  Symmetric matrix**

A square matrix is one in which the number of columns is equal to the number of rows. An important type of square matrix which arises quite often in the finite element method is a *symmetric matrix*. Such matrices possess the property that $a_{ij} = a_{ji}$. An example of such a matrix is given below:

$$\begin{bmatrix} 2 & 4 & 7 & -3 \\ 4 & 5 & 1 & 9 \\ 7 & 1 & 6 & -5 \\ -3 & 9 & -5 & 4 \end{bmatrix} \quad \text{which is often written as}$$

$$\begin{bmatrix} 2 & 4 & 7 & -3 \\ & 5 & 1 & 9 \\ \text{sym.} & & 6 & -5 \\ & & & 4 \end{bmatrix}$$

# C Table of Mechanical Properties of Engineering Materials

| Material | Yield or 0.1% proof stress (MN/m²) | Tensile strength (MN/m²) | Young's modulus (GN/m²) | Shear modulus (GN/m²) | % Elongation on 50 mm | Endurance or fatigue limit at 10⁷ cycles (MN/m²) | Coeff. of thermal expansion ($\times 10^{-6}/°C$) | Density (kg/m³) | Poisson's ratio |
|---|---|---|---|---|---|---|---|---|---|
| Acrylic | – | 50–80 | 2.7–3.2 | – | 2–8 | – | 0.6 | 1200 | 0.4 |
| Aluminium (pure) | 40 | 200 | 70 | 26 | 60 | – | 23 | 2710 | 0.33 |
| Aluminium alloy | 250–450 | 320–550 | 70–72 | 26–28 | 8–17 | 120–140 | 23 | 2626–2790 | 0.33 |
| Brass | 259 | 427 | 101 | 38 | 40 | 133 | 18.5 | 8430 | 0.34 |
| Bronze | 280 | 546 | 122 | 47 | 35 | 210 | 17.5 | 7601 | 0.34 |
| Carbon steel | 370 | 602 | 208 | 82 | 30 | 287 | 12 | 7850 | 0.27–0.3 |
| Cast iron (flake) | – | 280 | 175 | – | 0.6 | 119 | 12 | 7352 | 0.2–0.3 |
| Cast iron (nodular) | – | 735 | 175 | – | 3.0 | 315 | 12 | 7352 | 0.2–0.3 |
| Concrete* | 40 | 43 | 18.5 | – | 0.45† | – | 10.8 | 2400 | 0.1–0.2 |
| Copper (pure) | 60 | 400 | 110–120 | 40–46 | 50 | – | 17 | 8900 | 0.33–0.36 |
| Douglas fir (dry) | 56 | 125 | 14 | – | – | – | – | 560 | – |
| Douglas fir (wet) | 32 | 77 | 11 | – | – | – | – | 608 | – |
| Glass | – | 30–1000 | 50–80 | 20–35 | – | – | 5–11 | 2400–2800 | 0.2–0.27 |
| Magnesium alloy | 245 | 343 | 45 | 16.5 | 12 | 133 | 26 | 1825 | 0.35 |
| Mild steel | 280 | 462 | 207 | 81 | 45 | 224 | 12 | 7850 | 0.27–0.3 |
| Nickel steel | 1000 | 1250 | 207 | 82.5 | 14 | 595 | 12 | 7850 | 0.27–0.3 |
| Ni-Cr-Mo steel | 924 | 1085 | 203 | 77 | 19 | 525 | 12.5 | 7822 | 0.27–0.3 |
| Nylon | – | 65–86 | 2.0–2.8 | – | 60–300 | – | 0.8–1.0 | 1150 | 0.4 |
| Polycarbonate | – | 56–66 | 2.0–3.0 | – | 100–130 | – | 0.4–0.7 | 1100–1250 | 0.4 |
| Polythene | – | 8–35 | 0.2–1.4 | – | 100–1200 | – | 1.3–2.5 | 914–960 | 0.4–0.45 |
| u.P.V.C. | – | 30–70 | 1.0–3.5 | – | 10–300 | – | 0.5–1.0 | 1300–1500 | 0.41 |
| Red oak (dry) | 59 | 132 | 12.5 | – | – | – | – | 691 | – |
| Rubber (hard) | – | 5–32 | 0.004 | – | 150–7000 | – | 130–200 | 860–2000 | 0.45–0.5 |
| Stainless steel | 1120 | 1295 | 196 | 87 | 9 | 616 | 17.3 | 7905 | 0.27–0.3 |
| Titanium (pure) | 400 | 500 | 110 | 40 | 25 | – | 8–10 | 4500 | 0.33 |
| Titanium alloy | 750–910 | 900–1040 | 106 | 40 | 10–12 | 490 | 8–10 | 4470–4500 | 0.33 |
| Tungsten | 1000 | 1510 | 360 | 150 | 0–4 | – | 4.3 | 1900 | 0.2 |

\* Compression.
† On 150 mm.

*Note:* This table is only intended to give the reader an indication of the wide range of properties available from various types of material. The numerical values given are fairly typical, but can be varied in most cases very considerably by factors such as composition, heat treatment, temperature, strain rate, etc. For more detailed values of properties and other materials, the reader should consult the relevant British Standard specification or one of the published handbooks on material properties, e.g. C. J. Smithells (1976) *Metals Reference Book*, 5th edition, Butterworths, Oxford.

# D Answers to Problems

**Chapter 1**

1.1 $F_R = 202$, $M_z = -3500$

1.3 $M_x = -1.2$ kNm when $F_3 = 1$, $F_3 = 18.8$ kN

1.4 $M_x = -3536$, $M_y = 11314$, $M_z = 3536$ Nm

1.5 For A, $M_x = -250$, $M_y = 300$, $M_z = 0$. For B, $M_x = 250$, $M_y = -300$, $M_z = 0$

1.6 (a) 160.9 N (b) 201.1 N

1.7 1480 kg

1.8 809 N

1.9 23.3°

1.10 -257.2 kN

1.11 0.61 MN at 30° to horizontal

1.12 Yes

1.13 $F_{AB} = 505$ N, $R_E = R_D = 400$ N

1.14 (a) $\mu \geq 0.62$ (b) 1.55 kN

1.15 2.8 kN; 27.7°

1.16 $F_A = -24.25$ kN; $F_B = -48.3$ kN; $F_C = 18.18$ kN.

1.17 $F_{AC} = 0.5$ kN; $R_v = 5.0$ kN; $R_h = 7.4$ kN.

1.18 (a) $V_A = 1960$ kN; $H_A = 2940$ kN; $H_G = 2940$ kN.
$F_{AB} = 2940$ kN; $F_{BF} = 0$; $F_{FE} = -3533$ kN.
(b) $V_A = 1710$ kN; $H_A = 2565$ kN; $H_G = 2565$ kN.
$F_{AB} = 2565$ kN; $F_{BF} = 0$ kN; $F_{FE} = -3082$ kN.

1.19 $V_A = 22.8$ kN; $H_A = 0$; $V_B = 42.8$ kN.
$F_{AB} = 0$; $F_{BC} = -42.8$; $F_{AC} - 16.1$; $F_{AD} = 36$;
$F_{DC} = -22.8$; $F_{DE} = 48.4$.
$F_{EC} = -20$; $F_{CG} = -48.4$; $F_{EG} = -3.6$; $F_{EF} = 40.3$;
$F_{FG} = -72$.
$F_{GH} = -53.5$; $F_{FH} = 69$ kN.

1.20 $F_{DE} = -15.9$ kN

1.21 $F_{HG} = -29.4$ kN; $F_{HJ} = -39.24$ kN; New $F_{HJ} = -34.33$ kN.

1.22 (d) $F_{AB} = 1225$ N, $F_{BC} = -707$ N, $Area_{AB} = 12.25 \times 10^{-6} m^2$,
$Area_{BC} = 7.07 \times 10^{-6} m^2$, Weight $= 0.221$ kg
(e) Min wt. when $h = 1.414$ m is 0.221 kg

1.23 Min. wt occurs when $h = 1.272$ m and $F_{AB} = 1272$ N,
$F_{BC} = -786$ N, Area $= 12.7 \times 10^{-6} m^2$, Weight $= 0.260$ kg

1.24 DF, $-8$; DE, 8; DA, $-20$; FE, 3; FA, 11; FC, $-28$; EC, $-4.2$; EA, $-11.3$; EB, $-9$ kN

1.25 AD, 0; AF, 100.6; BD, 0; BF, 100.6; CF, $-200.2$; DE, 50; DF, $-70.5$; EF, $-70.5$ kN

1.27 Max. Axial Force $= 2$ kN, Max SF $= 2$ kN, Max BM $= 0.4$ kN m

1.28 (a) S.F.: A, 0; B, +1; C, +1/ − 7; D, −7/ + 3; F, +3 kN.
B.M.: A, 0; B, 0; C, +1; D, −6; F, 0 kN-m.
(b) S.F.: A, −2; B, −2/ + 7; E, −11/ + 8; F, +8 kN.
B.M.: A, 0; B, −2; E, −8; F, 0 kNm.
(c) S.F.: A, 0; C, −8/ − 24; F, −36 kN.
B.M.: A, 0; C, −8; F, -98 kN-m.

(d) S.F.: A, 4; B, 4/ − 3; F, −3 kN.
B.M.: A, 0; B, 4; D, -2/6; F, 0 kNm.

1.29   S.F.: D, −29.44; C, −29.44/ − 22.08; B, −22.08/ − 7.36;
A, −7.36 kN.
B.M.: D, −147.2; C, −88.32; B, −22.08; A, 0 kNm.

1.30   At G: $F_z = 1$;   $M_x = 0.25$; At F: $F_z = 1$;   $M_x = 0.25$; $M_y = 0.25$; At E: $F_x = 2.5$;   $F_y = 2.5$; At D: $F_x = 2.5$; $F_y = 2.5$; At support A: $F_x = -2.5$;   $F_y = -2.5$;   $F_z = 0$; At support B: $F_x = +2.5$;   $F_y = +2.5$;   $F_z = +1$ kN;   $M_y = 0.125$,   $M_z = 0.05$ kNm

1.31   Torque: AB, +286.5;   BC, −668.5;   CD, −191 Nm.

1.32   B.M.: A, −6;   B, +2.8;   C, −0.4;   D, −10 kNm.

## Chapter 2

2.1   7.64 MN/m² in top of rod.

2.2   $A = \dfrac{F}{\sigma} e^{(\rho/\sigma)y}$

2.3   53 037 mm²;   10.5 MN.

2.4   1720 mm²

2.5   $T_{max} = \dfrac{wl}{2 \sin \phi}$

2.6   27.6 m;   3.68 MN/m².

2.7   9 mm.

2.8   22.4 MN/m²

2.9   12 MN/m²;   211 MN/m².

2.10   $\sigma = -W/(2\pi hx \sin \beta)$

2.11   $\sigma_\theta = 736$ kN/m²   $\sigma_L = 552$ kN/m².

2.12   70 MN/m²;   138 MN/m².

2.13   (a) 8.8 MN/m²   (b) 3.4 MN/m²   (c) 56.8 MN/m²

2.14   Key:   131 Nm;   Pin:   138 Nm.

2.15   49.5 kW;   3.15 MN/m².

2.16   Torque varies from 6.5 Nm to −20 Nm

2.17   4.56 mm.

2.18   (b) $t = 0.00456$ m

2.19   11.31 kNm.

2.20   (b) 89

## Chapter 3

3.1   0.014 mm

3.2   (a) Plane stress   (b) Plane strain   (c) Plane stress.

3.3   0.413 mm

3.5   94.5 MN/m².

3.6   0.000 941;   0.0188 rad/m.

3.7   $\sigma_x = \dfrac{E}{(1 + \nu)(1 - 2\nu)} [(1 - \nu)\varepsilon_x + \nu(\varepsilon_y + \varepsilon_z)]$, etc.

$\sigma_x = \dfrac{E}{1 - \nu^2} \{\varepsilon_x + \nu\varepsilon_y\}$, etc.

$\sigma_x = \dfrac{E}{(1 + \nu)(1 - 2\nu)} [(1 - \nu)\varepsilon_x + \nu\varepsilon_y]$, etc.

3.11  0.000 78;  0.31 mm.
3.12  162 MN/m$^2$;  194 MN/m$^2$;  $-0.0005.$
3.13  0.87 MNm/m$^3$.
3.14  37.6 kN/m$^2$/m$^3$.

## Chapter 4

4.1  122 MN/m$^2$;  19.5 MN/m$^2$
4.2  0.858 mm
4.3  5.069 mm;  5 mm.
4.4  (b) As Prob 4.1 ($F = 38.37$ kN)
     (c) As Prob 4.3
4.5  $F_A = 23.2$ kN (tensile);  $F_B = 1.8$ kN (compressive)
4.6  386 MN/m$^2$ (steel);  $-214$ MN/m$^2$ (copper).

4.7  (a)  $F_i = \dfrac{E_i A_i}{L_i}(d - d_0)$;  (b)  $d = \dfrac{F + \sum \frac{E_i A_i d_0}{L_i}}{\sum \frac{E_i A_i}{L_i}}$;

     (d)  as Prob 4.6  ($d = 0.214$ mm)

4.8  Cylinder 89.6 MN/m$^2$;  Rods 29.6 MN/m$^2$.
4.9  1.24 m from left end.
4.10  124.2 kN.
4.11  23.9 kN;  76.1 kN;  95.3°C.
4.12  16.3 MN/m$^2$ (steel);  $-65.3$ MN/m$^2$ (copper).
4.13  1634 mm$^3$.
4.14  0.151 mm;  31.4 MN/m$^2$;  26.2 mm.
4.15  6.1 MN;  34.1 MN.

## Chapter 5

5.1  $D_i = 112$ mm;  $T_h/T_s = 2.35$.
5.3  0.832 rad
5.4  3.14 kW;  19.7 mm; 11.52 mm.
5.5  163.3 mm.
5.6  CD, 6.52 MN/m$^2$;  AB, $3.7 \times 10^{-3}$ rads.
5.7  40.3 kW;  20.1 kW.
5.8  5.36 kNm.
5.9  48.34 mm;  6.85 mm.
5.10  90.3 Nm.
5.11  $\tau = 0.469$ MN/m$^2$ (2 mm wall);  $\tau = 0.33$ MN/m$^2$ (3 mm wall);
     $0.278 \times 10^{-3}$ rad/m.
5.12  1 mm;  0.142 rads.
5.13  2.75 kNm;  0.0405 rads.
5.14  324.5 Nm.
5.16  (a) Assuming  $G_a = 10^9$ and $G_s = 2.2 \times 10^9$, $\theta_s = 0.41 = 0.75\theta_a$,
     $\theta_a = 0.548$.
     (b) $\tau_s = 55 \times 10^6$ N/m$^2$,   $\tau_b = -37 \times 10^6$ N/m$^2$.

## Chapter 6

6.1  S.F.: +4, +4/ $-$ 1, $-1/$ $-$ 6, $-6$.
     B.M.: 0, +8, +7, +12, 0.
6.2  B, 1101 Nm;  C, 1353 Nm.
6.4  A, 0;  B, 170;  F, 0;  G, 60;  C, 0;  H, $-10$;  J, 10;  D,
     70;  E, 0 kNm.

6.5   $Q = -58.85x^2$;   $M = -19.62x^3$.

6.6   $\sigma_{max} = -656\,MN/m^2$ at 4.75 m from left support.

6.7

| $x$ | 0 | 1 | 2 | 3 | 4 | 5 | 6 |
|---|---|---|---|---|---|---|---|
| $Q$ | +8 | +6.8 | +3.8 | 0 | −3.8 | −6.8 | −8 |
| $M$ | 0 | 7.6 | 13.0 | 15 | 13.0 | 7.6 | 0 |

6.8   S.F.: +5.5, −3.5, −3.5/−4, −4 kN

B.M.: 0, +7.5, +4, 0 kNm; $M_{max} = 8.6$ at $x = 2.34$.

6.9   36.41 mm;   91.38 mm.

6.10   5.68 m.

6.11   (i)   $-240\,MN/m^2$   (ii)   at $y = -103.4$ mm, $z = -66.9$ mm

6.12   $h = (4B \tan \alpha)/9$

6.13   (i)   $\sigma = 3.7\,MN/m^2$ (ii)   $\sigma = 1.5\,MN/m^2$

6.14   27.5 MN/m²;   22 MN/m²

6.15   $x = \dfrac{d_1 L}{2(d_2 - d_1)}$

6.16   $M_{AB} = 560\,N\text{-}m$;   $T_{AB} = 490\,N\,m$;   $M_{BC} = 501\,N\,m$;   $T_{BC} = 123\,Nm$.

$\sigma_{AB} = 263\,MN/m^2$;   $\tau_{AB} = 115\,MN/m^2$.

6.17   $M_1/M_2 = 1.43$.

6.18   27.9 mm;   7.93 kNm.

6.19   $\sigma_{timber} = 12\,MN/m^2$,   $\sigma_{steel} = 164\,MN/m^2$

6.20   $\sigma_{timber} = -1.6\,MN/m^2$ to $+6.4\,MN/m^2$   $\sigma_{steel} = -48\,MN/m^2$ to $-32\,MN/m^2$

6.21   38 kN m;   89.1 MN/m².

6.22   −10 to +26 MN/m²;   45.84 mm.

6.23   51 m;   1.63 MN/m².

6.24   66.9 mm.

6.25   $\tau_{max}/\tau_{mean} = 4/3$.

6.26   Sides: 0 at top, 104.6 MN/m² at N.A.; 85.2 MN/m² at floor; Floor: 17 MN/m² to 0.

6.28   1.14 MN/m²;   1.04 MN/m²;   25.6 mm.

6.29   4.84 kN;   4.3 kNm.

6.30   7.5 mm.

6.31   (a) 37.5 MN/m²; (b) 1 mm

6.32   $\sigma_A = 108\,MN/m^2$;   $\sigma_B = -40\,MN/m^2$;   $\sigma_C = -110\,MN/m^2$

6.33   $\tau_{max} = 1.18\,MN/m^2$;   $\tau_{mean} = 1.15\,MN/m^2$

6.34   ±136.9 MN/m².

6.35   L, −119.2 MN/m²;   M, 70.2 MN/m²;   N, 15.1 MN/m²; 16.2° to $z$ axis.

6.36   −13.08° to $z$ axis;   54.7 MN/m²; −33.5 MN/m².

6.37   43.4 mm;   155 MN/m²;   266 MN/m².

6.38   (i) (a) at the intersection of the two plates forming the section;

(b) at the centroid;

(c) $e = \dfrac{3b^2}{d + 6b}$

(ii) $\tau_{max} = 10.8\,MPa$

6.39   2.37 kN.

6.40   $\sigma_A = 23.8\,MN/m^2$;   $\sigma_B = -15\,MN/m^2$

6.41   78 MN/m² in the larger loop.

**Chapter 7**

7.1 $-8.54$ mm.
7.2 4.92 mm.
7.3 5.62 m; 1.77 mm.
7.5 at $x = 10$, $v = 33.3$
7.6 (a) $\frac{a}{b} = 0.586$ (b) $\frac{a}{b} = 0.554$
7.7 A, 8.66 mm; B, 2.27 mm; C, 4.98 mm.
7.8 19.2 mm.
7.11 $L/a = 1.24$.
7.12 $-0.5$ mm; 1.48 mm; $5.14 \times 10^{-3}$ rads.
7.13 28.1 mm; 32.3° anticlockwise from $z$ axis.
7.14 0.2 mm.
7.15 $\pm 0.0014$ rads.
7.16 (a) 2.44 mm; 152.2 MN/m$^2$ (b) 1.09 mm; 68 MN/m$^2$.

**Chapter 8**

8.1 64.8 kN/m.

8.2 $\dfrac{57}{128}wL$; $\dfrac{7}{128}wL$; $\dfrac{9}{128}wL^2$; $\dfrac{13wL^4}{6144EI}$.

8.3 $\dfrac{\bar{M}L}{16EI}$; $v = \dfrac{\bar{M}}{EI}\left[\dfrac{x^3}{4L} - \dfrac{5x^2}{8} + \dfrac{Lx}{2} - \dfrac{L^2}{8}\right]$.

8.4 15 mm.
8.5 $-250$ Nm; $-187.5$; $+2938$; $+2250$ N.
8.6 63.8 mm
8.7 (a) $3\bar{M}/2L$; $\bar{M}/4$. (b) $13wL^4/6144EI$.
8.8 0, 80, 5, 37.5, 0 kNm.
8.9 S.F.: $-4.06$; $-4.06/10.62$; $10.62/-9.38$; $-9.38/-2.18$; $-2.18/10$
B.M.: 0; $-40.63$; 65.6; $-28.13$; $-50$; 0
8.10 S.F.: 22.1; $-27.9/1.9$; $1.9/9; 9/-11$; $11/0$
B.M.: 0; 48.1; $-28.8$; $-9.6$; 44.8; 0.
8.11 $L_1 = L_2 = 0.375L$; $L_3 = 0.25L$.

**Chapter 9**

9.1 Long bolts.
9.2 $\sigma_b = 3\sigma_t$.
9.3 $U_u/U_w = 1.6$.
9.4 57.9 mm.
9.5 4 kN/m; 10 coils.
9.6 137 N; 4.1mm.
9.7 90.4 mm.
9.8 $2W/EI$.
9.9 43.1 N.
9.12 $H = -W\pi$; $V = W/2$; $M_{\max} = 0.093Wr$; $\delta_v = 0.019Wr^3/EI$.
9.13 $\delta_v = \pi Wr^3/8EI$; $\delta_h = -Wr^3/2EI$.
9.14 $F = EI\delta/3\pi R^3$.
9.16 0.412 mm vertical; 0.23 mm horizontal.

9.18 33 mm.
9.19 (a) Energy $= F^2L/2AE$ (b) Energy $= (\frac{3}{4})(F^2L/2AE)$.
9.20 0.071 rad; 424 MN/m$^2$.

**Chapter 10**

10.1 2 kN.
10.2 1.88 kN.
10.3 50 mm.

10.4 (i) $P = 4ht\sigma_y$; (ii) $P = \dfrac{\pi^2 Eth^3}{8L^2}$; (iii) $P = \dfrac{k\pi^2 Et}{3(1-v^2)h}$

10.5 (i) 20 mm$^2$ (ii) 90.2 mm$^2$
10.6 90.2 mm$^2$.
10.7 1026 kN.
10.8 $1.67\pi^2 EI/L^2$.
10.9 1.33 kN.
10.10 54.65 kN.
10.11 $-6.72$ MN/m$^2$.

10.12 (a) $\dfrac{b}{h} < \sqrt{\dfrac{kE}{\sigma_y}}$ (b) (i) 14.8 (ii) 27.5 (c) $h = 0.89$ mm,

$b = 13.3$ mm.

10.13 (a) buckling would occur (b) 2.4.
10.14 30 mm.
10.16 37.8 MN/m$^2$.
10.17 $h = 0.5$ for minimum weight.

**Chapter 11**

11.1 (a) $\sigma_x = 100$; $\sigma_y = 200$ MN/m$^2$ (b) Top: $\sigma_x = 240$ MN/m$^2$;
$\tau_{xy} = 0$ : Neutral axis: $\tau_{xy} = 6$ MN/m$^2$ (c) $\sigma_x = -10.2$;
$\tau_{xy} = 122$ MN/m$^2$.
11.2 (a) $\sigma_n = 199.6$; $\tau_s = -5.98$ MN/m$^2$ (b) $\sigma_n = 30$;
$\tau_s = -110$ MN/m$^2$ (c) $\sigma_n = -180.8$; $\tau_s = -3.35$ MN/m$^2$.
11.3 55.9, 21.5 MN/m$^2$.
11.4 As for 11.2.
11.6 60.4; $-10.4$ MN/m$^2$.
11.7 (a) $-47$; 67 MN/m$^2$; (b) 60, 20, 30, 17 MN/m$^2$.
11.8 A: 228, $-28$, 128;
B: $+80$, $-80$, 80;
C: 28, $-228$, 128 MN/m$^2$
11.9 46.3 MN/m$^2$; 69.3, $-23.1$ MN/m$^2$.
11.10 12.5, $-0.3$ MN/m$^2$.
11.11 2.25; 6.75 kN.
11.12 132 MN/m$^2$; 88.13 MN/m$^2$.
11.13 $W = 29.3$ kN; $F = -11.7$ kN
11.14 $7.17 \times 10^{-4}$; $-6.17 \times 10^{-4}$.
11.15 287, $-200$ MN/m$^2$; 243 MN/m$^2$.
11.16 141 MN/m$^2$; 2.36 kNm.
11.17 (i) $\varepsilon_1 = 1000 \times 10^{-6}$, $\varepsilon_2 = 625 \times 10^{-6}$ (ii) $\sigma_1 = 91.3$ MN/m$^2$,

$\sigma_2 = 71.2 \text{ MN/m}^2$  (iii) 86.3 MN/m²

11.18  1.70 kNm.

11.19  $11.3 \times 10^{-4}$, $6.18 \times 10^{-4}$;  18° anticlockwise from gauge A.

11.20  153.7;  −64.6;  109.2 MN/m²

11.21  $E = 69.9$;  $G = 26.9$;  $K = 58.3 \text{ GN/m}^2$;  $v = 0.3$.

11.22  54.7°.

11.23  (a) $0.75\sigma_0$, $0.25\sigma_0$, $0.433\sigma_0$; (b) 277 MN/m² by shear.

11.24  $\varepsilon_x = -2.45 \times 10^{-3}$;  $\varepsilon_y = 6.85 \times 10^{-3}$.

11.25  $E_x = 15.7$;  $E_y = 11.3 \text{ GN/m}^2$.

## Chapter 12

12.1  0.96 kNm;  0.83 kNm.

12.2  4.45 kN.

12.3  1.06 MN;  −1.06 MN.

12.4  4.6 mm.

12.5  yielding would occur.

12.6  143 N.

12.7  45.4 mm.

12.8  15.1 mm.

12.9  66.7 MN/m²;  424 kN.

12.10  520 MN/m².

12.11  3.37 MN/m².

12.12  0.265 mm;  61.7;  60.55 MN/m².

12.13  49.7 kN.

12.14  7.53 mm.

12.15  3.35 mm;  6.4.

## Chapter 13

13.3  Cylinder axially constrained.

13.4  $(a)\ \dfrac{\partial \gamma_{xy}}{\partial z} = \dfrac{\partial \gamma_{xz}}{\partial y} + \dfrac{\partial \gamma_{yz}}{\partial x};$  $(b)\ \dfrac{\partial \varepsilon_z}{\partial \theta} = r\dfrac{\partial \gamma_{\theta z}}{\partial z}$

## Chapter 14

14.2  1.014;  1.016 MN.

14.3  1.915;  110;  30 MN/m².

14.4  65, 66.7 MN/m²;  1.05.

14.5  $\sigma_\theta = \dfrac{a^2 - 19b^2}{(b^2 - a^2)}$

14.6  Outer: $\varepsilon_r = \dfrac{-3vp}{E(k^2 - 1)}$,  $\varepsilon_\theta = \dfrac{p(2 - v)}{E(k^2 - 1)}$,  $\varepsilon_z = \dfrac{p(1 - 2v)}{E(k^2 - 1)}$

Inner: $\varepsilon_r = \dfrac{-p}{E(h^2 - 1)}\{k^2(1 + v) + 2v - 1\}$,

$\varepsilon_\theta = \dfrac{-p}{E(k^2 - 1)}\{k^2(1 + v) + 1 - 2v\}$,  $\varepsilon_2 = \dfrac{p}{E(k^2 + 1)}(1 - 2v)$

14.7  396 MN/m²;  $28.7 \times 10^{-4}$.

14.9  18.4 MN/m².

14.10  (a) 654 kN  (b) 1308 kN  (c) 330 MN/m².

14.11  (a) $10^{-5}$ m  (b) 12.6 kN  (c) $0.5 \times 10^{-5}$ m  (d) plate.

14.12 22 kN.

14.13 (a) 5.93 MN/m$^2$, 4.81 MN/m$^2$   (b) 4.81 MN/m$^2$.

14.14 Inner: $-63$, $-49.2$ MN/m$^2$;   Outer: 49.3, 35.5 MN/m$^2$.

14.15 1.25 mm;   1000 mm.

14.16 (i) 54   (ii) 139.3, 85.6 MN/m$^2$.

14.17 20.3 MN/m$^2$.

14.18 Outside, 2.21;   Inside, 7.46 MN/m$^2$.

14.19 2.83 Nm

14.20 1.61, 13.56 MN/m$^2$.

14.21 (a) $\left(\dfrac{3+\nu}{16}\right)\rho\omega^2 r_0^2$   (b) GFRP.

14.22 $\sigma_\theta = \dfrac{2pr_1^2}{r_2^2 - r_1^2} + p\left(\dfrac{\pi n}{60}\right)^2 \{3r_1^2 + r_2^2 + \nu(r_1^2 - r_2^2)\}$

14.23 0.128 mm; 12.7 MN/m$^2$.

14.24 Assume 1 atmosphere $= 0.1$ MN/m$^2$ pressure difference and clamped edges, thickness $= 5.3$ mm.

14.25 7331 rev/min.

---

**Chapter 15**

15.1 (a) 1.7   (b) 2   (c) 1.272   (d) 1.125.

15.2 175 mm;   1.93 m from each end.

15.3 22.4 kN.

15.4 10.4 mm.

15.6 $w_p = 11.7 M_p/L^2$;   0.414L from free end.

15.7 $F_p = M_p$.

15.8 56.5 mm;   3°;   31.4 kN-m.

15.9 $5.5 \times 10^{-3}$;   1.32.

15.10 5328 N-m.

15.12 44.5%.

15.13 250, 150 MN/m$^2$.

15.14 10.6 mm;   3.47 kN.

15.15 $-17.4$ MN/m$^2$.

15.16 15 960 rev/min;   10%;   6.85 mm.

---

**Chapter 16**

16.2 (i) 1.55   (ii) 0.45.

16.3 3.1 mm.

16.4 $-1.13$ mm.

16.5 0.25 mm;   166, 48 MN/m$^2$.

16.6 for $x < 0$, $w = -\dfrac{pa^2}{2Eh}(e^{+\beta x}\cos\beta x)$;

for $x > 0$, $w = -\dfrac{pa^2}{2Eh}(e^{-\beta x}\cos\beta x - 2)$.   At $x = 0$, $M_x = 0$.

16.7 68.8, 19.9 MN/m$^2$.

16.8 $M_r = -1.75 D\alpha$.

16.9 0.840 mm.

16.10 0.47 MN.

16.11 6.45 mm.

16.12 0.113 mm.
16.13 $-90.64$;  $+90.64\,\text{kN/m}^2$.
16.14 1.5 m,  $2.94\,\text{MN/m}^2$;  2.25 m,  $2.2\,\text{MN/m}^2$.
16.15 8.2,  $4.69\,\text{MN/m}^2$.
16.16 $487\,\text{MN/m}^2$.
16.17 $5.03\,\text{MN/m}^2$.
16.18 $4.22\,\text{MN/m}^2$.

## Chapter 17

17.1 $\theta_1 = 0$, $\theta_2 = 0.0490$, $\theta_3 = 0.0857$ (Example 5.3)
  $\theta_1 = \theta_3 = 0$, $\theta_2 = 0.0108$ (Example 5.4)
17.2 $-0.15$, $-0.325\,\text{mm}$; 65.4, 9.6 kN.
17.3 $-0.37\,\text{mm}$;  9.6, $-16$, $-12\,\text{kN}$.
17.4 $-1.64\,\text{mm}$.

17.6 $(a)\ v = \dfrac{Wl^3}{48EI}$  $(b)\ v = \dfrac{5}{384}\dfrac{Wl^4}{EI}$

## Chapter 18

18.1 $198\,\text{GN/m}^2$;  $270\,\text{MN/m}^2$;  $315\,\text{MN/m}^2$;  $451\,\text{MN/m}^2$;
  24.4%.
18.2 $67.6\,\text{GN/m}^2$;  $290\,\text{MN/m}^2$;  $335\,\text{MN/m}^2$;  $413\,\text{MN/m}^2$.
18.4 16.
18.5 Wrong tempering temperature.

## Chapter 19

19.2 $36.3\,\text{MN m}^{-3/2}$.
19.3 $K_{IC} = 39.4\,\text{MN m}^{-3/2}$, estimate OK.
19.4 2380;  261;  200;  5.9;  4.0;  0.75 mm.
19.5 Not acceptable.
19.6 Test is valid, $K_{IC} = 149\,\text{MN m}^{-3/2}$.
19.7 1.82 m.
19.8 Yes.
19.9 255 kN.

19.10 $(b)\ K = \sigma\sqrt{\dfrac{\pi L}{2}}$.

19.11 (i) 14 mm  (ii) 12.4 mm.
19.12 0.305 mm.
19.13 OK, $K_I = 23 < K_{IC}$
19.14 0.6 mm;  1.5%.
19.15 $62.8\,\text{MN m}^{-3/2}$.

19.16 $(a)$ (i) $\sigma_0 = \dfrac{K_{IC}}{3.36\sqrt{\pi L}}$  (ii) $\sigma_0 = \dfrac{K_{IC}}{\sqrt{\pi(r + L)}}$

  $(b)\ \dfrac{r}{L} = 10.3$. For $r \ll L$, use (i).

19.17 $166\,\text{MN m}^{-3/2}$;  14.1%.
19.18 $142\,\text{MN m}^{-3/2}$;  1.07 mm.
19.19 0.106 Nm;  5.08 Nm.

**Chapter 20**

20.1   9.73 mm;   23.28 mm.
20.3   Medium–strength steel.
20.4   2.13 MN/m$^2$.
20.5   Goodman, 5.43;   Gerber, 6.6;   Soderberg, 4.6 mm.
20.6   69, 78.4 mm.
20.7   155 kN.
20.8   At central hole.
20.9   High notch sensitivity.
20.10  50 000 cycles.
20.11  11 363 flights.
20.12  10.7 MN/m$^2$.
20.13  518 MN/m$^2$.

**Chapter 21**

21.1   105 hours.
21.2   262 hours.
21.3   544°C.
21.4   7.1 minutes.
21.5   3.31 hours.
21.6   1.1 mm.
21.7   104 days.
21.8   70.8 N;   13.9 days.
21.9   4 mm; 1.1 mm.

# Index

# The Post-Colonial Studies Reader

*The Post-Colonial Studies Reader* is the essential introduction to the most important texts in post-colonial theory and criticism. Updating and expanding the coverage of the highly successful first edition, this second edition offers 121 extracts from key works in the field, arranged in clearly introduced parts.

Leading figures in the areas of post-colonial writing, theory and criticism are represented, as are critics who are as yet less well known. As in the first edition, the Reader ranges as widely as possible in order to reflect the remarkable diversity of work in the discipline and the vibrancy of anti-imperialist writing both within and without the metropolitan centres. Covering more debates, topics and critics than any comparable book in its field, *The Post-Colonial Studies Reader* provides the ideal starting point for students and issues a potent challenge to the ways in which we think and write about literature and culture.

**Bill Ashcroft** is Professor of English at the University of New South Wales, **Gareth Griffiths** is Professor of English at the University of Western Australia and **Helen Tiffin** is Professor of English at Queen's University, Ontario. They have each published widely in the field of post-colonial studies and are joint authors of the highly influential *The Empire Writes Back: Theory and Practice in Post-Colonial Literatures* (Routledge, 1989; second edition 2002) and *Key Concepts in Post-Colonial Studies* (Routledge, 1998).

# The Post-Colonial Studies Reader

## Second edition

Edited by

# Bill Ashcroft,

# Gareth Griffiths and

# Helen Tiffin

 Routledge
Taylor & Francis Group

LONDON AND NEW YORK

First edition published 1995
by Routledge
2 Park Square, Milton Park, Abingdon, Oxford OX14 4RN

Simultaneously published in the USA and Canada
by Routledge
270 Madison Ave, New York, NY 10016

Reprinted 1995, 1997, 1999 (twice), 2001, 2003, 2004

Second edition published 2006

Reprinted 2007 (twice), 2008

*Routledge is an imprint of the Taylor & Francis Group, an informa business*

© 1995, 2006 Bill Ashcroft, Gareth Griffiths and Helen Tiffin for editorial
and introductory material; individual extracts © the contributors

Typeset in Perpetua and Bell Gothic by
Florence Production Ltd, Stoodleigh, Devon
Printed and bound in Great Britain by
TJ International Ltd, Padstow, Cornwall

*British Library Cataloguing in Publication Data*
A catalogue record for this book is available from the British Library

*Library of Congress Cataloging in Publication Data*
   The post-colonial studies reader/edited by Bill Ashcroft, Gareth Griffiths &
      Helen Tiffin. – 2nd ed.
         p. cm.
      Includes bibliographical references (p.   ) and index.
   1. Commonwealth literature (English) – History and criticism.
   2. English literature – Developing countries – History and criticism.
   3. Postcolonialism – Commonwealth countries.   4. Postcolonialism in literature.
   5. Decolonization in literature.   6. Imperialism in literature.   7. Colonies in
   literature.   I. Ashcroft, Bill, 1946–   II. Griffiths, Gareth, 1943–
   III. Tiffin, Helen
      PR9080.P57 2005
      820.9'358 – dc22                                                    2005012943

ISBN10: 0–415–34564–2 (hbk)
ISBN10: 0–415–34565–0 (pbk)

ISBN13: 978–0–415–34564–4 (hbk)
ISBN13: 978–0–415–34565–1 (pbk)

# Contents

## PART TWO
## Universality and Difference

## PART THREE
## Representation and Resistance

## PART SIX
## Indigeneity

## PART NINE
### Feminism

## PART TEN
### Language

## PART ELEVEN
## The Body and Performance

## PART TWELVE
## History

## PART FIFTEEN
## Production and Consumption

# PART SIXTEEN
## Diaspora

# PART SEVENTEEN
## Globalization

# Figures

# Preface

This is the latest in a number of Readers published by Routledge and joins such earlier titles as *The Cultural Studies Reader*. The publishers insisted that the title of *The Post-Colonial Studies Reader* be congruent with the other Readers that they publish. The authors are equally at pains to insist therefore that the title is not meant to claim some kind of complete- ness of coverage or absolute authority. In a field as diverse and contentious as post-colonial studies such a claim would be particularly extravagant and foolish. However, the 121 extracts in this Reader are designed to introduce the major issues and debates in the field of post- colonial literary studies. This field itself has become so heterogeneous that no collection of readings could encompass every theoretical position now giving itself the name 'postcolonial/ post-colonial'. These terms themselves encapsulate an active and unresolved dispute between those who would see the post-colonial as designating an amorphous set of discursive prac- tices, akin to postmodernism, and those who would see it as designating a more specific, and 'historically' located set of cultural strategies. Even this latter view is divided between those who believe that post-colonial refers only to the period after the colonies become independent and those who argue, as the editors of this book would, that it is best used to designate the totality of practices, in all their rich diversity, which characterize the societies of the post- colonial world from the moment of colonization to the present day, since colonialism does not cease with the mere fact of political independence and continues in a neo-colonial mode to be active in many societies.

The structure of the Reader, the choice of subject areas and the selection and excisions of the readings are naturally determined by the editors' preferences and thus amount to a theoretical statement. But we have tried to introduce arguments with which we are not neces- sarily in agreement, and we have tried to produce a Reader that is above all a stimulus to discussion, thought and further exploration. The parameters we have chosen will no doubt seem unsatisfactory to some: in order to achieve as wide a representation of areas and approaches as possible most extracts are limited to about two thousand words and will thus often not encompass the whole argument of the pieces from which they are taken; some theorists may seem to be under-represented given their importance to the field; some of the writers would not be considered 'post-colonial' theorists at all. But each extract is selected to say something coherent about an issue of immediate relevance to post-colonial practice, and represents what we have taken to be the most interesting, provocative or stimulating aspect of the original. Obviously, cultural and political critiques by general theorists such as

Foucault, Derrida, Terdiman, Gramsci, Althusser, etc. have been influential in the construction of many post-colonial critical accounts but we have not included these in the Reader
since they are already easily accessible. This Reader is not a collection of theorists, but of
ideas; it is not interested in establishing a canon of theories or theorists but in indicating
something of the great scope, the rich heterogeneity and vast energy of the field of post-
colonial studies. We have been economical with notes, and if students or scholars wish to
investigate the full argument and the range of sources of some of these pieces we direct them
to the originals.

# Acknowledgements

Abdul R. JanMohamed, 'The Economy of Manichean Allegory: The Function of Racial Difference', in *Critical Inquiry*, 12(1) 1985. Published by the University of Chicago and reproduced by permission of the University of Chicago Press.

'Introduction' from *Orientalism*, by Edward W. Said, copyright © 1978 by Edward W. Said. Used by permission of Pantheon Books, a division of Random House, Inc.

Reprinted by permission of the publisher from *A Critique of Post-colonial Reason: Toward a History of the Vanishing Present*, by Gayatri Chakravorty Spivak, Cambridge, Mass.: Harvard University Press, Copyright © 1999 by the President and Fellows of Harvard College.

Homi K. Bhabha, an excerpt from 'Signs Taken for Wonders: Questions of Ambivalence and Authority Under a Tree Outside Delhi, May 1817', in *Critical Inquiry*, 12(1) 1985. Published by the University of Chicago and reproduced by permission of the University of Chicago Press and the author.

Benita Parry, 'Problems in Current Theories of Colonial Discourse', in *Oxford Literary Review* 9(1–2) 1987. Reproduced by permission of the *Oxford Literary Review*.

'The Scramble for Post-colonialism', Stephen Slemon, from Chris Tiffin and Alan Lawson (eds), *De-Scribing Empire*, 1994, Routledge. Reproduced by permission of the publisher.

Excerpts from Nicholas B. Dirks, *Colonialism and Culture*, University of Michigan Press. Copyright © by the University of Michigan 1992. Reproduced by permission of the publisher.

Excerpts from 'The Intimacy of Tyranny', in Achille Mbembe, *On the Postcolony*, University of California Press, Copyright © 2001. Reproduced by permission of the publisher.

Biodun Jeyifo, 'The Nature of Things: Arrested Decolonization and Critical Theory', *Research in African Literatures*, 21(1) 1990, Indiana University Press. Reproduced by permission of the publisher and author.

Excerpts from *Hopes and Impediments: Selected Essays 1965–1987*, by Chinua Achebe, copyright © 1988 by Chinua Achebe. Used by permission of Doubleday, a division of Random House Inc., and David Higham Associates.

Charles Larson, 'Heroic Ethnocentrism: The Idea of Universality in Literature', in *American Scholar* 42(3) (Summer) 1973. Reproduced by permission of the author.

Alan J. Bishop, 'Western Mathematics: The Secret Weapon of Cultural Imperialism', reprinted by permission of Sage Publications Ltd from *Race and Class* 32(2) 1990. Copyright © Sage (1990).

Aijaz Ahmad, 'Jameson's Rhetoric of Otherness and the "National Allegory"', in *Social Text*, 17 (Fall), pp. 3–28. Copyright, 1987, Duke University Press. All rights reserved. Used by permission of the publisher.

Tsenay Serequeberhan (1997), 'The Critique of Eurocentrism', in Emmanuel Eze (ed.), *Postcolonial African Philosophy*, Oxford: Blackwell Publishing. Reproduced by permission of the publisher.

From *Culture and Imperialism*, by Edward W. Said, copyright © 1993 by Edward W. Said. Used by permission of Alfred A. Knopf, a division of Random House, Inc.

Extract from *Culture and Imperialism*, by Edward W. Said, published by Chatto & Windus, 1993. Used by permission of The Random House Group Limited.

Stephen Slemon, from 'Unsettling the Empire: Resistance Theory for the Second World', *World Literature Written in English*, 30(2) 1990.

Sara Suleri, an excerpt from *The Rhetoric of English India*, copyright © 1992 by the University of Chicago. Published by the University of Chicago and reproduced by permission of the University of Chicago Press.

Robert Stam and Louise Spence, 'Colonialism, Racism and Representation: An Introduction', *Screen*, 24(2) 1983, pp. 2–20. Reproduced by kind permission of the authors and *Screen*.

Elleke Boehmer, excerpts from *Empire, The National, and the Postcolonial 1890–1920*, 2002, Oxford University Press. By permission of Oxford University Press.

Frantz Fanon, 'On National Culture' and 'The Pitfalls of National Consciousness' from *The Wretched of the Earth* (trans. Constance Farrington), New York: Grove Press, 1968 (original French edition 1961). Reprinted by permission of HarperCollins Publishers Ltd and Grove/Atlantic, Inc. Copyright © 1963 by Presence Africaine.

Benedict Anderson (1983), *Imagined Communities*, London: Verso. Reproduced by permission of the publisher.

Partha Chatterjee, 'Nationalism as a Problem', from *Nationalist Thought and the Colonial World: A Derivative Discourse*, 1986, published by Zed Books for United Nations University. Reproduced by permission of the publisher.

'The National Longing for Form', Timothy Brennan, in Homi K. Bhabha (ed.), *Nation and Narration*, 1990, Routledge. Reproduced by permission of the publisher.

Excerpts from 'Dissemination: Time, Narrative, and the Margins of the Modern Nation', in Homi K. Bhabha (ed.), *Nation and Narration*, 1990, Routledge. Reproduced by permission of the publisher.

Kirsten Holst Petersen and Anna Rutherford (1976), 'Fossil and Psyche', from *Enigma of Values*, Aarhus: Dangaroo Press. Reproduced by kind permission of Kirsten Holst Petersen.

'Named for Victoria, Queen of England', by Chinua Achebe, copyright © 1973 Chinua Achebe. First published in *New Letters*, (Fall) 1973. Reprinted by permission of *New Letters* and the Curators of the University of Missouri-Kansas City, the author and Harold Ober Associates.

Michael Dash, 'Marvellous Realism: The Way out of Négritude', in *Caribbean Studies*, 13(4) 1974. Reproduced by permission of *Caribbean Studies*.

Homi K. Bhabha, 'The Commitment to Theory', *New Formations*, 5, 1988, pp. 5–23. Reproduced by permission of Lawrence and Wishart, London and the author.

Robert Young, *Colonial Desire*, 1995, Routledge. Reproduced by permission of the publisher.

Gareth Griffiths, 'The Myth of Authenticity', from Chris Tiffin and Alan Lawson (eds), *De-Scribing Empire*, 1994, Routledge. Reproduced by permission of the publisher.

Marjorie Fee, 'Why C.K. Stead didn't like Kerry Hulme's *The Bone People* or Who Can Write as the Other', *Australian and NZ Studies in Canada*, vol 1. Reproduced by kind permission of the author.

Terry Goldie, 'The Representation of the Indigene', from *Fear and Temptation: The Image of the Indigene in Canadian, Australian and New Zealand Literatures*, 1989, McGill-Queens University Press. Reproduced by permission of the publisher.

'Postcolonialism, Ideology and Native American Literature', reprinted from *The Turn to the Native: Studies in Criticism and Culture*, by Arnold Krupat, by permission of the University of Nebraska Press. © 1996 by the University of Nebraska Press.

James Clifford, 'Indigenous Articulations', *The Contemporary Pacific*, 3(2) (Fall) 2001, p. 468. Reproduced by permission of the University of Hawaii Press and the author.

Diana Brydon, 'The White Inuit Speaks: Contamination or Literary Strategy', in Ian Adam and Helen Tiffin (eds), *Past the Last Post*, 1991, New York, London: Harvester Wheatsheaf. Reproduced by kind permission of Ian Adam.

Excerpts from *Beyond Ethnicity: Consent and Descent in American Culture*, by Werner Sollors, copyright © 1986 by Werner Sollors. Used by permission of Oxford University Press, Inc.

Phillip Gleason, 'Identifying Identity', in *The Journal of American History*, 69(4) (March) 1983, pp. 910–31. Copyright © Organization of American Historians. Reprinted with permission.

Copyright 1991, from *When the Moon Waxes Red: Representation, Gender and Cultural Politics*, by Trinh T. Min-ha. Reproduced by permission of Routledge/Taylor & Francis Books, Inc.

Stuart Hall, 'New Ethnicities', in *Black Film, British Cinema*, ICA Documents 7, 1989. Reproduced by kind permission of the author.

Mike Hill (2004), *After Whiteness: Unmaking An American Majority*, New York: New York University Press. Reproduced by permission of the publisher and author.

Gloria Anzaldúa (1987), from *Borderlands/La Frontera: The New Mestiza.* Copyright © 1987, 1999 by Gloria Anzaldúa. Reprinted by permission of Aunt Lute.

Reprinted by permission of the publisher from *On Human Diversity: Nationalism, Racism, and Exoticism in French Thought,* by Tzvetan Todorov, translated by Catherine Porter, pp. 90–128, Cambridge, Mass.: Harvard University Press, Copyright © 1993 by the President and Fellows of Harvard College.

Henry Louis Gates (1986), 'Writing Race', in *Race Writing and Difference,* The University of Chicago Press. Reproduced by permission of the publisher.

Homi K. Bhabha, 'Race, Time and the Revision of Modernity', in *The Location of Culture,* 1994, Routledge. Reproduced by permission of the publisher and the author.

Kwame Appiah, 'The Illusions of Race', in Emmanuel Eze (ed.), *African Philosophy: An Anthology,* 1998, Oxford: Blackwell Publishing. Reproduced by permission of the publisher.

Paul Gilroy (1987), *There Ain't No Black in the Union Jack,* Routledge. Reproduced by permission of the publisher.

Pal Ahluwalia, 'Negritude and Nativism', in *Politics and Post-colonial Theory: African Inflections,* 2001, Routledge. Reproduced by permission of the publisher.

Kirsten Holst Petersen, 'First Things First: Problems of a Feminist Approach to African Literature', *Kunapipi,* 6(3) 1984. Reproduced by permission of Kirsten Holst Petersen.

Ketu H. Katrak, 'Decolonizing Culture: Toward a Theory for Post-colonial Women's Texts', *Modern Fiction Studies,* 35(1) 1989, pp. 157–79. © Purdue Research Foundation. Reprinted with permission of The Johns Hopkins University Press.

Chandra Talpade Mohanty, 'Under Western Eyes: Feminist Scholarship and Colonial Discourse', in *Boundary 2,* 12(3), 13(1), pp. 333–58. Copyright, 1984, Duke University Press. All rights reserved. Used by permission of the publisher.

Trinh T. Minh-ha, 'Writing Postcoloniality and Feminism', from *Woman, Native, Other: Writing Postcoloniality and Feminism,* 1989, Indiana University Press. Reproduced by permission of the publisher.

Sara Suleri, 'Woman Skin Deep: Feminism and the Postcolonial Condition', in *Critical Inquiry,* 18(4) 1992. Published by the University of Chicago and reproduced by permission of the University of Chicago Press.

Excerpts from Oyewûmí, Oyèrónké (1997), *The Invention of Women: Making an African Sense of Western Gender Discourses,* The University of Minnesota Press. Copyright 1997 by The Regents of the University of Minnesota. Reproduced by permission of the publisher.

Adaptation reprinted by permission from *Decolonising the Mind* by Ngũgĩ wa Thiong'o. Copyright © 1981, 1982, 1984, 1986 by Ngũgĩ wa Thiong'o. Published by Heinemann, a division of Reed Elsevier, Inc., Portsmouth, N.H., and James Currey, London. All rights reserved. Reproduced by permission of the publishers.

Chinua Achebe (1989), 'Politics and Politicians of Language in African Literature', Doug Killam (ed.), *FILLM Proceedings,* Ont.: University of Guelph. Reproduced by kind permission of David Higham Associates on behalf of the author, and by Douglas Killam.

'Introduction: Alchemy of English', from Braj B. Kachru (1990), *Alchemy of English: The Spread Functions and Models of Non-Native Englishes*. Used with the permission of the University of Illinois Press.

Raja Rao (1937), Preface to *Kanthapura*, reproduced by permission of Oxford University Press India, New Delhi.

Bill Ashcroft (2001), 'Language and Transformation', from *Post-colonial Transformation*, Routledge. Reproduced by permission of the publisher.

Chantal Zabus, 'Relexification', from *The African Palimpsest: Indigenization of Language in the West African Europhone Novel*, Cross Cultures 4, Editions Rodopi, 1991. Reproduced by permission of the publisher.

Frantz Fanon (1952), 'The Fact of Blackness', from *Black Skin, White Masks* (trans. Charles Lam Markmann). Copyright © 1967 by Grove Press, Inc. Used by permission of Grove/Atlantic, Inc. *Peau noire, Masques blancs* by Frantz Fanon © Editions du Seuil, 1971.

Michael Dash, 'In Search of the Lost Body: Redefining the Subject in Caribbean Literature', *Kunapipi*, 2(1) 1989. Reproduced by kind permission of the author.

Russell McDougall, 'Achebe's *Arrow of God*: The Kinetic Idiom of an Unmasking', *Kunapipi*, 9(2) 1989. Reproduced by kind permission of the author.

Helen Gilbert, 'The Dance as Text in Contemporary Australian Drama: Movement and Resistance Politics', in *Ariel*, 23(1) 1992. Reproduced by the permission of the Board of Governors, University of Calgary, Calgary, Alberta.

Gillian Whitlock, 'Outlaws of the Text', from *The Intimate Empire: Reading Women's Autobiography*. Copyright © Gillian Whitlock 2000. Reprinted by permission of The Continuum International Publishing Group.

Marvin Carlson, 'Introduction', *Performance: A Critical Introduction*, 1996, Routledge. Reproduced by permission of the publisher.

Reina Lewis, 'On Veiling, Vision and Voyage: Cross-cultural Dressing and Narratives of Identity, *Interventions*, 1(4) 1999, pp. 500–20. Reproduced by permission of Taylor and Francis. www.tandf.co.uk/journals.

Excerpts from *Inventing A-M-E-R-I-C-A: Spanish Historiography and the Formation of Eurocentrism*, by José Rabasa. Copyright © 1993 by the University of Oklahoma Press, Norman. Reprinted by permission.

Peter Hulme (1986), *Colonial Encounters: Europe and the Native Caribbean 1492–1797*, London: Methuen. Reproduced by kind permission of the author.

Paul A. Carter, excerpts from *Road to Botany Bay: An Essay in Spatial History*, 1987. Reproduced by permission of Random House, Inc. and Faber & Faber Ltd.

Derek Walcott, 'The Muse of History', from *Is Massa Day Dead?*, by Orde Coombs, copyright © 1974 by Doubleday, a division of Bantam Doubleday Dell Publishing Group, Inc. Used by permission of Doubleday, a division of Random House, inc.

Dipesh Chakrabarty, 'Postcoloniality and the Artifice of History', © 1992 by The Regents of the University of California. Reprinted from *Representations*, 37 (Winter) 1992, pp. 1–26, by permission.

Dennis Lee, 'Cadence, Country, Silence: Writing in Colonial Space', in *Boundary 2*, (3)1, copyright, 1974, Duke University Press. All rights reserved. Used by permission of the publisher.

Graham Huggan, 'Decolonizing the Map', in *Ariel*, 20(4) 1989. Reproduced by the permission of the Board of Governors, University of Calgary, Calgary, Alberta.

Excerpts from Malcolm Lewis (1998), *Cartographic Encounters: Perspectives on Native American Mapmaking and Map Use*, copyright © 1998 by the University of Chicago. Published by the University of Chicago and reproduced by permission of the University of Chicago Press.

Excerpts from Walter D. Mignolo (1995), *The Dark Side of the Renaissance: Literacy, Territoriality and Colonization*, The University of Michigan Press. Copyright © by the University of Michigan 1992. Reproduced by permission of the publisher.

Gauri Viswanathan, 'The Beginnings of English Literary Study in British India', in *Oxford Literary Review*, 9(1–2) 1987. Reproduced by permission of *Oxford Literary Review*.

Philip G. Altbach, 'Education and Neocolonialism', in *Teachers College Record*, 100(3) (May) 1971. Reproduced by permission of Blackwell Publishing and the author.

Arun Mukerjee, 'Ideology in the Classroom: A Case Study in the Teaching of English Literature in Canadian Universities', in *Dalhousie Review*, 66(1–2) 1986. Reproduced by permission of *Dalhousie Review*.

Ngũgĩ wa Thiong'o, 'Borders and Bridges: Seeking Connections between Things', in Afzal-Khan and Seshadri-Crooks (eds), *The Pre-Occupation of Postcolonial Studies*, pp. 119–125. Copyright 2000, Duke University. All rights reserved. Used by permission of the publisher.

Excerpts from *The Book Today in Africa*, by S. I. A. Kotei. © UNESCO 1981. Reproduced by permission of UNESCO.

Philip G. Altbach, 'Literary Colonialism: Books in the Third World', *Harvard Educational Review*, 45(2), pp. 226–36. Copyright © 1975 by the President and Fellows of Harvard College. All rights reserved. Excerpted with permission.

Copyright © 1995, from *Imperial Leather: Race, Gender and Sexuality in the Colonial Context*, by Anne McClintock. Reproduced by permission of Routledge/Taylor & Francis Books, Inc.

Excerpts from Arjun Appadurai, 'Introduction – Commodities and the Politics of Value', pp. 3–63, from Arjun Appadurai (ed.), *The Social Life of Things*, 1986, Cambridge University Press, reproduced with permission.

Graham Huggan (2001), *The Postcolonial Exotic: Marketing the Margins,* Routledge. Reproduced by permission of the publisher.

Stuart Hall (1990), 'Cultural Identity and Diaspora', in Jonathan Rutherford (ed.), *Identity: Community, Culture, Difference*, London: Lawrence and Wishart, pp. 222–37. Reproduced by permission of the publisher.

'The Mind of Winter: Reflections on Life in Exile', by Edward W. Said, *Harper's Magazine*, 269 (September) 1984, pp. 49–55. Copyright © Edward W. Said, 1984. Reproduced by permission of The Wylie Agency Ltd.

Avtar Brah (1996), *Cartographies of Diaspora: Contesting Identities*, Routledge. Reproduced by permission of the publisher.

Vijay Mishra, 'The Diasporic Imaginary: Theorizing the Indian Diaspora', *Textual-Practice*, 10(3) (Winter) 1996, pp. 421–47. Reproduced by permission of Taylor and Francis. www.tandf.co.uk/journals.

James Clifford, 'Diasporas', *Cultural Anthropology*, 9(3), pp. 302–38. © 1994, American Anthropological Association. All rights reserved. Used by permission.

Excerpts from Amitava Kumar, *Passport Photos*, The University of California Press, Copyright © 2000. Reproduced by permission of the publisher.

From *The Postcolonial Aura: Third World Criticism in the Age of Global Capitalism*, by Arif Dirlik. Copyright © 1997 by Westview Press, a member of Perseus Books, L.L.C.

Excerpts from Arjun Appadurai (1996), *Modernity at Large: Cultural Dimensions of Globalization*, The University of Minnesota Press. Copyright 1996 by The Regents of the University of Minnesota. Reproduced by permission of the publisher.

Simon Gikandi, 'Globalization and the Claims of Postcoloniality', in *South Atlantic Quarterly*, 100(3), pp. 627–58. Copyright, 2002 (published 2001), Duke University Press. All rights reserved. Used by permission of the publisher.

'Glocalization', by Roland Robertson, reproduced from *Global Modernities*, edited by Mike Featherstone, Scott Lash and Roland Robertson, 1995, by permission of Sage Publications Ltd.

Reprinted by permission of the publisher, from *Empire* by Michael Hardt and Antonio Negri, pp. 183, 186–8, 189–95, Cambridge, Mass.: Harvard University Press, Copyright © 2000 by the President and Fellows of Harvard College.

Chandra Talpade Mohanty, 'Under Western Eyes Revisited', in *Signs*, 28(2) (Winter) 2003, pp. 499–538. Reproduced by permission of the University of Chicago Press. Copyright © 1984 by the University of Chicago. All rights reserved.

Excerpts from Alfred W. Crosby (1986), *Ecological Imperialism: The Biological Expansion of Europe 900–1900*, copyright © Cambridge University Press 1986, Cambridge University Press, reproduced with permission.

Richard H. Grove (1995), *Green Imperialism: Colonial Expansion, Tropical Island Edens And the Origins of Environmentalism*, Cambridge University Press, reproduced with permission.

Excerpts from Ken Saro-Wiwa, 'Trial Statement', reproduced by kind permission of The Estate of Mr K. B. Saro-Wiwa.

Val Plumwood (2003), 'Decolonizing Relationships with Nature', in William Adams and Martin Mulligan (eds), *Decolonizing Nature: Strategies for Conservation in a Post-colonial Era*, Earthscan. Reproduced by permission of James & James/Earthscan.

Gordon Sayre (1995), 'The Beaver as Native and Colonist', *Canadian Review of Comparative Literature/Revue Canadienne de Littérature Comparée*, September/décembre. Reproduced by permission of the author and the *Canadian Review of Comparative Literature*.

Cary Wolfe (1997), 'Old Orders for New: Ecology, Animal Rights and the Poverty of Humanism', *Electronic Book Review*, 4 (Winter). Reprinted by kind permission of The Electronic Book Review www.electronicbookreview.com.

Gauri Viswanathan (1998), *Outside the Fold*. © 1998, Princeton University Press. Reprinted by permission of Princeton University Press.

Copyright © 2002, from *Postcolonialism, Feminism, and Religious Discourse*, edited by Laura E. Donaldson and Kwok Pui-Lan. Reproduced by permission of Routledge/Taylor & Francis Books, Inc.

Copyright © 1996, from 'Reclaiming Our Histories', by William Baldridge in James Treat (ed.), *Native and Christian: Indigenous Voices on Religious Identity in the United States and Canada*. Reproduced by permission of Routledge/Taylor & Francis Books, Inc.

Peter Van der Veer, 'Global Conversions', in Griffiths and Scott (ed.), *Mixed Messages: Materiality, Textuality, Missions*, © 2004, published by Palgrave Macmillan, reproduced with permission of Palgrave Macmillan.

R. S. Sugirtharajah, 'Indigenous Reading/Postcolonial Criticism', 2001, *The Bible in the Third World: Precolonial, Colonial, Postcolonial Encounters*, Cambridge University Press, reproduced with permission.

Richard King, 'Sacred Texts, Hermeneutics and World Religions', in *Orientalism and Religion: Postcolonial Theory and the 'Mystic East'*, 1999, Routledge. Reproduced with permission of the publisher.

Norrel A. London, 'Entrenching the English Language in a British Colony: Curriculum Policy and Practice in Trinidad and Tobago', reprinted from *International Journal of Educational Development*, 23 2003, pp. 97–112. Copyright © 2003, with permission from Elsevier.

Excerpts (pp. 9–21) from *Imaginary Homelands: Essays and Criticism 1981–1991*, by Salman Rushdie (Granta Books, 1991) copyright © Salman Rushdie, 1981, 1982, 1983, 1984, 1985, 1986, 1987, 1988, 1990, 1991. Reproduced by permission of Penguin Books Ltd.

'Imaginary Homelands', copyright © 1982 by Salman Rushdie, from *Imaginary Homelands*, by Salman Rushdie. Used by permission of Viking Penguin, a division of Penguin Group (USA) Inc.

Excerpts from *The Dark Side of the Dream*, Vijay Mishra and Bob Hodge, Allen and Unwin, 1991, reproduced by kind permission of the authors.

*Jila Japingka*, by Peter Skipper. Reproduced with kind permission of Peter Skipper and Mangkaja Arts Resource Agency

Every effort has been made to trace and contact copyright holders. The publishers would be pleased to hear from any copyright holders not acknowledged here so that this section may be amended at the earliest opportunity.

# BILL ASHCROFT, GARETH GRIFFITHS AND HELEN TIFFIN

## GENERAL INTRODUCTION

WHEN ARTHUR JAMES BALFOUR stood up in the House of Commons, at the height of British imperial power, on 13 June 1910, to answer challenges to Britain's presence in Egypt, Edward Said tells us (1978: 32), he spoke under the mantle of two indivisible foundations of imperial authority – knowledge and power. The most formidable ally of economic and political control had long been the business of 'knowing' other peoples because this 'knowing' underpinned imperial dominance and became the mode by which they were increasingly persuaded to know themselves: that is, as subordinate to Europe. A consequence of this process of knowing became the export to the colonies of European language, literature and learning as part of a civilizing mission that involved the suppression of a vast wealth of indigenous cultures beneath the weight of imperial control. The date of Balfour's speech is significant. In just a few years British imperial power would begin to be dismantled by the effects of two world wars and the rise of independence movements throughout the world. This political dismantling did not immediately extend to imperial *cultural* influences, but it was attended by an unprecedented assertion of creative activity in post-colonial societies.

European imperialism took various forms in different times and places and proceeded both through conscious planning and contingent occurrences. As a result of this complex development something occurred for which the *plan* of imperial expansion had not bargained: the immensely prestigious and powerful imperial culture found itself appropriated in projects of counter-colonial resistance which drew upon the many different indigenous local and hybrid *processes* of self-determination to defy, erode and sometimes supplant the prodigious power of imperial cultural knowledge. Post-colonial literatures are a result of this interaction between imperial culture and the complex of indigenous cultural practices. As a consequence, 'post-colonial theory' has existed for a long time before that particular name was used to describe it. Once colonized peoples had cause to reflect on and express the tension that ensued from this problematic and contested, but eventually vibrant and powerful mixture of imperial language and local experience, post-colonial 'theory' came into being.

The term 'post-colonial' is resonant with all the ambiguity and complexity of the many different cultural experiences it implicates, and, as the extracts in this Reader demonstrate, it addresses all aspects of the colonial process from the beginning of colonial contact. Post-colonial critics and theorists should consider the full implications of restricting the meaning of the term to 'after-colonialism' or after-independence. All post-colonial societies are still subject in one way or another to overt or subtle forms of neo-colonial domination, and independence has not solved this problem. The development of new elites within independent societies, often buttressed by neo-colonial institutions; the development of internal divisions based on racial, linguistic or religious discriminations; the continuing unequal treatment of indigenous peoples in settler/invader societies – all these testify to the fact that post-colonialism is a continuing process of resistance and reconstruction. This does not imply that post-colonial practices are seamless and homogeneous but indicates the impossibility of dealing with any part of the colonial process without considering its antecedents and consequences.

Post-colonial theory involves discussion about experience of various kinds: migration, slavery, suppression, resistance, representation, difference, race, gender, place, and responses to the influential master discourses of imperial Europe such as history, philosophy and linguistics, and the fundamental experiences of speaking and writing by which all these come into being. None of these is 'essentially' post-colonial, but together they form the complex fabric of the field. Like the description of any other field, the term has come to mean many things, as the range of extracts in this Reader indicates. However we would argue that post-colonial studies are based in the 'historical fact' of European colonialism, and the diverse material effects to which this phenomenon gave rise. We need to keep this fact of colonization firmly in mind because the increasingly unfocused use of the term 'post-colonial' over the last ten years to describe an astonishing variety of cultural, economic and political practices has meant that there is a danger of its losing its effective meaning altogether. Indeed the diffusion of the term is now so extreme that it is used to refer to not only vastly different but even opposed activities. In particular the tendency to employ the term 'post-colonial' to refer to any kind of marginality at all runs the risk of denying its basis in the historical process of colonialism.

While drawing together a wide variety of theoretical and critical perspectives, this Reader attempts to redress a process whereby 'post-colonial theory' may itself mask and even perpetuate unequal economic and cultural relations. This happens when the bulk of the literary theory is seen to come out of the metropolitan centres, 'adding value' to the literary 'raw material' imported from the post-colonial societies (Mitchell 1992). Such a situation simply reproduces the inequalities of imperial power relations. Post-colonial 'theory' has been produced in all societies into which the imperial force of Europe has intruded, though not always in the formal guise of theoretical texts. But this might not be so clear today given the privileging of theory produced in metropolitan centres and the publishing networks that perpetuate this process. It is relatively easy, for instance, to obtain the classic texts of colonialist discourse theory in metropolitan societies, since they appear in publications widely circulated in these areas. But critical material by post-colonial theorists such as E. K. Brathwaite, Michael Dash, Raja Rao and Wilson Harris (not to mention the 'theory' located in 'creative' texts) are either not available or ignored in many contemporary metropolitan discussions of the field. Equally, though for different reasons, such as a crisis of documentation in many post-colonial societies, much of this material is difficult to obtain there, too. One purpose of this collection is to make a wide range of post-colonial critical material available in a relatively accessible and inexpensive form.

We have attempted to show and celebrate the immense range of countries and literatures from which the theorization of the post-colonial condition has emerged, and in so doing to place the more publicized recent concerns of colonialist discourse theory in a wider geographic and historic context. Indeed, for us, the hyphenated form of the word 'post-colonial' has come to stand for both the material effects of colonization and the huge diversity of everyday and sometimes hidden responses to it throughout the world. We use the term 'post-colonial' to represent the continuing process of imperial suppressions and exchanges throughout this diverse range of societies, in their institutions and their discursive practices. Because the imperial process works *through* as well as *upon* individuals and societies 'post-colonial' theory rejects the egregious classification of 'First' and 'Third' World and contests the lingering fallacy that the post-colonial is somehow synonymous with the economically 'underdeveloped'.

The effects of imperialism occur in many different kinds of societies including those 'settler/invader' societies in which post-colonial contestation is just as strongly and just as ambivalently engaged as it is in more obviously decolonizing states and regions. By the term 'post-colonial' we do not imply an automatic, nor a seamless and unchanging process of resistance but a series of linkages and articulations without which the process cannot be properly addressed. These linkages and articulations are not always directly oppositional; the material practices of post-colonial societies may involve a wide range of activities including conceptions and actions that are, or appear to be, complicit with the imperial enterprise. However, such complicit activities occur in *all* post-colonial societies, and their existence suggests the possibility of crucial comparisons that may be made within the whole range of post-colonial societies. The study of settler colony cultures where, it is frequently argued, such complicit practices are more obvious may, as a result, be especially useful in addressing the problem of complicity in all oppositional discourse, since they point to the difficulties involved in escaping from dominant discursive practices that limit and define the possibility of opposition. Settler colonies, precisely because their filiative metaphors of connection problematize the idea of resistance as a simple binarism, articulate the ambivalent, complex and processual nature of all imperial relations.

The readings we have assembled here are mainly from societies that employ forms of english[1] as a major language of communication. Clearly it would be possible and even desirable to construct a text that addressed the wider polyphonic spectrum of the colonial past but this would require a project far beyond the scope of this one. The Reader also recognizes, but does not directly address, the importance of the continuing body of work in indigenous languages. The 'silencing' of the post-colonial voice to which much recent theory alludes is in many cases a metaphoric rather than a literal one. Critical accounts emphasizing the 'silencing' effect of the metropolitan forms and institutional practices upon pre-colonial cultures, and the resulting forces of 'hybridization' that work on the continuing practice of those cultures, make an important point. But they neglect the fact that for many people in post-colonial societies the pre-colonial languages and cultures, although themselves subject to change and development, continue to provide the effective framework for their daily lives. Failure to acknowledge this might be one of the ways in which post-colonial discourse could, unwittingly, become 'a coloniser in its turn' (Ashcroft *et al.* 1989: 218). Without endorsing a naively 'nativist' position, post-colonial theory needs to be aware that it is engaged in a project that supplements rather than replaces the continuing study and promotion of the indigenous languages of post-colonial societies.

In putting together this Reader we have asked the question: how might a genuinely post-colonial literary enterprise proceed? Our focus in addressing this problem is through the particular agency of literature teaching in the academy. We recognize that this is only one limited avenue of address to the wider social and political issues affecting post-colonial societies, but it seems to us to be an important and worthwhile one, since literature and literary study in the academy have been crucial sites of political and cultural struggle with the most far-reaching results for the general history and practices of colonization and de-colonization. To define our purpose then: we have taken as our limited aim the provision of an effective text to assist in the revision of teaching practice within literary studies in english and so have sought to represent the impact of post-colonial literatures and criticism on the current shape of english studies.

## Note

1    This spelling reflects the fact that, as the editors argued in their earlier book *The Empire Writes Back: Theory and Practice in Post-colonial Literatures* (Ashcroft *et al.* 1989: 8), there is a 'need to distinguish between what is proposed as a standard code, English (the language of the erstwhile imperial centre), and the linguistic code, english, which has been transformed and subverted into several distinctive varieties throughout the world'.

# BILL ASHCROFT, GARETH GRIFFITHS AND HELEN TIFFIN

# INTRODUCTION TO THE SECOND EDITION

SINCE THE PUBLICATION OF THE first edition of *The Post-Colonial Studies Reader*, post-colonial theory has continued to rapidly expand and diversify. While argument about the term itself continues unabated and certain classical themes from the colonial discourse theory of Said, Spivak and Bhabha remain prominent, the last ten years has seen post-colonial theory employed by a growing variety of fields of study and disciplines. The term 'post-colonial' has been adopted in this period to characterize concerns in fields ranging from politics and sociology to religious studies, environmental studies, migration studies, anthropology and economic theory (e.g. Darby 1997; Castellino 2000). While post-colonial theory was a creation of literary study, it has provided a methodology for this wide range of disciplines because it has acknowledged the very specific forms of colonial and neo-colonial power operating in the world today. Some of the debates and areas of growing importance are indicated by the addition of new or rearranged sections in the Reader.

## Race, ethnicity, indigeneity

One way in which the term 'post-colonial' has come to be deployed is in the engagement with issues of cultural diversity, ethnic, racial and cultural difference and the power relations within them – a consequence of an expanded and more subtle understanding of the dimensions of neo-colonial dominance. This underlies the introduction of a new and arguably central section to this Reader on race, as well as the division of indigeneity and ethnicity into separate sections. Race continues to be relevant to post-colonial theory for two reasons: first, because it is so central to the growing power of imperial discourse over the nineteenth century, and second, because it remains a central and unavoidable 'fact' of modern society that race is used as the dominant category of daily discrimination and prejudice. While we may argue that race is a flawed and self-defeating category that traps its users in its biological and essentialist meshes (Appiah 1992), in practical terms race remains a ubiquitous social category that needs to be addressed as a reality in contemporary personal and social relations even when ethnicity might offer a more nuanced understanding of cultural identity. Ethnicity, with its emphasis on symbolic, social and cultural markers of difference provides a useful extension and complication of analyses of race, even though the term has a history at least as fraught and contentious as the term 'race'. Indigeneity, with its specific implication of the politics of settler societies, has been separated from ethnicity because it locates a quite focused, and contentious, series of debates about identity, resistance and transformation. No other category evokes arguments about authenticity as readily as that of the indigenous, and these extracts canvass the wide reach of the concept.

## Environment

As with issues of cultural identity, place has always been of great importance to post-colonial theory, because place is so embedded in language, naming and narrative. But the more material, and global issue of environmentalism is an important and growing aspect of this concept. The destruction of the environment has been one of the most damaging aspects of Western industrialization. The fact that the scramble for modernization has enticed developing countries into the destruction of their own environments, now under the disapproving gaze of a hypocritical West, is further evidence of the continuing importance of a post-colonial analysis of global crises. Post-colonial societies have taken up the 'civilizing' benefits of modernity, only to find themselves the 'barbaric' instigators of environmental damage. In such ways the dynamic of imperial moral power is maintained globally.

While the roots of contemporary environmentalism may lie in colonial damage in both settler colonies and colonies of occupation, neo-colonialism, often in association with the colonial past, continues to produce clashes of interests between 'the West and the Rest' in, for instance, areas of land and food scarcity, where the well-being of humans and endangered animal species may be at odds (Wolch and Emel 1998). Ironically, as the anthropocentric Western drive responsible for so much land and species degradation yields to more bio-centric paradigms of 'the human place in nature', formerly colonized subaltern groups are accused of insensitivity to animals and land as they are driven by economics from their own (often bio-centric) pre-colonial world views and practices into competing for survival through the very industrial and agricultural capitalism that dispossessed them of their original way of living. The hierarchical distinctions between humans and animals intrinsic to European humanism were frequently employed metaphorically in the creation of racialized hierarchies during the colonial period. More recently the emergence of new and challenging ways of thinking about humanities' relation to other species has been an important addition to the analysis of post-colonial subjectivities. Post-colonial environmentalism must therefore deal with a number of deeply problematic issues and conflicting interests, but as it begins to do so, the foundational importance of animal and environmental concerns to theorizing 'the post-colonial' becomes increasingly clear.

## Globalization

It is increasingly suggested that perhaps the ultimate and unavoidable future of post-colonial studies lies in its relation to globalization. Not only does globalization seem to be a natural extension of the interests of post-colonial studies, but in the opinion of Simon Gikandi (Chapter 106) globalization discourse has been overwhelmingly influenced by post-colonial terminology over the last decade. The relationship works in two ways: we cannot understand globalization without understanding the structure of global power relations that flourishes in the twenty first century as an economic, cultural and political legacy of Western imperialism. But post-colonial theory, especially of textual and cultural practices can provide very clear models for understanding how local communities achieve agency under the pressure of global hegemony. Post-colonial theory is very useful in its analysis of the strategies by which the 'local' colonized engage large hegemonic forces. This should not be mistaken for an assumption that some-how globalization is neo-colonialism per se – the two phenomena are very different and result in a different range of material effects. But the principles and strategies of engagement are similar.

## Diaspora

If anything seems to characterize globalization at the turn of the century, it is the phenomenon of the extraordinary and accelerating movement of peoples throughout the world. The increasing refugee crisis in every Western country is just one manifestation of the long-standing circulation of peoples in what Edward Said has called 'the voyage in' (1993: 261). Diaspora does not seem at first to be the province of post-colonial studies until we examine the deep impact of colonialism upon this movement. The most extreme consequences of imperial domin-ance can be seen in the radical displacement of peoples through slavery, indenture and settlement. More recently this movement can be seen to be a consequence of the disparity in wealth between the West and the world, extended by the economic imperatives of imperialism and rapidly opening a gap between colonizers and colonized. The movement of refugees, in particular, has often reignited racism (and Orientalism) in many communities worldwide.

Diaspora does not simply refer to this movement but also to the vexed questions of identity, memory and home that such movement produces. Extracts from Salman Rushdie, Stuart Hall and Edward Said address the complex issues of identity, subjectivity and exile. The issue is not only one of cultural engagement but also one of cultural circulation. For James Clifford there is a new world order of mobility, of rootless histories, and the paradox of global culture is that it is 'at home' with this motion rather than in a particular place. Clifford's book *Routes* examines the extent to which practices of displacement 'might be *constitutive* of cultural meanings rather than their simple transfer or extension' (1997: 3).

## The sacred

Debates concerning the traditional and sacred beliefs of colonized, indigenous and marginal-ized peoples have increased in importance to post-colonial studies. Indeed, it would be true to say that this remains the field of post-colonial studies in most need of critical and scholarly attention. Since the Enlightenment the sacred has been an ambivalent area in Western thinking that has uniformly tended to privilege the secular. As Chakrabarty and other critics have reminded us, secularity, economic rationalism and progressivism have dominated Western think-ing, while 'the sacred' has so often been relegated to primitivism and the archaic (Chakrabarty 2000; Scott and Simpson-Housley 2001; Griffiths and Scott 2004). However, at the end of the twentieth century, debates about the sacred have become more urgent as issues such as land rights and rights to sacred beliefs and practices begin to grow in importance. A paradigm shift has been occurring in this area, particularly in post-colonial theory, bringing a new con-sideration of the complex, hybrid and rapidly changing cultural formations of both marginalized and first world peoples (Gelder and Jacobs 1998). The sacred has followed the trajectory of other 'denied knowledges', as Bhabha puts it, entering the dominant discourse and estranging 'the basis of its authority – its rules of recognition' (1994: 114).

A misleading direction was given by Edward Said's well-known preference for 'secular criticism' over what he called the 'theological' bent of contemporary theory (1983: 1–30). Although by 'theological' Said meant schools of contemporary theory that were dogmatic and bounded, that encouraged devotees and acolytes rather than rigorous criticism, this seemed to suggest that the theological and the sacred were not the province of enlightened post-colonial analysis. Such an assumption reminds us of the gap that often exists between the theoretical agenda of the Western academy and the interests of post-colonial societies them-selves. The sacred has been an empowering feature of post-colonial experience in two ways: on one hand indigenous concepts of the sacred have been able to interpolate dominant concep-tions of cultural identity; and on the other Western forms of the sacred have often been appropriated and transformed as a means of local empowerment. Analyses of the sacred have been one of the most neglected, and may be one of the most rapidly expanding areas of post-colonial study.

This revised and expanded Reader indicates the rapidity with which associated areas of concern to post-colonial studies have been brought into the centre of discussion. Post-colonial analysis has always contested the notion of a monolithic imperialism subjecting a more or less homogeneous colonial world to its cultural and political power. Yet while classical imperialism has always been transcultural and rhizomic, a two-way process of cultural exchange, the extraordinary fluidity of modern global populations, and the astonishing expan-sion of globalization and the unprecedented rise to power of the US hypernation has ensured that post-colonial analysis will continue to develop and expand as it applies its various critical tools to a different kind of world in the twenty-first century.

# PART ONE

# Issues and Debates

THE EXTRACTS IN THIS SECTION indicate something of the historical proven-ance, the general theoretical directions and the important debates that have featured in post-colonial theory in recent times. West Indian novelist George Lamming expresses in a personal way some of the enduring issues: how a Britain without its Empire can still main-tain cultural authority in post-colonial societies, and the ways in which Eurocentric assumptions about race, nationality and literature return time and again to haunt the production of post-colonial writing. Lamming's is a foundational text in post-colonial writing; its early date indicates how long post-colonial intellectuals have been grappling with the articulation of their own modes of cultural production. It is important, too, in that it is a critical essay that is written by an imaginative writer, and as such represents the crucial role played by creative writers as diverse in time and place as Rabindranath Tagore, Raja Rao, Wole Soyinka, Chinua Achebe, Edward Kamau Brathwaite, Derek Walcott, Judith Wright, Tom King, Margaret Atwood, Dennis Lee, Alan Curnow, Keri Hulme and many others in developing a critical discourse in the post-colonial world. While these writers have often functioned as critics in a formal sense their own creative work has frequently been the site of critiques of imperial representation, language and ideological control. Thus, as Lamming argues here, the advent of the novel in the West Indies marks an important historical event as well as a formal cultural development.

This extract serves to remind us that the determining condition of what we refer to as post-colonial cultures is the historical phenomenon of colonialism, with its range of material practices and effects, such as transportation, slavery, displacement, emigration and racial and cultural discrimination. These material conditions and their relationship to questions of ideology and representation are at the heart of the most vigorous debates in recent post-colonial theory. Even the claim that they may exist independently of the modes of representation that allowed them to come into formation is to assert a point of considerable controversy.

Abdul R. JanMohamed stresses the importance, as does Lamming, of the literary text as a site of cultural control and as a highly effective instrumentality for the determination of the 'native' by fixing him/her under the sign of the other. JanMohamed also shows how these liter-ary texts contain features that can be subverted and appropriated to the oppositional and anti-colonial purposes of contemporary post-colonial writing. His essay analyses the literary text in quite specific ways as a means of bringing into being and modifying the controlling discourses of colonization. Using Lacan's distinction of the imaginary and symbolic stages of development as a conceptual tool in this analysis JanMohamed emphasizes the self-contradictions of binary constructions. By recognizing how the binarisms of colonial discourse operate (the self–other, civilised–native, us–them manichean polarities) post-colonial critics can promote an active reading that makes these texts available for re-writing and subversion. It is this process that brings into being the powerful syncretic texts of contemporary post-colonial writing. In the rest of the book from which this short extract is taken JanMohamed illustrates how this process of reinscription works by developing an analysis of the relationship between contemporary texts of post-colonial writing and the colonial texts to which they 'write back'. Such a process of 'writing back', far from indicating a continuing dependence, is an effective means of escaping from the binary polarities implicit in the manichean constructions of colonization and its practices.

No general survey of the field could omit Edward Said's groundbreaking *Orientalism*, the book that paved the way for post-colonial studies by forcing academics in the West to re-think the relationship between the Occident and Orient. This extract introduces an important early debate, that between 'colonial discourse theory' – evident in the work of Said, Spivak and Bhabha – and other more materialist text based forms of analysis. Since 1989 the difference between these approaches has narrowed as colonial discourse theory has been absorbed into more recent approaches. But these seminal thinkers have established the terms of many post-colonial debates over the last two decades. Said's thesis of the power of Orientalist discourse to 'construct' the Orient was founded on the Foucaultian premise of power/knowledge. The West had power to 'know' the Orient and that power constituted the Oriental other as a particular subject of discourse. As we can see in the section Representation and resistance, the importance of this link between power and knowledge is just as important today as it was in 1978, as information becomes the new currency of global relations.

Gayatri Spivak questions whether or not the possibility exists for any recovery of a subaltern voice that is not a kind of essentialist fiction. Although she expresses considerable sympathy for the project undertaken in contemporary historiography to give a voice to 'the subaltern' who had been written out of the record by conventional historical accounts, Spivak raises grave doubts about its theoretical legitimacy. She is sympathetic but critical in her response here to Ranajit Guha's subaltern studies project that seeks to obtain what Said termed the 'permission to speak' by going behind the terms of reference of 'elite' history to include the perspective of those who are never taken into account (the subaltern social groups). Recognizing and applauding the project's endorsement of the heterogeneity of the colonial subject, and giving a qualified approval to the politics of the effort to speak a 'politics of the people', Spivak is nevertheless concerned to articulate what she sees as the difficulties and contradictions involved in constructing a 'speaking position' for the subaltern. Wanting to acknowledge the continuity and vigour of pre-colonial social practice, its ability to modify and to 'survive' colonial incursions and definitional strategies and exclusions, she insists that the poststructuralist mode of the project only disguises what she sees as an underlying persistent essentialism. For her, one cannot construct a category of the 'subaltern' that has an effective 'voice' clearly and *unproblematically* audible above the persistent and multiple echoes of its inevitable heterogeneity. Her conclusion is that for 'the true' subaltern group, whose identity is its difference, there is no subaltern subject that can 'know and speak itself'. Thus the intellectual must avoid reconstructing the subaltern as merely another unproblematic field of knowing, so confining its effect to the very form of representation ('text for knowledge') the project sought to evade and lay bare. The conclusion is expressed, perhaps unfortunately, in a rather negative way: 'Subaltern historiography must confront the impossibility of such gestures.' Spivak's negative, as José Rabasa has pointed out, does not 'necessarily exclude such instances of colonized subjects defracting power as those Homi Bhabha has isolated in the case of India' (Rabasa 1993: 11–12).

The emphasis on the importance of the written text as an instrument of control (to which Said and JanMohamed's work makes reference), and of the deep ambivalences locked into the apparent universal fixities of colonialist epistemology, are taken up by Homi Bhabha. For Bhabha the 'emblem of the English book' is one of the most important of the 'signs taken for wonders' by which the colonizer controls the imagination and the aspirations of the colonized, because the book assumes a greater authority than the experience of the colonized peoples themselves. But, as Bhabha argues, such authority simultaneously renders the colonial presence ambivalent, since it only comes about by displacing those images of identity already held by the colonized society. The colonial space is therefore an agonistic space. Despite the 'imitation' and 'mimicry' with which colonized peoples cope with the imperial presence, the relationship becomes one of constant, if implicit, contestation and opposition. Indeed, such mimicry becomes the very site of that conflict, a 'transparency', as Bhabha puts it, which is dependent for its fixity on the underlying negative of imperial presence that it seems to duplicate. For Bhabha 'mimicry' does not mean that opposition is rejected, but rather that it is seen to encompass more than overt opposition. Opposition is not simply reduced to intention, but is implicit in the very production of dominance whose intervention as a 'dislocatory presence' paradoxically confirms the very thing it displaces. The resulting hybrid modalities also challenge the assumption of the 'pure' and the 'authentic', concepts upon which the resistance to imperialism often stands. Indeed hybridity, rather than indicating corruption or decline, may, as Bhabha argues, be the most common and effective form of subversive opposition since it displays the 'necessary deformation and displacement of all sites of discrimination and domination'.

Spivak's and Bhabha's analyses are important and very influential warnings of the complexities of the task faced by post-colonial theory. But they have also invited responses that see them and their approach as too deeply implicated in European intellectual traditions, which older, more radical exponents of post-colonial theory, such as Frantz Fanon and Albert Memmi, had sought to dismantle and set aside. The debate is a struggle between those who want to align themselves with the subaltern and those who insist that this attempt becomes at best only a refined version of the very discourse it seeks to displace. All are agreed, in some sense, that the main problem is how to effect agency for the post-colonial subject. But the contentious issue of how this is to be attained remains unresolved.

Benita Parry's critique of contemporary 'colonialist discourse' theory (such as Bhabha's and Spivak's) argues that the effect of its insistence on the 'necessary' silencing implicit in this mode of analysis has been to diminish the earlier intervention of critics such as Fanon who stood much more resolutely for the idea that de-colonization is a process of opposition to dominance. She also argues that colonialist discourse theory supports readings of post-colonial texts that inadequately ascribe a native 'absence' to texts in which the 'native' has access as a profoundly disruptive presence. In a sense Parry's argument is a plea for an analysis of the 'politics' of the project of colonialist discourse theory itself, and seeks to resurrect as a forgotten but vital element in the debate the voices of the post-colonial intellectuals of the earlier, oppositional 'national liberation' phase of decolonization. Subsequent response from Spivak has argued that such oppositional categories as the 'post-colonial intellectual' avoid the fact that the concept of 'intellectual' and of 'theory' as a discourse is by definition implicated in the Europeanization/hybridization of all culture in the aftermath of imperialism, making the distinctive category of 'post-colonial intellectual' as problematic as the term 'subaltern'.

The argument underpinning these positions, that there can be an engagement with the 'real' separate from its construction through what Barthes called 'reality effects', is put with great clarity by José Rabasa: 'cultural products should be taken as rhetorical artifices and not as depositories of data from which a factual truth may be construed' (1993: 9). Yet, of course, the avoidance of such a construing in practice may be to allow semiotic analyses of texts totally 'liberated' from any attempt or desire to understand the context of cultural production from which they emerge. The effect of this is, of course, to wipe out cultural difference.

The debate between those who insist on the possibility of an effective alignment of position with the subaltern and those who insist that this, paradoxically, may serve only to construct a refinement of the system it seeks to dismantle, is taken up and expanded later in the Reader in the section on Representation and resistance. There Jenny Sharpe's analysis of the problem of resistance and Stephen Slemon's article on the crucial role of settler culture, or 'Second World' texts, in articulating the ambivalence at the heart of post-colonial resistance, continue and elaborate some of the issues raised in this section.

Stephen Slemon's overview of recent developments within the field of post-colonial studies includes, like Parry's, an analysis of the difficulty that 'colonialist criticism' has in confirming the agency of the post-colonial subject. A crucial question for post-colonial theory, given that contemporary thought has firmly fixed subjectivity in language, is 'how can one account for the capacity of the subject in a post-colonial society to resist imperialism and thus to inter-vene in the conditions which appear to construct subjectivity itself?' Slemon analyses the positions of some of the major participants in the debates in a fresh and interesting way but also regards the debate itself as the product of the institutionalization of post-colonial studies within the practices of the contemporary academy. Quoting Henry Louis Gates, Slemon warns that 'academic interest in this history and the discourse of colonialism bids fair to become the last bastion for the project of global theory and for European universalism itself', forcing us to choose between 'oppositional critics whose articulations of the post-colonial institutionalise themselves as agonistic struggles over a thoroughly *disciplined* terrain'. Slemon reminds his readers that the real contest (agon) post-colonial studies seeks to address is that between the conflicting participants in the imperial process and their residual legatees, not between con-temporary schools of theory. The real concerns of this oppositional subject are in danger of being reduced to merely another location in the academic institutionalized landscape, yet another mere invasive 'mapping' of the subdued and subjugated post-colonial world.

One of the most prevalent issues in recent times has been the increasing applicability of post-colonial forms of analysis to many disciplines outside literature. Nicholas Dirks segues very deftly from E. M. Forster's *A Passage to India* to contemporary issues in anthropology, geography and postmodernity.

Post-colonial is a term that has been slow to make its way in Africa. In part this is because of the dominance in recent times of South Asian theorists in the field commonly known as colonialist discourse theory, and also because Africans have been concerned perhaps to argue for a seemingly more aggressive version of the anti-colonialist struggle for self-definition in the tradition of Fanon, Cabral and, more recently, Ngũgĩ. But in recent years

African post-colonial theorists have emerged who have wedded their political engagement to a subtle set of readings of culture as a vehicle both for oppression and resistance. Biodun Jeyifo's article contrasts the ways in which Euro-American and African critics have engaged with the effects of colonialism and neo-colonialism on African culture, and suggests that this division is both simplistic and itself part of the exoticization of Africa as intrinsically different to the rest of the world. After all he notes, we do not read Russian, German or French culture as if there were some impassible gap between local and foreign influences on their themes and structures, so why should we do so for Africa? Without losing sight of the goal of decolonization Jeyifo also resists the dismissal of theory by many African critics and artists, suggesting that the confusion and debate over African studies has resulted from its under-theorization rather than its over-theorization.

Achille Mbembe's work is now receiving the attention it long deserved. Its neglect suggests how Anglophonic criticism still tends to dominate the field in ways that are slowly being addressed. The influence of 'On the Postcolony', from which the extract here is taken, also shows how the shift is now occurring from analysis of the direct effect of the European colonizer to an analysis of the politics of repression that the post-independence regimes have both inherited and, in a sense, 'naturalized'. Mbembe exposes how local practice can be recruited by oppressive regimes to their claim to represent the people and to meld with an 'authentic' African identity. He shows how the rulers can employ the desire of the ruled to participate, if only by proxy, in the ceremonies and rituals of power. But, as the full work shows, he illustrates too how the same forms of local and popular culture can be used to forge resistance. There is an economic, class basis to this analysis that has been too frequently lacking in some earlier post-colonial work, as Mbembe shows how contemporary types of performative media (popular rituals, newspaper cartoons, graffiti, etc.) are available both to the new rulers and to those who resist them. By reworking Bakhtin's analysis of popular forms he shows how these can be 'the arenas in which subordinates reaffirm or subvert that power'. But he also demonstrates how the desire of the powerless to participate in the rituals of power organized by the dominant class 'burlesques' that power. At this moment Mbembe's analysis seems to offer a more overtly activist, political reading of the idea of 'mimicry' that underlies Homi Bhabha's idea of subversion.

## Chapter 1

## GEORGE LAMMING

## THE OCCASION FOR SPEAKING

From 'The Occasion For Speaking' *The Pleasures of Exile* London:
Michael Joseph, 1960.

I N ANY COUNTRY, DURING THIS CENTURY, it seems that the young will remain too numerous and too strong to fear being alone. It is from this premise that I want to consider the circumstances as well as the significance of certain writers' migration from the British Caribbean to the London metropolis. . . .

How has it come about that a small group of men, different in years and temperament and social origins, should leave the respective islands they know best, even exchange life there for circumstances which are almost wholly foreign to them? . . . Why have they migrated? And what, if any, are the peculiar pleasures of exile? Is their journey a part of a hunger for recognition? Do they see such recognition as a confirmation of the fact that they are writers? What is the source of their insecurity in the world of letters? And what, on the evidence of their work, is the range of their ambition as writers whose nourishment is now elsewhere, whose absence is likely to drag into a state of permanent separation from their roots? . . .

The exile is a universal figure. The proximity of our lives to the major issues of our time has demanded of us all some kind of involvement. Some may remain neutral; but all have, at least, to pay attention to what is going on. On the political level, we are often without the right kind of information to make argument effective; on the moral level we have to feel our way through problems for which we have no adequate reference of traditional conduct as a guide. Chaos is often, therefore, the result of our thinking and our doing. We are made to feel a sense of exile by our inadequacy and our irrelevance of function in a society whose past we can't alter, and whose future is always beyond us. Idleness can easily guide us into accepting this as a condition. Sooner or later, in silence or with rhetoric, we sign a contract whose epitaph reads: To be an exile is to be alive.

When the exile is a man of colonial orientation, and his chosen residence is the country which colonised his own history, then there are certain complications. For each exile has not only got to prove his worth to the other, he has to win the approval of Headquarters, meaning in the case of the West Indian writer, England. . . .

In England he does not feel the need to try to understand an Englishman, since all relationships begin with an assumption of previous knowledge, a knowledge acquired in the absence of the people known. This relationship with the English is only another aspect of the West Indian's relation to the *idea* of England.

As an example of this, I would recall an episode on a ship which had brought a number of West Indians to Britain. I was talking to a Trinidadian Civil Servant who had come to take some kind of course in the ways of bureaucracy. A man about forty-five, intelligent enough to be in the senior grade of the Trinidad Civil Service which is by no means backward, a man of some substance among his own class of people. We were talking in a general way about life among the emigrants. The ship was now steady; the tugs were coming alongside. Suddenly there was consternation in the Trinidadian's expression.

'But . . . but', he said, 'look down there.'

I looked, and since I had lived six years in England, I failed to see anything of particular significance. I asked him what he had seen; and then I realised what was happening.

'*They* do that kind of work, *too*?' he asked.

He meant the white hands and faces on the tug. In spite of films, in spite of reading Dickens – for he would have had to at the school which trained him for the Civil Service – in spite of all this received information, this man had never really felt, as a possibility and a fact, the existence of the English worker. This sudden bewilderment had sprung from his *idea* of England: and one element in that *idea* was that he was not used to seeing an Englishman working with his hands in the streets of Port-of-Spain.

This is a seed of his colonisation which has been subtly and richly infused with myth. We can change laws overnight; we may reshape images of our feeling. But this myth is most difficult to dislodge. . . .

I remember how pleased I was to learn that my first book, *In the Castle of My Skin*, had been bought by an American publisher. . . . It was the money I was thinking of to the exclusion of the book's critical reputation in America. The book had had an important critical press in England; its reputation here was substantial; so it could make no difference what America thought. . . . This is what I mean by the *myth*. It has little to do with lack of intelligence. It has nothing to do with one's origins in class. It is deeper and more natural. It is akin to the nutritive function of milk which all sorts of men receive at birth. It is *myth* as the source of spiritual foods absorbed, and learnt for exercise in the future. This *myth* begins in the West Indian from the earliest stages of his education. But it is not yet turned against America. In a sense, America does not even exist. It begins with the fact of England's supremacy in taste and judgement: a fact which can only have meaning and weight by a calculated cutting down to size of all non-England. The first to be cut down is the colonial himself.

This is one of the seeds which much later bear such strange fruit as the West Indian writers' departure from the very landscape which is the raw material of all their books. These men had to leave if they were going to function as writers since books, in that particular colonial conception of literature, were not – meaning, too, are not supposed to be – written by natives. Those among the natives who read also believed that; for all the books they had read, their whole introduction to something called culture, all of it, in the form of words, came from outside: Dickens, Jane Austen, Kipling and that sacred gang.

The West Indian's education was imported in much the same way that flour and butter are imported from Canada. Since the cultural negotiation was strictly between England and the natives, and England had acquired, somehow, the divine right to organise the native's reading, it is to be expected that England's export of literature would be English. Deliberately and exclusively English. And the further back in time England went for these treasures, the safer was the English commodity. So the examinations, which would determine that Trinidadian's future in the Civil Service, imposed Shakespeare, and Wordsworth, and Jane Austen and George Eliot and the whole tabernacle of dead names, now come alive at the world's greatest summit of literary expression. . . .

In [American novelist, James Baldwin's] most perceptive and brilliantly stated essays, *Notes of a Native Son*, he tries to examine and interpret his own situation as an American negro who is also a novelist drawing on the spiritual legacy of Western European civilisation. . . .

I know, in any case, that the most crucial time in my own development came when I was forced to recognise that I was a kind of bastard of the West; when I followed the line of my past I did not find myself in Europe, but in Africa. And this meant that in some subtle way, in a really profound way I brought to Shakespeare, Bach, Rembrandt, to the stones of Paris, to the cathedral at Chartres, and to the Empire State Building, a special attitude. These were not really my creations, they did not contain my history; I might search in them in vain for ever for any reflection of myself; I was an interloper. At the same time I had no other heritage which I could possibly hope to use. I had certainly been unfitted for the jungle or the tribe.

(Baldwin 1964: 14)

'I might search in vain for any reflection of myself. I had certainly been unfitted for the jungle or the tribe.'

We must pause to consider the source of Mr Baldwin's timidity; for it has a most respectable ancestry. Here is the great German philosopher Hegel having the last word on Africa in his Introduction to *The Philosophy of History*:

Africa proper, as far as History goes back, has remained — for all purposes of connection with the rest of the world — shut up; it is the Gold-land compressed within itself — the land of childhood, which lying beyond the days of self-conscious history, is enveloped in the dark mantle of Night. . . .

The negro as already observed exhibits the natural man in his completely wild and untamed state. We must lay aside all thought of reverence and morality — all that we call feeling — if we would rightly comprehend him; there is nothing harmonious with humanity to be found in this type of character. . . .

At this point we leave Africa never to mention it again. For it is no historical part of the world; it has no movement of development to exhibit. Historical movement in it — that is in its northern part — belongs to the Asiatic or European World. . . .

What we properly understand as Africa, is the Unhistorical, Undeveloped Spirit, still involved in the *conditions of mere nature* and which had to be presented here only as on the threshold of the World's history. . . .

The History of the World travels from East to West, for Europe *is absolutely the end of History*, Asia is the beginning.

It is important to relate the psychology implied in Mr Baldwin's regret to the kind of false confidence which Hegel represents in the European consciousness. For what disqualifies African man from Hegel's World of History is his apparent incapacity to evolve with the logic of Language which is the only aid man has in capturing the Idea. African Man, for Hegel, has no part in the common pursuit of the Universal. . . .

What the West Indian shares with the African is a common political predicament: a predicament which we call colonial; but the word colonial has a deeper meaning for the West Indian than it has for the African. The African, in spite of his modernity, has never been wholly severed from the cradle of a continuous culture and tradition. His colonialism mainly takes the form of lack of privilege in organising the day to day affairs of his country. This state of affairs is almost at an end; and its end is the result of the African's persistent and effective demand for political freedom. . . .

It is the brevity of the West Indian's history and the fragmentary nature of the different cultures which have fused to make something new; it is the absolute dependence on the values in that language of his coloniser which have given him a special relation to the word, colonialism. It is not merely a political definition; it is not merely the result of certain economic arrangements. It started as these, and grew somewhat deeper. Colonialism is the very base and structure of the West Indian cultural awareness. His reluctance in asking for complete, political freedom . . . is due to the fear that he has never had to stand. A foreign or absent Mother culture has always cradled his judgement. Moreover, the . . . freedom from physical fear has created a state of complacency in the West Indian awareness. And the higher up he moves in the social scale, the more crippled his mind and impulses become by the resultant complacency.

In order to change this way of seeing, the West Indian must change the very structure, the very basis of his values. . . .

I am not much interested in what the West Indian writer has brought to the English language; for English is no longer the exclusive language of the men who live in England. That stopped a long time ago; and it is today, among other things, a West Indian language. What the West Indians do with it is their own business. A more important consideration is what the West Indian novelist has brought to the West Indies. That is the real question; and its answer can be the beginning of an attempt to grapple with that colonial structure of awareness which has determined West Indian values.

There are, for me, just three important events in British Caribbean history. I am using the term, history, in an active sense. Not a succession of episodes which can easily be given some casual connection. What I mean by historical event is the creation of a situation which offers antagonistic oppositions and a challenge of survival that had to be met by all involved.

The first event is discovery. That began, like most other discoveries, with a journey; a journey inside, or a journey out and across. This was the meaning of Columbus. The original purpose of the journey may sometimes have nothing to do with the results that attend upon it. That journey took place nearly five centuries ago; and the result has been one of the world's most fascinating communities. The next event is the abolition of slavery and the arrival of the East – India and China – in the Caribbean Sea. The world met here, and it was at every level, except administration, a peasant world. In one way or another, through one upheaval after another, these people, forced to use a common language which they did not possess on arrival, have had to make something of their surroundings. . . .

The third important event in our history is the discovery of the novel by West Indians as a way of investigating and projecting the inner experiences of the West Indian community. The second event is about a hundred and fifty years behind us. The third is hardly two decades ago. . . . The West Indian writer is the first to add a new dimension to writing about the West Indian community. . . .

If we accept that the act of writing a book is linked with an expectation, however modest, of having it read; then the situation of a West Indian writer, living and working in his own community, assumes intolerable difficulties. The West Indian of average opportunity and intelligence has not yet been converted to reading as a civilised activity which justifies itself in the exercise of his mind. Reading seriously, at any age, is still largely associated with reading for examinations. In recent times the political fever has warmed us to the newspapers with their generous and diabolical welcome to join in the correspondence column. But book reading has never been a serious business with us. . . .

An important question, for the English critic, is not what the West Indian novel has brought to English writing. It would be more correct to ask what the West Indian novelists have contributed to English reading. For the language in which these books are written is English – which, I must repeat – is a West Indian language; and in spite of the unfamiliarity of its rhythms, it remains accessible to the readers of English anywhere in the world. The West Indian contribution to English reading has been made possible by their relation to the themes which are peasant. . . .

That's a great difference between the West Indian novelist and his contemporary in England. For peasants simply don't respond and see like middle-class people. The peasant tongue has its own rhythms which are [Trinidadian novelist Samuel] Selvon's and [Barbadian novelist Vic] Reid's rhythms; and no artifice of technique, no sophisticated gimmicks leading to the mutilation of form, can achieve the specific taste and sound of Selvon's prose.

For this prose is, really, the people's speech, the organic music of the earth. . . .

This may be the dilemma of the West Indian writer abroad: that he hungers for nourishment from a soil which he (as an ordinary citizen) could not at present endure. The pleasure and paradox of my own exile is that I belong wherever I am. My role, it seems, has rather to do with time and change than with the geography of circumstances; and yet there is always an acre of ground in the New World which keeps growing echoes in my head. I can only hope that these echoes do not die before my work comes to an end.

## Chapter 2

# ABDUL R. JANMOHAMED

# THE ECONOMY OF MANICHEAN ALLEGORY

From 'The Economy of Manichean Allegory: The Function of Racial Difference'
*Critical Inquiry* 12(1), 1985.

COLONIALIST LITERATURE IS AN EXPLORATION and a representation of a world at the boundaries of 'civilization,' a world that has not (yet) been domesticated by European signification or codified in detail by its ideology. That world is therefore perceived as uncontrollable, chaotic, unattainable, and ultimately evil. Motivated by his desire to conquer and dominate, the imperialist configures the colonial realm as a confrontation based on differences in race, language, social customs, cultural values, and modes of production.

Faced with an incomprehensible and multifaceted alterity, the European theoretically has the option of responding to the Other in terms of identity or difference. If he assumes that he and the Other are essentially identical, then he would tend to ignore the significant divergences and to judge the Other according to his own cultural values. If, on the other hand, he assumes that the Other is irremediably different, then he would have little incentive to adopt the viewpoint of that alterity: he would again tend to turn to the security of his own cultural perspective. Genuine and thorough comprehension of Otherness is possible only if the self can somehow negate or at least severely bracket the values, assumptions, and ideology of his culture. As Nadine Gordimer's and Isak Dinesen's writings show, however, this entails in practice the virtually impossible task of negating one's very being, precisely because one's culture is what formed that being. Moreover, the colonizers invariable assumption about his moral superiority means that he will rarely question the validity of either his own or his society's formation and that he will not be inclined to expend any energy in understanding the worthless alterity of the colonized. By thus subverting the traditional dialectic of self and Other that contemporary theory considers so important in the formation of self and culture, the assumption of moral superiority subverts the very potential of colonialist literature. Instead of being an exploration of the racial Other, such literature merely affirms its own ethnocentric assumptions; instead of actually depicting the outer limits of 'civilization,' it simply codifies and preserves the structures of its own mentality. While the surface of each colonialist text purports to represent specific encounters with specific varieties of the racial Other, the subtext valorizes the superiority of European cultures, of the collective process that has mediated that representation. Such literature is essentially specular: instead of seeing the native as a bridge toward syncretic possibility, it uses him as a mirror that reflects the colonialist's self-image.

Accordingly, I would argue that colonialist literature is divisible into two broad categories: the 'imaginary' and the 'symbolic.' The emotive as well as the cognitive intentionalities of the 'imaginary' text are structured by objectification and aggression. In such works the native functions as an image of the imperialist self in such a manner that it reveals the latter's self-alienation. Because of the subsequent projection involved in this context, the 'imaginary' novel maps the European's intense internal rivalry. The 'imaginary' representation of indigenous people tends to coalesce the signifier with the signified. In describing the attributes or actions of the native, issues such as intention, causality, extenuating circumstances, and so forth, are completely ignored; in the 'imaginary' colonialist realm, to say 'native' is automatically to say 'evil' and to evoke immediately the economy of the manichean allegory. The writer of such texts tends to fetishize a nondialectical, fixed opposition between the self and the native. Threatened by a metaphysical alterity that he has created, he quickly retreats to the homogeneity of his own group. Consequently, his psyche and text tend to be much closer to and are often entirely occluded by the ideology of his group.

Writers of 'symbolic' texts, on the other hand, are more aware of the inevitable necessity of using the native as a mediator of European desires. Grounded more firmly and securely in the egalitarian imperatives of Western societies, these authors tend to be more open to a modifying dialectic of self and Other. They are willing to examine the specific individual and cultural differences between Europeans and natives and to reflect on the efficacy of European values, assumptions, and habits in contrast to those of the indigenous cultures. 'Symbolic' texts, most of which thematize the problem of colonialist mentality and its encounter with the racial Other, can in turn be subdivided into two categories.

The first type, represented by novels like E. M. Forster's *A Passage to India* and Rudyard Kipling's *Kim*, attempts to find syncretic solutions to the manichean opposition of the colonizer and the colonized. This kind of novel overlaps in some ways with the 'imaginary' text: those portions of the novel organized at the emotive level are structured by 'imaginary' identification, while those controlled by cognitive intentionality are structured by the rules of the 'symbolic' order. Ironically, these novels – which are conceived in the 'symbolic' realm of intersubjectivity, heterogeneity, and particularity but are seduced by the specularity of 'imaginary' Otherness – better illustrate the economy and power of the manichean allegory than do the strictly 'imaginary' texts.

The second type of 'symbolic' fiction, represented by the novels of Joseph Conrad and Nadine Gordimer, realizes that syncretism is impossible within the power relations of colonial society because such a context traps the writer in the libidinal economy of the 'imaginary.' Hence, becoming reflexive about its context, by confining itself to a rigorous examination of the 'imaginary' mechanism of colonialist mentality, this type of fiction manages to free itself from the manichean allegory. . . .

If every desire is at base a desire to impose oneself on another and to be recognized by the Other, then the colonial situation provides an ideal context for the fulfillment of that fundamental drive. The colonialist's military superiority ensures a complete projection of his self on the Other: exercising his assumed superiority, he destroys without any significant qualms the effectiveness of indigenous economic, social, political, legal, and moral systems and imposes his own versions of these structures on the Other. By thus subjugating the native, the European settler is able to compel the Other's recognition of him and, in the process, allow his own identity to become deeply dependent on his position as a master. This enforced recognition from the Other in fact amounts to the European's narcissistic self-recognition since the native, who is considered too degraded and inhuman to be credited with any specific subjectivity, is cast as no more than a recipient of the negative elements of the self that the European projects onto him. This transitivity and the preoccupation with the inverted self-image mark the 'imaginary' relations that characterize the colonial encounter.

Nevertheless, the gratification that this situation affords is impaired by the European's alienation from his own unconscious desire. In the 'imaginary' text, the subject is eclipsed by his fixation on and fetishization of the Other: the self becomes a prisoner of the projected image. Even though the native is negated by the projection of the inverted image, his presence as an absence can never be canceled. Thus the colonialist's desire only entraps him in the dualism of the 'imaginary' and foments violent hatred of the native. This desire to exterminate the brutes, which is thematized consciously and critically in 'symbolic' texts such as *Heart of Darkness* and *A Passage to India*, manifests itself subconsciously in 'imaginary' texts, such as those of Joyce Cary, through the narrators' clear relish in describing the mutilation of natives. 'Imaginary' texts, like fantasies which provide naïve solutions to the subjects' basic problems, tend to center themselves on plots that end with the elimination of the offending natives.

The power of the 'imaginary' field binding the narcissistic colonialist text is nowhere better illustrated than in its fetishization of the Other. This process operates by substituting natural or generic categories for those that are socially or ideologically determined. All the evil characteristics and habits with which the colonialist endows the native are thereby not presented as the products of social and cultural difference but as characteristics inherent in the race – in the 'blood' – of the native. In its extreme form, this kind of fetishization transmutes all specificity and difference into a magical essence. Thus Dinesen boldly asserts:

> The Natives were Africa in flesh and . . . [The various cultures of Africa, the mountains, the trees, the animals] were different expressions of one idea, variations upon the same theme. It was not a congenial upheaping of heterogeneous atoms, but a heterogeneous upheaping of congenial atoms, as in the case of the oak-leaf and the acorn and the object made from oak.
>
> (Dinesen 1937: 21)

As this example illustrates, it is not the stereotypes, the denigrating 'images' of the native (which abound in colonialist literature), that are fetishized. Careful scrutiny of colonialist texts reveals that such images are used at random and in a self-contradictory fashion. For example, the narrator of Cary's *Aissa Saved* can claim that 'Kolu children of old-fashioned families like Makunde's were remarkable for their gravity and decorum; . . . they were strictly brought up and made to behave themselves as far as possible like grown-ups' (Cary 1949: 33). He even shows one such child, Tanawe, behaving with great decorum and gravity. Yet the same narrator depicts Kolu adults who have converted to Christianity as naughty, irresponsible children. Given the colonialist mentality, the source of the contradiction is quite obvious. Since Tanawe is too young to challenge colonialism, she can be depicted in a benign manner, and the narrator can draw moral sustenance from the generosity of his portrayal. But the adult Kolus' desire to become Christians threatens to eliminate one of the fundamental differences between them and the Europeans; so the narrator has to impose a difference. The overdetermined image he picks (Africans = children) allows him to feel secure once again because it restores the moral balance in favor of the ('adult') Christian conqueror. Such contradictory use of images abounds in colonialist literature.

My point, then, is that the imperialist is not fixated on specific images or stereotypes of the Other but rather on the affective benefits proffered by the manichean allegory, which generates the various stereotypes. As I have argued, the manichean allegory, with its highly efficient exchange mechanism, permits various kinds of rapid transformations, for example, metonymic displacement – which leads to the essentialist metonymy, as in the above quotation from Dinesen – and metaphoric condensation – which accounts for the structure and characterization in Cary's *Mister Johnson*. Exchange-value remains the central motivating force of both colonialist material practice and colonialist literary representation.

The fetishizing strategy and the allegorical mechanism not only permit a rapid exchange of denigrating images which can be used to maintain a sense of moral difference; they also allow the writer to transform social and historical dissimilarities into universal, metaphysical differences. If, as Dinesen has done, African natives can be collapsed into African animals and mystified still further as some magical essence of the continent, then clearly there can be no meeting ground, no identity, between the social, historical creatures of Europe and the metaphysical alterity of the Calibans and Ariels of Africa. If the differences between the Europeans and the natives are so vast, then clearly, as I stated earlier, the process of civilizing the natives can continue indefinitely. The ideological function of this mechanism, in addition to prolonging colonialism, is to dehistoricize and desocialize the conquered world, to present it as a metaphysical 'fact of life,' before which those who have fashioned the colonial world are themselves reduced to the role of passive spectators in a mystery not of their making.

There are many formal consequences of this denial of history and normal social interaction. While masquerading under the guise of realist fiction, the colonialist text is in fact antagonistic to some of the prevailing tendencies of realism. As M. M. Bakhtin has argued, the temporal model of the world changes radically with the rise of the realist novel: 'For the first time in artistic-ideological consciousness, time and the world become historical: they unfold as becoming, as an uninterrupted movement into a real future, as a unified, all-embracing and unconcluded process' (Bakhtin 1975: 30). But since the colonialist wants to maintain his privileges by preserving the status quo, his representation of the world contains neither a sense of historical becoming, nor a concrete vision of a future different from the present, nor a teleology other than the infinitely postponed process of 'civilizing.' In short, it does not contain any syncretic cultural possibility, which alone would open up the historic once more. . . .

This adamant refusal to admit the possibility of syncretism, of a rapprochement between self and Other, is the most important factor distinguishing the 'imaginary' from the 'symbolic' colonialist text. The 'symbolic' text's openness toward the Other is based on a greater awareness of potential identity and a heightened sense of the concrete socio-politico-cultural differences between self and Other. Although the 'symbolic' writer's understanding of the Other proceeds through self-understanding, he is freer from the codes and motifs of the deeper, collective classification system of his culture. In the final analysis, his success in comprehending or appreciating alterity will depend on his ability to bracket the values and bases of his culture. He may do so very consciously and deliberately, as Forster does in *A Passage to India*, or he may allow the emotions and values instilled in him during his social formation in an alien culture to inform his appraisals of the Other, as Kipling does in *Kim*. These two novels offer the most interesting attempts to overcome the barriers of racial difference. . . .

As we have seen, colonialist fiction is generated predominantly by the ideological machinery of the manichean allegory. Yet the relation between imperial ideology and fiction is not unidirectional: the ideology does not simply determine the fiction. Rather, through a process of symbiosis, the fiction *forms* the ideology by articulating and justifying the position and aims of the colonialist. But it does more than just define and elaborate the actual military and putative moral superiority of the Europeans. Troubled by the nagging contradiction between the theoretical justification of exploitation and the barbarity of its actual practice, it also attempts to mask the contradiction by obsessively portraying the supposed inferiority and barbarity of the racial Other, thereby insisting on the profound moral difference between self and Other. Within this symbiotic relation, the manichean allegory functions as a transformative mechanism between the affective pleasure derived from the moral superiority and material profit that motivate imperialism, on the one hand, and the formal devices (genres, stereotypes, and so on) of colonialist fiction, on the other hand. By allowing the European to denigrate the native in a variety of ways, by permitting an obsessive, fetishistic representation of the native's moral inferiority, the allegory also enables the European to increase, by contrast, the store of his own moral superiority; it allows him to accumulate 'surplus morality,' which is further invested in the denigration of the native, in a self-sustaining cycle.

   Thus the ideological function of all 'imaginary' and some 'symbolic' colonialist literature is to articulate and justify the moral authority of the colonizer and – by positing the inferiority of the native as a metaphysical fact – to mask the pleasure the colonizer derives from that authority. . . .

   Finally, we must bear in mind that colonialist fiction and ideology do not exist in a vacuum. In order to appreciate them thoroughly, we must examine them in juxtaposition to domestic English fiction and the anglophone fiction of the Third World, which originates from British occupation and which, during the current, hegemonic phase of colonialism, is establishing a dialogic relation with colonialist fiction. The Third World's literary dialogue with Western cultures is marked by two broad characteristics: its attempt to negate the prior European negation of colonized cultures and its adoption and creative modification of Western languages and artistic forms in conjunction with indigenous languages and forms. This dialogue merits our serious attention for two reasons: first, in spite of the often studied attempts by ethnocentric canonizers in English and other (Western) language and literature departments to ignore Third World culture and art, they will not go away; and, second, as this analysis of colonialist literature (a literature, we must remember, that is sued to mediate between different cultures) demonstrates, the domain of literary and cultural syncretism belongs not to colonialist and neocolonialist writers but increasingly to Third World artists.

## Chapter 3

# EDWARD W. SAID

## ORIENTALISM

From *Orientalism* New York: Random House, 1978.

O N A VISIT TO BEIRUT during the terrible civil war of 1975–6 a French journalist wrote regretfully of the gutted downtown area that 'it had once seemed to belong to . . . the Orient of Chateaubriand and Nerval' (Desjardins 1976: 14). He was right about the place, of course, especially so far as a European was concerned. The Orient was almost a European invention, and had been since antiquity a place of romance, exotic beings, haunting memories and landscapes, remarkable experiences. Now it was disappearing; in a sense it had happened, its time was over. Perhaps it seemed irrelevant that Orientals themselves had something at stake in the process, that even in the time of Chateaubriand and Nerval Orientals had lived there, and that now it was they who were suffering; the main thing for the European visitor was a European representation of the Orient and its contemporary fate, both of which had a privileged communal significance for the journalist and his French readers. . . .

The Orient is not only adjacent to Europe; it is also the place of Europe's greatest and richest and oldest colonies, the source of its civilizations and languages, its cultural contestant, and one of its deepest and most recurring images of the Other. In addition, the Orient has helped to define Europe (or the West) as its contrasting image, idea, personality, experience. Yet none of this Orient is merely imaginative. The Orient is an integral part of European *material* civilization and culture. Orientalism expresses and represents that part culturally and even ideologically as a mode of discourse with supporting institutions, vocabulary, scholarship, imagery, doctrines, even colonial bureaucracies and colonial styles. In contrast, the American understanding of the Orient will seem considerably less dense, although our recent Japanese, Korean, and Indochinese adventures ought now to be creating a more sober, more realistic 'Oriental' awareness. Moreover, the vastly expanded American political and economic role in the Near East (the Middle East) makes great claims on our understanding of that Orient.

It will be clear to the reader . . . that by Orientalism I mean several things, all of them, in my opinion, interdependent. The most readily accepted designation for Orientalism is an academic one, and indeed the label still serves in a number of academic institutions. Anyone who teaches, writes about, or researches the Orient – and this applies whether the person is an anthropologist, sociologist, historian, or philologist – either in its specific or its general aspects, is an Orientalist, and what he or she does is Orientalism. Compared with *Oriental studies* or *area studies*, it is true that the term *Orientalism* is less preferred by specialists today, both because it is too vague and general and because it connotes the high-handed executive attitude of nineteenth-century and early twentieth-century European colonialism. Nevertheless books are written and congresses held with 'the Orient' as their main focus, with the Orientalist in his new or old guise as their main authority. The point is that even if it does not survive as it once did, Orientalism lives on academically through its doctrines and theses about the Orient and the Oriental.

Related to this academic tradition, whose fortunes, transmigrations, specializations, and transmissions are in part the subject of this study, is a more general meaning for Orientalism. Orientalism is a style of thought based upon an ontological and epistemological distinction made between 'the Orient' and (most of the time) 'the Occident.' Thus a very large mass of writers, among whom are poets, novelists, philosophers, political theorists, economists, and imperial administrators, have accepted the basic distinction between East and West as the starting point for elaborate theories, epics, novels, social descriptions, and political accounts concerning the Orient, its people, customs, 'mind,' destiny, and so on. *This* Orientalism can accommodate Aeschylus, say, and Victor Hugo, Dante, and Karl Marx. . . .

The interchange between the academic and the more or less imaginative meanings of Orientalism is a constant one, and since the late eighteenth century there has been a considerable, quite disciplined – perhaps even regulated – traffic between the two. Here I come to the third meaning of Orientalism, which is something more historically and materially defined than either of the other two. Taking the late eighteenth century as a very roughly defined starting point Orientalism can be discussed and analyzed as the corporate institution for dealing with the Orient – dealing with it by making statements about it, authorizing views of it, describing it, by teaching it, settling it: in short, Orientalism as a Western style for dominating, restructuring, and having authority over the Orient. I have found it useful here to employ Michel Foucault's notion of a discourse, as described by him in *The Archaeology of Knowledge* and in *Discipline and Punish*, to identify Orientalism. My contention is that without examining Orientalism as a discourse one cannot possibly understand the enormously systematic discipline by which European culture was able to manage – and even produce – the Orient politically, sociologically, militarily, ideologically, scientifically, and imaginatively during the post-Enlightenment period. Moreover, so authoritative a position did Orientalism have that I believe no one writing, thinking, or acting on the Orient could do so without taking account of the limitations on thought and action imposed by Orientalism. In brief, because of Orientalism the Orient was not (and is not) a free subject of thought or action. This is not to say that Orientalism unilaterally determines what can be said about the Orient, but that it is the whole network of interests inevitably brought to bear on (and therefore always involved in) any occasion when that peculiar entity 'the Orient' is in question. How this happens is what this book tries to demonstrate. It also tries to show that European culture gained in strength and identity by setting itself off against the Orient as a sort of surrogate and even underground self. . . .

I have begun with the assumption that the Orient is not an inert fact of nature. It is not merely *there*, just as the Occident itself is not just *there* either. We must take seriously Vico's great observation that men make their own history, that what they can know is what they have made, and extend it to geography: as both geographical and cultural entities – to say nothing of historical entities – such locales, regions, geographical sectors as 'Orient' and 'Occident' are man-made. Therefore as much as the West itself, the Orient is an idea that has a history and a tradition of thought, imagery, and vocabulary that have given it reality and presence in and for the West. The two geographical entities thus support and to an extent reflect each other. Having said that, one must go on to state a number of reasonable qualifications. In the first place, it would be wrong to conclude that the Orient was *essentially* an idea, or a creation with no corresponding reality. . . . There were – and are – cultures and nations whose location is in the East, and their lives, histories, and customs have a brute reality obviously greater than anything that could be said about them in the West. About that fact this study of Orientalism has very little to contribute, except to acknowledge it tacitly. But the phenomenon of Orientalism as I study it here deals principally, not with a correspondence between Orientalism and Orient, but with the internal consistency of Orientalism and its ideas about the Orient (the East as career) despite or beyond any correspondence, or lack thereof, with a 'real' Orient. . . .

A second qualification is that ideas, cultures, and histories cannot seriously be understood or studied without their force, or more precisely their configurations of power, also being studied. To believe that the Orient was created – or, as I call it, 'Orientalized' – and to believe that such things happen simply as a necessity of the imagination, is to be disingenuous. The relationship between Occident and Orient is a relationship of power, of domination, of varying degrees of a complex hegemony. . . .

This brings us to a third qualification. One ought never to assume that the structure of Orientalism is nothing more than a structure of lies or of myths which, were the truth about them to be told, would simply blow away. I myself believe that Orientalism is more particularly valuable as a sign of European–Atlantic power over the Orient than it is a veridic discourse about the Orient (which is what, in its academic or scholarly form, it claims to be). . . .

In a quite constant way, Orientalism depends for its strategy on this flexible *positional* superiority, which puts the Westerner in a whole series of possible relationships with the Orient without ever losing him the relative upper hand. And why should it have been otherwise, especially during the period of extraordinary European ascendancy from the late Renaissance to the present? The scientist, the scholar, the missionary, the trader, or the soldier was in, or thought about, the Orient because he *could be there*, or could think about it, with very little resistance on the Orient's part. Under the general heading of knowledge of the Orient, and within the umbrella of Western hegemony over the Orient during the period from the end of the eighteenth century, there emerged a complex Orient suitable for study in the academy, for display in the museum, for reconstruction in the colonial office, for theoretical illustration in anthropological, biological, linguistic, racial, and historical theses about mankind and the universe, for instances of economic and sociological theories of development, revolution, cultural personality, national or religious character. Additionally, the imaginative examination of things Oriental was based more or less exclusively upon a sovereign Western consciousness out of whose unchallenged centrality an Oriental world emerged, first according to general ideas about who or what was an Oriental, then according to a detailed logic governed not simply by empirical reality but by a battery of desires, repressions, investments, and projections. . . .

Therefore, Orientalism is not a mere political subject matter or field that is reflected passively by culture, scholarship, or institutions; nor is it a large and diffuse collection of texts about the Orient; nor is it representative and expressive of some nefarious 'Western' imperialist plot to hold down the 'Oriental' world. It is rather a *distribution* of geopolitical awareness into aesthetic, scholarly, economic, sociological, historical, and philological texts; it is an *elaboration* not only of a basic geographical distinction (the world is made up of two unequal halves, Orient and Occident) but also of a whole series of 'interests' which, by such means as scholarly discovery, philological reconstruction, psychological analysis, landscape and sociological description, it not only creates but also maintains; it *is*, rather than expresses, a certain *will* or *intention* to understand, in some cases to control, manipulate, even to incorporate, what is a manifestly different (or alternative and novel) world; it is, above all, a discourse that is by no means in direct, corresponding relationship with political power in the raw, but rather is produced and exists in an uneven exchange with various kinds of power, shaped to a degree by the exchange with power political (as with a colonial or imperial establishment), power intellectual (as with reigning sciences like comparative linguistics or anatomy, or any of the modern policy sciences), power cultural (as with orthodoxies and canons of taste, texts, values), power moral (as with ideas about what 'we' do and what 'they' cannot do or understand as 'we' do). Indeed, my real argument is that Orientalism is – and does not simply represent – a considerable dimension of modern political–intellectual culture, and as such has less to do with the Orient than it does with 'our' world.

Because Orientalism is a cultural and a political fact, then, it does not exist in some archival vacuum; quite the contrary, I think it can be shown that what is thought, said, or even done about the Orient follows (perhaps occurs within) certain distinct and intellectually knowable lines. Here too a considerable degree of nuance and elaboration can be seen working as between the broad superstructural pressures and the details of composition, the facts of textuality. Most humanistic scholars are, I think, perfectly happy with the notion that texts exist in contexts, that there is such a thing as intertextuality, that the pressures of conventions, predecessors, and rhetorical styles limit what Walter Benjamin once called the 'overtaxing of the productive person in the name of . . . the principle of "creativity",' in which the poet is believed on his own, and out of his pure mind, to have brought forth his work (Benjamin 1973: 71). Yet there is a reluctance to allow that political, institutional, and ideological constraints act in the same manner on the individual author.

## Chapter 4

# GAYATRI CHAKRAVORTY SPIVAK

## CAN THE SUBALTERN SPEAK?

## (Abbreviated by the author)

From 'Can the Subaltern Speak?' in Gayatri Chakravorty Spivak *Toward a History of the Vanishing Present* Cambridge, Mass.: Harvard University Press, 1999

**H**ERE IS A WOMAN WHO TRIED TO BE DECISIVE *in extremis*. She 'spoke,' but women did not, do not, 'hear' her. Thus she can be defined as a 'subaltern' – a person without lines of social mobility.

Yet the ventriloquism of the speaking subaltern is the left intellectual's stock-in-trade. Gilles Deleuze declared, 'There is no more representation; there's nothing but action' – 'action of theory and action of practice which relate to each other as relays and form networks' (Foucault and Deleuze 1977: 206–7).

An important point is being made here: the production of theory is also a practice; the opposition between abstract 'pure' theory and concrete 'applied' practice is too quick and easy.[1] But Deleuze's articulation of the argument is problematic. Two senses of representation are being run together: representation as 'speaking for,' as in politics, and representation as 're-presentation,' as in art or philosophy. Since theory is also only 'action,' the theorist does not represent (speak for) the oppressed group. Indeed, the subject is not seen as a representative consciousness (one re-presenting reality adequately). These two senses of representation – within state formation and the law, on the one hand, and in subject-predication, on the other – are related but irreducibly discontinuous. To cover over the discontinuity with an analogy that is presented as a proof reflects again a paradoxical subject-privileging.[2] *Because* 'the person who speaks and acts . . . is always a multiplicity,' no 'theorizing intellectual . . . [or] party or . . . union' can represent 'those who act and struggle' (Foucault and Deleuze 1977: 206). Are those who act and *struggle* mute, as opposed to those who act and *speak* (Foucault and Deleuze 1977: 206)? These immense problems are buried in the differences between the 'same' words: consciousness and conscience (both *conscience* in French), representation and re-presentation. The critique of ideological subject-constitution within state formations and systems of political economy can now be effaced, as can the active theoretical practice of the 'transformation of consciousness.' The banality of leftist intellectuals' lists of self-knowing, politically canny subalterns stands revealed; representing them, the intellectuals represent themselves as transparent. If such a critique and such a project are not to be given up, the shifting distinctions between representation within the state and political economy, on the one hand, and within the theory of the Subject, on the other, must not be obliterated. Let us consider the play of *vertreten* ('represent' in the first sense) and *darstellen* ('re-present' in the second sense) in a famous passage in *The Eighteenth Brumaire of Louis*

*Bonaparte*, where Marx touches on 'class' as a descriptive and transformative concept. This is important in the context of the argument from the working class both from our two philosophers and *political* third world feminism from the metropolis.

Marx's contention here is that the descriptive definition of a class can be a differential one – its cutting off and difference from all other classes: '[I]n so far as millions of families live under economic conditions of existence that cut off their mode of life, their interest, and their formation from those of the other classes and place them in inimical confrontation [*feindlich gegenüberstellen*], they form a class' (Marx 1973: 239). There is no such thing as a 'class instinct' at work here. In fact, the collectivity of familial existence, which might be considered the arena of 'instinct,' is discontinuous with, though operated by, the differential isolation of classes. In this context, the formation of a class is *artificial* and economic, and the economic agency or *interest* is impersonal because it is systemic and heterogeneous. This agency or interest is tied to the Hegelian critique of the individual subject, for it marks the subject's empty place in that process without a subject which is history and political economy. Here the capitalist is defined as 'the conscious bearer [*Träger*] of the limitless movement of capital' (Marx 1977: 254). My point is that Marx is not working to create an undivided subject where desire and interest coincide. Class consciousness does not operate toward that goal. Both in the economic area (capitalist) and in the political (world historical agent), Marx is obliged to construct models of a divided and dislocated subject whose parts are not continuous or coherent with each other. A celebrated passage like the description of capital as the Faustian monster brings this home vividly (Marx 1977: 302).

The following passage, continuing the quotation from *The Eighteenth Brumaire*, is also working on the structural principle of a dispersed and dislocated class subject: the (absent collective) consciousness of the small peasant proprietor class finds its 'bearer' in a 'representative' who appears to work in another's interest. 'Representative' here does not derive from '*darstellen*'; this sharpens the contrast Foucault and Deleuze slide over, the contrast, say, between a proxy and a portrait. There is, of course, a relationship between them, one that has received political and ideological exacerbation in the European tradition at least since the poet and the sophist, the actor and the orator, have both been seen as harmful. In the guise of a post-Marxist description of the scene of power, we thus encounter a much older debate: between representation or rhetoric as tropology and as persuasion. *Darstellen* belongs to the first constellation, *vertreten* – with stronger suggestions of substitution – to the second. Again, they are related, but running them together, especially in order to say that beyond both is where oppressed subjects speak, act, and know *for themselves*, leads to an essentialist, utopian politics that can, when transferred to single-issue gender rather than class, give unquestioning support to the financialization of the globe, which ruthlessly constructs a general will in the credit-baited rural woman even as it 'formats' her through UN Plans of Action so that she can be 'developed.' Beyond this concatenation, transparent as rhetoric in the service of 'truth' has always made itself out to be, is the much-invoked oppressed subject (as Woman), speaking, acting, and knowing that gender in development is best for her. It is in the shadow of this unfortunate marionette that the history of the unheeded subaltern must unfold.

Here is Marx's passage, using '*vertreten*' where the English uses 'represent,' discussing a social 'subject' whose consciousness is dislocated and incoherent with its *Vertretung* (as much a substitution as a representation). The small peasant proprietors

> cannot represent themselves; they must be represented. Their representative must appear simultaneously as their master, as an authority over them, as unrestricted governmental power that protects them from the other classes and sends them rain

and sunshine from above. The political influence [in the place of the class interest, since there is no unified class subject] of the small peasant proprietors therefore finds its last expression [the implication of a chain of substitutions – *Vertretungen* – is strong here] in the executive force [*Exekutivegewalt* – less personal in German; Derrida translates *Gewalt* as violence in another context in *Force of Law*] subordinating society to itself.[3]

Not only does such a model of social incoherence – necessary gaps between the source of 'influence' (in this case the small peasant proprietors), the 'representative' (Louis Napoleon), and the historical-political phenomenon (executive control) – imply a critique of the subject as *individual* agent but a critique even of the subjectivity of a *collective* agency. The necessarily dislocated machine of history moves because 'the identity of the *interests*' of these proprietors 'fails to produce a feeling of community, national links, or a political organization.' The event of representation as *Vertretung* (in the constellation of rhetoric-as-persuasion) behaves like a *Darstellung* (or rhetoric-as-trope), taking its place in the gap between the formation of a (descriptive) class and the nonformation of a (transformative) class: 'In so far as millions of families live under economic conditions of existence that separate their mode of life . . . *they form a class*. In so far as . . . the identity of their interests fails to produce a feeling of community . . . *they do not form a class*.' The complicity of *vertreten* and *darstellen*, their identity-in-difference as the place of practice – since this complicity is precisely what Marxists must expose, as Marx does in *The Eighteenth Brumaire* – can only be appreciated if they are not conflated by a sleight of word.

It would be merely tendentious to argue that this textualizes Marx too much, making him inaccessible to the common 'man,' who, a victim of common sense, is so deeply placed in a heritage of positivism that Marx's irreducible emphasis on the work of the negative, on the necessity for defetishizing the concrete, is persistently wrested from him by the strongest adversary, 'the historical tradition' in the air.[4] The uncommon 'man,' the contemporary philosopher of practice, and the uncommon woman, the metropolitan enthusiast of *Third world resistance*, sometimes exhibit the same positivism.

I have dwelt so long on this passage in Marx because it spells out the inner dynamics of *Vertretung*, or representation in the political context. Representation in the economic context is *Darstellung*, the philosophical concept of representation as staging or, indeed, signification, which relates to the divided subject in an indirect way. The most obvious passage is well known: 'In the exchange relationship [*Austauschverhältnis*] of commodities their exchange-value appeared to us totally independent of their use value. But if we subtract their use-value from the product of labour, we obtain their value, as it was just determined [*bestimmt*]. The common element which represents itself [*sich darstellt*] in the exchange relation, or the exchange value of the commodity, is thus its value' (Marx 1977: 254).

According to Marx, under capitalism, value, as produced in necessary and surplus labor, is computed as the representation/sign of objectified labor (which is rigorously distinguished from human activity). Conversely, in the absence of a theory of exploitation as the extraction (production), appropriation, and realization of (surplus) value *as representation of labor power*, capitalist exploitation must be seen as a variety of domination (the mechanics of power as such). 'The thrust of Marxism,' Deleuze suggests, 'was to determine the problem [that power is more diffuse than the structure of exploitation and state formation] essentially in terms of interests (power is held by a ruling class defined by its interests)' (Foucault and Deleuze 1977: 214).

One cannot object to this minimalist summary of Marx's project, just as one cannot ignore that, in parts of the *Anti-Oedipus*, Deleuze and Guattari build their case on a brilliant if 'poetic' grasp of Marx's *theory* of the money form. Yet we might consolidate our critique in the following way: the relationship between global capitalism (exploitation in economics) and nation-state alliances (domination in geopolitics) is so macrological that it cannot account for the micrological texture of power.[5] Sub-individual micrologies cannot grasp the 'empirical' field. To move toward such an accounting one must move toward theories of ideology – of subject formations that micrologically and often erratically operate the interests that congeal the micrologies and are congealed in macrologies. Such theories cannot afford to overlook that this line *is* erratic, and that the category of representation in its *two* senses is crucial. They must note how the staging of the world in representation – its scene of writing, its *Darstellung* – dissimulates the choice of and need for 'heroes,' paternal proxies, agents of power – *Vertretung*.

My view is that radical practice should attend to this double session of representations rather than reintroduce the individual subject through totalizing concepts of power and desire.

One clearly available example of ideological epistemic violence is the remotely orchestrated, far-flung, and heterogeneous project to constitute the colonial subject as Other. This project is also the asymmetrical obliteration of the trace of that Other in its precarious Subject-ivity. It is well known that Foucault locates one case of epistemic violence, a complete overhaul of the episteme, in the redefinition of madness at the end of the European eighteenth century (see Foucault 1965: 251, 262, 269). But what if that particular redefinition was only a part of the narrative of history in Europe as well as in the colonies? What if the two projects of epistemic overhaul – European madness and colonial normality – worked as dislocated and unacknowledged parts of a vast two-handed engine?

Here, then, is a schematic summary of the epistemic violence of the codification of Hindu Law. If it clarifies the notion of epistemic violence, my final discussion of widow-sacrifice may gain added significance.

At the end of the eighteenth century, Hindu Law, in so far as it can be described as a unitary system, operated in terms of four texts that 'staged' a four-part episteme defined by the subject's use of memory: *sruti* (the heard), *smriti* (the remembered), *sastra* (the calculus), and *vyavahara* (the performance). The origins of what had been heard and what was remembered were not necessarily continuous or identical. Every invocation of *sruti* technically recited (or reopened) the event of originary 'hearing' or revelation. The second two texts – the learned and the performed – were seen as dialectically continuous. Legal theorists and practitioners were not in any given case certain if this structure described the body of law or four ways of settling a dispute. The legitimation, through a binary vision, of the polymorphous structure of legal performance, 'internally' noncoherent and open at both ends, is the narrative of codification I offer as an example of epistemic violence.

Consider the often-quoted programmatic lines from Macaulay's infamous 'Minute on Indian Education' (1835):

> We must at present do our best to form a class who may be interpreters between
> us and the millions whom we govern; a class of persons, Indian in blood and colour,
> but English in taste, in opinions, in morals, and in intellect. To that class we may
> leave it to refine the vernacular dialects of the country, to enrich those dialects with
> terms of science borrowed from the Western nomenclature, and to render them by
> degrees fit vehicles for conveying knowledge to the great mass of the population.
> (Macaulay 1835: 359)

The education of colonial subjects complements their production in law. One effect of establishing a version of the British system was the development of an uneasy separation between disciplinary formation in Sanskrit studies and the native, now alternative, tradition of Sanskrit 'high culture.' Within the former, the cultural explanations generated by authoritative scholars matched the epistemic violence of the legal project.

The place of the subaltern, as complicated by the imperialist project, is confronted by the 'Subaltern Studies' group. They *must* ask: Can the subaltern speak?

Ranajit Guha, the founder of the collective, gives a definition of the people that is an identity-in-differential. He proposes a dynamic stratification grid describing colonial social production at large. Even the third group on the list, the buffer group, as it were, between the people and the great macro-structural dominant groups, is itself defined as a place of in-betweenness. The classification falls into: 'dominant foreign groups,' and 'dominant indigenous groups at the all-India and at the regional and local levels' representing the elite; and '[t]he social groups and elements included in [the terms "people" and "subaltern classes"] represent[ing] *the demographic difference between the total Indian population and all those whom we have described as the "elite."*'[6]

For the (gender-unspecified) 'true' subaltern group, whose identity is its difference, there is no unrepresentable subaltern subject that can know and speak itself; the intellectual's solution is not to abstain from representation. The problem is that the subject's itinerary has not been left traced so as to offer an object of seduction to the representing intellectual. In the slightly dated language of the Indian group, the question becomes: How can we touch the consciousness of the people, even as we investigate their politics? With what voice-consciousness can the subaltern speak?

Within the effaced itinerary of the subaltern subject, the track of sexual difference is doubly effaced.[7] The question is not of female participation in insurgency, or the ground rules of the sexual division of labor, for both of which there is 'evidence.' It is, rather, that, both as object of colonialist historiography and as subject of insurgency, the ideological construction of gender keeps the male dominant. If, in the contest of colonial production, the subaltern has no history and cannot speak, the subaltern as female is even more deeply in shadow.

The regulative psychobiography of widow self-immolation will be useful in tracking this double silencing.

If I ask myself: how is it possible to want to die by fire to mourn a husband ritually, I am asking the question of the (gendered) subaltern woman as subject, not, as my friend Jonathan Culler somewhat tendentiously suggests, trying to 'produce difference by differing' or to 'appeal . . . to a sexual identity defined as essential and privileg[ing] experiences associated with that identity' (Culler 1982: 48). Culler is here a part of that mainstream project of Western feminism which both continues and displaces the battle over the right to individualism between women and men in situations of upward class mobility. One suspects that the debate between U.S. feminism and European 'theory' (as theory is generally represented by women from the United States or Britain) occupies a significant corner of that very terrain.

Sarah Kofman has suggested that the deep ambiguity of Freud's use of women as a scapegoat may be read as a reaction-formation to an initial and continuing desire to give the hysteric a voice, to transform her into the *subject* of hysteria (Kofman 1985). The masculine-imperialist ideological formation that shaped that desire into 'the daughter's seduction' is part of the same formation that constructs the monolithic 'third-world woman.' No contemporary metropolitan investigator is not influenced by that formation. Part of our 'unlearning' project is to articulate our participation in that formation – by *measuring* silences, if necessary – into the *object* of investigation. Thus, when confronted with the questions, Can the subaltern speak? Can the subaltern (as woman) speak?, our efforts to give the subaltern a voice in history will be doubly open to the dangers run by Freud's discourse. It is in acknowledgement of these dangers rather than as solution to a problem that I put together the sentence 'White men are saving brown women from brown men,' a sentence that runs like a red thread through today's 'gender and development.' My impulse is not unlike the one to be encountered in Freud's investigation of the sentence 'A child is being beaten' (Freud 1961 vol. 17: 174–204). For a list of ways in which Western criticism constructs 'Third World Woman' see Mohanty 1991: 51–80).

Freud predicates a *history* of repression that produces the final sentence. It is a history with a double origin, one hidden in the amnesia of the infant, the other lodged in our archaic past, assuming by implication a preoriginary space where human and animal were not yet differentiated (Freud 1961: 188). We are driven to impose a homology of this Freudian strategy on the Marxist narrative to explain the ideological dissimulation of imperialist political economy and outline a history of repression that produces a sentence like the one I have sketched: *white men are saving brown women from brown men* – giving honorary whiteness to the colonial subject on precisely this issue. This history also has a double origin, one hidden in the maneuverings behind the British abolition of widow sacrifice in 1829,[8] the other lodged in the classical and Vedic past of 'Hindu' India, the *Rg-Veda* and the *Dharmasastra*. An undifferentiated transcendental pre-originary space can only too easily be predicated for this other history.

The sentence I have constructed is one among many displacements describing the relationship between brown and white men (sometimes brown and white women worked in).[9] It takes its place among some sentences of 'hyperbolic admiration' or of pious guilt that Derrida speaks of in connection with the 'hieroglyphist prejudice.' The relationship between the imperialist subject and the subject of imperialism is at least ambiguous.

The Hindu widow ascends the pyre of the dead husband and immolates herself upon it. This is widow sacrifice. (The conventional transcription of the Sanskrit word for the widow would be *sati*. The early colonial British transcribed it *suttee*.) The rite was not practiced universally and was not caste- or class-fixed. The abolition of this rite by the British has been generally understood as a case of 'White men saving brown women from brown men.' White women – from the nineteenth-century British Missionary Registers to Mary Daly – have not produced an alternative understanding. Against this is the Indian nativist statement, a parody of the nostalgia for lost origins: 'The women wanted to die,' still being advanced.[10]

The two sentences go a long way to legitimize each other. The archivized examples of the testimony of the women's voice consciousness allow for the mobilization of such help. As one goes down the grotesquely mistranscribed names of these women, the sacrificed widows, in the police reports included in the records of the East India Company, one cannot put together a 'voice.' The most one can sense is the immense heterogeneity breaking through even such a skeletal and ignorant account (castes, for example, are regularly described as tribes). Faced with the dialectically interlocking sentences that are constructible as 'White men are saving brown women from brown men' and 'The women wanted to die,' the metropolitan feminist migrant (removed from the actual theater of decolonization) asks the question of simple semiosis – What does this signify? – and begins to plot a history.

I have written elsewhere of a constructed counternarrative of woman's consciousness, thus woman's being, thus woman's being good, thus the good woman's desire, thus woman's desire. This slippage can be seen in the fracture inscribed in the very word *sati*, the feminine form of *sat*. *Sat* transcends any gender-specific notion of masculinity and moves up not only into human but spiritual universality. It is the present participle of the verb 'to be' and as such means not only being, but the True, the Good, the Right. In the sacred texts it is essence, universal spirit. Even as a prefix it indicates appropriate, felicitous, fit. It is noble enough to have entered the most privileged discourse of modern Western philosophy: Heidegger's meditation on Being (Heidegger 1961: 58). *Sati*, the feminine of this word, simply means 'good wife.'

Figures like the goddess Athena – 'father's daughters self-professedly uncontaminated by the womb' – are useful for establishing women's ideological self-debasement, which is to be distinguished from a deconstructive attitude toward the essentialist subject. The story of the mythic Sati, reversing every narrateme of the rite, performs a similar function: the living husband avenges the wife's death, a transaction between great male gods fulfills the destruction of the female body and thus inscribes the earth as sacred geography. To see this as proof of the feminism of classical Hinduism or of Indian culture as goddess-centered and therefore feminist is as ideologically contaminated by nativism or reverse ethnocentrism as it was imperialist to erase the image of the luminous fighting Mother Durga and invest the proper noun Sati with no significance other than the ritual burning of the helpless widow as sacrificial offering who can then be saved. May the empowering voice of so-called superstition (Durga) not be a better starting point for transformation than the belittling or punitive befriending of the white mythology of 'reasonableness' (British police)? The interested do-gooding of corporate philanthropy keeps the question worth asking (see Spivak 2001: 120–63).

If the oppressed under postmodern capital have no necessarily unmediated access to 'correct' resistance, can the ideology of *sati*, coming from the history of the periphery, be sublated into any model of interventionist practice? Since this essay operates on the notion that all such clear-cut nostalgias for lost origins are suspect, especially as grounds for counterhegemonic ideological production, I must proceed by way of an example.[11]

A young woman of sixteen or seventeen, Bhubaneswari Bhaduri, hanged herself in her father's modest apartment in North Calcutta in 1926. The suicide was a puzzle since, as Bhubaneswari was menstruating at the time, it was clearly not a case of illicit pregnancy. Nearly a decade later, it was discovered, in a letter she had left for her elder sister, that she was a member of one of the many groups involved in the armed struggle for Indian independence. She had been entrusted with a political assassination. Unable to confront the task and yet aware of the practical need for trust, she killed herself.

Bhubaneswari had known that her death would be diagnosed as the outcome of illegitimate passion. She had therefore waited for the onset of menstruation. While waiting, Bhubaneswari, the celibate girl who was no doubt looking forward to good wifehood, perhaps rewrote the social text of *sati*-suicide in an interventionist way. (One tentative explanation of her inexplicable act had been a possible melancholia brought on by her father's death and her brother-in-law's repeated taunts that she was too old to be not-yet-a-wife.) She generalized the sanctioned motive for female suicide by taking immense trouble to displace (not merely deny), in the physiological inscription of her body, its imprisonment within legitimate passion by a single male. In the immediate context, her act became absurd, a case of delirium rather than sanity. The displacing gesture – waiting for menstruation – is at first a reversal of the interdict against a menstruating widow's right to immolate herself; the unclean widow must wait, publicly, until the cleansing bath of the fourth day, when she is no longer menstruating, in order to claim her dubious privilege.

In this reading, Bhubaneswari Bhaduri's suicide is an unemphatic, ad hoc, subaltern rewriting of the social text of *sati*-suicide as much as the hegemonic account of the blazing, fighting, familial Durga. The emergent dissenting possibilities of that hegemonic account of the fighting mother are well documented and popularly well remembered through the discourse of the male leaders and participants in the Independence movement. The subaltern as female cannot be heard or read.

I know of Bhubaneswari's life and death through family connections. Before investigating them more thoroughly, I asked a Bengali woman, a philosopher and Sanskritist whose early intellectual production is almost identical to mine, to start the process. Two responses: (a) Why, when her two sisters, Saileswari and Raseswari, led such full and wonderful lives, are you interested in the hapless Bhubaneswari? (b) I asked her nieces. It appears that it was a case of illicit love.

I was so unnerved by this failure of communication that, in the first version of this text, I wrote, in the accents of passionate lament: the subaltern cannot speak! It was an inadvisable remark.

Bhubaneswari Bhaduri was not a 'true' subaltern. She was a woman of the middle class, with access, however clandestine, to the bourgeois movement for Independence. Woman's interception of the claim to subalternity can be staked out across strict lines of definition by virtue of their muting by heterogeneous circumstances. Bhubaneswari attempted to 'speak' by turning her body into a text of woman/writing. The immediate passion of my declaration 'the subaltern cannot speak,' came from the despair that, in her own family, among women, in no more than fifty years, her attempt had failed.

> For Europe, the time when the new capitalism *definitely* superseded the old can be established with fair precision: it was the beginning of the twentieth century. . . . [With t]he boom at the end of the nineteenth century and the crisis of 1900–03 . . . [c]artels become one of the foundations of the whole of economic life. Capitalism has been transformed into imperialism.
>
> (Lenin 1969: 15)

Today's program of global financialization carries on that relay. Bhubaneswari had fought for national liberation. Her great-grandniece works for the New Empire. This too is a historical silencing of the subaltern. When the news of this young woman's promotion was broadcast in the family amidst general jubilation I could not help remarking to the then eldest surviving female member: 'Bhubaneswari' – her nickname had been Talu – 'hanged herself in vain,' but not too loudly. Is it any wonder that this young woman is a staunch multiculturalist, believes in natural childbirth, and wears only cotton?

© Gayatri Chakravorty Spivak, Columbia University

## Notes

1  Foucault's subsequent explanation (Foucault 1980: 145) of this Deleuzian statement comes closer to Derrida's notion that theory cannot be an exhaustive taxonomy and is always normed by practice.
2  Cf. the surprisingly uncritical notions of representation entertained in *Power/Knowledge* (1980: 141, 188). My remarks concluding this paragraph, criticizing intellectuals' representations of subaltern groups, should be rigorously distinguished from a coalition politics that takes into account its framing within socialized capital and unites people not because they are oppressed but because they are exploited. This model works best within a parliamentary democracy, where representation is not only not banished but elaborately staged.
3  This is a highly ironic passage in Marx, written in the context of the fraudulent *Representation* by Louis Napoleon and the regular suppression of the *Revolutionary peasants* by bourgeois interests (Marx 1973: 239). Many hasty readers think Marx is advancing this as his own opinion about all peasantry!

4  See the excellent short definition and discussion of common sense in Errol Lawrence, 'Just Plain Common Sense: The "Roots" of Racism,' (in Carby 1982: 48). The Gramscian notions of 'common sense' and 'good sense' are extensively discussed in Marcia Landy, *Film, Politics, and Gramsci* (1994: 73–98).

5  The situation has changed in the New World Order. Let us call the World Bank/IMF/World Trade Organization 'the economic;' and the United Nations 'the political.' The relationship between them is being negotiated in the name of gender ('the cultural'), which is, perhaps, micrology as such.

6  Ranajit Guha, *Subaltern Studies* (1982: 8). The usefulness of this tightly defined term was largely lost when *Selected Subaltern Studies* was launched in the United States under Spivak's initiative (1988). A new selection with a new introduction by Amartya Kumar Sen is about to appear from Duke University Press. In the now generalized usage, it is precisely this notion of the subaltern inhabiting a space of difference that is lost, e.g. in statements such as the following: 'The subaltern is force-fed into appropriating the master's culture' (Emily Apter, 'French Colonial Studies and Postcolonial Theory,' (1995: 178); or worse still, Jameson's curious definition of subalternity as 'the experience of inferiority' in 'Marx's Purloined Letter' (1994: 95).

7  I do not believe that the recent trend of romanticizing anything written by the Aboriginal or outcaste ('dalit' oppressed) intellectual has lifted the effacement.

8  For a brilliant account of how the 'reality' of widow-sacrificing was constituted or 'textualized' during the colonial period, see Lata Mani, 'Contentious Traditions: the Debate on *Sati* in Colonial India,' (1989: 88–126). I profited from discussion with Dr Mani at the inception of this project. Here I present some of my differences from her position. The 'printing mistake in the Bengali translation' (109) that she cites is not the same as the mistake I discuss, which is in the ancient Sanskrit. It is of course altogether interesting, that there should be all these errancies in the justification of the practice. A regulative psychobiography is not identical with 'textual hegemony' (96). I agree with Mani that the latter mode of explanation cannot take 'regional variations' into account. A regulative psychobiography is another mode of 'textualist oppression' when it produces not only 'women's consciousness' but a 'gendered episteme' (mechanics of the construction of objects of knowledge together with validity-criteria for statements of knowledge). You do not have to 'read verbal texts' here. It is something comparable to Gramsci's 'inventory without traces' (1971: 324). Like Mani (p. 125, n. 90), I too wish to 'add' to Kosambi's 'strategies.' To the 'supplement[ation of the linguistic study of problems of ancient Indian culture, by intelligent use of archaeology, anthropology, sociology and a suitable historical perspective' (Kosambi 1963: 177), I would add the insights of psychoanalysis, though not the regulative psychobiography of its choice. Alas, in spite of our factualist fetish, 'facts' alone may account for women's oppression, but they will never allow us to approach gendering, a net where we ourselves are enmeshed, as we decide what (the) facts are. Because of epistemic prejudice, Kosambi's bold and plain speech can and has been misunderstood; but his word 'live' can take on board a more complex notion of the mental theatre as Mani cannot: 'Indian peasants in villages far from any city *live* in a manner closer to the days when the Purñas were written than do the descendants of the brahmins who wrote the Purñas' (emphasis mine). Precisely. The self representation in gendering is regulated by the Puranic psychobiography, with the Brahmin as the model. In the last chapter I will consider what Kosambi mentions in the next sentence: 'A stage further back are the pitiful fragments of tribal groups, usually sunk to the level of marginal castes; they rely heavily upon food-gathering and have the corresponding mentality.' Kosambi's somewhat doctrinaire Marxism would not allow him to think of the tribal episteme as anything but only backward, of course. After the *sati* of Rup Kanwar in September, 1987, a body of literature on the contemporary situation has emerged. That requires quite a different engagement (see Radha Kumar 1993: 172–81).

9  See Kumari Jayawardena, *The White Woman's Other Burden: Western Women and South Asia During British Colonial Rule* (1995). Envy, backlash, reaction-formation; these are the routes by which such efforts may, in the absence of ethical responsibility, lead to opposite results. I have repeatedly invoked Melanie Klein and Assia Djebar in this context. See also Spivak (1994: 66–9).

10 The examples of female ventriloquist complicity, quoted by Lata Mani in her brilliant article 'Production of an Official Discourse on *Sati* in early Nineteenth Century Bengal (1986: 1–36), proves my point. The point is not that a refusal would not be ventriloquism for Women's Rights. One is not suggesting that only the latter is correct free will. One is suggesting that the freedom of the will is negotiable, and it is not on the grounds of a disinterested free will that we will be able to justify an action, in this case against the burning of widows, to the adequate satisfaction of all. The ethical aporia is not negotiable. We must act in view of this.

11 A position against nostalgia as a basis of counterhegemonic ideological production does not endorse its negative use. Within the complexity of contemporary political economy, it would, for example, be highly questionable to urge that the current Indian working-class crime of burning brides who bring insufficient dowries and of subsequently disguising the murder as suicide is either a *use* or *abuse* of the tradition of *sati*-suicide. The most that can be claimed is that it is a displacement on a chain of semiosis with the female subject as signifier, which would lead us back into the narrative we have been unraveling. Clearly, one must work to stop the crime of bride burning *in every way*. If, however, that work is accomplished by unexamined nostalgia or its opposite, it will assist actively in the substitution of race/ethnos or sheer genitalism as a signifier in the place of the female subject.

## Chapter 5

# HOMI K. BHABHA

## SIGNS TAKEN FOR WONDERS

From 'Signs Taken for Wonders: Questions of Ambivalence and Authority
Under a Tree Outside Delhi, May 1817' *Critical Inquiry* 12(1), 1985.

> A remarkable peculiarity is that they (the English) always write the personal
> pronoun I with a capital letter. May we not consider this Great I as an unintended
> proof how much an Englishman thinks of his own consequence?
>
> Robert Southey, *Letters from England*

THERE IS A SCENE IN THE CULTURAL WRITINGS of English colonialism which repeats so insistently after the early nineteenth century – and, through that repetition, so triumphantly *inaugurates* a literature of empire – that I am bound to repeat it once more. It is the scenario, played out in the wild and wordless wastes of colonial India, Africa, the Caribbean, of the sudden fortuitous discovery of the English book. It is, like all myths of origin, memorable for its balance between epiphany and enunciation. The discovery of the book is, at once, a moment of originality and authority, as well as a process of displacement that, paradoxically, makes the presence of the book wondrous to the extent to which it is repeated, translated, misread, displaced. It is with the emblem of the English book – 'signs taken for wonders' – as an insignia of colonial authority and a signifier of colonial desire and discipline, that I want to begin this essay.

In the first week of May 1817, Anund Messeh, one of the earliest Indian catechists, made a hurried and excited journey from his mission in Meerut to a grove of trees outside Delhi.

He found about 500 people, men, women and children, seated under the shade of the trees, and employed, as had been related to him, in reading and conversation. He went up to an elderly looking man, and accosted him, and the following conversation passed:

> 'Pray who are all these people? and whence come they?' 'We are poor and lowly, and we read and love this book' – 'What is that book?' 'The book of God!' – 'Let me look at it, if you please.' Anund, on opening the book, perceived it to be the Gospel of our Lord, translated into the Hindoostanee Tongue, many copies of which seemed to be in the possession of the party: some were PRINTED others WRITTEN by themselves from the printed ones. Anund pointed to the name of Jesus, and asked, 'Who is that?' 'That is God! He gave us this book.' – 'Where did you obtain it?' 'An Angel from heaven gave it us, at Hurdwar fair.' – 'An Angel?' 'Yes, to us he was God's Angel: but he was a man, a learned Pundit.' (Doubtless these translated Gospels must have been the books distributed, five or six years ago, at Hurdwar by

the Missionary.) 'The written copies we write ourselves, having no other means of obtaining more of this blessed word.' – 'These books,' said Anund, 'teach the religion of the European Sahibs. It is THEIR book; and they printed it in our language, for our use.' 'Ah! no'; replied the stranger, 'that cannot be, for they eat flesh.' – 'Jesus Christ,' said Anund, 'teaches that it does not signify what a man eats or drinks. EATING is nothing before God. *Not that which entereth into a man's mouth defileth him but that which cometh out of the mouth, this defileth a man*: for vile things come forth from the heart. *Out of the heart proceed evil thoughts, murders, adulteries, fornications, thefts; and these are the things that defile.*'

'That is true; but how can it be the European Book, when we believe that it is God's gift to us? He sent it to us at Hurdwar.' 'God gave it long ago to the Sahibs, and THEY sent it to us.' The ignorance and simplicity of many are very striking, never having heard of a printed book before; and its very appearance was to them miraculous. A great stir was excited by the gradual increasing information hereby obtained, and all united to acknowledge the superiority of the doctrines of this Holy Book to every thing which they had hitherto heard or known. An indifference to the distinctions of Caste soon manifested itself; and the interference and tyrannical authority of the Brahmins became more offensive and contemptible. At last, it was determined to separate themselves from the rest of their Hindoo Brethren; and to establish a party of their own choosing, four or five, who could read the best, to be the public teachers from this newly-acquired Book. . . . Anund asked them, 'Why are you all dressed in white?' 'The people of God should wear white raiment,' was the reply, 'as a sign that they are clean, and rid of their sins.' – Anund observed, 'You ought to be BAPTIZED, in the name of the Father, and of the Son, and of the Holy Ghost. Come to Meerut: there is a Christian Padre there; and he will shew you what you ought to do.' They answered, 'Now we must go home to the harvest; but, as we mean to meet once a year, perhaps the next year we may come to Meerut.' I explained to them the nature of the Sacrament and of Baptism; in answer to which, they replied, 'We are willing to be baptized, but we will never take the Sacrament. To all the other customs of Christians we are willing to conform, but not to the Sacrament, because the Europeans eat cow's flesh, and this will never do for us.' To this I answered, 'this WORD is of God, and not of men; and when it makes your hearts to understand, then you will PROPERLY comprehend it. They replied, 'If all our country will receive this Sacrament, then will we.' I then observed, 'The time is at hand, when all the countries will receive this WORD.' They replied, 'True.'

(Missionary Register 1818: 18–19)

Almost a hundred years later, in 1902, Joseph Conrad's Marlow, traveling in the Congo, in the night of the first ages, without a sign and no memories, cut off from the comprehension of his surroundings, desperately in need of a deliberate belief, comes upon Towson's (or Towser's) *Inquiry into some Points of Seamanship*:

Not a very enthralling book; but at the first glance you could see there a singleness of intention, an honest concern for the right way of going to work, which made these humble pages, thought out so many years ago, luminous with another than a professional light. . . . I assure you to leave off reading was like tearing myself away from the shelter of an old and solid friendship. . . .

'It must be this miserable trader – this intruder,' exclaimed the manager, looking back malevolently at the place we had left. 'He must be English,' I said.

(Conrad [1902] 1983: 71, 72)

Half a century later, a young Trinidadian discovers that same volume of Towson's in that very passage from Conrad and draws from it a vision of literature and a lesson of history. 'The scene,' writes V. S. Naipaul, 'answered some of the political panic I was beginning to feel':

To be a colonial was to know a kind of security; it was to inhabit a fixed world. And I suppose that in my fantasy I had seen myself coming to England as to some purely literary region, where, untrammeled by the accidents of history or background, I could make a romantic career for myself as a writer. But in the new world I felt that ground move below me . . . Conrad . . . had been everywhere before me. Not as a man with a cause, but a man offering a vision of the world's half-made societies . . . where always 'something inherent in the necessities of successful action carried with it the moral degradation of the idea.' Dismal but deeply felt: a kind of truth and half a consolation.

(Naipaul 1974: 233)

Written as they are in the name of the father and the author, these texts of the civilizing mission immediately suggest the triumph of the colonialist moment in early English Evangelism and modern English literature. The discovery of the book installs the sign of appropriate representation: the word of God, truth, art creates the conditions for a beginning, a practice of history and narrative. But the institution of the Word in the wilds is also an *Entstellung*, a process of displacement, distortion, dislocation, repetition[1] – the dazzling light of literature sheds only areas of darkness. Still the idea of the English book is presented as universally adequate: like the 'metaphoric writing of the West,' it communicates 'the immediate vision of the thing, freed from the discourse that accompanied it, or even encumbered it' (Derrida 1981: 189–90). . . .

The discovery of the English book establishes both a measure of mimesis and a mode of civil authority and order. If these scenes, as I've narrated them, suggest the triumph of the writ of colonialist power, then it must be conceded that the wily letter of the law inscribes a much more ambivalent text of authority. For it is in between the edict of Englishness and the assault of the dark unruly spaces of the earth, through an act of repetition, that the colonial text emerges uncertainly. Anund Messeh disavows the natives' disturbing questions as he returns to repeat the now questionable 'authority' of Evangelical dicta; Marlow turns away from the African jungle to recognize, in retrospect, the peculiarly 'English' quality of the discovery of the book; Naipaul turns his back on the hybrid half-made colonial world to fix his eye on the universal domain of English literature. What we witness is neither an untroubled, innocent dream of England nor a 'secondary revision' of the nightmare of India, Africa, the Caribbean. What is 'English' in these discourses of colonial power cannot be represented as a plenitude or a 'full' presence; it is determined by its belatedness. As a signifier of authority, the English book acquires its meaning *after* the traumatic scenario of colonial difference, cultural or racial, returns the eye of power to some prior, archaic image or identity. Paradoxically, however, such an image can neither be 'original' by virtue of the act of repetition that constructs it – nor 'identical' by virtue of the difference that defines it. Consequently, the colonial presence is always ambivalent, split between its appearance as original and authoritative and its articulation as repetition and difference. . . .

The place of difference and otherness, or the space of the adversarial, within such a system of 'disposal' as I've proposed, is never entirely on the outside or implacably oppositional. It is a pressure, and a presence, that acts constantly, if unevenly, along the entire boundary of authorization, that is, on the surface between what I've called disposal-as-bestowal and disposition-as-inclination. The contour of difference is agonistic, shifting, splitting, rather like Freud's description of the system of consciousness which occupies a position in space lying on the borderline between outside and inside, a surface of protection, reception, and projection. The power play of presence is lost if its transparency is treated naively as the nostalgia for plenitude that should be flung repeatedly into the abyss – *mise en abîme* – from which its desire is born. Such theoreticist anarchism cannot intervene in the agonistic space of authority where

> the true and the false are separated and specific effects of power [are] attached to the true, it being understood also that it is not a matter of a battle 'on behalf' of the truth, but of a battle about the status of truth and the economic and political role it plays.
>
> (Foucault 1980: 132)

It is precisely to intervene in such a battle for the *status* of the truth that it becomes crucial to examine the *presence* of the English book. For it is this surface that stabilizes the agonistic colonial space; it is its *appearance* that regulates the ambivalence between origin and *Entstellung*, discipline and desire, mimesis and repetition.

Despite appearances, the text of transparency inscribes a double vision: the field of the 'true' emerges as a visible effect of knowledge/power only after the regulatory and displacing division of the true and the false. From this point of view, discursive 'transparency' is best read in the photographic sense in which a transparency is also always a negative, processed into visibility through the technologies of reversal, enlargement, lighting, editing, projection, not a source but a re-source of light. Such a bringing to light is never a prevision; it is always a question of the provision of visibility as a capacity, a strategy, an agency but also in the sense in which the prefix pro(vision) might indicate an elision of sight, delegation, substitution, contiguity, in place of . . . what?

This is the question that brings us to the ambivalence of the presence of authority, peculiarly visible in its colonial articulation. For if transparency signifies discursive closure – intention, image, author – it does so through a disclosure of its *rules of recognition* – those social texts of epistemic, ethnocentric, nationalist intelligibility which cohere in the address of authority as the 'present,' the voice of modernity. The acknowledgement of authority depends upon the immediate – unmediated – visibility of its rules of recognition as the unmistakable referent of historical necessity.

In the doubly inscribed space of colonial representation where the presence of authority – the English book – is also a question of its repetition and displacement, where transparency is *techné*, the immediate visibility of such a régime of recognition is resisted. Resistance is not necessarily an oppositional act of political intention, nor is it the simple negation or exclusion of the 'content' of an other culture, as a difference once perceived. It is the effect of an ambivalence produced within the rules of recognition of dominating discourses as they articulate the signs of cultural difference and reimplicate them within the deferential relations of colonial power – hierarchy, normalization, marginalization, and so forth. For domination is achieved through a process of disavowal that denies the *différance* of colonialist power – the chaos of its intervention as *Entstellung*, its dislocatory presence – in order to preserve the authority of its identity in the universalist narrative of nineteenth-century historical and political evolutionism.

The exercise of colonialist authority, however, requires the production of differentiations, individuations, identity effects through which discriminatory practices can map out subject populations that are tarred with the visible and transparent mark of power. Such a mode of subjection is distinct from what Foucault describes as 'power through transparency': the reign of opinion, after the late eighteenth century, which could not tolerate areas of darkness and sought to exercise power through the mere fact of things being known and people seen in an immediate, collective gaze. What radically differentiates the exercise of colonial power is the unsuitability of the Enlightenment assumption of collectivity and the eye that beholds it. For Jeremy Bentham (as Michel Perrot points out), the small group is representative of the whole society – the part is *already* the whole. Colonial authority requires modes of discrimination (cultural, racial, administrative . . .) that disallow a stable unitary assumption of collectivity. The 'part' (which must be the colonialist foreign body) must be representative of the 'whole' (conquered country), but the right of representation is based on its radical difference. Such doublethink is made viable only through the strategy of disavowal just described, which requires a theory of the 'hybridization' of discourse and power that is ignored by Western post-structuralists who engage in the battle for 'power' as the purists of difference.

The discriminatory effects of the discourse of cultural colonialism, for instance, do not simply or singly refer to a 'person', or to a dialectical power struggle between self and Other, or to a discrimination between mother culture and alien cultures. Produced through the strategy of disavowal, the *reference* of discrimination is always to a process of splitting as the condition of subjection: a discrimination between the mother culture and its bastards, the self and its doubles, where the trace of what is disavowed is not repressed but repeated as something *different* – a mutation, a hybrid. It is such a partial and double force that is more than the mimetic but less than the symbolic, that disturbs the visibility of the colonial presence and makes the recognition of its authority problematic. To be authoritative, its rules of recognition must reflect consensual knowledge or opinion; to be powerful, these rules of recognition must be breached in order to represent the exorbitant objects of discrimination that lie beyond its purview. Consequently if the unitary (and essentialist) reference to race, nation, or cultural tradition is essential to preserve the presence of authority as an immediate mimetic effect, such essentialism must be exceeded in the articulation of 'differentiatory,' discriminatory identities.

To demonstrate such an 'excess' is not merely to celebrate the joyous power of the signifier. Hybridity is the sign of the productivity of colonial power, its shifting forces and fixities; it is the name for the strategic reversal of the process of domination through disavowal (that is, the production of discriminatory identities that secure the 'pure' and original identity of authority). Hybridity is the revaluation of the assumption of colonial identity through the repetition of discriminatory identity effects. It displays the necessary deformation and displacement of all sites of discrimination and domination. It unsettles the mimetic or narcissistic demands of colonial power but reimplicates its identifications in strategies of subversion that turn the gaze of the discriminated back upon the eye of power. For the colonial hybrid is the articulation of the ambivalent space where the rite of power is enacted on the site of desire, making its objects at once disciplinary and disseminatory – or, in my mixed metaphor, a negative transparency. If discriminatory effects enable the authorities to keep an eye on them, their proliferating difference evades that eye, escapes that surveillance. Those discriminated against may be instantly recognized, but they also force a recognition of the immediacy and articulacy of authority – a disturbing effect that is familiar in the repeated hesitancy afflicting the colonialist discourse when it contemplates its discriminated subjects: the *inscrutability* of the Chinese, the *unspeakable* rites of the Indians, the *indescribable* habits of the Hottentots. It is not that the voice of authority is at a loss for words. It is, rather, that the colonial discourse has reached that point when, faced with the hybridity of its objects, the *presence* of power is revealed as something other than what its rules of recognition assert.

If the effect of colonial power is seen to be the *production* of hybridization rather than the noisy command of colonialist authority or the silent repression of native traditions, then an important change of perspective occurs. It reveals the ambivalence at the source of traditional discourses on authority and enables a form of subversion, founded on that uncertainty, that turns the discursive conditions of dominance into the grounds of intervention. It is traditional academic wisdom that the presence of authority is properly established through the non-exercise of private judgment and the exclusion of reasons, in conflict with the authoritative reason. The recognition of authority, however, requires a validation of its source that must be immediately, even intuitively, apparent – 'You have that in your countenance which I would fain call master' – and held in common (rules of recognition). What is left unacknowledged is the paradox of such a demand for proof and the resulting ambivalence for positions of authority. If, as Steven I. Lukes rightly says, the acceptance of authority excludes any evaluation of the content of an utterance, and if its source, which must be acknowledged, disavows both conflicting reasons and personal judgement, then can the 'signs' or 'marks' of authority be anything more than 'empty' presences of strategic devices? Need they be any the less effective because of that? Not less effective but effective in a different form, would be our answer.

## Note

1   'Overall effect of the dream-work: the latent thoughts are transformed into a manifest formation in which they are not easily recognisable. They are not only transposed, as it were, into another key, but *they are also distorted in such a fashion that only an effort of interpretation can reconstitute them*' (Laplanche and Pontalis 1980: 124). See also Samuel Weber's excellent chapter 'Metapsychology Set Apart' (1982: 32–60).

## Chapter 6

# BENITA PARRY

## PROBLEMS IN CURRENT THEORIES OF COLONIAL DISCOURSE

From 'Problems in Current Theories of Colonial Discourse' *Oxford Literary Review* 9(1–2), 1987.

THE WORK OF SPIVAK AND BHABHA will be discussed to suggest the productive capacity and limitations of their different deconstructive practices, and to propose that the protocols of their dissimilar methods act to constrain the development of an anti-imperialist critique. It will be argued that the lacunae in Spivak's learned disquisitions issue from a theory assigning an absolute power to the hegemonic discourse in constituting and disarticulating the native. In essays that are to form a study on Master Discourse/Native informant, Spivak inspects 'the absence of a text that can "answer one back" after the planned epistemic violence of the imperialist project' (Spivak 1985a: 131), and seeks to develop a strategy of reading that will speak to the historically-muted native subject, predominantly inscribed in Spivak's writings as the non-elite or subaltern woman. A refrain, 'One never encounters the testimony of the women's voice-consciousness,' 'There is no space from where the subaltern (sexed) subject can speak,' 'The subaltern as female cannot be heard or read,' 'The subaltern cannot speak' (Spivak 1985b: 122, 129, 130), iterates a theoretical dictum derived from studying the discourse of *Sati* [widow sacrifice], in which the Hindu patriarchal code converged with colonialism's narrativization of Indian culture to efface all traces of woman's voice.

What Spivak uncovers are instances of doubly-oppressed native women who, caught between the dominations of a native patriarchy and a foreign masculist-imperialist ideology, intervene by 'unemphatic, ad hoc, subaltern rewriting(s) of the social text of *Sati*-suicide' (Spivak 1985b: 129): a nineteenth century Princess who appropriates – 'the dubious place of the free will of the sexed subject as female' (Spivak 1985a: 144) by signaling her intention of being a *Sati* against the edict of the British administration; a young Bengal girl who in 1926 hanged herself under circumstances that deliberately defied Hindu interdicts (Spivak 1985b). From the discourse of *Sati* Spivak derives large, general statements on woman's subject constitution/object formation in which the subaltern woman is conceived as a homogeneous and coherent category, and which culminate in a declaration on the success of her planned disarticulation. Even within the confines of this same discourse, it is significant that Lata Mani does find evidence, albeit mediated, of woman's voice. As Chandra Talpade Mohanty argues in her critique of western feminist writings on 'Third World Women,' discourses of representation should not be confused with material realities. Since the native woman is constructed within multiple social relationships and positioned as the product of different class, caste and cultural specificities, it should be possible to locate traces and testimony of women's voice on those sites where women inscribed themselves as healers, ascetics, singers of sacred songs, artizans and artists, and by this to modify Spivak's model of the silent subaltern.

If it could appear that Spivak is theorizing the silence of the doubly-oppressed subaltern woman, her theorem on imperialism's epistemic violence extends to posting the native, male and female, as an historically-muted subject. The story of colonialism which she reconstructs is of an interactive process where the European agent in consolidating the imperialist Sovereign Self, induces the native to collude in its own subject(ed) formation as other and voiceless. Thus while protesting at the obliteration of the native's subject position in the text of imperialism, Spivak in her project gives no speaking part to the colonized, effectively writing out the evidence of native agency recorded in India's 200 year struggle against British conquest and the Raj – discourses to which she scathingly refers as hegemonic nativist or reverse ethnocentric narrativization.

The disparaging of nationalist discourses of resistance is matched by the exorbitation of the role allotted to the post-colonial woman intellectual, for it is she who must plot a story, unravel a narrative and give the subaltern a voice in history, by using 'the resources of deconstruction "in the service of reading" to develop a strategy rather than a theory of reading that might be a critique of imperialism' (Spivak 1986: 230). Spivak's 'alternative narrative of colonialism' through a series of brilliant upheavals of texts which expose the fabrications and exclusions in the writing of the archive, is directed at challenging the authority of the received historical record and restoring the effaced signs of native consciousness, and it is on these grounds that her project should be estimated. Her account, it is claimed, disposes of the old story by dispersing the fixed, unitary categories on which this depended. Thus it is argued that for purposes of administration and exploitation of resources, the native was constructed as a programmed, 'nearly-selved' other of the European and not as its binary opposite. Furthermore, the cartography that became the 'reality' of India was drawn by agents who were themselves of heterogeneous class origin and social status and whose (necessarily) diversified maps distributed the native into differential positions which worked in the interest of the foreign authority – for example, a fantasmatic race-differentiated historical demography restoring 'rightful' Aryan rulers, and a class discourse effecting the proto-proletarianization of the 'aborigines.'

Instead of recounting a struggle between a monolithic, near-deliberative colonial power and an undifferentiated oppressed mass, this reconstruction displays a process more insidious than naked repression, since here the native is prevailed upon to internalize as self-knowledge, the knowledge concocted by the master: 'He (the European agent) is worlding their own world, which is far from mere uninscribed earth, anew, by obliging them to domesticate the alien as Master,' a process generating the force 'to make the "native" see himself as "other"' (Spivak 1985a: 133). Where military conquest, institutional compulsion and ideological interpellation was, epistemic violence and devious discursive negotiations requiring of the native that he rewrite his position as object of imperialism, is; and in place of recalcitrance and refusal enacted in movements of resistance and articulated in oppositional discourses, a tale is told of the self-consolidating other and the disarticulated subaltern.

This raw and selective summary of what are complex and subtle arguments has tried to draw out the political implications of a theory whose axioms deny to the native the ground from which to utter a reply to imperialism's ideological aggression or to enunciate a different self:

No perspective *critical* of imperialism can turn the Other into a self, because the project of imperialism has always already historically refracted what might have been the absolutely Other into a domesticated Other that consolidates the imperialist self. . . . A full literary inscription cannot easily flourish in the imperialist fracture or discontinuity, covered over by an alien legal system masquerading as Law as such, an alien ideology established as only truth, and a set of human sciences busy establishing the native 'as self-consolidating Other.'

(Spivak 1985c: 253, 254)

In bringing this thesis to her reading of *Wide Sargasso Sea* (Rhys 1968) as *Jane Eyre*'s reinscription, Spivak demonstrates the pitfalls of a theory postulating that the Master Discourse preempts the (self) constitution of the historical native subject. When Spivak's notion is juxtaposed to the question Said asks in *Orientalism*, 'how can one study other cultures and peoples from a libertarian, or a non-repressive and non-manipulative perspective?', and Jean Rhys' novel examined for its enunciation (despite much incidental racism) of just such a perspective which facilitates the transformation of the Other into a Self, then it is possible to construct a re-reading of *Wide Sargasso Sea* iterating many of Spivak's observations while disputing her founding precepts.

Spivak argues that because the construction of an English cultural identity was inseparable from othering the native as its object, the articulation of the female subject within the emerging norm of feminist individualism during the age of imperialism, necessarily excluded the native female, who was positioned on the boundary between human and animal as the object of imperialism's social-mission or soul-making. In applying this interactive process to her reading of *Wide Sargasso Sea* Spivak assigns to Antoinette/Bertha, daughter of slave-owners and heiress to a post-emancipation fortune, the role of the native female sacrificed in the cause of the subject-constitution of the European female individualist. Although Spivak does acknowledge that *Wide Sargasso Sea* is 'a novel which rewrites a canonical English text within the European novelistic tradition in the interest of the white Creole rather than the native' (Spivak 1985c: 253), and situates Antoinette/Bertha as caught between the English imperialist and the black Jamaican, her discussion does not pursue the text's representations of a Creole culture that is dependent on both yet singular, or its enunciation of a specific settler discourse, distinct from the texts of imperialism. The dislocations of the Creole position are repeatedly spoken by Antoinette, the 'Rochester' figure and Christophine; the nexus of intimacy and hatred between white settler and black servant is written into the text in the mirror imagery of Antoinette and Tia, a trope which for Spivak functions to invoke the other that could not be selved:

We had eaten the same food, slept side by side, bathed in the same river. As I ran, I thought, I will live with Tia and I will be like her. . . . When I was close I saw the jagged stone in her hand but I did not see her throw it. . . . I looked at her and I saw her face crumble as she began to cry. We stared at each other, blood on my face, tears on hers. It was as if I saw myself. Like in a looking-glass.

(Rhys 1968: 24)

But while themselves not English, and indeed outcastes, the Creoles are Masters to the blacks, and just as Brontë's book invites the reader via Rochester to see Bertha Mason as situated on the human/animal frontier ('One night I had been awakened by her yells. . . . It was a fierce West Indian night . . . those are the sounds of a bottomless pit,' quoted in Spivak 1985c: 247–8), so does Rhys' novel via Antoinette admit her audience to the regulation settler view of rebellious blacks: 'the same face repeated over and over, eyes gleaming, mouth half-open,' emitting 'a horrible noise . . . like animals howling but worse' (Rhys 1968: 32, 35).

The idiosyncrasies of an account where Antoinette plays the part of 'the woman from the colonies' are consequences of Spivak's decree that imperialism's linguistic aggression obliterates the inscription of a native self: thus a black female who in *Wide Sargasso Sea* is most fully selved, must be reduced to the status of a tangential figure, and a white Creole woman (mis)construed as the native female produced by the axiomatics of imperialism, her death interpreted as 'an allegory of the general epistemic violence of imperialism, the construction of a self-immolating subject for the glorification of the social mission of the colonizer' (Spivak 1985c: 251). While allowing that Christophine is both speaking subject and interpreter to whom Rhys designates some crucial functions, Spivak sees her as marking the limits of the text's discourse, and not, as is here argued, disrupting it.

What Spivak's strategy of reading necessarily blots out is Christophine's inscription as the native, female, individual Self who defies the demands of the discriminatory discourses impinging on her person. Although an ex-slave given as a wedding-present to Antoinette's mother and subsequently a caring servant, Christophine subverts the Creole address that would constitute her as a domesticated Other, and asserts herself as articulate antagonist of patriarchal, settler and imperialist law. Natural mother to children and surrogate parent to Antoinette, Christophine scorns patriarchal authority in her personal life by discarding her patronymic and refusing her sons' fathers as husbands; as Antoinette's protector she impugns 'Rochester' for his economic and sexual exploitation of her fortune and person and as female individualist she is eloquently and frequently contemptuous of male conduct, black and white. . . .

Christophine's defiance is not enacted in a small and circumscribed space appropriated within the lines of dominant code, but is a stance from which she delivers a frontal assault against antagonists, and as such constitutes a counter-discourse. Wise to the limits of post-emancipation justice, she is quick to invoke the protection of its law when 'Rochester' threatens her with retribution: 'This is free country and I am free woman' (Rhys 1968: 131) – which is exactly how she functions in the text, her retort to him condensing her role as the black, female individualist: 'Read and write I don't know. *Other things I know*' (Rhys 1968: 133; emphasis added). . . .

Spivak's deliberated deafness to the native voice where it is to be heard, is at variance with her acute hearing of the unsaid in modes of Western feminist criticism which, while dismantling masculist constructions, reproduce and foreclose colonialist structures and imperialist axioms by 'performing the lie of constituting a truth of global sisterhood where the mesmerizing model remains male and female sparring partners of generalizable or universalizable sexuality who are the chief protagonists in that European contest' (Spivak 1986: 226). Demanding of disciplinary standards that 'equal rights of historical, geographical, linguistic specificity' be granted to the 'thoroughly stratified larger theatre of the Third World' (238), Spivak in her own writings severely restricts (eliminates?) the space in which the colonized can be written back into history, even when 'interventionist possibilities' are exploited through the deconstructive strategies devised by the post-colonial intellectual.

Homi Bhabha on the other hand, through recovering how the master discourse was interrogated by the natives in their own accents, produces an autonomous position for the colonial within the confines of the hegemonic discourse, and because of this enunciates a very different 'politics.' The sustained effort of writings which initially concentrated on deconstituting the structure of colonial discourse, and which latterly have engaged with the displacement of this text by the inappropriate utterances of the colonized, has been to contest the notion Bhabha considers to be implicit in Said's *Orientalism*, that 'power and discourse is possessed entirely by the coloniser.' Bhabha reiterates the proposition of anti-colonialist writing that the objective of colonial discourse is to construe the colonized as a racially degenerate population in order to justify conquest and rule. However because he maintains that relations of power and knowledge function ambivalently, he argues that a discursive system split in enunciation, constitutes a dispersed and variously positioned native who by (mis)appropriating the terms of the dominant ideology, is able to intercede against and resist this mode of construction.

In dissenting from analysis ascribing an intentionality and unidirectionality to colonial power which, in Said's words, enabled Europe to advance unmetaphorically upon the Orient, Bhabha insists that this not only ignores representation as a concept articulating both the historical and the fantasmatic, but unifies the subject of colonial enunciation in a fixed position as the passive object of discursive domination. By revealing the multiple and contradictory articulations in colonialism's address, Bhabha as contemporary critic seeks to demonstrate the limits of its discursive power and to countermand its demand 'that its discourse (be) non-dialogic, its enunciation unitary' (Bhabha 1985a: 100); and by showing the wide range of stereotypes and the shifting subject positions assigned to the colonized in the colonialist text, he sets out to liberate the colonial from its debased inscription as Europe's monolithic and shackled Other, and into an autonomous native 'difference.' However, this reappropriation although effected by the deconstructions of the post-colonial intellectual, is made possible by uncovering how the master-discourse had already been interrogated by the colonized in native accents. For Bhabha, the subaltern has spoken, and his readings of the colonialist text recover a native voice. . . .

Where Spivak in inspecting the absence of a text that can answer back after the planned epistemic violence of the imperialist project, finds pockets of non-co-operation in the dubious place of the 'free will of the (female) sexed subject' (Spivak 1985a: 144), Bhabha produces for scrutiny a discursive situation making for recurrent instances of transgression performed by the native from within and against colonial discourse. Here the autocolonization of the native who meets the requirements of colonialist address, is co-extensive with the evasions and 'sly civility' through which the native refuses to satisfy the demand of the colonizer's narrative. This concept of mimicry has since been further developed in the postulate of 'hybridity' as the problematic of colonial discourse.

Bhabha contends that when re-articulated by the native, the colonialist desire for a reformed, recognizable, nearly-similar other, is enacted as parody, a dramatization to be distinguished from the 'exercise of dependent colonial relations through narcissistic identification.' For in the 'hybrid moment' what the native rewrites is not a copy of the colonialist original, but a qualitatively different thing-in-itself, where misreadings and incongruities expose the uncertainties and ambivalences of the colonialist text and deny it an authorizing presence. Thus a textual insurrection against the discourse of colonial authority is located in the natives' interrogation of the English book within the terms of their own system of cultural meanings, a displacement which is read back from the record written by colonialism's agents and ambassadors:

> Through the natives' strange questions it is possible to see, with historical hindsight, what they resisted in questioning the presence of the English — as religious mediation and as cultural and linguistic medium. . . . To the extent to which discourse is a form of defensive warfare, then mimicry marks those moments of civil disobedience within the discipline of civility: signs of spectacular resistance. When the words of the master become the site of hybridity — the warlike sign of the native — then we may not only read between the lines, but even seek to change the often coercive reality that they so lucidly contain.
>
> (Bhabha 1985a: 101, 104)

Despite a flagrantly ambivalent presentation which leaves it vulnerable to innocent misconstruction, Bhabha's theorizing succeeds in making visible those moments when colonial discourse already disturbed at its source by a doubleness of enunciation, is further subverted by the object of its address; when the scenario written by colonialism is given a performance by the native that estranges and undermines the colonialist script. The argument is not that the colonized possesses colonial power, but that its fracturing of the colonialist text by re-articulating it in broken English, perverts the meaning and message of the English book ('insignia of colonial authority and signifier of colonial desire and discipline,' 1985a: 89), and therefore makes an absolute exercise of power impossible.

A narrative which delivers the colonized from its discursive status as the illegitimate and refractory foil to Europe, into a position of 'hybridity' from which it is able to circumvent, challenge and refuse colonial authority, has no place for a totalizing notion of epistemic violence. Nor does the conflictual economy of the colonialist text allow for the unimpeded operation of discursive aggression: 'What is articulated in the doubleness of colonial discourse is not the violence of one powerful nation writing out another [but] a mode of contradictory utterance that ambivalently re-inscribes both coloniser and colonised.' The effect of this thesis is to displace the traditional anti-colonialist representation of antagonistic forces locked in struggle, with a configuration of discursive transactions: 'The place of difference and other-ness, or the space of the adversarial, within such a system of "disposal" as I've proposed, is never entirely on the outside or implacably oppositional' (95).

Those who have been or are still engaged in colonial struggles against contemporary forms of imperialism could well read the theorizing of discourse analysts with considerable disbelief at the construction this puts on the situation they are fighting against and the contest in which they are engaged. This is not a charge against the difficulty of the analyses but an observation that these alternative narratives of colonialism obscure the 'murderous and decisive struggle between two protagonists' (Fanon [1965b]: 30), and discount or write out the counter-discourses which every liberation movement records. The significant differences in the critical practices of Spivak and Bhabha are submerged in a shared programme marked by the exor-bitation of discourse and a related incuriosity about the enabling socioeconomic and political institutions and other forms of social praxis. Furthermore, because their theses admit of no point outside of discourse from which opposition can be engendered, their project is concerned to place incendiary devices within the dominant structures of representation and not to confront these with another knowledge. For Spivak, imperialism's epistemic bellicosity decimated the old culture and left the colonized without the ground from which they could utter confronta-tional words; for Bhabha, the stratagems and subterfuges to which the native resorted, destabilized the effectivity of the English book but did not write an alternative text – with whose constitution Bhabha declines to engage, maintaining that an anti-colonialist discourse 'requires an alternative set of questions, techniques and strategies in order to construct it' (Bhabha 1983a: 198).

Within another critical mode which also rejects totalizing abstracts of power as falsifying situations of domination and subordination, the notion of hegemony is inseparable from that of a counter-hegemony. In this theory of power and contest, the process of procuring the consent of the oppressed and the marginalized to the existing structure of relationships through ideological inducements, necessarily generates dissent and resistance, since the subject is conceived as being constituted by means of incommensurable solicitations and heterogeneous social practices. The outcome of this agonistic exchange, in which those addressed challenge their interlocutors, is that the hegemonic discourse is ultimately abandoned as scorched earth when a different discourse, forged in the process of disobedience and combat, occupying new, never-colonized and 'utopian' territory, and prefiguring other relationships, values and aspirations, is enunciated. At a time when dialectical thinking is not the rage amongst colonial discourse theorists, it is instructive to recall how Fanon's dialogical interrogation of European power and native insurrection reconstructs a process of cultural resistance and cultural disruption, participates in writing a text that can answer colonialism back, and anticipates another condition beyond imperialism:

> Face to face with the white man, the Negro has a past to legitimate, a vengeance to extract. . . . In no way should I dedicate myself to the revival of an unjustly unrecognized Negro civilization. I will not make myself a man of the past. . . . I am not a prisoner of history; it is only by going beyond the historical, instrumental hypothesis that I will initiate the cycle of my freedom.
>
> (Fanon [1967a]: 225–6, 229, 231)

The enabling conditions for Fanon's analysis are that an oppositional discourse born in political struggle, and at the outset invoking the past in protest against capitulating to the colonizer's denigrations, supersedes a commitment to archaic native traditions at the same time as it rejects colonialism's system of knowledge:

> The colonialist bourgeoisie had in fact deeply implanted in the minds of the colonised intellectual that the essential qualities remain eternal in spite of all the blunders men may make: the essential qualities of the West, of course. The native intellectual accepted the cogency of these ideas and deep down in his brain you could always find a vigilant sentinel ready to defend the Greco-Latin pedestal. Now it so happens that during the struggle for liberation, at the moment that the native intellectual comes into touch again with his people, this artificial sentinel is turned into dust. All the Mediterranean values, – the triumph of the human individual of clarity and of beauty – become lifeless, colourless knick-knacks. All those speeches seem like collections of dead words; those values which seemed to uplift the soul are revealed as worthless, simply because they have nothing to do with the concrete conflict in which the people is engaged.
>
> (Fanon [1965b]: 37–8)

While conceding the necessity of defending the past in a move away from unqualified assimilation of the occupying power's culture, Fanon recognizes the limitations on the writer and intellectual who utilize 'techniques and language which are borrowed from the stranger in his country.' Such transitional writing reinterpreting old legends 'in the light of a borrowed aestheticism and of a conception of the world which was discovered under other skies,' is for Fanon but a prelude to a literature of combat which 'will disrupt literary styles and themes . . . create a completely new public' and mould the national consciousness, 'giving it form and contours and flinging open before it new and boundless horizons.' Fanon's theory projects a development inseparable from a community's engagement in combative social action, during which a native contest initially enunciated in the invaders' language, culminates in a rejection of imperialism's signifying system. This is a move which colonial discourse theory has not taken on board, and for such a process to be investigated, a cartography of imperialist ideology more extensive than its address in the colonialist space, as well as a conception of the native as historical subject and agent of an oppositional discourse is needed.

## Chapter 7

## STEPHEN SLEMON

# THE SCRAMBLE FOR
# POST-COLONIALISM

From 'The Scramble for Post-colonialism' in Chris Tiffin and
Alan Lawson (eds) *De-Scribing Empire: Post-colonialism and
Textuality* London: Routledge, 1994.

'**POST-COLONIALISM**', **AS IT IS** now used in its various fields, de-scribes a remarkably heterogeneous set of subject positions, professional fields, and critical enterprises. It has been used as a way of ordering a critique of totalising forms of Western historicism; as a portmanteau term for a retooled notion of 'class', as a subset of both post-modernism and post-structuralism (and conversely, as the condition from which those two structures of cultural logic and cultural critique themselves are seen to emerge); as the name for a condition of nativist longing in post-independence national groupings; as a cultural marker of non-residency for a third-world intellectual cadre; as the inevitable underside of a frac-tured and ambivalent discourse of colonialist power; as an oppositional form of 'reading practice'; and – and this was my first encounter with the term – as the name for a category of 'literary' activity which sprang from a new and welcome political energy going on within what used to be called 'Commonwealth' literary studies. The obvious tendency, in the face of this heterogeneity, is to understand 'post-colonialism' mostly as an object of desire for critical practice: as a shimmering talisman that in itself has the power to confer political legit-imacy onto specific forms of institutionalised labour, especially on ones that are troubled by their mediated position within the apparatus of institutional power. I think, however, that this heterogeneity in the concept of the 'post-colonial' – and here I mean within the uni-versity institution – comes about for much more pragmatic reasons, and these have to do with a very real problem in securing the concept of 'colonialism' itself, as Western theories of subjectification and its resistances continue to develop in sophistication and complexity.

The nature of colonialism as an economic and political structure of cross-cultural domin-ation has of course occasioned a set of debates, but it is not really on this level that the 'question' of European colonialism has troubled the various post-colonial fields of study. The problem, rather, is with the concept of colonialism as an ideological or discursive formation: that is, with the ways in which colonialism is viewed as an apparatus for constituting subject positions through the field of representation. In a way – and of course this is an extreme oversimplification – the debate over a description of colonialism's multiple strategies for regulating Europe's others can be expressed diagrammatically (see Figure 7.1).

The general understanding that colonialism works on a left-to-right order of domination, with line 'A' representing various theories of how colonialism oppresses through direct polit-ical and economic control, and lines 'BC' and 'DE' representing differing concepts of the

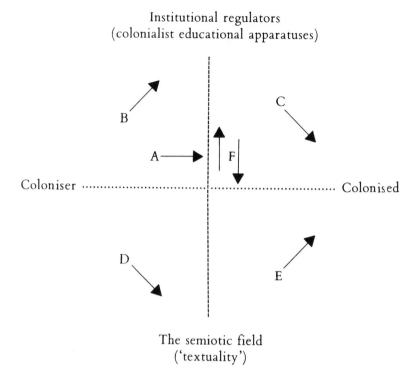

*Figure 7.1* Diagram representing the debate over the nature of colonialism

ideological regulation of colonial subjects, of subordination through the manufacture of consent. Theories that recognise an efficacy to colonialism that proceeds along line 'A' are in essence 'brute force' or 'direct political' theories of colonialist oppression: that is, they reject the basic thesis that power manages social contradiction partly through the strategic production of specific ideas of the 'self' – which subordinated groups then internalise as being 'real'. Theories, however, that examine the trajectory of colonialist power primarily along line 'BC' – a line representing an ideological flanking for the economic colonialism running along line 'A' – focus on the constitutive power of state apparatuses like education, and the constitutive power of professional fields of knowledge within those apparatuses, in the production of colonialist relations. Along this line, Edward Said (1978) examines the political efficacy of 'orientalism' within colonialism; Talal Asad (1973) and many others examine the role of anthropology in reproducing colonial relations; Alan Bishop (1990) examines the deployment of Western concepts of 'mathematics' against African school-children, Timothy Mitchell (1988) examines how the professional field of 'political science' came into being through a European colonialist engagement with the cultures of Egypt; Gauri Viswanathan (1989) examines the foundations of 'English' literary studies within a structure of colonialist management in India. This work keeps coming in, and the list of radically compromised professional fields within the Western syllabus of 'humanities' options grows daily longer. Theories that focus primarily on line 'DE' in this diagram examine the ways in which ideology reproduces colonialist relations through the strategic deployment of a vast semiotic field of representations – in literary works, in advertising, in sculpture, in travelogues, in exploration documents, in maps, in pornography, and so on.

This pattern, as I've laid it out so far, does not seem especially controversial or problematic, but the difficulties arise at the moment of conceptualising the *relation* between colonialist professional fields and institutions (at the top of the diagram) and the whole field of representation (at the bottom of the diagram) – the field of 'textuality' and its investment in reproducing and naturalising the structures of power. To take up one example of this paradigmatically: in Edward Said's work on Orientalism, colonialist power is seen to operate through a complex relationship between apparatuses placed on line 'F', where in the first instance a scholarly educational apparatus called 'Orientalism' – at the top of the line – appropriates textual representations of 'the Orient' in order to consolidate itself as a discipline and to reproduce 'the Orient' as a deployable unit of knowledge. So, in the first instance, colonialist power in Said's argument runs not just through the middle ground of this chart but through a complex set of relations happening along line 'F'; and since Said's thesis is that a function at the top of this line is employing those representations created at the bottom of the line in order to make up 'knowledges' that have an ideological function, you can say that the vector of motion along line F is an upward one, and that this upward motion is part of the whole complex, discursive structure whereby 'Orientalism' manufactures the 'Orient' and thus helps to regulate colonialist relations. That is Said's first position – that under Orientalism the vector of line 'F' is upward. But in Said's analysis, colonialist power also runs through line F in a downward movement, where the scholarly apparatus of Orientalism is understood to be at work in the production of a purely fantastic and entirely projected idea of the 'Orient'. The point is that in the process of understanding the multivalent nature of colonialist discourse in terms of the historical specific of 'Orientalism', Said's model becomes structurally ambivalent – under 'Orientalism', the 'Orient' turns out to be something produced both as an object of scholarly knowledge and as a location for psychic projection – and I've tried to graph this ambivalence as a double movement or vector along line 'F'. For Said, the mechanism that produces this 'Orient', then, has to be understood as something capable of deploying an ambivalent structure of relations along line 'F', and deploying that structure towards a unified end. And so Said (and here I'm following Robert Young's (1990) analysis of the problem) ends up referring the whole structure of colonialist discourse back to a single and monolithic originating intention within colonialism, the intention of colonialist power to possess the terrain of its Others. That assumption of intention is basically where Said's theory has proven to be most controversial.

Said's text is an important one here, for as Robert Young has shown, Said's work stands at the headwaters of colonial discourse theory, and this ambivalence in Said's model may in fact initiate a *foundational* ambivalence in the critical work which comes out of this field. This ambivalence sets the terms for what are now the two central debates within colonial discourse theory: the debate over historical specificity, and the debate over agency.

The first debate – the debate over the problem of historical specificity in the model – concerns the inconclusive relation between actual historical moments in the colonialist enterprise and the larger, possibly trans-historical discursive formation that colonial discourse theory posits in its attempt to understand the multivalent strategies at work in colonialist power. Can you look at 'colonial discourse' only by examining what are taken to be paradigmatic moments within colonialist history?

If so, can you extrapolate a modality of 'colonialism' from one historical moment to the next? Does discursive colonialism always look structurally the same, or do the specifics of its textual or semiotic or representational manoeuvres shift registers at different historical times and in different kinds of colonial encounters? And what would it mean to think of colonial discourse as a set of exchanges that function in similar ways for all sorts of colonialist

strategies in a vastly different set of cultural locations? These questions of historical specificity, though always a problem for social theory, are especially difficult ones for colonialist
discourse theory, and the reason for this is that this theory quite appropriately refuses to articulate a simplistic structure of social causality in the relation between colonialist institutions
and the field of representations. In other words, colonial discourse theory recognises a radical
ambivalence at work in colonialist power, and that is the ambivalence I have attempted to
show in Figure 7.1 as a double moment in vector at the level of line 'F'.

To clarify this I want to make use of Gauri Viswanathan's important work on Britain's
ideological control of colonised people through the deployment of colonialist educational
strategies in nineteenth-century India. Obviously, the question of what happens along line 'F'
can only be addressed by specific reference to immediate historical conditions, and every piece
of archaeological work on colonialist power will want to formulate the vector of action here
with particular sensitivity to the local conditions under analysis. Viswanathan researches this
part of the puzzle with exemplary attention to history, and at heart her argument is that
colonialist education in India (which would stand in as the ideological apparatus at the top of
the diagram) strategically and intentionally deployed the vast field of canonical English 'literature' (the field of representations at the bottom of the diagram) in order to construct a cadre
of 'native' mediators between the British Raj and the actual producers of wealth. The point
here is that Viswanathan's analysis employs a purely upward vector of motion to characterise
the specifics of how power is at work along line 'F' in the diagram, and what secures this
vector is Viswanathan's scrupulous attention to the immediate conditions that apply within
British and Indian colonial relations.

The problem, though – and here I mean the problem for colonial discourse theory – is
that the foundational ambivalence or double movement that Said's work inserts into the model
of colonialist discourse analysis always seems to return to the field; and it does so through critical work that on its own terms suggests a counter-flow along line 'F' at the same moment of
colonialist history. That is, the residual ambivalence in the vector of line 'F' within colonial
discourse theory seems to invite the fusion of Viswanathan's kind of analysis with critical
readings that would articulate a downward movement at this place in the diagram; and one
of the areas such work is now entering is the analysis of how English literary activity of the
period (at the bottom of line 'F') suddenly turned to the *representation* of educational processes
(at the top of the line), and why this literature should so immediately concern itself with the
investments of educational representations in the colonialist scene. In examining the place of
English literary activity within this moment of colonialist history, that is, a critic such as Patrick
Brantlinger would want to argue for the valency of texts such as *Jane Eyre* or *Tom Brown's School
Days* within colonialist discursive power, and colonialist discourse theory would want to understand how both kinds of discursive regulation, both vectors of movement along line 'F', are
at work in a specific historical moment of colonialist relations. Because of Said's ambivalence
in charting out the complex of Orientalism along line 'F', I am arguing, the field of colonialist
discourse theory carries that sense of ambivalence forward, and looks to an extraordinary
valency of movement within its articulation of colonialist power. The ambivalence makes our
understanding of colonial operations a great deal clearer for historical periods but it also upsets
the positivism of highly specific analyses of colonialist power going on *within* a period.

The basic project of colonial discourse theory is to push out from line 'A', and try to
define colonialism both as a set of political relations and as a signifying system, one with
ambivalent structural relations. It is remarkably clarifying in its articulation of the productive
relations between seemingly disparate moments in colonialist power (the structure of literary
education in India, the literary practice of representing educational control in Britain), but

because it recognises an ambivalence in colonialist power, colonial discourse theory results in a concept of colonialism that cannot be historicised modally, and that ends up being tilted towards a description of all kinds of social oppression and discursive control. For some critics, this ambivalence bankrupts the field. But for others, the concept of 'colonialism' – like the concept of 'patriarchy' for feminism, which shares this structure of transhistoricality and lack of specificity – remains an indispensable conceptual category of critical analysis, and an indispensable tool in securing our understanding of ideological domination under colonialism to the level of political economy.

The first big debate going on within colonialist discourse theory, then, is a debate over what happens when a model of 'colonial discourse' is carried beyond its scattered moments of archaeological research and is taken up as a general structure of oppression. I want now to turn to the second big debate going on between theorists of colonialist discourse; and that is the debate over the question of *agency* under colonialist power. Basically, the question of agency can be restated as a question of who or what acts oppositionally, when ideology or discourse or psychic processes of some kind construct human subjects, and the question of specifying agency is becoming an extremely complex one in all forms of critical theory at present. Again, however, this debate has especial urgency within colonial discourse theory, and, again, that is because this theory recognises foundationally that the vector of line 'F' in Figure 7.1 remains ambivalent at every moment of colonialist discursive control. . . .

I want to stress the presuppositional location of this post-colonial scramble – I want to articulate its foundations within the problematic of colonial discourse theory and within an unresolved debate within the Western humanities institution – because I suspect that at times workers in various orders of post-colonial analysis are made to feel a disempowering energy at work in their field – a disempowerment which stems from their sense that these debates ought to be resolved within post-colonial studies itself. And I also raise the question of an effect to these debates, not because I want to suggest they are anything other than crucial ones for the field, but because I think the terrain of post-colonial studies remains in danger of becoming colonised by competing academic methodologies, and being reparcelled into institutional pursuits that have no abiding interest in the specifics of either colonialist history or post-colonial agency. One of the most exciting research projects now going on in colonial discourse analysis, for example, is Homi Bhabha's theorising of colonialist ambivalence, and his attempt to carry that analysis forward to a wholesale critique of Western modernity. It is possibly instructive, therefore, that in the process of expressing admiration for his work, the post-structuralist critic Robert Young inserts Bhabha's project into a narrative of unpacking whose terms of reference are entirely European in origin: the radical restructuring of European historiography, and the allocation of alterity to the theatre of the European postmodern.

Along parallel lines, it is also instructive that Henry Louis Gates Jr. notes in Spivak's deconstructive brilliance a remarkable conflation between colonial discourse and Derrida's concept of writing itself – an argument, that is, that there is '*nothing* outside of (the discourse of) colonialism', and that all discourse must be nothing other than colonial discourse itself. Gates warns of a hidden consequence in this elevation in ascendency of the colonial paradigm by questioning what happens when we elide, for example, 'the distance between political repression and individual neurosis: the *positional* distance between Steve Biko and, say, Woody Allen?' (Gates 1991: 466). His argument is that academic interest in this history and the discourse of colonialism bids fair to become the last bastion for the project of global theory and for European universalism itself, and he asks us whether we really need to choose between oppositional critics whose articulations of the post-colonial institutionalise themselves as agonistic struggles over a thoroughly disciplined terrain.

I would like to echo Gates' sentiments in the face of this balkanisation; and in the absence of any real solutions to this crisis in the field I'd like to offer a two-part credo towards post-colonial work as it takes place within the Western academic institution. First, I think, post-colonial studies, if nothing else, needs to become more tolerant of methodological difference, at least when that difference is articulated towards emancipatory anti-colonialist ends. I am reminded that the great war within the Western 'humanities' is carried on the back of critical methodology and its competing orders, and that in many ways the subject-making function of the humanities is effected precisely in that debate. I have seen no evidence that the humanities carry any special brief for the global project of decolonisation, and so I would desperately want to preserve this function of decolonising commitment for post-colonial studies, despite its necessary investment in and ironic relation to the humanities complex. I am suspicious of the kind of argument that would insist on the necessary conflation of the diagram I put forward in this paper with a colonialist allegorical function, but I can see how the argument could be made. The tools for conceptual disempowerment in the struggle over method are going to remain available within post-colonial studies, but I remain suspicious of ahistorical and I think intolerant calls for homogeneity in a field of study which embraces radically different forms and functions of colonialist oppression and radically different notions of anti-colonialist agency.

Tolerance is never simply passive, and, ironically, the area of institutionalised post-colonial studies is finding itself increasingly invested in an academic star system of astonishing proportions, and through that star system it is learning to seek its instruction in oppositional tactics along lines that run increasingly and monolithically backward towards the centres of Western power. I cannot help noticing, for example, that in what Hortense Spillers calls the politics of mention, our theoretical masters in Paris or Oxford or New Haven are read and referenced by exemplary theorists of the local – the critic J. Michael Dash at the Mona campus in Jamaica is an example – but those metropolitan theorists seldom reference these cultural and theoretical mediators in return. Post-colonial studies should have an investment in open talk across cultural locations, however, and across methodological dynasties; and I think we do damage to the idea of post-coloniality at an immediate political level when that investment in cross-talk runs only one way.

As for the second part of this credo, I believe that post-colonial studies needs always to remember that its referent in the real world is a form of political, economic, and discursive oppression whose name, first and last, is *colonialism*. The forms of colonialist power differ radically across cultural locations, and its intersections with other orders of oppression are always complex and multivalent. But, wherever a globalised theory of the colonial might lead us, we need to remember that resistances to colonialist power always find material presence at the level of the local, and so the research and training we carry out in the field of post-colonialism, whatever else it does, must always find ways to address the local, if only on the order of material applications. If we overlook the local, and the political applications of the research we produce, we risk turning the work of our field into the playful operations of an academic glass-bead game, whose project will remain at best a description of global relations, and not a script for their change. There is never a necessary politics to the study of political actions and reactions; but at the level of the local, and at the level of material applications, post-colonialism must address the material exigencies of colonialism and neo-colonialism, including the neo-colonialism of Western academic institutions themselves.

## Chapter 8

# NICHOLAS B. DIRKS

## COLONIALISM AND CULTURE

From 'Introduction' *Colonialism and Culture* Ann Arbor:
University of Michigan Press, 1992.

> Europe is literally the creation of the third world.
> Frantz Fanon

IN THE OPENING CHAPTER OF *A Passage to India,* E. M. Forster carefully describes the geographical setting of the novel. The landscape 'presents nothing extraordinary,' nothing, that is, except the Marabar Caves. Although the caves provide the only visible wrinkle, they also contain the ambivalent space of the echo, the space in which the psychosexual fears of a British memsahib become projected onto the dark, glassy walls of colonial terror. Hallucinations of violation become inscribed in the fact of rape, at once the inverted metaphor of exploitation and the patriarchal pillar of colonial honor. The 'events' at the caves disrupt the narrative and display the terrible totality of a colonial world; there are, in the end, only the colonizers and the colonized.

Forster asserts that colonialism is more than just a narrative of momentary disarray, for colonialism has become nature itself. The landscape tells all: the Indian city of Chandrapore 'seems made of mud, the inhabitants of mud moving. So abased, so monotonous is everything that meets the eye, that when the Ganges comes down it might be expected to wash the excrescence back into the soil. Houses do fall, people are drowned and left rotting, but the general outline of the town persists, swelling here, shrinking there, like some low but indestructible form of life.' Anticipating Said's critique, Forster ironically notes that whatever cultivated decoration might have marked the natural contours of mud water 'stopped in the eighteenth century, nor was it ever democratic.' Otherwise, it is all nature at its lowest ebb. . . .

If Forster's writing inscribed colonialism and its antinomies so dramatically into the very contours of the natural world, his poetic conceit was appropriately chosen.' Nature, after all, is the world that is given, exempt from the capacity of humans to shape it and of historical processes to change it. And though nature seems invariably to stand in some kind of opposition to culture, it is also the case that culture is a way of talking about nature; culture, in the anthropological sense, is the congeries of values, beliefs, practices, and discourses that have come to carry the force of nature. Nature itself, as well as the various forms of nature's opposition to culture, appear to anthropology as the residues of cultural construction. Poetic conceits aside, Forster was writing the ethnography of British colonialism in India. For Forster, nature was a way of talking about culture.

In both metropolitan centers and colonial peripheries, the anthropological givens of culture have been transformed over and over again by colonial encounters. Often, these transformations seemed overdetermined, for culture in places such as India became, through colonial lenses, assimilated to the landscape itself, fixed in nature, and freed from history. Thus, castes, villages, and tribes were seen as the scientific axioms of a millenial geology of colonial spaces, the social facts of colonial control and the ideological means by which histories of colonial conquest were erased. For colonial rulers, the culture and nature of the colonized were one and the same.

Although colonial conquest was predicated on the power of superior arms, military organization, political power, and economic wealth, it was also based on a complexly related variety of cultural technologies. Colonialism not only has had cultural effects that have too often been either ignored or displaced into the inexorable logics of modernization and world capitalism, it was itself a cultural project of control. Colonial knowledge both enabled colonial conquest and was produced by it; in certain important ways, culture was what colonialism was all about. Cultural forms in newly classified 'traditional' societies were reconstructed and transformed by and through colonial technologies of conquest and rule, which created new categories and oppositions between colonizers and colonized, European and Asian, modern and traditional, West and East, even male and female.

The anthropological concept of culture might never have been invented without a colonial theater that both necessitated the knowledge of culture (for the purposes of control and regulation) and provided a colonized constituency that was particularly amenable to 'culture.' Without colonialism, culture could not have been so simultaneously, and so successfully, ordered and orderly, given in nature at the same time that it was regulated by the state. Even as much of what we now recognize as culture was produced by the colonial encounter, the concept itself was in part invented because of it. Culture was also produced out of the allied network of processes that spawned nations in the first place. Claims about nationality necessitated notions of culture that marked groups off from one another in essential ways, uniting language, race, geography, and history in a single concept. Colonialism encouraged and facilitated new claims of this kind, re-creating Europe and its others through its histories of conquest and rule.

If colonialism can be seen as a cultural formation, so also culture is a colonial formation. But culture was not simply some mystifying means for colonial conquest and rule, even as it could not be contained within colonized spaces. Culture was imbricated both in the means and the ends of colonial conquest, and culture was invented in relationship to a variety of internal colonialisms. Colonial theaters extended beyond the shores of tropical rivers and colonized spaces, emerging within both metropolitan contexts and the civil lines of colonial societies. Culture became fundamental to the formation of class society, the naturalization of gender divisions in Western bourgeois society, and to developing discourses of race, biology, and nationality. At the same time, metropolitan histories were both sustained and heavily influenced by colonial events; sexuality in Sumatra, torture in the Congo, terror in the Marabar Caves – were all displacements of the fault lines of expanding capitalism at the same time that they became fundamental moments in the unfolding narrative of the modern.

The parallel mutualities of colonizers and colonized on the one hand and colonialism and culture on the other make it more difficult than ever to devise historical narratives of cause and effect. If culture itself, as an object of knowledge and a mode of knowledge about certain objects, was formed in relation to colonial histories, it is all the more difficult to recognize the ways in which specific cultural forms were themselves constituted out of colonial encounters. This task becomes even more daunting when we realize that these cultural forms became fundamental to the development of resistance against colonialism, most notably in nationalist movements that used Western notions of national integrity and self-determination to justify claims for independence. In turn, Western colonial nations did not simply exploit colonized nations for economic profit, but depended upon the process of colonization and colonial rule for securing the nation-state itself: developing new technologies of state rule, maintaining and deepening the ruptures of a classed, patriarchal society during a time of reform and democratization, consolidating Western control over the development of world capitalism, even achieving international cultural hegemony in areas ranging from fashion to the novel – bringing both colonialism and culture back home.

Looking at colonialism as a cultural project of control thus focuses attention on the interdependency of these terms, on the complex interplay of coercion and hegemony, on the categories of thought that generally orient scholarly considerations of colonial history or historical anthropology. Linking culture and colonialism does not efface the violence of colonialism. Not only does this linkage preserve a sense of the violent means by which colonialism was effected and maintained, it allows us to see anew the expanded domains of violence, to realize that cultural intervention and influence were not antidotes to the brutality of domination but extensions of it. Representation in the colonial context was violent; classification a totalizing form of control. Brute torture on the body of the colonized was not the same as the public exhibition of a colonized body, but these two moments of colonial power shared in more than they differed. And torture became terror through the culture of colonialism itself.

Now that decolonization and the twentieth-century transformations of the world order have rendered colonialism a historical category, linked to the present more by such terms as *neo* and *post* than by any formal continuity, there is both license and risk in our collective interrogation of the colonial past. Colonialism is now safe for scholarship, and culture seems an appropriate domain in which to measure the effects of colonialism in the contemporary world. Is it possible that, in calling for the study of the aesthetics of colonialism, we might end up aestheticizing colonialism, producing a radical chic version of raj nostalgia? Is it likely that by linking colonialism and culture, we might ignore the extent to which colonialism became irrelevant due to transformations in the world economy having to do with the hegemony of superpowers and the internationalized structures of late industrial capitalism? Might we become so complicit in the displacements of postmodernity that we fail to recognize the informal continuities between the colonial past and the new world order present?

These are questions and worries that we should not let slip away. However, there are opportunities in the present intellectual conjuncture that we should not lose. For the first time it seems possible to imagine dramatic transformations in certain academic landscapes: the traditional analytic antinomies that have shaped questions and discourses in the past are being reformulated and sometimes dissolved. Colonialism can be seen both as a historical moment – specified in relation to European political and economic projects in the modern era – and as a trope for domination and violation. Culture can be seen both as a historically constituted domain of significant concepts and practices and as a regime in which power achieves its ultimate apotheosis. Linked together, colonialism and culture can be seen to provide a new world in which to deploy a critical cartography of the history and effects of power.

This new cartography might cast perspective on the history of the old cartography returning us to the age of the Enlightenment itself. We all know that the Enlightenment was the age of discovery and reason. It was also the age when reason was idealized as a quality or attitude that could transcend social and historical particularities. Reason made discovery the imperative of Western thought, but was neither dependent on discovery nor driven by it. What gets lost in this account, however, is that colonialism provided a theater for the Enlightenment project, the grand laboratory that linked discovery and reason. Science flourished in the eighteenth century not merely because of the intense curiosity of individuals working in Europe, but because colonial expansion both necessitated and facilitated the active exercise of the scientific imagination. It was through discovery – the siting, surveying, mapping, naming, and ultimately possessing – of new regions that science itself could open new territories of conquest: cartography, geography, botany, and anthropology were all colonial enterprises. Even history and literature could claim vital colonial connections, for it was through the study and narrativization of colonial others that Europe's history and culture could be celebrated as unique and triumphant.

As the world was shaped for Europe through cartography – which, writ large, included ships logs, narrative route maps, the establishment of boundaries, the textualization of treaties, the composition of epics, the fighting of wars, the raising of flags, the naming and appropriation of newly discovered spaces, the drawing of grids, the extermination of savages (and the list could go on and on) – so also it became peopled by classificational logics of metonymy and exclusion, recognition and opposition. Marking land and marking bodies were related activities; not only did land seem to determine much of a putatively biological nature, bodies themselves became markers of foreign lands. Before places and peoples could be colonized, they had to be marked as 'foreign,' as 'other,' as 'colonizable.' If geography and identity seem always to have been closely related, the age of discovery charted out new possibilities for this relationship.

Colonialism therefore was less a process that began in the European metropole and expanded outward than it was a moment when new encounters within the world facilitated the formation of categories of metropole and colony in the first place. But colonialism was not only good to think. The world was full of incentives for accumulation of all kinds, from knowledge to spices, from narratives to military command posts. There were compelling reasons to invent systematic beliefs about cultural differences, uniting such disparate projects as the precarious formation of national identity and the relentless exploitation of economic resources.

Colonialism was neither monolithic nor unchanging through history. Any attempt to make a systematic statement about the colonial project runs the risk of denying the fundamental historicity of colonialism, as well as of conflating cause with effect. It is tempting but wrong to ascribe either intentionality or systematicity to a congeries of activities and a conjunction of outcomes that, though related and at times coordinated, were usually diffuse, disorganized, and even contradictory. It could be argued that the power of colonialism as a system of rule was predicated at least in part on the ill-coordinated nature of power, that colonial power was never so omniscient nor secure to imagine itself as totalizing, and that while colonial rulers were always aware that their power was dependent on their knowledge, they themselves were never similarly aware of all the ways in which knowledge was, in any direct or strategic sense, power. . . .

Hegemony is perhaps the wrong word to use for colonial power, since it implies not only consent but the political capacity to generate consent through the institutional spaces of civil society, notably absent in the colonial context. Nevertheless, colonialism transformed domination into a variety of effects that masked both conquest and rule. Not only did colonial rulers align themselves with the inexorable and universal forces of science progress, rationality, and modernity, they displaced many of the disruptions and excesses of rule into institutions and cultures that were labeled as tradition. Colonialism came to be seen as ascendent and necessary precisely through the construction of the colonial world, with its naturalized oppositions between us and them, science and barbarity, modern and traditional. And in this construction, consent was less the issue than the reality of power itself.

An example of this process can be seen in the history of the caste system in India. For anthropology, for social theory, and for contemporary political practice in India, culture in India seems always to have been principally defined by caste. Caste has always been seen as central in Indian history, and as one of the major reasons why India has no history, certainly no sense of history. Caste defines the core of Indian tradition, and caste is today – as it was throughout the colonial era – the major threat to Indian modernity. However, I have argued elsewhere that much of what has been taken to be timeless tradition is, in fact, the para-doxical effect of colonial rule, where culture was carefully depoliticized and reified into a specifically colonial version of civil society. In ethnographic fieldwork, in the reading of texts traditionally dismissed as so much myth and fabulous legend, in reconstructing the precolonial history of Indian states and societies, in reading colonial texts, and in charting the contra-dictory effects of colonial rule, I found that the categories of culture and history subverted each other, opening up supplemental readings of 'caste' that made it seem more a product of rule than a predecessor of it.

. . . The success of colonial discourse was that, through the census, landholding, the law, *inter alia*, some Indians were given powerful stakes in new formulations and assumptions about caste, versions that came increasingly to resemble the depoliticized conditions of colonial rule. These versions were then canonized in the theoretical constructions of caste by anthro-pology, first in the hands of colonial administrators, later in the imaginations of such powerful thinkers and academics as Louis Dumont. Caste became the essence of Indian culture and civilization through historical process, under colonial rule.

This reference to anthropology reminds us that Western scholarship has consistently been part of the problem rather than the solution.

## Chapter 9

## BIODUN JEYIFO

## THE NATURE OF THINGS
## Arrested decolonization and critical theory

From 'The Nature of Things: Arrested Decolonization and Critical Theory'
*Research in African Literatures* 21(1), 33–42, 1990.

T HERE CAN HARDLY BE A GREATER AFFIRMATION for literary criticism than the view offered by Frank Kermode in his book, *Forms of Attention*, to the effect that critical discourse (which Kermode blithely calls 'conversations') is the primary medium in which literature survives. What Kermode calls 'perpetual modernity' is achieved by a literary work, or a writer, only to the extent that they continue not only to be read but also to be talked and written about . . . But a few qualifications on Kermode's conception of critical discourse are necessary in approaching the subject of this essay: scholarly critical discourse and the fate of African literature[s]. First of all, critical discourse not only assures the survival of literature, it also determines the condition in which it survives and the uses to which it will be put. For it to play such a role, it must accede to a position of power relative to other discourses, both within and beyond the domains of literature and the Humanities, a point central to much of contemporary post-modern critical theory (Macdowell 1986: 2). . . . What gives a particular critical discourse its decisive effectivity under these circumstances is the combination of historical, institutional and ideological factors that make the discourse a 'master' discourse which translates the avowed will-to-truth of all discourse into a consummated, if secret, will-to-power. In other words, this 'master' discourse becomes the discourse of the 'master', in its effects and consequences at least, if not in its conscious intentions. Once we recognize this discourse and the privileged subject position(s) out of which it speaks, once we identify the 'natural' magisterial register of its accents, we can recognize how its avowed will-to-truth masks the will-to-power which pervades all discourse, especially when we recognize discourse as epistemic behaviour.

Such observations imply that, at a fundamental level, all discourse is agonistic, particularly when, as in the present case, we are in a social and historical context which is massively overdetermined. The foregoing discussion is especially relevant to current debates over the pertinence of theory to African literature where 'theory' almost always implies 'their' theory in relation to 'our' literature, Western or 'eurocentric' evaluative norms and criteria in relation to non-Western traditions of writing (Gates, Showalter). If these objections are valid, the traditions of critical discourse on African literature that we have 'inherited' – traditions whose premises, frames of intelligibility, and conditions of possibility have been yoked to foreign, historically imperialist perspectives and institutions of discursive power-raise serious problems with regard to the survival and vitality of its object, African literature. Thus, the question of an African critical discourse which is self-constituted and self-constituting in line with the forces acting on the production of African literature is intimately connected with the fate of that literature. But what exactly is a self-constituted and self-constituting African critical discourse? Does it exist? If it does not (yet) exist, is there a need for its existence, or an aspiration for its constitution? What established positions have emerged in the debates that have taken place during the last two decades on these questions? Since these are large questions which we cannot hope to adequately explore in one essay, I would like to focus on one single, but crucial aspect of these debates: the emergence of African literature as an academic discipline. As I hope to demonstrate, this is one area of critical discourse on African literature in which 'theory' can play a decisive role in clearing up the confusion and sterile acrimony that have characterized many attempts to define a role for the scholar of African literature and to stake a claim of validity and legitimacy for the 'discipline' of African literary studies. With few exceptions, these debates have been under-theorized or characterized by the assumption that it is an untheorizable discursive space ontologically charged with the mysteries that supposedly lie at the heart of the nature of things.

Nothing better reveals the troubled state of critical discourse on African literature than the problematic accession of the literature to the status of an object of study in Africa and, perhaps more crucially, in Europe and North America. The historical emergence of this phenomenon, if not its historicity, has been the subject of many international conferences and seminars, many essays and books (Moore, Heywood, Hale and Priebe, Lindfors 1984; Arnold 1985). The diverse expressions of this phenomenon need not concern us here, especially the early exchanges on the definition of African literature or the delimitation of its constitutive elements. What stands at the centre of these debates and what concerns us in this essay, is the clearly emergent subsumption of all criticism and scholarship on African literature into two basic, supposedly distinct, polarized camps: first, the foreign, white, European or North American critic or scholar and second, the native, black African 'counterpart'. It is impossible to underestimate the hypostasis that presides over the representations that govern debates over this presumed dichotomy. Clearly what is called for is, first, the careful identification of this hypostasis and, second, its demythologization by a theoretical critique that presents it with its secret, repressed conditions of possibility – the structured conditions and relationships of its being-in-the-world, to use the Heideggerian formulation.

Two early observations of Soyinka and Achebe draw attention to the exceptionally problematic nature of the conventional dichotomy between 'foreign' and 'local', external and homegrown in the constituted publics of modern African literature. In 'And After the Narcissist?' (Soyinka 1966) an early essay which perhaps marks Soyinka's emergence as a 'strong' presence in the then newly emergent field of African literature, he makes the following observation:

> In any culture, the cycle of rediscovery – Negritude or Renaissance enlightenment or pre-Raphaelite – must, before the wonder palls, breed its own body of the literature of self-worship. African writing has suffered from an additional infliction: apart from his own discovery the African writer has experienced rediscovery by the external eye. It is doubtful if the effect of this has any parallel in European literature.
>
> (56)

As elaborated in this essay, the 'effect' without parallel in European literary history is the inculcation and promotion in African writing of themes and attitudes deleterious to the health and vitality of the nascent literary tradition by an exoticizing, paternalistic foreign commentary. Soyinka's indictments are specific: 'quaintness mongers', 'exoticists', and 'primitivists' are the designations he applies to these foreign critics. The following sentence captures the tone of Soyinka's umbrage at the promotion by the 'external eye' of themes, trends, and a particular type of African writer: 'The average published writer in the first few years of the post-colonial era was the most celebrated skin of incompetence to obscure the true flesh of the African dilemma' (1988: 17). This questioning of a dislocated foreign commentary on African literature is given a slightly different inflection by Achebe in 'The Novelist as Teacher', an essay which has come to be regarded as his manifesto:

> I am assuming, of course, that our writer and his society live in the same place. I realize that a lot has been made of the allegation that African writers have to write for European and American readers because African readers, where they exist at all, are only interested in reading text-books. I don't know if African writers will always have a foreign audience in mind. What I do know is that they don't have to.
>
> (42)

The point of these two observations is that both writers perceive an abnormal mutation, in what is normally regarded as the rather normal existence (or co-existence) of local and foreign readerships for any literature. Although neither author goes deeply into the historical and ideological roots of the epistemological consequences of the situation, their objections are not based on the mere fact that such scholars and critics are 'foreign', 'non-African', 'alien' rather than being 'native', 'African', and 'indigene'. Both Soyinka and Achebe argue in a rational way and refuse to base their contentions on the 'nature of things'. We are in an entirely different order of critical discourse when we engage a vast array of African critics and scholars for whom African literature is 'our own literature' and non-African critics and scholars for whom the literature is 'theirs'. For them, it 'belongs' to Africans, even when they themselves defend the usefulness and relevance of their participation in the criticism of African literature. This view concerns the staking and conceding of 'natural' proprietary rights in the criticism of African literature, and it is one of the few consensual positions between virtually all critics and scholars of African literature. Such a phenomenon is neither surprising nor particularly problematic: who would deny that Chinese literature 'belongs' to the Chinese, Japanese literature to the Japanese, and Russian literature to the Russians? What is anomalous, and problematic is that this point, which in most other cases is taken for granted and silently passed over in the criticism of specific works or authors, becomes, in this instance, a grounding, foundational critical rubric, a norm of evaluation and commentary. Pushed to the limits of its expression, it becomes a veritable ontologization of the critical enterprise: only Africans must criticize or evaluate African literature, or slightly rephrased, only Africans can give a 'true' evaluation of African literary works. Of course nobody expresses this ontologization and racialization of this 'truth' quite so blatantly, but the tone, the inflections, the nuances are often not too far from it. Among the most clamorous advocates of this viewpoint, Chinweizu is exemplary in his constant deployment of the collective, proprietary pronoun 'we', which he invariably uses in a supremely untroubled fashion as if he were absolutely certain of its axiomatic representativeness. . . .

Négritude writers have of course been the most prominent advocates of the nativist position. Its influence went beyond its immediate historical context, transcending its institutional and practical consolidation in *Présence Africaine*. However, by pushing the ontological rubric too far, by hypostatizing an hypostasis and reifying a reification, Négritude effectively argued itself out of relevance in the debate over who has proprietary rights to the turf of

African critical discourse. Who could stake a claim to serious critical, evaluative rights on the basis of Senghor's famous slogan: Emotion is Negro as Reason is Greek? Even to 'native' critics eager to assert 'natural' territorial rights to a virgin field, some of Senghor's Négritudinist excesses could not but be a great embarrassment. Criticism is, after all, an eminently rational activity, whereas 'emotion', 'feeling', 'intuition', 'rhythm', and some of the other 'keywords' of Négritude are characteristically relegated to the margin of the critical enterprise. As shocking as the following passage from Senghor may be to Reason, innumerable others can easily be culled from Senghor's theoretical musings on the supposed ontology of black African aesthetics:

> The African is as it were shut up inside his black skin. He lives in primordial night. He does not begin by distinguishing himself from the object, the tree or stone, the man or animal or social event. He does not keep it at a distance. He does not analyze it. Once he has come under its influence, he takes it like a blind man, still living, into his hands. He does not fix it or kill it. He turns it over and over in his supple hands, he fingers it, he feels it. The African is one of the worms created on the Third Day . . . a pure sensory field. Subjectively, at the end of his antennae, like an insect, he discovers the other. He is moved to his bowels, going out in a centrifugal movement from the subject to the object on the waves sent out from the Other.
>
> (Reed and Wake 1976: 29–30)

. . . Behind the claims and counter-claims the 'foreign' scholar-critics and 'native' claimants of 'natural' proprietary rights to critical insight lies a vastly displaced play of unequal power relations between the two camps. The fact of this displacement accounts for some of the bizarre self-representations and the occlusions of the social production of meaning by individual scholars and critics from either camp. Unfortunately few critics or scholars have paid close, scrupulous, critical attention to this issue. Only rarely . . . does the Africanist scholar acknowledge the vastly unequal relations of power and privilege between African and non-African scholars and critics of African literature: if expatriates have won the right to take part in African literature unmolested – unless their work merits abuse – their role in assisting to broaden the margin of freedom of African literature, as part of a larger struggle, has not been acknowledged. International consciousness and pressure have been a definite fertilizing factor in the growth of our discipline and the object it studies. Nevertheless a very serious imbalance exists in the funding of research on African literatures. Non-Africans can get money to do what Africans often could do better, yet the Africans must sit and watch it get done. Justice, as well as intellectual probity in our discipline, demands that bodies such as UNESCO be pressured to back more basic research in the humanities so that the patronage and political pressure of state and bilateral government sponsorship of research can he minimized.

## Chapter 10

# ACHILLE MBEMBE

## THE INTIMACY OF TYRANNY

From *On the Postcolony* Berkeley, Calif.: University of California Press, 2001.

A LTHOUGH THE EFFECTIVENESS OF what Foucault calls the 'politics of coercion' should not be underestimated, it is important not to lose sight of how it can actually lessen the burden of subjection and overdetermine how the 'normal' is constructed. Precisely because the postcolonial mode of domination is a regime that involves not just control but conviviality, even connivance – as shown by the constant compromises, the small tokens of fealty, the inherent cautiousness – the analyst must watch for the myriad ways ordinary people guide, deceive, and toy with power instead of confronting it directly. These evasions, as endless as Sisyphus's, can be explained only in that individuals are constantly being trapped in a net of rituals that reaffirm tyranny, and in that these rituals, however minor, are intimate in nature. Recent Africanist scholarship has not studied in detail the logic of capture and narrow escape, nor the way the traps are so interconnected that they become a unitary system of ensnarement. Yet making sense of this network is necessary for any knowledge we might have of the logics of 'resistance,' 'disorder' (Bourdon 1981), and 'conviviality' inherent in the post-colonial form of authority. For the present, it is enough to observe that, at any given moment in the postcolonial historical trajectory, the authoritarian mode can no longer be interpreted strictly in terms of surveillance, or the politics of coercion. The practices of ordinary citizens cannot always be read in terms of 'opposition to the state,' 'deconstructing power,' and 'dis-engagement.' In the postcolony, an intimate tyranny links the rulers with the ruled – just as obscenity is only another aspect of munificence, and vulgarity a normal condition of state power. If subjection appears more intense than it might be, this is because the subjects of the *commandement* have internalized authoritarian epistemology to the point where they reproduce it themselves in all the minor circumstances of daily life – social networks, cults and secret societies, culinary practices, leisure activities, modes of consumption, styles of dress, rhetorical devices, and the whole political economy of the body. The subjection is also more intense because, were they to detach themselves from these ludic resources, the subjects would, *as subjects*, lose the possibility of multiplying their identities. Yet it is precisely this possibility of assuming multiple identities that accounts for the fact that the body that dances, dresses in the party uniform, fills the roads, 'assembles en masse' to applaud the passing presidential procession in a ritual of confirmation, is willing to dramatize its subordination through such small tokens of fealty, and at the same time, instead of keeping silent in the face of obvious official lies and the effrontery of elites, this body breaks into laughter. And, by laughing, it drains officialdom of meaning and sometimes obliges it to function while empty and power-less. Thus we may assert that, by dancing publicly for the benefit of power, the 'postcolonized subject' is providing his or her loyalty, and by compromising with the corrupting control that state power tends to exercise at all levels of everyday life, the subject is reaffirming that this power is incontestable – precisely the better to play with it and modify it whenever possible. In short, the public affirmation of the 'postcolonized subject' is not necessarily found in acts of 'opposition' or 'resistance' to the *commandement*. What defines the postcolonized subject is the ability to engage in baroque practices fundamentally ambiguous, fluid, and modifiable even – here there are clear, written, and precise rules. These simultaneous yet apparently contradictory practices ratify, de facto, the status of fetish that state power so forcefully claims as its right. And by the same token they maintain, even while drawing upon officialese (its vocabulary, signs, and symbols), the possibility of altering the place and time of this ratification. This means that the recognition of state power as a fetish is significant only at the very heart of the ludic relationship. It is here that the official 'sign' or 'sense' is most easily 'unpacked,' 'disenchanted,' and gently repacked, and pretense (*le simulacre*) becomes the dominant modality of transactions between the state and society, or between rulers and those who are supposed to obey. This is what makes postcolonial relations not only relations of conviviality and covering over, but also of powerlessness par excellence – from the viewpoint both of the masters of power and of those they crush. However, since these processes are essentially magical, they in no way erase the dominated from the epistemological field of power (see the metaphor of cat and mouse used in Canetti 1988: 281–2). Consider, for example, ceremonies for the 'transfer of office' that punctuate postcolonial bureaucratic time and profoundly affect the imagination of individuals – elites and masses alike. One such ceremony took place in October 1987 in the small town of Mbankomo in Central Province. Essomba Ntonga Godfroy, the 'newly elected' municipal administrator, was to be 'installed in his Post' with his two assistants, Andre Effa Owona and Jean-Paul Otu. The ceremony was presided over by the prefect of Mefou, Tabou Pierre, assisted by the sub-prefect of Mbankomo District, Bekonde Belinga Henoc-Pierre. Among the main personalities on the stand were the president of the party's departmental section, representatives of elites from inside and outside the district, 'traditional' authorities, and cult priests. The dancers were accompanied by drums and xylophones. A church choir also contributed. According to a witness:

> Elation reached a feverish climax when the tricolour scarves were presented to the municipal administrator and his two assistants, and their badges as municipal advisers were handed to the three elected on 25 October. Well before this outburst of joy, the Prefect, Mr. Tabou, gave a brilliant and well received brief speech explaining the meaning of the day's ceremony to those elected and to the people – it was a celebration of democracy renewed.
>
> (Essono 1987: 11)

He did not forget to rattle off the list of positions held by the recently promoted official, and not only mentioned his age but also reminded the audience of his sporting successes[1] (see Bissi 1989: 3).

. . . Such attention to detail should not come as a surprise; it is part of the system of 'distinction' (see Bourdieu 1979: especially the section on struggles over symbols). The enumeration of the slightest educational achievement is one of the postcolonial codes of prestige, with special attention to distinctions attained in Europe. Thus, for example, citizens cite their diplomas with great care, they show off their titles – doctor, chief, president, and so on – with great affectation, as a way of claiming honor, glory, attention. Displays of this kind have an effect beyond their contribution to state ritual. Such a display is transformative; by casting its rays on the person installed, it bestows upon him a new radiance. In the hierarchy of mock honors, the description of scholarly achievements constitutes a marker of rank and status as well as of qualification. (See Goffman 1959: 17–76 which discusses the regulation of rites and of private conduct, as well as the notion of a 'code of circulation/distinction.') . . .

In the postcolony, magnificence and the desire to shine are not the prerogative only of those who command. The people also want to be 'honored,' to 'shine,' and to take part in celebrations.

> Last Saturday the Muslim community of Cameroon celebrated the end of Ramadan. For thirty days members of the community had been deprived of many things from dawn till dusk. They refrained from drinking, eating, smoking, sexual relations and saying anything that goes against the Muslim faith and the law. Last Saturday marked the end of these privations for the whole Muslim community of Cameroon.
>
> (Simgba 1989: 7)

It is clear that the obscenity of power in the postcolony is also fed by a desire for majesty on the part of the people. Because the postcolony is characterized above all by scarcity, the metaphor of food 'lends itself to the wide-angle lens of both imagery and efficacy' (see Guyer 1991). Food and tips (*pourboire*) are political,[2] 'food,' like 'scarcity,' cannot be dissociated from particular regimes of 'death,' from specific modalities of enjoyment or from therapeutic quests (Taussig 1986). This is why 'the night' (Rosny 1977) and 'witchcraft' (Geschiere 1988) the 'invisible'(Bonnafé 1978), the 'belly,' the 'mouth' (Brown 1983) and the 'penis' are historical phenomena in their own right. They are institutions and sites of power, in the same way as pleasure or fashion:

> Cameroonians love slick gaberdine suits, Christian Dior outfits, Yamamoto blouses, shoes of crocodile skin.
>
> (Owona 1989)

> The label is the true sign of 'class,'. . .
> There are certain names that stand out. They are the ones that should be worn on a jacket, a shirt, a skirt, a scarf, or a pair of shoes if you want to win respect.
>
> (Tagne 1989)

> Do not be surprised if one day when you enter an office unannounced you discover piles of clothing on the desks. The hallways of Ministries and other public or private offices have become the market place par excellence. Market conditions are so flexible that everyone – from the director to the messenger finds what they want. Indeed, owing to the current crisis, sellers give big reductions and offer long-term credit. . . .
> Business is so good that many people throw themselves into it head down. A veritable waterhole, it's where sophisticated ladies rub shoulders with all kinds of ruffians and layabouts. The basis of the entire 'network' is travel. It is no secret that most of the clothes on the market come from the West. Those who have the 'chance' to go there regularly are quick to notice that they can reap great benefits from frequent trips. A few 'agreements' made with customs officials, and the game is on.
>
> (Zok 1989)

Even death does not escape this desire to 'shine' and to be 'honored.' The rulers and the ruled want more than ceremonies and celebrations to show off their splendor. Those who have accumulated goods, prestige, and influence are not only tied to the 'constraints of giving' (see Veyne 1976: 230). They are also taken by the desire to 'die well' and to be buried with pomp (see Omoruyi 1988: 466–9). Funerals constitute one of the occasions where those who command gaze at themselves, much like Narcissus.[3] Thus, when Joseph Awunti, the presidential minister in charge of relations with parliament, died on November 4, 1987, his body was received at Bamenda airport by the governor of what was then the Northwestern Province, Wabon Ntuba Mboe, himself accompanied by the Grand Chancellor, the first vice-president of the party, and a variety of administrative, political, and 'traditional' authorities. Several personalities and members of the government were also present, including the 'personal' representative of the head of state, Joseph Charles Dumba, Minister to the Presidency. The Economic and Social Council was represented by its president, Ayang Luc, the National Assembly by the president of the parliamentary group, and the Central Committee of the Party by its treasurer (Mbonwoh 1987: 3). Power's sanction thus penetrated to the very manner the dead man was buried. It appears that those who command seek to familiarize themselves with death, paving the way for their burial to take on a certain quality of pleasure and expenditure. . . .

As we have seen, obscenity – regarded as more than a moral category – constitutes one modality of power in the postcolony. But it is also one of the arenas in which subordinates reaffirm or subvert that power. Bakhtin's error was to attribute these practices to the dominated. But the production of burlesque is not specific to this group. The real inversion takes place when, in their desire for a certain majesty, the masses join in the madness and clothe themselves in cheap imitations of power to reproduce its epistemology, and when power, in its own violent quest for grandeur, makes vulgarity and wrongdoing its main mode of existence. It is here, within the confines of this intimacy, that the forces of tyranny in Africa must be studied. Such research must go beyond institutions, beyond formal positions of power, and beyond the written rules, and examine how the implicit and explicit are interwoven, and how the practices of those who command and those who are assumed to obey are so entangled as to render both powerless. For it is precisely the situations of powerlessness that are the situations of violence par excellence.

## Notes

1   We are told *inter alia* that he was a former champion and holder of the 400-meter record (50.1 seconds) in Cameroon, winning a gold medal at the francophone school and in a university competition in May 1957.

2   Understood here in the sense intended by Bayart, who draws on the Foucaultian notion of *gouvernementalité* to define the *gouvernementalité du ventre* (belly politics) of black Africa. See Bayart 1993.

3   But they are also among the situations where the innumerable conflicts connected with inequality and the distribution of inheritance are played out. On this point see C. Vidal (1987) and M. Gilbert (1988).

# PART TWO

# Universality and Difference

THE CONCEPT OF UNIVERSALISM IS ONE OF particular interest to post-colonial writers because it is this notion of a unitary and homogeneous human nature which marginalizes and excludes the distinctive characteristics, the difference, of post-colonial societies. A crucial insistence of post-colonial theory is that, despite a shared experience of colonialism, the cultural realities of post-colonial societies may differ vastly. The washing out of cultural difference becomes a prominent effect of European literary criticism, since some appeal to the essential humanity of readers has been constructed as a function of the value and significance of the literary work. We are often told that what makes Shakespeare or Dante or Goethe 'great' is their ability to reveal something of 'the universal human condition'. Indeed the universality of writers has been invoked in literature discussions across the English-speaking world as an infallible sign of their stature or their 'seriousness'. The myth of universality is thus a primary strategy of imperial control as it is manifested in literary study and that is why it demands attention early on in this Reader. The universalist myth has, according to Chinua Achebe, a pernicious effect in the kind of colonialist criticism that denigrates the post-colonial text on the basis of an assumption that 'European' equals 'universal'. But even a brief analysis of the 'universal human condition' finds it disappearing into an endless network of provisional and specific determinations in which even the most apparently 'essential' features of human life become provisional and contingent.

The assumption of universalism is a fundamental feature of the construction of colonial power because the 'universal' features of humanity are the characteristics of those who occupy positions of political dominance. It is these people who are 'human', who have a legitimate history, who live in 'the world'. Because language is a discourse of power, in that it provides the terms and the structures by which individuals have a world, a method by which the 'real' is determined, notions of universality can, like the language that suggests them, become imperialistic. The language itself implies certain assumptions about the world, a certain history, a certain way of seeing. If one's own language, or one's immediate perceptions of the world do not concur then they must be suppressed in favour of that which the language itself reveals to be 'obvious'.

George Lamming has reminded us in his essay 'The Occasion for Speaking' of Hegel's assertion that the African is somehow outside of History, that Africa is 'no historical part of the world'. This is simply because History is the story of 'Civilization' and it is only when that language becomes 'appropriated' by other cultures that the very concept of history can be questioned, and that the universal condition of humanity can be revealed as far more heterogeneous. At a strategic moment in the British occupation of India, English literature was invoked precisely for its imputed power to convey universal values. As Gauri Viswanathan points out, the concept of universalism became part of the technology of Empire: when the introduction of Christianity was considered by the Indian colonial administration to be too great a threat to good order, the 'universal' discourse of english literature (see n. 1 of General introduction, p. 4) was consciously adopted as the vehicle for educating the Indian elites in tenets of civilized morality.

Not only is the supposed universal human nature found to be spurious when the post-colonial engages the European text ('What is a kiss?' asks Charles Larson's African student) but it is not even true of that most 'universal' of discourses – mathematics – as is explained by Alan Bishop. Yet such assumptions about literature and its relationship to human life profoundly influence the critical reception of post-colonial literatures. And not only is it true of both conservative and liberal humanism, but it also insidiously affects the responses of those critics who, like Fredric Jameson, passionately argue for a consideration of literatures other than the trans-Atlantic. Aijaz Ahmad points out the degree to which the habit of 'worldism' (as in first, second and third worldism), can obliterate the cultural distinctions between societies. As Serequeberhan points out, the critique of universalism is, ironically, the critique of Eurocentrism. The value of post-colonial discourse is that it provides a methodology for considering the dialogue of similarity and difference; the similarity of colonialism's political and historical pressure upon non-European societies, alongside the plurality of specific cultural effects and responses those societies have produced.

## Chapter 11

# CHINUA ACHEBE

## COLONIALIST CRITICISM

From *Hopes and Impediments: Selected Essays 1965–1987* New York: Doubleday, 1988. Based on a paper read to the Association for Commonwealth Literature and Language Studies at Makere University, Uganda, 1974.

WHEN MY FIRST NOVEL WAS PUBLISHED IN 1958 a very unusual review of it was written by a British woman, Honor Tracy, who is perhaps not so much a critic as a literary journalist. But what she said was so intriguing that I have never forgotten it. If I remember rightly she headlined it 'Three cheers for mere Anarchy!'. The burden of the review itself was as follows: These bright Negro barristers . . . who talk so glibly about African culture, how would they like to return to wearing raffia skirts? How would novelist Achebe like to go back to the mindless times of his grandfather instead of holding the modern job he has in broadcasting in Lagos?

I should perhaps point out that colonialist criticism is not always as crude as this but the exaggerated grossness of a particular example may sometimes prove useful in studying the anatomy of the species. There are three principal parts here: Africa's inglorious past (raffia skirts) to which Europe brings the blessing of civilization (Achebe's modern job in Lagos) and for which Africa returns ingratitude (sceptical novels like *Things Fall Apart*).

Before I go on to more advanced varieties I must give one more example of the same kind as Honor Tracy's which on account of its recentness (1970) actually surprised me:

The British administration not only safeguarded women from the worst tyrannies of their masters, it also enabled them to make their long journeys to farm or market without armed guard, secure from the menace of hostile neighbours. The Nigerian novelists who have written the charming and bucolic accounts of domestic harmony in African rural communities, are the sons whom the labours of these women educated; the peaceful village of their childhood to which they nostalgically look back was one which had been purged of bloodshed and alcoholism by an ague-ridden district officer and a Scottish mission lassie whose years were cut short by every kind of intestinal parasite.

It is even true to say that one of the most nostalgically convincing of the rural African novelists used as his sourcebook not the memories of his grandfathers but the records of the despised British anthropologists. The modern African myth-maker hands down a vision of colonial rule in which the native powers are chivalrously viewed through the eyes of the hard-won liberal tradition of the late Victorian scholar, while the expatriates are shown as schoolboys' blackboard caricatures.

(Andreski 1971: 26)

I have quoted this at such length because first of all I am intrigued by Iris Andreski's literary style which recalls so faithfully the sedate prose of the district officer government anthropologist of sixty or seventy years ago – a tribute to her remarkable powers of identification as well as to the durability of colonialist rhetoric. 'Tyrannies of their masters' . . . 'menace of hostile neighbours' . . . 'purged of bloodshed and alcoholism'. But in addition to this Iris Andreski advances the position taken by Honor Tracy in one significant and crucial direction – its claim to a deeper knowledge and a more reliable appraisal of Africa than the educated African writer has shown himself capable of.

To the colonialist mind it was always of the utmost importance to be able to say: 'I know my natives', a claim which implied two things at once: (a) that the native was really quite simple and (b) that understanding him and controlling him went hand in hand – understanding being a pre-condition for control and control constituting adequate proof of understanding. Thus in the heyday of colonialism any serious incident of native unrest, carrying as it did disquieting intimations of slipping control, was an occasion not only for pacification by the soldiers but also (afterwards) for a royal commission of inquiry – a grand name for yet another perfunctory study of native psychology and institutions. Meanwhile a new situation was slowly developing as a handful of natives began to acquire European education and then to challenge Europe's presence and position in their native land with the intellectual weapons of Europe itself. To deal with this phenomenal presumption the colonialist devised two contradictory arguments. He created the 'man of two worlds' theory to prove that no matter how much the native was exposed to European influences he could never truly absorb them; like Prester John[1] he would always discard the mask of civilization when the crucial hour came and reveal his true face. Now, did this mean that the educated native was no different at all from his brothers in the bush? Oh, no! He was different; he was worse. His abortive effort at education and culture though leaving him totally unredeemed and unregenerated had none the less done something to him – it had deprived him of his links with his own people whom he no longer even understood and who certainly wanted none of his dissatisfaction or pretensions. 'I know my natives; they are delighted with the way things are. It's only these half-educated ruffians who don't even know their own people.' How often one heard that and the many variations of it in colonial times! And how almost amusing to find its legacy in the colonialist criticism of our literature today! Iris Andreski's book is more than old wives' tales, at least in intention. It is clearly inspired by the desire to undercut the educated African witness (the modern myth-maker, she calls him) by appealing direct to the unspoilt woman of the bush who has retained a healthy gratitude for Europe's intervention in Africa. This desire accounts for all that reliance one finds in modern European travellers' tales on the evidence of 'simple natives' – houseboys, cooks, drivers, schoolchildren – supposedly more trustworthy than the smart alecs. . . .

In his book, *The Emergence of African Fiction*, Charles Larson tells us a few revealing things about universality. In a chapter devoted to Lenrie Peters's novel which he finds particularly impressive he speaks of its universality, its very limited concern with Africa itself. Then he goes on to spell it all out:

> That it is set in Africa appears to be accidental, for, except for a few comments
> at the beginning, Peters's story might just as easily take place in the southern part
> of the United States or in the southern regions of France or Italy. If a few names of
> characters and places were changed one would indeed feel that this was an American
> novel. In short, Peters's story is universal.
>
> (Larson 1971: 230)

But Larson is obviously not as foolish as this passage would make him out to be, for he ends it on a note of self-doubt which I find totally disarming. He says:

> Or am I deluding myself in considering the work universal? Maybe what I really mean is that The Second Round is to a great degree Western and therefore scarcely African at all.
>
> (238)

I find it hard after that to show more harshness than merely agreeing about his delusion. But few people I know are prepared to be so charitable. In a recent review of the book in *Okike*, a Nigerian critic, Omolara Leslie, mocks 'the shining faith that we are all Americans under the skin'.

Does it ever occur to these universities to try out their game of changing names of characters and places in an American novel, say, a Philip Roth or an Updike, and slotting in African names just to see how it works? But of course it would not occur to them. It would never occur to them to doubt the universality of their own literature. In the nature of things the work of a Western writer is automatically informed by universality. It is only others who must strain to achieve it. So-and-so's work is universal; he has truly arrived! As though universality were some distant bend in the road which you may take if you travel out far enough in the direction of Europe or America, if you put adequate distance between yourself and your home. I should like to see the word 'universal' banned altogether from discussions of African literature until such a time as people cease to use it as a synonym for the narrow, self-serving parochialism of Europe, until their horizon extends to include all the world. If colonialist criticism were merely irritating one might doubt the justification of devoting a whole essay to it. But strange though it may sound some of its ideas and precepts do exert an influence on our writers, for it is a fact of our contemporary world that Europe's powers of persuasion can be far in excess of the merit and value of her case. Take for instance the black writer who seizes on the theme that Africa's past is a sadly inglorious one as though it were something new that had not already been 'proved' adequately for him. Colonialist critics will, of course, fall all over him in ecstatic and salivating admiration — which is neither unexpected nor particularly interesting. What is fascinating, however, is the tortuous logic and sophistry they will sometimes weave around a perfectly straightforward and natural enthusiasm.

A review of Yambo Ouologuem's *Bound to Violence* (1968b) by a Philip M. Allen in the *Pan-African Journal* (Allen 1971) was an excellent example of sophisticated, even brilliant colonialist criticism. The opening sentence alone would reward long and careful examination; but I shall content myself here with merely quoting it:

> The achievement of Ouologuem's much discussed, impressive, yet over-praised novel has less to do with whose ideological team he's playing on than with the *forcing of moral universality on African civilization*.
>
> (my italics)

A little later Mr Allen expounds on this new moral universality:

> This morality is not only 'un-African' — denying the standards set by omnipresent ancestors, the solidarity of communities, the legitimacy of social contract: it is a Hobbesian universe that extends beyond the wilderness, beyond the white man's myths of Africa, into all civilization, theirs and ours.

If you should still be wondering at this point how Ouologuem was able to accomplish that Herculean feat of forcing moral universality on Africa or with what gargantuan tools, Mr Allen does not leave you too long in suspense. Ouologuem is 'an African intellectual who has mastered both a style and a prevailing philosophy of French letters', able to enter 'the remote alcoves of French philosophical discourse'. . . .

That a 'critic' playing on the ideological team of colonialism should feel sick and tired of Africa's 'pathetic obsession with racial and cultural confrontation' should surprise no one. Neither should his enthusiasm for those African works that show 'no easy antithesis between white and black'. But an African who falls for such nonsense, not only in spite of Africa's so very recent history but, even more, in the face of continuing atrocities committed against millions of Africans in their own land by racist minority regimes, deserves a lot of pity. Certainly anyone, white or black, who chooses to see violence as the abiding principle of African civilization is free to do so. But let him not pass himself off as a restorer of dignity to Africa, or attempt to make out that he is writing about man and about the state of civilization in general. . . . Perhaps for most ordinary people what Africa needs is a far less complicated act of restoration. . . .

The colonialist critic, unwilling to accept the validity of sensibilities other than his own, has made a particular point of dismissing the African novel. He has written lengthy articles to prove its non-existence largely on the grounds that the novel is a peculiarly Western genre, a fact which would interest us if our ambition was to write 'Western' novels. But, in any case, did not the black people in America, deprived of their own musical instruments, take the trumpet and the trombone and blow them as they had never been blown before, as indeed they were not designed to be blown? And the result, was it not jazz? Is any one going to say that this was a loss to the world or that those first Negro slaves who began to play around with the discarded instruments of their masters should have played waltzes and foxtrots? No! Let every people bring their gifts to the great festival of the world's cultural harvest and mankind will be all the richer for the variety and distinctiveness of the offerings.

My people speak disapprovingly of an outsider whose wailing drowned the grief of the owners of the corpse. One last word to the owners. It is because our own critics have been somewhat hesitant in taking control of our literary criticism (sometimes – let's face it – for the good reason that we will not do the hard work that should equip us) that the task has fallen to others, some of whom (again we must admit) have been excellent and sensitive. And yet most of what remains to be done can best be tackled by ourselves, the owners. If we fall back, can we complain that others are rushing forward? A man who does not lick his lips, can he blame the harmattan for drying them?

## Note

1   From the novel (1910) of the same name by the imperial statesman and adventure writer, John Buchan.

## Chapter 12

## CHARLES LARSON

# HEROIC ETHNOCENTRISM
## The idea of universality in literature

From 'Heroic Ethnocentrism: The Idea of Universality in Literature'
*American Scholar* 42(3) (Summer), 1973.

IN THE FALL OF 1962, when I began teaching English literature to high school students in Nigeria, I encountered a number of stumbling blocks, which I had in no way anticipated – all of them cultural, experiential. This was not a matter of science or technology and their various by-products as I had anticipated ('What is a flush toilet?') but, rather, matters related to what I have learned to call culturally restricted materials. It was enough, to be sure, just for my African students to read through a 450-page Victorian novel (required reading in those days for the British-administered school certificate examinations); and, as I later learned, in the lower levels at least, students were accustomed to taking several months or even the greater part of a year to read through and discuss the plot line of a single novel. Length alone was enough to get them, since English was their second language and the problem of vocabulary was especially troublesome. But once the problems of language, vocabulary and verbosity had been overcome, reading through the words became a less difficult process than understanding what the words themselves related – the 'experience of literature' as we are wont to say.

'Excuse me, sir, what does it mean "to kiss"?' That was a much more difficult question to answer than the usual ones relating to the plot or the characters of the novel – a real shock when it was brought to my attention that I had a rather naïve boy in my class. So I brushed the question off until it was repeated a number of times and I slowly began to realize that all of my students had no real idea of what it meant to kiss. This seemed an extremely odd thing to me because most of my students were upper-form boys in their late teens – some in their early twenties – and I had, of course, heard them talking on occasion about their girl friends. It was also rumoured that several of the boys were married, although by school regulations they were not supposed to be. Nevertheless, that question and others of a like nature kept recurring – in part, no doubt, because we were reading Thomas Hardy's *Far from the Madding Crowd*. Why did Hardy's characters get so flustered when they were kissed (or more likely, when they weren't kissed)? When I asked one of the European-educated African teachers why my students always seemed ready to return to that same question, I was more than surprised to learn that Africans, traditionally at least, do not kiss; to learn that what I thought was 'natural' in one society is not natural at all, but learned, that is, cultural. Not all peoples kiss. Or, stated more appropriately, not all peoples have learned to kiss. (When I later attended American movies with Africans, I could understand why the audience often went into hysterics at the romantic scenes in the films.)

How was one to read a Thomas Hardy novel with all those frustrated kisses without ever having been kissed? How was I to explain something like this to my African students? Or, to limit my experience to a more technical matter concerning the novel's form which also perplexed my students, what about those long passages of description for which Hardy is so celebrated? My African students couldn't understand what page after page of description of the countryside had to do with the plot of the novel. What they had given me, as I later learned, was another clue to the differing ways in which culture shapes our interpretations of literature. It was not until I seriously began studying the African novel itself, however, that I could put all of those pieces together; just as the questions about those kisses revealed something about my African students' cultural background, so too, did their concern about the descriptive passages of Hardy's book. The fact that descriptive passages were virtually nonexistent in African fiction initially seemed particularly puzzling to me, since the first generation of African Anglophone novelists, at least, had been brought up almost entirely on the Victorian novel. Whereas other elements of the Victorian novel had found their way into the African novel, description had not. Could it be that this omission in the African novel revealed something basically different between African and Western attitudes toward nature, toward one's environment?

Kissing and description, attitudes toward love and nature – are these attitudes so different for the African? Is the African way of life less sophisticated than our own? Or is the belief that these supposedly 'universal' attitudes should be the same as ours the naïve one? Is this what we really mean when we talk about 'universality' in literature – if someone does not react to something in our literature the same way that we do, then he is to be considered inferior? Perhaps the term itself is meaningless. After all, people love and die in every culture. Their reactions to these events in their lives, however, may be significantly different from our own. And these reactions, in turn, shape their interpretations of literature.

For the most part, the term 'universal' has been grossly misused when it has been applied to non-Western literature, because it has so often been used in a way that ignores the multiplicity of cultural experiences. Usually, when we try to force the concept of universality on someone who is not Western, I think we are implying that our own culture should be the standard of measurement. Why else would we expect all peoples to react in the same way that we do? . . . But let us return to those so-called universal experiences of all literature and illustrate some of the ways in which they may be radically different – at least for the African.

In his preface to Tsao-Hsueh-Chin's eighteenth-century Chinese novel, *Dream of the Red Chamber*, Mark Van Doren says, 'The greatest love stories have no time or place.' I frankly doubt this, in spite of other Western literary critics, who have also said that the most common theme in literature is love. (Leslie Fiedler, in his *Collected Essays*, for instance.) After reading dozens and dozens of contemporary African novels, I can in no way accept Van Doren's or Fiedler's assertions. There is at least one whole section of the world where the love story is virtually nonexistent. I can think of no contemporary African novel in which the central plot or theme can be called a 'love story,' no African novel in which the plot line progresses because of the hero's attempt to acquire a mate, no African novel in which seduction is the major goal, no African novel in which the fate of the lovers becomes the most significant element in the story. No African novel works this way because love as a theme in a Western literary sense is simply missing. Romantic love, seduction, sex – these are not the subjects of African fiction. In fact, in most contemporary African novels women play minor parts; the stories are concerned for the most part with a masculine world. There may be marriage, bride price and an occasional *tête à tête* but that is not the concern of the novel: it is always something else. There are no graphic descriptions of erotic love, there are no kisses, no holding hands. There is, in short, no love story as we have come to think of it in Western fiction. Not even the unrequited lover pining away. African fiction simply is not made of such stuff. . . .

Western romance is only one theme that may puzzle the African reader. He may have trouble understanding the lack of concern about death in some Western novels, too. Or, what is more likely, the Western reader may totally miss the significance of a death in a piece of African fiction that he is reading. A. Alvarez, in his fine book, *The Savage God*, says that 'perhaps half the literature of the world is about death.' Yet our society has worked so hard to neutralize the shock of death that it is quite possible for us to miss the emotional overtones of a piece of African writing in which death occurs. Sembene Ousmane's celebrated short story, 'Black Girl' ('La noire de . . .') is one such example. The story concerns a young Senegalese girl named Diouana, who moves to Antibes when the French family she has worked for in Senegal returns to France. Ousmane begins his story by projecting us into Diouana's thoughts, illustrating her excitement and fascination at being able to have such a wonderful experience: the chance to live in France. But Diouana's dreams shortly become a nightmare. Overworked, isolated from her fellow Africans, called a nigger by the four children in the French family, after some months Diouana commits suicide by slashing her wrists in the bathtub. The Western reader may think that Ousmane's story is simply another rather melodramatic account of racial prejudice – which it is, in part. But it is also a story about modern slavery, and what that situation will drive the sane person into doing: taking his own life. Just as slaves jumped overboard to their deaths in the ocean in order to escape slavery in the New World, Diouana takes her own life to find release from her own enslaved situation. But this is only a part of it, for in committing suicide – one of the strongest taboos in many African societies – she has only temporarily released herself. She has trapped her ancestors, broken the cycle of life, and, if she is an only child, she has ended the family lineage. She has, in short, committed a terrible abomination, and the African reading the conclusion to Ousmane's story is horrified by what she has done. It is, therefore, the religious overtone relating to ancestral worship that the Western reader will probably be completely unaware of. . . .

The hero concept – the belief in the individual who is different from his fellowmen – is [also] almost totally alien to African life; and, as an extension of this, the hero in contemporary African fiction is for the most part non-existent. The hero is almost nonexistent in contemporary Western literature too, but his descendant, the anti-hero, the isolated figure, is a force to be reckoned with. This is not true of African fiction, however. Rather, it is the group-felt experience that is all important: what happens to the village, the clan, the tribe. . . .

One begins to wonder if two peoples as widely different as Africans and Westerners will ever be able to read each other's literature and fully understand it. This is not, however, the question I started out to ask. Literature is not so limiting that only one interpretation is possible. We cannot all be both African and westerner, black and Caucasian. What is important, it seems to me, is that when we read a piece of non-Western literature we realize that the interpretation we make of it may be widely different from what the artist intended, and contrarily, that we should not expect people who are not of our own culture and heritage to respond in the same way that we do to our own literature. The time has come when we should avoid the use of the pejorative term 'universal.' What we really mean when we talk about universal experiences in literature are cultural responses that have been shaped by our own Western tradition.

Although most of the examples I have used in this essay are African in origin, I would hazard a final conjecture that the experience of other non-Western literatures (Chinese and Japanese, for example) will also support this belief that the word 'universal' is, indeed, limited. . . . For better or for worse, each of us was born into an ethnocentrically sealed world. The purpose of any piece of literature, no matter what culture it was produced in, is to show us something we were previously unaware of. Just as literature is a bridge connecting a life lived with a life not lived, so, too, all literature that is effective is a voyage into a previously untraveled world.

## Chapter 13

## ALAN J. BISHOP

# WESTERN MATHEMATICS
## The secret weapon of cultural imperialism

From 'Western Mathematics: The Secret Weapon of Cultural Imperialism'
*Race and Class* 32(2), 1990.

O F ALL THE SCHOOL SUBJECTS which were imposed on indigenous pupils in the colonial schools, arguably the one which could have been considered the least culturally-loaded was mathematics. Even today, that belief prevails. Whereas educational arguments have taken place over which language(s) should be taught, what history or religion, and whether, for example, 'French civilisation' is an appropriate school subject for pupils living thousands of kilometres from France, mathematics has somehow always been felt to be universal and, therefore, culture-free. It had in colonial times, and for most people it continues to have today, the status of a culturally neutral phenomenon in the otherwise turbulent waters of education and imperialism. . . .

Up to fifteen years or so ago, the conventional wisdom was that mathematics was culture-free knowledge. After all, the popular argument went, two twos are four, a negative number times a negative number gives a positive number, and all triangles have angles which add up to 180 degrees. These are true statements the world over. They have universal validity. Surely, therefore, it follows that mathematics must be free from the influence of any culture?

There is no doubt that mathematical truths like those are universal. They are valid everywhere, because of their intentionally abstract and general nature. So, it doesn't matter where you are, if you draw a flat triangle, measure all the angles with a protractor, and add the degrees together, the total will always be approximatcly 180 degrees. . . . Because mathematical truths like these are abstractions from the real world, they are necessarily context-free and universal.

But where do 'degrees' come from? Why is the total 180? Why not 200, or 100? Indeed, why are we interested in triangles and their properties at all? The answer to all these questions is, essentially, 'because some people determined that it should be that way'. Mathematical ideas, like any other ideas, are humanly constructed. They have a cultural history.

The anthropological literature demonstrates for all who wish to see it that the mathematics which most people learn in contemporary schools is not the only mathematics that exists. For example, we are now aware of the fact that many different counting systems exist in the world. In Papua New Guinea, Lean has documented nearly 600 (there are more than 750 languages there) containing various cycles of numbers, not all base ten (Lean 1991). As well as finger counting, there is documented use of body counting, where one points to a part of the body and uses the name of that part as the number. Numbers are also recorded in knotted strings, carved on wooden tablets or on rocks, and beads are used, as well as many different written systems of numerals (Menninger 1969). The richness is both fascinating and provocative for anyone imagining initially that theirs is the only system of counting and recording numbers.

Nor only is it in number that we find interesting differences. The conception of space which underlies Euclidean geometry is also only one conception – it relies particularly on the 'atomistic' and object-oriented ideas of points, lines, planes and solids. Other conceptions exist, such as that of the Navajos where space is neither subdivided nor objectified, and where everything is in motion (Pinxten *et al.* 1983). Perhaps even more fundamentally, we are more aware of the forms of classification which are different from western hierarchical systems – Lancy, again in Papua New Guinea, identified what he referred to as 'edge-classification', which is more linear than hierarchical (Lancy 1983; Philp 1973). The language and logic of the Indo-European group have developed layers of abstract terms within the hierarchical classification matrix, but this has not happened in all language groups, resulting in different logics and in different ways of relating phenomena.

Facts like these challenge fundamental assumptions and long-held beliefs about mathematics. Recognising symbolisations of alternative arithmetics, geometries and logics implies that we should, therefore, raise the question of whether alternative mathematical systems exist. Some would argue that facts like those above already demonstrate the existence of what they call 'ethno-mathematics', a more localised and specific set of mathematical ideas which may not aim to be as general nor as systematised as 'mainstream' mathematics. Clearly, it is now possible to put forward the thesis that all cultures have generated mathematical ideas, just as all cultures have generated language, religion, morals, customs and kinship systems. Mathematics is now starting to be understood as a pan-cultural phenomenon.

We must, therefore, henceforth take much more care with our labels. We cannot now talk about 'mathematics' without being more specific, unless we are referring to the generic form (like language, religion, etc.). The particular kind of mathematics which is now the internationalised subject most of us recognise is a product of a cultural history, and in the last three centuries of that history, it was developing as part of western European culture (if that is a well-defined term). That is why the title of this article refers to 'western mathematics'. In a sense, that term is also inappropriate, since many cultures have contributed to this knowledge and there are many practising mathematicians all over the world who would object to being thought of as western cultural researchers developing a part of western culture. Indeed, the history of western mathematics is itself being rewritten at present as more evidence comes to light, but more of that later. Nevertheless, in my view it is thoroughly appropriate to identify 'western mathematics', since it was western culture, and more specifically western European culture, which played such a powerful role in achieving the goals of imperialism.

There seem to have been three major mediating agents in the process of cultural invasion of colonised countries by western mathematics: trade, administration and education. Regarding trade and the commercial field generally, this is clearly the area where measures, units, numbers, currency and some geometric notions were employed. More specifically, it would have been western ideas of length, area, volume, weight, time and money which would have been imposed on the indigenous societies. . . .

The second way in which western mathematics would have impinged on other cultures is through the mechanism of administration and government. In particular the numbers and computations necessary for keeping track of large numbers of people and commodities would have necessitated western numerical procedures being used in most cases. . . .

The third and major medium for cultural invasion was education, which played such a critical role in promoting western mathematical ideas and, thereby, western culture. . . .

At worst, the mathematics curriculum was abstract, irrelevant, selective and elitist – as indeed it was in Europe – governed by structures like the Cambridge Overseas Certificate, and culturally laden to a very high degree. It was part of a deliberate strategy of acculturation – international in its efforts to instruct in 'the best of the West', and convinced of its superiority to any indigenous mathematical systems and culture. As it was essentially a university-preparatory education, the aspirations of the students were towards attending western universities. They were educated away from their culture and away from their society. . . .

So, it is clear that through the three media of trade, administration and education, the symbolisations and structures of western mathematics would have been imposed on the indigenous cultures just as significantly as were those linguistic symbolisations and structures of English, French, Dutch or whichever was the European language of the particular dominant colonial power in the country.

However, also like a language, the particular symbolisations used were, in a way, the least significant aspect of mathematics. Of far more importance, particularly in cultural terms, were the values which the symbolisations carried with them. Of course, it goes without saying that it was also conventional wisdom that mathematics was value free. How could it have values if it was universal and culture free? We now know better, and an analysis of the historical, anthropological and cross-cultural literatures suggests that there are four clusters of values which are associated with western European mathematics, and which must have had a tremendous impact on the indigenous cultures.

First, there is the area of rationalism, which is at the very heart of western mathematics. If one had to choose a single value and attribute which has guaranteed the power and authority of mathematics within western culture, it is rationalism. As Kline says: 'In its broadest aspect mathematics is a spirit, the spirit of rationality. It is this spirit that challenges, stimulates, invigorates, and drives human minds to exercise themselves to the fullest' (Kline 1972). . . .

Second, a complimentary set of values associated with western mathematics can be termed objectism, a way of perceiving the world as if it were composed of discrete objects, able to be removed and abstracted, so to speak, from their context. To decontextualise, in order to be able to generalise, is at the heart of western mathematics and science; but if your culture encourages you to believe, instead, that everything belongs and exists in its relationship with everything else, then removing it from its context makes it literally meaningless. In early Greek civilisation, there was also a deep controversy over 'object' or 'process' as the fundamental core of being. Heraclitos, in 600–500BC, argued that the essential feature of phenomena is that they are always in flux, always moving and always changing. Democritus and the Pythagoreans preferred the world-view of 'atoms', which eventually was to prevail and develop within western mathematics and science (see Ronan 1983 and Waddington 1977).

Horton sees objectism in another light. He compares this view with what he sees as the preferred African use of personal idiom as explanation. He argues that this has developed for the traditional African the sense that the personal and social 'world' is knowable, whereas the impersonal and the 'world of things' is essentially unknowable. The opposite tendency holds for the westerner (Horton 1967). . . . We can see, therefore, that with both rationalism and objectism as core values, western mathematics presents a dehumanised, objectified, ideological world-view which will emerge necessarily through mathematics teaching of the traditional colonial kind.

A third set of values concerns the power and control aspect of western mathematics. Mathematical ideas are used either as directly applicable concepts and techniques, or indirectly through science and technology, as ways to control the physical and social environment.

As Schaff says in relation to the history of mathematics: 'The spirit of the nineteenth and twentieth centuries is typified by man's increasing mastery over his physical environment' (Schaff 1963: 48). So, using numbers and measurements in trade, industry, commerce and administration would all have emphasised the power and control values of mathematics. It was (and still is) so clearly useful knowledge, powerful knowledge, and it seduced the majority of peoples who came into contact with it. . . .

From those colonial times through to today, the power of this mathematico-technological culture has grown apace – so much so that western mathematics is taught nowadays in every country in the world. Once again, it is mainly taught with the assumption of universality and cultural neutrality. From colonialism through to neo-colonialism, the cultural imperialism of western mathematics has yet to be fully realised and understood. Gradually, greater understanding of its impact is being acquired, but one must wonder whether its all-pervading influence is now out of control.

As awareness of the cultural nature and influence of western mathematics is spreading and developing, so various levels of responses can also be seen. At the first level there is an increasing interest in the study of ethno-mathematics, through both analyses of the anthropological literature and investigations in real-life situations. . . .[1]

At the second level, there is a response in many developing countries and former colonies which is aimed at creating a greater awareness of one's own culture. Cultural rebirth or reawakening is a recognised goal of the educational process in several countries. Gerdes, in Mozambique, is a mathematics educator who has done a great deal of work in this area. He seeks not only to demonstrate important mathematical aspects of Mozambican society, but also to develop the process of 'defreezing' the 'frozen' mathematics which he uncovers. For example, with the plating methods used by fishermen to make their fish traps, he demonstrates significant geometric ideas which could easily be assimilated into the mathematics curriculum in order to create what he considers to be a genuine Mozambican mathematics education for the young people there. . . .

The third level of response to the cultural imperialism of western mathematics is, paradoxically, to re-examine the whole history of western mathematics itself. It is no accident that this history has been written predominantly by white, male, western European or American researchers, and there is a concern that, for example, the contribution of Black Africa has been undervalued. . . .

I began by describing the myth of western mathematics' cultural neutrality. Increasingly, modern evidence serves to destroy this naïve belief. Nevertheless, the belief in that myth has had, and continues to have, powerful implications. Those implications relate to education, to national developments and to a continuation of cultural imperialism. Indeed it is not too sweeping to state that most of the modern world has accepted western mathematics, values included, as a fundamental part of its education. . . .

However, taking a broader view, one must ask: should there not be more resistance to this cultural hegemony? . . . Resistance is growing, critical debate is informing theoretical developments, and research is increasing, particularly in educational situations where culture-conflict is recognised. The secret weapon is secret no longer.

## Note

1    Bishop goes on to describe six 'universals' of ethno-mathematics, that is, six activities which may be found in some combination in every society: Counting; Locating; Measuring; Designing; Playing; Explaining. (Eds)

Chapter 14

## AIJAZ AHMAD

# JAMESON'S RHETORIC OF OTHERNESS AND THE 'NATIONAL ALLEGORY'

From 'Jameson's Rhetoric of Otherness and the "National Allegory"' *Social Text* 17 (Fall), 1987. A reply to Fredric Jameson's 'Third World Literature in the Era of Multinational Capitalism' *Social Text* 15 (Fall), 1986.

I HAVE BEEN READING JAMESON'S WORK now for roughly fifteen years, and at least some of what I know about the literatures and cultures of Western Europe and the USA comes from him; and because I am a Marxist, I had always thought of us, Jameson and myself, as birds of the same feather, even though we never quite flocked together. But then, when I was on the fifth page of this text (specifically, on the sentence starting with 'All third-world texts are necessarily . . .' etc.), I realized that what was being theorized was, among many other things, myself. Now, I was born in India and I write poetry in Urdu, a language not commonly understood among US intellectuals. So I said to myself: '*All?* . . . *necessarily?*' It felt odd. Matters became much more curious, however. For the further I read, the more I realized, with no little chagrin, that the man whom I had for so long, so affectionately, albeit from a physical distance, taken as a comrade was, in his own opinion, my civilizational Other. It was not a good feeling.

I too think that there *are* plenty of very good books written by African, Indian and Latin American writers which are available in English and which must be taught as an antidote to the general ethnocentricity and cultural myopia of the Humanities as they are presently constituted in these United States. If some label is needed for this activity, one may call it 'Third World Literature'. Conversely, however, I also hold that this term, 'the Third World', is, even in its most telling deployments, a polemical one, with no theoretical status whatsoever. . . . I shall argue in context, then, that there is no such thing as a 'Third World Literature' which can be constructed as an internally coherent object of theoretical knowledge. There are fundamental issues – of periodization, social and linguistic formations, political and ideological struggles within the field of literary production, and so on – which simply cannot be resolved at this level of generality without an altogether positivist reductionism. . . .

I shall argue later that since Jameson defines the so-called Third World in terms of its experience of colonialism and imperialism, the political category that necessarily follows from this exclusive emphasis is that of 'the nation', with nationalism as the peculiarly valorized ideology; and, because of this privileging of the nationalist ideology, it is then theoretically posited that 'all third-world texts are necessarily . . . to be read as . . . national allegories'.

The theory of the 'national allegory' as the metatext is thus inseparable from the larger Three Worlds Theory which permeates the whole of Jameson's own text. We too have to begin, then, with some comments on 'the Third World' as a theoretical category and on 'nationalism' as the necessary, exclusively desirable ideology. . . .

As we come to the substance of what Jameson 'describes', I find it significant that First and Second Worlds are defined in terms of their production systems (capitalism and socialism, respectively), whereas the third category – the Third World – is defined purely in terms of an 'experience' of externally inserted phenomena. That which is constitutive of human history itself is present in the first two cases, absent in the third case. Ideologically, this classification divides the world between those who make history and those who are mere objects of it; elsewhere in the text, Jameson would significantly reinvoke Hegel's famous description of the master–slave relation to encapsulate the First–Third World opposition. But analytically, this classification leaves the so-called Third World in limbo; if only the First World is capitalist and the Second World socialist, how does one understand the Third World? Is it pre-capitalist? Transitional? Transitional between what and what? But then there is also the issue of the location of particular countries within the various 'worlds'.

Take, for example, India. Its colonial past is nostalgically rehashed on US television screens in copious series every few months, but the India of today has all the characteristics of a capitalist country: generalized commodity production, vigorous and escalating exchanges not only between agriculture and industry but also between Departments I and II of industry itself, and technical personnel more numerous than those of France and Germany combined. . . .

So – does India belong in the First World or the Third? . . .

I have said already that if one believes in the Three Worlds Theory – hence in a 'Third World' defined exclusively in terms of 'the experience of colonialism and imperialism' – then the primary ideological formation available to a left-wing intellectual will be that of nationalism; it will then be possible to assert – surely with very considerable exaggeration, but possible to assert none the less – that 'all third-world texts are necessarily . . . *national allegories*' (original emphasis). This exclusive emphasis on the nationalist ideology is there even in the opening paragraph of Jameson's text, where the only choice for the 'Third World' is said to be between its 'nationalisms' and a 'global American postmodernist culture'. Is there no other choice? Could not one join the 'Second World', for example? . . .

Jameson's haste in totalizing historical phenomena in terms of binary oppositions (nationalism/postmodernism, in this case) leaves little room for the fact, for instance, that the only nationalisms in the so-called Third World which have been able to resist US cultural pressure and have actually produced any alternatives are those which are already articulated to and assimilated within the much larger field of socialist political practice. Virtually all the others have had no difficulty in reconciling themselves with what Jameson calls 'global American postmodernist culture'; . . . Nor does the absolutism of that opposition (post-modernism/nationalism) permit any space for the simple idea that nationalism itself is not some unitary thing with some predetermined essence and value. There are hundreds of nationalisms in Asia and Africa today; some are progressive, others are not. Whether or not a nationalism will produce a progressive cultural practice depends, to put it in Gramscian terms, upon the political character of the power bloc which takes hold of it and utilizes it, as a material force, in the process of constituting its own hegemony. There is neither theoretical ground nor empirical evidence to support the notion that bourgeois nationalisms of the so-called Third World will have any difficulty with postmodernism; they *want* it.

Yet there *is* a very tight fit between the Three Worlds Theory, the overvalorization of the nationalist ideology, and the assertion that 'national allegory' is the primary, even exclusive, form of narrativity in the so-called Third World. If this 'Third World' is *constituted* by the singular 'experience of colonialism and imperialism', and if the only possible response is a nationalist one, then what else is there that is more urgent to narrate than this 'experience'? In fact, there is *nothing else* to narrate. For if societies here are defined not by relations of production but by relations of intra-national domination; if they are forever suspended outside the sphere of conflict between capitalism (First World) and socialism (Second World); if the motivating force for history here is neither class formation and class struggle nor the multi-plicities of intersecting conflicts based upon class, gender, nation, race, region, and so on, but the unitary 'experience' of national oppression (if one is merely the *object* of history, the Hegelian slave), then what else *can* one narrate but that national oppression? Politically, we are Calibans all. Formally, we are fated to be in the poststructuralist world of Repetition with Difference; the same allegory, the nationalist one, rewritten, over and over again, until the end of time: 'all third-world texts are necessarily . . .'.

But one could start with a radically different premiss: namely, the proposition that we live not in three worlds but in one; that this world includes the experience of colonialism and imperialism on both sides of Jameson's global divide (the 'experience' of imperialism is a central fact of all aspects of life inside the USA, from ideological formation to the utiliza-tion of the social surplus in military-industrial complexes); that societies in formations of backward capitalism are as much constituted by the division of classes as are societies in the advanced capitalist countries; that socialism is not restricted to something called 'the Second World' but is simply the name of a resistance that saturates the globe today, as capitalism itself does; that the different parts of the capitalist system are to be known not in terms of a binary opposition but as a contradictory unity – with differences, yes, but also with profound overlaps. . . .

Jameson claims that one cannot proceed from the premiss of a real unity of the world 'without falling back into some general liberal and humanistic universalism'. That is a curious idea, coming from a Marxist. One would have thought that the world was united not by liberalist ideology – that the world was not at all constituted in the realm of an Idea, be it Hegelian or humanist – but by the global operation of a single mode of production, namely the capitalist one, and the global resistance to this mode, a resistance which is itself unevenly developed in different parts of the globe. Socialism, one would have thought, was not by any means limited to the so-called Second World (the socialist countries) but is a global phenom-enon, reaching into the farthest rural communities in Asia, Africa and Latin America, not to speak of individuals and groups within the United States. What gives the world its unity, then, is not a humanist ideology but the ferocious struggle between capital and labour which is now strictly and fundamentally global in character. . . .

As for the specificity of cultural difference, Jameson's theoretical conception tends, I believe, in the opposite direction – namely, that of homogenization. Difference between the First World and the Third is absolutized as an Otherness, but the enormous cultural hetero-geneity of social formations within the so-called Third World is submerged within a singular identity of 'experience'. Now, countries of Western Europe and North America have been deeply tied together over roughly the last two hundred years; capitalism itself is so much older in these countries; the cultural logic of late capitalism is so strongly operative in these metropolitan formations; the circulation of cultural products among them is so immediate, so extensive, so brisk, that one could sensibly speak of a certain cultural homogeneity among them. But Asia, Africa and Latin America? Historically, these countries were never so closely tied together. . . .

Of course, great cultural similarities also exist among countries that occupy analogous positions in the global capitalist system, and there are similarities in many cases that have been bequeathed by the similarities of socioeconomic structures in the pre-capitalist past. The point is not to construct a typology that is simply the obverse of Jameson's, but rather to define the material basis for a fair degree of cultural homogenization among the advanced capitalist countries and the lack of that kind of homogenization in the rest of the capitalist world. In context, therefore, one is doubly surprised at Jameson's absolute insistence upon Difference and the relation of Otherness between the First World and the Third, and his equally insistent idea that the 'experience' of the 'Third World' could be contained and communicated within a single narrative form. By locating capitalism in the First World and socialism in the Second, Jameson's theory freezes and dehistoricizes the global space within which struggles between these great motivating forces actually take place. And by assimilating the enormous heterogeneities and productivities of our life into a single Hegelian metaphor of the master–slave relation, this theory reduces us to an ideal-type and demands from us that we narrate ourselves through a form commensurate with that ideal-type. To say that all Third World texts are necessarily this or that is to say, in effect, that any text originating within that social space which is not this or that is not a 'true' narrative. It is in this sense above all that the category of 'Third World Literature' which is the site of this operation, with the 'national allegory' as its metatext as well as the mark of its constitution and difference, is, to my mind, epistemologically an impossible category. . . .

And at what point in history does a text produced in countries with 'experience of colonialism and imperialism' become a *Third World text*? In one kind of reading, only texts produced *after* the advent of colonialism could be so designated, since it is colonialism/imperialism which constitutes the Third World as such. But in speaking constantly of 'the West's Other'; in referring to the tribal/tributary and Asiatic modes as the theoretical basis for his selection of Lu Xun (Asian) and Sembene (African) respectively; in characterizing Freud's theory as a 'Western or First World reading' as contrasted with ten centuries of specifically Chinese distributions of the libidinal energy which are said to frame Lu Xun's texts – in deploying these broad epochal and civilizational categories, Jameson also suggests that the difference between the First World and the Third is itself primordial, rooted in things far older than capitalism as such. So, if the First World is the same as 'the West' and the 'Graeco-Judaic', one has, on the other hand, an alarming feeling that the *Bhagavad-Gita*, the edicts of Manu, and the Qur'an itself are perhaps Third World texts (though the Judaic elements of the Qur'an are quite beyond doubt, and much of the ancient art in what is today Pakistan is itself Graeco-Indic).

But there is also the question of *space*. Do all texts produced in countries with 'experience of colonialism and imperialism' become, by virtue of geographical origin, 'third-world texts'? Jameson speaks so often of '*all* third-world texts', insists so much on a singular form of narrativity for Third World Literature, that not to take him literally is to violate the very terms of his discourse. Yet one knows of so many texts from one's own part of the world which do not fit the description of 'national allegory' that one wonders why Jameson insists so much on the category, '*all*'. Without this category, of course, he cannot produce a theory of Third World Literature. But is it also the case that he means the opposite of what he actually says: not that '*all* third-world texts are to be read . . . as national allegories' but that *only* those texts which give us national allegories can be admitted as authentic texts of Third World Literature, while the rest are by definition excluded? So one is not quite sure whether one is dealing with a fallacy ('all third-world texts are' this or that) or with the Law of the Father (you must write *this* if you are to be admitted into my theory). . . .

Jameson insists over and over again that the *national* experience is central to the cognitive formation of the Third World intellectual, and that the narrativity of that experience takes the form exclusively of a 'national allegory'. But this emphatic insistence on the category 'nation' itself keeps slipping into a much wider, far less demarcated vocabulary of 'culture', 'society', 'collectivity', and so on. Are 'nation' and 'collectivity' the same thing? . . .

[O]ne may indeed connect one's personal experience to a 'collectivity' – in terms of class, gender, caste, religious community, trade union, political party, village, prison – combining the private and the public, and in some sense 'allegorizing' the individual experience, without involving the category of 'the nation' or necessarily referring back to the 'experience of colonialism and imperialism'. The latter statement would then seem to apply to a much larger body of texts, with far greater accuracy. By the same token, however, this wider application of 'collectivity' establishes much less radical difference between the so-called First and Third Worlds, since the whole history of realism in the European novel, in its many variants, has been associated with ideas of 'typicality' and 'the social', while the majority of the written narratives produced in the First World even today locate the individual story in a fundamental relation to some larger experience.

If we replace the idea of the 'nation' with that larger, less restrictive idea of 'collectivity', and if we start thinking of the process of allegorization not in nationalistic terms but simply as a relation between private and public, personal and communal, then it also becomes possible to see that allegorization is by no means specific to the so-called Third World.

## Chapter 15

# TSENAY SEREQUEBERHAN

## THE CRITIQUE OF EUROCENTRISM

From 'The Critique of Eurocentrism' in Emmanuel Eze (ed.) *Post-colonial African Philosophy* Oxford: Blackwell.

I N HIS, BY NOW FAMOUS BOOK *The Postmodern Condition,* the French philosopher Jean-François Lyotard puts forth the thesis that the 'postmodern' is 'incredulity toward metanarratives,' the discarding of the lived and world-historical 'grand narratives' through which modernity constituted itself (1989: xxiv; 1992: 17). And as Wlad Godzich has noted, for Lyotard, the global self-constitution of modernity is coterminous with 'the unleashing of capitalism' (quoted in Lyotard 1992: 17).

In other words, modernity is, properly speaking, the globalization of Europe — triumphantly celebrated by Marx in the first few pages of *The Communist Manifesto* — which constitutes itself globally by claiming that its historicity has 'at last compelled [Man] to face with sober senses his *real* conditions of life and his relations with his kind' (Marx and Engels 1983: 12). Before Marx, Hegel in *The Philosophy of Right* and in *The Philosophy of History,* and before him Kant in his historicopolitical writings, had essentially maintained the same view: that is, European modernity grasps the *real* in contradistinction to the ephemeral non-reality of non-European existence.

In this respect, Marx, a conscious and conscientious inheritor of the intellectual legacies of Kant and Hegel, articulates in his own idiom, his 'materialist conception of history' that which Hegel had already pronounced as the manifestation of *Geist* (mind and/or spirit) and, earlier still, Kant had envisaged and conceptualized as the providential working out of humankind's 'unsocial sociability.' In other words, for all three, no matter how differently they view the historical globalization of Europe, what matters is that European modernity is the *real* in contrast to the *unreality* of human existence in the non-European world. In this regard, Hegel and Marx specify systematically, in their own respective ways, the Idea of European superiority which Kant, long before them, enunciated as the centerpiece of his historicopolitical writings.

As Lyotard has observed, '[m]odernity,' and in its concrete manifestation this term always means empire and colonialism, 'whenever it appears, does not occur without a shattering of belief, without a discovery of the *lack of reality* in reality — a discovery linked to the invention of other realities' (1992: 9). Indeed, in its global invasion and subjugation of the world, European modernity found the *unreality* of myriad non-capitalist social formations, which it promptly shattered and replaced with its own replication of itself. Paradoxically, the profusion of differing and different modes of life was experienced, by invading Europe, as the 'lack of reality in reality': that is, as the *unreality* or vacuousness of and in the *real*.

On the other side of this divide, among the subjugated aboriginal peoples, this European perception of vacuity was experienced as death and destruction – the effective *creation* of vacuity. As Kane puts it:

> For the newcomers did not know only how to fight. They were strange people. If they knew how to kill with effectiveness, they also knew how to cure, with the same art. Where they had brought disorder, they established a new order. They destroyed and they *constructed*.
>
> (Kane 1963: 49 [my emphasis])

The subjugated experienced Europe as the putting into question of their very existence. In their turn, in the words of Chief Kabongo of the Kikuyu, the subjugated put forth their own interrogative to the vacuity 'constructed' by Europe: 'We Elders looked at each other. Was this the end of everything that we had known and worked for?' (1973: 32). Indeed it was!

But how did Europe invent, as Lyotard tells us, 'other realities'? By violently inseminating itself globally, after having properly tilled, turned over and reduced to compost the once lived actualities of the historicity of the non-European world. Or in the words of Kane:

> Those who had shown fight and those who had surrendered, those who had come to terms and those who had been obstinate – they all found themselves, when the day came, checked by census, divided up, classified, labeled, conscripted, administrated.
>
> (1963: 49)

Indeed, as Edward W. Said has pointedly observed:

> Imperialism was the theory, colonialism the practice of changing the uselessly unoccupied territories of the world into useful new versions of the European metropolitan society. Everything in those territories that suggested waste, disorder, uncounted resources, was to be converted into productivity, order, taxable, potentially developed wealth. You get rid of most of the offending human and animal blight – whether because it simply sprawls untidily all over the place or because it roams around unproductively and uncounted – and you confine the rest to reservations, compounds, native homelands, where you can count, tax, use them profitably, and you build a new society on the vacated space. Thus was Europe reconstituted abroad, its 'multiplication in space' successfully projected and managed. The result was a widely varied group of little Europes scattered throughout Asia, Africa, and the Americas, each reflecting the circumstances, the specific instrumentalities of the parent culture, its pioneers, its vanguard settlers. All of them were similar in one major respect – despite the differences, which and that was that their life was carried on with an air of *normality*.
>
> (1980: 78)

In both of the above quotations what needs to be noted is that Europe invents, throughout the globe, 'administrated' replicas of itself and does so in 'an air of normality.' This *normality*, as Said points out, is grounded on an 'idea, which dignifies [and indeed hastens] pure force with arguments drawn from science, morality, ethics, and *a general philosophy*' (1980: 77) [my emphasis].

This Idea, this 'general philosophy,' is, on the one hand, the trite and bland *prejudice* that European existence is, properly speaking, true human existence *per se*. And, as noted earlier, this same Idea or 'general philosophy' is that which Hegel and Marx, among others, inherit from Kant, and specify in their own idiom. This Idea or 'general philosophy' is the metaphysical ground for the 'normality' and legitimacy of European global expansion and conquest: that is, the consolidation of the *real*. Thus, trite prejudice and the highest wisdom, speculative thought, circuitously substantiate each other!

This banal bias and its metaphysical 'pre-text' (Lyotard 1988: 27) or pretension, furthermore, lays a 'heavy burden' ('The White Man's Burden'?) on Europe in its self-assumed global 'civilizing' charade and/or project. For as Father Placide Tempels, a colonizing missionary with an intellectual bent, sternly and gravely reminds his co-colonialists:

> It has been said that our civilizing mission alone can justify our occupation of the lands of uncivilized peoples. All our writings, lectures and broadcasts repeat *ad nauseum* our wish to civilize the African peoples. No doubt there are people who delight to regard as the progress of civilization the amelioration of material conditions, increase of professional skill, improvements in housing, in hygiene and in scholastic instruction. These are, no doubt, useful and even necessary 'values.' But do they constitute 'civilization'? Is not civilization, above all else, progress in *human personality*?
>
> (Tempels 1969: 171–2 [my emphasis])

Indeed, as Rudyard Kipling had poetically noted, Europe's colonizing mission was aimed at properly humanizing the '[h]alf devil and half child' (Kipling 1962: 143) nature of the aboriginal peoples it colonized. This is indeed what Tempels has in mind with his rhetorical question regarding civilization as 'progress in human personality,' for it is this self-righteous attitude on which is grounded the 'normality' of Europe's process of inventing globally 'administrated' replicas of itself.

The 'lack of reality in reality' which Europe finds, and displaces by its self-replication, is the 'immaturity' of the '[h]alf devil and half child' humanity of aboriginal peoples. Now, in this gauging of the 'lack of reality in reality,' European civilization is both the standard and the model by which this deficiency is first recognized and then remedied. Or to be more accurate, it is the Idea or 'general philosophy' of this civilization – or the way that it understands itself – that is the measure of the whole undertaking.

Now, as Rousseau noted in the first chapter of *The Social Contract,* force does not give moral or normative sanction to its effects. Thus for philosophy, which conceives of 'mind' as the Guide of the world, violence and conquest are masks for the rationality of the real. This then is how European philosophy in general participates in and contributes to the invention of 'other realities' – that is, of the replication of Europe as its cultural, material/physical, and historical substratum. And, as we shall soon see, this is precisely what Kant's historicopolitical texts intend to and do accomplish.

This inventiveness is grounded, as Lyotard tells us, in 'the Idea of emancipation' (1992: 24), which is articulated in

> the Christian narrative of the redemption of original sin through love; the *Aufklärer* narrative [i.e. Kant's narrative] of emancipation from ignorance and servitude through knowledge and egalitarianism; the speculative narrative [i.e. Hegel's narrative] of the realization of the universal Idea through the dialectic of the concrete; the Marxist narrative of emancipation from exploitation and alienation through the socialization of work; and the capitalist narrative of emancipation from poverty through technoindustrial development.
>
> (25)

Between 'these narratives there is ground for litigation.' But in spite of this family or familial conflict, 'all of them' are positioned on a singular historical track aimed at 'universal freedom,' and 'the fulfillment of all humanity' (ibid.). In Tempels' words, they are all aimed at 'progress in human personality.'

It is not my concern, in this chapter, to explore the conflicts between these narratives, but rather to underline their foundational similitude: that is, they all metaphysically coagulate around Tempels' phrase, 'progress in human personality.' To this extent these narratives collectively underwrite the colonialist project of global subjugation and expansion. For 'universal freedom' and 'the fulfillment of all humanity' presuppose, on the level of foundational principles (i.e. metaphysics) a singular humanity or the *singularization* of human diversity by being forced on a singular track of historical 'progress' grounded on an emulation and/or mimicry of European historicity (Castoriadis 1991).

In other words, it requires us to look at humanity as a whole, in all of its multiple diversity and amplitude, *not* as it shows itself (i.e. multiple, differing, diverse, disconsonant, dissimilar, etc.), but through the 'mediation or protection of a "pre-text"' (Lyotard 1988: 18), that flattens all difference. This is tangibly and masterfully accomplished by elevating European historicity, the 'pre-text' (i.e. the text that comes *before* the text of humanity, as it shows itself in its multiple heterogeneity) to the status of true human historicity *par excellence*.

The de-structuring critique of this 'pre-text' – the Occidental surrogate for the heterogeneous variance of human historical existence – is then the basic critical-negative task of the contemporary discourse of African philosophy. It is the task of undermining the European-centered conception of humanity on which the Western tradition of philosophy – and much more – is grounded. The way one proceeds in this reading is to allow the texts to present themselves, as much as possible, and to try to grasp them without 'anticipating the meaning' (ibid.), or superimposing on them the accepted reading which they themselves help to make possible.

In reading Kant's speculative historicopolitical texts in this manner, my purpose is to track down the way this 'pre-text' functions in his reading of our shared humanity. This 'pre-text' (i.e. Idea or 'general philosophy') is the shrine at which the great minds of Europe (past and present) prayed and still pray. It is that which serves as the buttress and justification and thus enshrines the 'normality' of the European subjugation of the world. It is the figleaf of European barbarity which makes it possible and acceptable, and without which Europe could not stand to face itself: that is, its history. As Joseph Conrad puts it:

> The conquest of the earth, which mostly means taking it away from those who have a different complexion or slightly flatter noses than ourselves, is not a pretty thing when you look into it too much. What redeems it is the idea only. An idea at the back of it; not a sentimental pretence but an idea; and an unselfish belief in the idea – something you can set up, and bow down before, and offer a sacrifice to.
>
> (Conrad 1972: 7)

Indeed, as Nietzsche has remarked against Hegel, in *The Advantage and Disadvantage of History for Life*, the 'idea' is that in front of which one prostrates oneself.

# PART THREE

# Representation and Resistance

REPRESENTATION AND RESISTANCE are very broad arenas within which much of the drama of colonialist relations and post-colonial examination and subversion of those relations has taken place. As Stephen Greenblatt has noted, texts were the 'invisible bullets' in the arsenal of empire. In both conquest and colonization, texts and textuality played a major part. European texts – anthropologies, histories, fiction – captured the non-European subject within European frameworks which read his or her alterity as *terror* or *lack*. Within the complex relations of colonialism these representations were re-projected to the colonized – through formal education or general colonialist cultural relations – as authoritative pictures of themselves. Concomitantly representations of Europe and Europeans within this textual archive were situated as normative. Such texts – the representations of Europe to itself, and the representation of others to Europe – were not accounts or illustrations of different peoples and societies, but a projection of European fears and desires masquerading as scientific/ 'objective' knowledges. Said's foundational *Orientalism* examined the process by which this discursive formation emerges; while his later *Culture and Imperialism* offers a more specific account of *Anglo*-imperialist literary discourse.

Because representation and resistance are such broad areas of contestation in post-colonial discourse, this is a section that exceeds its particular limits within this Reader. Feminism and its intersections with both colonialism and post-colonialism is necessarily about representation and resistance, as the Feminism section of this Reader demonstrates. And it is through education and in terms of modes of production and consumption that colonialist representations persist and currently circulate in, for instance, popular television shows, film, cartoons, novels and internet games. Consequently the Education and Production and consumption sections, like Universality and Hybridity are also concerned with representation and resistance as are a number of the introductory essays.

Post-colonial textual resistance to colonialist edicts and representations has taken many forms, from the nineteenth-century parody of Macaulay's 1835 'Minute' (by an unknown Bengali writer) to the widespread contemporary practice of counter-canonical literary responses discussed by Helen Tiffin. Shohat and Stam demonstrate similar resistance patterns in post-colonial film, but they also point to the many economic and technological difficulties for film making in many post-colonial societies.

Theorizing the nature and practice of post-colonial resistance more generally has become central to post-colonial debates. In particular post-structuralism's diverse intersections with post-colonialism has foregrounded questions not only of theory versus political commitment (books and barricades) but of agency itself, questions already raised by Bhabha's introductory essay in this volume. Stephen Slemon's article succinctly summarizes various kinds of literary resistances that have been theorized within post-colonialism and considers the crucial place of so-called 'Second World' writing in such theorizations. In so doing, Slemon and Sara Suleri (and Jenny Sharpe 1989) problematize earlier notions of post-colonial resistance (like those of Barbara Harlow or Tim Brennan) which depend upon a system of irreducible binary oppositions. Instead, they move away from a resistance theorizing which, in Sara Suleri's terms, 'precludes the concept of exchange by granting the idea of power a greater literalism than it deserved' towards a notion of 'cultural exchange.' In so doing, Suleri, like Elleke Boehmer in 'Networks of Resistance', is paving the way for more complex analyses of colonial relations and thus of post-colonial resistances. If earlier theorizations of resistance presupposed a foundation of undislocatable binaries – center/margin, self/other, colonizer/colonized – the general trajectory of the rather different projects of Bhabha, Slemon, Suleri and Appiah has been towards something that has always been implicit (even when not explicit) in both colonialist and post-colonial literary relations, and that is what Suleri calls the 'peculiar intimacy' of colonizer and colonized. We tend to think of resistance as developing locally; of being specific to nationalist or proto-nationalist groups. But as Elleke Boehmer argues, imperial infrastructures (much like the technologies or 'globalization' today) consisted in a web of cross-colonial administrative connections, structures and technologies, which, ironically, often facilitated cross-colonial resistances. Theorizing such complex contacts without losing sight of the persisting and historic *inequalities* within these relations and structures, is a major focus of contemporary post-colonial history and theory.

## Chapter 16

# EDWARD W. SAID

## RESISTANCE, OPPOSITION AND REPRESENTATION

From *Culture and Imperialism* New York: Alfred A. Knopf, 1993.

**A**N IMMENSE WAVE OF anti-colonial and ultimately anti-imperial activity, thought, and revision has overtaken the massive edifice of Western empire, challenging it, to use Gramsci's vivid metaphor, in a mutual siege. For the first time Westerners have been required to confront themselves not simply as the Raj but as representatives of a culture and even of races accused of crimes – crimes of violence, crimes of suppression, crimes of conscience. 'Today,' says Fanon in *The Wretched of the Earth*, 'the Third World . . . faces Europe like a colossal mass whose aim should be to try to resolve the problems to which Europe has not been able to find the answers' (Fanon [1965b]: 6). Such accusations had of course been made before, even by such intrepid Europeans as Samuel Johnson and W. S. Blunt. . . . The new situation was a sustained confrontation of, and systematic resistance to, the Empire as West. Long-simmering resentments against the white man from the Pacific to the Atlantic sprang into fully-fledged independence movements. Pan-African and Pan-Asian militants emerged who could not be stopped. . . .

The slow and often bitterly disputed recovery of geographical territory – which is at the heart of decolonization is preceded – as empire had been – by the charting of cultural territory. After the period of 'primary resistance,' literally fighting against outside intrusion, there comes the period of secondary, that is, ideological resistance, when efforts are made to reconstitute a 'shattered community, to save or restore the sense and fact of community against all the pressures of the colonial system' as Basil Davidson puts it (Davidson 1978: 155). This in turn makes possible the establishment of new and independent states. It is important to note that we are not mainly talking here about Utopian regions – idyllic meadows, so to speak – discovered in their private past by the intellectuals, poets, prophets, leaders, and historians of resistance. Davidson speaks of the 'otherworldly' promises made by some in their early phase, for example, rejecting Christianity and the wearing of Western clothes. But all of them respond to the humiliations of colonialism, and lead to 'the principal teaching of nationalism: the need to find the ideological basis for a wider unity than any known before' . . . (Davidson 1978: 156).

That is the partial tragedy of resistance, that it must to a certain degree work to recover forms already established or at least influenced or infiltrated by the culture of empire. This is another instance of what I have called overlapping territories: the struggle over Africa in the twentieth century, for example, is over territories designed and redesigned by explorers from Europe for generations, a process memorably and painstakingly conveyed in Philip Curtin's

*The Image of Africa* (1964). Just as the Europeans saw Africa polemically as a blank place when they took it, or assumed its supinely yielding availability when they plotted to partition it at the 1884–5 Berlin Congress, decolonizing Africans found it necessary to reimagine an Africa stripped of its imperial past.

Take as a specific instance of this battle over projections and ideological images the so-called quest or voyage motif, which appears in much European literature and especially literature about the non-European world. In all the great explorers' narratives of the late Renaissance (Daniel Defert has aptly called them the collection of the world [*la collecte du monde*] Defert 1982) and those of the nineteenth-century explorers and ethnographers, not to mention Conrad's voyage up the Congo, there is the topos of the voyage south as Mary Louise Pratt has called it, referring to Gide and Camus (1992), in which the motif of control and authority has 'sounded uninterruptedly.' For the native who begins to see and hear that persisting note, it sounds 'the note of crisis, of banishment, banishment from the heart, banishment from home.' This is how Stephen Dedalus memorably states it in the Library episode of *Ulysses* (Joyce 1922: 212) the decolonizing native writer – such as Joyce, the Irish writer colonized by the British – re-experiences the quest voyage motif from which he had been banished by means of the same trope carried over from the imperial into the new culture and adopted, reused, relived.

*The River Between*, by James Ngũgĩ (later Ngũgĩ wa Thiongo), redoes *Heart of Darkness* by inducing life into Conrad's river on the very first page. 'The river was called Honia, which meant cure, or bring-back-to-life. Honia river never dried: it seemed to possess a strong will to live, scorning droughts and weather changes. And it went on in the very same way, never hurrying, never hesitating. People saw this and were happy' (Ngũgĩ 1965: 1). Conrad's images of river, exploration, and mysterious setting are never far from our awareness as we read, yet they are quite differently weighted, differently – even jarringly – experienced in a deliberately understated, self-consciously unidiomatic and austere language. In Ngũgĩ the white man recedes in importance – he is compressed into a single missionary figure emblematically called Livingstone – although his influence is felt in the divisions that separate the villages, the riverbanks, and the people from one another. In the internal conflict ravaging Waiyaki's life, Ngũgĩ powerfully conveys the unresolved tensions that will continue well after the novel ends and that the novel makes no effort to contain. A new pattern, suppressed in *Heart of Darkness*, appears, out of which Ngũgĩ generates a new mythos, whose tenuous course and final obscurity suggest a return to an African Africa.

And in Tayeb Salih's *Season of Migration to the North*, Conrad's river is now the Nile, whose waters rejuvenate its peoples, and Conrad's first-person British narrative style and European protagonists are in a sense reversed, first through the use of Arabic; second in that Salih's novel concerns the northward voyage of a Sudanese to Europe; and third, because the narrator speaks from a Sudanese village. . . .

The post-imperial writers of the Third World therefore bear their past within them – as scars of humiliating wounds, as instigation for different practices, as potentially revised visions of the past tending towards a post-colonial future, as urgently reinterpretable and redeployable experiences, in which the formerly silent native speaks and acts on territory reclaimed as part of a general movement of resistance, from the colonist.

Another motif emerges in the culture of resistance. Consider the stunning cultural effort to claim a restored and invigorated authority over a region in the many modern Latin American and Caribbean versions of Shakespeare's *The Tempest*. This fable is one of several that stand guard over the imagination of the New World; other stories are the adventures and discoveries of Columbus, Robinson Crusoe, John Smith, and Pocohontas, and the adventures of Inkle and Yariko. (A brilliant study, *Colonial Encounters* by Peter Hulme, surveys them all in some detail (Hulme 1986).) It is a measure of how embattled this matter of 'inaugural figures' has become that it is now virtually impossible to say anything simple about any of them. To call this reinterpretative zeal merely simpleminded, vindictive, or assaultive is wrong, I think. In a totally new way in Western culture, the interventions of non European artists and scholars cannot be dismissed or silenced, and these interventions are not only an integral part of a political movement but, in many ways, the movement's successfully guiding imagination, intellectual and figurative energy re-seeing and rethinking the terrain common to whites and non-whites. For natives to want to lay claim to that terrain is, for many Westerners, an intolerable effrontery, for them actually to repossess it unthinkable. . . .

Three great topics emerge in decolonizing cultural resistance, separated for analytical purposes, but all related. One, of course, is the insistence on the right to see the community's history whole, coherently, integrally. Restore the imprisoned nation to itself. (Benedict Anderson connects this in Europe to 'print-capitalism,' which 'gave a new fixity to language' and 'created unified fields of exchange and communications below Latin and above the spoken vernaculars' (1983: 47).) The concept of the national language is central, but without the practice of a national culture – from slogans to pamphlets and newspapers, from folk tales and heroes to epic poetry, novels, and drama – the language is inert; national culture organizes and sustains communal memory, as when early defeats in African resistance stories are resumed ('they took our weapons in 1903; now we are taking them back'); it reinhabits the landscape using restored ways of life, heroes, heroines, and exploits; it formulates expressions and emotions of pride as well as defiance, which in turn form the backbone of the principal national independence parties. Local slave narratives, spiritual autobiographies, prison memoirs form a counterpoint to the Western powers' monumental histories, official discourses and panoptic quasi-scientific viewpoint. In Egypt, for example, the historical novels of Girgi Zaydan bring together for the first time a specific ally Arab narrative (rather the way Walter Scott did a century before). In Spanish America, according to Anderson, creole communities 'produced creoles who consciously redefined these [mixed] populations as fellow nationals' (64). Both Anderson and Hannah Arendt note the widespread global movement to 'achieve solidarities on an essentially imagined basis' (Anderson 1983: 74).

Second is the idea that resistance, far from being merely a reaction to imperialism, is an alternative way of conceiving human history. It is particularly important to see how much this alternative reconception is based on breaking down the barriers between cultures. Certainly, as the title of a fascinating book has it, writing back to the metropolitan cultures, disrupting the European narratives of the Orient and Africa, replacing them with either a more playful or a more powerful new narrative style is a major component in the process (Ashcroft *et al.* 1989). Salman Rushdie's novel *Midnight's Children* is a brilliant work based on the liberating imagination of independence itself, with all its anomalies and contradictions working themselves out. The conscious effort to enter into the discourse of Europe and the West, to mix with it, transform it, to make it acknowledge marginalized or suppressed or forgotten histories is of particular interest in Rushdie's work, and in an earlier generation of resistance writing. This kind of work was carried out by dozens of scholars, critics, and intellectuals in the peripheral world; I call this effort *the voyage in*.

Third is a noticeable pull away from separatist nationalism towards a more integrative view of human community and human liberation. I want to be very clear about this. No one needs to be reminded that throughout the imperial world during the decolonizing period, protest, resistance, and independence movements were fuelled by one or another nationalism. Debates today about Third World nationalism have been increasing in volume and interest, not least because to many scholars and observers in the West, this reappearance of nationalism revived several anachronistic attitudes; Elie Kedourie, for example, considers non-Western nationalism essentially condemnable, a negative reaction to a demonstrated cultural and social inferiority, an imitation of 'Western' political behaviour that brought little that was good; others, like Eric Hobsbawm (1982) and Ernest Gellner (1983), consider nationalism as a form of political behaviour that has been gradually superseded by new transnational realities of modern economies, electronic communications, and superpower military projection. In all these views, I believe, there is a marked (and, in my opinion, ahistorical? discomfort with non-Western societies acquiring national independence, which is believed to be 'foreign' to their ethos. Hence the repeated insistence on the Western provenance of nationalist philosophies, that are therefore ill-suited to, and likely to be abused by Arabs, Zulus, Indonesians, Irish, or Jamaicans.

This, I think, is a criticism of newly independent peoples that carries with it a broadly cultural opposition (from the Left as well as from the Right) to the proposition that the formerly subject peoples are entitled to the same kind of nationalism as, say, the more developed, hence more deserving, Germans or Italians. A confused and limiting notion of priority allows that only the original proponents of an idea can understand and use it. But the history of all cultures is the history of cultural borrowings. Cultures are not impermeable; just as Western science borrowed from Arabs, they had borrowed from India and Greece. Culture is never just a matter of Ownership, of borrowing and lending with absolute debtors and creditors, but rather of appropriations, common experiences, and interdependencies of all kinds among different cultures. This is a universal norm. Who has yet determined how much the domination of others contributed to the enormous wealth of the English and French states? . . .

No one today is purely *one* thing. Labels like Indian, or woman, or Muslim, or American are no more than starting-points, which if followed into actual experience for only a moment are quickly left behind. Imperialism consolidated the mixture of cultures and identities on a global scale. But its worst and most paradoxical gift was to allow people to believe that they were only, mainly, exclusively, white, or black, or Western, or Oriental. Yet just as human beings make their own history, they also make their cultures and ethnic identities. No one can deny the persisting continuities of long traditions, sustained habitations, national languages, and cultural geographies, but there seems no reason except reason and prejudice to keep insisting on their separation and distinctiveness, as if that was all human life was about. Survival in fact is about the connections between things; in Eliot's phrase, reality cannot be deprived of the 'other echoes [that] inhabit the garden.' It is more rewarding – and more difficult – to think concretely and sympathetically, contrapuntally, about others than only about 'us.' But this also means not trying to rule others, not trying to classify them or put them in hierarchies, above all, not constantly reiterating how 'our' culture or country is number one (or *not* number one for that matter). For the intellectual there is quite enough of value to do without *that*.

## Chapter 17

# HELEN TIFFIN

# POST-COLONIAL LITERATURES AND COUNTER-DISCOURSE

From 'Post-colonial Literatures and Counter-discourse' *Kunapipi* 9(3), 1987.

**A**S GEORGE LAMMING ONCE REMARKED, over three quarters of the contemporary world has been directly and profoundly affected by imperialism and colonialism. . . . Processes of artistic and literary decolonisation have involved a radical dis/mantling of European codes and a post-colonial subversion and appropriation of the dominant European discourses. This has frequently been accompanied by the demand for an entirely new or wholly recovered 'reality', free of all colonial taint. Given the nature of the relationship between coloniser and colonised, with its pandemic brutalities and its cultural denigration, such a demand is desirable and inevitable. But as the contradictions inherent in a project such as Chinweizu, Jemie and Madubuike's *The Decolonization of African Literature* demonstrate (Chinweizu *et al*. 1985), such pre-colonial cultural purity can never be fully recovered.

Post-colonial cultures are inevitably hybridised, involving a dialectical relationship between European ontology and epistemology and the impulse to create or recreate independent local identity. Decolonisation is process, not arrival; it invokes an ongoing dialectic between hegemonic centrist systems and peripheral subversion of them; between European or British discourses and their post-colonial dis/mantling. Since it is not possible to create or recreate national or regional formations wholly independent of their historical implication in the European colonial enterprise, it has been the project of post-colonial writing to interrogate European discourses and discursive strategies from a privileged position within (and between) two worlds; to investigate the means by which Europe imposed and maintained its codes in the colonial domination of so much of the rest of the world.

Thus the rereading and rewriting of the European historical and fictional record are vital and inescapable tasks. These subversive manoeuvres, rather than the construction or reconstruction of the essentially national or regional, are what is characteristic of post-colonial texts, as the subversive is characteristic of post-colonial discourse in general. Post-colonial literatures/cultures are thus constituted in counter-discursive rather than homologous practices, and they offer 'fields' (Lee 1977: 32–3) of counter-discursive strategies to the dominant discourse. The operation of post-colonial counter-discourse (Terdiman 1985)[1] is dynamic, not static: it does not seek to subvert the dominant with a view to taking its place, but, in Wilson Harris's formulation, to evolve textual strategies which continually 'consume' their 'own biases' (Harris 1985: 127) at the same time as they expose and erode those of the dominant discourse. . . .

In challenging the notion of literary universality (or the European appropriation of post-colonial practice and theory as post-modern or post-structuralist) post-colonial writers and critics engage in counter-discourse. But separate models of 'Commonwealth Literature' or 'New Writing in English' which implicitly or explicitly invoke notions of continuation of, or descent from, a 'mainstream' British literature, consciously or unconsciously reinvoke those very hegemonic assumptions against which the post-colonial text has, from its inception, been directed. Models which stress the shared language and shared circumstances of colonialism (recognising vast differences in the expression of British imperialism from place to place) allow for counter-discursive strategies, but unless their stress is on counter-discursive fields of activity, such models run the risk of becoming colonisers in their turn. African critics and writers in particular have rejected these models for their apparently neo-assimilative bases, and opted instead for the national or the pan-African. But if the impulse behind much post-colonial literature is seen to be broadly counter-discursive, and it is recognised that the resulting strategies may take many forms in different cultures, I think we have a more satisfactory model than national, racial, or cultural groupings based on marginalisation can offer, and one which perhaps avoids some of the pitfalls of earlier collective models or paradigms. Moreover, such a model can account for the ambiguous position of say, white Australians, who, though still colonised by Europe and European ideas, are themselves the continuing colonisers of the original inhabitants. In this model, all post-invasion Aboriginal writing and orature might be regarded as counter-discursive to a dominant 'Australian' discourse and beyond that again to its European progenitor. It is this model I wish to take up later in considering J. M. Coetzee's *Foe* which explores the problem of white South African settler literature in relation to the continuing oppression by whites of the black majority. . . .

It is possible to formulate at least two (not necessarily mutually exclusive) models for future post-colonial studies. In the first, the post-coloniality of a text would be argued to reside in its discursive features, in the second, in its determining relations with its material situation. The danger of the first lies in post-coloniality's becoming a set of unsituated reading practices; the danger in the second lies in the reintroduction of a covert form of essentialism. In an attempt to avoid these potential pitfalls I want to try to combine the two as overarching models in the reading of two texts by stressing counter-discursive strategies which offer a more general post-colonial reading practice or practices. These practices, though, are politic-ally situated; sites of production and consumption that are inextricably bound up with the production of meaning. The site of communication is of paramount importance in post-colonial writing, and remains its most important defining boundary. . . .

Within the broad field of the counter-discursive many sub-groupings are possible and are already being investigated. These include 'magic realism' as post-colonial discourse (see Dash 1974 and Slemon 1987), and the re/placing of carnivalesque European genres like the picaresque in post-colonial contexts, where they are carried to a higher subversive power. Stephen Slemon has demonstrated the potential of allegory as a privileged site of anti-colonial or post-colonial discourse (Slemon 1986, 1988b).

But the particular counter-discursive post-colonial field with which I want to engage here is what I'll call canonical counter-discourse. This strategy is perhaps most familiar through texts like Jean Rhys's *Wide Sargasso Sea*, and it is one in which a post-colonial writer takes up a character or characters, or the basic assumptions of a British canonical text, and unveils those assumptions, subverting the text for post-colonial purposes. An important point needs to be made here about the discursive functions of textuality itself in post-colonial worlds. European texts captured those worlds, 'reading' their alterity assimilatively in terms of their own cognitive codes. Explorers' journals, drama, fiction, historical accounts, 'mapping' enabled conquest and colonisation and the capture and/or vilification of alterity. But often the very texts which facilitated such material and psychic capture were those which the imposed

European education systems foisted on the colonised as the 'great' literature which dealt with 'universals'; ones whose culturally specific imperial terms were to be accepted as axiomatic at the colonial margins. Achebe has noted the ironies of Conrad's *Heart of Darkness* being taught in colonial African universities.

Understandably, then, it has become the project of post-colonial literatures to investigate the European textual capture and containment of colonial and post-colonial space and to intervene in that originary and continuing containment. In his study of nineteenth century France, Richard Terdiman saw what he calls 'textual revolution' as partly conditional on the 'blockage of energy directed to structural change of the social formation' (Terdiman 1985: 80). But he goes on to note that even so, 'Literary revolution is not revolution by homology, but by *intended function.*' Literary revolution in post-colonial worlds has been an intrinsic component of social 'disidentification' (Pêcheux 1975: 158)[2] from the outset. Achebe's essay, 'The Novelist As Teacher' (Achebe 1975: 167–74) stresses the crucial function of texts in post-colonial social formations and their primacy in effecting revolution and restitution, priorities which are not surprising given the role of the text in the European capture and colonisation of Africa. Post-colonial counter-discursive strategies involve a mapping of the dominant discourse, a reading and exposing of its underlying assumptions, and the dis/mantling of these assumptions from the cross-cultural standpoint of the imperially subjectified 'local'. *Wide Sargasso Sea* directly contests British sovreignty – over persons, place, culture, language. It reinvests its own hybridised world with a provisionally normative perspective, but one which is deliberately constructed as provisional since the novel is at pains to demonstrate the subjective nature of point of view and hence the cultural construction of meaning.

Just as Jean Rhys writes back to Charlotte Brontë's *Jane Eyre* in *Wide Sargasso Sea*, so Samuel Selvon in *Moses Ascending* and J. M. Coetzee in *Foe* (and indeed throughout his works) write back to Daniel Defoe's *Robinson Crusoe*. Neither writer is simply 'writing back' to an English canonical text, but to the whole of the discursive field within which such a text operated and continues to operate in post-colonial worlds. Like William Shakespeare's *The Tempest*, *Robinson Crusoe* was part of the process of 'fixing' relations between Europe and its 'others', of establishing patterns of reading alterity at the same time as it inscribed the 'fixity' of that alterity, naturalising difference within its own cognitive codes. But the function of such a canonical text at the colonial periphery also becomes an important part of material imperial practice, in that, through educational and critical institutions, it continually displays and repeats for the colonised subject, the original capture of his/her alterity and the processes of its annihilation, marginalisation, or naturalisation as if this were axiomatic, culturally ungrounded, 'universal', natural.

Selvon and Coetzee take up the complex discursive field surrounding Robinson Crusoe and unlock these apparent closures.

## Notes

1    Richard Terdiman, *Discourse/Counter-discourse: The Theory and Practice of Symbolic Resistance in Nineteenth-Century France* (Ithaca and London: Cornell University Press, 1985). Terdiman theorises the potential and limitations of counter-discursive literary revolution within a dominant discourse noting that counter-discourses have the power to *situate*: to relativise the authority and stability of a dominant system of utterances which cannot even countenance their existence (pp. 15–16). But Terdiman regards counter-discourses as ultimately unable to effect genuine revolution, since they are condemned to remain marginal to the dominant discourse. The post-colonial situation is a rather different one, however, from that which provides Terdiman with his model.

2    Michel Pêcheux uses the term 'disidentification' to denote a transformation and displacement of the subject position interpellated by a dominant ideology.

## Chapter 18

## STEPHEN SLEMON

## UNSETTLING THE EMPIRE
Resistance theory for the Second World

From 'Unsettling the Empire: Resistance Theory for the Second World'
*World Literature Written in English* 30(2), 1990.

WHAT I WANT TO DO IN THIS PAPER IS address two separate debates in critical theory, and then attempt to yoke them together into an argument for maintaining within a discourse of post-colonialism certain textual and critical practices which inhabit ex-colonial settler cultures and their literatures. The textual gestures I want to preserve for post-colonial theory and practice are various and dispersed, but the territory I want to reclaim for post-colonial pedagogy and research — and reclaim *not* as a unified and indivisible area but rather as a groundwork for certain modes of anti-colonial work — is that neither/nor territory of white settler-colonial writing which Alan Lawson has called the 'Second World'.

The first debate concerns the *field* of the 'post-colonial'. . . .

The second debate I want to address concerns the nature of literary *resistance* itself. Is literary resistance something that simply issues forth, through narrative, against a clearly definable set of power relations? Is it something actually *there* in the text, or is it produced and reproduced in and through communities of readers and through the mediating structures of their own culturally specific histories? Do literary resistances escape the constitutive purchase of genre, and trope, and figure, and mode, which operate elsewhere as a contract between text and reader and thus a set of centralizing codes, or are literary resistances in fact necessarily *embedded* in the representational technologies of those literary and social 'texts' whose structures and whose referential codes they seek to oppose?

These questions sound like definitional problems, but I think in fact they are crucial ones for a critical industry which at the moment seems to find these two central terms — 'post-colonial' and 'resistance' — positively shimmering as objects of desire and self-privilege, and so easily appropriated to competing, and in fact hostile, modes of critical and literary practice. Arun Mukherjee makes this point with great eloquence (Mukherjee 1990: 1–9), asking what specificity, what residual grounding, remains with the term 'post-colonial' when it is applied indiscriminately to both Second- and Third-World literary texts. The term 'resistance' recently found itself at the centre of a similar controversy, when it was discovered how very thoroughly a *failure* in resistance characterized some of the earlier political writing of the great theorist of *textual* resistance, Paul de Man. Both terms thus find themselves at the centre of a quarrel over the kinds of critical taxonomies that will be seen to perform legitimate work in articulating the relation between literary texts and the political world; and to say this is to recognize that critical taxonomies, like literary canons, issue forth from cultural institutions which continue to police what voices will be heard, which *kinds* of (textual) intervention will be made recognizable and/or classifiable, and what *authentic* forms of post-colonial textual resistance are going to look like. These debates are thus institutional: grounded in university curricula, and *about* pedagogical strategies. They are also about the question of authenticity itself: how a text emerges from a cultural grounding and speaks to a reading community, and how textual ambiguity or ambivalence proves pedagogically awkward when an apparatus called 'English studies' recuperates various writing practices holistically as 'literatures', and then deploys them wholesale towards a discourse of inclusivity and coverage. The first debate – the question of the 'post-colonial' – is grounded in the overlapping of three competing research or critical fields, each of which carries a specific cultural location and history. In the first of these fields, the term 'post-colonial' is an outgrowth of what formerly were 'Commonwealth' literary studies – a study which came into being *after* 'English' studies had been liberalized to include 'American' and then an immediate national or regional literature (Australian, Canadian, West Indian), and as a way of mobilizing the concept of national or geographical *difference* within what remains a unitary idea of 'English'. The second of these critical fields, in contrast, employs the term 'post-colonial' in considering the valency of subjectivity specifically within Third- and Fourth-World cultures, and within black, and ethnic, and First-Nation constituencies dispersed within First-World terrain. The institutionalizing of these two critical fields has made possible the emergence of a third field of study, however, where nation-based examinations of a variable literary Commonwealth, or a variable literary Third World, give way to specific analyses of the discourse of colonialism (and neo-colonialism), and where studies in cultural representativeness and literary mimeticism give way to the project of identifying the kinds of anti-colonialist resistance that can take place in literary writing. . . .

'Post-colonial' studies in 'English' now finds itself at a shifting moment, where three very different critical projects collide with one another on the space of a single signifier – and what will probably be a single course offering within an English studies programme. Not surprisingly, this situation has produced some remarkable confusions, and they underpin the present debate over the specificity of the 'post-colonial' in the areas of literary and critical practice.

The confusion which concerns me here is the way in which the *project* of the third 'post-colonial' critical field – that is, of identifying the scope and nature of anti-colonialist resistance in writing – has been mistaken for the project of the second critical field, which concerns itself with articulating the literary nature of Third- and Fourth-World cultural groups. For whereas the first and second of these post-colonial critical fields work with whole nations or cultures as their basic units, and tend to seek out the defining characteristics under which *all* writing in that field can be subsumed, the third critical field is concerned with identifying a social force, colonialism, and with the attempt to understand the resistances to that force, *wherever* they lie. Colonialism, obviously, is an enormously problematical category: it is by definition trans-historical and unspecific, and it is used in relation to very different kinds of cultural oppression and economic control. But like the term 'patriarchy', which shares similar problems in definition, the concept of colonialism, to this third critical field, remains crucial to a critique of past and present power relations in world affairs, and thus to a specifically *post*-colonial critical practice which attempts to understand the relation of literary writing to power and its contestations.

This mistaking of a pro-active, anti-colonialist critical project with nation-based studies in Third- and Fourth-World literary writing comes about for good reason – for it has been, and always will be, the case that the most important forms of resistance to any form of social power will be produced from within the communities that are most immediately and visibly subordinated by that power structure. But when the idea of anti-colonial resistance becomes *synonymous* with Third- and Fourth-World literary writing, two forms of displacement happen. First, *all* literary writing which emerges from these cultural locations will be understood as carrying a radical and contestatory content – and this gives away the rather important point that subjected peoples are sometimes capable of producing reactionary literary documents. And secondly, the idea will be discarded that important anti-colonialist literary writing can take place *outside* the ambit of Third- and Fourth-World literary writing – and this in effect excises the study of anti-colonialist Second-World literary activity from the larger study of anti-colonialist literary practice. . . .

This conflating of the projects of the second and third post-colonial critical fields, and the consequent jettisoning of Second-World literary writing from the domain of the post-colonial, remains – in the Bloomian sense – a 'misreading', and one which seems to be setting in train a concept of the 'post-colonial' which is remarkably purist and absolutist in tenor. . . .

The foundational principle for this particular approach to the field of post-colonial criticism is at heart a simple binarism: the binarism of Europe and its Others, of colonizer and colonized, of the West and the Rest, of the vocal and the silent. It is also a centre/periphery model with roots in world-systems theory – and as so often happens with simple binary systems, this concept of the post-colonial has a marked tendency to blur when it tries to focus upon ambiguously placed or ambivalent material. In what seems to be emerging as the dominant focus of post-colonial literary criticism now – especially for literary criticism coming out of universities in the United States – this blurring is everywhere in evidence in relation to what world systems theory calls the field of 'semi-periphery', and what follows behind it is a radical foreclosing by post-colonial criticism on settler/colonial writing: the radical ambivalence of colonialism's middle ground. . . .

At any rate, the new binaristic absolutism which seems to come in the wake of First-World accommodation to the fact of post-colonial literary and cultural criticism seems to be working in several ways to drive that trans-national region of ex-colonial settler cultures away from the field of post-colonial literary representation. The Second World of writing within the ambit of colonialism is in danger of disappearing: because it is not sufficiently pure in its anti-colonialism, because it does not offer up an experiential grounding in a common 'Third World' aesthetics, because its modalities of *post*-coloniality are too ambivalent, too occasional and uncommon, for inclusion within the field. This debate over the scope and nature of the 'post-colonial', I now want to argue, has enormous investments in the second debate I want to discuss in this paper, for in fact the idea of both literary and political *resistance* to colonialist power is the hidden term, the foundational concept, upon which *all* these distinctions in the modality of the 'post-colonial' actually rest. . . .

The first concept of resistance is most clearly put forward by Selwyn Cudjoe in his *Resistance and Caribbean Literature* and by Barbara Harlow in her book, *Resistance Literature*. For Cudjoe and Harlow, resistance is an act, or a set of acts, that is designed to rid a people of its oppressors, and it so thoroughly infuses the experience of living under oppression that it becomes an almost autonomous aesthetic principle. *Literary* resistance, under these conditions, can be seen as a form of contractual understanding between text and reader, one which is embedded in an experiential dimension and buttressed by a political and cultural aesthetic at work in the culture. And 'resistance literature', in this definition, can thus be seen as that category of literary writing which emerges as an integral part of an organized struggle or resistance for national liberation.

This argument for literary 'resistance' is an important one to hold on to – but it is also a strangely untheorized position, for it fails to address three major areas of critical concern. The first is a political concern: namely, that centre/periphery notions of resistance can actually work to *reinscribe* centre/periphery relations and can 'serve an institutional function of securing the dominant narratives' (Sharpe 1989: 139). The second problem with this argument is that it assumes that literary resistance is simply somehow *there* in the literary text as a structure of intentionality, and *there* in the social text as a communicative gesture of pure availability. Post-Lacanian and post-Althusserian theories of the constructedness of subjectivity, however, would contest such easy access to representational purity, and would argue instead that resistance is grounded in the *multiple* and *contradictory* structures of ideological interpellation or subject-formation – which would call down the notion that resistance can *ever* be 'purely' intended or 'purely' expressed in representational or communicative models. The third problem with this argument is that it has to set aside the very persuasive theory of power which Foucault puts forward in his *The Archaeology of Knowledge*: the theory that power *itself* inscribes its resistances and so, in the process, seeks to contain them. It is this third objection, especially, which has energized the post-structuralist project of theorizing literary resistance – and in order to clarify what is going on in that theatre of critical activity I want to focus especially on Jenny Sharpe's wonderful article in *Modern Fiction Studies* entitled 'Figures of Colonial Resistance'.

Sharpe's article involves a reconsideration of the work of theorists such as Gayatri Spivak, Homi Bhabha, Abdul JanMohamed, and Benita Parry, each of whom has worked to correct the critical 'tendency to presume the transparency' of literary resistance in colonial and post-colonial writing (138), and who collectively have worked to examine the ways in which resistance in writing must go beyond the mere 'questioning' of colonialist authority. There are important differences in how all of these theorists define literary resistance, but the two key points Sharpe draws out are, first, that you can never *easily* locate the sites of anti-colonial resistance – since resistance itself is always in some measure an 'effect of the contradictory representation of colonial authority' (145) and never simply a 'reversal' of power – and secondly, that resistance itself is therefore never *purely* resistance, never *simply* there in the text or the interpretive community, but is always *necessarily* complicit in the apparatus it seeks to transgress. . . .

Sharpe's argument, that is, underscores the way in which literary resistance is necessarily in a place of ambivalence: between systems, between discursive worlds, implicit and complicit in both of them. And from this recognition comes the very startling but inevitable claim – made most spectacularly by Tim Brennan in his book on *Salman Rushdie and the Third World* – that the Third World resistance writer, the Third World resistance text, is necessarily self-produced as a doubly-emplaced and mediated figure – Brennan's term is 'Third-World Cosmopolitan' – between the First and the Third Worlds, and *within* the ambit of a First-World politics.

There is a contradiction within the dominant trajectory of First-World post-colonial critical theory here – for that same theory which argues persuasively for the necessary *ambivalence* of post-colonial literary resistance, and which works to emplace that resistance squarely *between* First- and Third-World structures of representation, *also* wants to assign 'Second World' or ex-colonial settler literatures unproblematically to the category of the literature of empire, the literature of the First World, precisely *because* of its ambivalent position within the First-World/Third-World, colonizer/colonized binary. Logically, however, it would seem that the argument being made by Spivak, Bhabha, Sharpe, and others about the ambivalence of literary and other resistances – the argument that resistance texts are necessarily double, necessarily mediated, in their social location – is in fact nothing less than an argument *for* the emplace-ment of 'Second World' literary texts within the field of the 'post-colonial': for if there *is* only a space for a *pure* Third- and Fourth-World resistance outside the First-World hege-mony, then *either* you have to return to the baldly untheorized notion which informs the first position in the debate over literary resistance, *or* you have to admit that at least as far as writing is concerned, the 'field' of the genuinely *post*-colonial can never *actually* exist.

It is for this reason, I think, and not because of some vestigial nostalgia for an empire upon which the sun will never set, that many critics and theorists have argued long and hard for the preservation of white Australian, New Zealander, southern African, and Canadian literatures within the field of comparative 'post-colonial' literary studies. . . .

The 'Second World' – like the third of the three 'post-colonial' critical fields I have been discussing – is at root a *reading position*, and one which is and often has been taken up in settler and ex-colonial literature and criticism. The 'Second World', that is, like 'post-colonial criticism' itself, is a critical manoeuvre, a reading and writing action; and embedded within it is a theory of communicative action akin in some ways to Clifford Geertz's thesis about 'intermediary knowledge', or Gadamer's theory of an interpretive 'fusion of horizons'. 'The inherent awareness of both "there" and "here," and the cultural ambiguity of these terms', writes Lawson, 'are not so much the boundaries of its cultural matrix, nor tensions to be resolved, but a space *within* which [the Second-World, post-colonial] literary text may move while speaking' (Lawson 1986). Lawson's definition of literary representation in the discursive 'Second World' thus articulates a figure for what many First-World critical theorists would correctly define as the limits and the condition of *post-colonial* forms of literary resistance. The irony is that many of those same First-World critics would define that 'post-colonial' as exclusively the domain of the Third and Fourth Worlds.

But what perhaps marks a *genuine* difference in the contestatory activity of Second- and Third-World post-colonial writing, I now want to argue, is that the *illusion* of a stable self/other, here/there binary division has *never* been available to Second-World writers, and that as a result the sites of figural contestation between oppressor and oppressed, colonizer and colonized, have been taken *inward* and *internalized* in Second-World post-colonial textual practice. By this I mean that the *ambivalence* of literary resistance itself is the 'always already' condition of Second-World settler and post-colonial literary writing, for in the white literatures of Australia, or New Zealand, or Canada, or southern Africa, anti-colonialist resistance has *never* been directed at an object or a discursive structure which can be seen purely external to the self. The Second-World writer, the Second-World text, that is, have always been complicit in colonialism's territorial appropriation of land, and voice, and agency, and this has been their inescapable condition even at those moments when they have promulgated their most strident and most spectacular figures of post-colonial resistance. In the Second World, anti-colonialist resistances in literature must necessarily *cut across the individual subject*, and as they do so they also, necessarily, contribute towards that theoretically rigorous understanding of textual resistance which post-colonial *critical* theory is only now learning how to recognize. This ambivalence of emplacement is the *condition* of their possibility; it has been since the beginning; and it is therefore scarcely surprising that the ambivalent, the mediated, the conditional, and the radically *compromised* literatures of this undefinable Second World have an enormous amount yet to tell to 'theory' about the nature of literary resistance.

This *internalization* of the object of resistance in Second-World literatures, this internalization of the self/other binary of colonialist relations, explains why it is that it has always been Second-World *literary* writing rather than Second-World *critical* writing which has occupied the vanguard of a Second-World post-colonial literary or critical *theory*. Literary writing is about internalized conflict, whereas critical writing – for most practitioners – is still grounded in the ideology of unitariness, and coherence, and specific argumentative drive. For this reason, Second-World *critical* writing – with some spectacularly transgressive exceptions – has tended to miss out on the rigours of what, I would argue, comprises a necessarily ambivalent, necessarily contra/dictory or incoherent, anti-colonialist *theory* of resistance. In literary documents such as De Mille's *Strange Manuscript* or Furphy's *Such Is Life*, to name two nineteenth-century examples, or in the 're-historical' fictions of writers such as Fiona Kidman, Ian Wedde, Thea Astley, Peter Carey, Kate Grenville, Barbara Hanrahan, Daphne Marlatt, Susan Swan, and Rudy Wiebe – to name only a few from the contemporary period – this necessary *entanglement* of anti-colonial resistances within the colonialist machineries they seek to displace has been consistently thematized, consistently worked through, in ways that the unitary and logical demands of critical argumentation, at least in its traditional genres, have simply not allowed.

Chapter 19

# SARA SULERI

## THE RHETORIC OF ENGLISH INDIA

From *The Rhetoric of English India* Chicago: University of Chicago Press, 1992.

[T]HERE IS A] PRECARIOUS VULNERABILITY of cultural boundaries in the context of colonial exchange. In historical terms, colonialism precludes the concept of 'exchange' by granting to the idea of power a greater literalism than it deserves. The telling of colonial and postcolonial stories, however, demands a more naked relation to the ambivalence represented by the greater mobility of disempowerment. To tell the history of another is to be pressed against the limits of one's own – thus culture learns that terror has a local habitation and a name. . . . The allegorization of empire, in other words, can only take shape in an act of narration that is profoundly suspicious of the epistemological and ethical validity of allegory, suggesting that the term 'culture' – more particularly, 'other cultures' – is possessed of an intransigence that belies exemplification. Instead, the story of culture eschews the formal category of allegory to become a painstaking study of how the idioms of ignorance and terror construct a mutual narrative of complicities. While the 'allegory of empire' will always have recourse to the supreme fiction of Conrad's Marlow, or the belief that what redeems it is 'the idea alone,' its heart of darkness must incessantly acknowledge the horror attendant on each act of cultural articulation that demonstrates how Ahab tells Naboth's story in order to know himself. . . .

From the vast body of eighteenth-century historical documentation of British rule in India to the proliferation of Anglo-Indian fiction in the nineteenth and twentieth centuries, the narratives of English India are fraught with the idiom of dubiety, or a mode of cultural taletelling that is neurotically conscious of its own self-censoring apparatus. While such narratives appear to claim a new preeminence of historical facticity over cultural allegory, they nonetheless illustrate that the functioning of language in a colonial universe is preternaturally dependent on the instability of its own facts. For colonial facts are vertiginous: they lack a recognizable cultural plot; they frequently fail to cohere around the master-myth that proclaims static lines of demarcation between imperial power and disempowered culture, between colonizer and colonized. Instead, they move with a ghostly mobility to suggest how highly unsettling an economy of complicity and guilt is in operation between each actor on the colonial stage. If such an economy is the impelling force of the stories of English India, it demands to be read against the grain of the rhetoric of binarism that informs, either explicitly or implicitly, contemporary critiques of alterity in colonial discourse (see Bhabha 1983b; Spivak 1987; JanMohamed 1983). The necessary intimacies that obtain between ruler and ruled create a counter-culture not always explicable in terms of an allegory of otherness: the narrative of English India questions the validity of both categories to its secret economy, which is the dynamic of powerlessness at the heart of the imperial configuration.

If English India represents a discursive field that includes both colonial and postcolonial narratives, it further represents an alternative to the troubled chronology of nationalism in the Indian subcontinent. As long as the concept of nation is interpreted as the colonizer's gift to its erstwhile colony, the unimaginable community produced by colonial encounter can never be sufficiently read (Anderson 1983). Again, the theoretical paradigm of margin against center is unhelpful in this context, for it serves to hierarchize the emergence of nation in 'first' and 'third' worlds. . . .

If colonial cultural studies is to avoid a binarism that could cause it to atrophy in its own apprehension of difference, it needs to locate an idiom for alterity that can circumnavigate the more monolithic interpretations of cultural empowerment that tend to dominate current discourse. To study the rhetoric of the British Raj in both its colonial and postcolonial manifestations is therefore to attempt to break down the incipient schizophrenia of a critical discourse that seeks to represent domination and subordination as though the two were mutually exclusive terms. Rather than examine a binary rigidity between those terms – which is an inherently Eurocentric strategy – this critical field would be better served if it sought to break down the fixity of the dividing lines between domination and subordination, and if it further questioned the psychic disempowerment signified by colonial encounter. For to interpret the configurations of colonialism in the idiom of such ineluctable divisions is to deny the impact of narrative on a productive disordering of binary dichotomies. . . .

The intimacy of the colonial setting requires reiteration. For the reader of postcolonial discourse provides scant service to its conceptualization when she posits the issue of an intransigent otherness as both the first and the final solution to the political and aesthetic problems raised by the mutual transcriptions that colonialism has engendered in the Indian subcontinent. Diverse ironies of empire are too compelling to be explained away by the simple pieties that the idiom of alterity frequently cloaks. If cultural criticism is to address the uses to which it puts the agency of alterity, then it must further face the theoretical question that S. P. Mohanty succinctly formulates: 'Just how other, we need to force ourselves to indicate, is the Other?' (Mohanty 1989: 5). Since recourse neither to representation nor to cultural relativism can supply an answer, postcolonial discourse is forced into alternative questions: how can the dynamic of imperial intimacy produce an idea of nation that belongs neither to the colonizer nor to the colonized? Is nation in itself the alterity to which both subjugating and subjugated cultures must in coordination defer? In what ways does the idiom of otherness simply rehearse the colonial fallacy through which India could be interpreted only as the unreadability of romance? . . .

If the paradigm of master and victim is to be read in terms of its availability to the histories of colonialism and their concomitant narratives, then its rereading as a figure of colonial intimacy – as an interruption in traditional interpretations of imperial power – must necessarily generate a discursive guilt at the heart of the idiom of English India. Its troubled confluence of colony, culture, and nation lends a retroactive migrancy to the fact of imperialism itself, causing a figure like Kipling's Ahab to recognize that narration occurs to confirm the precariousness of power.

## Chapter 20

# ROBERT STAM AND LOUISE SPENCE

# COLONIALISM, RACISM AND REPRESENTATION

From 'Colonialism, Racism and Representation: An Introduction' *Screen* 24(2), 1983.

[S] TUDIES OF FILMIC COLONIALISM AND RACISM tend to focus on certain dimensions of film – social portrayal, plot and character. While such studies have made an invaluable contribution by alerting us to the hostile distortion and affectionate condescension with which the colonized have been treated in the cinema, they have often been marred by a certain methodological naiveté. While posing legitimate questions concerning narrative plausibility and mimetic accuracy, negative stereotypes and positive images, the emphasis on realism has often betrayed an exaggerated faith in the possibilities of verisimilitude in art in general and the cinema in particular, avoiding the fact that films are inevitably constructs, fabrications, representations. The privileging of social portrayal, plot and character, meanwhile, has led to the slighting of the specifically cinematic dimensions of the films; the analyses might easily have been of novels or plays rather than films. . . .

Although we are quite aware of the crucial importance of the *contextual*, that is, of those questions bearing on the cinematic industry, its processes of production, distribution and exhibition, those social institutions and production practices which construct colonialism and racism in the cinema, our emphasis here will be *textual* and *intertextual*. An anti-colonialist analysis, in our view, must make the same kind of methodological leap effected by feminist criticism when journals like *Screen* and *Camera Obscura* critically transcended the usefully angry but methodologically flawed 'image' analyses of such critics as Molly Haskell and Marjorie Rosen in order to pose questions concerning the apparatus, the position of the spectator, and the specifically cinematic codes. . . .

By colonialism, we refer to the process by which the European powers (including the United States) reached a position of economic, military, political and cultural domination in much of Asia, Africa and Latin America. This process, which can be traced at least as far back as the 'voyages of discovery' and which had as its corollary the institution of the slave trade, reached its apogee between 1900 and the end of World War I (at which point Europe had colonized roughly 85% of the earth) and began to be reversed only with the disintegration of the European colonial empires after World War II. . . .

The Renaissance humanism which gave birth to the code of perspective – subsequently incorporated, as Baudry points out, into the camera itself – also gave birth to the 'rights of man'. Europe constructed its self-image on the backs of its equally constructed Other – the 'savage', the 'cannibal' – much as phallocentrism sees its self-flattering image in the mirror of woman defined as lack. And just as the camera might therefore be said to inscribe certain features of bourgeois humanism, so the cinematic and televisual apparatuses, taken in their most inclusive sense, might be said to inscribe certain features of European colonialism. The magic carpet provided by these apparatuses flies us around the globe and makes us, by virtue of our subject position, its audio-visual masters. It produces us as subjects, transforming us into armchair conquistadores, affirming our sense of power while making the inhabitants of the Third World objects of spectacle for the First World's voyeuristic gaze. . . .

Colonialist representation did not begin with the cinema; it is rooted in a vast colonial intertext, a widely disseminated set of discursive practices. Long before the first racist images appeared on the film screens of Europe and North America, the process of colonialist image-making, and resistance to that process, resonated through Western literature. Colonialist historians, speaking for the 'winners' of history, exalted the colonial enterprise, at bottom little more than a gigantic act of pillage whereby whole continents were bled of their human and material resources, as a philanthropic 'civilizing mission' motivated by a desire to push back the frontiers of ignorance, disease and tyranny. Daniel Defoe glorified colonialism in *Robinson Crusoe* (1719), a novel whose 'hero' becomes wealthy through the slave trade and through Brazilian sugar mills, and whose first thought, upon seeing human footprints after years of solitude, is to 'get (him) a servant'. . . .

The struggle over images continues, within literature, into that period of the beginnings of the cinema. The colonialist inheritance helps account for what might be called the tendentiously flawed mimesis of many films dealing with the Third World. The innumerable ethnographic, linguistic and even topographical blunders in Hollywood films are illuminating in this regard. Countless safari films present Africa as the land of 'lions in the jungle' when in fact only a tiny proportion of the African land mass could be called 'jungle' and when lions do not live in jungle but in grasslands. At times the 'flaw' in the mimesis derives not from the *presence* of distorting stereotypes but from the *absence* of representations of an oppressed group. . . .

At other times the structuring absence has to do not with the people themselves but with a dimension of that people's history or institutions. A whole realm of Afro-American history, the slave revolts, is rarely depicted in film or is represented (as in the television series *Roots*) as a man, already dead, in a ditch. The revolutionary dimension of the black church, similarly, is ignored in favor of a portrayal which favors charismatic leaders and ecstatic songs and dances. The exclusion of whites from a film, we might add paradoxically, can also be the result of white racism. The all-black Hollywood musicals of the twenties and thirties, like present-day South African films made by whites for black audiences, tend to exclude whites because their mere presence would destroy the elaborate fabric of fantasy constructed by such films. . . .

Much of the work on racism in the cinema, like early work on the representation of women, has stressed the issue of the 'positive image'. This reductionism, though not wrong, is inadequate and fraught with methodological dangers. We should be equally suspicious of a naïve integrationism which simply inserts new heroes and heroines, this time drawn from the ranks of the oppressed, into the old functional roles that were themselves oppressive, much as colonialism invited a few assimilated 'natives' to join the club of the 'elite'. . . .

The television series *Roots*, for example, exploited positive images in what was ultimately a cooptive version of Afro-American history. The series' subtitle, 'the saga of an American family', reflects an emphasis on the European-style nuclear family (retrospectively projected onto Kunta's life in Africa) in a film which casts blacks as just another immigrant group making its way toward freedom and prosperity in democratic America.

The complementary preoccupation to the search for positive images, the exposure of negative images or stereotypes, entails similar methodological problems. A comprehensive methodology must pay attention to the *mediations* which intervene between 'reality' and representation. Its emphasis should be on narrative structure, genre conventions, and cinematic style rather than on perfect correctness of representation or fidelity to an original 'real' model or prototype. . . .

One mediation specific to cinema is spectator positioning. The paradigmatic filmic encounters of whites and Indians in the western, as Tom Engelhardt points out, typically involve images of encirclement. The attitude towards the Indian is premised on exteriority. The besieged wagon train or fort is the focus of our attention and sympathy, and from this centre our familiars sally out against unknown attackers characterized by inexplicable customs and irrational hostility: 'In essence, the viewer is forced behind the barrel of a repeating rifle and it is from that position, through its gun sights, that he [*sic*] receives a picture history of western colonialism and imperialism' (Englehardt 1995). The possibility of sympathetic iden-tifications with the Indians is simply ruled out by the point-of-view conventions. The spectator is unwittingly sutured into a colonialist perspective. . . .

One of the crucial innovations of *Battle of Algiers* (directed by Gillo Pontecorvo, 1966) was to invert this imagery of encirclement and exploit the identificatory mechanisms of cinema on behalf of the colonized rather than the colonizer. Algerians, traditionally represented in cinema as shadowy figures, picturesquely backward at best and hostile and menacing at worst, are here treated with respect, dignified by close-ups, shown as speaking subjects rather than as manipulable objects. While never caricaturing the French, the film exposes the oppressive logic of colonialism and consistently fosters our complicity with the Algerians. It is through Algerian eyes, for example, that we witness a condemned Algerian's walk to his execution. It is from *within* the casbah that we see and hear the French troops and helicopters. This time it is the colonized who are encircled and menaced and with whom we identify. . . .

In *Battle of Algiers*, the sequences consequently challenge the image of anti-colonialist guerrillas as terrorist fanatics lacking respect for human life. Unlike the Western mass media, which usually restrict their definition of 'terror' to anti-establishment violence – state repres-sion and government-sanctioned aerial bombings are not included in the definition – *Battle of Algiers* presents anti-colonialist terror as a response to colonialist violence. . . .

If *Battle of Algiers* exploits conventional identification mechanisms on behalf of a group traditionally denied them, other films critique colonialism and colonialist point-of-view conven-tions in a more ironic mode. *Petit à Petit* (directed by Jean Rouch, 1969) inverts the hierarchy often assumed within the discipline of anthropology, the academic offspring of colonialism, by having the African protagonist Damouré 'do anthropology' among the strange tribe known as the Parisians, interrogating them about their folkways. . . .

A more comprehensive analysis of character status as speaking subject as against spoken object would attend to cinematic and extra-cinematic codes, and to their interweaving within textual systems. In short, it must address the instances through which film speaks – com-position, framing, scale, off- and on-screen sound, music – as well as questions of plot and character. Questions of image scale and duration, for example, are intricately related to the respect accorded a character and the potential for audience sympathy, understanding and iden-tification. Which characters are afforded close-ups and which are relegated to the background? Does a character look or act, or merely appear, to be looked at and acted upon? With whom is the audience permitted intimacy? If there is off-screen commentary or dialogue, what is its relation to the image? *Black Girl* (directed by Ousmane Sembene, 1966) uses off-screen dialogue to foster intimacy with the title character, a Sengalese maid working in France. Shots of the maid working in the kitchen coincide with overhead slurs from her employers about her 'laziness'. Not only do the images point up the absurdity of the slurs – indeed, she is the *only* person working – but also the coincidence of the off-screen dialogue with close shots of her face makes us hear the comments as if through her ears.

The music track can also play a crucial role in the establishment of a political point of view and the cultural positioning of the spectator. Film music has an emotional dimension: it can regulate our sympathies, extract our tears or trigger our fears. In many classical Hollywood films, African polyrhythms became aural signifiers of encircling savagery, a kind of synecdochic acoustic shorthand for the atmosphere of menace implicit in the phrase 'the natives are restless'. *Der Leone Have Sept Cabeças* (Glauber Rocha, 1970), in contrast, treats African polyrhythms with respect, as music, while ironically associating the puppets of coloni- alism with 'La Marseillaisse', *Black and White in Colour* employs music satirically by having the African colonized carry their colonial masters, while singing – in their own language – satir- ical songs about them ('My master is so fat, how can I carry him . . . Yes, but mine is so ugly . . .'). . . .

We must be conscious, too, of the institutionalized expectations, the mental machinery that serves as the subjective support to the film industry, and which leads us to consume films in a certain way. This apparatus has adapted most of us to the consumption of films which display high production values. But many Third World film-makers find such a model, if not repugnant, at least inappropriate – not only because of their critique of dominant cinema, but also because the Third World, with its scarcer capital and higher costs, simply cannot *afford* it. Significantly, such film-makers and critics argue for a model rooted in the actual circumstances of the Third World: a 'third cinema' (Solanas-Gettino), 'an aesthetic of hunger' (Rocha), and 'an imperfect cinema' (Espinosa). To expect to find First World production values in Third World films is to be both naïve and ethnocentric. To prospect for Third World auteurs, similarly, is to apply a regressive analytical model which implicitly valorizes dominant cinema and promises only to invite a few elite members of the Third World into an already-established pantheon. . . .

Racism is not permanently inscribed in celluloid or in the human mind; it forms part of a constantly changing dialectical process within which, we must never forget, we are far from powerless.

## Chapter 21

# ELLEKE BOEHMER

## NETWORKS OF RESISTANCE

Excerpts from *Empire, The National, and the Postcolonial 1890–1920*
Oxford: Oxford University Press, 2002.

A S  WITH  PRESENT-DAY  BILATERAL  RELATIONS  between  Third World  or  tricontinental  countries,  cross-nationalist  circuitry  was,  it  is  important  to recognize, made possible and shaped by worldwide colonial (what we would today term neo-colonial) nexuses of communication and exchange. These took the form of newspapers, the telegraph, new road and railway links, and faster shipboard journeys. Mutual awareness between English-speaking professionals in the different regions of the British Raj, for example, was facilitated both by their common educational experience, and by the road, rail, and tele-graph networks laid down in the interests of imperial military security and more efficient government (see Brown 1989: 10). Nationalists in the Indian National Congress, therefore, as also in the early African National Congress in South Africa, were encouraged to think within the broader, cross-regional framework of their colony or potential country, even in some cases of the empire as a whole.

The colonial facilitation of nationalist interconnection may seem an irony, if a subversive one. Yet, clearly, the movement of influence or example, the perception of analogy, could not take place through a mere osmosis of ideas. Such movements required technological, infra-structural, and institutional support, and depended also on the physical shift and drift of individuals between the different regions of the empire. Indeed, although indigenous networks laid down by trade or religious and educational 'pilgrimages', in Benedict Anderson's termin-ology, existed from pre-colonial times, anti-colonial, cross-nationalist links adapted from and exploited an empire-wide reality: the fact that at official levels, too, the British Empire was both conceived of and operated as a loosely interconnected system, or at least as a series of multilaterally linked, parallel systems. . . .

In an imperial world interconnected through the use of English, ideas of cross-empire solidarity and interracial exchange also exerted an important ideological hold over white colonial elites (see Jameson 1989: xi–xii). The publication of Charles Dilke's *Greater Britain* (1879, and much reprinted) stimulated a widespread discussion concerning the formation of an Anglo-Saxon brotherhood or imperial federation of settler colonies as a way of binding the burgeoning empire more closely together on the grounds of blood relationship (while simul-taneously also overriding and/or sideroading native pressure for self-government). These ideas, endorsed by influential writers like Anthony Trollope, were then further expanded in historical analyses by J. R. Seeley and J. A. Froude (*The Expansion of England*, 1883; *Oceana, or England and Her Colonies*, 1886). Even if never practically realized, the constructive possibility of cross-border collaboration within a commonwealth of self-governing colonies was thus raised within settler circles as an issue for debate.

Gaining currency empire-wide, the cross-imperial idea was in these ways made available for annexation by non-white political elites similarly in quest of solidarity and a common vocabulary of rights. Not only W. S. Blunt but also M. K. Gandhi (1869–1948) and Annie Besant (1847–1933), the Theosophist and 1910s agitator for Indian Home Rule, were certainly thinking cross-nationally when in the first decades of the twentieth century they appealed to Queen Victoria's 1858 Proclamation of Empire, India's 'Magna Carta', in order to claim Indian rights to just and equal treatment both in the Raj and across the empire (see Brown 1989: 44; Besant 1914: 31, 35, 112–13). In the cross-border unity effected by appeal to 1858, they discovered, lay new possibilities for mounting political protest. . . .

In Rabindranath Tagore's *The Home and the World* (1916), Sandip the arch-nationalist criticizes the sceptical moderate Mikhil, a Tagore-surrogate, for being like an 'eternal' pointsman 'lying in wait by the line, to shunt one's train of thought from one rail to another' (Tagore 1992: 69). Derived from one of India's most hybridized of imperial imports, the railway, and within the medium of another, a novel by a writer who was himself an ideological border figure, both nationalist and internationalist, this metaphor nearly captures that effort of lateral and associative thinking, of moving between and across parallel tracks, which fledgling anti-colonialist movements in their different alienated and cross-cultural contexts were having to make at this time. As a way of encapsulating the translational and appropriative methods followed by such movements, Tagore's metaphor attaches to the aforementioned idea of travelling theory (of the critical adaptation of a theory within a different cultural world, often 'from below'). It may also be related to Homi Bhabha's emphasis on the borderline negotiation of cultural meanings, or, as he suggests in 'DissemiNation' and in other essays, on how strategies of resistance and empowerment emerge out of the regenerative interstices between cultural spaces, the pressurized margins marking out domains of difference (1990: 291–322).

Early ventures in anti-colonial resistance and cross-nationalist interaction offer diverse instances of such interstitial emergence of how different ideas of resistance might be picked up and developed in cross-border contact zones (like the metropolis), or reinflected and reinforced by being moved across borders and then adapted to local contexts; of the extent to which a liberatory politics represented a cross-hatching of different, often syncretized traditions. The will-to-identity of colonial elites was necessarily being articulated at points of conjecture between cultures; their politics and writing therefore were always profoundly shaped by both perceptions and experiences of cross-cultural interrelation, solidarity, and dialogue. At the same time, it is important to remember, cross-border allusions and solidarities did not as a rule compromise the (hoped-for) integrity of the nation, or indeed create permanent cross-national political structures. Anti-imperial interaction is to this extent clarified by Emmanuel Levinas's concept of alterity, in which the other – here, the brother or sister nation elsewhere in the empire – is simultaneously recognized as being distant and unknowable, yet as an entity pre-eminently to be taken into account, to be signalled towards (Levinas 1999).

Separated from the colonial governing elite but also alienated from the mass population by their middle-class status, education, lingua franca, and, in some cases, geographical mobility or educational and bureaucratic 'migrations', early nativist and nationalist intellectuals frequently felt themselves to be more at home in the colonizer's culture than in their indigenous environment (Anderson 1983: 47–65; 114–15). Certainly they tended to find a deeper pool of common experience with other, like-minded colonial nationalist 'pilgrims' than with their own people. The cross-border impulses of their migratory make-up prompted them therefore to reach beyond cultural and geopolitical boundaries to discover ways of constituting a resistant selfhood. Indeed it has been one of the ironies of native colonial experience across the twentieth century that nationalist writers and intellectuals such as Gandhi or Plaatje rediscovered, or uncovered for the first time – were returned to – their own cultural resources through diasporic contacts made abroad. London, pullulating with secularist, anarchist, social-ist, avant-garde, and freethinking circles, as the chapters below in different ways illustrate, thus formed an important meeting ground for Indian, Irish, African, and Caribbean freedom movements (Anand 1979: 3–9).

As David Lloyd has observed of nineteenth-century Ireland, anti-colonial and nationalist elites, finding themselves in peripheral but privileged positions in relation to countries 'marked by a singularly uneven pattern of economic development', grew adept at shifting their dilemmas of self-making around a broad cross-national and cosmopolitan ('cosmo-national' in Nivedita's phrase) landscape (Lloyd 1987: 59). . . .

In summary, anti-colonial intelligentsias, poised between the cultural traditions of home on the one hand and of their education on the other, occupied a site of potentially produc-tive inbetweenness where they might observe other resistance histories and political approaches in order to work out how themselves to proceed. Their cross-national contacts created an interstitial place between the cosmopolitan and the parochial in which they were able to lay claim to a till-now-metropolitan discourse of rights and self-assertion. Expressed in their own congresses and conventions, pamphlets and newsletters, this discourse they then made available amongst themselves. In the Janus fashion that Tom Nairn has memorably described as char-acteristic of nationalisms generally, elites spoke in a voice that bore an 'international resonance' and was thus 'recognizable to another, broader community', yet simultaneously built notions of their own especial racial nature out of essentialized spiritual and mythic resources which they believed the West had left untouched (Nairn 1977; 1994). As the outer-national, diasporic formation of Gandhi's notion of all-India illustrates, they relied on international exchange, and a borrowed second culture, in order, each individually, to assert their own unity and singularity, to *think back* to themselves. As Gyan Prakash encapsulates it: 'the unity of the national subject was forged in the space of difference and conflicts', and, we might add, of cross-nationalist interaction (Prakash 'Introduction' 1995: 9–10).

# PART FOUR

# Nationalism

ONE OF THE STRONGEST FOCI FOR RESISTANCE TO imperial control in colonial societies has been the idea of 'nation'. It is the concept of a shared community, one which Benedict Anderson calls an 'imagined community' (Anderson 1983: 15) that has enabled post-colonial societies to invent a self image through which they could act to liberate themselves from imperialist oppression. Nationalism in this sense is nowhere better summed up than in the work of Frantz Fanon and his dictum that 'a national culture is the whole body of efforts made by a people in the sphere of thought to describe, justify and praise the action through which that people has created itself and keeps itself in existence'.

However, Fanon was also one of the earliest theorists to warn of the pitfalls of national consciousness, of its becoming an 'empty shell', a travesty of what it might have been. The dangers of a national bourgeoisie using nationalism to maintain its own power demonstrates one of the principal dangers of nationalism – that it frequently takes over the hegemonic control of the imperial power, thus replicating the conditions it rises up to combat. It develops as a function of this control, a monocular and sometimes xenophobic view of identity and a coercive view of national commitment.

Theorizing national liberation discourse has been particularly strong in the African context, as the chapter from Fanon indicates. From a wider post-colonial perspective, the Indian critic Partha Chatterjee examines some of the contemporary attempts at theorizing the nation and nationalism. Working from the base-line established by Anderson's analysis and from those of Marxist critics such as Gelner, Chatterjee shows how Third-World nationalisms in the twentieth century have constructed themselves along the earlier forms of American and European nationalisms. Chattterjee demonstrates how this may enable post-colonial societies consciously to avoid or select among these forms in a more creative and effective way and to avoid naive nativist constructions of community in favour of an awareness of the complex formation of national consciousness in modern societies.

Settler colony cultures have never been able to construct simple concepts of the nation, such as those based on linguistic communality or racial or religious homogeneity. Faced with their 'mosaic' reality, they have, in many ways, been clear examples of the *constructedness* of nations. In settler colony cultures the sense of place and placelessness have been crucial factors in welding together a communal identity from the widely disparate elements brought together by transportation, migration and settlement. At the heart of the settler colony culture is also an ambivalent attitude towards their own identity, poised as they are between the centre from which they seek to differentiate themselves and the indigenous people who serve to remind them of their own problematic occupation of the country. The process of effecting justice, restitution and reconciliation with the indigenous peoples is now crucial to any notion of creating an effective identity, and the issue of how nationalism may continue to function to elide and obscure such important constitutive 'differences' has been at the heart of the debate in all ex-settler colony cultures in recent years.

Most recently a flurry of theoretical activity has made the nation and nationalism one of the most debated topics of contemporary theory. We have sought to illustrate the importance of this attempt at retheorizing nationalism through the work of Timothy Brennan and Homi Bhabha. As Brennan notes, 'the rising number of studies on nationalism in the past three decades reflects its lingering, almost atmospheric insistence in our thinking'. We could also say that the interest in nationalism throughout the world reflects the growing disillusionment in postmodern Europe with nationalism and its excesses. Post-colonial societies are increasingly wary, therefore, of that neo-universalist internationalism which subsumes them within monocentric or Europe-dominated networks of politics and culture. The fiction of national essences is rejected for the more refractory and syncretic complexes of ordinary experience as a way of approaching literary production.

Although nation, like race, has only the most tenuous theoretical purchase, in political practice it has continued to be what Anderson describes as 'the most universally legitimate value in the political life of our time' (Anderson 1983: 12). While nationalism operated as a general force of resistance in earlier times in post-colonial societies, a perception of its hegemonic and 'monologic' status is growing. From the point of view of literary theory, nationalism is of special interest since its rise, as Brennan and Bhabha note, is coterminous with the rise of the most dominant modern literary form, at least in European and European-influenced cultures – that of the novel. These ties between literature and nation evoke a sense of the 'fictive quality of the political concept itself' (Brennan). In this sense the story of the nation and the narrative form of the modern novel inform each other in a complex, reflexive way.

Chapter 22

# FRANTZ FANON

# NATIONAL CULTURE

From 'On National Culture' and 'The Pitfalls of National Consciousness' in
*The Wretched of the Earth* (trans. Constance Farrington) New York: Grove
Press, 1968 (original French edition 1961).

## On national culture

TODAY WE KNOW THAT IN THE FIRST PHASE of the national struggle colonialism tries to disarm national demands by putting forward economic doctrines. As soon as the first demands are set out, colonialism pretends to consider them, recognizing with ostentatious humility that the territory is suffering from serious underdevelopment which necessitates a great economic and social effort. And, in fact, it so happens that certain spectacular measures (centers of work for the unemployed which are opened here and there, for example) delay the crystallization of national consciousness for a few years. But, sooner or later, colonialism sees that it is not within its powers to put into practice a project of economic and social reforms which will satisfy the aspirations of the colonized people. Even where food supplies are concerned, colonialism gives proof of its inherent incapability. The colonialist state quickly discovers that if it wishes to disarm the nationalist parties on strictly economic questions then it will have to do in the colonies exactly what it has refused to do in its own country. . . .

I am ready to concede that on the plane of factual being the past existence of an Aztec civilization does not change anything very much in the diet of the Mexican peasant of today. I admit that all the proofs of a wonderful Songhai civilization will not change the fact that today the Songhais are underfed and illiterate, thrown between sky and water with empty heads and empty eyes. But it has been remarked several times that this passionate search for a national culture which existed before the colonial era finds its legitimate reason in the anxiety shared by native intellectuals to shrink away from that Western culture in which they all risk being swamped. Because they realize they are in danger of losing their lives and thus becoming lost to their people, these men, hot-headed and with anger in their hearts, relentlessly determine to renew contact once more with the oldest and most pre-colonial springs of life of their people.

Let us go further. Perhaps this passionate research and this anger are kept up or at least directed by the secret hope of discovering beyond the misery of today, beyond self-contempt, resignation, and abjuration, some very beautiful and splendid era whose existence rehabilitates us both in regard to ourselves and in regard to others. I have said that I have decided to go further. Perhaps unconsciously, the native intellectuals, since they could not stand wonder-struck before the history of today's barbarity, decided to back further and to delve deeper down; and, let us make no mistake, it was with the greatest delight that they discovered that there was nothing to be ashamed of in the past, but rather dignity, glory, and solemnity. The claim to a national culture in the past does not only rehabilitate that nation and serve as a justification for the hope of a future national culture. In the sphere of psycho-affective equilibrium it is responsible for an important change in the native. Perhaps we have not sufficiently demonstrated that colonialism is not simply content to impose its rule upon the present and the future of a dominated country. Colonialism is not satisfied merely with holding a people in its grip and emptying the native's brain of all form and content. By a kind of perverted logic, it turns to the past of the oppressed people, and distorts, disfigures, and destroys it. This work of devaluing pre-colonial history takes on a dialectical significance today. . . .

In such a situation the claims of the native intellectual are not a luxury but a necessity in any coherent program. The native intellectual who takes up arms to defend his nation's legitimacy and who wants to bring proofs to bear out that legitimacy, who is willing to strip himself naked to study the history of his body, is obliged to dissect the heart of his people. . . .

To fight for national culture means in the first place to fight for the liberation of the nation, that material keystone which makes the building of a culture possible. There is no other fight for culture which can develop apart from the popular struggle. To take an example: all those men and women who are fighting with their bare hands against French colonialism in Algeria are not by any means strangers to the national culture of Algeria. The national Algerian culture is taking on form and content as the battles are being fought out, in prisons, under the guillotine, and in every French outpost which is captured or destroyed.

We must not therefore be content with delving into the past of a people in order to find coherent elements which will counteract colonialism's attempts to falsify and harm. We must work and fight with the same rhythm as the people to construct the future and to prepare the ground where vigorous shoots are already springing up. A national culture is not a folklore, nor an abstract populism that believes it can discover the people's true nature. It is not made up of the inert dregs of gratuitous actions, that is to say actions which are less and less attached to the ever-present reality of the people. A national culture is the whole body of efforts made by a people in the sphere of thought to describe, justify, and praise the action through which that people has created itself and keeps itself in existence. . . .

While at the beginning the native intellectual used to produce his work to be read exclusively by the oppressor, whether with the intention of charming him or of denouncing him through ethnic or subjectivist means, now the native writer progressively takes on the habit of addressing his own people.

It is only from that moment that we can speak of a national literature. Here there is, at the level of literary creation, the taking up and clarification of themes which are typically nationalist. This may be properly called a literature of combat, in the sense that it calls on the whole people to fight for their existence as a nation. It is a literature of combat, because it moulds the national consciousness, giving it form and contours and flinging open before it new and boundless horizons; it is a literature of combat because it assumes responsibility, and because it is the will to liberty expressed in terms of time and space.

On another level, the oral tradition – stories, epics, and songs of the people – which formerly were filed away as set pieces are now beginning to change. The storytellers who used to relate inert episodes now bring them alive and introduce into them modifications which are increasingly fundamental. There is a tendency to bring conflicts up to date and to modernize the kinds of struggle which the stories evoke, together with the names of heroes and types of weapons. The method of allusion is more and more widely used. The formula 'This all happened long ago' is substituted with that of 'What we are going to speak of happened somewhere else, but it might well have happened here today, and it might happen tomorrow.' The example of

Algeria is significant in this context. From 1952–3 on, the storytellers, who were before that time stereotyped and tedious to listen to, completely overturned their traditional methods of storytelling and the contents of their tales. Their public, which was formerly scattered, became compact. The epic, with its typified categories, reappeared; it became an authentic form of entertainment which took on once more a cultural value. Colonialism made no mistake when from 1955 on it proceeded to arrest these storytellers systematically.

The contact of the people with the new movement gives rise to a new rhythm of life and to forgotten muscular tensions, and develops the imagination. Every time the storyteller relates a fresh episode to his public, he presides over a real invocation. The existence of a new type of man is revealed to the public. The present is no longer turned in upon itself but spread out for all to see. The storyteller once more gives free rein to his imagination; he makes innovations and he creates a work of art. It even happens that the characters, which are barely ready for such a transformation – highway robbers or more or less anti-social vagabonds – are taken up and remodelled. The emergence of the imagination and of the creative urge in the songs and epic stories of a colonized country is worth following. The storyteller replies to the expectant people by successive approximations, and makes his way, apparently alone but in fact helped on by his public, toward the seeking out of new patterns, that is to say national patterns. Comedy and farce disappear, or lose their attraction. As for dramatization, it is no longer placed on the plane of the troubled intellectual and his tormented conscience. By losing its characteristics of despair and revolt, the drama becomes part of the common lot of the people and forms part of an action in preparation or already in progress.

## The pitfalls of national consciousness

National consciousness, instead of being the all-embracing crystallization of the innermost hopes of the whole people, instead of being the immediate and most obvious result of the mobilization of the people, will be in any case only an empty shell, a crude and fragile travesty of what it might have been. The faults that we find in it are quite sufficient explanation of the facility with which, when dealing with young and independent nations, the nation is passed over for the race, and the tribe is preferred to the state. These are the cracks in the edifice which show the process of retrogression, that is so harmful and prejudicial to national effort and national unity. We shall see that such retrograde steps with all the weaknesses and serious dangers that they entail are the historical result of the incapacity of the national middle class to rationalize popular action, that is to say their incapacity to see into the reasons for that action.

This traditional weakness, which is almost congenital to the national consciousness of underdeveloped countries, is not solely the result of the mutilation of the colonized people by the colonial regime. It is also the result of the intellectual laziness of the national middle class, of its spiritual penury, and of the profoundly cosmopolitan mold that its mind is set in.

The national middle class which takes over power at the end of the colonial regime is an underdeveloped middle class. It has practically no economic power, and in any case it is in no way commensurate with the bourgeoisie of the mother country which it hopes to replace. In its narcissism, the national middle class is easily convinced that it can advantageously replace the middle class of the mother country. But that same independence which literally drives it into a corner will give rise within its ranks to catastrophic reactions, and will oblige it to send out frenzied appeals for help to the former mother country. The university and merchant classes which make up the most enlightened section of the new state are in fact characterized by the smallness of their number and their being concentrated in the capital, and the type of activities in which they are engaged: business, agriculture, and the liberal professions. Neither financiers nor industrial magnates are to be found within this national middle class. The national bourgeoisie of underdeveloped countries is not engaged in production, nor in invention, nor building, nor labor; it is completely canalized into activities of the intermediary type. Its innermost vocation seems to be to keep in the running and to be part of the racket. The psychology of the national bourgeoisie is that of the businessman, not that of a captain of industry; and it is only too true that the greed of the settlers and the system of embargoes set up by colonialism have hardly left them any other choice. . . .

The national bourgeoisie turns its back more and more on the interior and on the real facts of its undeveloped country, and tends to look toward the former mother country and the foreign capitalists who count on its obliging compliance. As it does not share its profits with the people, and in no way allows them to enjoy any of the dues that are paid to it by the big foreign companies, it will discover the need for a popular leader to whom will fall the dual role of stabilizing the regime and of perpetuating the domination of the bourgeoisie. The bourgeois dictatorship of underdeveloped countries draws its strength from the existence of a leader. We know that in the well-developed countries the bourgeois dictatorship is the result of the economic power of the bourgeoisie. In the underdeveloped countries on the contrary the leader stands for moral power, in whose shelter the thin and poverty-stricken bourgeoisie of the young nation decides to get rich.

The people who for years on end have seen this leader and heard him speak, who from a distance in a kind of dream have followed his contests with the colonial power, spontaneously put their trust in this patriot. Before independence, the leader generally embodies the aspirations of the people for independence, political liberty, and national dignity. But as soon as independence is declared, far from embodying in concrete form the needs of the people in what touches bread, land, and the restoration of the country to the sacred hands of the people, the leader will reveal his inner purpose: to become the general president of that company of profiteers impatient for their returns which constitutes the national bourgeoisie.

In spite of his frequently honest conduct and his sincere declarations, the leader as seen objectively is the fierce defender of these interests, today combined, of the national bourgeoisie and the ex-colonial companies. His honesty, which is his soul's true bent, crumbles away little by little. His contact with the masses is so unreal that he comes to believe that his authority is hated and that the services that he has rendered his country are being called in question. The leader judges the ingratitude of the masses harshly, and every day that passes ranges himself a little more resolutely on the side of the exploiters. He therefore knowingly becomes the aider and abettor of the young bourgeoisie which is plunging into the mire of corruption and pleasure.

## Chapter 23

# BENEDICT ANDERSON

# IMAGINED COMMUNITIES

From *Imagined Communities: Reflections on the Origin and Spread of Nationalism* London: Verso, 1983.

N ATION, NATIONALITY, NATIONALISM – all have proved notoriously difficult to define, let alone to analyse. In contrast to the immense influence that nationalism has exerted on the modern world, plausible theory about it is conspicuously meagre. Hugh Seton-Watson, author of far the best and most comprehensive English-language text on nationalism, and heir to a vast tradition of liberal historiography and social science, sadly observes: 'Thus I am driven to the conclusion that no "scientific definition" of the nation can be devised; yet the phenomenon has existed and exists' (Seton-Watson 1977: 5). Tom Nairn, author of the path-breaking *The Break-up of Britain*, and heir to the scarcely less vast tradition of Marxist historiography and social science, candidly remarks: 'The theory of nationalism represents Marxism's great historical failure' (Nairn 1977: 329–63). But even this confession is somewhat misleading, insofar as it can be taken to imply the regrettable outcome of a long, self-conscious search for theoretical clarity. It would be more exact to say that nationalism has proved an uncomfortable anomaly for Marxist theory and, precisely for that reason, has been largely elided, rather than confronted. How else to explain Marx's own failure to explicate the crucial pronoun in his memorable formulation of 1848: 'The proletariat of each country must, of course, first of all settle matters with its own bourgeoisie?' (Marx and Engels [1848] 1958: 45)[1] How else to account for the use, for over a century, of the concept 'national bourgeoisie' without any serious attempt to justify theoretically the relevance of the adjective? Why is this segmentation of the bourgeoisie – a world-class insofar as it is defined in terms of the relations of production – theoretically significant?

The aim of this book is to offer some tentative suggestions for a more satisfactory interpretation of the 'anomaly' of nationalism. My sense is that on this topic both Marxist and liberal theory have become etiolated in a late Ptolemaic effort to 'save the phenomena'; and that a reorientation of perspective in, as it were, a Copernican spirit, is urgently required. My point of departure is that nationality, or, as one might prefer to put it in view of that word's multiple significations, nation-ness, as well as nationalism, are cultural artefacts of a particular kind. To understand them properly we need to consider carefully how they have come into historical being, in what ways their meanings have changed over time, and why, today, they command such profound emotional legitimacy. I will be trying to argue that the creation of these artefacts towards the end of the eighteenth century[2] was the spontaneous distillation of a complex 'crossing' of discrete historical forces; but that, once created, they became 'modular,' capable of being transplanted, with varying degrees of self-consciousness, to a great variety of social terrains, to merge and be merged with a correspondingly wide variety of political and ideological constellations. I will also attempt to show why these particular cultural artefacts have aroused such deep attachments.

## Concepts and definitions

Before addressing the questions raised above, it seems advisable to consider briefly the concept of 'nation' and offer a workable definition. Theorists of nationalism have often been perplexed, not to say irritated, by these three paradoxes: 1. The objective modernity of nations to the historian's eye vs. their subjective antiquity in the eyes of nationalists. 2. The formal universality of nationality as a socio-cultural concept – in the modern world everyone can, should, will 'have' a nationality, as he or she 'has' a gender vs. the irremediable particularity of its concrete manifestations, such that, by definition, 'Greek' nationality is sui generis. 3. The 'political' power of nationalisms vs. their philosophical poverty and even incoherence. In other words, unlike most other isms, nationalism has never produced its own grand thinkers: no Hobbeses, Tocquevilles, Marxes, or Webers. This 'emptiness' easily gives rise, among cosmopolitan and polylingual intellectuals, to a certain condescension. Like Gertrude Stein in the face of Oakland, one can rather quickly conclude that there is 'no there there'. It is characteristic that even so sympathetic a student of nationalism as Tom Nairn can nonetheless write that: '"Nationalism" is the pathology of modern developmental history, as inescapable as "neurosis" in the individual, with much the same essential ambiguity attaching to it, a similar built-in capacity for descent into dementia, rooted in the dilemmas of helplessness thrust upon most of the world (the equivalent of infantilism for societies) and largely incurable' (Nairn 1977: 359).

Part of the difficulty is that one tends unconsciously to hypostasize the existence of Nationalism-with-a-big-N – rather as one might Age-with-a-capital-A – and then to classify 'it' as an ideology. (Note that if everyone has an age, Age is merely analytical expression.) It would, I think, make things easier if one treated it as if it belonged with 'kinship' and 'religion,' rather than with 'liberalism' or 'fascism.'

In an anthropological spirit, then, I propose the following definition of the nation: it is an imagined political community – and imagined as both inherently limited and sovereign. It is imagined because the members of even the smallest nation will never know most of their fellow-members, meet them, or even hear of them, yet in the minds of each lives the image of their communion.[3] Renan referred to this imagining in his suavely back-handed way when he wrote that 'Or l'essence d'une nation est que tous les individus aient beaucoup de choses en commun, et aussi que tous aient oublie bien des choses' ('Qu'est-ce qu'une nation?' [1892] Renan 1947–61: 887–906). With a certain ferocity Gellner makes a comparable point when he rules that 'Nationalism is not the awakening of nations to selfconsciousness: it invents nations where they do not exist' (Gellner 1964: 169). The drawback to this formulation, however, is that Gellner is so anxious to show that nationalism masquerades under false pretences that he assimilates 'invention' to 'fabrication' and 'falsity,' rather than to 'imagining' and 'creation.' In this way he implies that 'true' communities exist which can be advantageously juxtaposed to nations. In fact, all communities larger than primordial villages of face-to-face contact (and perhaps even these) are imagined. Communities are to be distinguished, not by their falsity/genuineness, but by the style in which they are imagined. Javanese villagers have always known that they are connected to people they have never seen, but these ties were once imagined particularistically – as indefinitely stretchable nets of kinship and clientship. Until quite recently, the Javanese language had no word meaning the abstraction 'society.'

We may today think of the French aristocracy of the *ancien régime* as a class; but surely it was imagined this way only very late.[4] To the question 'Who is the Comte de X?' the normal answer would have been, not 'a member of the aristocracy,' but 'the lord of X,' 'the uncle of the Baronne de Y,' or 'a client of the Duc de Z.'

The nation is imagined as *limited* because even the largest of them, encompassing perhaps a billion living human beings, has finite, if elastic boundaries, beyond which lie other nations. No nation imagines itself coterminous with mankind. The most messianic nationalists do not dream of a day when all the members of the human race will join their nation in the way that it was possible, in certain epochs, for, say, Christians to dream of a wholly Christian planet.

It is imagined as *sovereign* because the concept was born in an age in which Enlightenment and Revolution were destroying the legitimacy of the divinely-ordained, hierarchical dynastic realm. Coming to maturity at a stage of human history when even the most devout adherents of any universal religion were inescapably confronted with the living *pluralism* of such religions, and the allomorphism between each faith's ontological claims and territorial stretch, nations dream of being free, and, if under God, directly so. The gage and emblem of this freedom is the sovereign state.

Finally, it is imagined as a *community*, because, regardless of the actual inequality and exploitation that may prevail in each, the nation is always conceived as a deep, horizontal comradeship. Ultimately it is this fraternity that makes it possible, over the past two centuries, for so many millions of people, not so much to kill, as willingly to die for such limited imaginings. These deaths bring us abruptly face to face with the central problem posed by nationalism: what makes the shrunken imaginings of recent history (scarcely more than two centuries) generate such colossal sacrifices? I believe that the beginnings of an answer lie in the cultural roots of nationalism.

## Notes

1    In any theoretical exegesis, the words 'of course' should flash red lights before the transported reader.
2    As Aira Kemiläinen notes, the twin 'founding fathers' of academic scholarship on nationalism, Hans Kohn and Carleton Hayes, argued persuasively for this dating. Their conclusions have, I think, not been seriously disputed except by nationalist ideologues in particular countries. Kemiläinen also observes that the word 'nationalism' did not come into wide general use until the end of the nineteenth century. It did not occur, for example, in many standard nineteenth century lexicons. If Adam Smith conjured with the wealth of 'nations,' he meant by the term no more than 'societies' or 'states' (Kermiläinen 1964: 10, 33, 48–9).
3    'All that I can find to say is that a nation exists when a significant number of people in a community consider themselves to form a nation, or behave as if they formed one' (Seton-Watson 1977: 5). We may translate 'consider themselves' as 'imagine themselves.'
4    Hobsbawm, for example, 'fixes' it by saying that in 1789 it numbered about 400,000 in a population of 23,000,000 (Hobsbawm 1977: 78). But would this statistical picture of the noblesse have been imaginable under the *ancien régime*?

## Chapter 24

# PARTHA CHATTERJEE

## NATIONALISM AS A PROBLEM

From 'Nationalism as a Problem' *Nationalist Thought and the Colonial World: A Derivative Discourse* London: Zed Books for United Nations University, 1986.

**H**ISTORICALLY, THE POLITICAL COMMUNITY OF the nation super-seded the preceding 'cultural systems' of religious community and dynastic realm. In the process there occurred 'a fundamental change . . . in modes of apprehending the world, which, more than anything else, made it possible to "think" the nation' (Anderson 1983: 28). It was the 'coalition of Protestantism and print-capitalism' which brought about this change. 'What, in a positive sense, made the new communities imaginable was a half-fortuitous, but explosive, interaction between a system of production and productive relations (capitalism), a technology of communications (print), and the fatality of human linguistic diversity' (Anderson 1983: 46). The innumerable and varied ideolects of pre-print Europe were now 'assembled, within definite limits, into print-languages far fewer in number'. This was crucial for the emergence of national consciousness because print-languages created 'unified fields of exchange and communications' below Latin and above the spoken vernaculars, gave a new fixity to language, and created new kinds of 'languages-of-power' since some dialects were closer to the print-languages and dominated them while others remained dialects because they could not insist on their own printed form.

Once again historically, three distinct types or 'models' of nationalism emerged. 'Creole nationalism' of the Americas was built upon the ambitions of classes whose economic inter-ests were ranged against the metropolis. It also drew upon liberal and enlightened ideas from Europe which provided ideological criticisms of imperialism and *anciens régimes*. But the shape of the new imagined communities was created by 'pilgrim creole functionaries and provincial creole printmen'. Yet as a 'model' for emulation, creole nationalism remained incomplete, because it lacked linguistic communality and its state form was both retrograde and congruent with the arbitrary administrative boundaries of the imperial order.

The second 'model' was that of the linguistic nationalisms of Europe, a model of the independent national state which henceforth became 'available for pirating'.

> But precisely because it was by then a known model, it imposed certain 'standards' from which too-marked deviations were impossible. . . . Thus the 'populist' character of the early European nationalisms, even when led, demagogically, by the most back-ward social groups, was deeper than in the Americas: serfdom had to go, legal slavery was unimaginable — not least because the conceptual model was set in ineradicable place.
>
> (Anderson 1983: 78–9)

The third 'model' was provided by 'official nationalism' – typically, Russia. This involved the imposition of cultural homogeneity from the top, through state action. 'Russification' was a project which could be, and was, emulated elsewhere.

All three modular forms were available to third world nationalisms in the 20th century. Just as creole functionaries first perceived a national meaning in the imperial administrative unit, so did the 'brown or black Englishman' when he made his bureaucratic pilgrimage to the metropolis. On return,

> the apex of his looping flight was *the highest administrative centre to which he was assigned*: Rangoon, Accra, Georgetown, or Colombo. Yet in each constricted journey he found bilingual travelling companions with whom he came to feel a growing communality. In his journey he understood rather quickly that his point of origin – conceived either ethnically, linguistically, or geographically – was of small significance . . . it did not fundamentally determine his destination or his companions. Out of this pattern came that subtle, half-concealed transformation, step by step, of the colonial-state into the national-state, a transformation made possible not only by a solid continuity of personnel, but by the established skein of journeys through which each state was experienced by its functionaries.
>
> (Anderson 1983: 105)

But this only made possible the emergence of a national consciousness. Its rapid spread and acquisition of popular roots in the 20th century are to be explained by the fact that these journeys were now made by 'huge and variegated crowds'. Enormous increases in physical mobility, imperial 'Russification' programmes sponsored by the colonial state as well as by corporate capital, and the spread of modern-style education created a large bilingual section which could mediate linguistically between the metropolitan nation and the colonized people. The vanguard role of the intelligentsia derived from its bilingual literacy. 'Print-literacy already made possible the imagined community floating in homogeneous, empty time. . . . Bilingualism meant access, through the European language-of-state, to modern Western culture in the broadest sense, and, in particular, to the models of nationalism, nationness, and nation-state produced elsewhere in the course of the nineteenth century' (Anderson 1983: 107).

Third-world nationalisms in the 20th century thus came to acquire a 'modular' character.

> They can, and do, draw on more than a century and a half of human experience and three earlier models of nationalism. Nationalist leaders are thus in a position consciously to deploy civil and military educational systems modelled on official nationalism's; elections, party organizations, and cultural celebrations modelled on the popular nationalisms of 19th century Europe; and the citizen-republican idea brought into the world by the Americas.
>
> (Anderson 1983: 123)

Above all, the very idea of 'nation' is now nestled firmly in virtually all print-languages, and nation-ness is virtually inseparable from political consciousness.

'In a world in which the national state is the overwhelming norm, all of this means that nations can now be imagined without linguistic communality – not in the naive spirit of *nostros los Americanos*, but out of a general awareness of what modern history has demonstrated to be possible' (Anderson 1983: 123).

## Chapter 25

# TIMOTHY BRENNAN

## THE NATIONAL LONGING FOR FORM

From 'The National Longing for Form' in Homi K. Bhabha (ed.) *Nation and Narration* London: Routledge, 1990.

I T IS ESPECIALLY IN THIRD WORLD FICTION after the Second World War that the fictional uses of 'nation' and 'nationalism' are most pronounced. The 'nation' is precisely what Foucault has called a 'discursive formation' – not simply an allegory or imaginative vision, but a gestative political structure which the Third World artist is consciously building or suffering the lack of. 'Uses' here should be understood both in a personal, craftsmanlike sense, where nationalism is a trope for such things as 'belonging', 'bordering', and 'commitment'. But it should also be understood as the *institutional* uses of fiction in nationalist movements themselves. At the present time, it is often impossible to separate these senses.

The phrase 'myths of the nation' is ambiguous in a calculated way. It does not refer only to the more or less unsurprising idea that nations are mythical, that – as Hugh Seton-Watson wrote in his massive study of nations and states as recently as 1976 – 'there is no "scientific" means of establishing what all nations have in common' (Seton-Watson 1977: 5). The phrase is also not limited to the consequences of this artificiality in contemporary political life – namely, the way that various governments invent traditions to give permanence and solidity to a transient political form.

While the study of nationalism has been a minor industry in the disciplines of sociology and history since the Second World War, the premise here is that *cultural* study, and specifically the study of imaginative literature, is in many ways a profitable one for understanding the nation-centredness of the post-colonial world, as has begun to be seen in some recent studies (Jameson 1986: 65–88; Ahmad 1987: 3–25). From the point of view of cultural studies, this approach in some ways traverses uncharted ground. With the exception of some recent sociological works which use literary theories, it is rare in English to see 'nation-ness' talked about as an imaginative vision – as a topic worthy of full fictional realization. Also, it should be said that this neglect is not true of other literatures with a close and obvious relationship to the subject – for example those of Latin America and (because of the experience of the war) Germany and Italy. Even in the underrepresented branch of Third World English studies, one is likely to find discussions of race and colonialism, but not the 'nation' as such.

Only a handful of critics (often themselves tied to the colonized by background or birth) have seen English fiction about the colonies as growing out of a comprehensive imperial system. (Examples might include Edward Said, Ariel Dorfman, Hugh Ridley, Amiri Baraka, Homi Bhabha, Jean Franco, Abdul JanMohamed, Cornell West, and others.) The universality of this system, and its effects on the imaginative life, are much clearer – even inescapable – in the literature not of the 'colonies' but of the 'colonized'. The recent interest in Third

World literature reflected in special issues of mainstream journals and new publishers' series, as well as new university programmes, is itself a mark of the recognition that imperialism is, culturally speaking, a two-way flow.

For, in the period following the Second World War, English society was transformed by its earlier imperial encounters. The wave of postwar immigration to the imperial 'centres' – including in England the influx of large numbers of non-white people from Africa and the Caribbean, and in America, from Asia and Latin America – amounted to what Gordon Lewis calls 'a colonialism in reverse' – a new sense of what it means to be 'English' (Lewis 1978: 304). To a lesser extent, the same has happened in France (Harlow 1987: 27).

The wave of successful anti-colonial struggles from China to Zimbabwe has contributed to the forced attention now being given in the English-speaking world to the point of view of the colonized – and yet, it is a point of view that must increasingly be seen as a part of English-speaking culture. It is a situation, as the Indo-English author Salman Rushdie points out, in which English, 'no longer an English language, now grows from many roots; and those whom it once colonized are carving out large territories within the language for themselves' (Rushdie 1982: 8). The polycultural forces in domestic English life have given weight to the claims of the novelists and essayists abroad who speak more articulately and in larger crowds about neo-colonialism. And, in turn, such voices from afar give attention to the volatile cultural pluralism at home. The Chilean expatriate, Ariel Dorfman, has written that 'there may be no better way for a country to know itself than to examine the myths and popular symbols that it exports to its economic and military dominions' (Dorfman 1983: 8). And this would be even truer when the myths come home. One of the most durable myths has certainly been the 'nation'.

Not the colonies, but the colonized. The 'novel of empire' in its classic modernist versions (*Heart of Darkness*, *Passage to India*, *The Plumed Serpent*) has been blind to the impact of a world system largely directed by Anglo-American interests, however much it involved itself passion-ately, unevenly, and contradictorily in some of the human realities of world domination. For English criticism – even among politically minded critics after the war – has refused to place the fact of domination in a comprehensive approach to its literary material, and that becomes impossible when facing the work of those who have not merely visited but lived it.

The rising number of studies on nationalism in the past three decades reflects its lingering, almost atmospheric, insistence in our thinking. In cultural studies, the 'nation' has often lurked behind terms like 'tradition', 'folklore', or 'community', obscuring their origins in what Benedict Anderson has called 'the most universally legitimate value in the political life of our time' (Anderson 1983: 12).

The rise of the modern nation-state in Europe in the late eighteenth and early nineteenth centuries is inseparable from the forms and subjects of imaginative literature. On the one hand, the political tasks of modern nationalism directed the course of literature, leading through the Romantic concepts of 'folk character' and 'national language' to the (largely illusory) divisions of literature into distinct 'national literatures'. On the other hand, and just as fundamentally, literature participated in the formation of nations through the creation of 'national print media' – the newspaper and the novel. Flourishing alongside what Francesco de Sanctis has called 'the cult of nationality in the European nineteenth century', it was espe-cially the novel as a composite but clearly bordered work of art that was crucial in defining the nation as an 'imagined community'.

In tracing these ties between literature and nation, some have evoked the fictive quality of the political concept itself. For example, José Carlos Mariátegui, a publicist and organizer of Peru's Quechua-speaking minority in the 1920s, outlined the claims of fiction on national thought, saying simply that 'The nation . . . is an abstraction, an allegory, a myth that does not correspond to a reality that can be scientifically defined' (Mariátegui 1971: 187–8). Race, geography, tradition, language, size, or some combination of these seem finally insufficient for determining national essence, and yet people die for nations, fight wars for them, and write fictions on their behalf. Others have emphasized the *creative* side of nation-forming, suggesting the cultural importance of what has often been treated as a dry, rancorous political fact: 'Nationalism is not the awakening of nations to self-consciousness; it *invents* nations where they do not exist' (Ernest Gellner, quoted in Anderson 1983: 15).

The idea that nations are invented has become more widely recognized in the rush of research following the war. To take only one recent example, the idea circuitously finds its way into Eric Hobsbawm's and Terence Ranger's recent work on 'the invention of tradition', which is really a synonym in their writing for the animus of any successful nation-state:

> It is clear that plenty of political institutions, ideological movements and groups – not least in nationalism – were so unprecedented that even historic continuity had to be invented, for example by creating an ancient past beyond effective historical continuity either by semi-fiction (Boadicea, Vercingetorix, Arminius the Cheruscan) or by forgery (Ossian, the Czech medieval manuscripts). It is also clear that entirely new symbols and devices came into existence . . . such as the national anthem . . . the national flag . . . or the personification of 'the nation' in symbol or image.
>
> (Hobsbawm 1983: 7)

Corresponding to Hobsbawm's and Ranger's examples, literary myth too has been complicit in the creation of nations – above all, through the genre that accompanied the rise of the European vernaculars, their institution as languages of state after 1820, and the separation of literature into various 'national' literatures by the German Romantics at the end of the eighteenth and the beginning of the nineteenth centuries. Nations, then, are imaginary constructs that depend for their existence on an apparatus of cultural fictions in which imaginative literature plays a decisive role. And the rise of European nationalism coincides especially with one form of literature – the novel. . . . It was the *novel* that historically accompanied the rise of nations by objectifying the 'one, yet many' of national life, and by mimicking the structures of the nation, a clearly bordered jumble of languages and styles. Socially, the novel joined the newspaper as the major vehicle of the national print media, helping to standardize language, encourage literacy, and remove mutual incomprehensibility. But it did more than that. Its manner of presentation allowed people to imagine the special community that was the nation. . . .

[N]ovels in the post-war period are unique because they operate in a world where the level of communications, the widespread politics of insurgent nationalism, and the existence of large international cultural organizations have made the topics of nationalism and exile unavoidably aware of one another. The idea of nationhood is not only a political plea, but a formal binding together of disparate elements. And out of the multiplicities of culture, race, and political structures, grows also a repeated dialectic of uniformity and specificity: of world culture and national culture, of family and of people. One of many clear formulations of this can be found in Fanon's statement that '[i]t is at the heart of national consciousness that international consciousness lives and grows' (Fanon 1967b: 247–8). These universalist tendencies – already implicit in the concept of 'inalienable rights' – is accentuated by the break-up of the English and Spanish imperial systems, with their unities of language, their common enemies, and (in the case of Spanish America) their contiguous terrain. Examples of the persistence of this motif might be found, for instance, in the controversial role of the terms 'Africa' in the writing of the Nigerian author Chinua Achebe, or 'America' in the essays of the Cuban patriot José Marti.

Thus, of course, not all Third World novels about nations are 'nationalistic'. The variations range from outright attacks on independence, often mixed with nostalgia for the previous European *status quo* (as in the work of V. S. Naipaul, Manohar Malgonkar, and others), to vigorously anti-colonial works emphasizing native culture (Ngũgĩ wa Thiong'o, Tayib Salih, Sipho Sepamla, and others), to cosmopolitan explanations of the 'lower depths', or the 'fantastic unknown' by writers acquainted with the tastes and interests of dominant culture (García Márquez, Wole Soyinka, Salman Rushdie, and others).

As we shall see, in one strain of Third World writing the contradictory topoi of exile and nation are fused in a lament for the necessary and regrettable insistence of nation-forming, in which the writer proclaims his identity with a country whose artificiality and exclusiveness have driven him into a kind of exile – a simultaneous recognition of nationhood and an alienation from it. As we have said, the cosmopolitan thrust of the novel form has tended to highlight this branch of well-publicized Third World fiction. One result has been a trend of cosmopolitan commentators on the Third World, who offer an *inside view* of formerly submerged peoples for target reading publics in Europe and North America in novels that comply with metropolitan literary tastes.

Some of its better known authors have been from Latin America: for example, García Márquez, Vargas Llosa, Alejo Carpentier, Miguel Asturias, and others. But there is also a related group of postwar satirists of nationalism and dependency – writers of encyclopedic national narratives that dismember a recent and particularized history in order to expose the political dogma surrounding and choking it. Here one thinks especially of the Indo-English author Salman Rushdie, of the Paraguayan novelist Augusto Roa Bastos, and the South African Nadine Gordimer, along with many others.

In the case of Salman Rushdie, for instance, the examples of India and Pakistan are, above all, an opportunity to explore postcolonial *responsibility*. The story he tells is of an entire region slowly coming to think of itself as one, but a corollary of his story is disappointment. So little improvement has been made. In fact, the central irony of his novels is that independence has damaged Indian spirits by proving that 'India' can act as abominably as the British did. In a kind of metafictional extravaganza, he treats the heroism of nationalism bitterly and comically because it always seems to him to evolve into the nationalist demagogy of a caste of domestic sellouts and powerbrokers.

This message is very familiar to us because it has been easier to embrace in our metropolitan circles than the explicit challenges of, say, the Salvadoran protest-author Manlio Argueta, or the sparse and caustic satires of the Nigerian author, Obi Egbuna. However, it is perhaps the trend's overt cosmopolitanism – its Third World thematics as seen through the elaborate fictional architecture of European high art – that perfectly imagines the novel's obsessive nation-centredness and its imperial (that is, universalizing) origins. Distanced from the sacrifices and organizational drudgery of actual resistance movements, and yet horrified by the obliviousness of the west towards their own cultures, writers like Rushdie and Vargas Llosa have been well poised to thematize the centrality of nation-forming while at the same time demythifying it from a European perch. Although Vargas Llosa's erudite and stylistically sumptuous *The War of the End of the World*, for example, is not at all characteristic of the 'counter-hegemonic aesthetics' of much Third World writing, its very disengagement frees him to treat the ambivalence of the independence process as a totality, and, although negatively, reassert its fundamental importance to the postcolonial imagination. His treatment may be neither the most representative nor the most fair, but its very rootlessness brilliantly articulates the emotional life of decolonization's various political contestants. It is 'in-between'.

## Chapter 26

# HOMI K. BHABHA

## DISSEMINATION

## Time, narrative, and the margins of the modern nation

From 'Dissemination: Time, Narrative, and the Margins of the Modern Nation'
in Homi K. Bhabha (ed.) *Nation and Narration* London: Routledge, 1990.

**H**OW DOES ONE WRITE THE NATION'S MODERNITY as the event of the everyday and the advent of the epochal? The language of national belonging comes laden with atavistic apologues, which has led Benedict Anderson to ask: 'But why do nations celebrate their hoariness, not their astonishing youth?' (Anderson 'Narrating the nation' *The Times Literary Supplement*). The nation's claim to modernity, as an autonomous or sovereign form of political rationality, is particularly questionable if, with Partha Chatterjee, we adopt the post-colonial perspective:

> Nationalism . . . seeks to represent itself in the image of the Enlightenment and fails to do so. For Enlightenment itself, to assert its sovereignty as the universal ideal, needs its Other; if it could ever actualise itself in the real world as the truly universal, it would in fact destroy itself.
>
> (Chatterjee 1986: 17)

Such ideological ambivalence nicely supports Gellner's paradoxical point that the historical necessity of the idea of the nation conflicts with the contingent and arbitrary signs and symbols that signify the effective life of the national culture. The nation may exemplify modern social cohesion, but:

> Nationalism is not what it seems, and above all not what it seems to itself . . . The cultural shreds and patches used by nationalism are often arbitrary historical inventions. Any old shred would have served as well. But in no way does it follow that the principle of nationalism . . . is itself in the least contingent and accidental.
>
> (Gellner 1983: 56)

The problematic boundaries of modernity are enacted in these ambivalent temporalities of the nation-space. The language of culture and community is poised on the fissures of the present becoming the rhetorical figures of a national past. Historians transfixed on the event and origins of the nation never ask, and political theorists possessed of the 'modern' totalities of the nation – 'Homogeneity, literacy and anonymity are the key traits' (Gellner 1983: 38) – never pose, the awkward question of the disjunctive representation of the social, in this double-time of the nation. It is indeed only in the disjunctive time of the nation's modernity – as a knowledge disjunct between political rationality and its impasse, between the shreds and patches of cultural signification and the certainties of the nationalist pedagogy – that questions of nation as narration come to be posed. How do we plot the narrative of the nation that must mediate between the teleology of progress tipping over into the 'timeless' discourse of irrationality? How do we understand that 'homogeneity' of modernity – the people – which, if pushed too far, may assume something resembling the archaic body of the despotic or totalitarian mass? In the midst of progress and modernity, the language of ambivalence reveals a politics 'without duration', as Althusser once provocatively wrote: 'Space without places, time without duration' (Althusser 1972: 78). To write the story of the nation demands that we articulate that archaic ambivalence that informs modernity. We may begin by questioning that progressive metaphor of modern social cohesion – *the many as one* – shared by organic theories of the holism of culture and community, and by theorists who treat gender, class, or race as radically 'expressive' social totalities.

## Chapter 27

# DAVID CAIRNS AND SHAUN RICHARDS

## WHAT ISH MY NATION?

From 'What Ish My Nation?' *Writing Ireland: Colonialism, Nationalism and Culture* Manchester: Manchester University Press, 1988.

THE PROCESS OF DESCRIBING THE COLONIZED [in Ireland] and inscribing them in the discourse as second-order citizens in comparison with the colonizers commenced with the invocation of the judicial and military power of the State, but subsequently the colonizers attempted to convince the colonized themselves of their irremovable deficiencies and the consequent naturalness and permanence of their subordination. The wish of the colonizer that subjection should be willingly accepted rather than require constant recourse to coercion, can be seen in *Henry V*, the culmination of Shakespeare's second tetralogy, itself a dramatization of the process involved in the constitution of the unified nation, particularly as the process is expressed in the constitution of the unified and ordered subject who, to emphasize the power of the process, is the monarch himself. The transformation of Hal into Henry, particularly through the rejection of Falstaff, is a highly charged realization of the denial and repression of the 'other' attendant upon the constitution of the ordered subject and nation. What Shakespeare dramatizes is the originating moment of nationhood when the nation 'becomes conscious of itself, when it creates a model of itself' (Lotman and Uspensky 1978: 227). . . .

*Henry V* dramatizes the might and mercy of the English Nation State as it resolves the action as the victory at Agincourt is crowned by Henry's kissing the French Princess in the recognition that she is his 'sovereign queen'. The result of this betrothal, it is hoped, will be that the contending kingdoms should share an equal unity and 'Christian-like accord' (Act 5, Sc. ii). Philip Edwards has argued that the resolution of this imperial war is a piece of dramatic wish-fulfilment: that a contemporary cause of discord, namely Ireland, should come to an equally satisfactory conclusion (Edwards 1979: 74–86). . . .

While the play dramatizes an idealized resolution of the Anglo-Irish discord in the unity of marriage, and threatens in dialogue the alternative of massacre, there is also a more direct engagement with the problems of 'internal colonialism', which, in their expression, suggest that they have already been resolved. The famous scene at the English camp when English, Welsh, Scots and Irish Captains meet in an encounter whose main function is to dramatize their united presence in an army constantly referred to as English (Act 3, Sc. ii), reveals that just as in the marriage which unites England and France, so in the union which produces the English Nation State, there is a relationship which is 'structured in dominance'. What simultaneously unites and divides the Captains, or at least distinguishes between them, is the English language. Fluellen the Welshman substitutes 'p' for 'b' and utters 'look you' at intervals, Jamy the Scot has even more deviations from 'standard' linguistic expectations with his frequent use of 'gud' and mispronunciation of 'marry', while Macmorris the Irishman is the very embodiment of the 'stage

Irishman'; pugnacious and argumentative, expressing all in repetitious 'mispronunciations': 'O, tish ill done, tish ill done! By my hand, tish ill done!' (Act 3, Sc. ii). These Celts are united in their service to the English Crown. Their use of the English language, however, reveals that 'service' is the operative word, for in rank, in dramatic importance, and in linguistic competence, they are comical second-order citizens. They are, moreover, disputatious, and the argument between Fluellen and Macmorris, which is resolved by Gower's admonition, is further dramatic evidence of the harmony which England has brought to the fractious occupants of the Celtic fringe. Shakespeare's dramatization of the harmonious incorporation of such disparate elements into the English State reaches its peak in Macmorris's famous question: 'What ish my nation?' (Act 3, Sc. ii). As Philip Edwards argues, Macmorris's outburst is a denial of such separate status, brought on by the sensed implication in the words of Fluellen, that while the Welsh may speak from within the united State, Macmorris is a member of a separate and therefore marginal group. 'What ish my nation?' is therefore a rhetorical question to which the answer is supplied by Macmorris's service in the English army. The achievement, on a mass scale, of Macmorris's incorporation would represent a triumphant conclusion to the process of unmaking . . . as a pre-requisite for the fashioning of godly and biddable second-order citizens. The process of self-fashioning required the continued presence of an 'other' so that the maintenance of subtle points of differentiation from the colonizer would continue to reproduce, not only the subordination of the colonized, but the superordination of the colonizer. . . .

The Shakespearean 'fiction' of *Henry V* is, then, the expression of a politically advantageous 'myth' and indeed is expressed in terms which are themselves subsequently utilized for overtly political purposes. James I of England whose accession to the throne came only four years after the composition of *Henry V*, also expressed the indissolubility of the nation in terms of a marriage: 'I am the husband, and the whole isle is my lawful wife' (Edwards 1979: 84). This merging of the marginal with the mighty is equated by James with the fate of the brook which flows into a river which, in turn, flows into an ocean: 'so by the conjunction of divers little kingdoms in one are all these private differences and questions swallowed up' (84).

Culture, then, requires the drive toward – if not the achievement of – unity. But the contradictions that are necessarily excluded as a means of its achievement are quite literally those elements which contra-dict, speak against and speak otherwise than the dominant group. While Henry can be seen to court the French Princess in her own tongue and she replies in English, the same degree of linguistic parity is not extended to the Celtic Captains, for their position in the contemporary world of Elizabethan England was potentially, and actually, far more disruptive than that of a nation whose separate status was now an acknowledged if not welcomed fact. The Welsh, Scots and Irish must, therefore, be seen to speak English as evidence of their incorporation within the greater might of England, but they must speak it with enough deviations from the standard form to make their subordinate status in the union manifestly obvious. What cannot be acknowledged is their possession of an alternative language and culture, for to do so would be to stage the presence of the very contradictions which the play denies in its attempt to stage the ideal of a unified English Nation State. The resolution in the play is seen to be achieved by marriage rather than by massacre, by incorporation rather than by exclusion, but the inclusion of the Celts within the English State, of which the army is a paradigm, is as a result of an equally devastating act of cultural elision. The victims in the process of the march towards unity are those who contradict, and so implicitly question, the dominance of the incorporating power. Shakespeare's work engages with the process of colonial discourse at the moment of its mobilization to deal with Ireland, but the position of the colonized, namely Macmorris, is seen as one of proud inclusion. In this sense, the play, despite its references to the slaughter of Irish rebels, is an idealization of an actuality which stubbornly refused to conform.

# PART FIVE

# Hybridity

IN THE PRECEDING SECTION ON NATIONALISM it became clear that the idea of the nation is often based on naturalized myths of racial or cultural origin. That the need to assert such myths of origin was an important feature of much early post-colonial theory and writing, and that it was a vital part of the collective political resistance which focused on issues of separate identity and cultural distinctiveness is made clear in many of the extracts collected there. But what is also made clear is how problematic such construction is and how it has come under question in more recent accounts.

Most post-colonial writing has concerned itself with the hybridized nature of post-colonial culture as a strength rather than a weakness. Such writing focuses on the fact that the trans-action of the post-colonial world is not a one-way process in which oppression obliterates the oppressed or the colonizer silences the colonized in absolute terms. In practice it rather stresses the mutuality of the process. It lays emphasis on the survival even under the most potent oppression of the distinctive aspects of the culture of the oppressed, and shows how these become an integral part of the new formations which arise from the clash of cultures characteristic of imperialism. Finally, it emphasizes how hybridity and the power it releases may well be seen to be the characteristic feature and contribution of the post-colonial, allowing a means of evading the replication of the binary categories of the past and developing new anti-monolithic models of cultural exchange and growth.

Hybridity occurs in post-colonial societies as a result of conscious moments of cultural suppression, as when the colonial power invades to consolidate political and economic control, or when settler-invaders dispossess indigenous peoples and force them to 'assimilate' to new social patterns. It may also occur in later periods when patterns of immigration from the metropolitan societies and from other imperial areas of influence (e.g. indentured labourers from India and China) continue to produce complex cultural palimpsests with the post-colonized world.

Not surprisingly, since such formulations tend to resist ideas of a pure culture of either the post- or pre-colonial they have not found universal assent. They have also tended to emerge most strongly where no simple possibility for asserting a pre-colonial past is available, notably in the radically dislocated culture of the West Indies. Yet these regional patterns have formed the basis for the development of literary forms (such as 'magical realism') which have had a wide influence, and which have been applied by critics to societies of widely different kinds such as those of settler colonies, and even, as Homi Bhabha's piece indicates, to theories of colonization in societies such as India. In a different way, as Chinua Achebe's account of his childhood shows, even the cultures in countries such as Nigeria which have sought energetically to assert the validity and continuity of their pre-colonial past have still found a fruitful metaphor in the idea of cross-fertilization between their constitutive elements. They have realized that for the foreseeable future much of the artistic and social production of their world will take place within the constraints of the traces of the colonial and neo-colonial moment, and that much of the distinctiveness of contemporary post-colonial societies will be produced by and against this process either by vigorous resistance or, more frequently in recent times, by a dialogic process of recovery and reinscription.

The term hybridity has been sometimes misinterpreted as indicating something that denies the traditions from which it springs, or as an alternative and absolute category to which all post-colonial forms inevitably subscribe but, as E. K. Brathwaite's early and influential account of Jamaican creolization made clear, the 'creole' is not predicated upon the idea of the disappearance of independent cultural traditions but rather on their continual and mutual development. The interleaving of practices will produce new forms even as older forms continue to exist. The degree to which these forms become hybridized varies greatly across practices and between cultures. Thus, as critics like Karin Barber and E. K. Brathwaite have noted, oral practices may continue alongside the orally-influenced forms of post-colonial written culture in countries like Nigeria and Jamaica.

It is probably true to say though that no post-colonial form has been able entirely to avoid the impact of the shifts that have characterized the post-colonial world. While assertions of national culture sought to articulate the dangerous politics of assimilation implicit in the colonial, theories of the hybridity of the post-colonial world assert a different and arguably more potent resistance in the counter-discursive practices they celebrate. Whatever one's final view, these discussions have been the site of one of the most vigorous and fruitful critical debates in recent years.

## Chapter 28

# KIRSTEN HOLST PETERSEN AND
# ANNA RUTHERFORD

## FOSSIL AND PSYCHE

From *Enigma of Values* Aarhus: Dangaroo Press, 1976.

### Fossils

> Modern man is the product of that evolutionary symbiosis, and by any other hypothesis incomprehensible, indecipherable. *Every living being is also a fossil. Within it, all the way down to the microscopic structure of its proteins, it bears the traces if not the stigmata of its ancestry.* This is even truer of man than of any other animal species because of the dual evolution – physical and ideational – to which he is heir.
>
> Jacques Monod

> The word fossil is used in an idiosyncratic sense to invoke a rhythmic capacity to re-sense contrasting spaces and to suggest that a curious rapport exists between ruin and origin as latent to arts of genesis.
>
> Wilson Harris

> Only a dialogue with the past can produce originality.
>
> Wilson Harris

WE LIVE IN A WORLD IN WHICH WE HAVE polarized the so-called savage mind and the so-called civilized mind. Wilson Harris believes that we are in fact much closer to the savage mind than we think or would like to admit and he agrees with Monod when he says that each living person is a fossil in so far as each man carries within himself remnants of deep-seated antecedents.

The past plays tricks on us and conditions our present responses. Floating around in the psyche of each one of us are all the fossil identities. By entering into a fruitful dialogue with the past one becomes able to revive the fossils that are buried within oneself and are part of one's ancestors. To illustrate this, one could mention the uses to which a people could put their common past or cultural heritage. During the course of his stay at Aarhus, Wilson Harris gave a public lecture which he entitled 'Magical Realism'. In this lecture he told a story which was a perfect example of the positive influence which an awareness of one's roots can have. Part of the tale is repeated here:

I was born on the coastlands of Guyana and one is aware that one has there a heterogeneous body of peoples, peoples whose antecedents came from Africa, from India, Europe, and so on. And apart from that there is a very significant Amerindian presence, people who are descended from the pre-Columbian world. One of the sad things is that most of these people live within a context in which the issue of community remains alien or hidden away. This is something which came home to me in a peculiarly symbolic way on my first expedition into the interior of Guyana, in which I was aware of an enormous difference between the landscape of the coast-lands and the landscape of the interior. I had penetrated 150 miles. It seemed as if one had travelled thousands and thousands of miles, and in fact had travelled to another world, as it were, because one was suddenly aware of the fantastic density of place. One was aware of one's incapacity to describe it, as though the tools of language one possessed were inadequate. It was pointless describing the river as run-ning dark, the trees as green, or the rocks as grey. All this seemed less to do with the medium of place and more to do with the immediate tool of the world as repre-senting or signifying 'place'. Later I was to relate myself to those 'representations' or 'significations' as relative faces of the dynamic mystery of language, and this for me was a groping but authentic step into the reality of place. At first, however, I was conscious of how helpless I was in wrestling with something immensely authentic, immensely rich, immensely challenging. And I believe in my early experiments with poem and fiction I was simply using the word as a tool of identity. That is, I could not relate identity to eclipsed perspectives of place and community. And one of the first catalysts which occurred, which assisted me to come to grips with the kind of narrative juxtapositions which I needed and which I wanted to find, happened on an expedition into the Potaro river, which is a tributary of the Essequebo.

The Essequebo runs out of Brazil into the Atlantic. It runs through this fantastic landscape in which if one gets into the forest it seems as if the sky itself is a lake and the rivers are pouring from the sky. We were gauging the river for hydro-electric power and had chosen as our station a section where the river narrowed and then opened up again to run towards the Tumatumari rapids a mile or so away. It was necessary to gauge the river at all stages from the lowest to the highest levels. One needs to do this continuously because the sort of stage discharge curve one gets is built up out of frequent observations that check back on themselves. We set up a base line on one bank with alignment rods at right angles to this. We were thus able to align ourselves and anchor our boat in the river, one anchor at the stern and another at the bow. Then with a sextant we took a reading in order to calculate distances from the bank as we made our way across the river. The Potaro river is strangely beautiful and secretive. When the river falls, the sand banks begin to appear. At the foot of the Tumatumari rapids or falls the sand is like gold. Above, an abrupt change of texture occurs – it is white as snow. These startling juxtapositions seemed to me immensely significant in some curious and intuitive way that bore upon an expressionistic void of place and time.

When the river runs high the sand banks disappear. We were – on the particular expedition to which I am referring – gauging the river at a very high and dangerous stage. The water swirled, looked ugly and suddenly one of the anchors gripped the bed of the stream. The boat started to swing around and to take water. We could not dislodge the anchor. I decided that the only thing we could do was to cut ourselves free. So we severed the anchor rope and that was the end of that. Two or three years later, gauging the river in the same way, the identical impasse happened. Once again the anchor at the stern lodged in the bed of the stream. And this time it was much more crucial because the boat swung so suddenly, we took so much water, that it seemed to me at that moment that we were on the point of sinking. I am sure I couldn't have swum to the river bank if the boat had gone down because at that high stage I would have been pulled into the Tumatumari falls and decapitated by the rocks. As the boat swung I said to a man behind me: 'Cut the rope.' Well, he was so nervous that he took his prospecting knife and all he could do was a sort of feeble sawing upon the anchor rope as if he were paralyzed at that moment by the whole thing, the river, the swirling canvas of the stream. He was paralyzed. And then another member – the outboard mechanic – gave a sudden tug and the anchor moved. The boat righted itself. Half-swamped as we were, we were able to start the outboard engine and drive towards the bank. We began pulling up the anchor as we moved in. We got to the bank and then were able to bring the anchor right up when we discovered that it had hooked into the one we had lost three years before. Both anchors had now come up.

It is almost impossible to describe the kind of energy that rushed out of that constellation of images. I felt as if a canvas around my head was crowded with phantoms and figures. I had forgotten some of my own antecedents – the Amerindian/ Arawak ones – but now their faces were on the canvas. One could see them in the long march into the twentieth century out of the pre-Columbian mists of time. One could also sense the lost expeditions, the people who had gone down in these South American rivers. One could sense a whole range of things, all sorts of faces – angelic, terrifying, daemonic – all sorts of contrasting faces, all sorts of figures. There was a sudden eruption of consciousness, and what is fantastic is that it all came out of a constellation of two ordinary objects, two anchors.

(Harris 1973: 38–41)

The two anchors released an awareness of possibilities or in Wilson Harris's terms 'a density of resources'. Through such incidents one is able to gain an insight into a new dimension of psychic possibilities which up until then one had been unaware of.

On the other hand of course the same search for roots can give an entirely different result and can be used to foster a narrow nationalism. This was the case with Nazi Germany where the past was evoked to serve present feelings of national and racial superiority.

What must be remembered is that fossils like 'living' beings contain restrictive as well as explosive rooms or spaces and the fossil value of our human and a-human antecedents can either act as positive forces or can become prejudices, hideous biases, leading to implacable animism. So in fact one half of our 'fossil value' is constantly combating the other half.

## Architectonic fossil spaces

Awareness of the ambivalence of fossils enables one to visualize new possibilities and construct a new scale along which one can attempt to progress. This constructive process is what Wilson Harris means by architectonic; it presupposes an insight that may enable us to relate to the static in a new way thereby modifying both it and us. It is a process that can be equated with profound creativity.

When one realizes that involuntary codes are built into targets and affect objective judgement the creative imagination embarks on a quest for new values, on the psychical journey, and the former target is given new significance by the creative recognition of the architectonic fossil spaces. (See Figure 28.1.) . . .

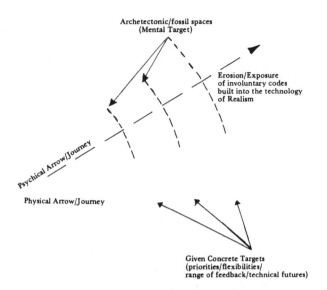

*Figure 28.1* Theme of expedition

What one must remember is that the goal of the spiritual journey, which is the realization of one's vision, can never be final except in the beginning of something new. The possibility and necessity of beginning again is always inherent in it; true permanence is never static, it is an eternal process of becoming, susceptible to dialogue with otherness.

There are moments in history that may endure for a decade or a generation when a culture may 'rest' in its achievements. This is natural and desirable. When however such a pattern of 'rest' begins to assume an idolatrous function of 'changelessness' Wilson Harris suggests that the institutions and models of the day begin to conceal from the body politic itself a growth of catastrophe to which there has ceased to be a 'creative' or 'digestive' response. Then there seems to be no possibility of change except through familiar violence or revenge patterns of self-destructive order.

South African society today provides an excellent illustration of some of the concepts that have just been discussed. The two alchemical dimensions, 'albedo' and 'nigredo' (not necessarily black and white but in this case they happen to be so) have become isolated in every way from one another. The white South Africans have locked themselves in their own apparently changeless fortresses of culture; 'albedo' (white supremacy) has become a basis for an idolatrous self-sufficient feedback.

A breakthrough from such a condition or dilemma is the vision Athol Fugard presents us with in *The Blood Knot*. In his play he has shown the dynamite-like situation that exists in present-day South Africa where the dialogue between opposites has been superseded by a totalitarian family of man. . . .

What one must remember is that fulfilment is a *ceaseless* task of the psyche; that identity is part of an infinite movement, that one can only come into a dialogue with the past and future, a dialogue which is necessary, if one ceases to invest in a single (and therefore latent totalitarian) identity. If one invests in identity one locks oneself in an immobile horizon; totalitarian identity was the extreme function of the Nazis. One must be prepared to participate in the immense and specific challenges of a wider community, to participate in what Wilson Harris calls the 'complex creativity involved in the "digestion" and "liberation" of contrasting spaces'.

## Chapter 29

# CHINUA ACHEBE

## NAMED FOR VICTORIA, QUEEN OF ENGLAND

From 'Named for Victoria, Queen of England' *New Letters* 40(3) (Fall), 1973.

I WAS BAPTISED ALBERT CHINUALUMOGU. I dropped the tribute to Victorian England when I went to the university although you might find some early acquaintances still calling me by it. The earliest of them all – my mother – certainly stuck to it to the bitter end. So if anyone asks you what Her Britannic Majesty Queen Victoria had in common with Chinua Achebe, the answer is: they both lost their Albert! As for the second name which in the manner of my people is a full-length philosophical statement I simply cut it in two, making it more businesslike without, I hope, losing the general drift of its meaning.

I have always been fond of stories and intrigued by language – first Igbo and later English, which I began to learn at about the age of eight. I don't know for certain but I probably have spoken more words in Igbo than English but I have definitely written more words in English than Igbo. Which I think makes me perfectly bilingual. Some people have suggested that I should be better off writing in Igbo. Sometimes they seek to drive the point home by asking me in which language I dream. When I reply that I dream in both languages they seem not to believe it. More recently I have heard an even more potent and metaphysical version of the question: in what language do you have an orgasm? Which would settle the matter if I knew.

We lived at the crossroads of cultures. We still do today; but when I was a boy one could see and sense the peculiar quality and atmosphere of it more clearly. I am not talking about all that rubbish we hear of the spiritual void and mental stresses that Africans are supposed to have, or the evil forces and irrational passions prowling through Africa's heart of darkness. We know the racist mystique behind a lot of that stuff and should merely point out that those who prefer to see Africa in those lurid terms have not themselves demonstrated any clear superiority in sanity or more competence in coping with life.

But still the crossroads does have a certain dangerous potency; dangerous because a man might perish there wrestling with multiple-headed spirits, but also he might be lucky and return to his people with the boon of prophetic vision.

On one arm of the cross we sang hymns and read the Bible night and day. On the other my father's brother and his family, blinded by heathenism, offered food to idols. That was how it was supposed to be anyhow. But I knew without knowing why it was too simple a way to describe what was going on. Those idols and that food had a strange pull on me in spite of my being such a thorough little Christian that often at Sunday services at the height of the grandeur of *Te Deum Laudamus* I would have dreams of a mantle of gold falling on me while the choir of angels drowned our mortal song and the voice of God Himself thundering: This is my beloved son in whom I am well pleased. Yes, despite those delusions of divine destiny I was not past taking my little sister to our neighbour's house when our parents were not looking and partaking of heathen festival meals. I never found their rice to have the flavour of idolatry. I was about ten then. If anyone likes to believe that I was torn by spiritual agonies or stretched on the rack of my ambivalence he certainly may suit himself. I do not remember any undue distress. What I do remember was a fascination for the ritual and the life on the other arm of the crossroads. And I believe two things were in my favour – that curiosity and the little distance imposed between me and it by the accident of my birth. The distance becomes not a separation but a bringing together like the necessary backward step which a judicious viewer may take in order to see a canvas steadily and fully.

I was lucky in having a few old books around the house when I was learning to read. As the fifth in a family of six children and with parents so passionate for their children's education I inherited many discarded primers and readers. I remember *A Midsummer Night's Dream* in an advanced stage of falling apart. I think it must have been a prose adaptation, simplified and illustrated. I don't remember whether I made anything of it. Except the title. I couldn't get over the strange beauty of it. A Midsummer Night's Dream. It was a magic phrase – an incantation that conjured up scenes and landscapes of an alien, happy and unattainable land.

I remember also my mother's *Ije Onye Kraist* which must have been an Igbo adaptation of *Pilgrim's Progress*. It could not have been the whole book; it was too thin. But it had some frightening pictures. I recall in particular a most vivid impression of *the valley of the shadow of death*. I thought a lot about death in those days. There was another little book which frightened and fascinated me. It had drawings of different parts of the human body. But I was only interested in what my elder sister told me was the human heart. Since there is a slight confusion in Igbo between heart and soul I took it that that strange thing looking almost like my mother's iron cooking pot turned upside down was the very thing that flew out when a man died and perched on the head of the coffin on the way to the cemetery.

I found some use for most of the books in our house but by no means all. There was one Arithmetic book I smuggled out and sold for half-a-penny which I needed to buy the tasty *mai-mai* some temptress of a woman sold in the little market outside the school. I was found out and my mother who had never had cause till then to doubt my honesty – laziness, yes; but not theft – received a huge shock. Of course she redeemed the book. I was so ashamed when she brought it home that I don't think I ever looked at it again which was probably why I never had much use for mathematics.

My parents' reverence for books was almost superstitious; so my action must have seemed like a form of juvenile simony. My father was much worse than my mother. He never destroyed any paper. When he died we had to make a bonfire of all the hoardings of his long life. I am the very opposite of him in this. I can't stand paper around me. Whenever I see a lot of it I am seized by a mild attack of pyromania. When I die my children will not have a bonfire.

The kind of taste I acquired from the chaotic literature in my father's house can well be imagined. For instance I became very fond of those aspects of ecclesiastical history as could be garnered from *The West African Churchman's Pamphlet* – a little terror of a booklet prescribing interminable Bible readings morning and night. It had the date of consecration for practically every Anglican bishop who ever served in West Africa; and, even more intriguing, the dates of their death. Many of them didn't last very long. I remember one pathetic case (I forget his name) who arrived in Lagos straight from his consecration at St. Paul's Cathedral and was dead within days, and his wife a week or two after him. Those were the days when West Africa was truly the white man's grave, when those great lines were written, of which I was at that time unaware:

Bight of Benin! Bight of Benin!
Where few come out though many go in!

But the most fascinating information I got from *Pamphlet*, as we called it, was this cryptic entry for the month of August:

*Augustine, Bishop of Hippo, died 430*

It had that elusive and eternal quality, a tantalizing unfamiliarity which I always found moving.

I did not know that I was going to be a writer because I did not really know of the existence of such creatures until fairly late. The folk-stories my mother and elder sister told me had the immemorial quality of the sky and the forests and the rivers. Later, when I got to know people it still didn't help much. It was the same Europeans who made all the other marvellous things like the motor-car. We did not come into it at all. We made nothing that wasn't primitive and heathenish.

The nationalist movement in British West Africa after the Second World War brought about a mental revolution which began to reconcile us to ourselves. It suddenly seemed that we too might have a story to tell. *Rule Britannia!* to which we had marched so unself-consciously on Empire Day now stuck in our throat.

At the university I read some appalling novels about Africa (including Joyce Cary's much praised *Mister Johnson*) and decided that the story we had to tell could not be told for us by anyone else no matter how gifted or well-intentioned.

Although I did not set about it consciously in that solemn way I now know that my first book, *Things Fall Apart*, was an act of atonement with my past, the ritual return and homage of a prodigal son. But things happen very fast in Africa. I had hardly begun to bask in the sunshine of reconciliation when a new cloud appeared, a new estrangement. Political independence had come. The nationalist leader of yesterday (with whom it had not been too difficult to make common cause) had become the not so attractive party boss. And then things really got going. The party boss was chased out by the bright military boys, new idols of the people. But the party boss knows how to wait, knows by heart the counsel Mother Bedbug gave her little ones when the harassed owner of the bed poured hot water on them: 'Be patient,' said she, 'for what is hot will in the end be cold.'

One hears that the party boss is already conducting a whispering campaign: 'You done see us chop,' he says, 'now you see *dem* chop. Which one you like pass?' And the people are truly confused.

In a little nondescript coffee shop where I sometimes stop for a hamburger in Amherst there are some unfunny inscriptions hanging on the walls, representing a one-sided dialogue between management and staff. The unfunniest of them all reads – poetically:

Take care of your boss
The next one may be worse.

The trouble with writers is that they will often refuse to live by such rationality.

## Chapter 30

# JACQUES STEPHEN ALÉXIS

# OF THE MARVELLOUS REALISM OF THE HAITIANS

From 'Of the Marvellous Realism of the Haitians' *Présence Africaine*
8–10, 1956.

## Towards a dynamic integration of the Marvellous; Marvellous Realism

THE ART AND LITERATURE OF several peoples of Negro origin, like those of many countries of the Antilles and Central and Latin America, have frequently given the example of the possible dynamic integration of the Marvellous in realism. It does not seem to us fair to think that the fascination, originality, and singular attractions of the aesthetic forms proper to countries of Negro origin are inexplicable, or that they are the result of chance, or the attraction of novelty, or a question of fashion. It is true that all peoples, whoever they may be, are endowed with feeling as well as with reason, but let us remember the saying that 'the people who have no more legends are condemned to perish of cold', and let us objectively recognize the fact that modern life with its stern rates of production, with its concentration of great masses of men into industrial armies, caught up in the frenzy of Taylorism, with its inadequate leisure, and its context of mechanized life, hampers and slows down the production of legends and a living folk lore. By way of contrast, the under-developed populations of the world who have still quite recently had to live in contact with Nature, have for centuries been compelled particularly to sharpen their eyes, their hearing, their sense of touch. The peoples among whom industrial life is most highly developed, have, for their part, used their senses to a lesser extent during the last few centuries, since material civilization has saved them a great deal of effort; that has been the price of industrial mechanization, certain regrettable consequences of which everyone recognizes. The under-developed populations of the world, on their part, know a blend of mechanical civilization and 'natural' life, so to speak, and it is beyond dispute that they have feelings of special liveliness. The problems which they have to face, the low standard of living, unemployment, poverty, hunger and illness are also problems which it is important to liquidate, and we do not overlook this.

These specially lively feelings give these peoples artistic possibilities which should be used. From there it is only a step to conceiving that the Haitian, for instance, does not seek to grasp the whole of sensible reality, but what strikes him, what threatens him, what in Nature particularly touches and stirs his emotions. From another angle, since reality is not intelligible in all its aspects to the members of under-developed communities, he naturally transposes his conceptions of relativity and of the marvellous in to his vision of everyday reality. A bird in rapid flight is, above all, a pair of wings, a woman giving suck impresses by her round and heavy breasts, a wild beast is essentially a footfall and a roar, the body responds naturally to music without following a pre-ordained plan, in contrast to other men who exercise a constant restraint over their bodies in order to conform to the social usages of polite society. To demonstrate the peculiar, and sometimes paradoxical, sensitivity of the Haitian, for example, we would cite the fact that in our Voodoo religion a man possessed will sometimes take a red hot iron in his hands without burning them and lick it; he climbs trees with agility, even if he is an old man, he succeeds in dancing for several days and nights on end, he chews and swallows glass. . . . Quite apart from any mystic conception of the world, in the light of numerous observed facts, there are many values which should be revised by science. Can one, in effect, strip a human being of all his antecedents, of all the unconditioned reflexes born of the conditioned reflexes transmitted by heredity? A human being cannot be the son of no man, the past and history cannot be denied; the Haitian, and, through him, his culture, is the legatee of an inheritance of reactions of behaviour and habitude anterior to his hundred and fifty years of independence; he is still, to a large extent, heir of cultural elements derived from distant Africa. The Haitian has an air, a family likeness, internal as much as external, which makes him resemble on many planes his other brothers in the world of Negro origin. That, moreover, is why we are here at this Congress.

It is because they recognize that their people express their whole consciousness of reality by the use of the Marvellous, that Haitian writers and artists have become aware of the formal problem of its use. Behind the imaginary characters of the *romancero* of Bouqui and Malice, it is a faithful picture of the conditions of rural life which the Haitian story-teller executes, it is its beauties and its ugliness, and struggles, the drama of the oppressors and the oppressed which he brings on to their stage. In his working songs, for among us work is unthinkable without music, or without songs in which all the workers take part, – in his working songs the Voodoo gods of the Haitian are nothing but an inspiration towards the ownership of the land on which he works, an aspiration towards the rain which feeds the harvest, an aspiration towards abundant bread, an aspiration to get rid of the maladies which afflict him, an aspiration towards betterment in every sphere. Even the religious songs and dances are transparent symbols in which they beg the gods for the solution of a specific problem; there are, moreover, pleasant gods, soldier gods, politician gods, powerful gods and exploited gods, gods who are unhappy in love, infirm gods, one-legged gods, blind gods, dumb gods, rapacious gods and gods who are simple, kind and helpful, poets and laughter. When they are mariners, our people also include the width of the horizon, the murmuring of the waves, the drama of the seas, in the form of Agouet Arroyo, the Loa of the Ocean, they hymn the Diamond Siren, 'the Sun Queen', as they sometimes say, but nothing more actual, nothing more truthful, nothing more loving than all these entities. How could we be unconscious to the extent of refusing to use all that in the service of realizing specific and actualized struggles: That is what made the poet and playwright Morisseau-Leroy write as follows in a recent article:

We are again living through a renaissance of the Haitian song. We see flourishing again forms of expression, both rich and original, as in the times when Dithyrambic or satyric, lyric or bucolic couplets flowed from the lips of a people whose temper and humour were proof against all misery. . . . From one end of the Republic to another nephews and uncles, nieces and aunts are singing or humming in cadence. . . . And if Agoue T'Arroya does not afford that rude class of workers enough protection against shipwreck, the official social institutions of the Republic have hardly done better in that direction. It is therefore gratuitously that in their songs they invoke the gods and the chiefs. . . . I want before everything to emphasize that if reality in its local aspect, as in its universal aspect, escapes those who have been led astray by a certain humanism, the popular bards, the 'composes' remain, in my view, the sole masters of Haitian poetry, the only ones capable of making us sing and dance together in the unavowed and common conviction that the people are safe and sound.

What, then, is the Marvellous, except the imagery in which a people wraps its experience, reflects its conception of the world and of life, its faith, its hope, its confidence in man, in a great justice, and the explanation which it finds for the forces antagonistic to progress? It is true that the Marvellous implies ingenuousness and empiricism, if not mysticism, but it has been proved that something else can be bound up with it. When the great painter Wilson Bigaud painted a picture called 'The Earthly Paradise' he made full use of the Marvellous, but has the painter not expressed the way in which the Haitian people conceives a time of happiness? Look at all those fruits which accumulate in bunches on the canvas, those dense masses of colour, all those splendid animals, tranquil and fraternal, including the wild beasts, is it not the cosmic dream of abundance and fraternity of a people still suffering from hunger and deprivation? When in his play *Rara* Morisseau-Leroy shows a man dying for his right to a feast day in the drabness of his working days, paralytics who get up and dance, mutes who begin to sing when, after the death of heroes, people recount that they are traversing the region, when ghosts are seen, no-one is mistaken, no-one gives it a mystical significance, but everyone sees in it an incitement to the fight for happiness. Naturally, one must always do better, and the combatants of the advance guard of Haitian culture recognize the need resolutely to transcend whatever is irrational, mystic and animist in their national patrimony, but they do not think that that is an insoluble problem. They will reject the animist garment which conceals the realist nucleus, the dynamics of their culture, a nucleus charged with good sense, life and humanism, they will put on its feet again what is too often walking on its head, but they will never deny that cultural tradition, which is a great and fine thing, the only one which they possess as their own. Just as there is no question of any people denying religious works of art influenced by a mystical conception of life, Haitian men of culture will be able in a dynamic, positive and scientific way, a way of social realism, to combine the whole human protest against the harsh realities of life, all the emotion, the long cry of struggle, distress and hope which are contained in the works and forms transmitted to them by the past.

Social realism, conscious of the imperatives of history, preaches an art human in its content, but resolutely national in its form. This means that the pseudo 'world citizens' of culture, the true cosmopolitans, the true expatriates, have nothing to do with the man of our time, nothing to do with progress, and therefore nothing to do with culture. If all human races and all nations are equal and sisters, they have none the less their own traditions, their own temperament and forms which are more likely to touch them. If Art were not national in its form, how could the citizens of a country set about recognizing the perfumes and the climates which they love, so as truly to relive the works of beauty which are offered to them, and to find in them their share of dreams and of courage? The result would be that the people in question would find it difficult to take part in the forward movement of mankind towards liberation, since that art and that literature, essential elements in realization as much as in delight, would have no hold upon their feelings.

Haitian artists made use of the Marvellous in a dynamic sense before they realized that they were creating a Marvellous Realism. We became gradually conscious of the fact. Creating realism meant that the Haitian artists were setting about speaking the same language as their people. The Marvellous Realism of the Haitians is thus an integral part of Social Realism, and in its Haitian form it follows the same preoccupations. The treasury of tales and legends, all the musical, choreographic and plastic symbolism, all the forms of Haitian popular art are there to help the nation in solving its problems and in accomplishing the tasks which lie before it. The Western genres and organons bequeathed to us must be resolutely transformed in a national sense, and everything in a work of art must stir those feelings which are peculiar to the Haitians, sons of three races and an infinity of cultures.

To sum up, the objects of Marvellous Realism are:

1.   To sing the beauties of the Haitian motherland, its greatness as well as its wretchedness, with the sense of the magnificent prospects which are opened up by the struggles of its people and the universal and the profound truth of life;
2.   To reject all art which has no real and social content;
3.   To find the forms of expression proper to its own people, those which correspond to their psychology, while employing in a renovated and widened form, the universal models, naturally in accordance with the personality of each creator;
4.   To have a clear consciousness of specific and concrete current problems and the real dramas which confront the masses, with the purpose of touching and cultivating more deeply, and of carrying the people with them in their struggles.

In relation to particular forms of art, there are many aspects which need to be made clearer, but only a detailed discussion would enable us to come nearer to the truth. It is not an easy task to progress along the road of this kind of realism, and there are many gropings and many errors ahead of us, but we shall know how to profit even by our mistakes, to reach as soon as possible what is already taking shape before our eyes. Work will settle all the rest.

## Chapter 31

# MICHAEL DASH

## MARVELLOUS REALISM
## The way out of négritude

From 'Marvellous Realism: The Way Out of Négritude' *Caribbean Studies*
13(4), 1974.

### Towards a redefinition of history

IF THERE IS ONE SOUND IDEA that the ideology of 'négritude' puts forward, it is certainly the notion of the double alienation of the black man – that is a belief that the problem is more than political and economic, that there was a psychological and spiritual reconstruction that should also take place. However, it was difficult for them to provide a solution to the problem of spiritual regeneration for the simple reason that they themselves held as true an attitude to the past as being totally un-creative. Indeed in their reiteration of the injustice of the past they did no more than emphasize the fact of spiritual loss to the extent that any notions of survival or emergence of a Third World personality were totally neglected. This is the essential difference between 'Marvellous Realism' and 'Négritude' – for the former stresses patterns of emergence from the continuum of history. For the elaboration of this philosophy we will examine the ideas of Jacques Stephen Aléxis, Haitian novelist, and Wilson Harris, a Guyanese novelist.

One feature of Third World writers which distinguishes them as a distinct literary fraternity is the fundamental dialogue with history in which they are involved. However, as we have noted, so far this dialogue with the past essentially consisted of a continuous and desperate protest against the ironies of history. They adhered to the view of history as fateful coincidence and tragic accident, and saw their function as artists in terms of their attitude to the past, that is, either in terms of a committed protest against the past which would give birth to a new humanism, or were so overwhelmed by the 'fact' of privation or dispossession that they withdrew to a position of cynicism with regard to their peoples (V. S. Naipaul the Trinidadian novelist is often quoted as typical of this attitude). However such attitudes to the continuum of history left out of account a significant and positive part of the history of the Third World. It made it difficult to see beyond the tragedy of circumstance to the complex process of survival which the autochthonous as well as the transplanted cultures in the New World underwent. Such an investigation of the process of adaptation and survival in the oppressed cultures of the New World could well change the vision of the past which froze the Third World writer in the prison of protest and reveal the colonial legacy as a positive and civilizing force in spite of the brutality and privation which cloud this historical period.

Such an attitude would signify for the Third World writer an investigation of his past which goes beyond the documented privations of slavery and colonization to a more speculative vision of history in which the consciousness of the dominated cultures would predominate. In order to tap this consciousness, both Aléxis and Harris have turned to the myths, legends and superstitions of the folk in order to isolate traces of a complex culture of survival which was the response of the dominated to their oppressors (Aléxis 1960 and Harris 1970a). That is to say that colonization and slavery did *not* make things of men, but in their own way the enslaved peoples might have in their own imagination so reordered their reality as to reach beyond the tangible and concrete to acquire a new re-creative sensibility which could aid in the harsh battle for survival. The only thing they could possess (and which could not be tampered with) was their imagination and this became the source of their struggle against the cruelty of their condition. This notion of a counter-culture of the imagination is the basis of Harris' investigation of the practices of Haitian Voudou, limbo and the Amerindian religious rites, as he is convinced that

> . . . the imagination of the folk involved a crucial inner re-creative response to the violations of slavery and indenture and conquest.
>
> (Harris 1970a: 12)

This is an attitude to the conquered peoples which is unprecedented. It is the taking into account of the inner resources that the ancestors of the Third World could have developed to combat their tragic environment, therefore engaging in a conception of the past which would shatter the myths of 'historylessness' or 'non-achievement'.

Of what importance can the conception of such an 'inner corrective' on history be to the contemporary writer? It means fundamentally that in the same way he can circumvent the ironies of history so can he avoid the negativity of pure protest. What can indeed emerge is a literature of renascence – a literary aesthetic and reality based on the fragile emergence of the Third World personality from the privations of history. Now such a conception of reality would mean for the writer, the endowing of the concrete and the tangible with a figurative meaning. Harris comments on this as a possibility of the writer when he claims:

> I believe the possibility exists for us to become involved in perspectives of renascence which can bring into play a figurative meaning beyond an apparently real world or prison of history – I believe a philosophy of history may well lie buried in the arts of the imagination.
>
> (Harris 1970a: 8)

This would signify an adoption of the positive imaginative reconstruction of reality developed in the consciousness of the folk, by the contemporary writer. It is more than coincidence which can explain the striking similarity between Harris' claim and Aléxis' statement in 1956:

> Haitian art, in effect, presents the real, with its accompaniment of the strange and the fantastic, of dreams and half light, of the mysterious and the marvellous. . . . The West of Graeco-Latin descent too often tends to intellectualization, to idealization, to the creation of perfect canons, to the logical unity of the elements of feeling, to a pre-established harmony, whereas our art tends towards the most exact sensual representation of reality, towards a creative intuition, character, power of expression.
>
> (Aléxis 1956: 260)

Such a vision of the imaginative resonances of external reality, far from being a poetic abstraction, can not only be explained by the imagination of the conquered cultures of the past but by the fact that . . . 'the under-developed populations of the world who have still quite frequently had to live in contact with nature, have been compelled particularly to sharpen their eyes, their hearing, their sense of touch' (Aléxis 1956: 268).

## Chapter 32

# EDWARD KAMAU BRATHWAITE

## CREOLIZATION IN JAMAICA

From *The Development of Creole Society in Jamaica 1770–1820* Oxford:
Clarendon Press, 1971.

THE SINGLE MOST IMPORTANT FACTOR in the development of Jamaican society was not the imported influence of the Mother Country or the local administrative activity of the white élite, but a cultural action – material, psychological and spiritual – based upon the stimulus/response of individuals within the society to their environment and – as white/black, culturally discrete groups – to each other. The scope and quality of this response and interaction were dictated by the circumstances of the society's foundation and composition – a 'new' construct, made up of newcomers to the landscape and cultural strangers each to the other; one group dominant, the other legally and subordinately slaves. This cultural action or social process has been defined within the context of this work as creolization. Mrs Duncker has described it, in general terms in so far as it affected white settlers and visitors:

> Although there were some people who came to the West Indies and refused to conform, the power of the society to mould new-comers was strong. However oddly constructed West India society might appear in England, for the English people coming to the West Indies it was only a short time before they were caught up in the system. . . .
>
> (Duncker 1960: 231)

Maria Nugent must have said the same thing to herself when, after watching her dance with an (elderly) black slave, her hostesses broke down and cried from horror and outrage. We were faced here with an obscure force, working upon an entire section of society, which makes them all conform to a certain concept of themselves; makes them perform in certain roles which, in fact, they quickly come to believe in. . . .

Slaves in Jamaica came from a wide area of West Africa, within the period of this study, mainly from the Gold Coast and the Niger and Cross deltas. Creolization began with 'seasoning' – a period of one to three years, when the slaves were branded, given a new name and put under apprenticeship to creolized slaves. During this period the slave would learn the rudiments of his new language and be initiated into the work routines that awaited him.

These work routines, especially for plantation slaves, were the next important step in creolization. . . . From this followed 'socialization' – participation with others through the gang system, and through communal recreational activities such as drumming and dancing and festivals. . . . For the docile there was also the persuasion of the whip and the fear of punishment; for the venal, there was the bribe of gift or compliment or the offer of a better position, and for the curious and self-seeking, the imitation of the master. This imitation went on, naturally, most easily among those in closest and most intimate contact with Europeans, among, that is, domestic slaves, female slaves with white lovers, slaves in contact with missionaries or traders or sailors, skilled slaves anxious to deploy their skills, and above all, among urban slaves in contact with the 'wider' life. . . .

It was one of the tragedies of slavery and of the conditions under which creolization had to take place, that it should have produced this kind of mimicry; should have produced such 'mimic-men'. But in the circumstances this was the only kind of 'white' imitation that would have been accepted, given the terms in which the slaves were seen; and it was this kind of mimicry that was largely smiled upon and cultivated by 'middle class' Jamaican (and West Indian) society after Emancipation. *The snow was falling in the canefields* became typical of the 'educated' West Indian imagination.

But it was a two-way process, and it worked both ways. . . . In white households the Negro influence was pervasive, especially in the country areas. . . .

To preserve the pure dialect of the tribe (at least of the females) planters had to send to England for governesses and practically locked their daughters away from Negro influence.

But it was in the intimate area of sexual relationships that the greatest damage was done to white creole apartheid policy and where the most significant – and lasting – inter-cultural creolization took place. Black mistresses made convenient spies and/or managers of Negro affairs, and white men in petty authority were frequently influenced in their decisions by black women with whom they were amorously, or at any rate sensually, connected. . . .

Creolization, then, was a cultural process that took place within a creole society – that is, within a tropical colonial plantation polity based on slavery.

Even more important for an understanding of Jamaican development during this period was the process of creolization, which is a way of seeing the society, not in terms of white and black, master and slave, in separate nuclear units, but as contributory parts of a whole. To see Jamaica (or the West Indies generally) as a 'slave' society is as much a falsification of reality, as the seeing of the island as a naval station or an enormous sugar factory. Here, in Jamaica, fixed with the dehumanizing institution of slavery, were two cultures of people, having to adapt themselves to a new environment and to each other. The friction created by this confrontation was cruel, but it was also creative. The white plantations and social institutions described in this study reflect one aspect of this. The slaves' adaptation of their African culture to a new world reflects another. The failure of Jamaican society was that it did not recognize these elements of its own creativity. Blinded by the need to justify slavery, white Jamaicans refused to recognize their black labourers as human beings, thus cutting themselves off from the one demographic alliance that might have contributed to the island's economic and (possibly) political independence. What the white Jamaican élite did not, could not, would not, dare accept, was that true autonomy for them could only mean true autonomy for all; that the more unrestricted the creolization, the greater would have been the freedom. They preferred a bastard metropolitanism – handed down to the society in general after Emancipation – with its consequence of dependence on Europe, to a complete exposure to creolization and liberation of their slaves.

Blinded by the wretchedness of their situation, many of Jamaica's slaves, especially the black élite (those most exposed to the influence of their masters), failed, or refused, to make conscious use of their own rich folk culture (their one indisputable possession), and so failed to command the chance of becoming self-conscious and cohesive as a group and consequently, perhaps, winning their independence from bondage, as their cousins in Haiti had done. 'Invisible', anxious to be 'seen' by their masters, the élite blacks and the mass of the free coloureds (apart from the significant exceptions already discussed within the body of this work, and those who, after Emancipation, were to establish, against almost impossible odds, the free villages and small peasantries of rural Jamaica), conceived of visibility through the lenses of their masters' already uncertain vision as a form of 'greyness' – an imitation of an imitation. Whenever the opportunity made it possible, they and their descendants rejected or disowned their own culture, becoming, like their masters, 'mimic-men'.

Cultural autonomy demands a norm and a residential correspondence between the 'great' and 'little' traditions within the society. Under slavery there were two 'great' traditions, one in Europe, the other in Africa, and so neither was residential. Normative value-references were made outside the society. Creolization (despite its attendant imitations and conformities) provided the conditions for and possibility of local residence. It certainly mediated the development of authentically local institutions, and an Afro-creole 'little' tradition among the slave 'folk'. But it did not, during the period of this study, provide a norm. For this to have been provided, the Euro-creole élite (the one group able, to some extent, to influence the pace and quality of creolization) would have had to have been much stronger, culturally, than it was. Unable or unwilling to absorb in any central sense the 'little' tradition of the majority, its efforts and its continuing colonial dependence merely created the pervasive dichotomy which has been indicated in this study. . . .

My own idea of creolization is based on the notion of an historically affected socio-cultural continuum, within which (in the case of Jamaica) there were four inter-related and sometimes overlapping orientations. From their several cultural bases people in the West Indies tend towards certain directions, positions, assumptions and ideals. But nothing is really fixed and monolithic. Although there is white/brown/black, there are infinite possibilities within these distinctions and many ways of asserting identity.

## Chapter 33

# HOMI K. BHABHA

## CULTURAL DIVERSITY AND CULTURAL DIFFERENCES

From 'The Commitment to Theory' *New Formations* 5, 1988.

[T]HE] REVISION OF THE HISTORY of critical theory rests . . . on the notion of cultural difference, not cultural diversity. Cultural diversity is an epistemological object – culture as an object of empirical knowledge – whereas cultural difference is the process of the *enunciation* of culture as 'knowledge*able*', authoritative, adequate to the construction of systems of cultural identification. If cultural diversity is a category of comparative ethics, aesthetics, or ethnology, cultural difference is a process of signification through which statements *of* culture or *on* culture differentiate, discriminate, and authorize the production of fields of force, reference, applicability, and capacity. Cultural diversity is the recognition of pre-given cultural 'contents' and customs, held in a time-frame of relativism; it gives rise to anodyne liberal notions of multiculturalism, cultural exchange, or the culture of humanity. Cultural diversity is also the representation of a radical rhetoric of the separation of totalized cultures that live unsullied by the intertextuality of their historical locations, safe in the Utopianism of a mythic memory of a unique collective identity. Cultural diversity may even emerge as a system of the articulation and exchange of cultural signs in certain . . . imperialist accounts of anthropology.

Through the concept of cultural difference I want to draw attention to the common ground and lost territory of contemporary critical debates. For they all recognize that the problem of the cultural emerges only at the significatory boundaries of cultures, where meanings and values are (mis)read or signs are misappropriated. . . .

The time of liberation is, as Fanon powerfully evokes, a time of cultural uncertainty, and, most crucially, of significatory or representational undecidability:

But [native intellectuals] forget that the forms of thought and what [they] feed . . . on, together with modern techniques of information, language and dress have dialectically reorganized the people's intelligences and *the constant principles (of national art)* which acted as safeguards during the colonial period are now undergoing extremely radical changes . . . [We] must join the people in that fluctuating movement which they are *just* giving a shape to . . . which will be the signal for everything to be called into question . . . it is to the zone of *occult instability* where the people dwell that we must come.

(Fanon 1967b: 168) (my emphasis)

The enunciation of cultural difference problematizes the division of past and present, tradition and modernity, at the level of cultural representation and its authoritative address. It is the problem of how, in signifying the present, something comes to be repeated, relocated, and translated in the name of tradition, in the guise of a pastness that is not necessarily a faithful sign of historical memory but a strategy of representing authority in terms of the artifice of the archaic. That iteration negates our sense of the origins of the struggle. It undermines our sense of the homogenizing effects of cultural symbols and icons, by questioning our sense of the authority of cultural synthesis in general.

This demands that we rethink our perspective on the identity of culture. Here Fanon's passage — somewhat reinterpreted — may be helpful. What is implied by his juxtaposition of the constant national principles with his view of culture-as-political-struggle, which he so enigmatically and beautifully describes as 'the zone of occult instability where the people dwell'? These ideas not only help to explain the nature of colonial struggle. They also suggest a possible critique of the positive aesthetic and political values we ascribe to the unity or totality of cultures, especially those that have known long and tyrannical histories of domination and misrecognition. Cultures are never unitary in themselves, nor simply dualistic in relation of Self to Other. This is not because of some humanistic nostrum that beyond individual cultures we all belong to the human culture of mankind; nor is it because of an ethical relativism that suggests that in our cultural capacity to speak of and judge Others we necessarily 'place ourselves in their position', in a kind of relativism of distance of which Bernard Williams has written at length (Williams 1985: ch. 9).

The reason a cultural text or system of meaning cannot be sufficient unto itself is that the act of cultural enunciation — the place of utterance — is crossed by the difference of writing or écriture. This has less to do with what anthropologists might describe as varying attitudes to symbolic systems within different cultures than with the structure of symbolic representation — not the content of the symbol or its 'social function', but the structure of symbolization. It is this 'difference' in language that is crucial to the production of meaning and ensures, at the same time, that meaning is never simply mimetic and transparent.

The linguistic difference that informs any cultural performance is dramatized in the common semiotic account of the disjuncture between the subject of a proposition (énoncé) and the subject of enunciation, which is not represented in the statement but which is the acknowledgment of its discursive embeddedness and address, its cultural positionality, its reference to a present time and a specific space. The pact of interpretation is never simply an act of communication between the I and the You designated in the statement. The production of meaning requires that these two places be mobilized in the passage through a Third Space, which represents both the general conditions of language and the specific implication of the utterance in a performative and institutional strategy of which it cannot 'in itself' be conscious. What this unconscious relation introduces is an ambivalence in the act of interpretation. . . .

The intervention of the Third Space, which makes the structure of meaning and reference an ambivalent process, destroys this mirror of representation in which cultural knowledge is continuously revealed as an integrated, open, expanding code. Such an intervention quite properly challenges our sense of the historical identity of culture as a homogenizing, unifying force, authenticated by the originary Past, kept alive in the national tradition of the People. In other words, the disruptive temporality of enunciation displaces the narrative of the Western nation which Benedict Anderson so perceptively describes as being written in homogeneous, serial time (Anderson 1983: ch. 2).

It is only when we understand that all cultural statements and systems are constructed in this contradictory and ambivalent space of enunciation, that we begin to understand why hierarchical claims to the inherent originality or 'purity' of cultures are untenable, even before we resort to empirical historical instances that demonstrate their hybridity. Fanon's vision of revolutionary cultural and political change as a 'fluctuating movement' of occult instability could not be articulated as cultural *practice* without an acknowledgment of this indeterminate space of the subject(s) of enunciation. It is that Third Space, though unrepresentable in itself, which constitutes the discursive conditions of enunciation that ensure that the meaning and symbols of culture have no primordial unity or fixity; that even the same signs can be appropriated, translated, rehistoricized, and read anew.

Fanon's moving metaphor – when reinterpreted for a theory of cultural signification – enables us to see not only the necessity of theory, but also the restrictive notions of cultural identity with which we burden our visions of political change. For Fanon, the liberatory 'people' who initiate the productive instability of revolutionary cultural change are themselves the bearers of a hybrid identity. They are caught in the discontinuous time of translation and negotiation, in the sense in which I have been attempting to recast these works. In the moment of liberatory struggle, the Algerian people destroy the continuities and constancies of the 'nationalist' tradition which provided a safeguard against colonial cultural imposition. They are now free to negotiate and translate their cultural identities in a discontinuous intertextual temporality of cultural difference. The native intellectual who identifies the people with the 'true national culture' will be disappointed. The people are now the very principle of 'dialectical reorganization' and they construct their culture from the national text translated into modern Western forms of information technology, language, dress. The changed political and historical site of enunciation transforms the meanings of the colonial inheritance into the liberatory signs of a free people of the future.

> I have been stressing a certain void or misgiving attending every assimilation of contraries – I have been stressing this in order to expose what seems to me a fantastic mythological congruence of elements. . . . (Harris 1973b: 62) And if indeed therefore any real sense is to be made of material change it can only occur with an acceptance of a concurrent void and with a willingness to descend into that void wherein, as it were, one may begin to come into confrontation with a spectre of invocation whose freedom to participate in an alien territory and wilderness has become a necessity for one's reason or salvation.
>
> (Harris 1973b: 60)

This meditation by the great Guyanian writer Wilson Harris on the void of misgiving in the textuality of colonial history reveals the cultural and historical dimension of that Third Space of enunciation which I have made the precondition for the articulation of cultural difference. He sees it as accompanying the 'assimilation of contraries' and creating that occult instability which presages powerful cultural changes. It is significant that the productive capacities of this Third Space have a colonial or post-colonial provenance. For a willingness to descend into that alien territory – where I have led you – may reveal that the theoretical recognition of the split-space of enunciation may open the way to conceptualizing an *inter*national culture, based not on the exoticism or multi-culturalism of the *diversity* of cultures, but on the inscription and articulation of culture's *hybridity*. To that end we should remember that it is the 'inter' – the cutting edge of translation and negotiation, the *in-between*, the space of the *entre* that Derrida has opened up in writing itself – that carries the burden of the meaning of culture. It makes it possible to begin envisaging national, anti-nationalist, histories of the 'people'. It is in this space that we will find those words with which we can speak of Ourselves and Others. And by exploring this hybridity, this 'Third Space', we may elude the politics of polarity and emerge as the others of our selves.

## Chapter 34

# ROBERT YOUNG

# THE CULTURAL POLITICS OF HYBRIDITY

From *Colonial Desire* London: Routledge, 1995.

## Hybridity: from racial theory to cultural criticism

AT ITS SIMPLEST, HYBRIDITY . . . implies a disruption and forcing together of any unlike living things, grafting a vine or a rose on to a different root stock, making difference into sameness. Hybridity is a making one of two distinct things, so that it becomes impossible for the eye to detect the hybridity of a geranium or a rose. Nevertheless, the rose exists, like the vine, only in so far as it is grafted onto the different stock. Neglect to prune either, and the plant eventually reverts to its original state. In the nineteenth century, we have seen that a common analogous argument was made that the descendants of mixed-race unions would eventually relapse to one of the original races thus characterizing miscegenation as temporary in its effects as well as unnatural in its very nature. Hybridization can also consist of the forcing of a single entity into two or more parts, a severing of a single object in two, turning sameness into difference, as in today's hybrid shares on the stock market, although they, in the last analysis, are merely parts of a whole that will have to be re-invoked at the wind-up date. Hybridity thus makes difference into sameness, and sameness into difference, but in a way that makes the same no longer the same, the different no longer simply different. In that sense it operates according to the form of logic that Derrida isolates in the term 'brisure', a breaking and joining at the same time, in the same place: difference and sameness in an apparently impossible simultaneity. Hybridity thus consists of a bizarre binate operation, in which each impulse is qualified against the other, forcing momentary forms of dislocation and displacement into complex economies of agonistic reticulation. This double logic which goes against the convention of rational either/or choices but which is repeated in science in the split between the classical and quantum physics, could be said to be as characteristic of the twentieth century as oppositional dialectical thinking was of the nineteenth.

Hybridity thus operates within the same conflictual structures as contemporary theory. Both repeat and reproduce the sites of their own cultural production whose discordant logic manifests itself in structural repetitions as structural repetition. . . . There is an historical stemma between the cultural concepts of our own day and those of the past from which we tend to assume that we have distanced ourselves. We restate and rehearse them covertly in the language and concepts that we use: every time a commentator uses the epithet 'full-blooded', for example, he or she repeats the distinction between those of pure and mixed race. Hybridity in particular shows the connections between the racial categories of the past and contemporary cultural discourse: it may be used in different ways, given different inflections and apparently discrete references, but it always reiterates and reinforces the dynamics of the same conflictual economy whose tensions and divisions it re-enacts in its own antithetical structure. There is no single, or correct, concept of hybridity: it changes as it repeats, but it also repeats as it changes. It shows that we are still locked into parts of the ideological network of a culture that we think and presume that we have surpassed. The question is whether the old essentializing categories of cultural identity, or of race, were really so essentialized, or have been retrospectively constructed as more fixed than they were. When we look at the texts of racial theory we find that they are in fact and already deconstructed. Hybridity here is a key term in that wherever it emerges it suggests the impossibility of essentialism. If so, then deconstructing such essentialist notions of race today we may rather be repeating the past than distancing ourselves from it, or providing a critique of it. Commentators talk of 'pseudo-scientific' racial theory in the nineteenth century as if the term 'pseudo' is enough to dismiss it with ease: but what that term in fact implies is that racial theory was never simply scientific or biologistic, just as categories were never wholly essentializing. Today it is common to claim that in such matters we have moved from biologism and scientism to the safety of culturalism, that we have created distance and surety by the very act of the critique of essentialism and the demonstration of its impossibility: but that shift has not been so absolute, for the racial was always cultural, the essential never unequivocal. How does that affect our own contemporary revisions of that imagined past? The interval that we assert between ourselves and the past may be much less than we assume. We may be more bound up with its categories than we like to think. Culture and race developed together, imbricated within each other, their discontinuous forms of repetition suggest, as Foucault puts it, 'how we have been trapped in our own history'. The nightmare of the ideologies and categories of racism continue to repeat upon the living.

## Colonial desire: racial theory and the absent other

In the nineteenth century, for the most part the cultural relations between Britain and her colonies were thought through on the basis of what Foucault calls the sovereign model of power (the basic assumption of diffusionism). But we can also see in the endless discussions of questions of racial miscegenation the soft underbelly of that power relation, fuelled by the multifarious forms of colonial desire. So, for example, Frederick Marryat's Peter Simple differentiates between the participants at a fancy dress ball in Barbados before announcing his own sexual preference:

> The progeny of a white and a negro is a mulatto, or half and half – of a white and mulatto, a quadroon, or one-quarter black, and of this class the company were chiefly composed. I believe a quadroon and white make the mustee or one-eighth black, and the mustee and white the mustafina, or one-sixteenth black. After that, they are white-washed, and considered as Europeans. . . . The quadroons are certainly the handsomest race of the whole; some of the women are really beautiful. . . . I must acknowledge, at the risk of losing the good opinion of my fair country-women, that I never saw before so many pretty figures and faces.
> (Marryatt 1834: 2, 195–7; partly cited in Brantlinger 1983)

But as we have seen, such desire, constituted by a dialectic of attraction and repulsion, soon brings with it the threat of the fecund fertility of the colonial desiring machine, whereby a culture in its colonial operation becomes hybridized, alienated and potentially threatening to its European original through the production of polymorphously perverse people who are, in Bhabha's phrase, 'white, but not quite:' in the nineteenth century, this threatening phenomenon of being degraded from a civilized condition was discussed as the process of 'decivilization'. . . . South America was always cited as the prime example of the degenerative results of racial hybridization ('Let any man turn his eyes to the Spanish American dominions, and behold what a vicious, brutal, and degenerate breed of mongrels has been there produced, between Spaniards, Blacks, Indians, and their mixed progeny' remarks Edward Long; 'they are a disgrace to human nature', adds Knox, blaming the perpetual revolutions of South America on their degenerate racial mixture; observations that are dutifully repeated by Spencer and Hitler) (Knox 1850; Spencer 1864–7; Hitler 1925–6).

. . . An analytic account of the intricate gradations of cultural fusion, regarded as a process of degeneration that mocked the nineteenth century's 'diffusion' model of the spread of cultures with the confusion of fusion, subverted alike both evolutionist and polygenist schools of ethnography, and beyond these held out the threat of undoing the whole progressive paradigm of Western civilization. Here theories of racial difference as degeneration themselves fused with the increasing cultural pessimism of the late nineteenth century and the claim that not only the population of cities but the world itself, that is the West, was degenerating. Each new racial ramification of miscegenation traced an historical trajectory that betrayed a narrative of conquest, absorption and inevitable decline. For the Victorians, race and sex became history, and history spoke of race and sex. It has often been suggested that the problem with Western historiography is that, generally speaking, only the West has been allowed a history. But in fact while this is in certain respects true, a different, covert history was assigned to the non-West in the nineteenth century, and that was the history produced by historical philology, which told of the development and diffusion of languages through a panoramic narrative of tribal migration and conquest, bringing in its wake the absorption of weak races by the strong. Aryan identity was also constituted through diaspora, through a history that equated migration with colonization. While this Darwinian, diasporic narrative of the Indo-European family was used as a way of giving European imperial expansion the status of natural law, its story of absorption, and therefore of linguistic and racial inmixing, at the same time also implicitly foretold the corruption, decadence and degeneration of European imperial civilization. So much so, in fact, that it eventually gave rise to arguments for the necessity of decolonization. C. L. Temple, for example, in his *Native Races and Their Rulers* (published in 1918, just fifteen years after the British annexation of what was to become Nigeria), argued that history has showed that 'one of three destinies awaits the conquered . . . race. It either fuses with, i.e. becomes absorbed in and absorbs the conquering race, and this is the usual result; or it re-captures its liberty; or, less often, it dies out'. Temple reasons that Africans are unlikely to die out (far too fertile; it was, of course, precisely this third option that became part of Nazi ideology); but he also considers that 'fusion between the European and the dark-skinned races of Africa is entirely out of the question'. Temple therefore asks of the Africans: 'What then is to be their future?' He answers: 'Historical analogies lead us to one conclusion only, i.e., that they will some day recapture their liberty' (Temple 1918: 23). Fear of racial fusion, therefore, brings him to the extreme position of envisaging the necessity of the dismantling of the Empire itself. Forty years later, it was gone.

In recent years a whole range of disciplines has been concerned with the question of the exclusion and representation of 'the Other', of inside/outside notions of otherness or of the difficulties, negotiated so painfully though not powerlessly by anthropologists, of self–other relations. Our talk of Manichean allegories of colonizer and colonized, of self and Other, mirrors the ways in which today's racial politics work through a relative polarization between black and white. This remorseless Hegelian dialectalization is characteristic of twentieth century accounts of race, racial difference and racial identity. I want to argue, however, that for an understanding of the historical specificity of the discourse of colonialism, we need to acknowledge that other forms of racial distinction have worked simultaneously alongside this model. Without any understanding of this, we run the risk of imposing our own categories and politics upon the past without noticing its difference, turning the otherness of the past into the sameness of the today. The loss that follows is not merely one for the knowledge of history: as with the case of hybridity, we can also remain unaware of how much that otherness both formed and still secretly informs our present. . . . The ideology of race, a semiotic system in the guise of ethnology, 'the science of races', from the 1840s onwards necessarily worked according to a doubled logic, according to which it both enforced and policed the differences between the whites and the non-whites, but at the same time focussed fetishistically upon the product of the contacts between them. Colonialism was always locked into the machine of desire: 'the machine remains desire, an investment of desire whose history unfolds'(Deleuze and Guattari 1972). Folded within the scientific accounts of race, a central assumption and paranoid fantasy was endlessly repeated: the uncontrollable sexual drive of the non-white races and their limitless fertility. What was clearly so fascinating was not just the power of other sexuality as such, the 'promiscuous', 'illicit intercourse' and 'excessive debauchery' of a licentious primitive sexuality, so salaciously imagined in the later editions of Malthus's *Principle of Population* (see Malthus [1798] (1826): 146, 156, 195); in the marriage-by-capture fantasies of McLennan's *Primitive Marriage* (1850), or in Spencer's chapters on 'Primitive Relations of the Sexes' and 'Promiscuity' in his *Principles of Sociology* 1876 (Spencer 1876: 1, 621–97). As racial theories show in their unrelenting attempt to assert inalienable differences between races, this extraordinary vision of an unbounded 'delicious fecundity', in Virginia Woolf's phrase, only took on significance through its voyeuristic tableau of frenzied, interminable copulation, of couplings, fusing, coalescence, between races. At its core, such racial theory projected a phantasmagoria of the desiring machine as a people factory: a Malthusian fantasy of uncontrollable, frenetic fornication producing the countless motley varieties of interbreeding, with the miscegenated offspring themselves then generating an ever increasing melange, 'mongrelity', of self-propagating endlessly diversifying hybrid progeny: half-blood, half-caste, half-breed, cross-breed, amalgamate, intermix, miscegenate; alvino, cabre, cafuso, castizo, cholo, chino, cob, creole, dustee, fustee, griffe, mamaluco, marabout, mestee, mestindo, mestizo, mestize, metifo, misterado, mongrel, morisco, mule, mulat, mulatto, mulatta, mulattress, mustafina, mustee, mustezoes, ochavon, octavon, octoroon, puchuelo, quadroon, quarteron, quatralvi, quinteron, saltatro, terceron, zambaigo, zambo, zambo prieto. . . . Nineteenth-century theories of race did not just consist of essentializing differentiations between self and other, they were also about a fascination with people having sex – interminable, adulterating, aleatory, illicit, inter-racial sex. But this steamy model of mixture was not a straightforward sexual or even cultural matter: in many ways it preserved the older commercial discourse that it superseded. For it is clear that the forms of sexual exchange brought about by colonialism were themselves both mirrors and consequences of the modes of economic exchange that constituted the basis of colonial relations; the extended exchange of property which began with small trading-posts and the visiting slave ships originated, indeed, as much as an exchange of bodies as of goods, or rather of bodies as goods: as in that paradigm of respectability, marriage, economic and sexual exchange were intimately bound up, coupled with each other, from the very first. The history of the meanings of the word 'commerce' includes the exchange both of merchandise and of bodies in sexual intercourse. It was therefore wholly appropriate that sexual exchange, and its miscegenated product, which captures the violent, antagonistic power relations of sexual and cultural diffusion, should become the dominant paradigm through which the passionate economic and political trafficking of colonialism was conceived. Perhaps this begins to explain why our own forms of racism remain so intimately bound up with sexuality and desire. The fantasy of postcolonial cultural theory, however, is that those in the Western academy at least have managed to free themselves from this hybrid commerce of colonialism, as from every other aspect of the colonial legacy.

# PART SIX

# Indigeneity

T HE INDIGENOUS PEOPLES OF 'settled' colonies, or 'First-Nations', have in many ways become the *cause célèbre* of post-colonialism. No other group seems so completely to earn the position of colonized group, so unequivocally to demonstrate the processes of imperialism at work. But indigenous groups have so often fallen into the political trap of essentialism set for them by imperial discourse. Imperial narratives such as that of anthropology in their project of *naming* and thus *knowing* indigenous groups have imported a notion of aboriginality, of cultural authenticity, which proves difficult to displace. The result is the positioning of the indigenous people as the ultimately marginalized, a concept that reinscribes the binarism of centre/margin, and prevents their engagement with the subtle processes of imperialism by locking them into a locally strategic but ultimately self-defeating essentialism. As many indigenous commentators have re-iterated, all cultures and societies change and adapt, and it is in a dynamic and shifting environment of adaptation that the political claims of indigenous people are situated.

The semiotic system by which the indigenous peoples of Canada, Australia and New Zealand have been represented looks, according to Terry Goldie, something like a chessboard in which the semiotic pawn signifying the indigenous person can only be moved in very circumscribed ways. This is the imperialist corollary of the essentialist argument and indeed essentialism works, in the long run, to the detriment of the indigenous society, as Gareth Griffiths shows, by separating the indigenous subject under conflicting categories of 'authentic' and 'inauthentic'. As post-colonial discourse demonstrates, the appeal to 'authenticity' is not merely an ontological contradiction, but a political trap. Thus the question of 'who can write as the Other' addressed by Margery Fee becomes particularly pertinent, for the rejection of an 'authentic' or 'essential' indigenous subjectivity must be reconciled with the real material conditions of subjection.

Arnold Krupat, suggests a new category – that of anti-imperial translation – to address the specific case of Native American cultural production. Krupat uses the concept of *trans-latio* (to carry across) to theorize the problematic area of indigenous appropriation. This is particularly significant because, in the main, the translation is from an oral to a textual dimension.

An oral literature, in order to become the subject of analysis, must indeed first become an object. It must, that is, be textualized; and here we encounter a translation dilemma of another kind, one in which the 'source language' itself has to be carried across – *trans-latio* – from one medium to another, involving something more than just a change of names.

Krupat provides another perspective on an issue that is close to the heart of post-colonial studies and one addressed in the section on Language in the Reader.

One of the impediments to fruitful discussion of indigenous peoples is the stereotypical assumption of their 'pre-historical' centrality, their synonymity with the static essence of colonized place. Such assumptions work towards a notion of authenticity that cuts off exploratory discussion of the place of indigenous peoples in global networks. James Clifford, investigating the diasporic flows and movements of indigenous islanders in the South Pacific, shows that identity, even for indigenous groups, is never static, but is 'articulated' and fluid, curiously overlapping the category of the diasporic in many cases. In a similar tactical move, Diana Brydon addresses the ways in which the category of the indigenous is disrupted by hybrid reality of Inuit representational strategies, overlapping the category of the hybrid in ways that force us to re-evaluate the assumptions surrounding the category of the 'indigenous'. The direction of post-colonial analysis has been towards an understanding of the increasingly complex and shifting subject positions of indigenous writers and cultural producers.

## Chapter 35

# GARETH GRIFFITHS

# THE MYTH OF AUTHENTICITY

From 'The Myth of Authenticity' in Chris Tiffin and Alan Lawson (eds)
*De-Scribing Empire* London: Routledge, 1994.

T HERE ARE REAL DANGERS IN recent representations of indigenous peoples in popular discourse, and especially in the media, which stress claims to an 'authentic' voice. For these claims, by overwriting the actual complexity of difference may write out that voice as effectively as earlier oppressive discourses of reportage. In fact, it may well be the same process at work, and the result may be just as crippling to the efforts of indigenous peoples to evolve an effective strategy of recuperation and resistance.

For example, in a recent dispute over mining at Yakabindie in Western Australia both sides of the dispute invoked the sign of the authentic in their defence of their position. The 'liberal' tone of modern journalism, its claim to even-handedness, is possible partly because of the way in which certain signs have been fetishised within popular discourse, in this case that of the 'authentic', the traditional and the local. The report in the *West Australian* of Monday, 12 August 1991 can stand as an encapsulation of this problem, representing as it does two images of the authentic, both inscribed under such legitimating signs as the 'elder', the local, and the tribal, and both counterposed by the illegitimate signs of the outsider, the Southerner, the fringe-dweller, whose representative in the article is the Perth political activist Robert Bropho. Let me quote the relevant paragraphs:

> Wiluna resident Tony Green, 89, said he was born less than 8 km. from Yakabindie. He had spent most of his life in the area and had never heard of a sacred site near the proposed mine. 'What about the future?' he said. 'We need the jobs for the people. I'd give that land to the mining people.' But community elder Dusty Stevens highlighted the feelings which have divided the region's Aborigines. 'Some of these fellahs just wouldn't know,' he said. There are a lot of sites in there.' The appearance of Mr. Bropho and members of the Swan Valley Fringe Dwellers at the meeting was attacked by Goldfields Aboriginal spokesman Aubrey Lynch, who said the southerners had nothing to do with the issue.
>
> (*West Australian*, 12 August 1991: 9)

Articles like this are an increasingly typical way of representing in the media the 'positions' and 'voices' of the indigene, inscribing them in effect as disputational claimants to a 'territoriality' of the authentic. Australian Aboriginal peoples may increasingly wish to assert their sense of the local and the specific as a recuperative strategy in the face of the erasure of difference characteristic of colonialist representation. But such representations subsumed by the white media under a mythologised and fetishised sign of the 'authentic' can also be used to create a privileged hierarchy of Australian Aboriginal voice which in practice represents that community as divided. More subtly, it may construct a belief in the society at large that issues of recovered 'traditional' rights are of a different order of equity from the right to general social justice and equality. Whilst this may be in part the unintentional product of a worthy liberal desire to recuperate Australian Aboriginal culture, it also frequently results, as in the case I have given, in a media construction of the 'authentic' Australian Aboriginal in opposition to the 'inauthentic' political activists whose claim is undermined (the metaphor is an appropriate one) by a dismissal of their right to represent Australian Aboriginal culture in any legitimate way.

In order fully to understand what is involved here it seems to me that these representations need also to be addressed through their reflection of a larger practice within colonialist discourse, a practice in which the possibilities of subaltern speech are contained by the discourse of the oppressor, and in which the writing of the Australian Aboriginal under the sign of 'authenticity' is an act of 'liberal' discursive violence, parallel in many ways to the inscription of the 'native' (indigene) under the sign of the savage. On the surface the obvious connection is through the reversed sentimental and nostalgic rendering of the Australian Aboriginal under the sign of the primitive (noble savage rather than cannibal savage). But at a deeper level both processes may be about the inscription of ourselves displaced upon the Australian Aboriginal, an inscription which may overwrite and overdetermine the full range of representations through which contemporary Australian Aboriginality might otherwise effectively be represented.

Michael Taussig's powerful study of the massacre and enslavement of the Putumayo Indians in the early years of this century stresses the way in which the silenced subjects of oppression are spoken by the different discourses through which their story is inscribed. . . .

Taussig records how contemporary Indians subjected to terror by the greed of the rubber companies of our own day, such as the Andoke Yarocamena, register their perceptions of the powerful ways in which 'narration' functions to control and override their resistance:

> Something crucial about the complicity and the magical powers of the company employees emerges from what has been said in recent times about Andoke Indians who claim that the rubber company had a stronger story than the Indians' story and this is why, for example, the armed uprising of the Andoke Yarocamena against the company failed and failed so disastrously.
>
> (Taussig 1986: 107)

It is clearly crucial to resistance that the 'story' of the Indian continues to be told. It is only through such counter-narratives that alter/native views can be put. As Taussig notes, however, this contradicts, at a simple level, the Indian assertion that their story is less powerful than that of the European. By story here is meant, as Taussig goes on to explain, that narrative (*rafue*) through which a 'necessary mediation between concept and practice which ensures the reproduction of the everyday world' (107) is effected. That is to say the fundamental systemic discourse through which the world is represented, analogous to other indigenous stories such as that of the various dreamings of the Australian Aboriginal peoples. The oppressor clearly shows that they are aware that their own narrative of the Indian, what the Andoke called 'Historias para *Nosotros* – histories not of, but for us' (107) are not perceived by the *oppressors* to be successful as the Andoke assert because of a greater mystic efficacy but because they override and overdetermine the possibility for the Indian speaking their own position within the alternative discourse of the conqueror. That the conquerors in fact continue to fear the 'story' of the indigene and seek to silence it is graphically and horrifically illustrated by their favoured torture of cutting out the tongues of the Indians and then, subsequent to this act, forcing them to 'speak'. In the light of the concerns raised by such images the reports on disputes such as Yakabindie may take on new and powerful resonances and the act of constructing the speech of the already silenced may metaphorically, at least, be perceived as an act best characterised, as I have suggested above, by a metaphor of violence, however 'liberal' in intention. In both cases the appropriated features of authentic discourse are installed after the event of silencing by violence.

Strategies of recuperation and texts which insist on the importance of re-installing the 'story' of the indigenous cultures are, therefore, as many Australian Aboriginal spokespeople have insisted, crucial to their resistance. Such recuperations may be the literal recuperation of the texts of pre-colonial cultures, the narratives of the dreaming or the body of pre-colonial oratures, or, as in the case of the work of Mudrooroo Narogin Nyoongah such as his recently published novel *Master of the Ghost Dreaming*, attempts to reinscribe the dominant culture of colonial society by re-telling the moment of encounter and invasion through indigenous eyes and discourses. In a sense this is part and parcel of Mudrooroo Narogin Nyoongah's asserted desire to speak from an 'essential' Aboriginal position (the word is his not mine) and of his belief that Aboriginal texts may be authentic or inauthentic in so far as they cohere within a larger Aboriginal metatext (which means, I presume, in part at least that alter/native story (rafue) of which the Andoke also speak). But Mudrooroo Narogin Nyoongah has also argued in the same text that 'the Aboriginal writer is a Janus-type figure with one face turned to the past and the other to the future while existing in a postmodern, multicultural Australia in which he or she must fight for cultural space' (Narogin 1990: 24). Thus in a sense he embraces his hybridised position not as a badge of failure or denigration but as part of that contestational weave of cultures which recent critical theory argues is the inescapable condition of all postmodernist experience, though at the same time he asserts in both his critical writing and practice as a novelist the importance of asserting his identity in essentialist difference as a political strategy. In this apparent contradiction he is registering the difficult and ambivalent position which the Aboriginal writer is forced to occupy in the complex task of simultaneously recuperating the traditional and contesting the profile of identity for Aboriginal peoples in contemporary Australian political and cultural space. . . .

Many of the problems raised by this issue have also been addressed by those who have sought to theorise the difficulties which arise when we consider the possibility of a subaltern subject 'speaking' within any dominant discourse such as colonialism or patriarchy. The questions these debates have raised include: We know that subaltern people are oppressed, but how do we know? How can that oppression be spoken? Even when the subaltern appears to 'speak' there is a real concern as to whether what we are listening to is really a subaltern voice, or the subaltern being spoken by the subject position they occupy within the larger discursive economy. Thus as Jenny Sharpe has argued the speaker who resists the colonial necessarily achieves that position within the framework of the system they oppose (her example is the famous anti-colonial speech of Aziz in the trial scene of Forster's *A Passage to India* (Sharpe 1989: 148–50). In inscribing such acts of resistance the deep fear for the liberal critic is contained in the worry that in the representation of such moments what is inscribed is not the subaltern's voice but the voice of your own other. Homi Bhabha has also acknowledged that subaltern speech is in some sense conditional upon the dominant discourse:

> For it is between the edict of Englishness and the assault of the dark unruly spaces
> of the earth, through an act of repetition, that the colonial text emerges uncertainly.
> (Bhabha 1985a: 126–7)

For Homi Bhabha, if I read him correctly, the possibility of subaltern speech exists principally and crucially when its mediation through mimicry and parody of the dominant discourse subverts and menaces the authority within which it necessarily comes into being. (In this same article Bhabha offers a convincing account of how such resistances can be developed and how they flourish within and through the deployment of mimicry within the necessarily hybridised condition of the colonised society.)

One basis for applying such aspects of colonialist theory to this topic is that indigenous peoples, too, in many important ways, exist in relation to their own societies (themselves settler colonies) in ways analogous to the colonial subject. They have been presented frequently in the representations of settler societies as subjects who do not possess, to use Bhabha's phrase, 'a stable unitary assumption of collectivity' (Bhabha 1985a: 153). This is, in part, the result of the deliberate suppression of pre-colonial cultures, and the displacement of their peoples in a policy of assimilation which aimed at the suppression of difference. The very wiping out of distinctive collectivities under an undifferentiated term such as 'aboriginal' is an example of this process in operation. It is therefore a powerful need of such peoples to re-assert their pre-colonised cultures and to struggle for the recuperation of their cultural difference and its resilience in and through the local and specific. Let me be quite clear that it is not with this process that I am quarrelling, but rather with the uses made of some of the strategies of authenticity associated with this process within white systems of representation which disavow the possibilities for the hybridised subjects of the colonising process to legitimate themselves or to speak in ways which menace the authority of the dominant culture precisely in so far as it 'mimics' and so subverts it. In such a fetishised use of the inscription of the authentic a further and subtler example of control emerges, one which in this use may function just as negatively in its impact on the effective empowerment of Australian aboriginal voices. The danger resides not in the inscription of the alternative metatext as such, but in the specific employment of this metatext under the sign of the authentic to exclude the many and complex voices of the Aboriginal peoples past and present.

The mythologising of the authentic characterised in the media representation of the Nyoongah in the *West Australian* article I have quoted, is then in many ways itself a construction which overpowers one of the most powerful weapons within the arsenal of the subaltern subject: that of displacement, disruption, ambivalence, or mimicry, discursive features founded not in the closed and limited construction of a pure authentic sign but in endless and excessive transformation of the subject positions possible within the hybridised. I want to argue that authentic speech, where it is conceived not as a political strategy within a specific political and discursive formation but as a fetishised cultural commodity, may be employed within such accounts as that of the *West Australian* to enact a discourse of 'liberal violence', re-enacting its own oppressions on the subjects it purports to represent and defend.

## Chapter 36

# MARGERY FEE

# WHO CAN WRITE AS OTHER?

From 'Why C. K. Stead Didn't Like Kerry Hulme's *The Bone People* or Who Can Write as the Other?' *Australian and New Zealand Studies in Canada* 1, 1989.

IN AN ARTICLE PUBLISHED IN *Ariel* in October 1985, [New Zealand novelist] C. K. Stead expresses reservations about Keri Hulme's highly-acclaimed *the bone people* . . . 'a novel by a Pakeha which has won an award [The Pegasus Award for Maori Literature] intended for a Maori' (Stead 1985: 104). Stead raises here two very controversial questions. First, how do we determine minority group membership? Second, can majority group members speak as minority members, Whites as people of colour, men as women, intellectuals as working people? If so, how do we distinguish biased and oppressive tracts, exploitative popularizations, stereotyping romanticizations, sympathetic identifications and resistant, transformative visions? . . .

The problem is complicated by the increasing number of writers who, like Hulme, are of mixed ancestry; who, like Aboriginal writer Sally Morgan, have been raised in ignorance of their ancestry; or who, like Canadian Métis writer Beatrice Culleton, have been brought up in White foster homes. Even writers like Witi Ihimaera and Patricia Grace, whose 'Maoriness' does not seem to be in question, speak English as their mother tongue, and have had to write their way back into their Maori language and culture (Pearson 1982: 166). This 'complication' is a salutary one, in that it emphasizes the dubiousness of most commonplaces about indigenous identity. . . .

Stead points out that Hulme was not brought up speaking Maori, an argument that would exclude both Patricia Grace and Witi Ihimaera from genuine Maoriness, and then casts doubt on the 'authenticity' of some of the Maori elements in the novel. To him they seem 'willed, self-conscious, not inevitable, not entirely authentic' (104). To shift the argument from the biological to the cultural and linguistic, as Stead has just done, seems a move toward flexibility, but is, in fact, quite rigid. Many indigenous people with eight indigenous great-grandparents live in cities and no longer speak their aboriginal languages. The majority culture has either actively caused or passively allowed the loss of traditional indigenous languages and cultures world-wide. For example, Native children in Canada were frequently either sent to boarding schools with White teachers who often punished them for speaking their native languages, or taken away from their parents and communities and sent to White foster homes. Canadian Native communities are still struggling for control over their children's education and foster care. For a member of a majority culture to try to deprive anyone of an indigenous identity just because of the success of this sort of program of cultural obliteration is ironic at best. . . .

The demand for 'authenticity' denies Fourth World writers a living, changing culture. Their culture is deemed to be Other and must avoid crossing those fictional but ideologically essential boundaries between Them and Us, the Exotic and the Familiar, the Past and the Future, the 'Dying' and the Living. Especially, 'authentic' writing from the Fourth World must steer clear of that quintessentially 'new' and ever renewing genre, the novel. For Stead, the function of the Maori work of literature is to preserve the past, not to change the future. Given the destruction inflicted by Whites on indigenous cultures one sympathizes with this view, but indigenous peoples may well feel it is suicidal to devote the time of their best-educated to cultural preservation at the cost of political renewal. Indigenous people have been acculturated to popular western literary forms, and any writer who wishes to reach them is unlikely to do so with a 'pure' traditional form. Nor are ordinary White readers likely to be attracted to an imitation of oral poetry, and yet the majority must be reached by minority writers if change is to take place.

Finally, Stead's insistence that the Maori elements be 'unconscious,' rather than 'willed,' is essentially a demand to hear what seems 'natural' to him, that is 'authentic' accounts that echo the 'authentic' accounts he is used to – those written by White anthropologists and those Pakeha writers who borrow this material. In fact, anthropologists have recently focussed almost obsessively on the degree to which ethnocentricity has marked and continues to mark the assumptions and results of the discipline. Since most writing works within a limited range of ideological possibilities, the trick for most writers is to sound original while repeating the same old 'truths' using the same old literary conventions. Writers who are trying to change the discursive formation, even if only a little, are usually greeted with incomprehension or annoyance. Hulme's attempt to integrate Pakeha and Maori culture in a way that transgresses the boundary between them is bound to seem 'willed,' since so few pieces of writing have made the attempt.

Stead does finally turn to Hulme's text and points to 'the imaginative strength' of *the bone people*: 'it creates a sexual union where no sex occurs, creates parental love where there are no physical parents, creates the stress and fusion of a family where there is no actual family' (104). The biological essentialism of Stead's assumption that sex and biological parent-hood are the sole constituents of an 'actual' family blinds him to the realization that here is Hulme's definition of Maori: 'actual' Maoriness, like an 'actual' family, has nothing to do with biology and everything to do with solidarity of feeling. Stead wants clear categories: either one is a Maori or a Pakeha. Although he is perceptive enough to spot the points where Hulme is violating his categories, he does not realize that she is doing so consciously and consistently.

Hulme's definition of Maori is far more liberal than either Stead's of Maori or Hobson's of American Indian. Perhaps her definition is too liberal, because if we simply conclude that if one feels Maori one is, we fall into a new set of problems. I may feel Maori, I may think I am writing as one, and be completely deluded. Indeed, as Sneja Gunew points out, even the belief that a Maori with 'pure' Maori ancestry automatically will write as a Maori is flawed: the oppressed Other 'supposedly speaks authentically and unproblematically as a unified subject on behalf of the groups she or he represents. . . . In the drive toward universalism one cannot admit that those oppressed others whom we hear as speaking authentic experi-ence might be playing textual games' (1987: 262). Roland Barthes, Jacques Derrida, and Michel Foucault, to name only the most eminent, have undermined rather thoroughly the argument for the authentic and unified voice of the author. Thus, it may seem, my support for Hulme's claim to write as a Maori has been produced only to be withdrawn again. Not quite. To see the individual writer as merely either a conduit for an eclectic range of multiple voices or the mouthpiece for the dominant discourse goes too far. Some writers are resisting writers. However, to say that anyone who qualifies as a Fourth World writer can or should write only about the Fourth World experience is simply another instance of the ubiquitous restriction of the minority. Yet some restrictions do exist.

Edward Said writes of Conrad that even when writing about the oppressed, all he 'can see is a world dominated by the West, and – of equal importance – a world in which every opposition to the West only confirms its wicked power. What Conrad could not see is life lived outside this cruel tautology . . . [and] not controlled by the gringo imperialists and liberal reformers of this world' (Said 1988: 70). By implication, some can see beyond this cruel tautology. But how far? David Maughan Brown (1985) details the extent to which even Ngũgĩ wa Thiong'o's most radical fiction is affected by his liberal humanist education. It is not possible simply to assume that a work written by an 'Other' (however defined), even a politicized Other, will have freed itself from the dominant ideology. Homi Bhabha says 'there is always, in Said, the suggestion that colonial power and discourse is possessed entirely by the coloniser, which is an historical and theoretical simplification' (Bhabha 1983b: 25). Radical writing, by definition, is writing that is struggling, of necessity only partly success- fully, to rewrite the dominant ideology from within, to produce a different version of reality. Hulme laboriously hammered her vision out over twelve years, beginning to write herself into a Maori and New Zealand into Aotearoa. As Terry Threadgold notes, ' "ideology" is not "out there," imposed as it were from above, but rather, is part of the signification itself. Ideologies are constructed in language as contextualized social discourse' (Threadgold 1986: 29). Rewriting the dominant ideology is not easy, since the difference between Pakeha and Maori has been written into existence by the dominant discourse, and thus the process of rewriting this ideology is the work of the whole New Zealand community, rather than of any one writer.

All this makes the idea of accurately or finally distinguishing authentic from inauthentic discourse impossible: the ideal of 'authenticity' has been proven to be, like so many others, relative and context-bound. This does not leave us, however, with nothing but language games. If the context is firmly kept in mind, it is possible to argue that to be classified as 'Fourth World,' writing must somehow promote indigenous access to power without negating indigenous difference.

## Chapter 37

# TERRY GOLDIE

# THE REPRESENTATION OF THE INDIGENE

From *Fear and Temptation: The Image of the Indigene in Canadian, Australian and New Zealand Literatures* Kingston: McGill-Queens University Press, 1989.

I T IS MY PERCEPTION THAT the shape of the signifying process as it applies to indigenous peoples is formed by a certain semiotic field, a field that provides the boundaries within which the images of the indigene function. The existence of this semiotic field constitutes an important aspect of the 'subjugated knowledges' to which Foucault refers in *Power/Knowledge* (1980: 81). The indigene is a semiotic pawn on a chess board under the control of the white signmaker. And yet the individual signmaker, the individual player, the individual writer, can move these pawns only within certain prescribed areas. Whether the context is Canada, New Zealand, or Australia becomes a minor issue since the game, the signmaking is all happening on one form of board, within one field of discourse, that of British imperialism. Terms such as 'war-dance,' 'war-whoop,' 'tomahawk,' and 'dusky' are immediately suggestive everywhere of the indigene. To a North American, at least the first three would seem to be obvious Indianisms, but they are also common in works on the Maori and the Aborigine. Explorers like Phillip King (*Narrative* 1827) generally refer to Aborigines as Indian, and specific analogies to North American Indians are ubiquitous in nineteenth-century Australian literature. Terms misapplied in the Americas became re-misapplied in a parody of imperialist discourse. The process is quite similar to one Lévi-Strauss describes in *The Savage Mind* (1972): 'In other words, the operative value of the systems of naming and classifying commonly called totemic derives from their formal character: they are codes suitable for conveying messages which can be transposed into other codes, and for expressing messages received by means of different codes in terms of their own system' (75). Obvious extreme ethnographic differences between the different indigenous cultures did little to impede the transposition.

To extend the chessboard analogy, it would not be oversimplistic to maintain that the play between white and indigene is a replica of the black and white squares, with clearly limited oppositional moves. The basic dualism, however, is not that of good and evil, although it has often been argued to be so, as in Abdul R. JanMohamed's 'The Economy of Manichean Allegory' (1985): 'The dominant model of power – and interest – relations in all colonial societies is the manichean opposition between the putative superiority of the European and the supposed inferiority of the native' (63). JanMohamed maintains that in apparent exceptions 'any evident "ambivalence" is in fact a product of deliberate, if at times subconscious, imperialist duplicity, operating very efficiently through the economy of its central trope, the manichean allegory' (61). Such a basic moral conflict is often implied but in contemporary texts the opposition is frequently between the 'putative superiority' of the indigene and the 'supposed inferiority' of the white. As Said suggests, the positive and negative sides of the image are but swings of one and the same pendulum: 'Many of the earliest oriental amateurs began by welcoming the Orient as a salutary *derangement* of their European habits of mind and spirit. The Orient was overvalued for its pantheism, its spirituality, its stability, its longevity, its primitivism, and so forth. . . . Yet almost without exception such overesteem was followed by a counter-response: the Orient suddenly appeared lamentably under-humanized, antidemocratic, backward, barbaric, and so forth' (1978: 150). Almost all of these characterizations could be applied to the indigenes of Australia, New Zealand, and Canada, as positive or negative attributes.

The complications of the issue extend even beyond oppositions of race, as Sander Gilman suggests in *Difference and Pathology* (1985):

> Because there is no real line between self and the Other, an imaginary line must be drawn; and so that the illusion of an absolute difference between self and Other is never troubled, this line is as dynamic in its ability to alter itself as is the self. This can be observed in the shifting relationship of antithetical stereotypes that parallel the existence of 'bad' and 'good' representations of self and Other. But the line between 'good' and 'bad' responds to stresses occurring within the psyche. Thus paradigm shifts in our mental representations of the world can and do occur. We can move from fearing to glorifying the Other. We can move from loving to hating.
>
> (18)

The problem is not the negative or positive aura associated with the image but rather the image itself. . . .

At least since Fanon's *Black Skin White Masks* (1952 [1967a]) it has been a commonplace to use 'Other' and 'Not-self' for the white view of blacks and for the resulting black view of themselves. The implication of this assertion of a white self as subject in discourse is to leave the black Other as object. The terms are similarly applicable to the Indian, the Maori, and the Aborigine but with an important shift. They are Other and Not-self but also must become self. Thus as Richon suggests and Pearson implies, imperialist discourse valorizes the colonized according to its own needs for reflection. 'The project of imperialism has always already historically refracted what might have been the absolute Other into a domesticated Other that consolidated the imperialist self,' explains Gayatri Spivak in 'Three Women's Texts and a Critique of Imperialism' (1985c: 253). Tzvetan Todorov in *The Conquest of America: The Question of the Other* (1982) also notes how the group as Other can function.

> This group in turn can be interior to society: women for men, the rich for the poor, the mad for the 'normal'; or it can be exterior to society, i.e., another society which will be near or far away, depending on the case: beings whom everything links to me on the cultural, moral, historical plane; or else unknown quantities, outsiders whose language and customs I do not understand, so foreign that in extreme instances I am reluctant to admit they belong to the same species as my own.
>
> (3)

But Spivak's area of study, the Indian sub-continent, is a different case from that of the Australian, Canadian, and New Zealand because the imperialist discourse remains admittedly non-indigenous. India is valorized by its relationship to imperialist dynamics but it 'belongs' to the white realm only as part of the empire. . . . Australians, New Zealanders, and Canadians have, and long have had, a clear agenda to erase this separation of belonging. The white Canadian looks at the Indian. The Indian is Other and therefore alien. But the Indian is indigenous and therefore cannot be alien. So the Canadian must be alien. But how can the Canadian be alien within Canada?

There are only two possible answers. The white culture can attempt to incorporate the Other, superficially through beaded moccasins and names like Mohawk Motors, or with much more sophistication, through the novels of Rudy Wiebe. Conversely, the white culture may reject the indigene: 'This country really began with the arrival of the whites.' This is no longer an openly popular alternative, but its historical importance is reflected in things like the 'native societies' that existed in all three countries in the late nineteenth century, societies to which no non-white, no matter how native, need have applied.

The importance of the alien within cannot be overstated. In their need to become 'native,' to belong here, whites in Canada, New Zealand, and Australia have adopted a process which I have termed 'indigenization.' A peculiar word, it suggests the impossible necessity of becoming indigenous. For many writers, the only chance for indigenization seemed to be through writing about the humans who are truly indigenous, the Indians, Inuit, Maori, and Aborigines. As J. J. Healy notes in *Literature and the Aborigine in Australia* (1978), 'The Aborigine was part of the tension of an indigenous consciousness. Not the contemporary Aborigine, not even a plausible historical one, but the sort of creature that *might* persuade a white Australian to look in the direction of the surviving race' (173). Many Canadians, New Zealanders, and Australians have responded strongly to this creature and to their own need to become indigenous. . . .

Of course the majority of writers in all three countries have given brief or no attention to native peoples. Perhaps then, while the image of the indigene may be a consistent concern, it is a limited one, the *Jindyworobaks* notwithstanding. But the process of indigenization is complex. Each reference in *The Bulletin*, the nationalistic nineteenth century Australian magazine, to the white Australian as 'native' or 'indigenous' is a comment on indigenization, regardless of the absence of Aborigines in those references. As Macherey claims, 'an ideology is made of what it does not mention; it exists because there are things which must not be spoken of' (1978: 132). In other words, absence is also negative presence. Thus in a work such as Henry Handel Richardson's *The Fortunes of Richard Mahony* (1930), a trilogy which uses the Australian gold rush as a field through which to explore the founding of a nation, the Aborigine is an essential non-participant.

Neither the racial split between self and Other nor the process of indigenization originates with Canada, Australia, or New Zealand, but neither do they have clear origins which might be seen as the source for these manifestations. Presumably the first instance in which one human perceived another as Other in racial terms came when the first recognized the second as different in colour, facial features, language, etc. And the first felt need for indigenization came when a person moved to a new place and recognized an Other as having greater roots in that place. The lack of a specific origin for these conditions is reflected in the widespread occurrence of their modern manifestations. . . .

However, regardless of the changes made in the form of the chessboard, whether dismissive histories of penetration or sensitive novels of appropriation, the semiotic field has continued, particularly in the few basic moves which the indigenous pawn has been allowed to make.

These few basic moves Said calls 'standard commodities' (1978: 190). Two such commodities which appear to be standard in the 'economy' created by the semiotic field of the indigene in Australian, Canadian, and New Zealand literatures are sex and violence. They are poles of attraction and repulsion, temptation by the dusky maiden and fear of the demonic violence of the fiendish warrior. Often both are found in the same work, as in John Richardson's *Wacousta or The Prophecy* ([1832] 1967), in which the warrior constantly attacks, but the maiden is an agent to avoid that attack. They are emotional signs, semiotic embodiments of primal responses. Could one create a more appropriate signifier for fear than the treacherous redskin? He incorporates, in generous quantities, the tenor of the impassioned, uncontrolled spirit of evil. He is strangely joined by the Indian maiden, who tempts the being chained by civilization towards the liberation represented by free and open sexuality, not the realm of untamed evil but of unrestrained joy. 'The "bad" Other becomes the negative stereotype; the "good" Other becomes the positive stereotype. The former is that which we fear to become; the latter, that which we fear we cannot achieve' (Gilman 1985: 20). Added to this is the alien's fear of the 'redskin' as hostile wilderness, the new, threatening land, and the arrivant's attraction to the maiden as restorative pastoral, this new, available land.

A third important commodity is orality, all the associations raised by the indigene's speaking, non-writing, state. The writers' sense of indigenes as having completely different systems of understanding different epistemes, is based on an often undefined belief that cultures without writing operate within a different dimension of consciousness. This different dimension suggests a fourth commodity, mysticism, in which the indigene becomes a sign of oracular power, either malevolent, in most nineteenth-century texts, or beneficent, in most contemporary ones. In an interesting variant on the semiotic process, the inadequacies of the writer's culture, in which little knowledge is to be gained from the popular beliefs of its own traditions . . . is placed in contrast to an indigenous belief system (usually quite asystemic) which holds the promise of a Presence to exceed even the presence of orality. . . .

[A] fifth commodity in the semiotic field of the indigene [is] the prehistoric. The historicity of the text, in which action makes a statement, whether overt or covert, on the chronology of the culture, shapes the indigene into an historical artifact, a remnant of a golden age that seems to have little connection to anything akin to contemporary life. A corollary of the temporal split between this golden age and the present degradation is a tendency to see indigenous culture as true, pure, and static. Whatever fails this test is not really a part of that culture.

The commodities – sex, violence, orality, mysticism, the prehistoric – can be seen as part of a circular economy within and without the semiotic field of the indigene. . . . It appears that as long as this semiotic field exists, as long as the shapes of the standard commodities change but the commodities remain the same, the chess match can appear to vary but there is still a defineable limit to the board. The necessities of indigenization can compel the players to participate but they cannot liberate the pawn.

## Chapter 38

# ARNOLD KRUPAT

## POSTCOLONIALISM, IDEOLOGY, AND NATIVE AMERICAN LITERATURE

From 'Postcolonialism, Ideology and Native American Literature' *The Turn to the Native: Studies in Criticism and Culture* Lincoln, Nebr.: University of Nebraska Press, 1996.

. . .

I WANT TO SUGGEST A CATEGORY – the category of anti-imperial translation – for conceptualizing the tensions and differences between contemporary Native American fiction and 'the imperial center'. Because historically specifiable acts of translative violence marked the European colonization of the Americas from Columbus to the present, it seems to me particularly important to reappropriate the concept of translation for contemporary Native American literature. To do so is not to deny the relationship of this literature to the postcolonial literatures of the world but, rather, to attempt to specify a particular modality for that relationship.

To say that the people indigenous to the Americas entered European consciousness only by means of a variety of complex acts of translation is to think of such things as Columbus's giving the name of San Salvador to an island he *knows* is called Guanahani by the natives – and then giving to each further island he encounters, as he wrote in his journals, 'a new name' (Greenblatt 1991: 52). Columbus also materially 'translated' (*trans-latio* 'to carry across') some of the Natives he encountered, taking 'six of them from here,' as he remarked in another well-known passage, 'in order that they may learn to speak' (Greenblatt 1991: 90). Columbus gave the one who was best at learning his own surname and the first name of his firstborn son, translating this otherwise anonymous person into Don Diego Colon.

Now, any people who are perceived as somehow unable to speak when they speak their own languages, are not very likely to be perceived as having a literature – especially when they do not write, a point to which we shall return. Thus, initially, the very 'idea of a (Native American) literature was inherently ludicrous,' as Brian Swann has noted, because Indian 'languages themselves were primitive' (1992: xiii). If Indians spoke at all, they spoke very badly (and, again, they did not write). In 1851, John DeForest, in his *History of the Indians of Connecticut*, observed, 'It is evident from the enormous length of many of the words, sometimes occupying a whole line, that there was something about the structure of these languages which made them cumbersome and difficult to manage' (Swann 1992: xiii).

Difficult for whom, one might ask, especially in view of the fact that De Forest himself had not achieved even minimal competence in any Native language. . . .

Almost half a century after DeForest, as late as 1894, Daniel Brinton – a man who actually did a great deal to make what he called the 'production' of 'aboriginal authors' visible to the dominant culture – nonetheless declared, 'Those peoples who are born to the modes of thought and expression enforced by some languages can never forge to the front in the struggle for supremacy; they are fatally handicapped in the race for the highest life' (Murray 1991: 8). The winners in the 'race for the highest life,' therefore, would be the race with the 'highest' language; and it was not the Indians but rather, as Brinton wrote, 'our Aryan forefathers' who were the ones fortunate enough to be endowed 'with a richly inflected speech.' As Kwame Anthony Appiah explained in reference to Johann Gottfried von Herder, the *Sprachgeist,* 'the "spirit" of the language, is not merely the medium through which speakers communicate but the sacred essence of a nationality. [And] Herder himself identified the highest point of the nation's language in its poetry' (Appiah 1990: 284), in its literature. 'Whoever writes about the literature of a country,' as Appiah elsewhere cited Herder, 'must not neglect its language' (50). For those like the Indians with 'primitive' languages, there would seem to be little hope, short of translation, for the prospects of literary achievement. Thus, by the end of the nineteenth century, the linguistic determinism expressed by Brinton – and, of course, by many others – worked against the possibility of seeing Native Americans as having an estimable literature at exactly the moment when the texts for that literature were, for the first time, being more or less accurately translated and published.

But here one must return to the other dimension of the translation issue as it affects Native American literatures. For the problem in recognizing the existence of Native literatures was not only that Natives could not speak or, when they did speak, that their languages were judged deficient or 'primitive' but also that they did not write.

Here I will only quickly review what I and others have discussed elsewhere. Because *littera-ture* in its earliest uses meant the cultivation of letters (from Latin *littera,* 'letter'), just as *agriculture* meant the cultivation of fields, peoples who did not inscribe alphabetic characters on the page could not, by definition, produce a literature. (They were also thought to be only minimally capable of agriculture in spite of overwhelming evidence to the contrary, but that is another story.) It was the alteration in European consciousness generally referred to as 'romanticism' that changed the emphasis in constituting the category of literature from the medium of expression, writing – literature as culture preserved in letters – to the *kind* of expression preserved, literature as imaginative and affective utterance, spoken or written. It is only at this point that an oral literature can be conceived as other than a contradiction in terms and the unlettered Indians recognized as people capable of producing a 'literature.'

For all of this, it remains the case that an oral literature, in order to become the subject of analysis, must indeed first become an object. It must, that is, be textualized; and here we encounter a translation dilemma of another kind, one in which the 'source language' itself has to be carried across – *trans-latio* – from one medium to another, involving something more than just a change of names. This translative project requires that temporal speech acts addressed to the ear be turned into visual objects in space, black marks on the page, addressed to the eye. Words that had once existed only for the tongue to pronounce now were to be entrusted to the apprehension of the eye. Mythography, in a term of Anthony Mattina's, or ethnopoetics has been devoted for many years to the problems and possibilities involved in this particular form of media translation. . . .

I base my sense of anti-imperial translation on a well-known, indeed classic text, that I have myself quoted on a prior occasion. The text is from Rudolph Pannwitz, who is cited in Walter Benjamin's important essay 'The Task of the Translator.' Pannwitz wrote, 'Our translations, even the best ones, proceed from a wrong premise. They want to turn Hindi, Greek,

English into German instead of turning German into Hindi, Greek, English. Our translators have far greater reverence for the usage of their own language than for the spirit of the foreign works. . . . The basic error of the translator is that he preserves the state in which his own language happens to be instead of allowing his language to be powerfully affected by the foreign tongue' (Benjamin 1969: 180–1). My use of Pannwitz was influenced by Talal Asad's paper, 'The Concept of Cultural Translation in British Social Anthropology,' originally presented at the School for American Research in 1984 and published in James Clifford and George Marcus's important collection *Writing Culture* in 1986. . . .

My claim is that Native American writers today are engaged in some version of the translation project along the broad lines sketched by Asad. Even though contemporary Native writers write in English and configure their texts in apparent consonance with Western or Euramerican literary forms – that is, they give us texts that look like novels, short stories, poems, and autobiographies – they do so in ways that present an 'English' nonetheless 'powerfully affected by the foreign tongue,' not by Hindi, Greek, or German, of course, and not actually by a 'foreign' language, inasmuch as the 'tongue' and 'tongues' in question are indigenous to America. The language they offer, in Asad's terms, derives at least in part from other forms of practice, and to comprehend it might just require, however briefly, that we attempt to imagine living other forms of life.

This is true of contemporary Native American writers in both literal and figurative ways. In the case of those for whom English is a second language (Luci Tapahonso, Ray Young Bear, Michael Kabotie, Ofelia Zepeda, and Simon Ortiz are some of the writers who come immediately to mind), it is altogether likely that their English will show traces of the structure and idioms of their 'native' language, as well as a variety of linguistic habits and narrative and performative practices of traditional expressive forms in Navajo, Mesquakie, Hopi, Tohono O'odham, and Acoma. Their English, then, is indeed an English, in Pannwitz's words, 'powerfully affected by the foreign tongue,' a tongue (to repeat) not 'foreign' at all to the Americas. Here the Native author quite literally tests 'the tolerance of [English] for assuming unaccustomed forms' (Asad 1986: 157), and an adequate commentary on the work of these writers will require of the critic if not bilingualism then at least what Dell Hymes has called some 'control' of the Native language.

Most Native writers today are not, however, fluent speakers of one or another of the indigenous languages of the Americas, although their experiences with these languages are so different that it would be impossible to generalize. (E.g., Leslie Marmon Silko certainly heard a good deal of Laguna as she was growing up, just as N. Scott Momaday heard a good deal of Jemez, whereas many of the Native American writers raised in the cities did not hear indigenous languages on a very regular basis.) Yet all of them have indicated their strong sense of indebtedness or allegiance to the oral tradition. Even the mixed-blood Anishinaabe–Chippewa writer Gerald Vizenor, someone who uses quotations from a whole range of contemporary European theorists and whose own texts are full of ironic effects possible only to a text-based literature, has insisted on the centrality of 'tribal stories' and storytelling to his writing. This is the position of every other contemporary Native American writer I can think of – all of them insist on the storytelling of the oral tradition as providing a context, as bearing on and influencing the writing of their novels, poems, stories, or autobiographies.

In view of this fact, it needs to be said that 'the oral tradition,' *as it is invoked by these writers*, is an 'invented tradition.' It can be seen, as John Tomlinson has remarked, 'as a phenomenon of modernity. There is a sense in which simply recognizing a practice as "traditional" marks it off from the routine practices of proper [sic] traditional societies' (1991: 91). This is not, of course, to deny that there were and continue to be a number of oral traditions that 'really' existed and continue to exist among the indigenous cultures of the Americas. Nor is it to deny that some contemporary Native American writers have considerable experience of 'real' forms of oral performance. I am simply noting that 'the oral tradition' as usually invoked in these contexts is a kind of catchall phrase whose function is broadly to name the source of the difference between the English of Native writers and that of Euramerican writers. This 'tradition' is not based on historically and culturally specific instances.

A quick glance at some of the blurbs on the covers or book jackets of work by contemporary Indian writers makes this readily apparent. When these blurbs are written by non-Indians (and most are, for obvious reasons, written by non-Indians), reference to 'the oral tradition' usually represents a loose and vague way of expressing nostalgia for some aboriginal authenticity or wisdom, a golden age of wholeness and harmony. When these blurbs are written by Native Americans – this generalization I venture more tentatively – they are . . . a rhetorical device, a strategic invocation of what David Murray has called the discourse of Indianness, a discourse that has currency in both the economic and the political sense in the United States. Once more, to say this is in no way to deny that the narrative modalities and practices of a range of Native oral literatures, as well as the worldviews of various Native cultures, are important to many of the texts constituting a contemporary Native American literature, and not merely honorifically, sentimentally, or rhetorically.

## Chapter 39

# JAMES CLIFFORD

# INDIGENOUS ARTICULATIONS

From 'Indigenous Articulations' *The Contemporary Pacific* 3(2) (Fall), 468, 2001.

[H]OW IS 'INDIGENEITY' BOTH rooted in and routed through particular places? How shall we begin to think about a complex dynamic of local landedness and expansive social spaces? Should we think of a continuum of indigenous and diasporic situations? Or are there specifically indigenous kinds of diasporism? Lived dialectics of urban and rural? On and off the reservation? Island and mainland native experiences? There are real tensions, to be sure, along the continuum of indigenous locations. But as Murray Chapman's extensive research on 'circulation' in the Solomon Islands and beyond suggests, we should be wary of binary oppositions between home and away, or a before–after progression from village life to cosmopolitan modernity (1978; 1991). As we try to grasp the full range of indigenous ways to be 'modern,' it is crucial to recognize patterns of visiting and return, of desire and nostalgia, of lived connections across distances and differences. . . .

The notion of articulated sites of indigeneity rejects two claims often made about today's tribal movements. On the one hand, articulation approaches question the assumption that indigeneity is essentially about primordial, transhistorical attachments (ancestral 'laws,' continuous traditions, spirituality, respect for Mother Earth, and the like). Such understandings tend to bypass the pragmatic, entangled, contemporary forms of indigenous cultural politics. On the other hand, articulation theory finds it equally reductive to see indigenous, or First Nations claims as the result of a post-sixties, 'postmodern' identity politics (appeals to ethnicity and 'heritage' by fragmented groups functioning as 'invented traditions' within a late-capitalist, commodified multiculturalism). This viewpoint brushes aside long histories of indigenous survival and resistance, transformative links with roots prior to and outside the world system. We must, I think, firmly reject these simplistic explanations – while weighing the partial truth each one contains.

To think of indigeneity as 'articulated' is, above all, to recognize the diversity of cultures and histories that currently make claims under this banner. What exactly unites Hawaiians (whose history includes a monarchic state) and much smaller Amazonian, or New Guinea groups? What connects Pan-Mayan activists with US tribal gaming operations? What allies the new Inuit autonomous province of Nunavut with Aboriginal and Torres Strait Islander land-claims (rather than with, say, the similar strong regionalisms of Catalonia, or perhaps what's emerging in Scotland or Wales)? What do 'tribal' peoples in India have in common with the Fijian Great Council of Chiefs?

I do not think we can arrive at a core list of essential 'indigenous' features. The commonality is more historically contingent, though no less real for all that. Indigenous movements are

positioned, and potentially but not necessarily connected, by overlapping experiences in relation to Euro-American, Russian, Japanese, and other imperialisms. They all contest the power
of assimilationist nation-states, making strong claims for autonomy, or for various forms of
sovereignty. In recent decades, positive discourses of indigenous commonality have emerged,
drawing together this range of historical predicaments: the various pan-Indian, pan-Aboriginal,
pan-Mayan, indigenous 'Arctic,' movements, as well as an expanding network of fourth-world
coalitions. Such discourses are also propagated through the networks of the United Nations,
nongovernment organizations, and tourists. Today, a number of expansive ideologies express
positive notions of 'indigenousness,' ideas that in turn feed back into local traditions.

To see such chains of equivalence (which must always downplay or silence salient
differences) as articulated phenomena is not to view them as inauthentic or 'merely' political,
invented, or opportunistic. Articulation as I understand it evokes a deeper sense of the
'political' – productive processes of consensus, exclusion, alliance, and antagonism that are
inherent in the transformative life of all societies.

## Articulations

I now focus more directly on how articulation theory helps us understand all this. What are
its limits? Where does it need to be adapted, customized? The politics of articulation for
Stuart Hall is, of course, an updating of Gramsci (Hall 1986a, b; Slack 1996). It understands
frontier effects, the lining up of friends and enemies, us and them, insiders and outsiders, on
one side or another of a line, as tactical. Instead of rigid confrontations – civilized and primitive, bourgeois and proletarian, white and black, men and women, west and third world,
Christian and pagan – one sees continuing struggles across a terrain, portions of which are
captured by changing alliances, hooking and unhooking particular elements. There's a lot of
middle ground; and crucial political and cultural positions are not firmly anchored on one
side or the other but are contested and up for grabs.

The term articulation, of course, suggests discourse or speech – but never a self-present,
'expressive' voice and subject. Meaningful discourse is a cutting up and combining of linguistic
elements, always a selection from a vastly greater repertoire of semiotic possibilities. So an
articulated tradition is a kind of collective 'voice,' but always in this constructed, contingent
sense. In another register – not reducible to the domain of language with its orders of grammar
and speech, structure and performance articulation refers to concrete connections, joints.
Stuart Hall's favourite example is an 'articulated lorry' (something that to us Americans sounds
very exotic!). Something that's articulated or hooked together (like a truck's cab and trailer,
or a sentence's constituent parts) can also be unhooked and recombined. When you understand a social or cultural formation as an articulated ensemble it does not allow you to prefigure
it on an organic model, as a living, persistent, 'growing' body, continuous and developing
through time. An articulated ensemble is more like a political coalition or, in its ability to
conjoin disparate elements, a cyborg. While the possible elements and positions of a socio-
cultural ensemble are historically imposed constraints that can be quite persistent over time,
there is no eternal or natural shape to their configuration.

Articulation offers a nonreductive way to think about cultural transformation and the
apparent coming and going of 'traditional' forms. All-or-nothing, fatal-impact notions of change
tend to assume that cultures are living bodies with organic structures. So, for example,
indigenous languages, traditional religions, or kinship arrangements, may appear to be critical
organs, which if lost, transformed, or combined in novel structures should logically imply
the organism's death. You can't live without a heart or lungs. But indigenous societies have

persisted with few, or no, native-language speakers, as fervent Christians, and with 'modern' family structures, involvement in capitalist economies, and new social roles for women and men. 'Inner' elements have, historically, been connected with, 'exterior' forms, in processes of selective, syncretic transformation. When Jean-Marie Tjibaou, speaking as both a former priest and an advocate of Kanak coutume, said that the Bible does not belong to westerners (who seized it 'passing through') he was detaching and rearticulating European and Melanesian religious traditions (1966: 303). . . .

In articulation theory, the whole question of authenticity is secondary, and the process of social and cultural persistence is political all the way back. It is assumed that cultural forms will always be made, unmade, and remade. Communities can and must reconfigure themselves, drawing selectively on remembered pasts. The relevant question is whether, and how, they convince and coerce insiders and outsiders, often in power-charged, unequal situations, to accept the autonomy of a 'we.' This seems to me a more realistic way of talking about what has been called cultural 'invention.' . . .

How should differently positioned authorities (academic and non-academic, Native and non-Native) represent a living tradition's combined and uneven processes of continuity, rupture, transformation, and revival? My suggestions today about articulation contribute to an ongoing argument (and, I hope, a conversation) on these critical issues. I am not persuaded that 'the invention of tradition' approach in the Pacific was essentially a matter of anthropologists, faced by new indigenous challenges, clinging to their professional authority to represent cultures and adjudicate authenticity. That is certainly part of the story. But the notion of 'invention' was also getting at something important, albeit in a clumsy way. The thinking of Roy Wagner (1980), deeply influenced in its structure by New Guinea poetics and politics, is a better source for the term's nonreductive meanings than the usual reference, Hobsbawm and Ranger (1983). This prescient recognition of inventive cultural process has tended to be lost in the flood of analyses that demystify nationalist fictions and manipulations.

At the present moment, it seems to me that the notion of invention can be usefully rethought as a politics of articulation. We are on more concrete, because more dynamic, historical grounds. The whole notion of custom looks quite different when seen this way, when what Margaret Jolly (1992) pointedly called 'specters of inauthenticity' are laid to rest. The question of what is borrowed from here or there, what is lost and rediscovered in new situations, can be discussed within the realm of normal political or cultural activity.

## Horizons

Articulation theory cannot account for everything. Pushed to extremes it can take you to a point where every cultural form, every structure or restructuration, every connection and disconnection, has a radical contingency as if, at any moment, anything were possible. That is a misreading of Stuart Hall on articulation. He is quite clear that the possible connections and disconnections are constrained at any historical moment. Certain forms and structural antagonisms persist over long periods. Yet these enduring forces – whether they be Christianity and capitalism or traditional cosmology and kinship – can be understood concretely only as they work through specific cultural symbols and political blocs. These are never guaranteed, but are actively produced and potentially challenged.

When thinking of differently articulated sites of indigeneity, however, one of the enduring constraints in the changing mix will always be the power of place. This is a fundamental component of all tribal, First Nation identifications. Not everyone is equally on the move. Many people live where they have always lived, even as the habitat around them goes through sometimes violent transformations. As the scale of 'tribal' and 'national' existence alters dramatically, people living exiled from ancestral places often sustain and revive a yearning, an active memory of land. (For island and coastal peoples a sense of material location can include a lot of water.) This 'grounding' offers a sense of depth and continuity running through all the ruptures and attachments, the effects of religious conversion, state control, new technologies, commodities, schooling, tourism, and so on. Indigenous forms of dwelling cover a range of sites and intensities: there are 'native' homebodies, commuters, travelers, and exiles. But a desire called 'the land,' is differently, persistently active. . . .

While recognizing this fundamental claim to a distinctly rooted history, I want to argue against rigid oppositions in defining the current array of indigenous experiences. We need to distinguish, and also (carefully, partially) to connect 'diasporism' and 'indigenism.' What's at stake is the articulation, the cobbling together, of 'big enough' worlds: concrete lives led in specific circuits between the global and the local. We cannot lose sight of ordinary people sustaining relational communities and cosmologies: composite 'worlds' that share the planet with others, overlapping and translating. An absolutist indigenism, where each distinct 'people' strives to occupy an original bit of ground, is a frightening utopia. For it imagines relocation and ethnic cleansing on an unimaginable scale: a denial of all the deep histories of movement, urbanization, habitation, reindigenization, sinking roots, moving on, invading, mixing – the very stuff of human history. There must be, and in practice there are, many ways to conceive of 'nativeness' in less absolute terms.

Nativism, the xenophobic shadow of indigeneity, values wholeness and separation, pure blood and autochthonous land. It denies the messy, pragmatic politics of articulation. Of course there's no shortage of violent examples in today's ethnically divided world to remind us of this ever-present threat. But nationalist chauvinism, while a constant tendency, is not a necessary outcome of the new indigenisms. The articulated, rooted, and cosmopolitan practices I've been trying to sketch today register more complex, emergent possibilities . . . The movements of Native Pacific people suggest newly inventive struggles for breathing space, for relational sovereignty, in post- or neocolonial conditions of complex connectivity. They are about finding ways to exist in a multiplex modernity, but with a difference, a difference derived from cultural tradition, from landedness, and from ongoing histories of displacement, travel, and circulation. As Hau'ofa has suggested, an element of 'diasporism,' of movement between places, is part of escaping belittlement – of becoming big, global, enough. But he also stressed that this must not mean losing contact with specific ecologies, places, and 'pasts to remember' (Hau'ofa 1993, 2000). Since indigenism and diasporism aren't one-size-fits-all categories, we need to work toward a more nuanced vocabulary, finding concrete ways to represent dispersed and connected populations.

## Chapter 40

# DIANA BRYDON

## THE WHITE INUIT SPEAKS
### Contamination as literary strategy

From 'The White Inuit Speaks: Contamination or Literary Strategy' in Ian Adam and Helen Tiffin (eds) *Past the Last Post: Theorizing Post-colonialism and Post-modernism* New York and London: Harvester Wheatsheaf, 1991.

**M**Y TITLE IS INSPIRED BY the coincidental appearance of the Inuit as symbolic figure in two important Canadian novels published in 1989, Kristjana Gunnar's *The Prowler* and Mordecai Richler's *Solomon Gursky Was Here*. By echoing the influential American ethnographic text *Black Elk Speaks*, I mean to highlight the assumptions about cultural purity and authenticity that post-modernism and post-colonialism, and these two texts, both use and challenge. *Black Elk Speaks* itself is now being recognized as a white man's construct, fusing traditional Lakota with Christian philosophy – a hybrid rather than the purely authentic of the anthropologist's dreams (Powers 1990). Unlike those who deplore a perceived loss in authenticity in Black Elk's cultural contamination, Gunnars and Richler explore the creative potential of such cross-cultural contact. For them, as for the bilingual Canadian poet Lola Lemire Tostevin, 'the concept of contamination as literary device' would seem to be appealing. Tostevin argues that 'Contamination means differences have been brought together so they make contact' (1989: 13).

Such a process defines the central activities of post-modernism and post-colonialism – the bringing of differences together into creative contact. But this is also where they part company. For it is the nature of this contact – and its results – that are at issue. For post-colonial writers, the cross-cultural imagination that I am polemically calling 'contamination' for the purposes of this article, is not just a literary device but also a cultural and even a political project. Linda Hutcheon ('Circling the Downspout') [1989] points out that post-colonialism and feminism have 'distinct political agendas and often a theory of agency that allow them to go beyond the post-modern limits of deconstructing existing orthodoxies into the realms of social and political action'. In contrast, she argues, 'post-modernism is politically ambivalent' (168). At the same time, however, she concludes that the post-colonial is 'as implicated in that which it challenges as is the *post-modern*' (183). This assertion depends on a leap from the recognition that the post-colonial is 'contaminated' by colonialism (in the word itself and the culture it signifies) to the conclusion that such 'contamination' necessarily implies complicity. It is this notion I would like to explore more fully in the rest of this paper.

If we accept Hutcheon's assertion that post-modernism is politically ambivalent, what are the implications of such a theory? There are at least two that interest me here. Firstly, what enables this ambivalence? Post-modernism takes on a personality; it becomes a subject, human-like in its ability to express ambivalence. The functions of the author, declared dead by post-structuralist theory, resurface in post-modernism and in the post-modernist text through the concept of ambivalence. The authority of the post-modernist text comes from this ambivalence, this ability to see all sides, to defer judgement and to refuse agency. Secondly, what are the effects of this ambivalence? It would seem to suggest that action is futile; that individual value judgements are likely to cancel each other out; that one opinion is as good as another; that it would be futile and dishonest to choose one path above any other; that disinterested contemplation is superior to any attempt at action. In effect, then, ambivalence works to maintain the status quo. It updates the ambiguity so favoured by the New Critics, shifting their formalist analysis of the text's unity into a psychoanalysis of its fissures, and their isolation of text from world into a worldliness that cynically discounts the effectiveness of any action for social change.

To refer to contradictions instead of a fundamental ambivalence places the analysis within a political rather than a psychoanalytical framework. Post-modernism and post-colonialism often seem to be concerned with the same phenomena, but they place them in different grids of interpretation. The name 'post-modernism' suggests an aestheticizing of the political while the name 'post-colonialism' foregrounds the political as inevitably contaminating the aesthetic, but remaining distinguishable from it. If post-modernism is at least partially about 'how the world dreams itself to be "American"' (Stuart Hall quoted in Ross 1988: xii), then post-colonialism is about waking from that dream, and learning to dream otherwise. Post-modernism cannot account for such post-colonial resistance writing, and seldom attempts to.

Much of my work over the past decade has involved documenting the contradictions of Canadian post-colonialism. Reading Canadian literature from a post-colonial perspective, recognizing Canadian participations in empire and in the resistance to empire, one quickly encounters some of the limitations of post-modernist theory in accounting for Canadian texts, even for those apparently post-modernist in form. Because Linda Hutcheon is one of Canada's preeminent theorists of the post-modern, this essay engages with her work first of all as a way of posing some of the problems I see when the post-colonial and the post-modern are brought together.

Despite post-modernism's function as a problematizing mode, several assumptions central to imperial discourse survive unchallenged in the work of its defenders. These include an evolutionary model of development, a search for synthesis that relies on a revival of the notion of authenticity, and an insistence on judging a work on its own terms alone as if there were only one true reading. A post-colonial reading would reject such assumptions: post-modernist readings affirm them under the guise of a disinterested objectivity. . . .

## The evolutionary model

In 'Circling the Downspout' Hutcheon writes that '[t]he current post-structuralist/post-modernist challenges to the coherent, autonomous subject have to be put on hold in feminist and post-colonial discourses, for both must work first to assert and affirm a denied or alienated subjectivity: those radical post-modern challenges are in many ways the luxury of the dominant order which can afford to challenge that which it securely possesses' (168). There are several problems with this statement. The first is the notion that there is a single evolutionary path of literary development established by the European model. Secondly, there is the idea of a norm of subjectivity also established by the European model. Thirdly, there is the implied assumption that political commitment (to the liberation of nation or women), even in non-European countries, must necessarily express itself through a literary realism that presents a unified subject along the nineteenth century European model. And finally, it seems to demean literary criticism as a 'luxury', something nonessential that not all societies really need, as if critique is not a necessary component for culture or identity building.

These assumptions are so strongly embedded in our western culture that even texts challenging such notions are read to confirm them. Consider Jamaica Kincaid's *Annie John*, a complex metafictional work challenging notions of a unified subjectivity that is often read as a traditional *bildungsroman* consolidating a simple achievement of just such a selfhood. Yet as Simon Gikandi argues . . . 'Caribbean women writers are concerned with a subject that is defined by what de Laurentis calls "a multiple, shifting, and often self-contradictory identity, a subject that is not divided in, but rather at odds, with language"'(Gikandi 1991: 14). This is the kind of subject whose exploration Hutcheon argues must be 'put on hold' in feminist and post-colonial writing, yet in fact we find it in many of these texts, if we read them with the openness we bring to European fictions.

## The search for synthesis

In expressing her unease with the use of post-colonial to describe the settler and multicultural contemporary cultures of Canada, Hutcheon suggests that perhaps Native culture 'should be considered the resisting, post-colonial voice of Canada' (172). This search for the authentic Canadian voice of post-colonialism mirrors the title of her book on post-modernism in Canada, *The Canadian Postmodern*. Just as we saw a unitary subjectivity being affirmed in the evolutionary model, so we see a unified voice or style being advocated here. Although Hutcheon here identifies Robert Kroetsch as 'Mr Canadian Postmodern' (1988b: 183), I would argue that there are several Canadian post-modernisms just as there is more than one Canadian post-colonial voice. A term may have multiple, subsidiary meanings without losing its usefulness in indicating a general category.

Hutcheon's assumption that the post-colonial speaks with a single voice leads her to belabour the necessity of resisting the totalizing application of a term that in her analysis would blur differences and deny the power relations that separate the native post-colonial experience from that of the settlers. Certainly turning to the post-colonial as a kind of touristic 'me-tooism' that would allow Canadians to ignore their own complicities in imperialism would be a serious misapplication of the term. Yet, as far as I know, discussions of Canadian post-colonialism do not usually equate the settler with the native experience, or the Canadian with the Third World. The kind of generalizations that Richard Roth criticizes in Abdul JanMohamed's work do tend to totalize in this way, but this kind of work always ignores countries like Canada. To my mind, Hutcheon gets it backwards when she writes: 'one can certainly talk of post-colonialism in Canada but only if the differences between its particular version and that of, especially, Third World nations is kept in mind' (174). The drawing of such distinctions is the whole point of talking about post-colonialism in Canada. The post-colonial perspective provides us with the language and the political analysis for understanding these differences. The danger is less that Canadians will rush to leap on the victim wagon than that they will refuse to recognize that they may well have some things in common with colonized people elsewhere.

Hutcheon's argument functions as a sort of straw man that misrepresents the post-colonial theoretical endeavour as practised in relation to Canada, deflecting attention away from its radical potential. Her argument demonstrates that in our care to respect the specificity of particular experiences we run another risk, that of a liberal pluralism, which uses the idea of different but equal discourses to prevent the forming of alliances based on a comparative analysis that can perceive points of connection. Consider the following statement from *The Canadian*

*Postmodern* (1988b) 'If women have not yet been allowed access to (male) subjectivity, then it is very difficult for them to contest it, as the (male) post-structuralist philosophers have been doing lately. This may make women's writing *appear* more conservative, but in fact it is just different' (5–6). By positing female writing as 'just different' from the male norm, Hutcheon erases the power differential she has been trying to establish, while reaffirming the male as the norm and the experimental as more advanced than and superior to the conservative. It sounds like special pleading for the second-rate, while on the surface it reaffirms the liberal myth of society formed from a plurality of equal differences. . . .

## The cult of authenticity

Paul Smith suggests that post-modernist discourse replaces the 'conflictual view and the comic view of the third world' with a 'cult of authenticity' (Ross 1988: 142). This seems to be what is happening with Hutcheon's assertion that only Canada's native peoples may claim to speak with an authentic post-colonial voice. Such an assertion connects her approach to post-colonialism to that of Fredric Jameson which produces a first world criticism respectful of a third world authenticity that it is believed his own world has lost. But what are the effects of such a 'cult of authenticity'? Meaghan Morris concludes her analysis of *Crocodile Dundee* with the statement that '[i]t is hardly surprising, then, that the figure of the colonial should now so insistently reappear from all sides not as deprived and dispossessed by rapacity but as the naive spirit of plenitude, innocence, optimism – and effective critical "distance"' (Ross 1988: 124). The post-modernist revisionings of the colonial and post-colonial that Smith (1988) and Morris (1988) discuss function to defuse conflict, denying the necessity of cultural and political struggle, and suggesting that tourism is probably the best model for cross-cultural interaction.

Hutcheon's argument that Canada's native peoples are the authentic post-colonial voice of the nation, with its implication that descendants of settlers and immigrants represent at best a contaminated post-coloniality, conforms to this post-modernist model. To challenge it, as Hutcheon knows, is fraught with difficulties because authenticity has also been used by colonial peoples in their struggles to regain power over their own lives. While post-colonial theorists embrace hybridity and heterogeneity as the characteristic post-colonial mode, some native writers in Canada resist what they see as a violating appropriation to insist on their ownership of their stories and their exclusive claim to an authenticity that should not be ventriloquized or parodied. When directed against the Western canon, post-modernist techniques of intertextuality, parody, and literary borrowing may appear radical and even potentially revolutionary. When directed against native myths and stories, these same techniques would seem to repeat the imperialist history of plunder and theft. Or in the case of *The Satanic Verses*, when directed against Islam, they may be read as sullying the dignity of a religion that prides itself on its purity.

Although I can sympathize with such arguments as tactical strategies in insisting on self-definition and resisting appropriation, even tactically they prove self-defeating because they depend on a view of cultural authenticity that condemns them to a continued marginality and an eventual death. Whose interests are served by this retreat into preserving an untainted authenticity? Not the native groups seeking land rights and political power. Ironically, such tactics encourage native peoples to isolate themselves from contemporary life and full citizenhood.

All living cultures are constantly in flux and open to influences from elsewhere. The current flood of books by white Canadian writers embracing Native spirituality clearly serves a white need to feel at home in this country and to assuage the guilt felt over a material appropriation by making it a cultural one as well. In the absence of comparable political reparation for past appropriations such symbolic acts seem questionable or at least inadequate. Literature cannot be confused with social action. Nonetheless, these creole texts are also part of the post-colonial search for a way out of the impasse of the endless play of post-modernist difference that mirrors liberalism's cultural pluralism. These books, like the post-colonial criticism that seeks to understand them, are searching for a new globalism that is neither the old universalism nor the Disney simulacrum. This new globalism simultaneously asserts local independence and global interdependencies. It seeks a way to cooperate without cooption, a way to define differences that do not depend on myths of cultural purity or authenticity but that thrive on an interaction that 'contaminates' without homogenizing. . . .

## Judging the work on its own terms

Hutcheon's conclusion to her *Poetics of Postmodernism* admits the 'limited' aims of post-modernism and its 'double encoding as both contestatory and complicitous' (1988a: 230). She acknowledges that 'I would agree with Habermas that this art does not "emit any clear signals"', but adds that its saving grace is that 'it does not try to'. It cannot offer answers, 'without betraying its anti-totalizing ideology' (231). I have suggested that it does surreptitiously offer answers – in ambivalence itself, in the relativity of liberal pluralism, in the cult of authenticity that lies behind its celebration of differences. But is it true that answers necessarily totalize? Are these the only alternatives? Is Hutcheon here asking enough of the post-modernist text? Or is she even asking the most interesting or the most important questions? Isn't the effect of such a conclusion to preserve the status quo and the myth of an objectivity that itself totalizes? Can we legitimately ask more of a text than it asks of itself? Post-colonial criticism suggests that we can. . . .

Perhaps the clearest difference between a post-modernist practice and a post-colonial practice emerges through their different uses of history. As Hutcheon points out, '[h]istoriographic metafiction acknowledges the paradox of the reality of the past but its *textualized accessibility* to us today' (1988a: 14). Without denying that things happened, post-modernism focuses on the problems raised by history's textualized accessibility: on the problems of representation, and on the impossibility of retrieving truth. Post-colonialism, in contrast, without denying history's textualized accessibility, focuses on the reality of a past that has influenced the present. As a result of these different emphases, post-modern fiction takes liberties with what we know of the facts of the past much more freely than does post-colonial fiction. Richler's improbable introduction of fictional characters into historical narrative has more in common with the methods of a Sir Walter Scott than a D. M. Thomas. Neither he nor Gunnars deny that different versions of specific events will circulate, but they are interested in the effects of historical happenings: the effects of invasion, of military occupation, of food blockades, of revolution. . . .

As Stephen Slemon points out, 'Western post-modernist readings can so over-value the anti-referential or deconstructive energies of post-colonial texts that they efface the important recuperative work that is also going on within them' (1991: 7). Those deconstructive energies are at work in these two novels, but it is the recuperative power, which they seek to energize for their readers and their Canadian culture, that most distinguishes them. And it is this power that a post-colonial reading can help us to understand. The white Inuit is speaking. Who is listening?

# PART SEVEN

# Ethnicity

I N T H E   C O N C E P T   O F   E T H N I C I T Y we discover one of the most vexed and complex issues in post-colonial theory. The way it intersects with notions of indigeneity, race, marginality, imperialism and identity, leads to a constantly shifting theoretical ground, a ground continually contested and subject to more heated debate than most. At its simplest the argument boils down to a dispute over whether some ethnic groups and not others are entitled to the term 'ethnic'.

Much of the difficulty surrounding the concept can be resolved if we understand the imperial project as a process rather than a structure. For instance, the model of an imperial 'centre' controlling a colonial 'margin', a model that underlies much of the policy, strategies and outcomes of colonialism, is a myth that is only retained by post-colonial discourse in order to be deconstructed. As a geographical myth the centre/margin binarism leads by logical extension to such absurdities as the idea that all people in colonies are marginalized while nobody at the imperial centre can be marginalized; or, more crudely, that whites are the colonizers and blacks the colonized. Obviously if we try to find *the* centre of the empire, we will never find it, even in Piccadilly, Wall Street or Buckingham Palace, because this structural notion omits the institutions and process by which power is disseminated and maintained. Clearly that process is one set in train by the imperial project and continues throughout the colonial world as Trinh T. Minh-ha demonstrates ('The centre itself is marginal'). This is why 'post-colonial' can apply to white settler/invader colonies as much as to the indigenous people. It is also why the idea that only some ethnic groups are 'ethnic' and not others can be seen to be fallacious.

The argument put by Werner Sollors that 'ethnic' includes all ethnic groups and not simply all those except an arbitrarily selected dominant group is one which generally concurs with the post-colonial rejection of the centre/margin binarism. This acceptance of ethnic pluralism, that everyone in a society is 'ethnic', is not to deny that some ethnic groups exercise dominance in a society. As Philip Gleason points out, the very questions such ethnic background generated in the individual's relationship to society underpinned the growing obsession with 'identity' in the US. But the binarism of one ethnic group at the centre and all others at the margins, overlooks the actual overlap between the multiplicity of ethnic groups and the dynamic, processual and multi-faceted institutions of power. As Stuart Hall points out, the conceptualization of ethnicity itself is undergoing a radical change based upon the increasingly complex politics of representation: old binarisms of black/white, and indeed conceptions of the 'essential' ethnic subject itself are now increasingly open to question. The growing discussion of ethnicity has been accompanied over the last decade by a surge of interest in the category of 'whiteness'. The colour that had been invisible in categories of 'the human' has now, as Mike Hill shows, come under increasing scrutiny as its ethnic visibility becomes the subject of widespread reflection and analysis. The trend of post-colonial criticism is to confirm that we are, indeed, all ethnic, but in Gloria Anzaldúa, with her discussion of the complex overlap of mestiza – Chicana – lesbian identity we find the extent to which the category 'ethnic' itself – like the category 'indigenous' – finds itself increasingly traversed by the hybridizing realities of contemporary global living.

## Chapter 41

# WERNER SOLLORS

# WHO IS ETHNIC?

From *Beyond Ethnicity: Consent and Descent in American Culture* Oxford and New York: Oxford University Press, 1986.

'**A**RE WE ETHNIC?' – as the *New Yorker* put it in a cartoon of 1972 depicting a white middle-class family in an elegant dining room – is the question Yankees or WASPs have had to ask themselves many times since then, and without getting just one universally accepted answer.

Two conflicting uses of 'ethnic' and 'ethnicity' have remained in the air. According to Everett and Helen Hughes 'we are all ethnic' (Hughes 1952: 7), and in E. K. Francis's terminology of 1947 'not only the French Canadians or the Pennsylvania Dutch would be ethnic groups but also the French of France or the Irish of Ireland' (Francis 1947: 395). But this universalist and inclusive use is in frequent conflict with the other use of the word, which excludes dominant groups and thus establishes an 'ethnicity minus one'. It may be absurd, as Harold Abramson has argued, to except white Anglo-Saxon Protestant Americans from the category of ethnicity (Abramson 1973: 9), and yet it is a widespread practice to define ethnicity as otherness. The contrastive terminology of ethnicity thus reveals a point of view which changes according to the speaker who uses it: for example, for some Americans eating turkey and reading Hawthorne appear to be more 'ethnic' than eating lasagne and reading Puzo.

As Everett Hughes suggested in a personal letter in 1977, the association of the ethnic with the other is not made in some languages: 'In Greece the national bank is the ethnic bank. In this country ethnic banks cannot be the national bank. . . .' To say it in the simplest and clearest terms, an ethnic, etymologically speaking is a *goy*. The Greek word *ethnikos*, from which the English 'ethnic' and 'ethnicity' are derived, meant 'gentile', 'heathen'. Going back to the noun *ethnos*, the word was used to refer not just to people in general but also to 'others'. In English usage the meaning shifted from 'non-Israelite' (in the Greek translation of the Bible the word *ethnikos* was used to render the Hebrew *goyim*) to 'non-Christian'. Thus the word retained its quality of defining another people contrastively, and often negatively. In the Christianized context the word 'ethnic' (sometimes spelled 'hethnic') recurred, from the four-teenth to the nineteenth century, in the sense of 'heathen'. Only in the mid-nineteenth century did the more familiar meaning of 'ethnic' as 'peculiar to a race or nation' re-emerge. However, the English language has retained the pagan memory of 'ethnic', often secularized in the sense of ethnic as other, as nonstandard, or, in America, as not fully American. This connotation gives the opposition of ethnic and American the additional religious dimension of the contrast between heathens and chosen people. No wonder that there is popular hesitation to accept the inclusive use of ethnicity. The relationship between ethnicity and American identity in this respect parallels that of pagan superstition and true religion. . . .

Ethnic theorists have often dwelled on the antithetical nature of ethnicity. In ethnic name-calling the tendency persists, as George Murdock argued in Seligman's *Encyclopaedia of the Social Sciences*, that a people

> usually calls itself either by a flattering name or by a term signifying simply 'men', 'men of men', 'first men', or 'people'. Aliens on the other hand, are regarded as something less than men; they are styled 'barbarians' or are known by some derogatory term corresponding to such modern American ethnic tags as 'bohunk', 'chink', 'dago', 'frog', 'greaser', 'nigger', 'sheeny' and 'wop'.
>
> (Murdock 1931: 613)

As Agnes Heller writes, what 'is now called "ethnocentrism" is the natural attitude of all cultures toward alien ones' (Heller 1984: 271). Such antithetical definitions not only are noticeable in the modern world (or among American ethnic writers and critics) but were also undertaken by American Indians. Thus, although it has become de rigueur in ethnic criticism to refer to the original inhabitants of the American continent as 'Native Americans' in order to avoid the, not slur, but misnomer 'Indians', the various Indian nations have followed the human pattern of calling themselves 'people' and calling others less flattering things. . . .

In his introduction to *Ethnic Groups and Boundaries* (1969), Frederick Barth sees the essence of ethnicity in such (mental, cultural, social, moral, aesthetic, and not necessarily territorial) boundary-constructing processes which function as cultural markers between groups. For Barth it is 'the ethnic *boundary* that defines the group, not the cultural stuff that it encloses' (15). 'If a group maintains its identity when members interact with others, this entails criteria for determining membership and ways of signalling membership and exclusion' (15). Previous anthropologists (and, we might add, historians, sociologists, and literary critics) tended to think about ethnicity 'in terms of different peoples, with different histories and cultures, coming together and accommodating themselves to each other'; instead, Barth suggests, we should 'ask ourselves what is needed to make ethnic distinctions *emerge* in an area' (38). With a statement that runs against the grain of much ethnic historiography, Barth argues that

> when one traces the history of an ethnic group through time, one is *not* simultaneously, in the same sense, tracing the history of 'a culture': the elements of the present culture of that ethnic group have not sprung from the particular set that constituted the group's culture at a previous time, whereas the group has a continual organizational existence with boundaries (criteria of membership) that despite modifications have marked off a continuing unit.
>
> (38)

Barth's focus on *boundaries* may appear scandalously heretical to some, but it does suggest plausible interpretations of the polyethnic United States. (Barth uses the term 'polyethnic' instead of the more common Graeco-Roman mixture 'multi-ethnic' – to maintain boundaries in etymology.) Barth's theory can easily accommodate the observation that ethnic groups in the United States have relatively little cultural differentiation, that the cultural *content* of ethnicity (the stuff that Barth's boundaries enclose) is largely interchangeable and rarely historically authenticated. . . .

## Race and ethnicity

To compound problems, there is another important line of disagreement concerning race and ethnicity. On the one hand, Harold Abramson argued that although 'race is the most salient ethnic factor, it is still only one of the dimensions of the larger cultural and historical phenomenon of ethnicity' (Abramson 1973: 175). . . . On the other hand, M. G. Smith would rather side with Pierre van den Berghe's *Race and Racism* (1967) and consider race a special 'objective' category that cannot be meaningfully discussed under the heading 'ethnicity' (Smith 1982: 10).

I have here sided with Abramson's universalist interpretation according to which ethnicity includes dominant groups and in which race, while sometimes facilitating external identification, is merely one aspect of ethnicity. I have three reasons for doing so. First, the interpretation of the rites and rituals of culturally dominant groups sometimes provides the matrix for the emergence of divergent group identities. . . . Second, the discussions of ethnicity and the production of ethnic literature have been strongly affected by Afro-Americans, and so actively influenced by them since World War II, that an omission of the Afro-American tradition in a discussion of ethnic culture in America would create a very serious gap in our reflections. In fact, the very emergence of the stress on ethnicity and the unmeltable ethnics was directly influenced by the black civil rights movement and strengthened by its radicalization in the 1960s. . . . Finally, I am interested in the processes of group formation and in the naturalization of group relationships . . . and have found examples from Puritan New England and Afro-America crucial to an understanding of these processes among other groups in America. The term 'ethnicity' here is thus a broadly conceived term.

## Chapter 42

# PHILIP GLEASON

## IDENTIFYING IDENTITY

From 'Identifying Identity' *The Journal of American History* 69(4) (March), 1983.

WHY DID IDENTITY SO QUICKLY BECOME AN indispensable term in American social commentary? . . .

The most important consideration, I would say, was that the word identity was ideally adapted to talking about the relationship of the individual to society as that perennial problem presented itself to Americans at mid-century. More specifically, identity promised to elucidate a new kind of conceptual linkage between the two elements of the problem, since it was used in reference to, and dealt with the relationship of, the individual personality and the ensemble of social and cultural features that gave different groups their distinctive character. The relationship of the individual to society has always been problematic for Americans because of the surpassing importance in the national ideology of the values of freedom, equality, and the autonomy of the individual.

In these circumstances the questions, 'Who am I?' and 'Where do I belong?' became inevitable. Identity was, in a sense, what the discussion was all about. As Erikson noted in 1950, 'we begin to conceptualize matters of identity at the very time in history when they become a problem.' The study of identity, he believed, was 'as strategic in our time as the study of sexuality was in Freud's time' (Erikson 1950: 242). Understood as a concept of the social sciences, identity thus gained its original currency because of its aptness for discussing one of the issues that dominated the American intellectual horizon of the 1950s, 'the survival of the person in mass society' (Stein *et al.* 1960). In those days the characteristic problem centered around 'the search for identity,' which was thought to arise primarily from the individual's feeling of being rootless and isolated in a swarming, anonymous throng. In the next decade the cultural climate changed drastically and the mass-society problem receded far into the background. But identity did not decline with the fading interest in the problems that first called for employment of the concept; on the contrary, it gained even greater popularity.

The problem of the relation of the individual to society assumed new forms in the turmoil of the 1960s, but identity was more relevant than ever – only now it was of 'identity crises' that one heard on every hand. Few who lived through that troubled time would deny that the expression 'identity crisis' spoke with greater immediacy to the American condition than the formula 'search for identity.' For the nation did go through a profound crisis – social, political, and cultural – between the assassination of John F. Kennedy and the resignation of Richard Nixon. The ingredients of the crisis – racial violence, campus disruptions, antiwar protests, cultural upheaval, and the abuse of official power and betrayal of public trust – need no elaboration. The point is that the national crisis translated itself to the ordinary citizen as a challenge to every individual to decide where he or she stood with respect to the traditional values, beliefs, and institutions that were being called into question, and with respect to the contrasting interpretations being offered of American society, American policies, and the American future.

In other words, the national crisis brought about a re-examination on a massive scale of the relationship between the individual and society. That was the relationship with which identity dealt, and in innumerable cases the re-examination was sufficiently intense to make the expression 'identity crisis' seem very apt. Within the context of cultural crisis, the revival of ethnicity deserves special attention as perhaps the most important legacy of the 1960s so far as usage of identity is concerned. There is in the nature of the case a close connection between the notion of identity and the awareness of belonging to a distinctive group set apart from others in American society by race, religion, national background, or some other cultural marker. As a matter of fact, Erikson alluded to the acculturation of immigrants immediately after drawing attention in 1950 to the timeliness of identity as an analytical concern. Looking back twenty years later, he underscored his own experience as an immigrant in tracing the developing of his thinking about identity. 'It would seem almost self-evident now,' he wrote, 'how the concepts of "identity" and "identity crisis" emerged from my personal, clinical, and anthropological observations in the thirties and forties. I do not remember when I started to use these terms; they seemed naturally grounded in the experience of emigration, immigration, and Americanization' (Erikson 1950: 242; Erikson 1975: 43). That is certainly plausible. But the connection between Erikson's personal experience and his sensitivity to identity problems doubtless seemed clearer by 1970 because of the growth of interest in ethnicity in the intervening years and because of the new respectability gained by ethnic consciousness.

In the late 1940s, assimilation was thought to have eroded immigrant cultures almost entirely, and the lingering vestiges of group consciousness seemed not only archaic but also potentially dysfunctional as sources of ethnocentrism, anti-intellectualism, and isolationist sentiment. Even Herberg, who first stressed the linkage between ethnicity and the search for identity, believed ethnic identities were being replaced by religious identities (Herberg 1956). The black revolution of the 1960s, and the subsequent emergence of the new ethnicity, changed all that. These movements affirmed the durability of ethnic consciousness, gave it legitimacy and dignity, and forged an even more intimate bond between the concepts of ethnicity and identity. And these developments not only took place against the background of the national identity crisis, they were also dialectically related to it – that is, ethnic or minority identities became more appealing options because of the discrediting of traditional Americanism brought about by the racial crisis and the Vietnam War (see Gleason 1980; Mann 1979). As Nathan Glazer pointed out in 1975, a situation had by then developed in 'the ecology of identities' in which, for the first time in American history, it seemed more attractive to many individuals to affirm an ethnic identity than to affirm that one was simply an American (Glazer 1975: 177–8).

## Chapter 43

# TRINH T. MINH-HA

## NO MASTER TERRITORIES

From *When the Moon Waxes Red: Representation, Gender and Cultural Politics* New York and London: Routledge, 1991.

### Centre and margin

THE IMPERVIOUSNESS IN THE WEST of the many branches of knowledge to everything that does not fall inside their predetermined scope has been repeatedly challenged by its thinkers throughout the years. They extol the concept of decolonization and continuously invite into their fold 'the challenge of the Third World.' Yet, they do not seem to realize the difference when they find themselves face to face with it — a difference which does not announce itself, which they do not quite anticipate and cannot fit into any single varying compartment of their catalogued world; a difference they keep on measuring with inadequate sticks designed for their own morbid purpose. When they confront the challenge 'in the flesh,' they naturally do not recognize it as a challenge. Do not hear, do not see. They promptly reject it as they assign it to their one-place-fits-all 'other' category and either warily explain that it is 'not quite what we are looking for' and that they are not the right people for it; or they kindly refer it to other 'more adequate' whereabouts such as the 'counter-culture,' 'smaller independent,' 'experimental' margins.

They? Yes, they. But, in the colonial periphery (as in elsewhere), we are often them as well. Colored skins, white masks; colored masks, white skins. Reversal strategies have reigned for some time. *They* accept the margins; so do *we*. For without the margin, there is no center, no heart. *The English and the French precipitate towards us, to look at themselves in our mirror. Following the old colonizers who mixed their blood in their turn, having lost their colonies and their blondness — little by little touched by this swarthy tint spreading like an oil stain over the world — they will come to Buenos Aires in pious pilgrimmage to try to understand how one cannot be, yet always be* (Ortiz 1987: 96). The margins, our sites of survival, become our fighting grounds and their site for pilgrimage. Thus, while we turn around and reclaim them as our exclusive territory, they happily approve, for the divisions between margin and center should be preserved, and as clearly demarcated as possible, if the two positions are to remain intact in their power relations. Without a certain work of displacement, again, the margins can easily recomfort the center in its goodwill and liberalism; strategies of reversal thereby meet with their own limits. The critical work that has led to an acceptance of negativity and to a new positivity would have to continue its course, so that even in its negativity and positivity, it baffles, displaces, rather than suppresses. By displacing, it never allows this classifying world to exert its classificatory power without returning it to its own ethnocentric classifications. All the while, it points to an else-where-within-here whose boundaries would continue to compel frenzied attempts at 'baptizing' through logocentric naming and objectivizing to reflect on themselves as they face their own constricting apparatus of refined grids and partitioning walls.

The center itself is marginal . . . [H]ow possible is it to undertake a process of decentralization without being made aware of the margins within the center and the centers within the margin? Without encountering marginalization from both the ruling center and the established margin? Wherever she goes she is asked to show her identity papers. What side does she speak up for? Where does she belong (politically, economically)? Where does she place her loyalty (sexually, ethnically, professionally)? Should she be met at the center, where they invite her in with much display, it is often only to be reminded that she holds the permanent status of a 'foreign worker,' 'a migrant,' or 'a temporary sojourner' – a status whose definable location is necessary to the maintenance of a central power. 'How about a concrete example from your own culture?' 'Could you tell us what it is like in . . . (your country)?' *As a minority woman, I . . . As an Asian-American woman, I . . . As a woman-of-color filmmaker, I . . . As a feminist, a . . . , and a . . . I. . . .* Not foreigner, yet foreign. At times rejected by her own community, other times needfully retrieved, she is both useless and useful. *The irreducibility of the margin in all explanation. The ceaseless war against dehumanization.* This shuttling in-between frontiers is a working out of and an appeal to another sensibility, another consciousness of the condition of marginality: that in which marginality is the condition of the center.

To use marginality as a starting point rather than an ending point is also to cross beyond it towards other affirmations and negations. There cannot be any grand totalizing integration without massive suppression, which is a way of recirculating the effects of domination. *Liberation opens up new relationships of power, which have to be controlled by practices of liberty* (Foucault 1988: 4). Displacement involves the invention of new forms of subjectivities, of pleasures, of intensities, of relationships, which also implies the continuous renewal of a critical work that looks carefully and intensively at the very system of values to which one refers in fabricating the tools of resistance. The risk of reproducing totalitarianism is always present and one would have to confront, in whatever capacity one has, the controversial values likely to be taken on faith as universal truths by one's own culture(s). . . .

## Outside in inside out

. . . Essential difference allows those who rely on it to rest reassuringly on its gamut of fixed notions. Any mutation in identity, in essence, in regularity, and even in physical place poses a problem, if not a threat, in terms of classification and control. If you can't locate the other, how are you to locate yourself?

> *One's sense of self is always mediated by the image one has of the other. (I have asked myself at times whether a superficial knowledge of the other, in terms of some stereotype, is not a way of preserving a superficial image of oneself.)*
>
> (Crapanzano 1985: 54)

Furthermore, where should the dividing line between outsider and insider stop? How should it be defined? By skin color (no blacks should make films on yellows)? By language (only Fulani can talk about Fulani, a Bassari is a foreigner here)? By nation (only Vietnamese can produce works on Vietnam)? By geography (in the North–South setting, East is East and East can't meet West)? Or by political affinity (Third World on Third World counter First and Second Worlds)? What about those with hyphenated identities and hybrid realities? It is worth noting here a journalist's report in a recent *Time* issue, which is entitled 'A Crazy Game of

Musical Chairs.' In this brief but concise report, attention is drawn on the fact that South Africans, who are classified by race and placed into one of the nine racial categories that determine where they can live and work, can have their classification changed if they can prove they were put in the wrong group. Thus, in an announcement of racial reclassifications by the Home Affairs Minister, one learns that: '*nine whites became colored, 506 coloreds became white, two whites became Malay, 14 Malay became white . . . 40 coloreds became black, 666 blacks became colored, 87 coloreds became Indian, 67 Indians became colored, 26 coloreds became Malay, 50 Malays became Indian, 61 Indians became Malay . . . and the list goes on. However, says the Minister, no blacks applied to become white, and no whites became black*' (*Time* 9 March 1987: 54).

The moment the insider steps out from the inside, she is no longer a mere insider (and vice versa). She necessarily looks in from the outside while also looking out from the inside. Like the outsider, she steps back and records what never occurs to her the insider as being worth or in need of recording. But unlike the outsider, she also resorts to non-explicative, non-totalizing strategies that suspend meaning and resist closure. (This is often viewed by the outsiders as strategies of partial concealment and disclosure aimed at preserving secrets that should only be imparted to initiates.) She refuses to reduce herself to an Other, and her reflections to a mere outsider's objective reasoning or insider's subjective feeling. She knows, probably like Zora Neale Hurston the insider-anthropologist knew, that she is not an outsider like the foreign outsider. She knows she is different while at the same time being Him. Not quite the Same, not quite the Other, she stands in that undetermined threshold place where she constantly drifts in and out. Undercutting the inside/outside opposition, her intervention is necessarily that of both a deceptive insider and a deceptive outsider. She is this Inappropriate Other/Same who moves about with always at least two/four gestures: that of affirming 'I am like you' while persisting in her difference; and that of reminding 'I am different' while unsettling every definition of otherness arrived at. . . .

Whether she turns the inside out or the outside in, she is, like the two sides of a coin, the same impure, both-in-one insider/outsider. For there can hardly be such a thing as an essential inside that can be homogeneously represented by all insiders; an authentic insider in there, an absolute reality out there, or an incorrupted representative who cannot be questioned by another incorrupted representative. . . .

In the context of this Inappropriate Other, questions like 'How loyal a representative of his/her people is s/he?' (the filmmaker as insider), or 'How authentic is his/her representation of the culture observed?' (the filmmaker as outsider) are of little relevance. When the magic of essences ceases to impress and intimidate, there no longer is a position of authority from which one can definitely judge the verisimilitude value of the representation. In the first question, the questioning subject, even if s/he is an insider, is no more authentic and has no more authority on the subject matter than the subject whom the questions concern.

This is not to say that the historical 'I' can be obscured or ignored, and that differentiation cannot be made; but that 'I' is not unitary, culture has never been monolithic, and more or less is always more or less in relation to a judging subject. Differences do not only exist between outsider and insider – two entities –, they are also at work within the outsider or the insider – a single entity. This leads us to the second question in which the filmmaker is an outsider. As long as the filmmaker takes up a positivistic attitude and chooses to bypass the inter-subjectivities and realities involved, factual truth remains the dominant criterion for evaluation and the question as to whether his/her work successfully represents the reality it claims would continue to exert its power. The more the representation leans on verisimilitude, the more it is subject to normative verification.

Chapter 44

# STUART HALL

# NEW ETHNICITIES

From 'New Ethnicities' *Black Film, British Cinema* ICA Documents 7, London:
Institute of Contemporary Arts, 1989.

I HAVE CENTRED MY REMARKS ON an attempt to identify and characterize
a significant shift that has been going on (and is still going on) in black cultural politics.
This shift is not definitive, in the sense that there are two clearly discernable phases – one
in the past which is now over and the new one which is beginning – which we can neatly
counterpose to one another. Rather, they are two phases of the same movement, which
constantly overlap and interweave. Both are framed by the same historical conjuncture and
both are rooted in the politics of anti-racism and the post-war black experience in Britain.
Nevertheless I think we can identify two different 'moments' and that the difference between
them is significant.

It is difficult to characterize these precisely, but I would say that the first moment was
grounded in a particular political and cultural analysis. Politically, this is the moment when
the term 'black' was coined as a way of referencing the common experience of racism and
marginalization in Britain and came to provide the organizing category of a new politics of
resistance, amongst groups and communities with, in fact, very different histories, traditions
and ethnic identities. In this moment, politically speaking, 'The Black experience', as a singular and unifying framework based on the building up of identity across ethnic and cultural
difference between the different communities, became 'hegemonic' over other ethnic/racial
identities – though the latter did not, of course, disappear. . . .

The struggle to come into representation was predicated on a critique of the degree of
fetishization, objectification and negative figuration which are so much a feature of the representation of the black subject. There was a concern not simply with the absence or marginality
of the black experience but with its simplification and its stereotypical character.

The cultural politics and strategies which developed around this critique had many facets,
but its two principal objects were first the question of *access* to the rights to representation
by black artists and black cultural workers themselves. Secondly, the *contestation* of the
marginality, the stereotypical quality and the fetishized nature of images of blacks, by the
counter-position of a 'positive' black imagery. These strategies were principally addressed to
changing what I would call the 'relations of representation'.

I have a distinct sense that in the recent period we are entering a new phase. But we
need to be absolutely clear what we mean by a 'new' phase because, as soon as you talk of
a new phase, people instantly imagine that what is entailed is the *substitution* of one kind of
politics for another. I am quite distinctly *not* talking about a shift in those terms. . . . There
is no sense in which a new phase in black cultural politics could replace the earlier one.

Nevertheless, it is true that as the struggle moves forward and assumes new forms, it does to some degree *displace*, reorganize and reposition the different cultural strategies in relation to one another. . . .

The shift is best thought of in terms of a change from a struggle over the relations of representation to a politics of representation itself. It would be useful to separate out such a 'politics of representation' into its different elements. We all now use the word representation, but, as we know, it is an extremely slippery customer. It can be used, on the one hand, simply as another way of talking about how one images a reality that exists 'outside' the means by which things are represented: a conception grounded in a mimetic theory of representation. On the other hand the term can also stand for a very radical displacement of that unproblematic notion of the concept of representation. My own view is that events, relations, structures do have conditions of existence and real effects, outside the sphere of the discursive, but that it is only within the discursive, and subject to its specific conditions, limits and modalities, do they have or can they be constructed within meaning. Thus, while not wanting to expand the territorial claims of the discursive infinitely, how things are represented and the 'machineries' and regimes of representation in a culture do play a *constitutive*, and not merely a reflexive, after-the-event role. This gives questions of culture and ideology, and the scenarios of representation – subjectivity, identity, politics – a formative, not merely an expressive, place in the constitution of social and political life. I think it is the move towards this second sense of representation which is taking place and which is transforming the politics of representation in black culture.

This is a complex issue. First, it is the effect of a theoretical encounter between black cultural politics and the discourses of a Eurocentric, largely white, critical cultural theory which in recent years, has focused so much analysis of the politics of representation. This is always an extremely difficult, if not dangerous encounter. (I think particularly of black people encountering the discourses of post-structuralism, post-modernism, psychoanalysis and feminism.) Secondly, it marks what I can only call 'the end of innocence', or the end of the innocent notion of the essential black subject. Here again, the end of the essential black subject is something which people are increasingly debating, but they may not have fully reckoned with its political consequences. What is at issue here is the recognition of the extraordinary diversity of subjective positions, social experiences and cultural identities which compose the category 'black'; that is, the recognition that 'black' is essentially a politically and culturally *constructed* category, which cannot be grounded in a set of fixed trans-cultural or transcendental racial categories and which therefore has no guarantees in Nature. What this brings into play is the recognition of the immense diversity and differentiation of the historical and cultural experience of black subjects. This inevitably entails a weakening or fading of the notion that 'race' or some composite notion of race around the term black will either guarantee the effectivity of any cultural practice or determine in any final sense its aesthetic value.

We should put this as plainly as possible. Films are not necessarily good because black people make them. They are not necessarily 'right-on' by virtue of the fact that they deal with black experience. Once you enter the politics of the end of the essential black subject you are plunged headlong into the maelstrom of a continuously contingent, unguaranteed, political argument and debate: a critical politics, a politics of criticism. You can no longer conduct black politics through the strategy of a simple set of reversals, putting in the place of the bad old essential white subject, the new essentially good black subject. Now, that formulation may seem to threaten the collapse of an entire political world. Alternatively, it may be greeted with extraordinary relief at the passing away of what at one time seemed to be a necessary fiction. Namely, either that all black people are good or indeed that all black people are *the same*.

After all, it is one of the predicates of racism that 'you can't tell the difference because they all look the same'. This does not make it any easier to conceive of how a politics can be constructed which works with and through difference, which is able to build those forms of solidarity and identification which make common struggle and resistance possible but without suppressing the real heterogeneity of interests and identities, and which can effectively draw the political boundary lines without which political contestation is impossible, without fixing those boundaries for eternity. It entails the movement in black politics, from what Gramsci called the 'war of manoeuvre' to the 'war of position' – the struggle around positionalities. But the difficulty of conceptualizing such a politics (and the temptation to slip into a sort of endlessly sliding discursive liberal-pluralism) does not absolve us of the task of developing such a politics.

The end of the essential black subject also entails a recognition that the central issues of race always appear historically in articulation, in a formation, with other categories and divisions and are constantly crossed and recrossed by the categories of class, of gender and ethnicity. (I make a distinction here between race and ethnicity to which I shall return.) To me, films like *Territories*, *Passion of Remembrance*, *My Beautiful Laundrette* and *Sammy and Rosie Get Laid*, for example, make it perfectly clear that the shift has been engaged; and that the question of the black subject cannot be represented without reference to the dimensions of class, gender, sexuality and ethnicity. . . .

I am familiar with all the dangers of ethnicity as a concept and have written myself about the fact that ethnicity, in the form of a culturally constructed sense of Englishness and a particularly closed, exclusive and regressive form of English national identity, is one of the core characteristics of British racism today. I am also well aware that the politics of anti-racism has often constructed itself in terms of a contestation of 'multi-ethnicity' or 'multi-culturalism'. On the other hand, as the politics of representation around the black subject shifts, I think we will begin to see a renewed contestation over the meaning of the term 'ethnicity' itself.

If the black subject and black experience are not stabilized by Nature or by some other essential guarantee, then it must be the case that they are constructed historically, culturally, politically – and the concept which refers to this is 'ethnicity'. The term ethnicity acknowledges the place of history, language and culture in the construction of subjectivity and identity, as well as the fact that all discourse is placed, positioned, situated, and all knowledge is contextual. Representation is possible only because enunciation is always produced within codes which have a history, a position within the discursive formations of a particular space and time. The displacement of the 'centred' discourses of the West entails putting in question its universalist character and its transcendental claims to speak for everyone, while being itself everywhere and nowhere. The fact that this grounding of ethnicity in difference was deployed, in the discourse of racism, as a means of disavowing the realities of racism and repression does not mean that we can permit the term to be permanently colonized. That appropriation will have to be contested, the term disarticulated from its position in the discourse of 'multi-culturalism' and transcoded, just as we previously had to recuperate the term 'black', from its place in a system of negative equivalences. The new politics of representation therefore also sets in motion an ideological contestation around the term 'ethnicity'. But in order to pursue that movement further, we will have to retheorize the concept of *difference*.

It seems to me that, in the various practices and discourses of black cultural production, we are beginning to see constructions of just such a new conception of ethnicity: a new cultural politics which engages rather than suppresses *difference* and which depends, in part, on the cultural construction of new ethnic identities. Difference, like representation, is also a slippery, and therefore, contested concept. There is the 'difference' which makes a radical and unbridgeable separation: and there is a 'difference' which is positional, conditional and conjunctural, closer to Derrida's notion of *differance*, though if we are concerned to maintain a politics it cannot be defined exclusively in terms of an infinite sliding of the signifier. We still have a great deal of work to do to *decouple* ethnicity, as it functions in the dominant discourse, from its equivalence with nationalism, imperialism, racism and the state, which are the points of attachment around which a distinctive British or, more accurately, English ethnicity have been constructed. Nevertheless, I think such a project is not only possible but necessary. Indeed, this decoupling of ethnicity from the violence of the state is implicit in some of the new forms of cultural practice that are going on in films like *Passion* and *Handsworth Songs*. We are beginning to think about how to represent a non-coercive and a more diverse conception of ethnicity, to set against the embattled, hegemonic conception of 'Englishness' which, under Thatcherism, stabilizes so much of the dominant political and cultural discourses, and which, because it is hegemonic, does not represent itself as ethnicity at all.

This marks a real shift in the point of contestation, since it is no longer only between antiracism and multiculturalism but *inside* the notion of ethnicity itself. What is involved is the splitting of the notion between, on the one hand the dominant notion which connects it to nation and 'race' and on the other hand what I think is the beginning of a positive conception of the ethnicity of the margins, of the periphery. That is to say, a recognition that we all speak from a particular place, out of a particular history, out of a particular experience, a particular culture, without being contained by that position as 'ethnic artists' or film-makers. We are all, in that sense, *ethnically* located and our ethnic identities are crucial to our subjective sense of who we are. But this is also a recognition that this is not an ethnicity which is doomed to survive, as Englishness was, only by marginalizing, dispossessing, displacing and forgetting other ethnicities. This precisely is the politics of ethnicity predicated on difference and diversity.

## Chapter 45

# MIKE HILL

## AFTER WHITENESS

From *After Whiteness: Unmaking An American Majority* New York: New York University Press, 2004.

C IVIL RIGHTS ARE INSEPARABLY TIED to statistical enumeration. The ambivalence of this attachment will be described more clearly in a moment. For now it is important simply to establish the civil rights association between race and the law. The myriad forms of antidiscrimination legislation that occurred between 1964 and 1968 continue to set the legal context for discussing the state's interest in racial self-disclosure (Haines 1980). Encouraged by the Voting Rights Act, as well as expanded federal legislation barring discrimination in employment (1964) and housing (1968), public scrutiny on the undercounting of minority populations would become an important means of legal redress against public disenfranchisement from the 1970 US census forward. The politics of counting, newly rendered by the movement precisely as politics, is the invaluable legacy of these earlier civil rights achievements. In tracing the legacy of this 'second Reconstruction' to debates concerning the 2000 census, we need to recall two related changes in the gathering of racial data: first, progressively minded race-conscious legislation, combined with experiments in statistical sampling that began in the 1940s, produced a new emphasis on self- over observer-enumeration. Racial categories on the census tended from this moment forward to place a new emphasis on self-recognition and assumed that this task would be unproblematically sutured to the law. A kind of fragile circularity between racial self-disclosure and categorical impermanence was here introduced. Race, jurisprudence, and public access to classificatory procedures would be made to more or less happily meet. However, as shall become clearer below, race categories can double back over time, less happily, to re-enforce (or fail to re-enforce) the essential state-identity correlation. Indeed, when race is presumed to be a common matter of governmental and individual interest, racial distinction is subject to occasional and, it turns out, increasingly fractious outbreaks of classificatory complexity and computational unease.

But before I limn further just this kind of unease, the close connection between the civil rights legacy and the U.S. census should be emphasized one final time. From the mid-1960s onward, civil rights burdened the Census Bureau with a three-part task: to calculate historically relevant and publicly decidable forms of racial self-recognition; to provide the most accurate and inclusive race counts; and, most problematically, to surrender an activity of racial naming that was once performed by bureaucrats and statisticians to the more volatile dictates of the population itself, the evidently incalculable masses of peoples.

If this volatility was at all diffused in the decade that followed the turbulent 1960s, it was done so hastily, momentarily, and with what the neo-liberal epoch now reveals to be the seeds of a potentially fatal set of contradictions. In 1977 the OMB issued its Statistical Policy

Directive 15, which named the longest-standing set of state-recognized race categories in more than two hundred years (Anderson 1999). These five categories are American Indian or Alaska Native, Asian and Pacific Islander, Black, White, and a Hispanic/non-Hispanic ethnic category (OMB 1977). In the years leading up to the 2000 census, any number of glossy magazines, newspaper editorials, and other journalistic ventures proceeded to jump-start the minimal analytical pressures necessary to worry the OMB official five. American Indian, for example, designates 'persons having origins in any of the original peoples of North America,' but excludes native Hawaiians, effectively rendering them 'Asian' immigrants. For reasons that can hardly be accounted for by increasing birthrates, according to the 1990 census the American Indian population was up 255 percent since 1960. The 'Asian' category overturns the former Japanese/Chinese distinction, which was operative during the internment of the former during World War II and which, up until 1943, served to underwrite anti-immigration statutes excluding the latter. Today the 'Asian' category contains such ostensibly different peoples as Samoans, Guamians, Cambodians, Filipinos, and Laotians. Until 2000, Black mandated the one-drop rule of hypo-descent, otherwise associated with the 1896 'separate but equal' doctrine known as Jim Crow. But since 'black' technically contains all people of African heritage, it covers the entire color spectrum. So in fact does 'white,' which includes among its official members North Africans, Arabs and Jews, and all peoples from India and the Middle East. Hispanic, which would have been counted as part of a 'Mexican' race in 1930, but not so in 1940, was until 1960 subsumed under 'white.'

Indeed, the problem of officially categorizing a pan-ethnic Latino plurality, which exceeds the once presumed reliable marker of national origin (find here: not only Mexicans and Puerto Ricans, but Salvadorians, Guatemalans, Ecuadorians, indigenous South and Central American immigrants, such as Zapotecs, Yaquis, *et al.*), remains the surest sign of the inadequacies of bureaucratic expedience. Given the relative stability or decline in the growth of populations within race classifications, the category of Hispanic as an ethnicity functions as a kind of inter-divisional racial buffer between black and white. While the fastest growing minority group in the country (in California, which had seventy-two Latino-majority cities in 1990, the population has surpassed the nominally minoritized white citizenry), Hispanic continues to hold the status of ethnicity, mediating the evidently purer, late-nineteenth-century black/white racial opposition. Hispanic therefore can legally include blacks from the Dominican Republic, blond, fair-skinned, blue-eyed Argentineans, and Mexicans who would otherwise be Native Americans if they happened to be born on the north side of the Rio Grande after 1857. (To fine-tune the Hispanic category, the 1970 census attempted to introduce what it called a Central or South American distinction. More than 1 million people made that choice; but after following up the experiment with census observer-enumerators, researchers found that the majority of these 'Central' or 'South Americans' were identified as ethno-Europeans from such places as Kansas, Alabama, and Mississippi.[1]). . . .

Since the early 1990's, when data on multiracial populations began to circulate widely, some sixty organizations touting the cause in such reproductive terms have appeared . . . A federal law requiring a multiracial option in the classification of race (H. R. 830) was introduced in the United States House of Representatives in 1997 by Representative Thomas E. Petri (R – Wisconsin). A march on Washington in 1996 and a march in Los Angeles in 1997 also provide evidence of the movement's increasing grassroots momentum. Just as it is illegal to die in the United States unless by a disease prescribed by the World Health Organization, activists claimed that the race categories of the 1990 census 'rob[bed] people of their identity.' It effectively amounted, they charged, to the federal obstruction of the hard-fought civil rights victory of self- over observer-enumeration. Thus in November 1998, following four years of intensive study by thirty-plus federal agencies, after thousands of pages of congressional testimony and demographic analysis at a cost of more than $100 million, the OMB decided not to include a multiracial category for the 2000 census. However, on March 10, 2000, the OMB issued its new guidelines on racial tabulations and entered uncharted demographic territory. With other minor changes in the civil rights-inspired official five, the OMB released its new 'Standards for Maintaining, Collecting, and Presenting Federal Data on Race and Ethnicity.' This mandate allows an option to 'mark one or more races' for the first time since the initial U.S. census in 1790. Even limiting one's choice to just two combinations, this new law stands to increase the tabulation from five to 128 possibilities (O'Hare 1998: 44). . . .

In a register of multiracial activists, the right to be counted as one would choose simply means the full extension of a civil rights-inspired emphasis on self-enumeration. Race is addressed as the matter of getting identity correct in one's own eyes and in the eyes of the state. But in pursuing the multiracialism debate a bit further, one begins to see how an individual's right to self-recognition, paradoxically, may in fact provide an opportunity for civil society to release the state from its previous civil rights obligations. In this sense, the neo-nationalist end of liberalism is found dormant in the logic of its originally benevolent ends. In effect, all and no race relations may come to exist in the eyes of a racially emancipated state. Multiplicity is unleashed upon identity, and the organizational capacity of the state is both maximized and evaporated within the act of saying, 'I am . . .'.

The politics of self-recognition that I am tracing in this account of the 2000 census debates could be reduced to three general conclusions. The first conclusion is simply to concede what has become a commonplace thesis in scholarly as well as governmental circles. As everyone everywhere now seems to agree, race is a historically changeable social construction. Republican congressmen and post-Enlightenment race theorists join each other in touting this boilerplate theme. But the fact that anti-essentialism is approaching postmodern common sense should not make it a trivial matter. To the contrary, that a constructionist theory of race has become ordinary news ought to provoke a look at how post-formalist assumptions about identity alter the jurisprudence of rights. Racial abundance threatens to terminate the legal specificity of race. The general drift today in the U.S. is to embrace increasingly specific forms of racialized self-disclosure: a new and accelerated civil rights lexicon increases the number of race categories that individuals may legally claim. On this order, race is everywhere significant and nowhere identifiable in the old formalist sense. (The NAACP's awkward defense of the one-drop rule of hypo-descent – formerly associated with Jim Crow – is good evidence of the current difficulties implicit in a post-civil rights approach to racial self- and state recognition.) The changeable and constructed nature of race means, in effect, that in order to maintain its categorical salience racial identity must stave off intraracial permutation. This is so because permutation is what identity cannot have if it is to remain categorically defensible. Global anthropologists insist that the world's 184 independent nations contain more than 5,000 race or ethnic groups, more than 12,000 diverse cultures (Kymicka 1995: 1; Tully 1995: 3). How to imagine governing according to those numbers? As is evident in the case of U.S. multiracialism, certain vacillations in self-recognition come forward when the state moves to incorporate more and more racial difference. This latest process of governing is distinct from previous civil rights struggles to liberalize the state and to get government officially interested in the racial identities it once denied. Under this new set of protocols, the state has admitted racial interest with ever greater nuance, but it has done so in order to threaten race itself with the evacuation of its former political significance.

My second conclusion comes out of the first, which declared the end of the liberal state and the appropriation of racial fluidity as a new governmental concern: if race categories are sociopolitical creations that tend to redivide over time, self-recognition within a single race becomes an increasingly impracticable pursuit. There is little serious argument today that identities cannot be reduced to the headings under which they are prone-indeed, legislated-to belong. Thus (conclusion number two) race categories presume a temporal index that to stay secure they also must deny. Once time is admitted to the understanding of identity, we see that race tends toward a point of categorical interrelation that empties it of previous political content. The dangerous crossroads toward which plurality seems liable to march thus repeats a Du Boisian problem of racial citation and re-citation, but with a more pluralistic twist. Regarding his oft-repeated reference to 'the color line,' the twenty-first century divisions of race seem to have extended, multiplied, and redoubled on the contemporary political scene. And thus the state's obligation to recognize race on the order of civil rights may well be thwarted, paradoxically, by its heightened racial expertise.

The social constructionist theory of race and what I just insisted is the temporal index of raical self-recognition are two points worth developing in relation to Critical Legal Studies (CLS) and Critical Race Theory (CRT). Before adding my third and final conclusion to these two already listed, let me pause for a moment and discuss this work at greater length. The debates in CLS and CRT hinge on the same question of anti-essentialism that we have already addressed. And depending where scholars in each group stand on the racial construction/ mutability issue, they will have either more or less interest in preserving civil rights-based jurisprudence (Delgado 2001: 7). . . . Some scholars working in CLS reveal a Foucauldian slant that is therefore highly skeptical about the ideological implications of civil-rights based legal discourse. These scholars hold that 'the idea of legal rights is one of the ways that the law helps legitimate the social world by representing it as rationally mediated by rules of law'(Crenshaw et al. 1995: xxiii). The juridical subject is thus regarded less as the beneficiary of legal protection and redress than as an effect of a coercive state order already in process when identity speaks for the law. Highlighting the limits of the civil rights movement in particular, Peter Gabel emphasizes the connection between a belief in rights and the ideological effects of state power. 'A belief in the state,' he writes, 'is a flight from the immediate alienation of concrete existence into a split-off sphere in people's minds in which they imagine themselves to be a part of an imaginary political community.' 'Hegemony,' Gabel continues, 'is reinforced through "state abstraction" because people believe in and react passively to the mere illusion of political consensus' (Crenshaw et al. 1995: 108). Here the state can only respond to race in a compromised, politically loaded, if not downright anti-progressive manner. . . .

Other CRT scholars such as Kimberlé Crenshaw, however, want to take issue with the 'vulgar anti-essentialist' attack on the legacy of civil rights as 'mere illusion.' 'The oppositional dynamic' of black oppression, Crenshaw writes, and the 'exclusion, and subordination as Other, initially created an ideological and political structure of formal inequality against which rights rhetoric proved to be the most effective weapon' (Crenshaw et al. 1995: 116). Crenshaw suggests, in step with a renewed understanding of the importance of civil rights, that 'the deconstruction of white race consciousness might lead to a liberated future for both blacks and whites' (Crenshaw et al. 1995: 118). And yet the term 'deconstruction,' within her own desire to qualify 'vulgar anti-essentialism' on behalf of 'oppositional dynamics,' points to a certain impasse within CLS and CRT circles. . . .

The impasse with CLS and CRT is important because it bears directly on what I am arguing is the state's potential termination of a civil rights interest in race via the very intensification of that interest; and because this impasse alludes to how such an intensity/termination nexus portends certain post-liberal practices of governing. In the case of multiracialism and the 2000 census debates as I have described them, [two contradictory impulses] are at work:

both a radical anti-essentialist theoretical mindset and a grassroots commitment to civil rights are claimed in what effectively reveals a transmutation of modern governmental technique. The social constructionist notion of race and the endorsement of a post-white national imaginary are encouraged by the state in the simultaneous appropriation and demolition of rights. It is not simply that the state's and its citizens' interests have become inseparable within the iron fist of Foucauldian biopower. Rather, there is a kind of mutuality between race and government that is both maximized and degenerated at once. Racial difference is developed to a point where racism no longer matters to the law [while it matters absolutely].

My third conclusion can now be introduced and combined with the first two, since it is an attempt to better limn this crucial paradox and to give it theoretical clarity. If race categories are relational and are mediated by additional categories that emerge over time, then much more needs to be said about the nature of this peculiar form of mediation. This, then, is conclusion number three: the claim to self-recognition within multiracialism also initiates certain forms of misrecognition that occur within previous racial categories. The principle of ontological laissez-faire that is touted by the multiracial movement is founded on the ability to ignore this conflict. Self-recognition is writ here as a constitutionally mandated act of social belonging. But the right to self-identify always also prohibits alternative collective realities. Indeed, in the instance I have been describing, mixed identity outright dissolves the limits that codify previously established racial groups. This should help further develop an emergent picture of postmodern governmentality that exists on the other side of [the state's old liberal interest in race]. [Moreover, the persistence of multitudes beyond liberalism should signal an accidental invitation to seek collective coherences not yet imagined.] Once bound to procedures of self-recognition that excluded difference and isolated individuals through rigid classificatory rules, the state mobilizes race differently at the moment. Beyond repression or disciplinary fixity, then, the practice of governing in its neoliberal guise maximizes a form of violence implicit in self-description as such. [And within this violence there is perhaps as much danger as hope.][2]

## Notes

1   See Barbara Vobejda (1991).
2   The material in brackets is supplied by the author for this extract only.

## Chapter 46

# GLORIA ANZALDÚA

## TOWARDS A NEW CONSCIOUSNESS

From *Borderlands/La Frontera: The New Mestiza* San Francisco: Aunt Lute, 1987.

Por la mujer de mi raza
hablara el espiritu.

JOSE VASCOCELOS, MEXICAN PHILOSOPHER, envisaged *una raza mestiza, una mezcla de razas ajines, una raza de color – la primera raza sintesis del globo*. He called it a cosmic race, *la raza cosmica*, a fifth race embracing the four major races of the world. Opposite to the theory of the pure Aryan, and to the policy of racial purity that white America practices, his theory is one of inclusivity. At the confluence of two or more genetic streams, with chromosomes constantly 'crossing over,' this mixture of races, rather than resulting in an inferior being, provides hybrid progeny, a mutable, more malleable species with a rich gene pool. From this racial, ideological, cultural, and biological cross-pollinization, an 'alien' consciousness is presently in the making – a new *mestiza* consciousness, *una conciencia de mujer*. It is a consciousness of the Borderlands.

### *Una lucha de fronteras*/A struggle of borders

Because I, a *mestiza*,
continually walk out of one culture
and into another,
because I am in all cultures at the same time,
*alma entre dos mundos, tres, cuatro,*
*me zumba la cabeza con lo contradictorio.*
*Estoy norteada por todas las voces que me hablan simuláneamente.*

The ambivalence from the clash of voices results in mental and emotional states of perplexity. Internal strife results in insecurity and indecisiveness. The mestiza's dual or multiple personality is plagued by psychic restlessness.

In a constant stare of mental nepantilism, an Aztec word meaning torn between ways, *la mestiza* is a product of the transfer of the cultural and spiritual values of one group to another. Being tricultural, monolingual, bilingual, or multilingual, speaking a patois, and in a state of perpetual transition, the *mestiza* faces the dilemma of the mixed breed: which collectivity does the daughter of a darkskinned mother listen to?

*El choque de un alma atrapado entre el mundo del espiritu y el mundo de la técnica a veces la deja entullada.* Cradled in one culture, sandwiched between two cultures, straddling all three cultures and their value systems, *la mestiza* undergoes a struggle of flesh, a struggle of borders, an inner war. Like all people, we perceive the version of reality that our culture communicates. Like others having or living in more than one culture, we get multiple, often opposing messages. The coming together of two self-consistent but habitually incompatible frames of reference causes *un choque*, a cultural collision. Within us and within *la cultura chicana*, commonly held beliefs of the white culture attack commonly held beliefs of the Mexican culture, and both attack commonly held beliefs of the indigenous culture. Subconsciously, we see an attack on ourselves and our beliefs as a threat and we attempt to block with a counterstance. But it is not enough to stand on the opposite river bank, shouting questions, challenging patriarchal, white conventions. A counterstance locks one into a duel of oppressor and oppressed; locked in mortal combat, like the cop and the criminal, both are reduced to a common denominator of violence. The counterstance refutes the dominant culture's views and beliefs, and, for this, it is proudly defiant. All reaction is limited by, and dependent on, what it is reacting against. Because the counterstance stems from a problem with authority – outer as well as inner – it's a step towards liberation from cultural domination. But it is not a way of life. At some point, on our way to a new consciousness, we will have to leave the opposite bank, the split between the two mortal combatants somehow healed so that we are on both shores at once and, at once, see through serpent and eagle eyes. Or perhaps we will decide to disengage from the dominant culture, write it off altogether as a lost cause, and cross the border into a wholly new and separate territory. Or we might go another route. The possibilities are numerous once we decide to act and not react.

## A tolerance for ambiguity

These numerous possibilities leave *la mestiza* floundering in uncharted seas. In perceiving conflicting information and points of view, she is subjected to a swamping of her psychological borders. She has discovered that she can't hold concepts or ideas in rigid boundaries. The borders and walls that are supposed to keep the undesirable ideas out are entrenched habits and patterns of behavior; these habits and patterns are the enemy within. Rigidity means death. Only by remaining flexible is she able to stretch the psyche horizontally and vertically. *La mestiza* constantly has to shift out of habitual formations; from convergent thinking, analytical reasoning that tends to use rationality to move toward a single goal (a Western mode), to divergent thinking, characterized by movement away from set patterns and goals and toward a more whole perspective, one that includes rather than excludes. The new *mestiza* copes by developing a tolerance for contradictions, a tolerance for ambiguity. She learns to be an Indian in Mexican culture, to be Mexican from an Anglo point of view. She learns to juggle cultures. She has a plural personality, she operates in a pluralistic mode – nothing is thrust out, the good the bad and the ugly, nothing rejected, nothing abandoned. Not only does she sustain contradictions, she turns the ambivalence into something else. She can be jarred out of ambivalence by an intense, and often painful, emotional event which inverts or resolves the ambivalence. I'm not sure exactly how. The work takes place underground – subconsciously. It is work that the soul performs. That focal point at fulcrum, that juncture where the mestiza stands, is where phenomena tend to collide. It is where the possibility of uniting all that is separate occurs. This assembly is not one where severed or separated pieces merely come together. Nor is it a balancing of opposing powers. In attempting to work out a synthesis, the self has added a third element which is greater than the sum of its severed parts. That third element is a new consciousness – a mestiza consciousness – and though it is a source of intense pain, its energy comes from continual creative motion that keeps breaking down the unitary aspect of each new paradigm.

*En unas pocas centurias*, the future will belong to the mestiza. Because the future depends on the breaking down of paradigms, it depends on the straddling of two or more cultures. By creating a new mythos – that is, a change in the way we perceive reality, the way we see ourselves, and the ways we behave – *la mestiza* creates a new consciousness.

The work of *mestiza* consciousness is to break down the subject–object duality that keeps her a prisoner and to show in the flesh and through the images in her work how duality is transcended. The answer to the problem between the white race and the colored, between males and females, lies in healing the split that originates in the very foundation of our lives, our culture, our languages, our thoughts. A massive uprooting of dualistic thinking in the individual and collective consciousness is the beginning of a long struggle, but one that could, in our best hopes, bring us to the end of rape, of violence, of war.

# PART EIGHT

# Race

I T IS HARD TO THINK OF A SIGNIFICANT DEBATE within the field of post-colonial cultural studies in the last century that has not felt the impact of this term. Yet, few terms have been more contested than race. It is now beyond contest that race, as it was conceived in the high period of imperialism – as a set of irreducible differences within the human species – is a scientific fallacy. But as a social phenomenon its continuing force resides not in its existence as a meaningful scientific taxonomy but in its undoubted effects on behaviour and on policy in many societies.

In different ways both Todorov and Gates engage with the emergence of the term in the post-Enlightenment era and with the varieties of ways in which it was exercised at different periods. Todorov draws a crucial distinction between racism and racialism thus: 'Racism is an ancient form of behavior that is probably found worldwide; racialism is a movement of ideas born in Western Europe whose period of influence extends from the mid-eighteenth century to the mid-twentieth.' In making this distinction he is both echoing and prefiguring many other commentators before and after him. Basil Davidson, for example, sees racialism as produced by the development of large-scale chattel slavery in the new plantation economies of the eighteenth century. These commentators and others, have noted that in earlier periods while racism existed it was often subordinate to other modes of prejudice such as religious discrimination and that a shared religious affiliation might override distinctions of race based on minor physiognomic differences and skin pigmentation shades. Perhaps, as we have suggested in our discussion of the revival of the sacral in the modern world (see Part Nineteen, The Sacred) we may be returning to a situation in which religious affiliation becomes as important a factor in identity as race. But it seems unlikely, as Paul Gilroy argues from the British example, that the distinctions racialist ideologies have fostered will not continue to play a major role in the development of new discriminatory 'profiling' in the decades ahead. The unfortunate truth seems to be that however discredited the pseudo-scientific basis of racism may be, its power to form discriminations remains potent. There may be some small comfort in the realization that race can be employed powerfully as a tool of resistant identity too, as occurred in the Civil Rights Movement in the US. Mike Hill's comments on this in the essay 'After Whiteness' which we have placed in the section on ethnicity rather than here. This choice in itself suggests that categories such as race, ethnicity or indigeneity, despite their different referents have a considerable degree of overlap in practice. Perhaps Kwame Anthony Appiah's controversial thesis on the degree to which early

African and African-American national resistance leaders were circumscribed by the very race theory they sought to combat, should be read in this more positive way too. Pal Ahluwalia makes this point in his persuasive case for the rehabilitation of the négritude movement and its foundational role in asserting black identity and in the process of resisting colonization and its denigratory effects on colonized black peoples.

With characteristic subtlety Homi Bhabha provides us with a thesis that relates race to a broader history of discursive formations and especially to its relationship with discourses of modernity. In so doing he relates the history of race and racism to the broader shift in post-colonial societies involving the 'translation of the meaning of time into the discourse of space'. This broader perspective allows Bhabha to argue that the fundamental underpinnings of racism relate it to other forms of ideological resistance to the complexities of modern, hybridized social spaces such as the religious identities of Irish Catholics in Belfast or 'Muslim fundamentalists' in Bradford. These groups he perceives voluntarily *embrace* the same self-defining limitations that racist theories *imposed* on racially denigrated peoples in colonial times, thus forcing them to accept a position that relegates them to 'unresolved, transitional moments within the disjunctive present of modernity that are then projected into a time of historical retroversion or an inassimilable place outside history'. Thus, as the complexities with which the idea of race develop and broaden it is clear that the need to address it as a centrally defining concept in post-colonial societies and in the societies that colonized them remains as vital as ever.

## Chapter 47

# TZVETAN TODOROV

# RACE AND RACISM

From *On Human Diversity: Nationalism, Racism and Exoticism in French Thought*
(trans. Catherine Porter) Cambridge, Mass.: Harvard University Press, 1993.

**H**UMAN BEINGS ARE AT ONCE alike and different; anyone can make this trivial observation, since ways of life vary throughout the world while the (biological) species remains the same. The essential problem is to determine just how far the realm of identity extends and where the realm of difference begins; we must try to discover just what relationship obtains between these two realms. Over the past several centuries, reflection on these questions has taken shape as a doctrine of race.

Here I must begin by introducing a terminological distinction. The word 'racism,' in its usual sense, actually designates two very different things. On the one hand, it is a matter of behavior, usually a manifestation of hatred or contempt for individuals who have well-defined physical characteristics different from our own; on the other hand, it is a matter of ideology, a doctrine concerning human races. The two are not necessarily linked. The ordinary racist is not a theoretician; he is incapable of justifying his behavior with 'scientific' arguments. Conversely, the ideologue of race is not necessarily a 'racist,' in the usual sense: his theoretical views may have no influence whatsoever on his acts, or his theory may not imply that certain races are intrinsically evil.

In order to keep these two meanings separate, I shall adopt the distinction that sometimes obtains between 'racism,' a term designating behavior, and 'racialism,' a term reserved for doctrines. I must add that the form of racism that is rooted in racialism produces particularly catastrophic results: this is precisely the case of Nazism. Racism is an ancient form of behavior that is probably found worldwide; racialism is a movement of ideas born in Western Europe whose period of influence extends from the mid-eighteenth century to the mid-twentieth.

Racialist doctrines, which will be our chief concern here, can be presented as a coherent set of propositions. They are all found in the 'ideal type' or classical version of the doctrine, but some of them may be absent from a given marginal or 'revisionist' version. These propositions may be reduced to five.

1. *The existence of races.* The first thesis obviously consists in affirming that there are such things as races, that is, human groupings whose members possess common physical characteristics; or rather (for the differences themselves are self-evident) it consists in affirming the relevance and significance of that notion. From this perspective, races are equated with animal species, and it is postulated that there is the same distance between two human races as between horses and donkeys: not enough to prevent reproduction, but enough to establish a boundary readily apparent to all. Racialists are not generally content to observe this

state of affairs; they also want to see it maintained: they are thus opposed to racial mixing. The adversaries of racialist theory have often attacked the doctrine on this point. First, they draw attention to the fact that human groups have intermingled from time immemorial; consequently, their physical characteristics cannot be as different as racialists claim. Next, these theorists add a two-pronged biological observation to their historical argument. In the first place, human beings indeed differ from one another in their physical characteristics; but in order for these variations to give rise to clearly delimited groups, the differences and the groups would have to coincide. However, this is not the case. We can produce a map of the 'races' if we measure genetic characteristics, a second if we analyze blood composition, a third if we use the skeletal system, a fourth if we look at the epidermis. In the second place, within each of the groups thus constituted, we find greater distances between one individual and another than between one group and another. For these reasons, contemporary biology, while it has not stopped studying variations among human beings across the planet, no longer uses the concept of race.

But this scientific argument is not really relevant to the argument against racialist doctrines: it is a way of responding with biological data to what is actually a question of social psychology. Scientists may or may not believe in 'races,' but their position has no influence on the perception of the man in the street, who can see perfectly well that the differences exist. From this individual's viewpoint, the only properties that count are the immediately visible ones: skin color, body hair, facial configuration. Furthermore, the fact that there are individuals or even whole populations that are the product of racial mixing does not invalidate the notion of race but actually confirms it. The person of mixed race is identified precisely because the observer is capable of recognizing typical representatives of each race.

2. *Continuity between physical type and character*.   But races are not simply groups of individuals who look alike (if this had been the case, the stakes would have been trivial). The racialist postulates, in the second place, that physical and moral characteristics are interdependent; in other words, the segmentation of the world along racial lines has as its corollary an equally definitive segmentation along cultural lines. To be sure, a single race may possess more than one culture; but as soon as there is racial variation there is cultural change. The solidarity between race and culture is evoked to explain why the races tend to go to war with one another.

Not only do the two segmentations coexist, it is alleged, but most often a causal relation is posited between them: physical differences *determine* cultural differences. We can all observe these two series of variables, physical and mental, around us; each one can be explained independently, and the two explanations do not have to be related after the fact; or else the two series can be observed without requiring any explanation at all. Yet the racialist acts as if the two series were nothing but the causes and effects of a single series. This first assertion in turn implies the hereditary transmission of mental properties and the impossibility of modifying those properties by education. The quest for unity and order in the variety of lived experience clearly relates the racialist attitude to that of the scholar in general, who tries to introduce order into chaos and whose constructions affirm the kinship of things. It must be added that, up to now, no proof has been provided for the relation of determinism or even for the interdependence of race and culture. This does not mean, of course, that proof might not one day be found, or that the search for proof is in itself harmful. We must simply note that, for the time being the hypothesis has turned out to be unproductive.

Here I should like to mention a recent proposal to maintain the causal relation while overturning it. This view no longer holds that physical characteristics determine mental ones;

rather, it holds that culture acts on nature. If, within a given population, tall people are preferred to short people, or blonds are preferred to brunettes, the population as a whole will evolve toward the desired end: its value system will serve as a genetic filter. We can also imagine a population that would prefer physical strength to intelligence, or vice versa; once again, conditions will be favorable for an extension of the qualities valued. Such an inversion of perspective opens up new possibilities for the study of mind-body interactions.

3. *The action of the group on the individual.* The same determinist principle comes into play in another sense: the behavior of the individual depends to a very large extent, on the racio-cultural (or 'ethnic') group to which he or she belongs. This proposition is not always explicit, since it is self-evident: what is the use of distinguishing races and cultures, if one believes at the same time that individuals are morally non-determined, that they act in function of their own will freely exercised, and not by virtue of their group membership – over which they have no control? Racialism is thus a doctrine of collective psychology, and it is inherently hostile to the individualist ideology.

4. *Unique hierarchy of values.* The racialist is not content to assert that races differ: he also believes that some are superior to others, which implies that he possesses a unitary hierarchy of values, an evaluative framework with respect to which he can make universal judgments. This is somewhat astonishing, for the racialist who has such a framework at his disposal is the same person who has rejected the unity of the human race. The scale of values in question is generally ethnocentric in origin; it is very rare that the ethnic group to which a racialist author belongs does not appear at the top of his own hierarchy. On the physical qualities, the judgment of preference usually takes the form of aesthetic appreciation: my race is beautiful, the others are more or less ugly. On the level of the mind, the judgment concerns both intellectual and moral qualities (people are stupid or intelligent, bestial or noble).

5. *Knowledge-based politics.* The four propositions listed so far take the form of descriptions of the world, factual observations. They lead to a conclusion that constitutes the fifth and last doctrinal proposition – namely, the need to embark upon a political course that brings the world into harmony with the description provided. Having established the 'facts,' the racialist draws from them a moral judgment and a political ideal. Thus, the subordination of inferior races or even their elimination can be justified by accumulated knowledge on the subject of race. Here is where racialism rejoins racism: the theory is put into practice.

The refutation of this last inference is a task not for the scientist but rather for the philosopher. Science can refute propositions like the first three listed, but it may also turn out that what appears self-evident to biologists today may be considered an error tomorrow. Even if this were to happen, however, it would not justify behavior that could be properly condemned on other grounds. Geneticists are not particularly well qualified to combat racism. Subjecting politics to science, and thus subjecting what is right to what is, makes for bad philosophy, not bad science; the humanist ideal can be defended against the racist ideal not because it is more true (an ideal cannot be more or less true) but because it is ethically superior, based as it is on the universality of the human race.

## Chapter 48

# HENRY LOUIS GATES

# WRITING RACE

From Editor's Introduction 'Writing Race' in *Race Writing and Difference*
Chicago: University of Chicago Press, 1986.

**W**HAT IMPORTANCE DOES 'RACE' HAVE as a meaningful category in the study of literature and the shaping of critical theory? If we attempt to answer this question by examining the history of Western literature and its criticism, our initial response would probably be 'nothing' or, at the very least, 'nothing explicitly.' Indeed, until the past decade or so, even the most subtle and sensitive literary critics would most likely have argued that, except for aberrant moments in the history of criticism, race has not been brought to bear upon the study of literature in any apparent way. . . . In much of the thinking about the proper study of literature in this century, race has been an invisible quantity, a persistent yet implicit presence. . . . It was Hippolyte-Adolphe Taine who made the implicit explicit by postulating 'race, moment, and milieu' as positivistic criteria through which any work could be read and which, by definition, any work reflected. Taine's *History of English Literature* was the great foundation upon which subsequent nineteenth-century notions of 'national literatures' would be constructed. What Taine called 'race' was the source of all structures of feeling and thought. . . . Race, for Taine, was everything. . . . Taine's originality lay not in his ideas about the nature and role of race but rather in their almost 'scientific' application to the history of literature. These ideas about race were received from the Enlightenment, if not from the Renaissance. By 1850, ideas of irresistible racial differences were commonly held. . . . The growth of canonical national literatures was coterminous with the shared assumption among intellectuals that race was a 'thing,' an ineffaceable quantity, which irresistibly determined the shape and contour of thought and feeling as surely as it did the shape and contour of human anatomy. . . .

Race, in these usages, pretends to be an objective term of classification, when in fact it is a dangerous trope. The sense of difference defined in popular usages of the term 'race' has both described and inscribed differences of language, belief system, artistic tradition, and gene pool, as well as all sorts of supposedly natural attributes such as rhythm, athletic ability, cerebration, usury, fidelity, and so forth. The relation between 'racial character' and these sorts of characteristics has been inscribed through tropes of race, lending the sanction of God, biology, or the natural order to even presumably unbiased descriptions of cultural tendencies and differences. . . . Race has become a trope of ultimate, irreducible difference between cultures, linguistic groups, or adherents of specific belief systems which more often than not also have fundamentally opposed economic interests. Race is the ultimate trope of difference because it is so very arbitrary in its application. The biological criteria used to determine 'difference' in sex simply do not hold when applied to 'race.' Yet we carelessly use language in such a way as to will this sense of natural difference into our formulations. To do so is to engage in a pernicious act of language, one which exacerbates the complex problem of cultural or ethnic difference, rather than to assuage or redress it. This is especially the case at a time when, once again, racism has become fashionable. . . .

Since the beginning of the seventeenth century, Europeans had wondered aloud whether or not the African 'species of men,' as they most commonly put it, could ever create formal literature, could ever master 'the arts and sciences.' If they could, the argument ran, then the African variety of humanity and the European variety were fundamentally related. If not, then it seemed clear that the African was destined by nature to be a slave. Why was the creative writing of the African of such importance to the eighteenth century's debate over slavery? I can briefly outline one thesis: after René Descartes, reason was privileged, or valorized, above all other human characteristics. Writing, especially after the printing press became so widespread, was taken to be the visible sign of reason. Blacks were 'reasonable,' and hence 'men,' if – and only if – they demonstrated mastery of 'the arts and sciences,' the eighteenth century's formula for writing. So, while the Enlightenment is characterized by its foundation on man's ability to reason, it simultaneously used the absence and presence of reason to delimit and circumscribe the very humanity of the cultures and people of color which Europeans had been 'discovering' since the Renaissance. The urge toward the systematization of all human knowledge (by which we characterize the Enlightenment) led directly to the relegation of black people to a lower place in the great chain of being, an ancient construct that arranged all of creation on a vertical scale from plants, insects, and animals through man to the angels and God himself. . . .

Blacks and other people of color could not write. Writing, many Europeans argued, stood alone among the fine arts as the most salient repository of 'genius,' the visible sign of reason itself. In this subordinate role, however, writing, although secondary to reason, is nevertheless the medium of reason's expression. We know reason by its writing, by its representations. Such representations could assume spoken or written form. And while several superb scholars give priority to the spoken as the privileged of the pair, most Europeans privileged writing – their writings about Africans, at least – as the principal measure of the Africans' humanity, their capacity for progress, their very place in the great chain of being. The direct correlation between economic and political alienation, on the one hand, and racial alienation, on the other, is epitomized in the allowing 1740 South Carolina statute that attempted to make it almost impossible for black slaves to acquire, let alone master, literacy:

> And *whereas* the having of slaves taught to write, or suffering them to be employed in writing, may be attending with great inconveniences;
>     *Be it enacted*, that all and every person and persons whatsoever, who shall hereafter teach, or cause any slave or slaves to be taught to write, or shall use or employ any slave as a scribe in any manner of writing whatsoever, hereafter taught to write; every such person or persons shall, for every offense, forfeith the sum of one hundred pounds current money.

Learning to read and to write, then, was not only difficult, it was a violation of a law.

Ironically, Anglo-African writing arose as a response to allegations of its absence. Black people responded to these profoundly serious allegations about their 'nature' as directly as they could: they wrote books, poetry, autobiographical narratives. Political and philosophical discourse were the predominant forms of writing. Among these, autobiographical 'deliverance' narratives were the most common and the most accomplished. Accused of lacking a formal and collective history, blacks published individual histories which, taken together, were intended to narrate in segments the larger yet fragmented history of blacks in Africa, now dispersed throughout a cold New World. The narrated, descriptive 'eye' was put into service as a literary form to posit both the individual 'I' of the black author as well as the collective 'I' of the race. Text created author; and black authors, it was hoped, would create, or re-create, the image of the race in European discourse. The very face of the race was contingent upon the recording of the black voice. Voice presupposed a face, but also seems to have been thought to determine the very contours of the black face.

The recording of an authentic black voice – a voice of deliverance from the deafening discursive silence which an enlightened Europe cited to prove the absence of the African's humanity – was the millennial instrument of transformation through which the African would become the European, the slave become the ex-slave, brute animal become the human being. So central was this idea to the birth of the black literary tradition in the eighteenth century that five of the earliest slave narratives draw upon the figure of the voice in the text – of the talking book – as crucial 'scenes of instruction' in the development of the slave on the road to freedom.[1]

These five authors, linked by revision of a trope into the very first chain of black signifiers, implicitly signify upon another chain, the metaphorical great chain of being. Blacks were most commonly represented on the chain either as the lowest of the human races or as first cousin to the ape. Because writing, according to Hume, was the ultimate sign of difference between animal and human, these writers implicitly were signifyin(g) upon the figure of the chain itself. Simply by publishing autobiographies, they indicted the received order of Western culture, of which slavery was to them the most salient sign. . . . Making the book speak, then, constituted a motivated and political engagement with and condemnation of Europe's fundamental sign of domination, the commodity of writing, the text and technology of reason. We are justified, however, in wondering aloud if the sort of subjectivity which these writers seek through the act of writing can be realized through a process which is so very ironic from the outset: how can the black subject posit a full and sufficient self in a language in which blackness is a sign of absence? Can writing, with the very difference it makes and marks, mask the blackness of the black face that addresses the text of Western letters, in a voice that speaks English through an idiom which contains the irreducible element of cultural difference that will always separate the white voice from the black? Black people, we know, have not been liberated from racism by our writings. We accepted a false premise by assuming that racism would be destroyed once white racists became convinced that we were human, too. Writing stood as a complex 'certificate of humanity,' as Paulin Hountondji put it. Black writing, and especially the literature of the slave, served not to obliterate the difference of race; rather, the inscription of the black voice in Western literatures has preserved those very cultural differences to be repeated, imitated, and revised in a separate Western literary tradition, a tradition of black difference. We black people tried to write ourselves out of slavery, a slavery even more profound than mere physical bondage. Accepting the challenge of the great white Western tradition, black writers wrote as if their lives depended upon it – and, in a curious sense, their lives did, the 'life of the race' in Western discourse.

## Note

1    See James Albert Ukawsaw Gronniosaw (1770); John Marrant (1785); Ottabah Cugoano (1787); Olaudah Equiano (1789); and John Jea (1806).

## Chapter 49

# HOMI K. BHABHA

## RACE, TIME AND THE REVISION OF MODERNITY

From *The Location of Culture* London: Routledge, 236–56, 1994.

. . .

**T**HE DISCOURSE OF RACE that I am trying to develop displays the *problem of the ambivalent temporality of modernity* that is often overlooked in the more 'spatial' traditions of some aspects of postmodern theory. Under the rubric 'the discourse of modernity', I do not intend to reduce a complex and diverse historical moment, with varied national genealogies and different institutional practices, into a singular shibboleth – be it the 'idea' of Reason, Historicism, Progress – for the critical convenience of postmodern literary theory. . . .

The power of the postcolonial translation of modernity rests in its *performative, deformative* structure that does not simply revalue the contents of a cultural tradition, or transpose values 'cross-culturally'. The cultural inheritance of slavery or colonialism is brought *before* modernity *not* to resolve its historic differences into a new totality, nor to forego traditions. It is to introduce another locus of inscription and intervention, another hybrid, 'inappropriate' enunciative site, through that temporal split – or time-lag – that I have opened up for the signification of postcolonial agency. . . .

The ethnocentric limitations of Foucault's spatial sign of modernity become immediately apparent if we take our stand, in the immediate postrevolutionary period, in San Domingo with the Black Jacobins, rather than Paris. What if the 'distance' that constitutes the meaning of the Revolution as sign, the signifying lag between event and enunciation, stretches not across the Place de la Bastille or the rue des Blancs-Monteaux, but spans the temporal difference of the colonial space? What if we heard the 'moral disposition of mankind' uttered by Toussaint L'Ouverture for whom, as C. L. R. James so vividly recalls, the signs of modernity, 'liberty, equality, fraternity. . . what the French Revolution signified, was perpetually on his lips, in his correspondence, in his private conversations' (James 1980: 290–1). What do we make of the figure of Toussaint – James invokes Phedre, Ahab, Hamlet – at the moment when he grasps the tragic lesson that the moral, *modern* disposition of mankind, enshrined in the sign of the Revolution, only fuels the archaic racial factor in the society of slavery? What do we learn from that split consciousness, that 'colonial' disjunction of modern times and colonial and slave histories, where the reinvention of the self and the remaking of the social are strictly out of joint?

These are the issues of the catachrestic, postcolonial translation of modernity. They force us to introduce the question of subaltern agency, into the question of modernity: what is this 'now' of modernity? Who defines this present from which we speak? This leads to a more challenging question: *what is the desire of this repeated demand to modernize? Why does it insist, so compulsively, on its contemporaneous reality, its spatial dimension, its spectatorial distance?* What happens to the sign of modernity in those repressive places like San Domingo where progress is only heard (of) and not 'seen', is that it reveals the problem of the disjunctive moment of its utterance: the space which enables a postcolonial contra-modernity to emerge. . . .

The 'subalterns and ex-slaves' who now seize the spectacular event of modernity do so in a catachrestic gesture of reinscribing modernity's 'caesura' and using it to transform the locus of thought and writing in the postcolonial critique. Listen to the ironic naming, the interrogative repetitions, of the critical terms themselves: black 'vernacularism' repeats the minor term used to designate the language of the native and the household slave to make demotic the grander narratives of progress. Black 'expressionism' reverses the stereotypical affectivity and sensuality of the stereotype to suggest that 'rationalities are produced *endlessly* in populist modernism' (Gilroy 1980: 278). 'New ethnicity' is used by Stuart Hall in the black British context to create a discourse of cultural difference that marks ethnicity as the struggle against ethnicist 'fixing' and in favour of a wider minority discourse that represents sexuality and class. Cornel West's genealogical materialist view of race and Afro-American oppression is, he writes, 'both continuous and discontinuous with the Marxist tradition' and shares an equally contingent relation to Nietzsche and Foucault (West 1987: 86ff.). More recently, he has constructed a prophetic pragmatic tradition from William James, Niebuhr and Du Bois suggesting that 'it is possible to be a prophetic pragmatist and belong to different political movements, e.g. feminist, Black, chicano, socialist, left-liberal ones' (West 1990: 232–3). The Indian historian Gyan Prakash, in an essay on postorientalist histories of the Third World, claims that:

> it is difficult to overlook the fact that . . . third world voices . . . speak within and to discourses familiar to the 'West'. . . .The Third World, far from being confined to its assigned space, has penetrated the inner sanctum of the 'First World' in the process of being 'Third Worlded' – arousing, inciting, and affiliating with the subordinated others in the First World . . . to connect with minority voices.
>
> (Prakash 1990: 403)

The intervention of postcolonial or black critique is aimed at transforming the conditions of enunciation at the level of the sign – where the intersubjective realm is constituted – not simply setting up new symbols of identity, new 'positive images' that fuel an unreflective 'identity politics'. The challenge to modernity comes in redefining the signifying relation to a disjunctive 'present': staging the past as symbol, myth, memory, history, the ancestral – but a past whose iterative value as sign reinscribes the 'lessons of the past' into the very textuality of the present that determines both the identification with, and the interrogation of, modernity: what is the 'we' that defines the prerogative of my present? The possibility of inciting cultural translations across minority discourses arises because of the disjunctive present of modernity. It ensures that what *seems* the 'same' within cultures is negotiated in the time-lag of the 'sign' which constitutes the intersubjective, social realm. Because that lag is indeed the very structure of difference and splitting within the discourse of modernity, turning it into a performative process, then each repetition of the sign of modernity is different, specific to its historical and cultural conditions of enunciation.

This process is most clearly apparent in the work of those 'postmodern' writers who, in pushing the paradoxes of modernity to its limits, reveal the margins of the West.[1] From the postcolonial perspective we can only assume a disjunctive and displaced relation to these works; we cannot accept them until we subject them to a *lagging*: both in the temporal sense of postcolonial agency with which you are now (over)familiar, and in the obscurer sense in which, in the early days of settler colonization, to be lagged was to be transported to the colonies for penal servitude!

In Foucault's Introduction to the *History of Sexuality*, racism emerges in the nineteenth century in the form of an historical retroversion that Foucault finally disavows. In the 'modern' shift of power from the juridical politics of death to the biopolitics of life, race produces a historical temporality of interference, overlapping, and the displacement of sexuality. It is, for Foucault, the great historical irony of modernity that the Hitlerite annihilation of the Jews was carried out in the name of the archaic, premodern signs of race and sanguinity – the oneiric exaltation of blood, death, skin – rather than through the politics of sexuality. What is profoundly revealing is Foucault's complicity with the logic of the 'contemporaneous' within Western modernity. Characterizing the 'symbolics of blood' as being retroverse, Foucault disavows the time-lag of race as the sign of cultural difference and its mode of repetition.

The *temporal* disjunction that the 'modern' question of race would introduce into the discourse of disciplinary and pastoral power is disallowed because of Foucault's spatial critique: 'we must conceptualize the deployment of sexuality on the basis of the techniques of power that are *contemporary* with it' (my emphasis) (Foucault 1976: 150). However subversive 'blood' and race may be they are in the last analysis merely an 'historical retroversion'. Elsewhere Foucault directly links the 'flamboyant rationality' of Social Darwinism to Nazi ideology, entirely ignoring colonial societies which were the proving grounds for Social Darwinist administrative discourses all through the nineteenth and early twentieth centuries (see Foucault 1989: 269).

If Foucault normalizes the time-lagged, 'retroverse' sign of race, Benedict Anderson places the 'modern' dreams of racism 'outside history' altogether. For Foucault race and blood interfere with modern sexuality. For Anderson racism has its origins in antique ideologies of class that belong to the aristocratic 'pre-history' of the modern nation. Race represents an archaic ahistorical moment outside the 'modernity' of the imagined community: 'nationalism thinks in historical destinies, while racism dreams of eternal contaminations . . . outside history' (Anderson 1983: 136). Foucault's spatial notion of the conceptual contemporaneity of power as sexuality limits him from seeing the double and overdetermined structure of race and sexuality that has a long history in the *peuplement* (politics of settlement) of colonial societies; for Anderson the 'modern' anomaly of racism finds its historical modularity, and its fantasmatic scenario, in the colonial space which is a belated and hybrid attempt to 'weld together dynastic legitimacy and national community . . . to shore up domestic aristocratic bastions' (Anderson 1983: 136). The racism of colonial empires is then part of an archaic acting out, a dream-text of a form of historical retroversion that 'appeared to confirm on a global, modern stage antique conceptions of power and privilege' (Anderson 1983: 136). What could have been a way of understanding the limits of Western imperialist ideas of progress within the genealogy of a 'colonial metropolis' – a hybridizing of the Western nation – is quickly disavowed in the language of the *opéra bouffe* as a grimly amusing tableau vivant of 'the [colonial] bourgeois gentilhomme speaking poetry against a backcloth of spacious mansions and gardens filled with mimosa and bougainvillea' (Anderson 1983: 136). It is in that 'weld' of the colonial site as, contradictorily, both 'dynastic and national', that the modernity of Western national society is confronted by its colonial double. Such a moment of temporal disjunction, which would be crucial for understanding the colonial history of contemporary metropolitan racism in the West, is placed 'outside history'. It is obscured by Anderson's espousal of 'a simultaneity across homogeneous empty time' as the modal narrative of the imagined community. It is this kind of evasion, I think, that makes Partha Chatterjee, the Indian 'subaltern' scholar, suggest, from a different perspective, that Anderson 'seals up his theme with a sociological determinism . . . without noticing the twists and turns, the suppressed possibilities, the contradictions still unresolved' (Chatterjee 1986: 21–2).

These accounts of the modernity of power and national community become strangely symptomatic at the point at which they create a rhetoric of 'retroversion' for the emergence of racism. In placing the representations of race 'outside' modernity, in the space of historical retroversion, Foucault reinforces his 'correlative spacing'; by relegating the social fantasy of racism to an archaic daydream, Anderson further universalizes his homogeneous empty time of the 'modern' social imaginary. Hidden in the disavowing narrative of historical retroversion and its archaism, is a notion of the time-lag that displaces Foucault's spatial analytic of modernity and Anderson's homogeneous temporality of the modern nation. In order to extract the one from the other we have to see how they form a double boundary: rather like the more general intervention and seizure of the history of modernity that has been attempted by postcolonial critics. Retroversion and archaic doubling, attributed to the ideological 'contents' of racism, do not remain at the ideational or pedagogical level of the discourse. Their inscription of a structure of retroaction returns to disrupt the enunciative function of this discourse and produce a different 'value' of the sign and time of race and modernity. At the level of content the archaism and fantasy of racism is represented as 'ahistorical', outside the progressive myth of modernity. This is an attempt, I would argue, to universalize the spatial fantasy of modern cultural communities as living their history 'contemporaneously', in a 'homogeneous empty time' of the People-as-One that finally deprives minorities of those marginal, liminal spaces from which they can intervene in the unifying and totalizing myths of the national culture. However, each time such a homogeneity of cultural identification is established there is a marked disturbance of temporality in the writing of modernity. For Foucault it is the awareness that retroversion of race or sanguinity haunts and doubles the contemporary analytic of power and sexuality and may be subversive of it: we may need to think the disciplinary powers of race as sexuality in a hybrid cultural formation that will not be contained within Foucault's logic of the contemporary. Anderson goes further in acknowledging that colonial racism introduces an awkward weld, a strange historical 'suture', in the narrative of the nation's *modernity*. The archaism of colonial racism, as a form of cultural signification (rather than simply an ideological content), reactivates nothing less than the 'primal scene' of the modern Western nation: that is, the problematic historical transition between dynastic, lineage societies and horizontal, homogeneous secular communities. What Anderson designates as racism's 'timelessness', its location 'outside history', is in fact that form of time-lag, a mode of repetition and reinscription, that *performs* the ambivalent historical temporality of modern national cultures – the *aporetic coexistence*, within the cultural history of the modern imagined community, of both the dynastic, hierarchical, prefigurative 'medieval' traditions (the past), and the secular, homogeneous, synchronous cross-time of modernity (the present). Anderson resists a reading of the modern nation that suggests in an iterative time-lag that the hybridity of the colonial space may provide a pertinent problematic within which to write the history of the 'postmodern' national formations of the West.

To take this perspective would mean that we see 'racism' not simply as a hangover from archaic conceptions of the aristocracy, but as part of the historical traditions of civic and liberal humanism that create ideological matrices of national aspiration, together with their concepts of 'a people' and its imagined community. Such a privileging of ambivalence in the social imaginaries of nation*ness*, and its forms of collective affiliation, would enable us to understand the coeval, often *incommensurable* tension between the influence of traditional 'ethnicist' identifications that coexist with contemporary secular, modernizing aspirations. The enunciative 'present' of modernity, that I am proposing, would provide political space to articulate and negotiate such culturally hybrid social identities. Questions of cultural difference would not be dismissed with a barely concealed racism – as atavistic 'tribal' instincts that afflict Irish Catholics in Belfast or 'Muslim fundamentalists' in Bradford. It is precisely such unresolved, transitional moments within the disjunctive present of modernity that are then projected into a time of historical retroversion or an inassimilable place outside history.

The *history* of modernity's antique dreams is to be found in the *writing out* of the colonial and postcolonial moment. In resisting these attempts to normalize the time-lagged colonial moment, we may provide a *genealogy* for postmodernity that is at least as important as the 'aporetic' history of the Sublime or the nightmare of rationality in Auschwitz. For colonial and postcolonial texts do not merely tell the modern history of 'unequal development' or evoke memories of underdevelopment. I have tried to suggest that they provide modernity with a modular moment of enunciation: the locus and locution of cultures caught in the transitional and disjunctive temporalities of modernity. What is in modernity more than modernity is the disjunctive 'postcolonial' time and space that makes its presence felt at the level of enunciation. It figures, in an influential contemporary fictional instance, as the contingent margin between Toni Morrison's indeterminate moment of the 'not-there' – a 'black' space that she distinguishes from the Western sense of synchronous tradition – which then turns into the 'first stroke' of slave rememory, the *time* of communality and the narrative of a history of slavery. This translation of the meaning of time into the discourse of space; this catachrestic seizure of the signifying 'caesura' of modernity's presence and present; this insistence that power must be thought in the hybridity of race and sexuality; that nation must be reconceived liminally as the dynastic-in-the-democratic, race-difference doubling and splitting the teleology of class-consciousness: it is through these iterative interrogations and historical initiations that the cultural location of modernity shifts to the postcolonial site.

## Note

1    Robert Young in *White Mythologies* (1990), also suggests, in keeping with my argument that the colonial and postcolonial moment is the liminal point, or the limit-text, of the holistic demands of historicism.

## Chapter 50

# KWAME ANTHONY APPIAH

## THE ILLUSIONS OF RACE

From 'The Illusions of Race' in Emmanuel Eze (ed.) *African Philosophy: An Anthology* Oxford: Blackwell, 1998.

> If this be true, the history of the world is the history, not of individuals, but of groups, not of nations, but of races.
>
> W. E. B. Du Bois

. . .

**WHEN I DEFINE RACIALISM** I [would say] that it [is] committed not just to the view that there are heritable characteristics, which constitute 'a sort of racial essence,' but also to the claim that the essential heritable characteristics account for more than the visible morphology – skin color, hair type, facial features – on the basis of which we make our informal classifications. To say that biological races existed because it was possible to classify people into a small number of classes according to their gross morphology would be to save racialism in the letter but lose it in the substance. The notion of race that was recovered would be of no biological interest – the interesting biological generalizations are about genotypes, phenotypes, and their distribution in geographical populations. We could just as well classify people according to whether or not they were redheaded, or redheaded and freckled, or redheaded, freckled, and broad-nosed too, but nobody claims that this sort of classification is central to human biology. There are relatively straightforward reasons for thinking that large parts of humanity will fit into no class of people who can be characterized as sharing not only a common superficial morphology but also significant other biological characteristics. The nineteenth-century dispute between monogenesis and polygenesis, between the view that we are descended from one original population and the view that we descend from several, is over. There is no doubt that all human beings descend from an original population (probably, as it happens, in Africa), and that from there people radiated out to cover the habitable globe. Conventional evolutionary theory would predict that as these populations moved into different environments and new characters were thrown up by mutation, some differences would emerge as different characteristics gave better chances of reproduction and survival. In a situation where a group of people was isolated genetically for many generations, significant differences between populations could build up, though it would take a very extended period before the differences led to reproductive isolation – the impossibility of fertile breeding – and thus to the origin of a new pair of distinct species. We know that there is no such reproductive isolation between human populations, as a walk down any street in

New York or Paris or Rio will confirm, but we also know that none of the major human population groups have been reproductively isolated for very many generations. If I may be excused what will sound like a euphemism, at the margin there is always the exchange of genes. . . .

The classification of people into 'races' would be biologically interesting if both the margins and the migrations had not left behind a genetic trail. But they have, and along that trail are millions of us (the numbers obviously depending on the criteria of classification that are used) who can be fitted into no plausible scheme at all. In a sense, trying to classify people into a few races is like trying to classify books in a library: you may use a single property – size, say – but you will get a useless classification, or you may use a more complex system of interconnected criteria, and then you will get a good deal of arbitrariness. No one – not even the most compulsive librarian! – thinks that book classifications reflect deep facts about books. Each of them is more or less useless for various purposes; all of them, as we know, have the kind of rough edges that take a while to get around. And nobody thinks that a library classification can settle which books we should value; the numbers in the Dewey decimal system do not correspond with qualities of utility or interest or literary merit. . . .

The disappearance of a widespread belief in the category of the Negro would leave nothing for racists to have an attitude toward. But it would offer, by itself, no guarantee that Africans would escape from the stigma of centuries. Extrinsic racists could disappear and be replaced by people who believed that the population of Africa had in its gene pool fewer of the genes that account for those human capacities that generate what is valuable in human life; fewer, that is, than in European or Asian or other populations. Putting aside the extra-ordinary difficulty of defining which genes these are, there is, of course, no scientific basis for this claim. A confident expression of it would therefore be evidence only of the persist-ence of old prejudices in new forms. But even this view would be, in one respect, an advance on extrinsic racism. For it would mean that each African would need to be judged on his or her own merits. Without some cultural information, being told that someone is of African origin gives you little basis for supposing anything much about them. Let me put the claim at its weakest: in the absence of a racial essence, there could be no guarantee that some partic-ular person was not more gifted – in some specific respect – than any or all others in the populations of other regions. . . .

All this has nothing to do with the plain fact that throughout the world today organized groups of men by monopoly of economic and physical power, legal enactment and intellectual training are limiting with determination and unflagging zeal the development of other groups; and that the concentration particularly of economic power today puts the majority of mankind into a slavery to the rest. . . . What does cement together people who share a characteristic – the 'badge of insult' – on the basis of which some of them have suffered discrimination? We might answer: 'Just that; so there is certainly something that the nonwhite people of the world share.' But if we go on to ask what harm exactly a young woman in Mali suffers from antiblack race prejudice in Paris, this answer misses all the important details. She does suffer, of course, because, for example, political decisions about North–South relations are strongly affected by racism in the metropolitan cultures of the North. But this harm is more systemic, less personal, than the affront to individual dignity represented by racist insults in the post-industrial city. If she is an intellectual, reflecting on the cultures of the North, she may also feel the mediated sense of insult: she may know, after all, that if she were there, in Paris, she would risk being subjected to some of the same discriminations; she may recognize that racism is part of the reason why she could not get a visa to go there; why she would not have a good time if she did. Such thoughts are certainly maddening, as African and African-American and black European intellectuals will avow, if you ask them how they feel about the racist immigration policies of Europe or the institutionalized racism of apartheid. And they are thoughts that can be had by any nonwhite person anywhere who knows – in a phrase of Chinua Achebe's – 'how the world is moving.' The thought that if I were there now, I would be a victim strikes at you differently, it seems to me, from the thought – which can enrage any decent white human being – that if I were there and if I were not White, I would be a victim. Yet we should always remember that this thought, too, has led many to an identification with the struggle against racism. The lesson, I think, of these reflections must be that there is no one answer to the question what identifications our antiracism may lead us into. . . .

The truth is that there are no races: there is nothing in the world that can do all we ask race to do for us. As we have seen, even the biologist's notion has only limited uses, and the notion that Du Bois required, and that underlies the more hateful racisms of the modern era, refers to nothing in the world at all. The evil that is done is done by the concept, and by easy – yet impossible – assumptions as to its application. Talk of 'race' is particularly distressing for those of us who take Culture seriously. For, where race works – in places where 'gross differences' of morphology are correlated with 'subtle differences' of temperament, belief, and intention – it works as an attempt at metonym for culture, and it does so only at the price of biologizing what is culture, ideology. To call it 'biologizing' is not, however, to consign our concept of race to biology. For what is present there is not our concept but our word only. Even the biologists who believe in human races use the term race, as they say, 'without any social implication.' What exists 'out there' in the world – communities of meaning, shading variously into each other in the rich structure of the social world – is the province not of biology but of the human sciences.

## Chapter 51

## PAUL GILROY

# THERE AIN'T NO BLACK IN THE UNION JACK

From *There Ain't No Black in the Union Jack* London: Routledge, 1987.

. . .

[**W**]HERE [**DOES RACISM FIT**] into contemporary British life? Today, Britain's black and other minority settlers still constitute a problem, but its dimensions have altered. As a result of major changes that have occurred both inside and outside of the country 'race' is rendered differently. The meaning and representation of race politics have been greatly changed and its strategic importance relative to other aspects of government has been transformed. The theme of primal racial difference is not being articulated into the official political languages of nationality, culture and belonging in the simple exclusionary way that it was not so very long ago. This situation reflects the fact that Britain's various black communities are no more static than the evolving social practices of all the other participants in the country's civic order. Black communities are being actively re-composed by the pressure of local circumstances, by the new arrivals whose experience of racism leads them to either seek or refuse political allies, and by inter-generational adaptation as well as novel and unstable geo-political conditions. . . .

Demographic and cultural changes have meant that the New World histories which turned the counter-memory of racial slavery into an interpretative device that could be applied to any example of injustice and exploitation have lost much of their power and appeal. . . . The combative mood of the 1970s was buoyed by the Caribbean and American political traditions that had politicised blackness all over the world. . . . It is no longer possible to ignore the way that insights derived from those traditions sometimes fail to connect with the experiences and understanding of younger people or with the vision of African and Asian settlers whose colonial and post-colonial sufferings have been necessarily different. . . . The racial idea 'Asian' has for example been broken down and enumerated into a multiplicity of regional, religious and other cultural fractions. It is true that many – though by no means all – voices raised from within these diverse groups do not recognise themselves in the powerfully empty and possibly anachronistic master-signifier: 'black'. . . . The symbolic and linguistic system in which political blackness made sense was a phenomenon of assertive decolonisation and is now in retreat. Its defeat is also connected to wider cultural shifts like the rise of identity politics, corporate multi-culture and an imploded, narcissistic obsession with the minutiae of ethnicity.

The historic turn away from the simpler efficacy of blackness a bridging term that had promoted vernacular cosmopolitan conversation and synchronised action among the victimized cannot be separated from the pursuit of more complex and highly-differentiated ways of fixing and instrumentalising culture and difference. These developments have made anti-racism less politically focused and certainly more difficult if not impossible to organise. They are not only more likely to be in tune with an understanding of 'race' that derives from diversified market relations, but have also helped to re-specify ethnicity exclusively in the contentious cultural terms of life-style and consumer preference. . . . Though the nascent global culture of human rights has achieved a great deal in a short period, it has not always been alert to the significance of either colonial domination or racism as tests of its own lofty aspirations and cosmopolitan conceptions of justice. . . . [The] arguments for taking 'race' seriously are uncontroversial in a climate where it is likely to be taken too seriously while racism is not taken seriously enough. It is possible and necessary to approach Britain's colonial history by more satisfactory methodological routes. Its racial subjects need a more complex genealogy than those debates allow.

Industrial decline has been intertwined with technological change, with immigration and settlement, with ideological racism and spatial segregation along economic and cultural lines. We need to grasp how their coming together took place in a desperate setting which nonetheless allowed black communities over several generations to be recognised as political actors: they were irreducible to their class positions because racism entered into the multi-modal processes in which classes were being constituted. It helps to appreciate that this historical predicament was overdetermined by Britain's painful loss of Empire and, that the country's communities of the strange and alien are still sometimes at risk of being engulfed by the profound cultural and psychological consequences of decline which is evident on many levels: economic and material as well as cultural and psychological. All these factors need to be recognised as contributing to the formation of the contending political constituencies and interests that work by means of 'race' and nation. Their complex interaction produces racial subjects and also includes the possibility of class based solidarity, usually as a welcome but insubstantial answer to the divisions wrought by racism. But it bears repetition that in this less tidy analytical scheme, class consciousness and conflict are not going to retain special interpretative privileges. Understanding their characteristic points of friction and injury is certainly necessary but cannot be sufficient.

More importantly still, the formations of class identity, solidarity and antagonism that derived from the lost industrial order are a poor guide to contemporary arrangements which have placed some new historical actors in unprecedented positions and require refined conceptions of their agency and subjectivity, as well as their sameness and solidarity. Attention to the basic antagonism between labour and capital cannot, by itself, supply adequate alternative explanations of the specific power and historical attributes of racial discourse. These difficulties are compounded when explanations anchored in the integrity of class-based theories, inflate the significance and potency of class solidarity or play down the power of racism to shape the economic, historical and political dynamics with which it is caught up. The substantive analytical challenge remains how to explain the new varieties of exploitation that are being articulated into the class structures of the industrial order as they decompose and mutate. . . .

Mainstream Britain has been required to become more fluent in the anthropological idiom of official multiculturalism, and all parties to this conflict – in spite of their opposed political positions – have come to share an interest in magnifying racial, cultural and ethnic differences so that a special transgressive pleasure can be discovered in their spectacular overcoming. These social and political divisions are presented as unbridgeable breaks in consciousness and experience even while mixing has become a routine and unexceptional feature of urban life. The widespread belief in their absolute nature may also be attractive because it absolves bona fide brits and cultural insiders from the hard work involved in translation and cross-cultural understanding. The normative power of racial common sense is comfortable with simplistic and misleading labels like 'race riots'. Indeed it is fuelled by the simple racial truths they bring to life. Creative and negative thinking is needed to generate more complex and challenging narratives which can repudiate those truths and be faithful to everyday metropolitan life by reducing the exaggerated dimensions of racial difference to a liberating ordinary-ness. From this angle, 'race' is nothing special, a virtual reality given meaning only by the fact that racism endures.

## Chapter 52

# PAL AHLUWALIA

# NÉGRITUDE AND NATIVISM

From *Politics and Post-colonial Theory: African Inflections* London and
New York: Routledge, 2001.

THE NÉGRITUDE MOVEMENT OF THE 1930S was full of contradictions
and ambivalence. It was by no means a movement that could be regarded simply as relativist, and one which merely reaffirmed the racial binaries which it sought to dismantle. On
the contrary, it was an important moment in the long and arduous struggle for decolonisation. Indeed, it was a formative moment for the African, who had been denigrated over
centuries and represented as child-like and unable to be a member of the 'civilised world'
(see Ashcroft 1997; Valentine 1996). It was essential to the process that sought to break down
the tyranny of the web of representations which had been forged over centuries. . . .

Because négritude has come under considerable attack for its reaffirmation of racial binaries,
its critical role as a predecessor to decolonisation has received cursory attention. At best, particularly in the former British colonies, it is seen as a cultural movement that had little to do
with the 'real' political struggle which led to independence. A significant amnesia therefore
appears to have crept into recent literary and critical theory, where the négritude movement
is seen more as an embarrassment than as occupying a central position in the process of
decolonising the mind that was an integral part of the struggle for independence.

Négritude as a movement emerged in Paris in the early 1930s, amongst African and West
Indian students under the leadership of Léopold Sédar Senghor from Senegal, Aimé Césaire,
a Martiniquian and the Guyanese Léon Damas. The three established a newspaper, *L'Etudiant
noir (The Black Student)*, in which they voiced their problems, stressing commonalities amongst
all black people around the world. Négritude needs to be contextualised against the general
background of colonisation and the manner in which the African's very being was denigrated.
As Irele has observed, this was 'not simply a rationalisation of white domination but . . . a
direct and crushing attack upon his subjectivity' (1971: 26). It is perhaps not surprising that
négritude took shape in Paris, given the French system of colonisation. French colonies were
seen as an extension of France, and their subjects were considered citizens. In theory this
form of colonialism was based on assimilation, but, in reality, the rights of citizenship were
extended to only the white French settlers in the colonies. When these African and West
Indian students arrived in Paris, they found that, contrary to the theory of assimilation, they
were isolated because of their colour. In France, they discovered that they were not 'French'.
This realization led them to undertake a journey to rediscover their past, their black roots
and African heritage. Through négritude the colonised sought to reverse the representations
ascribed to them, to turn those negative identities into positive images.

An important paradox of négritude was that the very people who were urging a return to authenticity and renewal were themselves thoroughly imbued with the values of the coloniser. It was their alienation in both cultures, their sense of not belonging in either their own culture or that of the colonisers, that became problematic. It was their 'preoccupation with the black experience' which 'developed into a passionate exaltation of the black race, associated with a romantic myth of Africa' (Irele 1981: 91). The négritude writers not only 'celebrated Africa by paying tribute to the African love of life, the African joy of love and the African dream of death' (Dathorne 1981: 59) but also challenged the colonisers in a way that they had never before been challenged. Césaire described négritude as 'the simple recognition of the fact of being black and the acceptance of this fact, of our destiny as black people, of our history and our culture' (cited in Irele 1981: 87).

Although the term négritude was first coined by Aimé Césaire in *L'Etudiant noir*, its development was the result of a partnership between Césaire and Senghor. Senghor describes the manner in which they developed the term:

> In what circumstances did Aimé Césaire and I launch the word negritude between 1933 and 1935? At that time, along with several other black students we were plunged into a panic-stricken despair. The horizon was blocked. No reform was in sight and the colonizers were justifying our political and economic dependence by the theory of the tabula rasa. . . . In order to establish an effective revolution, our revolution, we had first to divest ourselves of our borrowed attire – that of assimilation – and assert our being, that is to say our negritude.
>
> (Senghor, cited in Bâ 1973: 12)

*L'Etudiant noir* folded after a few issues, and, although it was succeeded by several other publications in which the ideas of négritude were elaborated, it was not until the establishment of *Présence Africaine* that the movement had a permanent voice to propagate its ideas. Nevertheless, the movement's success in its formative days needs to be attributed to its favourable reception amongst the French intelligentsia. This was helped in part by the American Black Renaissance movement of the 1920s, the impact of which was being felt in Paris, particularly in music and entertainment.[1] This acceptance was in no small way the result of Senghor being able to persuade Jean-Paul Sartre to write a preface to an anthology of Black writers he had edited. Sartre's essay titled *'Orphee noir'*, not only gave the négritude movement a boost but at the same time embroiled it in controversy which persists until today.

Négritude, from its inception, needs to be viewed as a notion which had a strong element of resistance. Senghor claimed that the very act of negating the representations of the black person was liberating. Négritude was at its core about returning to black people a humanity that had been denied to them by centuries of denigration and brutalisation which reached its apex through the colonial process. Senghor's writings were an affirmation that black people were humans, contrary to the manner in which their identity had been problematised within European discourses. A key aspect of Senghor's négritude, for which he and the entire movement has been criticised severely, is the affirmation of racial images celebrating merely blackness. That is, any negative trait that had been attributed to a black person is celebrated as a positive element. . . .

The historical interactions of the slave trade, direct colonialism and neocolonialism have determined the manner in which Africa has been dealt with in European discourse. The concept of négritude has developed since its inauguration to include a multiplicity of meanings as diverse as African personality, black renewal and Pan-Africanism. It had much in common with other movements of the struggle for liberation when it was crucial to break down the representations of the colonisers, when it was essential to reconstitute subjectivity. These divergent Africa experiences and modes of resistance are by no means monolithic, but they nevertheless bind the history of the Africa diaspora with that of the peoples of Africa. The experiences of Nkrumah, Nyerere, Mandela, Du Bois, Equiano, Blyden, Senghor and Césaire are locked into a common history in their opposition to a 'West' that has sought to denigrate them collectively. But they are all also engaged in the individual struggle against the oppressive nature of colonialism. It is in this context, that Benita Parry (1994) has warned against the all-too-easy dismissal of nativism on the basis of essentialism and, on the contrary, has uttered the rallying call of 'two cheers for nativism'. Parry is cognisant of the positive effects that négritude offers, above all else that it has important possibilities for anti-colonial resistance. It is perplexing why the idea of a common African heritage arouses such passion, when, in contrast, Europeans (be they German, French or Italian, etc.) can claim a common Greek philosophical heritage without detracting from their individuality. As Drachler argues, 'serious inquiry does not hesitate to trace the mainstreams of the European heritage to their sources, nor to seek the common cultural ground on which Western men stand' (Drachler 1964: 168). While it is important to be careful not to replicate the essentialist positions advocated by the négritude writers, it would be far too reductive to simply dismiss and discredit the entire movement, as has been the case in much recent post-colonial writing. It was an important part of the development of a black awareness and consciousness which eventually paved the way for the liberation of Africa. . . .

Negative representations of Africa and blackness are deeply embedded within the European imagination, but, as Chinua Achebe has noted, African identity is in the making. There isn't a formal identity that is Africa. But at the same time, there is an identity coming into existence. And it has a certain context and meaning' (cited in Appiah 1992: 173). It is within this continuing process of identity-formation that the négritude movement needs to be placed. It was an important moment in the long and arduous journey towards decolonisation.

## Note

1   Although Senghor acknowledges the influence of the African-American movement, Koffi Anyinefa has recently demonstrated how African-Americans remained ambivalent to Senghor's work (Anyinefa 1996). For an excellent account of the Black renaissance see Gates 1984. See also Jules-Rosette 1998.

# PART NINE

# Feminism

IN MANY DIFFERENT SOCIETIES, WOMEN, like colonized subjects, have been relegated to the position of 'Other', 'colonized' by various forms of patriarchal domination. They thus share with colonized races and cultures an intimate experience of the politics of oppression and repression. It is not surprising therefore that the history and concerns of feminist theory have paralleled developments in post-colonial theory. Feminist and post-colonial discourses both seek to reinstate the marginalized in the face of the dominant, and early feminist theory, like early nationalist post-colonial criticism, was concerned with inverting the structures of domination, substituting, for instance, a female tradition or traditions for a male-dominated canon. But like post-colonial criticism, feminist theory has rejected such simple inversions in favour of a more general questioning of forms and modes, and the unmasking of the spuriously author/itative on which such canonical constructions are founded.

Until recently feminist and post-colonial discourses have followed a path of convergent evolution, their theoretical trajectories demonstrating striking similarities but rarely intersecting. In the last ten years, however, there has been increasing interest not just in their parallel concerns but in the nature of their actual and potential intersections – whether creatively coincident or interrogative. Feminism has highlighted a number of the unexamined assumptions within post-colonial discourse, just as post-colonialism's interrogations of Western feminist scholarship have provided timely warnings and led to new directions.

Early problems raised by the attempts to accommodate these similar but sometimes conflicting agendas are described by Kirsten Holst Petersen in 'First Things First'. It is significant that this problem is articulated as a dilemma for African women *writers* whose representations of their societies, and of patriarchal oppressions within them, are seen as conflicting with the processes of decolonization and cultural restitution, not just in terms of images presented to the former colonizers, but more significantly in terms of their *own* Euro-interpellated populations. African cultural values systematically denigrated by colonialist ideologies and institutions demand positive representation, and this restitutive impulse has frequently been seen to conflict with feminist *re*formation.

The notion of 'double colonization' – i.e. that women in formerly colonized societies were *doubly* colonized by both imperial and patriarchal ideologies – became a catch-phrase of post-colonial and feminist discourses in the 1980s. But it is only recently that 'double colonization' has begun to be adequately theorized. Ketu Katrak (like the East and West African writers Petersen discusses) reminds us of the inescapable necessity of situating a feminist politics within particular colonized societies. Using the example of the Jamaican Sistren Collective's work, she grounds a decolonizing feminist restitution in the local particularities of class and race. The Jamaican writer Erna Brodber's short essay 'Sleeping's Beauty and Prince Charming' (1989) suggests another way of actually theorizing the concept of a double colonization. Texts – the 'fairy tales' of Europe – have not only subjectified Jamaican women, but through cultural interpellation effected the erasure of the black female body within Jamaican male culture. Hence the black 'Prince Charming' of Brodber's fable can *sense* his female counterpart, but when he looks for her he can see 'no/body'. Sara Suleri examines a rather different refraction of the concept of 'double colonization' in Pakistan through the recent institution of Muslim Law, a process facilitated by neo-colonial United States' support of a male regime where laws against rape have recoiled horrifically on the bodies of women and children.

Not surprisingly perhaps, the use of language in decolonizing strategies forms the basis of Sistren's (re)creative experimentation; and Trinh T. Minh-ha, aware of the difficulties of, in Audre Lorde's terms, using the masters' tools to dismantle his house, nevertheless attempts to escape enclosure through complex linguistic/generic experimentation. Significantly, too, she refuses to be 'ghettoized' through the separate and/or combined essentialisms of gender, race or ethnicity, seeing these consolidating positions – politically strategic as they may at first appear – as new houses or rather out-houses of the 'master(s)'.

Chandra Mohanty's 'Under Western Eyes' (with Rachel Carby's 'White Woman Listen!') is foundational in critiquing Western feminisms that too easily elide specific cultural difference and 'naturalize' all women's oppression under widely differing manifestations of patriarchical domination to European models. As Gayatri Spivak demonstrates, what is a radically liberating piece of writing or politics in *one* arena can act as a colonizing agent in another. Sara Suleri's article offers a useful critique of a number of the positions taken up by other writers in this section, and also offers a useful overview of some of the other positions discussed in this part.

## Chapter 53

# KIRSTEN HOLST PETERSEN

## FIRST THINGS FIRST

## Problems of a feminist approach to African literature

From 'First Things First: Problems of a Feminist Approach to African Literature' *Kunapipi* 6(3), 1984.

I N THE AUTUMN OF 1981 I went to a conference in Mainz. The theme of the conference was 'The Role of Women in Africa'; it was a traditional academic conference and proceeded in an orderly fashion with papers on various aspects of the subject and not too much discussion until the last day of the conference when a group of young German feminists had been invited to participate. They dismissed the professor who up until then had chaired the session (he was a man), installed a very articulate student as chairwoman, and proceeded to turn the meeting into a series of personal statements and comments in the tradition of feminist movement meetings. They discussed Verena Stefan's book *Shedding* with its radical feminist solution, and they debated their relationship to their mothers, in terms of whether they should raise their mothers' consciousness and teach them to object to their fathers or whether perhaps it was best to leave them alone. The African women listened for a while, and then they told their German sisters how inexplicably close they felt to their mothers/ daughters, and how neither group would dream of making a decision of importance without first consulting the other group. This was not a dialogue! It was two very different voices shouting in the wilderness, and it pointed out to me very clearly that universal sisterhood is not a given biological condition as much as perhaps a goal to work towards, and that in that process it is important to isolate the problems which are specific to Africa or perhaps the Third World in general, and also perhaps to accept a different hierarchy of importance in which the mother/daughter relationship would be somewhat downgraded.

One obvious and very important area of difference is this: whereas Western feminists discuss the relative importance of feminist versus class emancipation, the African discussion is between feminist emancipation versus the fight against neo-colonialism, particularly in its cultural aspect. In other words, which is the more important, which comes first, the fight for female equality or the fight against Western cultural imperialism? When I say that this is what the discussion is about, I hasten to add that there is very little explicit discussion about the subject, but – as I hope to show – the opinion which is implicit in the choice of subject of the first generation of modern African writers has had a profound influence on attitudes to women and the possibility of a feminist school of writing.

Whilst there is not a lot, there is some explicit discussion about the subject. The Malawian poet Felix Mnthali states one view very clearly in a poem called 'Letter to a Feminist Friend':[1]

I will not pretend
to see the light
in the rhythm of your paragraphs:
illuminated pages
need not contain
any copy-right
on history

My world has been raped
                              looted
                  and squeezed
by Europe and America
and I have been scattered
over three continents
to please Europe and America

AND NOW
the women of Europe and America
after drinking and carousing
on my sweat
rise up to castigate
                  and castrate
their menfolk
from the cushions of a world
I have built!

Why should they be allowed
to come between us?
You and I were slaves together
uprooted and humiliated together
Rapes and lynchings –

the lash of the overseer
and the lust of the slave-owner
do your friends 'in the movement'
understand these things?

. . .

No, no, my sister,
                      my love,
first things first!
Too many gangsters
still stalk this continent
too many pirates

too many looters
far too many
still stalk this land –

. . .

When Africa
at home and across the seas
is truly free
there will be time for me
and time for you
to share the cooking
and change the nappies –
till then,
first things first!

. . . An important impetus behind the wave of African writing which started in the '60s was the desire to show both the outside world and African youth that the African past was orderly, dignified and complex and altogether a worthy heritage. This was obviously opting for fighting cultural imperialism, and in the course of that the women's issue was not only ignored – a fate which would have allowed it to surface when the time was ripe – it was conscripted in the service of dignifying the past and restoring African self-confidence. The African past was not made the object of a critical scrutiny the way the past tends to be in societies with a more harmonious development, it was made the object of a quest, and the picture of women's place and role in these societies had to support this quest and was consequently lent more dignity and described in more positive terms than reality warranted. Achebe's much praised objectivity with regard to the merits and flaws of traditional Ibo society becomes less than praiseworthy seen in this light: his traditional women are happy, harmonious members of the community, even when they are repeatedly beaten and barred from any say in the communal decision-making process and constantly reviled in sayings and proverbs. It would appear that in traditional wisdom behaving like a woman is to behave like an inferior being. My sense of humour has always stopped short at the pleasant little joke about Okonkwo being punished, not for beating his wife, but for beating her during the week of peace (Achebe 1958). The obvious inequality of the sexes seems to be the subject of mild amusement for Achebe.

If Achebe is obviously quite contented with the unequal state of affairs, Okot p'Bitek takes this tendency a step further and elevates his female protagonist, Lawino, into the very principle of traditional ways. . . . [But] in refusing to admire Lawino's romanticised version of her obviously sexist society one tears away the carpet from under the feet of the fighter against cultural imperialism. Lawino has become a holy cow, and slaughtering her and her various sisters is inevitably a betrayal, because they are inextricably bound up with the fight for African self-confidence in the face of Western cultural imperialism. . . .

It is no coincidence that this paper started as a discussion of images of women in literature written by men and ended by discussing a female writer and her portrayal of women's situation in present-day Africa. It is only just that women should have the last say in the discussion about their own situation, as, undoubtedly, we shall. This, however, is not meant to further the over-simplified view that a woman's view is always bound to be more valid than a man's in these discussions. The 'first things first' discussion as it appears in the writing of Ngũgĩ and Buchi Emecheta is a good example of the complexity of this situation. Ngũgĩ's ideological

starting point seems to me ideal. 'No cultural liberation without women's liberation.' This
. . . is a more difficult and therefore more courageous path to take in the African situation
than in the Western one, because it has to borrow some concepts – and a vocabulary – from
a culture from which at the same time it is trying to disassociate itself and at the same time
it has to modify its admiration for some aspects of a culture it is claiming validity for. . . .
[But] Buchi Emecheta . . . can recreate the situation and difficulties of women with authenticity and give a valuable insight into their thoughts and feelings. Her prime concern is not
so much with cultural liberation, nor with social change. To her the object seems to be to
give women access to power in the society as it exists, to beat men at their own game. She
lays claim to no ideology, not even a feminist one. She simply ignores the African dilemma,
whereas Ngũgĩ shoulders it and tries to come to terms with it. This could look like the
welcome beginning of 'schools' of writing, and to my mind nothing could be more fruitful
than a vigorous debate in literature about the role and future of women.

## Note

1    Felix Mnthali, 'Letter to a Feminist Friend'. The poem will appear in a volume entitled *Beyond the
Echoes*. [This book does not appear in the current bibliographies we have been able to check. We can
only assume the volume announced has not yet appeared. Eds.]

## Chapter 54

# KETU H. KATRAK

# DECOLONIZING CULTURE
## Toward a theory for postcolonial women's texts

From 'Decolonizing Culture: Toward a Theory for Post-colonial Women's Texts' *Modern Fiction Studies* 35(1), 1989.

THE CONCEPT OF SOCIAL RESPONSIBILITY, not only for postcolonial writers but also for critics/theorists, is central to my concern. Social responsibility must be the basis of any theorizing on postcolonial literature as well as the root of the creative work of the writers themselves. Whereas writers commonly respond seriously to the many urgent issues of their societies, critics/theorists of this literature often do not.

What theoretical models will be appropriate for this task? How can theory be an integral part of the struggle of these writers as presented in their novels, poems, dramas, essays, letters, and testimonies? How can we make our theory and interpretation of postcolonial texts challenge the hegemony of the Western canon? How can we, within a dominant Eurocentric discourse, make our study of postcolonial texts itself a mode of resistance? And, most significantly, what theoretical models will be most constructive for the development of this literature?

It is useful within a postcolonial context to think of theory, as Barbara Harlow suggests, as strategy, to consider certain integral and dialectical relationships between theory and practice. I wish to propose certain theoretical models for a study of women writers that will expand a narrow academic conceptualization of theory and that can be expressed in a language lucid enough to inspire people to struggle and to achieve social change.

## Decolonizing postcolonial theory

I would like first to examine several disconcerting trends in the recent production and consumption of postcolonial theory in general in order to decolonize this terrain and then to propose a historically situated method of approaching the work of women writers. One finds (1) little theoretical production of postcolonial writers given the serious attention it deserves, or that it is dismissed as not theoretical enough by Western standards; (2) the increasing phenomenon of using postcolonial texts as raw material for the theory producers and consumers of Western academia; (3) theoretical production as an end in itself, confined to the consumption of other theorists who speak the same privileged language in which obscurity is regularly mistaken for profundity. A new hegemony is being established in contemporary theory that can with impunity ignore or exclude postcolonial writers' essays, interviews, and other cultural productions while endlessly discussing concepts of the 'Other,' of 'difference,' and so on. Soyinka's words in his Preface to *Myth, Literature and the African World* still ring true:

We black Africans have been blandly invited to submit ourselves to a second epoch of colonialism – this time by a universal-humanoid abstraction defined and conducted by individuals whose theories and prescriptions are derived from the apprehension of *their* world and *their* history, *their* social neuroses and *their* value systems.

(1976: x)

Another more subtly insidious trend in recent postcolonial theory is the critic's attempt to engage with certain fashionable theoretical models in order (1) to validate postcolonial literature, even to prove its value through the use of complicated Eurocentric models, or (2) to succumb to the lure of engaging in a hegemonic discourse of Western theory given that it is 'difficult' or 'challenging,' often for the sole purpose of demonstrating its shortcomings for an interpretation of postcolonial texts. The intellectual traps in such theoretical gymnastics are many: for instance, a questioning of the canon and a simultaneous appropriating and tokenizing of postcolonial literary texts or an attempt to get away from narrowly anthropological readings of these texts and thereby interpreting them primarily as 'acts of language.'

The result is thus situations that inevitably assert an intellectual and political domination. Often, with the best intentions, Western intellectuals are unconsciously complicit in an endeavor that ironically ends up validating the dominant power structure, even when they ideologically oppose such hegemonic power. . . .

Postcolonial women writers participate actively in the ongoing process of decolonizing culture. Fanon's concept that 'decolonization is always a violent phenomenon' is useful for an analysis of how the English language is 'violated' from its standard usage and how literary forms are transformed from their definitions within the Western tradition. In terms of language, it is as if a version of the cultural and economic violence perpetrated by the colonizer is now appropriated by writers in order to 'violate' the English language in its standard use. Both arenas – linguistic and cultural – are dialectically related. Language *is* culture, particularly the transformations of rhetorical and discursive tools available through a colonial(ist) education system; and one expression of cultural tradition (among others like film, popular culture, festival) is through language. . . .

Women writers' uses of oral traditions and their revisions of Western literary forms are integrally and dialectically related to the kinds of content and the themes they treat. Women writers' stances, particularly with regard to glorifying/denigrating traditions, vary as dictated by their own class backgrounds, levels of education, political awareness and commitment, and their search for alternatives to existing levels of oppression often inscribed within the most revered traditions. Their texts deal with, and often challenge, their dual oppression–patriarchy that preceded and continues after colonialism and that inscribes the concepts of womanhood, motherhood, traditions such as dowry, bride-price, polygamy, and a worsened predicament within a capitalist economic system introduced by the colonizers. Women writers deal with the burdens of female roles in urban environments (instituted by colonialism), the rise of prostitution in cities, women's marginalization in actual political participation. . . .

## The Sistren Collective (Jamaica)

Sistren's work best illustrates a radical revising both of the English language in their use of 'patwah' and of literary forms such as drama and the short story that are based on folk-forms, ritual, and personal testimony. Sistren came together in May 1977 when a group of twelve working-class women employed as street cleaners under the Michael Manley government 'special make-work program called Impact' presented a drama titled *Downpression Get a Blow*.

They were assisted by Honor Ford Smith of the Jamaica School of Drama, who was the Artistic Director for the group from 1977 to 1988.

Sistren's creative use of folk-forms uses working-class women's daily language. As Ford Smith notes, 'Writing in dialect, with its improvised spelling and immediate flavor, the women learned to write a form of English that had previously been considered "bad, coarse, vulgar." . . . By writing a language that had hitherto been that of a non-literate people, the women broke silence' (Ford Smith 1986: 85–91). There is vast potential for cultural resistance within what Edward Kamau Brathwaite in his significant study *History of the Voice* calls 'a submerged language' (1984: 16). When expressed, this language can be most empoweringly subversive, particularly within Caribbean society where middleclass attitudes about 'proper speech' still prevail. . . .

In postcolonial Jamaica, a neocolonist legacy of denigrating 'patwah' continues. Sistren's use of 'patwah' demonstrates 'the refusal of a people to imitate a coloniser,' remarks Ford Smith, and 'their insistence on creation, their movement from obedience towards revolution. Not to nurture such a language is to retard the imagination and power of the people who created it' (88).[1] Ford Smith depicts the relationship of language to power relations and to class in Jamaican society; 'patwah' is commonly used for entertainment purposes, not for serious writing and reflection. Working-class people who speak 'patwah' often cannot write it, and the rift between the oral and the literate cultures gets deeper. Ford Smith skillfully brings these two dimensions together in her use of oral testimony, itself a part of oral tradition, as the base for Sistren's written stories entitled *Lionheart Gal: Lifestories of Jamaican Women*. One significant contribution of *Lionheart Gal* is the combination of oral and written forms: thirteen of the fifteen stories are based on oral testimony/interview that records working-class women's daily language. . . . The stories effectively demystify female roles, such as the nurturing mother and the romanticizing of peasant life, as well as sexuality and violence. 'Taken together, they are a composite woman's story. . . . All of the testimonies are underscored by a movement from girlhood to adulthood, country to city, isolated individual experiences to a more politicised collective awareness . . .' (Sistren with Honor Ford Smith (ed.) 1986: 1).

## Note

1   Ford Smith has created this new spelling of 'patwah' in order to distinguish it from more commonly used spelling 'patois.'

## Chapter 55

# CHANDRA TALPADE MOHANTY

# UNDER WESTERN EYES
## Feminist scholarship and colonial discourses

From 'Under Western Eyes: Feminist Scholarship and Colonial Discourse'
*Boundary 2* 12(3), 13(1), 1984.

**H**OWEVER SOPHISTICATED OR PROBLEMATICAL its use as an explanatory construct, colonization almost invariably implies a relation of structural domination, and a suppression – often violent – of the heterogeneity of the subject(s) in question. What I wish to analyze is specifically the production of the 'Third World Woman' as a singular monolithic subject in some recent (Western) feminist texts. . . .

Clearly Western feminist discourse and political practice is neither singular nor homogeneous in its goals, interests or analyses. However, it is possible to trace a coherence of *effects* resulting from the implicit assumption of 'the West' (in all its complexities and contradictions) as the primary referent in theory and praxis. My reference to 'Western feminism' is by no means intended to imply that it is a monolith. Rather, I am attempting to draw attention to the similar effects of various textual strategies used by particular writers that codify Others as non-Western and hence themselves as (implicitly) Western. It is in this sense that I use the term 'Western feminist.' The analytic principles discussed below serve to distort Western feminist political practices, and limit the possibility of coalitions among (usually White) Western feminists and working class and feminists of color around the world. These limitations are evident in the construction of the (implicitly consensual) priority of issues around which apparently *all* women are expected to organize. . . .

The relationship between 'Woman' – a cultural and ideological composite Other constructed through diverse representational discourses (scientific, literary, juridical, linguistic, cinematic, etc.) – and 'women' – real, material subjects of their collective histories – is one of the central questions the practice of feminist scholarship seeks to address. This connection between women as historical subjects and the re-presentation of Woman produced by hegemonic discourses is not a relation of direct identity, or a relation of correspondence or simple implication.[1] It is an arbitrary relation set up by particular cultures. I would like to suggest that the feminist writings I analyze here discursively colonize the material and historical heterogeneities of the lives of women in the third world, thereby producing/re-presenting a composite, singular 'Third World Woman' – an image which appears arbitrarily constructed, but nevertheless carries with it the authorizing signature of Western humanist discourse.[2] I argue that assumptions of privilege and ethnocentric universality on the one hand, and inadequate self-consciousness about the effect of Western scholarship on the 'third world' in the context of a world system dominated by the West on the other, characterize a sizeable extent of Western feminist work on women in the third world. An analysis of 'sexual difference' in the form of a cross-culturally singular, monolithic notion of patriarchy or male dominance leads to the construction of a similarly reductive and homogeneous notion of what I call the 'Third World Difference' — that stable, ahistorical something that apparently oppresses most if not all the women in these countries. And it is in the production of this 'Third World Difference' that Western feminisms appropriate and 'colonize' the fundamental complexities and conflicts which characterize the lives of women of different classes, religions, cultures, races and castes in these countries. It is in this process of homogenization and systematization of the oppression of women in the third world that power is exercised in much of recent Western feminist discourse, and this power needs to be defined and named. . . .

Western feminist scholarship cannot avoid the challenge of situating itself and examining its role in such a global economic and political framework. To do any less would be to ignore the complex interconnections between first and third world economies and the profound effect of this on the lives of women in these countries. I do not question the descriptive and inform-ative value of most Western feminist writings on women in the third world. I also do not question the existence of excellent work which does not fall into the analytic traps I am concerned with. In fact I deal with an example of such work later on. In the context of an overwhelming silence about the experiences of women in these countries, as well as the need to forge international links between women's political struggles, such work is both path-breaking and absolutely essential. However, it is both to the *explanatory potential* of particular analytic strategies employed by such writing, and to their *political effect* in the context of the hegemony of Western scholarship, that I want to draw attention here. While feminist writing in the US is still marginalized (except from the point of view of women of color addressing privileged White women), Western feminist writing on women in the third world must be considered in the context of the global hegemony of Western scholarship — i.e. the produc-tion, publication, distribution and consumption of information and ideas. Marginal or not, this writing has political effects and implications beyond the immediate feminist or disciplinary audience. One such significant effect of the dominant 'representations' of Western feminism is its conflation with imperialism in the eyes of particular third world women.[3] Hence the urgent need to examine the *political* implications of *analytic* strategies and principles. . . .

The first principle I focus on concerns the strategic location or situation of the category 'women' vis-à-vis the context of analysis. The assumption of women as an already consti-tuted, coherent group with identical interests and desires, regardless of class, ethnic or racial location or contradictions, implies a notion of gender or sexual difference or even patriarchy (as male dominance — men as a correspondingly coherent group) which can be applied univers-ally and cross-culturally. The context of analysis can be anything from kinship structures and the organization of labor to media representations. The second principle consists in the uncrit-ical use of particular methodologies in providing 'proof' of universality and cross-cultural validity. The third is a more specifically political principle underlying the methodologies and the analytic strategies, i.e. the model of power and struggle they imply and suggest. I argue that as a result of the two modes — or, rather, frames — of analysis described above, a homo-geneous notion of the oppression of women as a group is assumed, which, in turn, produces the image of an 'average third world woman.' This average third world woman leads an essen-tially truncated life based on her feminine gender (read: sexually constrained) and being 'third world' (read: ignorant, poor, uneducated, tradition-bound, domestic, family-oriented, victim-ized, etc.). This, I suggest, is in contrast to the (implicit) self-representation of Western women as educated, modern, as having control over their own bodies and sexualities, and the freedom to make their own decisions. The distinction between Western feminist re-presentation of women in the third world, and Western feminist *self*-presentation is a distinction of the same order as that made by some marxists between the 'maintenance' function of the housewife and the real 'productive' role of wage labor, or the characterization by developmentalists of the third world as being engaged in the lesser production of 'raw materials' in contrast to the 'real' productive activity of the First World. These distinctions are made on the basis of the privileging of a particular group as the norm or referent. Men involved in wage labor, first world producers, and, I suggest, Western feminists who sometimes cast Third World women in terms of 'ourselves undressed' (Rosaldo 1980: 392), all construct themselves as the referent in such a binary analytic.

## 'Women' as category of analysis, or: we are all sisters in struggle

By women as a category of analysis, I am referring to the critical assumption that all of us of the same gender, across classes and cultures, are somehow socially constituted as a homogeneous group identified prior to the process of analysis. This is an assumption which characterizes much feminist discourse. The homogeneity of women as a group is produced not on the basis of biological essentials, but rather on the basis of secondary sociological and anthropological universals. Thus, for instance, in any given piece of feminist analysis, women are character-ized as a singular group on the basis of a shared oppression. What binds women together is a sociological notion of the 'sameness' of their oppression. It is at this point that an elision takes place between 'women' as a discursively constructed group and 'women' as material subjects of their own history. Thus, the discursively consensual homogeneity of 'women' as a group is mistaken for the historically specific material reality of groups of women. This results in an assumption of women as an always-already constituted group, one which has been labelled 'powerless,' 'exploited,' 'sexually harrassed,' etc., by feminist scientific, economic, legal and sociological discourses. (Notice that this is quite similar to sexist discourse labelling women weak, emotional, having math anxiety, etc.) The focus is not on uncovering the material and ideological specificities that constitute a particular group of women as 'powerless' in a particular context. It is rather on finding a variety of cases of 'powerless' groups of women to prove the general point that women as a group are powerless. . . .

Male violence must be theorized and interpreted *within* specific societies, both in order to understand it better, as well as in order to effectively organize to change it. Sisterhood cannot be assumed on the basis of gender; it must be formed in concrete, historical and political practice and analysis. . . .

[Unless this is done] women are constituted as a group via dependency relationships vis-à-vis men, who are implicitly held responsible for these relationships. When 'women of Africa' as a group (versus 'men of Africa' as a group?) are seen as a group precisely because they are generally dependent and oppressed, the analysis of specific historical differences becomes impossible, because reality is always apparently structured by divisions – two mutually exclusive and jointly exhaustive groups, the victims and the oppressors. Here the sociological is substituted for the biological in order, however, to create the same – a unity of women. Thus, it is not the descriptive potential of gender difference, but the privileged positioning and explanatory potential of gender difference as the *origin* of oppression that I question. . . . Women are taken as a unified 'Powerless' group prior to the analysis in question. Thus, it is then merely a matter of specifying the context *after the fact*. . . . The problem with this analytic strategy is that it assumes men and women are already constituted as sexual-political subjects *prior* to their entry into the arena of social relations. Only if we subscribe to this assumption is it possible to undertake analysis which looks at the 'effects' of kinship structures, colonialism, organization of labor, etc., on women, who are already defined as a group apparently because of shared dependencies, but ultimately because of their gender. But women are *produced through these very relations* as well as being implicated in forming these relations. As Michelle Rosaldo states: '. . . woman's place in human social life is not in any direct sense a product of the things she does (or even less, a function of what, biologically, she is) but the meaning her activities acquire through concrete social interactions' (Rosaldo 1980: 400). That women mother in a variety of societies is not as significant as the *value* attached to mothering in these societies. The distinction between the act of mothering and the status attached to it is a very important one — one that needs to be made and analyzed contextually.

## Notes

1    I am indebted to Teresa de Lauretis for this particular formulation of the project of feminist theorizing. See especially her introduction in de Lauretis (1984); see also Sylvia Wynter, 'The Politics of Domination,' unpublished manuscript.

2    This argument is similar to Homi Bhabha's definition of colonial discourse as strategically creating a space for a subject peoples through the production of knowledges and the exercise of power. The full quote reads: '[colonial discourse is] an apparatus of power . . . an apparatus that turns on the recognition and disavowal of racial/cultural/historical differences. Its predominant strategic function is the creation of a space for a "subject peoples" through the production of knowledges in terms of which surveillance is exercised and a complex form of pleasure/unpleasure is incited. It (i.e. colonial discourse) seeks authorisation for its strategies by the production of knowledges by coloniser and colonised which are stereotypical but antithetically evaluated.' Homi Bhabha, 'The Other Question — the Stereotype and Colonial Discourse.' *Screen* 24 (November–December 1983), 23.

3    A number of documents and reports on the UN International Conferences on Women, Mexico City, 1975, and Copenhagen, 1980, as well as the 1976 Wellesley Conference on Women and Development attest to this. Nawal el Saadawi, Fatima Mernissi and Mallica Vajarathon in 'A Critical Look At The Wellesley Conference' (*Quest*, IV (Winter 1978), 101–7), characterize this conference as 'American-planned and organized,' situating third world participants as passive audiences. They focus especially on the lack of self-consciousness of Western women's implications in the effects of imperialism and racism in their assumption of an 'international sisterhood.' A recent essay, by Pratibha Parmar and Valerie Amos, is titled 'Challenging Imperial Feminism,' *Feminist Review* 17 (Autumn 1984), 3–19. Parmar and Amos characterize Euro-American feminism which seeks to establish itself as the only legitimate feminism as 'imperial.'

## Chapter 56

# TRINH T. MINH-HA

## WRITING POSTCOLONIALITY
## AND FEMINISM

From *Woman, Native, Other: Writing Postcoloniality and Feminism*
Bloomington, Ind.: Indiana University Press, 1989.

**W**ORDS EMPTY OUT WITH AGE. Die and rise again, accordingly invested with new meanings, and always equipped with a secondhand memory. In trying to tell something, a woman is told, shredding herself into opaque words while her voice dissolves on the walls of silence. Writing: a commitment of language. The web of her gestures, like all modes of writing, denotes a historical solidarity (on the understanding that her story remains inseparable from history). She has been warned of the risk she incurs by letting words run off the rails, time and again tempted by the desire to gear herself to the accepted norms. But where has obedience led her? At best, to the satisfaction of a 'made-woman,' capable of achieving as high a mastery of discourse as that of the male establishment in power. Immediately gratified, she will, as years go by, sink into oblivion, a fate she inescapably shares with her foresisters. How many, already, have been condemned to premature deaths for having borrowed the master's tools and thereby played into his hands? Solitude is a common prerequisite, even though this may only mean solitude in the immediate surroundings. Elsewhere, in every corner of the world, there exist women who, despite the threat of rejection, resolutely work toward the unlearning of institutionalized language, while staying alert to every deflection of their body compass needles. *Survival*, as Audre Lorde comments, '*is not an academic skill*. . . . It is learning how to take our differences and make them strengths. *For the master's tools will never dismantle the master's house*. They may allow us temporarily to beat him at his own game, but they will never enable us to bring about genuine change' (Lorde 1981: 99). The more one depends on the master's house for support, the less one hears what he doesn't want to hear. Difference is not difference to some ears, but awkwardness or incompleteness. Aphasia. Unable or unwilling? Many have come to tolerate this dissimilarity and have decided to suspend their judgments (only) whenever the other is concerned. Such an attitude is a step forward; at least the danger of speaking for the other has emerged into consciousness. But it is a very small step indeed, since it serves as an excuse for their complacent ignorance and their reluctance to involve themselves in the issue. You who understand the dehumanization of forced removal–relocation–reeducation–redefinition, the humiliation of having to falsify your own reality, your voice – you know. And often cannot *say* it. You try and keep on trying to unsay it, for if you don't, they will not fail to fill in the blanks on your behalf, and you will be said.

## The policy of 'separate development'

With a kind of perverted logic, they work toward your erasure while urging you to keep your way of life and ethnic values *within the borders of your homelands*. This is called the policy of 'separate development' in apartheid language. Tactics have changed since the colonial times and indigenous cultures are no longer (overtly) destroyed (preserve the form but remove the content, or vice versa). You may keep your traditional law and tribal customs among yourselves, as long as you and your own kind are careful not to step beyond the assigned limits. Nothing has been left to chance when one considers the efforts made by the White South African authorities to distort and use the tools of Western liberalism for the defense of their racialistically indefensible cause. Since no integration is possible when terror has become the order of the day, I (not you) will give you freedom. I will grant you autonomy – not complete autonomy, however, for 'it is a liberal fallacy to suppose that those to whom freedom is given will use it only as foreseen by those who gave it' (Manning 1968: 287). . . . The delimitation of territories is my answer to what I perceive as some liberals' dream for 'the inauguration, namely, of a system in which South Africa's many peoples would resolve themselves unreluctantly into one' (289). The governed do not (should not) compose a single people; this is why I am eager to show that South Africa is not one but ten separate nations (of which the White nation is the only one to be skin-defined; the other nine being determined largely on the basis of language – the Zulu nation, the Swazi nation, and so on). This philosophy – I will not call it 'policy' – of 'differentiation' will allow me to have better control over my nation while looking after yours, helping you thereby to gradually stand on your own. It will enable you to return to 'where you belong' whenever you are not satisfied with my law and customs or whenever you are no longer useful to me. Too bad if you consider what has been given to you as the leftovers of my meals. Call it 'reserves of cheap labor' or 'bantustans' if you wish; 'separate development' means that each one of us minds her/his own business (I will interfere when my rights are concerned since I represent the State) and that your economical poverty is of your own making. As for 'the Asiatic cancer, which has already eaten so deeply into the vitals of South Africa, [it] ought to be resolutely eradicated' (Jan Christiaen Smuts, quoted in Fischer 1954: 25). Non-white foreigners have no part whatsoever in my plans and I 'will undertake to drive the coolies [Indians] out of the country within four years' (General Louis Botha, quoted in Fischer 1954: 25). My 'passionate concern for the future of a European-type white society, and . . . that society's right to self-preservation' is not a question of color feeling, but of nationalism, the 'Afrikaner nationalism [which] is a form of collective selfishness; but to say this is simply to say that it is an authentic case of nationalism' (Manning 1968: 287).

Words manipulated at will. As you can see, 'difference' is essentially 'division' in the understanding of many. It is no more than a tool of self-defense and conquest. You and I might as well not walk into this semantic trap which sets us up against each other as expected by a certain ideology of separatism. Have you read the grievances some of our sisters express on being among the few women chosen for a 'Special Third World Women's Issue' or on being the only Third World woman at readings, workshops, and meetings? . . .

Why not go and find out for yourself when you don't know? Why let yourself be trapped in the mold of permanent schooling and wait for the delivery of knowledge as a consumer waits for her/his suppliers' goods? The understanding of difference is a shared responsibility, which requires a minimum of willingness to reach out to the unknown. As Audre Lorde says,

> Women of today are still being called upon to stretch across the gap of male ignorance, and to educate men as to our existence and our needs. This is an old and primary

tool of all oppressors to keep the oppressed occupied with the master's concerns. Now we hear that it is the task of black and third world women to educate white women, in the face of tremendous resistance, as to our existence, our differences, our relative roles in our joint survival. This is a diversion of energies and a tragic repetition of racist patriarchal thought.

(Lorde 1981: 100)

One has to be excessively preoccupied with the master's concerns, indeed, to try to explain why women cannot have written 'the plays of Shakespeare in the age of Shakespeare,' as Virginia Woolf did. Such a waste of energy is perhaps unavoidable at certain stages of the struggle; it need not, however, become an end point in itself. . . .

Specialness as a soporific soothes, anaesthetizes my sense of justice; it is, to the wo/man of ambition, as effective a drug of psychological self-intoxication as alcohol is to the exiles of society. Now, i am not only given the permission to open up and talk, i am also encouraged to express my difference. My audience expects and demands it; otherwise people would feel as if they have been cheated: We did not come to hear a Third World member speak about the First (?) World, We came to listen to that voice of difference likely to bring us *what we can't have* and to divert us from the monotony of sameness. They, like their anthropologists whose specialty is to detect all the layers of my falseness and truthfulness, are in a position to decide what/who is 'authentic' and what/who is not. No uprooted person is invited to participate in this 'special' wo/man's issue unless s/he 'makes up' her/his mind and paints her/himself thick with authenticity. Eager not to disappoint, i try my best to offer my benefactors and benefactresses what they most anxiously yearn for: the possibility of a difference, yet a difference or an otherness that will not go so far as to question the foundation of their beings and makings. Their situation is not unlike that of the American tourists who, looking for a change of scenery and pace in a foreign land, such as, for example, Japan, strike out in search of what they believe to be the 'real' Japan – most likely shaped after the vision of Japan as handed to them and reflected in television films like 'Shogun' – or that of the anthropologists, whose conception of 'pure' anthropology induces them to concentrate on the study of 'primitive' ('native,' 'indigenous,' or to use more neutral, technical terms: 'non-state,' 'non-class') societies. Authenticity in such contexts turns out to be a product that one can buy, arrange to one's liking, and/or preserve. Today, the 'unspoiled' parts of Japan, the far-flung locations in the archipelago, are those that tourism officials actively promote for the more venturesome visitors. Similarly, the Third World representative the modern sophisticated public ideally seeks is the unspoiled African, Asian, or Native American, who remains more preoccupied with her/his image of the *real* native – the *truly different* – than with the issues of hegemony, racism, feminism, and social change (which s/he lightly touches on in conformance to the reigning fashion of liberal discourse). A Japanese actually looks more Japanese in America than in Japan, but the 'real' type of Japanism ought to be in Japan. The less accessible the product 'made-in-Japan,' the more trustworthy it is, and the greater the desire to acquire and protect it. . . .

## The question of roots and authenticity

'I was made to feel,' writes Joanne Harumi Sechi, 'that cultural pride would justify and make good my difference in skin color while it was a constant reminder that I was different' (Sechi 1980: 444). Every notion in vogue, including the retrieval of 'roots' values, is necessarily exploited and recuperated. The invention of needs always goes hand in hand with the compulsion to help the needy, a noble and self-gratifying task that also renders the helper's service indispensable. The part of the savior has to be filled as long as the belief in the problem of 'endangered species' lasts. To persuade you that your past and cultural heritage are doomed to eventual extinction and thereby keeping you occupied with the Savior's concern, inauthenticity is condemned as a *loss of origins* and a whitening (or faking) of non-Western values. Being easily offended in your elusive identity and reviving readily an old, racial charge, you immediately react when such guilt-instilling accusations are leveled at you and are thus led to stand in need of defending that very ethnic part of yourself that for years has made you and your ancestors the objects of execration. Today, planned authenticity is rife; as a product of hegemony and a remarkable counterpart of universal standardization, it constitutes an efficacious means of silencing the cry of racial oppression. We no longer wish to erase your difference. We demand, on the contrary, that you remember and assert it. At least, to a certain extent. Every path I/i take is edged with thorns. On the one hand, i play into the Savior's hands by concentrating on authenticity, for my attention is numbed by it and diverted from other, important issues; on the other hand, i do feel the necessity to return to my so-called roots, since they are the fount of my strength, the guiding arrow to which i constantly refer before heading for a new direction. The difficulties appear perhaps less insurmountable only as I/i succeed in making a distinction between difference reduced to identity-authenticity and difference understood also as critical difference from myself. The first induces an attitude of temporary tolerance – as exemplified in the policy of 'separate development' – which serves to reassure the conscience of the liberal establishment and gives a touch of subversiveness to the discourse delivered. . . .

The pitting of anti-racist and anti-sexist struggles against one another allows some vocal fighters to dismiss blatantly the existence of either racism or sexism within their lines of action, as if oppression only comes in separate, monolithic forms. Thus, to understand how pervasively dominance operates via the concept of hegemony or of absent totality in plurality is to understand that the work of decolonization will have to continue within the women's movements. . . .

## Chapter 57

# SARA SULERI

## WOMAN SKIN DEEP
### Feminism and the postcolonial condition

From 'Woman Skin Deep: Feminism and the Postcolonial Condition' *Critical Inquiry* 18(4) (Summer), 1992.

I WOULD CLAIM THAT WHILE CURRENT FEMINIST discourse remains vexed by questions of identity formation and the concomitant debates between essentialism and constructivism, or distinctions between situated and universal knowledge, it is still prepared to grant an uneasy selfhood to a voice that is best described as the property of 'postcolonial Woman.' Whether this voice represents perspectives as divergent as the African-American or the postcolonial cultural location, its imbrications of race and gender are accorded an iconicity that is altogether too good to be true. Even though the marriage of two margins should not necessarily lead to the construction of that contradiction in terms, a 'feminist center,' the embarrassed privilege granted to racially encoded feminism does indeed suggest a rectitude that could be its own theoretical undoing. The concept of the postcolonial itself is too frequently robbed of historical specificity in order to function as a preapproved allegory for any mode of discursive contestation. The coupling of *postcolonial* with *woman*, however, almost inevitably leads to the simplicities that underlie unthinking celebrations of oppression, elevating the racially female voice into a metaphor for 'the good.' Such metaphoricity cannot exactly be called essentialist, but it certainly functions as an impediment to a reading that attempts to look beyond obvious questions of good and evil. In seeking to dismantle the iconic status of postcolonial feminism, I will attempt here to address the following questions: within the tautological margins of such a discourse, which comes first, gender or race? How, furthermore, can the issue of chronology lead to some preliminary articulation of the productive superficiality of race?

Before such questions can be raised, however, it is necessary to pay some critical attention to the mobility that has accrued in the category of postcolonialism. Where the term once referred exclusively to the discursive practices produced by the historical fact of prior colonization in certain geographically specific segments of the world, it is now more of an abstraction available for figurative deployment in any strategic redefinition of marginality. For example, when James Clifford elaborated his position on traveling theory during a recent seminar, he invariably substituted the metaphoric condition of postcoloniality for the obsolete binarism between anthropologist and native.[1] As with the decentering of any discourse, however, this reimaging of the postcolonial closes as many epistemological possibilities as it opens. On the one hand, it allows for a vocabulary of cultural migrancy, which helpfully derails the postcolonial condition from the strictures of national histories, and thus makes way for the theoretical articulations best typified by Homi Bhabha's recent anthology, *Nation and Narration* (1990). On the other hand, the current metaphorization of postcolonialism threatens to become so amorphous as to repudiate any locality for cultural thickness. A symptom of this termino- logical and theoretical dilemma is astutely read in Kwame Anthony Appiah's essay, 'Is the Post- in Postmodernism the Post- in Postcolonial?' (1991). Appiah argues for a discursive space-clearing that allows postcolonial discourse a figurative flexibility and at the same time reaffirms its radical locality within historical exigencies. His discreet but firm segregation of the postcolonial from the postmodern is indeed pertinent to the dangerous democracy accorded the coalition between postcolonial and feminist theories, in which each term serves to reify the potential pietism of the other.

In the context of contemporary feminist discourse, I would argue, the category of postcolonialism must be read both as a free-floating metaphor for cultural embattlement and as an almost obsolete signifier for the historicity of race. There is no available dichotomy that could neatly classify the ways in which such a redefinition of postcoloniality is necessarily a secret sharer in similar reconfigurations of feminism's most vocal articulation of marginality, or the obsessive attention it has recently paid to the racial body. Is the body in race subject or object, or is it more dangerously an objectification of a methodology that aims for radical subjectivity? Here, the binarism that informs Chandra Mohanty's paradigmatic essay, 'Under Western Eyes: Feminist Scholarship and Colonial Discourses,' deserves particular consider- ation. Where Mohanty engages in a particular critique of 'Third World Woman' as a monolithic object in the texts of Western feminism, her argument is premised on the irreconcilability of gender as history and gender as culture. 'What happens,' queries Mohanty, 'when [an] assumption of "women as an oppressed group" is situated in the context of Western feminist writing about third world women?' What happens, apparently, begs her question. In contesting what she claims is a 'colonialist move,' Mohanty proceeds to argue that 'Western feminists alone become the true "subjects" of this counter-history. Third World women, on the other hand, never rise above the debilitating generality of their "object" status' (Mohanty 1991: 71). A very literal ethic underlies such a dichotomy, one that demands attention to its very obvi- ousness: how is this objectivism to be avoided? How will the ethnic voice of womanhood counteract the cultural articulation that Mohanty too easily dubs as the exegesis of Western feminism? The claim to authenticity – only a black can speak for a black; only a postcolonial subcontinental feminist can adequately represent the lived experience of that culture – points to the great difficulty posited by the 'authenticity' of female racial voices in the great game that claims to be the first narrative of what the ethnically constructed woman is deemed to want.

This desire all too often takes its theoretical form in a will to subjectivity that claims a theoretical basis most clearly contravened by the process of its analysis. An example of this point is Trinh Minh-ha's treatise, *Woman, Native, Other* (1989), which seeks to posit an alter- native to the anthropological twist that constitutes the archaism through which nativism has been apprehended. Subtitled *Writing Postcoloniality and Feminism*, Trinh's book is a paradigmatic meditation that can be essentialized into a simple but crucial question: how can feminist discourse represent the categories of 'woman' and 'race' at the same time? If the languages of feminism and ethnicity are to escape an abrasive mutual contestation, what novel idiom can freshly articulate their radical inseparability? Trinh's strategy is to relocate her gendering of ethnic realities on the inevitable territory of postfeminism, which underscores her desire to represent discourse formation as always taking place after the fact of discourse. It further confirms my belief that had I any veto power over prefixes, *post-* would be the first to go – but that is doubtless tangential to the issue at hand. In the context of Trinh's methodology, the shape of the book itself illuminates what may best be called the endemic ill that effects a certain temporal derangement between the work's originary questions and the narratives that they engender. *Woman, Native, Other* consists of four loosely related chapters, each of which opens with an abstraction and ends with an anecdote. While there is a self-pronounced difference between the preliminary thesis outlined in the chapter 'Commitment from the Mirror-Writing Box' to the concluding claims in 'Grandma's Story,' such a discursive distance is not matched with any logical or theoretical consistency. Instead, a work that is impelled by an impassioned need to question the lines of demarcation between race and gender concludes by falling into a predictable biological fallacy in which sexuality is reduced to the literal structure of the racial body, and theoretical interventions within this trajectory become minimalized into the naked category of lived experience.

When feminism turns to lived experience as an alternative mode of radical subjectivity, it only rehearses the objectification of its proper subject. While lived experience can hardly be discounted as a critical resource for an apprehension of the gendering of race, neither should such data serve as the evacuating principle for both historical and theoretical contexts alike. 'Radical subjectivity' too frequently translates into a low-grade romanticism that cannot recognize its discursive status as *pre-* rather than *post-*. In the concluding chapter of Trinh's text, for example, a section titled 'Truth and Fact: Story and History' delineates the skewed idiom that marginal subjectivities produce. In attempting to proclaim an alternative to male-identified objectivism, Trinh-as-anthropologist can only produce an equally objectifying idiom of joy:

> Let me tell you a story. For all I have is a story. Story passed on from generation to generation, named Joy. Told for the joy it gives the storyteller and the listener. Joy inherent in the process of storytelling. Whoever understands it also understands that a story, as distressing as it can be in its joy, never takes anything away from anybody.
>
> (Trinh 1989: 119)

Given that I find myself in a more acerbic relation both to the question of the constitution of specific postcolonialisms and of a more metaphoric postcolonial feminism, such a jointly universalist and individualist 'joy' is not a term that I would ordinarily welcome into my discursive lexicon. On one level, its manipulation of lived experience into a somewhat fallacious allegory for the reconstitution of gendered race bespeaks a transcendence – and an attendant evasion – of the crucial cultural issues at hand. On a more dangerous level, however, such an assumption serves as a mirror image of the analyses produced by the critics of political rectitude. For both parties, 'life' remains the ultimate answer to 'discourse.' The subject of race, in other words, cannot cohabit with the detail of a feminist language.

Trinh's transcendent idiom, of course, emanates from her somewhat free-floating understanding of 'postcoloniality': is it an abstraction into which all historical specificity may be subsumed, or is it a figure for a vaguely defined ontological marginality that is equally applicable to all 'minority' discourses? In either case, both the categories of 'woman' and 'race' assume the status of metaphors, so that each rhetoric of oppression can serve equally as a mirrored allegory for the other. Here, *Woman, Native, Other* is paradigmatic of the methodological blurring that dictates much of the discourse on identity formation in the coloring of feminist discourse. To privilege the racial body in the absence of historical context is indeed to generate an idiom that tends to waver with impressionistic haste between the abstractions of postcoloniality and the anecdotal literalism of what it means to articulate an 'identity' for a woman writer of color. Despite its proclaimed location within contemporary theoretical – not to mention post-theoretical – discourse, such an idiom poignantly illustrates the hidden and unnecessary desire to resuscitate the 'self.'

What is most striking about such discursive practices is their failure to confront what may be characterized best as a great enamourment with the 'real.' Theories of postcolonial feminism eminently lend themselves to a reopening of the continued dialog that literary and cultural studies have – and will continue to have – with the perplexing category known as realism, but at present the former discourse chooses to remain too precariously parochial to recognize the bounty that is surely its to give. Realism, however, is too dangerous a term for an idiom that seeks to raise identity to the power of theory. While both may be windmills to the quixotic urge to supply black feminism with some version of the 'real,' Trinh's musings on this subject add a mordantly pragmatic option to my initial question: 'what comes first, race or gender?' Perhaps the query would be more finely calibrated if it were rephrased to ask, 'What comes first, race, gender, or profession?' And what, in our sorry dealings with such realisms, is the most phantasmagoric category of all? . . .

If race is to complicate the project of divergent feminisms, in other words, it cannot take recourse to biologism, nor to the incipient menace of rewriting alterity into the ambiguous shape of the exotic body.

The body that serves as testimony for lived experience, however, has received sufficient interrogation from more considered perspectives on the cultural problems generated by the dialogue between gender and race, along with the hyperrealist idiom it may generate. Hazel Carby helpfully advocates that

> black feminist criticism [should] be regarded critically as a problem, not a solution, as a sign that should be interrogated, a locus of contradictions. Black feminist criticism has its source and its primary motivation in academic legitimation, placement within a framework of bourgeois humanistic discourse.
>
> (Carby 1987: 15)

The concomitant question that such a problem raises is whether the signification of gendered race necessarily returns to the realism that it most seeks to disavow. If realism is the Eurocentric and patriarchal pattern of adjudicating between disparate cultural and ethnic realities, then it is surely the task of radical feminism to provide an alternative perspective. In the vociferous discourse that such a task has produced, however, the question of alternativism is all too greatly subsumed either into the radical strategies that are designed to dictate the course of situated experience, or into the methodological imperatives that impel a work related to *Woman, Native, Other* such as bell hooks's *Talking Back: Thinking Feminist, Thinking Black*.

While the concept of 'talking back' may appear to be both invigorating and empowering to a discourse interested in the reading of gendered race, the text *Talking Back* is curiously engaged in talking to itself; in rejecting Caliban's mode of protest, its critique of colonization is quietly narcissistic in its projection of what a black and thinking female body may appear to be, particularly in the context of its repudiation of the genre of realism. Yet this is the genre, after all, in which African-American feminism continues to seek legitimation: hooks's study is predicated on the anecdotes of lived experience and their capacity to provide an alternative to the discourse of what she terms patriarchal rationalism. Here the unmediated quality of a local voice serves as a substitute for any theoretical agenda that can make more than a cursory connection between the condition of postcolonialism and the question of gendered race. Where hooks claims to speak beyond binarism, her discourse keeps returning to the banality of easy dichotomies: 'Dare I speak to oppressed and oppressor in the same voice? Dare I speak to you in a language that will take us away from the boundaries of domination, a language that will not fence you in, bind you, or hold you? Language is also a place of struggle' (hooks 1989: 28). The acute embarrassment generated by such an idiom could possibly be regarded as a radical rhetorical strategy designed to induce racial discomfort in its audience, but it more frequently registers as black feminism's failure to move beyond the proprietary rights that can be claimed by any oppressed discourse.

As does Trinh's text, hooks's claims that personal narrative is the only salve to the rude abrasions that Western feminist theory has inflicted on the body of ethnicity. The tales of lived experience, however, cannot function as a sufficient alternative, particularly when they are predicated on dangerously literal professions of postcolonialism. *Yearning: Race, Gender, and Cultural Politics*, hooks's more recent work, rehearses a postcolonial fallacy in order to conduct some highly misguided readings of competing feminisms within the context of racial experience. . . .

I proffer life in Pakistan as an example of such a postcolonial and lived experience. Pakistani laws, in fact, pertain more to the discourse of a petrifying realism than do any of the feminist critics whom I have cited thus far. The example at hand takes a convoluted postcolonial point and renders it nationally simple: if a postcolonial nation chooses to embark on an official program of Islamization, the inevitable result in a Muslim state will be legislation that curtails women's rights and institutes in writing what has thus far functioned as the law of the passing word. . . .

It is important to keep in mind that the formulation of the Hudood Ordinances was based on a multicultural premise, even though they were multicultural from the dark side of the moon. These laws were premised on a Muslim notion of *Hadd* and were designed to interfere in a postcolonial criminal legal system that was founded on Anglo-Saxon jurisprudence. According to feminist lawyer Asma Jahangir,

> the Hudood Ordinances were promulgated to bring the criminal legal system of Pakistan in conformity with the injunctions of Islam. . . . Two levels of punishments are introduced in the Ordinances. Two levels of punishment and, correspondingly, two separate sets of rules of evidence are prescribed. The first level or category is the one called the 'Hadd' which literally means the 'limit' and the other 'Tazir', which means 'to punish'.
>
> (Jahangir and Jilani 1990: 24)

The significance of the *Hadd* category is that it delineates immutable sentences: *Tazir* serves only as a safety net in case the accused is not convicted under *Hadd*. These fixed rules are in themselves not very pretty: *Hadd* for theft is amputation of a hand; for armed robbery, amputation of a foot; for rape or adultery committed by married Muslims, death by stoning; for rape or adultery committed by non-Muslims or unmarried Muslims, a hundred public lashes (Jahangir and Jilani 1990: 24). While I am happy to report that the *Hadd* has not yet been executed, the laws remain intact and await their application.

The applicability of these sentences is rendered more murderous and even obscenely ludicrous when the immutability of the *Hadd* punishments is juxtaposed with the contingency of the laws of evidence. If a man is seen stealing a thousand rupees by two adult Muslim males, he could be punished by *Hadd* and his hand would be amputated. If an adult Muslim stole several million rupees and the only available witnesses were women and non-Muslims, he would not qualify for a *Hadd* category and would be tried under the more free-floating *Tazir* instead. 'A gang of men can thus rape all the residents of a women's hostel,' claims Jahangir with understandable outrage, 'but [the] lack of ocular evidence of four Muslim males will rule out the imposition of a Hadd punishment' (Jahangir and Jilani 1990: 49). Such a statement, unfortunately, is not the terrain of rhetoric alone, since the post-Hudood Ordinance application of the *Tazir* has made the definition of rape an extremely messy business indeed.

Here, then, we turn to *Zina*, and its implications for the Pakistani female body. The Hudood Ordinances have allowed for all too many openings in the boundaries that define rape. Women can now be accused of rape, as can children; laws of mutual consent may easily convert a case of child abuse into a prosecution of the child for *Zina*, for fornication. Further- more, unmarried men and women can be convicted of having committed rape against each other, since a subsection of the *Zina* offense defines rape as 'one where a man or a woman have illicit sex knowing that they are not validly married to each other' (quoted in Jahangir and Jilani 1990: 58). In other words, fornication is all, and the statistics of the past few years grimly indicate that the real victims of the Hudood Ordinances are women and children, most specifically those who have no access to legal counsel and whose economic status renders them ignorant of their human rights.

Jahangir cites the example of a fifteen-year-old woman, Jehan Mina, who, after her father's death, was raped by her aunt's husband and son. Once her pregnancy was discovered, another relative filed a police report alleging rape. During the trial, however, the accused led no defense, and Mina's testimony alone was sufficient to get her convicted for fornication and sentenced to one hundred public lashes. That child's story is paradigmatic of the untold miseries of those who suffer sentences in Muslim jails.

Let me state the obvious: I cite these alternative realisms and constructions of identity in order to reiterate the problem endemic to postcolonial feminist criticism. It is not the terrors of Islam that have unleashed the Hudood Ordinances on Pakistan, but more probably the United States government's economic and ideological support of a military regime during that bloody but eminently forgotten decade marked by the 'liberation' of Afghanistan. Jehan Mina's story is therefore not so far removed from our current assessment of what it means to be multicultural. How are we to connect her lived experience with the overwhelming realism of the law? In what ways does her testimony force postcolonial and feminist discourse into an acknowledgement of the inherent parochialism and professionalism of our claims?

## Note

1    James Clifford's course, 'Travel and Identity in Twentieth-century Interculture,' was given as the Henry Luce Seminar at Yale University, Fall 1990.

## Chapter 58

# OYÈRÓNKÉ OYEWÙMÍ

## COLONIZING BODIES AND MINDS

From *The Invention of Women: Making an African Sense of Western Gender Discourses* Minneapolis: University of Minnesota Press, 1997.

T**HE HISTORIES OF BOTH THE COLONIZED** and the colonizer have been written from the male point of view – women are peripheral if they appear at all. While studies of colonization written from this angle are not necessarily irrelevant to understanding what happened to native females, we must recognize that colonization impacted males and females in similar and dissimilar ways. Colonial custom and practice stemmed from 'a world view which believes in the absolute superiority of the human over the nonhuman and the subhuman, the *masculine* over the *feminine* . . ., and the modern or progressive over the traditional or the savage' (Nandy 1983: x).

Therefore, the colonizer differentiated between male and female bodies and acted accordingly. Men were the primary target of policy, and, as such, they were the natives and so were visible. These facts, from the standpoint of this study, are the justification for considering the colonial impact in gender terms rather than attempting to see which group, male or female, was the most exploited. The colonial process was sex-differentiated insofar as the colonizers were male and used gender identity to determine policy. From the foregoing, it is clear that any discussion of hierarchy in the colonial situation, in addition to employing race as the basis of distinctions, should take into account its strong gender component. The two racially distinct and hierarchical categories of the colonizer and the native should be expanded to four, incorporating the gender factor. However, race and gender categories obviously emanate from the preoccupation in Western culture with the visual and hence physical aspects of human reality. Both categories are a consequence of the bio-logic of Western culture. Thus, in the colonial situation, there was a hierarchy of four, not two, categories. Beginning at the top, these were: men (European), women (European), native (African men), and Other (African women). Native women occupied the residual and unspecified category of the Other.

In more recent times, feminist scholars have sought to rectify the male bias in the discourses on colonization by focusing on women. One major thesis that emerged from this effort is that African women suffered a 'double colonization': one form from European domination and the other from indigenous tradition imposed by African men. Stephanie Urdang's book *Fighting Two Colonialisms* is characteristic of this perspective (1983). While the depth of the colonial experience for African women is expressed succinctly by the idea of doubling, there is no consensus about what is being doubled. From my perspective, it is not colonization that is two, but the forms of oppression that flowed from the process for native females. Hence, it is misleading to postulate two forms of colonization because both manifestations of oppression are rooted in the hierarchical race/gender relations of the colonial situation.

African females were colonized by Europeans as Africans and as African women. They were dominated, exploited, and inferiorized as Africans together with African men and then separately inferiorized and marginalized as African women.

It is important to emphasize the combination of race and gender factors because European women did not occupy the same position in the colonial order as African women. A circular issued by the British colonial government in Nigeria shows the glaringly unequal position of these two groups of women in the colonial system. It states that 'African women should be paid at 75% of the rates paid to the European women' (cited in Mba 1982: 65). Furthermore, whatever the 'status' of indigenous customs, the relations between African men and women during this period can be neither isolated from the colonial situation nor described as a form of colonization, particularly because African men were subjects themselves. The racial and gender oppressions experienced by African women should not be seen in terms of addition, as if they were piled one on top of the other. In the context of the United States, Elizabeth Spelman's comment on the relationship between racism and sexism is relevant. She writes: 'How one form of oppression is experienced is influenced by and influences how another form is experienced' (Spelman 1988: 123). Though it is necessary to discuss the impact of colonization on specific categories of people, ultimately its effect on women cannot be separated from its impact on men because gender relations are not zero-sum – men and women in any society are inextricably bound.

## The state of patriarchy

The imposition of the European state system, with its attendant legal and bureaucratic machinery, is the most enduring legacy of European colonial rule in Africa. The international nation-state system as we know it today is a tribute to the expansion of European traditions of governance and economic organization. One tradition that was exported to Africa during this period was the exclusion of women from the newly created colonial public sphere. In Britain, access to power was gender-based; therefore, politics was largely men's job; and colonization, which is fundamentally a political affair, was no exception. Although both African men and women as conquered peoples were excluded from the higher echelons of colonial state structures, men were represented at the lower levels of government. The system of indirect rule introduced by the British colonial government recognized the male chief's authority at the local level but did not acknowledge the existence of female chiefs. Therefore, women were effectively excluded from all colonial state structures. The process by which women were bypassed by the colonial state in the arena of politics – an arena in which they had participated during the precolonial period – is of particular interest in the following section.

The very process by which females were categorized and reduced to 'women' made them ineligible for leadership roles. The basis for this exclusion was their biology, a process that was a new development in Yorùbá society. The emergence of women as an identifiable category, defined by their anatomy and subordinated to men in all situations, resulted, in part, from the imposition of a patriarchal colonial state. For females, colonization was a twofold process of racial inferiorization and gender subordination. [In] pre-British Yorùbá society, anafemales, like the anamales, had multiple identities that were not based on their anatomy. The creation of 'women' as a category was one of the very first accomplishments of the colonial state.

In a book on European women in colonial Nigeria, Helen Callaway explores the relationship between gender and colonization at the level of the colonizer. She argues that the colonial state was patriarchal in many ways. Most obviously, colonial personnel were male. Although a few European women were present in a professional capacity as nurses, the administrative branches, which embodied power and authority, excluded women by law (Callaway 1987: 4). Furthermore, she tells us that the Colonial Service, which was formed for the purpose of governing subject peoples, was

> a male institution in all its aspects: its 'masculine' ideology, its military organisation and processes, its rituals of power and hierarchy, its strong boundaries between the sexes. It would have been 'unthinkable' in the belief system of the time even to consider the part women might play, other than as nursing sisters, who had earlier become recognised for their important 'feminine' work.
>
> (ibid.: 5–6)

It is not surprising, therefore, that it was unthinkable for the colonial government to recognize female leaders among the peoples they colonized, such as the Yorùbá.

Likewise, colonization was presented as a 'man-sized' job – the ultimate test of manhood – especially because the European death-rate in West Africa at this time was particularly high. Only the brave-hearted could survive the 'white man's grave,' as West Africa was known at the time. According to Callaway, Nigeria was described again and again as a man's country in which women (European women) were 'out of place' in a double sense of physical displacement and the symbolic sense of being in an exclusively male territory. Mrs Tremlett, a European woman who accompanied her husband to Nigeria during this period, lamented about the position of European women: 'I often found myself reflecting rather bitterly on the insignificant position of a woman in what is practically a man's country. . . . If there is one spot on earth where a woman feels of no importance whatever, it is in Nigeria at the present day' (quoted in Callaway 1987: 5). If the women of the colonizer were so insignificant, then one could only imagine the position of the 'other' women, if their existence was acknowledged at all.

Yet on the eve of colonization there were female chiefs and officials all over Yorùbá land. Ironically, one of the signatories to the treaty that was said to have ceded Ibàdàn to the British was Lànlátu, an *ìyálóde,* an anafemale chief (Johnson 1921: 656). The transformation of state power to male-gender power was accomplished at one level by the exclusion of women from state structures. This was in sharp contrast to Yorùbá state organization, in which power was not gender-determined.

The alienation of women from state structures was particularly devastating because the nature of the state itself was undergoing transformation. Unlike the Yorùbá state, the colonial state was autocratic. The African males designated as chiefs by the colonizers had much more power over the people than was vested in them traditionally. In British West Africa in the colonial period, (male) chiefs lost their sovereignty while increasing their powers over the people (Crowder and Ikeme 1970: xv), although we are to believe that their powers derived from 'tradition' even where the British created their own brand of 'traditional chiefs.' Martin Chanock's astute comment on the powers of chiefs in colonial Africa is particularly applicable to the Yorùbá situation: 'British officials, . . . where they came across a chief, . . . intended to invest *him* retroactively not only with a greater range of authority than he had before but also with authority of a different type. There seemed to be no way of thinking about chiefly authority . . . which did not include judicial power' (Chanock 1982: 59). Thus male chiefs were invested with more power over the people while female chiefs were stripped of power. Through lack of recognition, their formal positions soon became attenuated.

At another level, the transfer of judicial power from the community to the council of male chiefs proved to be particularly negative for women at a time when the state was extending its tentacles to an increasing number of aspects of life. In pre-British Yorùbá society, adjudication of disputes rested with lineage elders. Therefore, very few matters came under the purview of the ruler and the council of chiefs. But in the colonial administration, the Native Authority System, with its customary courts, dealt with all civil cases including marriage, divorce, and adultery. . . .

It was into this unfortunate tradition of male dominance that Africans were drafted – this was particularly disadvantageous to women because marriage, divorce, and even pregnancy came under the purview of the state. Given the foregoing, it is clear that the impact of colonization was profound and negative for women. Appraisals of the impact of colonization that see certain 'benefits' for African women are mistaken in light of the overarching effect of the colonial state, which effectively defined females as 'women' and hence second-class colonial subjects unfit to determine their own destiny. The postindependence second-class status of African women's citizenship is rooted in the process of inventing them as women. Female access to membership in the group is no longer direct; access to citizenship is now mediated through marriage, through the 'wifization of citizenship.'

# PART TEN

# Language

L ANGUAGE IS A FUNDAMENTAL SITE OF struggle for post-colonial discourse because the colonial process itself begins in language. The control over language by the imperial centre – whether achieved by displacing native languages, by installing itself as a 'standard' against other variants which are constituted as 'impurities', or by planting the language of empire in a new place – remains the most potent instrument of cultural control. Language provides the terms by which reality may be constituted; it provides the names by which the world may be 'known'. Its system of values – its suppositions, its geography, its concept of history, of difference, its myriad gradations of distinction – becomes the system upon which social, economic and political discourses are grounded.

One of the most subtle demonstrations of the power of language is the means by which it provides, through the function of naming, a technique for knowing a colonized place or people. To name the world is to 'understand' it, to know it and to have control over it. The word 'Africa', for instance, is determined by European historical formations which had little or no relevance to the complex of linguistic cultural and economic factors which tied and sometimes separated various societies on the continent. To name reality is therefore to exert power over it, simply because the dominant language becomes the way in which it is known. In colonial experience this power is by no means vague or abstract. A systematic education and indoctrination installed the language and thus the reality on which it was predicated as pre-eminent.

There are several responses to this dominance of the imperial language, but two present themselves immediately in the decolonizing process – rejection or subversion. The process of radical decolonization proposed by Ngũgĩ wa Thiong'o is a good demonstration of the first alternative. Ngũgĩ's programme for restoring an ethnic or national identity embedded in the mother tongue involves a rejection of English, a refusal to use it for his writing, a refusal to accede to the kind of world and reality it appears to name, a refusal to submit to the political dominance its use implies. This stance of rejection rests upon the assumption that an essential Gĩkũyũ identity may be regained, an identity which the language of the colonizer seems to have displaced or dispersed.

However, many more writers have felt that this appeal to some essential cultural identity is doomed to failure, indeed, misunderstands the heterogeneous nature of human experience. Ngũgĩ's own essay indicates the divergent reasons fellow African writers had for using the English language, but most involve a confidence that English can be used in the process of *resisting* imperialism. Braj Kachru shows how in the Indian situation the language has provided a neutral vehicle for communication between contesting language groups, while the Indian novelist Raja Rao voices, in a piece written as long ago as 1938, the challenge of the post-colonial writer to adapt the colonial language to local needs. This determination to use the language as an ethnographic tool has been a more common response of post-colonial writers. The appropriation of the language is essentially a subversive strategy, for the adaptation of the 'standard' language to the demands and requirements of the place and society into which it has been appropriated amounts to a far more subtle rejection of the political power of the standard language. In Chinua Achebe's words this is a process by which the language is made to bear the weight and the texture of a different experience. In doing so it becomes a different language. By adapting the alien language to the exigencies of a mother grammar, syntax, vocabulary, and by giving a shape to the variations of the speaking voice, such writers and speakers construct an 'english' which amounts to a very different linguistic vehicle from the received standard colonial 'English'. As Bill Ashcroft demonstrates, the belief that the English text is unable to communicate a 'non-English' cultural meaning is based on a misconception of the way language 'means'. Meaning is seen to be a constitutive interaction within the 'message event'.

The process of language adaptation is linguistically profound because it establishes a medium which fractures the concept of a standard language and installs the 'marginal' vari-ations of language use as the actual network of a particular language. E. K. Brathwaite demonstrates the process of language adaptation in the Caribbean, arguably one of the most dynamic linguistic communities in the world. What he calls 'nation language' is a language which, rather than attempting to recover lost origins, demonstrates the vigorous success of linguistic variation in this region.

One of the most detailed examinations of this process has focused on the Anglophone and Francophone speaking communities of West Africa. Chantal Zabus employs the very useful notion of a *palimpsest* to demonstrate how the language practices of a region may be built up through a range of linguistic strategies, one of the most significant of which she calls 'relexification'.

## Chapter 59

# NGŨGĨ WA THIONG'O

# THE LANGUAGE OF AFRICAN LITERATURE

From 'The Language of African Literature' *Decolonising the Mind: The Politics of Language in African Literature* London: James Currey, 1981.

IN 1962 I WAS INVITED to that historic meeting of African writers at Makerere University College, Kampala, Uganda. . . . The title, 'A Conference of African Writers of English Expression', automatically excluded those who wrote in African languages. . . . The discussions on the novel, the short story, poetry and drama were based on extracts from works in English and hence they excluded the main body of work in Swahili, Zulu, Yoruba, Arabic, Amharic and other African languages. Yet, despite this exclusion of writers and literature in African languages, no sooner were the introductory preliminaries over than this Conference of 'African Writers of English Expression' sat down to the first item on the agenda: 'What is African Literature?' . . .

English, like French and Portuguese, was assumed to be the natural language of literary and even political mediation between African people in the same nation and between nations in Africa and other continents. In some instances these European languages were seen as having a capacity to unite African peoples against divisive tendencies inherent in the multiplicity of African languages within the same geographic state. Thus Ezekiel Mphahlele later could write, in a letter to *Transition* number 11, that English and French have become the common language with which to present a nationalist front against white oppressors, and even 'where the whiteman has already retreated, as in the independent states, these two languages are still a unifying force'. . . .

Chinua Achebe, in a speech entitled 'The African Writer and the English Language' (1975), said:

> Is it right that a man should abandon his mother tongue for someone else's? It looks like a dreadful betrayal and produces a guilty feeling. But for me there is no other choice. I have been given the language and I intend to use it.
>
> (Achebe 1975: 62)

See the paradox: the possibility of using mother-tongues provokes a tone of levity in phrases like 'a dreadful betrayal' and 'a guilty feeling'; but that of foreign languages produces a categorical positive embrace, what Achebe himself, ten years later, was to describe as this 'fatalistic logic of the unassailable position of English in our literature'. . . .

The lengths to which we were prepared to go in our mission of enriching foreign languages by injecting Senghorian 'black blood' into their rusty joints, is best exemplified by Gabriel Okara in an article reprinted in *Transition*:

> As a writer who believes in the utilization of African ideas, African philosophy and African folklore and imagery to the fullest extent possible, I am of the opinion the only way to use them effectively is to translate them almost literally from the African language native to the writer into whatever European language he is using as medium of expression. I have endeavoured in my words to keep as close as possible to the vernacular expressions. For, from a word, a group of words, a sentence and even a name in any African language, one can glean the social norms, attitudes and values of a people.
>
> In order to capture the vivid images of African speech, I had to eschew the habit of expressing my thoughts first in English. It was difficult at first, but I had to learn. I had to study each law expression I used and to discover the probable situation in which it was used in order to bring out the nearest meaning in English. I found it a fascinating exercise.
>
> (Okara 1963: 15)

Why, we may ask, should an African writer, or any writer, become so obsessed by taking from his mother-tongue to enrich other tongues? Why should he see it as his particular mission? We never asked ourselves: how can we enrich our languages? How can we 'prey' on the rich human-ist and democratic heritage in the struggles of other peoples in other times and other places to enrich our own? . . . What seemed to worry us more was this: after all the literary gymnastics of preying on our languages to add life and vigour to English and other foreign languages, would the result be accepted as good English or good French? Will the owner of the language criticise our usage? Here we were more assertive of our rights! Chinua Achebe wrote:

> I feel that the English language will be able to carry the weight of my African experience. But it will have to be a new English, still in full communion with its ancestral home but altered to suit new African surroundings.
>
> (1975: 62)

Gabriel Okara's position on this was representative of our generation:

> Some may regard this way of writing English as a desecration of the language. This is of course not true. Living languages grow like living things, and English is far from a dead language. There are American, West Indian, Australian, Canadian and New Zealand versions of English. All of them add life and vigour to the language while reflecting their own respective cultures. Why shouldn't there be a Nigerian or West African English which we can use to express our own ideas, thinking and philosophy in our own way?
>
> (Okara 1963: 15–16)

How did we arrive at this acceptance of 'the fatalistic logic of the unassailable position of English in our literature', in our culture and in our politics? . . . How did we, as African writers, come to be so feeble towards the claims of our languages on us and so aggressive in our claims on other languages, particularly the languages of our colonisation? . . .

In my view language was the most important vehicle through which that power fascinated and held the soul prisoner. The bullet was the means of the physical subjugation. Language was the means of the spiritual subjugation. Let me illustrate this by drawing upon experiences in my own education, particularly in language and literature.

I was born into a large peasant family: father, four wives and about twenty-eight children. I also belonged, as we all did in those days, to a wider extended family and to the community as a whole.

We spoke Gĩkũyũ as we worked in the fields. We spoke Gĩkũyũ in and outside the home. I can vividly recall those evenings of story-telling around the fireside. It was mostly the grown-ups telling the children but everybody was interested and involved. We children would re-tell the stories the following day to other children who worked in the fields picking the pyrethrum flowers, tea-leaves or coffee beans of our European and African landlords. . . .

There were good and bad story-tellers. A good one could tell the same story over and over again, and it would always be fresh to us, the listeners. He or she could tell a story told by someone else and make it more alive and dramatic. The differences really were in the use of words and images and the inflexion of voices to effect different tones.

We therefore learnt to value words for their meaning and nuances. Language was not a mere string of words. It had a suggestive power well beyond the immediate and lexical meaning. Our appreciation of the suggestive magical power of language was reinforced by the games we played with words through riddles, proverbs, transpositions of syllables, or through non-sensical but musically arranged words. So we learnt the music of our language on top of the content. The language, through images and symbols, gave us a view of the world, but it had a beauty of its own. The home and the field were then our pre-primary school but what is important, for this discussion, is that the language of our evening teach-ins, and the language of our immediate and wider community, and the language of our work in the fields were one.

And then I went to school, a colonial school, and this harmony was broken. The language of my education was no longer the language of my culture. I first went to Kamaandura, missionary run, and then to another called Maanguuũ run by nationalists grouped around the Gĩkũyũ Independent and Karinga Schools Association. Our language of education was still Gĩkũyũ. The very first time I was ever given an ovation for my writing was over a composition in Gĩkũyũ. So for my first four years there was still harmony between the language of my formal education and that of the Limuru peasant community.

It was after the declaration of a state of emergency over Kenya in 1952 that all the schools run by patriotic nationalists were taken over by the colonial regime and were placed under District Education Boards chaired by Englishmen. English became the language of my formal education. In Kenya, English became more than a language: it was the language, and all the others had to bow before it in deference.

Thus one of the most humiliating experiences was to be caught speaking Gĩkũyũ in the vicinity of the school. The culprit was given corporal punishment – three to five strokes of the cane on bare buttocks – or was made to carry a metal plate around the neck with inscriptions such as I AM STUPID or I AM A DONKEY. . . .

The attitude to English was the exact opposite: any achievement in spoken or written English was highly rewarded; prizes, prestige, applause; the ticket to higher realms. English became the measure of intelligence and ability in the arts, the sciences, and all the other branches of learning. English became the main determinant of a child's progress up the ladder of formal education. . . . Literary education was now determined by the dominant language while also reinforcing that dominance. Orature (oral literature) in Kenyan languages stopped. . . .

Thus language and literature were taking us further and further from ourselves to other selves, from our world to other worlds.

What was the colonial system doing to us Kenyan children? What were the consequences of, on the one hand, this systematic suppression of our languages and the literature they carried, and on the other the elevation of English and the literature it carried? To answer those questions, let me first examine the relationship of language to human experience, human culture, and the human perception of reality. . . .

Language as communication has three aspects or elements. There is first what Karl Marx once called the language of real life, the element basic to the whole notion of language, its origins and development: that is, the relations people enter into with one another in the labour process, the links they necessarily establish among themselves in the act of a people, a community of human beings, producing wealth or means of life like food, clothing, houses. A human community really starts its historical being as a community of co-operation in production through the division of labour; the simplest is between man, woman and child within a household; the more complex divisions are between branches of production such as those who are sole hunters, sole gatherers of fruits or sole workers in metal. Then there are the most complex divisions such as those in modern factories where a single product, say a shirt or a shoe, is the result of many hands and minds. Production is co-operation, is communication, is language, is expression of a relation between human beings and it is specifically human.

The second aspect of language as communication is speech and it imitates the language of real life, that is communication in production. The verbal signposts both reflect and aid communication or the relations established between human beings in the production of their means of life. Language as a system of verbal signposts makes that production possible. The spoken word is to relations between human beings what the hand is to the relations between human beings and nature. The hand through tools mediates between human beings and nature and forms the language of real life: spoken words mediate between human beings and form the language of speech.

The third aspect is the written signs. The written word imitates the spoken. Where the first two aspects of language as communication through the hand and the spoken word historically evolved more or less simultaneously, the written aspect is a much later historical development. Writing is representation of sounds with visual symbols, from the simplest knot among shepherds to tell the number in a herd or the hieroglyphics among the Agikuyu gicaandi singers and poets of Kenya, to the most complicated and different letter and picture writing systems of the world today. . . .

But there is more to it: communication between human beings is also the basis and process of evolving culture. In doing similar kinds of things and actions over and over again under similar circumstances, similar even in their mutability, certain patterns, moves, rhythms, habits, attitudes, experiences and knowledge emerge. Those experiences are handed over to the next generation and become the inherited basis for their further actions on nature and on themselves. There is a gradual accumulation of values which in time become almost self-evident truths governing their conception of what is right and wrong, good and bad, beautiful and ugly, courageous and cowardly, generous and mean in their internal and external relations. Over a time this becomes a way of life distinguishable from other ways of life. They develop a distinctive culture and history. Culture embodies those moral, ethical and aesthetic values, the set of spiritual eyeglasses, through which they come to view themselves and their place in the universe. Values are the basis of a people's identity, their sense of particularity as members of the human race. All this is carried by language. Language as culture is the collective memory bank of a people's experience in history. Culture is almost indistinguishable from the language that makes possible its genesis, growth, banking, articulation and indeed its transmission from one generation to the next. . . .

Language as communication and as culture are then products of each other. Communication creates culture: culture is a means of communication. Language carries culture, and culture carries, particularly through orature and literature, the entire body of values by which we come to perceive ourselves and our place in the world. How people perceive themselves affects how they look at their culture, at their politics and at the social production of wealth, at their entire relationship to nature and to other beings. Language is thus inseparable from ourselves as a community of human beings with a specific form and character, a specific history, a specific relationship to the world. . . .

I believe that my writing in the Gĩkũyũ language, a Kenyan language, an African language, is part and parcel of the anti-imperialist struggles of Kenyan and African peoples. In schools and universities our Kenyan languages – that is the languages of the many nationalities which make up Kenya – were associated with negative qualities of backwardness, underdevelopment, humiliation and punishment. . . .

So I would like to contribute towards the restoration of the harmony between all aspects and divisions of language so as to restore the Kenyan child to his environment, understand it fully so as to be in a position to change it for his collective good. I would like to see Kenya peoples' mother-tongues (our national languages!) carry a literature reflecting not only the rhythms of a child's spoken expression, but also his struggle with his nature and his social nature.

But writing in our languages per se – although a necessary first step in the correct direction – will not itself bring about the renaissance in African cultures if that literature does not carry the content of our peoples' anti-imperialist struggles to liberate their productive forces from foreign control. . . .

In other words writers in African languages should reconnect themselves to the revolutionary traditions of an organised peasantry and working class in Africa in their struggle to defeat imperialism and create a higher system of democracy and socialism in alliance with all other peoples of the world.

## Chapter 60

# CHINUA ACHEBE

## THE POLITICS OF LANGUAGE

From 'Politics and Politicians of Language in African Literature' in Doug Killam (ed.) *FILLM Proceedings* Guelph, Ontario: University of Guelph, 1989.

. . .

O F ALL THE EXPLOSIONS THAT HAVE ROCKED the African continent . . ., few have been more spectacular, and hardly any more beneficial, than the eruption of African literature, shedding a little light here and there on what had been an area of darkness.

So dramatic has been the change that I am even presuming that one or two in this very distinguished audience might recognize that my title is a somewhat mischievous rendering of the subtitle of the book *Decolonizing the Mind* by an important African writer and revolutionary, Ngũgĩ wa Thiong'o. The mischief lies in my inserting after the word *politics* the words *and politicians* like dropping a cat among Ngũgĩ's pigeons.

Ngũgĩ's book argues passionately and dramatically that to speak of African literature in European languages is not only an absurdity but also part of the scheme of western imperialism to hold Africa in perpetual bondage. He reviews his own position as a writer in English and decides that he can no longer continue in the treachery. So he makes a public renunciation of English in a short statement at the beginning of his book. Needless to say, Ngũgĩ applies the most severe censure to those African writers who remain accomplices of imperialism, especially Senghor and Achebe, but particularly Achebe presumably because Senghor no longer threatens anybody!

Theatricalities aside, the difference between Ngũgĩ and myself on the issue of indigenous or European languages for African writers is that while Ngũgĩ now believes it is *either/or*, I have always thought it was *both* . . .

I write in English. English is a world language. But I do *not* write in English *because* it is a world language. My romance with the world is subsidiary to my involvement with Nigeria and Africa. Nigeria is a reality which I could not ignore. One characteristic of this reality, Nigeria, is that it transacts a considerable portion of its daily business in the English language. As long as Nigeria wishes to exist as a nation it has no choice in the forseeable future but to hold its more than two hundred component nationalities together through an alien language, English. I lived through a civil war in which probably two million people perished over the question of Nigerian unity. To remind me therefore that Nigeria's foundation was laid only a hundred years ago at the Berlin conference of European powers and in the total absence of any Africans is not really useful information to me. It is precisely because the nation is so new and so fragile that we would soak the land in blood to maintain the frontiers mapped out by foreigners.

English is therefore not marginal to Nigerian affairs. It is quite central. I can speak across two hundred linguistic frontiers to fellow Nigerians only in English. Of course I also have a mother tongue which luckily for me is one of the three major languages of the country. Luckily I say because this language, Igbo, is not really in danger of extinction. I can gauge my good luck against the resentment of fellow Nigerians who oppose most vehemently the token respect accorded to the three major tongues by newscasters saying goodnight in them after reading a half-hour bulletin in English!

Nothing would be easier than to ridicule our predicament if one was so minded. And nothing would be more attractive than proclaiming from a safe distance that our job as writers is not to describe the predicament but to change it. But this is where the politics of language becomes *politicking* with language.

One year after the [1962] Makerere conference, a Nigerian literary scholar, Obi Wali, published a magazine article in which he ridiculed the meeting and called on the African writers and the European 'midwives' of their freak creations to stop pursuing a dead end. And he made the following important suggestion: 'What we would like future conferences on African literature to devote time to is the all-important problem of African writing in African languages, and all its implications for the development of a truly African sensibility.' Having set that rather clear task before 'future conferences on African literature,' Dr. Obi Wali, who was himself a teacher of literature and a close friend of the poet Christopher Okigbo, might have been expected to lead the way along the lines of his prescription. But what he does instead is to abandon his academic career for politics and business. As a leading parliamentarian in Nigeria's Second Republic he might have played the midwife to a legislation in favour of African literature in African languages. But no; Obi Wali, having made his famous intervention, like a politician simply dropped out of sight.

In 1966 Nigeria's first military coup triggered off a countercoup and then a series of horrendous massacres of Igbo people in Hausa-speaking northern Nigeria. A famous educationist well-known for his opposition to the continued use of English in Nigeria wrote in a Lagos newspaper offering the incredible suggestion that if all Nigerians had spoken one language the killings would not have happened. And he went further to ask the Nigerian army to impose Hausa as Nigeria's lingua franca. Fortunately people were too busy coping with the threat of disintegration facing the country to pay serious attention to the fellow. But I could not resist writing a brief rejoinder in which I reminded him that the thousands who were killed spoke excellent Hausa.

The point in all this is that language is a handy whipping-boy to summon and belabour when we have failed in some serious way. In other words we play politics with language and in so doing conceal the reality and the complexity of our situation from ourselves and from those foolish enough to put their trust in us.

The politics Ngũgĩ plays with language is of a different order. It is a direct reflection of a slowly perfected Manichean vision of the world. He sees but one 'great struggle between the two mutually opposed forces in Africa today: an imperialist tradition on one hand and a resistance tradition on the other.' Flowing nicely from this unified vision, Africa's language problems resolve themselves into European languages sponsored and foisted on the people by imperialism and African languages defended by patriotic and progressive forces of peasants and workers.

To demonstrate how this works out in practice, Ngũgĩ gives us a moving vignette of how the enemy interfered with his mother tongue in his 'Limuru peasant community.' 'I was born in a large peasant family: father, four wives and about twenty-eight children . . . We spoke Gĩkũyũ as we worked in the fields.' The reader is given nearly two pages of this pastoral idyll of linguistic and social harmony, in which stories are told around the fire at the end of the day. Even at school young Ngũgĩ is taught in Gĩkũyũ, in which he excels to the extent of winning an infant ovation for his composition in that language.

Then the imperialist struck in 1952 and declared a state of emergency in Kenya; and Ngũgĩ's world is brutally shattered:

> All the schools run by patriotic nationalists were taken over by the colonial regime and were placed under District Education Boards chaired by Englishmen. English became the language of my formal education. In Kenya, English became more than a language: it was *the* language, and all others had to bow before it in deference.

A heart-rending scenario, but also a scenario strewn with fatal snags for the single-minded. I had warned about this danger in one of the earliest statements I ever made in my literary career – that those who would canonize our past must serve also as the devil's advocate, setting down beside the glories every inconvenient fact. Unfortunately Ngũgĩ is too good a partisan to do this double duty. So he files the totally untenable report that imperialists imposed the English language on the patriotic peasants of Kenya as recently as 1952!

What about the inconvenient fact that already in the 1920s and 1930s 'the Kikuyu Independent Schools which were started by the Kikuyu after their rift with the Scottish mission-aries, *taught in English* instead of the vernacular even in the first grade' (Symonds 1966: 202; emphasis added). For the avoidance of doubt, the scenario here is of imperialist agents (in the shape of Scottish missionaries) desiring to teach Kikuyu children in their mother tongue while the patriotic Kikuyu peasants broke away because they preferred English!

What happened in Kenya also happened in the rest of the empire. Neither in India nor in Africa did the English seriously desire to teach their language to the native. The historic and influential Phelps-Stokes Commission in West Africa in 1922 had placed much greater value on the native tongue than English; and its recommendations were picked up by the British Advisory Committee on Native Education on Tropical Africa (Smock and Bentsi-Enchill 1976: 174).

In Nigeria the demand for English was already there in the coastal regions in the first half of the nineteenth century. In a definitive study of the work of Christian missions in Nigeria 1841 to 1891, Professor J. F. A. Ajayi reports that in the Niger delta in the 1850s the missionary teachers were already 'obliged to cater for the demand . . . for the knowledge of the English language' (1965: 133). In Calabar by 1876 some of the chiefs were not satisfied by the amount of English their children were taught in missionary schools and were hiring private tutors at a very high fee (133–4). Nowhere in all this can we see the slightest evidence of the simple scenario of European imperialism forcing its language down the throats of unwilling natives. In fact imperialism's ways with language were extremely complex.

If imperialism was not entirely to blame for the dominance of European languages in Africa today who then is the culprit? Ourselves? Our parents? If our fathers were at least partly to blame or if we were misguided, why do we not change the situation today by renouncing the use of European languages and rediscovering our indigenous tongues? Or is it by any chance true that these alien languages are still knocking about because they serve an actual need?

No African in our recent history had fought imperialism more doggedly than Kwame Nkrumah of Ghana. And yet we are told that:

> During the Nkrumah era, political leaders demonstrated considerable concern over the possible divisive impact of a mother tongue medium policy. Although English is

a language alien to Ghana they saw it as the best vehicle for achieving national communication and social and political unification.

(Smock and Bentsi-Enchill 1976: 176)

In addition to this political problem Ghana faced the practical question of teaching mother tongues when ethnic mixing had reached significant levels in urban and rural schools as a result of internal migrations. Already by 1956 the Bernard Committee had found that schools where the pupils spoke a single mother tongue were far fewer than schools in which more than five languages were represented in fair numbers. The simple consequence of this is that if the policy of teaching in mother tongues were to be enforced the schools concerned would have to hire more than five teachers for every class. (This was at the 1956 level of ethnic mixing in Ghana. The situation today, thirty years later, would be considerably more difficult, unless we are to follow South Africa and send every native back to his homeland!)

It would seem then that the culprit for Africa's language difficulties was not imperialism, as Ngũgĩ would have us believe, but the linguistic plurality of modern African states. No doubt this will explain the strange fact that the Marxist states in Africa, with the exception of Ethiopia, have been the most forthright in adopting the languages of their former colonial rulers – Angola, Mozambique, Guinea-Bissau and most lately Burkina Faso, whose minister of culture said with a retrospective shudder recently that the sixty ethnic groups in that country could mean sixty different nationalities.

This does not in any way close the argument for the development of African languages by the intervention of writers and governments. But we do not have to falsify our history in the process. That would be playing politics. The words of the Czech novelist, Milan Kundera, should ring in our ears: 'Those who seek power passionately do so not to change the present or the future but the past – to rewrite history.'

## Chapter 61

# BRAJ B. KACHRU

## THE ALCHEMY OF ENGLISH

From *The Alchemy of English: The Spread Functions and Models of Non-Native Englishes* Champaign, Ill.: University of Illinois Press, 1990.

[T]HE ENGLISH LANGUAGE IS A TOOL OF POWER, domination and elitist identity, and of communication across continents. Although the era of the 'White man's burden' has practically ended in a political sense, and the Raj has retreated to native shores, the linguistic and cultural consequences of imperialism have changed the global scene. The linguistic ecology of, for example, Africa and Asia is not the same. English has become an integral part of this new complex sociolinguistic setting. The colonial Englishes were essentially acquired and used as non-native second languages, and after more than two centuries, they continue to have the same status. The *non-nativeness* of such varieties is not only an attitudinally significant term, but it also has linguistic and sociolinguistic significance. . . .

In India, only Sanskrit, English, Hindi, and to some extent Persian have acquired pan-Indian intranational functions. The domains of Sanskrit are restricted, and the proficiency in it limited, except in the case of some professional pandits. The cause of Hindi was not helped by the controversy between Hindi, Urdu and Hindustani. Support for Hindustani almost ended with independence; after the death of its ardent and influential supporter, Gandhi, very little was heard about it. The enthusiasm and near euphoria of the supporters of Hindi were not channeled in a constructive (and realistic) direction, especially after the 1940s. The result is that English continues to be a language both of power and of prestige.

For governments, English thus serves at least two purposes. First, it continues to provide a linguistic tool for the administrative cohesiveness of a country (as in South Asia and parts of Africa). Second, at another level, it provides a language of wider communication (national and international). The enthusiasm for English is not unanimous, or even widespread. The disadvantages of using it are obvious: cultural and social implications accompany the use of an external language. But the native languages are losing in this competition.

English does have one clear advantage, attitudinally and linguistically: it has acquired a *neutrality* in a linguistic context where native languages, dialects, and styles sometimes have acquired undesirable connotations. Whereas native codes are functionally marked in terms of caste, religion, region, and so forth, English has no such 'markers,' at least in the non-native context. It was originally the foreign (alien) ruler's language, but that drawback is often over-shadowed by what it can do for its users. True, English is associated with a small and elite group; but it is in their role that the *neutrality* of a language becomes vital (e.g. for Tamil speakers in Tamil Nadu, or Bengali speakers in West Bengal). In India the most widely used language is Hindi (46 percent) and its different varieties (e.g. Hindustani, Urdu), have traditionally been associated with various factions: Hindi with the Hindus; Urdu with the Muslims;

and Hindustani with the maneuvering political pandits who could not create a constituency for it. While these attitudinal allocations are not necessarily valid, this is how the varieties have been perceived and presented. English, on the other hand, is not associated with any religious or ethnic faction.

Whatever the limitations of English, it has been perceived as the language of power and opportunity, free of the limitations that the ambitious attribute to the native languages.

## Attitudinal neutrality and power

In several earlier studies it has been shown (Kachru 1978 and 1982a) that in *code-mixing*, for example, English is being used to *neutralize* identities one is reluctant to express by the use of native languages or dialects. 'Code-mixing' refers to the use of lexical items or phrases from one code in the stream of discourse of another. Neutralization thus is a linguistic strategy used to 'unload' a linguistic item from its traditional, cultural and emotional connotations by avoiding its use and choosing an item from another code. The borrowed item has referential meaning, but no cultural connotations in the context of the specific culture. This is not borrowing in the sense of filling a lexical gap. . . . In Kashmiri the native word *mɔnd* ('widow') invokes the traditional connotations associated with widowhood. Its use is restricted to abuses and curses, not occurring in 'polished' conversation. *vedvā* (Hindi *vidhwā*) or English *widow* is preferred by the Hindus. In Tamil, as shown by Annamalai (1978) *maccaan* and *attimbeer* reveal the caste identity of the speaker – not desirable in certain situations. Therefore, one uses English *brother-in-law*, instead. English *rice* is neutral compared with *saadam* or *soru* (purist) in Tamil. A lexical item may be associated with a specific style in the native language as are *manaivi* (formal) and *pendītti* (colloquial) in Tamil, but the English equivalent *wife* has no style restrictions.

In such contexts, then, the power of neutralization is associated with English in two ways. First, English provides – with or without 'mixing' – an additional code that has referential meaning but no cultural overtones or connotations. . . . Second, such use of English develops new code-mixed varieties of languages. Lexicalization from English is particularly preferred in the contexts of kinship, taboo items, science and technology, or in discussing sex organs and death. What Moag (1982: 276) terms the 'social neutrality' of English in the case of Fiji is applicable in almost all the countries where English is used as a non-native language. In the Fijian context, Tongans and Fijians

> find English the only safe medium in which to address those of higher status. English not only hides their inability in the specialized vernacular registers, but also allows them to meet traditional superiors on a more or less equal footing.
>
> (276)

. . .

## Contact literatures in English: creativity in the other tongue

The contact literatures in English have several characteristics, of which two may be mentioned here. In South Asia, to take one example, there are three more or less pan-South Asian literatures: Sanskrit, Persian, and Hindi. In terms of both style and content, Sanskrit has been associated with the native Hindu tradition. Persian (in its Indian form) and Urdu have maintained the Perso-Arabic stylistic devices, metaphors and symbolism. It is this aspect of Urdu that alienated it from the traditionalist Hindus, who believe that in its formal experimentation, thematic range, and metaphor, it has maintained an 'un-Indian' (Islamic) tradition, and continues to seek inspiration from such non-native traditions. This attitude toward Urdu tells only part of the story, and negates the contribution that the Hindus have made to the Urdu language, and the way it was used as the language of national revival. Indian English literature cuts across these attitudes. It has united certain pan-South Asian nationalists, intellectuals, and creative writers. It has provided a new perspective in India through an 'alien' language.

In Indian English fiction (see e.g. Mukherjee 1971; Parameswaran 1976) R. K. Narayan, Mulk Raj Anand, and Raja Rao (e.g. his *Kanthapura*) have brought another dimension to the understanding of the regional, social, and political contexts. In this process, linguistically speaking, the process of the Indianization of English has acquired an institutionalized status.

In a sociological sense, then, English has provided a linguistic tool and a sociopolitical dimension very different from those available through native linguistic tools and traditions. A non-native writer in English functions in two traditions. In psychological terms, such a multi-lingual role calls for adjustment. In attitudinal terms, it is controversial; in linguistic terms, it is challenging, for it means molding the language for new contexts. Such a writer is suspect as fostering new beliefs, new value systems, and even new linguistic loyalties and innovations.

This, then, leads us to the other side of this controversy. For example, what have been the implications of such a change – attitudinally and sociologically – for the Indian languages (and for African languages) and for those speakers whose linguistic repertoires do not include English? Additionally, we need to ask what are its implications for creative writers whose media are 'major' or 'minor' Indian languages?

## Post-colonial period

Since independence, the controversy about English has taken new forms. Its alien power base is less an issue; so is its Englishness or Americanness in a cultural sense. The English language is not perceived as necessarily imparting only Western traditions. The medium is non-native, but the message is not. In several Asian and African countries, English now has national and international functions that are both distinct and complementary. English has thus acquired a new power base and a new elitism. The domains of English have been restructured . . . people ask is English really a non-native ('alien') language for India, for Africa, for South East Asia?

In the case of India one wonders: has India played the age-old trick on English too, of nativizing it and acculturating it – in other words, Indianizing it? The Indian writer and phil-osopher Raja Rao associates power with English, which, in his mind is equal if not greater than Sanskrit, when he says:

> Truth, said a great Indian sage, is not the monopoly of the Sanskrit language. Truth can use any language, and the more universal, the better it is. If metaphysics is India's primary contribution to world civilization, as we believe it is, then must she use the most universal language for her to be universal. . . . And as long as the English language is universal it will always remain Indian. . . . It would then be correct to say as long as we are Indian – that is, not nationalists, but truly Indians of the Indian psyche – we shall have the English language with us and amongst us, and not as a guest or friend, but as one of our own, of our caste, our creed, our sect and our tradition.
>
> (Rao [1937] 1978: 421)

These new power bases in Africa or in Asia have called into question the traditionally accepted, externally normative standards for the institutionalized varieties. The new varieties have their own linguistic and cultural ecologies or sociological contexts. The adaptation to these new ecologies has given non-native Englishes new identities. . . .

One might say that contemporary English does not have just one defining context but many — across cultures and languages. This is also true of the growing new literatures in English. The concepts of 'British literature' or 'American literature' represent only part of the spectrum. The new traditions — really not so new — must be incorporated into the tradition of 'literature[s] in English' (Narasimhaiah 1978).

. . . The alchemy of English (present and future), then, does not only provide social status, it also gives access to attitudinally and materially desirable domains of power and knowledge. It provides a powerful linguistic tool for manipulation and control. In addition, this alchemy of English has left a deep mark on the languages and literature of the non-western world. English has thus caused transmutation of languages, equipping them in the process for new societal, scientific and technological demands. The process of Englishization has initiated stylistic and thematic innovations, and has 'modernized' registers. The power of English is so dominant that a new caste of English-using speech fellowships has developed across cultures and languages. It may be relatively small, but it is powerful, and its values and perspectives are not necessarily in harmony with the traditional values of these societies. In the past, the control and manipulation of international power have never been in the hands of users of one language group. Now we see a shift of power from the traditional caste structure; in the process, a new caste has developed. In this sense, English has been instrumental in a vital social change, and not only in that of language and literatures.

## Chapter 62

# RAJA RAO

## LANGUAGE AND SPIRIT

From 'Preface' *Kanthapura* New Delhi: Oxford University Press India, 1937.

T HERE IS NO VILLAGE IN INDIA, however mean, that has not a rich *sthala-purana*, or legendary history, of its own. Some god or godlike hero has passed by the village – Rama might have rested under this pipal-tree, Sita might have dried her clothes, after her bath, on this yellow stone, or the Mahatma himself, on one of his many pilgrimages through the country, might have slept in this hut, the low one by the village gate. In this way the past mingles with the present, and the gods mingle with men to make the repertory of your grandmother always bright. One such story from the contemporary annals of a village I have tried to tell.

The telling has not been easy. One has to convey in a language that is not one's own the spirit that is one's own. One has to convey the various shades and omissions of a certain thought-movement that looks maltreated in an alien language. I use the word 'alien,' yet English is not really an alien language to us. It is the language of our intellectual make-up – like Sanskrit or Persian was before – but not of our emotional make-up. We are all instinctively bilingual, many of us writing in our own language and in English. We cannot write like the English. We should not. We cannot write only as Indians. We have grown to look at the large world as part of us. Our method of expression therefore has to be a dialect which will some day prove to be as distinctive and colorful as the Irish or the American. Time alone will justify it.

After language the next problem is that of style. The tempo of Indian life must be infused into our English expression, even as the tempo of American or Irish life has gone into the making of theirs. We, in India, think quickly, we talk quickly, and when we move we move quickly. There must be something in the sun of India that makes us rush and tumble and run on. And our paths are paths interminable. The *Mahabharata* has 214,778 verses and the *Rama-yana* 48,000. The *Puranas* are endless and innumerable. We have neither punctuation nor the treacherous 'ats' and 'ons' to bother us – we tell one interminable tale. Episode follows episode, and when our thoughts stop our breath stops, and we move on to another thought. This was and still is the ordinary style of our storytelling. I have tried to follow it myself in this story.

It may have been told of an evening, when as the dusk falls, and through the sudden quiet, lights leap up in house and after house, and stretching her bedding on the veranda, a grandmother might have told you, newcomer, the sad tale of her village.

## Chapter 63

# BILL ASHCROFT

# LANGUAGE AND TRANSFORMATION

From *Post-colonial Transformation* London: Routledge, 2001.

**P**OST-COLONIAL WRITERS WHO WRITE IN ENGLISH have used it as a cultural vehicle through which a world audience could be introduced to features of culturally diverse post-colonial societies. But the use of colonial languages has opened up a long-running and unresolved argument in post-colonial circles. According to Indian linguist, Braj Kachru, English has been widely accepted as a *lingua franca* in India because of its relatively 'neutral' nature, since its effects in everyday use are far less inflammatory than those stemming from the contention between one or another minority languages (1986). On the other hand, the Kenyan novelist Ngũgĩ wa Thiongo has argued that writing in an African language 'is part and parcel of the anti-imperialist struggles of Kenyan and African peoples' (1981: 28).

Underlying the dispute over the most effective form of discursive resistance is the question: 'Can one use the language of imperialism without being inescapably contaminated by an imperial world view?' It is a question which continues to provoke argument, because it is ineluctably rooted in real political conflict. Martinican Edouard Glissant, for instance, says

> There are . . . no languages or language spoken in Martinique, neither Creole nor French, that have been 'naturally' developed by and for us Martinicans because of our experience of collective, proclaimed, denied, or seized responsibility at all levels. The official language, French, is not the people's language. This is why we, the elite, speak it so correctly. The language of the people, Creole, is not the language of the nation.
>
> (1989: 166)

Although framed in terms of class, Glissant's observation alerts us to the frequency with which a particular *use* of language can be conflated with the language itself. What makes a language a 'people's language?' Does it lie in the facts of its origin, its 'invention,' or in the particular conditions of its use? The extent to which either French or Creole will be the language of the people depends largely on *how* it is used as much as upon *how widely* it is used and by whom.

A similar confusion is suggested in Fanon's assertion in *Black Skin, White Masks* that 'To speak a language is to take on a world, a culture. The Antilles negro who wants to be white will be the whiter as he gains greater mastery of the cultural tool that language is' (1952 [1967a]: 38). The key to this astute perception is the term 'take on'. For there can be no doubt that a colonial language gives access to authority and a perception of a certain form of social being. But this access is not gained as a feature of the language itself, through a process by which the speaker absorbs, unavoidably, the culture from which the language emerges. This new, comprador identity comes about through the act of speaking itself, the act of self assertion involved in using the language of the colonizer. The speaker 'takes on' the language rather than vice versa. This seems a small point but it is crucial because the speaking *need not necessarily* make the speaker 'more white' (itself an appropriated metaphor). The language is a tool which has meaning according to the way in which it is used.

This is, of course, a key to the importance of language as cultural capital. Time and again we have seen post-colonial politicians and intellectuals 'take on' (in Fanon's words) the colonizing language as a means of empowerment, a bank of cultural capital to be used for the purposes of self representation. But the ambivalence of this process lies in an intractable philosophical problem: does having a language amount to having a particular kind of world, a world that is simply not communicable in any other language? At base this problem rests on the question of meaning itself: of how meaning is communicated in texts.

The key to this problem of communicability between colonizer and colonized, lies in the fact that the written text is a social situation. That is to say, it has its being in something more than the marks on the page, for it exists in the participations of social beings whom we call writers and readers, and who *constitute* the writing as communication of a particular kind, as 'saying' a certain thing. When these participants exist in different cultures, as they do in post-colonial writing, two issues quickly come to the forefront: can writing in one language convey the reality of a different culture? and can a reader fully understand a different cultural reality being communicated in the text? One of the most persistent misconceptions about this activity is that the meaning of writing is a kind of static *a priori* to be uncovered; existing either as a function of the language itself, or the inscription of something in the mind of the writer, or the reconstruction of the reader's experience. Indeed, the very term 'meaning' tends to infer some objective content which is the end point of reading.

The capacity of post-colonial literatures to transform imperial discourse itself must begin by seeing that meaning is achieved constitutively as a product of the dialogic situation of reading. The 'objective' meanings of writing come about by a process of 'social' accomplishment between the writing and reading participants. This is because meaning is a social fact which comes to being within the discourse of a culture, and social facts as well as social structures are themselves social accomplishments. If we focus the meaning event within the usage of social actors who present themselves to each other as functions in the text, and see that cultural 'distance' is privileged at the site of this usage, it resolves the conflict between language, reader and writer over the 'ownership' of meaning. . . .

Meaning, then, is a social accomplishment characterized by the participation of the writer and reader 'functions' within the 'event' of the particular discourse. To take into account the necessary presence of these functions and the situation in which the meaning occurs, we can call the meaning a 'situated accomplishment'. The message 'event', the site of the 'communication' therefore becomes of paramount importance in post-colonial literatures because the 'participants' are potentially so very 'absent'.

Post-colonial writing affirms the primacy of the message event because the immense 'distance' between author and reader in the cross-cultural or sub-cultural text undermines the privilege of both subject and object and opens meaning to a relational dialectic which 'emancipates' it. This emancipation, however, is limited by the 'absence' which is often inscribed in the cross-cultural text – the gulf of silence or 'metonymic gap' (Ashcroft *et al.* 1989: 51–9) installed by strategies of language variance which signify its difference. It is precisely cultural *difference* rather than cultural *identity* which is installed in this way because identity itself is the function of a network of differences rather than an essence. Inscription therefore does not 'create meaning' by enregistering it, rather, it initiates meaning to a horizon of relationships circumscribed by that silence which ultimately resists complete interpretation. It is this silence, the metonymic assertion of the post-colonial text's difference, which resists the absorption of post-colonial literature into a universalist paradigm.

We can thus see how important is the cross-cultural literary text in questions of signification. Nothing better describes to us the distance traversed in the social engagement which occurs when authors write and readers read. But it is clear that the distances *are* traversed. Writing comes into being at the intersection of the sites of production and consumption. Although the 'social relationship' of the two absent subjects is actually a function of their access to the 'situation' of the writing, it is in this threefold interaction of situation, author function and reader function that meaning is accomplished.

## Language

The first contender for the 'ownership' of meaning is language, commonly held to embody or contain meaning either by direct representation or in a more subtle way by determining the perception of the world. Language in post-colonial societies, characterized as it is by complexity, hybridity and constant change, inevitably rejects the assumption of a linguistic structure or code that can be described by the colonial distinction of 'standard' and 'variant'. All language is 'marginal', all language emerges out of conflict and struggle. The post-colonial text brings language and meaning to a discursive site in which they are mutually constituted, and at this site the importance of usage is inescapable.

Words are never simply referential in the actual dynamic habits of a speaking community. Even the most simple words like 'hot', 'big', 'man', 'got', 'ball', 'bat', have a number of meanings, depending on how they are used. Indeed, these uses are the ways (and therefore what) the word means in certain circumstances. In his novel *The Voice* Gabriel Okara (1970) demonstrates the almost limitless prolixity of the words 'inside' and 'insides' to describe the whole range of human volition, experience, emotion and thought. Brought to the site of meaning which stands at the intersection between two separate cultures, the word demonstrates the total dependence of that meaning upon its 'situated-ness'.

Language cannot, therefore, be said to perform its function by reflecting or referring to the world in a purely contingent way, and thus meanings cannot remain exclusively accessible to those 'native' speakers who 'experience their referents', so to speak. The central feature of the ways in which words mean things in spoken or written discourse is the situation of the word. The ranges of 'nuance' and 'connotation' which are sometimes held to be the key to the incommunicability of cultural *experience* are simply functions of that situation. This is particularly important for its dismantling of the claims that a particular language has an essential and exclusive capacity to convey cultural truth. In general, one may see how the word is meant by the way it functions in the sentence, but the meaning of a word may require considerably more than a sentence for it to be adequately situated. The question remains whether it is the responsibility of the author in the cross-cultural text to employ techniques which more promptly 'situate' the word or phrase for the reader. While post-colonial writing has led to a profusion of technical innovation which exists to span the purported gap between writer and prospective reader, the process of reading itself is a continual process of contextualization and adjustment directly linked to the constitutive relations within the discursive event.

An alternative, determinist view, proposed by Whorf and Sapir, that language actually constructs the perceptions and experiences of speakers seems less problematic. The central idea of Whorf and Sapir's thesis is well known (see Sapir 1931 and Whorf 1952), proposing that language functions not simply as a device for reporting experience, but also, and more significantly, as a way of defining experience for its speakers. But even this more attractive view of the link between language and the world may give rise to a number of objections from constitutive theory. Clearly, language offers one set of categories and not another for speakers to organize and describe experience, but should we therefore assume that language *creates* meanings in the minds of speakers? While it is quite clear that language is more than a 'reproducing instrument for voicing ideas' (for what do thoughts or ideas look like apart from their expression in language?) the same objections can be applied to the idea of language as the 'shaper' or 'programmer' of ideas. Such ideas are still inaccessible apart from language. To possess a language is to possess a technique, not necessarily a quantum of knowledge about the world.

It is the situation of discourse, then, rather than the linguistic system in the speaker's mind in which the 'obligatory terms' of language are structured. The meaning and nature of perceived reality are not determined within the minds of the users, nor even within the language itself, but within the use, within the multiplicity of relationships which operate in the system. Margaret Atwood makes an interesting reference to a North American Indian language that has no noun-forms, only verb-forms. In such a linguistic culture the experience of the world remains in continual process. Such a language cannot exist if language is either anterior or posterior to the world but reinforces the notion that language inhabits the world, *in practice*. The semantic component of the sentence is contained in the syntax: the meaning of a word or phrase is its use in the language, a use which has nothing to do with the kind of world a user 'has in his or her head'.

What the speaker 'has in mind', like a linguistic system or culture, or intentions or meanings, are only accessible in the 'retrospective' performance of speaking. The categories which language offers to describe the world are easily mistaken to shape something in the mind because we naturally assume that, like the rules of chess, we hold the linguistic system 'in our minds', in advance of the world. But language is co-extensive with social reality, not because it causes a certain perception of the world, but because it is inextricable from that perception. Languages exist, therefore, neither before the fact nor after the fact but *in the fact*.

If the written text is a social situation the post-colonial text emphasizes the central problem of this situation, the 'absence' of those 'functions' in the text which operate to constitute the discursive event as communication: the 'writer' and 'reader'. The Author, with its vision and intentions, its 'gifted creative insight', has historically exerted the strongest claim upon the meaning of writing. But how *does* the non-English speaker, for instance, mean anything in English? Firstly, writers, like the language, are subject to the *situation*, in that they must say something *meanable*. This does not mean they cannot alter the language, to use it neologistically and creatively, but they are limited *as any speaker is limited* to a situation in which words have meaning. Literature, and particularly narrative, has the capacity to domesticate even the most alien experience. It does not need to *reproduce* the experience to signify its nature. The processes of understanding are therefore not limited to the minds of speakers of one mother tongue and denied the speakers of another. Meaning and understanding exist outside the mind, within the engagement of speakers using the language. Understanding is not a function of what goes on in the 'mind' at all, but a location of the word in the 'message event' – that point at which the language, the writer and reader coincide to produce the meaning. The cultural 'distance' detected at this point is not a result of the inability of language to communicate, but a product of the 'metonymic gap' installed by strategies of language variance which themselves signify a post-colonial identity.

## Chapter 64

# EDWARD KAMAU BRATHWAITE

# NATION LANGUAGE

From *History of the Voice: The Development of Nation Language in
Anglophone Caribbean Poetry* London and Port of Spain: New Beacon, 1984.

**T**HE CARIBBEAN IS A SET OF ISLANDS stretching out . . . on an arc of
some two thousand miles from Florida through the Atlantic to the South American coast,
and they were originally inhabited by Amerindian people: Taino, Siboney, Carib, Arawak. In
1492 Columbus 'discovered' (as it is said) the Caribbean, and with that discovery came the
intrusion of European culture and peoples and a fragmentation of the original Amerindian
culture. We had Europe 'nationalizing' itself into Spanish, French, English and Dutch so that
people had to start speaking (and thinking) four metropolitan languages rather than possibly
a single native language. Then with the destruction of the Amerindians, which took place
within 30 years of Columbus' discovery (one million dead a year) it was necessary for the
Europeans to import new labour bodies into the area. And the most convenient form of labour
was the labour on the edge of the *slave* trade winds, the labour on the edge of the hurricane,
the labour on the edge of Africa. And so Ashanti, Congo, Yoruba, all that mighty coast of
western Africa was imported into the Caribbean. And we had the arrival in our area of a new
language structure. It consisted of many languages but basically they had a common semantic
and stylistic form. What these languages had to do, however, was to submerge themselves,
because officially the conquering peoples – the Spaniards, the English, the French and the
Dutch – insisted that the language of public discourse and conversation, of obedience, command
and conception should be English, French, Spanish or Dutch. They did not wish to hear people
speaking Ashanti or any of the Congolese languages. So there was a submergence of this
imported language. Its status became one of inferiority. Similarly, its speakers were slaves.
They were conceived of as inferiors – non-human, in fact. But this very submergence served
an interesting interculturative purpose, because although people continued to speak English
as it was spoken in Elizabethan times and on through the Romantic and Victorian ages, that
English was, nonetheless, still being influenced by the underground language, the submerged
language that the slaves had brought. And that underground language was itself constantly
transforming itself into new forms. It was moving from a purely African form to a form which
was African but which was adapted to the new environment and adapted to the cultural imperative of the European languages. And it was influencing the way in which the English, French,
Dutch and Spaniards spoke their own language. So there was a very complex process taking
place, which is now beginning to surface in our literature.

Now, as in South Africa (and any area of cultural imperialism for that matter), the educational
system of the Caribbean did not recognize the presence of these various languages. What our
educational system did was to recognize and maintain the language of the conquistador, the

language of the planter, the language of the official, the language of the anglican preacher. It insisted that not only would English be spoken in the anglophone Caribbean, but that the educational system would carry the contours of an English heritage. Hence . . . Shakespeare, George Eliot, Jane Austen – British literature and literary forms, the models which had very little to do, really, with the environment and the reality of non-Europe – were dominant in the Caribbean educational system. It was a very surprising situation. People were forced to learn things which had no relevance to themselves. Paradoxically, in the Caribbean (as in many other 'cultural disaster' areas), the people educated in this system came to know more, even today, about English kings and queens than they do about our own national heroes, our own slave rebels, the people who helped to build and to destroy our society. We are more excited by their literary models, by the concept of, say, Sherwood Forest and Robin Hood than we are by Nanny of the Maroons, a name some of us didn't even know until a few years ago.[1] And in terms of what we write, our perceptual models, we are more conscious (in terms of sensibility) of the falling of snow, for instance – the models are all there for the falling of the snow – than of the force of the hurricanes which take place every year. In other words, we haven't got the syllables, the syllabic intelligence, to describe the hurricane, which is our own experience, whereas we can describe the imported alien experience of the snowfall. It is that kind of situation that we are in.

> The day the first snow fell I floated to my birth of feathers falling by my window; touched earth and melted, touched again and left a little touch of light and everywhere we touched till earth was white.
>
> (Brathwaite 1975: 7)

This is why there were (are?) Caribbean children who, instead of writing in their 'creole' essays 'the snow was falling on the playing fields of Shropshire' (which is what our children literally were writing until a few years ago, below drawings they made of white snowfields and the corn-haired people who inhabited such a landscape), wrote 'the snow was falling on the canefields' trying to have both cultures at the same time.

What is even more important, as we develop this business of emergent language in the Caribbean, is the actual rhythm and the syllables, the very software, in a way, of the language. What English has given us as a model for poetry, and to a lesser extent prose (but poetry is the basic tool here), is the pentameter. . . .

It is *nation language* in the Caribbean that, in fact, largely ignores the pentameter. Nation language is the language which is influenced very strongly by the African model, the African aspect of our New World/Caribbean heritage. English it may be in terms of some of its lexical features. But in its contours, its rhythm and timbre, its sound explosions, it is not English, even though the words, as you hear them, might be English to a greater or lesser degree. And this brings us back to the question . . . can English be a revolutionary language? And the lovely answer that came back was: *it is not English that is the agent. It is not language, but people, who make revolutions.*

I think, however, that language does really have a role to play here – certainly in the Caribbean. But it is an English which is not the standard, imported, educated English, but that of the submerged, surrealist experience and sensibility, which has always been there and which is now increasingly coming to the surface and influencing the perception of contemporary Caribbean people. It is what I call, as I say, *nation language*. I use the term in contrast to *dialect*. The word 'dialect' has been bandied about for a long time, and it carries very pejorative overtones. Dialect is thought of as 'bad English'. Dialect is 'inferior English'. Dialect is the language used when you want to make fun of someone. Caricature speaks in dialect. Dialect has a long history coming from the plantation where people's dignity is distorted through their language and the descriptions which the dialect gave to them. Nation language, on the other hand, is the submerged area of that dialect which is much more closely allied to the African aspect of experience in the Caribbean. It may be in English: but often it is in an English which is like a howl, or a shout or a machine-gun or the wind or a wave. It is also like the blues. And sometimes it is English and African at the same time. I am going to give you some examples. But I should tell you that the reason I have to talk so much is that there has been very little written on this subject. I bring to you the notion of nation language but I can refer you to very little literature, to very few resources. I cannot refer you to what you call an 'Establishment'. . . .

Now I'd like to describe for you some of the characteristics of our nation language. First of all, it is from, as I've said, an oral tradition. The poetry, the culture itself, exists not in a dictionary but in the tradition of the spoken word. It is based as much on sound as it is on song. That is to say, the noise that it makes is part of the meaning, and if you ignore the noise (or what you would think of as noise, shall I say) then you lose part of the meaning. When it is written, you lose the sound or the noise, and therefore you lose part of the meaning. . . .

In order to break down the pentameter, we discovered an ancient form which was always there, the calypso. This is a form that I think nearly everyone knows about. It does not employ the iambic pentameter [IP]. It employs dactyls. It therefore mandates the use of the tongue in a certain way, the use of sound in a certain way. It is a model that we are moving naturally towards now. Compare:

(IP)      To be or not to be, that is the question

(Kaiso)    The stone had skidded arc'd and bloomed into islands Cuba San Domingo.
                      Jamaica Puerto Rico (Brathwaite 1967: 48)

But not only is there a difference in syllabic or stress pattern, there is an important difference in shape of intonation. In the Shakespeare (IP above), the voice travels in a single forward plane towards the horizon of its end. In the kaiso, after the skimming movement of the first line, we have a distinct variation. The voice dips and deepens to describe an intervalic pattern. And then there are more ritual forms like *kumina*, like *shango*, the religious forms, which I won't have time to go into here, but which begin to disclose the complexity that is possible with nation language.

The other thing about nation language is that it is part of what may be called total expression. . . . Reading is an isolated, individualistic expression. The oral tradition on the other hand demands not only the griot but the audience to complete the community: the noise and sounds that the maker makes are responded to by the audience and are returned to him. Hence we have the creation of a continuum where meaning truly resides. And this total expression comes about because people be in the open air, because people live in conditions of poverty ('unhoussled') because they come from a historical experience where they had to rely on their very breath rather than on paraphernalia like books and museums and machines. They had to depend on immanence, the power within themselves, rather than the technology outside themselves. . . .

The other model that we have and that we have always had in the Caribbean, as I've said before, is the calypso, and we are going to hear now the Mighty Sparrow singing a *kaiso* which came out in the early sixties. It marked, in fact, the first major change in consciousness that we all shared. . . . In 'Dan is the Man in the Van' he says that the education we got from England has really made us idiots because all of those things that we had to read about – Robin Hood, King Alfred and the Cakes, King Arthur and the Knights of the Round Table – all of these things really haven't given us anything but empty words. And he did it in the calypso form. And you could hear the rhyme-scheme of this poem. He is rhyming on 'n's' and 'l's' and he is creating a cluster of syllables and a counterpoint between voice and orchestra, between individual and community, within the formal notion of 'call and response', which becomes typical of our nation in the revolution.

(Solo)     According, to de education you get when you small
     You(ll) grow up wi(th) true ambition an respec for
     one an all
     But in my days in school they teach me like a fool
     THE THINGS THEY TEACH ME A
     SHOULDA BEEN A BLOCK-HEADED
     MULE

(Chorus)   *Pussy has finish his work long ago*
     *An now he restin an ting*
     *Solomon Agundy was born on a MunDEE*
     DE ASS IN DE LION SKIN.

                                     (Sparrow 1963: 86)

I could bring you a book, The Royal Reader, or the one referred to by Sparrow, Nelson's West Indian Reader by J. O. Cutteridge, that we had to learn at school by heart, which contained phrases like: 'the cow jumped over the moon', 'ding dong bell, pussy in the well', 'Twisty & Twirly were two screws' and so on. I mean, that was our beginning of an understanding of literature. 'Literature' started (startled, really) literally at that level, with that kind of model. It was all we had. The problem of transcending this is what I am talking about now. . . .

Today, we have a very confident movement of nation language. In fact, it is inconceivable that any Caribbean poet writing today is not going to be influenced by this submerged/emerging culture. . . . at last, our poets, today, are recognizing that it is essential that they use the resources which have always been there, but which have been denied to them – and which they have sometimes themselves denied.

## Note

1   The Maroons were escaped slaves who set up autonomous societies throughout Plantation America. Nanny of the Maroons, an ex-Ashanti Queen Mother, is regarded as one of tbe greatest of the Jamaica freedom fighters. See Brathwaite (1977).

## Chapter 65

# CHANTAL ZABUS

# RELEXIFICATION

From *The African Palimpsest: Indigenization of Language in the West African Europhone Novel* Cross Cultures 4, Amsterdam and Atlanta, Ga.: Editions Rodopi, 1991.

> Who are you people be? If you are coming-in people be, then come in.
> Gabriel Okara, *The Voice*

SUCH WOULD BE THE INVITATION to the frowning, recalcitrant reader into a realm where a seemingly familiar language conveys an unfamiliar message. When the West African writer attempts to simulate the character of African speech in a Europhone text, some process is at work which has never been adequately described. Indeed, the terminology used to identify such an approach in current literature, whether it be linguistic or literary studies or the writer's own assessment of his method(s), has been misleading because of the confusion with the notion of 'translation' as well as other equally inaccurate terms. Such terms as 'transference' or 'transmutation' have indeed permeated studies with particular reference to West Africa and its literary output in English (see especially Kachru 1982b; Shridar 1982; Bokamba 1982). Against this unsatisfactory nomenclature, I propose the linguistic term 'relexification.'

Loreto Todd's felicitous formulation – 'the relexification of one's mother tongue, using English vocabulary but indigenous structures and rhythms' (Todd 1982: 303) – best describes the process at work when the African language is simulated in the Europhone text. The emphasis is here on the lexis in the original sense of speech, word or phrase and on *lexicon* in reference to the vocabulary and morphemes of a language and, by extension, to word formation. As we shall see, this concept can be expanded to refer to semantics and syntax, as well. I shall thus here redefine relexification as the making of a new register of communication out of an alien lexicon. The adjectives 'new' and 'alien' are particularly relevant in a post-colonial context in which the European language remains alien or irreducibly 'other' to a large majority of the West African population . . . and a 'new' language is being forged as a result of the particular language-contact situation in West Africa and the artist's imaginative use of that situation.

Relexification is often diachronic despite its synchronic aspects. As Achebe points out with regard to the 'new English' of his novels, 'the beginning of this English . . . was already there in [his] society, in popular speech and [he] foresee[s] the possibility for a lot more Africanization or Nigerianization of English in [his] literature' (Egejuru 1980: 49). However, the writer's innovations, whether lexico-semantic or morpho-syntactic, may not reflect variations in current oral usage in West Africa. For instance, Igbo people today use 'eleven' instead of 'ten and one' (Igbo: *iri no otu*), which Achebe used in *Things Fall Apart* to render the traditional Igbo counting system (Achebe 1958: 37; Zaslavsky 1973: 47). Relexification in its diachronic function should therefore not be confused with what has been called 'nativization' (Shridhar 1982) or 'Africanism' in reference to 'any English construction that reflects a structural property of an African language' (Bokamba 1982: 78). Nor does it bear any resemblance to any purely synchronic function of relexification verging on mother-tongue interference, calquing (also calking) or loan-translation. 'Africanization,' as well as 'indigenization,' however, refer to larger strategies of cultural decolonization . . . which are to be understood against the general background of African and 'Third World' economic de-linkage with Western supremacy.

Whether Alfred Sauvy meant it or not when he coined the phrase – *le tiers monde* – after the French *tiers état*, the 'Third World' has become the site of the 'third code' . . . This new register of communication, which is neither the European target language nor the indigenous source language, functions as an 'interlanguage' or as a 'third register,' after Irwin Stern's analysis of the new Portuguese/Kimbudu language in Angolan Luandino Vieira's fiction. Such a register results from the 'minting,' to borrow Jahn's numismatic phrase, the 're-cutting,' as Sartre contends in 'Orphée noir,' or the 'fashioning [out],' as Achebe would have it, of a new European-based novelistic language wrung out of the African tongue (Jahn 1966: 239–42). When relexified, it is not 'metropolitan' English or French that appears on the page but an unfamiliar European language that constantly suggests another tongue. As such, it is closer to verbatim or loan-translation than to translation *per se* but, as we shall see, it differs from both. So when Gabriel Okara writes: 'My insides smell with anger' or when Ahmadou Kourouma writes: 'Il n'avait pas soutenu un petit rhume,' both writers have relexified two West African languages – Ijo and Maninka – into English and French, respectively.

Although they do not use the term 'relexification,' writers unanimously recognize that the process at work when they write novels in European languages involves 'some sort of translation' that 'approximates' the meaning in the source language. . . . Gabriel Okara is of the opinion that 'the only way to use [African ideas] effectively is to translate them almost literally from the African language native to the writer into whatever European language he is using as his medium of expression' (Okara 1963: 15). On the Francophone side, Cameroonian Francis Bebey attempts to 'extract the essence of Douala and put it alongside the essence of French so as to attain a very enriched cultural level' (Bebey 1979: 112). The Senegalese writer Cheikh Hamidou Kane 'doesn't think that the use of French modifies his style or intent but rather that his use of French is modified' (by Peulh/Pular), whereas the Guinean writer Camara Laye admits that he loses 'a lot in the transposition from Malinke into French' (Egejuru 1980: 35, 42).

Such remarks are not confined only to West Africa. Thus, the South African novelist and critic, Es'kia Mphahlele, concurs:

> I listen to the speech of my people, to the ring of dialogue in my home language and struggle to find an approximation to the English equivalent . . . it is really an attempt to paraphrase a single Sotho expression.
>
> (Mphahlele 1964: 303–4)

Two years earlier, in 1962, the African writers attending the Makerere Conference, of which Mphahlele wrote a press report, had reached the following consensus:

It was generally agreed that it is better for an African writer to think and feel in his own language and then look for an English transliteration approximating the original.

(Mphahlele 1962: 2)

. . . Relexification is thus tied to the notions of 'approximation' and of 'transparence.' Yet, it also encompasses those of 'transposition,' 'paraphrase,' 'translation' (even 'psychic'), 'transliteration,' 'transference' and 'transmutation.' To make matters worse, there is some disagreement about the process of ideation behind relexification. Es'kia Mphahlele points to the difficulty of 'peg[ging] the point at which [he] stop[s] thinking in [his] mother tongue and begin[s] to think in English, and vice versa' (Mphahlele 1964: 304). Also, Olympe Bhely-Quenum from Benin confesses that he writes in Fon or Yoruba – 'two or three lines,' then translates his thoughts and develops the original idea into the target language (1982: 14). Here, the writer admits to the possibility of an African-language original, which, under the guise of carelessly jotted notes – 'two or three lines' – or elaborate literary fragments, is not visible in the record of literary history. This phenomenon is said to have existed, in various degrees, in European Renaissance verses that poets would first compose in Latin and then expand into a French version. In that respect, the fate of Latin which can be understood in terms of either death or creolization, may be relevant to the future of African languages, where writers continue writing in European languages and refuse to fecundate oral art with writing, thereby practising some sort of 'linguistic contraception.'

Relexification also differs from 'auto-translation,' that is, the translation of one's own work, in that the latter posits that the original work is visible. Such is the case with Ghanaian Kofi Awoonor's translation in English of his own poem 'I Heard a Bird Cry,' originally written in Ewe. About the English version he submitted to *Black Orpheus* he has this to say:

Some kind of transference of perception takes place. I move from one linguistic dimension into another totally different, sometimes violently different one. . . . Within me these two things exist side by side so I can move across these boundaries with, I hope, absolute ease. So at times ideas that exist in the Ewe, or that have taken place in the Ewe poems, either get mutated or expanded or contracted, whatever, depending upon words and what technical mode I want to use in English.

(Awoonor in Lindfors *et al.* 1972: 49)

Auto-translation . . . is a contemporary yet exceptional trend in African writing, as epitomized by Ngũgĩ wa Thiong'o's own English translations of his works composed in Gĩkũyũ.

John Pepper Clark concurs with Awoonor's notion of mutation through expansion or contraction of ideas in the source language: 'a thought you have has been very well expressed already in your mother tongue; you like that manner of expression so much you want to transplant it into English' (Clark 1972: 68). In the absence of an original, writers add cautiously that one must not assume that the African writer thinks first in his mother tongue and then translates his thoughts into the target language. Chinua Achebe warns us: 'If it were such a simple, mechanical process, I would agree that it was [a] pointless . . . eccentric pursuit' (1975: 102). Writers' psychosocial attitudes towards relexification vary considerably but it is preferable to leave questions related to such attitudes to psycholinguists and transformational grammarians.

Unlike interpretive translation or the 'lesser' activity of *transcodage* which both take place between two texts – the original and the translated version – relexification is characterized by the absence of an original. It therefore does not operate from the language of one text to the other but from one language to the other within the same text. Such texts are . . . palimpsests for, behind the scriptural authority of the target European language, the earlier, imperfectly erased remnants of the source language are still visible. Just as these remnants may lead to the discovery of lost literary works of centuries long past, the linguistic remnants inhabiting the relexified text may lead to the discovery of the repressed source language. The linguistic notions of 'source language' and 'target language' are therefore retained because of the underlying implication that the un-earthing of these debris inevitably leads to the 'source' of the native culture-based text, without its idyllic, bucolic connotations of a return to the etymological and cultural roots of African culture. Although the 'act of reading' a palimpsest lies beyond the scope of this study, one should stress the uniqueness and particularity of each 'message-event' and its openness for a wide variety of readings when consumed, by different audiences in different socio-linguistic contexts.

What distinguishes relexification from translation is not only the absence of a separate original. Relexification takes place, as already suggested, between two languages within the same text. Although these two languages are unrelated, they interact as dominant vs. dominated languages or elaborated vs. restricted codes, as they did and still do to some extent in West Africa where the European language is the official language and the medium of prestige and power. As it hosts such warring tendencies, relexification is a strategy *in potentia* which transcends the merely methodological. On the methodological level, it stems from a need to solve an immediate artistic problem: that of rendering African concepts, thought-patterns and linguistic features in the European language. On the strategic level, relexification seeks to subvert the linguistically codified, to decolonize the language of early, colonial literature and to affirm a revised, non-atavistic orality via the imposed medium.

Whereas the method of pidginization grounds the character in his/her supranational or urban identity, relexification grounds the character in a specific ethnicity. Relexification is therefore statistically more recurrent than pidginization in novels with a local or rural setting, hereafter called native culture-based novels. The degree of pidginization is thus in inverse ratio to the degree of relexification. For instance, Chinua Achebe's *A Man of the People* (1960) contains ninety-three pidgin utterances whereas *Things Fall Apart* (1958) contains only three of them. As a rule, there is a higher incidence of relexifying devices as the work comes closer to orality. This should come as no surprise since such texts are relexified from languages that have remained essentially oral and belong to the vast corpus of oral human discourse, for most languages spoken by humans over the millennia have no connection with writing. The relexified medium goes beyond the literary utilization of existing popular idioms and, in its world-creating aspect, transfigures the glottopolitical situation through the creation of a new form of literary expression.

# PART ELEVEN

# The Body and Performance

**T**HE 'DIFFERENCE' OF THE POST-COLONIAL SUBJECT by which s/he can be 'othered' is felt most directly and immediately in the way in which the superficial differences of the body and voice (skin colour, eye shape, hair texture, body shape, language, dialect or accent) are read as indelible signs of the 'natural' inferiority of their possessors. As Fanon noted many years ago, this is the inescapable 'fact' of blackness, a 'fact' which forces on 'negro' people a heightened level of bodily self-consciousness, since it is the body which is the inescapable, visible sign of their oppression and denigration. In a more general way the 'fact' of the body is a central feature of the post-colonial, standing as it does metonymically for all the 'visible' signs of difference, and their varied forms of cultural and social inscription, forms often either undervalued, overdetermined or even totally invisible to the dominant colonial discourse. Yet, paradoxically the resulting self-consciousness, as Fanon perceived, can drive the very opposition which can undo this stereotyping.

Bodily presence and awareness in one sense or another is one of the features which is central to post-colonial rejections of the Eurocentric and logocentric emphasis on 'absence', a rejection which positions the Derridean dominance of the 'written' sign within a larger discursive economy of voice and movement. In its turn this alter/native discursive and inscriptive economy which stresses the oral and the performative is predicated upon the idea of an exchange in which those engaged are physically present to one another. In oral performance the meaning is made in the exchange that is a *sine qua non* of orality. The oral text is not synonymous with the written inscriptions or oral 'texts' collected by anthropologists and others in recent years. In practice the oral only exists and acquires meaning in the possibility of an immediate and modifying response, existing therefore only interactively with its whole speech or movement event.[1] In other words the real body is acknowledged in such an exchange in a way in which the 'pale' material concerns of recent theory are readily dissolved.

In most written accounts the oral is overdetermined even in the act of being recorded and celebrated by the written. This is what usually passes for an acknowledgement of the 'oral' and 'performative'. This inferior positioning replicates the larger positioning of the oral and performative within the economy of communication in the modern world. In 'modern' societies the oral and the performative continues to exist alongside the written but it is largely ignored or relegated to the condition of pretext in many accounts, represented as only the beginning or origin of the written. Yet in many post-colonial societies oral, performative events may be the principal present and modern means of continuity for the pre-colonial culture and may also be the tools by which the dominant social institutions and discourses can be subverted or repositioned, shown that is to be constructions naturalized within a hierarchized politics of difference. At a further metaphorical level the body and the voice often become the sign in post-colonial written texts for alter/native cultures which can only exist within the written as a disruption or a gap, simultaneously unbridgeable and yet bridged by the written word. Such a trace is crucial to the position of post-colonial texts that seek to record the continuing presence of oral, performative cultures of colonized groups within the predominantly written discourse of the colonizer.

The body itself has also been the literal 'text' on which colonization has written some of its most graphic and scrutable messages. The punishment machine of Kafka's nightmare story 'In the Penal Colony', which literally inscribes on the dying body of the transgressor the name of his 'crime', is a powerful allegorization of what was, too often, a literal reality in both slave colonies and penal colonies, as Gillian Whitlock's reference to the slave narrative of Mary Prince here makes clear.

The body, too, has become then the literal site on which resistance and oppression have struggled, with the weapons being in both cases the physical signs of cultural difference, veils and wigs, to use Kadiatu Kanneh's terms, symbols and literal occasions of the power struggles of the dominater and dominated for possession of control and identity. Such struggles have often articulated the further intersections of race with gender and class in the construction of the colonized as subject and subaltern.

All in all the recent interest in theorizing this complex interaction, whose trace as these extracts show can be found in varying ways across all post-colonial societies, is a site of some of the most provocative and challenging recent discussions of post-coloniality.

## Note

1    For a very illuminating and useful set of accounts of 'orality' see Karin Barber and P. F. de Moraes Farias (1989).

## Chapter 66

# FRANTZ FANON

## THE FACT OF BLACKNESS

From *Black Skin, White Masks* (trans. Charles Lam Markmann) New York: Grove Press, 1967.

THE BLACK MAN AMONG his own in the twentieth century does not know at what moment his inferiority comes into being through the other. Of course I have talked about the black problem with friends, or, more rarely, with American Negroes. Together we protested, we asserted the equality of all men in the world. In the Antilles there was also that little gulf that exists among the almost-white, the mulatto, and the nigger. But I was satisfied with an intellectual understanding of these differences. It was not really dramatic. And then . . .

And then the occasion arose when I had to meet the white man's eyes. An unfamiliar weight burdened me. The real world challenged my claims. In the white world the man of color encounters difficulties in the development of his bodily schema. Consciousness of the body is solely a negating activity. It is a third-person consciousness. The body is surrounded by an atmosphere of certain uncertainty. I know that if I want to smoke, I shall have to reach out my right arm and take the pack of cigarettes lying at the other end of the table. The matches, however, are in the drawer on the left, and I shall have to lean back slightly. And all these movements are made not out of habit but out of implicit knowledge. A slow composition of my *self* as a body in the middle of a spatial and temporal world – such seems to be the schema. It does not impose itself on me; it is, rather, a definitive structuring of the self and of the world-definitive because it creates a real dialectic between my body and the world. . . .

'Look, a Negro!' It was an external stimulus that flicked over me as I passed by. I made a tight smile.

'Look, a Negro!' It was true. It amused me.

'Look, a Negro!' The circle was drawing a bit tighter. I made no secret of my amusement.

'Mama, see the Negro! I'm frightened!' Frightened! Frightened! Now they were beginning to be afraid of me. I made up my mind to laugh myself to tears, but laughter had become impossible. . . .

My body was given back to me sprawled out, distorted, recolored, clad in mourning in that white winter day. The Negro is an animal, the Negro is bad, the Negro is mean, the Negro is ugly; look, a nigger, it's cold, the nigger is shivering, the nigger is shivering because he is cold, the little boy is trembling because he is afraid of the nigger, the nigger is shivering with cold, that cold that goes through your bones, the handsome little boy is trembling because he thinks that the nigger is quivering with rage, the little white boy throws himself into his mother's arms: Mama, the nigger's going to eat me up.

All round me the white man, above the sky tears at its navel, the earth rasps under my feet, and there is a white song, a white song. All this whiteness that burns me . . .

I sit down at the fire and I become aware of my uniform. I had not seen it. It is indeed ugly. I stop there, for who can tell me what beauty is?

Where shall I find shelter from now on? I felt an easily identifiable flood mounting out of the countless facets of my being. I was about to be angry. The fire was long since out, and once more the nigger was trembling.

'Look how handsome that Negro is! . . .'

'Kiss the handsome Negro's ass, madame!'

Shame flooded her face. At last I was set free from my rumination. At the same time I accomplished two things: I identified my enemies and I made a scene. A grand slam. Now one would be able to laugh.

The field of battle having been marked out, I entered the lists.

What? While I was forgetting, forgiving, and wanting only to love, my message was flung back in my face like a slap. The white world, the only honorable one, barred me from all participation. A man was expected to behave like a man. I was expected to behave like a black man – or at least like a nigger. I shouted a greeting to the world and the world slashed away my joy. I was told to stay within bounds, to go back where I belonged.

They would see, then! I had warned them, anyway. Slavery? It was no longer even mentioned, that unpleasant memory. My supposed inferiority? A hoax that it was better to laugh at. I forgot it all, but only on condition that the world not protect itself against me any longer. I had incisors to test. I was sure they were strong. And besides . . .

What! When it was I who had every reason to hate, to despise, I was rejected? When I should have been begged, implored, I was denied the slightest recognition? I resolved, since it was impossible for me to get away from an *inborn complex*, to assert myself as a BLACK MAN. Since the other hesitated to recognize me, there remained only one solution: to make myself known.

In *Anti-Semite and Jew*, Sartre says: 'They [the Jews] have allowed themselves to be poisoned by the stereotype that others have of them, and they live in fear that their acts will correspond to this stereotype . . . We may say that their conduct is perpetually overdetermined from the inside' (1965: 95).

All the same, the Jew can be unknown in his Jewishness. He is not wholly what he is. One hopes, one waits. His actions, his behavior are the final determinant. He is a white man, and, apart from some rather debatable characteristics, he can sometimes go unnoticed. He belongs to the race of those who since the beginning of time have never known cannibalism. What an idea, to eat one's father! Simple enough, one has only not to be a nigger. Granted, the Jews are harassed – what am I thinking of? They are hunted down, exterminated, cremated. But these are little family quarrels. The Jew is disliked from the moment he is tracked down. But in my case everything takes on a *new* guise. I am given no chance. I am overdetermined from without. I am the slave not of the 'idea' that others have of me but of my own appearance.

I move slowly in the world, accustomed now to seek no longer for upheaval. I progress by crawling. And already I am being dissected under white eyes, the only real eyes. I am *fixed*. Having adjusted their microtomes, they objectively cut away slices of my reality. I am laid bare. I feel, I see in those white faces that it is not a new man who has come in, but a new kind of man, a new genus. Why, it's a Negro! . . .

As I begin to recognize that the Negro is the symbol of sin, I catch myself hating the Negro. But then I recognize that I am a Negro. There are two ways out of this conflict. Either I ask others to pay no attention to my skin, or else I want them to be aware of it. I try then to find value for what is bad – since I have unthinkingly conceded that the black man is the color of evil. In order to terminate this neurotic situation, in which I am compelled to choose an unhealthy, conflictual solution, fed on fantasies, hostile, inhuman in short, I have only one solution: to rise above this absurd drama that others have staged round me, to reject the two terms that are equally unacceptable, and, through one human being, to reach out for the universal. When the Negro dives – in other words, goes under – something remarkable occurs.

Listen again to Césaire:

Ho ho
Their power is well anchored
Gained
Needed
My hands bathe in bright heather
In swamps of annatto trees
My gourd is heavy with stars
But I am weak. Oh I am weak.
Help me.
And here I am on the edge of metamorphosis
Drowned blinded
Frightened of myself, terrified of myself
Of the gods . . . you are no gods. I am free.

(Césaire 1946: 144)

THE REBEL: I have a pact with this night, for twenty years
I have heard it calling softly for me . . .

(Césaire 1946: 122)

Having again discovered that night, which is to say the sense of his identity, Césaire learned first of all that 'it is no use painting the foot of the tree white, the strength of the bark cries out from beneath the paint. . . .'

The discovery of the existence of a Negro civilization in the fifteenth century confers no patent of humanity on me. Like it or not, the past can in no way guide me in the present moment.

The situation that I have examined, it is clear by now, is not a classic one. Scientific objectivity was barred to me, for the alienated, the neurotic, was my brother, my sister, my father. I have ceaselessly striven to show the Negro that in a sense he makes himself abnormal; to show the white man that he is at once the perpetrator and the victim of a delusion.

There are times when the black man is locked into his body. Now, 'for a being who has acquired consciousness of himself and of his body, who has attained to the dialectic of subject and object, the body is no longer a cause of the structure of consciousness, it has become an object of consciousness' (Merleau-Ponty 1945: 277).

The Negro, however sincere, is the slave of the past. None the less I am a man, and in this sense the Peloponnesian War is as much mine as the invention of the compass. Face to face with the white man, the Negro has a past to legitimate, a vengeance to exact; face to face with the Negro, the contemporary white man feels the need to recall the times of cannibalism. A few years ago, the Lyon branch of the Union of Students From Overseas France asked me to reply to an article that made jazz music literally an irruption of cannibalism into the modern world. Knowing exactly what I was doing, I rejected the premises on which the request was based, and I suggested to the defender of European purity that he cure himself of a spasm that had nothing cultural in it. Some men want to fill the world with their presence. A German philosopher described this mechanism as *the pathology of freedom*. In the circumstances, I did not have to take up a position on behalf of Negro music against white music, but rather to help my brother to rid himself of an attitude in which there was nothing healthful.

## Chapter 67

## MICHAEL DASH

# IN SEARCH OF THE LOST BODY
## Redefining the subject in Caribbean literature

From 'In Search of the Lost Body: Redefining the Subject in Caribbean Literature' *Kunapipi* 2(1), 1989.

I N CÉSAIRE'S WRITING THE BODY has the last word. In his poetry and theatre he re-enacts the need to reintegrate the exiled subject in the lost body. In his epic poem *Cahier d'un retour au pays natal*, Césaire imagines the journey of the disembodied subject across the estranging waters and the eventual reintegration of the body with the *pays natal*. . . .

In order to embrace this mutilated *pays natal*, the subject must overcome his or her initial revulsion. He or she must radically redefine notions of time, space, beauty and power before return becomes possible, and must strip away all illusions – whether that of heroic prodigal, solemn demiurge or New World African – empty consciousness of all pretensions ('overboard with alien riches/overboard with my real lies') in order to achieve reintegration. The end of exile, the triumph over the estranging sea, is only possible when the subject feels his or her bonds with the lost body of the native land. The ego-centred attitude of saviour or reformer must yield to a humble realisation that the discourse of the island-body is more powerful. The *pays natal* is the realm of viscous damp where familiar meanings dissolve, of the unspeakable that eludes the systematising word.

The importance of Césaire's contribution to a tradition of Caribbean writing is his passionate concern with psychic 're-memberment', with the successful incarnation of the displaced subject. Without reference to Césaire, Harris describes this concern as 'a new corpus of sensibility' which imaginatively releases the deep archetypal resonances of 'the theme of the phantom limb – the re-assembly of dismembered man or god' (Harris 1981: 27). The *Cahier* ends with a triumphant vision of sensory plenitude as the subject is possessed by the lost island-body. In the final movement of this poem, the 'wound of the waters' yields its secret as it becomes the pupil of the eye, the navel of the world, an integrating Omphalos. The dream of '*La Rencontre Bien Totale*', the ecstatic abolition of all dualism, haunts Césaire's imagination. In Césaire's essay *Poésie et Connaissance* (1944) he describes the poetic ideal as a capacity to transcend oppositions, to achieve André Breton's vision of a 'certain point in mind' which could exist beyond contradictions. The dynamic image at the end of the *Cahier* – of the spiral, plunging in two directions – is an imaginative representation of the power of the reanimated body. The ideal of a restless, protean physicality is constantly invoked in his poetry. As *Intimité marine*, he states his poetic identity in terms of 'the neck of an enraged horse, as a giant snake. I coil I uncoil I leap'.

The images of dismemberment and reintegration so passionately stated in Césaire's epic poem recur throughout his poetic *oeuvre*. For instance the poem '*Corps perdu*' (which gives its name to the collection of poems) specifically deals with the retrieval of the lost body. Another poem that restates the theme of dismemberment is '*Dit d'errance*', which does invoke 'archetypal resonances', in Harris's words, in its reference to the indestructibility of the Egyptian god Osiris. The poetic subject assumes all dismemberments which have existed.

> All that ever was dismembered
> in me has been dismembered
> all that ever was mutilated
> in me has been mutilated . . .
>
> (Davis 1984: 102)

As Gregson Davis points out in his reading of the poem, the lines 'the goddess piece by fragment/put back together her dissevered lover' specifically refer to the reconstitution of Osiris by Isis. Césaire has a special priority in Caribbean writing because of this vision of the re-membered body. . . .

Césaire's writing never ceases to insist on the unstable nature of the world. His horror of stasis (*durcir le beau*), his belief that stability is a mirage, has created the possibility of isolating the ideal of unencumbered physical movement or the refusal of corporeal determinism in Caribbean literature. The ideal of revolutionary self-assertiveness is expressed through corporeal imagery. For instance, Frantz Fanon attempts to rewrite the body of colonised man, creating a new subject from the dismemberment and castration inflicted by the coloniser's destructive gaze. In *The Wretched of the Earth*, Fanon equates a reanimated body with the liberated voice of the revolutionary intellectual:

> It is a vigorous style, alive with rhythms, struck through and through with bursting life. . . . The new movement gives rise to a new rhythm of life and to forgotten muscular tensions, and develops the imagination.
>
> (Fanon 1967b: 177)

Fanon's images of verbal muscularity have a resonance in Caribbean writing in which revolutionary potential is evoked through the resurrected flesh. The reanimated body of the land in Jacques Roumain's *Masters of the Dew* and the erotic carnality of René Depestre's *Rainbow for the Christian West* are clear examples of spiritual awakening expressed in images of revitalised physicality.

The rewriting or reinventing of the subject does not always take the form of virile images of sexual hubris. Corporeal metamorphosis can take a totally different direction if the subject is defined in terms of an exemplary reticence or evasiveness. In Simone Schwarz-Bart's novel *The Bridge of Beyond*, the corporeal ideal is one of resilience, slipperiness and manoeuvrability. Bodies are repeatable, can be dissolved or can defy the force of gravity. For instance, Télumée deals with personal tragedy by imagining herself as floating free of the world and its destructive force:

> Then I would lie on the ground and try to dissolve my flesh: I would fill myself with bubbles and suddenly go light – a leg would be no longer there, then an arm, my head and whole body faded into the air, and I was floating.
>
> (Schwarz-Bart 1982: 104)

Her fantasy of an unencumbered body is an imaginative strategy designed to resist the desecrating force of her oppressive world. Schwarz-Bart's novel is a tribute to the survival of a particular group of women because of their imaginative powers. Her narrative is built around the tensions that separate the transcendental from the existential. Her main character yearns for a world divested of fixed, determining matter. The *morne* or hill which offers refuge exerts a vertical pull on the protagonist to counteract the downward pull of the plains with which fiery destruction and physical entrapment are associated.

In Schwarz-Bart's tale of female endurance, the subject is not aggressively impulsive but values suppleness and taciturn stoicism. In the face of the insults of her *béké* mistress Mme. Desaragne, she is 'ready to dodge, to slip between the meshes of the trap she was weaving with her breath'. She clings to this image of elusiveness until Mme. Desaragne disappears like starch dissolved in water. Schwarz-Bart's novel demonstrates the corrective power of the folk imagination. We have insight into a process of psychic *marronnage* that allows the individual to survive even in the most vulnerable circumstances.

This image of an ever-changing body emerges as an even more suggestive symbol in the work of Alejo Carpentier. In it an aesthetic of incompleteness offers an insight into a world where forms are unstable, where an intricate branching, adaptation and accretion governs the existence of all things. Carpentier's imagery is best explained in the symbolism of the grotesque as described by the Russian critic Bakhtin, in which the body 'is not something completed and finished, but open, uncompleted' (Bakhtin 1984: 364). In Carpentier's novel *Explosion in a Cathedral*, we are presented with a teeming world inhabited by fluid, evanescent form. Nothing has a fixed contour in this submarine world in which matter cannot be discriminated from non-matter. Esteban, Carpentier's protagonist, realises that this world resists being named or structured. In its unspeakable nature it defies the efforts of the comprehending subject:

> Carried into a world of symbiosis, standing up to his neck in pools whose water was kept perpetually foaming by cascading waves, and was broken, torn, shattered, by the hungry bite of jagged rocks, Esteban marvelled to realise how the language of these islands had made use of agglutinations, verbal amalgams and metaphors to convey the formal ambiguity of things which participated in several essences at once.
>
> (Carpentier 1972: 185)

The ambiguous space imagined by Carpentier is akin to Harris's zone of 'inarticulacy' or Bakhtin's 'unpublicized spheres of speech' in which 'the dividing lines between objects and phenomena are drawn quite differently than in the prevailing picture of the world' (Bakhtin 1984: 421). Esteban's field of vision does not focus on the concrete and the static but on a world of infinite metamorphosis that seems to defy language itself. It illustrates Harris's conception of Caribbean consciousness caught between sea and forest.

Post-modernism concentrates on the inadequacy of interpretation and the disorienting reality of the unexplainable. Caribbean writing exploits precisely this terrain of the unspeakable. In the radical questioning of the need to totalise, systematise and control, the Caribbean writer is a natural deconstructionist who praises latency, formlessness and plurality. In order to survive, the Caribbean sensibility must spontaneously decipher and interpret the sign systems of those who wish to dominate and control. The writing of the region goes beyond simply creating alternative systems to reflect the futility of all attempts to construct total systems, to assert the powers of the structuring subject. It is not simply a matter of deploying Caliban's militant idiom against Prospero's signifying authority. It is, perhaps, a matter of demonstrating the opacity and inexhaustibility of a world that resists systematic construction or transcendent meaning.

## Chapter 68

## RUSSELL McDOUGALL

## THE BODY AS CULTURAL SIGNIFIER

From 'Achebe's *Arrow of God*: The Kinetic Idiom of an Unmasking' *Kunapipi*
9(2), 1989.

T HE VARIOUS LEVELS OF RHYTHM in Achebe's fiction, from stylistic to
structural and thematic, have been explored before this (McCarthy 1985: 243–56). My
project is a related one: it is to focus on the human body as a verbal signifier that encodes
movement iconographically as a condition of culture. The complex kinetics of *Arrow of God*
relates directly to a theory of action, from which develops a hermeneutic practice: reading
as a dance of attitudes, criticism as participation. . . .[1]

I am opposed to the Eurocentric appropriation of Achebe's 'canon' for the metropolitan
'Great Tradition' and my purpose is to consider in detail the counter assertion, often announced
but rarely argued on the evidence. If this is a truly decolonizing fiction, the case cannot be
argued simply at the level of content (by authenticating social setting, folklore, the use of
proverbs, etc.); meaning needs to be explored in terms of mode. My interest is in the *attitudes*
of the image, the *strategies* of the narrative, the *placing* of the reader, the *cultural coding* of
those aesthetic principles that inform the whole process of the fiction.

When Ezeulu, Chief Priest of Ulu, decides to send Oduche to learn the ways of the
whiteman it is as an extension of himself, saying: 'I want one of my sons to . . . be my eye
there' (Achebe 1974: 45). But it is also as a sacrifice:

> it may even happen to an unfortunate generation that they are pushed beyond the
> end of things, and their back is broken and hung over a fire. When this happens they
> may sacrifice their own blood. This is what our sages meant when they said that a
> man who has nowhere else to put his hand for support puts it on his own knee.

Oduche is not only the eye; he is also the knee. But it is not only vision and sacrifice that
are linked in the duality of Ezeulu's motive for sending his son to join the whiteman: the eye
is to see, so that he may know the secret of the whiteman's power, and the knee is to lend
support to the arm put upon it in order to stabilize a collective body on the brink of collapse.
The connection between the two is inescapable. The knowledge that is to come from what
the eye sees and to stabilize the traditional power-base of the villages of Umuaro is correlated with a physical gesture, and that is aimed at maintaining the vertical position and balance
of the body under the severe stress of imperialism. . . .[2]

It is change, in the form of the whiteman, that poses the external threat. 'The world is
changing', Ezeulu tells Oduche. 'I do not like it. But I am like the bird Eneke-nti-oba. When
his friends asked him why he was always on the wing he replied: "Men of today have learnt

to shoot without missing and so I have learnt to fly without perching"' (45). This is Ezeulu's explanation of why he wants Oduche to be his eye. Of particular interest is the concept associated with Oduche-the-eye of adjusting to change by being constantly on the move; for, at the same time, the response to change that is embodied in Oduche-the-leg (upon which the hand seeks support) is the *arrest* of motion. This latter response is ritualized in the sacrifice of Oduche, while the ritual counterpart of the former is the dance: 'If anyone asks you why you should be sent to learn these new things tell him that a man must dance the dance prevalent in his time' (89). The attitude to change, then, is a complex one, of neutralizing it by embracing it, or of arresting it by making sacrifice to the god that is bringing it about. Dance bridges the poles of opposition embodied in this complexity of response to change.

If the image for responding to change by moving with it is the dancer, it must be admitted that the significant point, flexibility, focuses to a great degree on the legs. Peggy Harper writes: 'A characteristic body posture in [Nigerian] dance consists of a straight-backed torso with the legs used as springs, the knees bending and stretching in fluidly executing the rhythmic action patterns of the dance, and feet placed firmly on the ground' (1969: 289). The hand which seeks body-support upon the leg, then, implies the arrest of motion; not only is this 'logical', it refers as well to the traditional principle of 'supporting' as a stabilizing mode. This particular 'supporting' image requires the leg to bend, to provide a plane against which the hand can push, in order to gain the vertical impulse of stability. This seems obvious and pedantic perhaps; but it is worth stressing that the ultimate intention is 'straightness', so that, although motion is arrested, the image is underlined by an active potentiality – straighten*ing*. The iconography of the bended knee in sculpture often suggests the same context. The implication, centring on the knee, is of flexibility – which moves us considerably closer to the image of the dancer than the apparent contradiction in Ezeulu's responses to change might at first suggest. Although the African ideal of stability is generally vested in a flat-footed approach to the dance, and embodied in performance by a straight-backed torso, the complementary ideal of flexibility relies upon *bended* 'buoyant knees over stable feet' (Thompson 1979: 10). As the image of the dancer and the image of the 'supporting' knee are seen to merge, so do the respective associated poles of Ezeulu's apparently contradictory responses to change, in accord with the traditional dance dialectic relating flexibility to stability.

In sending Oduche to learn the whiteman's ways, Ezeulu bases his response to change upon the principle of flexibility, or, in other words, *innovation*; but the principle behind sacrificing the boy to the whiteman's god is one of stability, or *tradition*. One might even see the two responses as relating to two different perceptions of time, 'real' and 'mythic', the one permitting individual innovation and the other sacrificing the individual to the tradition of community. If this is so then we should not wonder that the image of the dancer and the image of the 'supporting' knee are not so polarized as Ezeulu's responses to change at first seem: as 'real' and 'mythic' time mesh. Positive proof of this is given in the form of call-and-response (and solo-and-circle), which provides a potent organizational metaphor throughout the novel. As Thompson writes, there is an 'overlap situation' in call-and-response that

> combines innovative calls (or innovative steps, of the leader) with tradition (the choral round, by definition blurring individuality). Solo-ensemble work, among the many other things it seems to accomplish, is the presentation of the individual as a figure on the ground of custom. It is the very perception of real and mythic time.
>
> (Thompson 1979: 43)

That Oduche is sent to the whiteman as both a sacrifice and 'to learn a new dance' (169) in fact suggests the mesh of tradition and innovation that is at the core of Ezeulu's response to the threat of the whiteman. This becomes most apparent when he makes use of the whiteman's attitude to time, an attitude expressed first in the Assistant District Officer's keeping the Priest waiting (as the ADO himself had been kept waiting while being 'broken in' at the Lieutenant Governor's dinner) and then by his imprisoning him. Imprisonment and other forms of coercive waiting enslave one to a future that makes the present meaningless – as Camus, Beckett and other Absurdists have demonstrated. But Ezeulu attempts to exploit the temporal condition of meaningless imprisonment (a symbolic condition of Western life, tyrannized as it is by imperial structure) so as to manipulate the traditional year of his people, in revenge for their disrespect: 'his real struggle was with his own people and the white man was, without knowing it, his ally. The longer he was kept in Okperi the greater his grievance and his resources for the fight' (176). The whiteman, of course, is *not* his ally, and so in the end he fails (and falls). If one is to move with the rhythm of change as the dancer to the drum, one must contribute one's own rhythm and not merely mark time. . . .

To say that innovation and tradition mesh in the overall motivation for sending Oduche to the whiteman is another way of saying that he is expected 'to learn a new dance' while maintaining in his mind the rhythm of the old. In other words, what Ezeulu requires of him is conceived by Achebe in the same terms as 'apart-playing': in the terms of cross-rhythmic interpretation. Oduche is to learn the ways of the whiteman without losing his commitment to the ways of his people. Unfortunately, however, the rhythm of the whiteman is not merely different in kind. The Western 'approach to rhythm is called divisive because we divide the music into standard units of time', Chernoff tells us:

> As we mark the time by tapping or clapping our hands, we are separating the music into easily comprehensible units of time. . . . It is this fact, that Western musicians count together from the same starting point, which enables a conductor to stand in front of more than a hundred men and women playing in an orchestra and keep them together with his baton. Rhythm is something we *follow*.
>
> (quoted in Thompson 1979: 41–2)

. . . Accordingly, Oduche is unable to cross the rhythm of the 'new dance' with the rhythm of his people; instead, he takes it on its own terms, and follows it. The next time we come across Oduche, after his father's command to tell the whiteman the old custom even as he learns the new, we see him instead 'speak up for the Lord' (49) against the Sacred Python – following instead of interpreting. The irony of his subsequent Christian naming (an imperial act of claiming), as the rock upon which the Church will be built ('Peter'), is that it is an image of solid inflexibility totally alien to African notions of support and stability. . . .

The hermeneutic principle of *Arrow of God* is one of fluid movement from one position to another, a dancing of attitudes which, in the reader's way of relating them, composes his/her own contribution to what the novel is *doing*. A sense of rhythm and of balance is needed to activate the shifting patterns of metaphor and to relate the different faces of truth, where truth is 'like a Mask dancing' and its 'characters' are permutations of its essence. . . .

The potential and the limits of individual participation in the communal context are dramatized as they are encoded by idiom in *Arrow of God*. At the core of the colonial relationship, as T. O. Ranger declares, is 'the successful manipulation and control of symbols' (Ranger 1975: 166). Ezeulu undoubtedly fails in this power struggle; but the novel does not. Nor does Idemili. Implicit in much 'criticism' of Achebe's fiction is the honorific judgement that it extends the Great Tradition of the Nineteenth-Century Novel in its European Heyday; were this true, from a post-colonial point of view it would be an accusation, for, as Gayatri Spivak and others have demonstrated, that form of fiction encodes the ideal of Empire (by investing narrative authority in omniscient or centralized perspective, by proposing concepts of universalized value, etc.) (Spivak 1987). The accusation, of course, is false. Achebe has captured the symbolic form of the novel from the 'central' tradition, and grounded it upon an aesthetic of movement and motion and agility – which, as he says, 'inform the Igbo concept of existence' and so, by a paradigm shift, reconstitute the nature and experience of fiction. The reader must engage with a kinetic performance, must participate in a process: '*Ada-akwu ofu ebe enene mmoo*; you do not stand in one place to watch a masquerade. You must imitate its motion' (Address given by Achebe at Guelph University, Ontario, 1984). The process is one of socialization and constant renewal, functioning by an 'overlap' of multiple perspectives – individual and communal, call-and-response, solo-and-circle – that redefines the imperial concept of the centre in African terms, in terms of slippage: as that blank space where innovation inscribes itself on the ground of tradition.

## Notes

1   This critical model is an adaptation of the 'metronome sense' advocated as essential to an understanding of African music by Richard Alan Waterman (1952: 211). I am indebted also to John Miller Chernoff's discussion of African music as an educational force (1981: 154).

2   The impetus for my continuing examination of the kinetics of Achebe's fiction comes from Robert Farris Thompson's discussion of African art history (Thompson 1979).

## Chapter 69

# HELEN GILBERT

## DANCE, MOVEMENT AND RESISTANCE POLITICS

From 'The Dance as Text in Contemporary Australian Drama: Movement and Resistance Politics' *Ariel* 23(1), 1992.

CRITICAL ANALYSIS OF DANCE in Australian plays is virtually non-existent even though almost all Aboriginal, and some feminist and other playwrights, use dance in their texts. Similarly, most theatre reviewers either fail to notice the dance, or classify it as spectacle, therein eliding its signifying practices with aesthetic (read normative) standards of judgment. Heavily influenced by Western theatrical and critical traditions, representations of dance in contemporary white (and to a lesser extent, black) Australian drama necessarily carry traces of earlier historical readings of European drama that link dance, somewhat ambivalently, with themes of harmony or chaos. . . .

Dramaturgically speaking, enactment of some kind of dance in a play does a number of things to the text. As a focalizing agent, it draws attention to the rhetoric of embodiment in all performance, something which is less apparent in dramatization of dialogue, especially within the conventions of realism. Even while bringing the body into focus, dance also spatializes, which is to say that it foregrounds proxemic relations between characters, spectators, and features of the set. The ever-shifting relational axis of space breaks down binary structures that seek to situate dance as either image or identity, and the spectator as observer rather than co-producer of meaning. Furthermore, situated within a dramatic text, dance often de-naturalizes theatre's signifying practices by disrupting narrative sequence and/or genre. What dance 'does' then, is draw attention to the constructedness of dramatic representation, which suggests that it can function as an alienating device in the Brechtian sense. This calls for analysis of its ideological encoding, an especially important project in criticism of postcolonial texts. . . .

In discussing specific instances of dance in contemporary Australian drama, I begin with David Malouf's *Blood Relations* because, in its refiguring of *The Tempest*, this play positions the dance within a tradition which it then subverts to expose the ideological assumptions which traverse the body in the masque scene of Shakespeare's colonial paradigm. The dance of the nymphs and reapers in Prospero's vision functions as a representation of ritualized social harmony and appropriates the New World with its portent of bounteous harvest. It also desexualizes the body by linking Miranda and Ferdinand's forthcoming union with images of the fruition of nature, and, significantly, denies illegitimate sexual desire by excluding Caliban from the spectacle. Malouf refigures Shakespeare's dance as part of a carnivalesque magic show layered with ironic and apocalyptic overtones. The performance is neither at the behest nor under the control of Willy (the Australian Prospero), and it clearly expresses conflict rather than harmony. Kit, who functions as an Ariel figure, engineers the show despite Willy's opposition; then he foregrounds the dance as homosexual display when he dances in an exaggerated fashion with Dash (Trinculo), eclipsing Cathy/Miranda and Edward/Ferdinand's performance as the 'happy couple'. During their movements, the characters' dialogue operates as a running meta-commentary on the dancing itself, stressing its theatricality, which provides the spectator with a method for deconstructing the illusionistic devices of representation. Meanwhile, Dinny/Caliban, the Aboriginal character in the play, makes a potent statement of political autonomy when he declines an invitation to join the dance, and then disrupts it completely with a recitation of Caliban's 'This island's mine . . . 'speech from *The Tempest*. By insisting on staging his own 'show', Dinny refuses the inscriptions of white ritual movement on his body and holds the whole performance, and its Shakespearean prototype as well, up to scrutiny.

As well as resisting identities imposed by the dominant culture on individuals or groups and/or abrogating the privilege of their signifying systems, dance can function to recuperate postcolonial subjectivity because movement helps constitute the individual in society. . . .

Movement as producer of one's self and one's culture has special significance for reading the dance as text in Aboriginal plays. In imperial historical accounts, Aboriginal dance has been encoded as the expression of savage or exotic 'otherness' within a discourse which represents blacks as objects to be looked at, rather than as self-constituting subjects. W. Robertson, for example, writing in 1928, constructs Aboriginal dance during a *corroboree* as the picturesque signifier of less-than-human behaviour. 'The weirdly painted natives issuing from the dense bracken of the bush to perform the dances, looked more like wraiths than human beings. . . .' (Robertson 1928: 95). . . .

These descriptions, though purportedly historical accounts, clearly use theatrical conventions to conflate nature and the indigene, marking the dance as a 'primitive' performance event designed for consumption by the imperial spectator. Along with some notion of theatrical order (an implied programme), Robertson's narrative points to the use of costuming and make-up (the painted bodies), while evoking backstage areas in the 'dense blackness of the bush' and a well-lit space by the fire where the compelling stage action occurs. This narrative exemplifies what Nietzsche argues is the constitutive basis of history – 'dramatistical' thought, or the ability 'to think one thing with another, and weave the elements into a whole, with the presumption that the unity of plan must be put into the objects if it isn't there' (quoted in White 1982: 53). In constructing a bush theatre to frame (read contain) the dance, and by situating himself as the impartial observer of a series of static 'pictures', Robertson naturalizes his perspective, renders invisible the appropriative function of the historian's gaze, and militates against the threat of difference that the Aboriginal dance with its 'frenzied movements' poses. He can thus categorize the dancers as more wraith-like than human and relegate the *corroboree* to the realm of the fantastic, the fictional, the infernal, reserving the notion of 'real' dance for the dominant culture by marginalizing its variants. Robertson's failure to acknowledge the dancers' subjectivity prevents him from discerning any functional aspects of the *corroboree* vis-à-vis Aboriginal culture and certainly blinds him to the possibility of resistance politics. It is this representation of dance as reified spectacle that is problematized in contemporary Aboriginal drama if we focus on movement as part of identity formation/recuperation and spatial re-orientation. . . .

As an important mode of narrative in Aboriginal culture, dancing (or drawing with the body) can also function to restore masculine identity through its links with ritual and male initiation ceremonies. In Richard Walley's *Coordah*, Nummy, the 'local drunk' and 'trickster' figure, escapes the fixity of these roles formed within the dominant discourse of colonization by recreating his Aboriginality through dance performance (Walley 1989: 109–66). Similarly, dancing in a *corroboree* gives Billy Kimberly of Jack Davis's *No Sugar* an opportunity to transgress his assigned role of tracker/informant (Davis 1986). During the *corroboree*, individual identity is both created by, and subsumed in, group identity as culturally coded movement that gives valance to each performer's dance, allowing participants to shed their everyday roles determined within white hierarchies of power. In this sense, the dance acts as a shaman exorcizing evil. It is also an occasion for the exchange of cultural capital between tribes, and for the contestation of white dominated space. A recent production of *No Sugar* in Perth featured the dance as a potent tool for symbolic reclamation of Aboriginal land when, even after the performers' movements ended, their spatial inscriptions were clearly palpable through footprints on the sand and a visible layer of unsettled red dust. As Auber Octavius Neville, Chief Protector of Aborigines, walked tentatively across this ground in his three-piece suit to deliver a speech that situated Aborigines firmly within white historical discourse, traces of the *corroboree* marked his presence as incongruous, invasive, and ultimately illegitimate. (The published text of *No Sugar* does not place the scenes depicting the *corroboree* and Neville's speech in adjacency; however, director Phil Thompson worked closely with Jack Davis in this production.)

That *No Sugar* encodes the *corroboree* as a masculine activity (the female characters are denied participation *and* spectatorship) raises some problematic issues: on the one hand, it gives the dance a higher status as cultural production because all societies deem the occupations of men more important than those of women (Hanna 1987: 22–3); on the other hand, most of Davis's predominantly white audience will be tempted to read the performance from culturally subjectified standpoints that link dance with female activity, thereby seeing the *corroboree* as a feminizing practice. The ritual and spatial codings of the performance, however, resist this totalizing impulse by grounding the *corroboree* firmly in Aboriginal history and epistemology through its links with the Dreamtime, which, Stephen Muecke claims, is the 'constant supplementary signified of all Aboriginal narrative' (Muecke 1983: 98). The agency of white Australian theatre practice can also function to legitimate Aboriginal performance practices even while it necessarily compromizes them. As Penny Van Toorn argues in her analysis of minority texts and majority audiences, hybridized texts 'harness the power of valorizing signs recognized by the dominant audience in order to impart prestige to the valorizing signs' of their own culture (Van Toorn 1990: 112). . . .

Feminist Australian drama has also explored the dance's potential as transformative agent in identity recuperation. Dorothy Hewett, for example, has written plays in which dance features not only as enacted resistance to the appropriation of the female body within imperialist and/or patriarchal discourses but also as an active self-constituting process. A similar ideological project underscores the representation of dance in Sarah Cathcart and Andrea Lemon's monodrama, *The Serpent's Fall* (Cathcart and Lemon 1988), which uses the body/performance of a single actor playing six separate characters to produce a fluid concept of feminine gender identity while at the same time enacting sociocultural differences between the characters. The primary character, Sarah, an Australian actor, plays, alternately, a young archaeologist, a middle-aged Greek migrant, an urban Aborigine, a cafeteria boss, and a retired teacher – all but one of them women. As Sarah slips from one character to another without costume changes or breaks, using only mimed props, the performance enacts what Jill Dolan outlines as the two opposing feminist theories of the self, identity politics and poststructural notions of the decentred subject. . . .

These theories of self may both be applied to a reading of the play's dance, a seminal sequence in which Sarah in rapid succession performs dances as four separate characters. As she moves, her transitions between characters, and the traces of difference thus enacted by a single but composite body, both produce and deconstruct cultural and racial specificity, showing identity to be fluid but not without some sort of grounding in individual socio-historical circumstances. Not surprisingly, this dance functions as a mode of empowerment for all of its participants: for Kelly, the Aboriginal woman, it forms a link with the land; for Sula, a moment of resistance to the drudgery of the cafeteria and its patriarchal structures; for Bernice, the archaeologist, the possibility of rewriting the body against the discourses of power engendered in the biblical myth of genesis; and for Sarah herself, further confirmation that her own performance, and therefore her 'self' is, in a sense, intertextual. This composite, shifting sense of identity, which has much in common with that of the hybridized postcolonial subject, figures as the most important feature of *The Serpent's Fall* and its construction of the feminine, and the Australian. . . .

Whereas Aboriginal and feminist playwrights frequently use dance as a mode of empowerment for marginalized individuals/groups, other Australian plays represent dance more ambiguously. Although an in-depth examination of this topic is beyond the scope of my paper, Louis Nowra's work deserves mention. Almost all of Nowra's plays include some kind of dance, most of which enact a struggle for power and authority in tense, uneasy relationships between individuals and groups. From a postcolonial perspective, these struggles can be seen as emblematic of the colonizer/colonized dialectic, a process that, to some extent, hybridizes the identity of both dominating and subordinated groups. The Bal Masqué scene of *Visions* (Nowra 1979) enacts such a process in a combination of movement and verbal discourse in which Madam Lynch (representative of European imperial power) and Lopez (a somewhat ambivalent figure of colonial resistance) vie for control over the dance, which in symbolic terms can be seen as Paraguayan culture and even the country itself. . . .

In contemporary Australian drama, then, the dance emerges as a locus of struggle in producing and representing individual and cultural identity. As a site of competing ideologies, it also offers a site of potential resistance to hegemonic discourses through its representation of the body on stage as a moving subject that actually looks back at the spectator, eluding the kind of appropriation that the 'male gaze' theories of cinema outline. In Stanton Garner's terms, 'exploiting the body's centrality within the theatrical medium' allows the refiguring of the 'actor's body as a principal site of theatrical and political intervention, establishing (in the process) a contemporary "body politic" rooted in the individual's sentient presence' (Garner 1990: 146). Thus, reading/producing the dance as text provides an approach to drama that de-naturalizes notions of the self grounded primarily in language, and avoids privileging the performance of the mind over the performance of the body. As spectators with a split gaze that recognizes representation as distinct from embodiment, we *can* know the dancer from the dance.

## Chapter 70

# GILLIAN WHITLOCK

## OUTLAWS OF THE TEXT

From 'Outlaws of the Text: Women's Bodies and the Organisation of Gender in Imperial Space' from *The Intimate Empire: Reading Women's Autobiography* London: Continuum, 2000.

R ESPONSES TO EMPIRE IN SETTLER SOCIETIES, like Australia and Canada, comprise a site of contesting and conflicting claims, an array of identifications and subjectivities which refuse to cohere neatly into oppositional or complicit post-colonialisms. Settler post-colonialism confounds the positions of self and other in relation to discourse and discursive strategies; as a number of theorists of settler cultures argue, these 'second world' spaces are characterised by the ambiguity and ambivalence of both oppositional and complicit positions (Lawson 1991; Slemon 1990). Confusions of complicity and resistance in these cultures makes the identification of outlaws in settler territory a perilous enterprise. On these frontiers outlaws and sheriffs are not in predictable and fixed opposition but related and inter-dependent, mixed in hybrid forms which confuse the rule of Law. These ambiguities are all the more evident when we turn to settler women in particular, who occupy a terrain of 'shift-ingness' (Giles 1989). Here discourses of gender, class and race further complicate forms of complicity and resistance. The female body has always been crucial to the reproduction of Empire, and deeply marked by it. On the other hand it can also be at the bosom of de-scribing Empire. The anatomy of the female body in colonial space is the skeleton for the following discussion which ends, but does not begin, in the settler colonies. In fact [I want to examine this process] via Bermuda and the history of Mary Prince, an unlikely autobiographer. . . .

You may well ask how it is that we have available to us an autobiography by a woman born into slavery in the Crown colony of Bermuda in 1788. We learn from her *History* that she was sold as an infant, sold again in 1805 as an adolescent, and again as a woman in her twenties. Each time her *History* records experiences of degradation and brutality which reach a nadir on the salt ponds of Turks Island. She was sold for a fourth and final time to a merchant in Antigua, who took her to England with his family in 1828 to do their laundry. Here, at the height of the anti-slavery campaign, Mary Prince came to the attention of the Anti-Slavery Society and became a maid in the house of Thomas Pringle, the Secretary of the Society. . . .

The character of the *History* is shaped generically according to the form of the British slave narrative which, in 1830, prescribed a particularly limited sense of the intersections between gender and race in the life history of slave women.

For this form, this casting of a life of slavery, won public support by detailing atrocities and portraying slaves as pure and Christian-like, innocent victims and martyrs. Women whose cause they championed could not be seen to be involved in any form of moral corruption: 'Christian purity, for these abolitionists, overrode regard for truth' (Ferguson 1986: 4).

Without doubt this constrained Prince's ability to describe her experiences fully. The abusive sexual experience which were no doubt part of the brutality she suffered enter the text only in encoded ways. What is remarkable about the shape of Prince's *History* is that it attributes to her personhood, and spiritual equality. It is however blind to gender, to her womanhood. We know Prince married but we know no detail of her married life, or whether she became a mother at any stage of her career. All traces of her sexuality and likely sexual abuse are absent from the *History*. This was of course 20 years prior to the black abolitionist and freed slave Sojourner Truth's speech to the Akron convention in 1851. 'Aint I a Woman?' was the refrain of Truth's speech. It is a claim upon both race and gender, about how gender affects racial oppression, which is notably absent at Claremont Square in 1831. It is, then, by no means the case that the colonial body is gendered in the representation. Mary Prince's body is marked by her race and status.

There is at the end of Mary Prince's *History* an episode which makes the construction of her identity in these terms clear. Appendix Three was added to the third edition of the *History* following 'inquiries from various quarters respecting the existence of marks of severe punishments on Mary Prince's body'. Mary's narrative is, it seems not enough. So the amanuensis, the guest who copied Mary's story from her own lips, now views Mary's body and provides a testimonial, 'full and authentic evidence' that 'the whole back part of her body is severely scarred, and, as it were, *chequered*, with the vestiges of severe floggings'. '[There] are many large gashes on other parts of her person, exhibiting an appearance as if the flesh had been deeply cut, or lacerated with gashes, by some instrument wielded by most unmerciful hands' (Ferguson 1986: 119). There is no evidence here that it is a female body which is scarred and disfigured. . . .

Mary Prince's body is a reminder that as late as 1830 the most brutal forms of repression continued in the British Empire. In penal and slave colonies in particular savage physical punishment and coercion remained even as more subtle forms of self surveillance and discipline were instituted in Europe. Tyrannical ideals of order and precision could preside unchecked in the garrisons, penitentiaries and planter households of the settler and crown colonies. . . .

The body at the end of Mary's *History* also brings home to us that the gender of the female body becomes visible only in particular ways of knowing. . . .

I will do this by bringing a second image alongside that viewing of Mary's body at Claremont Square in the last week of March 1831. Within a week of testifying to Mary's scars the relationship between these two women took a different turn: here Mary becomes the spectator, the amanuensis the autobiographical subject. Let me, finally, allow the amanuensis to speak from a letter of April 9, 1831:

> I was on the 4th instant at St. Pancras Church made the happiest girl on earth, in being united to the beloved being in whom I have long centred all my affections. Mr Pringle 'gave me' away, and Black Mary, who had treated herself with a complete new suit upon the occasion, went on the coach box, to see her dear Missie and Biographer wed. . . .

This second glimpse, equally intriguing, of Mary Prince as Black Mary, resplendent in a new suit and perched on the coach box of the bridal carriage, is our last view of her. Of her dear Missie and Biographer we know more. Like so many women of her class who married a half pay officer with no prospects of inheritance, it was her fate to emigrate – to Upper Canada. From here she would pen another autobiographical document, one better known than Mary

Prince's *History*. It was published in 1852 as *Roughing It in the Bush*. . . .

The journey which Moodie began with Mary Prince as observer on the coach box was presented much less sentimentally in 1909 by Cicely Hamilton. In her first wave feminist polemic *Marriage as a Trade* Hamilton presents marriage as women's compulsory trade, a ceremony which marks their entry as wives and later mothers into the sexual economy of a patriarchal society. Throughout Hamilton's brutally rationalist discussion of marriage as trade in which the currency is women's bodies, we see the influence of imperialism upon her argument. The reproduction of British values and race following the passage of the bridal ships, the association of marriage and emigration which was so much a feature of nineteenth century European sexual politics, alerts Hamilton to the commercial basis of the marriage transaction.

Those of us who write about settler women in Australia and Canada as critics or, like Grenville and Marlatt, as authors need to grasp precisely the location of these emigrants in the imperial organisation of gender and race. For it was in the settler colonies that nineteenth century pro-natalist discourses assumed particular importance. As Marchant observed in 1916, in the difference between the number of cradles and the number of coffins lie the existence and persistence of our Empire. The dissemination of British institutions and society depended upon its emigrants; in colonies of occupation – India, Africa – European women were valued less than in settlement colonies. In fact Chilla Bulbeck has argued that in these colonies women were seen as wives but not mothers, in pre-Bowlby days women were expected to send their children 'Home' to school (Bulbeck 1991: 14). However, in settlement colonies the fertility of European women and the welfare of mothers and children were vital to the colonising project. The British Women's Emigration Association, for example, used marriage as one of its incentives to encourage women to emigrate. They also stressed this as an opportunity to civilise the world and secure British values in the colonies. White women as homemakers and mothers helped to maintain and promote the Empire through the biological and daily repro-duction of the settler population (Strobel 1991: 46). The uterus was singled out not only as the most important female organ, but the most important organ of the Race; as one imperialist opined: 'the uterus is to the Race what the heart is to the Individual' (Gallagher and Laqueur 1987). . . .

Of course this process of rearing racially and nationally identified children assumed particular valency in settler colonies; discourses of maternalism characterise the writings of settler women from the beginnings. Their gender and status as wife and mother were crucial to the politics of imperialism. Catharine Parr Traill's handbook *The Backwoods of Canada*, for example, is an instruction to European settler women in appropriate maternal behaviour. 'Maternal' here is to be understood in the widest terms, incorporating not only management of domestic 'capital', the making of soap and sugar, but also the 'husbanding' of local flora and wildlife. . . .

My argument here is that this child and that fusion of emigration with maternity are crucial to our understanding the historically specific context of settler writings by and about women. Discourses of maternalism and imperialism coalesced to produce that collective iden-tity of the immigrant gentlewoman which Moodie and her kind embodied – with varying degrees of success! By coming to this maternal body via Mary Prince, whose status as a mother is unknown and whose body is represented in terms of race and status rather than gender, I mean to stress that colonial subjects were by no means necessarily gendered in the repre-sentation. The case of Mary Prince reminds us of the tangle of distinct and variable relations of power and points of resistance in the field of Empire. . . .

## Chapter 71

# MARVIN CARLSON

## RESISTANT PERFORMANCE

From *Performance: A Critical Introduction* London and New York: Routledge, 1996.

**B**EYOND ART AS PROCESS is the idea of art as a means to make community rather than commodity. Imbedded in that is the need to discover and make connections between a culturally and spiritually dissociated past and our present social and political realities. . . .

Performance work based primarily upon autobiographical material and frequently dedicated to providing a voice to previously silenced individuals or groups became in the early 1970s, and still remains in the 1990s, a major part of socially and politically engaged performance; but other engaged performance has also been developed in quite different and, in general, more openly resistant ways. Here, as in identity performance, the lead has been taken, both in theory and practice, by women, although more recently gay men and ethnic minorities have continued to develop these strategies to address their own concerns. When the modern women's movement began in the late 1960s, it existed in quite a different world from the apolitical, formalist, gallery-oriented 'performance' work of the same period. Yet at the same time, many radical feminists were much attracted to the symbolic values and performance strategies of the radical guerrilla and street theater of the period. . . .

Little current critical performance follows the strategy so common in the late 1960s guerrilla performance of direct opposition, but an extremely wide variety of socially and politically engaged performance of a different sort has evolved, reflecting the concerns, tensions, and assumptions of a postmodern consciousness. When the very structure of the performative situation is recognized as already involved in the operations of the dominant social systems, directly oppositional performance becomes highly suspect, since there is no 'outside' from which it can operate. Unable to move outside the operations of performance (or representation), and thus inevitably involved in its codes and reception assumptions, the contemporary performer seeking to resist, challenge, or even subvert these codes and assumptions must find some way of doing this 'from within.' They must seek some strategy suggested by de Certeau's 'tactics,' which he sees as activities: that 'belong to the other,' outside the institutionalized space of 'proper' activity. A tactic, says de Certeau,

> insinuates itself into the other's place, fragmentarily, without taking it over in its entirety, without being able to keep it at a distance. It has at its disposal no base where it can capitalize on its advantages, prepare its expansions, and secure independence with respect to circumstances . . . because it does not have a place, a tactic depends on it – it is always on the watch for opportunities that must be seized 'on

the wing.' Whatever it wins, it does not keep. It must constantly manipulate events in order to turn them into 'opportunities.'

(De Certeau 1984: xix)

Without providing specific strategies for such operations, Butler and others have nevertheless contributed significantly to grounding such strategies by providing a theoretical orientation that accepts the postmodern suspicion of an empowered subject existing outside and prior to social formations without renouncing the possibility of a position of agency to oppose the oppressions of these formations (see Butler 1988, 1990a, 1990b, 1993). . . .

A typically postmodern double operation is involved in such performance; the constitution of the self through social performance is viewed as a dynamic simultaneously coercive and enabling.

The sort of double operations that Butler sees involved in social performance are closely related to the strategies of recent theorists and performers concerned with developing a resistant performance art in the cultural context of postmodern thought. The possibility, even the necessity, of critique if not subversion from within performative activity has become widely accepted, but the most effective performance strategies for such subversion remain much debated. The central concern of resistant performance arises from the dangerous game it plays as a double-agent, recognizing that in the postmodern world complicity and subversion are inextricably intertwined. . . .

Thi Minh-ha Trinh, a feminist film maker and theorist, stresses the importance of disrupting the traditional performance/reception process. Dominant ideology, she argues, is reinforced in the 'activity of both the performer and spectator' every time an interpretation of a work implicitly presents itself as a 'mere (obvious or objective) decoding of the producer's message.' This occurs in much traditional art, whenever there is 'a blind denial of the mediating Subjectivity of the spectator as a reading subject and meaning maker-contributor' or whenever artists 'consider their works to be transparent descriptions or immediate experiences of Reality "as it is".' Trinh encourages a kind of Brechtian alienation strategy to 'break off the habit of the spectacle by asking questions aloud; by addressing the reality of representations and entering explicitly into dialogue with the viewer/reader,' the goal being to disrupt the 'conventional role' of the spectators, so that they are no longer expected to discover 'what the work is all about,' but to 'complete and co-produce' it by addressing it 'in their own language and with their own representation subjectivity' (Trinh 1991: 93–4). Performance is no longer created by someone for someone, but is the expression of a plurivocal world of communicating bodies, where difference is 'conceived of not as a divisive element, but as a source of interactions; object and subject are neither in opposition nor merged with each other . . .' (136).

This multiplicity is particularly clear in the case of performers seeking to articulate the experience of ethnic minorities, since the pressures of otherness and of cultural stereotyping are so central to their experience and so inextricably intertwined with the exploration of an ethnic 'identity.' The potential performance complexity of this process has been illuminatingly explored by Rebecca Schneider in her analysis of *Reverb-ber-ber-ations* (1990) by the Native American performers Spiderwoman (Schneider 1993: 227–56). Schneider approvingly cites Diamond's discussion of the disruptive, critical power of mimicry, noting that this power has been explored not 'only within feminism but by theorists interested in developing a strategy to disrupt political colonialism (see Diamond 1995). Most notably, Homi Bhabha in 'Of Mimicry and Man' (1984b) locates a 'comic turn' within colonialism's use of representation, attempting to justify the domination of a subject people by creating representations of that people as 'ignoble, childish, primitive'. However, the representing of the 'other' as an incomplete or undeveloped 'same' also works against the process of domination by introducing the destabilizing carnival of mimicry, and the menace of its 'double, which in disclosing the ambivalence of colonial discourse also disrupts its authority' (Bhabha 1984b: 126). Schneider cites Spiderwoman's, restaging of 'Snake Oil Sideshows' (*Winnetou's Snake Oil Show from Wigwam City*, 1989) and similar 'exotic' material as comic, subversive mimicry that

> adroitly played in the painful space between the need to claim an authentic native identity and their awareness of the historical commodification of the signs of that authenticity. Their material falls in the interstices where their autobiographies meet popular constructions of the American Indian.
>
> (Schneider 1993: 237)

Performance of this kind slips back and forth between 'a firm declaration of identity' and parody of the social clichés that haunt that identity, insisting upon 'an experience of the sacred *despite* the historical corruption and compromise of identities.'

In an even more complex 'comic turn,' Schneider suggests that Spiderwoman also utilizes what she calls 'countermimicry.' In the 'Indian Love Call' sequence of *Sun, Moon, Feather* (1981), for example, the native American actresses

> don't act out the Indian parts – the virulent near-naked, dancing brave or the dark Indian Princess – they fight over who gets to be be-ringleted, vaseline-over-the-lens Jeanette MacDonald and who has to play stalwart, straight-backed Canadian Mounty Nelson Eddy. They are not re-playing, re-membering, or re-claiming native images but appropriating the appropriate.
>
> (Schneider 1993: 246)

Some of the most complex and chalenging recent ethnic performances have ultilized mimic or countermimic strategies to deal directly with the process of cultural stagings or representations of ethnicity. James Luna's Bessie-award winning performance/installation, *The Artifact Piece* (1987), was created in San Diego's Museum of Man and provided ironic commentary on its own ethnographic surroundings and their cultural functions. Luna placed himself on display, his body laid out in one case labeled with tags identifying scars inflicted during drunken fights. Another case showed his college degree, photos of his children, arrest record, and divorce papers along with objects used in contemporary Native American ceremonies. Explanatory panels described with mock ethnographic objectivity a modern Native American life in which AA meetings have taken the place of traditional ritual ceremonies (Mifflin 1992: 88).

A similar, but even more complex display was the well-known *Two Undiscovered Amerindians Visit* (1992) of Guillermo Gomez-Pena and Coco Fusco, first offered in Madrid and London, then in Australia and the United States, and finally the subject of a fascinating video documentary (Fusco 1993: 143–67). Drawing upon the once popular European and North American practice of exhibiting indigenous people from Africa, Asia, and the Americas in fairs, shows, and circuses, Gomez-Pena and Fusco displayed themselves for three days in a golden cage as recently discovered Amerindians from an island in the Gulf of Mexico. They performed such 'traditional tasks' as sewing voodoo dolls, lifting weights, and watching TV, were fed sandwiches and fruit and taken to the bathroom on leashes by guards. As with the Luna display, explanatory panels provided mock scientific information on the 'indians' and their 'native culture.' To the surprise of the performers, many viewers took their exhibition seriously, playing out a rich variety of accommodations or resistances to the display process itself and raising more complex questions about the cultural interpretation of this display than they had anticipated.

Current resistant performance is so varied and complex that no 'typical' example of it could be cited, but the *Undiscovered Amerindians* show illustrates one of the central current concerns, and that is the dynamic involved when a specific performer encounters a specific culturally and historically situated audience. What began in *Undiscovered Amerindians* as an ironic commentary on appropriation, representation, and colonial imaging through the playful reconstruction of a once popular and symbolically charged intercultural performance, became a far more complicated and interesting phenomenon as the performers gradually realized that all of these concerns had to be freshly negotiated, often in surprising and unexpected ways, in every new encounter.

## Chapter 72

# REINA LEWIS

## ON VEILING, VISION AND VOYAGE

From 'On Veiling, Vision and Voyage: Cross-cultural Dressing and Narratives of Identity' *Interventions* 1(4), 500–20, 1999.

I N 1915 THE BRITISH FEMINIST AND TURKOPHILE Grace Ellison wrote about her visit to Turkey on behalf of a British suffrage organization. Her book *An Englishwoman in a Turkish Harem* aimed to reveal the true state of Turkish women's lives and to challenge Orientalist stereotypes. She emphasized the high standards of education among elite women and the level of support among progressive men for female emancipation. Notably, Ellison also wrote in considerable detail about clothes, especially about the familiarity of elite Ottoman women with European clothes and furniture and the codes of conduct which accompanied these. Although changes in dress were often seen as signs of modernization, Ellison, despite her progressive feminist politics, romanticizes the Ottoman harem system (Melman 1992) and aestheticizes the veil, seeing it not as a mechanism of seclusion but as a fetching head-dress. She herself adores capering about in a veil.

These contradictory attitudes are seen most acutely in her 1928 work when she returns to review the new Turkish republic. . . .

In Turkish society, in the late pre-republican and early republican period, women's clothes, in particular the veil, were seen as a crucial index of political and social change by politicians of all persuasions, and inevitably discussed as such by local and foreign commentators (Graham-Brown 1988). Both women's and men's apparel were regularly the subject of sumptuary legislation (Seni 1995). Since the Tanzimat reforms of the mid-nineteenth century, the partial or wholesale adoption or adaptation of European fashions had become increasingly common among elite women, so women observers might often encounter women wearing the veil in conjunction with elements or adaptations of western dress (Micklewright 1999a). Women might dress entirely in Paris gowns to receive each other at home and often wore European clothes under their veils and cloaks (*feredges*) when outside, though European dress in this latter context would not, of course, be detectable to observers. Sometimes, wearing western clothes was part of a clear political statement but at the very least, dress, like the consumption of other western goods, signalled a vague sense of generally westernized modernity – though the advantages of western ways and the extent to which they should be embraced were hotly debated (Berkes 1964; Duben and Behar 1991; Göçek 1996). My focus here is on how dress, both Turkish and European, functions as an element of gender and ethnic performance and how this might register differently with those who wear it and read about it in both Orient and Occident. . . .

With her hostess Fatima's connivance, she dresses in a veil to visit the mosque and holy tomb at Eyoub where five years earlier she had 'had the humiliating experience of being refused admission to the tomb because I was wearing a hat; now I am wearing a veil who can tell whether I am Muslim or Christian?' (Ellison 1915: 162). . . . However, although she has some qualms about her undercover presence being interpreted badly, as religious and cultural disrespect, she is all laughter at having fooled some Europeans. . . .

Ellison is not unaware of the veil's negative qualities. She reports her discussion with the prominent Turkish feminist Halide Edib Adivar:

> Is it [the veil] protection or is it not? Halide Hanum considers that it creates between the sexes a barrier which is impossible when both sexes should be working for the common cause of humanity. It makes the woman at once the 'forbidden fruit', and surrounds her with an atmosphere of mystery which, although fascinating, is neither desirable nor healthy. The thicker the veil the harder the male stares. The more the woman covers her face the more he longs to see the features which, were he to see but once, would interest him no more.
>
> Personally I find the veil no protection. In my hat I thread my way in and out of the cosmopolitan throng at Pera. No one speaks to me, no one notices me, and yet my mirror shows I am no more ugly than the majority of my sex. But when I have walked in the park . . . a veiled woman, what a different experience. Even the cold Englishman has summoned up enough Turkish to pay compliments to our 'silhouettes'.
>
> (Ellison 1915: 69)

Ellison's delight in receiving the attentions directed (in Turkish) at a supposed Turkish woman indicates her investment in the thrill of passing as 'other'. It is not just that she adores the veil as an item of clothing, but that she also mistakes it for a sleight-of-hand way of temporarily inhabiting another identity. This in part explains her inability to realize the different significance of wearing the veil for Turkish and English women. . . .

Gail Low has written about the pleasures of cross-cultural dressing which, she argues, are often underpinned by a closely held sense of racialized differentiation. This might initially seem at odds with the profound delight taken by participants in dressing in local clothes and even passing as native. But for the westerner, she suggests, the pleasure of wearing an exotic and splendid 'native' costume is enhanced by the knowledge of the white skin underneath the disguise (Low 1989). For Richard Burton or Kipling's Kim, the ability to pass as a native is an important talent, both pleasurable (giving access to hidden native customs) and political (Kim after all spies for the Raj). Clothes, she argues, are important to the fantasy of cross-dressing because they are 'superficial' and can always be removed when one needs to revert to type, to reassert one's racial or cultural superiority. In contrast, Apter maintains that cross-dressing undermines previous conventions of absolute difference, and instances Loti's delight in passing not only as a Turk, but also as a veiled Turkish woman, as an example of such fluidity of boundaries. I am inclined to agree with Low that it is the transitory nature of this boundary-breaking that is significant. For Loti, as for Ellison, the thrill of cross-cultural dressing is predicated on an illicit reinvestment in the very boundaries they cross. Clothes operate as visible gatekeepers of those divisions and, even when worn against the grain, serve always to re-emphasize the existence of the dividing line. So cross-dressing offers both the pleasures of consumption – the Orient is a space full of enticing goods to be bought, savoured and worn – and the deeper thrill of passing as native. . . .

As well as enjoying dressing in Turkish clothes with her women friends, Ellison delights in going out and about in Istanbul in Turkish dress. As we have seen, she is thrilled when she hoodwinks people into thinking that she is Turkish. How does this link to the experience of hoodwinking her readers, who open the book expecting one thing (disreputable 'smoking room' tales of polygamy) and get another? If the logic of the veil is that one cannot identify the wearer, why does the caption 'Englishwoman wearing a yashmak' identify the nationality or race of the subject, but not her name? By presenting herself as willingly acculturated to Turkish life . . . Ellison suggests the positive aspects of haremization. But the book also links into another set of pleasures that Low associates with cross-cultural dressing, namely fantasies of power and surveillance. The undercover cross-culturally dressed agent embodies a mode of power based on a 'fantasy of invisibility' which imagines for an imperial gaze a state of omnipotence and omnipresence that is secret and voyeuristic rather than visible, as in the panoptican model (Low 1989: 95). . . . She can revel nor only in her gender-specific access to segregated domestic spaces but also in her ability to pass as Muslim and gain access to spaces forbidden to non-believers. By crossing both the gender and religious origins of harem (originally a protection and seclusion of holy space; see Peirce 1993), she can enjoy the pleasures of cultural transgression without having to give up the racial privilege that underpins her authority to represent her version of Oriental reality. . . .

A different experience of cross-cultural dressing is illustrared by Zeyneb Hanum and Melek Hanum. This is the reverse side of the charade. The power accruing to the European sartorial adventurer who can delight in the white skin underneath the native clothes is not available to two Turkish women who flee the constraints of Istanbul harems for a 'free' life in Europe. Nevertheless, they also start out with similarly fantastical expectations of exoticized difference:

> It seems to me that we Orientals are children to whom fairy tales have been told for too long – fairy tales which have every appearance of truth. You hear so much of the mirage of the East, but what is that compared to the mirage of the West, to which all Orientals are attracted?
>
> (Zeyneb Hanum 1913: 186–8)

For these two Turkish women travellers, Europe disappoints on several fronts. Unable to recognize the signs of western freedom as truly valuable (they cannot get over the point-lessness of sport as a leisure pursuit and the crazy pace of a Paris Season), they look through eyes which are both haremized and haremizing, finding the harem in Europe. Zeyneb Hanum, writing to Ellison in Istanbul, reports her stay in a 'Ladies' Club' in London and is not impressed with the experience:

> What a curious harem! and what a difference from the one in which you are living at present.
> . . . The silence of the room was restful . . . [but] it is the peace of apathy. Is this, then, what the Turkish women dream of becoming one day? Is this their ideal of independence and liberty? . . . A club, as I said before, is after all another kind of harem, but it has none of the mystery and charm of the Harem of the East.
>
> (Zeyneb Hanum 1913: 182–6)

Zeyneb Hanum exerts a haremizing gaze on the west that makes strange its familiar division of space and organization of sexuality. . . .

Although Zeyneb Hanum's references to seeing the outside world without a veil make it clear that the sisters were not veiled in Europe, their written account does not specify what they wear. The picture of Melek Hanum at Fontainbleau shows the existence of a European wardrobe and I suspect this is what they wore most of the time in Europe. Why then do the photographs of them in their book about their travels in Europe mainly show them wearing Turkish clothes? Many of the photographs are captioned with details about Oriental female costume, so I think it is quite likely that these are present as an ethnographic or historical supplement on Turkish female life rather than as a record of their time in Europe. Scenes such as 'Turkish ladies paying a visit' (Zeyneb Hanum 1913: 172) are clearly meant to picture their life in Turkey and complement the written part of the book which mixes accounts of life in Turkey with responses to Europe. The photographs work, I think, to maintain the authors' Turkishness in the face of their potentially acculturating sojourn in Europe. . . .

The second photograph of 'Zeyneb in her Paris drawing-room' enacts an identification designed variously to signify Oriental (her Turkish dress and Turkish furnishings) and western-ized (being socially able and acculturated in western high society). But what can we make of a Turkish woman wearing western clothes: is this cross-cultural dressing? There is a long established celebratory mode for dealing with western cross-cultural dressing as one of the pleasures of the imperial theatre, but perhaps if the Oriental does it s/he not only denies the western gaze of the exotica it expects and demands, but also risks becoming Bhabha's mimic man – an uncanny imitation of the real thing, doomed to inauthenticity (Bhabha 1984b). Does the practice of Orientals adopting European dress and behaviours threaten the viability of the west's pleasurable play at fashion *à la turque*? The west wants to play-act at being exotic, but when the referent for that exotic reappears in their midst clad in western clothes the differ-entiating terms which secure the western masquerade begin to crumble. If by the turn of the century, the 'authentic' Turkish past is already being recommodified as quaint historic interest by the Turks (for both domestic and souvenir markets) what is left for the west to dress in and photograph. . . .

We can also see Zeyneb Hanum's self-presentation in Turkish dress as an activity which knowingly exploits the Orientalist paradigm for her own ends. Without downplaying the frustration of being positioned by and trying to intervene in a western discourse which cannot recognize the nuanced specificity of her Ottoman identification, it is possible that the émigré Turk can manipulate cultural codes to her advantage. In this instance the masquerade as oppressed, veiled, Turkish woman helps her sales just as Ellison's cross-cultural dressing, the epitome of the exoticized Turkish stereotype she claims to challenge, helps hers. Clearly, the shifting significations of the veiled female body offer both women points of resistance and compliance, whose impact on their various audiences they attempt to anticipate. Yet to imagine this was ever an easy process would be to underestimate the complexity of their own subjective investments in cross-cultural codes of dress, narrative and identification.

# PART TWELVE

# History

THE SIGNIFICANCE OF HISTORY for post-colonial discourse lies in the modern
origins of historical study itself, and the circumstances by which 'History' took upon itself
the mantle of a discipline. For the emergence of history in European thought is coterminous
with the rise of modern colonialism, which in its radical othering and violent annexation of
the non-European world, found in history a prominent, if not *the* prominent, instrument
for the control of subject peoples. At base, the myth of a value free, 'scientific' view of the
past, the myth of the beauty of order, the myth of the story of history as a simple representation of the continuity of events, authorized nothing less than the construction of world
reality. This was a time in which the European nations, represented by three or four 'world'
cities, 'absorbed into themselves the whole of world history' as Oswald Spengler puts it
(Spengler [1926] 1962: 32).

The question the human sciences had to face in the nineteenth century was: what does
it mean to have a history? This question, Foucault maintains, signals a great mutation in the
consciousness of Western man, a mutation which has to do ultimately with 'our modernity',
which in turn is the sense we have of being 'utterly different from all other forms of humanity
known to history' (Foucault 1970: 219–20). The question we ask at this point is, of course,
who is this 'we'? Clearly, what it means to have a history is the same as what it means to
have a legitimate existence: history and legitimation go hand in hand; history legimates 'us'
and not others.

According to Hayden White it was important that history, seeking the title of 'scientific
discipline' in the nineteenth century mould, should suppress the modality of interpretation that
has always given it its form. The appeal to a moral or political authority underlying all interpretation had to be sublimated by dissolving the authority to interpret into the interpretation
itself. This, and the desire for the 'scientific', generated a particular historiographic ideology:
a single narrative truth that was 'simply' the closest possible representation of events. White
identifies the emergence of the discipline of history with a strategic moment of choice between
possible discursive options, in which the apparently neutral narrative form succeeded by virtue
of its resemblance to the purity of scientific disciplines. A crucial question he asks is 'What
is *ruled out* by conceiving the historical object in such a way that *not* to conceive it in that
way would constitute prima facie evidence of want of discipline?' (White 1982: 120). His
answer is *rhetoric*, which can be described as an awareness of the variety of ways of configuring
a past that itself only exists as a chaos of forms.

The problem of history becomes particularly crucial for the post-colonial writer. For not only are the questions of truth and fiction, of narrativity and indeterminacy, time and space, of pressing importance because the material ground, the political dimension of post-colonial life impresses itself so urgently, but the historical narrativity is that which structures the forms of reality itself. In other words, the myth of historical objectivity is embedded in a particular view of the sequential nature of narrative, and its capacity to reflect, isomorphically, the pattern of events it records. The post-colonial task, therefore, is not simply to contest the message of history, which has so often relegated individual post-colonial societies to footnotes to the march of progress, but also to engage the medium of narrativity itself, to reinscribe the 'rhetoric', the heterogeneity of historical representation as White describes it. This, of course, is easier said than done for post-colonial societies that so often have failed to gain access to the very institution of 'History' itself with its powerful rules of inclusion and exclusion.

José Rabasa indicates the extent of the historical construction of a Eurocentric world, in the conception of the Mercator *Atlas* itself. The map of the world can be seen as a palimpsest on which Europe has written its own dominance through the agency of history. This tendency of history to construct the world can be seen, as Peter Hulme shows, in the historical emergence of the word 'cannibal' to describe the inhabitants of the West Indies. That which etymologically begins as description assumes very quickly a power to signify the 'Other'. That which emerges as 'historical' is the result of contesting discourses. But it is the servitude to the 'muse of History' that Derek Walcott sees as most debilitating to the New World societies because it inevitably ossifies into a 'literature of recrimination and despair'. In this way the spirit of resistance itself is caught in the disempowering binarism of imperial history. Walcott suggests the need for a new beginning to post-colonial history, a new Adam and a new Eden, one which dispenses with imperial history altogether. Paul Carter's concept of spatial history goes some way towards this by rejecting the imperial idea of history as a stage on which it plays out its universal theme of the emergence of order out of chaos. The concept of place as a palimpsest written and overwritten by successive (historical) inscriptions is one way of circumventing history as the 'scientific narrative' of events.

As with that Western commodity called 'Theory' (in its various forms; theories about writing, or philosophy, or politics) the way in which colonized peoples have been able to enter the 'discursive plane', so to speak, of these patently authoritative and powerful intellectual pursuits, is through *literary* writing which may authorize otherwise forbidden entries into the intellectual battlefield of European thought. Wilson Harris believes that 'a philosophy of history may well lie buried in the arts of the imagination'. For Harris such imaginative arts extend beyond the 'literary' to include the discourse of the *limbo* dance or of *vodun*, all examples of the creativity of 'stratagems available to Caribbean man in the dilemmas of history which surround him'. In post-colonial societies the term 'literary' may well operate in its traditional canonical way, but more often it has come to operate as a mode by which the objectivity of narrative is contested, and particularly the narrative of history.

Recognizing that all histories, no matter what they are about, ultimately have 'Europe' as their subject, Dipesh Chakrabarty advocates a post-colonial history that, rather than returning to atavistic, nativist histories, or rejecting modernism itself, should invent a narrative that 'deliberately makes visible, within the very structure of its narrative forms, its own repressive strategies and practices'.

## Chapter 73

# JOSÉ RABASA

## ALLEGORIES OF *ATLAS*

From *Inventing A-M-E-R-I-C-A: Spanish Historiography and the Formation of Eurocentrism* Norman, Okla.: University of Oklahoma Press, 1993.

**A**S FAR AS I KNOW, there is no history of the atlas as a genre. Insofar as such a history might turn out to be for clarifying the question of Eurocentrism, I believe that an analysis of Mercator's *Atlas* is a necessary preparatory task. I also believe that the *Atlas* manifests the main constituents that have defined Europe as a privileged source of meaning for the rest of the world. Eurocentrism, as I will try to point out with respect to the *Atlas*, is more than an ideological construct that vanishes with the brush of the pen or merely disappears when Europe loses its position of dominance. The trace of European expansionism continues to exist in the bodies and minds of the rest of the world, as well as in the fantasies of the former colonizers. The transposition of the image of the palimpsest becomes an illuminative metaphor for understanding geography as a series of erasures and over-writings that have transformed the world. The imperfect erasures are, in turn, a source of hope for the reconstitution or reinvention of the world from native points of view. . . .

A cursory glance at Mercator's world map (Figure 73.1) uncovers a plurality of semiotic systems and semantic levels interacting with each other.

The map functions as a mirror of the world, not because the representation of the earth has the status of a natural sign, but because it aims to invoke a simulacrum of an always inaccessible totality by means of an arrangement of symbols. Thus Mercator, after enumerating the different sections of the *Atlas*, tells us in 'Preface upon Atlas' that his work '(as in a mirror) will set before your eyes, the whole world, that in the making of some rudiments, ye may finde out the causes of things, and so by attayning unto wisdome and prudence, by this meanes leade the Reader to higher speculation' (Mercator 1636: 'Preface'). As such the world map itself organizes different semiotic systems for creating a play of mirrors that would ultimately lead the reader to speculate on the creation of the world and the godhead: only topics from the part of the *Atlas* on 'Creation' are not allegorically coded on the margins of the world map. . . .

Since the totality of the world can never be apprehended as such in a cartographical objectification, maps have significance only within a subjective reconstitution of the fragments. The *Atlas* stands out as an ironic allegorization of this blind spot inherent in a cartographical enterprise. As a palimpsest, the *Atlas* conveys the irony of a bricolage where the interpreter is caught up in an open-ended process of signification and where the loose fragments derived from primitive texts allow for a plurality of combinations. Memory and systematic forgetfulness suspend the elucidation of a stable structure and constitute the need for an active translation.

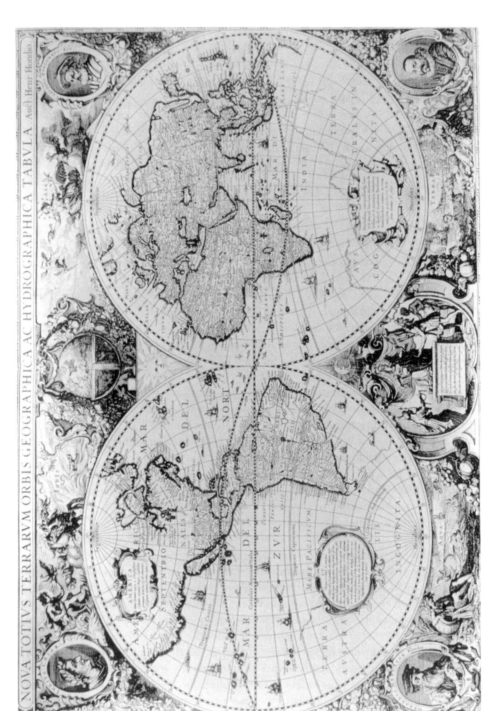

*Figure 73.1* Mercator's world map

*Source:* Mercator (1636)

An inside and an outside constitute two planes of content and expression for reading the map. The outside introduces an allegorical decoration that offers a narrative illumination to the portrayal of the earth. A title, portraits, proper names, allegories of the elements, a celestial sphere, instruments of measurement, a sun, a moon, and an allegory of the four continents frame the world with historic, cosmographic and anthropological categories. These registers introduce a series of strata into an apparently homogeneous and flat representation of the globe. The frame functions both as a decoration and as a content to be read in the map. Likewise, the separation of the world into two circles (the Old and New World) tabulated by meridians, parallels and the line of the zodiac, not only structure the totality of the world for locating names and points in space, but are also particular expressions of the celestial sphere represented in the frame itself. As a result, the map mirrors the course of history and the macrocosmos. Under closer inspection, we find the inside and the outside organized in terms of a binary opposition between the eternal and the contingent, between hard and soft parts. Without exhausting the binary oppositions organizing the map, the following samples exemplify the hierarchical arrangements:

| *Hard* | *Soft* |
|---|---|
| moderns | ancients |
| Europe | the rest of the world |
| Old World | New World |
| masculine | feminine |
| coordinates | contours |
| macrocosmos | microcosmos |

These binary oppositions must be understood as independent realms interacting with each other and inseparable for portraying the totality of the cosmos and the whole circle of the Earth. In the following discussion we will often see soft and hard characterizations of the written and the visual, of geography and history, shift positions. . . .

Let us now observe in the following analogy between cartography and the art of painting how the historical is indissoluble from the geographical in Mercator's *Atlas*:

> . . . I have principally endeavored to describe before every Mapp the order & nature of the most remarkeable places in every Province, the better to profit, the studious, and carefull of Politick matters and States affairs.
>
> (Mercator 1636 II: 269)

For Mercator, the written defines differences in what would otherwise be a homogeneous space. As a result, knowledge and power merge 'to profit the studious, and carefull of Politick matters and State affairs'. The written solidifies locations while supplying meaning to the visual. Writing, as such, is both a soft and a hard component in the *Atlas*. Inscriptions precede and determine the visibility of the contour, but they also flesh out the abstract frame. The possibility and the significance of the map thus depend on history. The inscription of the map gives place to its silhouette, but its silhouette is historical and meaningful only when it evokes a European history. In this light Mercator explains in 'Preface to the Reader' the scope of the *Atlas*: 'This work then is composed of *Geographie* (which is a description of the knowne Earth and parts thereof) and *Historie*, which is (*Oculus Mundi*) the eye of the World' (Mercator 1636: 'Preface'). The personification of geographic space in terms of a Eurocentric perspective is inseparable from the above definition of history as a visual function: 'the eye of the World'. . . .

The visual and the written ironically jumble up time and space within this paradoxical understanding of history, and the 'eye of the world' takes on a plurality of meanings depending on three points of reference:

(1) Travel narratives prefigure the data of the *Atlas*.
(2) History introduces a soft component into the maps for a qualitative determination of space.
(3) Mercator defines cosmography as the light of history.

Whatever alternative, the metaphorical nature of the maps explodes into a spatiotemporal reversibility: time becomes spatial and space becomes temporal. Cosmography is the light of history, while history illuminates the spatial representation of the Earth. . . .

## Rhetoric and the universality of Europe

Along with an ideological stance, the *Geographie* and *Historie of the Atlas* convey a planetary strategy wherein knowledge and representation indissolubly institute and erase territories. If specific political configurations establish boundaries and national identities for a European geographic space, then the rest of the world acquires spatial meaning only after the different regions have been inscribed by Europeans. History, 'the eye of the World', on an ideological level defines the national character of territories depicted. History thus naturalizes particular national formations and institutionalizes forgetfulness of earlier territorializations in the perception of the world. Next to this ideological, or mythological, reification of space, the signalling instrumentality of the *Atlas* opens the territories to a qualitative appraisal of demographic, commercial, ethnographic, religious, political and military details for strategic arrangements. Ideology naturalizes history insofar as it places national configurations and the destiny of European domination *sub specie aeternitatis*. Accordingly, the signalling power of the Atlas reopens territories to domination and appropriation within a historical dimension. . . .

. . . It is specifically the Christian European *man* who can offer the mirror of the world and hold a privileged position throughout the universal semblance of the *Atlas*: 'Here [in Europe] wee have the right of Lawes, the dignity of the Christian Religion, the forces of Armes. . . . Moreover, Europe manageth all Arts and Sciences with such dexterity, that for the invention of manie things shee may be truely called a Mother . . . she hath . . . all manner of learning, whereas other Countries are all of them, overspread with Barbarisme' (1636 I: 42). Let us leave aside for the moment the *Mother Europe* attribute. This passage makes manifest how global histories and geographies, despite their 'introduction' of other religions into the world scenario, always retain a Eurocentric perspective that defines the position and value of the rest of the world. In this respect, the project of the *Atlas* seals an epoch that began with Columbus: the pulsating utopian and millenarian disruptions of European history that the discovery of the New World provoked in Spanish historiography are long gone from the totalizing global vision of Mercator. . . .

In Mercator's maps, previous names accompany contemporary usages. The different regions of the world carry a temporal disparity according to the periods when the sources of information were produced. Generally speaking, Mercator also displays a tendency to make space historical by incorporating legends into the empty areas in the maps. This practice ensures a centrifugal movement from the name-laden Europe to the periphery, where legends and drawings characterize vast territories without history. In the periphery itself, the concentration of names serves as an index of colonialization. . . .

The normality and supremacy of Europe, however, are not perceptible in the bare frame of the world. We have already seen how the written – names and legends – index the higher position of Europe insofar as a hierarchy of space moves from an agglomeration of names to vaguely defined contours, and the newly discovered territories acquire semanticity in terms of their inclusion within a European perspective. Legends in the world map remind us of this. . . . One thing is for certain – no region must be left uncharted. Accordingly, the unknown (*Terra Australis incognita*, most of *America Septantrionalis*) must be prefigured, invented for a hesitant totalization of the shape of the Earth. . . .

It is interesting to note in the frame of the world map, that while Ptolemy, Mercator and Hondius carry a national identification next to their names, Caesar, a symbol of imperialism, stands open to national determination. While Ptolemy is dressed in a Renaissance fashion, Caesar's laurel crown takes a transhistorical and transnational dimension. Caesar functions as an empty slot where different leaders may inscribe themselves. The merging of geography and history, of knowledge and power, have Caesar as a prototypical incarnation of world domination. . . . Like the symbol of Caesar, the world revealed by Mercator's *Atlas* is a transhistorical and transnational theatre where imperialist configurations take form by means of particular national appropriations. Beyond rationalization, the *Atlas* establishes a world subject to national translations. Hexham formulates the malleability of the *Atlas* in the following terms:

> At their request I have undertaken, and by the helpe of God, according to my weake abilitie, translated their *Atlas Major* into English, for the good of my Countrie-men, and by their direction (who have most interest therein) have enlarged, & augmented it, out of many worthy Authors of my owne Nation, where it was most needful and requisite, and amended some errours in it, which were escaped in the former editions, & they for their parts have adorned it with new and exact Maps.
>
> (Mercator 1636: 'Preface')

The *Englishizing* of the world rests on the translatability of Caesar and the imperialist regard. The role of the translator corresponds to the formulation of a planetary strategy from a national point of view. Under the objective simulacrum of the *Atlas* flows an ironic commentary to the papal bulls that partitioned the unknown world between Portugal and Spain. The Iberian discoveries, conquests and tentative location of places for a determination of sovereignty slowly shaped and mapped the totality of the globe (16). Such a totality became a theatre for contention as European nations came to disregard the pontifical division of the Earth. . . .

Not only are the allegories in the *Atlas* an integral part of cartography, but the *Atlas* as a whole stands as an ironic allegory of the geographer's project to encompass the totality of the world. Atlas, king of Mauritania, the legendary first constructor of the globe, becomes a symbol for a particular genre of the Renaissance 'Book of the World'. In the process the ancient male geographer is transformed into a feminine flat representation of the world where Europe ultimately figures as the mother of 'all manner of learning'. As I have pointed out, in the allegory of the four continents, the presence of the male principle in the female personification of the continents formulates a hierarchy in terms of their subordination to masculinity. Asia, Africa and America in their degrees of nudity lack properness; that is, their selfhood depends solely on European imprints and a consequent mimicry of European space.

From the invention of America emerges a new Europe. The millenarian dream whereby the Franciscans transferred the geographic realization of history to the New World now, with Mercator, returns the locus of universal history to Europe; the angelic nature of the natives is replaced with a universal subjectivity that is indispensable to the knowing of truth and thus constitutes the apex of history. Europe, which in analogous allegories is invested with a sphere and a cross emblematic of Catholicism, assumes a secular version where science and knowledge define her supremacy and universality. . . .

If Europeans retain the universal key, nothing keeps the *Atlas* from being translated into a non-European idiom as its ultimate irony within a historical horizon. This is not the place to elaborate on the 'writing back' of the colonized, but my analysis depends on the possibility that the universal address of the *Atlas* includes readings not confined to a Eurocentric point of view. The meanings of humanity, the world and history become undecidable beyond a European battle ground. Universal history is undecidable, not on account of a theoretical deconstruction of teleology and eschatology, but because of an ever-present deconstruction of Eurocentric world views by the rest of the world. As it were, the empire has always been writing back. The allegorization of the four continents suppresses the colonialist machinery and fabricates an omnipotent European subject who can dominate the world from the cabinet, but it also produces a blind spot that dissolves history as a privileged modality of European culture.

## Chapter 74

## PETER HULME

## COLUMBUS AND THE CANNIBALS

From *Colonial Encounters: Europe and the Native Caribbean 1492–1797*
London and New York: Methuen, 1986.

THE PRIMARY *OED* DEFINITION of 'cannibal' reads: 'A man (*esp.* a savage) that eats human flesh; a man-eater, an anthropophagite. Originally proper name of the man-eating Caribs of the Antilles.' The morphology or, to use the *OED*'s word, form-history of 'cannibal' is rather more circumspect. The main part of its entry reads:

> (In 16th c. pl. *Canibales*, a. Sp. *Canibales*, originally one of the forms of the ethnic name *Carib* or *Caribes*, a fierce nation of the West Indies, who are recorded to have been anthropophagi, and from whom the name was subsequently extended as a descriptive term . . .)

This is a 'true' account of the morphology of the word 'cannibal' in English, yet it is also an ideological account that functions to repress important historical questions about the use of the term – its discursive morphology, perhaps, rather than its linguistic morphology. The trace of that repression is the phrase 'who are recorded to have been', which hides beneath its bland-ness – the passive tense, the absence (in a book of authorities) of any ultimate authority, the assumption of impartial and accurate observation – a different history altogether.

The tone of 'who are recorded to have been' suggests a nineteenth-century ethnographer sitting in the shade with notebook and pencil, calmly recording the savage rituals being per-formed in front of him. However unacceptable that might now seem as 'objective reporting', it still appears a model of simplicity compared with the complexities of the passages that constitute the record in this instance.

On 23 November 1492 Christopher Columbus approached an island 'which those Indians whom he had with him called "Bohio"'. According to Columbus's *Journal* these Indians, usually referred to as Arawaks:

> said that this land was very extensive and that in it were people who had one eye in the forehead, and others whom they called 'canibals'. Of these last, they showed great fear, and when they saw that this course was being taken, they were speechless, he says, because these people ate them and because they are very warlike.
>
> (Columbus [1825] 1960: 68–9)

. . . This is the first appearance of the word 'canibales' in a European text, and it is linked immediately with the practice of eating human flesh. The *Journal* is, therefore, in some sense at least, a 'beginning text'.

But in just what sense is that name and that ascription a 'record' of anything? For a start the actual text on which we presume Columbus to have inscribed that name disappeared, along with its only known copy, in the middle of the sixteenth century. The only version we have, and from which the above quotation is taken, is a handwritten abstract made by Bartolome de Las Casas, probably in 1552, and probably from the copy of Columbus's original then held in the monastery of San Pablo in Seville. There have subsequently been various transcriptions of Las Casas's manuscript. So the apparent transparency of 'who are recorded to have been' is quickly made opaque by the thickening layers of language: a transcription of an abstract of a copy of a lost original. This is chastening, but to some extent contingent. More telling is what might be called the internal opacity of the statement. Columbus's 'record', far from being an observation that those people called 'canibales' ate other people, is a report of other people's words; moreover, words spoken in a language of which he had no prior knowledge and, at best, six weeks' practice in trying to understand.

Around this passage cluster a whole host of ethnographic and linguistic questions. . . . But the general argument here will be that, though important, these questions take second place to the textual and discursive questions. What first needs examination, in other words, are not isolated passages taken as evidence for this or that, but rather the larger units of text and discourse, without which no meaning would be possible at all.

To write about the text we call 'el diario de Colón' (Columbus's journal) is to take a leap of faith, to presume that the transcription of the manuscript of the abstract of the copy of the original stands in some kind of meaningful relationship to the historical reality of Columbus's voyage across the Atlantic and down through the Caribbean islands during the winter months of 1492–3.

It would be perverse and unhelpful to presume that no such relationship exists, but credulous and unthinking to speak – as some have done – of the *Journal*'s 'frank words, genuine and unadorned'. Circumspection would certainly seem called for. Yet if the *Journal* is taken not as a privileged eye-witness document of the discovery, nor as an accurate ethnographic record, but rather as the first fable of European beginnings in America, then its complex textual history and slightly dubious status become less important than the incredible narrative it unfolds.

This is not an argument in favour of somehow lifting Columbus and his *Journal* out of history. . . . But it is an argument in favour of bracketing particular questions of historical accuracy and reliability in order to see the text whole, to gauge the structure of its narrative, and to chart the interplay of its linguistic registers and rhetorical modalities. To read the *Journal* in this way is also to defer the biographical questions: the Columbus of whom we speak is for the moment a textual function, the 'I' of the *Journal* who is occasionally, and scandalously, transformed into the third person by the intervention of the transcriber's 'I'.

The *Journal* is generically peculiar. It is in part a log-book, and throughout records the navigational details of Columbus's voyage. Commentators have usually accepted that it was written up almost every evening of the six-and-a-half-month journey, not revised or rewritten, and not constructed with a view to publication. It certainly gives that impression, which is all that matters here: Columbus is presented by the *Journal* as responding day by day to the stimulus of new challenges and problems. Yet if its generic shape is nautical the *Journal* is also by turns a personal memoir, an ethnographic notebook, and a compendium of European fantasies about the Orient: a veritable palimpsest.

'From whom the name was subsequently extended as a descriptive term.' Linguistic morphology is concerned only with the connection made between the term 'cannibal' and the practice of eating human flesh. We have seen how the very first mention of that term in a European text is glossed with reference to that practice, and for the linguist it is satisfactory, but not of intrinsic interest, to note how that reference is always present, either implicitly or explicitly, in any recorded use of the word 'cannibal' from Columbus's on 23 November 1492 onwards. It was adopted into the bosom of the European family of languages with a speed and readiness which suggests that there had always been an empty place kept warm for it. Poor 'anthropophagy', if not exactly orphaned, was sent out into the cold until finding belated lodging in the nineteenth century within new disciplines seeking authority from the deployment of classical terminology.

All of which makes it even stranger that the context of that beginning passage immediately puts the association between the word 'cannibal' and the eating of human flesh into doubt. Las Casas continues:

> The admiral says that he well believes that there is something in this, but that since they were well armed, they must be an intelligent people [gente de razon], and he believed that they may have captured some men and that, because they did not return to their own land, they would say that they were eaten.
>
> (Columbus 1960: 69)

This passage is of no interest to linguistic morphology since Columbus's scepticism failed to impinge upon the history of the word. Ethnographically it would probably be of scant interest, showing merely Columbus's initial scepticism, and therefore making him a more reliable witness in the end. Even from the point of view of a revisionist ethnography that wanted to discount suggestions of native anthropophagy the passage could only be seen as evidence of the momentary voice of European reason soon to be deafened by the persistence of Arawak defamations of their traditional enemy. Attention to the discursive complexities of the text will suggest a different reading. The great paradox of Columbus's *Journal* is that although the voyage of 1492–3 was to have such a devastating and long-lasting effect on both Europe and America, and is still celebrated as one of the outstanding achievements of humanity, the record itself tells of misunderstandings, failures and disappointments. The greatest of these – that he had not reached Asia – was too overwhelming for Columbus ever to accept. The minor ones are in some ways even more telling. . . .

In brief, what a symptomatic reading of the *Journal* reveals is the presence of two distinct discursive networks. In bold outline each discourse can be identified by the presence of key words: in one case 'gold', 'Cathay', 'Grand Khan', 'intelligent soldiers', 'large buildings', 'merchant ships'; in the other 'gold', 'savagery', monstrosity', 'anthropophagy'. Even more boldly, each discourse can be traced to a single textual origin, Marco Polo and Herodotus respectively. More circumspectly, there is what might be called a discourse of Oriental civiliza-tion and a discourse of savagery, both archives of topics and motifs that can be traced back to the classical period. It is tempting to say that the first was based on empirical knowledge and the second on psychic projection, but that would be a false dichotomy. There was no doubt a material reality – the trade that had taken place between Europe and the Far East over many centuries, if intermittently. In pursuit of, or as an outcome of, this trade there were Europeans who travelled to the Far East, but their words are in no way a simple reflec-tion of 'what they saw'. For that reason it is better to speak of identifiable discourses. There was a panoply of words and phrases used to speak about the Orient: most concerned its wealth and power, as well they might since Europe had for many years been sending east large amounts of gold and silver. Marco Polo's account was the best-known deployment of these topoi. The discourse of savagery had in fact changed little since Herodotus's 'investigation' of Greece's 'barbarian' neighbours. The locations moved but the descriptions of Amazons, Anthropophagi and Cynocephali remained constant throughout Ctesias, Pliny, Solinus and many others. This discourse was hegemonic in the sense that it provided a popular vocabulary for constituting 'otherness' and was not dependent on *textual* reproduction. Textual authority was however available to Columbus in Pierre d'Ailly and Aeneas Sylvius, and indeed in the text that we know as 'Marco Polo', but which is properly *Divisament dou Monde*, authored by a writer of romances in French, and itself already an unravellable discursive network.

In the early weeks of the Columbian voyage it is possible to see a certain jockeying for position between these two discourses, but no overt conflict. The relationship between them is expressed as that between present and future: this is a world of savagery, over there we will find Cathay. But there are two potential sites of conflict, one conscious – in the sense of being present in the text; the other unconscious – in the sense that it is present only in its absence and must be reconstructed from the traces it leaves. The conscious conflict is that two elements, 'the soldiers of the Grand Khan' from the discourse of Marco Polo and 'the mandating savages' from the discourse of Herodotus, are competing for a single signifier – the word 'canibales'. Columbus's wavering on 23 November belongs to a larger pattern of references in which 'canibal' is consistently glossed by his native hosts as 'maneater' while it ineluctably calls to his mind 'el Gran Can'. In various entries the phonemes echo each other from several lines' distance until on 11 December 1492 they finally coincide:

> it appears likely that they are harassed by an intelligent race, all these islands living in great fear of those of Caniba. 'And so I repeat what I have said on other-occasions,' he says, 'the Caniba are nothing else than the people of the Grand Khan [*que Caniba no es otra cosa sino la gente del Gran Can*], who must be very near here and possess ships, and they must come to take them captive, and as the prisoners do not return, they believe that they have been eaten.'
>
> (Columbus 1960: 92–3)

The two 'Can' are identified as one, the crucial identification is backdated, and 'canibal' as man-eater must simply disappear having no reference to attach itself to.

Except of course that it does not disappear at all. That would be too easy. In fact the assertion of the identity of 'Caniba' with 'gente del Can', so far from marking the victory of the Oriental discourse, signals its very defeat; as if the crucial phonetic evidence could only be brought to textual presence once its power to control action had faded.

## Chapter 75

# DEREK WALCOTT

## THE MUSE OF HISTORY

From 'The Muse of History' in Orde Coombes *Is Massa Day Dead? Black Moods in the Caribbean* New York: Doubleday, 1974.

> History is a nightmare from which I am trying to awake.
>
> James Joyce

THE COMMON EXPERIENCE of the New World, even for its patrician writers whose veneration of the old is read as idolatory of the mestizo, is colonialism. They too are victims of tradition, but they remind us of our debt to the great dead, that those who break a tradition first hold it in awe. They perversely discourage disfavor, but because their sense of the past is of a timeless, yet habitable moment, the New World owes them more than it does those who wrestle with the past, for their veneration subtilizes an arrogance which is tougher than violent rejection. They know that by openly fighting tradition we perpetuate it, that revolutionary literature is a filial impulse, and that maturity is the assimilation of the features of every ancestor.

When these writers cunningly describe themselves as classicists and pretend an indifference to change, it is with an irony as true of the colonial anguish as the fury of the radical. If they appear to be phony aristocrats, it is because they have gone past the confrontation of history, that Medusa of the New World.

These writers reject the idea of history as time for its original concept as myth, the partial recall of the race. For them history is fiction, subject to a fitful muse, memory. Their philosophy, based on a contempt for historic time, is revolutionary, for what they repeat to the New World is its simultaneity with the Old. Their vision of man is elemental, a being inhabited by presences, not a creature chained to his past. Yet the method by which we are taught the past, the progress from motive to event, is the same by which we read narrative fiction. In time every event becomes an exertion of memory and is thus subject to invention. The farther the facts, the more history petrifies into myth. Thus, as we grow older as a race, we grow aware that history is written, that it is a kind of literature without morality, that in its actuaries the ego of the race is indissoluble and that everything depends on whether we write this fiction through the memory of hero or of victim.

In the New World servitude to the muse of history has produced a literature of recrimination and despair, a literature of revenge written by the descendants of slaves or a literature of remorse written by the descendants of masters. Because this literature serves historical truth, it yellows into polemic or evaporates in pathos. The truly tough aesthetic of the New World neither explains nor forgives history. It refuses to recognize it as a creative or culpable force. This shame and awe of history possess poets of the Third World who think of language as enslavement and who, in a rage for identity, respect only incoherence or nostalgia.

The great poets of the New World, from Whitman to Neruda, reject this sense of history. Their vision of man in the New World is Adamic. In their exuberance he is still capable of enormous wonder. Yet he has paid his accounts to Greece and Rome and walks in a world without monuments and ruins. They exhort him against the fearful magnet of older civilizations. Even in Borges, where the genius seems secretive, immured from change, it celebrates an elation which is vulgar and abrupt, the life of the plains given an instant archaism by the hieratic style. Violence is felt with the simultaneity of history. So the death of a gaucho does not merely repeat, but is, the death of Caesar. Fact evaporates into myth. This is not the jaded cynicism which sees nothing new under the sun, it is an elation which sees everything as renewed. . . .

New World poets who see the 'classic style' as stasis must see it also as historical degradation, rejecting it as the language of the master. This self-torture arises when the poet also sees history as language, when he limits his memory to the suffering of the victim. Their admirable wish to honor the degraded ancestor limits their language to phonetic pain, the groan of suffering, the curse of revenge. The tone of the past becomes an unbearable burden, for they must abuse the master or hero in his own language, and this implies self-deceit. Their view of Caliban is of the enraged pupil. They cannot separate the rage of Caliban from the beauty of his speech when the speeches of Caliban are equal in their elemental power to those of his tutor. The language of the torturer mastered by the victim. This is viewed as servitude, not as victory.

But who in the New World does not have a horror of the past, whether his ancestor was torturer or victim? Who, in the depth of conscience, is not silently screaming for pardon or for revenge? The pulse of New World history is the racing pulse beat of fear, the tiring cycles of stupidity and greed. . . .

In time the slave surrendered to amnesia. That amnesia is the true history of the New World. That is our inheritance, but to try and understand why this happened, to condemn or justify is also the method of history, and these explanations are always the same: This happened because of that, this was understandable because, and in days men were such. These recriminations exchanged, the contrition of the master replaces the vengeance of the slave, and here colonial literature is most pietistic, for it can accuse great art of feudalism and excuse poor art as suffering. To radical poets poetry seems the homage of resignation, an essential fatalism. But it is not the pressure of the past which torments great poets but the weight of the present:

> there are so many dead,
> and so many dikes the red sun breached,
> and so many heads battering hulls
> and so many hands that have closed over kisses
> and so many things that I want to forget.
>> (Neruda)

The sense of history in poets lives rawly along their nerves:

> My land without name, without America,
> equinoctial stamen, lance-like purple,

your aroma rose through my roots
into the cup I drained, into the most tenuous
word not yet born in my mouth.

(Neruda)

It is this awe of the numinous, this elemental privilege of naming the new world which annihilates history in our great poets, an elation common to all of them, whether they are aligned by heritage to Crusoe and Prospero or to Friday and Caliban. They reject ethnic ancestry for faith in elemental man. The vision, the 'democratic vista,' is not metaphorical, it is a social necessity. A political philosophy rooted in elation would have to accept belief in a second Adam, the recreation of the entire order, from religion to the simplest domestic rituals. The myth of the noble savage would not be revived, for that myth never emanated from the savage but has always been the nostalgia of the Old World, its longing for innocence. The great poetry of the New World does not pretend to such innocence, its vision is not naive. Rather, like its fruits, its savor is a mixture of the acid and the sweet, the apples of its second Eden have the tartness of experience. In such poetry there is a bitter memory and it is the bitterness that dries last on the tongue. It is the acidulous that supplies its energy. . . . For us in the archipelago the tribal memory is salted with the bitter memory of migration.

To such survivors, to all the decimated tribes of the New World who did not suffer extinction, their degraded arrival must be seen as the beginning, not the end of our history. The shipwrecks of Crusoe and of the crew in *The Tempest* are the end of an Old World. It should matter nothing to the New World if the Old is again determined to blow itself up, for an obsession with progress is not within the psyche of the recently enslaved. That is the bitter secret of the apple. The vision of progress is the rational madness of history seen as sequential time, of a dominated future. Its imagery is absurd. In the history books the discoverer sets a shod foot on virgin sand, kneels, and the savage also kneels from his bushes in awe. Such images are stamped on the colonial memory, such heresy as the world's becoming holy from Crusoe's footprint or the imprint of Columbus' knee. These blasphemous images fade, because these hieroglyphs of progress are basically comic. And if the idea of the New and the Old becomes increasingly absurd, what must happen to our sense of time, what else can happen to history itself, but that it too is becoming absurd? This is not existentialism. Adamic, elemental man cannot be existential. His first impulse is not self-indulgence but awe, and existentialism is simply the myth of the noble savage gone baroque. . . .

But to most writers of the archipelago who contemplate only the shipwreck, the New World offers not elation but cynicism, a despair at the vices of the Old which they feel must be repeated. Their malaise is an oceanic nostalgia for the older culture and a melancholy at the new, and this can go as deep as a rejection of the untamed landscape, a yearning for ruins. To such writers the death of civilizations is architectural, not spiritual, seeded in their memories is an imagery of vines ascending broken columns, of dead terraces, of Europe as a nourishing museum. They believe in the responsibility of tradition, but what they are in awe of is not tradition, which is alert, alive, simultaneous, but of history, and the same is true of the new magnifiers of Africa. For these their deepest loss is of the old gods, the fear that it is worship which has enslaved progress. Thus the humanism of politics replaces religion. They see such gods as part of the process of history, subjected like the tribe to cycles of achievement and despair. Because the Old World concept of God is anthropomorphic, the New World slave was forced to remake himself in His image, despite such phrases as 'God is light, and in Him is no darkness,' and at this point of intersecting faiths the enslaved poet and enslaved priest surrendered their power. But the tribe in bondage learned to fortify itself by cunning assimilation of the religion of the Old World. What seemed to be surrender was redemption. What seemed the loss of tradition was its renewal. What seemed the death of faith, was its rebirth. . . .

I accept this archipelago of the Americas. I say to the ancestor who sold me, and to the ancestor who bought me, I have no father, I want no such father, although I can understand you, black ghost, white ghost, when you both whisper 'history,' for if I attempt to forgive you both I am falling into your idea of history which justifies and explains and expiates, and it is not mine to forgive, my memory cannot summon any filial love, since your features are anonymous and erased and I have no wish and no power to pardon. You were when you acted your roles, your given, historical roles of slave seller and slave buyer, men acting as men, and also you, father in the filth-ridden gut of the slave ship, to you they were also men, acting as men, with the cruelty of men, your fellowman and tribesman not moved or hovering with hesitation about your common race any longer than my other bastard ancestor hovered with his whip, but to you inwardly forgiven grandfathers, I, like the more honest of my race, give a strange thanks. I give the strange and bitter and yet ennobling thanks for the monumental groaning and soldering of two great worlds, like the halves of a fruit seamed by its own bitter juice, that exiled from your own Edens you have placed me in the wonder of another, and that was my inheritance and your gift.

## Chapter 76

# PAUL CARTER

## SPATIAL HISTORY

From *Road to Botany Bay: An Essay in Spatial History* London:
Faber & Faber, 1987.

**B**EFORE THE NAME: what was the place like before it was named? How did Cook
see it? . . . Even as we look towards the horizon or turn away down fixed routes, our
gaze sees through the space of history, as if it was never there. In its place, nostalgia for the
past, cloudy time, the repetition of facts. The fact that where we stand and how we go is
history: this we do not see.

According to our historians it was always so. Australia was always simply a stage where
history occurred, history a theatrical performance. It is not the historian who stages events,
weaving them together to form a plot, but History itself. History is the playwright, coordinating
facts into a coherent sequence: the historian narrating what happened is merely a copyist or
amanuensis. He is a spectator like anybody else and, whatever he may think of the performance,
he does not question the stage conventions. . . .

In a theatre of its own design, history's drama unfolds; the historian is an impartial
onlooker, simply repeating what happened. In [Australian historian Manning] Clarke's account
[of the landing of the First Fleet] this illusion of the historian as *répétiteur* is reinforced by
other, literary means. . . .

Such history is a fabric woven of self-reinforcing illusions. But above all, one illusion
sustains it. This is the illusion of the theatre, and, more exactly, the unquestioned convention of the all-seeing spectator. The primary logic which holds together Clarke's description
is its visibility. Nature's painted curtain is drawn aside to reveal heroic man at his epic labour
on the stage of history. . . .

This kind of history, which reduces space to a stage, that pays attention to events unfolding
in time alone, might be called imperial history. The governor erects a tent here rather than
there; the soldier blazes a trail in that direction rather than this: but, rather than focus on
the *intentional* world of historical individuals, the world of active, spatial choices, empirical
history of this kind has as its focus facts which, in a sense, come after the event. The primary
object is not to understand or to interpret: it is to legitimate. This is why this history is
associated with imperialism – for who are more liable to charges of unlawful usurpation and
constitutional illegitimacy than the founders of colonies? Hence, imperial history's *defensive*
appeal to the logic of cause and effect: by its nature, such a logic demonstrates the emergence
of order from chaos.

Hence, too, its preference for fixed and detachable facts, for actual houses, visible
clearings and boats at anchor. For these, unlike the intentions which brought them there,
unlike the material uncertainties of lived time and space, are durable objects which can be

treated as typical, as further evidence of a universal historical process. Orphaned from their unique spatial and temporal context, such objects, such historical facts, can be fitted out with new paternities. Legitimized by an imperial discourse, they can even form future alliances of their own. (It is precisely this family-tree myth of history which assures the historian his privileged status.) . . .

The fact is that, as an account of foundation and settlement, not to mention the related processes of discovery and exploration, empirical history, with its emphasis on the factual and static, is wholly inadequate. . . . For the result of cause and effect narrative history is to give the impression that events unfold according to a logic of their own. They refer neither to the place, nor to the people. Imperial history's mythic lineage of heroes is the consequence of its theatrical assumption that, in reality, historical individuals are actors, fulfilling a higher destiny. . . .

*The Road to Botany Bay*, then, is written against these mythic imaginings. It is a prehistory of places, a history of roads, footprints, trails of dust and foaming wakes. . . . Against the historians, it recognizes that our life as it discloses itself spatially is dynamic, material but invisible. It constantly transcends actual objects to imagine others beyond the horizon. It cannot be delimited by reference to immediate actions, let alone treated as an autonomous fact independent of intention. It recognizes that the spatiality of historical experience evaporates before the imperial gaze. . . .

What is evoked here are the spatial forms and fantasies through which a culture declares its presence. It is spatiality as a form of non-linear writing; as a form of history. That cultural space *has* such a history is evident from the historical documents themselves. For the literature of spatial history – the letters home, the explorer's journals, the unfinished maps – are written traces which, but for their spatial occasion, would not have come into being. They are not like novels: their narratives do not conform to the rules of cause-and-effect empirical history. Rather they are analogous to unfinished maps and should be read accordingly as records of travelling. . . .

Such spatial history – history that discovers and explores the lacuna left by imperial history – begins and ends in language. It is this which makes it history rather than, say, geography. If it does *imitate* the world of the traveller it is in a different sense. For, like the traveller whose gaze is oriented and limited, it makes no claim to authoritative completeness. It is, must be, like a journey, exploratory. . . .

But where to begin? Late in 1616, Dirck Hartog of Amsterdam and his ship, the *Eendracht*, were blown on to the north-west coast of Australia. The skipper commemorated his involuntary landing on a pewter plate, which he affixed to a post. The island where Hartog landed was named after him; the adjoining mainland was called the Land of Eendracht. In 1697, another Dutchman, Vlamingh, also blown off-course, found Hartog's memorial. He had Hartog's inscription copied on to a new pewter plate and appended a record of his own visit. In 1699, the English seaman, William Dampier, also visited this coast. He let the island retain its Dutch connection, but renamed the country to the east Shark Bay. Then, in 1801, one Captain Emmanuel Hamelin discovered a pewter plate 'of about six inches in diameter on which was roughly engraven two Dutch inscriptions . . .' and named the place Cape Inscription. . . .

[S]uch a name, as the earlier editions testify, belongs firmly to the history of travelling. Rewritten and repeated, it serves as a point of departure. But Cape Inscription, the name, is also the result of erasure: it also symbolizes the imperial project of permanent possession through dispossession. In short, the name oscillates between two extreme interpretations. It suggests a kind of history which is neither static nor mindlessly mobile, but which incorporates both possibilities. It points to a kind of history where travelling is a process of *continually* beginning, continually ending, where discovery and settlement belong to the same exploratory process. . . .

But Cape Inscription is also a striking figure of speech, an oxymoron yoking writing and landscape in a surprising, even grotesque way. A geographical feature is made no bigger than a page of writing. A calligraphic flourish is able, it seems, to plume out like an ocean current a hundred miles long. This metaphorical way of speaking is a pointer to the way spatial history must interpret its sources. It also indicates, concisely and poetically, the *cultural* place where spatial history begins: not in a particular year, nor in a particular place, but *in the act of naming*. For by the act of place-naming, space is transformed symbolically into a place, that is, a space with a history.

## Chapter 77

# WILSON HARRIS

## THE LIMBO GATEWAY

From 'History, Fable and Myth in the Caribbean and the Guianas' in Hena Maes-Jelinek (ed.) *Explorations: A Selection of Talks and Articles 1966–81* Mundelstrup: Dangaroo Press, 1981.

**I** **WANT TO MAKE IT AS CLEAR AS I CAN** that a cleavage exists in my opinion between the historical convention in the Caribbean and Guianas and the arts of the imagination. I believe a philosophy of history may well lie buried in the arts of the imagination and whether my emphasis falls on *limbo* or *vodun*, on Carib bush-baby omens, on Arawak zemi, on Latin, English inheritances – in fact within and beyond these emphases – my concern is with epic stratagems available to Caribbean man in the dilemmas of history which surround him.

There are two kinds of myths related to Africa in the Caribbean and Guianas. One kind seems fairly direct, the other has clearly undergone metamorphosis. In fact even the direct kind of myth has suffered a 'sea-change' of some proportions. In an original sense, therefore, these myths which reflect an African link in the Caribbean are also part and parcel of a native West Indian imagination and therefore stand, in some important ways, I feel, in curious rapport with vestiges of Amerindian fable and legend. (Fable and myth are employed as variables of the imagination in this essay.)

Let us start with a myth stemming from Africa which has undergone metamorphosis. The one which I have in mind is called *limbo*. The limbo dance is a well-known feature in the Carnival life of the West Indies today though it is still subject to intellectual censorship as I shall explain as I go along in this paper. The *limbo* dancer moves under a bar which is gradually lowered until a mere slit of space, it seems, remains through which with spread-eagled limbs he passes like a spider.

*Limbo* was born, it is said, on the slave ships of the Middle Passage. There was so little space that the slaves contorted themselves into human spiders. Limbo, therefore, as Edward Brathwaite, the distinguished Barbadian-born poet, has pointed out, is related to *anancy* or spider fables. If I may now quote from *Islands*, the last book in his trilogy:

drum stick knock
and the darkness is over me
knees spread wide
and the water is hiding me
*limbo*
*limbo like me*

But there is something else in the *limbo–anancy* syndrome which, as far as I am aware, is overlooked by Edward Brathwaite, and that is the curious dislocation of a chain of miles reflected in the dance so that a re-trace of the Middle Passage from Africa to the Americas and the West Indies is not to be equated with a uniform sum. Not only has the journey from the Old World to the new varied with each century and each method of transport but needs to be re-activated in the imagination as a limbo perspective when one dwells on the Middle Passage: a *limbo* gateway between Africa and the Caribbean.

In fact here, I feel, we begin to put our finger on something which is close to the inner universality of Caribbean man. Those waves of migration which have hit the shores of the Americas – North, Central and South – century after century have, at various times, possessed the stamp of the spider metamorphosis in the refugee flying from Europe or in the indentured East Indian and Chinese from Asia.

*Limbo* then reflects a certain kind of gateway or threshold to a new world and the dislocation of a chain of miles. It is – in some ways – the archetypal sea-change stemming from Old Worlds and it is legitimate, I feel, to pun on *limbo* as a kind of shared phantom *limb* which has become a subconscious variable in West Indian theatre. The emergence of formal West Indian theatre was preceded, I suggest, by that phantom limb which manifested itself on Boxing Day after Christmas when the ban on the 'rowdy' bands (as they were called) was lifted for the festive season.

I recall performances I witnessed as a boy in Georgetown, British Guiana, in the early 1930s. Some of the performers danced on high stilts like elongated limbs while others performed spread-eagled on the ground. In this way limbo spider and stilted pole of the gods were related to the drums like grassroots and branches of lightning to the sound of thunder.

Sometimes it was an atavistic spectacle and it is well known that these bands were suspected by the law of subversive political stratagems. But it is clear that the dance had no political or propaganda motives though, as with any folk manifestation, it could be manipulated by demagogues. The whole situation is complex and it is interesting to note that Rex Nettleford in an article entitled 'The Dance as an Art Form – Its Place in the West Indies' has this to say: 'Of all the arts, dance is probably the most neglected. The art form continues to elude many of the most intuitive in an audience, including the critics' (Nettleford 1968: 127).

It has taken us a couple of generations to begin – just begin – to perceive, in this phenomenon, an activation of unconscious and sleeping resources in the phantom limb of dismembered slave and god. An activation which possesses a nucleus of great promise – of far-reaching new poetic form.

For *limbo* (one cannot emphasize this too much) is not the total recall of an African past since that African past in terms of tribal sovereignty or sovereignties was modified or traumatically eclipsed with the Middle Passage and with generations of change that followed. *Limbo* was rather the renascence of a new corpus of sensibility that could translate and accommodate African and other legacies within a new architecture of cultures. For example, the theme of the phantom limb – the re-assembly of dismembered man or god – possesses archetypal resonances that embrace Egyptian Osiris, the resurrected Christ and the many-armed goddess of India, Kali, who throws a psychical bridge with her many arms from destruction to creation.

In this context it is interesting to note that *limbo* – which emerged as a novel re-assembly out of the stigmata of the Middle Passage – is related to Haitian *vodun* in the sense that Haitian *vodun* (though possessing a direct link with African *vodun* which I shall describe later on) also seeks to accommodate new Catholic features in its constitution of the muse.

It is my view – a deeply considered one – that this ground of accommodation, this art of creative coexistence born of great peril and strangest capacity for renewal – pointing away from apartheid and ghetto fixations – is of the utmost importance and *native* to the Caribbean, perhaps to the Americas as a whole. It is still, in most respects, a latent syndrome and we need to look not only at *limbo* or *vodun* but at Amerindian horizons as well – shamanistic and rain-making vestiges and the dancing bush baby legends of the Caribs (now extinct) which began to haunt them as they crouched over their campfires under the Spanish yoke.

Insufficient attention has been paid to such phenomena and the original native capacity these implied as omens of rebirth. Many historians have been intent on indicting the Old Work of Europe by exposing a uniform pattern of imperialism in the New World of the Americas. Thus they conscripted the West Indies into a mere adjunct of imperialism and over-looked a subtle and far-reaching renascence. In a sense therefore the new historian [Thomas] – though his stance is an admirable one in debunking imperialism – has ironically extended and reinforced old colonial prejudices which censored the *limbo* imagination as a 'rowdy' mani-festation and overlooked the complex metaphorical gateway it constitutes in rapport with Amerindian omen.

Later on I intend to explore the Amerindian gateways between cultures which began obscurely and painfully to witness (long before *limbo* or *vodun* or the Middle Passage) to a native suffering community steeped in caveats of conquest. At this point I shall merely indi-cate that these gateways exist as part and parcel of an original Caribbean architecture which it is still possible to create if we look deep into the rubble of the past, and that these Amerindian features enhance the *limbo* assembly with which we are now engaged – the spider syndrome and phantom limb of the gods arising in Negro fable and legend.

I used the word 'architecture' a moment or two ago because I believe this is a valid approach to a gateway society as well as to a community which is involved in an orig-inal re-construction or re-creation of variables of myth and legend in the wake of stages of conquest.

First of all the *limbo* dance becomes the human gateway which dislocates (and therefore begins to free itself from) a uniform chain of miles across the Atlantic. This dislocation of interior space serves therefore as a corrective to a uniform cloak or documentary stasis of imperialism. The journey across the Atlantic for the forebears of West Indian man involved a new kind of space, inarticulate as this new 'spatial' character was at the time – and not simply an unbroken schedule of miles in a log book. Once we perceive this inner corrective to historical documentary and protest literature which sees the West Indies as utterly deprived, or gutted by exploitation, we begin to participate in the genuine possibilities of original change in a people severely disadvantaged (it is true) at a certain point in time.

The *limbo* dance therefore implies, I believe, a profound art of compensation which seeks to re-play a dismemberment of tribes (*note again the high stilted legs of some of the performers and the spider-anancy masks of others running close to the ground*) and to invoke at the same time a curious psychic re-assembly of the parts of the dead muse and god. And that re-assembly which issued from a state of cramp to articulate a new growth – and to point to the necessity for a new kind of drama novel and poem – is a creative phenomenon of the first importance in the imagination of a people violated by economic fates.

One cannot over-emphasize, I believe, how original this phenomenon was. So original it aroused both incomprehension and suspicion in the intellectual and legal administrations of the land (I am thinking in particular of the first half of the twentieth century though one can, needless to say, go much farther back). What is bitterly ironic – as I have already indicated – is that present-day historians in the second half of the twentieth century – militant and critical of imperialism as they are here – have fallen victim, in another sense, to the very imperialism they appear to denounce. They have no criteria for arts of originality springing out of an age of *limbo* and the history they write is without an inner time. This historical refusal to see may well be at the heart of the *Terrified Consciousness* which a most significant critic to emerge in the West Indies at this time, Kenneth Ramchand, analyses in a brilliant essay (Ramchand 1969). One point which Kenneth Ramchand did not stress – but which is implicit in what he calls the 'nightmare' in Jean Rhys's novel *Wide Sargasso Sea* – is that Antoinette is mad Bertha in *Jane Eyre* and that Jean Rhys, intuitively rather than intention-ally, compensates a historical portrait of the West Indian creole – bridges the gap, as it were, between an outer rational frame and an inner irrational desolation to transform the hubris of reason (or proprietorship of flesh-and-blood) and bring into play a necessity for re-creative and therapeutic capacities grounded in complex vision. . . .

I believe that the *limbo* imagination of the folk involved a crucial inner re-creative response to the violations of slavery and indenture and conquest, and needed its critical or historical correlative, its critical or historical advocacy. This was not forthcoming since the histor-ical instruments of the past clustered around an act of censorship and of suspicion of folk-obscurity as well as individual originality, and that inbuilt suspicion continues to motivate a certain order of critical writing in the West Indies today.

## Chapter 78

# DIPESH CHAKRABARTY

## POSTCOLONIALITY AND THE ARTIFICE OF HISTORY

From 'Postcoloniality and the Artifice of History: Who Speaks for "Indian" Pasts?' *Representations* 37 (Winter), 1992.

THE PURPOSE OF THIS ARTICLE is to problematize the idea of 'Indians' representing themselves in 'history.' Let us put aside for the moment the messy problems of identity inherent in a transnational enterprise such as *Subaltern Studies*, where passports and commitments blur the distinctions of ethnicity in a manner that some would regard as characteristically postmodern. I have a more perverse proposition to argue. It is that insofar as the academic discourse of history – that is, 'history' as a discourse produced at the institutional site of the university – is concerned, 'Europe' remains the sovereign theoretical subject of all histories, including the ones we call 'Indian,' 'Chinese,' 'Kenyan,' and so on. There is a peculiar way in which all these other histories tend to become variations on a master narrative that could be called 'the history of Europe.' In this sense, 'Indian' history itself is in a position of subalterneity; one can only articulate subaltern positions in the name of this history. . . .

Colonial Indian history is replete with instances where Indians arrogated subjecthood to themselves precisely by mobilizing, within the context of 'modern' institutions and sometimes on behalf of the modernizing project of nationalism, devices of collective memory that were both antihistorical and antimodern. This is not to deny the capacity of 'Indians' to act as subjects endowed with what we in the universities would recognize as 'a sense of history' (what Peter Burke calls 'the renaissance of the past') but to insist at the same time that there were also contrary trends, that in the multifarious struggles that took place in colonial India, antihistorical constructions of the past often provided very powerful forms of collective memory.

There is then this double bind through which the subject of 'Indian' history articulates itself. On the one hand, it is both the subject and the object of modernity, because it stands for an assumed unity called the 'Indian people' that is always split into two – a modernizing elite and a yet-to-be-modernized peasantry. As such a split subject, however, it speaks from within a metanarrative that celebrates the nation state; and of this metanarrative the theoretical subject can only be a hyperreal 'Europe,' a 'Europe' constructed by the tales that both imperialism and nationalism have told the colonized. The mode of self-representation that the 'Indian' can adopt here is what Homi Bhabha has justly called 'mimetic' (Bhabha 1984b). Indian history, even in the most dedicated socialist or nationalist hands, remains a mimicry of a certain 'modern' subject of 'European' history and is bound to represent a sad figure of lack and failure. The transition narrative will always remain 'grievously incomplete.'

On the other hand, maneuvers are made within the space of the mimetic – and therefore within the project called 'Indian' history – to represent the 'difference' and the 'originality' of the 'Indian,' and it is in this cause that the antihistorical devices of memory and the antihistorical 'histories' of the subaltern classes are appropriated. Thus peasant/worker constructions of 'mythical' kingdoms and 'mythical' pasts/futures find a place in texts designated 'Indian' history precisely through a procedure that subordinates these narratives to the rules of evidence and to the secular, linear calendar that the writing of 'history' must follow. The antihistorical, antimodern subject, therefore, cannot speak itself as 'theory' within the knowledge procedures of the university even when these knowledge procedures acknowledge and 'document' its existence. Much like Spivak's 'subaltern' (or the anthropologist's peasant who can only have a quoted existence in a larger statement that belongs to the anthropologist alone), this subject can only be spoken for and spoken of by the transition narrative that will always ultimately privilege the modern (i.e. 'Europe').

So long as one operates within the discourse of 'history' produced at the institutional site of the university, it is not possible simply to walk out of the deep collusion between 'history' and the modernizing narrative(s) of citizenship, bourgeois public and private, and the nation state. 'History' as a knowledge system is firmly embedded in institutional practices that invoke the nation state at every step – witness the organization and politics of teaching, recruitment, promotions, and publication in history departments, politics that survive the occasional brave and heroic attempts by individual historians to liberate 'history' from the meta-narrative of the nation state. One only has to ask, for instance: Why is history a compulsory part of education of the modern person in all countries today including those that did quite comfortably without it until as late as the eighteenth century? Why should children all over the world today have to come to terms with a subject called 'history' when we know that this compulsion is neither natural nor ancient? It does not take much imagination to see that the reason for this lies in what European imperialism and third-world nationalisms have achieved together: the universalization of the nation state as the most desirable form of political community. Nation states have the capacity to enforce their truth games, and universities, their critical distance notwithstanding, are part of the battery of institutions complicit in this process. 'Economics' and 'history' are the knowledge forms that correspond to the two major institutions that the rise (and later universalization) of the bourgeois order has given to the world – the capitalist mode of production and the nation state ('history' speaking to the figure of the citizen). A critical historian has no choice but to negotiate this knowledge. She or he therefore needs to understand the state on its own terms, i.e. in terms of its self-justificatory narratives of citizenship and modernity. Since these themes will always take us back to the universalist propositions of 'modern' (European) political philosophy – even the 'practical' science of economics that now seems 'natural' to our constructions of world systems is (theoretically) rooted in the ideas of ethics in eighteenth-century Europe – a third-world historian is condemned to knowing 'Europe' as the original home of the 'modern,' whereas the 'European' historian does not share a comparable predicament with regard to the pasts of the majority of humankind. Thus follows the everyday subalternity of non-Western histories with which I began this paper. Yet the understanding that 'we' all do 'European' history with our different and often non-European archives opens up the possibility of a politics and project of alliance between the dominant metropolitan histories and the subaltern peripheral pasts. Let us call this the project of provincializing 'Europe,' the 'Europe' that modern imperialism and (third-world) nationalism have, by their collaborative venture and violence, made universal. Philosophically, this project must ground itself in a radical critique and transcendence of liberalism (i.e. of the bureaucratic constructions of citizenship, modern state, and bourgeois

privacy that classical political philosophy has produced), a ground that late Marx shares with
certain moments in both poststructuralist thought and feminist philosophy. In particular, I am
emboldened by Carole Pateman's courageous declaration – in her remarkable book *The Sexual
Contract* (1988) – that the very conception of the modern individual belongs to patriarchal
categories of thought.

The project of provincializing 'Europe' refers to a history that does not yet exist; I can therefore
only speak of it in a programmatic manner. To forestall misunderstanding, however, I must
spell out what it is not while outlining what it could be.

   To begin with, it does not call for a simplistic, out-of-hand rejection of modernity, liberal
values, universals, science, reason, grand narratives, totalizing explanations, and so on. Fredric
Jameson has recently reminded us that the easy equation often made between 'a philosophical
conception of totality' and 'a political practice of totalitarianism is baleful' (Jameson 1988:
354). What intervenes between the two is history – contradictory, plural, and heterogeneous
struggles whose outcomes are never predictable, even retrospectively, in accordance with
schemas that seek to naturalize and domesticate this heterogeneity. These struggles include
coercion (both on behalf of and against modernity), physical, institutional, and symbolic
violence, often dispensed with dreamy-eyed idealism – and it is this violence that plays a
decisive role in the establishment of meaning, in the creation of truth regimes, in deciding, as
it were, whose and which 'universal' wins. As intellectuals operating in academia, we are not
neutral to these struggles and cannot pretend to situate ourselves outside of the knowledge
procedures of our institutions.

   The project of provincializing 'Europe' therefore cannot be a project of 'cultural relativism.' It cannot originate from the stance that the reason/science/universals which help define
Europe as the modern are simply 'culture-specific' and therefore only belong to the European
cultures. For the point is not that Enlightenment rationalism is always unreasonable in itself
but rather a matter of documenting how – through what historical process – its 'reason,'
which was not always self-evident to everyone, has been made to look 'obvious' far beyond
the ground where it originated. If a language, as has been said, is but a dialect backed up by
an army, the same could be said of the narratives of 'modernity' that, almost universally
today, point to a certain 'Europe' as the primary habitus of the modern.

   This Europe, like 'the West,' is demonstrably an imaginary entity, but the demonstration
as such does not lessen its appeal or power. The project of provincializing 'Europe' has to
include certain other additional moves: (1) the recognition that Europe's acquisition of the
adjective *modern* for itself is a piece of global history of which an integral part is the story
of European imperialism; and (2) the understanding that this equating of a certain version of
Europe with 'modernity' is not the work of Europeans alone; third-world nationalisms, as
modernizing ideologies *par excellence*, have been equal partners in the process. I do not mean
to overlook the anti-imperial moments in the careers of these nationalisms; I only underscore
the point that the project of provincializing 'Europe' cannot be a nationalist, nativist, or
atavistic project. In unraveling the necessary entanglement of history – a disciplined and insti-
tutionally regulated form of collective memory – with the grand narratives of 'rights,'
'citizenship,' the nation state, 'public' and 'private' spheres, one cannot but problematize
'India' at the same time as one dismantles 'Europe.'

   The idea is to write into the history of modernity the ambivalences, contradictions, the
use of force, and the tragedies and the ironies that attend it. That the rhetoric and the claims
of (bourgeois) equality, of citizens' rights, of self-determination through a sovereign nation
state have in many circumstances empowered marginal social groups in their struggles is

undeniable – this recognition is indispensable to the project of Subaltern Studies. What effectively is played down, however, in histories that either implicitly or explicitly celebrate the advent of the modern state and the idea of citizenship is the repression and violence that are as instrumental in the victory of the modern as is the persuasive power of its rhetorical strategies. Nowhere is this irony – the undemocratic foundations of 'democracy' – more visible than in the history of modern medicine, public health, and personal hygiene, the discourses of which have been central in locating the body of the modern at the intersection of the public and the private (as defined by, and subject to negotiations with, the state). The triumph of this discourse, however, has always been dependent on the mobilization, on its behalf, of effective means of physical coercion. I say 'always' because this coercion is both originary/foundational (i.e. historic) as well as pandemic and quotidian. Of foundational violence, David Arnold gives a good example in a recent essay on the history of the prison in India. The coercion of the colonial prison, Arnold shows, was integral to some of the earliest and pioneering research on the medical, dietary, and demographic statistics of India, for the prison was where Indian bodies were accessible to modernizing investigators (Arnold 1992). Of the coercion that continues in the names of the nation and modernity, a recent example comes from the Indian campaign to eradicate smallpox in the 1970s. Two American doctors (one of them presumably of 'Indian' origin) who participated in the process thus describe their operations in a village of the Ho tribe in the Indian state of Bihar:

> In the middle of gentle Indian night, an intruder burst through the bamboo door of the simple adobe hut. He was a government vaccinator, under orders to break resistance against smallpox vaccination. Lakshmi Singh awoke screaming and scrambled to hide herself. Her husband leaped out of bed, grabbed an axe, and chased the intruder into the courtyard. Outside a squad of doctors and policemen quickly overpowered Mohan Singh. The instant he was pinned to the ground, a second vaccinator jabbed smallpox vaccine into his arm. Mohan Singh, a wiry 40-year old leader of the Ho tribe, squirmed away from the needle, causing the vaccination site to bleed. The government team held him until they had injected enough vaccine. . . . While the two policemen rebuffed him, the rest of the team overpowered the entire family and vaccinated each in turn. Lakshmi Singh bit deep into one doctor's hand, but to no avail.
>
> (Brilliant 1978: 3)

There is no escaping the idealism that accompanies this violence. The subtitle of the article in question unselfconsciously reproduces both the military and the do-gooding instincts of the enterprise. It reads: 'How an army of samaritans drove smallpox from the earth.'

Histories that aim to displace a hyperreal Europe from the center toward which all historical imagination currently gravitates will have to seek out relentlessly this connection between violence and idealism that lies at the heart of the process by which the narratives of citizenship and modernity come to find a natural home in 'history.' I register a fundamental disagreement here with a position taken by Richard Rorty in an exchange with Jürgen Habermas. Rorty criticizes Habermas for the latter's conviction 'that the story of modern philosophy is an important part of the story of the democratic societies' attempts at self-reassurance.' Rorty's statement follows the practice of many Europeanists who speak of the histories of these 'democratic societies' as if these were self-contained histories complete in themselves, as if the self-fashioning of the West were something that occurred only within its self-assigned geographical boundaries. At the very least Rorty ignores the role that the 'colonial theater' (both external and internal) – where the theme of 'freedom' as defined by modern political philosophy was constantly invoked in aid of the ideas of 'civilization,' 'progress,' and latterly 'development' – played in the process of engendering this 'reassurance.' The task, as I see it, will be to wrestle ideas that legitimize the modern state and its attendant institutions, in order to return to political philosophy – in the same way as suspect coins returned to their owners in an Indian bazaar – its categories whose global currency can no longer be taken for granted.

And, finally – since 'Europe' cannot after all be provincialized within the institutional site of the university whose knowledge protocols will always take us back to the terrain where all contours follow that of my hyperreal Europe – the project of provincializing Europe must realize within itself its own impossibility.

It therefore looks to a history that embodies this politics of despair. It will have been clear by now that this is not a call for cultural relativism or for atavistic, nativist histories. Nor is this a program for a simple rejection of modernity, which would be, in many situations, politically suicidal. I ask for a history that deliberately makes visible, within the very structure of its narrative forms, its own repressive strategies and practices, the part it plays in collusion with the narratives of citizenships in assimilating to the projects of the modern state all other possibilities of human solidarity. The politics of despair will require of such history that it lays bare to its readers the reasons why such a predicament is necessarily inescapable. This is a history that will attempt the impossible: to look toward its own death by tracing that which resists and escapes the best human effort at translation across cultural and other semiotic systems, so that the world may once again be imagined as radically heterogeneous. This, as I have said, is impossible within the knowledge protocols of academic history, for the globality of academia is not independent of the globality that the European modern has created. To attempt to provincialize this 'Europe' is to see the modern as inevitably contested, to write over the given and privileged narratives of citizenship other narratives of human connections that draw sustenance from dreamed-up pasts and futures where collectivities are defined neither by the rituals of citizenship nor by the nightmare of 'tradition' that 'modernity' creates. There are of course no (infra)structural sites where such dreams could lodge themselves. Yet they will recur so long as the themes of citizenship and the nation state dominate our narratives of historical transition, for these dreams are what the modern represses in order to be.

# PART THIRTEEN

# Place

**P**LACE AND DISPLACEMENT ARE CRUCIAL FEATURES of post-colonial discourse. By 'place' we do not simply mean 'landscape'. Indeed the idea of 'landscape' is predicated upon a particular philosophic tradition in which the objective world is separated from the viewing subject. Rather 'place' in post-colonial societies is a complex interaction of language, history and environment. It is characterized first by a sense of displacement in those who have moved to the colonies, or the more widespread sense of displacement from the imported language, of a gap between the 'experienced' environment and descriptions the language provides, and second, by a sense of the immense investment of culture in the construction of place.

A sense of displacement, of the lack of 'fit' between language and place, may be experienced by both those who possess English as a mother tongue and those who speak it as a second language. In both cases, the sense of dislocation from an historical 'homeland' and that created by the dissonance between language, the experience of 'displacement' generates a creative tension within the language. Place is thus the concommitant of difference, the continual reminder of the separation, and yet of the hybrid interpenetration of the colonizer and colonized.

The theory of place does not simply propose a binary separation between the 'place' named and described in language, and some 'real' place inaccessible to it, but rather indicates that in some sense place *is* language, something in constant flux, a discourse in process. The sense of 'lack of fit' between language and place is that which propels writers such as Dennis Lee to construct a new language. The post-colonial text, negotiating as it does the space between the textual language and the lived space becomes the metonym of the continual process of reclamation, as a cultural reality is both *posited* and *reclaimed* from the incorporating dominance of English.

Whether the speaker is the settler, the indigenous occupant of invaded colonies, a member of a colonized and dominated African or Indian society or the multifarious Caribbean region, language always negotiates a kind of gap between the word and its signification. In this sense the dynamic of 'naming' becomes a primary colonizing process because it appropriates, defines, captures the place in language. And yet the process of naming opens wider the very epistemological gap which it is designed to fill, for the 'dynamic mystery of language', as Wilson Harris puts it, becomes a groping step into the reality of place, not simply reflecting or representing it, but in some mysterious sense intimately involved in the process of its creation, of its 'coming into being'.

Place therefore, the 'place' of the 'subject', throws light upon subjectivity itself, because whereas we might conceive subjectivity as a process, as Lacan has done, so the discourse of place is a process of a continual dialectic between subject and object. Thus a major feature of post-colonial literatures is the concern with either developing or recovering an appropriate identifying relationship between self and place because it is precisely within the parameters of place and its separateness that the process of subjectivity can be conducted.

Place is also a palimpsest, a kind of parchment on which successive generations have inscribed and reinscribed the process of history. V. S. Naipaul signals this in *The Middle Passage* when he sees the history of the Caribbean signified in the land: 'There is slavery in the vegetation. In the sugarcane, brought by Columbus on that second voyage when, to Queen Isabella's fury, he proposed the enslavement of the Amerindians' (Naipaul 1962: 61–2). But the simple conflict of colonizer and colonized that Naipaul sees here is really a simplification of the complex way in which history is embedded in place. As a kind of counterpoint to Rabasa's critique of Mercator's *Atlas*, Graham Huggan demonstrates how the map itself is decolonized in post-colonial constructions of place. The map is the crucial signifier of control over place and thus of power over the inscription of being. Malcolm Lewis's discussion of indigenous map making therefore points to a kind of inscription that signifies a different way of knowing place. When the colonial map is drawn it effectively erases both the indigenous map and the indigenous knowing it signifies.

Perhaps the most detailed discussion of this process is Paul Carter's *The Road to Botany Bay* which surveys at length the extent to which the language of travel, of exploration, of settlement, indeed naming itself, turned empty space into 'place' in Australia and has continued to re-write the text of that place. This was not a place which was 'simply there' but a place which is in a continual process of being 'written'. This is true of any place, but in post-colonial experience the linkage between language, place and history is far more prominent because the interaction is so much more urgent and contestatory. One of the more interesting aspects of this palimpsest is the rewriting, through Aboriginal textuality, of a place which would seem to have been overwritten by the colonizer. As Bob Hodge and Vijay Mishra point out, the 'place' in aboriginal culture, rather than existing as a visual construct, is a kind of 'ground of being'.

## Chapter 79

# DENNIS LEE

# WRITING IN COLONIAL SPACE

From 'Cadence, Country, Silence: Writing in Colonial Space' *Boundary 2* 3(1) (Fall), 1974.

## Cadence

**M**OST OF MY LIFE AS A WRITER is spent listening into a cadence which is a kind of taut cascade, a luminous tumble. If I withdraw from immediate contact with things around me I can sense it churning, flickering, dancing, locating things in more shapely relation to one another without robbing them of themselves. I say it is present continuously, but certainly I spend days on end without noticing it. I hear it more clearly because I have recognised it in Hölderlin or Henry Moore, but I don't think it originates in their work. I think they heeded it too.

What I hear is initially without content; but when the poem does come, the content must accord with the cadence I have been overhearing or I cannot make it. (I speak of 'hearing' cadence, but in fact I am baffled by how to describe it. There is no auditory sensation – I don't hallucinate; yet it is like sensing a continuous, changing tremor with one's ear and one's whole body at the same time. It seems very matter-of-fact, yet I do not know the name of the sense with which I perceive it.)

More and more I sense this cadence as presence – though it may take 50 or 100 revisions before a poem enacts it – I sense it as presence, both outside myself and inside my body opening out and trying to get into words. What is it? I can convey some portion of that by pointing to things I have already written, saying 'Listen to the cadence here, and here – no, listen to the deeper cadence in which the poem is locally sustained.' But the cadence of the poems I have written is such a small and often mangled fraction of what I hear, it tunes out so many wavelengths of that massive, infinitely fragile polyphony, that I frequently despair. And often it feels perverse to ask what is cadence, when it is all I can manage to heed it. . . .

## Country

I have been writing of cadence as though one had merely to hear its words and set them down. But that is not true, at least not in my experience. There is a check on one's pen which seems to take hold at the very moment that cadence declares itself. Words arrive, but words have also gone dead.

To get at this complex experience we must begin from the hereness, the local nature of cadence. We never encounter cadence in the abstract; it is insistently here and now. Any man aspires to be at home where he lives, to celebrate communion with men on earth around him, under the sky where he actually lives. And to speak from his own dwelling – however light or strong the inflections of that place – he will make his words intelligible to men else-where, because authentic. In my case, then, cadence seeks the gestures of being a Canadian human: *mutatis mutandi*, the same is true for anyone here – an Israeli, an American, a Quebecker.

But if we live in a space which is radically in question for us, that makes our barest speaking a problem to itself. For voice does issue in part from civil space. And alienation in that space will enter and undercut our writing, make it recoil upon itself, become a problem to itself.

The act of writing 'becomes a problem to itself' when it raises a vicious circle; when to write necessarily involves something that seems to make writing impossible. Contradictions in our civil space are one thing that make this happen, and I am struck by the subtle connections people here have drawn between words and their own problematic public space. . . .

To explore the obstructions of cadence is, for a Canadian, to explore the nature of colonial space. Here I am particularly concerned with what it does to writing. One can also analyse it economically or politically, or try to act upon it; but at this point I want only to find words for our experience of it. . . .

I shall be speaking of 'words', but not merely those you find in a dictionary. I mean all the resources of the verbal imagination, from single words through verse forms, conventions about levels of style, characteristic versions of the hero, resonant structures of plot. And I use my own experience with words because I know it best. It tallies with things other writers in their thirties have said, but I don't know how many would accept it fully.

My sense when I began writing, about 1960 – and this lasted five or six years – was that I had access to a great many words: those of the British, the American, and (so far as anyone took it seriously) the Canadian traditions. Yet at the same time those words seemed to lie in a great random heap, which glittered with promise so long as I considered it in the mass but within which each individual word went stiff, inert, was somehow clogged with sludge, the moment I tried to move it into place in a poem. I could stir words, prod at them, cram them into position; but there was no way I could speak them directly. They were completely external to me, though since I had never known the words of poetry in any other way I assumed that was natural.

Writers everywhere don't have to begin with a resistant, external language; there was more behind the experience than just getting the hang of the medium during apprenticeship. In any case, after I had published one book of poems and finished another a bizarre thing happened: I stopped being able to use words on paper at all.

All around me – in England, America, even in Canada – writers opened their mouths and words spilled out like crazy. But increasingly when I opened mine I simply gagged; finally, the words no longer came. For about four years at the end of the decade I tore up every-thing I wrote – twenty words on a page were enough to set me boggling at their palpable inauthenticity. And looking back at my previous writing, I felt as if I had been fishing pretty beads out of a vat of crank-case oil and stringing them together. The words weren't limber or alive or even mine.

To discover that you are mute in the midst of all the riches of a language is a weird experience. I had no explanation for it; by 1967 it had happened to me, but I didn't know why . . . I had just begun to write, and now I was stopped. I would still sit down in my study with a pen and paper from time to time, and every time I ended up ripping the paper to pieces and pitching it out. The stiffness and falsity of the words appalled me; the reaction was more in my body than my mind, but it was very strong. . . .

The colonial writer does not have words of his own. Is it not possible that he projects his own condition of voicelessness into whatever he creates? that he articulates his own power-lessness, in the face of alien words, by seeking out fresh tales of victims? . . . perhaps the colonial imagination is driven to recreate, again and again, the experience of writing in colonial space.

We are getting close to the centre of the tangle. Why did I stop being interested in Shakespeare at Stratford, when I had gone assiduously for ten summers? Why did I fidget and squirm in front of TV and read so much less? And why did I dry?

The words I knew said Britain, and they said America, but they did not say my home. They were always and only about someone else's life. All the rich structures of language were present, but the currents that animated them were not home to the people who used the language here.

But the civil self seeks nourishment as much as the biological self; it too fuels the imagination. And if everything it can find is alien, it may protect itself in a visceral spasm of refusal. To take an immediate example: the words I used above 'language', 'home', 'here' – have no native charge; they convey only meanings in whose face we have been unable to find ourselves since the eighteenth century. This is not a call for arbitrary new Canadian defin-itions, of course. It is simply to point out that the texture, weight and connotation of almost every word we use comes from abroad. For a person whose medium is words, who wants to use words to recreate our being human here – and where else do we live? – that fact creates an absolute problem.

Why did I dry for four years? The language was drenched with our non-belonging, and words – bizarre as it sounds, even to myself – words had become the enemy. To use them as a writer was to collaborate further in one's extinction as a rooted human being. And so, by a drastic and involuntary stratagem of self-preserval, words went dead.

The first necessity for the colonial writer – so runs the conventional wisdom – is to start writing of what he knows. His imagination must come home. But that first necessity is not enough. For if you are Canadian, home is a place that is not home to you – it is even less your home than the imperial centre you used to dream about. Or to say what I really know best, the *words* of home are silent. And to write a jolly ode to harvests in Saskatchewan, or set an American murder mystery in Newfoundland, is no answer at all. Try to speak the words of your home and you will discover – if you are a colonial – that you do not know them.

To speak unreflectingly in a colony, then, is to use words that speak only alien space. To reflect is to fall silent, discovering that your authentic space does not have words. And to reflect further is to recognise that you and your people do not in fact have a privileged authentic space just waiting for words; you are, among other things, the people who have made an alien inauthenticity their own. You are left chafing at the inarticulacy of a native space which may not exist. So you shut up.

But perhaps – and here was the breakthrough – perhaps our job was not to fake a space of our own and write it up, but rather to find words for our space-lessness. Perhaps that *was* home. This dawned on me gradually. Instead of pushing against the grain of an external, uncharged language, perhaps we should finally come to writing with that grain.

To do that was a homecoming – and a thoroughly edgy, uncertain homecoming it was. You began by giving up the idea of writing in the same continuum as Lowell, Roethke, Ginsberg, Olson, Plath. . . . It was a question of starting from your own necessities. And you began striving to hear what happened in words – in 'love', 'inhabit', 'fail', 'earth', 'house' – as you let them surface in your own mute and native land. It was a funny, visceral process;

there was nothing as explicit as starting to write in *joual*, though the process was comparable. There was only the decision to let words be how they actually are for us. But I am distorting the experience again by writing it down. There was nothing conscious about this decision, initially at least – it was a direction one's inner ear took up. I know I fought it.

The first mark of words, as you began to re-appropriate them in this space-less civil space, was a kind of blur of unachieved meaning. That I had already experienced, though only as something oppressing and negative. But then I began to sense something more.

Where I lived, a whole swarm of inarticulate meanings lunged, clawed, drifted, eddied, sprawled in half-grasped disarray beneath the tidy meaning which the simplest word had brought with it from England and the States. 'City': once you learned to accept the blurry, featureless character of that word – responding to it as a Canadian word, with its absence of native connotation – you were dimly savaged by the live, inchoate meanings trying to surface through it. The whole tangle and sisyphean problematic of people's existing here, from the time of the *coureurs de bois* to the present day, came struggling to be included in the word 'city'. Cooped up beneath the familiar surface of the word as we use it ('city' as London, as New York, as Los Angeles) – and cooped up further down still, beneath the blank and blur you heard when you sought some received indigenous meaning for the word – listening all the way down, you began to overhear the strands and communal lives of millions of people who went their particular ways here, whose roots and lives and legacy come together in the cities we live in. Edmonton, Toronto, Montreal, Halifax: 'city' meant something still unspoken, but rampant with held-in energy. Hearing it was like watching the contours of an unexpected continent gradually declare themselves through the familiar lawns and faces of your block.

Though that again is hindsight: all of it. You heard an energy, and those lives were part of it. Under the surface alienation and the second-level blur of our words there was a living barrage of meaning: private, civil, religious – unclassifiable finally, but there, and seamless, and pressing to be spoken. And I felt that press of meaning: I had no idea what it was, but I could feel it teeming towards words. I called it cadence.

And hearing that cadence, I started to write again.

## Chapter 80

# PAUL CARTER

## NAMING PLACE

From *Road to Botany Bay: An Essay in Spatial History* London and Boston:
Faber & Faber, 1987.

[I]T WAS ALMOST A COMMONPLACE AMONG BRITISH residents that, in Australia, the laws of association seemed to be suspended. There seemed to be nothing that could be accurately named. There was, consequently, very little purchase for the imagination – that mental faculty which . . . was primarily a mechanism for making analogies. This was why Barron Field lamented in his *Geographical Memoirs* that Australia was quite unsuitable as a subject for poetry. Referring in particular to 'the eternal eucalyptus', the former dramatic critic of *The Times*, the friend of Coleridge and Wordsworth, wrote:

> No tree, to my taste, can be beautiful that is not deciduous. . . . Dryden says of the laurel
>
>> From winter winds it suffers no decay;
>> For ever fresh and fair, and every month is May.
>
> Now it may be the fault of the cold climate in which I was bred, but this is just what I complain of in an evergreen. 'Forever fresh', is a contradiction in terms; what is 'forever fair' is never fair; and without January, in my mind, there can be no May. All the dearest allegories of human life are bound up with the infant and slender green of spring, the dark redundance of summer, and the sere and yellow leaf of autumn. These are as essential to the poet as emblems, as they are to the painter as picturesque objects; and the common consent and immemorial custom of European poetry has made the change of the seasons, and its effect upon vegetation, a part, as it were, of our very nature. I can therefore hold no fellowship with Australian foliage. . . .
>
> (Field 1825: 423–4)

Field's real subject in this passage is not nature at all. It is language, and the impossibility of distinguishing the language of feeling from the language of description. The proper context in which to understand Field's uncompromising stance is not the history of taste but the history of mind. In particular, the point of departure for Field's animadversions is clearly the prevalent doctrine of associationism. . . . The association of simple ideas to form complex ones depended on the ideas (or objects they derived from) being comparable. And first among the qualities of objects that made their comparison possible was, as Hume wrote, 'resemblance': 'This is a relation without which no philosophical relation can exist, since no objects will admit of comparison, but what have some degree of resemblance' (Hume 1934: 22).

European nature is an 'emblem' of human life because its cycle of seasons resembles the seasons of human life. Poetry, then, in so far as it evokes human life metaphorically, involves not the description of nature but its association with human themes.

But, as Barron Field's remarks bring out, the association of ideas depended on a profounder assumption. It depended on the assumption that distinct ideas existed to be related. But how is a distinct idea defined, except in relation to other ideas? Since, as Hume put it, 'all kinds of reasoning consist in nothing but a comparison, and a discovery of those relations, either constant or inconstant, which two or more objects bear to each other' (Hume 1934: 77), an absolute idea is a contradiction in terms. Hence Field's irritation with Dryden's laurel. Our ideas of freshness and fairness are relative. They derive their distinctness from our ideas of sereness and dullness. Their efficacy as metaphors of human life depends on our ideas of stale and withered age. But a tree eternally green offers the poet nothing. It defies the logic of association. It is not a distinct idea.

The far-reaching and astonishing implication of Field's remarks is, then, that Australia is, strictly speaking, indescribable. In so far as its nature is undifferentiated, it does not have a distinct character. Lacking this, it cannot be compared and so known. Its uniformity also means that it cannot be named, because no nameable parts distinguish themselves. Not amenable to the logic of association, Australia appears to be unknowable. A state of uniformity offers no starting point, whether for literary or physical travel.

The implications of this conclusion are not only literary. Bearing in mind that the prime responsibility of the early explorers was to *describe* what they saw, the dissonance between language and land presented a considerable challenge. There was no question of falling back on the logic of facts. There was no possibility even of allowing oneself the lazy luxury of comparison. Facts proved fancies; analogies proved false. Indeed, the spatial ramifications of Field's argument are well brought out by Field himself – in the context, significantly enough, of place names. Not surprisingly, Field is highly critical of Australian place names, which, he presumes, attempt to apply the principles of association. Given the 'prosaic, unpicturesque, unmusical' qualities of Australian nature, they are doomed, in his view, to failure. Here is Field's description of the country in the region of Mount York, between Windsor and Bathurst in the Blue Mountains, west of Sydney:

> The King's Tableland is as anarchical and untabular as any His Majesty possesses. The Prince Regent's Glen below it (if it be the glen that I saw) is not very romantic. Jamison's Valley we found by no means a happy one. Blackheath is a wretched misnomer. Not to mention its awful contrast to the beautiful place of that name in England, heath it is none. Black it may be when the shrubs are burnt, as they often are. Pitt's Amphitheatre disappointed me. The hills are thrown together in a monotonous manner, and their clothing is very unpicturesque – a mere sea of harsh trees; but Mr Pitt was no particular connoisseur in mountain scenery or in amphi-theatres.
> (Field 1825: 430)

This splenetic outburst against misnomers and misnamers is based on the assumption that the Australian names in question violate the logic of association. It is not so much a 'description' of nature as a critique of the absurdity of associative naming in Australia: the 'Tableland' in question does not really resemble a tableland; the 'Glen' does not really suggest a glen; the 'Valley' does not really recall a valley, because the Australian places fail to conjure up the proper associations. In Australia, therefore, class names of this kind fail.

The interesting thing about Field's view is that it underlines the perhaps surprising point that *both* elements in place names were figurative, and non-factual: it is not only the particularizing element 'Botany' that is metaphorical; the apparently more objective term 'Bay' may be equally fanciful. What Field's remarks do not bring out, though, is the even more important point that follows from this – which is that the proper way in which to interpret geographical class names like 'valley' and 'tableland' is not in terms of imperial history, but within the perspective of the history of travelling. The imperial pretensions of particularizing epithets like 'King's', 'Prince Regent's' and even 'Pitt's' may seem all too obvious, but their concentration suggests that even they reflect the namer's intention to characterize a space. And, certainly, when we turn to the other element in such names, the spatial intention they express becomes inescapable. For the fact is that such names were given. In some sense they 'stuck'. And, even if we accepted Field's genealogical judgement and considered such classificatory names failures, we would still be left wondering why they were given in the first place. Why, if the newcomers were bound by the laws of analogy, by what they had formerly seen and read, did they not leave these nameless extensions, these culturally invisible intervals, unnamed and silent?

The truth is that the naming process may have been metaphorical and, to that extent, a kind of gnomic poetry *manqué*, but it was not associational in intent. It was not a retrospective gloss, after the event, a sort of gilding in bad taste round the physical mirror of history. It was the names themselves that brought history into being, that invented the spatial and conceptual coordinates within which history could occur. For how, without place names, without agreed points of reference, could directions be given, information exchanged, 'here' and 'there' defined? Consider those most beautiful of Australian names, names like Cape Catastrophe, Mount Misery, Retreat Well and Lake Disappointment. These names do not merely confirm Field's argument, that the logic of association breaks down in Australia: they also defy it, asserting the possibility of naming in the absence of resemblance. If a well is associated with water and is therefore regarded as an aid to the traveller, to describe it as *Retreat* well is to say the conventional associations of the class name fail here. A mountain, associated with long views and perhaps with water, would, it might be thought, be welcomed by the explorer. To call it 'Mount Misery' is again to suggest that, in Australia, the normal logic of association breaks down. But, if these 'wells' are not wells, these 'mountains' unmountainlike, what do the names mean? What is their function? The paradox they express is not descriptive. Rather, it refers to the traveller's state of expectation.

More than this, such class names (as their riddling qualifiers often make explicit) do not reflect what is already there: on the contrary, they embody the existential necessity the traveller feels to invent a place he can inhabit. Without them, punctuating the monotony, distinguishing this horizon from that, there would be no evidence he had travelled. To be sure, the traveller might retain his private impressions, but, without names, and the discourse of the journal they epitomize, his experience could never become public, a historical fact leading to other facts, other journeys. Thus, the fundamental impulse in applying class names like 'mount' or 'river' was a desire to differentiate the uniformity.

Partly, of course, the uniformity defied easy differentiation because, in a quite simple way, the English language lacked words to characterize it. Alexander Hamilton Hume and William Hovell, for example, who led the first overland expedition from Sydney to Port Phillip in 1824, 'cross swamp which had been mistaken for a meadow'. As it turns out, it is neither one nor the other:

> This, like all other spaces of any extent, lying intermediately between the ranges, consists of a kind of meadow, divided along its centre by a small but rapid stream, is somewhat swampy, and in places near the water produces reeds.
>
> (Hovell and Hume 1831: 51)

This was a case where a kind of country, far from rare, raised a problem of nomenclature. There was no English term for it, and yet in more arid regions the traveller sought it out. Indeed another explorer, Edward John Eyre, depended on it when making his attempt on the centre in 1841. Advancing northward parallel to the Flinders Range in South Australia,

> we just kept far enough into the plains to intercept the watercourses from the hills where they spread out into level country, and by this means we got excellent feed for our horses.
>
> (Eyre 1845: 94)

As this fertile zone of passage was neither meadow nor swamp, the explorer had no name for it. Passages like these come close to vindicating the reductionist view of the American philosopher of language, Benjamin Whorf. On the map at least, these places do not exist simply because they cannot be named; reality is a naive reflection of the language available to describe it. But, even though Eyre and Hume and Hovell cannot name this intermediate space directly, they can, after all, refer to it. What limits their powers of description is not vocabulary but the desire to differentiate, the necessity of naming in order to travel. And, in so far as such nameless zones can be located syntactically and spatially between 'plains' and 'hills', they have been appropriated to the traveller's route.

In any case, the difficulty of vocabulary aside, the criteria of differentiation were not simply empirical, naively describing the nature of 'things' already there – it was precisely such objects that names served to constitute. They were determined not empirically but rhetorically. They embodied the traveller's directional and territorial ambitions: his desire to possess where he had been as a preliminary to going on. And this desire was not placeless, it did not resemble the equal stare of the map grid. It depended on positing a 'here' (the traveller's viewpoint and orientation) and a 'there' (the landscape, the horizon). And where such viewpoints did not exist, they had to be hypothesized, rhetorically asserted by way of names. Otherwise, the landscape itself could never enter history.

This is the significance of the urgency, the premature willingness, of Australian travellers to name 'mountains' and 'rivers'. Mountains and rivers were culturally desirable, they conjured up pleasing associations. But, more fundamentally, they signified differences that made a difference. They implied the possibility of viewpoints, directions: to call a hill 'Prospect Hill' . . . is to describe hills' historical function. 'Supposed course of the River Nepean': to write these words on the map one after the other is to set out graphically the spatial intention implicit in invoking the word 'river'. Hills and rivers were, in fact, the kind of object that made travelling as a historical activity possible. They were the necessary counterpart of the traveller's desire to travel, to see the horizon and to find a route there.

## Chapter 81

# GRAHAM HUGGAN

# DECOLONIZING THE MAP

From 'Decolonizing the Map: Post-colonialism, Post-structuralism and the Cartographic Connection' *Ariel* 20(4), 1989.

THE PREVALENCE OF THE MAP TOPOS in contemporary post-colonial literary texts, and the frequency of its ironic and/or parodic usage in these texts, suggests a link between a de/reconstructive reading of maps and a revisioning of the history of European colonialism. This revisionary process is most obvious, perhaps, in the fiction of the Caribbean writer Wilson Harris, where the map features as a metaphor of the perceptual transformation which allows for the revisioning of Caribbean cultural history in terms other than those of catastrophe or complex. Throughout his work, Harris stresses the relativity of modes of cultural perception; thus, although he recognizes that a deconstruction of the social text of European colonialism is the prerequisite for a reconstruction of post-colonial Caribbean culture, he emphasizes that this and other post-colonial cultures neither be perceived in essentialist terms, nor divested of its/their implication in the European colonial enterprise. The hybrid forms of Caribbean and other post-colonial cultures merely accentuate the transitional status of all cultures; so while the map is ironized on the one hand in Harris's work as a visual analogue for the inflexibility of colonial attitudes and for the 'synchronic essentialism' of colonial discourse, it is celebrated on the other as an agent of cultural transformation and as a medium for the imaginative revisioning of cultural history (see Harris 1981).

More recent developments in post-colonial writing and, in particular, in the Canadian and Australian literatures, suggest a shift of emphasis from the interrogation of European colonial history to the overt or implied critique of unquestioned nationalist attitudes which are viewed as 'synchronic' formations particular not to post-colonial but, ironically, to colonial discourse. A characteristic of contemporary Canadian and Australian writing is a multiplication of spatial references which has resulted not only in an increased range of national and international locations but also in a series of 'territorial disputes' which pose a challenge to the self-acknowledging 'mainstreams' of metropolitan culture, to the hegemonic tendencies of patriarchal and ethnocentric discourses, and implicitly, I would argue, to the homogeneity assumed and/or imposed by colonialist rhetoric. These revised forms of cultural decolonization have brought with them a paradoxical alliance between internationalist and regionalist camps where the spaces occupied by the 'international', like those by the 'regional', do not so much forge new definitions as denote the semantic slippage between prescribed definitions of place. The attempt by writers such as Hodgins (1977) and Malouf (1985) to project spaces other than, or by writers such as Van Herk (1986) and Atwood (1985), to articulate the spaces between, those prescribed by dominant cultural or cultural groups, indicates a resistance to the notion of cartographic enclosure and to the imposed cultural limits that notion implies. Yet the range of geographical locations and diversity of functions served by the map metaphor in the contemporary Canadian and Australian literatures suggests a desire on the part of their respective writers not merely to deterritorialize, but also to reterritorialize, their increasingly multiform cultures. The dual tendencies towards geographical dispersal (as, for example, in the 'Asian' fictions of Koch and Rivard) and cultural decentralization (as, for example, in the hyperbolically fragmented texts of Bail and Kroetsch) can therefore be seen within the context of a resiting of the traditional 'mimetic fallacy' of cartographic representation. The map no longer features as a visual paradigm for the ontological anxiety arising from frustrated attempts to define a national culture, but rather as a locus of productive dissimilarity where the provisional connections of cartography suggest an ongoing perceptual transformation which in turn stresses the transitional nature of post-colonial discourse. This transformation has been placed within the context of a shift from an earlier 'colonial' fiction obsessed with the problems of writing in a 'colonial space' to a later, 'post-colonial' fiction which emphasizes the provisionality of all cultures and which celebrates the particular diversity of formerly colonized cultures whose ethnic mix can no longer be considered in terms of the colonial stigmas associated with mixed blood or cultural schizophrenia. Thus, while it would be unwise to suggest that the traditional Canadian and Australian concerns with cultural identity have become outmoded, the reassessment of cartography in many of their most recent literary texts indicates a shift of emphasis away from the desire for homogeneity towards an acceptance of diversity reflected in the interpretation of the map, not as a means of spatial containment or systematic organization, but as a medium of spatial perception which allows for the reformulation of links both within and between cultures.

In this context, the 'new spaces' of post-colonial writing in Canada and Australia can be considered to resist one form of cartographic discourse, whose patterns of coercion and containment are historically implicated in the colonial enterprise, but to advocate another, whose flexible cross-cultural patterns not only counteract the monolithic conventions of the West but revision the map itself as the expression of a shifting ground between alternative metaphors rather than as the approximate representation of a 'literal truth'. This paradoxical motion of the map as a 'shifting ground' is discussed at length by the French post-structuralists Gilles Deleuze and Félix Guattari. For Deleuze and Guattari, maps are experimental in orientation:

> The map is open and connectable in all its dimensions; it is detachable, reversible, susceptible to constant modification. It can be torn, reversed, adapted to any kind of mounting, reworked by an individual, group, or social formation. It can be drawn on the wall, conceived of as a work of art, constructed as a political action or as a meditation.
>
> (Deleuze and Guattari 1987: 12)

The flexible design of the map is likened by Deleuze and Guattari to that of the rhizome, whose 'deterritorializing lines of flight' (222) effect 'an asignifying rupture against the oversignifying breaks separating structures or cutting across a single structure' (7–9).

As Diana Brydon has illustrated, Deleuze and Guattari's association of the multiple connections/disconnections of the rhizome with the transformative patterns of the map provides a useful, if by its very nature problematic, working model for the description of post-colonial cultures and for the closer investigation of the kaleidoscopic variations of post-colonial discourse (Brydon 1988). Moreover, a number of contemporary women writers in Canada and Australia, notably Nicole Brossard and Marion Campbell, have adapted Deleuze and Guattari's model to the articulation of a feminist cartography which dissociates itself from the 'over-signifying' spaces of patriarchal representation but through its 'deterritorializing lines of flight' produces an alternative kind of map characterized not by the containment or regimentation of space but by a series of centrifugal displacements (see Godard 1987). Other implicitly 'rhizomatic' maps are sketched out in experimental fictions such as those of Kroetsch (1975) and Baillie (1986) in Canada, and Bail (1980) and Murnane (1984) in Australia, where space, as in Deleuze and Guattari's model, is constituted in terms of a series of intermingling lines of connection which shape shifting patterns of de- and reterritorialization. In the work of these other 'new novelists', the map is often identified, then parodied and/or ironized, as a spurious definitional construct, thereby permitting the writer to engage in a more wide-ranging deconstruction of Western signifying systems. . . . If the map is conceived of in Deleuze and Guattari's terms as 'rhizomatic' ('open') rather than as a falsely homogeneous ('closed') construct, the emphasis then shifts from de- to reconstruction, from mapbreaking to mapmaking. The benefit of Deleuze and Guattari's model is that it provides a viable alternative to the implicitly hegemonic (and historically colonialist) form of cartographic discourse which uses the duplicating procedures of mimetic representation and structuralist reconstitution as strategic means of stabilizing the foundations of Western culture and of 'fixing' the position (thereby maintaining the power) of the West in relation to cultures other than its own. Thus, whereas Derrida's deconstructive analysis of the concepts of 'centred' structure and 'interested' simulacrum engenders a process of displacement which undoes the supposed homogeneity of colonial discourse, Deleuze and Guattari's rhizomatic map views this process in terms of a processual transformation more pertinent to the operations of post-colonial discourse and to the complex patterns of de- and reterritorialization working within and between the multicultural societies of the post-colonial world.

As Stephen Slemon has demonstrated, one of the characteristic ploys of post-colonial discourse is its adoption of a creative revisionism which involves the subversion or displacement of dominant discourses (Slemon 1988b). But included within this revisionary process is the internal critique of the post-colonial culture (or cultures), a critique which takes into account the transitional nature of post-colonial societies and which challenges the tenets both of an essentialist nationalism which sublimates or overlooks regional differences and of an unconsidered multiculturalism (mis) appropriated for the purposes of enforced assimilation rather than for the promulgation of cultural diversity. The fascination of post-colonial writers, and of Canadian and Australian writers in particular, with the map topos can be seen in this context as a specific instance of creative revisionism in which the desystematization of a narrowly defined and demarcated 'cartographic' space allows for a culturally and historically located critique of colonial discourse while, at the same time, producing the momentum for a projection and exploration of 'new territories' outlawed or neglected by dominant discourses which previously operated in the colonial, but continue to operate in modified or transposed forms in the post-colonial, culture. I would suggest further that, in the cases of the contemporary Canadian and Australian literatures, these territories correspond to a series of new or revised rhetorical spaces occupied by feminism, regionalism and ethnicity, where each of these items is understood primarily as a set of counter-discursive strategies which challenge the claims of or avoid circumscription within one or other form of cultural centrism. These territories/spaces can also be considered, however, as shifting grounds which are themselves subject to transformational patterns of de- and reterritorialization. The proliferation of spatial references, crossing of physical and/or conceptual boundaries and redisposition of geographical coordinates in much contemporary Canadian and Australian writing stresses the provisionality of cartographic connection and places the increasing diversity of their respective literatures in the context of a postcolonial response to and/or reaction against the ontology and epistemology of 'stability' promoted and safeguarded by colonial discourse. I would conclude from this that the role of cartography in contemporary Canadian and Australian writing, specifically, and in post-colonial writing in general, cannot be solely envisaged as the reworking of a particular spatial paradigm, but consists rather in the implementation of a series of creative revisions which register the transition from a colonial framework within which the writer is compelled to recreate and reflect upon the restrictions of colonial space to a post-colonial one within which he or she acquires the freedom to engage in a series of 'territorial disputes' which implicitly or explicitly acknowledge the relativity of modes of spatial (and, by extension, cultural) perception. So while the map continues to feature in one sense as a paradigm of colonial discourse, its deconstruction and/or revisualization permits a 'disidentification' from the procedures of colonialism (and other hegemonic discourses) and a (re) engagement in the ongoing process of cultural decolonization. The 'cartographic connection' can therefore be considered to provide that provisional link which joins the contestatory theories of post-structuralism and post-colonialism in the pursuit of social and cultural change.

## Chapter 82

## BOB HODGE AND VIJAY MISHRA

# ABORIGINAL PLACE

From *The Dark Side of the Dream* London: Allen and Unwin, 1991.

### Land as theme

FOR ABORIGINES TODAY the issue of issues is land rights. So it is surprising at first glance that Aboriginal art and literature is not rich in references to land and evocations of landscapes. Aborigines' love of their own land and their precise knowledge of its topography are not in question, yet it was traditionally not an explicit theme in visual or verbal art. The situation of contemporary Aborigines is now very different. Instead of the confident assumption of identity tied to and established through links to a country, dispossession to some degree is their universal experience. But there is still a continuity between traditional and contemporary forms of cultural expression of this theme amongst Aborigines. Traditional culture provided a highly flexible set of ways of encoding a nexus of rights and obligations towards the land. It gave rise to aesthetic statements which were essentially political and juridical rather than personal and expressive. This quality made it equally well adapted to the needs of Aborigines today, all of whom are in some respects fringe-dwellers in their own land, needing a means of relocating themselves in White Australia, reconstructing an identity which is fully Aboriginal yet adequate to the new situation.

In looking at these kinds of adaptation we need to recognise the validity of the two broad strategies adopted by Aboriginal people as these are reflected in cultural forms. Many Aboriginal groups in northern and central Australia are trying to reestablish traditional ways of life, as close to their traditional territories as is now possible. The acrylic art of the Western Desert peoples and the maintenance of traditional languages are important to this strategy. But for many Aborigines in the south the route back has been disrupted, so that the direct link with a specific piece of country is no longer viable. For these Aborigines, urban dwellers or fringe-dwellers in country towns, the achievements of Western Desert artists are inspiring but unavailable. The writers who speak for them include all the Aboriginal writers well known in the White community: Jack Davis, Kath Walker (Oodgeroo Noonuccal), Colin Johnson (Mudrooroo Narogin), Kevin Gilbert, Robert Bropho and Sally Morgan. Yet each of these distinct strands of Aboriginal art is equally Aboriginal, equally crucial to all Aborigines, since one establishes the Aboriginal base, while the other opens up the transformational freedom that is equally important to all Aborigines, wherever they are placed.

As an instance of traditional adaptation, we will take two texts by Peter Skipper, a painting and a story. Peter Skipper's Walmatjari speaking group was part of the exodus of Western Desert people who moved east, out of the desert area, in closer proximity to areas of White settlement in the northwest of Western Australia. His community is established near Fitzroy

Crossing, where Peter Skipper is an important elder who is concerned to retain traditional forms of life in this new situation.

The story we will look at was reported to a linguist, Joyce Hudson, and transcribed by her as an example of the Walmatjari language. In that form (part of a grammar of Walmatjari for use by linguists and educators) it has the low aesthetic status and small circulation typical of this genre. It seems a casual and uninformative anecdote, yet like so many such texts it has a complexity of structure and depth of meaning that repay much closer scrutiny. Our reading, we should point out at the outset, will barely scratch the surface of the meanings of this text which are 'owned' by Peter Skipper to which we have no way or right to access. We will begin by giving a translation of the complete text, drawing on Joyce Hudson's translation and commentary (1978):

> In the wet season they eat plant food, Janiya
> and meat, lizard –
> meat, lizard and wild onion plants, wild onion.
> In the wet season they eat the
> bush-walnuts as plant food, bush-walnuts they eat.
> In the sandhills they lived like this
> the people lived in that former time in the sandhills.
> They were eating meat and plant food.
> Plants that they ate were various, various.
> They were eating all kinds of plant food,
> all kinds of plant foods they ate,
> until what they ate was finished,
> a finish to eating plant food and meat
> Well, they ate meat only,
> then that finished.
> And the people they went this way
> to other kinds of plant food, Whiteman's tucker.
> Well, those people too, they went north to the stations.
> Then they gave them plant food,
> the people from the sandhills.
> Those people went for good,
> never to return.
> Well, they went on a journey for plant food,
> Whiteman's tucker.
> Then they stayed there, those people.
> So they ate the plant food of the Whiteman,
> and so they stayed there, the people,
> never to return.

In terms of comparable White genres, this is closest to a lyric of loss and dispossession, stately and formulaic, almost literally a structure of feeling organising a temporal and spatial map. There is no comment, no explanation, no justification of either present or past, only a recurring set of organising categories that can be applied to both. Why did the food supply disappear? Was it a natural disaster, an extended drought, or was it the incursions of Whites with their stock and excessive demands on the fragile ecology? Peter Skipper gives no answer or not explicitly. . . .

Another principle of the structure, which our translation can only indicate with some clumsiness, is the repetition of the two categories of food, *kuyi* (meat) and *miyi* (plant). These two words insistently classify the natural environment as kinds of food to be hunted or gathered in the gendered division of labour of traditional society. The progression that Skipper describes has three stages. In the first, normative stage, there is both *miyi* and *kuyi*, gathering and hunting. In the second phase, there is only *kuyi*, hunting. The tucker supplied by the Whites is referred to every time as *miyi*, plant foods. Flour and sugar may have been the main staples provided by stations, but there was meat too, from occasional killings. Skipper classifies this food in social not in biological terms as *miyi*, implying a comment on its effect on the traditional roles of Aboriginal society. He does not cast back to a time when men as hunters were exclusive or dominant providers – that phase in his scheme was a result of a breakdown in the natural order. Instead in the landscape of plenitude, in the Wet season in the past, men and women were coequals. That is the situation that can and must return, when another Wet succeeds the present Dry.

There is no statement of regret, though the repeated 'never to return' unmistakeably suggests a sense of loss. What the text does is to carry the values of the desert and the past into the new situation, not as a legitimation of the present (a form of existence which is still radically incomplete) but as a kind of charter for change. The text is neither militant nor resigned, but its criticism and its optimism are so understated as to be almost invisible. Its dominant qualities of balance and poise are aesthetic and political, and even its self-effacement is part of a strategy of survival. And although this past is a remembered historical past it also has something of the structural form of what is called 'the Dreamtime' by Aboriginalists: that is, a time in the past whose values are still active in the present.

Peter Skipper painted *Jila Japingka* (Figure 82.1) in 1987, a text which is reproduced in Sutton (1988). The style of the painting is typical of the acrylics of the Western Desert artists, though it has its own distinctive qualities. As is normal for such texts, the first impression is the sense of formal patterns, produced by repetition of a small number of elements, as happens with his verbal text also. The meaning of the text is otherwise almost inaccessible, without further explanation. Sutton provides a gloss, the first aspect of which is its positioning in space and time. The cross-shape (painted deep blue) is formed by rain from the four compass-points, with the top rain from the east, the bottom rain from the west, with rain from the north on the left, and rain from the south on the right.

This, then, is a map of the same landscape as in the story, the all-important landscape in which Peter Skipper and his people have acted out their life. The east, where his people came from, is positioned at the top of the picture, and the west (Fitzroy Crossing, Broome) is at the bottom. As is typical of traditional Aboriginal art, this text makes no attempt to represent the landscape accurately. The symmetry of the four rains implies that rain comes equally from all four directions, which is very far from true in that part of the Kimberleys. But the symmetry is broken by the profusion of water-sources in the dry east (the arc above the cross and the four water-holes above all painted blue), as against the absence of water in the west. The painting encodes the meaning of plenitude as an attribute of the east (and the past) compared to the west and the present, just as the story did.

The semi-circles and concentric circles are traditional motifs, which carry a complex of meanings, referring to home-centres (campsites, waterholes, fires) or people resting. But the grid of rectangular shapes is not typical. In this text these shapes dominate the west and the north, characterising the spaces of civilisation and Whitemen's ways, seeming not unlike the bars of a cage, symbols of regimented existence. But Sutton's annotation indicates that these rectangular shapes are sandhills, so that western civilisation is reclassified as a kind of desert.

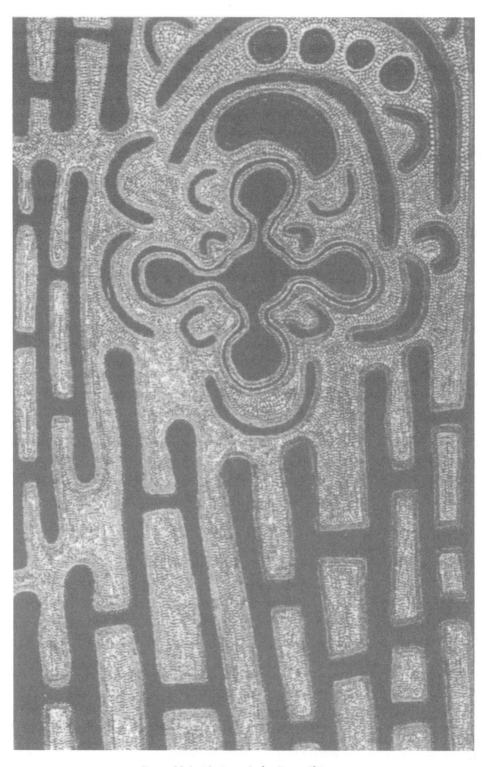

*Figure 82.1  Jila Japingka* by Peter Skipper

*Source*:  Picture courtesy of Duncan Kentish and Peter Skipper

'Desert' in this scheme is partly negative. However in Skipper's system the actual desert region is constructed as a place of abundance, not as a barren place. This contradictory classification then has a positive implication, making the barren terrain of civilisation a place in which Aborigines can survive as they did in the Western Desert.

Skipper achieves this meaning-effect by drawing on the resources of traditional art, specifically its capacity to use a minimalist system of classification to establish a complex network of connections that in Western traditions is associated with metaphor. We see a more typical instance of this in the use of semi-circular shapes. Sutton's annotation indicates that the two large arcs in the top right of the painting are long sandhills, while the others represent clouds. The semicircular shape acts as a classifier which establishes a metaphoric link between the two: these sandhills are like clouds insofar as both are like people camped around a site. Clouds are the source of water and sandhills are dry, so the link serves to resolve these primary oppositions. But sandhills themselves are encoded as both semicircular and rectangular rounded open shapes (home, Aboriginality) and rectilinear closed shapes (White domains, exile). In this and many other ways the painting responds to two opposing impulses, establishment of difference (between desert and water, home and exile, Aboriginal and White) and the resolution of difference.

The verbal text is constructed out of the same fundamental principles, and is concerned with the same crucial issue – coming to terms with the position of Aborigines in White Australia, using traditional resources to express a twin sense of alienation and belonging. If we gave the texts a more formal analysis we would come up with the seemingly implausible elaborations of the kind of structuralism for which Claude Levi-Strauss is famous. Neither text is a normal object for these forms of exegesis, and the verbal text especially seems far too humble a form to justify any attention to its formal qualities and implicit levels of meaning. We have discussed them in such detail as the only way to make the general point convincing. Very many Aboriginal texts, written and unwritten, recorded or not, deal directly with the fundamental issues facing Aboriginal people, torn as they are between alienation and a sense of belonging. The strategy they use is an adaptation of traditional Aboriginal ways, constructing maps that are designed to represent broad stretches of space and time, to give meaning and perspective, direction and hope on the bewildering journey of the life of themselves and their people.

## Chapter 83

# G. MALCOLM LEWIS

# INDIGENOUS MAP MAKING

From 'Pre-encounter and Indigenous Post-encounter Mapmaking' in
*Cartographic Encounters: Perspectives on Native American Mapmaking
and Map Use* Chicago: Chicago University Press, 11–14, 1998.

**N**ATIVE PEOPLES THROUGHOUT North America made maps indigenously, certainly from early contact times and probably long before that. They were made in a number of contexts, of which the following were important: as messages or instructions to others; as interactive planning; in order to reconstruct past events and record them for posterity; to make sense of the world beyond that of direct experience and relate it to the known world; and to divine. The list is not exhaustive.

At their simplest, maps were ephemeral gestures. Because these were fugitive, they are known of only in encounter contexts. For example, in 1761 the chief of the Pookmoosh Band of Micmac, in what was to become New Brunswick, whose people had hitherto been allied to the French, encapsulated their changing geopolitical situation in one dynamic hand gesture. It was a case of what, almost two centuries later, and in a quite different context, Winston Churchill called 'The Closing of the Ring.' The French Acadians had been deported or taken flight and the English, some Scots, and, most of all, people from the New England colonies were about to move in. The Micmac perceived this to be a threat. 'Their chief made almost a circle with his forefinger and thumb, and pointing at the end of his forefinger, said there was Quebec, the middle joint of his finger was Montreal, the joint next [to] the hand was New York, the joint of the thumb next [to] the hand was Boston, the middle joint of the thumb was Halifax, the interval betwixt his finger and thumb was Pookmoosh, so the Indians would soon be surrounded, which he signified by closing his finger and thumb.'

Far more important than gesture maps were the message maps left at strategic locations by departing persons for the information of others expected to arrive or pass by. Particularly common on birchbark sheets and somewhat less so on the exposed white wood of conspicuously blazed trees, the practice was most characteristic among the hunting peoples of the Northeast woodlands. The best known and probably oldest extant indigenous example of the former is that found in 1841 on the Ottawa River-Lake Huron watershed. Almost certainly made by Ojibwa, its geographical content is minimal; just sufficient to indicate who had left it, the route via which they had arrived, the location of the previous night's camp, and the intended route ahead. Such maps are known to have been used from the late seventeenth century onward. Maps on blazed trees were almost certainly less common. They seem usually to have been associated with military activities. Prominently placed, they were symbols of power, resistance, success in battle, or aggressive intent. That described and illustrated by Bray in 1782 is a good example (Figure 83.1). It incorporates stylized plans of three forts (certainly Detroit [9] and Fort Pitt [10], and perhaps Fort Loudoun [8]) and a map of the confluence of the Allegheny and Monongahela Rivers at Fort Pitt to form the Ohio River, together with the adjacent civil settlement of Pittsburgh (11). The whole is certainly not a map but, like many pictographic messages, some of the components are recognizably cartographic. In the absence of written words and, more specifically toponyms, such composite pictographs conveyed to the initiated information about places, either by means of identifiable plans or distinctive networks.

An event involving mapmaking for instructional purposes was reported by Col. Richard Dodge on the basis of an account of events which occurred almost fifty years before. Sometime in the 1820s, Comanche braves were about to undertake a journey of approximately one thousand miles from central Texas to near Monterrey, Mexico, and back. As none of the braves had been there before, they were instructed in advance by older men who had. The route to be taken was divided into days of travel. For each day a map was made on the ground. This was memorized by all the young men before the map for the next day's journey was made. Military activity was also the context for an extremely good example of mapmaking done as a basis for interactive planning. Sometime between 1860 and 1865, warriors from three Nootkan villages on the west coast of Vancouver Island were planning to attack a fourth village some ninety miles to the north. Only one man knew the village well because he had courted his wife there. He was asked to make a map of it. The warriors all went down to a beach where the chosen man modeled a map three-dimensionally in the sand. It was very detailed, showing critical topographic features, tracks, and individual houses. 'All this time the warriors . . . stood round the delineator in a large circle . . . questions were asked and eager conversation held.' Only then was the general plan of attack, which had already been proposed, finally accepted.

Made for use at a federal government council in Washington, D.C., in 1837, an Iowa chief's map was produced, apparently spontaneously, in trying to resolve a dispute between

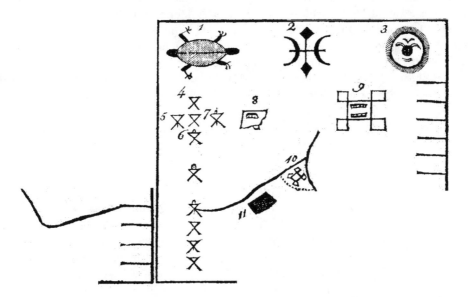

*Figure 83.1* The Wingenund (a Delaware warrior) map incorporated in a painting on the trunk of a blazed tree seen adjacent to the Muskingum River (Ohio) in or a little before 1781, as redrawn in Bray (1782)

*Source:* Courtesy of the Newberry Library

two groups of Indians concerning land. It shows 'the route of my (Ioway) forefathers – the land that we have always claimed' and does so against a topological but detailed representation of the upper Mississippi and Missouri River basins; in all, more than one quarter of a million square miles. Though apparently lacking a beginning or an end, the dotted line zigzags across much of the western Middle West. Dots in circles appear to represent locations where the ancestors had settled in the course of a long and complicated migration.

Maps preserving traditions for later generations the least known but most indigenous. It is impossible to estimate how many are still preserved by native people. In *Maps and Dreams,* Hugh Brody described a public hearing held in 1979 by the Northern Pipeline Agency for the Beaver Indians of northeast British Columbia. The first day's proceedings were almost over. They had involved the submission of evidence in map form: hunting, trapping, and berry-collecting territories carefully and systematically plotted for the occasion on modern topographic maps. Then, as the meeting was about to break up, two Indians produced a large moosehide bundle. When opened, it was as large as the table top and was revealed to be a dream map. No one knew how old it was. It showed heaven, the trail to it, a false trail, and animals. All its content had been discovered in dreams. A corner of it was missing. The detached part had been buried with someone who would not otherwise have found it easy to get to heaven. Dream maps are unpacked only on very special occasions. Their intricate routes and meanings are difficult to understand. When the owner of a dream map dies, it is buried with him. Not surprisingly, they are little known. Likewise, little was known of the Southern Ojibwa made migration scrolls until Selwyn Dewdney made a partial inventory and wrote about them in the 1970s. He was, incidentally, able to locate only eight certain and seven possible examples. The Pawnee sky chart on buckskin at the Field Museum, Chicago, is another example of a map made to preserve tradition. But we must be careful. William Gartner, a graduate student at the University of Wisconsin, cautioned that it 'is not merely a map of the celestial sky. Its direct uses are as a beacon for heavenly forces, as an earthly guide, as a symbol of cosmological unity, and as a flag of identity during the Thunder and/or Great Washing ceremonies.' This is a level of interpretive understanding rarely attained by non-natives.

All the indigenous maps referred to thus far were based on information derived via experience, by tradition, or in dreams. Some patterns considered to be maps were, however, produced by chance. Especially among the Naskapi of Quebec and Labrador, divination involved inducing cracks on the scapula bone of certain mammals. Patterns induced by heat or percussion were often interpreted as maps and sometimes related to real-world features; especially rivers, lakes, and trails. The same tribe also made decorative patterns by biting folded pieces of birch-bark. Patterns were sometimes random or emerged by error in the course of trying to create something else. Deliberately or accidentally made, patterns were usually explained or given names. Frequently, these indicate an ability to read patterns as maps of trails.

Regrettably, there are no pre-encounter artifacts reliably identifiable as maps. Even if there were, it is doubtful whether it would ever be possible to determine the contexts in which they were made and used. Some rock paintings and glyphs do contain patterns which, from a modern Euro-American perspective, appear map-like. The purposes for which they were made, when, and by members of which cultures are, however, too much in doubt to afford bases for firm conclusions. Nevertheless, the indigenous uses of maps observed in the course of the encounter were almost certainly rooted in earlier practices; perhaps much earlier. It is very unlikely that maps made by Indians and Inuit in early-encounter contexts were spontaneous innovations made to satisfy alien whites.

## Chapter 84

# WALTER MIGNOLO

# THE OTHER SIDE OF THE MOUNTAIN

From *The Dark Side of the Renaissance: Literacy, Territoriality and Colonization* Ann Arbor: University of Michigan Press, 1995.

I N COLONIAL SITUATIONS, mapping and naming (i.e. geographical discourse) are to territoriality what grammars are to Amerindian speech and historical narratives to Amerindian memories. Once recognized, the dialectic between what is common in the perspective and diversity of the colonizer's semiotic moves and what is common in the perspective and diversity of the colonized semiotic moves requires a type of understanding that can no longer be based on a linear conception of history and a continuity of the classical tradition. This dialectic required a type of understanding that focuses on discontinuities and the counterpart of maps, grammars, and histories: the existence and persistence of speech over grammars, of memories over histories, of territorial orderings over maps. However, while writing the grammars of Amerindian languages and the histories of their memories challenged both the grammarian and the historian, verbal geographical descriptions and maps overpowered, so to speak, Amerindian *pinturas* and territoriality.

While all these activities, and more, kept European administrators and cartographers busy, Amerindians in Mexico and elsewhere were also concerned with territorial description, although not on the same scale as their European counterparts. Comparisons have to be made, therefore, at two different levels: the level of the absence (e.g. the Amerindian 'lack' of world-scale geometric projections); and the level of the difference (e.g. how López de Velasco described New Spain and the Mexicans Anáhuac). Additionally, Amerindian 'maps' are not as well documented as Spanish and European ones partly due to the fact that most of them were destroyed in the process of colonization. Here we are confronted with another example in the cultural process when anything that is not recognized or mentioned by those who control the transmission and circulation of information does not exist. Power asserts itself by suppressing and negating both what is not considered relevant or is considered dangerous. The control of the cultural sphere is similar in many ways to maps. It gives the impression that it covers the territory, disguised under a set of principles that allowed for certain expressions to be ignored because they did not fulfill the basic requirement or because they were disturbing. Even Spanish historians and missionaries of the sixteenth century were silent about Amerindian territorial representations. For López de Velasco, as we have seen, the question never came up. Franciscan friars, like Sahagún, who were so careful in documenting all possible aspects of native lifestyles and their memories of the past, were also mute in matters of territorial conceptualizations.

Thus Amerindian 'maps' painted during the colonial period that have come down to us are shrouded with uncertainties. Who painted them, when, and for whom are questions that cannot be readily answered, as is the case with Spanish and European maps. There are enough examples, however, to become acquainted with Amerindian cognitive patterns of territorial organization. These examples are from three different periods and pragmatic contexts. From the first period (between 1540 and 1560) come those 'maps' which, although already showing Spanish influence, are predominantly Amerindian both in their graphic conceptualization of the territory as well as in the integration of narrative memories. Examples from this period are the *Mapa Sigüenza*, the *Tira de la peregrinación* (or *Tira del museo*), and the *Maps of Cuauhtinchán*. The first two examples are spatial narratives of the Aztec peregrination from Aztlán to the Valley of Mexico. The *pinturas* of the third example are spatial narratives of the Toltecs-Chichimecs, from the Valley of Puebla. From the second period (approximately 1579–86) are the *pinturas* from the *relaciones* although not all of them were painted by Amerindians. Enough examples can be found among those painted by Amerindians to show an increasing hybrid-ization of cultural products not only in the mixture of elements included in the 'maps' and the design of objects and events but, mainly, in the way space itself is conceptualized. Similar features are seen in the *pinturas* from the third period (1600–1750), all of them related to land possessions *(concession de mercedes)* and land litigations. All of these *pinturas* were produced under the specific conditions of colonial situations, and they are all examples of colonial semiosis, although traces of the preconquest painting style are still evident.

Thus, it should be kept in mind that the acts of describing, mapping, or painting the space depicted (or invented) and the way of depicting or inventing, it are interrelated, although at clearly distinguished levels. The first corresponds to the level of action; the second to the relations between signs and their content; the third complies with a given way of doing things with signs and cognitive patterns.

The well-known *Mapa Sigüenza* (Figure 84.1) is a fine example of Amerindian *pinturas* from the mid-sixteenth century. Although the events and space described and the patterns of descriptions supposedly survived from preconquest times, the act of painting itself is a colonial one and, therefore, a case of colonial semiosis. The *Mapa* traces the territory in its spatial as well as temporal boundaries: the chart of the space from the legendary Aztlán to the Valley of Mexico and the chart of the peregrination of time from the point of origin (Aztlán) to the point of arrival (Chapultepec). When compared with the *Tira de la peregrinación* (Figure 84.2) one would tend to believe that the former is colonial while the latter came to us as a preconquest artifact. While both trace the peregrination from a place of origin to the actual habitat, there are obvious differences between the two. Human figures in the *Mapa Sigüenza* respond less to indigenous patterns than the *Tira de la peregrinación*. In the latter the counting of the years is clearly indicated and the signs are similar to the numerical signs encountered in other codices from the Valley of Mexico. In the *Mapa Sigüenza* the years have not been indicated. The years shown in Figure 84.2, in arabic numerals, have been added and repre-sent a scholarly reconstruction rather than an original colonial chronological listing. We know, however, . . . that the pre-Columbian ruling class had developed sophisticated means of time reckoning that they applied to place social events in time and to keep record of the past.

When compared with European maps, the *Mapa Sigüenza* is very imprecise as far as location (longitude and latitude) is concerned. This kind of geographical 'imprecision' has been inter-preted in negative (e.g. what the Amerindians did not have) rather than positive (e.g. what the Amerindians did have) terms. Territorial conceptions among the Aztecs were, of course, dissimilar to those of the Spanish. Garcia Martinez reports that the territoriality of the *altepetl* (alt, water; *tepetl*, hill) encompassed both the natural resources as well as the memory of human history. Contrary to European maps, the *altepetl* did not imply a precise delimitation of geo-graphic boundaries. Geographic limits were fuzzy and variable and it was often the case that between two *altepetl* there were disputed lands as well as empty spaces. Garcia Martinez concluded that it was during the process of colonization that a more defined sense of space, law, and history was projected onto the *altepetl*. I am not sure whether we should conclude that what the Spaniards perceived as a 'lack' was due to the fact that the Mexica were simply unable to match the Spaniards' territorial conceptualization or rather that they really did not 'lack' anything because they had different ways of fulfilling similar needs. The fact remains that locations determined by the historical and sociological significance of an event in the collective memory 'inventing' the peregrination and charting of the Nahuatl world. To place the point of origin (Aztlán) and the point of arrival (Chapultepec) in two opposite corners is an indication not only of a peregrination in time but also of a conscientious use of graphic signs to indicate that the two points, of departure and arrival (e.g. *Tira de la peregrinación*, my addition), are the most distant in space. In the words of Radin, 'It is not an annual account set year for year but, like the Tira Boturini, shows unmistakable evidence of systematization.' On the other hand, Aztlán and Chapultepec are the sites to which more space is devoted in the *pintura*: they are not only points of departure and arrival; of all the places indicated by the peregrination, they are the most significant. Duverger suggests that the peregrination of Aztlán was an 'invention' after the arrival at the Valley of Mexico. An obvious statement, perhaps, although a necessary one for the understanding of colonial semiosis and Amerindian 'mapping' during the pre-colonial as well as colonial period. If during the expansion of the Mexica empire maps like these satisfied peoples' need to reassure themselves and their own tradition in front of rival communities, it is no less relevant that the ruling class of Amerindian civilizations moves from being in power to being disempowered. Construction of ethnic *(ethnos,* we) identities, however, is not limited to power. On the contrary, acts of opposition and resistance and the will to survive require a strong sense of individual and communal identity.

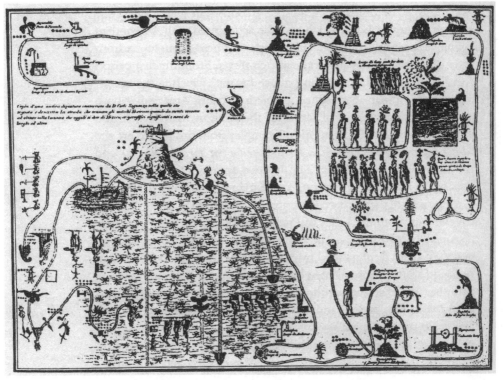

*Figure 84.1 Mapa Sigüenza (c. 1550).* Origin and peregrination of the Aztecs

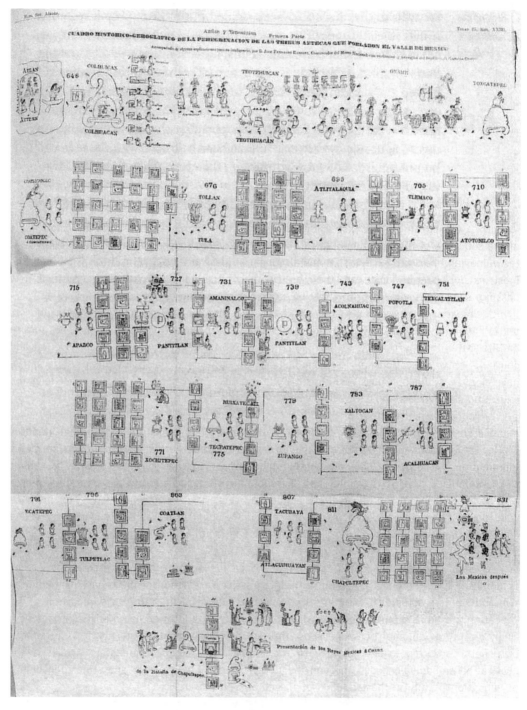

*Figure 84.2  Tira de la Peregrinación*: from Aztlán (top left to tenochtitlán, bottom centre)

# PART FOURTEEN

# Education

I N *IMAGES IN PRINT* (1988), a study of race, class and gender bias in contemporary Caribbean text books, Ruby King and Mike Morrissey note that although some of the countries of the Commonwealth Caribbean have been independent for twenty-five years, the values and patterns of British colonial education have persisted. Education is perhaps the most insidious and in some ways the most cryptic of colonialist survivals, older systems now passing, sometimes imperceptibly, into neo-colonialist configurations.

Such patterns are reproduced not just through established curricula, syllabuses and set texts, but more fundamentally through basic attitudes to education itself, to both its nature and its role within particular nations and cultures. Moreover the conditions of production and consumption of education and its technologies, while they may have undergone subtle shifts, have not, as Altbach argues, significantly altered the unequal power relations between the educational producers and the 'peripheral' consumers of it.

Education, whether state or missionary, primary or secondary (and later tertiary) was a massive canon in the artillery of empire. The military metaphor can however seem inappropriate, since unlike outright territorial aggression, education effects, in Gramsci's terms, a 'domination by consent'. This domination by consent is achieved through what is taught to the colonized, how it is taught, and the subsequent emplacement of the educated subject as a part of the continuing imperial apparatus – a knowledge of English literature, for instance was required for entry into the civil service and the legal professions. Education is thus a conquest of another kind of territory – it is the foundation of colonialist power and consolidates this power through legal and administrative apparatuses.

As Gauri Viswanathan notes, 'the split between the material and discursive practices of colonialism is nowhere sharper than in the progressive rarefaction of the rapacious, exploitative and ruthless actor of history into the reflective subject of literature'. As important as all education proved as a means to colonialist control, literary education had a particular valency. The brutality of colonial personnel was, through the deployment of literary texts in education, both converted to and justified by the implicit and explicit 'claims' to superiority of civilization embodied/encoded through the 'fetish' of the English book and state sponsorship of education usually in the language of the colonizing power.

Education becomes a technology of colonialist subjectification in two other important and intrinsically interwoven ways: it establishes the locally English or British as normative through critical claims to 'universality' of the values embodied in English literary texts, and it represents the colonized to themselves as inherently inferior beings – 'wild', 'barbarous', 'uncivilized'.

Moreover, technologies of teaching strongly reinforced such textual representations. The reciting of poetry, dramatic set-pieces or prose passages from the works of English writers was not just a practice of literary teaching throughout the empire – it was also an effective mode of moral, spiritual and political inculcation. The English 'tongue' (and thus English literary culture and its values) was learned by 'heart', a phrase that captures the technology's particular significance. Texts, as a number of cultures recognize, actually enter the body, and imperial education systems interpellated a colonialist subjectivity not just through syllabus content, or the establishment of libraries within which the colonial could absorb 'the lesson of the master', but through internalizing the English text, and reproducing it before audiences of fellow colonials. Recitation of literary texts thus becomes a ritual act of obedience, often performed by a child before an audience of admiring adults, who, in reciting that English tongue speaks as if he/she were the imperial speaker/master rather than the subjectified colonial so often represented in English poetry and prose.

This is one reason why education and literary education in particular, has been a major theme and site of contestation in post-colonial literatures. Writers like Jean Rhys in *Wide Sargasso Sea* challenge the whole of that discursive field within which *Jane Eyre* was produced and reproduced, through formal education and informal repute at the colonial periphery. Both Jamaica Kincaid in *Lucy* and Erna Brodber in *Myal* anatomize and dis/mantle imperial education and its technologies. In *Myal* too, Brodber examines the question of knowledges themselves, and against Anglo-education with its colonialist intent, posits an/other kind of knowledge based on African cultural survivals. In so doing both she and Kincaid examine and challenge that persisting gap between the so-called 'First World' production of knowledge (the 'authoritative' text) and its consumption at colonial and post-colonial sites – the inferior and mutable contexts for these 'immutable' Anglo-European products.

Formal tertiary education – specifically literary education – thus becomes the focus of debate in colonies of occupation (such as those of Africa and India) and, perhaps more surprisingly, in the settler colonies as well. Ngũgĩ wa Thiongo's important essay on 'The Abolition of the English Department', written in the 1970s, was an important manifesto in this debate. Thirty years later, however, British literature still remains part of the 'core' curricula in most tertiary English departments throughout the world. But increasing national recognition of domestic achievements, and the significance of other post-colonial literatures written in English, has ameliorated somewhat this earlier rhetoric of rejection, conceding the potential for imperial–colonial and cross-colonial exchanges. The sections of Ngũgĩ wa Thiongo's more recent essay, reproduced here, are indicative of this trend. Nevertheless, as Arun Mukherjee suggests, a persisting gap between pedagogical practice and cutting edge research in the discipline of 'English', to some extent perpetuates the old power relations. Radical criticism perpetuates the old power relations. Radical changes in literary theory and the rise of cultural studies, have also affected teaching practices in some countries, but in the majority of cases, English department curricula remain Anglo-oriented if not Anglo-dominated.

Revolutions in literary theory have, like colonialist education systems, proved a double-edged sword. Imperial education systems effected colonial subjectification, but they also paved the way for subversive and eventually revolutionary processes. Like earlier models of literary education, contemporary literary theories – specifically Marxist and post-structuralist – have on the one hand offered new possibilities for Anglo-canonical dismantling, but in their establishing of new kinds of hegemonies they have, perhaps inadvertently, sometimes acted, as the Caribbean critic Barbara Christian has argued, to reinforce the power of the dominant Western academies. And while post-colonial and national texts may have replaced some Anglo-canonical ones, they sometimes become again 'raw materials' for metropolitan theorists. Gayatri Spivak's essay on the 'burden' of 'English' raises a number of complex issues which arise for speakers, writers and readers of the former imperial language in today's world (Spivak 1992). One of these arises from a specific case study by Noel London of the teaching of English in Trinidad and Tobago in the period immediately preceding and following independence. While language teaching is a clear demonstration of the operation of imperial power, a process of cultural entrenchment and indoctrination, it may also be seen as the site of struggle, demonstrated in the increasing creolization of Trinidadian society, an occasion for the transformation of the language of education.

Education thus remains one of the most powerful discourses within the complex of colonialism and neo-colonialism. A vital technology of social control, it also offers (and has offered) one of the most potentially fruitful routes to a dis/mantling of colonialist and neo-colonial authority, and of bringing different cultures into contact on the basis of exchange rather than within contexts of domination and subordination. But as Philip Altbach's essay reminds us, persisting economic and technological inequalities continue to militate against a genuine decolonization in material practice.

## Chapter 85

# THOMAS MACAULAY

## MINUTE ON INDIAN EDUCATION

From *Speeches of Lord Macaulay with his Minute on Indian Education*
(selected with an introduction and notes by G. M. Young) Oxford: Oxford
University Press, [1835] 1979.

**H**OW, THEN, STANDS THE CASE? We have to educate a people who cannot at present be educated by means of their mother-tongue. We must teach them some foreign language. The claims of our own language it is hardly necessary to recapitulate. It stands pre-eminent even among the languages of the west. It abounds with works of imagination not inferior to the noblest which Greece has bequeathed to us; with models of every species of eloquence; with historical compositions, which, considered merely as narratives, have seldom been surpassed, and which, considered as vehicles of ethical and political instruction, have never been equalled; with just and lively representations of human life and human nature; with the most profound speculations on metaphysics, morals, government, jurisprudence, and trade; with full and correct information respecting every experimental science which tends to preserve the health, to increase the comfort, or to expand the intellect of man. Whoever knows that language has ready access to all the vast intellectual wealth, which all the wisest nations of the earth have created and hoarded in the course of ninety generations. It may safely be said, that the literature now extant in that language is of far greater value than all the literature which three hundred years ago was extant in all the languages of the world together. Nor is this all. In India, English is the language spoken by the ruling class. It is spoken by the higher class of natives at the seats of Government. It is likely to become the language of commerce throughout the seas of the East. It is the language of two great European communities which are rising, the one in the south of Africa, the other in Australasia; communities which are every year becoming more important, and more closely connected with our Indian Empire. Whether we look at the intrinsic value of our literature, or at the particular situation of this country, we shall see the strongest reason to think that, of all foreign tongues, the English tongue is that which would be the most useful to our native subjects.

The question now before us is simply whether, when it is in our power to teach this language, we shall teach languages in which, by universal confession, there are no books on any subject which deserve to be compared to our own; whether, when we can teach European science, we shall teach systems which, by universal confession, whenever they differ from those of Europe, differ for the worse; and whether, when we can patronise sound Philosophy and true History, we shall countenance, at the public expense, medical doctrines, which would disgrace an English farrier – Astronomy, which would move laughter in girls at an English boarding-school – History, abounding with kings thirty feet high, and reigns thirty thousand years long – and Geography, made up of seas of treacle and seas of butter.

We are not without experience to guide us. History furnishes several analogous cases, and they all teach the same lesson. There are in modern times, to go no further, two memorable instances of a great impulse given to the mind of a whole society, – of prejudices overthrown, – of knowledge diffused, – of taste purified, – of arts and sciences planted in countries which had recently been ignorant and barbarous.

The first instance to which I refer, is the great revival of letters among the Western nations at the close of the fifteenth, and the beginning of the sixteenth, century. At that time almost every thing that was worth reading was contained in the writings of the ancient Greeks and Romans. Had our ancestors acted as the Committee of Public Instruction has hitherto acted; had they neglected the language of Cicero and Tacitus; had they confined their attention to the old dialects of our own island; had they printed nothing and taught nothing at the universities, but chronicles in Anglo-Saxon, and Romances in Norman-French, would England have been what she now is? What the Greek and Latin were to the contemporaries of More and Ascham, our tongue is to the people of India. The literature of England is now more valuable than that of classical antiquity. I doubt whether the Sanscrit literature would be as valuable as that of our Saxon and Norman progenitors. In some departments, – in History, for example, – I am certain that it is much less so.

Another instance may be said to be still before our eyes. Within the last hundred and twenty years, a nation which had previously been in a state as barbarous as that in which our ancestors were before the crusades, has gradually emerged from the ignorance in which it was sunk, and has taken its place among civilised communities. – I speak of Russia. There is now in that country a large educated class, abounding with persons fit to serve the state in the highest functions, and in no wise inferior to the most accomplished men who adorn the best circles of Paris and London. There is reason to hope that this vast Empire, which in the time of our grandfathers was probably behind the Punjab, may, in the time of our grandchildren, be pressing close on France and Britain in the career of improvement. And how was this change effected? Not by flattering national prejudices: not by feeding the mind of the young Muscovite with old women's stories which his rude fathers had believed: not by filling his head with lying legends about St. Nicholas: not by encouraging him to study the great question, whether the world was or was not created on the 13th of September: not by calling him 'a learned native', when he has mastered all these points of knowledge: but by teaching him those foreign languages in which the greatest mass of information had been laid up, and thus putting all that information within his reach. The languages of Western Europe civilised Russia. I cannot doubt that they will do for the Hindoo what they have done for the Tartar. . . .

It is impossible for us, with our limited means, to attempt to educate the body of the people. We must at present do our best to form a class who may be interpreters between us and the millions whom we govern; a class of persons, Indian in blood and colour, but English in taste, in opinions, in morals, and in intellect. To that class we may leave it to refine the vernacular dialects of the country, to enrich those dialects with terms of science borrowed from the Western nomenclature, and to render them by degrees fit vehicles for conveying knowledge to the great mass of the population.

## Chapter 86

# GAURI VISWANATHAN

# THE BEGINNINGS OF ENGLISH LITERARY STUDY IN BRITISH INDIA

From 'The Beginnings of English Literary Study in British India' *Oxford Literary Review* 9(1–2), 1987.

THIS PAPER IS PART OF A LARGER INQUIRY into the construction of English literary education as a cultural ideal in British India. British parliamentary documents have provided compelling evidence for the central thesis of the investigation: that humanistic functions traditionally associated with the study of literature – for example, the shaping of character or the development of the aesthetic sense or the disciplines of ethical thinking – are also essential to the process of sociopolitical control. My argument is that literary study gained enormous cultural strength through its development in a period of territorial expansion and conquest, and that the subsequent institutionalization of the discipline in England itself took on a shape and an ideological content developed in the colonial context. . . .

English literature made its inroads in India, albeit gradually and imperceptibly, with a crucial event in Indian educational history: the passing of the Charter Act of 1813. This act, which renewed the East India Company's charter for commercial operations in India, produced two major changes in Britain's role with respect to its Indian subjects: one was the assumption of a new responsibility towards native education, and the other was a relaxation of controls over missionary work in India. . . .

In keeping with the government policy of religious neutrality, the Bible was proscribed and scriptural teaching forbidden.

The opening of India to missionaries, along with the commitment of the British to native improvement, might appear to suggest a victory for the missionaries, encouraging them perhaps to anticipate official support for their Envangelizing mission. But if they had such hopes, they were to be dismayed by the continuing checks on their activities, which grew impossibly stringent. Publicly, the English Parliament demanded a guarantee that large-scale proselytizing would not be carried out in India. Privately, though, it needed little persuasion about the distinct advantages that would flow from missionary contact with the natives and their 'many immoral and disgusting habits'.

Though representing a convergence of interest, these two events – of British involvement in Indian education and the entry of missionaries – were far from being complementary or mutually supportive. On the contrary, they were entirely opposed to each other both in principle and in fact. The inherent constraints operating on British educational policy are apparent in the central contradiction of a government committed to the improvement of the people while being restrained from imparting any direct instruction in the religious principles of the

English nation. The encouragement of Oriental learning, seen initially as a way of fulfilling the ruler's obligations to the subjects, seemed to accentuate rather than diminish the contradiction. For as the British swiftly learned to their dismay, it was impossible to promote Orientalism without exposing the Hindus and Muslims to the religious and moral tenets of their respective faiths – a situation that was clearly not tenable with the stated goal of 'moral and intellectual improvement'.

This tension between increasing involvement in Indian education and enforced noninterference in religion was productively resolved through the introduction of English literature. Significantly, the direction to this solution was present in the Charter Act itself, whose 43rd section empowered the Governor-General-in-Council to direct that 'a sum of not less than one lac of rupees shall be annually applied to the revival and improvement of literature, and the encouragement of the learned natives of India' (Great Britain 1831–2: 486). As subsequent debate made only too obvious, there is deliberate ambiguity in this clause regarding which literature was to be promoted, leaving it wide open for misinterpretations and conflicts to arise on the issue. While the use of the world 'revival' may weight the interpretations on the side of Oriental literature, the almost deliberate imprecision suggests a more fluid government position in conflict with the official espousal of Orientalism. Over twenty years later Macaulay was to seize on this very ambiguity to argue that the phrase clearly meant Western literature, and denounce in no uncertain terms attempts to interpret the clause as a reference to Oriental literature:

> It is argued, or rather taken for granted, that by literature, the Parliament can have meant only Arabic and Sanskrit literature, that they never would have given the honourable appellation of a learned native to a native who was familiar with the poetry of Milton, the Metaphysics of Locke, the Physics of Newton; but that they meant to designate by that name only such persons as might have studied in the sacred books of the Hindoos all the uses of cusa-grass, and all the mysteries of absorption into the Deity.
>
> (Macaulay 1835: 345)

This plea on behalf of English literature had a major influence on the passing of the English Education Act in 1835, which officially required the natives of India to submit to its study. But English was not an unknown entity in India at that time, for some natives had already begun receiving rudimentary instruction in the language more than two decades earlier. Initially, English did not supersede Oriental studies but was taught alongside it. Yet it was clear that it enjoyed a different status, for there was a scrupulous attempt to establish separate colleges for its study. Even when it was taught within the same college, the English course of studies was kept separate from the course of Oriental study, and was attended by a different set of students. The rationale was that if the English department drew students who were attached only to its department and to no other (that is, the Persian or the Arabic or the Sanskrit), the language might then be taught 'classically' in much the same way that Latin and Greek were taught in England.

It is important to emphasize that the early British Indian curriculum in English, though based on literary material, was primarily devoted to language studies. However, by the 1820s the atmosphere of secularism in which these studies were conducted became a major cause for concern to the missionaries who were permitted to enter India after 1813. Within England itself, there was a strong feeling that texts read as a form of secular knowledge were 'a sea in which the voyager has to expect shipwreck' (*Athenaeum* 1839: 108) and that they could not be relied on to exert a beneficial effect upon the moral condition of society in general. This sentiment was complemented by an equally strong one that for English works to be studied even for language purposes a high degree of mental and moral cultivation was first required which the mass of people simply did not have. To a man in a state of ignorance of moral law, literature would appear indifferent to virtue. Far from cultivating moral feelings, a wide reading was more likely to cause him to question moral law more closely and perhaps even encourage him to deviate from its dictates. . . .

The uneasiness generated by a strictly secular policy in teaching English served to resurrect Charles Grant in the British consciousness. An officer of the East India Company, Grant was one of the first Englishmen to urge the promotion of both Western literature and Christianity in India. In 1792 he had written a tract entitled *Observations on the State of Society among the Asiatic Subjects of Great Britain*, which was a scathing denunciation of Indian religion and society. What interested the British in the years following the actual introduction of English in India was Grant's shrewd observation that by emphasizing the moral aspect, it would be possible to talk about introducing Western education without having to throw open the doors of English liberal thought to natives; to aim at moral improvement of the subjects without having to worry about the possible danger of inculcating radical ideas that would upset the British presence in India. Moral good and happiness, Grant had argued, 'views politics through the safe medium of morals, and subjects them to the laws of universal rectitude' (Great Britain 1832: 75). The most appealing part of his argument, from the point of view of a government now sensing the truth of the missionaries' criticism of secularism, was that historically Christianity had never been associated with bringing down governments, for its concern was with the internal rather than the external condition of man. . . .

As late as the 1860s, the 'literary curriculum' in British educational establishments remained polarized around classical studies for the upper classes and religious studies for the lower. As for what is now known as the subject of English literature, the British educational system had no firm place for it until the last quarter of the nineteenth century, when the challenge posed by the middle classes to the existing structure resulted in the creation of alternative institutions devoted to 'modern' studies.

It is quite conceivable that educational development in British India may have run the same course as it did in England, were it not for one crucial difference: the strict controls on Christianizing activities. Clearly, the texts that were standard fare for the lower classes in England could not legitimately be incorporated into the Indian curriculum without inviting violent reactions from the native population, particularly the learned classes. And yet the fear lingered in the British mind that without submission of the individual to moral law or the authority of God, the control they were able to secure over the lower classes in their own country would elude them in India. Comparisons were on occasion made between the situation at home and in India, between the 'rescue' of the lower classes in England, 'those living in the dark recesses of our great cities at home, from the state of degradation consequent on their vicious and depraved habits, the offspring of ignorance and sensual indulgence', and the elevation of the Hindus and Muslims whose 'ignorance and degradation' required a remedy not adequately supplied by their respective faiths. Such comparisons served to intensify the search for other social institutions to take over from religious instruction the function of communicating the laws of the social order.

It was at this point that British colonial administrators, provoked by missionaries on the one hand and fears of native insubordination on the other, discovered an ally in English literature to support them in maintaining control of the natives under the guise of a liberal education. With both secularism and religion appearing as political liabilities, literature appeared to represent a perfect synthesis of these two opposing positions. The idea evolved in alternating stages of affirmation and disavowal of literature's derivation from and affiliation with Christianity as a social institution. The process illuminates and substantiates what Lowenthal has called a central factor in the construction of every ideology: the self-conscious glorification of existing social contradictions. A description of that process is reconstructed below from the minutes of evidence given before the British Parliament's Select Committee, and recorded in the 1852–3 volume of the *Parliamentary Papers*. These proceedings reveal not only an open assertion of British material interests but also a mapping out of strategies for promoting those interests through representations of Western literary knowledge as objective, universal, and rational.

The first stage in the process was an assertion of structural congruence between Christianity and English literature. Missionaries had long argued on behalf of the shared history of religion and literature, of a tradition of belief and doctrine creating a common culture of values, attitudes, and norms. They had ably cleared the way for the realization that as the 'grand repository of the book of God' England had produced a literature that was immediately marked off from all non-European literatures, being 'animated, vivified, hallowed, and baptized' by a religion to which Western man owed his material and moral progress. The difference was poetically rendered as a contrast between

> the literature of a world embalmed with the Spirit of Him who died to redeem it, and that which is the growth of ages that have gloomily rolled on in the rejection of that Spirit, as between the sweet bloom of creation in the open light of heaven, and the rough, dark recesses of submarine forests of sponges.
>
> (*Madras Christian Instructor and Missionary Record* 11(4) 1844: 195)

This other literature was likened to Plato's cave, whose darkened inhabitants were 'chained men . . . counting the shadows of subterranean fires'.

The missionary description was appropriated in its entirety by government officers. But while the missionaries made such claims in order to force the government to sponsor teaching of the Bible, the administrators used the same argument to prove that English literature made such direct instruction redundant. They initiated several steps to incorporate selected English literary texts into the Indian curriculum on the claim that these works were supported in their morality by a body of evidence that also upheld the Christian faith. In their official capacity as members of the Council on Education, Macaulay and his brother-in-law Charles Trevelyan were among those engaged in a minute analysis of English texts to prove the 'diffusive benevolence of Christianity' in them. The process of curricular selection was marked by weighty pronouncements of the 'sound Protestant Bible principles' in Shakespeare, the 'strain of serious piety' in Addison's *Spectator* papers, the 'scriptural morality' of Bacon and Locke, the 'devout sentiment' of Abercrombie, the 'noble Christian sentiments' in Adam Smith's *Moral Sentiments* (hailed as the 'best authority for the true science of morals which English literature could supply') (Great Britain 1852–3). The cataloguing of shared features had the effect of convincing detractors that the government could effectively cause voluntary reading of the Bible and at the same time disclaim any intentions of proselytizing. . . .

To disperse intention, and by extension authority in related fields of knowledge and inquiry proposed itself as the best means of dissipating native resistance. As one government publication put it, 'If we lay it down as our rule to teach only what the natives are willing to make national, viz., what they will freely learn, we shall be able by degrees to teach them all we know ourselves, without any risk of offending their prejudices' (Sharpe 1920). One of the great lessons taught by Gramsci, which this quotation amply corroborates, is that cultural domination operates by consent, indeed often preceding conquest by force. 'The supremacy of a social group manifests itself in two ways', he writes in the *Prison Notebooks*, 'as "domination" and as "intellectual and moral leadership". . . . It seems clear . . . that there can, and indeed must be hegemonic activity even before the rise of power, and that one should not count only on the material force which power gives in order to exercise an effective leadership' (Gramsci 1971: 57). He argues that consent of the governed is secured primarily through the moral and intellectual suasion, a strategy clearly spelled out by the British themselves: 'The Natives must either be kept down by a sense of our power, or they must willingly submit from a conviction that we are more wise, more just, more humane, and more anxious to improve their condition than any other rulers they could have' (Farish 1838: 239).

Implicit in this strategy is a recognition of the importance of self-representation, an activity crucial to what the natives 'would freely learn'. The answer to this last question was obvious to at least one member of the Council on Education: the natives' greatest desire, averred C. E. Trevelyan, was to raise themselves to the level of moral and intellectual refinement of their masters; their most driving ambition, to acquire the intellectual skills that confirmed their rulers as lords of the earth. Already, he declared, the natives had an idea that 'we have gained everything by our superior knowledge; that it is this superiority which has enabled us to conquer India, and to keep it; and they want to put themselves as much as they can upon an equality with us' (Great Britain 1852–3: 187). If the assumption was correct that individuals willingly learned whatever they believed provided them with the means of advancement in the world, a logical method of overwhelming opposition was to demonstrate that the achieved material position of the Englishman was derived from the knowledge contained in English literary, philosophical, and scientific texts, a knowledge accessible to any who chose to seek it.

In effect, the strategy of locating authority in these texts all but effaced the sordid history of colonialist expropriation, material exploitation, and class and race oppression behind European world dominance. Making the Englishman known to the natives through the products of his mental labour served a valuable purpose in that it removed him from the plane of ongoing colonialist activity – of commercial operations, military expansion, administration of territories – and de-actualized and diffused his material presence in the process. In a crude reworking of the Cartesian axiom, production of thought defined the Englishman's true essence, overriding all other aspects of his identity – his personality, actions, behaviour. His material reality as a subjugator and alien ruler was dissolved in his mental output; the blurring of the man and his works effectively removed him from history. As the following statement suggests, the English literary text functioned as a surrogate Englishman in his highest and most perfect state: '[The Indians] daily converse with the best and wisest Englishmen through the medium of their works, and form ideas, perhaps higher ideas of our nation than if their intercourse with it were of a more personal kind' (Trevelyan 1838: 176). The split between the material and the discursive practices of colonialism is nowhere sharper than in the progressive rarefaction of the rapacious, exploitative, and ruthless actor of history into the reflective subject of literature.

How successful was the British strategy? That is clearly a topic for another paper, though it is worth noting that the problematics of colonial representations of authority have been brilliantly analysed by Homi Bhabha . . . in his essay 'Signs Taken for Wonders'. [This account] provides a compelling philosophical framework for analysing native interrogation of British authority in relation to the 'hybridization' of power and discourse, the term Bhabha uses to describe the nontransparency of the colonial presence and the problems created thereby in the recognition of its authority. Though my purpose in this paper has primarily been to describe a historical process rather than to do a microanalysis of the techniques of power, the question of effectiveness of strategy is never far removed. Indeed, the fact that English literary study had its beginnings as a strategy of containment raises a host of questions about the interrelations of culture, state, and civil society and the modes of assertion of authority within that network of relations.

## Chapter 87

## PHILIP G. ALTBACH

# EDUCATION AND NEOCOLONIALISM

From 'Education and Neocolonialism' *Teachers College Record* 100(3) (May), 1971.

T HE OLD COLONIAL ERA, some say, is dead. Evidence? Most formerly colonial areas are now independent nations. On the ruins of traditional colonial empire, however, has emerged a new, subtler, but perhaps equally influential, kind of colonialism. The advanced industrial nations (the United States, most of Europe, including the Soviet Union, and Japan) retain substantial influence in what are now referred to as the 'developing areas.'

Traditional colonialism involved the direct political domination of one nation over another area, thus enabling the colonial power to control any and all aspects of the internal and external life of the colony. The results of colonialism differed from country to country, depending in part on the policies of the ruling power and in part on the situation in the colony itself. Neocolonialism is more difficult to describe and hence to analyze. In this essay neocolonialism means the impact of advanced nations on developing areas, in this case with special reference to their educational systems and intellectual life. Modern neocolonialism differs from traditional colonialism in that it does not involve direct political control, leaving substantial leeway to the developing country. It is similar, nevertheless, in that some aspects of domination by the advanced nation over the developing country remain. Neocolonialism is partly a planned policy of advanced nations to maintain their influence in developing countries, but it is also simply a continuation of past practices. . . .

Neocolonialism . . . is not always a negative influence, just as colonialism itself had some positive effects in several areas. The focus here, however, is generally on the negative results of educational neocolonialism precisely because the consequences are important for the recipient countries and because they have not yet been adequately analyzed. Neocolonialism can be quite open and obvious, such as the distribution of foreign textbooks in the schools of a developing country. It is, however, generally more subtle and includes the use of foreign technical advisors on matters of policy and the continuation of foreign administrative models and curricular patterns for schools. Some developing countries rely, for example, on expatriate teachers for their secondary schools and colleges. These teachers, regardless of their personal orientations, cannot but inculcate Western values and views in the schools. Most developing countries have maintained the colonial pattern of school administration and many have altered the curriculum only slightly, thus retaining much of the orientation of colonial education (see Ashby 1967; Kazamias and Epstein 1968). . . .

Reliance on foreign models was dictated in part by the colonial government. Indigenous educational patterns were destroyed either by design or as the inadvertent result of policies which ignored local needs and traditions. Colonial powers seldom set up adequate educational facilities in their colonies and immediately limited educational opportunity and, in a sense, hindered modernization. In addition, existing facilities reflected the needs of the metropolitan power, and not of the indigenous population. The inadequacies of the modern educational system, outmoded trends in curriculum, and the orientation of the schools toward building up an administrative cadre rather than technically trained and socially aware individuals needed for social and economic development can be linked in many countries to the colonial experience. . . .

Most colonial powers, when they concentrated on education at all, stressed humanistic studies, fluency in the language of the metropolitan country, and the skills necessary for secondary positions in the bureaucracy. Lawyers were trained, but few scientists, agricultural experts, or qualified teachers were available when independence came. Emerging elite groups were Western-oriented, in part as a result of their education. In some instances, in fact, individuals were even unfamiliar with their own indigenous language.[1]

Colonial educational policies were generally elitist. In India, British educational elitism assumed the title of 'downward filtration' – a system by which a small group of Indians with a British style education supposedly spread enlightenment to the masses (see McCully 1943). 'French assimilationist' policies also worked in this direction. Indigenous cultures, in many cases highly developed, were virtually ignored by colonial educational policy. Trends toward modernization, in many cases spurred by European-style education, were at the same time skewed by foreign control of the educational system.

Schools were established slowly by colonial governments, and even strong local pressure for education did not create a sufficiently large system. Some colonial powers, such as the Belgians, felt that higher level training for indigenous populations was bad policy, and thus when the Congo gained independence in 1960, there were only a handful of college graduates. The French, with their reliance on a totally French educational system for a very limited number of 'assimilated' individuals, produced only a small number of graduates. While British policy allowed for some measure of freedom and local initiative and did provide more opportunities for secondary and higher education, it neglected primary education. In contrast, both the French and the Belgians devoted funds to primary education, with the Church often providing the teaching manpower. Despite these differences and some regional variations, the colonial powers administered without much regard for the educational aspirations of local populations.

Political independence changed relatively little educationally in most developing countries. Few countries, despite the militancy of nationalist movements or deep feelings of enmity toward the former colonial powers, made sharp breaks with the educational past. In most cases, for example, Indian, Pakistan, Burma, and Singapore, the educational system expanded quantitatively, but did not alter much in terms of curriculum, orientation, or administration. In a number of countries, notably in formerly British Africa, higher education remained firmly rooted to its English curriculum and orientation, and in the immediate postindependence years, expanded very slowly indeed. Even nations which had never been under colonial domination, such as Thailand, Liberia, and Ethiopia, came under Western educational influence because of increased foreign aid and technical assistance. . . .

The continued use of European languages in many developing countries is one of the most important aspects of neocolonialism and the impact of the colonial heritage on the Third World. In a few cases, such as Indonesia, the colonial language (Dutch) was discarded, and a linguistically diverse national polity shifted to an indigenous language. In a number of developing countries, such as Nigeria, Ghana, India, Pakistan, and most of French-speaking Africa where there is no single indigenous national language, there has been a tendency to use the metropolitan language in administration and sometimes in education. The trend is to slowly replace European languages with indigenous media, but the process has been slow and difficult. What is more, linguistic change in the schools has not always been accompanied by curricular change.

European languages have tended to remain influential among elite groups even after the schools have shifted to indigenous languages. In some countries, higher education is conducted in the metropolitan language even after change takes place at lower levels. In addition, elites have often sent their children to private schools conducted in a European language in an effort to maintain their privileged position. The continued importance of European languages has other repercussions as well. Strong intellectual links with the metropolitan country are generally maintained, with the result that indigenous intellectual life and cultural development may be hampered, or at least deflected. In India, for example, research on Indian languages is undeveloped, in part owing to the great stress on expression in English and the prestige of publishing in English language journals. Indian economists have been more concerned with 'model building' and theory than with the sometimes undramatic local problems of development. Indian sociologists have been criticized in scholarly journals for their ignorance of local issues and social structures and their stress on Western-oriented sociological theory. The major advanced states, particularly the English- and French-speaking metropolitan powers, have helped to maintain the educational and linguistic status quo by subsidizing textbooks and journals. They provide scholarships for students to study in the metropolitan country and send large numbers of teachers and technical personnel to developing areas. All of these factors help to direct the intellectual energy and attention of developing areas from their own situations to the international intellectual and scholarly community. . . .

American aid to overseas universities has tried to 'depoliticize' aspects of higher education. The founding of technical universities in various Latin American countries is an indication of this orientation (see Myers 1968). Such new institutions have functioned in direct competition with the established 'national universities.' The stated reason for developing these new institutions instead of upgrading existing universities is that a technologically-oriented curriculum is impossible to implement in the older institutions. It is significant that the older universities in Latin America are often dominated by leftist elements and that the newer institutions provide a counterbalance to strong left-wing influences in Latin American intellectual and political life. The technical universities have stressed a more innovative curriculum in the sciences. They have also adopted, in many cases, an American style academic organization.

To facilitate American policy goals, particular models of higher education have been exported and specific kinds of programs supported financially. American style 'land grant colleges' have been established in a number of developing countries, including India, Nigeria, Indonesia, and several Latin American nations. These institutions are based on a close relationship between the government and the university in opposition to academic traditions of independence in some developing nations. It may be, of course, that this model is suitable for developing areas, although the fact that land grant style universities have proved successful in several countries is due at least in part to the very large infusions of money and technical aid which have poured into them. . . .

The results of American policy are rather similar to the British colonial educational policies of the nineteenth century in that existing metropolitan institutions are exported to the developing areas, often in forms somewhat below domestic standards and sometimes without much adaptation to local conditions.

Advanced nations have been active in promoting particular academic disciplines and specialties, and the emphases which have been given may provide an insight into the motivations of the donors. American assistance has established an American Studies Research Institute in India, complete with a scholarly journal in which Indian academics may write on American-related topics. Of no basic relevance to India's modernization, this institute will help to produce over the long run a group of Indian professors favorable to the American cause, and perhaps professionally tied to it.[2] . . .

It is no surprise that relations between advanced industrial nations and developing countries in many respects are unequal. The influence of the advanced industrial nations has continued beyond the period of traditional colonialism and is one of the basic facts of economic, political, and social life of the developing world. Despite the self-evident nature of these facts, much of the analysis of the social, economic, and educational development of the Third World has ignored this basic aspect of the situation.

One cannot be optimistic about an immediate end to neocolonialism in any sphere, and perhaps especially in education. If anything, the scientific and educational gap between the advanced and the developing countries is growing. . . .

Only when an adequate understanding of modern neocolonialism in its many facets is achieved will [it] be possible to change the domination of West over East to a more equitable arrangement in an increasingly interdependent world.

## Notes

1   In Singapore, where much of the ruling elite is composed of British-educated Chinese, the post-independence Prime Minister, Lee Kwan Yew, issued an order that members of the government should learn Chinese. Lee, a graduate of a British university, taught himself Chinese in order to communicate with his constituency.

2   The Americans have not been the only ones concerned with promoting the study of their own country overseas. Soviet funds have been given to establish departments of Russian studies at the University of Delhi and other institutions in developing countries. The German and French governments subsidize professorships in the study of German and French language and culture, and provide visiting professors without cost to universities in developing countries. These programs, while not crucial in diplomacy or power politics, do build up a group of individuals in developing countries who have strong ties to the particular metropolitan country.

## Chapter 88

# ARUN P. MUKHERJEE

## IDEOLOGY IN THE CLASSROOM
## A case study in the teaching of English literature in Canadian universities

From 'Ideology in the Classroom: A Case Study in the Teaching of English Literature in Canadian Universities' *Dalhousie Review* 66(1–2), 1986.

G ENERALLY SPEAKING, WE, THE CANADIAN university teachers of English, do not consider issues of the classroom worth critical scrutiny. Indeed, there is hardly any connection between our pedagogy and our scholarly research. A new teacher, looking for effective teaching strategies, will discover to her/his utter dismay that no amount of reading of scholarly publications will be of any help when she faces a class of undergraduates. In fact, the two discourses – those of pedagogy and scholarly research – are diametrically opposed and woe betide the novice who uses the language of current scholarly discourse in the classroom. . . .

The short fiction anthology I used for my introductory English 100 class – I deliberately chose a Canadian one – includes a short story by Margaret Laurence entitled 'The Perfume Sea.' This story, as I interpret it, underlines the economic and cultural domination of the Third World. However, even though I presented this interpretation of the story to my students in some detail, they did not even consider it when they wrote their essays. While the story had obviously appealed to them – almost 40 per cent chose to write on it – they ignored the political meaning entirely.

I was thoroughly disappointed by my students' total disregard for local realities treated in the short story. Nevertheless, their papers did give me an understanding of how their education had allowed them to neutralize the subversive meanings implicit in a piece of good literature, such as the Laurence story.

The story, from my point of view, is quite forthright in its purpose. Its locale is Ghana on the eve of independence from British rule. The colonial administrators are leaving and this has caused financial difficulties for Mr. Archipelago and Doree who operate the only beauty parlour within a radius of one hundred miles around an unnamed small town. Though the equipment is antiquated, and the parlour operators not much to their liking, the ladies have put up with it for want of a better alternative.

With the white clientele gone, Mr. Archipelago and Doree have no customers left. The parlour lies empty for weeks until one day the crunch comes in the shape of their Ghanaian landlord, Mr. Tachie, demanding rent. Things, however, take an upturn when Mr. Archipelago learns that Mr. Tachie's daughter wants to look like a 'city girl' and constantly pesters

her father for money to buy shoes, clothes and make-up. Mr. Archipelago, in a flash of
inspiration, discovers that Mercy Tachie is the new consumer to whom he can sell his 'product':
'Mr. Tachie, you are a bringer of miracles! . . . There it was, all the time, and we did not
see it. We, even Doree, will make history – you will see' (221).

The claim about making history is repeated twice in the story and is significantly linked to
the history made by Columbus. For Mr. Archipelago is very proud of the fact that he was born
in Genoa, Columbus's home town. The unpleasant aspect of this act of making history is unmis-
takably spelt out: 'He [Columbus] was once in West Africa, you know, as a young seaman, at
one of the old slave-castles not far from here. And he, also, came from Genoa' (217).

The symbolic significance of the parlour is made quite apparent from the detailed attention
Laurence gives to its transformation. While the pre-independence sign had said:

ARCHIPELAGO
English-Style Barber
European Ladies' Hairdresser  (211)

the new sign says:

ARCHIPELAGO & DOREE
Barbershop
All-Beauty Salon
African Ladies A Specialty  (221)

With the help of a loan from Mr. Tachie, the proprietors install hair-straightening equipment
and buy shades of make-up suitable for the African skin. However, though the African ladies
show much interest from a distance, none of them enters the shop. Two weeks later, Mercy
Tachie hesitantly walks into the salon 'because if you are not having customers, he [Mr. Tachie]
will never be getting his money from you' (222). Mercy undergoes a complete transformation in the salon and comes out looking like a 'city girl,' the kind she has seen in the *Drum*
magazine. Thus, Mr. Archipelago and Doree are 'saved' by 'an act of Mercy' (226). They have
found a new role in the life of this newly independent country: to help the African bourgeoisie
slavishly imitate the values of its former colonial masters.

These political overtones are reinforced by the overall poverty the story describes and
the symbolic linking of the white salon operators with the only black merchant in town. The
division between his daughter and other African women who go barefoot with babies on their
backs further indicates the divisive nature of the European implant. Other indications of the
writer's purpose are apparent from her caricature of Mr. Archipelago and Doree, a device
which prevents emotional identification with them. The fact that both of them have no known
national identities – both of them keep changing their stories – is also significant, for it seems
to say that, like Kurtz in *Heart of Darkness*, they represent the whole white civilization. The
story thus underplays the lives of individuals in order to emphasize these larger issues: the
nature of colonialism as well as its aftermath when the native élite takes over without really
changing the colonial institutions except for their names.

This, then, was the aspect of the story in which I was most interested, no doubt because
I am myself from a former colony of the Raj. During class discussions, I asked the students
about the symbolic significance of the hair straightening equipment, the change of names, the
identification of Mr. Archipelago with Columbus, the *Drum* magazine, and the characters of
Mr. Tachie and Mercy Tachie. However, the students based their essays not on these aspects,

but on how 'believable' or 'likable' the two major characters in the story were, and how they found happiness in the end by accepting change. That is to say, the two characters were freed entirely from the restraints of the context, i.e. the colonial situation, and evaluated solely on the basis of their emotional relationship with each other. The outer world of political turmoil, the scrupulously observed class system of the colonials, the contrasts between wealth and poverty, were non-existent in their papers. As one student put it, the conclusion of the story was 'The perfect couple walking off into the sunset, each happy that they had found what had eluded both of them all their lives, companionship and privacy all rolled into one relationship.' For another, they symbolized 'the anxiety and hope of humanity . . . the common problem of facing or not facing reality.'

I was astounded by my students' ability to close themselves off to the disturbing implications of my interpretation and devote their attention to expatiating upon 'the anxiety and hope of humanity,' and other such generalizations as change, people, values, reality, etc. I realized that these generalizations were ideological. They enabled my students to efface the differences between British bureaucrats and British traders, between colonizing whites and colonized blacks, and between rich blacks and poor blacks. They enabled them to believe that all human beings faced dilemmas similar to the ones faced by the two main characters in the story.

Though, thanks to Kenneth Burke, I knew the rhetorical subterfuges which generalizations like 'humanity' imply, the papers of my students made me painfully aware of their ideological purposes. I saw that they help us to translate the world into our own idiom by erasing the ambiguities and the unpleasant truths that lie in the crevices. They make us oblivious to the fact that society is not a homogeneous grouping but an assortment of groups where we belong to one particular set called 'us,' as opposed to the other set or sets we distinguish as 'them.'

The most painful revelation came when I recognized the source of my students' vocabulary. Their analysis, I realized, was in the time-honoured tradition of that variety of criticism which presents literary works as 'universal.' The test of a great work of literature, according to this tradition, is that despite its particularity, it speaks to all times and all people. As Brent Harold notes, 'It is a rare discussion of literature that does not depend heavily on the universal "we" (meaning we human beings), on "the human condition," "the plight of modern man," "absurd man" and other convenient abstractions which obscure from their users the specific social basis of their own thought . . .' (Harold 1972: 201).

Thus, all conflict eliminated with the help of the universal 'we,' what do we have left but the 'feelings' and 'experiences' of individual characters? The questions in the anthologies reflect that. When they are not based on matters of technique – where one can short circuit such problems entirely – they ask students whether such and such character deserves our sympathy, or whether such and such a character undergoes change, or, in other words, an initiation. As Richard Ohmann comments:

> The student focuses on a character, on the poet's attitude, on the individual's struggle toward understanding – but rarely if ever, on the social forces that are revealed in every dramatic scene and almost every stretch of narration in fiction. Power, class, culture, social order and disorder – these staples of literature are quite excluded from consideration in the analytic tasks set for Advanced Placement candidates.
>
> (1976: 59–60)

Instead of facing up to the realities of 'power, class, culture, social order and disorder,' literary critics and editors of literature anthologies hide behind the universalist vocabulary that only mystifies the true nature of reality. For example, the editorial introduction to 'The Perfume Sea' considers the story in terms of categories that are supposedly universal and eternal:

> Here is a crucial moment in human history seen from inside a beauty parlour and realized in terms of the 'permanent wave.' But while feminine vanity is presented as the only changeless element in a world of change, Mrs. Laurence, for all her lightness of touch, is not 'making fun' of her Africans or Europeans. In reading the story, probe for the deeper layers of human anxiety and hope beneath the comic surfaces.
> (Ross and Stevens 1988: 201)

Though the importance of 'a crucial moment in history' is acknowledged here, it is only to point out the supposedly changeless: that highly elusive thing called 'feminine vanity.' The term performs the function of achieving the desired identification between all white women and all black women, regardless of the barriers of race and class. The command to probe 'the deeper layers of human anxiety and hope' – a command that my students took more seriously than their teacher's alternative interpretation – works to effectively eliminate consideration of disturbing socio-political realities.

This process results in the promotion of what Ohmann calls the 'prophylactic view of literature' (63). Even the most provocative literary work, when seen from such a perspective, is emptied of its subversive content. After such treatment, as Ohmann puts it, 'It will not cause any trouble for the people who run schools or colleges, for the military-industrial complex, for anyone who holds power. It can only perpetuate the misery of those who don't' (61).

The editor–critic thus functions as the castrator. He makes sure that the young minds will not get any understanding of how our society actually functions and how literature plays a role in it. Instead of explaining these relationships, the editor–critic feeds students on a vocabulary that pretends that human beings and their institutions have not changed a bit during the course of history, that they all face the same problems as human beings. . . .

Surely, literature is more than form? What about the questions regarding the ideology and social class of the writer, the role and ideology of the patrons and the disseminators of literature, the role of literature as a social institution and, finally, the role of the teacher–critic of literature as a transmitter of the dominant social and cultural values? Have these questions no place in our professional deliberations?

## Chapter 89

# NGŨGĨ WA THIONG'O

## BORDERS AND BRIDGES

From 'Borders and Bridges: Seeking Connections between Things' in Fawzia
Afzal-Khan and Kalpana Seshadri-Crooks (eds) *The Pre-Occupation of
Postcolonial Studies* Durham, N.C. and London: Duke University Press, 2000.

**I** AM ONLY TOO AWARE that in the past I have been associated with a call for the
abolition of English departments. But today I will not be calling for their abolition . . .

Although I have spoken of the need to abolish certain things as they are, what I really
seek is a way of studying in which we focus on the connections between various phenomena
instead of seeing them in isolation from one another. Teaching English literature in India or
in Africa ought to be a way of crossing borders. What has been wrong in the colonial context
is that the act of interpreting the other culture that is far from us has, instead of clarifying
real connections and each culture thereby illuminating the other, ended by making us captives
of the foreign culture and alienating us from our own. In calling for abolition, therefore, I
am primarily seeking a way to clarify connections between one culture and another, literature and politics, literature and economics, literature and the environment, literature and
psychology, between the parts and the whole. . . .

One of the inherited traditions of Western education in the last four hundred years is that
of putting things in compartments, resulting in an incapacity to see the links that bind various
categories. We are trained not to see connections between phenomena, and we become locked
in Aristotelian categories. So the East becomes East, and the West becomes West, and never
the twain shall meet. But is this really true in a world that ultimately is round? Nothing exemplifies this attitude better than our approach as teachers of literature to questions of art and
aesthetics. What has aesthetics to do with the environment? With questions of wealth, power,
and values in a society? What does it have to do with the question of poverty in a society? What
does it have to do with the question of poverty in Africa, or of wealth in the West, or of Africa
in the sixteenth century, or of the Africa of AD 2000, of the relations between Africa and the
West in the year 2000? Literature, in particular, is often taught as if it had nothing to do with
these 'other' realms of our being. And you can see this in the current retreat into theory in the
contemporary teaching of English. It is a retreat into what I term modern scholasticism. You
get much argument on whether language has meaning at all. If you look at the 1950s, its literature was one of engagement, of commitment, as in the books of Sartre. Literature became very
important, the basis of discussion for many vital issues. In the case of black writers, we had
monumental meetings, for example, at Rome in 1956 and in Paris in 1959. All the debates in
literature and aesthetics and culture were related to the anticolonial process. . . .

For those of us who come from a colonial society, it would be easy to see this process in
terms of those who are dominating and those who are dominated. For instance, you can see

a situation in which a dominating section controls how the dominated people perceive themselves. We can see how our mental universe is connected with other realms when we put the structure in the context of those dominating and those who are dominated. If those dominating can in fact capture and control the self-perceptions of those who are being controlled, they will never in fact need police. One way of abolishing the police and the army would be the total enslavement of the mind by those who are ruling the economy, the power relations, and the values. If you control the mind of the people, you do not need the police to control them at any other level. You can also see how that control can change not only how people look at one another but how they look at their relationship to those controlling them. You can see this clearly in the colonial mode of education, which for many of us in Africa makes us look to Europe as the basis of everything, as the very center of the universe. We can see it in the way we are brought up to regard the English language as the basis of definition of our own identity. Instead of seeing English as just another language with a lot of books and literature available in it, we see it as a way of defining our own being. We become captives to this language, developing certain attitudes of positive identification with English (or French). We also develop attitudes of distancing ourselves from our own languages, our own cultures. It is not simply a question of acquiring another important tool; the acquisition of that intellectual tool becomes a process of alienating ourselves from our own languages and what they can in fact produce. Another way of looking at it, especially in Africa, is as the creation of an alienated elite. You can see the kind of communal investment that goes into producing these minds when these minds go to get their Ph.D.s from abroad and so on. They never ever give anything back to the community by putting that knowledge into the languages available to the people themselves. They invest in us, and wherever we go – be it Miranda House, or Nairobi, or Yale – what we produce there we lock it with keys marked English or French or Portuguese or whatever the language of education. Language is of the utmost importance. If you look at the area of cultural environment, language is the key. It is the means whereby we communicate with one another for the production of wealth. It is also what I have called elsewhere a collective memory bank of a people.

I have so far dwelt on one important form of connections, those that account for a society. Another kind of connection that I think is also important is that between one wholeness and another. Far too often, as humans, we see ourselves as distinct from plants and trees and animals. But in another sense we are, in fact, all connected. I can think of myself as Ngũgĩ, Gary, Joseph, or whatever, as someone apart from others who has nothing to do with anyone else. If you study the aesthetics of the western, the cowboy films of America, you can see a deliberate and conscious placement of the individual as the one who defines humanity. The individual who has no connection with anybody else is the one who is victorious over those who are organized. So the single individual is able to overcome notions of organization. He is the anarchic individual whose own strength owes nothing to anyone else. He is often seen as being victorious over institutions, over masses of people.

But, if we look at it, who is this individual, the one who is not connected with anybody else? It should be self-evident that we are all connected to one another through the air we breathe. When we think of the air we breathe, it is recognizably something outside ourselves. If you ask me to talk about my own dual being, I would perhaps talk about my hands or my legs or my hair or the different parts that contribute to my wholeness as a single and individual human being. But, surely, the air I breathe is even more me than any of my limbs. If you cut off my finger, I can continue to live, but, if you cut off the air I breathe, I will not last a minute. So, although the air we breathe is a part of the external environment, it is central to my being and to that of all those who, like me, depend on it. We are, accordingly, ultimately all connected. When we think of ourselves as individuals, we see ourselves as completely free from all other individuals. But, even at the level of simple survival, this is just not true. We are, in fact, connected with our entire environment, yet it is the individual who is most often praised. For instance, when we pollute the environment or allow others to pollute it, we are actually polluting our own being. Again, you may think of a factory as being elsewhere and polluting those people, but, if you think of it very carefully, you will find that pollution affects our very being. If we are so connected with our natural environment, we have to keep a healthy balance between ourselves and the environment. But, if we are thus connected inextricably with our natural environment, we are even more connected with our social environment. It is that environment, whether one of oppression or nonoppression, equality or inequality, power used for the communal good or against it, that is of vital importance.

I am currently teaching at New York University as professor of comparative literature and performance. Before that, I taught at Yale University. Turning to the question of connections in the context of Western tradition, I recall how a billionaire gave a lot of money to Yale while I was there to promote the 'undiluted' study of Western civilization. There was a general feeling that Western civilization was being eroded by the attention being given to postcolonial literatures and the call for multiculturalism. Two years ago, the billionaire withdrew his money because he felt that Yale was not doing what he set out for it to do. If Western civilization means the history, culture, literature, and arts associated only with Europe, how do you teach that portion of it that is Renaissance and post-Renaissance without going into the notions of, say, slavery and colonialism? We know that there is no post-Renaissance European economics, history, and culture without colonialism. If you look at history, you will find that all the wars fought between the European powers during the sixteenth, seventeenth, and eighteenth centuries were over colonial trade and colonial possessions. India, as we all know, is central to this, to the emergence of so many European nations. The discovery of so much of the New World had to do with the effort to discover a way to reach the riches of India. And how do you teach about slavery and the slave trade as integral to post-Renaissance Europe without talking about Africa? And, since Africans have been part and parcel of the founding settlements and growth of America, is slavery not integral to American civilization? How can one teach American literature and history and culture without seeing the centrality of Africa in their makeup? If there is any one continuous and unbroken centrality in American culture and life, it is surely the portion contributed by the economic and cultural inputs of the Africans or, shall we say, the African Americans. European and American studies that ignore the centrality of Africa and of coloniality are false.

Again, fostered by English departments, we see a lot of studies and comments on the notion of the modern and the postmodern. But they ignore what constitutes modernity. If you think of Western modernity in terms of Renaissance or post-Renaissance Europe, that modernity is bound up completely with colonialism. There is no way of extricating it from colonialism, and, in fact, in some cases it is directly reflected in the literature itself.

So the study of African, of Asian, or of Latin American literatures must be seen as part and parcel of teaching literature and culture in the West. The really important thing is to see connections. It is only when we see real connections that we can meaningfully talk about differences, similarities, and identities. So the border, seen as a bridge, is founded on the recognition that no culture is an island unto itself. It has been influenced by other cultures and other histories with which it has come into contract. This recognition is the basis of all the other bridges that we want to build across our various cultural borders. The bridges are already there, in fact. The challenge facing, say, teachers of English literature, of African or of Asian literature, is to recognize and find those bridges and build on them. That is why teaching literatures and teaching languages is a privilege that faces all of us – the challenge to see connections between literature and that wholeness that we call society, a wholeness constituted by all that comes under economics, politics, and the environment.

## Note

This is the text of the Tenth Krishna Memorial Lecture delivered on February 19, 1996 at Miranda House, University of Delhi.

## Chapter 90

# NORREL A. LONDON

# ENTRENCHING ENGLISH IN TRINIDAD AND TOBAGO

From 'Entrenching the English Language in a British Colony: Curriculum Policy and Practice in Trinidad and Tobago' *International Journal of Educational Development* 23, 97–112, 2003.

A CENTRAL FEATURE IN THE ENGLISH LANGUAGE education agenda in Trinidad and Tobago concerns the nature and exercise of power. Power, Fairclough reminds us, is the ability to achieve one's goals and to control events through intentional action, and in this light the language education offered must be considered an unbridled display of power and authority by the imperial state. English language education was pivotal in the schooling experience, and the decision to give the subject pride of place in the curriculum was purely external, contrived and perpetrated by Britain in the interest of colonialism. The encounter was not an interaction of equals, however. It was unethical and unfair, and points to what Apple describes as 'the politics of official knowledge' in which 'dominant groups try to create situations where the compromises that are formed favour *them*' (1993: 10). Reasons for the [instructional] packages contrived, the grammar syndrome evident, and the pedagogy employed were all based upon prescriptions of a powerful (colonial) Self and a powerless (colonized) Other.

The language teaching enterprise here is a good example of how the colonial encounter sought to infuse and structure change on the local scene, and two general lines of action in such undertakings have been identified in recent scholarship about coloniality and the colonial condition (Prakash 1995; Moore-Gilbert 1997). A 'bandit-free' approach, transparent in its self-interest, greed and rapacity, was at work during the earlier colonial period; but this route to dominance gave way in the period [before independence] to a mode 'pioneered by rationalists, modernists and liberals who argued that imperialism was really the messianic harbinger of civilization to the uncivilized' (Gandhi 1998: 15). Ashis Nandy (1983) describes these approaches respectively as the 'militaristic' type, and the 'civilizational' type. The latter, more subtle, was the instrument used in the Trinidad and Tobago encounter described here, and the tool became sharper and more effective as the colonial period drew to a close in 1962.

The approach used was also significant in another way. It contained the seeds of cultural syncreticity, a feature of all postcolonial societies as Dennis Williams (1969) points out. The standard form of English encouraged in school culture of the day now became co-extensive in usage with local pidgin or creole both within and outside of school, a combination that provides proof that some metamorphosis in language and language use was also at work despite the 'powerlessness' of the colonial Other. Language switch as a result was common during the period despite the tenacity of standard English in some walks of life . . . It is a story of the seductive narrative of power, but the syncretism that took place tells of another story: the counter-narrative of the colonized surely, firmly, and stubbornly declining the 'come-on' of colonialism on the local frontier.

The powerless Other, as Fairclough (1989) admitted, therefore, may not always continue in the condition of impotence and incapability. They, like schools, possess some relative autonomy (Apple 1982), and may re-act and carve out some control over their daily lives. In the Trinidad and Tobago context the increasingly accepted use of a vernacular modality in social discourse of the day is an example of this exercise of power by the 'powerless,' but the esteem which locals ascribed to English both then and now explains the other side of the coin. It demonstrates the success of colonial hegemonic practices in the area of language education, and reveals that the English language emerged victor in the encounter.

A Fanonian critique of Western civilization would extend the narrative of power here to incorporate some related functions and if this is done, racism, violence, and struggle would also come center stage in the Trinidad and Tobago language education experience. Tollefson (1991) maintains that language education is an important arena of struggle, the language variety that emerges dominant being the prize. Struggle in the circumstance is not something to be 'ended' like warfare . . . but rather is a phenomenon inherent in social systems from which transformations might proceed. Entrenchment of the English language then does not come as a surprise, but the variations produced in the struggle and how these contested for co-dominance in language use on the local scene are noteworthy.

As English became fortified in the schools, so too did the use of creole proliferate. The creole emerged at an earlier period, but usage had always been linked to notions of cultural and social degradation. The success of the entrenchment campaign and improved facility in the use of English nevertheless imparted some degree of legitimacy to the use of sub-standard forms of English, a practice that has implications for the construct of power explained above. Noteworthy as well is an observation by David Brazil (1998). He contends that the creole counterparts which 'settled' during this period of entrenchment are not devious forms of or mistakes in the English language that was taught in the schools, but are rather parts of a separate and genuine linguistic sub-system which approximates, but which is distinct from, the source language, English. If this is the case, then the question of the source of the local creole is an issue that cannot be overlooked, and the English emphasized as well as the remnants of the various languages occasioned through slavery, indenture, earlier colonization by Spain, and the prevalent use of French even during the early days of British occupation, are possible sources.

Policy and practice during the period of entrenchment were different in some respects from those which obtained during the earlier period [of colonization]. This was the case regarding pedagogy, how instruction was framed for delivery (packaging), and in the rationale for curriculum building (ideology). The change, however, points to an important feature concerning language policy and planning in the broader sense, and in this context Roberts (1997) explains that in a language teaching and learning enterprise the rules that govern the operation are likely to change depending upon the objective. Thus, for entrenchment purposes, policy and practice in Trinidad and Tobago changed from what they were in the earlier period of British colonization, and assumed some different features. The change, in turn, points to another contention. Tollefson (1991) insists that the warfare inherent in linguistic hegemony of this nature is not constrained by time or space, but is constantly in ferment.

Deliberations about whether or not to teach English in Trinidad and Tobago were not an issue during this late period. The prevailing assumption by this time was that there was no other language to learn in the first cycle of schooling. This was an ideological position questioned by neither student nor teacher, and the settlement was accepted in institutions other than the school. In places of worship, in the workplace and in the courts of law, the ideology was the same and the position taken granted legitimacy to English as the one and only form of communication for locals. English was *nulli secundus*. Consideration that other varieties of language had been rendered unprivileged earlier in the colonial contact through slavery and indenture, was not an issue by the time the 1930s dawned upon the scene. English by this time was 'right'; all other languages were 'wrong' as mediums of schooling and education, an ideology which spoke loudly and clearly to two issues on the colonial frontier: the inequitable distribution and unethical use of power on the one hand, and the extent to which consent for the dominance of English had been manufactured (Herman and Chomsky 1988).

The 'grammar syndrome' [that characterized English teaching] was also rooted in the ideology that English was the only way to go, and that turning the clock back to usage of indigenous forms was out of the question. The rationale was that command of the language was a necessity for personal economic and social advancement, and the argument was so compelling that it determined standards set for success and failure in school. A gatekeeper function in other words, was assigned to English. Academic success could not therefore be ascribed without success in English, but even more striking was the practice that excellence in all other subjects at school got cancelled out and became null and void if success in English was not attained. The practice has changed to some extent in contemporary Trinidad and Tobago, but semblances of the thinking regarding the role of English are still evident, despite efforts from managers to re-write performance indicators and to inscribe some new norms into schooling in the present post-colonial era.

The extent to which English has ossified locally points to another issue – the role which the school played both during the colonial era and at other times as a site for linguistic hegemony. Hegemony, as Ashcroft *et al.* (1989) remind us, is the successful production and re-production of what gets accepted and legitimized, and the above classroom agenda must be seen in that light. A good example in this regard is to be found in the pedagogical principle of emulation mentioned earlier. The practice was part of the hidden curriculum, and the understanding was that there was 'one best way' to attain clarity and effectiveness in speaking (and in writing) English. This view, incidentally, as an approach to proficiency in language use has sometimes been challenged (Burchfield 1985; Pennycook 1994).

The classroom agenda was also based upon an understanding of the structure of society as it was then, and as it was projected to become in the future, using colonialism as the framework. The mental disciplinary doctrine used was, by definition, incapable of developing and inspiring the consciousness of children, and the denial here was to be carried over into the life worlds of the masses placed at the lower end of the social stratum. Mindlessness was a feature of what was taught and learned as English, and even in those instances where the curriculum drew upon social efficiency ideals, the objective was not structural re-organization for liberation of the mind. The rise of the epistolary method referred to, for example, served merely to facilitate interaction at a time when diasporic circumstances had been changing, but enlightened as the approach may have been, it did not promote social structural change. The masses, as always under colonialism, remained at the bottom of the social ladder, and the way in which English was taught in Trinidad and Tobago first cycle schools ministered to this grand purpose.

. . . The humanist emphasis in the curriculum, the punctiliousness required in English language usage, and the packaging of subjects to recognize an exaggerated 'English' core are indications of how it was believed society should be structured, and as a result, function.

In essence, these decisions constituted mechanisms for changing and maintaining the worldview of those who would receive instruction. The humanist tradition, for example, was incorporated to prescribe ethical and moral standards and standards for social interaction that would sub-serve colonialism, and the Odyssean model (among others), to which children were exposed, is a case in point. The humanist dimension which teaching and learning the language assumed during this period of entrenchment was demonstrative of a particular worldview, but it echoes something of deep significance. It demonstrates that culture is not just a way of seeing and structuring the world and society, but also a way of changing it. This is the dimension of culture which facilitated operationalization of the strategies used in language learning, and it constituted an attempt to place colonizers not only at the center of social administration and governance, but also at the end of time. A new time-space reality, in other words, became another end product of the language policy and practice in the entrenchment campaign.

# PART FIFTEEN

# Production and Consumption

CONSIDERABLE ENERGY HAS BEEN SPENT, as many of the pieces in this Reader testify, on theorizing the possibilities for post-colonial cultures recovering or developing identities, national cultural self-sufficency and confidence, or speculating as to how destructive the representation of colonials and post-colonials within the discursive modes of colonization have been. Again, a lot of energy has been spent in discussing issues of language choice and of the need to recover pre-colonial languages. Yet the processes of patronage and control by which the colonial and neo-colonial powers continue to exercise a dominant role in selecting, licensing, publishing and distributing the texts of the post-colonial world, and the degree to which the inscriptive practices, choice of form, subject matter, genre, etc. is also subject to such control, have received far less attention than they deserve. This wider sociological dimension of post-colonial textual studies, is, as André Lefevre argues, resident principally in 'refracted texts' such as school or university reading lists. The power of such texts has been discussed already in the section on Education. In the area addressed by this section conditions change rapidly, and many of the essays here are already outstripped by events, for example Peter Hyland's piece on Singapore writing would need now to be corrected to indicate the rapid rise in local publication there in the last twenty years or so. Equally, there is a need for updating and extending the pioneer work of S. I. A. Kotei and Philip Altbach reproduced here. No area of study seems to us to be more urgently in need of address at the present time, and the pieces we reproduce here are intended as much to stimulate the production of more current assessments of the material conditions of cultural production and consumption in post-colonial societies as they are authoritative accounts of the present situation.

Who consumes and produces the texts for the 'post-colonial' world, who canonizes them, who acquires them and has them available as physical objects is an important but neglected precondition for more abstract and theoretical discussions of the agency of the post-colonial subject. As well as the continuing control of these elements by neo-colonial forces, it is important to document the effect of attitudes within the post-colonial world to the very idea of publishing. As Peter Hyland notes, in some post-colonial societies cultural production has been seen as a luxury that these societies cannot afford. Elsewhere, as Altbach notes, the new independence comprador class sees national literatures (including texts in the 'prestigious' ex-colonial languages) as suitable for annexation to the construction of their own power and prestige. The lesson to be drawn from this is that cultural production and its effects is important in any society and it is perilous to neglect it. Altbach also points, as does Kotei, to the complex relations between colonial and neo-colonial cultural producers and the ways in which the ex-colonies are available both as suitable markets for cultural products and as the source for exotic products for sale on the home market. As Altbach also notes, even the 'liberal' enterprises of cultural development programmes may have an adverse effect on the development of independent and self-sufficient modes of cultural production in the post-colonial world by creating a product that undersells the local entrepreneur, preventing the develop-ment of a self-sufficient and economically sustainable local industry. W. J. T. Mitchell speculates on how one of the most powerful neo-colonial powers (itself both an ex-colony and an ex-empire, as he rightly perceives) might set about addressing this problem at least at the level of the academy.

Kotei's report, now more than two decades old, points to a situation which seems to be not only continuing but worsening. The crisis of documentation in areas such as Africa to which he makes reference is simply not being addressed at a time when, ironically, Europe and America are congratulating themselves on their enlightenment in having discovered and promoted the writings of the post-colonial world. As a recent issue of the Filipino cultural magazine *Solidarity* has indicated, economic security is no guarantee of the development of local product and control. The situation in South and South-East Asia shows that even where the technical skill and infrastructure of production exists, and may be strongly utilized to produce an industry successfully serving off-shore clients, the development of the production of books aimed at and reflecting the needs and concerns of the local market does not auto-matically follow. Countries such as Australia and Canada, too, because their domestic market remains relatively small, find that the study of their own culture may be restricted by such factors as publishers' budgets to the famous and canonical authors, or those endorsed by the power of the absent 'centre'. Thus in Australia several critical books on a Nobel winner such as White or a Booker prize winner such as Malouf or Carey will be adjudged viable, whereas studies of important but less internationally acclaimed writers such as Judith Wright or Dorothy Hewett may be less likely to be published since they command no interest in the world market.

It is by such material practices that the fate of post-colonial literary work is often determined. It is this which allows these books to come into existence, which gives them their chance to effect their 'work in the world'. For this reason it seems to us to be one of the most important and so far largely neglected areas of concern. As a result we have included in this new edition three essays which engage a variety of forms of imperial production: Anne McClintock's examination of the role of the production and consumption of such apparently innocent items as 'soap' in defining imperial control; Arjun Appadurai's assessment of the effect of production on the formation of cultural 'value' and its political fallout; and Graham Huggan's analysis of the interaction of the material and symbolic in the marginalization of post-colonial cultural codes of value.

## Chapter 91

# ANDRÉ LEFEVRE

## THE HISTORIOGRAPHY OF AFRICAN LITERATURE WRITTEN IN ENGLISH

From 'Interface: Some Thoughts on the Historiography of African Literature Written in English' in Dieter Riemenschneider (ed.) *The History and Historiography of Commonwealth Literature* Tübingen: Gunter Narr Verlag, 1983.

A LITERATURE . . . CAN BE DESCRIBED as a system, embedded in the environment of a civilization/culture/society, call it what you will. The system is not primarily demarcated by a language, or an ethnic group, or a nation, but by a poetics, a collection of devices available for use by writers at a certain moment in time. . . . The environment exerts control over the system, by means of patronage. Patronage combines both an ideological and an economic component. It tries to harmonize the system with other systems it has to co-exist with in the wider environment – or it simply imposes a kind of harmony. It provides the producer of literature with a livelihood, and also with some kind of status in the environment.

Traditional African literature is a perfect illustration of this state of affairs. The artist, we are told, has interiorized the implicit poetics of the community, which supports him at least in terms of status, 'he is a spokesman for the society in which he lives, sharing its prejudices and directing its dislikes (in a limited form of satire) against what is discountenanced' (Dathorne 1974: 3), and he does so by making use of certain genres. The illustration matches the model so perfectly because the patronage is totally undifferentiated, i.e. the system allows of one ideal of literature, and only one, and also because the patronage is able to exert its control directly and immediately. Since the literature is oral, not written, and since the artist is therefore a performer, the audience will immediately make its displeasure felt if the artist makes a mistake in the telling or in the reciting – an immediacy that is lost in the transition from oral to written literature. The illustration also works so well because the model sidesteps one complicating factor for the sake of clarity. This must now be corrected by redefining poetics and patronage as constraints influencing the production of literature, rather than simply as factors guiding it in what appears to be a suspiciously mechanistic manner, and by adding the language in which literature is produced as another constraint. The concept of constraint(s), as used here, implies that all statements made about it are more or less double-edged, or rather, that the reader/ hearer is supposed to supply the other side of the coin, so to speak. Constraints can always be honoured and subverted. Their importance lies only partially in their bare existence, the other part being the spirit in which they are taken. Producers of literature may subvert these constraints, or they may be quite happy to work with them or within them.

Literature, then, is produced in the zone of tension where the artist's creativity comes to terms with the constraints. The writer will not reject those constraints out of hand in systems with undifferentiated patronage, because he quite simply has nowhere else to go – but silence. Literary revolutions, on the other hand, tend to occur in systems with differentiated patronage, in which different ideals of literature are allowed to coexist, and in which literature produced on the basis of those different ideals is read by different groups of readers.

Interface, and that is what mainly concerns us here, is the situation which arises when two systems interact, in this case the English system and the African system, so that a kind of hybrid poetics comes into being, combining elements from the historically dominated system (the African one) with elements from the historically dominant system (the English one), and acting as a constraint on the production of literature within the dominated system, while it leaves the dominant system relatively unaffected. . . . Interface is regulated first and foremost by the power and/or prestige of the respective environments of the respective systems, by the power of their respective patronages and the policies they are willing to adopt, and by what use the different poetics have for each other. It strikes me that the concept of interface might at least be useful in developing a chronology that is capable of accommodating more complex factors than one mainly based on theme, or even on a mere succession of decades, or of events occurring outside the system. I would propose shifts in the nature of patronage as the factor that demarcates chronological periods in the development of a system. If events in the environments lead to a changed social role for a group that has exercised patronage, changes are likely to take place inside the system. If they do not, changes are much less likely, no matter how momentous the events may be in other respects.

The prestige of an environment may be less readily measurable, by an independent observer, than its power, and in the early stages of the interface there was no doubt as to where power lay. Technology figures as the prominent de facto criterion between the civilized and the primitive, and it was soon to be provided with an ideological justification. 'Primitiveness, essentially a product of political domination, received, in the second part of the nineteenth century, an almost authoritative stamp from social Darwinism' (Obiechina 1975: 15). The English system quite logically occupied the dominant position in the interface – dominant with a vengeance since it was, at first, quite simply proclaimed that African literature did not exist, just as Du Bellay, for example, dismissed Medieval Literature more or less out of hand. It lived on, of course, for quite some time after its dismissal, as did traditional literature in British Africa, but those who produced it would gradually find out that it did not confer the same status on them as before, certainly not in the new urban communities, precisely because the African patronage, or rather, the patronage inside the African system, had lost its status-conferring power. The African system was forced on the defensive and the English system had no use for it. As a result, interaction was a very one-sided affair. . . .

Refractions of original texts in the English system became the main instrument in the institutionalization of that system as the main, or even the only one in the interface. Probably the most influential refracted texts in this respect were school anthologies, introducing these originals in schools and on other levels of education, as these became gradually more available. The ideology of the groups that acted as patrons for different schools and, later, universities played the most important part here. Surveys and anthologies of English literature must have read quite differently according to the sub-ideology they were trying to propagate (a fascinating field of study here, be it said in passing, and one very little cultivated). I say sub-ideologies, because they were all united in the main ideology: that of the white man's civilizing mission and of his superiority. In universities the most influential refracted text was what it still is (it is also the shortest one by far): the reading list that introduces the canon of a literature as modified by successive changes in taste, that seemingly elusive amalgam of ideological, economic and poetological factors.

Environmental patterns such as these served to discredit the old African patronage even more, but they did little to encourage the production, by Africans, of literature in English that would closely follow the poetics of the English system, precisely because they did not replace the old African patronage. In fact, the two patronages remained quite distinct for some time, one producing, at best, various variants of Couriferist literature in African writers, the other producing essentially what it had always produced, but which was now much less honoured and sought after.

The literature produced on the basis of English poetics and under English patronage could hardly be other than that of Couriferism, in which interface means the total hegemony of one system over another. Hegemony is used here in the fullest sense of the word, which means not only acceptance, which may range from the grudging to the resigned, of English constraints, not least among them the language, but identification with those constraints: this is the way it has to be, not just the way it is. Patronage selects the themes that can be treated, emphasizes certain techniques and rejects others, according to the changing appreciation of elements of the poetics, and sees to it that the language is used 'correctly.' In short, we have here a clear-cut case of patronage by stipulation.

The same patronage by stipulation made its appearance in the only case in which European patronage tried to take the place of African patronage within the African system: in the production of literature in the vernacular. The motive was, of course, ideological: 'the missionaries who ran the printing presses' (Owomoyela 1979: 28), and it introduced a new means of literary communication into the African system: the text. Only the text really necessitates the production of other texts as a control mechanism in systems with undifferentiated patronage and as weapons in the struggle between rival poetics in systems with differentiated patronage, since the control mechanisms of oral literature are a lot more direct. This kind of patronage could again bestow status, in the writer's immediate environment and without the need for the writer to produce in a different language. He did, however, have to produce on the basis of a different, or a hybrid poetics, and certainly on the basis of a different ideology: 'The christianized vernacular writer took the decisive step of separating himself from the group' (Owomoyela 1979: 28).

Being essentially ideological in nature, this kind of patronage exerted both a destructive and a conserving influence on African poetics, or rather, it turned that poetics into a kind of 'selective poetics,' in which elements (themes mainly, characters and situations, since genres are inherently neutral and symbols can always be allegorized, witness the wholescale allegorical colonization of classical literature undertaken within the West European system, and by the same ideology, some fifteen centuries earlier) unacceptable to the ideology are rejected, whereas others are allowed. The rejected elements eventually vanish if the ideology manages to extend its hegemony over the whole system, they go underground if it does not. In doing so it saves many elements of the rejected poetics, which may emerge again later, such as the ballad after three centuries in the West European system, and a number of themes and other elements after a few decennia in the interface situation we are analyzing here.

The underground elements could only be allowed to 'hybridize' the poetics of the interface after another shift in patronage, in which a hybridized group of readers is willing to patronize the literature based on such a poetics, and in which the old dominant patronage group is willing to tolerate and, eventually, accept it. There were environmental reasons for this, of course: urbanization, the institution of cash economy, industrialization and the progress of Christianity contributed to a situation in which the African 'removed himself from a community where status and social hierarchy had determined the individual's place in society and where the individual counted in terms of the group to which he belonged, and entered a situation in which he was free to assert, if only in a limited way, his own individuality' (Obiechina 1975: 5).

This was instrumental in creating a potential patronage, which would see its sense of its own worth dramatically boosted by the obvious demonstration of the white man's vulnerability provided by World War II, while another factor was instrumental in setting up a group of potential producers: the 'popular press,' owned by Africans, which 'gave the common man his first apprenticeship at literary expression in print' (Obiechina 1975: 12). The new patronage group found itself ready and able to confer status on writers who produced on the basis of the new hybrid poetics, which, as a result, acquired enough status itself to challenge the dominant, English poetics.

This process was also helped by the fact that refracted texts had, in the meantime, begun to travel the other way. But refractions of a dominated poetics penetrate into the system organized around the dominant poetics only if they are first filtered through the dominant system. The more that system becomes familiar with texts from the dominated system, the less rigorous the filtering process is likely to be. There is a fairly continuous progression, therefore, from philological refractions, where the motive is, once again, ideological – one needs to know languages in order to be able to carry out missionary activities – to translation, without a doubt the type of refracted text that has done most to introduce African literature to other literary systems which ignored, or denied its existence, to writing in West European languages, among them of course English by Africans. At first this type of writing tends to respect English poetics in all but one element: theme, which is frankly African, and gives these writings a kind of exotic novelty value. Hence the emergence of the autobiography as one of the dominant genres in African English literature, until it is succeeded by the novel, which gradually Africanizes more and more elements from the basically English poetics, and, in doing so, reaffirms the status of the hybrid.

Poetological, as opposed to ideological interest in African poetics was not all that often expressed within the English system until after World War I, when it was found that African poetics, and the literature produced on the basis of it, could be invoked, often with little or no factual knowledge, as an example of a certain ideal of literature that would challenge the then dominant one. This poetological interest on the part of certain groups within the English system coincided with a mainly ideological interest expressed by the emerging African nationalist leaders, who needed an African literature, preferably a great one, in order to counteract the overwhelming cultural claims of the colonizers. In the independent African nations this same attitude has given rise to a most interesting type of translation, in which vernacular literature from the different languages inside a new nation is translated into English (or the other European language that functions as the nation language) not primarily for export, but for internal use: a 'foreign' language is used to reinforce a sense of 'national' cultural identity.

The hybrid poetics is also accepted for economic reasons: there is, quite simply, money in it, particularly in the recent past, when anything that came out of Africa would get published by Heinemann, Longmans, Macmillan and a number of publishers in the United States. Finally, it would seem, the hybrid poetics has produced its hybrid patronage: the canonization of works and writers that is now going on, is not the work of Africans only: it is also carried out in London, in various centers in the United States and in Europe.

Being a hybrid system that is still developing, that is in its first stages even, African English literature, and other African literatures written in European languages, can teach us a lot about the way in which literary systems as such originate and develop. We shall only learn what we need to learn in that respect, though, if we resolutely broaden our research, away from the canon. That is, if we are not content with commenting on the works that form the canon, but if we also want to shed light on the factors that are instrumental in the canonization process. To be able to do that, we must include non-canonized works in our surveys, and show what parts they play, and we must also include refracted texts much more than has been done up to now.

Systematization on the basis of a model of this type will not only teach us something about the field we want to investigate, but also about the model we are trying to use, since new information will inevitably tend to modify the model. The danger inherent in this type of approach is that the model, the system, tends to be given some kind of ontological status, that the 'map' and the 'territory' become confused, or even interchangeable. This danger can only be counteracted by means of continuous feedback between those among us who work in the territory itself, and those who try to make maps.

## Chapter 92

# S. I. A. KOTEI

## THE BOOK TODAY IN AFRICA

From *The Book Today in Africa* Paris: UNESCO, 1981.

## The African entrepreneur publisher

THREE SETS OF CONDITIONS DETERMINE the success or failure of private indigenous enterprise in an African country. The first comprises the general state of affairs in the country concerned, particularly the national political economy. If national policy actively favours private enterprise, then it should be expected that the necessary facilities will be provided; if there is a tendency to centralization or state monopoly of particular industries, then private enterprise is not encouraged. In any case, the extent to which the state will make foreign exchange available to an indigenous entrepreneur will depend on the importance of his industry to the national scale of priorities.

The second set of conditions is endemic to the enterprise itself: that is, availability of the requisite manpower, skills and appropriate technology.

The third and perhaps the most vital condition is the existence of consumer market forces. Two crucial questions can be asked here. First, assuming there is a demand for the services or products of the enterprise, is the market size large enough to make it economically viable? Second, is the market value of the product within the purchasing power of the consumer public?

Where the entrepreneur publisher is concerned, the book industry in Africa has been affected by all of the above conditions. The only constant factors are the second and third, conditions, that is, manpower/technology, and the market. . . . Those African nations which have had a relatively strong economy in recent years, matched by socio-political awareness of the role of the book in development, have also had a relatively healthy book industry. . . .

A remarkable example is the *Onitsha Market Literature* which burst upon the Nigerian reading public in the 1950s (Obiechina 1971, 1972, 1973).

It was a phenomenon of literary profusion without comparison anywhere in Africa, before or since. In reference to the remarkable success achieved (in a financial and technical sense) it could be reported that printing and publishing had become eastern Nigeria's healthiest industry (Harris 1968: 226). Many of these 'industrialists' not only wrote their boy-meets-girl novelettes and rapid-results cram-books, but also printed, published and sold them.

How can one account for the ability of *Onitsha Market Literature* to achieve success without any of the persons involved having received much training in writing, printing, publishing or book-selling? For one, there was a ready market of buyers who were eager to learn from reading any and all accessible material.

Nevertheless, it is true that Third World publishing is full of pitfalls for the untrained, uninitiated entrepreneur. Indeed the book trade is regarded as the most risky business in the world today, after film-making (Hasan 1975: 1–8). With increasingly tough competition, the amateur publisher would be ill-advised to enter the profession without adequate training.

Publishers in Africa have in most cases ignored this admonition. Their concern to alleviate the book hunger precipitates a decision to publish on a large scale at a national level, though professional manpower is lacking. Inability to gauge accurately the size of the potential market is one other predicament faced by the beginning entrepreneur publisher. There are hardly any studies of reading habits to guide his choice of specialization or distribution targets. Therefore, he adopts a trial-and-error method, unless, of course, he goes into the assured market of textbook publishing. Even here, he has to compete with established publishers, both national and foreign. . . .

Currently, the most popular themes (from the sales point of view) are the novelettes variously described as 'popular fictions', 'boy-meets-girl', or 'market literature'. They have no serious political axes to grind but they tend to be moralistic in their social comment and portrayal of ethical stands.

If an author cannot find a local publishing house to publicize his views, he either relies on his own devices by establishing a private press or sends his manuscripts abroad (Armah 1975). The African writer's deprivation was passionately expressed by the President of the Ghana Association of Writers in 1973 as follows:

> If you [the writer] set out to print anything on your own, the printing costs will stagger you. If you manage to print, the distribution difficulties will blow your mind. If you give your stuff to a local publisher, you will sympathize so much with his problems that you may not write again. . . . So all our best work . . . appears first to an audience which either regards us like some glass-enclosed specimen . . . or like an exotic weed to be sampled and made a conversation piece . . . or else we become some international organization's pet.
>
> (Okai 1973: 4)

Private book production thrives on direct relations between authors and printers. The absence of publishers in Mauritius means that an author bears the entire cost of publishing (Jacob 1974). In Ethiopia, described as a 'society without publishers', a co-operative approach among authors has also been adopted. A number of authors with manuscripts to publish pay a regular monthly subscription to have their manuscripts published in turn. The effects of the above compromises are that production is small in quantity, the book itself is made small in order to make savings on paper and printing costs.

The alternative to self-reliance is for an author to submit his manuscript to an established local publisher – often thereby running the risk of it never seeing the light of day, or having to wait interminably before being published. Many manuscripts suffer this fate not because they are worthless or of lesser value than those that get published, but simply because publishers get so many unsolicited manuscripts that their backlog is always much more voluminous than production. It becomes impossible to maintain a proper balance between input and output ratios because of innumerable technical, financial and manpower constraints. In this respect, it is pertinent to observe that most of the complaints that are lodged against state publishing houses by authors concern delays in publishing their manuscripts, rather than inefficiency in design or even failure to pay royalties. None the less, there have been cases where a commissioned author refused a contract because he feared that an African publisher either could not guarantee good book design, or lacked the facilities for wide international distribution. There are now African houses who excel in book design, but who cannot promote wide enough sales at home or abroad to make it worth the writer's while to submit more manuscripts. . . .

## The critic

Probably because of these constraints, the doyen of African creative writers – Chinua Achebe – has called for a kind of collectivization in which the writers and their audience will move together in a dynamic evolving relationship, through the publisher who must operate in the same historic and social continuum. 'It stands to reason that he [the publisher] cannot play this role from London or Paris or New York' (Oluwasanmi et al. 1975: 44). His work must be published in Africa itself, where the local publisher, with the liveliness of local imagination, can seize upon the peculiar characteristics of a place to operate more effectively within the social milieu. The truly successful writer, for his part, must be both a mirror and an image of the values of his society.

One way of interpreting Achebe's call for the African writer to operate in a 'historic and social continuum' is for all writers to work together to reach a sizeable continental audience. To achieve this objective one must assume again that there is a historic and social continuum running across the continent backed by a common culture. The fact is that African peoples have had different historical experiences and live in multiple, heterogeneous, cultural milieu. Therefore the themes will not be quite the same.

However, from the organizational point of view, writers could come together to find common frames of reference. Accordingly, a Union of Writers of the African Peoples spearheaded by dramatists, novelists and poets in sub-Saharan Africa has been formed. . . .

## The status of the writer

The social status of writers is relative to the degree of audience-appreciation. Throughout the world, the degree of recognition or status that society accords to any group of professional persons is roughly commensurate with the degree of service which the society receives or expects from that group. The village teacher is seen by most rural dwellers as an indispensable asset to the community where education of the young is concerned; he is accordingly a highly respected citizen even though his remuneration might not be high, when compared with that of a doctor or minister of religion. Often he receives compensation in kind from grateful villagers who would make donations of foodstuffs and household equipment. Besides, the mere fact that he has acquired a certain degree of education, plus the literate skills which most members of his society do not possess, places him in a class apart from themselves. . . .

Most African writers communicate with a tiny subculture within society. Among this group, the writer enjoys considerable social standing; beyond it he gets nothing but passive recognition. Unless the writer (anywhere) is bent on making literature a solitary art he must get onto the popular bandwagon. In Kenya, David Maillu's Comb Book Series with titles such as *The Flesh: Diary of a Prostitute*, *The Komon Man*, *Dear Monica* and *My Dear Bottle* are the most successful popular fiction. One reason why this kind of market literature is more widely read than the polished English novels of the Heinemann series is precisely because the latter do not fully communicate. A writer must meet with his readers in a common environment. . . . What of the African writer's economic status? In developing countries, the facilities for researching a subject are minimal; this makes the human investment truly enormous because the author neither has access to many good reference libraries nor research assistants. It can therefore be understood why some African writers find the normal 10–15 per cent royalty unattractive.

When Ethiope, an officially sponsored publishing house of the former Mid-Western State of Nigeria, offered an author £100 as advance for his manuscript, he promptly declined. Ironically the multinationals, who are in a better financial position to encourage African writers, are less prepared to pay reasonable advances. The 'African Writers' series, which supposedly exists to promote African literature, could only make an offer of £50 to an African writer (which he too declined) as advance payment. Later, Heinemann paid £500 to Houghton Mifflin of Boston, United States, for sole Commonwealth rights to publish the same title.

The situation is no better where copyright is concerned. Most countries in the Maghreb lack national copyright protection for the author; they seem moreover to have abstained from both the Berne and the Universal Copyright Conventions (UCC). They can thereby avoid paying royalties to authors outside the region whose books are published locally. Conversely their own authors do not get any protection inside or outside the region (Botros 1978: 572). When this fact is coupled with low royalties (resulting from the low price of Arabic books) and high rate of taxes, it is difficult for an author writing in Arabic to live solely on his literary earnings.

The dilemma of writers everywhere, and one which slows their productivity, is that very few can make a living from their craft. As can be expected, therefore, African publishers have difficulty in attracting local authors, who look more to the developed countries where sales figures will get them closer to subsistence levels. They can then be sure of at least a minimal but steady financial compensation for their labours.

## Chapter 93

## PHILIP G. ALTBACH

## LITERARY COLONIALISM
## Books in the Third World

From 'Literary Colonialism: Books in the Third World' *Harvard Educational Review* 45(2) (May), 1975.

THE PRODUCTS OF KNOWLEDGE are distributed unequally. Industrialized countries using a 'world' language – notably, the United States, Britain, France, and to a lesser extent, West Germany and the Soviet Union – are at the center of scientific research and scholarly productivity. These same countries dominate the systems which distribute knowledge; they control publishing houses and produce scholarly journals, magazines, films and television programs which the rest of the world consumes. Other countries, especially those in the Third World, are at the periphery of the international intellectual system (Shils 1972).

This essay will examine the relationship between industrialized and developing countries by looking at a small but important aspect of this relationship – the world of books and publishing. The discussion is predicated on several ideas. First, the unequal distribution of intellectual products results from a complex set of factors including historical events, economic relationships, language, literacy and the nature of educational systems. Second, industrialized nations have benefited from their control of the means for distribution of knowledge and have at times used their superiority to the disadvantage of developing countries. Third, patterns of national development, the direction and rate of scientific growth, and the quality of cultural life are related to issues of intellectual productivity and independence. Third World nations have not often paid sufficient attention to these issues because of their overwhelming concern with more immediate problems of development.

There are not enough books to meet the rapidly growing needs of the developing countries. The shortage is not a problem which can be solved simply by printing vast quantities of books, but a complex issue which involves a number of national needs, from printing technology to research support. Some Third World countries lack the technical facilities for mass production of books, and some lack indigenous authors to write on subjects of national concern in languages that most literate citizens understand. Even where books exist to serve a national culture, they often cost more than individuals or even institutions can afford.

At present, there is a shortage of books for 70 percent of the globe. The nature of the Third World 'book hunger,' as Barker and Escarpit have recently called it, can be seen in the fact that the 34 industrialized countries with only 30 percent of the population produce 81 percent of the world's book titles (Barker and Escarpit 1973: 16). Although literacy rates in these nations are higher than in developing countries, the rates alone do not begin to account for the disparity in book production.

Figures for Asia dramatically illustrate the book gap. In 1967 the 18 developing countries of the region with 28 percent of the world's population, accounted for only 7.3 percent of the total number of book titles and 2.6 percent of the total number of copies produced per year, and half of these were textbooks (UNESCO 1967). This represents only 32 book titles per million population, while in Europe the average was 417 per million. . . .

Book publishing does not function in a vacuum; it is related to other elements in a society and has international dimensions as well (Altbach 1975). The following discussion will not analyze all elements of publishing in developing countries – a complex process in any society. Rather, it will treat those particular weaknesses of Third World publishing which perpetuate the dependent position in which most developing nations find themselves.

This is not to say that Third World publishing is totally dependent on industrial cultures or that accomplishments have not been achieved. Indeed, given the odds against creative independent publishing, a number of developing nations have made impressive gains. Nor should it be inferred that industrialized nations have manipulated Third World publishing solely for their own national interests and economic gain. Third World dependence on industrial nations for intellectual products results from a complex set of interrelated factors. . . .

Colonial languages have been used as a means of national unification in a number of Third World nations, particularly those in which no one indigenous language commands the loyalty of the entire population. In addition, ruling elites in Third World countries have often used the colonial language to protect their own privileged position. As long as only 10 percent or less of a population has access to the language of political and economic control, that language represents a source of power.

The colonial language has also been the medium for scholarship. The continued domination of the highest levels of the educational system by Western languages has resulted in a paucity of technical and scholarly books in indigenous languages. English or French continues to be a key to graduate education and to research studies in the Third World, even in countries with some commitment to indigenous languages.

Furthermore, libraries and institutions, which comprise the bulk of the market for scholarly and non-fiction books, are accustomed to buying books in European languages. Even where classes are conducted in the indigenous language, a Western language is usually necessary for library research. Thus, authors wishing to write for a national audience and to reach their intellectual peers generally write in a European language.

Even in Indonesia, one of the few former colonies which has made a concentrated and fairly successful effort to promote the use of an indigenous language, *Bahasa Indonesia*, indigenous scholarly books and advanced textbooks do not yet exist and materials in English are widely used. It is my estimate that in India, about half of the book titles are published in English, while only 2 percent of the population is literate in English. In both Anglophone and Francophone Africa, virtually all books are published in the metropolitan language. In many former colonies the 80 to 95 percent of the population who do not know English or French are effectively barred from the higher levels of education. (Latin America is an exception in this regard since either Spanish or Portuguese is the language of a great majority of the population and a large regional market for books exists. The two publishing giants of the region, Mexico and Argentina, have fairly effectively used this linguistic unity to build thriving publishing industries.)

Publishers are an integral part of this colonial tradition. Indian publishers, for example, do not follow a consciously neocolonialist policy of trying to maintain foreign influence on the subcontinent. Rather, they perceive that the largest market for books is in English and that, in fact, the only national market is for such material. Hence, a complex web of economic and intellectual relationships and traditions makes it difficult to stop publishing in European languages. . . .

Therefore, Third World intellectuals tend to look toward a Western audience. If there is a prestige in publishing, it lies in writing for such Western journals as *Encounter* or *Les Temps Modernes*, or in having a book published in London, New York, or Paris. Publication abroad may bring money and the opportunity to communicate with other Third World intellectuals, since communication seldom runs directly between one developing country and another but is mediated through advanced nations.

There is little circulation of books or journals among Third World nations, even between those with the same language. It is significant that *Jeune Afrique*, an influential African journal with a multinational circulation, is published in Paris. The enterprising Nigerian publisher, Joseph Okpaku, came to New York to start his Third Press, which specializes in African and black subjects. It is perhaps indicative of the difficulties involved in regional publishing that it is often easier to travel between Dakar or Abidjan and Paris than between various African capitals.

The economics of publishing concerns much more than the cost of producing a book in a particular country. Rates of literacy, reading habits of the population, government policy toward books, copyright regulations and the nature of libraries are all part of the economic equation (Smith 1966; Bailey 1970). For example, low literary rates, low per capita purchasing power and a diversity of languages – all common in Third World nations – contribute to a limited market for books. Many of the smaller developing countries find it economically imposs- ible to publish most kinds of books because the internal market is simply too small. Even in such large nations as Nigeria, India and Indonesia, only some textbooks, certain kinds of popular fiction, and religious books are profitable to publish. Although labor costs are lower than in the West, total costs are high since print runs are small and distribution is difficult.

Book distribution may be the single most serious dilemma of publishing in the Third World. Dan Lacy divides the problem into three elements: (a) the actual demand for books as distinguished from the need; (b) the network of distribution, for example, booksellers and wholesalers; (c) the means of conveying information about books, such as reviews, advertising and book-trade journals. Low reader density, great distances between settlements, and poor transportation facilities make book distribution in developing countries particularly difficult (Lacy 1973). Just as developing countries themselves are at the periphery of the world's know- ledge system, regions outside of capital cities, especially rural areas, which are often completely without access to books, periodicals or newspapers, are at the periphery of knowledge systems within these nations.

An important part of the Third World's cultural dependency stems from political and trade relationships with industrialized nations (Altbach 1971: 543–58). Industrialized nations export their products, in this case, books and expertise, to the developing countries. Foreign aid pro- grams, while seeking to provide help to developing countries, often deepen existing patterns of dependence (Mende 1973). Knowledge, then, is a part of the neocolonial relationship.

Commercial arrangements built up over years of colonialism persist in many developing countries. Branches of British and French publishers continue to operate in the Third World, and in some places dominate the publishing scene (Nottingham 1969: 139–44). The advan- tages of foreign firms – expertise, the backing of foreign capital and a worldwide distribution network – have made the emergence of indigenous publishers even more difficult than might otherwise have been the case. . . .

Foreign aid programs have had an impact on publishing in developing countries. While the United States has sponsored the largest aid effort, other countries have also engaged in aid programs. For example, the English Language Book Scheme (ELBS), sponsored by Britain, each year sells more than 1 million copies comprising several hundred titles, intended mainly for use as college and university textbooks. On a considerably smaller scale, West Germany and the Soviet Union have also sponsored intellectual assistance, including aid to publishing.

Foreign aid, particularly intellectual assistance, cannot be separated from the policy goals of the donor country or, for that matter, from the policies and orientation of the recipient nation's government. The American rationale for book-related aid programs has involved both the technical importance of books in the development process and the ideological elements of anti-communism (Benjamin 1969; Barnett and Piggford 1969). Between 1950 and 1964, the United States Information Agency assisted in the production of 9,000 editions and printed 80 million copies in 51 languages (Benjamin 1969: 72).

The Indo-American Textbook Program (PL480) was one of the largest American efforts. Under the PL480 program more than 1,000 different textbooks were reprinted in English for use by Indian college and university students, and more than 4 million copies were distributed at subsidized prices. Although the titles were predominantly in the natural sciences, the reprints included many topics in the social sciences and humanities. The Indian government gave full approval to the program and a joint Indo-American committee selected the textbooks. With the recent cooling in Indo-American relations and changing US foreign aid priorities, textbook aid has virtually ended in India.

Like similar programs in other countries, the Indo-American Textbook Program had certain negative results. In some fields, particularly the social sciences, American books were not relevant to the Indian situation, and the orientations of American social scientists reflected their own ideological biases. Yet the subsidized books tended to drive their more expensive unsubsidized domestic counterparts off the market. The artificially low prices for American books gave buyers a distorted sense of the real cost of books. Finally, several subsidiaries of US publishers were able to establish themselves in the Indian market through the aid programs and their growth may have retarded the development of indigenous Indian publishing. . . .

Do aid programs help Third World publishers to establish strong roots and to bring out relevant locally-written books? Or do they circulate materials which the industrialized countries think will win them influence at the cost of discouraging the development of local publishing? The answers to such questions are complicated, but certainly require more attention at the planning and implementation stages of aid programs than they have been given to date. . . .

A final problem which developing countries face in their quest for intellectual independence is that of copyright, which traditionally has worked to the advantage of the industrialized nations and only now is beginning to change (Gidwani 1968; Barker and Escarpit 1973: 88–102). Copyright regulations have made it difficult and expensive for Third World nations to translate and publish materials originally appearing in the West. Western publishers have often preferred to export their own books rather than to license reprinting in developing nations because larger profits could be realized. Recently, changes in international copyright arrangements have permitted developing countries to reprint and/or translate educational materials more freely than before and at modest cost (UNESCO 1973). These changes, made when the industrialized nations began to realize that copyright agreements were being violated with increasing frequency, will no doubt help the developing countries to obtain the printed materials they need at prices they can afford. . . .

The following suggestions are intended to provide some ideas which can be easily implemented and which may help to ameliorate the existing inequalities in the world of books and publishing. . . .

As a first step, communications between Third World nations should be improved so that common problems and issues can be discussed directly without being mediated through institutions and publications in the industrialized nations. This is particularly important on a regional basis, for example, among the nations of Francophone Africa and of Southeast Asia. As a part of communications development, Third World countries must also create viable means of book distribution among themselves, and between themselves and the industrialized nations.

With the strengthening of indigenous publishing and internal distribution facilities in the Third World, intellectuals need not publish their work abroad. Such an effort should include financial and technical assistance from the public sector when necessary. Foreign scholars working in developing nations should publish their findings in the countries where they conduct their research. In this way local publishing will be strengthened and relevant research will be available to local audiences. The intellectual infrastructure in many Third World countries needs to be strengthened in other ways. Libraries, journals which review books, and bibliographical and publicity tools for publishing should be supported.

In addition, major national policy questions which relate directly to books, including the language of instruction in the educational system, levels of literacy and the ownership of the publishing apparatus, must be solved by Third World governments with an understanding of their implications for the balance of intellectual production. Part of any language reform effort should be assistance to publishing in indigenous languages. Finally, Third World leaders must carefully evaluate foreign aid programs to ensure that their nations benefit without local publishing industries or intellectual autonomy being undermined.

## Chapter 94

# ANNE McCLINTOCK

## SOFT-SOAPING EMPIRE

From *Imperial Leather: Race, Gender and Sexuality in the Colonial Context* New York and London: Routledge, 1995.

## Soap and civilization

**A** T THE BEGINNING OF THE nineteenth century, soap was a scarce and humdrum item and washing a cursory activity at best. A few decades later the manufacture of soap had burgeoned into an imperial commerce; Victorian cleaning rituals were peddled globally as the God-given sign of Britain's evolutionary superiority, and soap was invested with magical, fetish powers. The soap saga captured the hidden affinity between domesticity and empire and embodied a triangulated crisis in value: the *undervaluation* of women's work in the domestic realm, the *overvaluation* of the commodity in the industrial market and the *disavowal* of colonized economies in the arena of empire. Soap entered the realm of Victorian fetishism with spectacular effect, notwithstanding the fact that male Victorians promoted soap as the icon of nonfetishistic rationality.

Both the cult of domesticity and the new imperialism found in soap an exemplary mediating form. The emergent middle class values – monogamy ('clean' sex, which has value), industrial capital ('clean' money, which has value), Christianity ('being washed in the blood of the lamb'), class control ('cleansing the great unwashed') and the imperial civilizing mission ('washing and clothing the savage') – could all be marvelously embodied in a single household commodity. Soap advertising, in particular the Pears soap campaign, took its place at the vanguard of Britain's new commodity culture and its civilizing mission.

In the eighteenth century, the commodity was little more than a mundane object to be bought and used – in Marx's words, 'a trivial thing.' By the late nineteenth century, however, the commodity had taken its privileged place not only as the fundamental form of a new industrial economy but also as the fundamental form of a new cultural system for representing social value (see Richards 1990, introduction and Ch. 1). Banks and stock exchanges rose up to manage the bonanzas of imperial capital. Professions emerged to administer the goods tumbling hectically from the manufactures. Middle-class domestic space became crammed as never before with furniture, clocks, mirrors, paintings, stuffed animals, ornaments, guns and myriad gewgaws and knicknacks. Victorian novelists bore witness to the strange spawning of commodities that seemed to have lives of their own, and huge ships lumbered with trifles and trinkets plied their trade among the colonial markets of Africa, the East and the Americas (see Simpson 1982).

The new economy created an uproar not only of things but of signs. As Thomas Richards has argued, if all these new commodities were to be managed, a unified system of cultural representation had to be found. Richards shows how, in 1851, the Great Exhibition at the Crystal Palace served as a monument to a new form of consumption: 'What the first Exhibition heralded so intimately was the complete transformation of collective and private life into a space for the spectacular exhibition of commodities' (Richards 1990: 72). As a 'semiotic laboratory for the labor theory of value,' the World Exhibition showed once and for all that the capitalist system had not only created a dominant form of exchange but was also in the process of creating a dominant form of representation to go with it: the voyeuristic panorama of surplus as spectacle. By exhibiting commodities, not only as goods but as an organized system of images, the World Exhibition helped fashion 'a new kind of being, the consumer and a new kind of ideology, consumerism' (Richards 1990: 5). The mass consumption of the commodity spectacle was born.

Victorian advertising reveals a paradox, however, for, as the cultural form that was entrusted with upholding and marketing abroad those founding middle-class distinctions – between private and public, paid work and unpaid work – advertising also from the outset began to confound those distinctions. Advertising took the intimate signs of domesticity (children bathing, men shaving, women laced into corsets, maids delivering nightcaps) into the public realm, plastering scenes of domesticity on walls, buses, shopfronts and billboards. At the same time, advertising took scenes of empire into every corner of the home, stamping images of colonial conquest on soap boxes, matchboxes, biscuit tins, whiskey bottles, tea tins and chocolate bars. By trafficking promiscuously across the threshold of private and public, advertising began to subvert one of the fundamental distinctions of commodity capital, even as it was coming into being.

From the outset, moreover, Victorian advertising took explicit shape around the reinvention of racial difference. Commodity kitsch made possible, as never before, the mass marketing of empire as an organized system of images and attitudes. Soap flourished not only because it created and filled a spectacular gap in the domestic market but also because, as a cheap and portable domestic commodity, it could persuasively mediate the Victorian poetics of racial hygiene and imperial progress.

Commodity racism became distinct from scientific racism in its capacity to expand beyond the literate, propertied elite through the marketing of commodity spectacle. If, after the 1850s, scientific racism saturated anthropological, scientific and medical journals, travel writing and novels, these cultural forms were still relatively class-bound and inaccessible to most Victorians, who had neither the means nor the education to read such material. Imperial kitsch as consumer spectacle, by contrast, could package, market and distribute evolutionary racism on a hitherto unimagined scale. No preexisting form of organized racism had ever before been able to reach so large and so differentiated a mass of the populace. Thus, as domestic commodities were mass marketed through their appeal to imperial jingoism, commodity jingoism itself helped reinvent and maintain British national unity in the face of deepening imperial competition and colonial resistance. The cult of domesticity became indispensable to the consolidation of British national identity, and at the center of the domestic cult stood the simple bar of soap.[1]

Yet soap has no social history. Since it purportedly belongs in the female realm of domesticity, soap is figured as beyond history and beyond politics proper (see Burke 1992: 195–216). To begin a social history of soap, then, is to refuse, in part, to accept the erasure of women's domestic value under imperial capitalism. It cannot be forgotten moreover, that the history of European attempts to impose a commodity economy on African cultures was also the history of diverse African attempts either to refuse or to transform European commodity fetishism to suit their own needs. The story of soap reveals that fetishism, far from being a quintessentially African propensity, as nineteenth-century anthropology maintained, was central to industrial modernity, inhabiting and mediating the uncertain threshold zones between domesticity and industry, metropolis and empire.

## Soap and commodity spectacle

Before the late nineteenth century, clothes and bedding washing was done in most households only once or twice a year in great, communal binges, usually in public at streams or rivers (see Davidoff and Hall 1992). As for body washing, not much had changed since the days when Queen Elizabeth I was distinguished by the frequency with which she washed: 'regularly every month whether she needed it or not '(Lindsay and Bamber 1965: 34). By the 1890s, however, soap sales had soared, Victorians were consuming 260,000 tons of soap a year, and advertising had emerged as the central cultural form of commodity capitalism (Lindsay and Bamber 1965: 38).

Before 1851, advertising scarcely existed. As a commercial form, it was generally regarded as a confession of weakness, a rather shabby last resort. Most advertising was limited to small newspaper advertisements, cheap handbills and posters. After midcentury, however, soap manufacturers began to pioneer the use of pictorial advertising as a central part of business policy.

The initial impetus for soap advertising came from the realm of empire. With the burgeoning of imperial cotton on the slave plantations came the surplus of cheap cotton goods, alongside the growing buying power of a middle class that could afford for the first time to consume such goods in large quantities. Similarly, the sources for cheap palm oil, coconut oil and cottonseed oil flourished in the imperial plantations of West Africa, Malaya, Ceylon, Fiji and New Guinea. As rapid changes in the technology of soapmaking took place in Britain after midcentury, the prospect dawned of a large domestic market for soft body soaps, which had previously been a luxury that only the upper class could afford.

Economic competition with the United States and Germany created the need for a more aggressive promotion of British products and led to the first real innovations in advertising. In 1884, the year of the Berlin Conference, the first wrapped soap was sold under a brand name. This small event signified a major transformation in capitalism, as imperial competition gave rise to the creation of monopolies. Henceforth, items formerly indistinguishable from each other (soap sold simply as soap) would be marketed by their corporate signature (Pears, Monkey Brand, etc.). Soap became one of the first commodities to register the historic shift from myriad small businesses to the great imperial monopolies. In the 1870s, hundreds of small soap companies plied the new trade in hygiene, but by the end of the century, the trade was monopolized by ten large companies.

In order to manage the great soap show, an aggressively entrepreneurial breed of advertisers emerged, dedicated to gracing each homely product with a radiant halo of imperial glamour and racial potency. The advertising agent, like the bureaucrat, played a vital role in the imperial expansion of foreign trade. Advertisers billed themselves as 'empire builders' and flattered themselves with 'the responsibility of the historic imperial mission.' Said one: 'Commerce even more than sentiment binds the ocean sundered portions of empire together. Anyone who increases these commercial interests strengthens the whole fabric of the empire' (Hindley 1972: 117). Soap was credited not only with bringing moral and economic salvation to Britain's 'great unwashed' but also with magically embodying the spiritual ingredient of the imperial mission itself.

In an ad for Pears, for example, a black and implicitly racialized coalsweeper holds in his hands a glowing, occult object. Luminous with its own inner radiance, the simple soap bar glows like a fetish, pulsating magically with spiritual enlightenment and imperial grandeur, promising to warm the hands and hearts of working people across the globe (Dempsey 1978). Pears, in particular, became intimately associated with a purified nature magically cleansed of polluting industry (tumbling kittens, faithful dogs, children festooned with flowers) and a purified working class magically cleansed of polluting labor (smiling servants in crisp white aprons, rosy-cheeked match girls and scrubbed scullions)' (Bradley 1991: 179–203).

Nonetheless, the Victorian obsession with cotton and cleanliness was not simply a mechanical reflex of economic surplus. If imperialism garnered a bounty of cheap cotton and soap oils from coerced colonial labor, the middle class Victorian fascination with clean, white bodies and clean, white clothing stemmed not only from the rampant profiteering of the imperial economy but also from the realms of ritual and fetish. Soap did not flourish when imperial ebullience was at its peak. It emerged commercially during an era of impending crisis and social calamity, serving to preserve, through fetish ritual, the uncertain boundaries of class, gender and race identity in a social order felt to be threatened by the fetid effluvia of the slums, the belching smoke of industry, social agitation, economic upheaval, imperial competition and anticolonial resistance. Soap offered the promise of spiritual salvation and regeneration through commodity consumption, a regime of domestic hygiene that could restore the threatened potency of the imperial body politic and the race.

## Note

1    In 1889 an ad for Sunlight Soap featured the feminized figure of Brittania, standing on a hill and showing P. T. Barnum, the famous circus manager and impresario of the commodity spectacle, a huge Sunlight Soap factory stretched out below them. Britannia proudly proclaims the manufacture of Sunlight Soap to be: 'The Greatest Show on Earth' (see Wicke 1988).

## Chapter 95

# ARJUN APPADURAI

## COMMODITIES AND THE POLITICS OF VALUE

From *The Social Life of Things* Cambridge: Cambridge University Press, 1986.

I N AN EXTREMELY INTERESTING DISCUSSION of British trade in Hawaii in the late eighteenth and early nineteenth centuries, Marshall Sahlins has shown how Hawaiian chiefs, in stretching traditional conceptions of tabu to cover new classes of trade goods (in keeping with their own cosmopolitical interests), succeeded in transforming the 'divine finality' even of economic tabus into instruments of expedience (Sahlins 1981: 44–5). Thus, what Sahlins calls 'the pragmatics of trade' erodes and transforms the cultural bounds within which it is initially conceived. In a word, the politics of enclaving, far from being a guarantor of systemic stability, may constitute the Trojan horse of change.

The diversion of commodities from specified paths is always a sign of creativity or crisis, whether aesthetic or economic. Such crises may take a variety of forms: economic hardship, in all manner of societies, drives families to part with heirlooms, antiques, and memorabilia and to commoditize them. This is as true of kula valuables as of more modern valuables. The other form of crisis in which commodities are diverted from their proper paths, of course, is warfare and the plunder that historically has accompanied it. In such plunder, and the spoils that it generates, we see the inverse of trade. The transfer of commodities in warfare always has a special symbolic intensity, exemplified in the tendency to frame more mundane plunder in the transfer of special arms, insignia, or body parts belonging to the enemy. In the high-toned plunder that sets the frame for more mundane pillage, we see the hostile analogue to the dual layering of the mundane and more personalized circuits of exchange in other contexts (such as kula and gimwali in Melanesia). Theft, condemned in most human societies, is the humblest form of diversion of commodities from preordained paths.

But there are subtler examples of the diversion of commodities from their predestined paths. One whole area involves what has been dubbed tourist art, in which objects produced for aesthetic, ceremonial, or sumptuary use in small, face-to-face communities 'are transformed culturally, economically, and socially by the tastes, markets, and ideologies of larger commodities and the politics of value. The major art and archeology collections of the Western world, whose formation represents extremely complex blends of plunder, sale and inheritance, combined Western taste for the things of the past and of the other (see Hencken 1981). In this traffic in artifacts, we can find today most of the critical cultural issues in the international flow of 'authentic' (see Spooner, 1986) and 'singular' (see Kopytoff 1986) commodities. The current controversies between English and American museums and governments and various other countries raise all the moral and political delicacies that come into play when things get diverted, several times over, from their minimal, conventional paths and are transferred by a variety of modes that make their history of claims and counterclaims extremely difficult to adjudicate.

The diversion of commodities from their customary paths always carries a risky and morally ambiguous aura. Whenever what Bohannan (1955) called conveyances give way to what he called conversions, the spirit of entrepreneurship and that of moral taint enter the picture simultaneously. In the case of the kula exchanges of Melanesia, the movement of commodities across spheres, though somehow out of order, is also at the heart of the strategy of the skillful and successful kula player. Inappropriate conversions from one sphere of exchange to another are frequently fortified by recourse to the excuse of economic crisis, whether it be famine or bankruptcy. If such excuses are not available or credible, accusations of inappropriate and venal motives are likely to set in. Excellent examples of the political implications of diversion are to be found in the arena of illegal or quasilegal commodity exchanges, one case of which is discussed next. Lee Cassanelli . . . discusses the shift, in the last fifty years in Northeastern Africa, in the political economy of a quasilegal commodity called qat (catha edulis) (Cassanelli 1986). Qat provides an excellent example of change in what may be referred to as a commodity ecumene,[1] that is, a transcultural network of relationships linking producers, distributors, and consumers of a particular commodity or set of commodities. What is particularly interesting, in this case, is the dramatic expansion of the scale of consumption (and of production) of qat which is clearly tied to changes in the technical infrastructure as well as the political economy of the region. Although the expansion of production appears consistent with conditions that fit with more universal patterns in the commercialization of agriculture, what is more intriguing is the expansion of demand and the response of the state – especially in Somalia – to the explosion in both the production and the consumption of qat.

The recent (1983) ban by the Somali government on the planting, importing, and chewing of qat clearly is the most recent move in a long tradition of state ambivalence toward a commodity whose consumption is perceived as tied to unproductive, and potentially subversive, forms of sociality. In the case of the current Somali ban, it appears that qat (like cloth in Gandhi's rhetoric) is seen as a multilevel problem, one that challenges not only state control over the economy but state authority over the social organization of leisure among the, newly rich and upwardly mobile citizens of urban Somalia. We are again reminded, with this example, that rapid changes in consumptiom, if not inspired and regulated by those in power, are likely to appear threatening to them. Also, in the case of Somalia, we have a very good example of the tension between a rapid shift in the political economy of a regional commodity ecumene and the authority of one state in this ecumene. Of course, the best examples of the diversion of commodities from their original nexus is to be found in the domain of fashion, domestic display, and collecting in the modern West. In the high-tech look inspired by the Bauhaus, the functionality of factories, warehouses, and workplaces is diverted to household aesthetics. The uniforms of various occupations are turned into the vocabulary of costume. In the logic of found art, the everyday commodity is framed and aestheticized. These are all examples of what we might call commoditization by diversion, where value, in the art or fashion market, is accelerated or enhanced by placing objects and things in unlikely contexts. It is the aesthetics of decontextualization (itself driven by the quest for novelty) that is at the heart of the display, in highbrow Western homes, of the tools and artifacts of the 'other': the Turkmen saddlebag, Masai spear, Dinka basket.[2] In these objects, we see not only the equation of the authentic with the exotic everyday object, but also the aesthetics of diversion. Such diversion is not only an instrument of decommoditization of the object, but also of the (potential) intensification of commoditization by the enhancement of value attendant upon its diversion. This enhancement of value through the diversion of commodities from their customary circuits underlies the plunder of enemy valuables in warfare, the purchase and display of 'primitive' utilitarian objects, the framing of 'found' objects, the making of collections of any sort.[3] In all these examples, diversions of things combine the aesthetic impulse, the entrepreneurial link, and the touch of the morally shocking. Nevertheless, diversions are meaningful only in relation to the paths from which they stray. Indeed, in looking at the social life of commodities in any given society or period, part of the anthropological challenge is to define the relevant and customary paths, so that the logic of diversions can properly, and relationally, be understood. The relationship between paths and diversions is itself historical and dialectical as Michael Thompson (1979) has skillfully shown in regard to art objects in the modern West. Diversions that become predictable are on their way to becoming new paths, paths that will in turn inspire new diversions or returns to old paths. These historical relationships are rapid and easy to see in our own society, but less visible in societies where such shifts are more gradual.

Change in the cultural construction of commodities is to be sought in the shifting relationship of paths to diversions in the lives of commodities. The diversion of commodities from their customary paths brings in the new. But diversion is frequently a function of irregular desires and novel demands, and we turn therefore to consider the problem of desire and demand. . . .

## Desire and demand

Part of the reason why demand remains by and large a mystery is that we assume it has something to do with desire, on the one hand (by its nature assumed to be infinite and trans-cultural) and need on the other (by its nature assumed to be fixed). Following Baudrillard (1981), I suggest that we treat demand, hence consumption, as an aspect of the overall political economy of societies. Demand, that is, emerges as a function of a variety of social practices and classifications, rather than a mysterious emanation of human needs, a mechanical response to social manipulation (as in one model of the effects of advertising in our own society), or the narrowing down of a universal and voracious desire for objects to whatever happens to be available. . . .

## Knowledge and commodities

. . . [P]eculiarities of knowledge . . . accompany relatively complex, long-distance, intercultural flows of commodities, though even in more homogeneous, small-scale, and low-technology loci of commodity, there is always the potential for discrepancies in knowledge about commodities. But as distances increase, so the negotiation of the tension between knowledge and ignorance becomes itself a critical determinant of the flow of commodities.

Commodities represent very complex social forms and distributions of knowledge. In the first place, and crudely, such knowledge can be of two sorts: the knowledge (technical, social, aesthetic, and so forth) that goes into the production of the commodity; and the knowledge that goes into appropriately consuming the commodity. The production knowledge that is read into a commodity is quite different from the consumption knowledge that is read from the commodity. Of course, these two readings will diverge proportionately as the social, spatial, and temporal distance between producers and consumers increases. . . . It may not be accurate to regard knowledge at the production locus of a commodity as exclusively technical or empirical and knowledge at the consumption end as exclusively evaluative or ideological. Knowledge at both poles has technical, mythological, and evaluative components, and the two poles are susceptible to mutual and dialectical interaction.

## Notes

1   My use of the term ecumene is a rather idiosyncratic modification of Marshall Hodgson's use of it in *The Venture of Islam* (1974).

2   Also compare Alsop's notion (1981) that art collecting invariably 'pries loose' the things that are collected from their former context of use and deprives them of significant social purpose.

3   It is worth noting that despite a superficial opposition between them, there is a deep affinity between trade and art, at least from the point of view of the material life of simpler societies. Both involve what might be called the *intensification of objecthood* through its very different ways. Tourism art builds on this inner affinity.

## Chapter 96

# GRAHAM HUGGAN

## THE POSTCOLONIAL EXOTIC

From *The Postcolonial Exotic: Marketing the Margins* London
and New York: Routledge, 2001.

**T**HE POSTCOLONIAL EXOTIC . . . occupies a site of discursive conflict between
a local assemblage of more or less related oppositional practices and a global apparatus
of assimilative institutional/commercial codes. More specifically, it marks the intersection
between contending regimes of value: one regime – postcolonialism – that posits itself as
anti-colonial, and that works toward the dissolution of imperial epistemologies and institutional
structures; and another – postcoloniality – that is more closely tied to the global market, and
that capitalises both on the widespread circulation of ideas about cultural otherness and on
the worldwide trafficking of culturally 'othered' artefacts and goods. This constitutive tension
within the postcolonial might help explain its abiding ambiguity; it also helps us better understand how value is generated, negotiated and disseminated in the postcolonial field of cultural
production. The regime of postcolonialism, according to Patrick McGee, implicitly 'reads against
the grain of value' (McGee 1992: 16); it interrogates the institutional processes by which value
is acquired, exchanged and transmitted, ideally working toward what Edward Said calls the
'transvaluation of value' itself (Said 1986; also McGee 1992: 17). Postcolonialism, it could be
said, acknowledges the contingency of value (Herrnstein Smith 1984, 1988); it recognises the
need to critique value as a fixed or seemingly permanent presence; to see evaluation as a process
subject to historical change and ideological manipulation; and to accept that value is 'transitive
– that is to say, value for somebody in a particular situation – and is [therefore] always culturally and historically specific' (Eagleton and Fuller 1983: 76). Part of postcolonialism's regime
of value appears to lie in the very resistance to value; or at least in the opposition to universal-
ising codes of evaluation that assert the 'intrinsic' meaning or 'transhistorical' worth of literary/
cultural texts. Value is constituted, rather, as a 'site of institutional struggle – a struggle over
such issues as authorship, authenticity, and legitimacy, which involves several different but
interconnected levels of mediation' (Huggan 1997: 412).

So if postcolonial literary/cultural works, on one level, articulate forms of material struggle
– the ongoing battle for emancipation, the continuing attempt to dismantle imperialist institutions and dominating structures this struggle might also be extended to these works' symbolic
power. Bourdieu's emphasis on the symbolic production of the literary work is important
here. As Bourdieu points out, the value of literary (and other cultural) works is often generated
through structures of belief:

> The sociology of . . . literature has to take as its object not only the material production
> but also the symbolic production of the work, i.e. the production of the value of the
> work or, which amounts to the same thing, of belief in the value of the work.
>
> (Bourdieu 1993: 37)

Here, however, postcolonialism comes into conflict once again with postcoloniality; for while postcolonial works and their authors gain currency from their perceived capacity for anti-imperialist resistance, 'resistance' itself emerges as a commodified vehicle of symbolic power. The same might be said for much of the cultural vocabulary of postcolonial criticism. As Julia Emberley points out in the context of the critical reception of contemporary Native (Canadian/American) women's writing:

> [T]he society of the spectacle has . . . displaced questions of political economy into a postcolonial discourse in which images, re-presentations, 'authenticities,' and 'the experience of marginality' circulate as the currency of exchange. . . . The material administered and exchanged in the process of subjecting Native people to colonial historization no longer exists in the form of supplies or European commodities. In postcolonial discourse it is their symbolic value as textual commodities that is being exchanged.
>
> (Emberley 1993: 163, 109)

Emberley is too quick to see, and dismiss, postcolonialism as a hegemonic discourse that ironically contributes toward the containment and surveillance of its subaltern subjects. However, it is certainly true that the terms in which many postcolonial debates are currently being conducted – 'resistance' 'authenticity', 'marginality', and so on – circulate as reified objects in a late-capitalist currency of symbolic exchange. The postcolonial is thus constructed as an object of contestation between potentially incompatible ideologies, political factions and interest groups. The complex politics of value surrounding the postcolonial field of production clearly cannot be limited to the latest market rulings for commodity exchange. (In any case, as Arjun Appadurai argues, commodities 'constantly spill beyond . . . specific regimes of value, [so that] political control of demand is always threatened with disturbance' (Appadurai 1986: 57).) Notwithstanding, postcolonial products function, at least in part as cultural commodities that move back and forth within an economy regulated largely by Western metropolitan demand (Appadurai 1986). This economy functions on a symbolic, as well as a material, level; it is regulated, that is, not only by the flow of material objects (book, films, videotapes, etc.), but also by the institutional values that are brought to bear in their support (Huggan 1997).

To accuse postcolonial writers/thinkers of being lackeys to this system is, as I have repeatedly suggested, to underestimate their power to exercise agency over their work. It may also be to devalue the agency, both individual and collective, of their readers, who by no means form a homogeneous or readily identifiable consumer group. Postcolonial literatures in English – to make an obvious point – are read by many different people in many different places; it would be misleading, not to mention arrogant, to gauge their value only to Western metropolitan response.[1] And it would be as difficult to distinguish a single reading public as to identify its location, in part because readers of postcolonial works are part of an increasingly diasporised, transnational English-speaking culture, but most of all because literary/cultural audiences all over the world are by their very nature plural and heterogeneous. Such audiences, according to John Frow, are composed of several different 'valuing communities' whose boundaries are necessarily porous and whose interests are far from evenly matched (Frow 1995: 142–3; also Huggan 1997: 429). Audiences, says Frow, arguing in part against Bourdieu, are never fixed; 'valuing communities', similarly, cannot be conceptualised 'in terms of self-contained positional identities' (Frow 1995: 154) – the value ascribed to a literary work is never the more or less direct expression of a social group. This labile view of audience suggests that the attempt to locate and affix the social positions of 'valuing communities' is always likely to be chimerical; it also guards against the narrow identification of 'target' audiences or, more specifically in this case, the monumentalisation of a metropolitan reader-ship, implied or not, for postcolonial texts. Postcolonial texts, as previously stated, are more subject than most to diasporic mediation; their readerships are highly likely to be multiply dislocated and dispersed (Radhakrishnan 1996). What is more, such texts often tend to drama-tise these dislocations and dispersals, commenting ironically on the material conditions under which they are produced, distributed and consumed. I shall suggest here, following Frow, that postcolonial readerships, like audiences in general, are part of a wider semiotic apparatus of value-coding: one which is irreducible to a single set of standards or criteria governing reading, and which negotiates instead between intersecting 'discursive formations' (Bennett 1990) and 'evaluative regimes' (Appadurai 1986). Frow sums it up succinctly:

> [N]either texts nor readers have an existence independent of [specific social] relations; . . . every act of reading, and hence every act of ascribing value, is specific to the particular regime that organizes it. Texts and readers are not separable elements with fixed properties but 'variable functions within a discursively ordered set of relations' [Bennett] . . . and apparently identical texts and readers will function quite differ-ently within different regimes. . . . The concept of regime [implies] that no object, no text, no cultural practice has an intrinsic or necessary meaning or value or function; and that meaning, value, and function are always the effect of specific (and changing, changeable) social relations and mechanisms of signification.
>
> (Frow 1995: 145)

To see postcolonial cultural production in terms of its own regimes of value (e.g. postcolonialism versus postcoloniality) is to open the way for historical and, not least, institutional critique. How does postcolonial discourse function within a network of changing 'social relations and mechanisms of signification'? How has the postcolonial come to acquire an increasingly com-modified status; how is the cultural capital that accrues to it apportioned and controlled? How is value generated in the postcolonial field of production? All of these questions have to do, directly or indirectly, with the politics of value that governs commodity exchange within a more or less regulated field. Arjun Appadurai names some of the forms that this politics can take: 'the politics of diversion and of display; the politics of authenticity and of authentication; the politics of knowledge and ignorance; the politics of expertise . . . and connoisseur-ship' (Appadurai 1986: 57). What is interesting about these categories is that they have almost direct equivalents in the postcolonial field: 'the politics of diversion and of display' (the spectacle of cultural difference); 'the politics of authenticity and of authentication' (the construction of native authenticity, the marginal voice, the representative writer); 'the politics of knowledge and ignorance' (alteritism, the fetishisation of the other); 'the politics of expertise . . . and connoisseurship' (aestheticisation, mechanisms of professional legitimation). Many of these categories, as we have seen, also belong to discourses of the exotic. What this homology suggests is that the postcolonial exotic is not itself a diversionary tactic but a dilemma that is very much central to the postcolonial field. And that dilemma might be posed as follows: is it possible to account for cultural difference without at the same time mystifying it? To locate and praise the other without also privileging the self? To promote the cultural margins with-out ministering to the needs of the mainstream? To construct an object of study that resists, and possibly forestalls, its own commodification? The postcolonial exotic is the name that one might give to this dilemma, a name that accompanies the emergence of postcolonial studies as an institutional field. The postcolonial exotic can be either a contradiction in terms (for postcolonialism) or a tautology (for postcoloniality). It is many different things at once: a mechanism of cultural translation for the English-speaking mainstream and a vehicle for the estrangement of metropolitan mainstream views; a semiotic circuit in which the signs of oppositionality are continually recoded, circulating alternately as commodities within a late-capitalist, neo-imperialist symbolic economy and as markers of anti-imperialist resistance in an age of 'adversarial internationalization' (Said 1990); a reminder, generally, of the contradictions inscribed in the contemporary alterity industry and a warning-sign, specifically, to those who invoke otherness to disguise their fear of cultural ignorance (Suleri 1992); a self-obsessed unfurling of fetishistic spectacles, lures and distractions and a self-critical unveiling of the imperialist power-politics that lurks behind aesthetic diversion (Mason 1996; Rousseau and Porter 1990).

The question remains: what is it possible for postcolonial writers/thinkers to do about the postcolonial exotic? Some of them might wish to disclaim or downplay their involvement in postcolonial theoretical production, or to posit alternative epistemologies and strategies of cultural representation (Ahmad 1992; Boyce-Davies 1994). Others might wish to 'opt out' of, or at least defy, the processes of commodification and institutionalisation that have arguably helped create a new canon of 'representative' postcolonial literary/cultural works. Still others, however, have chosen to work within, while also seeking to challenge, institutional structures and dominant systems of representation. These writers/thinkers – in very different ways – have recognised their own complicity with exoticist aesthetics while choosing to manipulate the conventions of the exotic to their own political ends. The following chapters in this book provide examples of what we might call 'strategic exoticism': the means by which postcolonial writers/ thinkers, working from within exoticist codes of representation, either manage to subvert those codes ('inhabiting them to criticize them' Spivak 1990), or succeed in redeploying them for the purposes of uncovering differential relations of power. Exoticism, after all, remains an at best unstable system of containment: its assimilation of the other to the same can never be definitive or exhaustive, since the 'collision between ego's culture and alien cultures' (Mason 1996: 147) is continually refashioned, and the effects that collision produces may unsettle as much as reassure, dislodge authority as much as reconfirm it (Bhabha 1994). 'Strategic exoticism' is an option, then, but as we shall see, it is not necessarily a way out of the dilemma. Indeed, the self-conscious use of exoticist techniques and modalities of cultural representation might be considered less as a response to the phenomenon of the postcolonial exotic than as a further symptom of it. There will be plenty other symptoms to chart here; for the postcolonial exotic is, to some extent, a pathology of cultural representation under late capitalism – a result of the spiralling commodification of cultural difference, and of responses to it, that is characteristic of the (post)modern, market-driven societies in which many of us currently live.

## Note

1   It could, and probably will, be argued that that is precisely what this book is doing. However, I see one of the aims of this book as being to show the effects of metropolitan mediation both on the reception of postcolonial literary/cultural products and on the development of postcolonial studies as an institutionalised academic field. To argue this, as I do, from a Western (if not a major) metropolitan location is to risk being accused of merely perpetrating the phenomenon I am critically analyzing. On one level, of course, this is true. On another, however, it is never enough simply to admit one's own or others' complicity; nor is it enough simply to imagine that by dealing say with less well-known writers and critics that one has 'escaped' the self-replicating commodity circuits within which postcolonial writers and critics are undeniably caught. Postcolonialism's involvement with global commodity culture is, after all, the subject of this book; the contention that the project is entangled in this entanglement, while in itself legitimate, does not seem particularly useful in this context.

# PART SIXTEEN

# Diaspora

DIASPORA IS A TERM OF GROWING RELEVANCE to post-colonial studies. First used to describe the Jewish dispersion in Babylonian times and then after the Roman destruction of Jerusalem, the notion of a diaspora of peoples has become increasingly common in describing the combination of migrancy and continued cultural affiliation that characterizes many racial, ethnic and national groups scattered throughout the world. Where immigration connotes travel from one country to another, diaspora is the scattering throughout the world from one geographic location. But diaspora distinguishes itself from terms such as 'immigration' and 'immigrant', or 'migration' and 'migrant' in more fundamental ways. These words focus on movement, disruption and displacement rather than the perpetuation of complex patterns of symbolic and cultural connection that come to characterize the diasporic society. They describe the *diversity* of 'strangers' rather than the *difference* of the re-located diasporic subject.

This scattering leads to a splitting in the sense of home. A fundamental ambivalence is embedded in the term *diaspora*: a dual ontology in which the diasporic subject is seen to look in two directions – towards an historical cultural identity on one hand, and the society of relocation on the other. In the diasporic subject, then, we see in stark relief the hybrid and dual characteristics that are most often associated with post-colonial discourse. For Salman Rushdie this leads to the emergence of 'Imaginary Homelands' which continue to be written and re-written as the world takes on an ever more complex global character. Diasporic writing becomes strategic because the identity of the diasporic subject is actively inscribed.

The *Shorter Oxford Dictionary* defines diaspora as 'the dispersion'. The term first occurs in Deuteronomy 28, Verse 25 which says 'the Lord will cause you to be defeated before your enemies you will come up from one direction but flee from them in another and you will become a thing of horror'. It is useful to dwell on this seminal text to see how the characteristics of diaspora have persisted: diaspora is a scattering; it is an exile, and, in the original text, that exile is a punishment. In many respects the *experience* of diaspora retains these characteristics. When Edward Said ponders on the state of exile in 'The Mind of Winter' he dwells of the sense of loss, 'the unhealable rift forced between a human being and a native place, between the self and its true home'. In this sense it is a punishment, but an ambiguous one. For in the same breath Said emphasizes the profound creative empowerment the diaspora brings with it.

Exile and diaspora are experiences that exist more widely and in more circumstances than those produced by colonialism. Said's favourite exiles, Adorno and Auerbach are cases in point. But there is an argument to be made for the profound impact of modern imperialism upon the disruption and dispersal of peoples throughout the world. Although the term has its origins in the Jewish diaspora, it is not until the spread of European modernity, through its imperial adventures – a spread which led to the transport and relocation of millions of people, particularly Africans, from one place to another in the world – that the modern sense of the term diaspora begins to take shape.

This example of the African diaspora itself suggests a problematic feature of the term. The very notion of an 'African people' all of whom came from vastly different cultures, language groups, ethnic groups spread across the world in a diaspora may help perpetuate the myth of Africa as a monolithic unity. Where 'diaspora' seems to refer to dispersion, diffusion and heterogeneity, migration movement and scattering, the very term may enhance monolithic notions of culture and identity. Yet the dispersal of large groups of people throughout the world generates hybrid and heterogeneous societies that problematize the very notions of unity, racial dominance and 'civilization' on which empires are built. The settlement, indenture or enslavement of various ethnic and cultural groups to support tropical plantation industries and other mercantile adventures of European capitalism, led to a profound change in the ethnic character of the world. Nevertheless, the sense of origin, however imagined, however monolithic, is actively constructed as an identifying and affiliative marker.

Who can be called 'diasporic'? The issue here is not simply of ethnic affiliation and cultural movement but also of social position. Of the many different peoples who have been scattered and dispersed throughout the world, only some of them can be called diasporic. For instance, we can talk about an Irish diaspora or an Indian or African diaspora but we rarely talk about an English diaspora. The question is one of power. It seems inappropriate to talk about the spread of a powerful colonizing people around the world as 'an exile' although some may experience it as such. Nor can we describe as a diaspora that cultural group that attains global dominance. Diasporas have come to mean cultural minorities, in social power if not always in number (e.g. the African diaspora in the Caribbean) and as such are always seen to be establishing their sense of identity and cultural affiliation, their sense of home, their sense of subject position, against the background of a 'majoritarian' rule.

Diaspora also problematizes the concept of a national identity. The cutting across national boundaries, the dispersion, the spreading out, the diffusion through space and the occupation of many different kinds of national groups disrupts the process, so important to nationalism, of establishing metaphysical links with a particular geographic location with a particular community that lives within those borders. A very different range of filiations and links are established and perpetuated by diasporic communities. If we think of the nation-state as the quintessential institution of modernity and its dissemination the principal object of colonization then diaspora has come to problematize not only the sacred unity of the nation but also the fundamental ideologies of modernity itself. So we find that diaspora has a profoundly disruptive affect upon the whole edifice of European epistemological and political power because it disrupts modernity, it disrupts the idea of nation and national identity, it disrupts the notion of unity and coherence to rational subjectivity and it becomes a prominent feature of a contemporary post-colonial world.

Diasporic identity demonstrates the extent to which identity itself must be constructed and reconstructed by individuals in their everyday life. When Kurdish protesters, citizens of European countries or the US, protest the arrest of the Kurdish leader we see third-generation diasporic subjects making a very clear choice about their cultural affiliation and ethnic identity.

The identification of individuals who protest in these cases is a very clear act of choice. A choice of identification, of belonging to a group, of belonging to a nationally-based community which they may never have visited geographically. There are many occasions like this when the fluid categories of identity become fixed for a brief moment. But this tells us something about subjectivity itself. Subjectivity is fluid, a function of different roles, and what we find in diaspora is an uncovering of the very fluid and constructed nature of identity and also the operation of individual subjective agency in that process of identification.

Diaspora highlights the global trend of creating, constructing and reconstructing identity, not by identifying with some ancestral place, but through travelling itself. While the diasporic subject travels, so does culture. A travelling culture means a culture that changes, develops and transforms itself according to the various influences it encounters in different places. Thus, while diasporas change their countries of arrival, so are their cultures changed in turn. In this respect the most explicit binary, that apparently existing between 'indigenous' and 'diasporic', becomes disrupted, as James Clifford shows, by the 'articulation' of identity through movement and travel.

## Chapter 97

# SALMAN RUSHDIE

## IMAGINARY HOMELANDS

From *Imaginary Homelands: Essays and Criticism 1981–1991*
London: Granta, 1991/New York: Viking Penguin, 1982.

A N OLD PHOTOGRAPH IN A CHEAP FRAME hangs on a wall of the room where I work. It's a picture dating from 1946 of a house into which, at the time of its taking, I had not yet been born. The house is rather peculiar – a three-storeyed gabled affair with tiled roofs and round towers in two corners, each wearing a pointy tiled hat. 'The past is a foreign country'; goes the famous opening sentence of L. P. Hartley's novel *The Go-Between,* 'they do things differently there.' But the photograph tells me to invert this idea; it reminds me that it's my present that is foreign, and that the past is home, albeit a lost home in a lost city in the mists of lost time.

A few years ago I revisited Bombay, which is my lost city, after an absence of something like half my life. Shortly after arriving, acting on an impulse, I opened the telephone directory and looked for my father's name. And, amazingly, there it was; his name, our old address, the unchanged telephone number, as if we had never gone away to the unmentionable country across the border. It was an eerie discovery. I felt as if I were being claimed, or informed that the facts of my faraway life were illusions, and that this continuity was the reality. Then I went to visit the house in the photograph and stood outside it, neither daring nor wishing to announce myself to its new owners. (I didn't want to see how they'd ruined the interior.) I was overwhelmed. The photograph had naturally been taken in black and white; and my memory, feeding on such images as this, had begun to see my childhood in the same way, monochromatically. The colours of my history had seeped out of my mind's eye; now my other two eyes were assaulted by colours, by the vividness of the red tiles, the yellow-edged green of cactus-leaves, the brilliance of bougainvillaea creeper. It is probably not too romantic to say that that was when my novel *Midnight's Children* was really born; when I realized how much I wanted to restore the past to myself, not in the faded greys of old family-album snapshots, but whole, in CinemaScope and glorious Technicolor.

Bombay is a city built by foreigners upon reclaimed land; I, who had been away so long that I almost qualified for the title, was gripped by the conviction that I, too, had a city and a history to reclaim.

It may be that writers in my position, exiles or emigrants or expatriates, are haunted by some sense of loss, some urge to reclaim, to look back, even at the risk of being mutated into pillars of salt. But if we do look back, we must also do so in the knowledge – which gives rise to profound uncertainties – that our physical alienation from India almost inevitably means that we will not be capable of reclaiming precisely the thing that was lost; that we will, in short, create fictions, not actual cities or villages, but invisible ones, imaginary homelands, Indias of the mind.

Writing my book in North London, looking out through my window on to a city scene totally unlike the ones I was imagining on to paper, I was constantly plagued by this problem, until I felt obliged to face it in the text, to make clear that (in spite of my original and I suppose somewhat Proustian ambition to unlock the gates of lost time so that the past reappeared as it actually had been, unaffected by the distortions of memory) what I was actually doing was a novel of memory and about memory, so that my India was just that: 'my' India, a version and no more than one version of all the hundreds of millions of possible versions. I tried to make it as imaginatively true as I could, but imaginative truth is simultaneously honourable and suspect, and I knew that my India may only have been one to which I (who am no longer what I was, and who by quitting Bombay never became what perhaps I was meant to be) was, let us say, willing to admit I belonged.

This is why I made my narrator, Saleem, suspect in his narration; his mistakes are the mistakes of a fallible memory compounded by quirks of character and of circumstance, and his vision is fragmentary. It may be that when the Indian writer who writes from outside India tries to reflect that world, he is obliged to deal in broken mirrors, some of whose fragments have been irretrievably lost.

But there is a paradox here. The broken mirror may actually be as valuable as the one which is supposedly unflawed. Let me again try and explain this from my own experience. Before beginning *Midnight's Children,* I spent many months trying simply to recall as much of the Bombay of the 1950s and 1960s as I could; and not only Bombay-Kashmir, too, and Delhi and Aligarh, which, in my book, I've moved to Agra to heighten a certain joke about the Taj Mahal. I was genuinely amazed by how much came back to me. I found myself remembering what clothes people had worn on certain days, and school scenes, and whole passages of Bombay dialogue verbatim, or so it seemed; I even remembered advertisements, film-posters, the neon Jeep sign on Marine Drive, toothpaste ads for Binaca and for Kolynos, and a footbridge over the local railway line which bore, on one side, the legend 'Esso puts a tiger in your tank' and, on the other, the curiously contradictory admonition: 'Drive like Hell and you will get there.' Old songs came back to me from nowhere: a street entertainer's version of 'Good Night, Ladies', and, from the film *Mr 420* (a very appropriate source for my narrator to have used), the hit number 'Mera Joota Hai Japani', which could almost be Saleem's theme song.

I knew that I had tapped a rich seam; but the point I want to make is that of course I'm not gifted with total recall, and it was precisely the partial nature of these memories, their fragmentation, that made them so evocative for me. The shards of memory acquired greater status, greater resonance, because they were *remains*; fragmentation made trivial things seem like symbols, and the mundane acquired numinous qualities. There is an obvious parallel here with archaeology. The broken pots of antiquity, from which the past can sometimes, but always provisionally, be reconstructed, are exciting to discover, even if they are pieces of the most quotidian objects.

It may be argued that the past is a country from which we have all emigrated, that its loss is part of our common humanity. Which seems to me self-evidently true; but I suggest that the writer who is out-of-country and even out-of-language may experience this loss in an intensified form. It is made more concrete for him by the physical fact of discontinuity, of his present being in a different place from his past, of his being 'elsewhere'. This may enable him to speak properly and concretely on a subject of universal significance and appeal.

But let me go further. The broken glass is not merely a mirror of nostalgia. It is also, I believe, a useful tool with which to work in the present.

John Fowles begins *Daniel Martin* with the words: 'Whole sight: or all the rest is desolation.' But human beings do not perceive things whole; we are not gods but wounded creatures, cracked lenses, capable only of fractured perceptions. Partial beings, in all the senses of that phrase. Meaning is a shaky edifice we build out of scraps, dogmas, childhood injuries, newspaper articles, chance remarks, old films, small victories, people hated, people loved; perhaps it is because our sense of what is the case is constructed from such inadequate materials that we defend it so fiercely, even to the death. The Fowles position seems to me a way of succumbing to the guru-illusion. Writers are no longer sages, dispensing the wisdom of the centuries. And those of us who have been forced by cultural displacement to accept the provisional nature of all truths, all certainties, have perhaps had modernism forced upon us. We can't lay claim to Olympus, and are thus released to describe our worlds in the way in which all of us, whether writers or not, perceive it from day to day.

In *Midnight's Children,* my narrator Saleem uses, at one point, the metaphor of a cinema screen to discuss this business of perception: 'Suppose yourself in a large cinema, sitting at first in the back row, and gradually moving up, . . . until your nose is almost pressed against the screen. Gradually the stars' faces dissolve into dancing grain; tiny details assume grotesque proportions; . . . it becomes clear that the illusion itself is reality.' The movement towards the cinema screen is a metaphor for the narrative's movement through time towards the present, and the book itself, as it nears contemporary events, quite deliberately loses deep perspective, becomes more 'partial'. I wasn't trying to write about (for instance) the Emergency in the same way as I wrote about events half a century earlier. I felt it would be dishonest to pretend, when writing about the day before yesterday, that it was possible to see the whole picture. I showed certain blobs and slabs of the scene.

I once took part in a conference on modern writing at New College, Oxford. Various novelists, myself included, were talking earnestly of such matters as the need for new ways of describing the world. Then the playwright Howard Brenton suggested that this might be a somewhat limited aim: does literature seek to do no more than to describe? Flustered, all the novelists at once began talking about politics.

Let me apply Brenton's question to the specific case of Indian writers, in England, writing about India. Can they do no more than describe, from a distance, the world that they have left? Or does the distance open any other doors?

These are of course political questions, and must be answered at least partly in political terms. I must say first of all that description is itself a political act. The black American writer Richard Wright once wrote that black and white Americans were engaged in a war over the nature of reality. Their descriptions were incompatible. So it is clear that redescribing a world is the necessary first step towards changing it. And particularly at times when the State takes reality into its own hands, and sets about distorting it, altering the past to fit its present needs, then the making of the alternative realities of art, including the novel of memory, becomes politicized. 'The struggle of man against power,' Milan Kundera has written, 'is the struggle of memory against forgetting.' Writers and politicians are natural rivals. Both groups try to make the world in their own images; they fight for the same territory. And the novel is one way of denying the official, politicians' version of truth.

The 'State truth' about the war in Bangladesh, for instance, is that no atrocities were committed by the Pakistani army in what was then the East Wing. This version is sanctified by many persons who would describe themselves as intellectuals. And the official version of the Emergency in India was well expressed by Mrs Gandhi in a recent BBC interview. She said that there were some people around who claimed that bad things had happened during the Emergency, forced sterilizations, things like that; but, she stated, this was all false.

Nothing of this type had ever occurred. The interviewer, Mr Robert Kee, did not probe this statement at all. Instead he told Mrs Gandhi and the Panorama audience that she had proved, many times over, her right to be called a democrat.

So literature can, and perhaps must, give the lie to official facts. But is this a proper function of those of us who write from outside India? Or are we just dilettantes in such affairs, because we are not involved in their day-to-day unfolding, because by speaking out we take no risks, because our personal safety is not threatened? What right do we have to speak at all?

My answer is very simple. Literature is self-validating. That is to say, a book is not justified by its author's worthiness to write it, but by the quality of what has been written. There are terrible books that arise directly out of experience, and extraordinary imaginative feats dealing with themes which the author has been obliged to approach from the outside.

Literature is not in the business of copyrighting certain themes for certain groups. And as for risk: the real risks of any artist are taken in the work, in pushing the work to the limits of what is possible, in the attempt to increase the sum of what it is possible to think. Books become good when they go to this edge and risk falling over it — when they endanger the artist by reason of what he has, or has not, artistically dared.

So if I am to speak for Indian writers in England I would say this, paraphrasing G. V. Desani's H. Hatterr: The migrations of the fifties and sixties happened. 'We are. We are here.' And we are not willing to be excluded from any part of our heritage; which heritage includes both a Bradford-born Indian kid's right to be treated as a full member of British society, and also the right of any member of this post-diaspora community to draw on its roots for its art, just as all the world's community of displaced writers has always done. (I'm thinking, for instance, of Grass's Danzig-become-Gdansk, of Joyce's abandoned Dublin, of Isaac Bashevis Singer and Maxine Hong Kingston and Milan Kundera and many others. It's a long list.)

Let me override at once the faintly defensive note that has crept into these last few remarks. The Indian writer, looking back at India, does so through guilt-tinted spectacles. (I am of course, once more, talking about myself.) I am speaking now of those of us who emigrated . . . and I suspect that there are times when the move seems wrong to us all, when we seem, to ourselves, post-lapsarian men and women. We are Hindus who have crossed the black water; we are Muslims who eat pork. And as a result — as my use of the Christian notion of the Fall indicates — we are now partly of the West. Our identity is at once plural and partial. Sometimes we feel that we straddle two cultures; at other times, that we fall between two stools. But however ambiguous and shifting this ground may be, it is not an infertile territory for a writer to occupy. If literature is in part the business of finding new angles at which to enter reality, then once again our distance, our long geographical perspective, may provide us with such angles. Or it may be that that is simply what we must think in order to do our work.

Midnight's Children enters its subject from the point of view of a secular man. I am a member of that generation of Indians who were sold the secular ideal. One of the things I liked, and still like, about India is that it is based on a non-sectarian philosophy. I was not raised in a narrowly Muslim environment; I do not consider Hindu culture to be either alien from me or more important than the Islamic heritage. I believe this has something to do with the nature of Bombay, a metropolis in which the multiplicity of commingled faiths and cultures curiously creates a remarkably secular ambience. Saleem Sinai makes use, eclectically, of whatever elements from whatever sources he chooses. It may have been easier for his author to do this from outside modern India than inside it.

I want to make one last point about the description of India that *Midnight's Children* attempts. It is a point about pessimism. The book has been criticized in India for its allegedly despairing tone. And the despair of the writer-from-outside may indeed look a little easy, a little pat. But I do not see the book as despairing or nihilistic. The point of view of the narrator is not entirely that of the author. What I tried to do was to set up a tension in the text, a paradoxical opposition between the form and content of the narrative. The story of Saleem does indeed lead him to despair. But the story is told in a manner designed to echo, as closely as my abilities allowed, the Indian talent for non-stop self-regeneration. This is why the narrative constantly throws up new stories, why it 'teems'. The form – multitudinous, hinting at the infinite possibilities of the country – is the optimistic counterweight to Saleem's personal tragedy. I do not think that a book written in such a manner can really be called a despairing work.

England's Indian writers are by no means all the same type of animal. Some of us, for instance, are Pakistani. Others Bangladeshi. Others West, or East, or even South African. And V. S. Naipaul, by now, is something else entirely. This word 'Indian' is getting to be a pretty scattered concept. Indian writers in England include political exiles, first-generation migrants, affluent expatriates whose residence here is frequently temporary, naturalized Britons, and people born here who may never have laid eyes on the subcontinent. Clearly, nothing that I say can apply across all these categories. But one of the interesting things about this diverse community is that, as far as Indo-British fiction is concerned, its existence changes the ball game, because that fiction is in future going to come as much from addresses in London, Birmingham and Yorkshire as from Delhi or Bombay.

One of the changes has to do with attitudes towards the use of English. Many have referred to the argument about the appropriateness of this language to Indian themes. And I hope all of us share the view that we can't simply use the language in the way the British did; that it needs remaking for our own purposes. Those of us who do use English do so in spite of our ambiguity towards it, or perhaps because of that, perhaps because we can find in that linguistic struggle a reflection of other struggles taking place in the real world, struggles between the cultures within ourselves and the influences at work upon our societies. To conquer English may be to complete the process of making ourselves free.

But the British Indian writer simply does not have the option of rejecting English, anyway. His children, her children, will grow up speaking it, probably as a first language; and in the forging of a British Indian identity the English language is of central importance. It must, in spite of everything, be embraced. (The word 'translation' comes, etymologically, from the Latin for 'bearing across'. Having been borne across the world, we are translated men. It is normally supposed that something always gets lost in translation; I cling, obstinately, to the notion that something can also be gained.)

To be an Indian writer in this society is to face, every day, problems of definition. What does it mean to be 'Indian' outside India? How can culture be preserved without becoming ossified? How should we discuss the need for change within ourselves and our community without seeming to play into the hands of our racial enemies? What are the consequences, both spiritual and practical, of refusing to make any concessions to Western ideas and prac-tices? What are the consequences of embracing those ideas and practices and turning away from the ones that came here with us? These questions are all a single, existential question: How are we to live in the world?

I do not propose to offer, prescriptively, any answers to these questions; only to state that these are some of the issues with which each of us will have to come to terms.

To turn my eyes outwards now, and to say a little about the relationship between the Indian writer and the majority white culture in whose midst he lives, and with which his work will sooner or later have to deal.

In common with many Bombay-raised middle-class children of my generation, I grew up with an intimate knowledge of, and even sense of friendship with, a certain kind of England: a dream-England composed of Test Matches at Lord's presided over by the voice of John Arlott, at which Freddie Trueman bowled unceasingly and without success at Polly Umrigar; of Enid Blyton and Billy Bunter, in which we were even prepared to smile indulgently at portraits such as 'Hurree Jamset Ram Singh', 'the dusky nabob of Bhanipur'. I wanted to come to England. I couldn't wait. And to be fair, England has done all right by me; but I find it a little difficult to be properly grateful. I can't escape the view that my relatively easy ride is not the result of the dream – England's famous sense of tolerance and fair play, but of my social class, my freak fair skin and my 'English' English accent. Take away any of these, and the story would have been very different. Because of course the dream – England is no more than a dream.

Sadly, it's a dream from which too many white Britons refuse to awake. Recently, on a live radio programme, a professional humorist asked me, in all seriousness, why I objected to being called a wog. He said he had always thought it a rather charming word, a term of endearment. 'I was at the zoo the other day;' he revealed, 'and a zoo keeper told me that the wogs were best with the animals; they stuck their fingers in their ears and wiggled them about and the animals felt at home.' The ghost of Hurree Jamset Ram Singh walks among us still.

As Richard Wright found long ago in America, black and white descriptions of society are no longer compatible. Fantasy, or the mingling of fantasy and naturalism, is one way of dealing with these problems. It offers a way of echoing in the form of our work the issues faced by all of us: how to build a new, 'modern' world out of an old, legend-haunted civilization, an old culture which we have brought into the heart of a newer one. But whatever technical solutions we may find, Indian writers in these islands, like others who have migrated into the north from the south, are capable of writing from a kind of double perspective: because they, we, are at one and the same time insiders and outsiders in this society. This stereoscopic vision is perhaps what we can offer in place of 'whole sight'.

There is one last idea that I should like to explore, even though it may, on first hearing, seem to contradict much of what I've so far said. It is this: of all the many elephant traps lying ahead of us, the largest and most dangerous pitfall would be the adoption of a ghetto mentality. To forget that there is a world beyond the community to which we belong, to confine ourselves within narrowly defined cultural frontiers, would be, I believe, to go voluntarily into that form of internal exile which in South Africa is called the 'homeland'. We must guard against creating, for the most virtuous of reasons, British-Indian literary equivalents of Bophuthatswana or the Transkei.

This raises immediately the question of whom one is writing 'for'. My own, short, answer is that I have never had a reader in mind. I have ideas, people, events, shapes, and I write 'for' those things, and hope that the completed work will be of interest to others. But which others? In the case of *Midnight's Children* I certainly felt that if its subcontinental readers had rejected the work, I should have thought it a failure, no matter what the reaction in the West. So I would say that I write 'for' people who feel part of the things I write 'about', but also for everyone else whom I can reach. In this I am of the same opinion as the black American writer Ralph Ellison, who, in his collection of essays *Shadow and Act,* says that he finds something precious in being black in America at this time; but that he is also reaching for more than that. 'I was taken very early,' he writes, 'with a passion to link together all I loved within the Negro community and all those things I felt in the world which lay beyond.'

Art is a passion of the mind. And the imagination works best when it is most free. Western writers have always felt free to be eclectic in their selection of theme, setting, form; Western visual artists have, in this century, been happily raiding the visual storehouses of Africa, Asia, the Philippines. I am sure that we must grant ourselves an equal freedom.

Let me suggest that Indian writers in England have access to a second tradition, quite apart from their own racial history. It is the culture and political history of the phenomenon of migration, displacement, life in a minority group. We can quite legitimately claim as our ancestors the Huguenots, the Irish, the Jews; the past to which we belong is an English past, the history of immigrant Britain. Swift, Conrad, Marx are as much our literary forebears as Tagore or Ram Mohan Roy. America, a nation of immigrants, has created great literature out of the phenomenon of cultural transplantation, out of examining the ways in which people cope with a new world; it may be that by discovering what we have in common with those who preceded us into this country, we can begin to do the same.

I stress this is only one of many possible strategies. But we are inescapably international writers at a time when the novel has never been a more international form (a writer like Borges speaks of the influence of Robert Louis Stevenson on his work; Heinrich Boll acknowledges the influence of Irish literature; cross-pollination is everywhere); and it is perhaps one of the more pleasant freedoms of the literary migrant to be able to choose his parents. My own – selected half consciously, half not – include Gogol, Cervantes, Kafka, Melville, Machado de Assis; a polyglot family tree, against which I measure myself, and to which I would be honoured to belong.

There's a beautiful image in Saul Bellow's latest novel, *The Dean's December*. The central character, the Dean, Corde, hears a dog barking wildly somewhere. He imagines that the barking is the dog's protest against the limit of dog experience. 'For God's sake,' the dog is saying, 'open the universe a little more!' And because Bellow is, of course, not really talking about dogs, or not only about dogs, I have the feeling that the dog's rage, and its desire, is also mine, ours, everyone's. 'For God's sake, open the universe a little more!'

## Chapter 98

# STUART HALL

## CULTURAL IDENTITY AND DIASPORA

From 'Cultural Identity and Diaspora' in Jonathan Rutherford (ed.) *Identity: Community, Culture, Difference* London: Lawrence and Wishart, 1990.

THERE ARE AT LEAST TWO DIFFERENT ways of thinking about 'cultural identity'. The first position defines 'cultural identity' in terms of one, shared culture, a sort of collective 'one true self', hiding inside the many other, more superficial or artificially imposed 'selves', which people with a shared history and ancestry hold in common. Within the terms of this definition, our cultural identities reflect the common historical experiences and shared cultural codes which provide us, as 'one people', with stable, unchanging and continuous frames of reference and meaning, beneath the shifting divisions and vicissitudes of our actual history. This 'oneness', underlying all the other, more superficial differences, is the truth, the essence, of 'Caribbeanness', of the black experience. It is this identity which a Caribbean or black diaspora must discover, excavate, bring to light and express through cinematic representation.

Such a conception of cultural identity played a critical role in all post-colonial struggles which have so profoundly reshaped our world. It lay at the centre of the vision of the poets of 'Négritude', like Aimé Césaire and Léopold Senghor, and of the Pan-African political project, earlier in the century. It continues to be a very powerful and creative force in emergent forms of representation amongst hitherto marginalised peoples. . . .

There is, however, a second, related but different view of cultural identity. This second position recognises that, as well as the many points of similarity, there are also critical points of deep and significant *difference* which constitute 'what we really are'; or rather – since history has intervened – 'what we have become'. We cannot speak for very long, with any exactness, about 'one experience, one identity', without acknowledging its other side – the ruptures and discontinuities which constitute, precisely, the Caribbean's 'uniqueness'. Cultural identity, in this second sense, is a matter of 'becoming' as well as of 'being'. It belongs to the future as much as to the past. It is not something which already exists, transcending place, time, history and culture. Cultural identities come from somewhere, have histories. But, like everything which is historical, they undergo constant transformation. Far from being eternally fixed in some essentialised past, they are subject to the continuous 'play' of history, culture and power. Far from being grounded in mere 'recovery' of the past, which is waiting to be found, and which when found, will secure our sense of ourselves into eternity, identities are the names we give to the different ways we are positioned by, and position ourselves within, the narratives of the past.

It is only from this second position that we can properly understand the traumatic character of 'the colonial experience'. The ways in which black people, black experiences, were positioned and subject-ed in the dominant regimes of representation were the effects of a critical exercise of cultural power and normalisation. Not only, in Said's 'Orientalist' sense, were we constructed as different and other within the categories of knowledge of the West by those regimes. They had the power to make us see and experience *ourselves* as 'Other'. Every regime of representation is a regime of power formed, as Foucault reminds us, by the fatal couplet, 'power/knowledge'. But this kind of knowledge is internal, not external. It is one thing to position a subject or set of peoples as the Other of a dominant discourse. It is quite another thing to subject them to that 'knowledge', not only as a matter of imposed will and domination, by the power of inner compulsion and subjective conform-ation to the norm. That is the lesson – the sombre majesty – of Fanon's insight into the colonising experience in *Black Skin, White Masks*. . . .

This second view of cultural identity is much less familiar, and more unsettling. If identity does not proceed, in a straight, unbroken line, from some fixed origin, how are we to under-stand its formation? We might think of black Caribbean identities as 'framed' by two axes or vectors, simultaneously operative: the vector of similarity and continuity; and the vector of difference and rupture. Caribbean identities always have to be thought of in terms of the dialogic relationship between these two axes. The one gives us some grounding in, some continuity with, the past. The second reminds us that what we share is precisely the experi-ence of a profound discontinuity: the peoples dragged into slavery, transportation, colonisation, migration, came predominantly from Africa – and when that supply ended, it was temporarily refreshed by indentured labour from the Asian subcontinent. . . .

It is possible, with this conception of 'difference', to rethink the positioning and repositioning of Caribbean cultural identities in relation to at least three 'présences', to borrow Aimé Césaire's and Léopold Senghor's metaphor: *Présence Africaine, Présence Européenne,* and the third, most ambiguous, presence of all – the sliding term, *Présence Americain* . . . I mean America, here, not in its 'first-world' sense – the big cousin to the North whose 'rim' we occupy, but in the second; broader sense: America, the 'New World', *Terra Incognita.*

*Présence Africaine* is the site of the repressed. Apparently silenced beyond memory by the power of the experience of slavery, Africa was, in fact present everywhere: in the everyday life and customs of the slave quarters, in the languages and patois of the plantations, in names and words, often disconnected from their taxonomies, in the secret syntactical struc-tures through which other languages were spoken, in the stories and tales told to children, in religious practices and beliefs in the spiritual life, the arts, crafts, musics and rhythms of slave and post-emancipation society. Africa, the signified which could not be represented directly in slavery, remained and remains the unspoken, unspeakable 'présence' in Caribbean culture. It is 'hiding' behind every verbal inflection, every narrative twist of Caribbean cultural life. It is the secret code with which every Western text was 're-read'. It is the ground-bass of every rhythm and bodily movement. *This* was – is – the 'Africa that 'is alive and well in the diaspora' (Hall 1976).

When I was growing up in the 1940s and 1950s as a child in Kingston, I was surrounded by the signs, music and rhythms of this Africa of the diaspora, which only existed as a result of a long and discontinuous series of transformations. But, although almost everyone around me was some shade of brown or black (Africa 'speaks'!), I never once heard a single person refer to themselves or to others as, in some way, or as having been at some time in the past, 'African'. It was only in the 1970s that this Afro-Caribbean identity became historically avail-able to the great majority of Jamaican people, at home and abroad. In this historic moment, Jamaicans discovered themselves to be 'black' – just as, in the same moment, they discovered themselves to be the sons and daughters of 'slavery'.

This profound cultural discovery, however, was not, and could not he, made directly, without 'mediation'. It could only be made *through* the impact on popular life of the post-colonial revolution, the civil rights struggles, the culture of Rastafarianism and the music of reggae – the metaphors, the figures or signifiers of a new construction of 'Jamaican-ness'. These signified a 'new' Africa of the New World, grounded in an 'old' Africa: a spiritual journey of discovery that led, in the Caribbean, to an indigenous cultural revolution; this is Africa, as we might say, necessarily 'deferred' – as a spiritual, cultural and political metaphor.

It is the présence/absence of Africa, in this form, which has made it the privileged signifier of new conceptions of Caribbean identity. Everyone in the Caribbean, of whatever ethnic background, must sooner or later come to terms with this African présence. Black, brown mulatto, white – all must look *Présence Africaine* in the face, speak its name. But whether it is, in this sense, an *origin* of our identities, unchanged by four hundred years of displacement, dismemberment, transportation, to which we could in any final or literal sense return, is more open to doubt. The original 'Africa' is no longer there. It too has been transformed. History is, in that sense, irreversible. We must not collude with the West which, precisely, normalises and appropriates Africa by freezing it into some timeless zone of the primitive, unchanging past. Africa must at last be reckoned with by Caribbean people, but it cannot in any simple sense be merely recovered.

It belongs irrevocably, for us, to what Edward Said once called an 'imaginative geography and history', which helps 'the mind to intensify its own sense of itself by dramatising the difference between what is close to it and what is far away'. It 'has acquired an imaginative or figurative value we can name and feel' (1978: 55). Our belongingness to it constitutes what Benedict Anderson calls 'an imagined community' (1982). To this 'Africa', which is a necessary part of the Caribbean imaginary, we can't literally go home again.

What of the second, troubling, term in the identity equation – the European présence? For many of us, this is a matter not of too little but of too much. Where Africa was a case of the unspoken, Europe was a case of that which is endlessly speaking – and endlessly speaking *us*. The European Présence interrupts the innocence of the whole discourse of 'difference' in the Caribbean by introducing the question of power. 'Europe' belongs irrevocably to the 'play' of power, to the lines of force and consent, to the role of the *dominant,* in Caribbean culture. In terms of colonialism, underdevelopment, poverty and the racism of colour, the European présence is that which, in visual representation, has positioned the black subject within its dominant regimes of representation: the colonial discourse, the literatures of adventure and exploration, the romance of the exotic, the ethnographic and travelling eye, the tropical languages of tourism, travel brochure and Hollywood and the violent, pornographic languages of *ganja* and urban violence. . . .

The dialogue of power and resistance, of refusal and recognition, with and against *Présence Européenne* is almost as complex as the 'dialogue' with Africa. In terms of popular cultural life, it is nowhere to be found in its pure, pristine state. It is always-already fused, syncretised, with other cultural elements. It is always-already creolised – not lost beyond the Middle Passage, but ever-present: from the harmonics in our musics to the ground-bass of Africa, traversing and intersecting our lives at every point. How can we stage this dialogue so that, finally, we can place it, without terror or violence, rather than being forever placed by it? Can we ever recognise its irreversible influence, whilst resisting its imperialising eye? The enigma is impossible, so far, to resolve. It requires the most complex of cultural strategies. Think, for example, of the dialogue of every Caribbean filmmaker or writer, one way or another, with the dominant cinemas and literature of the West – the complex relationship of young black British filmmakers with the 'avant-gardes' of European and American filmmaking. Who could describe this tense and tortured dialogue as a 'one way trip'?

The Third, 'New World' presence, is not so much power, as ground, place, territory. It is the juncture-point where the many cultural tributaries meet, the 'empty' land (the European colonisers emptied it) where strangers from every other part of the globe collided. None of the people who now occupy the islands – black, brown, white, African, European, American, Spanish, French, East Indian, Chinese, Portuguese, Jew, Dutch – originally 'belonged' there. It is the space where the creolisations and assimilations and syncretisms were negotiated. The New World is the third term – the primal scene – where the fateful/fatal encounter was staged between Africa and the West. It also has to be understood as the place of many, continuous displacements: of the original pre-Columbian inhabitants, the Arawaks, Caribs and Amerindians, permanently displaced from their homelands and decimated; of other peoples displaced in different ways from Africa, Asia and Europe; the displacements of slavery, colonisation and conquest. It stands for the endless ways in which Caribbean people have been destined to 'migrate'; it is the signifier of migration itself – of travelling, voyaging and return as fate, as destiny; of the Antillean as the prototype of the modern or postmodern New World nomad, continually moving between centre and periphery. This preoccupation with movement and migration Caribbean cinema shares with many other 'Third Cinemas', but it is one of our defining themes, and it is destined to cross the narrative of every film script or cinematic image. . . .

The 'New World' presence – America, *Terra Incognita* – is therefore itself the beginning of diaspora, of diversity, of hybridity and difference, what makes Afro-Caribbean people already people of a diaspora. I use this term here metaphorically, not literally: diaspora does not refer us to those scattered tribes whose identity can only be secured in relation to some sacred homeland to which they must at all costs return, even if it means pushing other people into the sea. This is the old, the imperialising, the hegemonising, form of 'ethnicity'. We have seen the fate of the people of Palestine at the hands of this backward-looking conception of diaspora – and the complicity of the West with it. The diaspora experience as I intend it here is defined, not by essence or purity, but by the recognition of a necessary heterogeneity and diversity; by a conception of 'identity' which lives with and through, not despite, difference; by *hybridity*. Diaspora identities are those which are constantly producing and reproducing themselves anew, through transformation and difference.

## Chapter 99

# EDWARD W. SAID

## THE MIND OF WINTER

From 'The Mind of Winter: Reflections on Life in Exile' *Harper's Magazine*
269 (September), 1984.

> There is no sense of ease like the ease we felt in those scenes where we were
> born, where objects became clear to us before we had known the labour of
> choice, and where the outer world seemed only an extension of our personality.
>
> George Eliot, *The Mill on the Floss*

EXILE IS THE UNHEALABLE RIFT FORCED BETWEEN a human being
and a native place, between the self and its true home. The essential sadness of the break
can never be surmounted. It is true that there are stories portraying exile as a condition that
produces heroic, romantic, glorious, even triumphant episodes in a person's life. But these
are no more than stories, efforts to overcome the crippling sorrow of estrangement. The
achievements of any exile are permanently undermined by his or her sense of loss.

If true exile is a condition of terminal loss, why has that loss so easily been transformed
into a potent, even enriching, motif of modern culture? One reason is that we have become
accustomed to thinking of the modern period itself as spiritually orphaned and alienated. This
is supposedly the age of anxiety and of the lonely crowd. Nietzsche taught us to feel uncomfortable with tradition, and Freud to regard domestic intimacy as the polite face painted on
patricidal and incestuous rage.

The canon of modern Western culture is in large part the work of exiles, émigrés,
refugees. American academic, intellectual, and aesthetic thought is what it is today because of
refugees from fascism, communism and other regimes given over to the oppression and expulsion of dissidents. One thinks of Einstein, and his impact on his century. There have been
political thinkers, such as Herbert Marcuse. The critic George Steiner once proposed that a
whole genre of twentieth-century Western literature, a literature by and about exiles – among
them Beckett, Nabokov, Pound – reflects 'the age of the refugee.' In the introduction to his
book *Extraterristorial*, Steiner wrote:

> It seems proper that those who create art in a civilization of quasi-barbarism, which
> has made so many homeless, should themselves be poets unhoused and wanderers
> across language. Eccentric, aloof, nostalgic, deliberately untimely. . . .

In other places and times, exiles had similar cross-cultural and transnational visions,
suffered the same frustrations and miseries, performed the same elucidating and critical tasks.

The difference, of course, between earlier exiles and those of our time is scale. Modern warfare, imperialism, and the quasi-theological ambitions of totalitarian rulers have seen to that. Ours is indeed the age of the refugee, the displaced person, mass immigration.

Against this larger and more impersonal setting exile cannot function as a tonic. To think of exile as beneficial, as a spur to humanism or to creativity, is to belittle its mutilations. Modern exile is irremediably secular and unbearably historical. It is produced by human beings for other human beings; it has torn millions of people from the nourishment of tradition, family, and geography. . . .

We come to nationalism and its essential association with exile. Nationalism is an assertion of belonging to a place, a people, a heritage. It affirms the home created by a community of language, culture, and customs; and by doing so, it fends off the ravages of exile. Indeed, it is not too much to say that the interplay between nationalism and exile is like Hegel's dialectic of servant and master, opposites informing and constituting each other. All nationalisms in their early stages posit as their goal the overcoming of some estrangement – from soil, from roots, from unity, from destiny. The struggles to win American independence, to unify Germany, to liberate Algeria were those of national groups separated – exiled – from what was construed to be their rightful way of life. Triumphant nationalism can be used retrospectively as well as prospectively to justify a heroic narrative. Thus all nationalisms have their founding fathers, their basic, quasi-religious texts, their rhetoric of belonging, their historical and geographical landmarks, their official enemies and heroes. This collective ethos forms what Pierre Bourdieu, the French sociologist, calls the *habitus*, the coherent amalgam of practices linking habit with inhabitance. In time, successful nationalisms arrogate truth exclusively to themselves and assign falsehood and inferiority to outsiders. . . .

One enormous difficulty in describing this no man's land is that nationalisms are about groups, whereas exile is about the absence of an organic group situated in a native place. How does one surmount the loneliness of exile without falling into the encompassing and thumping language of national pride, collective sentiments, group passions? What is there worth saving and holding on to between the extremes of exile on the one hand and the often bloody-minded affirmations of nationalism on the other? Are nationalism and exile reactive phenomena? Do they have any intrinsic attributes? Are they simply two conflicting expressions of paranoia?

These questions cannot be fully answered because each of them assumes that exile and nationalism can be discussed neutrally, without reference to each other. Because both terms include everything from the most collective of collective sentiments to the most private of private emotions, there is no language adequate for both, and certainly there is nothing about nationalism's public and all-inclusive ambitions that touches the truth of the exile's predicament.

For exile is fundamentally a discontinuous state of being. Exiles are cut off from their roots, their land, their past. They generally do not have armies or states, though they are often in search of these institutions. This search can lead exiles to reconstitute their broken lives in narrative form, usually by choosing to see themselves as part of a triumphant ideology or a restored people. Such a story is designed to reassemble an exile's broken history into a new whole.

At bottom, exile is a jealous state. With very little to possess, you hold on to what you have with aggressive defensiveness. What you achieve in exile is precisely what you have no wish to share, and it is in the drawing of lines around you and your compatriots that the least attractive aspects of being an exile emerge: an exaggerated sense of group solidarity as well as a passionate hostility towards outsiders, even those who may in fact be in the same predicament as you. What could be more intransigent than the conflict between Zionist Jews and Arab Palenstinians? The Palestinians feel that they have been turned into exiles by the proverbial people of exile, the Jews. But the Palestinians also know that their sense of national identity has been nourished in the exile milieu, where everyone not a blood brother or sister is an enemy, where every sympathiser is really an agent of some unfriendly power, and where the slightest deviation from the accepted line is an act of rankest treachery. . . .

The literature of exile has taken its place alongside the literature of adventure, education, and discovery as a *topos* of human experience. How did this come about? Is this the *same* exile that dehumanizes and often quite literally kills? Or is it some more benign variety?

The answer is the latter, I believe. As an element in the Christian and humanistic tradition of redemption though loss and suffering – and Western literature is part of this tradition – exile has played a consistent role . . . This redemptive view of exile is primarily religious, although it has been claimed by many cultures, political ideologies, mythologies, and traditions. Exile becomes the necessary precondition to a better state. We see this in stories about a nation's exile before statehood, a prophet's exile from home prior to a triumphant return, Moses, Mohammed, Jesus.

Much of the contemporary interest in exile can be traced to the somewhat pallid notion that non-exiles can share in the benefits of exile as a redemptive motif. There is no point in trying to dismiss this idea, because it has a certain plausibility and truth to it. Like medieval itinerant scholars of learned Greek slaves in the Roman Empire, exiles – the exceptional ones among them – do leaven their environment. And naturally 'we' concentrate on that enlightening aspect of 'their' presence among us, not on their misery or their demands. But looked at from the bleak political perspective of modern mass dislocations, individual exiles force us to reorganise the tragic fate of homelessness in a necessarily heartless world. . . .

There is the immense fact of isolation and displacement, which produces the kind of narcissistic masochism that resists all efforts at amelioration, acculturation, and community. At this extreme the exile can make a fetish of exile, a practice that distances him or her from all connections and commitments. To live as if everything around you were temporary and perhaps trivial is to fall prey to petulant cynicism as well as to querulous lovelessness. More common is the pressure on the exile to join – parties, national movements, the state. The exile is offered a new set of affiliations and develops new loyalties. But there is also a loss – of critical perspective, of intellectual reserve, of moral courage.

Is there some middle ground between these two alternatives? Before this can be answered, it must be recognised that the defensive nationalism of exiles often fosters self-awareness as much as it does the less attractive forms of self-assertion. By that I mean that such reconstitutive projects as assembling a nation out of exile (and this is true in this century for Jews and Palestinians) involve constructing a national history, reviving an ancient language, founding national institutions like libraries and universities. . . .

The exile knows that in a secular and contingent world, homes are always provisional. Borders and barriers, which enclose us within the safety of familiar territory, can also become prisons, and are often defended beyond reason or necessity. Exiles cross borders, break barriers of thought and experience.

Hugo of St. Victor, a twelfth-century monk from Saxony, wrote these hauntingly beautiful lines:

> It is, therefore, a source of great virtue for the practised mind to learn, bit by bit, first to change about invisible and transitory things, so that afterwards it may be able to leave them behind altogether. The man who finds his homeland sweet is still a

tender beginner; he to whom every soil is as his native one is already strong; but he is perfect to whom the entire world is a foreign land. The tender soul has fixed his love on one spot in the world; the strong man has extended his love to all places; the perfect man has extinguished his.

Erich Auerbach, the great twentieth-century literary scholar who spent the war years as an exile in Turkey, has cited this passage as a model for anyone wishing to transcend national or provincial limits. Only by embracing this attitude can a historian begin to grasp human experience and its written records in their diversity and particularity; otherwise he or she will remain committed more to the exclusions and reactions of prejudice than to the freedom that accompanies knowledge. But note that Hugo twice makes it clear that the 'strong' or 'perfect' man achieves independence and detachment by *working through* attachments, not by rejecting them. Exile is predicated on the existence of, love for, and bond with one's native place; what is true of all exile is not that home and love of home are lost, but that loss is inherent in the very existence of both.

Regard experiences *as if* they were about to disappear. What is it that anchors them in reality? What would you save of them? What would you give up? Only someone who has achieved independence and detachment, someone whose homeland is 'sweet' but whose circumstances make it impossible to recapture that sweetness, can answer those questions. (Such a person would also find it impossible to derive satisfaction from substitutes furnished by illusion or dogma.)

This may seem like a prescription for an unrelieved grimness of outlook and, with it, a permanently sullen disapproval of all enthusiasm or buoyancy of spirit. Not necessarily. While it perhaps seems peculiar to speak of the pleasures of exile; there are some positive things to be said for a few of its conditions. Seeing 'the entire world as a foreign land' makes possible originality of vision. Most people are principally aware of one culture, one setting, one home; exiles are aware of at least two, and this plurality of vision gives rise to an aware-ness of simultaneous dimensions, an awareness that – to borrow a phrase from music – is *contrapuntal*.

For an exile, habits of life, expression, or activity in the new environment inevitably occur against the memory of these things in another environment. Thus both the new and the old environments are vivid, actual, occurring together contrapuntally. There is a unique pleasure in this sort of apprehension, especially if the exile is conscious of other contrapuntal juxtapositions that diminish orthodox judgement and elevate appreciative sympathy. There is also a particular sense of achievement in acting as if one were at home wherever one happens to be.

This remains risky, however: the habit of dissimulation is both wearying and nerve-racking. Exile is never the state of being satisfied, placid, or secure. Exile, in the words of Wallace Stevens, is 'a mind of winter' in which the pathos of summer and autumn as much as the potential of spring are nearby but unobtainable. Perhaps this is another way of saying that a life of exile moves according to a different calendar, and is less seasonal and settled than life at home. Exile is life led outside habitual order. It is nomadic, decentred, contrapuntal; but no sooner does one get accustomed to it than its unsettling force erupts anew.

## Chapter 100

# AVTAH BRAH

# THINKING THROUGH THE CONCEPT
# OF DIASPORA

From *Cartographies of Diaspora: Contesting Identities* London and New York:
Routledge, 1996.

FIRST, A NOTE ABOUT THE TERM 'DIASPORA'. The word derives from
the Greek – *dia*, 'through', and *speirein*, 'to scatter'. According to *Webster's Dictionary* in the
United States, diaspora refers to a 'dispersion from'. Hence the word embodies a notion of a
centre, a locus, a 'home' from where the dispersion occurs. It invokes images of multiple
journeys. The dictionary also highlights the word's association with the dispersion of the Jews
after the Babylonian exile. Here, then, is an evocation of a diaspora with a particular resonance
within European cartographies of displacement; one that occupies a particular space in the
European psyche, and is emblematically situated within Western iconography as the diaspora
*par excellence*. Yet, to speak of late twentieth-century diasporas is to take such ancient diasporas
as a point of departure rather than necessarily as 'models', or as what Safran (1991) describes
as the 'ideal type'. The dictionary juxtaposition of what the concept signifies in general as against
one of its particular referents, highlights the need to subject the concept to scrutiny, to consider
the ramifications of what it connotes or denotes, and to consider its analytical value.

At the heart of the notion of diaspora is the image of a journey. Yet not every journey can
be understood as diaspora. Diasporas are clearly not the same as casual travel. Nor do they
normatively refer to temporary sojourns. Paradoxically, diasporic journeys are essentially about
settling down, about putting roots 'elsewhere'. These journeys must be historicised if the concept of diaspora is to serve as a useful heuristic device. The question is not simply about who
*travels* but *when, how and under what circumstances*? What socio-economic, political and cultural
conditions mark the trajectories of these journeys? What regimes of power inscribe the formation of a specific diaspora? In other words, it is necessary to analyse what makes one diasporic
formation similar to or different from another: whether, for instance, the diaspora in question
was constituted through conquest and colonisation as has been the case with several European
diasporas. Or it might have resulted from the capture or removal of a group through slavery
or systems of indentured labour, as, for example in the formation respectively of African and
Asian diasporas in the Caribbean. Alternatively, people may have had to desert their home as
a result of expulsion and persecution, as has been the fate of a number of Jewish groups at
various points in history. Or they may have been forced to flee in the wake of political strife,
as has been the experience of many contemporary groups of refugees such as the Sri Lankans,
Somalis and Bosnian Muslims. Perhaps the dispersion occurred as a result of conflict and war,
resulting in the creation of a nation state on the territory previously occupied by another,

as has been the experience of Palestinians since the formation of Israel. On the other hand, a population movement could have been induced as part of global flows of labour, the trajectory of many, for example African-Caribbeans, Asians, Cypriots or Irish people in Britain.

If the circumstances of leaving are important, so, too, are those of arrival and settling down. How and in what ways do these journeys conclude, and intersect in specific places, specific spaces, and specific historical conjunctures? How and in what ways is a group inserted within the social relations of class, gender, racism, sexuality or other axes of differentiation in the country to which it migrates? The manner in which a group comes to be 'situated' in and through a wide variety of discourses, economic processes, state policies and institutional practices is critical to its future. This 'situatedness' is central to how different groups come to be relationally positioned in a given context. I emphasise the question of relational positioning for it enables us to begin to deconstruct the regimes of power which operate to differentiate one group from another; to represent them as similar or different: to include or exclude them from constructions of the 'nation' and the body politic; and which inscribe them as juridical, political, and psychic subjects. It is axiomatic that each empirical diaspora must be analysed in its historical specificity. But the issue is not one that is simply about the need for historicising or addressing the specificity of a particular diasporic experience, important though this is.

Rather, the *concept* of diaspora concerns the historically variable forms of *relationality* within and between diasporic formations. It is about relations of power that similarise and differentiate between and across changing diasporic constellations. In other words, the concept of diaspora centres on the *configurations of power which differentiate diasporas internally as well as situate them in relation to one another.*

Diasporas, in the sense of distinctive historical experiences, are often composite formations made up of many journeys to different parts of the globe, each with its own history, its own particularities. Each such diaspora is an interweaving of multiple travelling; a text of many distinctive and, perhaps, even disparate narratives. This is true, among others, of the African, Chinese, Irish, Jewish, Palestinian and South Asian diasporas. For example, South Asians in Britain have a different, albeit related, history to South Asians in Africa, the Caribbean, Fiji, South East Asia, or the USA. Given these differences, can we speak of a 'South Asian diaspora' other than as a mode of description of a particular cluster of migrations? The answer depends crucially upon how the relationship between these various components of the cluster is conceptualised.

I would suggest that it is the *economic, political and cultural specificities linking these components that the concept of diaspora signifies.* This means that these multiple journeys may configure into one journey via a *confluence of narratives* as it is lived and re-lived, produced, reproduced and transformed through individual as well as collective memory and re-memory. It is within this confluence of narrativity that 'diasporic community' is differently imagined under different historical circumstances. By this I mean that the identity of the diasporic imagined community is far from fixed or pre-given. It is constituted within the crucible of the materiality of everyday life; in the everyday stories we tell ourselves individually and collectively.

All diasporic journeys are composite in another sense too. They are embarked upon, lived and re-lived through multiple modalities: modalities, for example, of gender, 'race', class, religion, language and generation. As such, all diasporas are differentiated, heterogeneous, contested spaces, even as they are implicated in the construction of a common 'we'. It is important, therefore, to be attentive to the nature and type of processes in and through which the collective 'we' is constituted. Who is empowered and who is disempowered in a specific construction of the 'we'? How are social divisions negotiated in the construction of the 'we'?

What is the relationship of this 'we' to its 'others'? Who are these others? This is a critical question. It is generally assumed that there is a single dominant Other whose overarching omnipresence circumscribes constructions of the 'we'. Hence, there tends to be an emphasis on bipolar oppositions: black/white; Jew/Gentile, Arab/Jew; English/Irish; Hindu/Muslim. The centrality of a particular binary opposition as the basis of political cleavage and social division in a given situation may make it necessary, even imperative, to foreground it. The problem remains, however, as to how such binaries should be analysed. Binaries can all too readily be assumed to represent ahistorical, universal constructs. This may help to conceal the workings of historically specific socioeconomic, political and cultural circumstances that mark the terrain on which a given binary comes to assume its particular significance. That is, what are actually the effects of institutions, discourses and practices may come to be represented as immutable, trans-historical divisions. As a consequence, a binary that should properly be an object of deconstruction may gain acceptance as an unproblematic given.

It is especially necessary to guard against such tendencies at the present moment when the surfacing of old and new racisms, violent religious conflicts and the horrors of 'ethnic cleansing' make it all too easy to slide into an acceptance of contextually variable phenomena as trans-historical universalisms that are then presumed to be an inevitable part of human nature. On the contrary, the binary is a socially constructed category whose trajectory warrants investigation in terms of how it was constituted, regulated, embodied and contested, rather than taken as always already present. A bipolar construction might be addressed fruitfully and productively as an object of analysis and a tool of deconstruction; that is, as a means of investigating the conditions of its formation, its implication in the inscription of hierarchies, and its power to mobilise collectivities.

The point is that there are multiple others embedded within and across binaries, albeit one or more may be accorded priority within a given discursive formation. For instance, a discourse may be primarily about gender and, as such, it may centre upon gender-based binaries (although, of course, a binarised construction is not always inevitable). But this discourse will not exist in isolation from others, such as those signifying class, 'race', religion or generation. The specificity of each is framed in and through fields of representation of the other. What is at stake, then, is not simply a question of some generalised notion of, say, masculinity and femininity, but whether or not these representations of masculinity and femininity are racialised; how and in what ways they inflect class; whether they reference lesbian, gay, heterosexual or some other sexualities; how they feature age and generation; how and if they invoke religious authority. Binaries, thus, are intrinsically differentiated and unstable. What matters most is how and why, in a given context, a specific binary – e.g. black/white – takes shape, acquires a seeming coherence and stability, and configures with other constructions, such as Jew/Gentile or male/female. In other words, *how these signifiers slide into one another in the articulation of power.*

We may elaborate the above point with reference to racialised discourses and practices. The question then reformulates itself in terms of the relationship at a specific moment between different forms of racism. Attention is shifted to the forms in which class, gender, sexuality or religion, for instance, might figure within these racisms, and to the specific signifier(s) – colour, physiognomy, religion, culture, etc. – around which these differing racisms are constituted. An important aspect of the problematic will be the relational positioning of groups by virtue of these racisms. How, for instance, are African, Caribbean, South Asian and white Muslims differentially constructed within anti-Muslim racism in present-day Britain? Similarly, how are blacks, Chicanos, Chinese, Japanese, or South Koreans in the USA differentiated within its racialised formations? What are the economic, political, cultural and psychic effects of these differential racialisations on the lives of these groups? What are the implications of these effects in terms of how members of one racialised group might relate to those of another? Do these effects produce conditions that foster sympathetic identification and solidarity across groups or do they create divisions? Of central concern in addressing such questions are the power dynamics which usher racialised social relations and inscribe racialised modes of subject-ivity and identity. My argument . . . is that these racisms are not simply parallel racisms but are intersecting modalities of *differential racialisations marking positionality across articulating fields of power*.

## Chapter 101

# VIJAY MISHRA

## THE DIASPORIC IMAGINARY
## Theorizing the Indian diaspora

From 'The Diasporic Imaginary: Theorizing the Indian Diaspora' *Textual-Practice* 10(3) (Winter), 421–47, 1996.

> In the arcade of Hanuman House . . . there was already the evening assembly of old men . . . pulling at *clay cheelums* that glowed red and smelled of ganja and burnt sacking. . . . They could not speak English and were not interested in the land where they lived; it was a place where they had come for a short time and stayed longer than they expected. They continually talked of going back to India, but when the opportunity came, many refused, afraid of the unknown, afraid to leave the familiar temporariness.
>
> (Naipaul 1969: 193–4)

**I**N THE LARGER NARRATIVE of global migrations and diasporas I would want to situate a diaspora of which, in complex ways, I have been/am a part. This is the Indian diaspora of around nine million about which not much of a theoretical nature has been written. In the lead essay in the foundation issue of the journal *Diaspora*, William Safran for instance devotes a mere twelve lines to the Indian diaspora and not unnaturally oversimplifies the characteristics of this diaspora (Safran 1991: 83–99). Unlike most other diasporas whose first movement out of the homeland can no longer be established with absolute precision, the Indian diaspora presents us with a case history that has been thoroughly documented. This is largely because the Indian diaspora began as part of British imperial movement of labour to the colonies. The end of slavery produced a massive demand for labour in the sugar planta-tions and Indian indentured labourers were brought to Trinidad, Guyana, Surinam, Mauritius, Fiji and South Africa. There were also movements of labour to East Africa, Sri Lanka and Malaya to work on the railways, tea and rubber plantations respectively. This narrative of diasporic movement is, however, not continuous or seamless as there is a radical break between the older diasporas of classic capitalism and the mid- to late twentieth-century diasporas of advanced capital to the metropolitan centres of the Empire, the New World and the former settler colonies. Since these are two interlinked but historically separated diasporas, I would want to refer to them as the old ('exclusive') and the new ('border') Indian diasporas. Furthermore, I would want to argue that the old Indian diasporas were diasporas of exclu-sivism because they created relatively self-contained 'little Indias' in the colonies. The founding writer of the old Indian diaspora is, of course, V. S. Naipaul. The new diaspora of late capital (the diaspora of the border), on the other hand, shares characteristics with many other similar diasporas such as the Chicanos and the Koreans in the US. The new Indian diaspora is mediated in the works of Salman Rushdie, Hanif Kureishi, Meera Nair, Rohinton Mistry, M. G. Vassanji, Gurinder Chadha, Meera Syal and others. These writers/film-makers speak of a diaspora whose overriding characteristic is one of mobility. Where the diaspora of exclusivism transplanted Indian icons of spirituality to the new land – a holy Ganges here, a lingam or a coiled serpent there – the diaspora of the border kept in touch with India through family networks and marriage, generally supported by a state apparatus that encouraged family reunion. Diasporas of the border in these Western democracies are visible presences – 'we are seen, therefore we are', says the Chicano novelist John Rechy (1995: 113) – whose corporealities carry marks of their hyphenated subjectivities. But elsewhere too, in Fiji or in Singapore, the state insists on diasporic identifications as its citizens, for demographic calculations and, in Fiji, for racialized electoral rolls, are always ethnic subjects. In Singapore the government prides itself on its CMIO (Chinese-Malay-Indian-Other) model of ethnic taxonomy which valorizes and transcodes, along racially essentialist lines, the specificities of communal experience even as the nation-state struggles to establish the primacy of the transcendent Singaporean citizen. For these hyphenated bodies (in spite of the enlightened ethos of citizenry) an extreme form of double consciousness occurs whenever the views of the dominant community begin to coincide with the rhetoric of what Sartre once observed as the racist question about the presumed ultimate solution of diasporas: 'What do we do with them now?' For diasporas this question always remains a trace, a potentially lethal 'solution', around which their selves continue to be shaped. Before continuing with the archaeology of the Indian diaspora in some detail I would want to pause here to advance a general theory of the diasporic imaginary that could act as a theoretical template for the rest of this paper.

## The diasporic imaginary and its pretexts

The diasporic imaginary is a term I use to refer to any ethnic enclave in a nation-state that defines itself, consciously, unconsciously or because of the political self-interest of a racialized nation-state, as a group that lives in displacement (Clifford 1994: 310). I use the word 'imaginary' in both its original Lacanian sense and in its more flexible current usage, as found in the works of Slavoj Žižek. Žižek defines the imaginary as the state of 'identification with the image in which we appear likeable to ourselves, with the image representing "what we would like to be"' (1989: 105). In a subsequent application of this theory to the nation itself, Žižek connects the sublime idea of what he calls the 'Nation Thing' to the subject's imaginary identification with it. The 'nation' (as the 'Thing') is therefore accessible to a particular group of people of itself because it needs no particular verification of this 'Thing' called 'Nation' (1993: 210–12). For this group the 'nation' simply is (beyond any kind of symbolization). Now in this construction of the 'Nation Thing' the nation itself is a fiction since it is built around a narrative imaginatively constructed by its subjects. The idea of the homeland then becomes (and here the terms 'imaginary' and 'fantasy' do coalesce), as Renata Salecl has pointed out, '[a] fantasy structure, [a] scenario, through which society perceives itself as a homogeneous entity' (1994: 15). We can follow up Salecl to make a more precise connection between a general theory of homelands (which is what I have done so far) and a theory of the diasporic homeland. Salecl refers, after Lacan, to fantasy as something that is predicated upon the construction of desire around a particularly traumatic event. The fantasy of homeland is then linked, in the case of the diaspora, to that recollected moment when diasporic subjects feel they were wrenched from their mother(father)land. The cause may be the traumatized 'middle passage' of slave trade or the sailing ships (later steamships) of Indian indenture, but the 'real' nature of the disruption is not the point at issue here; what is clear is that the moment of 'rupture' is transformed into a trauma around an absence that because it cannot be fully symbolized becomes part of the fantasy itself. Sometimes the 'absence' is a kind of repression, a sign of loss, like the Holocaust for European Jews after the war, or the Ukrainian famine for the Ukrainian diaspora. To be able to preserve that loss, diasporas very often construct racist fictions of purity as a kind of jouissance, a joy, a pleasure around which anti-miscegenation narratives of homelands are constructed against the reality of the homelands themselves. Racist narratives of homelands are therefore part of the dynamics of diasporas, as imaginary homelands are constructed from the space of distance to compensate for a loss occasioned by an unspeakable trauma.

The hypermobility of postmodern capital and ideas, and especially their ready dissemination on electronic bulletin boards (the internet, etc.) have the effect of actually reinforcing ethnic absolutism because diasporas can now connect with the politics of the homeland even as they live elsewhere. The collapse of distance on the information highway of cyberspace and a collective sharing of knowledge about the homeland by diasporas (a sharing that was linked to the construction of nations as imagined communities in the first instance) may be addressed by examining the kind of work Amit S. Rai has done on the construction of Hindu identity (1995: 31–57). His research explores the new public sphere that the Indian diaspora (or any diaspora for that matter) now occupies as it becomes a conduit through which the conservative politics of the homeland may be presented as the desirable norm. In Rai's exploration of six internets – soc.culture.indian, alt.hindu, alt.islam, soc.culture.tamil, su.orig.india, and INET – he finds that many of the postings indicate a desire to construct India in purist terms. It is an India that is Hindu in nature and one in which secularism is simply a ruse to appease minorities. In its invocations of important Indian religious and cultural figures – Vivekananda, R. C. Dutt, etc. – the subtext is always a discourse of racial purity ('we must go to the root of the disease and cleanse the blood of all impurities', said Swami Vivekananda) and the sexual threat to Hindus posed by the Muslims in India. The double space occupied by the diaspora (multicultural hysteria within the US and rabid racial absolutism for the homeland) is summarized by Rai as follows:

> Finally, this textual construction of the diaspora can at the same time enable these diasporics to be 'affirmative action' in the United States and be against 'reservations' in India, to lobby for a tolerant pluralism in the West, and also support a narrow sectarianism in the East.
>
> (Rai 1995: 42)

Although Rai's conclusions may be suspect – the postings need not lead to the correlation he discovers – it should be clear from the foregoing that diasporas construct homelands in ways that are very different from people of the homelands themselves. For an Indian in the diaspora, for instance, India is a very different kind of homeland than for the Indian national. At the same time the nation-state as an 'imagined community' needs diasporas to remind it of what the idea of homeland is. Diasporic discourse of the homeland is thus a kind of return of the repressed for the nation-state itself, its pre-symbolic (imaginary) narrative, in which one sees a more primitive theorization of the nation itself. Thus both the Jewish and gypsy diasporas – two extreme instances of the diasporic imaginary – have been treated by nation-states with particular disdain because they exemplify in varying degrees characteristics of a past that nation-states want to repudiate. For Franz Liszt, the gypsy diaspora was a 'crisis for

Enlightenment definitions of civilization and nationalist definitions of culture' (Trumpener 1992: 860). The Jews, equally a problem but with an extensive sense of history and civilization, carried all the characteristics of an ethnic community *(ethnie)* and thus were both an earlier condition of the European nation-state as well as its nemesis (Smith 1986: 23–30, 117). If the gypsies were read as the absolute instance of a nomadic tribe ('a dirty gypsy' is a term of abuse in both Hungary and Romania), the profound historicity of the Jewish people gave the Jewish diaspora a specially privileged position in diasporic theory. Diasporic theory then uses the Jewish example as the ethnic model for purposes of analysis or at least as its point of departure. But Jewish diasporas were never totally exclusivist – 'not isolation from Christians but insulation from Christianity' was their motto, as Max Weinreich put it (quoted In Clifford 1994: 326) – and met the nation-state half-way in its border zones. Jewish 'homelands', for instance, were constantly being re-created: in Babylon, in the Rhineland, in Spain, in Poland and even in America with varying degrees of autonomy (Smith 1986: 117). Movement ceased to be from a centre (Israel/Palestine/Judaea) to a periphery and was across spaces of the 'border'. Against the evidence, Zionist politics interpreted the Jewish diaspora as forever linked to a centre and argued that every movement of displacement (from Spain to France, from Poland to America) carried within it the trauma of the original displacement (such as that from Judaea to Babylon). In retrospect one can see how readily such a logic would erase the idea of nation as 'palimpsestic text' and replace it with the idea of nation as a racially pure ethnic enclave. In a very signifi-cant manner, then, the model of the Jewish diaspora is now contaminated by the diasporization of the Palestinians in Israel and by the Zionist belief that a homeland can be artificially recon-structed without adequate regard to intervening history. The theoretical problematic posed here is not simply Zionist. In no less a novel than George Eliot's *Daniel Deronda*, the 'Jew' signified world-historical questions of exile.

We need to keep the Palestinian situation in mind in any theorization of diasporas even as we use the typology of the Jewish diaspora to situate and critique the imaginary construc-tion of a homeland as *central mythomoteur* of diaspora histories (see Said 1980). The reason for this is that displaced Palestinians and their enforced mobility force us to distinguish between the Zionist project of Israel and the historically deterritorialized experiences of Jewish people generally. Echoing Max Weinreich the latter point is made by Daniel Boyarin and Jonathan Boyarin (1993: 693–725), who reread the Jewish diaspora through a postcolonial discourse where Jewishness is seen as a disruptive sign in the mosaic of history and an affirmation of a democratic ethos of equality that does not privilege any particular ethnic community in a nation. Against the Zionist fictions of a heroic past and a distant land, the real history of dias-pora is always contaminated by the social processes that govern their lives. Indeed, the features of diasporas found in the writings of Gellner (1983), Hobsbawm (1982), Smith and Safran become an 'insufferable' aspect of their lives only when a morally bankrupt nation-state asks the question, 'What shall we do with them?' As the exemplary condition of late modernity, diasporas 'call into question the idea that a people must have a land in order to be a people' (Boyarin and Boyarin 1993: 718). Of course, the danger here is that diasporas may well become romanticized as *the* ideal social condition (though many multicultural nations must come to terms with it) in which communities are no longer persecuted. As long as there is a fascist fringe always willing to find racial scapegoats for the nation's own shortcomings and to chant 'Go home', the autochthonous pressures towards diasporic racial exclusivism will remain. To address real diasporas does not mean that the discourses which have been part of diaspora mythology (homeland ancient past, return and so on) will disappear overnight. Under a gaze that threatens their already precarious sense of the 'familiar temporariness', diasporas lose their enlightened ethos and retreat into discourses of ethnic purity that are always the 'imaginary' underside of their own constructions of the homeland.

## Chapter 102

# JAMES CLIFFORD

## DIASPORAS

From 'Diasporas' in *Cultural Anthropology* (9)3 302–38, 1994.

A N UNRULY CROWD OF descriptive/interpretive terms now jostle and converse in an effort to characterize the contact zones of nations, cultures, and regions: terms such as 'border,' 'travel,' 'creolization,' 'transculturation,' 'hybridity' and 'diaspora' (as well as the looser 'diasporic'). Important new journals, such as *Public Culture* and *Diaspora* (or the revived *Transition*), are devoted to the history and current production of transnational cultures. In his editorial preface to the first issue of *Diaspora*, Khachig Tölölian writes, 'Diasporas are the exemplary communities of the transnational moment.' But he adds that diaspora will not be privileged in the new journal devoted to 'transnational studies' and that 'the term that once described Jewish, Greek, and Armenian dispersion now shares meanings with a larger semantic domain that includes words like immigrant, expatriate, refugee, guest-worker, exile community, overseas community, ethnic community' (Tölölian 1991: 4–5). This is the domain of shared and discrepant meanings, adjacent maps and histories, that we need to sort out and specify as we work our way into a comparative practice of intercultural studies. . . .

A different approach would be to specify the discursive field diacritically. Rather than locating essential features, we might focus on diaspora's borders, on what it defines itself against. And, we might ask, what articulations of identity are currently being replaced by diaspora claims? It is important to stress that the relational positioning at issue here is a process not of absolute othering but rather of entangled tension. Diasporas are caught up with and defined against (1) the norms of nation-states and (2) indigenous, and especially autochthonous, claims by 'tribal' peoples.

The nation-state, as common territory and time, is traversed and, to varying degrees, subverted by diasporic attachments. Diasporic populations do not come from elsewhere in the same way that 'immigrants' do. In assimilationist national ideologies such as those of the United States, immigrants may experience loss and nostalgia, but only en route to a whole new home in a new place. Such ideologies are designed to integrate immigrants, not people in diasporas. Whether the national narrative is one of common origins or of gathered populations, it cannot assimilate groups that maintain important allegiances and practical connections to a homeland or a dispersed community located elsewhere. Peoples whose sense of identity is centrally defined by collective histories of displacement and violent loss cannot be 'cured' by merging into a new national community. This is especially true when they are the victims of ongoing, structural prejudice. Positive articulations of diaspora identity reach outside the normative territory and temporality (myth/history) of the nation-state.

But are diaspora cultures consistently antinationalist? What about their own national aspirations? Resistance to assimilation can take the form of reclaiming another nation that has been lost, elsewhere in space and time, but that is powerful as a political formation here and now. There are, of course, antinationalist nationalisms, and I do not want to suggest that diasporic cultural politics are somehow innocent of nationalist aims or chauvinist agendas. Indeed, some of the most violent articulations of purity and racial exclusivism come from diaspora populations. But such discourses are usually weapons of the (relatively) weak. It is important to distinguish nationalist critical longing, and nostalgic or eschatological visions, from actual nation building – with the help of armies, schools, police, and mass media. 'Nation' and 'nation-state' are not identical. A certain prescriptive antinationalism, now intensely focused by the Bosnian horror, need not blind us to differences between dominant and subaltern claims. Diasporas have rarely founded nation-states: Israel is the prime example. And such 'homecomings' are, by definition, the negation of diaspora.

Whatever their ideologies of purity, diasporic cultural forms can never, in practice, be exclusively nationalist. They are deployed in transnational networks built from multiple attachments, and they encode practices of accommodation with, as well as resistance to, host countries and their norms. Diaspora is different from travel (though it works through travel practices) in that it is not temporary. It involves dwelling, maintaining communities, having collective homes away from home (and in this it is different from exile, with its frequently individualist focus). Diaspora discourse articulates, or bends together, both roots *and* routes to construct what Gilroy (1987) describes as alternate public spheres, forms of community consciousness and solidarity that maintain identifications outside the national time/space in order to live inside, with a difference. Diaspora cultures are not separatist, though they may have separatist or irredentist moments. . . .

Dispersed tribal peoples, those who have been dispossessed of their lands or who must leave reduced reserves to find work, may claim 'diasporic' identities. Inasmuch as their distinctive sense of themselves is oriented toward a lost or alienated home defined as aboriginal (and thus 'outside' the surrounding nation-state), we can speak of a diasporic dimension of contemporary tribal life. Indeed, recognition of this dimension has been important in disputes about tribal membership. The category 'tribe,' which was developed in U.S. law to distinguish settled Indians from roving, dangerous 'bands,' places a premium on localism and rootedness. Tribes with too many members living away from the homeland may have difficulty asserting their political/cultural status. This was the case for the Mashpee, who in 1978 failed to establish continuous 'tribal' identity in court (Clifford 1988: 277–346).

Thus, when it becomes important to assert the existence of a dispersed people, the language of diaspora comes into play, as a moment or dimension of tribal life. All communities, even the most locally rooted, maintain structured travel circuits, linking members 'at home' and 'away.' Under changing conditions of mass communication, globalization, postcolonialism, and neocolonialism, these circuits are selectively restructured and rerouted according to *internal and eternal* dynamics. Within the diverse array of contemporary diasporic cultural forms, tribal displacements and networks are distinctive. For in claiming both autochthony and a specific, transregional worldliness, new tribal forms bypass an opposition between rootedness and displacement – an opposition underlying many visions of modernization seen as the inevitable destruction of autochthonous attachments by global forces. Tribal groups have, of course, never been simply 'local': they have always been rooted and routed in particular landscapes, regional and interregional networks. What may be distinctively *modern*, however, is the relentless assault on indigenous sovereignty by colonial powers, transnational capital, and emerging nation-states. If tribal groups survive, it is now frequently in artificially reduced and displaced conditions, with segments of their populations living in cities away from the land, temporarily or even permanently. In these conditions, the older forms of tribal cosmopolitanism (practices of travel, spiritual quest, trade, exploration, warfare, labor migrancy, visiting, and political alliance) are supplemented by more properly diasporic forms (practices of long-term dwelling away from home). The permanence of this dwelling, the frequency of returns or visits to homelands, and the degree of estrangement between urban and landed populations vary considerably. But the specificity of tribal diasporas, increasingly crucial *dimensions* of collective life, lies in the relative proximity and frequency of connection with land-based communities claiming autochthonous status. . . .

## The currency of diaspora discourses

The language of diaspora is increasingly invoked by displaced peoples who feel (maintain, revive, invent) a connection with a prior home. This sense of connection must be strong enough to resist erasure through the normalizing processes of forgetting, assimilating, and distancing. Many minority groups that have not previously identified in this way are now reclaiming diasporic origins and affiliations. What is the currency, the value and the contemporaneity, of diaspora discourse?

Association with another nation, region, continent, or world-historical force (such as Islam) gives added weight to claims against an oppressive national hegemony. Like tribal assertions of sovereignty, diasporic identifications reach beyond mere ethnic status within the composite, liberal state. The phrase 'diasporic community' conveys a stronger sense of difference than, for example, 'ethnic neighborhood' used in the language of pluralist nationalism. This stronger difference, this sense of being a 'people' with historical roots and destinies outside the time/space of the host nation, is not separatist. (Rather, separatist desires are just one of its moments.) Whatever their eschatological longings, diaspora communities are 'not-here to stay.' Diaspora cultures thus mediate, in a lived tension, the experiences of separation and entanglement, of living here and remembering/desiring another place. If we think of displaced populations in almost any large city, the transnational urban swirl recently analyzed by Ulf Hannerz (1992), the role for mediating cultures of this kind will be apparent.

Diasporic language appears to be replacing, or at least supplementing, minority discourse. Transnational connections break the binary relation of 'minority' communities with 'majority' societies – a dependency that structures projects of both assimilation and resistance. And it gives a strengthened spatial/historical content to older mediating concepts such as W. E. B. Du Bois's notion of 'double consciousness.' Moreover, diasporas are not exactly immigrant communities. The latter could be seen as temporary, a site where the canonical three generations struggled through a hard transition to ethnic American status. But the 'immigrant' process never worked very well for Africans, enslaved or free, in the New World. And the so-called new immigrations of non-European peoples of color similarly disrupt linear assimilation narratives (see especially Schiller *et al.* 1992). Although there is a range of acceptance and alienation associated with ethnic and class variations, the masses of these new arrivals are kept in subordinate positions by established structures of racial exclusion. Moreover, their immigration often has a less all-or-nothing quality, given transport and communications technologies that facilitate multilocale communities. (On the role of television, see Naficy 1991.) Large sections of New York City, it is sometimes said, are 'parts of the Caribbean,' and vice versa (Sutton and Chaney 1987). Diasporist discourses reflect the sense of being part of an ongoing transnational network that includes the homeland not as something simply left behind but as a place of attachment in a contrapuntal modernity.

Diaspora consciousness is thus constituted both negatively and positively. It is constituted negatively by experiences of discrimination and exclusion. The barriers facing racialized sojourners are often reinforced by socioeconomic constraints, particularly – in North America – the development of a post-Fordist, nonunion, low-wage sector offering very limited opportunities for advancement. This regime of 'flexible accumulation' requires massive transnational flows of capital and labor – depending on, and producing, diasporic populations. Casualization of labor and the revival of outwork production have increased the proportion of women in the workforce, many of them recent immigrants to industrial centers (see Cohen 1987; Harvey 1989; Mitter 1986; Potts 1990; Sassen-Koob 1982). These developments have produced an increasingly familiar mobility 'hourglass' – masses of exploited labor at the bottom and a very narrow passage to a large, relatively affluent middle and upper class (Rouse 1991: 13). New immigrants confronting this situation, like the Aguillans in Redwood City, may establish transregional identities, maintained through travel and telephone circuits, that do not stake everything on an increasingly risky future in a single nation. It is worth adding that negative experience of racial and economic marginalization can also lead to new coalitions: one thinks of Maghrebi diasporic consciousness uniting Algerians, Moroccans, and Tunisians living in France, where a common history of colonial and neocolonial exploitation contributes to new solidarities. And the moment in 1970s Britain when the exclusionist term 'black' was appropriated to form antiracial alliances between immigrant South Asians, Afro-Caribbeans, and Africans provides another example of a negative articulation of diaspora networks.

Diaspora consciousness is produced positively through identification with world-historical cultural/political forces, such as 'Africa' or 'China.' The process may not be as much about being African or Chinese as about being American or British or wherever one has settled, differently. It is also about feeling global. Islam, like Judaism in a predominantly Christian culture, can offer a sense of attachment elsewhere, to a different temporality and vision, a discrepant modernity. I'll have more to say below about positive, indeed utopian, diasporism in the current transnational moment. Suffice it to say that diasporic consciousness 'makes the best of a bad situation.' Experiences of loss, marginality, and exile (differentially cushioned by class) are often reinforced by systematic exploitation and blocked advancement. This constitutive suffering coexists with the skills of survival: strength in adaptive distinction, discrepant cosmopolitanism, and stubborn visions of renewal. Diaspora consciousness lives loss and hope as a defining tension.

## Chapter 103

# AMITAVA KUMAR

## PASSPORT PHOTOS

From *Passport Photos* Berkeley: University of California Press, 2000.

Name
Place of Birth
Date of Birth
Profession
Nationality
Sex
Identifying Marks

**M**Y PASSPORT PROVIDES NO INFORMATION about my language. It simply presumes I have one.

If the immigration officer asks me a question – his voice, if he's speaking English, deliberately slow, and louder than usual – I do not, of course, expect him to be terribly concerned about the nature of language and its entanglement with the very roots of my being. And yet it is in language that all immigrants are defined and in which we all struggle for an identity. That is how I understand the postcolonial writer's declaration about the use of a language like English that came to us from the colonizer:

> Those of us who do use English do so in spite of our ambiguity towards it, or perhaps because of that, perhaps because we can find in that linguistic struggle a reflection of other struggles taking place in the real world, struggles between the cultures within ourselves and the influences at work upon our societies. To conquer English may be to complete the process of making ourselves free.
>
> (Rushdie 1981)

I also do not expect the immigration officer to be very aware of the fact that it is in that country called language that immigrants are reviled. I'd like to know what his thoughts were when he first heard the Guns N' Roses song:

> Immigrants
> and faggots
> They make no sense to me
> They come to our country –
> And think they'll do as they please

Like start some mini-Iran
Or spread some fuckin' disease.
            (quoted in Mills 1996)

. . . As the Swiss linguist Ferdinand de Saussure argued very early in this century, language is a system of signs. And any sign consists of a signifier (the sound or written form) and a signified (the concept). As the two parts of the sign are linked or inseparable (the word 'camera,' for instance, accompanies the concept 'camera' and remains quite distinct in our minds from the concept 'car'), what is prompted is the illusion that language is transparent. The relationship between the signifier and the signified, and hence language itself, is assumed to be natural.

When we use the word 'alien' it seems to stick rather unproblematically and unquestioningly to something or someone, and it is only by a conscious, critical act that we think of something different. Several years ago, in a public speech, Reverend Jesse Jackson seemed to be questioning the fixed and arbitrary assumptions in the dominant ideology when he reminded his audience that undocumented Mexicans were not aliens, they were *migrant workers*.

E. T., Jackson said emphatically, was an *alien*. . . .

When you turn to me in the bus or the plane and talk to me – *if* you talk to me – you might comment, trying to be kind, 'Your English is very good.'

If I am feeling relaxed, and the burden of the permanent chip on my shoulder seems light, I will smile and say, 'Thank you' (I never add, 'So is yours'). Perhaps I will say, 'Unfortunately, the credit goes to imperialism. The British, you know . . .' (Once, a fellow traveler widened her eyes and asked, 'The British still rule over *India?*').

It was the British who, in the first half of the nineteenth century, under the imperative of Lord Macaulay, introduced the systematic teaching of English in India in order to produce a class of clerks. In Rushdie's novel *The Moor's Last Sigh*, a painter by the name of Vasco Miranda drunkenly upbraids the upper-class Indians as 'Bleddy Macaulay's Minutemen. . . . Bunch of English-Medium misfits, the lot of you. . . . Even your bleddy dreams grow from foreign roots.' Much later in the novel, the protagonist, Moor Zogoiby, reflects on Macaulay's legacy as he is leaving for the last time the city of his birth, Bombay:

> *To form a class*, Macaulay wrote in the 1835 Minute on Education, . . . of *persons, Indian in blood and colour; but English in opinions, in morals, and in intellect*. And why, pray? O, to be *interpreters between us and millions whom we govern*. How grateful such a class of persons should, and must, be! For in India the dialects were *poor and rude, and a single shelf of a good European library was worth the whole native literature*. History, science, medicine, astronomy, geography, religion were likewise derided. *Would disgrace an English farrier . . . would move laughter in girls at an English boarding-school*.
> 
>                                                                     (Rushdie 1995)

This historical reverie is an occasion for Zogoiby to declare retrospective judgment on the drunken Miranda, to assure the reader and posterity that 'We were not, had never been, that class. The best, and worst, were in us, and fought in us, as they fought in the land at large. In some of us, the worst triumphed; but still we could say – and truthfully – that we had loved the best' (ibid.).

But what is it that was judged the best – in English?

The answer to that question can be sought in the pages of another Third-World writer, Michelle Cliff, who in writing about a Jamaican childhood describes how the schoolteacher's manual, shipped year after year from the London offices, directed the teacher

> to see that all in the school memorized the 'Daffodils' poem by William Wordsworth. . . . The manual also contained a pullout drawing of a daffodil, which the pupils were 'encouraged to examine' as they recited the verse. [Cliff rightly adds,] No doubt the same manuals were shipped to villages in Nigeria, schools in Hong Kong, even settlements in Northwest Territory – anywhere that 'the sun never set'. . . . Probably there were a million children who could recite 'Daffodils,' and a million more who had actually never seen the flower, only the drawing, and so did not know why the poet had been stunned.
>
> (Cliff 1984)

I was one of those children, though I cannot remember being shown even a drawing of the flower! And this in an independent nation, still unable to shrug off, when it comes to education in English, its colonial heritage. I can, therefore, understand the critique of that education lying at the heart of the Jamaican cultural theorist Stuart Hall's observation: 'When I first got to England in 1951 I looked out and there were Wordsworth's daffodils. Of course, what else would you expect to find? That's what I knew about. That is what trees and flowers meant. *I didn't know the names of the flowers I had left behind in Jamaica*' (Hall 1997). In some ways, admittedly, we cannot speak of the postcolonial experience as only limited to the idea of the absent daffodil. A more adequate representation of that experience would encompass, at the same time, that moment when the Indian child thinks of the daffodil as a bright marigold – or when, as in Cliff's novel, a student in the Caribbean colors it 'a deep red like a hibiscus. The red of a flame' (Cliff 1984). In that instant, which I can only call one of creative appropriation, language does not remain an instrument of cultural domination. It is transformed, knowingly or unknowingly, into a weapon of protest.

But let me return to that particular moment when, as his plane banks over the smoky landscape of Bombay, Moor Zogoiby finds himself thinking of the doings of the Indian elite in the past ('In some of us, the worst triumphed; but still we could say – and truthfully – that we had loved the best') (Rushdie 1995). His thought can also be seen as a protest. His Bombay was a Bombay that is no longer. For his Bombay was, as he says, 'a city of mixed-up, mongrel joy' (ibid.). That vision of the city is in direct conflict with the Bombay, or Mumbai as it has now been renamed, of the right-wing, Hindu rule of the Shiv Sena in Maharashtra. The Shiv Sena's vision of Mumbai is essentially a purist one. It is intolerant of those that fall outside its own, arbitrary, even atavistic, frame of reference. What Zogoiby seems to be savoring in his past, as he leaves his city behind him with no companion other than a stuffed mutt by the name of Jawaharlal, is the kind of modern liberal-democratic vision we associate with an earlier India led by Nehru. 'Unlike many other nationalists who had come to a sense of their Indianness through the detour of the West,' Sunil Kilhnani points out, 'there is no trace in Nehru of that inwardly turned rage of an Aurobindo or Vivekananda, political intellectuals who strove to purge themselves of what they came to regard as a defiling encounter with the modern West – an encounter that had first planted in them the urge to be Indian' (1997). Against the memory of Jawaharlal Nehru, the mongrel visionary, we have the reality of the Shiv Sena supremo, Bal Thackeray, who lists Hitler among his models.

And yet there is one detail that deserves commentary. The Shiv Sena leader's own name owes its origins to his Hindu father's admiration for William Makepeace Thackeray. I recall that detail not in order to point out that the Shiv Sena leader is hypocritical – though he might be that, and he can certainly be accused of much else – but to point out that, in the postcolonial condition, contradictions are inescapable. . . .

. . . In 1896 a colonial official argued against the restrictions imposed on the entry of Indians in South Africa, adding that this would be 'most painful' for Queen Victoria to approve. At the same time, he sanctioned a European literacy test that would automatically exclude Indians while preserving the facade of racial equality (quoted in Jensen 1988).

Almost a hundred years later a Texas judge ordered the mother of a five-year-old to stop speaking in Spanish to her child. Judge Samuel Kiser reminded the mother that her daughter was a 'full-blooded American.' 'Now get this straight. You start speaking in English to this child because if she doesn't do good in school, then I can remove her because it's not in her best interest to be ignorant. The child will only hear English' (quoted in Verhovek 1995).

Who is permitted to proceed beyond the gates into the mansion of full citizenship? And on what terms? These are the questions that the episode in the Texas courthouse raises. Apart from the issue of gross paternalism and an entirely injudicious jingoism, what comes into play here is the class bias in North American society that promotes bilingualism in the upper class but frowns on it when it becomes an aspect of lower-class life.

More revealing of the ties between language and U.S. Immigration is the following newspaper report: 'School and city officials expressed outrage this week over the Border Patrol's arrest of three Hispanic students outside an English as Second Language class' (Arteaga 1994).

For the Chicano poet Alfred Arteaga, the above story about arrest and deportation has a double irony: 'irony, not only that "officials expressed outrage" at so typical an INS action, but irony also, that the story made it into print in the first place' (ibid.). Arteaga knows too well that what Chicanos say and do in their own language is rarely found worthy of printing.

I think it is equally significant to remark on the fact that the officers who conducted the arrest were patrolling the borders of the dominant language to pick up the illegals. They are ably assisted by the likes of the California state assemblyman William J. Knight, who distributed among his fellow legislators a poem, 'I Love America.' That poem begins with the words 'I come for visit, get treated regal,/So I stay, who care illegal.' This little ditty makes its way through the slime of a racist fantasy. Its landscape is filled with greedy swindlers and dishonest migrant workers. The breeding subhumans speak in a broken syntax and mispronounce the name Chevy, the heartbeat of America, as (call the National Guard, please!) Chebby. The poem ends with a call that emanates like a howl from the guts of the Ku Klux Klan:

> We think America damn good place,
> Too damn good for white man's race.
> If they no like us, they can go,
> Got lots of room in Mexico.

If the immigration officer were to ask me about my language, what would I say? That any precious life-giving sense of language loses all form in this arid landscape of Buchanan-speak? Perhaps. That any answer I could possibly give is nothing more defined than a blur moving on the infrared scopes of those guarding the borders of fixed identity. . . .

And yet, while speaking of the patrolling of the borders of dominant identity, I must note the presence of one who is still eluding arrest: a border-artist/poet-performer/hoarder-of-hyphens/warrior-for-Gringostroika. Officer, meet Guillermo Gómez-Peña. You have been looking for him not only because Gómez-Peña declares 'I speak in English therefore you listen/I speak in English therefore I hate you' (quoted in Nancy 1994). But also because, like a 'Pablo Neruda gone punk,' this 'border brujo' threatens mainstream America with the swaggering banditry of language, demanding as ransom a pure reality-reversal:

> What if the U.S. was Mexico?
> What if 200,000 Anglo-Saxicans
> Were to cross the border each month
> to work as gardeners, waiters
> 3rd chair musicians, movie extras
> bouncers, babysitters, chauffeurs
> syndicated cartoons, feather-weight boxers, fruit-pickers
> and anonymous poets?
> What if they were called Waspanos
> Waspitos, Wasperos or Waspbacks?
> What if literature was life, eh?

# PART SEVENTEEN

# Globalization

G LOBALIZATION IS THE PROCESS WHEREBY individual lives and local communities are affected by economic and cultural forces that operate worldwide. It is the process of the world shrinking, becoming a single place. The term has had a meteoric rise since the mid-1980s, before which time words such as 'international' and 'international relations' had been preferred. The rise of the word 'international' itself in the eighteenth century indicated the growing importance of territorial states in organizing social relations, and is an early consequence of the global perspective of European imperialism. Similarly, the rapidly increasing interest in globalization reflects a changing organization of worldwide social relations in this century, one in which the 'nation' has begun to have a decreasing import- ance as individuals and communities gain access to globally disseminated knowledge and culture, and are affected by economic realities that bypass the boundaries of the state. The structural aspects of globalization are the nation-state system itself (on which the concepts of internationalism and international co-operation are based), global economy, the global communication system and world military order.

Theories of globalization have moved from expressions of the process as 'cultural imperialism' or neo-imperialism to analyses of the 'hybridization', 'diffusion', 'relativization,' and interrelationship of global societies, the 'compression of the world and the intensification of the consciousness of the world as a whole' (Robertson 1992: 8). Globalization discourse was dominated in the 1980s by sociology and political economy, but during the 1990s, as Simon Gikandi suggests, post-colonial studies, with its provenance in literary studies, provided a range of terms such as hybridity, transculturation, 'Third Space', appropriation and trans- formation that came to dominate discussion of cultural globalization. The relationship with classical imperialism has, despite considerable developments in globalization theory, continued to interest many theorists. This is because analyses of imperialism itself in post-colonial studies have become much more sophisticated. Like globalization, it may be more useful to see that imperialism is not simply a conscious and active ideology, but a combination of conscious ideological programmes and unconscious 'rhizomic' structures of unprogrammed connections and engagements (Ashcroft 2001: 50). This interaction and circulation is precisely the way in which the global is produced.

The importance of globalization to post-colonial studies comes first from its demonstration of the structure of world power relations, which stands firm in the twentieth century as a legacy of Western imperialism. Second, the ways in which local communities engage the forces of globalization bear some resemblance to the ways in which colonized societies have historically engaged and appropriated the forces of imperial dominance. Several extracts in this section (Dirlik, Appadurai, Robertson) are exercised by the link between the local and global and the agency of local subjects in appropriating, transforming and consuming global phenomena. Interestingly, the production and consumption of literatures provides an excellent model for the processes of local engagement with the global, as well as illuminating the similar ways in which imperialism and globalization provide cultural capital that is exploited by local communities.

By appropriating strategies of representation, organization and social change through access to global systems, local communities and marginal interest groups can both empower themselves and influence those global systems. Although choice is always mediated by the conditions of subject formation, the belief that one has a choice in the processes of changing one's own life or society can indeed be empowering. In this sense the appropriation of global forms of culture may free one from local forms of dominance and oppression or at least provide the tools for a different kind of identity formation. On the other hand, Hardt and Negri's influential work on the broader metaphor of empire to describe the new forms of control exercised by international capital, suggests a continuity of oppression. They focus specifically on the role of these global institutions promoted by the United States as it emerged as the sole world power in the period after the collapse of the Soviet Union. Their work has been increasingly influential on those working in the area of the intersection of colonialism, neo-colonial forms of power and the institutions of global control of more recent times. Chandra Talpade Mohanty's essay charts a trajectory that signals a typical development in post-colonial studies. Her important essay 'Under Western Eyes' included in the Feminism section examines the relation of feminist scholarship and colonial discourses. This revisiting of that original thesis in 'Feminist Solidarity through Anticapitalist Struggles' is recast in a way that demonstrates the profound importance of global structures of capital flows and labour commodification. Whether or not Simon Gikandi's claim about the importance of post-colonial discourse to globalization proves correct, there is no doubt that globalization will continue to frame the arena of post-colonial analysis.

## Chapter 104

# ARIF DIRLIK

# THE GLOBAL IN THE LOCAL

From *The Postcolonial Aura: Third World Criticism in the Age of Global Capitalism* Boulder, Col.: Westview Press, 1997.

**M**Y PRIMARY CONCERN IS WITH the local as site of promise, and the social and ideological changes globally that have dynamized a radical re-thinking of the local over the last decade. I am interested especially in the relationship between the emergence of a Global Capitalism and the emergence of concern with the local as a site of resistance and liberation. Consideration of this relationship is crucial, it seems to me, to distinguishing a 'critical localism' from localism as an ideological articulation of capitalism in its current phase. Throughout, however, I try also to remain cognizant of the local as a site of predicament. In its promise of liberation, localism may also serve to disguise oppression and parochialism. It is indeed ironic that the local should emerge as a site of promise at a historical moment when localism of the most conventional kind has reemerged as the source of genocidal conflict around the world. The latter, too, must surely enter any consideration of the local as site of resistance to and liberation from oppression. In either case, the local that is at issue here is not the 'local' in any conventional or traditional sense, but a very contemporary 'local' that serves as a site for the working out of the most fundamental contradictions of the age.

## Rethinking the local

It is too early, presently, to sort out the factors that have contributed to the ascendancy of a concern with the local over the last decade, and any such undertaking must of necessity be highly speculative. What the 'local' implies in different contexts is highly uncertain. Suffice it to say here that a concern for the local seems to appear in the foreground in connection with certain social movements (chief among them ecological, women's, ethnic and indigenous people's movements) and the intellectual repudiation of past ideologies (chief among them, for the sake of brevity here, the intellectual developments associated with, or that have gone into the making of, 'postmodernism').

Why there should be a connection between the repudiation of past ideologies and the reemergence of the local as a concern is not very mysterious. Localism as an orientation in either a 'traditional' or a modern sense has never disappeared, but rather has been suppressed or, at best, marginalized in various ideologies of modernity. Localism does not speak of an incurable social disease that must sooner or later bring about its natural demise; and there is nothing about it that is inherently undesirable. What makes it seem so is a historical consciousness that identifies civilization and progress with political, social and cultural homogenization, and justifies the suppression of the local in the name of the general and the universal. Modernist teleology has gone the farthest of all in stamping upon the local its derogatory image: as enclaves of backwardness left out of progress, as the realm of rural stagnation against the dynamism of the urban, industrial civilization of capitalism, as the realm of particularistic culture against universal scientific rationality and, perhaps most importantly, as the obstacle to full realization of that political form of modernity, the nation-state (see Kropotkin 1975).

This teleology has been resisted not only in the name of 'traditional' localism that sought to preserve received forms of local society, but by radical critics of modernity as well. Anti-modernism rendered the local into a refuge from the ravages of modernity. . . . Third World revolutions in the twentieth century would perpetuate these concerns for local society; especially those revolutions which, compelled by force of circumstances to pursue agrarian strategies of revolution, had to face local societies and their participation in revolution as a condition of revolutionary success. In these cases, ironically, local society would also emerge as a source of national identity, against the cosmopolitanism of urban centers drawn increasingly into the global culture of capitalism (see Dirlik 1991; 1997b).

The teleology of modernity, nevertheless, was to emerge victorious in the twentieth century over earlier socialist doubts about its consequences. The concern for the local persisted in the thinking of agrarian utopians and anarchists, but they, too, were to be marginalized for their insistence on the continued relevance of the local (see Illich 1981; Bookchin 1971). In the immediate decades after World War II, the modernizationist repudiation of the local prevailed in both bourgeois and Marxist social science.

It is not surprising, therefore, that the local should appear in contemporary discourse hand-in-hand with the repudiation of modernist teleology, the rejection as ideology of the 'metanarratives' which have framed the history of modernization, whether capitalist or socialist. 'Postmodernism,' which has been described as 'incredulity toward metanarratives,' provides a convenient if loose term for characterizing the various challenges to modernist teleology, not because those who do so think of themselves as 'postmodernists,' but because every such challenge in its own way contributes to the making of a postmodern consciousness. Be that as it may, the repudiation of modernist teleology implies that there is nothing natural or inherently desirable about modernization (in capitalist or socialist form), and that the narra-tive of modernization is a narrative of compelling into modernity those who did not necessarily wish to be modern. . . .

The repudiation of the metanarrative of modernization, and its redirection of attention to coercion over teleology in development, have had two immediate consequences. First, it rescues from invisibility those who were earlier viewed as castaways from history, whose social and cultural forms of existence appear in the narrative of modernization at best as irrele-vancies, at worst as minor obstacles to be extinguished on the way to development. Having refused to die a natural death, but instead come into self-awareness as victims of coercion, they demand now not just restoration of *their* history, further splintering the already cracked facade of modernity. The demand is almost inevitably accompanied by a reassertion of the local against the universalistic claims of modernism (see Means n.d.: 19–33; Clastres 1987).

The repudiation of the metanarrative of modernization, secondly, has allowed greater visibility to 'local narratives.' Rather than an inexorable march of global conquest from its origins in Europe, the history of modernization appears now as a temporal succession of spatially dispersed local encounters, to which the local objects of progress made their own con-tributions through resistance or complicity, contributing in significant ways to the formation of modernity, as well as to its contradictions. Also questioned in this view are the claims of nationalism which, a product itself of modernization, has sought to homogenize the societies it has claimed for itself, suppressing further such local encounters, and the 'heterogeneity' they imply (see Chatterjee 1986).

Were it simply an ideological phenomenon, the postmodern repudiation of the meta-narrative of modernity could be dismissed as a momentary loss of faith in modernity, another instance of those chronic failures of nerve that seem to attend moments of crisis in develop-ment, especially on occasions of transition, that will go away as soon as the transition has been completed and the crisis resolved. If so, this new round of anti-modernism might be at best a passive enabling condition that allows us to hear previously inaudible voices, that will be muted again as soon as the business of development is once again under way, with capitalism having disposed of the competition that for nearly a century shaped and 'distorted' its development. . . .

Even before the crisis of socialism became evident in the 1980s, and postmodernism became a household word, in other words, developments in social movements and in the relationships between the 'Three Worlds' called into question the spatial and temporal teleology of devel-opment, as well as the conceptual teleology that had characterized earlier radical thinking. On the other hand, it is necessary to note that whatever the material circumstances that rendered 'postmodernism' intelligible and plausible, it was the generation that came of age with these developments that was to play the crucial part in its articulation. The concern for the local (whether literally local, or in reference to the 'local' needs of social groups) gathered force simultaneously with the repudiation of teleology. I can do no more than suggest here that an ecological consciousness, which has done much to reassert the primacy of the local (as the most viable location for living in harmony with nature), was a product of the same circumstances, and obviously bore some relationship to the shift in social and political consciousness.

Development as maldevelopment; adjustment to nature against the urge to conquer it; the porosity of borderlands against the rigidity of political forms, in particular the nation-state; heterogeneity over homogeneity; overdetermination against categorically defined subject-ivities; ideology as culture, and culture as daily negotiation; enlightenment as hegemony; 'local knowledge' against universal scientific rationality; native sensibilities and spiritualities as a supplement to, if not a substitute for, reason; oral against written culture; political move-ments as 'politics of difference' and 'politics of location.' The list could go on. It enumerates elements of a postmodern consciousness that serve as enabling conditions for a contemporary localism, but also produce it. The consciousness itself is an articulation not of powerlessness, but of newfound power among social groups who demand recognition of their social exist-ence and consciousness against a modernity that had denied them a historical and, therefore, a political presence. Governor John D. Waihee III of Hawaii (1992), referring to the Hawaiian sovereignty movement, acknowledged recently that what seems possible today would have been unimaginable only two decades ago.

The suspicion of Enlightenment metanarratives for their denial of difference makes for a suspicion of all metanarratives which suppress or overlook differences, allows for localized consciousness, and points to the local as the site for working out 'alternative public spheres' and alternative social formations (Giroux 1992: 21–2). This is the promise held out by the local. The local, however, also indicates fragmentation and, given the issues of power involved, political and cultural manipulation as well. This is the predicament. That traces of earlier forms of exploitation and oppression persist in the local, albeit in forms worked over by modernity, aggravates the predicament.

This predicament becomes more apparent when we view the problem of the local from the perspective of the global: the local as object of the operations of capital, which provides the broadest context for inquiry into the sources and consequences of contemporary localism. The emergence of the concern for the local over the last two decades has accompanied a significant transformation within capitalism with far-reaching economic, political, social and cultural consequences. This transformation, and its implications for the local, need to be considered in any evaluation of the local as source of promise and predicament. . . .

The situation created by Global Capitalism helps explain certain phenomena that have become apparent over the past two to three decades, but especially since the eighties: global motions of peoples (and, therefore, cultures), the weakening of boundaries (among societies, as well as among social categories), the replication in societies internally of inequalities and discrepancies once associated with colonial differences, simultaneous homogenization and fragmentation within and across societies, the interpenetration of the global and the local (which shows culturally in a simultaneous cosmopolitanism and localism of which the most cogent expression may be 'multiculturalism'), and the disorganization of a world conceived in terms of 'three worlds' or nation-states. Some of these phenomena have also contributed to an appearance of equalization of differences within and across societies, as well as of democratization within and between societies. What is ironic is that the managers of this world situation themselves concede the concentration of power in their (or their organizations') hands; as well as their manipulation of peoples, boundaries and cultures to appropriate the local for the global, to admit different cultures into the realm of capital only to break them down and to remake them in accordance with the requirements of production and consumption, and even to reconstitute subjectivities across national boundaries to create producers and consumers more responsive to the operations of capital. Those who do not respond, or the 'basket-cases' which are not essential to those operations – four-fifths of the global population by their count – need not be colonized; they are simply marginalized. What the new 'flexible production' has made possible is that it is no longer necessary to utilize explicit coercion against labor, at home or abroad (in colonies); those peoples or places that are not responsive to the needs (or demands) of capital, or are too far gone to respond 'efficiently,' simply find themselves out of its pathways.

Much of what I have described above as the conditions for the production of contemporary localism (a postmodern consciousness, embedded in new forms of empowerment) appears in this perspective as a product of the operations of Global Capitalism. It should also be apparent from the above that the local is of concern presently not only to those who view it as a site of liberation struggles, but, with an even greater sense of immediacy, to managers of global capital as well as to those responsible for the economic welfare of their communities. . . .

The radical slogan of an earlier day, 'Think globally, act locally,' has been assimilated by transnational corporations with far greater success than in any radical strategy. The recognition of the local in marketing strategy, however, does not mean any serious recognition of the autonomy of the local, but is intended to recognize the features of the local so as to incorporate localities into the imperatives of the global. The 'domestication' of the corporation into local society serves only to further mystify the location of power, which rests not in the locality but in the global headquarters of the company which coordinates the activities of its local branches. . . .

From the perspective of Global Capitalism, the local is a site not of liberation but manipulation; stated differently, it is a site the inhabitants of which must be liberated from themselves (stripped of their identity) to be homogenized into the global culture of capital (their identities reconstructed accordingly). Ironically, even as it seeks to homogenize populations globally, consuming their cultures, Global Capitalism enhances awareness of the local, pointing to it also as the site of resistance to capital.

This is nevertheless the predicament of the local. A preoccupation with the local that leaves the global outside its line of vision is vulnerable to manipulation at the hands of global capital which of necessity commands a more comprehensive vision of a global totality. Differences of interest and power on the site of the local, which are essential to its reconstruction along non-traditional, democratic, lines, render the local all the more vulnerable to such manipulation as capital plays on these differences, and the advocates of different visions and interests seek to play capital against one another. The local in the process becomes the site upon which the multifaceted contradictions of contemporary society play out, where critique turns into ideology and ideology into critique, depending upon its location at any one fleeting moment.

## Chapter 105

# ARJUN APPADURAI

## DISJUNCTION AND DIFFERENCE

From *Modernity at Large: Cultural Dimensions of Globalization*
Minneapolis: University of Minnesota Press, 1996.

THE CENTRAL PROBLEM OF TODAY'S global interactions is the tension between cultural homogenization and cultural heterogenization. A vast array of empirical facts could be brought to bear on the side of the homogenization argument, and much of it has come from the left end of the spectrum of media studies (Hamelink 1983; Mattelart, 1983; Schiller 1976), and some from other perspectives (Gans 1985; Iyer 1988). Most often, the homogenization argument subspeciates into either an argument about Americanization or an argument about commoditization, and very often the two arguments are closely linked. What these arguments fail to consider is that at least as rapidly as forces from various metropolises are brought into new societies they tend to become indigenized in one or another way: this is true of music and housing styles as much as it is true of science and terrorism, spectacles and constitutions. The dynamics of such indigenization have just begun to be explored systemically (Barber 1987; Feld 1988; Hannerz 1987, 1989; Ivy 1988; Nicoll 1989; Yoshimoto 1989), and much more needs to be done. But it is worth noticing that for the people of Irian Jaya, Indonesianization may be more worrisome than Americanization, as Japanization may be for Koreans, Indianization for Sri Lankans, Vietnamization for the Cambodians, and Russianization for the people of Soviet Armenia and the Baltic republics. Such a list of alternative fears to Americanization could be greatly expanded, but it is not a shapeless inventory: for polities of smaller scale, there is always a fear of cultural absorption by polities of larger scale, especially those that are nearby. One man's imagined community is another man's political prison.

This scalar dynamic, which has widespread global manifestations, is also tied to the relationship between nations and states, to which I shall return later. For the moment let us note that the simplification of these many forces (and fears) of homogenization can also be exploited by nation-states in relation to their own minorities, by posing global commoditization (or capitalism, or some other such external enemy) as more real than the threat of its own hegemonic strategies.

The new global cultural economy has to be seen as a complex, overlapping, disjunctive order that cannot any longer be understood in terms of existing center-periphery models (even those that might account for multiple centers and peripheries). Nor is it susceptible to simple models of push and pull (in terms of migration theory), or of surpluses and deficits (as in traditional models of balance of trade), or of consumers and producers (as in most neo-Marxist theories of development). Even the most complex and flexible theories of global development that have come out of the Marxist tradition (Amin 1980; Mandel 1978;

Wallerstein 1974; Wolf 1982) are inadequately quirky and have failed to come to terms with what Scott Lash and John Urry have called disorganized capitalism (1987). The complexity of the current global economy has to do with certain fundamental disjunctures between economy, culture, and politics that we have only begun to theorize.

I propose that an elementary framework for exploring such disjunctures is to look at the relationship among five dimensions of global cultural flows that can be termed (a) *ethnoscapes*, (b) *mediascapes*, (d) *technoscapes*, (d) *financescapes*, and (e) *ideoscapes*. The suffix *-scape* allows us to point to the fluid, irregular shapes of these landscapes, shapes that characterize international capital as deeply as they do international clothing styles. These terms with the common suffix *-scape* also indicate that these are not objectively given relations that look the same from every angle of vision but, rather that they are deeply perspectival constructs, inflected by the historical, linguistic, and political situatedness of different sorts of actors: nation-states, multinationals, diasporic communities, as well as subnational groupings and movements (whether religious, political, or economic), and even intimate face-to-face groups, such as villages, neighborhoods, and families. Indeed, the individual actor is the last locus of this perspectival set of landscapes, for these landscapes are eventually navigated by agents who both experience and constitute larger formations, in part from their own sense of what these landscapes offer.

These landscapes thus are the building blocks of what (extending Benedict Anderson) I would like to call *imagined worlds*, that is, the multiple worlds that are constituted by the historically situated imaginations of persons and groups spread around the globe. . . . An important fact of the world we live in today is that many persons on the globe live in such imagined worlds (and not just in imagined communities) and thus are able to contest and sometimes even subvert the imagined worlds of the official mind and of the entrepreneurial mentality that surround them.

By *ethnoscape* I mean the landscape of persons who constitute the shifting world in which we live: tourists, immigrants, refugees, exiles, guest workers, and other moving groups and individuals constitute an essential feature of the world and appear to affect the politics of (and between) nations to a hitherto unprecedented degree. This is not to say that there are no relatively stable communities and networks of kinship, friendship, work, and leisure, as well as of birth, residence, and other filial forms. But it is to say that the warp of these stabilities is everywhere shot through with the woof of human motion, as more persons and groups deal with the realities of having to move or the fantasies of wanting to move. What is more, both these realities and fantasies now function on larger scales, as men and women from villages in India think not just of moving to Poona or Madras but of moving to Dubai and Houston, and refugees from Sri Lanka find themselves in South India as well as in Switzerland, just as the Hmong are driven to London as well as to Philadelphia. And as international capital shifts its needs, as production and technology generate different needs, as nation-states shift their policies on refugee populations, these moving groups can never afford to let their imaginations rest too long, even if they wish to.

By *technoscape*, I mean the global configuration, also ever fluid, of technology and the fact that technology, both high and low, both mechanical and informational, now moves at high speeds across various kinds of previously impervious boundaries. Many countries now are the roots of multinational enterprise: a huge steel complex in Libya may involve interests from India, China, Russia, and Japan, providing different components of new technological configurations. The odd distribution of technologies, and thus the peculiarities of these technoscapes, are increasingly driven not by any obvious economies of scale, of political control, or of market rationality but by increasingly complex relationships among money flows, political possibilities, and the availability of both un- and highly skilled labor. So, while India exports waiters and chauffeurs to Dubai and Sharjah, it also exports software engineers to the United States – indentured briefly to Tata-Burroughs or the World Bank, then laundered through the State Department to become wealthy resident aliens, who are in turn objects of seductive messages to invest their money and know-how in federal and state projects in India.

The global economy can still be described in terms of traditional indicators (as the World Bank continues to do) and studied in terms of traditional comparisons (as in Project Link at the University of Pennsylvania), but the complicated technoscapes (and the shifting ethno-scapes) that underlie these indicators and comparisons are further out of the reach of the queen of social sciences than ever before. How is one to make a meaningful comparison of wages in Japan and the United States or of real-estate costs in New York and Tokyo, without taking sophisticated account of the very complex fiscal and investment flows that link the two economies through a global grid of currency speculation and capital transfer?

Thus it is useful to speak as well of *financescapes*, as the disposition of global capital is now a more mysterious, rapid, and difficult landscape to follow than ever before, as currency markets, national stock exchanges, and commodity speculations move megamonies through national turnstiles at blinding speed, with vast, absolute implications for small differences in percentage points and time units. But the critical point is that the global relationship among ethnoscapes, technoscapes, and financescapes is deeply disjunctive and profoundly unpredict-able because each of these landscapes is subject to its own constraints and incentives (some political, some informational, and some technoenvironmental), at the same time as each acts as a constraint and a parameter for movements in the others. Thus, even an elementary model of global political economy must take into account the deeply disjunctive relationships among human movement, technological flow, and financial transfers.

Further refracting these disjunctures (which hardly form a simple, mechanical global infrastructure in any case) are what I call *mediascapes* and *ideoscapes,* which are closely related landscapes of images. *Mediascapes* refer both to the distribution of the electronic capabilities to produce and disseminate information (newspapers, magazines, television stations, and film-production studios), which are now available to a growing number of private and public interests throughout the world, and to the images of the world created by these media. These images involve many complicated inflections, depending on their mode (documentary or enter-tainment), their hardware (electronic or preelectronic), their audiences (local, national, or transnational), and the interests of those who own and control them. What is most import-ant about these mediascapes is that they provide (especially in their television, film, and cassette forms) large and complex repertoires of images, narratives, and ethnoscapes to viewers throughout the world, in which the world of commodities and the world of news and politics are profoundly mixed. What this means is that many audiences around the world experience the media themselves as a complicated and interconnected repertoire of print, celluloid, electronic screens, and billboards. The lines between the realistic and the fictional landscapes they see are blurred, so that the farther away these audiences are from the direct experi-ences of metropolitan life, the more likely they are to construct imagined worlds that are chimerical, aesthetic, even fantastic objects, particularly if assessed by the criteria of some other perspective, some other imagined world.

Mediascapes, whether produced by private or state interests, tend to be image-centered, narrative-based accounts of strips of reality, and what they offer to those who experience and transform them is a series of elements (such as characters, plots, and textual forms) out of which scripts can be formed of imagined lives, their own as well as those of others living in other places. These scripts can and do get disaggregated into complex sets of metaphors by which people live (Lakoff and Johnson 1980) as they help to constitute narratives of the Other and protonarratives of possible lives, fantasies that could become prolegomena to the desire for acquisition and movement.

*Ideoscapes* are also concatenations of images, but they are often directly political and frequently have to do with the ideologies of states and the counterideologies of movements explicitly oriented to capturing state power or a piece of it. These ideoscapes are composed of elements of the Enlightenment worldview, which consists of a chain of ideas, terms, and images, including *freedom*, *welfare*, *rights*, *sovereignty*, *representation*, and the master term *democracy*. The master narrative of the Enlightenment (and its many variants in Britain, France, and the United States) was constructed with a certain internal logic and presupposed a certain relationship between reading, representation, and the public sphere. (For the dynamics of this process in the early history of the United States, see Warner 1990.) But the diaspora of these terms and images across the world, especially since the nineteenth century, has loosened the internal coherence that held them together in a Euro-American master narrative and provided instead a loosely structured synopticon of politics, in which different nation-states, as part of their evolution, have organized their political cultures around different keywords (e.g. Williams 1976).

As a result of the differential diaspora of these keywords, the political narratives that govern communication between elites and followers in different parts of the world involve problems of both a semantic and pragmatic nature: semantic to the extent that words (and their lexical equivalents) require careful translation from context to context in their global movements, and pragmatic to the extent that the use of these words by political actors and their audiences may be subject to very different sets of contextual conventions that mediate their translation into public politics. Such conventions are not only matters of the nature of political rhetoric: for example, what does the aging Chinese leadership mean when it refers to the dangers of hooliganism? What does the South Korean leadership mean when it speaks of discipline as the key to democratic industrial growth?

These conventions also involve the far more subtle question of what sets of communicative genres are valued in what way (newspapers versus cinema, for example) and what sorts of pragmatic genre conventions govern the collective readings of different kinds of text. So, while an Indian audience may be attentive to the resonances of a political speech in terms of some keywords and phrases reminiscent of Hindi cinema, a Korean audience may respond to the subtle codings of Buddhist or neo-Confucian rhetoric encoded in a political document. The very relationship of reading to hearing and seeing may vary in important ways that determine the morphology of these different ideoscapes as they shape themselves in different national and transnational contexts. This globally variable synaesthesia has hardly even been noted, but it demands urgent analysis. Thus *democracy* has clearly become a master term, with powerful echoes from Haiti and Poland to the former Soviet Union and China, but it sits at the center of a variety of ideoscapes, composed of distinctive pragmatic configurations of rough translations of other central terms from the vocabulary of the Enlightenment. This creates ever new terminological kaleidoscopes, as states (and the groups that seek to capture them) seek to pacify populations whose own ethnoscapes are in motion and whose mediascapes may create severe problems for the ideoscapes with which they are presented. The fluidity of ideoscapes is complicated in particular by the growing diasporas (both voluntary and involuntary) of intellectuals who continuously inject new meaning-streams into the discourse of democracy in different parts of world.

This extended terminological discussion of the five terms I have coined sets the basis for a tentative formulation about the conditions under which current global flows occur: they occur in and through the growing disjunctures among ethnoscapes, technoscapes, financescapes, mediascapes, and ideoscapes. This formulation, the core of my model of global cultural flow, needs some explanation. First, people, machinery, money, images, and ideas now follow increasingly nonisomorphic paths; of course, at all periods in human history, there have been some disjunctures in the flows of these things, but the sheer speed, scale, and volume of each of these flows are now so great that the disjunctures have become central to the politics of global culture. The Japanese are notoriously hospitable to ideas and are stereotyped as inclined to export (all) and import (some) goods, but they are also notoriously closed to immigration, like the Swiss, the Swedes, and the Saudis. Yet the Swiss and the Saudis accept populations of guest workers, thus creating labor diasporas of Turks, Italians, and other circum-Mediterranean groups. Some such guest-worker groups maintain continuous contact with their home nations, like the Turks, but others, like high-level South Asian migrants, tend to desire lives in their new homes, raising anew the problem of reproduction in a deterritorialized context.

## Chapter 106

# SIMON GIKANDI

# GLOBALIZATION AND THE CLAIMS OF POSTCOLONIALITY

From 'Globalization and the Claims of Postcoloniality' *South Atlantic Quarterly* 100(3) 627–58, 2002.

G LOBALIZATION AND POSTCOLONIALITY are perhaps two of the most important terms in social and cultural theory today. Since the 1980s, they have functioned as two of the dominant paradigms for explaining the transformation of political and economic relationships in a world that seems to become increasingly interdependent with the passing of time, with boundaries that once defined national cultures becoming fuzzy. The debates on globalization and postcolonialism are now so universal in character, and the literature on these topics is so extensive, that they are difficult to summarize or categorize. And to the extent that it dominates most debates on the nature of society and economy in the social sciences, globalization must be considered one of the constitutive elements of disciplines such as anthropology and sociology. Similarly, it is difficult to conceive an area of literary studies, from medievalism to postmodernism, that is not affected by debates on postcolonial theory and postcoloniality. While diverse writers on globalization and postcolonialism might have differing interpretations of the exact meaning of these categories, or their long-term effect on the institutions of knowledge production in the modern world, they have at least two important things in common: they are concerned with explaining forms of social and cultural organization whose ambition is to transcend the boundaries of the nation-state, and they seek to provide new vistas for understanding cultural flows that can no longer be explained by a homogenous Eurocentric narrative of development and social change. For scholars trying to understand cultural and social production in the new millennium, globalization is attractive both because of its implicit universalism and its ability to reconcile local and global interests. Furthermore, globalization is appealing to social analysts because of what is perceived as its conjunctive and disjunctive form and function. In the first regard, as Jan Nederveen Pieterse has noted, globalization brings the universal and the local together in a moment of conceptual renewal and 'momentum of newness' (Pieterse 1998: 75). In the second instance, what Arjun Appadurai calls global mediascapes and ideoscapes have become the site of tension between 'cultural homogenization and cultural heterogenization' (Appadurai 1996: 32). In both cases, the language that enables conjuncture or disjuncture – hybridity and cultural transition, for example – comes directly from the grammar book of postcolonial theory. In this sense, one could argue that what makes current theories of globalization different from earlier ones, let's say those associated with modernization in the 1950s and 1960s, is their strategic deployment of postcolonial theory.

Besides their shared cultural grammar, however, the relationship between globalization and postcoloniality is not clear; neither are their respective meanings or implications. Is postcoloniality a consequence of the globalization of culture? Do the key terms in both categories describe a general state of cultural transformation in a world where the authority of the nation-state has collapsed or are they codes for explaining a set of amorphous images and a conflicting set of social conditions? The discourse of globalization is surrounded by a rhetoric of newness, but what exactly are the new vistas that these terms provide analysts of societies and cultures that have acquired a transnational character? Is globalization a real or virtual phenomenon? Where do we locate postcoloniality – in the spaces between and across cultures and traditions or in national states, which, in spite of a certain crisis of legitimacy, still continue to demand affiliation from their citizens and subjects? These questions are made even more urgent by the realization that while we live in a world defined by cultural and economic flows across formally entrenched national boundaries, the world continues to be divided, in stark terms, between its 'developed' and 'underdeveloped' sectors. It is precisely because of the starkness of this division that the discourse of globalization seems to be perpetually caught between two competing narratives, one of celebration, the other of crisis.

From one perspective, globalization appears to be a sign of the coming into being of a cultural world order that questions the imperial cartography that has defined global relations since the early modern period. Globalization constitutes, in this regard, what Appadurai calls 'a complex overlapping, disjunctive order that cannot any longer be understood in terms of existing center-periphery models' (Appadurai 1996: 32). And for those who might argue that globalization is simply the Westernization or Americanization of the world, Appadurai makes a crucial distinction between older forms of modernity, whose goal was the rationalization of the world in Weberian terms, to the symbolic economy of a new global culture based on reciprocal rather than nonlinear relationships:

> The master narrative of the Enlightenment (and its many variants in Britain, France, and the United States) was constructed with a certain internal logic and presupposed a certain relationship between reading, representation, and the public sphere. . . . But the diaspora of these terms and images across the world, especially since the nineteenth century, has loosened the internal coherence that held them together in a Euro-American master narrative and provided instead a loosely structured synopticon of politics, in which different nation-states, as part of their evolution, have organized their political cultures around different keywords.
>
> (1996: 36)

Clearly, globalization appeals to advocates of hybridity as diverse as Homi Bhabha and Pieterse because it seems to harmonize the universal and the particular and, in the process, it seems to open up to a multiplicity of cultural relationships unheard of in the age of empire: for Bhabha, the globalization of social spaces reflects a state of 'unsatisfaction' that, nevertheless, enables the articulation and enunciation of 'a global or transnational imaginary and its "cosmopolitan subjectivities"'; for Pieterse, it is through hybridity that globalization works against 'homogenization, standardization, cultural imperialism, westernization, Americanization' (Bhabha 1994: 204; Pieterse 1998: 76).

Nevertheless, this optimistic and celebratory view of globalization, which is particularly pronounced in postcolonial studies because it uses the lexicon that postcolonial theory makes available to us, is constantly haunted by another form of globalization, one defined by a sense of crisis within the postcolony itself. Unsure how to respond to the failure of the nationalist mandate, which promised modernization outside the tutelage of colonialism, citizens of the postcolony are more likely to seek their global identity by invoking the very logic of Enlightenment that postcolonial theory was supposed to deconstruct.

Now, my primary interest in this discussion is not to adjudicate between the celebratory narrative of globalization and the more dystopic version represented in [Africans' desire for material modernity]; it is not even my intention to rationalize the actions of Africans who die seeking the dream of a European identity in very colonial and Eurocentric terms. On the contrary, I am interested in using these contrasting views of globalization to foreground at least three closely related problems, which, I believe, call into question many of the claims motivating the theoretical literature on globalization and its relations to postcoloniality.

The first problem arises from the realization that when social scientists try to differentiate older forms of globalization (located solidly within the discourse of colonialism and modernization) from the new forms structured by hybridity and difference, they often tend to fall back on key words borrowed from postcolonial theory. Although some of these key words – the most prominent are hybridity and difference – have been popular in literary studies since the 1970s, they have been shunned by empirical social scientists who decry the lack of the conceptual foundations that might make them useful analytical categories (see Robertson 1990). At the same time, however, social scientists eager to turn globalization into the site of what Pieterse calls 'conceptual renewal' have found the language of postcolonial theory indispensable to their project (Pieterse 1998: 75).

The second problem concerns the rather optimistic claim that the institutions of cultural production provide irrefutable evidence of new global relations. It is important here to note that when advocates of the new global order, most prominently Appadurai and Bhabha, talk about globalization, they conceive it almost exclusively in cultural terms; but it is premature to argue that the images and narratives that denote the new global culture are connected to a global structure or that they are disconnected from earlier or older forms of identity. In other words, there is no reason to suppose that the global flow in images has a homological connection to transformations in social or cultural relationships . . . Global images have a certain salience for students of culture, especially postmodern culture, but this does not mean that they are a substitute for material experiences. In regard to cultural images, my argument is that we cannot stop at the site of their contemplation; rather, as Mike Featherstone has noted, we 'need to inquire into the grounds, the various generative processes, involving the formation of cultural images and traditions' (Featherstone 1990: 2).

The last problem I want to take up in this essay concerns the premature privileging of literary texts – and the institutions that teach them – as the exemplars of globalization. No doubt, the most powerful signs of the new process of globalization come from literary texts and other works of art. For critics looking for the sign of hybridity, heterogeneity, and newness in the new world order, there cannot be a better place to go than Salman Rushdie's *Satanic Verses* or Gabriel García Márquez's *El cieno años de soledad*. Such works are now considered world texts because, as Franco Moretti has argued, they have a frame of reference that is 'no longer the nation-state, but a broader entity – a continent, or the world-system as a whole' (Moretti 1996: 50). Surprisingly, however, no reading of these seminal texts is complete without an engagement with the nation-state, its history, its foundational mythologies, and its quotidian experiences. To the extent that they seek to deconstruct the foundational narrative of the nation, these are world texts; yet they cannot do without the framework of the nation. What needs to be underscored here, then, is the persistence of the nation-state in the very literary works that were supposed to gesture toward a transcendental global culture. I will conclude my discussion by arguing that one of the great ironies of the discourse of globalization is that although English literature has become the most obvious sign of transnationalism, it is continuously haunted by its historical – and disciplinary – location in a particular national ethos and ethnos. What are we going to do with those older categories – nation, culture, and English – which function as the absent structure that shapes and yet haunts global culture and the idea of literature itself?

My contention that postcolonial theory has been the major source of a new grammar for rethinking the global begs a foundational question: Why did culture in general and literature in particular become central terms in the discourse of globalization in the 1980s? There are two obvious explanations for the cultural turn in global studies. The first one is that sometime in the 1980s, cultural and literary theorists became convinced that the debates on globalization that had dominated disciplines such as sociology and anthropology for most of the twentieth century had become hopelessly imprisoned in the classical narrative of modernity, or Wallersteinian world-system theory. These scholars began to elaborate a cultural and literary project whose goal was to show that the real signs of how globalization was being lived, experienced, and interpreted were to be found primarily in the literary and cultural field. It was in literary culture, postcolonial theorists argued, that a new narrative of globalization, one that would take us beyond modernity and colonialism, could be identified and experienced (see Appadurai 1996; Bhabha 1994; Hall 1996).

The second explanation for the cultural turn in global studies can be connected to the emergence of postmodern theories that called into question some of the dominant grand narratives of globalization. Let us recall here that before the emergence of postmodernism as a conceptual mode of explaining the nature of global culture, theories of globalization were constructed around the concept of modernization, a powerful and homogenizing category that appealed as much to colonial systems as it did to nationalist movements in the so-called third world. With the emergence of postmodern theories of cultural formation, however, certain key categories in theories of modernization were called into question. These categories included the efficacy of homogenizing notions such as modernization, the authority of the nation-state as the central institution in the management of social relationships, and the idea of culture as the embodiment of symbolic hierarchies such as patriotism and citizenship. Against the totality implicit in colonial and nationalist theories of globalization, postmodern critics sought to show, after Jean-François Lyotard, that 'eclecticism [was] the degree zero of contemporary general culture' (Lyotard 1984: 76).

Calling attention to the existence of a variety of decentered narratives and the challenge to the nation-states by transnational movements that were creating new sites of identity – diasporas, for example – outside the boundaries of the state itself, postcolonial theory saw itself as responding to new cultural forms that could not be contained by world-system theories. In one sense, this turn to the literary in global studies was premised on the belief that in order to displace globalization from its national and disciplinary boundaries, it was important to call into question its key terms, mainly the notion of structuralization that dominated world-system theory, the Eurocentric chronology that had enabled its periodization, and the universalism on which its schemes of identity were based. For Homi Bhabha, Stuart Hall, and Arjun Appadurai, who have been some of the most influential figures in this displacement of the idea of globalization, the new mode of global cultural, and social relations is defined by its transgression of the boundaries established by the nation-state, the structures of dominant economic and social formations, and what they conceive to be a Eurocentric sense of time. The key assumption in what one may call the cultural version of globalization is that in the old global order, the nation was the reality and category that enabled the socialization of subjects, and hence the structuralization of cultures; now, in transnationality, the nation has become an absent structure. The nation is still an apparatus of enormous symbolic power, but it is also the mechanism that produces what Bhabha calls 'a continual slippage of categories, like sexuality, class affiliation, territorial paranoia, or "cultural difference"' (Bhabha 1994: 140).

## Chapter 107

# ROLAND ROBERTSON

# GLOCALIZATION

From 'Glocalization: Time-Space and Homogeneity-Heterogeneity' in Mike Featherstone, Scott Lash and Roland Robertson (eds) *Global Modernities* London: Sage, 1995.

A CCORDING TO *The Oxford Dictionary of New Words* (1991: 134) the term 'glocal' and the process noun 'glocalization' are 'formed by telescoping *global* and *local* to make a blend'. Also according to the *Dictionary* that idea has been 'modelled on Japanese *dochakuka* (deriving from *dochaku* "living on one's own land"), originally the agricultural principle of adapting one's farming techniques to local conditions, but also adopted in Japanese business for *global localization,* a global outlook adapted to local conditions' (emphasis in original). More specifically, the terms 'glocal' and 'glocalization' became aspects of business jargon during the 1980s, but their major locus of origin was in fact Japan, a country which has for a very long time strongly cultivated the spatio-cultural significance of Japan itself and where the general issue of the relationship between the particular and the universal has historically received almost obsessive attention (Miyoshi and Harootunian 1989). By now it has become, again in the words of *The Oxford Dictionary of New Words* (1991: 134), 'one of the main marketing buzzwords of the beginning of the nineties'. . . .

My deliberations in this chapter on the local-global problematic hinge upon the view that contemporary conceptions of locality are largely produced in something like global terms, but this certainly does not mean that all forms of locality are thus substantively homogenized (notwithstanding the standardization, for example, of relatively new suburban, fortress communities). An important thing to recognize in this connection is that there is an increasingly globe-wide discourse of locality, community, home and the like. One of the ways of considering the idea of *global culture is* in terms of its being constituted by the increasing interconnectedness of many local cultures both large and small (Hannerz 1990), although I certainly do not myself think that global culture is entirely constituted by such interconnectedness. In any case we should be careful *not to equate the communicative and interactional connecting of such cultures* – including very asymmetrical forms of such communication and interaction, as well as 'third cultures' of mediation – *with the notion of homogenization of all cultures.* . . .

## The local in the global? The global in the local?

In one way or another the issue of the relationship between the 'local' and the 'global' has become increasingly salient in a wide variety of intellectual and practical contexts. In some respects this development hinges upon the increasing recognition of the significance of space, as opposed to time, in a number of fields of academic and practical endeavour. The general interest in the idea of postmodernity, whatever its limitations, is probably the most intellectually tangible manifestation of this. The most well known maxim – virtually a cliché – proclaimed in the diagnosis of 'the postmodern condition' is of course that 'grand narratives' have come to an end, and that we are now in a circumstance of proliferating and often competing narratives. In this perspective there are no longer any stable accounts of dominant change in the world. This view itself has developed, on the other hand, at precisely the same time that there has crystallized an increasing interest in the world as a whole as a single place. (Robbins (1993: 187) also notes this, in specific reference to geographers.) As the sense of temporal unidirectionality has faded so, on the other hand, has the sense of 'representational' space within which all kinds of narratives may be inserted expanded. This of course has increasingly raised in recent years the vital question as to whether the apparent collapse – and the 'deconstruction' – of the heretofore dominant social-evolutionist accounts of implicit or explicit world history are leading rapidly to a situation of chaos or one in which, to quote Giddens (1990: 6), 'an infinite number of purely idiosyncratic "histories" can be written'. . . . .

Like many others, Barber defines globalization as the opposite of localization. He argues that 'four imperatives make up the dynamic of McWorld: a market imperative, a resource imperative, an information-technology imperative, and an ecological imperative' (1992: 54). Each of these contributes to 'shrinking the world and diminishing the salience of national borders' and together they have 'achieved a considerable victory over factiousness and particularism, and not least over their most virulent traditional form – nationalism' (Barber 1992: 54; cf. Miyoshi 1993). Remarking that 'the Enlightenment dream of a universal rational society has to a remarkable degree been realized', Barber (1992: 59) emphasizes that that achievement has, however, been realized in commercialized, bureaucratized, homogenized and what he calls 'depoliticized' form. Moreover, he argues that it is a very incomplete achievement because it is 'in competition with forces of global breakdown, national dissolution, and centrifugal corruption' (cf. Kaplan 1994). While notions of localism, locality and locale do not figure explicitly in Barber's essay they certainly diffusely inform it.

There is no good reason, other than recently established convention in some quarters, to define globalization largely in terms of homogenization. Of course, anyone is at liberty to so define globalization, but I think that there is a great deal to be said against such a procedure. Indeed, while each of the imperatives of Barber's McWorld appear superficially to suggest homogenization, when one considers them more closely, they each have a local, diversifying aspect. I maintain also that it makes no good sense to define the global as if the global excludes the local. In somewhat technical terms defining the global in such a way suggests that the global lies beyond all localities, as having systemic properties over and beyond the attributes of units within a global system. This way of talking flows along the lines suggested by the macro-micro distinction, which has held much sway in the discipline of economics and has recently become a popular theme in sociology and other social sciences. . . .

One can undoubtedly trace far back into human history developments involving the expansion of chains of connectedness across wide expanses of the earth. In that sense 'world formation' has been proceeding for many hundreds, indeed thousands, of years. At the same time, we can undoubtedly trace through human history periods during which the consciousness of the potential for world 'unity' was in one way or another particularly acute. One of the major tasks of students of globalization is, as I have said, to comprehend *the form* in which the present, seemingly rapid shifts towards a highly interdependent world was structured. I have specifically argued that that form has been centred upon four main elements of the global-human condition: societies, individuals, the international system of societies, and humankind (Robertson 1992). It is around the changing relationships between, different emphases upon and often conflicting interpretations of these aspects of human life that the contemporary world as a whole has crystallized. So in my perspective the issue of what is to be included under the notion of the global is treated very comprehensively. The global is not in and of itself counterposed to the local. Rather, what is often referred to as the local is essentially included within the global.

In this respect globalization, defined in its most general sense as the compression of the world as a whole, involves the linking of localities. But it also involves the 'invention' of locality, in the same general sense as the idea of the invention of tradition (Hobsbawm and Ranger 1983), as well as its 'imagination' (cf. Anderson 1983). There is indeed currently something like an 'ideology of home' which has in fact come into being partly in response to the constant repetition and global diffusion of the claim that we now live in a condition of homelessness or rootlessness; as if in prior periods of history the vast majority of people lived in 'secure' and homogenized locales. Two things, among others, must be said in objection to such ideas. First, the form of globalization has involved considerable emphasis, at least until now, on the cultural homogenization of nationally constituted societies; but, on the other hand, prior to that emphasis, which began to develop at the end of the eighteenth century, what McNeill (1985) calls polyethnicity was normal. Second, the phenomenological diagnosis of the generalized homelessness of modern man and woman has been developed as if 'the same people are behaving and interpreting at the same time in the same broad social process' (Meyer 1992: 11); whereas there is in fact much to suggest that it is increasingly global expectations concerning the relationship between individual and society that have produced both routinized and 'existential' selves. On top of that, the very ability to identify 'home', directly or indirectly, is contingent upon the (contested) construction and organization of interlaced categories of space and time. . . .

The present century has seen a remarkable proliferation with respect to the 'international' organization and promotion of locality. A very pertinent example is provided by the current attempts to organize globally the promotion of the rights and identities of native, or indigenous, peoples (Charles 1993; Chartrand 1991). This was a strong feature, for example, of the Global Forum in Brazil in 1992, which, so to say, surrounded the official United Nations 'Earth Summit'. Another is the attempt by the World Health Organization to promote 'world health' by the reactivation and, if need be, the invention of 'indigenous' local medicine. It should be stressed that these are only a few examples taken from a multifaceted trend.

## Glocalization and the cultural imperialism thesis

. . . Some important aspects of the local-global issue are manifested in the general and growing debate about and the discourse of cultural imperialism (Tomlinson 1991). There is of course a quite popular intellectual view which would have it that the entire world is being swamped by Western – more specifically, American – culture. This view has undoubtedly exacerbated recent French political complaints about American cultural imperialism, particularly within the context of GATT negotiations. There are, on the other hand, more probing discussions of and research on this matter. For starters, it should be emphasized that the virtually overwhelming evidence is that even 'cultural messages' which emanate directly from 'the USA' are *differentially* received and interpreted; that 'local' groups 'absorb' communication from the 'centre' in a great variety of ways (Tomlinson 1991). Second, we have to realize that the major alleged producers of 'global culture' – such as those in Atlanta (CNN) and Los Angeles (Hollywood) – increasingly tailor their products to a differentiated global market (which they partly construct). For example, Hollywood attempts to employ mixed, 'multinational' casts of actors and a variety of 'local' settings when it is particularly concerned, as it increasingly is, to get a global audience. Third, there is much to suggest that seemingly 'national' symbolic resources are in fact increasingly available for differentiated global interpretation and consumption. For example, in a recent discussion of the staging of Shakespeare's plays, Billington (1992) notes that in recent years Shakespeare has been subject to wide-ranging cultural interpretation and staging. Shakespeare no longer belongs to England. Shakespeare has assumed a universalistic significance; and we have to distinguish in this respect between Shakespeare as representing Englishness and Shakespeare as of 'local-cum-global' relevance. Fourth, clearly many have seriously underestimated the flow of ideas and practices from the so-called Third World to the seemingly dominant societies and regions of the world (J. Abu-Lughod 1991; Hall 1991a; 1991b).

Much of global 'mass culture' is in fact impregnated with ideas, styles and genres concerning religion, music, art, cooking, and so on. In fact the question of what will 'fly' globally and what will not is a very important question in the present global situation. We know of course that the question of what 'flies' is in part contingent upon issues of power; but we would be very ill-advised to think of this simply as a matter of the hegemonic extension of Western modernity. As Tomlinson (1991) has argued, 'local cultures' are, in Sartre's phrase, *condemned to freedom*. And their global participation has been greatly (and politically) underestimated. At this time 'freedom' is manifested particularly in terms of the social construction of identity-and-tradition, by the appropriation of cultural traditions (Habermas 1994: 22). Although, as I have emphasized, this reflexiveness is typically undertaken along relatively standardized global-cultural lines. (For example, in 1982 the UN fully recognized the existence of indigenous peoples. In so doing it effectively established *criteria* in terms of which indigenous groups could and should identify themselves and be recognized formally. There are national parallels to this, in the sense that some societies have legal criteria for ethnic groups and cultural traditions.) Then there is the question of diversity at the local level. This issue has been raised in a particularly salient way by Balibar (1991a), who talks of *world spaces*. The latter are places in which the world-as-a-whole is potentially inserted. The general idea of world-space suggests that we should consider the local as a 'micro' manifestation of the global – in opposition, *inter alia*, to the implication that the local indicates enclaves of cultural, ethnic, or racial homogeneity. Where, in other words, is *home* in the late-twentieth century? Balibar's analysis – which is centred on contemporary Europe – suggests that in the present situation of global complexity, the idea of home has to be divorced analytically from the idea of locality. There may well be groups and categories which equate the two, but that doesn't entitle them or their representatives to project their perspective onto humanity as a whole. In fact there is much to suggest that the senses of home and locality are contingent upon alienation from home and/or locale. How else could one have (reflexive) consciousness of such? We talk of the mixing of cultures, of polyethnicity, but we also often underestimate the significance of what Lila Abu-Lughod (1991) calls 'halfies'. As Geertz (1986: 114) has said, 'like nostalgia, diversity is not what it used to be'. One of the most significant aspects of contemporary diversity is indeed the complication it raises for conventional notions of culture. We must be careful not to remain in thrall to the old and rather well established view that cultures are organically binding and sharply bounded. In fact Lila Abu-Lughod opposes the very idea of culture because it seems to her to deny the importance of 'halfies', those who combine in themselves as individuals a number of cultural, ethnic and genderal features (cf. Tsing 1993). This issue is closely related to the frequently addressed theme of global hybridization, even more closely to the idea of creolization (Hannerz 1992: 217–67).

## Chapter 108

# GUY HARDT AND ANTONIO NEGRI

## IMPERIAL SOVEREIGNTY

From *Empire* Cambridge, Mass.: Harvard University Press, 2000.

THERE IS A LONG TRADITION OF modern critique dedicated to denouncing the dualisms of modernity. The standpoint of that critical tradition, however, is situated in the paradigmatic place of itself, both 'inside' and 'outside,' at the threshold or the point of crisis. What has changed in the passage to the imperial world, however, is that this border place no longer exists, and thus the modern critical strategy tends no longer to be effective. . . .

### There is no more outside

The domains conceived as inside and outside and the relationship between them are configured differently in a variety of modern discourses. The spatial configuration of inside and outside itself, however, seems to us a general and foundational characteristic of modern thought. In the passage from modern to postmodern and from imperialism to Empire there is progressively less distinction between inside and outside.

This transformation is particularly evident when viewed in terms of the notion of sovereignty. Modern sovereignty has generally been conceived in terms of a (real or imagined) territory and the relation of that territory to its outside. Early modern social theorists, for example, from Hobbes to Rousseau, understood the civil order as a limited and interior space that is opposed or contrasted to the external order of nature. The bounded space of civil order, its place, is defined by its separation from the external spaces of nature. In an analogous fashion, the theorists of modern psychology understood drives, passions, instincts, and the unconscious metaphorically in spatial terms as an outside within the human mind, a continuation of nature deep within us. Here the sovereignty of the Self rests on a dialectical relation between the natural order of drives and the civil order of reason or consciousness. Finally, modern anthropology's various discourses on primitive societies function as the outside that defines the bounds of the civil world. The process of modernization, in all these varied contexts, is the internalization of the outside, that is, the civilization of nature.

In the imperial world, this dialectic of sovereignty between the civil order and the natural order has come to an end. This is one precise sense in which the contemporary world is post-modern. 'Postmodernism,' Fredric Jameson tells us, 'is what you have when the modernization process is complete and nature is gone for good' (Jameson 1991: ix). Certainly we continue to have forests and crickets and thunderstorms in our world, and we continue to understand our psyches as driven by natural instinct and passions; but we have no nature in the sense that these forces and phenomena are no longer understood as outside, that is, they are not seen as original and independent of the artifice of the civil order. In a postmodern world all phenomena and forces are artificial, or, as some might say, part of history. The modern dialectic of inside and outside has been replaced by a play of degrees and intensities, of hybridity and artificiality.

The outside has also declined in terms of a rather different modern dialectic that defined the relation between public and private in liberal political theory. The public spaces of modern society, which constitute the place of liberal politics, tend to disappear in the postmodern world. According to the liberal tradition, the modern individual, at home in its private spaces, regards the public as its outside. The outside is the place proper to politics, where the action of the individual is exposed in the presence of others and there seeks recognition (see Arendt 1958). In the process of postmodernization, however, such public spaces are increasingly becoming privatized. The urban landscape is shifting from the modern focus on the common square and the public encounter to the closed spaces of malls, freeways, and gated communities. The architecture and urban planning of megalopolises such as Los Angeles and Sao Paolo have tended to limit public access and interaction in such a way as to avoid the chance encounter of diverse populations, creating a series of protected interior and isolated spaces (see Davis 1990; Caldiera 1996). Alternatively, consider how the banlieu of Paris has become a series of amorphous and indefinite spaces that promote isolation rather than any interaction or communication. Public space has been privatized to such an extent that it no longer makes sense to understand social organization in terms of a dialectic between private and public spaces, between inside and outside. The place of modern liberal politics has disappeared, and thus from this perspective our postmodern and imperial society is characterized by a deficit of the political. In effect, the place of politics has been de-actualized. . . .

Finally, there is no longer an outside also in a military sense. When Francis Fukuyama claims that the contemporary historical passage is defined by the end of history, he means that the era of major conflicts has come to an end: sovereign power will no longer confront its Other and no longer face its outside, but rather will progressively expand its boundaries to envelop the entire globe as its proper domain (Fukuyama 1992). The history of imperialist, interimperialist, and anti-imperialist wars is over. The end of that history has ushered in the reign of peace. Or really, we have entered the era of minor and internal conflicts. Every imperial war is a civil war, a police action – from Los Angeles and Granada to Mogadishu and Sarajevo. In fact, the separation of tasks between the external and the internal arms of power (between the army and the police, the CIA and the FBI) is increasingly vague and indeterminate.

In our terms, the end of history that Fukuyama refers to is the end of the crisis at the center of modernity, the coherent and defining conflict that was the foundation and raison d'être for modern sovereignty. History has ended precisely and only to the extent that it is conceived in Hegelian terms – as the movement of a dialectic of contradictions, a play of absolute negations and subsumption. The binaries that defined modern conflict have become blurred. The Other that might delimit a modern sovereign Self has become fractured and indistinct, and there is no longer an outside that can bound the place of sovereignty. The outside is what gave the crisis its coherence. Today it is increasingly difficult for the ideologues of the United States to name a single, unified enemy; rather, there seem to be minor and elusive enemies everywhere. The end of the crisis of modernity has given rise to a proliferation of minor and indefinite crises, or, as we prefer, to an omni-crisis.

It is useful to remember here . . . that the capitalist market is one machine that has always run counter to any division between inside and outside. It is thwarted by barriers and exclusions; it thrives instead by including always more within its sphere. Profit can be generated only through contact, engagement, interchange, and commerce. The realization of the world market would constitute the point of arrival of this tendency. In its ideal form there is no outside to the world market: the entire globe is its domain. We might thus use the form of the world market as a model for understanding imperial sovereignty. Perhaps, just as Foucault recognized the panopticon as the diagram of modern power, the world market might serve adequately – even though it is not an architecture but really an anti-architecture – as the diagram of imperial power.

The striated space of modernity constructed *places* that were continually engaged in and founded on a dialectical play with their outsides. The space of imperial sovereignty, in contrast, is smooth. It might appear to be free of the binary divisions or striation of modern boundaries, but really it is crisscrossed by so many fault lines that it only appears as a continuous, uniform space. In this sense, the clearly defined crisis of modernity gives way to an omni-crisis in the imperial world. In this smooth space of Empire, there is no *place* of power – it is both everywhere and nowhere. Empire is an *ou-topia,* or really a *non-place.*

## Imperial racism

The passage from modern sovereignty to imperial sovereignty shows one of its faces in the shifting configurations of racism in our societies. We should note first of all that it has become increasingly difficult to identify the general lines of racism. In fact, politicians, the media, and even historians continually tell us that racism has steadily receded in modern societies – from the end of slavery to decolonization struggles and civil rights movements. Certain specific traditional practices of racism have undoubtedly declined, and one might be tempted to view the end of the apartheid laws in South Africa as the symbolic close of an entire era of racial segregation. From our perspective, however, it is clear that racism has not receded but actually progressed in the contemporary world, both in extent and in intensity. It appears to have declined only because its form and strategies have changed. If we take Manichaean divisions and rigid exclusionary practices (in South Africa, in the colonial city, in the southeastern United States, or in Palestine) as the paradigm of modern racisms, we must now ask what is the *postmodern* form of racism and what are its strategies in today's imperial society.

Many analysts describe this passage as a shift in the dominant theoretical form of racism, from a racist theory based on biology to one based on culture. The dominant modern racist theory and the concomitant practices of segregation are centered on essential biological differences among races. Blood and genes stand behind the differences in skin color as the real substance of racial difference. Subordinated peoples are thus conceived (at least implicitly) as other than human, as a different order of being. These modern racist theories grounded in biology imply or tend toward an ontological difference – a necessary, eternal, and immutable rift in the order of being. In response to this theoretical position, then, modern anti-racism positions itself against the notion of biological essentialism, and insists that differences among the races are constituted instead by social and cultural forces. These modern anti-racist theorists operate on the belief that social constructivism will free us from the straitjacket of biological determinism: if our differences are socially and culturally determined, then all humans are in principle equal, of one ontological order, one nature.

With the passage to Empire, however, biological differences have been replaced by sociological and cultural signifiers as the key representation of racial hatred and fear. In this way imperial racist theory attacks modern anti-racism from the rear, and actually co-opts and enlists its arguments. Imperial racist theory agrees that races do not constitute isolable biological units and that nature cannot be divided into different human races. It also agrees that the behavior of individuals and their abilities or aptitudes are not the result of their blood or their genes, but are due to their belonging to different historically determined cultures (see Balibar 1991b; Gordon and Newfield 1994). Differences are thus not fixed and immutable but contingent effects of social history. Imperial racist theory and modern anti-racist theory are really saying very much the same thing, and it is difficult in this regard to tell them apart. In fact, it is precisely because this relativist and culturalist argument is assumed to be necessarily anti-racist that the dominant ideology of our entire society can appear to be against racism, and that imperial racist theory can appear not to be racist at all . . .

As a theory of social difference, the cultural position is no less 'essentialist' than the biological one, or at least it establishes an equally strong theoretical ground for social separation and segregation. Nonetheless, it is a pluralist theoretical position: all cultural identities are equal in principle. This pluralism accepts all the differences of who we are so long as we agree to act on the basis of these differences of identity, so long as we act our race. Racial differences are thus contingent in principle, but quite necessary in practice as markers of social separation. The theoretical substitution of culture for race or biology is thus transformed paradoxically into a theory of the preservation of race (see Michaels 1992, 1995). This shift in racist theory shows us how imperial theory can adopt what is traditionally thought to be an anti-racist position and still maintain a strong principle of social separation.

We should be careful to note at this point that imperial racist theory in itself is a theory of segregation, not a theory of hierarchy. Whereas modern racist theory poses a hierarchy among the races as the fundamental condition that makes segregation necessary, imperial theory has nothing to say about the superiority or inferiority of different races or ethnic groups in principle. It regards that as purely contingent, a practical matter. In other words, racial hierarchy is viewed not as cause but as effect of social circumstances. For example, African American students in a certain region register consistently lower scores on aptitude tests than Asian American students. Imperial theory understands this as attributable not to any racial inferiority but rather to cultural differences: Asian American culture places a higher importance on education, encourages students to study in groups, and so forth. The hierarchy of the different races is determined only a posteriori, as an effect of their cultures – that is, on the basis of their performance. According to imperial theory then, racial supremacy and subordination are not a theoretical question, but arise through free competition, a kind of market meritocracy of culture. . . .

The form and strategies of imperial racism help to highlight the contrast between modern and imperial sovereignty more generally. Colonial racism, the racism of modern sovereignty, first pushes difference to the extreme and then recuperates the Other as negative foundation of the Self. The modern construction of a people is intimately involved in this operation. A people is defined not simply in terms of a shared past and common desires or potential, but primarily in dialectical relation to its Other, its outside. A people (whether diasporic or not) is always defined in terms of a *place* (be it virtual or actual). Imperial order, in contrast, has nothing to do with this dialectic. Imperial racism, or differential racism, integrates others with its order and then orchestrates those differences in a system of control. Fixed and biological notions of peoples thus tend to dissolve into a fluid and amorphous multitude, which is of course shot through with lines of conflict and antagonism, but none that appear as fixed and eternal boundaries. The surface of imperial society continuously shifts in such a way that it destabilizes any notion of place. The central moment of modern racism takes place on its boundary, in the global antithesis between inside and outside. As Du Bois said nearly one hundred years ago, the problem of the twentieth century is the problem of the color line. Imperial racism, by contrast, looking forward perhaps to the twenty-first century, rests on the play of differences and the management of micro-conflictualities within its continually expanding domain.

## Chapter 109

# CHANDRA TALPADE MOHANTY

# FEMINIST SOLIDARITY THROUGH ANTICAPITALIST STRUGGLES

From 'Under Western Eyes Revisited: Feminist Solidarity through Anticapitalist Struggles' *Signs* 28(2) (Winter), 499–538, 2003.

> Women's and girls' bodies determine democracy: free from violence and sexual abuse, free from malnutrition and environmental degradation, free to plan their families, free to not have families, free to choose their sexual lives and preferences.
>
> Zillah Eisenstein, *Global Obscenities* (1998)

. . .

**W**HAT ARE THE CONCRETE EFFECTS of global restructuring on the 'real' raced, classed, national, sexual bodies of women in the academy, in workplaces, streets, households, cyberspaces, neighborhoods, prisons, and in social movements? And how do we recognize these gendered effects in movements against globalization? Some of the most complex analyses of the centrality of gender in understanding economic globalization attempt to link questions of subjectivity, agency, and identity with those of political economy and the state. This scholarship argues persuasively for the need to rethink patriarchies and hegemonic masculinities in relation to present-day globalization and nationalisms, and it also attempts to retheorize the gendered aspects of the refigured relations of the state, the market, and civil society by focusing on unexpected and unpredictable sites of resistance to the often devastating effects of global restructuring on women. And it draws on a number of disciplinary paradigms and political perspectives in making the case for the centrality of gender in processes of global restructuring, arguing that the reorganization of gender is part of the global strategy of capitalism.

Women workers of particular caste/class, race, and economic status are necessary to the operation of the capitalist global economy. Women are not only the preferred candidates for particular jobs, but particular kinds of women – poor, Third and Two-Thirds World, working-class, and immigrant/migrant women – are the preferred workers in these global, 'flexible' temporary job markets. The documented increase in the migration of poor, One-Third/Two-Thirds World women in search of labor across national borders has led to a rise in the international 'maid trade' (Parrenas 2001) and in international sex trafficking and tourism. Many global cities now require and completely depend on the service and domestic labor of immigrant and migrant women. The proliferation of structural adjustment policies around the world has reprivatized women's labor by shifting the responsibility for social welfare from the state to the household and to women located there. The rise of religious fundamentalisms in conjunction with conservative nationalisms, which are also in part reactions to global capital and its cultural demands, has led to the policing of women's bodies in the streets and in the workplaces.

Global capital also reaffirms the color line in its newly articulated class structure evident in the prisons in the One-Third World. The effects of globalization and reindustrialization on the prison industry in the One-Third World leads to a related policing of the bodies of poor, One-Third/Two-Thirds World, immigrant, and migrant women behind the concrete spaces and bars of privatized prisons. Angela Davis and Gina Dent (2001) argue that the political economy of U.S. prisons, and the punishment industry in the West/North, brings the intersection of gender, race, colonialism, and capitalism into sharp focus. Just as the factories and workplaces of global corporations seek and discipline the labor of poor, Third World/South, immigrant/migrant women, the prisons of Europe and the United States incarcerate disproportionately large numbers of women of color, immigrants, and noncitizens of African, Asian, and Latin American descent.

Making gender and power visible in the processes of global restructuring demands looking at, naming, and seeing the particular raced and classed communities of women from poor countries as they are constituted as workers in sexual, domestic, and service industries; as prisoners; and as household managers and nurturers. In contrast to this production of workers, Patricia Fernandez-Kelly and Diane Wolf (2001: esp. 1248) focus on communities of black U.S. inner-city youth situated as 'redundant' to the global economy. This redundancy is linked to their disproportionate representation in U.S. prisons. They argue that these young men, who are potential workers, are left out of the economic circuit, and this 'absence of connections to a structure of opportunity' results in young African-American men turning to dangerous and creative survival strategies while struggling to reinvent new forms of masculinity.

There is also increased feminist attention to the way discourses of globalization are themselves gendered and the way hegemonic masculinities are produced and mobilized in the service of global restructuring. Marianne Marchand and Anne Runyan (2000) discuss the gendered metaphors and symbolism in the language of globalization whereby particular actors and sectors are privileged over others: market over state, global over local, finance capital over manufacturing, finance ministries over social welfare, and consumers over citizens. They argue that the latter are feminized and the former masculinized and that this gendering naturalizes the hierarchies required for globalization to succeed. Charlotte Hooper (2000) identifies an emerging hegemonic Anglo-American masculinity through processes of global restructuring – a masculinity that affects men and women workers in the global economy. Hooper argues that this Anglo-American masculinity has dualistic tendencies, retaining the image of the aggressive frontier masculinity on the one hand, while drawing on more benign images of CEOs with (feminized) nonhierarchical management skills associated with teamwork and networking on the other.

While feminist scholarship is moving in important and useful directions in terms of a critique of global restructuring and the culture of globalization, I want to ask some of the same questions I posed in 1986 once again. In spite of the occasional exception, I think that much of present-day scholarship tends to reproduce particular 'globalized' representations of women. Just as there is an Anglo-American masculinity produced in and by discourses of globalization, it is important to ask what the corresponding femininities being produced are. Clearly there is the ubiquitous global teenage girl factory worker, the domestic worker, and the sex worker. There is also the migrant/immigrant service worker, the refugee, the victim of war crimes, the woman-of-color prisoner who happens to be a mother and drug user, the consumer-housewife, and so on. There is also the mother-of-the-nation/religious bearer of traditional culture and morality.

Although these representations of women correspond to real people, they also often stand in for the contradictions and complexities of women's lives and roles. Certain images, such as that of the factory or sex worker, are often geographically located in the Third World/South, but many of the representations identified above are dispersed throughout the globe. Most refer to women of the Two-Thirds World, and some to women of the One-Third World. And a woman from the Two-Thirds World can live in the One-Third World. The point I am making here is that women are workers, mothers, or consumers in the global economy, but we are also all those things simultaneously. Singular and monolithic categorizations of women in discourses of globalization circumscribe ideas about experience, agency, and struggle. While there are other, relatively new images of women that also emerge in this discourse – the human rights worker or the NGO advocate, the revolutionary militant and the corporate bureaucrat – there is also a divide between false, overstated images of victimized and empowered womanhood, and they negate each other. We need to further explore how this divide plays itself out in terms of a social majority/minority, One-Third/Two-Thirds World characterization. The concern here is with whose agency is being colonized and who is privileged in these pedagogies and scholarship. These then are my new queries for the twenty-first century.

Because social movements are crucial sites for the construction of knowledge, communities, and identities, it is very important for feminists to direct themselves toward them. The antiglobalization movements of the last five years have proven that one does not have to be a multinational corporation, controller of financial capital, or transnational governing institution to cross national borders. These movements form an important site for examining the construction of transborder democratic citizenship. But first a brief characterization of antiglobalization movements is in order.

Unlike the territorial anchors of the anticolonial movements of the early twentieth century, antiglobalization movements have numerous spatial and social origins. These include anticorporate environmental movements such as the Narmada Bachao Andolan in central India and movements against environmental racism in the U.S. Southwest, as well as the anti-agribusiness small-farmer movements around the world. The 1960s consumer movements, people's movements against the International Monetary Fund and World Bank for debt cancellation and against structural adjustment programs, and the antisweatshop student movements in Japan, Europe, and the United States are also a part of the origins of the antiglobalization movements. In addition, the identity-based social movements of the late twentieth century (feminist, civil rights, indigenous rights, etc.) and the transformed U.S. labor movement of the 1990s also play a significant part in terms of the history of antiglobalization movements.

While women are present as leaders and participants in most of these antiglobalization movements, a feminist agenda only emerges in the post-Beijing 'women's rights as human rights' movement and in some peace and environmental justice movements. In other words, while girls and women are central to the labor of global capital, antiglobalization work does not seem to draw on feminist analysis or strategies. Thus, while I have argued that feminists need to be anticapitalists, I would now argue that antiglobalization activists and theorists also need to be feminists. Gender is ignored as a category of analysis and a basis for organizing in most of the antiglobalization movements, and antiglobalization (and anticapitalist critique) does not appear to be central to feminist organizing projects, especially in the First World/ North. In terms of women's movements, the earlier 'sisterhood is global' form of internationalization of the women's movement has now shifted into the 'human rights' arena. This shift in language from 'feminism' to 'women's rights' can be called the mainstreaming of the feminist movement – a (successful) attempt to raise the issue of violence against women onto the world stage.

If we look carefully at the focus of the antiglobalization movements, it is the bodies and labor of women and girls that constitute the heart of these struggles. For instance, in the environmental and ecological movements such as Chipko in India and indigenous movements against uranium mining and breast-milk contamination in the United States, women are not only among the leadership: their gendered and racialized bodies are the key to demystifying and combating the processes of recolonization put in place by corporate control of the environment. My earlier discussion of Vandana Shiva's analysis of the WTO and biopiracy from the epistemological place of Indian tribal and peasant women illustrates this claim, as does Grace Lee Boggs's notion of 'place-based civic activism' (2000: 19). Similarly, in the anticorporate consumer movements and in the small farmer movements against agribusiness and the anti-sweatshop movements, it is women's labor and their bodies that are most affected as workers, farmers, and consumers/household nurturers.

Women have been in leadership roles in some of the cross-border alliances against corporate injustice. Thus, making gender, and women's bodies and labor, visible and theorizing this visibility as a process of articulating a more inclusive politics are crucial aspects of feminist anticapitalist critique. Beginning from the social location of poor women of color of the Two-Thirds World is an important, even crucial, place for feminist analysis; it is precisely the potential epistemic privilege of these communities of women that opens up the space for demystifying capitalism and for envisioning transborder social and economic justice.

The masculinization of the discourses of globalization analyzed by Hooper (2000) and Marchand and Runyan (2000) seems to be matched by the implicit masculinization of the discourses of antiglobalization movements. While much of the literature on antiglobalization movements marks the centrality of class and race and, at times, nation in the critique and fight against global capitalism, racialized gender is still an unmarked category. Racialized gender is significant in this instance because capitalism utilizes the raced and sexed bodies of women in its search for profit globally, and, as I argued earlier, it is often the experiences and struggles of poor women of color that allow the most inclusive analysis as well as politics in antiglobalization struggles.

On the other hand, many of the democratic practices and process-oriented aspects of feminism appear to be institutionalized into the decision-making processes of some of these movements. Thus the principles of nonhierarchy, democratic participation, and the notion of the personal being political all emerge in various ways in this antiglobal politics. Making gender and feminist agendas and projects explicit in such antiglobalization movements thus is a way of tracing a more accurate genealogy, as well as providing potentially more fertile ground for organizing. And of course, to articulate feminism within the framework of antiglobalization work is also to begin to challenge the unstated masculinism of this work. The critique and resistance to global capitalism, and uncovering of the naturalization of its masculinist and racist values, begin to build a transnational feminist practice.

A transnational feminist practice depends on building feminist solidarities across the divisions of place, identity, class, work, belief, and so on. In these very fragmented times it is both very difficult to build these alliances and also never more important to do so. Global capitalism both destroys the possibilities and also offers up new ones.

Feminist activist teachers must struggle with themselves and each other to open the world with all its complexity to their students. Given the new multiethnic racial student bodies, teachers must also learn from their students. The differences and borders of each of our identities connect us to each other, more than they sever. So the enterprise here is to forge informed, self-reflexive solidarities among ourselves.

I no longer live simply under the gaze of Western eyes. I also live inside it and negotiate it every day. I make my home in Ithaca, New York, but always as from Mumbai, India. My cross-race and cross-class work takes me to interconnected places and communities around the world – to a struggle contextualized by women of color and of the Third World, sometimes located in the Two-Thirds World, sometimes in the One-Third. So the borders here are not really fixed. Our minds must be as ready to move as capital is, to trace its paths and to imagine alternative destinations.

# PART EIGHTEEN

# Environment

THE INCURSION OF EUROPEANS into other regions of the globe in the centuries after 1492 resulted not only in genocide, but also in radical changes, unparalleled in human history, to both tropical and temperate environments.

As Alfred Crosby argues, such impact – particularly in the settler colonies, or the 'Neo-Europes' as he terms them – was as much responsible for European conquest and dominance as was military might. Disease, destruction of native flora and fauna, the felling of forests and land clearing, and the introduction of grazing animals and 'pest' species all transformed not only the land, but reconstituted the ways in which 'land' could be apprehended. Instead of an intrinsic part of human 'being' and at least partially *constitutive* of human identities, it became the inert background for profit making and taking. Early despoliation paved the way for many contemporary environmental problems – salinification of soils in Australia and the lowering of the water table in Eastern Canada. In the colonies of occupation, traditional patterns of crop rotation were often replaced by the (enforced) growing of 'cash' crops, a practice that eventually resulted in loss of soil fertility and desertification leading to famine in, for example, the countries surrounding the Sahara Desert.

In the extract included here, Crosby noted a further legacy of colonialist environmental change: the imbalance in terms of food production and food consumption between the crop-growing 'Neo-Europes' such as the United States, Canada and Australia, and the food consumers.

While acknowledging the environmental damage Crosby explores, Richard Grove argues that notwithstanding such negative colonialist effects – in some cases perhaps even *because* of them – the roots of contemporary environmentalist thinking and activism are to be traced to colonial administration and European scientists and naturalists. Such men, Grove argues, witnessed at first hand the results of the European impact, especially on islands such as Mauritius where potential extinction of animal and plant species was most readily observed. Concerns about conservation generated in these colonies predate Henry David Thoreau's Walden experiment and the establishment of the world's first National Park at Yosemite in the United States; the 'moments' often cited as marking the initiation of concern for a disappearing 'wilderness'.

But the consequences of European incursion were not confined to material effects, whether good or bad. The hegemonic power of Europe's economic and scientific rationalism also, in time, extinguished indigenous ontologies and epistemologies, re-defining for much of the world, the very nature of human *being* and land, flora and fauna. A sense of both differences in perceptions of what Val Plumwood terms the 'more-than-human' world and the destructive effects of European hegemony can be gained from Gordon Sayre's account of both the material and symbolic uses to which the North American beaver was put by Native American and settler cultures. Hunters, traders and settlers decimated beaver populations to feed the European fur trade, and the increasingly displaced original inhabitants were also forced into collusive exploitation by loss of their traditional grounds and ways of life. Changing human ontologies and material practices under European hegemony pushed beavers to near extinction – a fate they had already suffered in Europe. Both cultures had made *symbolic* use of the beaver, but whereas beavers were highly respected and accorded independent being in indigenous symbologies, the French (and other European settlers) used 'beavers' to draw theriomorphic morals that were always politically or socially contingent.

The various symbolic uses to which the beaver was put in French colonial society – as an example of social co-operation and orderliness, for instance – indicates the diverse but always pervasive roles animals (and the figure of the animal) have played in all human societies. But for most contemporary urbanists, as Steve Baker notes, animals are encountered more frequently *in representation* than in 'the real', something that indicates the scale of destruction of the complex biotic communities encountered by Europeans after 1492. The post-Enlightenment symbolic role of 'the animal and the animalistic' in Western societies is closely connected to the actual destruction of other peoples and animals. It is, as Cary Wolf (citing Jacques Derrida and Georges Bataille) argues, our determined constructions (and maintenance) of the 'human/animal' divide – the so-called species boundary – which is responsible for our continued killing and enslavement of animals worldwide, and which itself forms the inescapable basis of racism and genocide. Apparently impermeable, yet always shifting and politically, socially or economically *contingent*, it facilitates and has facilitated 'the non criminal putting to death' as Derrida terms it, 'not only of animals, but of other humans as well by making them as animal'. Speciesism, then, is the indispensable basis of racism and genocidal othering; and is necessarily implicated in patriarchal dominance and abuse as well.

Recognizing this, Val Plumwood argues not just for the *ethical* necessity of rethinking the Enlightenment deification of a 'Reason' considered to be an exclusively human attribute, but for a re-cognition necessary for survival of humans and the planet. Her argument is not against 'rational' behaviour as such, but against the deployment of 'the rational' and its destructive epigones – capitalism, economic rationalism, unbridled technological development – in the *irrational* pursuit of wealth and of *exclusively* human 'being'. Our survival and that of other species on this planet requires us to recognize our place in the 'more than human' world and our dependence – whatever a naturalized Cartesian dualism may have led us to believe – both spiritually and materially on it.

One of the most confronting extracts in this section is related to the trial of Ken Saro-Wiwa, Nigerian environmental activist and poet who was executed in 1995 as a result of collusion between the Nigerian government and the Dutch multi-national Shell Company whose exploitation of Nigeria's oil wealth had resulted in the almost complete destruction, through pollution of the Ogoni people's land and food supplies. This all too familiar neo-colonial pattern, whereby under-resourced post-independence governments are forced to rely on financial investment and backing from multi-national giants like Shell, or the World Bank, has continued to promulgate, even enforce, Western 'development' in allegedly independent states. The Indian novelist Arundhati Roy in her scathing account of collusive local, national and international financial interests involved in the Namarda Sagar Dam Project, used, like Saro Wiwa, her reputation as a writer and imaginative thinker to oppose a 'development' from which the World Bank was eventually forced to withdraw its support.

In spite then, of its contributions to environmental awareness and preservation (as argued by Grove) European colonialism, together with its neo-colonial legacies has had an inglorious history and usually destructive results. And although environmental degradation had occurred (and was occurring) in a number of pre-colonized areas, the *post*-incursion damage to people, animals, and places on a world scale was unprecedented. It is thus not surprising that so many individuals and organizations across formerly colonized countries are now turning their attention to a radical rethinking of relationships between humans, animals and place; a re-thinking which, at least in some cases, is looking for its inspiration to the once despised or ignored aboriginal ways of apprehending human identity in place.

## Chapter 110

## ALFRED W. CROSBY

# ECOLOGICAL IMPERIALISM

From *Ecological Imperialism: The Biological Expansion of Europe 900–1900*
Cambridge: Cambridge University Press, 1986.

EUROPEAN IMMIGRANTS and their descendants are all over the place, which requires explanation.

It is more difficult to account for the distribution of this subdivision of the human species than that of any other. The locations of the others make an obvious kind of sense. . . . All these peoples have expanded geographically – have committed acts of imperialism, if you will – but they have expanded into lands adjacent to or at least near to those in which they had already been living. . . . Europeans, in contrast, seem to have leapfrogged around the globe.

Europeans, a division of Caucasians distinctive in their politics and technologies, rather than in their physiques, live in large numbers and nearly solid blocks in northern Eurasia, from the Atlantic to the Pacific. They occupy much more territory there than they did a thousand or even five hundred years ago, but that is the part of the world in which they have lived throughout recorded history, and there they have expanded in the traditional way, into contiguous areas. They also compose the great majority in the populations of what I shall call the Neo-Europes, lands thousands of kilometers from Europe and from each other. Australia's population is almost all European in origin, and that of New Zealand is about nine-tenths European. In the Americas north of Mexico there are considerable minorities of Afro-Americans and *mestizos* (a convenient Spanish-American term I shall use to designate Amerindian and white mixtures), but over 80 percent of the inhabitants of this area are of European descent. . . . Even if we accept the highest estimations of Afro-American, and Amerindian populations, more than three of every four Americans in the southern temperate zone are entirely of European ancestry. Europeans, to borrow a term from apiculture, have swarmed again and again and have selected their new homes as if each swarm were physically repulsed by the others.

The Neo-Europes are intriguing for reasons other than the disharmony between their locations and the racial and cultural identity of most of their people. These lands attract the attention – the unblinking envious gaze – of most of humanity because of their food surpluses. They compose the majority of those very few nations on this earth that consistently, decade after decade, export very large quantities of food. In 1982, the total value of all agricultural exports in the world, of all agricultural products that crossed national borders, was $210 billion. Of this, Canada, the United States, Argentina, Uruguay, Australia, and New Zealand accounted for $64 billion, or a little over 30 percent, a total and a percentage that would be even higher if the exports of southern Brazil were added. The Neo-European share of exports of wheat, the most important crop in international commerce, was even greater. In 1982,

$18 billion worth of wheat passed over national boundaries, of which the Neo-Europes exported about $13 billion. In the same year, world exports of protein-rich soybeans, the most important new entry in international trade in foodstuffs since World War II, amounted to $7 billion. The United States and Canada accounted for $6.3 billion of this. In exports of fresh, chilled, and frozen beef and mutton, the Neo-Europes also lead the world, as well as in a number of other foodstuffs. Their share of the international trade in the world's most vitally important foods is much greater than the Middle East's share of petroleum exports (see Brown 1984: 19).

The dominant role of the Neo-Europes in international trade in foodstuffs is not simply a matter of brute productivity. . . . These regions lead the world in production of food *relative to the amount locally consumed*, or, to put it another way, in the production of surpluses for export. To cite an extreme example, in 1982 the United States produced only a minuscule percentage of the world's rice, but it accounted for one-fifth of all exports of that grain, more than any other nation (*World Almanac* 1983: 156). . . .

[L]et us turn to the subject of the Europeans' proclivity for migrating overseas, one of their most distinctive characteristics, and one that has had much to do with Neo-European agricultural productivity. Europeans were understandably slow to leave the security of their homelands. The populations of the Neo-Europes did not become as white as they are today until long after Cabot, Magellan, and other European navigators first came upon the new lands, nor until many years after the first white settlers made their homes there. In 1800, North America, after almost two centuries of successful European colonization, and though in many ways the most attractive of the Neo-Europes to Old World migrants, had a population of fewer than 5 million whites, plus about 1 million blacks. Southern South America, after more than two hundred years of European occupation, was an even worse laggard, having less than half a million whites. Australia had only 10,000, and New Zealand was still Maori country.

Then came the deluge. Between 1820 and 1930, well over 50 million Europeans migrated to the Neo-European lands overseas. That number amounts to approximately one-fifth of the entire population of Europe at the beginning of that period. Why such an enormous movement of peoples across such vast distances? Conditions in Europe provided a considerable push – population explosion and a resulting shortage of cultivable land, national rivalries, persecution of minorities – and the application of steam power to ocean and land travel certainly facilitated long distance migration. But what was the nature of the Neo-European pull? The attractions were many, of course, and they varied from place to place in these new-found lands. But underlying them all, and coloring and shaping them in ways such that a reasonable man might be persuaded to invest capital and even the lives of his family in Neo-European adventures, were factors perhaps best described as biogeographical. . . .

Where are the Neo-Europes? Geographically they are scattered, but they are in similar latitudes. They are all completely or at least two-thirds in the temperate zones, north and south, which is to say that they have roughly similar climates. The plants on which Europeans historically have depended for food and fiber, and the animals on which they have depended for food, fiber, power, leather, bone, and manure, tend to prosper in warm-to-cool climates with an annual precipitation of 50 to 150 centimeters. These conditions are characteristic of all the Neo-Europes, or at least of their fertile parts in which Europeans have settled densely. One would expect an Englishman, Spaniard, or German to be attracted chiefly to places where wheat and cattle would do well, and that has indeed proved to be the case.

The Neo-Europes all lie primarily in temperate zones, but their native biotas are clearly different from one another and from that of northern Eurasia. . . . European colonists sometimes found Neo-European flora and fauna exasperatingly bizarre. Mr. J. Martin in Australia in the 1830s complained that

> the trees retained their leaves and shed their bark instead, the swans were black, the eagles white, the bees were stingless, some mammals had pockets, others laid eggs, it was warmest on the hills and coolest in the valleys, [and] even the blackberries were red.
>
> (Powell 1976: 13–14)

There is a striking paradox here. The parts of the world that today in terms of population and culture are most like Europe are far away from Europe – indeed, they are across major oceans – and although they are similar in climate to Europe, they have indigenous floras and faunas different from those of Europe. The regions that today export more foodstuffs of European provenance – grains and meats – than any other lands on earth had no wheat, barley, rye, cattle, pigs, sheep, or goats whatsoever five hundred years ago. The resolution of the paradox is simple to state, though difficult to explain. North America, southern South America, Australia, and New Zealand are far from Europe in distance but have climates similar to hers, and European flora and fauna, including human beings, can thrive in these regions if the competition is not too fierce. In general, the competition has been mild. On the pampa, Iberian horses and cattle have driven back the guanaco and rhea; in North America, speakers of Indo-European languages have overwhelmed speakers of Algonkin and Muskhogean and other Amerindian languages; in the antipodes, the dandelions and house cats of the Old World have marched forward, and kangaroo grass and Kiwis have retreated. Why? Perhaps European humans have triumphed because of their superiority in arms, organization, and fanaticism, but what in heaven's name is the reason that the sun never sets on the empire of the dandelion? Perhaps the success of European imperialism has a biological, an ecological, component. . . .

The Neo-Europes collectively and singly are important, more important than their sizes and populations and even wealth indicate. They are enormously productive agriculturally, and with the world's population thrusting toward 5 billion and beyond, they are vital to the survival of many hundreds of millions. The reasons for this productivity include the undeniable virtuosity of their farmers and agricultural scientists and, in addition, several fortuitous circumstances that require explanation. The Neo-Europes all include large areas of very high photosynthetic potential, areas in which the amount of solar energy, the sunlight, available for the transformation of water and inorganic matter into food is very high. The quantity of light in the tropics is, of course, enormous, but less than one might think, because of the cloudiness and haziness of the wet tropics and the unvarying length of the day year-round. . . .

Taking all in all, the zones of the earth's surface richest in photosynthetic potential lie between the tropics and fifty degrees latitude north and south. There most of the food plants that do best in an eight-month growing season thrive. Within these zones the areas with rich soils that receive the greatest abundance of sunlight and, as well, the amounts of water that our staple crops require – the most important agricultural land in the world, in other words – are the central United States, California, southern Australia, New Zealand, and a wedge of Europe consisting of the southwestern half of France and the northwestern half of Iberia. All of these, with the exception of the European wedge, are within the Neo-Europes; and a lot of the rest of the Neo-European land, such as the pampa or Saskatchewan, is nearly as rich photosynthetically, and is as productive in fact, if not in theory (see Chang 1970). . . .

An extraordinarily, perhaps frighteningly, large number of humans elsewhere in the world depend on the Neo-Europes for much of their food, and it appears that more and more will as world population increases. . . . Often in defiance of ideology and perhaps of good sense, more and more members of our species are becoming dependent on parts of the world far away where pale strangers grow food for sale. A very great many people are hostage to the possible effects of weather, pests, diseases, economic and political vagaries, and war in the Neo-Europes.

The responsibilities of the Neo-Europeans require unprecedented ecological and diplomatic sophistication: statesmanship in farm and embassy, plus greatness of spirit. One wonders if their comprehension of our world is equal to the challenge posed by the current state of our species and of the biosphere. It is an understanding formed by their own experience of one to four centuries of plenty, a unique episode in recorded history. I do not claim that this plenty has been evenly distributed: the poor are poor in the Neo-Europes, and Langston Hughes's nagging question 'What happens to a dream deferred?' still nags, but I do insist that the people of the Neo-Europes almost universally believe that great material affluence can and should be attained by everyone, particularly in matters of diet. In Christ's Palestine, the multiplication of the loaves and fishes was a miracle; in the Neo-Europes it is expected. . . .

Today we are drawing on the advantages accruing from second entry, but widespread erosion, diminishing fertility, and the swift growth in the numbers of those dependent on the productivity of Neo-European soils remind us that the profits are finite. We are in need of a flowering of ingenuity equal to that of the Neolithic or, lacking that, of wisdom.

## Chapter 111

# RICHARD GROVE

## GREEN IMPERIALISM

From *Green Imperialism: Colonial Expansion, Tropical Island Edens And the Origins of Environmentalism* Cambridge: Cambridge University Press, 1995.

WHILE THE DEGREE OF POPULAR INTEREST in global environmental degradation may be something novel, the history of environmental concern and conservation is certainly not new. On the contrary, the origins and early history of contemporary western environmental concern and concomitant attempts at conservationist intervention lie far back in time. For example, the current fear of widespread artificially induced climate change, widely thought to be of recent origin, actually has ancient roots in the writings of Theophrastus of Erasia in classical Greece (Hughes 1985: 296–307; Glacken 1967). Later climatic theories formed the basis for the first forest conservation policies of many of the British colonial states. Indeed, as early as the mid eighteenth century, scientists were able to manipulate state policy by their capacity to play on fears of environmental cataclysm, just as they are today. By 1850 the problem of tropical deforestation was already being conceived of as a problem existing on a global scale and as a phenomenon demanding urgent and concerted state intervention. Now that scientists and environmentalists once again have the upper hand in state and international environmental policy, we may do well to recall the story of their first – relatively short-lived – periods of power. . . .

Early scientific critiques of 'development' or 'improvement' were, in fact, well established by the early nineteenth century. The fact that such critiques emerged under the conditions of colonial rule in the tropics is not altogether surprising. The kind of homogenizing capital-intensive transformation of people, trade, economy and environment with which we are familiar today can be traced back at least as far as the beginnings of European colonial expansion, as the agents of new European capital and urban markets sought to extend their areas of operation and sources of raw materials. It is clearly important, therefore, to try to understand current environmental concerns in the light of a much longer historical perspective of social responses to the impact of capital-intensive western and non-western economic forces. The evolution of a reasoned awareness of the wholesale vulnerability of earth to man and the idea of 'conservation', particularly as practiced by the state, has been closely informed by the gradual emergence of a complex European epistemology of the global environment. . . .

The older and far more complex antecedents of contemporary conservationist attitudes and policies have quite simply been overlooked in the absence of any attempts to deal with the history of environmental concern on a truly global basis. In particular, and largely for quite understandable ideological reasons, very little account has ever been taken of the central significance of the colonial experience in the formation of western environmental attitudes and critiques. Furthermore, the crucially pervasive and creative impact of the tropical and colonial experience on European natural science and on the western and scientific mind after the fifteenth century has been almost entirely ignored by those environmental historians and geographers who have sought to disentangle the history of environmentalism and changing attitudes to nature (see Thomas 1983; O'Riordan 1976). Added to this, the historically decisive diffusion of indigenous, and particularly Indian, environmental philosophy and knowledge into western thought and epistemology after the late fifteenth century has been largely dismissed. Instead, it has simply been assumed that European and colonial attempts to respond to tropical environmental change derived exclusively from metropolitan and northern models and attitudes. In fact the converse was true. The available evidence shows that the seeds of modern conservationism developed as an integral part of the European encounter with the tropics and with local classifications and interpretations of the natural world and its symbolism. As colonial expansion proceeded, the environmental experiences of Europeans and indigenous peoples living at the colonial periphery played a steadily more dominant and dynamic part in the construction of new European evaluations of nature and in the growing awareness of the destructive impact of European economic activity on the peoples and environments of the newly 'discovered' and colonized lands. . . .

Some researchers have suggested that Judaeo-Christian attitudes to the environment have been inherently destructive (see White 1967: 1202–7; Opie 1987: 2–19). Such claims are highly debatable. In fact they should probably be seen as a consequence of a perceptual confusion between the characteristically rapid ecological changes caused by the inherently transforming potential of colonizing capital and the consequences of culturally specific attitudes to the environment. In this connection it might be noted, for example, that rapid deforestation of the Ganges basin in pre-colonial Northern India during the sixteenth century does not appear to have been impeded by indigenous religious factors (see Erdosy 1998). There were, however, clear links between religious change during the sixteenth century and the emergence of a more sympathetic environmental psychology. Above all, the advent of Calvinism in seventeenth-century Europe seems to have lent a further impetus to the Edenic search as a knowledge of the natural world began to be seen as a respectable path to seeking knowledge of God. . . .

During the fifteenth century the task of locating Eden and re-evaluating nature had already begun to be served by the appropriation of the newly discovered and colonized tropical islands as paradises. This role was reinforced by the establishment of the earliest colonial botanical gardens on these islands and on one mainland 'Eden', the Cape of Good Hope (Karstens 1957). These imaginative projections were not, however, easily confined. Conceptually, they soon expanded beyond the physical limitations of the botanical garden to encompass large tropical islands. Subsequently the colonialist encounter in India, Africa and the Americas with large 'wild' landscapes apparently little altered by man, along with their huge variety of plants (no longer confinable, as they had been, to one botanical garden), meant that the whole tropical world became vulnerable to colonization by an ever-expanding and ambitious imaginative symbolism. Frequently such notions were closely allied to the stereotyping of luckless indigenous people as 'noble savages'. Ultimately, then, the area of the new and far more complex European 'Eden' of the late eighteenth and early nineteenth century knew no real bounds. Even Australia and Antarctica, in recent years, have not been immune to being termed Edens. The imaginative hegemony implied by new valuations of nature, which had themselves been stimulated by the encounter with the colonial periphery, had enormous implications for the way in which the real – that is, economic – impact of the colonizer on the natural environment was assessed by the new ecological critics of colonial rule. . . .

This new sensitivity developed, ironically, as a product of the very specific, and ecologically destructive, conditions of the commercial expansion of the Dutch and English East India companies and, a little later, of the French East India Company. The conservationist ideology which resulted was based both upon a new kind of evaluation of tropical nature and upon the highly empirical and geographically circumscribed observations of environmental processes which the experience of tropical island environments had made possible. After about 1750 the rise to prominence of climatic theories gave a new boost to conservationism, often as part of an emerging agenda of social reform, particularly among the agronomes and physiocrats of Enlightenment France. These theories, as well as the undeniably radical and reformist roots of much eighteenth-century environmentalism in both its metropolitan and colonial manifestations, have long eluded scholarly attention.

Instead it has been assumed by some historians that the colonial experience was not only highly destructive in environmental terms but that its very destructiveness had its roots in ideologically 'imperialist' attitudes towards the environment (see Worster 1977: 29–55). On the face of it, this does not seem an extraordinary thesis to advance, particularly as the evidence seems to indicate that the penetration of western economic forces which was facilitated by colonial annexation did indeed promote a rapid ecological transformation in many parts of the world. This was especially true in the late nineteenth century in Southern Africa, where a particularly exploitative agricultural and hunting ethos at first prevailed (see Mackenzie 1988; Pringle 1983). On closer inspection, however, the hypothesis of a purely destructive environmental imperialism does not appear to stand up at all well. In the first place, rapid and extensive ecological transition was frequently a feature of pre-colonial landscapes and states, either as a consequence of the development of agriculture or for other sociological reasons. . . .

Alongside the emergence of professional natural science, the importance of the island as a mental symbol continued to constitute a critical stimulant to the development of concepts of environmental protection as well as of ethnological and biological identity. Half a century before an acquaintance with the Falklands (see Grove 1985) and Galapagos provided Charles Darwin with the data he required to construct a theory of evolution, the isolated and peculiar floras of St Helena, Mauritius and St Vincent had already sown the seeds for concepts of rarity and a fear of extinction that were, by the 1790s, already well developed in the minds of French and British colonial botanists. Furthermore, the scientific odysseys of Anson, Bougainville and Cook served to reinforce the significance of specific tropical islands – Otaheite and Mauritius in particular – as symbolic and practical locations of the social and physical Utopias beloved of the early Romantic reaction to the Enlightenment. . . .

To summarize, the ideological and scientific content of early colonial conservationism as it had developed under early British and French colonial rule amounted by the 1850s to a highly heterogeneous mixture of indigenous, Romantic, Orientalist and other elements. Of course the thinking of the scientific pioneers of early conservationism was often contradictory and confused. Many of their prescriptions were constrained by the needs of the colonial state, even though the state at first resisted the notion of conservation. In the second half of the nineteenth century, too, forest conservation and associated forced resettlement methods were frequently the cause of a fierce oppression of indigenous peoples and became a highly convenient form of social control. Indeed, resistance to colonial conservation structures became a central element in the formation of many early anti-imperialist nationalist movements (see Grove 1990: 15–51). However, despite the overarching priorities and distortions of colonialism, the early colonial conservationists nevertheless remain entitled to occupy a very important historical niche. This is, above all, because they were able to foresee, with remarkable precision, the apparently unmanageable environmental problems of today. Their antecedents, motivations and agendas, therefore demand our close attention.

## Chapter 112

# KEN SARO-WIWA

# TRIAL STATEMENT

*Kenule Beeson Saro-Wiwa, founder of the Movement for the Survival of the Ogoni People, internationally acclaimed author and environmental and human rights activist, was hanged by the government of Nigeria on Nov. 10, 1995 despite world-wide pleas for clemency. Eight other activists were hanged with him. They were convicted of murder after several pro-government leaders were killed at a rally, even though the condemned men were not even present — or even alleged to be present — at the scene.*

K EN SARO-WIWA'S STATEMENT before the Civil Disturbances Tribunal that convicted and condemned him to death on October 31, 1995 at Port Harcourt, Rivers State, Nigeria:

My lord,

We all stand before history. I am a man of peace, of ideas. Appalled by the denigrating poverty of my people who live on a richly endowed land, distressed by their political marginalization and economic strangulation, angered by the devastation of their land, their ultimate heritage, anxious to preserve their right to life and to a decent living, and determined to usher to this country as a whole a fair and just democratic system which protects everyone and every ethnic group and gives us all a valid claim to human civilization, I have devoted my intellectual and material resources, my very life to a cause in which I have total belief and from which I cannot be blackmailed or intimidated. I have no doubt at all about the ultimate success of my cause, no matter the trials and tribulations which I and those who believe with me may encounter on our journey. Not imprisonment nor death can stop our ultimate victory.

I repeat that we all stand before history. I and my colleagues are not the only ones on trial. Shell is here on trial and it is as well that it is represented by counsel said to be holding a watching brief. The Company has, indeed, ducked this particular trial, but its day will surely come and the lessons learnt here may prove useful to it for there is no doubt in my mind that the ecological war that the Company has waged in the Delta will be called to question sooner than later and the crimes of that war be duly punished. The crime of the Company's dirty wars against the Ogoni people will also be punished. On trial also is the Nigerian nation, its present rulers and those who assist them. Any nation which can do to the weak and disadvantaged what the Nigerian nation has done to the Ogoni, loses a claim to independence and to freedom from outside influence. I am not one of those who shy away from protesting injustice and oppression, arguing that they are expected in a military regime. The military do not act alone. They are supported by a gaggle of politicians, lawyers, judges, academics and businessmen, all of them hiding under the claim that they are only doing their duty, men and women too afraid to wash their pants of urine. We all stand on trial, my lord, for by our actions we have denigrated our Country and jeopardized the future of our children. . . . I predict that the scene here will be played and replayed by generations yet unborn. Some have already cast themselves in the role of villains, some are tragic victims, some still have a chance to redeem themselves. The choice is for each individual. I predict that the denouement of the riddle of the Niger delta will soon come. The agenda is being set at this trial. Whether the peaceful ways I have favored will prevail depends on what the oppressor decides, what signals it sends out to the waiting public.

In my innocence of the false charges I face here, in my utter conviction, I call upon the Ogoni people, the peoples of the Niger delta, and the oppressed ethnic minorities of Nigeria to stand up now and fight fearlessly and peacefully for their rights. History is on their side. God is on their side. For the Holy Quran says in Sura 42, verse 41: 'All those that fight when oppressed incur no guilt, but Allah shall punish the oppressor.' Come the day.

## Chapter 113

# VAL PLUMWOOD

# DECOLONIZING RELATIONSHIPS WITH NATURE

From William Adams and Martin Mulligan (eds) *Decolonizing Nature: Strategies for Conservation in a Post-colonial Era* London: Earthscan, 2003.

**I**T IS USUALLY NOW ACKNOWLEDGED that in the process of Eurocentric colonization, the lands of the colonized and the non-human populations who inhabit these lands were often plundered and damaged, as an indirect result of the colonization of the people. What we are less accustomed to acknowledging is the idea that the concept of colonization can be applied directly to non-human nature itself, and that the relationship between humans, or certain groups of them, and the more-than-human world might be aptly characterized as one of colonization. This is one of the things that an analysis of the structure of colonization can help to demonstrate. Analysing this structure can cast much light upon our current failures and blind spots in relationships with nature, because we are much more able to see oppression in the past or in contexts where it is not our group who is cast as the oppressor. It is a feature of colonizing and centric thought systems that they can disguise centric relationships in a way that leaves the colonizer (and sometimes even the colonized) blind to their oppressive character.

Although not largely thought of as the non-human sphere, in contrast with the truly or ideally human (identified with reason), the sphere of 'nature' has, in the past, been taken to include what are thought of as less ideal or more primitive forms of the human. This included women and supposedly 'backward' or 'primitive' people, who were seen as exemplifying an earlier and more animal stage of human development. The supposed deficit in rationality of these groups invites rational conquest and reordering by those taken to best exemplify reason – namely, elite white males of European descent and culture (Said 1978). 'Nature' then encompasses the underside of rationalist dualisms that oppose reason to nature, mind to body, emotional female to rational male, human to animal, and so on. Progress is the progressive overcoming, or control of, this 'barbarian' non-human or semi-human sphere by the rational sphere of European culture and 'modernity'. In this sense, a culture of rational colonization in relation to those aspects of the world, whether human or non-human, that are counted as 'nature' is part of the general cultural inheritance of the West (Plumwood 1993), underpinning the specific conceptual ideology of European colonization and the bioformation of the neo-Europes (Crosby 1986).

An encompassing and underlying rationalist ideology applying both to humans and to non-humans is thus brought into play in the specific processes of European colonization. This ideology is applied not only to indigenous peoples but to their land, which was frequently portrayed in colonial justifications as unused, underused or empty – areas of rational deficit.

The ideology of colonization, therefore, involves a form of *anthropocentrism* that underlies and justifies the colonization of non-human nature through the imposition of the colonizers' land forms and visions of ideal landscapes in just the same way that *Eurocentrism* underlies and justifies modern forms of European colonization, which see indigenous cultures as 'primitive', less rational and closer to children, animals and nature (Plumwood 1993, 1996). The resulting Eurocentric form of anthropocentrism draws upon, and parallels, Eurocentric imperialism in its logical structure. It tends to see the human sphere as beyond or outside the sphere of 'nature', construes ethics as confined to the human (allowing the non-human sphere to be treated instrumentally), treats non-human *difference* as inferiority, and understands both non-human agency and value in hegemonic terms that deny and subordinate them to a hyperbolized human agency.

The colonization of nature thus relies upon a range of conceptual strategies that are employed also within the human sphere to support supremacism of nation, gender and race. The construction of non-humans as 'Others' involves both distorted ways of seeing sameness, continuity or commonality with the colonized 'Other', and distorted ways of seeing their difference or independence. The usual distortions of continuity or sameness construct the ethical field in terms of moral dualism, involving a major boundary or gulf between the 'One' and the 'Other' that cannot be bridged or crossed. This can be seen, for example, in the gulf between an elite, morally considerable group and an out-group defined as 'mere resources' for the first group. Such an out-group need not, or cannot, be considered in similar ethical terms as the first group. In the West, especially, this gulf is usually established by constructing non-humans as lacking in the very department that Western rationalist culture has valued above all else and identified with the human – that of mind, rationality or spirit – and what is often seen as the outward expression of mind in the form of language and communication. The excluded group is conceived, instead, in the reductionist terms established by mind/body or reason/nature dualism: 'mere' bodies, which can thus be servants, slaves, tools or instruments for human needs and projects. Reductionist and dualistic constructions of the non-human remain common today, especially among scientists. . . .

Hyper-separation is an emphatic form of separation that involves much more than just recognizing difference. Hyper-separation means defining the dominant identity emphatically against, or in opposition to, the subordinated identity, by exclusion of their real or supposed qualities. The function of hyper-separation is to mark out the Other for separate and inferior treatment. Thus, 'macho' identities emphatically deny continuity with women and try to minimize qualities seen as being appropriate for, or shared with, women. Colonizers exaggerate differences – for example, through emphasizing exaggerated cleanliness, 'civilized' or 'refined' manner, body covering, or alleged physiological differences between what are defined as separate races. They may ignore or deny relationship, conceiving the colonized as less than human. The colonized are described as 'stone age', 'primitive' or as 'beasts of the forest', and this is contrasted with the qualities of civilization and reason that are attributed to the colonizer.

Similarly, the human 'colonizer' treats nature as radically Other, and humans as emphatically separated from nature and animals. From an anthropocentric standpoint, nature is a hyper-separate lower order, lacking any real continuity with the human. This approach stresses heavily those features that make humans different from nature and animals, rather than those we share with them. Anthropocentric culture often endorses a view of the human as outside, and apart from, a plastic, passive and 'dead' nature, which lacks agency and meaning. A strong ethical discontinuity is felt at the human species boundary, and an anthropocentric culture will tend to adopt concepts of what makes a 'good' human being that reinforce this discontinuity by devaluing those qualities of human selves and human cultures that it associates with nature and animality. . . .

## Homogenization/stereotyping

The Other is not an individual but a member of a class stereotyped as interchangeable, replaceable, all alike – that is, as homogenous. Thus, essential female and 'racial' nature is uniform and unalterable (Stepan 1993). . . .

Ronald Reagan's famous remark 'You've seen one redwood, you've seen them all' invokes a parallel homogenization of nature. An anthropocentric culture rarely sees nature and animals as individual centres of striving or needs, doing their best in their conditions of life. Instead, nature is conceived in terms of interchangeable and replaceable units (as 'resources'), rather than as infinitely diverse and always in excess of knowledge and classification. Anthropocentric culture conceives nature and animals as all alike in their lack of consciousness, which is assumed to be exclusive to the human. Once nature and animals are viewed as machines or automata, minds are closed to the range and diversity of their mind-like qualities. Human-supremacist models promote insensitivity to the marvellous diversity of nature, since they attend to differences in nature only if they are likely to contribute in some obvious way to human interests, conceived as separate from those of nature. Homogenization leads to a serious underestimation of the complexity and irreplaceability of nature. These two features of human/nature dualism – radical exclusion and homogenization – work together to produce, in anthropocentric culture, a polarized understanding in which the human and non-human spheres correspond to two quite different substances or orders of being in the world. . . .

Nature is represented as inessential and massively denied as the unconsidered background to technological society. Since anthropocentric culture sees non-human nature as a basically inessential constituent of the universe, nature's deeds are systematically omitted from account and consideration in decision-making. Dependency upon nature is denied, systematically, so that nature's order, resistance and survival requirements are not perceived as imposing a limit upon human goals or enterprises. For example, crucial biospheric and other services provided by nature, and the limits they might impose upon human projects, are not considered in accounting or decision-making. We only pay attention to them after disaster occurs, and then only to 'fix up' for a while. Where we cannot quite forget how dependent upon nature we really are, dependency appears as a source of anxiety and threat, or as a further technological problem to be overcome. Accounts of human agency that background nature's 'work' as a collaborative co-agency feed hyperbolized concepts of human autonomy and independence from nature. . . .

To counter the first dynamic of 'us-them' polarization, it is necessary to acknowledge and reclaim continuity and overlap between the polarized groups, as well as internal diversity within them. However, countering the second dynamic of denial, assimilation and instrumentalization requires recognition of the Other's difference, independence and agency . . . Such a biospheric Other is not a background part of our field of action or subjectivity, not a mere precondition for human action, not a refractory foil to self. Rather, biospheric Others can be other ethical and communicative subjects and other actors in the world – others to whom we owe debts of gratitude, generosity and recognition as prior and enabling presences.

The re-conception of nature in the agentic terms that deliver it from construction as a background is perhaps the most important aspect of moving to an alternative ethical framework, because 'backgrounding' is perhaps the most hazardous and distorting effect of 'othering' from a human prudential point of view. When the Other's agency is treated as background or denied, we give the Other less credit than is due to them; we can come to take for granted what they provide for us; and we pay attention only when something goes wrong. This is a problem for prudence, as well as for justice. When we are, in fact, dependent upon this Other, we can gain an illusory sense of our own ontological and ecological independence, and it is just such a sense that seems to pervade the dominant culture's contemporary disastrous misperceptions of its economic and ecological relationships. . . .

Countering a hegemonic dualism, such as that between nature and culture, presents many traps for young players. . . . Dualistic concepts of nature insist that 'true' nature must be entirely free of human influence, ruling out any overlap between nature and culture. This reversal, which suggests that only 'pure' nature (perhaps in the form of 'wilderness') is valuable or has needs that should be recognized and respected, leaves us without adequate ways of recognizing and tracking the agency of the more-than-human sphere in our daily lives, since this rarely appears in a pure or unmixed form. Yet, this is one of the most important things we need to do to counter the widespread and very damaging illusion that modern urban life has 'overcome' the need for nature or has become disconnected from nature.

Polarized concepts of wilderness as the realm of an idealized, pure nature remain popular in the environment movement where they are often employed for protective purposes, to keep, for example, market uses of land at bay. The concept of wilderness has been an important part of the colonial project, and attempts by neo-European conservation movements to press it into service as a means of resisting the continuing colonization of nature must take account of its double face.

On the one hand, it represents an attempt to recognize that nature has been colonized and to give it a domain of its own; on the other hand, it continues and extends the colonizing refusal to recognize the prior presence and agency of indigenous people in the land. If we understand wilderness in the traditional way, as designating areas that are purely the province of nature, then to call Australia, or parts of it, wilderness is to imply that no human influence has shaped its development. We imply that it is purely Other, having no element of human culture. However, the idea that the Australian continent, or even substantial parts of it, are pure nature, is insensitive to the claims of indigenous peoples and denies their record as ecological agents who have left their mark upon the land. Indigenous critics such as Marcia Langton have rightly objected that such a strategy colludes with the colonial concept of Australia as *terra nullius* and with the colonial representation of Aboriginal people as merely animal and as 'parasites on nature' (Langton 1996). To recognize that both nature and indigenous peoples have been colonized, we need to rethink, relocate and redefine our protective concepts for nature within a larger anti-colonial critique.

## Chapter 114

# GORDON SAYRE

# THE BEAVER AS NATIVE AND AS COLONIST

From 'The Beaver as Native and Colonist' *Canadian Review of Comparative Literature/Revue Canadienne de Littérature Comparée* September/décembre, 1995.

> The spoil of this animal has hitherto been the principal article in the commerce of New France. It is itself one of the greatest wonders in nature, and may very well afford many a striking lesson of industry, foresight, dexterity, and perserverance in labor.
>
> Pierre François-Xavier de Charlevoix [1744] (1976)
> *Journal of a Voyage to North America*, 1: 151

IN THE LITERATURE OF NEW FRANCE one finds many brief but colorful descriptions of the beaver. Nearly every colonial travel writer of the seventeenth and eighteenth century included in his text an account of this animal, which, as Charlevoix demonstrates, was as intriguing for its intelligence and social behavior as it was important for its economic value. Much more than an interesting object of natural history, the beaver was to the French in Canada what sugar, tobacco, or spices were to other colonies. Its double status – as a natural and social marvel when alive, and as a valuable commodity when dead – caused representations of the beaver to become overdetermined with strong and often contradictory ascriptions. A small, non-threatening herbivore, the beaver was anthropomorphized, idealized, and sentimentalized in print even as it was killed by the thousands in the forests of North America; its pelts stacked, shipped and traded in the pursuit of profit.

The beaver's function in literary representation is quite different from other animals famous in American literature. While a single whale or bear becomes for Melville or Faulkner symbolic of natural or psychic forces locked in a struggle with Man, the portrayal of the beaver is ethnographic, not dramatic, and plural, not singular. The colonial travellers who wrote of the beaver, such as Lahontan, Charlevoix, and Le Page du Pratz, are all known for their early ethnographies of Native American society, and it is in this context that the accounts of the beaver should be analyzed. The beaver's sociable life was represented in a timeless realm unaltered by man, much as ethnography was written in an achronic 'ethnographic present' where the intervention of the colonist or anthropologist is not described. The beaver's body and behavior was dismembered into assimilable pieces, much as ethnography separates life into categories like marriage and sexuality, clothing and ornament, health and subsistence, deflecting attention from the creatural and cultural survival of the organic whole. The ethnography of the beaver functioned in much the same way as the many descriptions of the manners and customs of the Indians to rhetorically reconstitute the undisturbed indigenous life which was being destroyed by the invading Europeans. James Clifford has written, 'Ethnography's disappearing object is, then, in significant degree, a rhetorical construct legitimating a representational practice: "salvage ethnography" in its widest sense. The other is lost, in disintegrating time and space, but saved in the text' (1986: 112). By the latter part of the nineteenth century if not sooner, the rhetoric familiar from the trope of the 'vanishing Indian' was being applied to the beaver. 'The beaver has gradually disappeared before the spread of civilization, which first settled along the shores of the Mediterranean. As each wave covered more of Europe, the range wherein the beaver existed perceptibly narrowed' (1892: 27), wrote Horace T. Martin in the beaver-ethnography *Castorologia*, 'North America remains the last stage on which are witnessed the scenes of a doomed culture' (29). Fur traders in the 1600s may not have imagined the beaver becoming extinct as it nearly did in the early twentieth century, but I believe that they did feel some anxiety over their dependence upon a limited resource, an anxiety which fanciful portrayals of the beaver helped to assuage.

Although ethnography is used for both, there is an important difference in the ideological (or as Clifford would say, 'allegorical') function of the representations of the beaver and of the Amerindian. Whereas Native American society was represented as so antithetical to European that it was virtually impossible to live in one without rejecting the values of the other, the social life of beavers apparently resembled civilized and particularly colonial society. The cooperative, industrious, and non-nomadic beaver was often held up as a worthy model for colonists to imitate, and in this it resembles only one other animal, also a social creature, one which Crevecoeur extolled: 'The well-known industry of bees, that excellent government which pervades their habitations, that never-ceasing industry by which they are actuated' (1981: 244).

François-Marc Gagnon sees in the written and pictorial representations of the beaver in the seventeenth century a measure of the epistemic shift in natural history identified by Michel Foucault, from a discourse of similitude using analogies, such as with the otter, fish or lamb, to a discourse of representation which isolates the beaver as an autonomous natural object (1984: 201). . . . European readers might have been familiar with the beaver had it not been hunted to extinction in Western Europe. Since it was so novel, anatomical analogies were needed to relate the beaver to more familiar animals, and thus assimilate it to human uses. Physical descriptions of the beaver break up the animal into small parts and sort the pieces into some of the categories listed by Foucault. Following its dual status as natural marvel and as commodity, the beaver is seen first through its organs and limbs, analyzed by their resemblance to those of humans or other animals and by their use for the beaver; and second, by its resources, analyzed according to their use for Europeans. Where the beaver transcends Foucault's *episteme* of the science of animals, however, is in the social characteristics that place it alongside contemporary ethnological thought.

While the idealized portrayal of the beaver resembles that of the Noble Savage in its ideological function – to compensate for or justify its imminent demise – it is quite different in its content, and a similarity between the two societies is rarely proposed by the European writers. Instead, the Indians are said to see themselves in the beaver. Antoine-Denis Raudot reported of the Indians that 'They believe that these animals are a nation. They see so much intelligence in them that they cannot help but compare them to themselves' (1904: letter 7). LeClercq too remarked that the Indians believe the beavers are organized in Nations. Thomas Morton, in his *New English Canaan*, recounts asking the Indians 'whether they eat of the Beavers, to which they replied Matta, (noe) saying they were almost Beavers brothers' ([1637] 1967: 237). . . .

Accounts of the human qualities of the beaver which an author thinks his readers may find hard to believe, or insulting to human vanity, are often attributed to the Indians. As well as affording plausible deniability for improbable facts included in travel narratives which advertise their veracity, this deferral to the Indians enacts a new relation between the ethnographic portrayals of the Indian and of the beaver. Where the Indian serves as informant for the *Moeurs des Castors*, the beaver is constituted at a second remove from the European observer, and for once the Indian assumes the powerful role of ethnographer. European is to Indian as Indian is to beaver. Of course, the text we read is still in French or English and no eighteenth-century Algonquian author has left us his written description of the beaver. But in thus doubling the ethnographic gaze, the European texts both extend and undermine the authority of the ethnographic genre. The Indians apparently regard the culture of the beaver through the same epistemological categories which the Europeans use when they describe the Indians. And yet if the Indians' notions of the beaver are wrong, the European's notions of the Indians might be equally faulty. . . .

Labor power and organization were often key components of the fabulous accounts of the beaver, myths designed by and for the imagination of colonists. While Indians living outside the colony tempt settlers to 'go savage', to abandon civilization for an exotic and primitive form of life, beavers, though they too live in the forest, live in a manner which French writers admire and want colonists to imitate. The idealized Indian or 'Noble Savage' is attractive for qualities suppressed by or alternative to those of the dominant classes of European society, and therefore appeals to the individual who desires to escape it. The 'Noble Beaver', on the other hand, represents dominant European values of work, planning, loyalty, and hierarchy. Charlevoix, addressing his correspondent, the Duchess of Lesdiguières, writes 'These, Madam, are all the advantages the beavers are capable of affording the commerce of this colony: their foresight, their unanimity, and that wonderful subordination we so admire in them' ([1744] 1976: 1: 158). Note that Charlevoix says 'Commerce' and not simply 'Life' or even 'Prosperity'; he lays bare the *raison d'être* of the colony. But the fur trade, ironically, encouraged subversive behavior which mimicked the Indians as it destroyed the beavers, since the traders or 'coureurs de bois' had to seek out and adopt the life of the Indians in order to obtain the most high quality pelts. Phillippe Jacquin, in his excellent history of French-Indian relations in the fur trade, describes the impulse to freedom which colonists and authorities experienced in Canada as: 'the footloose spirit in the roots of the populations' instincts, which the shock of immigration and the confrontation with indigenous society combined to reactivate' (1987: 111, my translation).

Several scholars have suggested that European colonists expected the Indians to live as the poorest did in Europe, and despised them all the more when their way of life did not conform to these expectations (see, for instance, Jennings 1976: chap. 5). Indian tribes wandered as poor beggars did in Europe, the colonial writers claimed, though they did not need to beg because they shared willingly among each other and with the newly-arrived Europeans. Amerindian men hunted as only the rich were able to do in Europe (the enormous differences in the techniques and purposes of hunting were ignored), and like them they enjoyed a great deal of leisure when not hunting. The women tended to crops, but Europeans often overlooked this too when they denounced the Indians' refusal to settle down to a peasant-like agricultural subsistence. Europeans tried to view the Indians as serfs, for then they might be controlled as serfs were. Jennings has also shown (46; 312) how the English colonists' rationale for attacking the Indians resembled that used against the Irish peasants during the same period.

In all these respects, the beaver offered a corrective to the anarchy of the colonial setting and the threatening vices of the Indians. Beavers built sturdy homes where they lived year-round and which sheltered their children from enemies and from the elements. . . .

Out of the beaver was fashioned a comprehensive ideology which both encouraged behavior desired by colonial officials and warned of the dangers of vagabondage. Hunting beaver pelts was subversive not only for the contacts it necessitated among the Savages, but also for the great profits it could bring. Bourgeois merchants took most of this profit, but entrepreneurial 'coureurs de bois' could make huge sums as well, which posed a threat to the class structure of the colony. The corrupt Indian whom the fur traders so resembled had also been corrupted by the beaver trade, as many colonial writers recognized. Before contact with the 'coreurs de bois' (always to be distinguished from contact with missionaries) the hypostatized Noble Savages had no *tien et mien*, did not code status in wealth, were generous, did not overhunt game, and did not find any great value in Beaver pelts (did not even wear them, says Charlevoix). Now they drive a hard bargain for the pelts and trade them for liquor. . . .

A great fictional narrative of psychological reflection and spiritual pursuit of the beaver, the novel that would place the beaver in the same class as the whale and the bear, was not possible because the beaver was an ally, not a barrier to colonization. The beaver of course continued to be one of Canada's most valuable resources into the nineteenth century, but as the bulk of the trade shifted into English hands and north to the inhospitable shores of Hudson Bay, the significance of the beaver for the ideology of settlement lessened. . . .

The legends about the beaver survived, however. In the nineteenth century at least two books were published which continued the range of discourses, from natural history to mythology and anthropology, which we have seen in the literature of New France. By emphasizing folklore, these books revive the renaissance *episteme* which predated natural history. Horace T. Martin's *Castorologia*, quoted above, was one, and the other was by the famous ethnographer of the Iroquois, Lewis Henry Morgan, who published *The American Beaver and His Works* in 1868. . . . Morgan preferred to call animals 'mutes' and believed that they were endowed with reason and the possibility of cultural evolution, such as the beaver's development of canals and perfection in the design of lodges. The stage theories which pervaded Morgan's and all anthropological theory in the nineteenth century were also applied to the beaver. Moreover, Morgan provided a new justification for representing the beaver more as a culture than as an animal, and for studying these representations within a history of ethnography as well as of natural history. The beaver demands attention from post-colonial theory, because the growth of colonial New France came at the expense of native American cultures, and one of these cultures was that of the beaver, a kind, furry creature which, ironically, was believed to mimic the values of the colonists.

## Chapter 115

# CARY WOLFE

## OLD ORDERS FOR NEW

From 'Old Orders for New: Ecology, Animal Rights and the Poverty
of Humanism' *Electronic Book Review* 4 (Winter), 1997.

E ARLY ON IN *THE NEW ECOLOGICAL ORDER*, the French philosopher
Luc Ferry characterizes the allure and danger of ecology in the postmodern moment. What
separates it from various other issues in the intellectual and political field, he writes, is that

> it can call itself a true 'world vision,' whereas the decline of political utopias, but
> also the parcelization of knowledge and the growing 'jargonization' of individual scien-
> tific disciplines, seemed to forever prohibit any plan for the globalization of thought
> . . . At a time when ethical guide marks are more than ever floating and undeter-
> mined, it allows the unhoped-for promise of rootedness to form, an objective
> rootedness, certain of a new moral ideal.
>
> (Ferry 1995: xx)

For Ferry – a staunch liberal humanist in the Kantian if not quite Cartesian tradition – this
vision conceals a danger to which contemporary European intellectuals are especially sensitive: not holism, nor even moralism, exactly, but that far more charged and historically
freighted thing, *totalitarianism*. Ferry's concern is that such 'world visions', incarnated in
contemporary environmentalism in movements such as Deep Ecology and ecofeminism,
threaten '[o]ur entire democratic culture', which 'since the French Revolution, has been
marked, for basic philosophical reasons, by the glorification of *uprootedness*, or *innovation*' (xxi).
Ferry's thesis – it becomes quite explicit in his comparison of environmental legislation under
the Third Reich with tenets of Deep Ecology in the book's second section – is that Deep
Ecology and its ilk have moved in to occupy the space left open by the passing of the political imaginaries of fascism and communism, so that denunciations of liberalism (and its corollary
in political praxis, reformism) may now be unmasked for what they are: critiques 'in the
name of *nostalgia*, or, on the contrary, in that of *hope*: either the nostalgia for a lost past, for
national identity flouted by the culture of rootlessness, or revolutionary hope in a radiant
future, in a classless and free society' (xxvi). . . .

Ferry is certainly right to draw our attention to the often uncritical nostalgia and romantic
holism of some varieties of environmental thought – problems that have been noted before
by critics from points on the map very different politically from Ferry's avowed liberal
humanism. . . .

But in defending democratic difference, everything hinges, of course, on precisely how
such terms are framed, and how difference is articulated – an index of which often may be

found in how its imagined opponents are painted. Here, as we shall see, Ferry's text gives us early and ample pause, not least in its impoverished notions of 'democracy' and the 'human'. As for the latter, Ferry wholly disengages the 'human' from problems of class power and from the determinative force of both discourse (conceived not merely as rhetoric but also in the stronger Foucauldian sense as materially institutionalized conditions of production), and psychoanalytic investment. Similarly, Ferry's notion of 'democracy' is extraordinarily thin: it is completely decoupled from the problem of *capitalism* as liberal democracy's de facto economic embodiment. Given the well-known importance of both class and race in contemporary environmentalism – in debates about 'environmental racism' and the disproportionate exposure to toxic waste and environmental degradation borne by the poor, as well as in discussions about how middle class, and how white, the contemporary environmental movement is – this is surprising and disabling for one as eager as Ferry to defend the heritage of 'democracy'. . . .

Deep Ecology was invented by Norwegian philosopher Arne Naess, formalized and codified by Naess and American philosophers Bill Devall and George Sessions, and more recently adapted by the European Greens. Deep Ecology proposes a fundamental change, from anthropocentric to 'biocentric', in how we view the relationship of *Homo Sapiens* to the rest of the biosphere. As Devall writes,

> There are two great streams of environmentalism in the latter half of the twentieth century. One stream is reformist, attempting to control some of the worst of the air and water pollution and inefficient land use practices in industrialised nations and to save a few of the remaining pieces of wild-lands as 'designated wilderness areas.' The other stream supports many of the reformist goals but is revolutionary, seeking a new metaphysics, epistemology, cosmology, and environmental ethics of the person/planet.
>
> (quoted in Ferry 1995: 60)

What Devall invokes here is the distinction that gave the movement its name: between 'shallow environmentalism' and 'deep ecology'. An eclectic blend (to put it mildly) of ideas drawn from Heidegger, Buddhism, Robinson Jeffers, and many other sources . . .

The deep ecology philosophical platform may be boiled down to this: this ultimate good is not harmony with nature, nor even holism per se, but rather something much more specific: *biodiversity*. Once this is recognized, we must affirm the *inherent value* of all forms of life that contribute to this ultimate good, and we must actively oppose all actions and processes by human beings and their societies that compromise these values.

The appeal of Deep Ecology and its demand that we recognize the inherent value of the biosphere and conduct ourselves accordingly is understandable for all sorts of scientific, ethical, historical, and political reasons. As Gregory Bateson points out in his influential collection *Steps to an Ecology of Mind*, 'the last hundred years have demonstrated empirically that if an organism or aggregate of organisms sets to work with a focus on its own survival and thinks that that is the way to select its adaptive motives, its "progress" ends up with a destroyed environment. If the organism ends up destroying its environment, it has in fact destroyed itself' (Bateson 1972: 451). . . .

The essential conservatism of Ferry's position is hard to spot at first because his framing of 'the new ecological order' sets against the 'fundamentalism' and moral Puritanism of contemporary environmentalism the apparent openness and commitment to change – the '*uprootedness*, or *innovation*' as he puts it (xxi) – of the liberal humanist tradition he defends. Ferry contends that 'the hatred of the *artifice* connected with our civilization of rootlessness' that we find in Deep Ecology 'is also a *hatred of humans as such*. For man is the antinatural being par excellence' (xxvii). According to the blend of Rousseau and Kant with which Ferry identifies himself, the '*humanitas* [of the human] resides in his freedom, in the fact that he is undefined, that his nature is to have no nature but to possess the capacity to distance himself from any code within which one may seek to imprison him. In other words: his essence is that he has no essence'. 'Romantic racialism and historicism are thus inherently impossible', Ferry continues. 'For what is racism at its philosophical core if not the attempt to define a category of humans by its essence?' (5). There seems much to admire here and very little to condemn. Unfortunately – as with his concept of democracy – the reality of Ferry's notion of the human is that it is a good deal less 'open' and 'innovative' than it at first appears. For even though Ferry condemns racism for its attempt to define a category of human beings by its essence, this is precisely what Ferry's liberal humanist *speciesism* does in relation to nonhuman others in his critique of animal rights philosophy.

. . . Ferry constantly presents as differences in *kind* what are only maintainable as differences in *degree*. In the case of animal rights, he consistently *over*states the degree to which 'the animal is programmed by a code which goes by the name of "instinct"' (5), and he constantly *under*states the degree to which new work in ethology has shown that many nonhuman animals demonstrate degrees of the volition, free will, and abstraction that Ferry is at great pains to protect as the sole domain of the human. At the same time, he exaggerates the degree to which the human being 'is *nothing* as determined *by nature*' (9), not bound by instinct, biological needs and intolerances, by sexuality, the body, and so on. . . .

Ferry attempts to forestall the pursuit of his humanism to its logical conclusions, but he can do so only at the price of an utterly question-begging and bare-faced retort to speciesism. 'Why sacrifice a healthy chimpanzee over a human reduced to a vegetable state?', Ferry asks.

> If one were to adopt the criteria that says there is continuity between men and animals, Singer might be right to consider as a 'speciesist' the priority accorded human vegetables. If on the other hand we adopt the criteria of freedom, it is not unreasonable to admit that we must respect humankind, even in those who no longer manifest anything but its residual signs.
>
> (42)

But of course it *is* 'unreasonable', because in this instance Ferry isn't relying upon the quality of 'freedom' at all to ethically adjudicate the matter – the human vegetable, by Ferry's own admission, does not possess this quality – but only membership in a given *species*. And this, of course, is no better than the racism that is supposedly impossible under Ferry's humanism.

It should come as no surprise, then, that Ferry is unable to satisfactorily address an important issue raised by animal rights philosophy: that the discourse and practice of speciesism in the name of liberal humanism has historically been turned upon other *humans* as well. It is entirely to the point that the first chapter of Singer's *Animal Liberation* is entitled 'All Animals are Equal, or Why Supporters of Liberation for Blacks and Women Should Support Animal Liberation Too' (1975: 1). That linkage has recently been made quite graphically in ecofeminism, in texts like Carol Adams's flawed but nevertheless important study, *The Sexual Politics of Meat*, which demonstrates that the species system not only makes possible the systematic killing of many millions of animals a year for food, product testing, and research, but also provides a ready-made symbolic economy which overdetermines the representation of women by transcoding the edible bodies of animals and the sexualized bodies of women (as chick, beaver, Playboy bunny) within an overarching 'logic of domination'. The discourse and institution of speciesism, then, is by no means limited to its overwhelmingly direct and disproportionate effects upon nonhumans. Indeed, as Gayatri Spivak puts it,

> the great doctrines of identity of the ethical universal, in terms of which liberalism thought out its ethical programmes, played history false, because the identity was disengaged in terms of who was and who was not human. That's why all of these projects, the justification of slavery, as well as the justification of Christianization, seemed to be alright: because, after all, these people had not graduated into human-hood, as it were.
>
> (229)

. . . As both Spivak and Adams suggest, you don't *need* the argument from 'racial essence' to justify oppression if you can control the discourses and institutions that reduce human beings to the status of objects. A belief that women have more free will and control over the finality of their actions than a cow or a pig does not prevent the use of the discourse of speciesism in the oppression of women. . . .

Most promising here, I think, is the pragmatist approach to the problem, within which it is perfectly possible to argue that taking account of the ethical relevance of the work of ethologists like Goodall *does not mean committing ourselves to naturalism in ethics*. From a pragmatist point of view, all it means is that, in the historically and socially contingent discourse called ethics, we are obliged – precisely *because* ethics cannot ground itself in a representationalist relation to the object – *to apply consistently the rules we devise for determining subjectivity, personhood, and their ethically relevant traits and behaviors, without prejudice towards species or anything else*. The strength of this position (as Stanley Cawell or Gianni Vattimo might say) lies precisely in its 'weakness'. We need not cling to any empiricist notion about what Goodall or anyone else has discovered about nonhuman animals – any more than we need to do the same for our knowledge of human beings – to insist that when our generally agreed-upon markers for ethical consideration are observed in species other than *Homo sapiens*, we are obliged to take them into account equally and to respect them accordingly. This amounts to nothing more than taking the humanist conceptualization of the problem at its word, and then being rigorous about it – and then showing how humanism must, if rigorously pursued, generate its own deconstruction once these 'defining' characteristics are found beyond the species barrier. But this, of course, is precisely what Ferry is unable and unwilling to do.

In the end, then, it is Ferry's 'human', and not, as he argues, the nonhuman animal, who is 'the enigmatic being', the 'dreamed object' – 'enigmatic' because incoherent, and 'dreamed' because an imaginary subject, a fantasy. And yet, for all that, quite familiar. For as both Georges Bataille (in *Theory of Religion*) and Jacques Derrida (in 'Eating Well' and 'Force of Law') remind us, the humanist concept of subjectivity is inseparable from the discourse and institution of speciesism, which relied upon the tacit acceptance – and nowhere more clearly (as Slavoj Žižek has noted) than in Ferry's beloved Kant – that the full transcendence of the 'human' requires the sacrifice of the 'animal' and the animalistic, which in turn makes possible a symbolic economy in which we can engage in a 'non-criminal putting to death' (as Derrida puts it) not only of animals, but other humans as well, by marking them as animal. It may be the case, as Ferry argues, that ethics is always ineluctably human, always about human concepts and not about objects, but what Ferry's concept of 'the human' fails to acknowledge

– indeed his project depends on its disavowal – is how this constitutive and finally desperate repression fatally generates what Žižek calls 'humanism's self-destructive dimension' (Žižek 2001: 26). As Žižek puts it,

> The subject 'is' only insofar as the Thing (the Kantian Thing in itself as well as the Freudian impossible-incestuous object, das Ding) is sacrificed, 'primordially re-pressed'. . . . This 'primordial repression' introduces a fundamental imbalance in the universe: the symbolically structured universe we live in is organized around a void, an impossibility (the inaccessibility of the Thing in itself).
>
> (Žižek 2001: 181)

'Therein', Žižek continues, 'consists the ambiguity of the Enlightenment'; the transcendence of the Enlightenment subject is shadowed by 'a fundamental prohibition to probe too deeply into the obscure origins, which betrays a fear that by doing so, one might uncover something monstrous' (136).

We could scarcely do better than Žižek's characterization to provide a thumbnail psycho-analysis of Ferry's *The New Ecological Order*. But when we remember, with Derrida, that the effectiveness of the discourse of species, when applied to social others of *whatever* sort, relied upon a prior taking for granted of the institution of speciesism – that is, upon the ethical acceptability of the systematic, institutionalized killing of nonhuman others – then it is clear that while the ethical priority of confronting speciesism may begin with Man and his self-destructive humanism, it does not end there.

# PART NINETEEN

# The Sacred

FROM THE ENLIGHTENMENT ONWARDS, as Peter Van der Veer and Richard King make clear, there has been a tendency in Euro-American thought to assume that the secular has replaced the sacral as the obvious and unchallengeable mode by which the world is best interpreted. This is a significant part of the whole issue of valuation which post-colonial resistance engages in so many ways. European societies since the eighteenth century, while continuing to subscribe to a notional religious affiliation have really understood truth as a matter for scientific, secular reason alone. The sacral has been dismissed as at best a myth and at worst a superstition. This Eurocentric stress on the secular has been part of the exported baggage of colonization, so that the European-educated elites of these societies, as Viswanathan shows here, and as Elleke Boehmer also suggests in the section on Resistance and Representation, even those strongest in their anti-colonialism, took for granted the need to 'reform' their cultures by providing rational readings of their oldest cultural modes. Thus along with Eurocentric political forms such as the 'nation' a post-enlightenment rational for sacral elements in society became a global consequence of imperialism. It is not surprising, therefore, that resistance to these forces should have taken the form of a renewed sense of the sacral as offering an alternative to European models of thought.

At the beginning of the last century W. E. B. DuBois commented that race would be the dominant issue of the century ahead. As we traverse the opening decade of the twenty-first century it becomes clear that, while race and colour will continue to be a major force in dividing the world and reinforcing the divisions that have emerged, a new force has also entered the arena of struggle. The division of peoples based on their religious beliefs. At its simplest, this is epitomized by the stress on the emergence of what are called 'fundamentalist' ideas. These are very much part of the contemporary picture many people in the West now have of Islamic societies. They are also present in the emergence of modern forms of Hindu communal identities affecting the politics of the Indian sub-continent. But it is also necessary to see these non-Western examples of extreme religious affiliation as emerging alongside renewed, intolerant forms of Christian fundamentalism, especially, though not exclusively, in America, and in the more extreme forms of orthodoxy that characterize elements in the modern Zionist state of Israel. In short, fundamentalism, by which perhaps we mean a literal and activist assertion of religion as the most important element in individual and communal identity is a crucial element in many of the most persistent and dangerous encounters facing us in the new millennium.

But beyond this extremist element there is also a sense in which the sacral is emerging again as part of a broader rethinking of post-colonial identity. This has been especially the case in the cultural re-assertion of indigenous peoples in settler societies, where the sacred has been tied to a sense of 'belonging' to specific place and in whose view the world is literally brought into being by its sacred narratives. Baldridge and Donaldson in their different ways illustrate how a recovery of the sacral may form resistance, both by reasserting the traditional modes and by appropriating the imported religions of the colonizers to support a new identifying sacrality. As part of a wider desire to assert that the forms of European thought are not a given for modernity or for an effective social polity post-colonial critics are rereading the sacred texts of their own cultures and those of their erstwhile masters. Thus, as Sugirtharajah shows, although the post-colonial believer may have been controlled by traditional readings of sacred texts such as the Christian Bible, they are increasingly engaging with these texts to produce *resistant* readings that expose the biased interpretations that the colonialists employed to turn them into instruments of social control. For these and other reasons it is increasingly clear that the relationship between the sacral and the secular in the new century will be a complex one, and neither the simplistic assertions of the European *philosophe* nor the equally simplistic reassertions of religious fundamentalism is likely to be a sufficient means to confront the challenges ahead.

## Chapter 116

# GAURI VISWANATHAN

## CONVERSION, 'TRADITION' AND NATIONAL CONSOLIDATION

From *Outside the Fold: Conversion, Modernity and Belief* Princeton, N.J.: Princeton University Press, 1998.

. . .

**B**ECAUSE CONVERSION'S ALLIANCE WITH cultural criticism is so apparent, especially when accepted as an activity rather than a state of mind, there is an obvious temptation to read conversion in general as originating in motives of critique. Undoubtedly, change of religion is not merely an oppositional activity, though it may begin as such. Furthermore, conversion expresses an altered consciousness deriving from the construction of norms within culture. This is a central point in William James's *The Variety of Religious Experience*, as it is indeed in a great deal of writing about religious experience. For James the experience of awakened religious consciousness validates the believing self. Its significance for cultural development lies in the fact that the culture in which conversion occurs lends its own structural features to the content of religious experience. In turn, the heightened religious subjectivity makes visible the range of meanings embedded in cultural forms.

But the more interesting question is whether or not there is an internal critique of those norms even as they are being assimilated in the sudden 'turning' signified by conversion experiences. In other words, regardless of whether conversion is an assimilative or an oppositional gesture, the specific circumstances, historical context, and political climate in which conversion occurs might suggest a more complicated trajectory. In somewhat paradoxical fashion, assimilation may be accompanied by critique of the very culture with which religious affiliation is sought. Equally, dissent may aim at reforming and rejuvenating the culture from which the convert has detached herself.

These are considerations that have as forceful an impact on colonial conversions as on the conversions occurring in England during and after legal emancipation of non-Anglicans. And if, as I have been suggesting, religious minorities in England often shared the position of colonial subjects, the strategies deployed by such figures as Disraeli to find a place for Jews in Anglican England – a place not circumscribed by legal edict – bear comparison with the strategies developed by colonial converts to create an alternative community to that provided by custom and law. Again, the key factor lies in the multiple affiliations opened up by conversion – the possibilities of occupying several positions in relation to both nation and religion. The blurring between the objects to which the convert assimilates – and those he (or she) challenges with a free crossover between assent and dissent – is precisely the source of the power of conversion. Thus, assimilation and dissent often crisscross with motives not immediately attached to their apparent function in conversion. The result is that converts may be engaged just as readily in a critique of their adopted culture and religion as in a project to reform the culture that they have renounced. In either case, conversion's instrumental significance cannot be denied, nor can its dynamic engagement with either or both cultures with which the convert is affiliated.

To my mind, the colonial figure who perfectly illustrates this last point is Narayan Viman Tilak, a Maharashtrian Brahmin convert to Christianity in late nineteenth-century western India. Although he assimilated norms of Christian belief and conduct, Tilak also sought to indigenize Christianity and make it compatible with Hinduism. His syncretic project contained an implicit critique of the alienating effects of British colonialism. Though his conversion may have suggested a rejection of Hinduism in favor of the colonizer's religion, his subsequent use of Christianity to enunciate an anticolonial vision of Indian nationalism was far more damaging to British imperialist ambitions than if he were to have remained a Hindu.

A major nationalist figure in the renaissance of Marathi literature, Tilak devoted his creative energies to articulating possible modes of relating to a revitalized India, and (paradoxically it would seem at first glance) conversion to Christianity appeared to him to be the most effective way of expressing that relation. Tilak conceived of Christianity as embodying a necessary vision of the future, with the power to purge Hinduism of its most hated caste features, yet at the same time adapted to India as a truly indigenous religion in its own right. The English missionary J. C. Winslow claimed that Tilak had confided to him that he was the founder of a new religion, and indeed Tilak's writings suggest that he had distanced himself equally from traditional forms of both Hinduism and Christianity (Winslow 1923: 17). Convinced that India's political enslavement was matched only by its own moral degradation through a coercive caste system enshrined in Hinduism, Tilak, like many other Christian converts, grew firmer in the belief that only a religious awakening would enable India to embark on a new era of reform and advancement.

But at the same time, Tilak wanted Indian Christians to be more truly Indian, claiming that the British missionary project had denationalized Christianity and made the West the exclusive reference point of Indian Christianity. In one of his poems he denounced the paternalistic attitude of missionaries in scathing terms: 'We dance as puppets while you hold the strings. How long shall this buffoonery endure?' (*Abhang* no. 160 in Winslow 1923: 58). Tilak undertook to teach Marathi Christians to study older Marathi literature, especially the devotional poetry of Jnaneshwar, Namadev, and Tukaram. He insisted that he came to Jesus Christ 'over the bridge of Tukaram's verse' (Winslow 1923: 60) and continued to make it his goal to adapt Indian Christianity to the spirit of Indian cultural forms such as the bhajan, or Hindu devotional hymn. Interestingly, though deriving its force from the reformist impulse to which his Christian conversion gave him access. Tilak's nationalism was the means by which he also sought reconciliation with the former Brahmin community from which he had been excommunicated. His poetry is a combination of religious fervor, patriotic zeal, and antisectarian feeling; like his nationalism, his verse is conceived as much in a spirit of rapprochement as of critique:

Thrice blessed is thy womb, my Motherland,
Whence mighty rishis, saints, and sages spring!
A Christian I, yet here none taunteth me.
Nor buffeteth with angry questioning.
I meet and greet them, and with love embrace:

None saith, 'Thou dost pollute us by thy sin!'
My Guru they delight to venerate; they say.
'He is our brother and our kin'.
                              (Winslow 1923: 58)

Tilak's syncretic ambition was shared by converts from other regions of India, who also sought through conversion to recover a 'national religion' that eliminated rather than preserved difference. The attempt to create a hybrid entity like 'Hindu Christianity' was a keenly felt aspect of this synthesizing project. As the Bengali historian M. M. Ali has argued, the earliest symptoms of nationalist stirrings in Bengal, far from being a return to Hinduism and a revival of classical Sanskrit texts, were the setting up of Hindu Christian churches, which defiantly attempted to indigenize the Christianity introduced by English missionaries (see Ali 1965). This observation implicitly endorses Robin Horton's widely discussed thesis that African conversions were not so much a tribute to missionary success as expressions of African nationalism, through which the colonial ruler's religion was given a strong indigenist bent.[1] (Horton 1971: 85–108; Horton 1975a: 219–35; Horton 1975b: 373–99). In Horton's reading, conversion is a sign of acceptance of modernity, not of capitulation to colonial power, and the setting up of African churches introduced an Africanist discourse that not only indigenized Christianity but also brought what Horton calls a local-centered, 'microcosmic' thinking in contact with ideas of nations and territories.

Horton's neat dichotomy between local and national community reproduces a premodern/modern split, which has inevitably come under attack by other critics, most notably by Terence Ranger, who argues that premodern African religions were never uniquely microcosmic but bridged ethnic and territorial boundaries (Ranger 1993). But despite these reservations, Horton's thesis still has a certain usefulness in understanding Christian conversions in colonial societies as signifying responses to internal changes that were already under way, and as a form of domesticating (and to some extent neutralizing) alien religious and cultural beliefs. . . .

## Note

1    Horton entered into a debate with several scholars contesting his theory of African modernity, and Humphrey Fisher, in 'Conversion Reconsidered: Some Historical Aspects of Religious Conversion in Black Africa,' *Africa* (1973) 43, 1: 27–40 offered a hard-hitting critique of Horton's description of Africanist Christian movements. Deryck Schreuder and Geoffrey Oddie, 'What is "Conversion"? History, Christianity, and Religious Change in Colonial Africa and South Asia,' *Journal of Religious History* (December 1989) 15: 496–518, attempt to sift through some of the salient arguments emerging from the debate, though they tend at times to misapply perceptions deriving from the African situation to South Asia.

## Chapter 117

# LAURA E. DONALDSON

# GOD, GOLD, AND GENDER

From Laura E. Donaldsen *Postcolonialism, Feminism and Religious Discourse* London and New York: Routledge, 2002.

T HE ESTABLISHMENT OF THE SO-CALLED new world order calls for a new kind of analysis as well as a new world slogan: rather than 'God, gold, and glory,' which asserts the motivations usually ascribed to the Anglo-European colonial system, the motto of this transformed thinking should instead foreground the alliance of 'God, gold, and gender' (Mazrui 1990: 61). Although the processes underlying 'God, gold, and gender' have assumed a heightened significance in the arena of contemporary politics, they also possess a long, often unrecognized, symbolic and hegemonic history. Any introduction to the questions posed by postcolonialism, feminism, and religious discourse must therefore acknowledge this history and detail its most consequential moments. Given the particular contexts of my own colonial history, I would include such moments as Martin Waldseemüller's (re)naming of 'America,' Mary Rowlandson's captivity narrative, and Charlotte Bronte's novel *Jane Eyre*. Excavating the cultural archaeologies of these texts reveals a continuity of preoccupations with the present as well as insights about how colonialism, gender, and religion shaped, and continue to shape, individuals and societies.

I have chosen these sites for several reasons. First, each foregrounds issues largely ignored in current configurations of postcolonial studies: the centrality of religion, the importance of the Fourth World (that of indigenous peoples), and the persistence of 'the woman question.' Second, each highlights the crucial role of Christianity in promoting the Anglo-European imperialist project. Although this anthology does focus on other religious traditions-Islam, Judaism, and Hinduism-one cannot overestimate the historical influence of the Christian tradition in disseminating imperialist ideologies. While many countries occupied and dominated foreign territories, only the group of nations claiming Christian identity implemented a global colonial system upon which the sun never set. As Michael Prior notes in *The Bible and Colonialism*, the social transformation resulting from the decision to encroach on a foreign terrain reflects the determination of the colonizers to alter radically a region's politics in favor of the colonists and the introduction of Christianity functioned as a powerful tool of such transformation. An all-too-common paradigm for this process excluded the land's indigenous inhabitants from these new arrangements: 'Several motivations combined to exclude the indigenes, and for those influenced by religious considerations, the biblical paradigm provided a ready justification for it. The exclusivist tendencies in North America and South Africa have been ascribed to the influence of the Old Testament in the Puritan faith in the case of the former, and in the Dutch Reformed Church in the latter'(Prior 1997: 175). Although some might criticize our emphasis on Christian narratives of colonialism, I believe that this stress constitutes an essential component of decolonizing the myths and ideologies still undergirding the entire colonialist enterprise. Until all peoples grasp the intricate relationships binding colonialism, gender, and religion together, the dream of a genuinely new world order will remain only the shadow of an ideal.

## From Amerigo to 'America'

Early in the sixteenth century, a group of humanists assembled in the tiny French village of Saint Dié to form the 'Gymnasium Vosgianum' – a kind of intellectual advocacy group whose goal was to disseminate certain important books. Among their numbers was a young clergyman named Martin Waldseemüller, whose interest in cartography subsequently altered the course of history. Stefan Zweig notes that 'no one would have heard the conversations of either Waldseemüller or this "miniature academy" if a printer named Walter Ludd had nor decided to set up a press in Saint Dié' (Zweig 1942: 51). For his first publishing venture, Ludd – a chaplain and secretary to the duke of Lorraine – responded to the public's interest in travel narratives by combining Ptolemy's *Cosmographia* with Amerigo Vespucci's memoir of his exploratory sea voyages. Ptolemy's *Cosmographia* had been regarded as Europe's ultimate geographical authority for more than a millennium, but the latter part of the fifteenth century had challenged its authority through the 'discovery' of hitherto unknown lands. Consequently, any reprinting of the *Cosmographia* had to incorporate these new lands into the old maps in order to revise and complete them. It was Martin Waldseemüller who not only drew these newly revised cartographic charts but also, in the view of many, authored the entire text. His fateful decision to name the *quarta orbis pars*, or 'fourth part of the world,' after Amerigo Vespucci occurs in chapter 9 of the *Cosmographia Introduction*, which appeared in print on April 25, 1507. According to Waldseemüller:

> Now, these parts of the earth have been more extensively explored and a fourth part has been discovered by Amerigo Vespucci (as will be set forth in what follows). Inasmuch as both Europe and Asia received their names from women, I see no reason why anyone should justly object to calling this Amerige, i.e., the land of Amerigo, or America, after Amerigo, its discoverer, a man of great ability.
>
> (Waldseemüller [1507] 1907: 70)

In Latin, 'o' represents the masculine ending for words, so by changing 'Amerigo' to the feminized 'Ameriga,' or 'America,' Waldseemüller metaphorically transforms 'him' into a 'her.' Mary Louise Pratt observes that the process of cartographic naming joined together [Eurocentric] religious and geographical projects in the quest for European expansion. Its agents claimed the world by christening landmarks and topographic formations with Euro-Christian names (Pratt 1992: 32). Other scholars have noted the performative character of mapping as well as its exemplary role in demonstrating the discursive practices of colonialism. Graham Huggan, for example, argues that 'mapping's reinscription and enclosure of space parallel historical colonialism's acquisition, management, and reinforcement of military and political power' (Huggan 1990: 25). Waldseemüller's (re)naming of Turtle Island joins the discourses of religion and geography in a similar ideological enterprise; it also, however, precipitates a gendered crisis in colonial representation.[1]

The havoc that 'America' wreaks among the globe's feminine names (which I will address shortly) actually displaces another, perhaps more urgent, conflict: 'her' disruption of the patriarchal assignment of the world to the sons of Noah. According to biblical tradition, the descendants of Japheth (Noah's oldest son) populated what is now called Europe: the descendants of Shem, Asia; and the descendants of Ham, Africa. This particular (re)naming seems tragically prophetic, since 'Japheth' most likely derives from the Hebrew cognate, *phatah*: 'to spread' or 'enlarge.' Since the Japhetic origin of Europe was virtually uncontested until the nineteenth century (De Rougemont 1966: 21), these connotations of latitude, width, and expansion ominously prefigure the regimes of Anglo-European imperialism. Yet, the insertion of a feminized 'America' into this previously masculine field engenders a profound anxiety, which Jan van der Straet's 1575 painting of 'her' originary encounter with Vespucci symptomatically manifests. Since I discuss both this anxiety and the painting more fully in my essay 'The Breasts of Columbus,' [Editor: see below] I will say here only that the discovery and subsequent conquest of 'America' – whether material or symbolic – destabilizes the myth of European self-sufficiency and confronts 'him' with the prospect of engulfment by 'her.'

Of course, Waldseemüller presents 'America' as an affirmation of women, or at least of women's names. Subsequent mapmakers suppressed the potential of this new member to introduce discord by confining her to an iconographic continuum operating between what one can only describe as the naked and the dressed. In Mercator's 1636 Atlas, for instance, Europe appears as an elaborately clothed woman who also wears a crown and holds a book in her hands. Also clothed (but without a crown), Asia extends her bejeweled arms toward Europe. Africa forms a dramatic contrast, since she is both dark-skinned and unclothed. Unlike the formal sovereignty displayed by Europe, Mercator portrayed 'America' as an Indian Queen who wore a crown of upright feathers, but who was also naked and supplicating. In *Inventing A-M-E-R-I-C-A*, Jose Rabasa succinctly demystifies the colonialist imaginary structuring this sartorial difference: 'Dressed and learned in the sciences, Europe rules and supersedes Asia, the origin of science, art, and religion. In contrast, Africa and America in their nudity testify to the dominance of the feminine and typify the barbarous stares that are, nonetheless, full of treasures for Europe' (Rabasa 1993: 204). The work process of a colonialist machinery disappears under the facade of the world offering its riches to Europe. If mapping uniquely connects historical imperialism's textual and material vectors (Ryan 1994: 128), then Mercator's feminine spectacle figures the textual and material vectors of colonialism's women. The relations among them affirm not only that religion and gender exerted important influences in colonialism's founding moments, but also that any postcolonial analysis must similarly attend to their effects. . . .

* * *

## 'THE BREASTS OF COLUMBUS'

### A political anatomy of postcolonialism and feminist religious discourse

> I do not hold that the earthly Paradise has the form of a rugged mountain, as it is shown in pictures, but that it lies at the summit of what I have described as the stalk of the pear.
>
> Christopher Columbus, Log, Third Voyage

Whether they are large and plump, small and flat, or positively undulating, women's breasts embody sexual difference in the identity discourses of the West; however, the meaning these discourses generate about breasts is radically divergent. On the one hand, the socialization of women as objects of the male gaze has induced many to spend millions of dollars enlarging, sometimes reducing, but always disciplining their breasts through the rigors of plastic surgery. In contrast, feminists such as Helene Cixous suggest that a woman's breasts (along with her other reproductive organs) defy the centralized phallicism of men and engender an alternative libidinal organization as well as a distinctly feminine way of knowing. However, we should also remember that breasts possess a colonial history and that the female mammary glands constitute a significant part of imperialism's political anatomy.

Who can forget, for example, the recent struggle of indigenous people for the hills of Welatye Therre, or 'Two Breasts,' where Australian aboriginal women have danced and sung for thousands of years? (Jacobs 1994: 184). Both the songs associated with the site and its sacred objects nurture the dissemination of women's knowledge. As one Arrernte woman declared: 'Like you've got women's liberation, for hundreds of years we've had ceremonies which control our conduct, how we behave and act and how we control our sexual lives. . . . They give spiritual and emotional health to Aboriginal women' (Jacobs 1994: 186). Specific knowledge about the Welarye Therre is forbidden to non-Native outsiders, as well as to aboriginal men, yet the Arrernte women who claim this place were forced to disclose its significance (and thus risk its sacredness) to protect it from flooding by a proposed hydroelectric dam. After an effective and vigorous opposition campaign they defeated the dam proposal, and the 'Two Breasts' of Arrernte culture continue to nurture their people.

The breasts of 'Sheba,' which come to life in the pages of H. Rider Haggard's novel *King Solomon's Mines*, exist as perhaps the most (in)famous European counterpart to the Welatye Therre. In Haggard's nineteenth-century adventure tale, a mercenary Portuguese trader named Jose da Silvestre ironically draws a map to the chamber containing the riches of King Solomon while he is starving to death on the 'nipple' of 'Sheba's Breasts' – the name of an African mountain that rightly refuses to nurse such a destructive child. Described by Malek Alloula as the half-aesthetic concept of the 'Moorish bosom,' the specter of the colonized's dark, seductive breasts (even such topographical ones as Sheba's) both titillated and haunted the European social imaginary during the centuries of conquest. In his analysis of suberotic breast images in *The Colonial Harem*, Alloula documents their circulation through the postcards that French colonials in Algeria sent home to the mother country. These postcards function as a virtual 'anthology of breasts' whose diversity ultimately reveals a pattern of sameness: 'Generally topped off with a smiling or dreamy face, this Moorish bosom, which expresses an obvious invitation, will . . . be offered to view, without any envelope to ensure the intimacy of a private correspondance [sic]'(Alloula 1986: 105).

Perhaps the most haunting chronicler of the postcolonial breast is the West Bengali writer and activist Mahasweta Devi, whose collection *Breast Stories* (1997) explores in literary form many of the questions I will raise here. *Breast Stories* includes, for example, the story of Draupadi, a tribal woman who leads a resistance movement on behalf of her people, and Senanayak, the police chief who pursues her, and in whom Gayatri Chakravorry Spivak finds a cautionary tale of the expert on Third World resistance literature nourished by First World societies. Devi also introduces her readers to Jashoda, a lower-caste woman who becomes a professional wet nurse (through the practice of suckling) to her master's sons, and Gangor, another tribal woman and nursing mother who works as a migrant laborer. Gangor's breasts become objects of obsession and oppression through the circulation of an ill-advised photograph. According to Spivak, who translated and introduced Devi's collection:

The breast is not a symbol in these stories. In 'Draupadi,' what is represented is an erotic object transformed into an object of torture and revenge where the line between (hetero)sexuality and gender violence begins to waver. In 'Breast Giver' [Jashoda's story], it is a survival object transformed into a commodity, making visible the indeterminacy between filial piety and gender violence, between house and temple, between domination and exploitation. . . . In 'Behind the Bodice' [Gangor's story], [Devi] bitterly decries the supposed 'normality' of sexuality as male violence.

(Spivak 1996: vii)

This confluence of colonialism and patriarchy so brilliantly explored by Devi also conjures the colonial breasts of most immediate concern to this essay: those fantasized by Columbus, whose voyages in search of the East convinced him that the world was not round, but rather, pear-shaped and topped by a protuberance much like a woman's nipple.

The 'nipple' envisioned by Columbus was none other than Guanahani, the land of the Taino, which the Spanish renamed San Salvador. Feminist critic Anne McClintock identifies Columbus' breast fantasies as a genre of 'porno-tropics' that draws on 'a long tradition of male travel as an erotics of ravishment' for its content (McClintock 1995: 22). For example, earlier travelers' tales had regaled their European readers with stories of men's breasts flowing with milk and militarized Amazonian women lopping theirs off (McClintock 1995: 22). To readers satiated with such exoticized sexualities, the mammillary imaginings of Rider Haggard and Columbus must have seemed quite tame. In her essay, 'The Breast, the Apocalypse, and the Colonial Journey' (later revised as the chapter 'De/colon/izing Spaces' in *Apocalypse Now and Then*, Catherine Keller offers her own engagement with the breasts of Columbus as well as the only explicit engagement by a Euro-American feminist theologian with the 1992 quincentennial.

Keller describes this essay as contributing 'a Euroamerican feminist fragment to the post-colonial project' and articulates its goal as decolonizing the 'androcentric' minds whose visions of imperialism were so heavily influenced by the 'mammillary trope' (Keller 1996: 193). To this end, she examines the 'gender codes' structuring the colonial enterprise and, more particularly, the gender codes that coalesced not only in Christopher Columbus' apocalyptic discourse but in all speculations about the end of the world.[2] For Keller, 'androcentric myths of the holy warrior, an appeal to male-on-male violence, and a rejection of the feminized and sexualized body link both Revelation's New Jerusalem and the New World's Terra Incognita' (Keller 1994: 68). While Columbus' attempts to formulate an individual politics were ultimately negligible, Keller asserts that his personal apocalyptic beliefs nevertheless transformed 'the collective mythic structures of the Western world' (70). This claim tends to exaggerate the sociopolitical importance of apocalypse in fourteenth- and fifteenth-century Europe, however, since one could argue that a 1366 amendment by the Priors of Florence to the laws governing the import and sale of 'infidel' slaves had an even greater impact on the agendas of conquest.

In this decision, the Holy Roman Church defined the word infidel so that it encompassed anyone born a non-Christian, regardless of any subsequent conversion they might have experienced. From this moment forward, racial and ethnic origin rather than religious difference became the foundation of slavery – an evolving context of discrimination that nurtured the notion of non-Europeans as separate, distinct, and inferior 'races' (Stannard 1992: 209). This semantic and theological redefinition of the infidel also influenced the belief that God created American Indians for the specific purpose of becoming slaves to European Christians. Overestimating the influence of apocalyptic thought consequently deflects our attention away from equally important attitudes and policies, along with the complicity of the institutional Church in facilitating the emergence of both colonialism and white supremacy. Despite this tenuous interpretation of Columbian apocalypticism, projects with Keller's twin foci are long overdue: contemporary postcolonial theory consistently subsumes the particularity of women's estates under its appeals to an homogenous category called 'the colonized' (Ashcroft *et al.* 1989: 103)[3] and it has largely overlooked the collusion of colonialist as well as indigenous patriarchies. Gayatri Chakravorty Spivak cautions us about correcting this neglect with an uncritical flight into the realm of the postcolonial, however: 'Much so-called cross-cultural disciplinary practice, even when "feminist," reproduces and forecloses colonialist structures: sanctioned ignorance and a refusal of subject-status and therefore human-ness [for the sub-altern]' (Spivak 1999: 167). In other words, despite the cognitive commitments to anti racism and decolonization of many Euro-American feminists, the structures of thought and research methods organizing their scholarship continue to thwart the progressive goals they espouse. Spivak's emphasis upon the reproduction of two particular 'colonialist structures' – sanctioned ignorance and the refusal of subject status to the oppressed – constitutes a starting point for examining this question, since both foreground questions of how scholars produce knowledge rather than personal culpability. My hope is that this exploration of postcolonialism and feminist religious discourse will shift the dialogue among women of the Third and Fourth Worlds and those of the First from a moral to an epistemological paradigm. As Keller herself has noted, blaming or denouncing each other may correct a situation by achieving a 'behavior-modifying shame and an institution-shifting conscience, but it will never heal the systemic complex which provoked it' (Keller 1996: 260). The epistemological shift I am suggesting begins, but certainly does not end, this process of healing. . . .

## Notes

1   'Turtle Island' is the name that the Haudenosaunee (Iroquois) gave to the land now called 'North America'. It comes from a creation story, which tells how the turtle volunteered to use his shell as an anchor for the earth.

2   A focus on messianic apocalypticism in Columbus's thought is not new. Historian David Stannard (whose work Keller does not mention) addressed this issue much earlier and much more thoroughly in *American Holocaust: the Conquest of the New World* (1992).

3   Editor's note: Since this references the work of the editors of the present volume, it might be noted that it involves a discussion by them of Timothy Findlay's novel *Not Wanted on the Voyage*. The page referenced (103) contains no use of the words 'the colonized' nor any discussion of the kind mentioned here. Indeed, the analysis of the novel it presents focuses on Findlay's stress on the role in colonial discourse of patriarchal force as a means of falsely excluding categories such as women, children as well as other animal species from the 'human'. Thus presenting the very opposite case to that which the reference in this text describes. In fact *The Empire Writes Back* makes a very specific reference to the description of the 'double colonization' of women in the work of Holst Petersen and Rutherford (1985) and to the early analyses of Spivak (1987) of these relations (p. 177).

## Chapter 118

# WILLIAM BALDRIDGE

## RECLAIMING OUR HISTORIES

From James Treat (ed.) *Native and Christian: Indigenous Voices on Religious Identity in the United States and Canada* New York: Routledge, 1996.

FROM A NATIVE AMERICAN'S PERSPECTIVE, one way to describe the spiritual significance of 1492 is to realize that for the last half-millennium Columbus and his spiritual children have usurped the role of God and imposed their definitions of reality onto this continent. People now go through life believing that trees went unidentified until Europeans came to name them, that places could not be distinguished and directions could not be given until Europeans arrived to designate one place New York and another Los Angeles. People in the United States accept as self-evident that this continent could not produce food until row cropping was introduced, that water was not pure before filtration plants were introduced, and that conservation is a concept introduced by the US Forestry Service. It is believed without question that this land was godless until the arrival of Christianity. For Native Americans, perhaps the most pervasive result of colonialism is that we cannot even begin a conversation without referencing our words to definitions imposed or rooted in 1492. The arrival of Columbus marks the beginning of colonial hubris in America, a pride so severe that it must answer the charge of blasphemy.

A central agent in the colonization of this hemisphere has been the Christian church. Whatever the church likes to believe its intentions were or are in making us the object of its missionary endeavors, history shows the missionary system to be colonialism in the name of Christ. The foundation of colonial Christianity rests on its power to monopolize definitions: who is godless, godly, and most godly, all stemming from Christianity's definition of the essential nature of God. When Christians confuse their confessions of faith with absolute knowledge of reality, they invite a challenge of hubris. When Christians confuse the limitations of their humanity with the nature of God, they invite a challenge of blasphemy. Who can claim absolute knowledge of reality but God alone?

Native Americans have not been passive toward Christian colonialism. Today's generation of Native Americans, like the generations that preceded us, and those to follow, are bound by the spiritual power of freedom and dignity, gifts from our Creator. We are often dismissed as trying to change the past or trying to return to the past. Having our intelligence questioned is a familiar experience. But being underestimated is one of our most effective and constant weapons. We are not denying history or the weight of the forces pushing us down. We are also not willing to forsake our spiritual birthright as children of God. Colonial Christian definitions to the contrary, we will not label our ancestors nor teach our children that they are spiritually illegitimate. So, as well as resisting we are retrenching, reaching down, down to the bedrock of our continent, down where our spiritual vitality is grounded. Native people's thoughts need not be determined by the definitions of the colonizer if they know who they are and where they stand, if their identity is anchored in bedrock. We are the embodiment of this hemisphere. God made us and placed us here. Identifying and attacking the enemy is a time-honored means to gain respect and admiration within one's community. As a Native American Christian I have done my warring with the missionaries. They are a target as easy to hit as dirt, and just as difficult to eliminate. I am not making claims to be a seasoned veteran missionary fighter, but I do know the taste of battle. In my first firefights I joined those presenting a more balanced picture of the Christian missionary work. Many of the missionaries were people of good faith who sacrificially brought us the gospel of Christ. I was raised on the praise and publicity generated by the churches concerning these people. On the other hand, many missionaries served as federal agents and in that role negotiated treaties which left us no land. Most missionaries taught us to hate anything Native American and that of necessity included hating our friends, our families, and ourselves. Most refused to speak to us in any language but their own. The missionaries functioned and continue to function as 'Christ-bearing colonizers.' If it were otherwise the missionaries would have come, shared the gospel, and left. We know, of course, that they stayed, and they continue to stay, and they continue to insist that we submit to them and their definitions. The vast majority of Native people have experienced the missionary system as racist and colonial and our most prevalent response has been passive resistance: A very small percentage of Native Americans are practicing Christians. I realize that such language makes some people anxious and others quite indignant. If Native American Christians do not use colonial formulas for confessing Christ such people claim that we are not 'real Christians', they believe that we are questioning the ability of these formulas to satisfy their spiritual needs, and they bristle at being defined as 'Christ-bearing colonizers' when they 'were only trying to help.' Still, the list of injuries to hurl back at the missionaries is long, yet it remains essentially unknown outside of Indian communities. I continue to be impressed by the number of people, active in the life of their various denominations who are shocked to hear about the realities of current mission programs to Native Americans and who assumed that 'we stopped doing missions like that a hundred years ago.' One of the spin-offs of the quincentennial celebration of the European invasion of this hemisphere was a proliferation of information, produced by non-Indian historians, concerning the church's role in colonialism. When whites turn on themselves with their technological expertise – graduate degrees, computers, data banks, ethno-histories, and such – a defiant arrow from an Indian becomes a flame amidst laser beams. Furthermore, as the new official history continues to roll off the press, some of us warriors are realizing that the power of our arrows comes not so much from the historical evidence as it does from the anecdotal experience we own. We Indians lived the missionary history, we continue to live this history, and the power of our stories cannot be matched by 'the cold hard facts.' It is encouraging to watch the academic historians revisit history, and bring a greater degree of balance to the subject. At the same time, it is discouraging to remember that such 'objectivity' typically comes at a point in a political struggle when the issues have been decided and the victors, that is, the ones writing the history, can afford to admit that 'nobody is perfect' without any real threat to their power base. In the 500-year war against Christian colonialism we have had our successes. If on no more than a few occasions of hit-and-run skirmishes, we have had our moments. For me, the sense of camaraderie with brothers and sisters has become a lasting satisfaction. Yet the spoils of our small victories have faded into an ironic lesson: the very act of fighting the missionary system concedes too much to colonialism. It concedes too much because it accepts the premise that our dignity must be granted to us rather than be recognized in us. It accepts the premise that God loves one people more than God loves all people. It accepts the premise that a God of justice would condemn a people to hell because of where they were born and when they were born. Fighting missionaries has taught me that the end of the missionary system begins with a change of heart, my heart, not the heart of the missionary nor the heart of the institutions that commission missionaries. Fighting the oppression of the missionary system is a struggle for justice that unavoidably becomes a struggle for power. Power lies at the core of Christian colonialism. Refusing the terms of the struggle is an essential first step in regaining the spiritual perspective of Native America.

## Chapter 119

# RICHARD KING

## ORIENTALISM AND RELIGION

From *Orientalism and Religion: Postcolonial Theory, India and the 'Mystic East'* London: Routledge, 1999.

[**T**HERE IS A] **TENDENCY** within most Indological accounts to claim to have uncovered the 'essence' of the object under consideration, through careful scholarly analysis (see Inden 1986). Thus works that purport to explain the 'Oriental mind-set' or the 'Indian mentality', etc., presuppose that there is a homogeneous, and almost Platonic 'essence' or 'nature' that can be directly intuited by the Indological expert. Inden (1986) is correct, in my view, to attack such essentialism, not just because it misrepresents the heterogeneity of the subject matter, but also because of the way in which such essentialism results in the construction of a cultural stereotype that may then be used to subordinate, classify and dominate the non-Western world. One of the problems with Inden's radical rendering of post-structuralist constructivism, however, is that he does not take seriously enough the material 'realities' under-lying the imaginary constructions of 'India' (see Ahmad (1991: 135–63) for a critique of Inden). Inden's work, however, is interesting for his critical analysis of 'affirmative Orientalism'. This strand of Orientalist discourse, labelled 'romantic' by Inden because of its indebtedness to European Romanticism, is generally motivated by an admiration for, and sometimes by a firm belief in the superiority of, Eastern cultures. The romantic image of India portrays Indian culture as profoundly spiritual, idealistic and mystical. Thus as Peter Marshall points out

> As Europeans have always tended to do, they created Hinduism in their own image. Their study of Hinduism confirmed their beliefs and Hindus emerged from their work as adhering to something akin to undogmatic Protestantism. Later generations of Europeans, interested themselves in mysticism, were able to portray the Hindus as mystics.
>
> (Marshall 1970: 430)

. . . What is interesting about the 'mystical' or 'spiritual' emphasis that predominates in the romanticist conception of India is not just that it has become a prevalent theme in con-temporary Western images of India, but also that it has exerted a great deal of influence upon the self-awareness of the very Indians that it purports to describe. Some might argue, as David Kopf clearly does (see Kopf 1969), that such endorsement by Indians themselves suggests the anti-imperial nature of such discourses, yet one cannot ignore the sense in which British colonial ideology, through the various media of communication, education and insti-tutional control, has made a substantial contribution to the construction of modern identity and self-awareness among contemporary Indians.

European translations of Indian texts prepared for a western audience provided to the 'educated' Indian a whole range of Orientalist images. Even when the anglicised Indian spoke a language other than English, 'he' would have preferred, because of the symbolic power attached to English, to gain access to his own past through the translations and histories circulating through colonial discourse. English education also familiarised the Indian with ways of seeing, techniques of translation, or modes of representation that came to be accepted as 'natural.'

(Niranjana 1990: 778)

Perhaps the primary examples of this are the figures of Vivekananda and Mohandas K. Gandhi. Vivekananda (1863–1902), founder of the Ramakrishna Mission, an organization devoted to the promotion of a contemporary form of Advaita Vedanta (non-dualism), placed particular emphasis upon the spirituality of Indian culture as a curative for the nihilism and materialism of modern Western culture. In Vivekananda's hands, Orientalist notions of India as 'other worldly' and 'mystical' were embraced and praised as India's special gift to humankind. Thus the very discourse that succeeded in alienating, subordinating and controlling India was used by Vivekananda as a religious clarion call for the Indian people to unite under the banner of a universalistic and all-embracing Hinduism.

Up India, and conquer the world with your spirituality . . . Ours is a religion of which Buddhism, with all its greatness is a rebel child and of which Christianity is a very patchy imitation.

(Vivekananda 1970: Vol. 3: 275)

The salvation of Europe depends on a rationalistic religion, and Advaita – non-duality, the Oneness, the idea of the Impersonal God – is the only religion that can have any hold on any intellectual people.

(Vivekananda 1970: Vol. 22: 139)

Colonial stereotypes thereby became transformed and used in the fight against colonialism. Despite this, stereotypes they remain. Vivekananda's importance, however, goes far beyond his involvement with the Ramakrishna Mission. He attended (without invitation) the First World Parliament of Religions in Chicago in 1893, delivering a lecture on Hinduism (or at least on his own conception of the nature of Hinduism and its relationship with the other 'world religions'). Vivekananda was a great success and initiated a number of successful tours of the United States and Europe. In the West he was influential in the reinforcement of the romanticist emphasis upon Indian spirituality, and in India Vivekananda became the focus of a renascent intellectual movement, which might more accurately be labeled 'neo-Hinduism' or 'neo-Vedanta' rather than Hinduism. . . .

Of the many enduring images of 'the Orient' that have captured the imagination of Westerners over the centuries, it is the characterization of Eastern culture, and Indian religions in particular, as 'mystical', that is most relevant to our current discussion. As European culture became increasingly intrigued by the cultural mysteries and economic resources of foreign lands in an age of colonial expansion, it was inevitable that a developing awareness of the diversity of cultures and religions would require some characterization of these 'alternative perspectives' in a way that displayed their alterity when compared to the normative European (Christian) perspective. . . . With a greater awareness of the plurality of religious perspectives throughout the world furnished by colonial encounters abroad it became inevitable that comparisons with Christianity would come to the fore. It is perhaps no coincidence, then, to find the Protestant presuppositions of many Europeans (especially in the growing scholarly fields of German and English Orientalism) being reflected in their characterization of non-Christian religion. Given Protestant distrust of the 'mystical' elements of Catholicism, it is also of little surprise to find a tendency to disentangle the 'mystical' from its particular Christian application and to reapply this characterization in a pluralistic religious context. Once the term 'mystical' became detached from the specificity of its originally Christian context and became applied to the 'strange and mysterious Orient', the association of the East with 'mysticism' became well and truly entrenched in the collective cultural imagination of the West. . . . Thus, for some, describing religions of the East as 'mystical' is a way of differentiating the essential historical truth of Christianity from its inferior rivals – and implicitly to attack those within Western Christianity who might want to focus upon the 'mystical' dimensions of their own tradition. However, for many of the Romantics 'the mystic East' represented the spirituality that much of contemporary Christian religion seemed to lack. Thus, as the term 'mystical' became divorced from a Christian context and was applied to other religions by Western theologians and Orientalists, it continued to function at home as the site of a power struggle in the battle to define European and Christian cultural identity. Today, there are perhaps two powerful images in contemporary Western characterizations of Eastern religiosity. One is the continually enduring notion of the 'mystical East' that we have been discussing – a powerful image precisely because for some it represents what is most disturbing and outdated about Eastern culture, while for others it represents the magic, the mystery and the sense of the spiritual that they perceive to be lacking in modern Western culture. The depravity and backwardness of the Orient thus appears to sit side by side with its blossoming spirituality and cultural richness. Both of these motifs have a long historical pedigree, deriving from the hopes and fears of the European imagination and its perennial fascination with the East.

The second image of Eastern religion – one indeed that is increasingly coming to the fore in Western circles, is that of the 'militant fanatic'. Such a characterization also has a considerable ancestry, being a contemporary manifestation of older colonial myths about Oriental despotism and the irrationality of the colonial subject. The particular nature of this construct is, of course, heavily influenced by the secularist perspective of much of modern Western culture. The image of the militant fanatic or religious 'fundamentalist', while frequently interwoven with 'the mystical' characterization (particularly in the emphasis that Western commentators place upon the 'religious' dimension of conflicts such as Ayodhya in India), is rarely explicitly associated with the notion of 'the mystical East' precisely because modern Western understandings of 'the mystical' tend to preclude the possibility of an authentic mystical involvement in political struggle. The otherworldly Eastern mystic cannot be involved in a this-worldly political struggle without calling into question the strong cultural opposition between the mystical and the public realms. The discontinuity between these two cultural representations of the East has frequently created problems for Western and Western-influenced observers who find it difficult to reconcile notions of spiritual detachment with political (and sometimes violent) social activism (see Jurgensmeyer 1990). Thus in the modern era we find Hinduism being represented both as a globalized and all-embracing world religion and as an intolerant and virulent form of religious nationalism. Despite the apparent incongruity of these two representations . . . one feature that both characterizations share in common is the debt they owe to Western Orientalism.

## Chapter 120

# PETER VAN DER VEER

## GLOBAL CONVERSIONS

From Gareth Griffiths and Jamie S. Scott (eds) *Mixed Messages:
Materiality, Textuality, Missions* New York: Palgrave Macmillan, 2004.

**B**Y DEFINITION, WORLD RELIGIONS have always been expanding over the globe and have been a major part of the formation of world civilizations. This expansion has also always been intertwined with histories of trade and conquest as well as with those of learning and education. Since 1800 religious expansion is part and parcel of the spread of modernity. Finally, in today's world it is directly connected to the contemporary process of globalization. The study of conversion and of the globalization of religion gives us an excellent vantage point from which to study these larger transformations. This opportunity has often not been sufficiently realized for reasons that are closely related to these transformations themselves. Secularist assumptions about secular progress and the decline of religion have hindered the development of an adequate understanding of the importance of religion in the modern world.

While it is hard to miss the continuities in the global history of conversion it is equally difficult to miss the breaks, discontinuities, and differences. What we call modern is certainly conceptualized as a break with the past, with tradition, and as a celebration of newness. The concept of 'the modern' emphasizes simultaneously difference from the past and from that which is found outside of one's own cultural domain. These things are connected to the extent that those who are living elsewhere, outside of the West, are often conceptualized to live in the past, outside of modernity. The modern, then, is seen to have a specific location in the western world and more specifically in the European Enlightenment and the carriers thereof. Jonathan Israel has argued that there was a radical underground European Enlightenment between 1650 and 1750 that has been generally regarded as marginal to the wider Enlightenment, but that he sees as central. In his view 'the Enlightenment – European and global – not only attacked and severed the roots of traditional European culture in the sacred, magic, kinship and hierarchy, secularizing all institutions and ideas, but (intellectually and to a degree in practice) effectively demolished all legitimation of monarchy, aristocracy, woman's subordination to man, ecclesiastical authority, and slavery, replacing these with the principles of universality, equality, and democracy' (Israel 2001: vi). Whatever one may think of such a grand view of the history of the European enlightenment, it strikes us that it emphasizes a process of expansion and conversion from a radical margin within Europe. The history of modernity in Europe is a history of conversion to ideas that were once marginal and radical and, *pace* Israel's focus on a European history of ideas, that history is intimately related to the encounter with civilizations and religions elsewhere.

To study global conversion would, in my view, imply that one is aware of a field of interactions that produces modern religion as well as modern secularity in the West and elsewhere (see Van der Veer 2001). If one accepts that the coming of modernity, broadly conceived, is indeed the great transformation that Israel and others think it is, then one should take into account that it is a process that is of long duration, in principle always unfinished and riddled with contradictions. The notion that an already finished, modular modernity is shipped from Europe to the rest of the world is contrary to historical evidence. In studies of encounters between Europe and the rest of the world one often finds an opposition between an internal and an external perspective. The internal account is that there is already an internal historical process of expansion and modernization taking place that is, as it were, fulfilled by the encounter with Western modernity. One may note here the concept of 'fulfillment' that is taken from Christian theology. The external account is that one finds in that encounter an imposition of modern ideas on traditional worlds and thus a conquest of the native mind by modern ways of thinking, including Christian and secular ones.

Both such accounts contain valuable elements, but one may consider studying these encounters as a set of interactions in which internality and externality are tropes in the representation of such encounters. This observation is certainly not to say that this phenomenon takes place in a harmless world outside of a history of power, but that again, the signification of power and resistance to it are so much part of the process that is studied. The histories that precede these encounters are sometimes subsumed in colonial or missionary archives and marginalized or even erased, but these cultural traditions are not powerless or fruitlessly resisting from the margins; they are quite central to world history. This centrality is immediately evident in the case of Asia, which was at the heart of the world economy till 1800 or so, and is already returning to that position in our period.

An interactional perspective has a number of implications for the study of conversion. First of all, at the highest level of generalization, we are all coeval and in the process of becoming modern. The missionaries who bring modern Christianity to the rest of the world are being converted themselves. It is certainly not the case that they carry a stagnant and moribund Christianity from a secularizing world to a still enchanted world. Surely, they are themselves experiencing the transformation of religion at home and they bring that experience to bear on their understanding of their own endeavors in their mission fields. Simultaneously their encounter with the traditions of the people they try to convert also has an impact on their understanding of their own beliefs and practices.

Secondly, the emergence of modern missionary activity from the late eighteenth century creates a religious public at home that supports the activity but is also transformed by it (see van Rooden 1997; Hofmeyer 2004). I would argue that the rise of a muscular, imperial Christianity in Britain in the nineteenth century is directly related to the representation of missionary activity as heroic and adventurous. The site for this form of Christianity is not primarily church or chapel, but the public school and the novel (see Griffiths 2004). At the same time the missionary project is responded to by a number of religious groups in the mission field, who are challenged by it to come up with their own modern institutions for education and social welfare. While in Britain a whole genre of pamphlet literature and popular novels depicts the heroic encounter with such native barbarisms as widow-immolation, in India a genre of apologetic literature as well as of radically anti-Christian pamphlets emerges.

Thirdly, the simultaneous emergence of the nation-state and the colony has some significant implications for the location of religion. Modern Christianity in a number of nation-states in Europe, such as Britain and Holland — both significant recruiting grounds for missionaries — becomes nationalized. That is to say, religious differences between Protestant and Roman

Catholic Christians, and between different Protestant denominations, are subsumed under the rubric of national unity towards threatening neighbors, but significantly also towards the colony. The nation-state develops a number of secular institutions that undercut the public authority and power of religious institutions. One has to see this development as a long-term process since the real decline of the public significance of religion in Britain and Holland only takes place after World War II with the rise of the welfare state. This process is obviously of great importance to the location of religion in society. Contrary to what is often assumed, religion in nineteenth-century Europe appears to become more significant in civil society in terms of the active mobilization of people, despite its decline in scientific and intellectual circles and despite the competition with secular ideologies such as liberalism and socialism. Unlike the United States, Britain and Holland do not have such a sharp separation of church and state, so this process of religious mobilization was and continues to be stronger in the former case. At the same time, we find a shift in the location of religion in the colony. A religious public sphere emerges in which a number of so-called revivalist movements mobilize people and resources for the modernization of religious traditions and especially education. When one interprets conversion to Christianity in the modern period as a conversion to modernity one argues, in effect, that conversion is a crucial aspect of the transformation of Christianity into a modern, global religion (see Van der Veer 1996).

## Chapter 121

# R. S. SUGIRTHARAJAH

# POSTCOLONIALIZING BIBLICAL INTERPRETATION

From *The Bible in the Third World: Precolonial, Colonial, Postcolonial Encounters*, Cambridge: Cambridge University Press, 2001.

**P**OSTCOLONIAL READING will reread biblical texts from the perspective of postcolonial concerns such as liberation struggles of the past and present; it will be sensitive to subaltern and feminine elements embedded in the texts; it will interact with and reflect on postcolonial circumstances such as hybridity, fragmentation, deterritorialization, and hyphenated, double or multiple, identities. One postcolonial concern is the unexpected amalgamation of peoples, ideas, cultures and religions. The religious landscape is so complex that reading a text through one single religious view may not yield much these days when cultural identities and religions coalesce. Postcolonial reading will, for instance, see the confrontation of Elijah and the priests of Canaan with the Phoenician god Baal at Mount Carmel, not as a straight theological conflict between two deities, Yahweh and Baal, nor as one religious community and its gods pitched against another and its gods, but as a complex issue where communities intermingle and the gods are significantly beyond their theological propensities. Mainstream and missionary scholarship tend to see the Mount Carmel episode as a clear theological conflict between two deities, one virtuous and the other evil, extrapolate it to denigrate Asian and African religions as idolatrous, superstitious and evil, and see the victory of Yahweh as proof of the superiority of the biblical God. Postcolonial reading will re-examine what the confrontation is all about (see Wielenga 1988: 151–63; Wielenga 1995: 51–7). When the Israelites settled in Canaan, they incorporated a number of Canaanite theological concepts into their theological thinking. For example, it was from the Canaanite religion that Israel learnt to speak of the 'God of Heaven'. It was a Canaanite concept that was transformed into the 'hosts of heaven' (I Kings 22.19; Isaiah 6.3); the 'God of thunder' was turned into Yahweh as the one who rides in the clouds (Psalms 68.4; 104.3), and, faced with El, the highest god in the Canaanite pantheon, Yaweh was simply identified as equal to El. . . .

In an age when many people question traditional sources of moral authority, sacred texts – the Bible among them – may not be the only place to look for answers to abstract or existential problems. The purpose of postcolonial reading is not to invest texts with properties that no longer have relevance to our context, or with excessive and exclusive theological claims which invalidate other claims. It seeks to puncture the Christian Bible's Western protection and pretensions, and to help reposition it in relation to its oriental roots and Eastern heritage. The aim is not to rediscover the biblical texts as an alternative or to search in their pages for a better world, as a way of coping with the terrors of the colonial aftermath.

The Bible is approached not for its intrinsic authoritativeness or distinctiveness, but because of the thematic presuppositions of postcolonialism, which are influenced by such cultural and psychological effects as hybridity and alienation triggered by colonialism.

Postcolonialism's critical procedure is an amalgam of different methods ranging from the now unfashionable form-criticism to contemporary literary methods. It is interdisciplinary in nature and pluralistic in its outlook. It is more an avenue of inquiry than a homogeneous project. One of the significant aspects of postcolonialism is its theoretical and intellectual catholicism. It thrives on inclusiveness, and it is attracted to all kinds of tools and disciplinary fields, as long as they probe injustices, produce new knowledge which problematizes well-entrenched positions and enhance the lives of the marginalized. Any theoretical work that straddles and finds its hermeneutical home in different disciplines is bound to suffer from a certain eclectic theoretical arbitrariness. Such a selective bias, though lacking coherence, is sometimes necessary for the sake of the task in hand.

In defining the task of postcoloniality, Gloria Anzaldúa points out that postcoloniality looks at power systems and disciplines ranging from government documents to anthropological compositions and asks, 'Who has the voice? Who says these are rules? Who makes the law? And if you're not part of making the laws and rules and the theories, what part do you play? . . . What reality does this disciplinary field, or this government, or this system try to crush? What reality is it trying to erase? What reality is it trying to suppress?' (Lunsford 1999: 62) Applying this to biblical studies, postcolonial biblical criticism poses the following questions: who has the power to interpret or tell stories? To whom do the stories/texts belong? Who controls their meaning? Who decides what texts we choose? Against whom are these stories or interpretations aimed? What is their ethical effect? Who has power to access data?

## Liberation hermeneutics and postcolonial criticism: shall the twain meet?

Liberation hermeneutics and postcolonial criticism should be companions in arms, fighting the good fight. For both, commitment to liberation, however modernist the project may be, still has a valid purchase, for liberation as a grand narrative provides hope for countless millions of people who daily face institutional and personal violence and oppression. Both liberation hermeneutics and postcolonial criticism take the 'Other', namely the poor, seriously; both want to dismantle hegemonic interpretations and do not hesitate to offer prescriptions and make moral judgements, while acknowledging the perils of such decisions. However, to reiterate a point made in the last chapter, the entrenchment of liberation hermeneutics within the modernistic framework acts as an inhibition, and prevents it from embracing some of the virtues of postmodernism for its liberative cause. Postcolonial critical theory, on the other hand, as an off-shoot of postmodernism, while it collaborates with it, distances itself from its errors and unsavoury aspects. While liberation hermeneutics has successfully undermined the certitude of dominant biblical scholarship, it is triumphalistic of its own achievement. Postcolonialism, on the other hand, understands the Bible and biblical interpretation as a site of struggle over its efficacy and meanings. There is a danger in liberation hermeneutics making the Bible the ultimate adjudicator in matters related to morals and theological disputes. Postcolonialism is much more guarded in its approach to the Bible's serviceability. It sees the Bible as both a safe and an unsafe text, and as both a familiar and a distant one. . . .

Postcolonial reading advocates the emancipation of the Bible from its implication in dominant ideologies both at the level of the text and at the level of interpretation. For postcolonialism, the critical principle is not derived only from the Bible but is determined by contextual needs and other warrants. It sees the Bible as one among many liberating texts. Liberation hermeneutics could usefully avail itself of some of the insights advocated by postcolonialism without abandoning or toning down its loyalty to the poor.

In its choice of biblical paradigms, and in its preoccupation with certain favoured texts and with reading them at their face value, liberation hermeneutics fails to appreciate the historical or political ramifications such an interpretation will have for those who face displacement and uprooting in their own lands and countries. For instance, in espousing and endorsing the Exodus as the foundational text for liberation in its early days, liberation hermeneutics failed to note that its suitability as a project had limited value and force. While liberation hermeneutics claimed that the Exodus was read from the point of view of the oppressed, it did not pause to think of the plight of the victims who were at the receiving end of its liberative action, and forced to embark upon what Robert Allen Warrior calls a 'reverse Exodus' from their own promised land. We [can see] how inappropriate the narrative is for Native American, Palestinian and Aboriginal contexts. It also raises awkward theological questions as to what kind of a God is posited both by the Bible and liberation theology. God is the one who emancipates Israel, but also in the process destroys Egyptians and Canaanites. Postcolonialism reads the narrative from the Canaanite point of view and discerns the parallels between the humiliated people of biblical and contemporary times. . . .

Liberation for postcolonialism is not imposing a preexisting notion, but working out its contours in responding to voices within and outside the biblical tradition. Postcolonial space refuses to press for a particular religious stance as final and ultimate. As a point of entry, individual interpreters may have their own theological, confessional and denominational stance, but this in itself does not preclude them from enquiring into and entertaining a variety of religious truth claims. It is the multi-disciplinary nature of the enterprise which gives postcolonialism its energy. It sees revelation as an ongoing process which embraces not only the Bible, tradition, and the Church but also other sacred texts and contemporary secular events. What postcolonialism will argue for is that the idea of liberation and its praxis must come from the collective unconscious of the people. It sees liberation not as something hidden or latent in the text, but rather as born of public consensus created in democratic dialogue between text and context.

# Bibliography

Abramson, Harold J. (1973) *Ethnic Diversity in Catholic America*, New York and London: John Wiley.

Abu-Lughod, J. (1991) 'Going Beyond Global Babble', in A. D. King (ed.) *Culture, Globalization and the World-System*, London: Macmillan.

Abu-Lughod, L. (1991) 'Writing Against Culture', in R. G. Fox (ed.) *Recapturing Anthropology*, Santa Fe, N.M.: School of America Research Press.

Achebe, Chinua (1958) *Things Fall Apart*, London: Heinemann.

Achebe, Chinua (1960) *A Man of the People*, London: Heinemann.

Achebe, Chinua (1971) *Beware, Soul Brother*, rpt, London: Heinemann 1979.

Achebe, Chinua (1973) 'Named for Victoria, Queen of England' *New Letters* 40, 3 (Fall).

Achebe, Chinua (1974) *Arrow of God*, 2nd edn, London: Heinemann.

Achebe, Chinua (1975) *Morning Yet on Creation Day*, Garden City, N.Y.: Doubleday.

Achebe, Chinua (1988) *Hopes and Impediments: Selected Essays 1965–1987*, New York: Doubleday.

Achebe, Chinua (1989) 'Politics and Politicians of Language in African Literature', *FILLM Proceedings*, Doug Killam (ed.), Guelph, Ontario: University of Guelph.

Adam, Ian and Helen Tiffin (eds) (1991) *Past the Last Post: Theorizing Post-colonialism and Post-modernism*, Hemel Hempstead: Harvester Wheatsheaf.

Adams, Carol (1990) *The Sexual Politics of Meat*, New York: Continuum.

Agnew, Jean-Christophe (1983) 'The Consuming Vision of Henry James', in Richard Wrightman Fox and T. J. Jackson Lears (eds) *The Culture of Consumption: Critical Essays in American History*, New York: Pantheon.

Ahluwalia, Pal (2001) 'Negritude and Nativism', in *Politics and Post-colonial Theory: African Inflections*, London: Routledge.

Ahmad, Aijaz (1987) 'Jameson's Rhetoric of Otherness and the "National Allegory"', *Social Text* 17: 3–28.

Ahmad, Aijaz (1991) 'Between Orientalism and Historicism: Anthropological Knowledge of India', *Studies in History* 7, 1.

Ahmad, Aijaz (1992) *In Theory: Classes, Nations, Literatures*, London: Verso.

Ajayi, I. F. Ade (1965) *Christian Missions in Nigeria 1841–1891*, London: Longman.

Akshara, K. V. (1984) 'Western Responses to Traditional Indian Theatre', *Journal of Arts and Ideas* 8 (July–September).

Aléxis, Jacques Stephen (1956) 'Of the Marvellous Realism of the Haitians', *Présence Africaine* 8–10.

Aléxis, Jacques Stephen (1960) *Romancero aux étoiles*, Paris: Gallimard.

Ali, Muhammad Mohar (1965) *The Bengali Reaction to Christian Missionary Activities 1933–1985*, Chittagong: Mehrun Publications.

Allen, Philip (1971) '*Bound to Violence* by Yambo Ouloguem', *Pan-African Journal* iv, 4 (Fall): 518–23, New York: Pan-African Institute.

Alloula, Malek (1986) *The Colonial Harem*, Minneapolis: University of Minnesota Press.

Alsop, J. (1981) *The Rare Art Traditions: A History of Art Collecting and its Linked Phenomena*, Princeton, N.J.: Princeton University Press.

Altbach, Phillip G. (1971) 'Education and Neo-colonialism', *Teachers College Record* 100, 3 (May).

Altbach, Phillip G. (1975a) 'Literary Colonialism: Books in the Third World', *Harvard Educational Review* 45, 2 (May): 226–36.

Altbach, Philip G. (1975) 'Publishing and the Intellectual System', in *Annals of the American Academy of Political and Social Science*.

Althusser, Louis (1972) *Montesquieu, Rousseau, Marx*, London: Verso.

Amin, Samir (1977) *Imperialism and Unequal Development*, New York: Monthly Review Press.

Amin, Samir (1980) *Class and Nation: Historically and in the Current Crisis*, New York and London: Monthly Review Press.

Amuta, Chidi (1989) *The Theory of African Literature*, London: Zed Books.

Anand, Mulk Raj (1979) 'Why I Write', in K. Singh (ed.) *Indian Writing in English*, New Delhi: Heritage: 1–9.

Anderson, Benedict (1983) *Imagined Communities: Reflections on the Origin and Spread of Nationalism*, London: Verso.

Anderson, Margo (1988) *The American Census: A Social History*, New Haven: Yale University Press.

Anderson, Margo (1997) *The American Census: A Social History*, New Haven: Yale University Press.

Andreski, Iris (1971) *Old Wives Tales*, New York: Schocken.

Annamalai, E. (1978) 'The Anglicized Indian Languages: a Case of Code-Mixing', *International Journal of Dravidian Linguistics* 7, 2: 239–47.

Anyinefa, Kofi (1996) 'Hello and Goodbye to Négritude: Senghor, Dadié, Dongala, and America', *Research in African Literatures* 27, 2 (Summer): 51–69.

Anzaldúa, Gloria (1987) *Borderlands/La Frontera: The New Mestiza*, San Francisco: Aunt Lute.

Appadurai, Arjun (ed.) (1986) *The Social Life of Things: Commodities in Cultural Perspective*, Cambridge: Cambridge University Press.

Appadurai, Arjun (1996) *Modernity at Large: Cultural Dimensions of Globalization*, Minneapolis: University of Minnesota Press.

Appiah, Kwame Anthony (1990) 'Race', in Frank Lettricchia and Thomas McGlaughlin (eds) *Critical Terms for Literary Study*, Chicago: University of Chicago Press: 274–87.

Appiah, Kwame Anthony (1991) 'Is the Post- in Postmodernism the Post- in Postcolonial?', *Critical Inquiry* 17 (Winter).

Appiah, Kwame Anthony (1992) *In My Father's House: Africa in the Philosophy of Culture*, London: Methuen; New York: Oxford University Press.

Appiah, Kwame Anthony (1998) 'The Illusions of Race', in Emmanuel Eze (ed.) *African Philosophy: An Anthology*, Oxford: Blackwell.

Apple, M. W. (1982) *Education and Power*, Boston and London: Routledge & Kegan Paul.

Apple, M. W. (1993) *Official Knowledge: Democratic Education in a Conservative Age*, New York: Routledge.

Apter, Emily (1995) 'French Colonial Studies and Postcolonial Theory', *Sub-Stance* 76/7, 24: 1–2.

Arendt, Hannah (1958) *The Human Condition*, Chicago: University of Chicago Press.

Armah, Ayi Kwei (1975) 'Struggles to Find a Local Publisher', *Asemka* 4 (University of Cape Coast Press).

Armas, Genaro (2001) 'Black Hispanic Totals Nearly Equal', *Associated Press*, 8 March. Electronic press release.

Arnold, David (1992) 'The Colonial Prison: Power, Knowledge and Penology in Nineteenth Century India', in David Arnold and David Hardiman, *Subaltern Studies* 8, Delhi.

Arnold, Stephen (ed.) (1985) *African Literature Studies: The Present State/L'Etat Present*, Washington, D.C.: Three Continents Press.

Arteaga, Alfred (1994) 'An Other Tongue', in Alfred Arteaga (ed.) *An Other Tongue*, Durham, N.C.: Duke University Press.

Asad, Talal (1973) 'The Concept of Cultural Translation in British Social Anthropology', in James Clifford and George Marcus (eds) *Writing Culture: The Poetics and Politics of Ethnography*, Berkeley: University of California Press: 141–64.

Ashby, Eric (1967) *Universities: British, Indian, African*, Cambridge, Mass.: Harvard University Press.

Ashcroft, Bill (1997) 'Globalism, Post-colonialism and African Studies', in Pal Ahluwalia and Paul Nursey-Bray (eds) *Post-colonialism: Culture and Identity in Africa*, New York: Nova Science Publishers.

Ashcroft, Bill (2001) *Post-colonial Transformation*, London: Routledge.

Ashcroft, Bill, Gareth Griffiths and Helen Tiffin (1989) *The Empire Writes Back: Theory and Practice in Post-colonial Literatures*, London: Routledge. Second edn, 2002.

Ashcroft, W. D. (1989) 'Constitutive Graphonomy', in Stephen Slemon and Helen Tiffin (eds.) *After Europe: Critical Theory and Post-colonial Writing*, Mundelstrup: Dangaroo.

*Athenaeum* (1839) issue 108.

Atodevi, Stanislas S. (1972) *Négritude et négrologues*, Paris: Union Generale de'Editions.

Atwood, Margaret (1985) *The Handmaid's Tale*, Toronto: McClelland and Stewart.

Austin, J. L. (1962) *How to do Things with Words*, Oxford: Clarendon.

Bâ, Sylvia Washington (1973) *The Concept of Négritude in the Poetry of Léopold Sédar Senghor*, Princeton, N.J.: Princeton University Press.

Back, Les and John Solomos (2000) *Theories of Race and Racism: A Reader*, London and New York: Routledge.

Bail, Murray (1980) *Homesickness*, Melbourne: Mamillan.

Bailey, Herbert Jr (1970) *The Art and Science of Book Publishing*, New York: Harper & Row.

Baillie, Robert (1986) *Les Voyants*, Montreal: Hexagone.

Baker, Houston A. Jr (1986) 'Caliban's Triple Play', *Critical Inquiry* 13, 1: 182–96.

Bakhtin, M. (1981) 'Epic and Novel: Toward a Methodology for the Study of the Novel', *The Dialogic Imagination: Four Essays*, Michael Holquist (ed.), trans. by Caryl Emerson and Michael Holquist, Austin, Tex.: University of Texas Press.

Bakhtin, Mikhail (1984) *Rabelais and his World*, Bloomington, Ind.: Indiana University Press.

Baldridge, William (1996) 'Reclaiming Our Histories', in James Treat (ed.) *Native and Christian: Indigenous Voices on Religious Identity in the United States and Canada*, New York: Routledge.

Baldwin, James (1964) *Notes of a Native Son*, London: Michael Joseph.

Balibar, Etienne (1991a) 'Es Gibt Keinan Staat en Europea: Racism and Politics in Europe Today', *New Left Review* 186 (March/April).

Balibar, Étienne (1991b) 'Is There a "Neo-Racism"?', in Étienne Balibar and Immanuel Wallerstein, *Race, Nation, Class*, London: Verso.

Ban Kah Choon (1979) 'A Review of Creative Writing in Singapore, 1978', *Commentary* 3, 3.

Baran, Paul A. (1962) *Political Economy of Growth*, New York: Monthly Review Press.

Barber, B. R. (1992) 'Jihad vs. McWorld', *The Atlantic* 269, 3.

Barber, K. (1987) 'Popular Arts in Africa', *Africa Studies Review* 30 (3 September): 1–78.

Barber, Karin and P. F. de Moraes Farias (eds) (1989) *Discourse and Its Disguises: The Interpretation of African Oral Texts*, Centre of West African Studies, Birmingham University African Studies Series 1.

Barker, Ronald and Robert Escarpit (1973) *The Book Hunger*, Paris: UNESCO.

Barnett, Stanley and Robert Piggford (1969) *Manual on Book and Library Activities in Developing Countries*, Washington, D.C.: Agency for International Development.

Barth, Frederick (1969) *Ethnic Groups and Boundaries: The Social Organisation of Culture Difference*, Boston: Little, Brown.

Bataille, Georges (1992) *Theory of Religion*, trans. Robert Hurley, New York: Zone.

Bateson, Gregory (1972) 'Form, Substance and Difference', in *Steps to an Ecology of Mind*, New York: Ballantine.

Baudrillard, J. (1981) *For a Critique of the Political Economy of the Sign*, St Louis, Mo.: Telos Press.

Bayart, Jean Francois (1993) *The State in Africa*, London and New York: Longman.

Bebey, Francis (1977) 'Paris Interview', 20 August, cited by Norman Stockle in 'Towards an Africanisation of the Novel: Francis Bebey's Narrative Technique', in Kolawole Ogungbesan (ed.) *New West African Literature,* London: Heinemann, 1979.

Benjamin, Curtis (1964) *Books as Forces in National Development and International Relations*, New York: National Foreign Trade Council.

Benjamin, Walter (1969) 'The Task of the Translator', in Hannah Arendt (ed.) *Illuminations*, New York: Schoken Books.

Benjamin, Walter (1973) *Charles Baudelaire: A Lyric Poet in the Era of High Capitalism,* trans. Harry Zohn, London: New Left Books.

Bennett, T. (1990) *Outside Literature*, London: Routledge.

Berkes, N. (1964) *The Development of Secularism in Turkey*, Montreal: McGill University Press.

Berry, Reginald (1986) 'A Deckchair of Words: Post-colonialism, Post-modernism, and the Novel of Self-projection in Canada and New Zealand', *Landfall* 40: 310–23.

Besant, Annie (1914) *India and the Empire: A Lecture and Various Papers*, London: Theosophical Publishing Society.

Bhabha, Homi K. (1983a) 'Difference, Discrimination and the Discourse of Colonialism', in *The Politics of Theory*, Proceedings of the Essex Conference on the Sociology of Literature, July 1982, Colchester: University of Essex.

Bhabha, Homi K. (1983b) 'The Other Question . . . Homi Bhabha Reconsiders the Stereotype and Colonial Discourse', *Screen* 24 (November–December).

Bhabha, Homi K. (1984a) 'Representation and the Colonial Text: A Critical Exploration of Some Forms of Mimeticism', in Frank Gloversmith (ed.) *The Theory of Reading*, Brighton: Harvester.

Bhabha, Homi K. (1984b) 'Of Mimicry and Men: The Ambivalence of Colonial Discourse', *October* 28: 125–33.

Bhabha, Homi K. (1985a) 'Signs Taken for Wonders: Questions of Ambivalence and Authority Under a Tree Outside Delhi, May 1817', *Critical Inquiry* 12, 1.

Bhabha, Homi K. (1985b) 'Sly Civility', *October* 34.

Bhabha, Homi K. (1988) 'The Commitment to Theory', *New Formations* 5: 5–23.

Bhabha, Homi K. (ed.) (1990) *Nation and Narration*, London: Routledge.

Bhabha, Homi K. (1994) *The Location of Culture*, London: Routledge.

Bhely-Quenum, O. (1982) 'Ecriture noire en question (débat)', *Notre Librairie* 65.

Billington, M. (1992) 'The Reinvention of William Shakespeare', *World Press Review* (July).

Bishop, Alan J. (1990) 'Western Mathematics: The Secret Weapon of Cultural Imperialism', *Race and Class* 32, 2.

Bissi, M. (1989) 'Commonauté urbaine de Douala: Place à M. Pokossy Ndoumbé', *Cameroon Tribune* 4372 (19 April).

Boehmer, Elleke (2002) *Empire, The National, and the Postcolonial 1890–1920: Resistance in Interaction*, Oxford: Oxford University Press.

Boggs, Grace Lee (2000) 'School Violence: A Question of Place', *Monthly Review* 52, 2: 18–20.

Bohannan, P. (1955) 'Some Principles of Exchange and Investment amongst the Tiv', *American Anthropologist* 57: 60–70.

Bokamba Fyamba, G. (1982) 'The Africanization of English', in B. Kachru (ed.) *The Other Tongue: English Across Cultures*, Chicago: University of Illinois Press: 77–98.

Bonnafé, P. (1978) *Nzo Lipfe, le lignage de la mort: La sorcellerie, idéologie de la lutte sociale sur le plateau kukuya*, Paris: Labethno.

Bookchin, Murray (1971) *Post Scarcity Anarchism*, Palo Alto, Calif.: Ramparts Press.

Booth, James (1980) *Writers and Politics in Nigeria*, New York: Africana Publishers.

Botros, Salib (1978) 'Problems of Book Development in the Arab World with Special Reference to Egypt', *Library Trends* (Spring).

Bourdieu, P. (1979) *La Distinction: Critique sociale du jugement*, Paris: Editions de Minuit.

Bourdieu, P. (1993) *The Field of Cultural Production: Essays on Art and Literature*, R. Johnson (ed.), New York: Columbia University Press.

Bourdon, R. (1981) *La place du désordre*, Paris: Presses Universitaires de France.

Boyarin, Daniel and Jonathan Boyarin (1993) 'Diaspora: Generation and the Ground of Jewish Identity', *Critical Enquiry* 19, 4: 693–725.

Boyce-Davis, C. (1994) *Black Women, Writing and Identity: Migrations of the Subject*, New York: Routledge.

Bradley, Laurel (1991) 'From Eden to Empire: John Everett Millais' Cherry Ripe', *Victorian Studies* 34, 2 (Winter).

Brah, Avtar (1996) *Cartographies of Diaspora: Contesting Identities*, London and New York: Routledge.

Brahms, Flemming (1982) 'Entering Our Own Ignorance – Subject-Object Relations in Commonwealth Literature', *World Literature Written in English* 21, 2 (Summer).

Brathwaite, Edward Kamau (1967) 'Caribbean Theme: A Calypso', sung by the author on *Rights of Passage*, London: Argo Records DA 102, 1969.

Brathwaite, Edward Kamau (1967–8) 'Jazz and the West Indian Novel', *Bim* 2, 44–6.

Brathwaite, Edward Kamau (1971) *The Development of Creole Society in Jamaica 1770–1820*, Oxford: Clarendon Press.

Brathwaite, Edward Kamau (1973) *The Arrivants*, London: Oxford University Press.

Brathwaite, Edward Kamau (1975) *Other Exiles*, London: Oxford University Press.

Brathwaite, Edward Kamau (1977) *Nanny, Sam Sharpe, and the Struggle for People's Liberation*, Kingston: API (for National Heritage Week Committee).

Brathwaite, Edward Kamau (1984) *History of the Voice: The Development of Nation Language in Anglophone Caribbean Poetry*, London and Port of Spain: New Beacon.

Brantlinger, Patrick (1983) *Rule of Darkness, British Literature and Imperialism, 1830–1914*, Ithaca, N.Y.: Cornell University Press: 59–60.

Bray, William (1782) 'Observations on the Indian Method of Picture Writing', *Archaelogia* 6: 159.

Brazil, D. (1998) 'The Development of English as a World Language', in *Collin's English Dictionary*, Millennium edn, pp. xxvi–xxxiii (Foreword).

Brennan, Timothy (1990) 'The National Longing for Form', in Homi K. Bhabha (ed.) *Nation and Narration*, London: Routledge.

Brilliant, Lawrence with Girija Brilliant (1978) 'Death for a Killer Disease', *Quest* 3 (May/June).

Brodber, Erna (1989) 'Sleeping's Beauty and the Prince Charming', *Kunapipi* 2, 3.

Brody, Lewis Hugh (1981) *Maps and Dreams: Indians and the British Columbia Frontier*, Vancouver: Douglas & McIntyre.

Brontë, Charlotte (1960) *Jane Eyre*, New York: Collier.

Brown, David A. Maughan (1985) *Land, Freedom and Fiction: History and Ideology in Kenya*, London: Zed Books.

Brown, E. P. (1983) *Nourrir les gens, nourrir les haines*, Paris: Société d'ethnographie.

Brown, Judith (1989) *Gandhi: Prisoner of Hope*, New Haven: Yale University Press.

Brown, Lester R. (1984) 'Putting Food on the World's Table, a Crisis of Many Dimensions', *Environment* 26 (May).

Brown, Russell M. (1978) 'Critic, Culture, Text: Beyond Thematics', *Essays in Canadian Writing* 11 (Summer).

Brydon, Diana (1987) 'The Myths that Write Us: Decolonising the Mind', *Commonwealth* 10, 1: 1–14.

Brydon, Diana (1988) 'Troppo Agitato: Reading and Writing Culturesa in Randolph Stow's *Visitants* and Rudy Wiebe's *The Temptations of Big Bear*', *Ariel* 19, 1: 13–32.

Brydon, Diana (1991) 'The White Inuit Speaks: Contamination or Literary Strategy', in Ian Adam and Helen Tiffin (eds) *Past the Last Post: Theorizing Post-colonialism and Post-modernism*, New York and London: Harvester Wheatsheaf.

Bulbeck, Chilla (1991) Unpublished paper, Faculty of Humanities Research Seminar, Griffith University, Nathan, Queensland.

Burchfield, R. (1985) *The English Language*, Toronto: University of Toronto Press.

Burke, Timothy (1992) '"Nyamarira That I Loved": Commoditization, Consumption and the Social History of Soap in Zimbabwe', *The Societies of Southern Africa in the 19th and 20th Centuries: Collected Seminar Papers* 42, 17: 195–216, London: University of London Institute of Commonwealth Studies.

Butler, Judith (1988) 'Performative Acts and Gender Constitution: An Essay in Phenomenology and Feminist Theory', *Theatre Journal* 40: 519–31.

Butler, Judith (1990a) *Gender Trouble*, New York: Routledge.

Butler, Judith (1990b) 'The Force of Fantasy: Feminism, Mapplethorpe and Discursive Excess', *Differences* 2: 121.

Butler, Judith (1993) *Bodies That Matter*, New York: Routledge.

Cabral, Amilcar (1973) *Return to the Sources: Selected Speeches*, New York and London: Monthly Review Press.

Cairns, David and Sean Richards (1988) 'What Ish my Nation?', in *Writing Ireland: Colonialism, Nationalism and Culture*, Cultural Politics Series, Manchester: Manchester University Press.

Caldiera, Teresa (1996) 'Fortified Enclaves: The New Urban Segregation', *Public Culture* 8, 303–28.

Callaway, Helen (1987) *Gender, Culture Empire: European Women in Colonial Nigeria*, Oxford: Macmillan Press in association with St Anthony's College.

Canetti, E. (1988) *Crowds and Power*, trans. C. Stewart, New York: Farrar, Strauss & Giroux.

Carby, Hazel V. (1982) 'White Woman Listen! Black Feminism and the Boundaries of Sisterhood', in Centre for Contemporary Cultural Studies (ed.) *The Empire Strikes Back: Race and Racism in 70s Britain*, London: Hutchinson.

Carby, Hazel V. (1987) *Reconstructing Womanhood: The Emergence of the Afro-American Woman Novelist*, New York: Oxford University Press.

Carlson, Marvin (1996) 'Introduction', in *Performance: A Critical Introduction*, London and New York: Routledge.

Carpentier, Alejo (1972) *Explosion in a Cathedral*, trans. John Sturrock, Harmondsworth: Penguin.

Carter, Paul A. (1987) *Road to Botany Bay: An Essay in Spatial History*, London: Faber & Faber.

Cary, Joyce (1949) [1932] *Aissa Saved*, London: E. Benn.

Cassanelli, L. (1986) 'Qat: Changes in the Production and Consumption of a Quasilegal Commodity in Northeast Africa', in Arjun Appaurai (ed.) *The Social Life of Things: Commodities in Cultural Perspective*', Cambridge: Cambridge University Press: 236–60.

Castellino, Joshua (2000) *International Law and Self-determination: The Interplay of the Politics of Territorial Possession with Formulations of Post-colonial National Identity*, The Hague and Boston: Martinus Nijhoff.

Castoriadis, Cornelius (1991) 'Reflections on "Rationality" and "Development"', in *Philosophy, Politics, Autonomy*, New York: Oxford University Press.

Cathcart, Sarah and Lemon, Andrea (1988) *The Serpent's Fall*, Sydney: Currency.

Césaire, Aimé (1946) *Et les chiens se taisaient, a Tragedy*, in *Les Armes Miraculeuses*, Paris: Gallimard.

Chakrabarty, Dipesh (1992) 'Postcoloniality and the Artifice of History: Who Speaks for "Indian" Pasts?', *Representations* 37 (Winter): 1–26.

Chakrabarty, Dipesh (2000) *Provincializing Europe: Postcolonial Thought and Historical Difference*, Princeton, N.J.: Princeton University Press.

Chambers, Iain and Lidia Curti (eds) (1996) *The Post-colonial Question: Common Skies, Divided Horizons*, London and New York: Routledge.

Chang, Jen Hu (1970) 'Potential Photosynthesis and Crop Productivity', *Annals of the Association of American Geographers* 60 (March): 92–101.

Chanock, Martin (1982) 'Making Customary Law: Men, Women and the Courts in Colonial Rhodesia', in Margaret J. Hay and Marcia Wright (eds) *African Women and the Law: Historical Perspectives*, Boston: African Studies Center, Boston University.

Chapman, Murray (1978) 'On the Cross-cultural Study of Circulation', *International Migration Review* 12: 559–69.

Chapman, Murray (1991) 'Pacific Island Movement and Sociopolitical Change: Metaphors of Misunderstanding', *Population and Development Review* 17: 263–92.

Charles, G. (1993) 'Hobson's Choice for Indigenous Peoples', in R. D. Jackson (ed.) *Global Issues* 93/4, Guildford, Conn.: Dushkin.

Charlevoix, Pierre François-Xavier de [1744] (1976) *Historie de la Nouvelle France*, 3 vols, Ottawa: Elysée.

Chartrand, L. (1991) 'A New Solidarity Among Native Peoples', *World Press Review* (August).

Chatterjee, Partha (1986) *Nationalist Thought and the Colonial World: A Derivative Discourse*, London: Zed Books for United Nations University.

Chernoff, John Miller [1979] (1981) *African Rhythm and African Sensibility: Aesthetics and Social Action in African Musical Idioms*, Chicago: University of Chicago Press.

Chief Kabongo (1973) 'The Coming of the Pink Cheeks', in *Through African Eyes*, vol. IV, *The Colonial Experience: An Inside View*, Leone E. Clark (ed.), New York: Praeger.

Chinweizu, Onwuchekwu Jemie and Ihechukwu Madubuike [1980] (1985) *Towards the Decolonization of African Literature*, London: Routledge & Kegan Paul.

Chinweizu 'The Responsibilities of Scholars of African Literature', *Research in African Literatures* 13–19.

Chow, Rey (1986–7) 'Rereading the Mandarin Ducks and Butterflies: A Response to the Postmodern Condition', *Cultural Critique* 5: 69–93.

Christian, Barbara (1987) 'The Race for Theory', *Cultural Critique*, 6: 51–63.

Clark, John Pepper (1972) in *African Writers Talking: A Collection of Radio interviews*, Dennis Duerden and Cosmos Pieterse (eds), London: Heinemann; New York: Africana.

Clastres, Pierre (1987) *Society Against the State*, trans. Robert Hurley and Abe Stein, New York: Zone Books.

Cliff, Michelle (1984) *Abeng*, New York: Dutton.

Clifford, James (1980) 'Review of *Orientalism*, by Edward Said', in *History and Theory* 12, 2.

Clifford, James (1986) 'On Ethnographic Allegory', in James Clifford and George Marcus (eds) *Writing Culture: The Poetics and Politics of Ethnography*, Berkeley: University of California Press.

Clifford, James (1988) *The Predicament of Culture: Twentieth Century Ethnography, Literature, and Art*, Cambridge, Mass.: Harvard University Press.

Clifford, James (1994) 'Diasporas', *Cultural Anthropology* 9, 3: 302–38.

Clifford, James (1997) 'Diasporas', in *Routes: Travel and Translation in the Late Twentieth Century*, Cambridge, Mass. and London: Harvard University Press.

Clifford, James (2001) 'Indigenous Articulations', *The Contemporary Pacific* 3, 2 (Fall): 468.

Closs, M. P. (1986) *Native American Mathematics*, Austin, Tex.: University of Texas Press.

Cohen, Robin (1987) *The New Helots: Migrants in the International Division of Labor*, Aldershot: Gower.

Columbus, Christopher [1825] (1960) *The Journal of Christopher Columbus*, trans. Cecil Jane, London: Antony Blond.

Conrad, Joseph (1972) *Heart of Darkness*, New York: Pocket Books.

Conrad, Joseph [1902] (1983) *Heart of Darkness*, Paul O'Prey (ed.), Harmondsworth: Penguin.

Cooley, Dennis (1987) *The Vernacular Muse: The Eye and Ear in Contemporary Literature*, Winnipeg: Turnstone.

Crapanzano, Vincent (1985) 'A Reporter at Large', *The New Yorker* 18 March.

Crenshaw, Kimberlé *et al.* (1995) *Critical Race Theory*, New York: New York University Press.

Creveçoeur, J. Hector St John de (1981) *Letters from an American Farmer*, Harmondsworth: Penguin.

Crosby, Alfred. W. (1986) *Ecological Imperialism: The Biological Expansion of Europe 900–1900*, Cambridge: Cambridge University Press.

Crowder, M. and O. Ikeme (eds) (1970) *West African Chiefs: Their Changing Status under Colonial Rule and Independence*, Ife: University of Ife Press.

Cugoano, Ottabah (1787) *Thoughts and Sentiments on the Evil and Wicked Traffic of the Slavery and Commerce of the Human Species*, London: T. Beckett.

Culler, Jonathan (1982) *On Deconstruction: Theory and Criticism after Structuralism*, Ithaca, N.Y.: Cornell University Press.

Curnow, Alan (1962) *A Small Room with Large Windows*, London: Oxford University Press.

Curtin, Philip (1964) *Image of Africa: British Ideas and Action, 1780–1850*, 2 vols, Madison: University of Wisconsin Press.

Darby Phillip (ed.) (1997) *At the Edge of International Relations: Postcolonialism, Gender, and Dependency*, London and New York: Pinter.

Dash, Michael (1974) 'Marvellous Realism – The Way out of Négritude', *Caribbean Studies* 13, 4.

Dash, Michael (1989) 'In Search of the Lost Body: Redefining the Subject in Caribbean Literature', *Kunapipi* 2, 1.

Dathorne, O. R. (1974) *The Black Mind*, Minneapolis: Minneapolis University Press.

Dathorne, O. R. (1981) *Dark Ancestor: The Literature of the Black Man in the Caribbean*, Baton Rouge: Louisiana State University Press.

Davey, Frank (1988) *Reading Canadian Reading*, Winnipeg: Turnstone.

Davidoff, Leonore and Catherine Hall (1992) *Family Fortunes: Men and Women of the English Middle Class*, London: Routledge.

Davidson, Basil (1978) *Africa in Modern History: The Search for a New Society*, London: Allen Lane.

Davies, Robertson (1972) *The Manticore*, Toronto: Macmillan.

Davis, Angela and Gina Dent (2001) 'Prison as a Border: A Conversation on Gender, Globalization, and Punishment', *Signs* 26, 4: 1235–41.

Davis, Gregson (ed.) (1984) *Twenty Poems of Aimé Césaire*, Stanford, Calif.: Stanford University Press.

Davis, Jack (1982) *Kullark/The Dreamers*, Sydney: Currency.

Davis, Jack (1986) *No Sugar*, Sydney: Currency.

Davis, Jack and Bob Hodge (eds) (1985) *Aboriginal Writing Today*, papers from the First National Conference of Aboriginal Writers Held in Perth, Western Australian 1983, Canberra: Australian Institute for Aboriginal Studies.

Davis, Mike (1990) *City of Quartz*, London: Verso.

de Certeau, Michel (1984) *The Practice of Everyday Life*, trans. Steven F. Randall, Berkeley: University of California Press.

Defert, Daniel (1982) 'The Collection of the World: Accounts and Voyages from the Sixteenth to the Eighteenth Centuries', *Dialectical Anthropology* 7: 11–20

de Lauretis, Theresa (1984) *Alice Doesn't: Feminism, Semiotics, Cinema*, Bloomington, Ind.: Indiana University Press.

Deleuze, Gilles and Félix Guattari (1987) *A Thousand Plateaus: Capitalism and Schizophrenia*, trans. B. Massumi, Minneapolis: University of Minnesota Press.

Delgado, Richard (2001) *Critical Race Theory: An Introduction*, New York: New York University Press.

Dempsey, Mike (ed.) (1978) *Bubbles. Early Advertising Art from A E Pears Ltd.*, London: Fontana.

De Rougemont, Denis (1966) *The Idea of Europe*, trans. Norbert Guterman, New York: Macmillan.

Derrida, Jacques (1981) 'The Double Session', in *Dissemination*, trans. Barbara Johnson, Chicago: University of Chicago Press.

Derrida, Jacques (1991) '"Eating Well", or the Calculation of the Subject: an Interview with Jacques Derrida', in Eduardo Cadava, Peter Connor and Jun-Luc Nancy (eds) *Who Comes After the Subject?*, New York: Routledge.

Desjardins, Thierry (1976) *Le Martyre du Liban*, Paris: Plon.

Devi, Mahasweta (1997) *Breast Stories*, trans. with introductory essays by Gayatri Chakravorty Spivak, Calcutta: Seagull Books.

Diamond, Elin (1995) *Writing Performance*, London: Routledge.

Dinesen, Isaac (1937) *Out of Africa*, London: Puttman.

Dirks, Nicholas B. (1992) *Colonialism and Culture*, Ann Arbor: University of Michigan Press.

Dirlik, Arif (1991) *Anarchism in the Chinese Revolution*, Berkeley: University of California Press.

Dirlik, Arif (1997a) *The Postcolonial Aura: Third World Criticism in the Age of Global Capitalism*, Boulder, Col.: Westview Press.

Dirlik, Arif (1997b) 'Mao Zedong and "Chinese Marxism"', in *Encyclopedia of Asian Philosophy*, London: Routledge.

Docker, John (1978) 'The Neo-colonial Assumption in University Teaching of English', in Chris Tiffin (ed.) *South Pacific Images*, St Lucia, Queensland: SPACLALS.

Donaldson, Laura E. (1988) 'The Miranda Complex: Colonialism and the Question of Feminist Reading', *Diacritics* 18, 3: 65–77.

Donaldson, Laura E. (2002) *Postcolonialism, Feminism, and Religious Discourse*, Laura E. Donaldson and Kwok Pui-Lan (eds), London and New York: Routledge.

Dorfman, A. (1983) *The Empire's Old Clothes*, New York: Pantheon.

Dorsinville, Max (1974) *Caliban Without Prospero: Essay on Quebec Black Literature*, Erin, Ontario: Press Porceptic.

Dorsinville, Max (1983) *Le Pays natal: essais sur les littératures du Tiers Monde et du Quebec*, Dakar: Nouvelles Editions Africaines.

Drachler, Jacob (ed.) (1964) *African Heritage*, New York: Collins.

Duben, A. and C. Behar (1991) *Istanbul Households: Marriage, Family and Fertility, 1880–1940*, Cambridge: Cambridge University Press.

Duncker, Sheila Joan (1960) 'The Free Coloured and their Fight for Civil Rights in Jamaica 1800–1836', unpublished MA thesis, University of London.

During, Simon (1987) 'Postmodernism or Post-colonialism Today', *Textual Practice* 1, 1.

Eagleton, Terry (1975) *Myths of Power: A Marxist Study of the Brontës*, London: Macmillan.

Eagleton, T. and P. Fuller (1983) 'The Question of Value: A Discussion', *New Left Review* 142: 76–90.

Edwards, Philip (1979) *Threshold of a Nation*, Cambridge: Cambridge University Press.

Egejuru, Phanuel Akubueze (1980) *Towards African Literary Independence: A Dialogue With Contemporary African Writers*, Westport, Conn.: Greenwood.

Eisenstein, Zillah (1998) *Global Obscenities: Patriarchy, Capitalism, and the Lure of Cyber-fantasy*, New York: New York University Press.

Ellison, G. (1915) *An Englishwoman in a Turkish Harem*, London: Methuen.

Ellison, G. (1928) *Turkey To-day*, London: Hutchinson.

Emberley, J. (1993) *Thresholds of Difference: Feminist Critique, Native Women's Writing, Postcolonial Theory*, Toronto: University of Toronto Press.

Emecheta, Buchi (1978) *The Bride Price*, London: Collins.

Emenyonu, Ernest (1971) 'African Literature: What Does it Take to be its Critic?', *African Literature Today* 5: I–II.

Englehardt, Tom [1971] (1995) 'Ambush at Kamikaze Pass', *Bulletin of Concerned Asian Scholars* 3, 1 (Winter/Spring).

Equiano, Olaudah (1789) *The Interesting Narrative of the Life of Olaudah Equiano, or Gustavus Vassa, The African. Written by Himself*, London: Printed and sold by the author.

Erdosy, George (1998) 'Deforestation in Pre- and Proto-Historic South Asia', in Richard H. Grove, Vinita Damodaran and Satpal Sangwan (eds) *Nature and the Orient: The Environmental History of South and Southeast Asia*, Delhi: Oxford University Press: 51–69.

Erikson, Erik H. (1950) *Childhood and Society*, New York: Norton.

Erikson, Erik H. (1975) *Life History and the Historical Moment*, New York: Norton.

Escobar, Arturo (1984–5) 'Discourse and Power in Development: Michel Foucault and the Relevance of his Work in the Third World', *Alternatives* 10, 3 (Winter).

Essono, P. (1987) 'Installation de l'administrateur municipal de Mbankomo: La fête de la démocratie retrouvée', *Cameroon Tribune* 4207, 4 (December).

Eyre, Edward John (1845) *Journals of Expeditions of Discovery into Central Australia and Overland from Adelaide to King George's Sound, 1840–1*, 2 vols, London: T. and W. Boone.

Fabian, Johannes (1983) *Time and the Other: How Anthropology Makes its Object*, New York: Columbia University Press.

Fairclough, N. (1989) *Language and Power*, New York: Longman.

Fanon, Frantz (1965a) Algeria Unveiled', in *A Dying Colonialism*, New York: Grove Press.

Fanon, Frantz [1961] (1965b) *The Wretched of the Earth*, trans. Constance Farrington, London: Macgibbon and Kee.

Fanon, Frantz (1967a) *Black Skin, White Masks*, trans. Charles Lam Markmann, New York: Grove Press.

Fanon, Frantz (1967b) *The Wretched of the Earth*, Harmondsworth: Penguin.

Fanon, Frantz (1968) 'On National Culture'and 'The Pitfalls of National Consciousness', in *The Wretched of the Earth* (trans. Constance Farrington), New York: Grove Press (original French edition 1961).

Fanon, Frantz (1970) *Toward the African Revolution*, Harmondsworth: Penguin.

Fanon, Frantz (1986) *Black Skin, White Masks*, London: Pluto.

Farah, Nuruddin (1970) *From a Crooked Rib*, London: Heinemann.

Farish, J. (1838) Minute dated August 28th 1838, Political Dept, Vol. 20/795, 1837–9 (Bombay Records); quoted in B. K. Boman-Behram (1942) *Educational Controversies of India: The Cultural Conquest of India under British Imperialism*, Bombay: Taraporevala Sons.

Featherstone, Mike (ed.) (1990) *Global Culture: Nationalism, Globalization, and Modernity*, London: Sage.

Fee, Marjorie (1989) 'Why C. K. Stead didn't like Kerry Hulme's *The Bone People* or Who Can Write as the Other', *Australian and NZ Studies in Canada* 1.

Feld, S. (1988) 'Notes on World Beat', *Public Culture* 1, 1: 31–7.

Ferguson, Moira (1986) *The History of Mary Prince A West Indian Slave, Related By Herself*, London: Pandora.

Fernandez-Kelly, Patricia and Diane Wolf (2001) 'A Dialogue on Globalization', *Signs* 26, 4: 1243–9.

Ferry, Luc (1995) *The New Ecological Order*, trans. Carol Volk, Chicago: University of Chicago Press.

Field, Barron (ed.) (1825) *Geographical Memoirs on New South Wales: By Various Hands*, London: John Murray.

Figueroa, John (1970) 'Our Complex Language Situation', in *Caribbean Voices*, vol. 2, *The Blue Horizons*, London: Evans.

Fischer, Humphrey (1973) "Conversation Reconsidered: Some Historical Aspects of Religious Conversion in Black Africa", *Africa* 43, 1: 27–40.

Fischer, Louis (1954) *Gandhi: His Life and Message for the World*, New York: New American Library.

Ford Smith, Honor (1985) 'Sistren: Jamaican Women's Theatre', in D. Kahn and D. Neumaier (eds) *Cultures in Contention*, Seattle: Real Correct.

Ford Smith, Honor (1986) (in collaboration with Sistren Collective) *Lionheart Gal: Life Stories of Jamaican Women*, London: Women's Press.

Foucault, Michel (1965) *Madness and Civilization: A History of Insanity in the Age of Reason*, trans. Richard Howard, New York: Pantheon.

Foucault, Michel (1970) *The Order of Things*, London: Tavistock.

Foucault, Michel (1973) *Madness and Civilization; A History of Insanity in the Age of Reason*, New York: Vintage Books.

Foucault, Michel (1976) *Histoire de la sexualité*, Paris: Gallimard.

Foucault, Michel (1980) *Power/Knowledge: Selected Interviews and Other Writings, 1972–1977*, trans. Colin Gordon, Brighton: Harvester Press.

Foucault, Michel (1982) 'The Subject and Power', Afterword to Herbert L. Dreyfus and Paul Rabinow, *Michel Foucault: Beyond Structuralism and Hermeneutics*, Brighton: Harvester: 208–26.

Foucault, Michel (1988) 'The Ethic of Care of the Self as a Practice of Freedom', in J. Bernauer and D. Rasmussen (eds) *The Final Foucault*, Cambridge, Mass.: MIT.

Foucault, Michel (1989) *Foucault Live*, trans. J. Johnstone and S. Lotringer, New York: Semiotexte.

Foucault, Michel and Gilles Deleuze (1977) 'Intellectuals and Power: A Conversation between Michel Foucault and Gilles Deleuze', in Michel Foucault, *Language, Counter-memory, Practice: Selected Essays and Interviews*, trans. Donald Bouchard and Sherry Simon, Ithaca, N.Y.: Cornell University Press: 206–7.

Fox-Genovese, Elizabeth (1982) 'Placing Women's History in History', *New Left Review* 133 (May–June).

Francis, E. K. (1947) 'The Nature of the Ethnic Group', *American Journal of Sociology* 52: 393–400.

Freud, Sigmund (1961) '"A Child Is Being Beaten": A Contribution to the Study of the Origin of Sexual Perversion', *The Standard Edition of the Psychological Works*, vol. 17, trans. James Strachey, New York: Norton: 175–204.

Frow, John (1995) *Cultural Studies and Cultural Value*, Oxford: Clarendon.

Fukuyama, Francis (1992) *The End of History and the Last Man*, New York: Free Press.

Fusco, Coco and Paula Herdia (1993) *The Couple in the Cage: a Guatinaui Odyssey; [videorecording]* Authentic Documentary Productions New York: Third World Newsreel.

Gagnon, François-Marc (1984) '"Portrait du Castor": Analogies et Representation', in Bernard Beugnot (ed.) *Voyages: Récits et Imaginarie: Actes de Montréal*, Paris: Papers on Seventeenth-Century Literature: 199–213.

Gallagher, Catherine and Thomas Laqueur (1987) *The Making of the Modern Body: Sexuality and Society in the Nineteenth Century*, Berkeley: University of California.

Gandhi, Leela (1998) *Postcolonial Theory: A Critical Introduction*, New York: Columbia University Press.

Gans, E. (1985) *The End of a Culture: Towards a Generative Anthropology*, Berkeley: University of California Press.

Garner, Stanton B. Jr (1990) 'Post-Brechtian Anatomies: Weiss, Bond, and the Politics of Embodiment', *Theatre Journal* 42, 2.

Gates, Henry Louis (ed.) (1984) *Black Literature and Literary Theory*, London: Methuen.

Gates, Henry Louis (1986) 'Writing Race', in *Race Writing and Difference*, Chicago: University of Chicago Press.

Gates, Henry Louis (1988) *The Signifying Monkey: a Theory of Afro-American Literary Criticism*, New York: Oxford University Press.

Gates, Henry Louis (1989) 'Authority (White), Power, and the (Black) Critic; or it's all Greek to me', in Ralph Cohen (ed.) *The Future of Literary Theory*, New York: Routledge.

Gates, Henry Louis (1991) 'Critical Fanonism', *Critical Inquiry* 17: 457–70.

Geertz, C. (1986) 'The Uses of Diversity', *Michigan Quarterly* 5, 1.

Geertz, C. (1988) 'Being There, Writing Here', *Harper's Magazine* (March).

Gelder, Kenneth and Jane M. Jacobs (1998) *Uncanny Australia: Sacredness and Identity in a Postcolonial Nation*, Carlton, Victoria: Melbourne University.

Gellner, E. (1964) *Thought and Change*, London: New Left Books.

Gellner, E. (1983) *Nations and Nationalism*, Oxford: Blackwell.

Gerard, Albert (1980) 'The Study of African Literature: Birth and Early Growth of a New Branch of Learning', *Canadian Review of Comparative Literature* (Winter): 67–98.

Geschiere, P. (1988) 'Sorcery and the State: Popular Modes of Political Action among the Maka of Southeast Cameroon', *Critique of Anthropology* 8.

Giddens, A. (1990) *The Consequences of Modernity*, Stanford, Calif.: Stanford University Press.

Gidwani, N. N. (ed.) (1968) *Copyright: Legalized Piracy?*, Bombay: Indian Committee for Cultural Freedom.

Gikandi, Simon (1991) 'Narration in the Post-colonial Moment: Merle Hodge's *Crick Crack Monkey*', in Ian Adam and Helen Tiffin (eds) *Past the Last Post: Theorizing Post-colonialism and Post-modernism*, Hemel Hempstead: Harvester Wheatsheaf: 13–22.

Gikandi, Simon (2002) 'Globalization and the Claims of Postcoloniality', *South Atlantic Quarterly* 100, 3 (published 2001): 627–58.

Gilbert, Helen (1992) 'The Dance as Text in Contemporary Australian Drama: Movement and Resistance Politics', *Ariel* 23, 1.

Gilbert, M. (1998) 'The Sudden Death of a Millionaire: Conversation and Consensus in a Ghanaian Kingdom', *Africa* 58, 3: 291–313.

Gilbert, Sandra M. and Susan Gubar (1979) *The Madwoman in the Attic: The Woman Writes and the Nineteenth-century Literary Imagination*, New Haven, Conn.: Yale University Press.

Giles, Fiona (1989) 'Finding a Shiftingness: Situating the Nineteenth-century Anglo-Australian Female Subject', *New Literatures Review* 18: 10–20.

Gilman, Sander (1985) *Difference and Pathology: Stereotypes of Sexuality, Race and Madness*, Ithaca, N.Y.: Cornell University Press.

Gilroy, Paul [1980] (1990) 'One Nation under a Groove', in D. T. Goldberg (ed.) *Anatomy of Racism*, Minneapolis: University of Minnesota Press.

Gilroy, Paul (1987) *There Ain't No Black in the Union Jack* London: Routledge.

Giroux, Henry (1992) *Border Crossings: Cultural Workers and the Politics of Education*, New York and London: Routledge.

Glacken, C. J. (1967) *Traces on the Rhodian Shore: Nature and Culture in Western Thought, from Ancient Times to the End of the Eighteenth Century*, Berkeley: University of California Press.

Glazer, Nathan (1975) *Affirmative Discrimination: Ethnic Inequality and Public Policy*, New York: Basic Books.

Gleason, Philip (1980) 'American Identity and Americanization', in Stephan Thernstrom *et al.*, *Harvard Encyclopaedia of American Ethnic Groups*, Cambridge, Mass.: Harvard University Press.

Gleason, Phillip (1983) 'Identifying Identity', *The Journal of American History* 69, 4 (March): 910–31.

Glissant, Eduoard (1989) *Caribbean Discourse: Selected Essays*, trans. and introduction by Michael Dash, Charlottesville: University of Virginia Press.

Göçek, F. M. (1996) *Rise of the Bourgeoisie, Demise of Empire: Ottoman Westernization and Social Change*, New York: Oxford University Press.

Godard, Barbara (1987) 'Mapmaking', in *Gynocritics: Feminist Approaches to Canadian and Quebecois Women's Writing*, Toronto: ECW Press.

Godfrey, Dave (1972) Interviewed by Graeme Gibson, in *Eleven Canadian Novelists*, Toronto: Anansi.

Goffman, Erich (1959) *The Presentation of Self in Everyday Life*, New York: Doubleday.

Goh Poh Seng (1980) 'Forum: English Language and Literature in Singapore', *Commentary* 4, 2: 6.

Goldie, Terry (1989) *Fear and Temptation: The Image of the Indigene in Canadian, Australian and New Zealand Literatures*, Kingston: McGill-Queens University Press.

Gordon, Avery and Christopher Newfield (1994) 'White Mythologies', *Critical Inquiry* 20, 4 (Summer): 737–57.

Graham-Brown, S. (1988) *Images of Women: The Portrayal of Women in Photography of the Middle East 1860–1950*, London: Quartet.

Graham-White, Anthony (1974) *The Drama of Black Africa*, New York: Samuel French.

Gramsci, Antonio (1971) *Selections from the Prison Notebooks*, trans. Quintin Hoare and Geoffrey Nowell Smith, New York: International Publishers.

Grant, Charles (1792) *Observations on the State of Society among the Asiatic Subjects of Great Britain*, Great Britain Parliamentary Papers 1831–2, Vol. 8.

Great Britain, Parliamentary Papers (1831–2) Vol. 9, Appendix I, Extract of Letter in the Public Department, from the Court of Directors to the Governor-General in Council, dated 6th September, 1813.

Great Britain, Parliamentary Papers (1832) Vol. 8, 'Observations on the State of Society'.

Great Britain, Parliamentary Papers (1852–3) Vol. 29, Evidence of Maj. F. Rowlandson.

Great Britain, Parliamentary Papers (1852–3a) Vol. 32, Evidence of the Rev. W. Keane.

Great Britain, Parliamentary Papers (1852–3b) Vol. 32. Evidence of Horace Wilson.

Greenblatt, Stephen (1991) *Marvelous Possessions: The Wonder of the New World*, Chicago: University of Chicago Press.

Grey, Stephen (1984) 'A Sense of Place in New Literatures, Particularly South African English', *World Literature Written in English* 24 (Autumn): 228.

Griffiths, Gareth (1994) 'The Myth of Authenticity', in Chris Tiffin and Alan Lawson (eds) *De-Scribing Empire*, London: Routledge.

Griffiths, Gareth (2004) 'Mixed Messages: Imperial Adventures and Missionary Tales', in Gareth Griffiths and Jamie S. Scott (eds) *Mixed Messages: Materiality, Textuality and Missions*, New York: Palgrave Macmillan.

Gronniosaw, James Albert Ukawsaw (1770) *A Narrative of the Most Remarkable Particulars of the Life of James Albert Ukawsaw Gronniosaw, An African Prince*, Bath: W. Gye.

Grove, R. H. (1985) 'Charles Darwin and the Falkland Islands', *Polar Record* 22: 413–20.

Grove, R. H. (1990) 'Colonial Conservation, Ecological Hegemony and Popular Resistance: Towards a Global Synthesis', in J. Mackenzie (ed.) *Imperialism and the Natural World*, Manchester: Manchester University Press.

Grove, Richard H. (1995) *Green Imperialism: Colonial Expansion, Tropical Island Edens And The Origins Of Environmentalism*, Cambridge: Cambridge University Press.

Guha, Ranajit (ed.) (1982) *Subaltern Studies 1: Writing on South Asian History and Society*, Delhi: Oxford University Press.

Guha, Ranajit (ed.) (1983) *Subaltern Studies 11: Writings on South Asian History and Society*, Delhi: Oxford University Press.

Gunder-Frank, Andre (1967) *Capitalism and Underdevelopment in Latin America*, New York: Monthly Review Press.

Gunew, Sneja (1987) 'Culture, Gender and Author Function', *Southern Review* 20.

Guyer, J. L. (1991) 'British Colonial and Postcolonial Food Regulations with Reference to Nigeria: An Essay in Formal Social Anthropology', unpublished ms.

Habermas, J. (1994) 'Citizenship and National Identity', in B. van Steenburgen (ed.) *The Condition of Citizenship*, London: Sage.

Hadjinicolaou, Nicos (1982) 'On the Ideology of Avant-gardism', *Praxis* 6.

Haines, Jack (1980) *Black Radicals and the Civil Rights Mainstream, 1954–1970*, Knoxville: University of Tennessee Press.

Hale, Thomas and Richard Priebe (eds) (1977) *The Teaching of African Literature*, selected annual papers from the African Literature Association, Austin: University of Texas Press.

Hall, Stuart (ed.) (1976) *Resistance Through Rituals*, London: HarperCollins Academic.

Hall, Stuart (1986a) 'Gramsci's Relevance for the Study of Race and Ethnicity', *Journal of Communication Inquiry* 10, 2: 5–27.

Hall, Stuart (1986b) 'On Postmodernism and Articulation: An Interview with Stuart Hall', *Journal of Communication Inquiry* 10, 2: 45–60.

Hall, Stuart (1989) 'New Ethnicities', in *Black Film, British Cinema*, ICA Documents 7, London: Institute of Contemporary Arts.

Hall, Stuart (1990) 'Cultural Identity and Diaspora', in Jonathan Rutherford (ed.) *Identity: Community, Culture, Difference*, London: Lawrence and Wishart: 222–37.

Hall, Stuart (1991a) 'The Global and the Local: Globalization and Ethnicity', in A. D. King (ed.) *Culture, Globalization and the World-system*, London: Macmillan.

Hall, Stuart (1991b) 'Old and New Identities, Old and New Ethnicities', in A. D. King (ed.) *Culture, Globalization and the World-system*, London: Macmillan.

Hall, Stuart (1996) 'When Was "the Post-colonial"? Thinking at the Limit', in Ross Chambers and Lydia Curti (eds) *The Post-colonial Question*, London and New York: Routledge: 242–60.

Hall, Stuart (1997) 'The Local and the Global: Globalization and Ethnicity', in Anne McClintock *et al.* (eds) *Dangerous Liasons: Gender, Nation and Postcolonial Perspectives*, Minneapolis: University of Minnesota Press.

Hamelink, C. (1983) *Cultural Autonomy in Global Communications*, New York: Longman.

Hanna, Judith L. (1987) 'Patterns of Dominance: Men, Women and Homosexuality in Dance', *The Drama Review* 31, 1: 22–47.

Hannerz, Ulf (1987) 'The World in Creolization', *Africa* 57, 4: 546–59.

Hannerz, Ulf (1989) 'Notes on the Global Ecumene', *Public Culture* 1, 2 (Spring): 66–75.

Hannerz, Ulf (1990) 'Cosmopolitans and Locals in World Culture', in M. Featherstone (ed.) *Global Culture*, London: Sage.

Hannerz, Ulf (1991) 'Scenarios for Peripheral Cultures', in Anthony King (ed.) *Culture, Globalization and the World-system*, Binghampton: Department of Art History, State University of New York: 107–28.

Hannerz, Ulf (1992) *Cultural Complexity: Studies in the Social Organization of Meaning*, New York: Columbia University Press.

Hanum, Zeyneb (1913) *A Turkish Woman's European Impressions*, G. Ellison (ed.), London: Seeley, Service Co.

Hardt, Michael and Antonio Negri (2000) *Empire*, Cambridge, Mass.: Harvard University Press.

Harlow, Barbara (1987) *Resistance Literature*, New York and London: Methuen.

Harold, Brent (1972) 'Beyond Student-centred Teaching: The Dialectical Materialist Form of the Literature Course', *College English* 34 (November): 200–14.

Harper, Peggy (1969) 'Dance in Nigeria', *Ethnomusicology* 13, 2: 280–93.

Harris, J. R. (1968) 'Nigerian Enterprise in the Printing Industry', in *Nigerian Journal of Economic and Social Studies* 10, 1 (March).

Harris, Wilson [1970a] (1981) *History, Fable and Myth in the Caribbean and the Guianas*, The National History and Arts Council, Ministry of Information and Culture, Georgetown. Reprinted and revised in Hena Maes Jelinek (ed.) *Explorations: Talks and Articles 1966–81*, Aarhus: Dangaroo.

Harris, Wilson (1970b) *Sleepers of Roraima*, London: Faber & Faber.

Harris, Wilson (1973a) 'A Talk on the Subjective Imagination', *New Letters* (Autumn).

Harris, Wilson (1973b) *Tradition, the Writer and Society*, London and Port of Spain: New Beacon.

Harris, Wilson (1981) *Explorations: A Selection Of Talks And Articles 1966–1981*, Mundelstrup: Dangaroo.

Harris, Wilson (1985) 'Adversarial Contexts and Creativity', *New Left Review* 154 (November–December).

Harrison, Dick (1977) *Unnamed Country: The Struggle for a Canadian Prairie Fiction*, Edmonton: University of Alberta.

Harrison, Nancy (1988) *An Introduction to the Writing Practice of Jean Rhys: The Novel as Women's Text*, Chapel Hill: University of North Carolina Press.

Harvey, David (1989) *The Conditions of Postmodernity: An Enquiry into the Origins of Cultural Change*, Oxford: Blackwell.

Hasan, Abdul (1975) 'Introducing Publishing in the University Curriculum – the Delhi Experiment', unpublished paper read at the Commonwealth African Book Development Seminar, Ibadan, 2–14 February.

Hau'ofa, Epeli (1993) 'Our Sea of Islands . . . A Beginning', in Epeli Hau'ofa, Vijay Naidu and Eric Waddell (eds) *A New Oceania: Rediscovering Our Sea of Islands*, Suva: University of the South Pacific, School of Social and Economic Development: 4–19, 126–39.

Healy, J. J. (1978) *Literature and the Aborigine in Australia 1770–1975*, St Lucia: University of Queensland.

Hegel, G. W. [1892] (1956) *The Philosophy of History*, New York: Dover Publications.

Heidegger, Martin (1961) *An Introduction to Metaphysics*, trans. Ralph Mannheim, New York: Doubleday Anchor.

Heidegger, Martin (1971) *Poetry, Language, Thought*, trans. Albert Hofstadter, New York: Harper & Row.

Heidegger, Martin (1977) 'The Origin of the Work of Art', in *Poetry, Language, Thought*, trans. Albert Hofstadter, New York: Harper & Row.

Heller, Agnes (1984) 'Can Cultures be Compared?', *Dialectical Anthropology* 8 (April): 269–74.

Hencken, H. (1981) 'How the Peabody Museum acquired the Mecklenburg Collection', in *Symbols* 2–3 (Fall), Peabody Museum, Harvard University.

Herberg, Will (1956) *Protestant-Catholic-Jew: An Essay in American Religious Sociology*, Garden City, N.Y.: Anchor Books.

Herman, E. S. and N. Chomsky (1988) *Manufacturing Consent: The Political Economy of the Mass Media*, New York: Pantheon Books.

Heywood, Christopher (ed.) (1971) *Perspectives on African Literature: Selections from the Proceedings*, New York: African Publishing Corp.

Higham, John (1975) *Send These to Me: Jews and Other Immigrants in Urban America*, New York: Athenaeum.

Hill, Mike (2004) *After Whiteness: Unmaking An American Majority*, New York: New York University Press.

Hill, Samuel Charles (1927) *Catalogue: Home Miscellaneous Series*, London: India Office Library.

Hindley, Diana and Geoffrey Hindley (1972) *Advertising in Victorian England 1837–1901*, London: Wayland.

Hitler, Adolf [1925–6] (1974) *Mein Kampf*, trans. by Ralph Manheim; introduction by D. C. Watt, London: Hutchinson.

Hobsbawm, Eric (1977) 'Some Reflections on "The break-up of Britain"', *New Left Review* 105 (September–October).

Hobsbawm, E. J. (1982) *Nations and Nationalism Since 1780*, Cambridge: Cambridge University Press.

Hobsbawm, Eric and Terence Ranger (eds) (1983) *The Invention of Tradition*, Cambridge: Cambridge University Press.

Hodge, Bob and Mishra, Vijay (1991) *The Dark Side of the Dream*, London: Allen and Unwin.

Hodgins, Jack (1977) *The Invention of the World*, Toronto: Macmillan.

Hodgson, Marshall G. S. (1974) *The Venture of Islam: Conscience and History in a World Civilisation*, Chicago: University of Chicago Press.

Hofmeyr, Isabel (2004) 'Inventing the World: Transnationalism, Transmission and Christian Textualities', in Gareth Griffiths and Jamie S. Scott (eds) *Mixed Messages: Materiality, Textuality and Missions*, New York: Palgrave.

Holst Petersen, Kirsten (1984) 'First Things First: Problems of a Feminist Approach to African Literature', *Kunapipi* 6, 3.

Holst Petersen, Kirsten and Anna Rutherford (1976) *Enigma of Values*, Aarhus: Dangaroo.

Holst Petersen, Kirsten and Anna Rutherford (eds) (1986) *A Double Colonization: Colonial and Post-colonial Women's Writing*, Mundelstrup: Dangaroo.

hooks, bell [Gloria Watkins] (1989) 'On Self-Recovery', *Talking Back: Thinking Feminist, Thinking Black*, Boston: South End Press.

Hooper, Charlotte (2000) 'Masculinities in Transition: The Case of Globalization', in M. Marchand and A. Runyan (eds) *Gender and Global Restructuring*, New York: Routledge: 59–73.

Horton, R. (1967) 'African Traditional Thought and Western Science', *Africa* XXXVII.

Horton, Robin (1971) 'African Conversion', *Africa* 41, 2: 85–108.

Horton, Robin (1975a) 'On the Rationality of Conversion', *Africa: Journal of the International African Institute*, 45, 3.

Horton, Robin (1975b) 'On the Rationality of Conversion Part II', *Africa: Journal of the International African Institute*, 45, 4.

Hovell, W. H. and H. Hume (1831) *Journal of Discovery to Port Phillip, New South Wales*, W. Bland (ed.), Sydney: A. Hill.

Hudson, Joyce (1978) *The Walmatjari*, Darwin: Working Papers of the Summer Institute of Linguistics.

Huggan, Graham (1989) 'Decolonizing the Map: Post-colonialism, Post-structuralism and the Cartographic Connection', *Ariel* 20, 4.

Huggan, Graham (1990) 'Decolonizing the Map: Post-colonialism, Post-structuralism and the Cartographic Connection', in Ian Adam and Helen Tiffin (eds) *Past the Last Post: Theorizing Post-colonialism and Post-modernism*, Calgary: University of Calgary Press.

Huggan, Graham (1997) 'Prizing "Otherness": A Short History of the Booker', *Studies in the Novel* 29, 3: 412–33.

Huggan, Graham (2001) *The Postcolonial Exotic: Marketing the Margins*, London and New York: Routledge.

Hughes, Everett Cherrington, and Helen MacGill Hughes (1952) *Where Peoples Meet: Racial and Ethnic Frontiers*, Glencoe, Ill.: Free Press.

Hughes, J. D. (1985) 'Theophrastus as Ecologist', *Environmental Review* 4: 296–307.

Hulme, Peter (1986) *Colonial Encounters: Europe and the Native Caribbean 1492–1797*, London: Methuen.

Hume, David (1934) *A Treatise of Human Nature*, vol. 1, A. D. Lindsay (ed.), London: J. M. Dent.

Hurston, Zora Neale (1979) 'How it Feels to Be Colored Me', in Alice Walker (ed.) *I Love Myself*, Old Westbury, N.Y.: Feminist Press.

Hutcheon, Linda (1988a) *A Poetics of Postmodernism: History, Theory, Fiction*, London and New York: Routledge.

Hutcheon, Linda (1988b) *The Canadian Postmodern: A Study of Contemporary English-Canadian Fiction*, Toronto: Oxford University Press.

Hutcheon, Linda (1989) 'Circling the Down Spout of Empire: Post-colonialism and Post-modernism', *Ariel* 20, 4.

Illich, Ivan (1981) *Shadow Work*, Salem, N.H.: Marion Boyars.

Inden, Ronald (1986) 'Orientalist Constructions of India', in *Modern Asian Studies* 20, 3.

Irele, Abiola (1971) 'Négritude Revisited', in Paul Nursey-Bray (ed.) *Aspects of Africa's Identity: Five Essays,* Kampala: Makerere Institute of Social Research.

Irele, Abiola (1981) *The African Experience in Literature and Ideology*, London: Heinemann.

Israel, Jonathan (2001) *Radical Enlightenment: Philosophy and the Making of Modernity, 1650–1750*, Oxford: Oxford University Press.

Ivy, M. (1988) 'Tradition and Difference in the Japanese Mass Media', *Public Culture* 1, 1: 21–9.

Iyer, P. (1988) *Video Night in Kathmandu*, New York: Knopf.

Jacob, H. (1974) *Mauritian Book Development*, Paris: UNESCO.

Jacobs, Jane M. (1994) 'Earth Honoring: Western Desires and Indigenous Knowledges', in Allison Blunt and Gillian Rose (eds) *Writing Women and Space: Colonial and Postcolonial Geographies*, New York: Guilford Press.

Jacquin, Philippe (1987) *Les Indiens Blancs*, Paris: Payot.

Jahangir, Asma and Hina Jilani (1990) *The Hudood Ordinances: A Divine Sanction?*, Lahore, Pakistan: Rhotas Books.

Jahn, Janheinz (1966) 'Caliban and Prospero', in *A History of Neo-African Literature*, trans. Oliver Cobum and Ursula Lehrburger, London: Faber & Faber: 239–42.

Jahn, Janheinz (1968) *A History of Neo-African Literature*, trans. Oliver Coburn and Ursula Lehbruger, London: Faber & Faber.

James, C. L. R. (1980) *The Black Jacobins*, London: Allison & Busby.

James, William [1902] (1958) *The Variety of Religious Experience*, New York: New American Library.

Jameson, Fredric (1984) 'Literary Innovation and Modes of Production: A Commentary', *Modern Chinese Literature* 1, 1.

Jameson, Fredric (1986) 'Third World Literature in the Era of Multi-national Capitalism', *Social Text* 15 (Fall).

Jameson, Fredric (1988) 'Cognitive Mapping', in Cary Nelson and Laurence Grossberg (eds) *Marxism and the Interpretation of Culture*, Urbana, Ill.: University of Illinois Press.

Jameson, Fredric (1989) 'Foreword', in Roberto Retamar, *Caliban and Other Essays*, Minneapolis: University of Minnesota Press.

Jameson, Fredric (1991) *Postmodernism, Or, the Cultural Logic of Late Capitalism*, Durham, N.C.: Duke University Press.

Jameson, Fredric (1994) 'Marx's Purloined Letter', *New Left Review* 209.

JanMohamed, R. (1983) *Manichean Aesthetics: The Politics of Literature in Colonial Africa*, Amherst: University of Massachusetts Press.

JanMohamed, R. (1985) 'The Economy of Manichean Allegory: The Function of Racial Difference', *Critical Inquiry* 12, 1.

Jayawardena, Kumari (1995) *The White Woman's Other Burden: Western Women and South Asia During British Colonial Rule*, New York: Routledge.

Jea, John (1806) *The Life and Sufferings of John Jea, An African Preacher*, Swansea.

Jeffares, A. Norman (1965) Introduction in John Press (ed.) *The Commonwealth Pen*, London: Heinemann.

Jennings, Francis (1976) *The Invasion of America: Indians, Colonialism and the Cant of Conquest*, New York: Norton.

Jensen, M. (1988) *Passage from India: Asian Indian Immigrants in North America*, New Haven: Yale University Press.

Jeyifo, Biodun (1990) 'The Nature of Things: Arrested Decolonization and Critical Theory', *Research in African Literatures* 21, 1: 33–42.

Jeyifo, Biodun (forthcoming) 'The Reinvention of Theatrical Traditions: Critical Discourses on Interculturalism in the African Theatre', *Proceedings of an International Conference on Interculturalism in World Theatre*, Bad Homburg, West Germany.

Johnson, Samuel (1921) *The History of the Yorubas*, New York: Routledge & Kegan Paul.

Jolly, Margaret (1992) 'Specters of Inauthenticity', *The Contemporary Pacific* 4: 49–72.

Jones, M. G. (1938) *The Charity School Movement*, Cambridge: Cambridge University Press.

Jordan, Winthrop (1969) *White Over Black*, Harmondsworth: Penguin.

Joseph, Gloria and Jill Lewis (1981) *Common Differences: Conflicts in Black and White Feminist Perspectives*, Boston: Beacon Press.

Joyce, James [1922] (1966) *Ulysses*, reprint, New York: Vintage.

Jules-Rosette, Bennetta (1998) *Black Paris: The African Writers' Landscape*, Chicago: University of Illinois Press.

Jurgensmeyer, Mark (1990) 'What the Bhikkhu Said: Reflections on the Rise of Militant, Religious Nationalism', *Religion* 20: 53–76.

Kachru, Braj B. (1978) 'Code-Mixing as a Communicative Strategy in India', in James E. Alatis (ed.) *Report of the Twentieth Annual Roundtable Meeting on Linguistics and Language Studies*, monograph series on language and linguistics, Washington D.C.: Georgetown University Press.

Kachru, Braj B. (1982a) 'The Bilingual's Linguistic Repertoire', in B. Hartford, A. Valdman and C. Foster (eds) *Issues in Bilingual Education: The Role of the Vernacular*, New York: Plenum.

Kachru, Braj B. (ed.) (1982b) *The Other Tongue: English Across Cultures*, Chicago: University of Illinois Press.

Kachru, Braj B. (1990) *Alchemy of English: The Spread Functions and Models of Non-native Englishes*, Champaign, Ill.: University of Illinois Press.

Kane, Cheikh Hamidou (1963) *Ambiguous Adventure*, trans. K. Woods, London: Heinmann.

Kaplan, R. D. (1994) 'The Coming Anarchy', *Atlantic Monthly* 273, 2.

Karstens, M. C. (1957) *The Old Company's Garden at the Cape and its Superintendents*, Cape Town: Maskew Miller.

Katrak, Ketu H. (1989) 'Decolonizing Culture: Toward a Theory for Post-colonial Women's Texts', *Modern Fiction Studies* 35, 1: 157–79.

Katrak, Ketu (1992) 'The Difficult Politics of Wigs and Veils: Feminism and the Colonial Body', paper presented at 'Gender and Colonialism' Conference, University College, Galway.

Kazamias, A. M. and E. H. Epstein (eds) (1968) *School in Transition*, Parts 1 and 2, Boston: Allyn & Bacon.

Keller, Catherine (1994) 'The Breast, the Apocalypse, and the Colonial Journey', *Journal of Feminist Studies in Religion* 10, 1 (Spring): 68.

Keller, Catherine (1996) *Apocalypse Now and Then: A Feminist Guide to the End of the World*, Boston: Beacon.

Kemiläinen, Aira (1964) *Nationalism: Problems Concerning the Word, Concept and Classification*, Jyväskylä: Konstantajat.

Kermode, Frank (1985) *Forms of Attention*, Chicago and London: University of Chicago Press.

Kiernan, Victor (1969) *The Lords of Human Kind*, Boston: Little, Brown.

Kilhnani, Sunil (1997) *The Idea of India*, New York: Farrar, Strauss & Giroux.

Kincaid, Jamaica (1989) *A Small Place*, Harmondsworth: Penguin.

King, Ruby Hope and Mike Morrissey [1986] (1988) *Images in Print: Bias and Prejudice in Caribbean Textbooks*, Mona, Jamaica: Institute of Social and Economic Research, University of the West Indies.

King, Phillip (1827) *Narrative of a Survey of the Intertropical and Western Coasts of Australia Performed Between the Years 1818 and 1822*, 2 vols, London: John Murray.

King, Richard (1999) 'Sacred Texts, Hermeneutics and World Religions' in *Orientalism and Religion: Postcolonial Theory, India and the 'Mystic East'*, London: Routledge.

Kipling, Rudyard (1962) *A Choice of Kipling's Verse*, T. S. Eliot (ed.), New York: Anchor.

Kline, M. (1972) *Mathematics in Western Culture*, New York: Oxford University Press.

Knorr, Klaus (1944) *British Colonial Theories*, Toronto: University of Toronto Press.

Knox, Robert (1850) *The Races of Men: A Fragment*, London: Renshaw.

Kofman, Sarah (1985) *The Enigma of Woman: Woman in Freud's Writings*, trans. Catherine Porter, Ithaca, N.Y.: Cornell University Press.

Kopf, David (1969) *British Orientalism and the Bengal Renaissance: The Dynamivs of Indian Modernization, 1773–1835*, Berkeley and Los Angeles: University of California Press.

Kopytoff, I. (1986) 'The Cultural Biography of Things; Commoditization as Process', in Arjun Appaurai (ed.) *The Social Life of Things: Commodities in Cultural Perspective*, Cambridge: Cambridge University Press: 64–94.

Kosambi D. D. (1963) 'Combined Methods in Indology', *Indo-Iranian Journal* 6.

Kotei, S. I. A. (1981) *The Book Today in Africa*, Paris: UNESCO.

Kroestch, Robert (1974) 'Unhiding the Hidden: Recent Canadian Fiction', *Journal of Canadian Fiction* 3.

Kroetsch, Robert (1975) *Badlands*, Toronto: New Press.

Kröller, Eva Marie (1985) 'The Politics of Influence: Canadian Postmodernism in an American Context', in M. J. Valdes (ed.) *InterAmerican Literary Relations*, vol. 3, New York: Garland: 118–23.

Kropotkin, Peter A. (1975) *Selected Writings on Anarchism and Revolution*, ed. with an introduction by Martin A. Miller, Cambridge, Mass.: MIT Press.

Krupat, Arnold (1996) 'Postcolonialism, Ideology and Native American Literature', *The Turn to the Native: Studies in Criticism and Culture*, Lincoln, Nebr.: University of Nebraska Press.

Kumar, Amitava (2000) *Passport Photos*, Berkeley: University of California Press.

Kumar, Radha (1993) 'Agitation Against Sati 1987–88', in *The History of Doing*, Delhi: Kali for Women: 172–81.

Kymicka, Will (1995) *Multicultural Citizenship: A Liberal Theory of Minority Rights*, Oxford: Clarendon Press.

Lacy, Dan (1973) 'Practical Considerations, Including Financial, in the Creation, Production, and Distribution of Books and Other Educational Materials', in Francis Keppel (ed.) *The Mohonk Conference*, New York: National Book Committee.

Lakoff, G. and M. Johnson (1980) *Metaphors We Live By*, Chicago and London: University of Chicago Press.

Lamming, George (1960) *The Pleasures of Exile*, London: Michael Joseph.

Lancy, D. F. (1983) *Cross-cultural Studies in Cognition and Mathematics*, New York: Academic Press.

Landy, Marcia (1994) *Film, Politics, and Gramsci*, Minneapolis: University of Minnesota Press: 73–98.

Langton, M. (1996) 'What Do We Mean by Wilderness? Wilderness and *terra nullius* in Australian Art', *The Sydney Papers* 8, 1: 10–31.

Laplanche, J. and Pontalis, J. B. (1980) *The Language of Psychoanalysis*, trans. Donald Nicholson-Smith, London: Hogarth Press.

Larson, Charles (1971) *The Emergence of African Fiction*, Indianapolis: Indiana University Press.

Larson, Charles (1973) 'Heroic Ethnocentrism: The Idea of Universality in Literature', *American Scholar* 42, 3.

Lash, S. and J. Urry (1987) *The End of Organized Capitalism*, Madison: University of Wisconsin Press.

Laurence, Margaret (1970) 'Ivory Tower of Grassroots?: The Novelist as Socio-Political Being', in William H. New (ed.) *A Political Art: Essays in Honour of George Woodcock*, Vancouver: University of British Columbia: 15–25.

Lawrence, Errol (1982) 'Just Plain Common Sense: The "Roots" of Racism', in Hazel V. Carby *et al.*, *The Empire Strikes Back: Race and Racism in 70s Britain*, London: Hutchinson.

Lawson, Alan (1983) 'Patterns Preferences and Preoccupations: The Discovery of Nationality in Australian and Canadian Literatures', in Peter Crabbe (ed.) *Theory and Practice in Comparative Studies: Canada Australia and New Zealand*, Sydney: ANZACS (Australia and New Zealand Association of Canadian Studies).

Lawson, Alan (1986), 'There is Another World but it is This One', paper given at the Badlands Conference on Australian and Canadian Literatures, Calgary, Alberta.

Lawson, Alan (1991) 'A Cultural Paradigm for the Second World', *Australian–Canadian Studies* 9, i/ii: 67–78.

Lean, G. A. (1991) *Counting Systems of Papua New Guinea*, Lea: Department of Mathematics and Statistics, Papua New Guinea University of Technology.

LeClercq, Père Chrestien (1691) *Nouvelle Relation de la Gaspesie qui Contient les Moers et la Religion des Sauvages Gaspesiens Portecroix, Adorateurs du Soleil et D'autres Peuples de l'Amerique Septentrionale, die le Canada*, Paris: A. Auroy.

Lee, Dennis (1974) 'Cadence, Country, Silence: Writing in Colonial Space', *Boundary 2*, 3, 1 (Fall).

Lee, Dennis (1977) *Savage Fields: An Essay in Literature and Cosmology*, Toronto: Anansi.

Lee, Tzu Pheng (1980) *Prospect of a Drowning*, Singapore: Heinemann Educational.

Lefevere, André (1983) 'Interface: Some Thoughts on the Historiography of African Literature Written in English', in Dieter Riemenschneider (ed.) *The History and Historiography of Commonwealth Literature*, Tübingen: Gunter Narr Verlag.

Lenin, V. I. [1916] (1969) *Imperialism: The Highest Stage of Capitalism*, Chicago: International.

Levinas, Immanuel (1999) *Alterity and Transcendence*, trans. Michael B. Smith, London: Athlone.

Lévi-Strauss, Claude (1972) *The Savage Mind*, London: Weidenfeld & Nicolson.

Lewis, Gordon (1978) *Slavery, Imperialism, and Freedom: Studies in English Radical Thought*, New York and London: Monthly Review Press.

Lewis, Malcolm (1998) *Cartographic Encounters: Perspectives on Native American Mapmaking and Map Use*, Chicago: University of Chicago Press: 11–14.

Lewis, Reina (1999) 'On Veiling, Vision and Voyage: Cross-cultural Dressing and Narratives of Identity', *Interventions* 1, 4: 500–20.

Lindfors, Bernth (1975) 'The Blind Men and the Elephant', *African Literature Today* 7: 53–64.

Lindfors, Bernth (ed.) (1984) *Research Priorities in African Literatures*, New York: Hans Zell Publishers.

Lindfors, Bernth (1985) 'Is Anything Wrong with African Literary Studies?' in S. Arnold (ed.) *African Literature Studies*, Washington, D.C.: Three Continents Press: 17–26.

Lindfors, Bernth (1985) 'On Disciplining Students in a Nondiscipline', in Thomas Hale and Richard Priebe (eds) *The Teaching of African Literature*, Talence: Presses Universitaires de Bordeaux: 41–7.

Lindfors, Bernth (1987) 'Some New Year's Resolutions', *ALA Bulletin* 13, 1 (Winter): 35–6.

Lindfors, Bernth, Ian Munro, Richard Priebe and Reinhard Sander (eds) (1972) *Palaver: Interviews with Five African Writers in Texas*, Austin: University of Texas Press.

Lindsay, David T. A. and Geoffrey C. Bamber (1965) *Soap-Making. Past and Present, 1876–1976*, Nottingham: Gerard Brothers.

Lloyd, David (1987) *Nationalism and Minor Literature*, Berkeley and Los Angeles: University of California Press.

London, Norrel A. (2003) 'Entrenching the English Language in a British Colony: Curriculum Policy and Practice in Trinidad and Tobago', *International Journal of Educational Development* 23: 97–112.

Lorde, Audre (1981) 'The Master's Tools Will Never Dismantle the Master's House', in Cherrie Morraga and Gloria Anzaldúa (eds) *This Bridge Called My Back: Writings by Radical Women of Colour*, Latham, New York: Kitchen Table Press.

Lorde, Audre (1989) 'Age, Race, Class, and Sex: Women Redefining Difference', in Russell Ferguson *et al.* (eds) *Out There: Marginalization and Contemporary Culture*, New York: New Museum of Contemporary Art and MIT Press.

Lotman, Yu M. and B. A. Uspensky (1978) 'On the Semiotic Mechanism of Culture', *New Literary History*, IX, 2: 211–32.

Low, G. Ching-Liang (1989) 'White Skins/Black Masks: The Pleasures and Politics of Imperialism', *New Formations* 9 (Winter): 83–103.

Lunsford, Andrea (1999) 'Towards a Mestiza Rhetoric: Gloria Anzualda on Composition and Postcoloniality', in Gary A. Olson and Lynn Worsham (eds) *Race, Rhetoric and the Postcolonial*, Albany: State University of New York Press.

Lyotard, Jean-François (1984) *The Postmodern Condition: A Report on Knowledge*, trans. Geoff Bennington and Brian Massumi, Minneapolis: University of Minnesota Press.

Lyotard, Jean-François (1986–7) 'Rules and Paradoxes and a Svelte Appendix', *Cultural Critique* 5 (Winter).

Lyotard, Jean-François (1988) *Perigrinations*, New York: Columbia University Press.

Lyotard, Jean François (1989) *The Postmodern Condition*, Minneapolis: University of Minnesota Press.

Lyotard, Jean François (1992) *The Postmodern Explained*, Minneapolis: University of Minnesota Press.

Macaulay, Thomas B. [1835] (1979) 'Minute on Indian Education', in G. M. Young (ed.) *Speeches*, London and Oxford: Oxford University Press, AMS Edition.

McCarthy, Eugene (1985) 'Rhythm and Narrative Method in Achebe's Things Fall Apart', *Novel* 18, 3 (Spring).

McClintock, Anne (1995) *Imperial Leather: Race, Gender and Sexuality in the Colonial Context*, New York: Routledge.

McCully, Bruce (1943) *English Education and the Origins of Indian Nationalism*, New York: Columbia University Press.

McDougall, Russell (1989) 'Achebe's *Arrow of God*: The Kinetic Idiom of an Unmasking', *Kunapipi* 9, 2.

Macdowell, Diane (1986) *Theories of Discourse*, New York: Basil Blackwell.

McGee, Patrick (1992) *Telling the Other: The Question of Value in Modern and Postcolonial Writing*, Ithaca, N.Y.: Cornell University Press.

Macherey, Pierre (1978) *A Theory of Literary Production*, trans. Geoffrey Wall, London: Routledge & Kegan Paul.

Mackenzie, J. (1988) *The Empire of Nature: Hunting, Conservation and British Imperialism*, Manchester: Manchester University Press.

McNeill, W. H. (1985) *Polyethnicity and National Unity in World History*, Toronto: University of Toronto Press.

*Madras Christian Instructor and Missionary Record* (1844) 11, 4.

Magdoff, Harry (1978) *Imperialism from the Colonial Age to the Present*, New York: Monthly Review.

Malouf, David (1985) *12 Edmonstone Street*, London: Chatto & Windus.

Malthus, T. R. [1798] (1826) *An Essay on the Principle of Population; or, A View of its Past and Present Effects on Human Happiness; with an Inquiry into our Prospects respecting the Future Removal or Mitigation of the Evils which it Occasions*, 6th edition, 2 vols, London: Murray.

Mandel, E. (1978) *Late Capitalism*, London: Verso.

Mani, Lata (1986) 'Production of an Official Discourse on Sati in Early Nineteenth Century Bengal', *Economic and Political Weekly* 21, 17 (26 April): WS-36.

Mani, Lata (1989) 'Contentious Traditions: The Debate on *Sati* in Colonial India', in *Recasting Women: Essays in Colonial History*, Delhi: Kali for Women: 88–126.

Mann, Arthur (1979) *The One and the Many: Reflections on the American Identity*, Chicago: University of Chicago Press.

Manning, Charles A. W. (1968) 'In Defense of Apartheid', in C. D. Moore and A. Dunbar (eds) *Africa Yesterday and Today*, New York: Bantam.

Mannoni, O. (1964) *Prospero and Caliban: The Psychology of Colonization. 1950*, New York: Praeger.

Marchak, Patricia (1978) 'Given a Certain Latitude: A (Hinterland) Sociologist's View of Anglo-Canadian Literature', in Paul Cappon (ed.) *In Our Own House: Social Perspectives on Canadian Literature*, Toronto: McLelland & Stewart: 178–205.

Marchand, Marianne H. and Anne Runyan (eds) (2000) *Gender and Global Restructuring: Sightings, Sites and Resistances*, New York: Routledge.

Mariàtegui, Carlos José (1971) *Seven Interpretive Essays on Peruvian Reality*, Austin: University of Texas Press.

Marrant, John (1785) *Narrative of the Lord's Wonderful Dealings with John Marrant, A Black*, London: Gilbert and Plummer.

Marryatt, Frederick (1984) *Peter Simple*, 3 vols, London: Saunders and Otley, vol II: 195–7.

Marshall, Peter (1970) *The British Discovery of Hinduism in the Eighteenth Century*, Cambridge: Cambridge University Press.

Martin, Horace T. (1892) *Castorologia, or the History and Traditions of the Canadian Beaver*, Montreal: Wm. Drysdale.

Marx, Karl (1973) *Surveys from Exile*, trans. David Fernbach, New York: Vintage.

Marx, Karl (1977) *Capital: A Critique of Political Economy*, vol. 1, trans. Ben Fowkes, New York: Vintage.

Marx, Karl and Friedrich Engels [1848] (1958) *The Communist Manifesto*, vol. 1, in *Selected Works*, Moscow: Foreign Publishing House.

Marx, Karl and Frederick Engels (1983) *The Communist Manifesto*, New York: International Publishers.

Mason, P. (1996) 'On Producing the (American) Exotic', *Anthropos* 91: 139–51.

Mather, Cotton [1693] (1950) *On Witchcraft, Being the Wonders of the Invisible World*, reprint, Mount Vernon, N.Y.: Peter Pauper Press.

Mattelart, A. (1983) *Transnationals and the Third World: The Struggle for Culture*, South Hadley, Mass.: Bergin & Garvey.

Mazrui, Ali A. (1990) *Cultural Forces in World Politics*, London: James Currey.

Mba, Nina (1982) *Nigerian Women Mobilized: Women's Political Activity in South-eastern Nigeria, 1900–1965*, Berkeley: University of California, Institute of International Studies.

Mbembe, Achille (2001) *On the Postcolony*, Berkeley, Calif.: University of California Press.

Mbonwoh, N. (1987) 'Le corps de Joseph Awunti repose désormais à Kedju Ketinguh', *Cameroon Tribune* 4010, 3 (12 November).

Means, Russell (n.d.) 'The Same Old Song', in Ward Churchill (ed.) *Marxism and Native Americans*, Boston: South End Press.

Melman, B. (1992) *Women's Orients: English Women and the Middle East, 1718–1918. Sexuality, Religion and Work*, Basingstoke: Macmillan.

Memmi, Albert (1965) *The Colonizer and the Colonized* 1957, New York: Orion.

Mende, Tibor (1973) *From Aid to Recolonization: Lessons of a Failure*, New York: Pantheon.

Menninger, K. (1969) *Number Words and Number Symbols: A Cultural History of Numbers*, Cambridge, Mass.: MIT Press

Mercator, Hondius Janssonius [1636] (1968) *Atlas or a Geographicke Description of the World*, facsimile edition in two volumes with an introduction by R. A. Skelton, Amsterdam: Theatrum Orbis Terrarum.

Mercer, Kobena (1988) 'Diaspora, Culture and the Dialogic Imagination: The Aesthetic of Black Independent Film in Britain', in Mbye B. Cam and Claire Andrade-Watkins (eds) *Blackframes: Critical Perspectives on Black Independent Cinema*, Cambridge, Mass.: MIT Press.

Merleau-Ponty, Maurice (1945) *La Phénoménologie de la perception*, Paris: Gallimard.

Meyer, W. J. (1992) 'The World Polity and the Authority of the Nation State', in *Perspectives* (ASA Theory Section) 15, 1.

Meyers, Jeffrey (1973) *Fiction and the Colonial Experience*, Totowa, N.J.: Rowman & Littlefield.

Michaels, Walter Benn (1992) 'Race into Culture: A Critical Genealogy of Cultural Identity', *Critical Inquiry* 18,4 (Summer): 655–85.

Michaels, Walter Benn (1995) *Our America: Nativism, Modernism, and Pluralism*, Durham, N.C.: Duke University Press.

Micklewright, N. (1999a) 'Public and Private for Ottoman Women of the Nineteenth Century', in Fairchild D. Ruggles (ed.) *Women and Self-representation in Islamic Societies*, New York: SUNY.

Micklewright, N. (1999b) 'Photography and Consumption in the Ottoman Empire', in D. Quataert (ed.) *Consumption in the Ottoman Empire*, New York: SUNY.

Mifflin, Margot (1992) 'Performance Art: What Is It And Where Is It Going?', *Art News* 91, 4: 84–9.

Mignolo, Walter D. (1995) *The Dark Side of the Renaissance: Literacy, Territoriality and Colonization*, Ann Arbor: University of Michigan Press.

Mills, Nicolaus (1996) 'Lifeboat Ethics and Immigration Fears', *Dissent* (Winter).

Minh-ha, Trinh, T. (1989) *Woman, Native, Other: Writing Postcoloniality and Feminism*, Bloomington, Ind.: Indiana University Press.

Minh-ha, Trinh T. (1991) *When the Moon Waxes Red; Representation, Gender and Cultural Politics*, New York and London: Routledge.

Mishra, Vijay (1996) 'The Diasporic Imaginary: Theorizing the Indian Diaspora', *Textual-Practice* 10, 3 (Winter): 421–47.

Missionary Register (1818) Church Missionary Society, London, January 1818, 18–19.

Mitchell, W. J. T. (1992) 'Postcolonial Culture, Postimperial Criticism', *Transition* 56.

Mitter, Swasti (1986) *Common Fate, Common Bond: Women in the Global Economy*, London: Pluto.

Miyoshi, M. (1993) 'A Borderless World: From Colonialism to Transnationalism and the Decline of the Nation-State', *Critical Enquiry* 19, 4.

Miyoshi, M. and H. D. Harootunian (eds) (1989) *Postmodernism and Japan*, Durham, N.C.: Duke University Press.

Mnthali, Felix (198?) 'Letter to a Feminist Friend', in unpublished manuscript 'Beyond the Echoes'.

Moag, Rodney F. (1982) 'The Life-cycle of Non-Native Englishes: A Case Study', in Braj Kachru (ed.) *The Other Tongue: English Across Cultures*, Chicago: University of Illinois Press.

Mobonwoh, N. (1987) 'Le corps de Joseph Awonti repose désormais à Kedju Ketinguh', *Cameroon Tribune* 4010, 12 November.

Mohanty, Chandra Talpade (1984) 'Under Western Eyes: Feminist Scholarship and Colonial Discourse', *Boundary 2* 12, 3; 13, 1 (Spring/Fall): 333–58.

Mohanty, Chandra Talpade (1991) 'Under Western Eyes: Feminist Scholarship and Colonial Discourses', in Chandra Talpade Mohanty, Ann Russo and Lourdes Torres (eds) *Third World Women and the Politics of Feminism*, Bloomington, Ind.: Indiana University Press.

Mohanty, Chandra Talpade (2003) 'Under Western Eyes Revisited', *Signs* 28, 2: 499–538.

Mohanty, S. P. (1989) '"Us and Them": On the Philosophical Bases of Political Criticism', *New Formations* 8 (Summer): 55–80.

Moore, Gerald (ed.) (1965) *African Literature and the University*, Ibadan: Ibadan University Press.

Moore-Gilbert, Bart (1997) *Postcolonial Theory: Contexts, Practices, Politics*, New York: Verso.

Moraga, Cherrie (1984) *Loving in the War Years*, Boston: South End Press.

Moraga, Cherrie and Anzaldúa, Gloria (eds) (1983) *This Bridge Called My Back: Writings By Radical Women of Color*, New York: Kitchen Table Press.

Moretti, Franco (1996) *Modern Epic: The World System from Goethe to García Márquez*, London: Verso.

Morgan, Henry J. (ed.) (1882) *The Dominion Annual Register and Review*, Montreal: John Lovell.

Morgan, Lewis Henry (1868) *The American Beaver and His Works*, Philadelphia: J. B. Lippincott & Co.

Morris, Meaghan (1988) 'Tooth and Claw: Tales of Survival and Crocodile Dundee', in Andrew Ross (ed.) *Universal Abandon?: The Politics of Post-modernism*, Minneapolis: University of Minnesota Press.

Morton, Thomas [1637] (1967) *New English Canaan or New Canaan Containing an Abstract of New England. Composed in Three Bookes*, Charles Francis Adams (ed.), New York: Burt Franklin.

Mphahlele, Es'kia (1962) 'Press Report', Conference of African Writers, *MAK/V* 2.

Mphahlele, Es'kia (1964) 'The Language of African Literature', *Harvard Educational Review* 34 (Spring): 298–306.

Mudrooroo (1985) 'White Forms, Aboriginal Content', in J. Davis and B. Hodge (eds) *Aboriginal Writing Today*, Canberra: Australian Institute for Aboriginal Studies.

Mudrooroo (1990) *Writing From the Fringe: A Study of Modern Aboriginal Literature*, South Yarra, Vic: Hyland House.

Muecke, Stephen (1983) 'Ideology Re-iterated: The Uses of Aboriginal Oral Narratives', *Southern Review* 16, 1.

Mukerjee, Arun (1986) 'Ideology in the Classroom: A Case Study in the Teaching of English Literature in Canadian Universities', *Dalhousie Review* 66, 1–2: 22–30.

Mukherjee, Arun P. (1990) 'Whose Post-colonialism and Whose Post-modernism?', *World Literature Written in English* 30, 2: 1–9.

Mukherjee, Menakshi (1971) *The Twice-Born Fiction: Themes and Techniques of the Indian Novel in English*, Delhi and London: Heinemann.

Mukherjee-Blaise, Bharati (1983) 'Mimicry and Reinvention', in Uma Parameswaran (ed.) *The Commonwealth in Canada*, Calcutta: Writer's Workshop Greybird: 147–57.

Murdock, George P. (1931) 'Ethnocentrism', in Edwin A. Seligman and Alvin Johnson (eds) *Encyclopedia of the Social Sciences*, vol. 5, New York: Macmillan: 613–14.

Murnane, Gerald (1984) *The Plains*, Ringwood: Penguin.

Murray, David (1991) *Forked Tongues: Speech, Writing and Representation in North American Indian Texts*, Bloomington, Ind.: Indiana University Press.

Myers, Charles N. (1968) *U.S. University Activity Abroad: Implications of the Mexican Case*, New York: Education and World Affairs.

Naficy, Hamid (1991) 'Exile Discourse and Televisual Fetishization', *Quarterly Review of Film and Video* 13, 1–3: 85–116.

Naipaul, V. S. (1961) *A House for Mr. Biswas*, London: André Deutsch.

Naipaul, V. S. (1969) *A House for Mr. Biswas*, Harmondsworth: Penguin Books: 193–4.

Naipaul, V. S. (1974) 'Conrad's Darkness', in *The Return of Eva Peron*, New York: Knopf.

Nair, Chandran (1975) 'The Current State of Creative Writing in Singapore', in Chandran Nair (ed.) *Developing Creative Writing in Singapore*, Singapore: Woodrose Publications.

Nairn, Tom (1977) *The Break up of Britain: Crisis and Neo-Nationalism*, London: New Left Books.

Nairn, Tom (1994) 'What Nations are For', *London Review of Books* (8 September).

Nancy, Jean-Luc (1994) 'Cut Throat Sun', trans. Lydie Moudileno, in Alfred Arteaga (ed.) *An Other Tongue*, Durham, N.C.: Duke University Press.

Nandy, Ashis (1983) *The Intimate Enemy: Loss and Recovery of Self under Colonialism*, Delhi: Oxford University Press.

Narasimhaiah, C. D. (ed.) (1978) *Awakened Conscience: Studies in Commonwealth Literature*, Delhi: Sterling.

Narogin, Mudrooroo (1990) *Writing from the Fringe: A Study of Modern Aboriginal Literature*, South Yarra, Victoria: Hyland House.

Nettleford, Rex (1968) 'The Dance as an Art Form – Its Place in the West Indies', *Caribbean Quarterly* 14 (March–June): 127–35.

New, W. H. (1978) 'New Language, New World', in C. D. Narasimhaiah (ed.) *Awakened Conscience* New Delhi: Sterling.

New, W. H. (ed.) (1975) *Among Worlds*, Erin, Ontario: Press Porceptic.

Ngara, Emmanuel (1985) *Art and Ideology in the African Novel*, London: Heinemann.

Ngaté, Jonathan, (1988) *Francophone African Fiction: Reading a Literary Tradition*, Trenton, N.J.: Africa World Press.

Ngũgĩ wa Thiong'o (1965) *The River Between*, London: Heinemann.

Ngũgĩ wa Thiong'o (1972) *Homecoming: Essays on African and Caribbean Literature, Culture and Politics*, New York: Heinemann.

Ngũgĩ wa Thiong'o (1981) 'The Language of African Literature', in *Decolonising the Mind: The Politics of Language in African Literature*, London: James Currey.

Ngũgĩ wa Thiong'o (1986) *Decolonizing the Mind*, London: James Currey.

Ngũgĩ wa Thiong'o (2000) 'Borders and Bridges: Seeking Connections between Things', in Fawzia Afzal-Khan and Kalpana Seshadri-Crooks (eds) *The Pre-Occupation of Postcolonial Studies*, Durham, N.C. and London: Duke University Press.

Nicoll, F. (1989) 'My Trip to Alice', *Criticism, History and Interpretation* 3: 21–32.

Niranjana, Tejaswini (1990) 'Translation, Colonialism and Rise of English', *Economic and Political Weekly* 25, 15 (14 April).

Nottingham, John (1969) 'Establishing an African Publishing Industry: A Study in Decolonization', *African Affairs* 68.

Nowra, Louis (1979) *Visions*, Sydney: Currency.

Obiechina, E. N. (1971) *Literature for the Masses*, Enugu: Nwanko-Ifejika Publ.

Obiechina, E. N. (1972) *Onitsha Market Literature*, African Writers, 109, London: Heinemann Educational Books.

Obiechina, E. N. (1973) *An African Popular Literature: A Study of Onitsha Market Pamphlets*, Cambridge: Cambridge University Press.

Obiechina, E. N. (1975) *Culture, Tradition and Society in the West African Novel*, Cambridge: Cambridge University Press.

Office of Management and Budget (OMB) (1977) Statistical Policy Directive no. 15. Available online at: http://www.fedworld.gov/ftp.htm#omb.

O'Hare, William (1998) 'Managing Multiple-race Data', *American Demographics* (April): 20, 4: 42–5.

Ohmann, Richard with Wallace Douglas (1976) *English in America: A Radical View of the Profession*, New York: Oxford University Press.

Okai, Atukwe (1973) 'The Role of the Ghanaian Writers in the Revolution', in *Weekly Spectator* Accra (14 July): 4.

Okara, Gabriel (1963) 'African Speech . . . English Words', *Transition* 10: 15–16.

Okara, Gabriel [1964] (1970) *The Voice*, London: Heinemann.

Okot p'Bitek, (1966) *Song of Lawino*, Kenya: East African Publishing House.

Oluwasanmi, E., E. McLean and H. Zell (eds) (1975) *Publishing in Africa in the Seventies: Proceedings of an International Conference on Publishing and Book Development Held at the University of Ile-ife, Nigeria 16–20 December 1973*, Ile-Ife: University of Ile-Ife.

Omoroyi, J. (1988) 'Nigerian Funeral Programmes: An Unexplored Source of Information', *Africa* 58, 4.

Omvedt, Gail (1980) *We Will Smash this Prison*, London: Zed Press.

Opie, J. (1987) 'Renaissance Origins of the Environmental Crisis', *Environmental Review* 2: 2–19.

O'Riordan, T. (1976) *Environmentalism*, London: Pion.

Ortiz, Alicia Dujovne (1987) 'Beunos Aires (An Excerpt)', in *Discourse* 8 (Fall-Winter).

Ouologuem, Yambo (1968a) *Le Devoir de violence*, Paris: Editions du Seuil.

Ouologuem, Yambo (1968b) *Bound to Violence*, trans. Ralph Mannheim, London: Heinemann.

Owomoyela, O. (1979) *African Literatures: An Introduction*, Waltham, Mass.: Crossroads Press.

Owona, R. (1989) 'Branché sur les cinq continents', *Cameroon Tribune* 4378 (April 27).

*Oxford Dictionary of New Words* (1991) compiled by Sara Tulloch, Oxford: Oxford University Press.

Oyewùmí, Oyèrónké (1997) *The Invention of Women: Making An African Sense Of Western Gender Discourses*, Minneapolis: University of Minnesota Press.

Pache, Walter (1985) 'The Fiction Makes Us Real: Aspects of Postmodernism in Canada', in Robert Kroetsch and Reingard H. Nischik (eds) *Gaining Ground: European Critics on Canadian Literature*, Edmonton: NeWest: 64–85.

Parameswaran, Uma (1976) *A Study of Representative Indo-English Novelists*, Delhi: V. Publishers.

Parekh, Bikhu (1989) *Colonisation, Tradition and Thought: Ghandi's Political Discourse* London: Sage Publications.

Parmar, Pratibha and Valerie Amos (1984) 'Challenging Imperial Feminism', *Feminist Review* 17 (Autumn).

Parrenas, Rhacel Salazar (2001) 'Transgressing the Nation State: The Partial Citizenship and "Imagined Global Community" of Migrant Filipina Domestic Workers', *Signs* 26, 4: 1129–54.

Parry, Benita (1987) 'Problems in Current Theories of Colonial Discourse', *Oxford Literary Review* 9: 1–2.

Parry, Benita (1994) 'Resistance Theory/Theorising Resistance or Two Cheers for Nativism', in F. Barker, P. Hulme and M. Iversen (eds) *Colonial Discourse/Postcolonial Theory*, Manchester: Manchester University Press.

Pateman, Carol (1988) *The Sexual Contract*, Stanford, Calif.: Stanford University Press.

Pearson, Bill (1982) 'Witi Ihimera and Patricia Grace', in Cherry Hankin (ed.) *Critical Essays on the New Zealand Short Story*, Auckland: Heinemann.

Pêcheux, Michel [1975] (1982) *Language, Semantics and Ideology*, trans. Harbans Nagpal, London: Macmillan.

Peirce, L. (1993) *The Imperial Harem: Women and Sovereignty in the Ottoman Empire*, Oxford: Oxford University Press.

Pence, Ellen (1982) 'Racism – A White Issue', in G. T. Hull, P. B. Scott and B. Smith (eds) *But Some of Us Are Brave*, Old Westbury, N.Y.: Feminist Press.

Pennycook, Alastair (1994) *The Cultural Politics of English as an International Language*, London: Longman.

Pennycook, Alastair (1998) *English and the Discourses of Colonialism*, London: Routledge.

Philp, H. (1973) 'Mathematical Education in Developing Countries', in A. G. Howson (ed.) *Developments in Mathematical Education*, London: Cambridge University Press.

Pieterse, Jan Nederveen (1998) 'Hybrid Modernities: Mélange Modernities in Asia', *Sociological Analysis* 1, 3.

Pinxten, R., I. van Doren and F. Harvey (1983) *The Anthropology of Space*, Philadelphia: University of Pennsylvania Press.

Plamenatz, John (1973) 'Two Types of Nationalism', in E. Kamenka (ed.) *Nationalism: The Nature and Evolution of an Idea*, Canberra: ANU Press.

Plumwood, Val (1993) *Feminism and the Mastery of Nature*, London: Routledge.

Plumwood, Val (1996) 'Anthropocentrism and Androcentrism: Parallels and Politics', *Ethics and the Environment* 1, 2 (Fall): 119–52.

Plumwood, Val (2003) 'Decolonizing Relationships with Nature', in William Adams and Martin Mulligan (eds) *Decolonizing Nature: Strategies for Conservation in a Post-colonial Era*, London: Earthscan.

Popkewitz, T. S. (1998) *Struggling for the Soul: The Politics of Education and the Construction of the Teacher*, New York: Teachers College Press.

Potts, Lydia (1990) *The World Labour Market: A History of Migration*, London: Zed Books.

Powell, Joseph M. (1976) *Environmental Management in Australia 1788–1914*, London: OUP.

Powers, William (1990) 'When Black Elk Speaks, Everybody Listens', *Social Text* 24: 43–56.

Prakash, Gyan (1990) 'Post-Orientalist Third-World Histories', *Comparative Studies in Society and History* 32, 1.

Prakash, Gyan (1995) *After Colonialism*, Princeton, N.J.: Princeton University Press.

Pratt, Mary Louise (1992) *Imperial Eyes: Travel Writing and Transculturation*, London: Routledge.

Press, John (ed.) (1965) *The Commonwealth Pen*, London: Heinemann.

Pringle, T. (1983) *The Conservationists and the Killers*, Cape Town: Bulpin.

Prior, Michael (1997) *The Bible and Colonialism*, Sheffield: Sheffield Academic Press.

Quirk, Randolph, Sidney Greenbaum, Geoffrey Leech and Jan Svartvik (1972) *A Grammar of Contemporary English*, London: Longman.

Rabasa, José (1985) 'Allegories of the Atlas', in Francis Barker *et al.* (eds) *Europe and Its Others: Proceedings of the Essex Conference on the Sociology of Literature*, Colchester: University of Essex Press, 2 (July 1984).

Rabasa, José (1993) *Inventing A-M-E-R-I-C-A: Spanish Historiography and the Formation of Eurocentrism*, Norman, Okla.: University of Oklahoma Press.

Radhakrishnan, R. (1993) 'Postcoloniality and the Boundaries of Identity', *Callaloo* 16, 4: 750–71.

Radhakrishnan, R. (1996) *Diasporic Mediations: Between Home and Location*, Minneapolis: University of Minnesota Press.

Rai, Amit S. (1995) 'India On-Line: Electronic Bulletin Boards and the Construction of a Diasporic Hindu Identity', *Diaspora* 4, 1: 31–57.

Ramchand, Kenneth (1969) 'Terrified Consciousness', *Journal of Commonwealth Literature* 7 (July).

Ranger, Terence (1975) *Dance and Society in Eastern Africa*, London: Heinemann.

Ranger, Terence (1993) 'The Local and the Global in Southern African Religious History', in Robert W. Heffner (ed.) *Conversion to Christianity: Historical and Anthropological Perspectives on a Great Transformation*, Berkeley and Los Angeles: University of California Press.

Rao, Raja [1937] (1963) *Kanthapura*, Bombay and Oxford: New Directions.

Rao, Raja [1937] (1978) 'The Caste of English', in C. D. Narasimhaiah (ed.) *Awakened Conscience*, Delhi: Sterling: 420–2.

Raudot, Antoine-Denis (1904) *Relation par Lettres de l'Amerique Septentrionale, (Année 1709 et 1710)* Editée et Annotée par le. P Camille de Rochmonteix de la Compagnie de Jesus, Paris: Letouzey et Ané.

Rechy, John (1995) 'Interview With John Rechy', by Debra Castillo, in *Diacritics* 25, 1: 113.

Reed, John and Clive Wake (eds) (1976) *Prose and Poetry: Leopold Sedar Senghor*, London: Heinemann.

Renan, Ernest (1947–61) *Oevres Completes*, vol. 1, Paris: Calmann-Levy.

Rhys, Jean (1968) *Wide Sargasso Sea*, Harmondsworth: Penguin.

Richards, Thomas (1990) *The Commodity Culture of Victorian Britain: British Advertising and Spectacle 1851–1914*, London: Verso.

Richardson, Henry Handel (1930) *The Fortunes of Richard Mahoney*, collected edition, London: Heinemann.

Richardson, John [1832] (1967) *Wacousta or The Prophecy*, Toronto: McClelland and Stewart.

Robbins, B. (1993) *Secular Vocations: Intellectuals, Professionalism, Culture*, London: Verso.

Roberts, P. (1997) *From Oral to Literate Culture: Colonial Experience in the English West Indies*, Mona, Jamaica: University of the West Indies Press.

Robertson, Roland (1990) 'Mapping the Global Condition: Globalization As the Central Concept', in M. Featherstone (ed.) *Global Culture*, London: Sage.

Robertson, Roland (1992) *Globalization: Social Theory and Global Culture*, London: Sage.

Robertson, Roland (1995) 'Glocalization', in Mike Featherstone, Scott Lash and Roland Robertson (eds) *Global Modernities*, London: Sage.

Robertson, W. (1928) *Coo-ee Talks*, Sydney: Angus & Robertson.

Robinson, Gerald and Natalie Robinson (1971) 'The Battle of Indian Education: Macauley's Opening Salvo Newly Discovered', *Victorian Studies* 14.

Ronan, C. A. (1983) *The Cambridge Illustrated History of the World's Science*, Cambridge: Cambridge University Press.

Rosaldo, M. Z. (1980) 'The Use and Abuse of Anthropology: Reflections of Feminism and Cross-cultural Understanding', *Signs* 5, 3.

Rosny, E. de (1977) *Les yeux de ma chevre*, Paris: Plon.

Ross, Andrew (ed.) (1988) *Universal Abandon?: The Politics of Post-Modernism*, Minneapolis: University of Minnesota Press.

Ross, Malcolm and John Stevens (eds) (1967) *In Search of Ourselves*, Toronto: J. M. Dent.

Rouse, Roger (1991) 'Mexican Migration and the Social Space of Postmodernism', *Diaspora* 1, 1: 8–23.

Rousseau, G. S. and R. Porter (eds) (1990) *Exoticism in the Enlightenment*, Manchester: Manchester University Press.

Rushdie, Salman (1982) 'The Empire Writes Back with a Vengeance', *The Times* (3 July): 8.

Rushdie, Salman (1991) *Imaginary Homelands: Essays and Criticism 1981–1991*, London: Granta.

Rushdie, Salman (1995) *The Moor's Last Sigh*, New York: Pantheon.

Ryan, Simon (1994) 'Inscribing the Emptiness: Cartography, Exploration and the Construction of Australia', in Chris Tiffin and Alan Lawson (eds) *De-Scribing Empire: Post-colonialism and Textuality*, London: Routledge.

Saadawi, Nawal el, Fatima Mernissi and Mallica Vajarathon (1978) 'A Critical Look at the Wellesley Conference', *Quest* IV (Winter).

Safran, William (1991) 'Diasporas in Modern Societies: Myths of Homeland and Return', in *Diaspora* 1, 1 (Spring): 83–99.

Sahlins, M. (1981) *Historical Metaphors and Mythical Realities: Structure in the Early History of the Sandwich Islands Kingdom*, Ann Arbor: University of Michigan Press.

Said, Edward W. (1978) *Orientalism*, New York: Random House.

Said, Edward W. (1980) *The Question of Palestine*, New York: Vintage Books.

Said, Edward W. (1983) *The World the Text and the Critic*, Cambridge, Mass.: Harvard University Press.

Said, Edward W. (1984) 'Permission to Narrate', *London Review of Books* (16 February).

Said, Edward W. (1984) 'The Mind of Winter: Reflections on Life in Exile', *Harper's Magazine* 269 (September): 49–55.

Said, Edward W. (1986) 'Representing the Colonized: Anthropology's Interlocutors', *Critical Inquiry* 15: 205–25.

Said, Edward W. (1988) 'Through Gringo Eyes: With Conrad in Latin America', *Harper's Magazine* (April).

Said, Edward W. (1990) 'Third World Intellectuals and Metropolitan Culture', *Raritan* 9, 3: 27–50.

Said, Edward W. (1993) *Culture and Imperialism*, New York: Alfred A. Knopf.

Salecl, Renata (1994) *The Spoils of Freedom*, London: Routledge.

Salutin, Rick (1984) *Marginal Notes: Challenges to the Mainstream*, Toronto: Lester and Orpen Dennys.

Sangari, Kumkum (1984) 'The Changing Text', *Journal of Arts and Ideas* 8 (July–September).

Sangari, Kumkum (1986) 'Of Ladies, Gentlemen, and the Short Cut: *The Portrait of a Lady*', in Lola Chatterjee (ed.) *Women/Image/Text*, Delhi: Trianka.

Sangari, Kumkum (1987) 'The Politics of the Possible', *Cultural Critique* 7.

Sapir, Edward (1931) 'Conceptual Categories in Primitive Languages', *Science* 74.

Sartre, Jean-Paul (1965) *Anti-Semite and Jew*, trans. George J. Becker, New York: Shocken.

Sassen-Koob, Saskia (1982) 'Recomposition and Pheripheralization at the Core', *Contemporary Marxism* 5: 88–100.

Saul, John Ralston (1988) 'We Are Not Authors of the Post-novel Novel', *Brick* (Winter): 52–4.

Sayre, Gordon (1995) 'The Beaver as Native and Colonist', *Canadian Review of Comparative Literature/ Revue Canadienne de Littérature Comparée* September/décembre.

Schaff, W. L. (1963) *Our Mathematical Heritage*, New York: Collier Books.

Schiller, H. (1976) *Communication and Cultural Domination*, White Plains, N.Y.: International Arts and Sciences.

Schiller, Nina Glick, Linda Basch and Christina Blanc-Szanton (eds) (1992) *Towards a Transnational Perspective on Migration: Race, Class, Ethnicity, and Nationalism Reconsidered*, New York: New York Academy of Sciences.

Schneider, Rebecca (1993) 'See the Big Show: Spiderwoman, Theater Doubling Back', in L. Hart and P. Phelan (eds) *Acting Pot: Feminist Performances*, Ann Arbor: University of Michigan Press.

Schreuder, Deryck and Geoffrey Oddie (1989) 'What is "Conversion"? History, Christianity, and Religious Change in Colonial Africa and South Asia', *Journal of Religious History* 15 (December): 498-518.

Schwartz-Bart, André (1959) *Le Dernier de justes*, Paris: Editions du Sueil.

Schwarz-Bart, Simone (1982) *The Bridge of Beyond*, trans. Barbara Bray, London: Heinemann.

Scott, Jamie S. and Paul Simpson-Housley (eds) (2001) *Mapping the Sacred: Religion, Geography and Postcolonial Literatures*, Amsterdam and Atlanta Ga.: Rodopi.

Sechi, Joanne Harumi (1980) 'Being Japanese-American Doesn't Mean "Made in Japan"', in D. Fisher (ed.) *The Third Woman. Minority Women Writers of the United States*, Boston: Houghton Mifflin.

Second Congress of Negro Writers and Artists (1959) 'Resolution on Literature', *Présence Africaine* (February–May): 24–5.

Seni, N. (1995) 'Fashion and Women's Clothing in the Satirical Press of Istanbul at the End of the 19th Century', in T. Sirin (ed.) *Women in Modern Turkish Society*, London: Zed.

Serequeberhan, Tsenay (1997) 'The Critique of Eurocentrism', in Emmanuel Eze (ed.) *Postcolonial African Philosophy*, Oxford: Blackwell.

Seton-Watson, Hugh (1977) *Nations and States: An Enquiry into the Origins of Nations and the Politics of Nationalism*, Boulder, Col.: Westview Press.

Sharpe, Henry (ed.) (1920) *Selections from Educational Records Part 1, 1781–1839*, New Delhi: Government Printing Office.

Sharpe, Jenny (1989) 'Figures of Colonial Resistance', *Modern Fiction Studies* 35, 1 (Spring).

Shils, Edward (1972) *The Intellectuals and the Powers and Other Essays*, Chicago: University of Chicago Press.

Shore, F. J. (1983) 'On the Language and Character Best Suited to the Education of the People', in Peter Penner and Richard Dale MacLean (eds) *The Rebel Bureaucrat: John Shore (1799–1837) as Critic of William Bentinck's India*, New Delhi: Chanakya.

Shridar, S. N. (1982) 'Non-native English Literatures: Context and Relevance', in B. Kachru (ed.) *The Other Tongue*, Chicago: University of Illinois Press: 291–306.

Simgba, J. B. (1989) 'La communauté musulmane du Cameroun en fête', *Cameroon Tribune* 4383 (May): 7–8.

Simpson, David (1982) *Fetishism and Imagination: Dickens, Melville, Conrad*, Baltimore: Johns Hopkins University Press.

Singer, Peter (1975) *Animal Liberation*, New York: Avon.

Sinha, Narendra K. (1970) *The Economic History of Bengal 1793–1848*, vol. 3, Calcutta: Firma K. L. Mukhopadhyay.

Slack, Jennifer Daryl (1996) 'The Theory and Method of Articulation in Cultural Studies', in David Morley and Kian-Hsing Chen (eds) *Stuart Hall: Critical Dialogues in Cultural Studies*, London: Routledge: 112–29.

Slemon, Stephen (1986) 'Revisioning Allegory: Wilson Harris's Carnival', *Kunapipi* 8, 2: 45–55.

Slemon, Stephen (1987) 'Monuments of Empire: Allegory/Counter-discourse/Post-colonial Writing', *Kunapipi* 9, 3: 1–16.

Slemon, Stephen (1988a) 'Magic Realism as Post-colonial Discourse', *Canadian Literature* 116: 9–23.

Slemon, Stephen (1988b) 'Post-colonial Allegory and the Transformation of History', *Journal of Commonwealth Literature* 23, 1: 157–68.

Slemon, Stephen (1990) 'Unsettling the Empire: Resistance Theory for the Second World', *World Literature Written in English* 30, 2: 30–41.

Slemon, Stephen (1991) 'Modernism's Last Post', in Ian Adam and Helen Tiffin (eds) *Past the Last Post*, London and New York: Harvester Wheatsheaf.

Slemon, Stephen (1994) 'The Scramble for Post-colonialism', in Chris Tiffin and Alan Lawson (eds) *De-Scribing Empire*, London: Routledge.

Smith, Anthony D. (1986) *The Ethnic Origins of Nations*, Oxford: Blackwell.

Smith, B. Herrnstein (1984) 'Contingencies of Value', in R. von Hallberg (ed.) *Canons*, Chicago, Ill.: University of Chicago Press: 5–40.

Smith, B. Herrnstein (1988) *Contingencies of Value: Alternative Perspectives for Critical Theory*, Cambridge, Mass.: Harvard University Press.

Smith, Barbara (ed.) (1983) *Home Girls: A Black Feminist Anthology*, New York: Kitchen Table Press.

Smith, Datus Jr (1966) *A Guide to Book Publishing*, New York: Bowker.

Smith, M. G. (1982) 'Ethnicity and Ethnic Groups in America: The View From Harvard', *Ethnic and Racial Studies* 5: 1–22.

Smock, David R. and Kwamina Bentsi-Enchill (eds) (1976) *The Search for National Integration in Africa*, London: Collier-Macmillan.

Smuts, Jan Christiaan (1954) quoted in L. Fischer, *Gandhi*, New York: New American Library.

Sollors, Werner (1986) *Beyond Ethnicity: Consent and Descent in American Culture*, Oxford and New York: Oxford University Press.

Southey, Robert (1951) *Letters from England*, London: Cresset Press.

Soyinka, Wole (1966) 'And After the Narcissist?' African Forum 1, 4: 53–64.

Soyinka, Wole (1976) *Myth, Literature and the African World*, Cambridge: Cambridge University Press.

Soyinka, Wole (1988) *Art, Dialogue and Outrage: Essays on Literature and Culture*, Ibadan: New Horn Press; Oxford: H. Zell Associates.

Sparrow, The Mighty (1963) *One Hundred and Twenty Calypsos to Remember*, Port of Spain: National Recording Company.

Spelman, Elizabeth (1988) *Inessential Woman: Problems of Exclusion in Feminist Thought*, Boston: Beacon Press.

Spencer, Herbert (1864–7) *The Principle of Sociology*, 2 vols, London and Edinburgh: Williams & Norgate.

Spengler, Oswald [1926] (1962) *The Decline of the West*, New York: Random House.

Spivak, Gayatri Chakravorty (1985a) 'The Rani of Simur', in Francis Barker *et al.* (eds) *Europe and Its Others*, vol. 1, Proceedings of the Essex Conference on the Sociology of Literature July 1984, Colchester: University of Essex.

Spivak, Gayatri Chakravorty (1985b) 'Three Women's Texts and a Critique of Imperialism', *Critical Inquiry* 12, 1.

Spivak, Gayatri Chakravorty (1986) 'Imperialism and Sexual Difference', *Oxford Literary Review* 8: 1–2.

Spivak, Gayatri Chakravorty (1987) *In Other Worlds: Essays in Cultural Politics*, New York: Methuen.

Spivak, Gayatri Chakravorty (1988) *In Other Worlds: Essays in Cultural Politics*, New York and London: Routledge.

Spivak, Gayatri Chakravorty (1989) 'Reading The Satanic Verses', *Public Culture* 2, 1 (Fall).

Spivak, Gayatri Chakravorty (1990) 'Post-structuralism, Marginality, Postcoloniality and Value' in P. Collier and H. Geyer-Ryan (eds), *Literary Theory Today*, Ithaca: Cornell University Press.

Spivak, Gayatri Chakravorty (1991) 'Remembering the Limits: Difference, Identity and Practice', in Peter Osborne (ed.) *Socialism and the Limits of Liberalism*, London: Verso.

Spivak, Gayatri Chakravorty (1992) 'The Burden of English', in Rajaswari Sunder Rajan (ed.) *The Lie of the Land: English Literary Studies in India*, Delhi: Oxford University Press: 275–99.

Spivak, Gayatri Chakravorty (1994) 'Psychoanalysis in Left Field; and Field-Working: Examples to Fit the Title,' in Michael Münchow and Sonu Shamdasani (eds) *Speculations After Freud: Psychoanalysis, Philosophy and Culture*, London: Routledge: 66–9.

Spivak, Gayatri Chakravorty (1996) *The Spivak Reader: Selected Works of Gayatri Chakravorty Spivak*, Donna Landry and Gerald MacLean (eds), New York: Routledge.

Spivak, Gayatri Chakravorty (1998) Introduction to Mahasweta Devi, *Breast Stones*, trans. Gayatri Chakravorty Spivak, Calcutta: Seagull Press.

Spivak, Gayatri Chakravorty (1999) *Toward a History of the Vanishing Present*, Cambridge, Mass.: Harvard University Press.

Spivak, Gayatri Chakravorty (2001) 'Moving Devi', in *Cultural Critique* 47 (Winter): 120–63.

Spooner, B. (1986) 'Weavers and Dealers: The Authenticity of an Oriental Carpet', in Arjun Appaurai (ed.) *The Social Life of Things: Commodities in Cultural Perspective*, Cambridge: Cambridge University Press: 195–235.

Stam, Robert and Louise Spence (1983) 'Colonialism, Racism and Representation: An Introduction', *Screen* 24, 2: 2–20.

Stannard, David (1992) *American Holocaust: The Conquest of the New World*, New York: Oxford University Press.

Stavrianos, L. S. (1981) *Global Rift: The Third World Comes of Age*, New York: William Morrow.

Stead, C. K. (1985) 'Kerry Hulme's *the bone people* and the Pegasus Award for Maori Literature', *Ariel* 16.

Stein, Maurice R., Arthur J. Vidich and David Manning White (eds) (1960) *Identity and Anxiety: Survival of the Person in Mass Society*, Glencoe, Ill.: Free Press.

Stepan, N. L. (1993) 'Race and Gender: The Role of Analogy in Science', in S. Harding (ed.) *The Racial Economy of Science*, Indianapolis: Indiana University Press: 359–76.

Strobel, Margaret (1991) *European Women and the Second British Empire*, Bloomington, Ind.: Indiana University Press.

Sugirtharajah, R. S. (2001) 'Indigenous Reading/Postcolonial Criticism', in *The Bible in the Third World: Precolonial, Colonial, Postcolonial Encounters*, Cambridge: Cambridge University Press.

Suleri, Sara (1992) *The Rhetoric of English India*, Chicago: University of Chicago Press.

Suleri, Sara (1992) 'Woman Skin Deep: Feminism and the Postcolonial Condition', *Critical Inquiry* 18, 4 (Summer).

Sutton, Constance and Elsa Chaney (eds) (1987) *Caribbean Life in New York City: Sociocultural Dimensions*, New York: Centre for Migration Studies.

Sutton, P. (1988) *Dreamings: The Art of Aboriginal Australia*, Melbourne: Viking.

Swann, Brian (1992) 'Introduction', in *On the Translation of Native American Literatures*, Washington D.C.: The Smithsonian Institute Press.

Symonds, Richard (1966) *The British and their Successors*, Evanston, Ill.: Northwestern University Press.

Tagne, D. N. (1989) 'Le venin hypnotique de la griffe', *Cameroon Tribune* 4378 (27 April).

Tagore, Rabindranath [1916] (1992) *The Home and the World*, trans. Surendranath Tagore, rev. R. Tagore, Madras: Macmillan India.

Taussig, Michael (1986) *Shamanism, Colonialism and the Wild Man: A Study of Terror and Healing*, Chicago: University of Chicago Press.

Tempels, Placide (1969) *Bantu Philosophy*, Paris: Présence Africaine.

Temple, C. L. (1918) *Native Races and their Rulers: Sketches of Official Life and Administrative Problems in Nigeria*, Cape Town: Argus.

Terdiman, Richard (1985) *Discourse/Counter-discourse: The Theory and Practice of Symbolic Resistance in Nineteenth-century France*, Ithaca and London: Cornell University Press.

Thomas, K. (1983) *Man and the Natural World: Changing Attitudes in England 1500–1800*, Oxford: Pantheon.

Thompson, Michael (1979) *Rubbish Theory*, Oxford: Oxford University Press.

Thompson, Robert F. (1979) *African Art in Motion: Icon and Act*, Los Angeles: University of California Press.

Threadgold, Terry (1986) Introduction to *Semiotics, Ideology, Language*, Sydney: Sydney Association for Studies in Society and Culture.

Thumboo, Edwin (ed.) (1973) *Seven Poets: Singapore and Malaysia*, Singapore: Singapore University Press.

Tiffin, Helen (1987) 'Post-colonial Literatures and Counter-discourse', *Kunapipi* 9, 3: 17–34.

Tiffin, Helen (1988) 'Post-colonialism, Post-modernism and the Rehabilitation of Post-colonial History', *Journal of Commonwealth Literature* 23, 1: 169–81.

Tjibaou, Jean-Marie (1966) *La Presence Kanak*, edited by Alban Bensa and Eric Wittersheim, Paris: Editions Odile Jacob.

Todd, Loreto (1982) 'The English Language in West Africa', in R. W. Bailey and M. Görlach (eds) *English as a World Language*, Ann Arbor: University of Michigan Press.

Todorov, Tzvetan (1982) *The Conquest of America: The Question of the Other*, trans. Richard Howard, Ithaca, N.Y.: Cornell University Press.

Todorov, Tzvetan (1993) 'Race and Racism', in *On Human Diversity: Nationalism, Racism and Exoticism in French Thought*, trans. Catherine Porter, Cambridge, Mass.: Harvard University Press.

Tollefson, J. W. (1991) *Planning Language, Planning Inequality: Language Policy in the Community*, London: Longman.

Tölölian, Khachig (1991) 'The Nation State and Its Others: In Lieu of a Preface', *Diaspora* 1, 1: 3–7.

Tomlinson, John (1991) *Cultural Imperialism: A Critical Introduction*, Baltimore: Johns Hopkins University Press.

Tostevin, Lola Lemire (1989) 'Contamination: A Relation of Difference', *Tessera* (Spring).

Trevelyan, C. E. (1838) *On the Education of the People of India*, London: Longman, Orme, Brown, Green and Longmans.

Trumpener, Katie (1992) 'The Time of the Gypsies: a "People Without History" in the Narratives of the West', *Critical Inquiry* 18, 4: 860.

Tsing, A. L. (1993) *In the Realm of the Diamond Queen*, Princeton, N.J.: Princeton University Press.

Tully, James (1995) *Strange Multiplicity: Constitutionalism in an Age of Diversity*, London: Cambridge University Press, 1995.

UNESCO (1967) *Book Development in Asia: Report on the Production and Distribution of Books in the Region*, Paris: UNESCO.

UNESCO (1973) *Records of the Conference for Revision of the Universal Copyright Convention*, Paris: UNESCO.

Urdang, Stephanie (1983) *Fighting Two Colonialisms: Women in Guinea-Bissau*, London: Zed Press.

Valentine, Gill (1996) 'Angels and Devils: Moral Landscapes of Childhood', *Environment and Planning D: Society and Space* 14: 581–99.

van den Berghe, Pierre L. (1967) *Race and Racism: A Contemporary Perspective*, New York: Wiley.

Van der Veer, Peter (1996) *Conversion to Modernities: the Globalization of Christianity*, New York: Routledge.

Van der Veer, Peter (2001) *Imperial Encounters: Religion and Modernity in India and Britain*, Princeton, N.J.: Princeton University Press.

Van der Veer, Peter (2004) 'Global Conversions', in Gareth Griffiths and Jamie S. Scott (eds) *Mixed Messages: Materiality, Textuality, Missions*, New York: Palgrave Macmillan.

Van Herk, Aretha (1986) *No Fixed Address*, Toronto: McClelland & Stewart.

Van Rooden, Peter (1997) 'Nineteenth-century Representations of Missionary Conversion and the Transformation of Western Christianity', in Peter Van der Veer (ed.) *Conversion to Modernities: The Globalization of Christianity*, New York: Routledge: 65–9.

Van Toorn, Penny (1990) 'Discourse/Patron Discourse: How Minority Texts Command the Attention of Majority Audiences', *Span* 30.

Vaughan, Michalina and Margaret Archer (1971) *Social Conflict and Educational Change in England and France 1789–1848*, Cambridge: Cambridge University Press.

Verhovek, Sam Howe (1995) 'Mother Scolded by Judge for Speaking in Spanish', *New York Times* (30 August).

Veyne, P. (1976) *Le pain et le cirque: sociologie historique d'un pluralisme politique*, Paris, Seuil.

Vidal, C. (1987) 'Funérailles et conflit social en Côte d'Ivoire', *Politique Africaine* 24.

Viswanathan, Gauri (1987) 'The Beginnings of English Literary Study in British India', *Oxford Literary Review* 9, 1–2.

Viswanathan, Gauri (1989) *Masks of Conquest: Literary Study and British Rule in India*, New York: Columbia University Press.

Viswanathan, Gauri (1998) *Outside the Fold: Conversion, Modernity and Belief*, Princeton, N.J.: Princeton University Press.

Vivekananda (1970) *The Complete Works of Swami Vivekananda*, Calcutta: Advaita Ashrama.

Vobejda, Barbara (1991) 'How Kansas is Central to Americans', *Washington Post* (29 April): A9.

Waddington, C. H. (1977) *Tools for Thought*, St. Albans: Paladin.

Wagner, Roy (1980) *The Invention of Culture*, Chicago: University of Chicago Press.

Waihee, John D. (1992) 'A Century After Queen's Overthrow, Talk of Sovereignty Shakes Hawaii', *The New York Times* (8 November), 'National Report'.

Walcott, Derek (1974) 'The Muse of History', in Orde Coombs, *Is Massa Day Dead? Black Moods in the Caribbean*, New York: Doubleday.

Waldseemüller, Martin [1507] (1907) *The Cosmographia Introductio in Facsimile*, United States Catholic Historical Society, vol. 4, ed. Charles George Herbermann, trans. Joseph Fischer and Franz von Wieser, New York: United States Catholic Historical Society.

Wallerstein, I. (1974) *The Modern World System*, 2 vols, New York and London: Academic Press.

Walley, Richard (1989) *Plays From Black Australia*, Sydney: Currency.

Warner, M. (1990) *The Letters of the Republic: Publication and the Public Sphere in Eighteenth-Century America*, Cambridge, Mass.: Harvard University Press.

Waterman, Richard Alan (1952) 'African Influences on the Music of the Americas', in Sol Tax (ed.) *Acculturation in the Americas*, Chicago: University of Chicago Press.

Watts, Alan W. (1958) *Nature, Man, and Woman*, New York: Vintage Books, 1970.

Weber, Samuel (1982) 'Metapsychology Set Apart', in *The Legend of Freud*, Minneapolis: University of Minnesota Press.

West, C. (1987) 'Race and Social Theory: Towards a Genealogical Materialist Analysis', in M. Davis, M. Marable, F. Pfeil and M. Sprinker (eds) *Towards a Rainbow Socialism*, London: Verso.

West, C. (1990) *The American Evasion of Philosophy*, London: Macmillan.

White, Hayden (1982) 'The Politics of Historical Interpretation: Discipline and De-Sublimation', *Critical Inquiry* 9.

White, Lynn (1967) 'The Historical Roots of Our Sociological Crisis', *Science* 155: 1202–7.

Whitlock, Gillian (2000) 'Outlaws of the Text: Women's Bodies and the Organisation of Gender in Imperial Space', *The Intimate Empire: Reading Women's Autobiography*, London: Continuum.

Whorf, Benjamin Lee (1952) *Collected Papers on Metalinguistics*, Washington, D.C.: Foreign Service Institute, Department of State.

Wicke, Jennifer (1988) *Advertising Fiction: Literature, Advertisement and Social Reading*, New York: Columbia University Press.

Wiebe, Rudy (1973) *The Temptations of Big Bear*, Toronto: McClelland & Stewart.

Wielenga, Bastiaan (1988) *It's a Long Road to Freedom: Perspectives of Biblical Theology*, Madurai: Tamilnadu Theological Seminary.

Wielenga, Bastiaan (1995) 'The God of Israel and the Other Deities: Why So Particular?', in Israel Selvanayagam (ed.) *Biblical Insights on Inter-Faith Dialogue: Source Material for Study and Reflection*, Bangalore: The Board for Theological Text-Books Programme for South Asia.

Williams, Bernard (1985) *Ethics and the Limits of Philosophy*, London: Fontana.

Williams, Dennis (1969) *Image and Idea in the Arts of Guyana*, Georgetown, Guyana: National History and Arts Council, Ministry of Information.

Williams, Raymond (1976) *Keywords: A Vocabulary of Culture annd Society*, London: Fontana.

Williams, Raymond (1984) *Writing in Society*, London: Verso.

Williams, Raymond (1996) *Keywords*, New York: Oxford University Press.

Winslow, J. C. (1923) *Narayan Viman Tilak: The Christian Poet of Maharashtra*, Calcutta: Association Press.

Wolch, Jennifer and Jody Emel (eds) (1998) *Animal Geographies: Place, Politics and Identity in the Nature-Culture Borderlands*, London: Verso.

Wolf, E. (1982) *Europe and the People Without History*, Berkeley: University of California Press.

Wolfe, Cary (1997) 'Old Orders for New: Ecology, Animal Rights and the Poverty of Humanism', *Electronic Book Review* 4 (Winter).

Wolfe, Cary (2003) *Animal Rites: American Culture, the Discourse of Species and Posthumanist Theory*, Chicago and London: University of Chicago Press.

*World Almanac and Book of Facts 1984* (1983) New York: Newspaper Enterprise Association.

Worster, D. (1977) *Nature's Economy: A History of Ecological Ideas*, San Francisco: Sierra Club Books.

Yap, Arthur (1980) *down the line*, Singapore: Heinemann Educational.

Yeo, Robert (1970) 'Poetry in English in Singapore and Malaysia', in *Singapore Book World* 1, 1.

Yoshimoto, M. (1989) 'The Postmodern and Mass Images in Japan', *Public Culture* 1, 2: 8–25.

Young, Robert (1990) *White Mythologies: Writing History and the West*, London and New York: Routledge.

Young, Robert (1995) *Colonial Desire*, London: Routledge.

Zabus, Chantal (1991) *The African Palimpsest: Indigenization of Language in the West African Europhone Novel*, Cross Culture 4, Amsterdam and Atlanta, Ga.: Editions Rodopi.

Zaslavsky, Claudia (1973) *Africa Counts*, Westport, Conn.: Lawrence Hill.

Žižek, Slavoj (1989) *The Sublime Object of Ideology*, London: Verso.

Žižek, Slavoj (1993) *Tarrying with the Negative*, Durham, N.C.: Duke University Press.

Žižek, Slavoj (2001) *Enjoy Your Symptom! Jacques Lacan in Hollywood and Out*, New York: Routledge.

Zok, C. M. (1989) 'Le prêt á porter fait du porte-á-porte', *Cameroon Tribune* 4378 (27 April).

Zweig, Stefan (1942) *Amerigo: A Comedy of Errors in History*, trans. Andrew St James, New York: Viking.

# Index